Applied Multivariate Statistical Analysis

Applied Multivariate Statistical Analysis

RICHARD A. JOHNSON
University of Wisconsin—Madison

DEAN W. WICHERN
Texas A&M University

PEARSON
Prentice
Hall

Pearson Education International

Library of Congress Cataloging-in-Publication Data

Johnson, Richard A.
  Statistical analysis/Richard A. Johnson.—6th ed.
  Dean W. Wichern
    p. cm.
  Includes index.
  ISBN 0-13-514350-0
  1. Statistical Analysis

CIP Data Available

Executive Acquisitions Editor: *Petra Recter*
Vice President and Editorial Director, Mathematics: *Christine Hoag*
Project Manager: *Michael Bell*
Production Editor: *Debbie Ryan*
Senior Managing Editor: *Linda Mihatov Behrens*
Manufacturing Buyer: *Maura Zaldivar*
Associate Director of Operations: *Alexis Heydt-Long*
Marketing Manager: *Wayne Parkins*
Marketing Assistant: *Jennifer de Leeuwerk*
Editorial Assistant/Print Supplements Editor: *Joanne Wendelken*
Art Director: *Jayne Conte*
Director of Creative Service: *Paul Belfanti*
Cover Designer: *Bruce Kenselaar*
Art Studio: *Laserswords*

 © 2007 Pearson Education, Inc.
Pearson Prentice Hall
Pearson Education, Inc.
Upper Saddle River, NJ 07458

Printed in the United States of America

10 9 8 7 6 5

ISBN: 0-13-514350-0

Pearson Education LTD., *London*
Pearson Education Australia PTY, Limited, *Sydney*
Pearson Education Singapore, Pte. Ltd
Pearson Education North Asia Ltd, *Hong Kong*
Pearson Education Canada, Ltd., *Toronto*
Pearson Educación de Mexico, S.A. de C.V.
Pearson Education–Japan, *Tokyo*
Pearson Education Malaysia, Pte. Ltd

To the memory of my mother and my father.
                                          R. A. J.

To Dorothy, Michael, and Andrew.
                                          D. W. W.

Contents

Preface

This book originally grew out of our lecture notes for an "Applied Multivariate Analysis" course offered jointly by the Statistics Department and the School of Business at the University of Wisconsin–Madison. *Applied Multivariate Statistical Analysis*, Sixth Edition, is concerned with statistical methods for describing and analyzing multivariate data. Data analysis, while interesting with one variable, becomes truly fascinating and challenging when several variables are involved. Researchers in the biological, physical, and social sciences frequently collect measurements on several variables. Modern computer packages readily provide the numerical results to rather complex statistical analyses. We have tried to provide readers with the supporting knowledge necessary for making proper interpretations, selecting appropriate techniques, and understanding their strengths and weaknesses. We hope our discussions will meet the needs of experimental scientists, in a wide variety of subject matter areas, as a readable introduction to the statistical analysis of multivariate observations.

LEVEL

Our aim is to present the concepts and methods of multivariate analysis at a level that is readily understandable by readers who have taken two or more statistics courses. We emphasize the applications of multivariate methods and, consequently, have attempted to make the mathematics as palatable as possible. We avoid the use of calculus. On the other hand, the concepts of a matrix and of matrix manipulations are important. We do not assume the reader is familiar with matrix algebra. Rather, we introduce matrices as they appear naturally in our discussions, and we then show how they simplify the presentation of multivariate models and techniques.

The introductory account of matrix algebra, in Chapter 2, highlights the more important matrix algebra results as they apply to multivariate analysis. The Chapter 2 supplement provides a summary of matrix algebra results for those with little or no previous exposure to the subject. This supplementary material helps make the book self-contained and is used to complete proofs. The proofs may be ignored on the first reading. In this way we hope to make the book accessible to a wide audience.

In our attempt to make the study of multivariate analysis appealing to a large audience of both practitioners and theoreticians, we have had to sacrifice a consistency of level. Some sections are harder than others. In particular, we have summarized a voluminous amount of material on regression in Chapter 7. The resulting presentation is rather succinct and difficult the first time through. We hope instructors will be able to compensate for the unevenness in level by judiciously choosing those sections, and subsections, appropriate for their students and by toning them down if necessary.

ORGANIZATION AND APPROACH

The methodological "tools" of multivariate analysis are contained in Chapters 5 through 12. These chapters represent the heart of the book, but they cannot be assimilated without much of the material in the introductory Chapters 1 through 4. Even those readers with a good knowledge of matrix algebra or those willing to accept the mathematical results on faith should, at the very least, peruse Chapter 3, "Sample Geometry," and Chapter 4, "Multivariate Normal Distribution."

Our approach in the methodological chapters is to keep the discussion direct and uncluttered. Typically, we start with a formulation of the population models, delineate the corresponding sample results, and liberally illustrate everything with examples. The examples are of two types: those that are simple and whose calculations can be easily done by hand, and those that rely on real-world data and computer software. These will provide an opportunity to (1) duplicate our analyses, (2) carry out the analyses dictated by exercises, or (3) analyze the data using methods other than the ones we have used or suggested.

The division of the methodological chapters (5 through 12) into three units allows instructors some flexibility in tailoring a course to their needs. Possible sequences for a one-semester (two quarter) course are indicated schematically. Each instructor will undoubtedly omit certain sections from some chapters to cover a broader collection of topics than is indicated by these two choices.

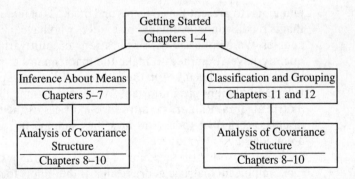

For most students, we would suggest a quick pass through the first four chapters (concentrating primarily on the material in Chapter 1; Sections 2.1, 2.2, 2.3, 2.5, 2.6, and 3.6; and the "assessing normality" material in Chapter 4) followed by a selection of methodological topics. For example, one might discuss the comparison of mean vectors, principal components, factor analysis, discriminant analysis and clustering. The discussions could feature the many "worked out" examples included in these sections of the text. Instructors may rely on diagrams and verbal descriptions to teach the corresponding theoretical developments. If the students have uniformly strong mathematical backgrounds, much of the book can successfully be covered in one term.

We have found individual data-analysis projects useful for integrating material from several of the methods chapters. Here, our rather complete treatments of multivariate analysis of variance (MANOVA), regression analysis, factor analysis, canonical correlation, discriminant analysis, and so forth are helpful, even though they may not be specifically covered in lectures.

CHANGES TO THE SIXTH EDITION

New material.   Users of the previous editions will notice several major changes in the sixth edition.

- Twelve new data sets including national track records for men and women, psychological profile scores, car body assembly measurements, cell phone tower breakdowns, pulp and paper properties measurements, Mali family farm data, stock price rates of return, and Concho water snake data.

- Thirty seven new exercises and twenty revised exercises with many of these exercises based on the new data sets.

- Four new data based examples and fifteen revised examples.

- Six new or expanded sections:

    1. Section 6.6 Testing for Equality of Covariance Matrices
    2. Section 11.7 Logistic Regression and Classification
    3. Section 12.5 Clustering Based on Statistical Models
    4. Expanded Section 6.3 to include "An Approximation to the Distribution of T^2 for Normal Populations When Sample Sizes are not Large"
    5. Expanded Sections 7.6 and 7.7 to include Akaike's Information Criterion
    6. Consolidated previous Sections 11.3 and 11.5 on two group discriminant analysis into single Section 11.3

Web Site.   To make the methods of multivariate analysis more prominent in the text, we have removed the long proofs of Results 7.2, 7.4, 7.10 and 10.1 and placed them on a web site accessible through *www.pearsonhighered.com/datasets*. In addition, all full data sets saved as ASCII files that are used in the book are available on the web site.

Instructors' Solutions Manual.   An Instructors Solutions Manual is available on the author's website accessible through *www.prenhall.com/statistics*. For information on additional for-sale supplements that may be used with the book or additional titles of interest, please visit the Prentice Hall web site at *www.prenhall.com*.

ACKNOWLEDGMENTS

We thank many of our colleagues who helped improve the applied aspect of the book by contributing their own data sets for examples and exercises. A number of individuals helped guide various revisions of this book, and we are grateful for their suggestions: Christopher Bingham, University of Minnesota; Steve Coad, University of Michigan; Richard Kiltie, University of Florida; Sam Kotz, George Mason University; Him Koul, Michigan State University; Bruce McCullough, Drexel University; Shyamal Peddada, University of Virginia; K. Sivakumar University of Illinois at Chicago; Eric Smith, Virginia Tech; and Stanley Wasserman, University of Illinois at Urbana-Champaign. We also acknowledge the feedback of the students we have taught these past 35 years in our applied multivariate analysis courses. Their comments and suggestions are largely responsible for the present iteration of this work. We would also like to give special thanks to Wai Kwong Cheang, Shanhong Guan, Jialiang Li and Zhiguo Xiao for their help with the calculations for many of the examples.

We must thank Dianne Hall for her valuable help with the Solutions Manual, Steve Verrill for computing assistance throughout, and Alison Pollack for implementing a Chernoff faces program. We are indebted to Cliff Gilman for his assistance with the multidimensional scaling examples discussed in Chapter 12. Jacquelyn Forer did most of the typing of the original draft manuscript, and we appreciate her expertise and willingness to endure cajoling of authors faced with publication deadlines. Finally, we would like to thank Petra Recter, Debbie Ryan, Michael Bell, Linda Behrens, Joanne Wendelken and the rest of the Prentice Hall staff for their help with this project.

R. A. Johnson
rich@stat.wisc.edu

D. W. Wichern
d-wichern@tamu.edu

Applied Multivariate Statistical Analysis

Aspects of Multivariate Analysis

1.1 Introduction

Scientific inquiry is an iterative learning process. Objectives pertaining to the explanation of a social or physical phenomenon must be specified and then tested by gathering and analyzing data. In turn, an analysis of the data gathered by experimentation or observation will usually suggest a modified explanation of the phenomenon. Throughout this iterative learning process, variables are often added or deleted from the study. Thus, the complexities of most phenomena require an investigator to collect observations on many different variables. This book is concerned with statistical methods designed to elicit information from these kinds of data sets. Because the data include simultaneous measurements on many variables, this body of methodology is called *multivariate analysis*.

The need to understand the relationships between many variables makes multivariate analysis an inherently difficult subject. Often, the human mind is overwhelmed by the sheer bulk of the data. Additionally, more mathematics is required to derive multivariate statistical techniques for making inferences than in a univariate setting. We have chosen to provide explanations based upon algebraic concepts and to avoid the derivations of statistical results that *require* the calculus of many variables. Our objective is to introduce several useful multivariate techniques in a clear manner, making heavy use of illustrative examples and a minimum of mathematics. Nonetheless, some mathematical sophistication and a desire to think quantitatively will be required.

Most of our emphasis will be on the *analysis* of measurements obtained without actively controlling or manipulating any of the variables on which the measurements are made. Only in Chapters 6 and 7 shall we treat a few experimental plans (designs) for generating data that prescribe the active manipulation of important variables. Although the experimental design is ordinarily the most important part of a scientific investigation, it is frequently impossible to control the generation of appropriate data in certain disciplines. (This is true, for example, in business, economics, ecology, geology, and sociology.) You should consult [6] and [7] for detailed accounts of design principles that, fortunately, also apply to multivariate situations.

It will become increasingly clear that many multivariate methods are based upon an underlying probability model known as the multivariate normal distribution. Other methods are ad hoc in nature and are justified by logical or commonsense arguments. Regardless of their origin, multivariate techniques must, invariably, be implemented on a computer. Recent advances in computer technology have been accompanied by the development of rather sophisticated statistical software packages, making the implementation step easier.

Multivariate analysis is a "mixed bag." It is difficult to establish a classification scheme for multivariate techniques that is both widely accepted and indicates the appropriateness of the techniques. One classification distinguishes techniques designed to study interdependent relationships from those designed to study dependent relationships. Another classifies techniques according to the number of populations and the number of sets of variables being studied. Chapters in this text are divided into sections according to inference about treatment means, inference about covariance structure, and techniques for sorting or grouping. This should not, however, be considered an attempt to place each method into a slot. Rather, the choice of methods and the types of analyses employed are largely determined by the objectives of the investigation. In Section 1.2, we list a smaller number of practical problems designed to illustrate the connection between the choice of a statistical method and the objectives of the study. These problems, plus the examples in the text, should provide you with an appreciation of the applicability of multivariate techniques across different fields.

The objectives of scientific investigations to which multivariate methods most naturally lend themselves include the following:

1. *Data reduction or structural simplification.* The phenomenon being studied is represented as simply as possible without sacrificing valuable information. It is hoped that this will make interpretation easier.

2. *Sorting and grouping.* Groups of "similar" objects or variables are created, based upon measured characteristics. Alternatively, rules for classifying objects into well-defined groups may be required.

3. *Investigation of the dependence among variables.* The nature of the relationships among variables is of interest. Are all the variables mutually independent or are one or more variables dependent on the others? If so, how?

4. *Prediction.* Relationships between variables must be determined for the purpose of predicting the values of one or more variables on the basis of observations on the other variables.

5. *Hypothesis construction and testing.* Specific statistical hypotheses, formulated in terms of the parameters of multivariate populations, are tested. This may be done to validate assumptions or to reinforce prior convictions.

We conclude this brief overview of multivariate analysis with a quotation from F. H. C. Marriott [19], page 89. The statement was made in a discussion of cluster analysis, but we feel it is appropriate for a broader range of methods. You should keep it in mind whenever you attempt or read about a data analysis. It allows one to maintain a proper perspective and not be overwhelmed by the elegance of some of the theory:

> If the results disagree with informed opinion, do not admit a simple logical interpretation, and do not show up clearly in a graphical presentation, they are probably wrong. There is no magic about numerical methods, and many ways in which they can break down. They are a valuable aid to the interpretation of data, not sausage machines automatically transforming bodies of numbers into packets of scientific fact.

1.2 Applications of Multivariate Techniques

The published applications of multivariate methods have increased tremendously in recent years. It is now difficult to cover the variety of real-world applications of these methods with brief discussions, as we did in earlier editions of this book. However, in order to give some indication of the usefulness of multivariate techniques, we offer the following short descriptions of the results of studies from several disciplines. These descriptions are organized according to the categories of objectives given in the previous section. Of course, many of our examples are multifaceted and could be placed in more than one category.

Data reduction or simplification

- Using data on several variables related to cancer patient responses to radiotherapy, a simple measure of patient response to radiotherapy was constructed. (See Exercise 1.15.)

- Track records from many nations were used to develop an index of performance for both male and female athletes. (See [8] and [22].)

- Multispectral image data collected by a high-altitude scanner were reduced to a form that could be viewed as images (pictures) of a shoreline in two dimensions. (See [23].)

- Data on several variables relating to yield and protein content were used to create an index to select parents of subsequent generations of improved bean plants. (See [13].)

- A matrix of tactic similarities was developed from aggregate data derived from professional mediators. From this matrix the number of dimensions by which professional mediators judge the tactics they use in resolving disputes was determined. (See [21].)

Sorting and grouping

- Data on several variables related to computer use were employed to create clusters of categories of computer jobs that allow a better determination of existing (or planned) computer utilization. (See [2].)

- Measurements of several physiological variables were used to develop a screening procedure that discriminates alcoholics from nonalcoholics. (See [26].)

- Data related to responses to visual stimuli were used to develop a rule for separating people suffering from a multiple-sclerosis-caused visual pathology from those not suffering from the disease. (See Exercise 1.14.)

- The U.S. Internal Revenue Service uses data collected from tax returns to sort taxpayers into two groups: those that will be audited and those that will not. (See [31].)

Investigation of the dependence among variables

- Data on several variables were used to identify factors that were responsible for client success in hiring external consultants. (See [12].)

- Measurements of variables related to innovation, on the one hand, and variables related to the business environment and business organization, on the other hand, were used to discover why some firms are product innovators and some firms are not. (See [3].)

- Measurements of pulp fiber characteristics and subsequent measurements of characteristics of the paper made from them are used to examine the relations between pulp fiber properties and the resulting paper properties. The goal is to determine those fibers that lead to higher quality paper. (See [17].)

- The associations between measures of risk-taking propensity and measures of socioeconomic characteristics for top-level business executives were used to assess the relation between risk-taking behavior and performance. (See [18].)

Prediction

- The associations between test scores, and several high school performance variables, and several college performance variables were used to develop predictors of success in college. (See [10].)

- Data on several variables related to the size distribution of sediments were used to develop rules for predicting different depositional environments. (See [7] and [20].)

- Measurements on several accounting and financial variables were used to develop a method for identifying potentially insolvent property-liability insurers. (See [28].)

- cDNA microarray experiments (gene expression data) are increasingly used to study the molecular variations among cancer tumors. A reliable classification of tumors is essential for successful diagnosis and treatment of cancer. (See [9].)

Hypotheses testing

- Several pollution-related variables were measured to determine whether levels for a large metropolitan area were roughly constant throughout the week, or whether there was a noticeable difference between weekdays and weekends. (See Exercise 1.6.)

- Experimental data on several variables were used to see whether the nature of the instructions makes any difference in perceived risks, as quantified by test scores. (See [27].)

- Data on many variables were used to investigate the differences in structure of American occupations to determine the support for one of two competing sociological theories. (See [16] and [25].)

- Data on several variables were used to determine whether different types of firms in newly industrialized countries exhibited different patterns of innovation. (See [15].)

The preceding descriptions offer glimpses into the use of multivariate methods in widely diverse fields.

1.3 The Organization of Data

Throughout this text, we are going to be concerned with analyzing measurements made on several variables or characteristics. These measurements (commonly called *data*) must frequently be arranged and displayed in various ways. For example, graphs and tabular arrangements are important aids in data analysis. Summary numbers, which quantitatively portray certain features of the data, are also necessary to any description.

We now introduce the preliminary concepts underlying these first steps of data organization.

Arrays

Multivariate data arise whenever an investigator, seeking to understand a social or physical phenomenon, selects a number $p \geq 1$ of *variables* or *characters* to record. The values of these variables are all recorded for each distinct *item, individual*, or *experimental unit*.

We will use the notation x_{jk} to indicate the particular value of the kth variable that is observed on the jth item, or trial. That is,

$$x_{jk} = \text{measurement of the } k\text{th variable on the } j\text{th item}$$

Consequently, n measurements on p variables can be displayed as follows:

	Variable 1	Variable 2	$\cdots$	Variable k	$\cdots$	Variable p
Item 1:	x_{11}	x_{12}	$\cdots$	x_{1k}	$\cdots$	x_{1p}
Item 2:	x_{21}	x_{22}	$\cdots$	x_{2k}	$\cdots$	x_{2p}
$\vdots$	$\vdots$	$\vdots$		$\vdots$		$\vdots$
Item j:	x_{j1}	x_{j2}	$\cdots$	x_{jk}	$\cdots$	x_{jp}
$\vdots$	$\vdots$	$\vdots$		$\vdots$		$\vdots$
Item n:	x_{n1}	x_{n2}	$\cdots$	x_{nk}	$\cdots$	x_{np}

Or we can display these data as a rectangular array, called $\mathbf{X}$, of n rows and p columns:

$$\mathbf{X} = \begin{bmatrix} x_{11} & x_{12} & \cdots & x_{1k} & \cdots & x_{1p} \\ x_{21} & x_{22} & \cdots & x_{2k} & \cdots & x_{2p} \\ \vdots & \vdots & & \vdots & & \vdots \\ x_{j1} & x_{j2} & \cdots & x_{jk} & \cdots & x_{jp} \\ \vdots & \vdots & & \vdots & & \vdots \\ x_{n1} & x_{n2} & \cdots & x_{nk} & \cdots & x_{np} \end{bmatrix}$$

The array $\mathbf{X}$, then, contains the data consisting of all of the observations on all of the variables.

Example 1.1 (A data array) A selection of four receipts from a university bookstore was obtained in order to investigate the nature of book sales. Each receipt provided, among other things, the number of books sold and the total amount of each sale. Let the first variable be total dollar sales and the second variable be number of books sold. Then we can regard the corresponding numbers on the receipts as four measurements on two variables. Suppose the data, in tabular form, are

Variable 1 (dollar sales):   42   52   48   58
Variable 2 (number of books):   4   5   4   3

Using the notation just introduced, we have

$$x_{11} = 42 \quad x_{21} = 52 \quad x_{31} = 48 \quad x_{41} = 58$$
$$x_{12} = 4 \quad x_{22} = 5 \quad x_{32} = 4 \quad x_{42} = 3$$

and the data array $\mathbf{X}$ is

$$\mathbf{X} = \begin{bmatrix} 42 & 4 \\ 52 & 5 \\ 48 & 4 \\ 58 & 3 \end{bmatrix}$$

with four rows and two columns. ∎

Considering data in the form of arrays facilitates the exposition of the subject matter and allows numerical calculations to be performed in an orderly and efficient manner. The efficiency is twofold, as gains are attained in both (1) *describing* numerical calculations as operations on arrays and (2) the *implementation* of the calculations on computers, which now use many languages and statistical packages to perform array operations. We consider the manipulation of arrays of numbers in Chapter 2. At this point, we are concerned only with their value as devices for displaying data.

Descriptive Statistics

A large data set is bulky, and its very mass poses a serious obstacle to any attempt to visually extract pertinent information. Much of the information contained in the data can be assessed by calculating certain summary numbers, known as *descriptive statistics*. For example, the arithmetic average, or sample mean, is a descriptive statistic that provides a measure of location—that is, a "central value" for a set of numbers. And the average of the squares of the distances of all of the numbers from the mean provides a measure of the spread, or variation, in the numbers.

We shall rely most heavily on descriptive statistics that measure location, variation, and linear association. The formal definitions of these quantities follow.

Let $x_{11}, x_{21}, \ldots, x_{n1}$ be n measurements on the first variable. Then the arithmetic average of these measurements is

$$\bar{x}_1 = \frac{1}{n} \sum_{j=1}^{n} x_{j1}$$

If the n measurements represent a subset of the full set of measurements that might have been observed, then $\bar{x}_1$ is also called the *sample mean* for the first variable. We adopt this terminology because the bulk of this book is devoted to procedures designed to analyze samples of measurements from larger collections.

The sample mean can be computed from the n measurements on each of the p variables, so that, in general, there will be p sample means:

$$\bar{x}_k = \frac{1}{n} \sum_{j=1}^{n} x_{jk} \qquad k = 1, 2, \ldots, p \tag{1-1}$$

this many means

A measure of spread is provided by the *sample variance*, defined for n measurements on the first variable as

$$s_1^2 = \frac{1}{n} \sum_{j=1}^{n} (x_{j1} - \bar{x}_1)^2$$

where $\bar{x}_1$ is the sample mean of the x_{j1}'s. In general, for p variables, we have

$$s_k^2 = \frac{1}{n} \sum_{j=1}^{n} (x_{jk} - \bar{x}_k)^2 \qquad k = 1, 2, \ldots, p \tag{1-2}$$

Two comments are in order. First, many authors define the sample variance with a divisor of $n - 1$ rather than n. Later we shall see that there are theoretical reasons for doing this, and it is particularly appropriate if the number of measurements, n, is small. The two versions of the sample variance will always be differentiated by displaying the appropriate expression.

Second, although the s^2 notation is traditionally used to indicate the sample variance, we shall eventually consider an array of quantities in which the sample variances lie along the main diagonal. In this situation, it is convenient to use double subscripts on the variances in order to indicate their positions in the array. Therefore, we introduce the notation s_{kk} to denote the same variance computed from measurements on the kth variable, and we have the notational identities

$$s_k^2 = s_{kk} = \frac{1}{n} \sum_{j=1}^{n} (x_{jk} - \bar{x}_k)^2 \qquad k = 1, 2, \ldots, p \tag{1-3}$$

The square root of the sample variance, $\sqrt{s_{kk}}$, is known as the *sample standard deviation*. This measure of variation uses the same units as the observations.

Consider n pairs of measurements on each of variables 1 and 2:

$$\begin{bmatrix} x_{11} \\ x_{12} \end{bmatrix}, \begin{bmatrix} x_{21} \\ x_{22} \end{bmatrix}, \ldots, \begin{bmatrix} x_{n1} \\ x_{n2} \end{bmatrix}$$

That is, x_{j1} and x_{j2} are observed on the jth experimental item ($j = 1, 2, \ldots, n$). A measure of linear association between the measurements of variables 1 and 2 is provided by the *sample covariance*

$$s_{12} = \frac{1}{n} \sum_{j=1}^{n} (x_{j1} - \bar{x}_1)(x_{j2} - \bar{x}_2)$$

or the average product of the deviations from their respective means. If large values for one variable are observed in conjunction with large values for the other variable, and the small values also occur together, s_{12} will be positive. If large values from one variable occur with small values for the other variable, s_{12} will be negative. If there is no particular association between the values for the two variables, s_{12} will be approximately zero.

The *sample covariance*

$$s_{ik} = \frac{1}{n} \sum_{j=1}^{n} (x_{ji} - \bar{x}_i)(x_{jk} - \bar{x}_k) \qquad i = 1, 2, \ldots, p, \quad k = 1, 2, \ldots, p \quad (1\text{-}4)$$

measures the association between the ith and kth variables. We note that the covariance reduces to the sample variance when $i = k$. Moreover, $s_{ik} = s_{ki}$ for all i and k.

The final descriptive statistic considered here is the *sample correlation coefficient* (or *Pearson's product-moment correlation coefficient*; see [14]). This measure of the linear association between two variables does not depend on the units of measurement. The sample correlation coefficient for the ith and kth variables is defined as

most useful

$$r_{ik} = \frac{s_{ik}}{\sqrt{s_{ii}}\ \sqrt{s_{kk}}} = \frac{\displaystyle\sum_{j=1}^{n} (x_{ji} - \bar{x}_i)(x_{jk} - \bar{x}_k)}{\sqrt{\displaystyle\sum_{j=1}^{n} (x_{ji} - \bar{x}_i)^2}\ \sqrt{\displaystyle\sum_{j=1}^{n} (x_{jk} - \bar{x}_k)^2}} \qquad (1\text{-}5)$$

for $i = 1, 2, \ldots, p$ and $k = 1, 2, \ldots, p$. Note $r_{ik} = r_{ki}$ for all i and k.

The sample correlation coefficient is a standardized version of the sample covariance, where the product of the square roots of the sample variances provides the standardization. Notice that r_{ik} has the same value whether n or $n - 1$ is chosen as the common divisor for s_{ii}, s_{kk}, and s_{ik}.

The sample correlation coefficient r_{ik} can also be viewed as a sample *covariance*. Suppose the original values x_{ji} and x_{jk} are replaced by *standardized* values $(x_{ji} - \bar{x}_i)/\sqrt{s_{ii}}$ and $(x_{jk} - \bar{x}_k)/\sqrt{s_{kk}}$. The standardized values are commensurable because both sets are centered at zero and expressed in standard deviation units. The sample correlation coefficient is just the sample covariance of the standardized observations.

Although the signs of the sample correlation and the sample covariance are the same, the correlation is ordinarily easier to interpret because its magnitude is bounded. To summarize, the sample correlation r has the following properties:

1. The value of r must be between -1 and $+1$ inclusive.

2. Here r measures the strength of the linear association. If $r = 0$, this implies a lack of linear association between the components. Otherwise, the sign of r indicates the direction of the association: $r < 0$ implies a tendency for one value in the pair to be larger than its average when the other is smaller than its average; and $r > 0$ implies a tendency for one value of the pair to be large when the other value is large and also for both values to be small together.

3. The value of r_{ik} remains unchanged if the measurements of the ith variable are changed to $y_{ji} = ax_{ji} + b, j = 1, 2, \ldots, n$, and the values of the kth variable are changed to $y_{jk} = cx_{jk} + d, j = 1, 2, \ldots, n$, provided that the constants a and c have the same sign.

The quantities s_{ik} and r_{ik} do not, in general, convey all there is to know about the association between two variables. Nonlinear associations can exist that are not revealed by these descriptive statistics. Covariance and correlation provide measures of *linear* association, or association along a line. Their values are less informative for other kinds of association. On the other hand, these quantities can be very sensitive to "wild" observations ("outliers") and may indicate association when, in fact, little exists. In spite of these shortcomings, covariance and correlation coefficients are routinely calculated and analyzed. They provide cogent numerical summaries of association when the data do not exhibit obvious nonlinear patterns of association and when wild observations are not present.

Suspect observations must be accounted for by correcting obvious recording mistakes and by taking actions consistent with the identified causes. The values of s_{ik} and r_{ik} should be quoted both with and without these observations.

The sum of squares of the deviations from the mean and the sum of cross-product deviations are often of interest themselves. These quantities are

$$w_{kk} = \sum_{j=1}^{n} (x_{jk} - \bar{x}_k)^2 \qquad k = 1, 2, \ldots, p \tag{1-6}$$

and

$$w_{ik} = \sum_{j=1}^{n} (x_{ji} - \bar{x}_i)(x_{jk} - \bar{x}_k) \qquad i = 1, 2, \ldots, p, \quad k = 1, 2, \ldots, p \tag{1-7}$$

The descriptive statistics computed from n measurements on p variables can also be organized into arrays.

Arrays of Basic Descriptive Statistics

Sample means

$$\bar{\mathbf{x}} = \begin{bmatrix} \bar{x}_1 \\ \bar{x}_2 \\ \vdots \\ \bar{x}_p \end{bmatrix}$$

Sample variances and covariances

$$\mathbf{S}_n = \begin{bmatrix} s_{11} & s_{12} & \cdots & s_{1p} \\ s_{21} & s_{22} & \cdots & s_{2p} \\ \vdots & \vdots & \ddots & \vdots \\ s_{p1} & s_{p2} & \cdots & s_{pp} \end{bmatrix} \tag{1-8}$$

Sample correlations

$$\mathbf{R} = \begin{bmatrix} 1 & r_{12} & \cdots & r_{1p} \\ r_{21} & 1 & \cdots & r_{2p} \\ \vdots & \vdots & \ddots & \vdots \\ r_{p1} & r_{p2} & \cdots & 1 \end{bmatrix}$$

The sample mean array is denoted by $\bar{\mathbf{x}}$, the sample variance and covariance array by the capital letter $\mathbf{S}_n$, and the sample correlation array by $\mathbf{R}$. The subscript n on the array $\mathbf{S}_n$ is a mnemonic device used to remind you that n is employed as a divisor for the elements s_{ik}. The size of all of the arrays is determined by the number of variables, p.

The arrays $\mathbf{S}_n$ and $\mathbf{R}$ consist of p rows and p columns. The array $\bar{\mathbf{x}}$ is a single column with p rows. The first subscript on an entry in arrays $\mathbf{S}_n$ and $\mathbf{R}$ indicates the row; the second subscript indicates the column. Since $s_{ik} = s_{ki}$ and $r_{ik} = r_{ki}$ for all i and k, the entries in symmetric positions about the main northwest–southeast diagonals in arrays $\mathbf{S}_n$ and $\mathbf{R}$ are the same, and the arrays are said to be *symmetric*.

Example 1.2 (The arrays $\bar{\mathbf{x}}$, $\mathbf{S}_n$, and R for bivariate data) Consider the data introduced in Example 1.1. Each receipt yields a pair of measurements, total dollar sales, and number of books sold. Find the arrays $\bar{\mathbf{x}}$, $\mathbf{S}_n$, and $\mathbf{R}$.

Since there are four receipts, we have a total of four measurements (observations) on each variable.

The sample means are

$$\bar{x}_1 = \frac{1}{4} \sum_{j=1}^{4} x_{j1} = \frac{1}{4}(42 + 52 + 48 + 58) = 50$$

$$\bar{x}_2 = \frac{1}{4} \sum_{j=1}^{4} x_{j2} = \frac{1}{4}(4 + 5 + 4 + 3) = 4$$

$$\bar{\mathbf{x}} = \begin{bmatrix} \bar{x}_1 \\ \bar{x}_2 \end{bmatrix} = \begin{bmatrix} 50 \\ 4 \end{bmatrix}$$

The sample variances and covariances are

$$s_{11} = \frac{1}{4} \sum_{j=1}^{4} (x_{j1} - \bar{x}_1)^2$$

$$= \frac{1}{4}((42 - 50)^2 + (52 - 50)^2 + (48 - 50)^2 + (58 - 50)^2) = 34$$

$$s_{22} = \frac{1}{4} \sum_{j=1}^{4} (x_{j2} - \bar{x}_2)^2$$

$$= \frac{1}{4}((4 - 4)^2 + (5 - 4)^2 + (4 - 4)^2 + (3 - 4)^2) = .5$$

$$s_{12} = \frac{1}{4} \sum_{j=1}^{4} (x_{j1} - \bar{x}_1)(x_{j2} - \bar{x}_2)$$

$$= \frac{1}{4}((42 - 50)(4 - 4) + (52 - 50)(5 - 4)$$

$$+ (48 - 50)(4 - 4) + (58 - 50)(3 - 4)) = -1.5$$

$$s_{21} = s_{12}$$

and

$$\mathbf{S}_n = \begin{bmatrix} 34 & -1.5 \\ -1.5 & .5 \end{bmatrix}$$

The sample correlation is

$$r_{12} = \frac{s_{12}}{\sqrt{s_{11}} \sqrt{s_{22}}} = \frac{-1.5}{\sqrt{34} \sqrt{.5}} = -.36$$

$$r_{21} = r_{12}$$

so

$$\mathbf{R} = \begin{bmatrix} 1 & -.36 \\ -.36 & 1 \end{bmatrix}$$ ∎

Graphical Techniques

Plots are important, but frequently neglected, aids in data analysis. Although it is impossible to simultaneously plot *all* the measurements made on several variables and study the configurations, plots of individual variables and plots of pairs of variables can still be very informative. Sophisticated computer programs and display equipment allow one the luxury of visually examining data in one, two, or three dimensions with relative ease. On the other hand, many valuable insights can be obtained from the data by constructing plots with paper and pencil. Simple, yet elegant and effective, methods for displaying data are available in [29]. It is good statistical practice to plot pairs of variables and visually inspect the pattern of association. Consider, then, the following seven pairs of measurements on two variables:

Variable 1 (x_1):	3	4	2	6	8	2	5
Variable 2 (x_2):	5	5.5	4	7	10	5	7.5

These data are plotted as seven points in two dimensions (each axis representing a variable) in Figure 1.1. The coordinates of the points are determined by the paired measurements: $(3, 5), (4, 5.5), \ldots, (5, 7.5)$. The resulting two-dimensional plot is known as a *scatter diagram* or *scatter plot*.

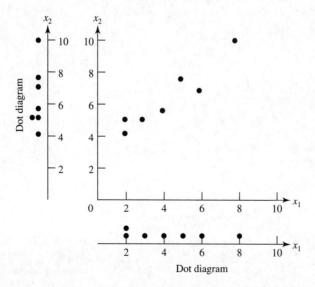

Figure 1.1 A scatter plot and marginal dot diagrams.

Also shown in Figure 1.1 are separate plots of the observed values of variable 1 and the observed values of variable 2, respectively. These plots are called (*marginal*) *dot diagrams*. They can be obtained from the original observations or by projecting the points in the scatter diagram onto each coordinate axis.

The information contained in the single-variable dot diagrams can be used to calculate the sample means $\bar{x}_1$ and $\bar{x}_2$ and the sample variances s_{11} and s_{22}. (See Exercise 1.1.) The scatter diagram indicates the orientation of the points, and their coordinates can be used to calculate the sample covariance s_{12}. In the scatter diagram of Figure 1.1, large values of x_1 occur with large values of x_2 and small values of x_1 with small values of x_2. Hence, s_{12} will be positive.

Dot diagrams and scatter plots contain different kinds of information. The information in the marginal dot diagrams is not sufficient for constructing the scatter plot. As an illustration, suppose the data preceding Figure 1.1 had been paired differently, so that the measurements on the variables x_1 and x_2 were as follows:

Variable 1 (x_1):	5	4	6	2	2	8	3
Variable 2 (x_2):	5	5.5	4	7	10	5	7.5

(We have simply rearranged the values of variable 1.) The scatter and dot diagrams for the "new" data are shown in Figure 1.2. Comparing Figures 1.1 and 1.2, we find that the marginal dot diagrams are the same, but that the scatter diagrams are decidedly different. In Figure 1.2, large values of x_1 are paired with small values of x_2 and small values of x_1 with large values of x_2. Consequently, the descriptive statistics for the individual variables $\bar{x}_1, \bar{x}_2, s_{11}$, and s_{22} remain unchanged, but the sample covariance s_{12}, which measures the association between pairs of variables, will now be negative.

The different orientations of the data in Figures 1.1 and 1.2 are not discernible from the marginal dot diagrams alone. At the same time, the fact that the marginal dot diagrams are the same in the two cases is not immediately apparent from the scatter plots. The two types of graphical procedures complement one another; they are not competitors.

The next two examples further illustrate the information that can be conveyed by a graphic display.

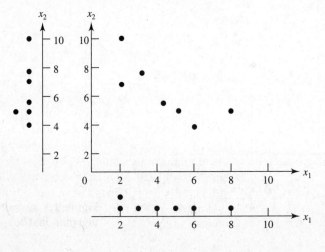

Figure 1.2 Scatter plot and dot diagrams for rearranged data.

Example 1.3 (The effect of unusual observations on sample correlations) Some financial data representing jobs and productivity for the 16 largest publishing firms appeared in an article in *Forbes* magazine on April 30, 1990. The data for the pair of variables x_1 = employees (jobs) and x_2 = profits per employee (productivity) are graphed in Figure 1.3. We have labeled two "unusual" observations. Dun & Bradstreet is the largest firm in terms of number of employees, but is "typical" in terms of profits per employee. Time Warner has a "typical" number of employees, but comparatively small (negative) profits per employee.

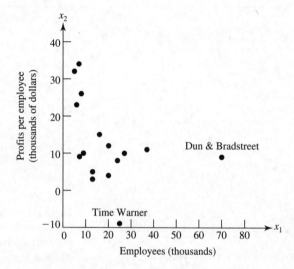

Figure 1.3 Profits per employee and number of employees for 16 publishing firms.

The sample correlation coefficient computed from the values of x_1 and x_2 is

$$r_{12} = \begin{cases} -.39 & \text{for all 16 firms} \\ -.56 & \text{for all firms but Dun \& Bradstreet} \\ -.39 & \text{for all firms but Time Warner} \\ -.50 & \text{for all firms but Dun \& Bradstreet and Time Warner} \end{cases}$$

It is clear that atypical observations can have a considerable effect on the sample correlation coefficient. ∎

Example 1.4 (A scatter plot for baseball data) In a July 17, 1978, article on money in sports, *Sports Illustrated* magazine provided data on x_1 = player payroll for National League East baseball teams.

We have added data on x_2 = won–lost percentage for 1977. The results are given in Table 1.1.

The scatter plot in Figure 1.4 supports the claim that a championship team can be bought. Of course, this cause–effect relationship cannot be substantiated, because the experiment did not include a random assignment of payrolls. Thus, statistics cannot answer the question: Could the Mets have won with $4 million to spend on player salaries?

Table 1.1 1977 Salary and Final Record for the National League East

Team	x_1 = player payroll	x_2 = won–lost percentage
Philadelphia Phillies	3,497,900	.623
Pittsburgh Pirates	2,485,475	.593
St. Louis Cardinals	1,782,875	.512
Chicago Cubs	1,725,450	.500
Montreal Expos	1,645,575	.463
New York Mets	1,469,800	.395

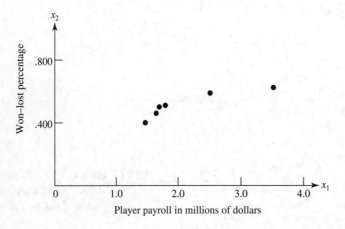

Figure 1.4 Salaries and won–lost percentage from Table 1.1.

To construct the scatter plot in Figure 1.4, we have regarded the six paired observations in Table 1.1 as the coordinates of six points in two-dimensional space. The figure allows us to examine visually the grouping of teams with respect to the variables total payroll and won–lost percentage. ∎

Example 1.5 (Multiple scatter plots for paper strength measurements) Paper is manufactured in continuous sheets several feet wide. Because of the orientation of fibers within the paper, it has a different strength when measured in the direction produced by the machine than when measured across, or at right angles to, the machine direction. Table 1.2 shows the measured values of

$$x_1 = \text{density (grams/cubic centimeter)}$$

$$x_2 = \text{strength (pounds) in the machine direction}$$

$$x_3 = \text{strength (pounds) in the cross direction}$$

A novel graphic presentation of these data appears in Figure 1.5, page 16. The scatter plots are arranged as the off-diagonal elements of a covariance array and box plots as the diagonal elements. The latter are on a different scale with this

Table 1.2 Paper-Quality Measurements

Specimen	Density	Strength	
		Machine direction	Cross direction
1	.801	121.41	70.42
2	.824	127.70	72.47
3	.841	129.20	78.20
4	.816	131.80	74.89
5	.840	135.10	71.21
6	.842	131.50	78.39
7	.820	126.70	69.02
8	.802	115.10	73.10
9	.828	130.80	79.28
10	.819	124.60	76.48
11	.826	118.31	70.25
12	.802	114.20	72.88
13	.810	120.30	68.23
14	.802	115.70	68.12
15	.832	117.51	71.62
16	.796	109.81	53.10
17	.759	109.10	50.85
18	.770	115.10	51.68
19	.759	118.31	50.60
20	.772	112.60	53.51
21	.806	116.20	56.53
22	.803	118.00	70.70
23	.845	131.00	74.35
24	.822	125.70	68.29
25	.971	126.10	72.10
26	.816	125.80	70.64
27	.836	125.50	76.33
28	.815	127.80	76.75
29	.822	130.50	80.33
30	.822	127.90	75.68
31	.843	123.90	78.54
32	.824	124.10	71.91
33	.788	120.80	68.22
34	.782	107.40	54.42
35	.795	120.70	70.41
36	.805	121.91	73.68
37	.836	122.31	74.93
38	.788	110.60	53.52
39	.772	103.51	48.93
40	.776	110.71	53.67
41	.758	113.80	52.42

Source: Data courtesy of SONOCO Products Company.

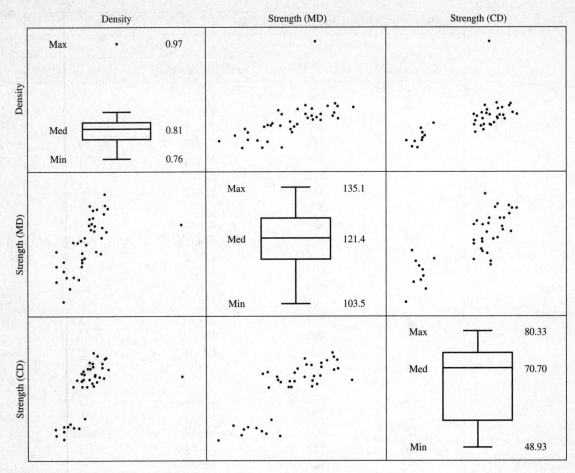

Figure 1.5 Scatter plots and boxplots of paper-quality data from Table 1.2.

software, so we use only the overall shape to provide information on symmetry and possible outliers for each individual characteristic. The scatter plots can be inspected for patterns and unusual observations. In Figure 1.5, there is one unusual observation: the density of specimen 25. Some of the scatter plots have patterns suggesting that there are two separate clumps of observations.

These scatter plot arrays are further pursued in our discussion of new software graphics in the next section. ∎

In the general multiresponse situation, p variables are simultaneously recorded on n items. Scatter plots should be made for pairs of important variables and, if the task is not too great to warrant the effort, for all pairs.

Limited as we are to a three-dimensional world, we cannot always picture an entire set of data. However, two further geometric representations of the data provide an important conceptual framework for viewing multivariable statistical methods. In cases where it is possible to capture the essence of the data in three dimensions, these representations can actually be graphed.

n Points in p Dimensions (p-Dimensional Scatter Plot). Consider the natural extension of the scatter plot to p dimensions, where the p measurements

$$(x_{j1}, x_{j2}, \ldots, x_{jp})$$

on the jth item represent the coordinates of a point in p-dimensional space. The coordinate axes are taken to correspond to the variables, so that the jth point is x_{j1}
units along the first axis, x_{j2} units along the second, ..., x_{jp} units along the pth axis.
The resulting plot with n points not only will exhibit the overall pattern of variability, but also will show similarities (and differences) among the n items. Groupings of
items will manifest themselves in this representation.

The next example illustrates a three-dimensional scatter plot.

Example 1.6 (Looking for lower-dimensional structure) A zoologist obtained measurements on $n = 25$ lizards known scientifically as *Cophosaurus texanus*. The
weight, or mass, is given in grams while the snout-vent length (SVL) and hind limb
span (HLS) are given in millimeters. The data are displayed in Table 1.3.

Although there are three size measurements, we can ask whether or not most of
the variation is primarily restricted to two dimensions or even to one dimension.

To help answer questions regarding reduced dimensionality, we construct the
three-dimensional scatter plot in Figure 1.6. Clearly most of the variation is scatter
about a one-dimensional straight line. Knowing the position on a line along the
major axes of the cloud of points would be almost as good as knowing the three
measurements Mass, SVL, and HLS.

However, this kind of analysis can be misleading if one variable has a much
larger variance than the others. Consequently, we first calculate the standardized
values, $z_{jk} = (x_{jk} - \bar{x}_k)/\sqrt{s_{kk}}$, so the variables contribute equally to the variation

Table 1.3 Lizard Size Data

Lizard	Mass	SVL	HLS	Lizard	Mass	SVL	HLS
1	5.526	59.0	113.5	14	10.067	73.0	136.5
2	10.401	75.0	142.0	15	10.091	73.0	135.5
3	9.213	69.0	124.0	16	10.888	77.0	139.0
4	8.953	67.5	125.0	17	7.610	61.5	118.0
5	7.063	62.0	129.5	18	7.733	66.5	133.5
6	6.610	62.0	123.0	19	12.015	79.5	150.0
7	11.273	74.0	140.0	20	10.049	74.0	137.0
8	2.447	47.0	97.0	21	5.149	59.5	116.0
9	15.493	86.5	162.0	22	9.158	68.0	123.0
10	9.004	69.0	126.5	23	12.132	75.0	141.0
11	8.199	70.5	136.0	24	6.978	66.5	117.0
12	6.601	64.5	116.0	25	6.890	63.0	117.0
13	7.622	67.5	135.0				

Source: Data courtesy of Kevin E. Bonine.

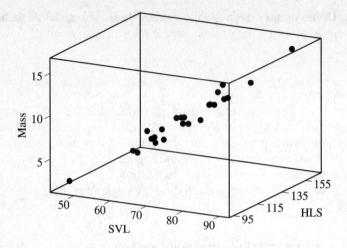

Figure 1.6 3D scatter plot of lizard data from Table 1.3.

in the scatter plot. Figure 1.7 gives the three-dimensional scatter plot for the standardized variables. Most of the variation can be explained by a single variable determined by a line through the cloud of points.

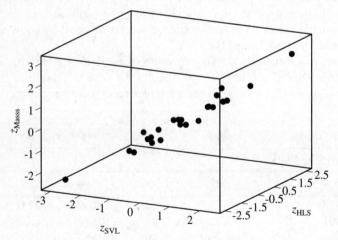

Figure 1.7 3D scatter plot of standardized lizard data.   ∎

A three-dimensional scatter plot can often reveal group structure.

Example 1.7 (Looking for group structure in three dimensions) Referring to Example 1.6, it is interesting to see if male and female lizards occupy different parts of the three-dimensional space containing the size data. The gender, by row, for the lizard data in Table 1.3 are

f m f f m f m f m f m f m

m m m f m m m f f m f f

Figure 1.8 repeats the scatter plot for the original variables but with males marked by solid circles and females by open circles. Clearly, males are typically larger than females.

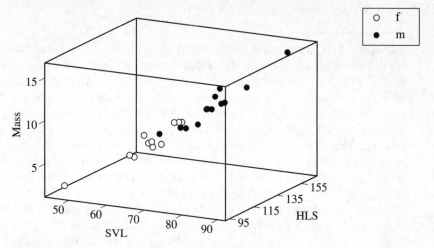

Figure 1.8 3D scatter plot of male and female lizards.

***p* Points in *n* Dimensions.**   The *n* observations of the *p* variables can also be regarded as *p* points in *n*-dimensional space. Each column of **X** determines one of the points. The *i*th column,

$$
\begin{bmatrix}
x_{1i} \\
x_{2i} \\
\vdots \\
x_{ni}
\end{bmatrix}
$$

consisting of all *n* measurements on the *i*th variable, determines the *i*th point.

In Chapter 3, we show how the closeness of points in *n* dimensions can be related to measures of association between the corresponding *variables*.

1.4 Data Displays and Pictorial Representations

The rapid development of powerful personal computers and workstations has led to a proliferation of sophisticated statistical software for data analysis and graphics. It is often possible, for example, to sit at one's desk and examine the nature of multidimensional data with clever computer-generated pictures. These pictures are valuable aids in understanding data and often prevent many false starts and subsequent inferential problems.

As we shall see in Chapters 8 and 12, there are several techniques that seek to represent *p*-dimensional observations in few dimensions such that the original distances (or similarities) between pairs of observations are (nearly) preserved. In general, if multidimensional observations can be represented in two dimensions, then outliers, relationships, and distinguishable groupings can often be discerned by eye. We shall discuss and illustrate several methods for displaying multivariate data in two dimensions. One good source for more discussion of graphical methods is [11].

Linking Multiple Two-Dimensional Scatter Plots

One of the more exciting new graphical procedures involves electronically connecting many two-dimensional scatter plots.

Example 1.8 (Linked scatter plots and brushing) To illustrate *linked* two-dimensional scatter plots, we refer to the paper-quality data in Table 1.2. These data represent measurements on the variables x_1 = density, x_2 = strength in the machine direction, and x_3 = strength in the cross direction. Figure 1.9 shows two-dimensional scatter plots for pairs of these variables organized as a 3 × 3 array. For example, the picture in the upper left-hand corner of the figure is a scatter plot of the pairs of observations (x_1, x_3). That is, the x_1 values are plotted along the horizontal axis, and the x_3 values are plotted along the vertical axis. The lower right-hand corner of the figure contains a scatter plot of the observations (x_3, x_1). That is, the axes are reversed. Corresponding interpretations hold for the other scatter plots in the figure. Notice that the variables and their three-digit ranges are indicated in the boxes along the SW–NE diagonal. The operation of marking (*selecting*), the obvious outlier in the (x_1, x_3) scatter plot of Figure 1.9 creates Figure 1.10(a), where the outlier is labeled as specimen 25 and the same data point is highlighted in all the scatter plots. Specimen 25 also appears to be an outlier in the (x_1, x_2) scatter plot but not in the (x_2, x_3) scatter plot. The operation of *deleting* this specimen leads to the modified scatter plots of Figure 1.10(b).

From Figure 1.10, we notice that some points in, for example, the (x_2, x_3) scatter plot seem to be disconnected from the others. Selecting these points, using the (dashed) rectangle (see page 22), highlights the selected points in all of the other scatter plots and leads to the display in Figure 1.11(a). Further checking revealed that specimens 16–21, specimen 34, and specimens 38–41 were actually specimens

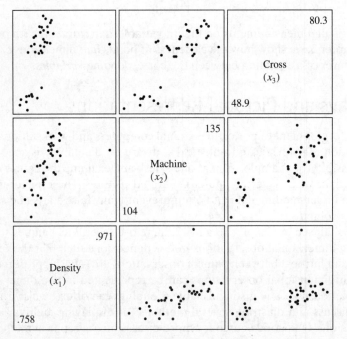

Figure 1.9 Scatter plots for the paper-quality data of Table 1.2.

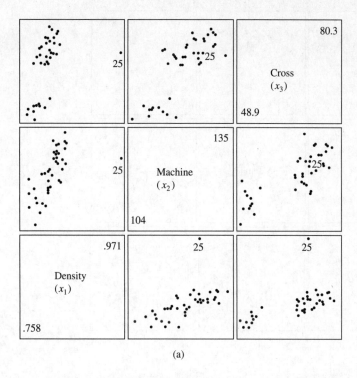

(a)

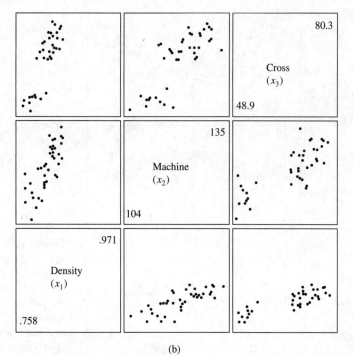

(b)

Figure 1.10 Modified scatter plots for the paper-quality data with outlier (25) (a) selected and (b) deleted.

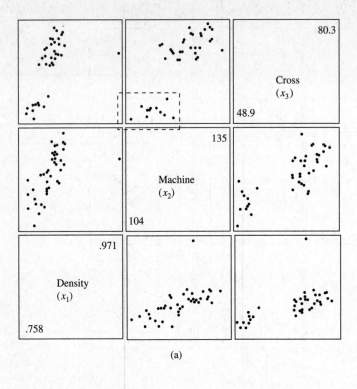

(a)

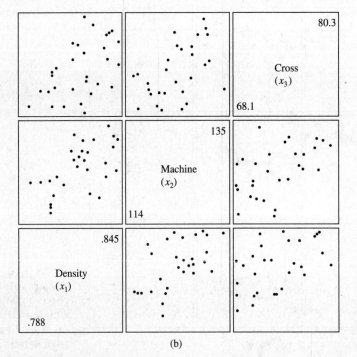

(b)

Figure 1.11 Modified scatter plots with (a) group of points selected and (b) points, including specimen 25, deleted and the scatter plots rescaled.

from an older roll of paper that was included in order to have enough plies in the cardboard being manufactured. Deleting the outlier and the cases corresponding to the older paper and adjusting the ranges of the remaining observations leads to the scatter plots in Figure 1.11(b).

The operation of highlighting points corresponding to a selected range of one of the variables is called *brushing*. Brushing could begin with a rectangle, as in Figure 1.11(a), but then the brush could be moved to provide a sequence of highlighted points. The process can be stopped at any time to provide a snapshot of the current situation. ∎

Scatter plots like those in Example 1.8 are extremely useful aids in data analysis. Another important new graphical technique uses software that allows the data analyst to view high-dimensional data as slices of various three-dimensional perspectives. This can be done dynamically and continuously until informative views are obtained. A comprehensive discussion of dynamic graphical methods is available in [1]. A strategy for on-line multivariate exploratory graphical analysis, motivated by the need for a routine procedure for searching for structure in multivariate data, is given in [32].

Example 1.9 (Rotated plots in three dimensions) Four different measurements of lumber stiffness are given in Table 4.3, page 186. In Example 4.14, specimen (board) 16 and possibly specimen (board) 9 are identified as unusual observations. Figures 1.12(a), (b), and (c) contain perspectives of the stiffness data in the x_1, x_2, x_3 space. These views were obtained by continually rotating and turning the three-dimensional coordinate axes. Spinning the coordinate axes allows one to get a better

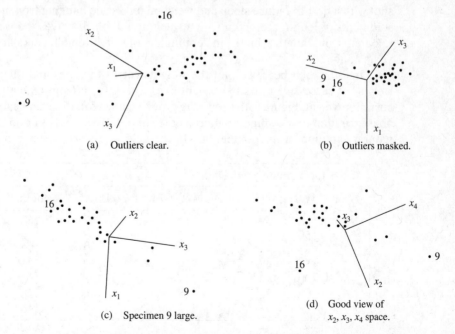

(a)   Outliers clear.

(b)   Outliers masked.

(c)   Specimen 9 large.

(d)   Good view of x_2, x_3, x_4 space.

Figure 1.12 Three-dimensional perspectives for the lumber stiffness data.

understanding of the three-dimensional aspects of the data. Figure 1.12(d) gives one picture of the stiffness data in x_2, x_3, x_4 space. Notice that Figures 1.12(a) and (d) visually confirm specimens 9 and 16 as outliers. Specimen 9 is very large in all three coordinates. A counterclockwiselike rotation of the axes in Figure 1.12(a) produces Figure 1.12(b), and the two unusual observations are masked in this view. A further spinning of the x_2, x_3 axes gives Figure 1.12(c); one of the outliers (16) is now hidden.

Additional insights can sometimes be gleaned from visual inspection of the slowly spinning data. It is this dynamic aspect that statisticians are just beginning to understand and exploit.   ∎

Plots like those in Figure 1.12 allow one to identify readily observations that do not conform to the rest of the data and that may heavily influence inferences based on standard data-generating models.

Graphs of Growth Curves

When the height of a young child is measured at each birthday, the points can be plotted and then connected by lines to produce a graph. This is an example of a *growth curve*. In general, repeated measurements of the same characteristic on the same unit or subject can give rise to a growth curve if an increasing, decreasing, or even an increasing followed by a decreasing, pattern is expected.

Example 1.10 (Arrays of growth curves) The Alaska Fish and Game Department monitors grizzly bears with the goal of maintaining a healthy population. Bears are shot with a dart to induce sleep and weighed on a scale hanging from a tripod. Measurements of length are taken with a steel tape. Table 1.4 gives the weights (wt) in kilograms and lengths (lngth) in centimeters of seven female bears at 2, 3, 4, and 5 years of age.

First, for each bear, we plot the weights versus the ages and then connect the weights at successive years by straight lines. This gives an approximation to growth curve for weight. Figure 1.13 shows the growth curves for all seven bears. The noticeable exception to a common pattern is the curve for bear 5. Is this an outlier or just natural variation in the population? In the field, bears are weighed on a scale that

Bear	Wt2	Wt3	Wt4	Wt5	Lngth 2	Lngth 3	Lngth 4	Lngth 5
1	48	59	95	82	141	157	168	183
2	59	68	102	102	140	168	174	170
3	61	77	93	107	145	162	172	177
4	54	43	104	104	146	159	176	171
5	100	145	185	247	150	158	168	175
6	68	82	95	118	142	140	178	189
7	68	95	109	111	139	171	176	175

Table 1.4 Female Bear Data

Source: Data courtesy of H. Roberts.

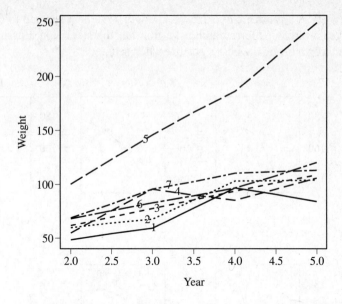

Figure 1.13 Combined growth curves for weight for seven female grizzly bears.

reads pounds. Further inspection revealed that, in this case, an assistant later failed to convert the field readings to kilograms when creating the electronic database. The correct weights are (45, 66, 84, 112) kilograms.

Because it can be difficult to inspect visually the individual growth curves in a combined plot, the individual curves should be replotted in an array where similarities and differences are easily observed. Figure 1.14 gives the array of seven curves for weight. Some growth curves look linear and others quadratic.

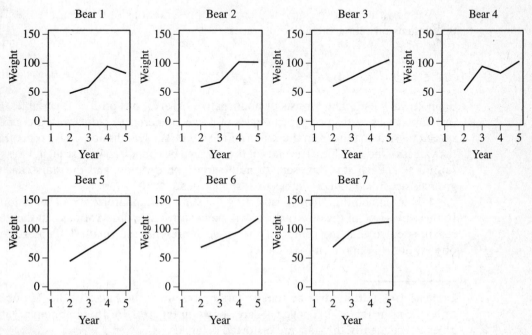

Figure 1.14 Individual growth curves for weight for female grizzly bears.

Figure 1.15 gives a growth curve array for length. One bear seemed to get shorter from 2 to 3 years old, but the researcher knows that the steel tape measurement of length can be thrown off by the bear's posture when sedated.

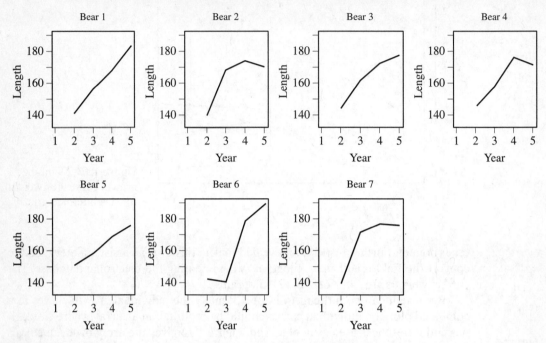

Figure 1.15 Individual growth curves for length for female grizzly bears.

We now turn to two popular pictorial representations of multivariate data in two dimensions: stars and Chernoff faces.

Stars

Suppose each data unit consists of nonnegative observations on $p \geq 2$ variables. In two dimensions, we can construct circles of a fixed (reference) radius with p equally spaced rays emanating from the center of the circle. The lengths of the rays represent the values of the variables. The ends of the rays can be connected with straight lines to form a star. Each star represents a multivariate observation, and the stars can be grouped according to their (subjective) similarities.

It is often helpful, when constructing the stars, to standardize the observations. In this case some of the observations will be negative. The observations can then be reexpressed so that the center of the circle represents the smallest standardized observation within the entire data set.

Example 1.11 (Utility data as stars) Stars representing the first 5 of the 22 public utility firms in Table 12.4, page 688, are shown in Figure 1.16. There are eight variables; consequently, the stars are distorted octagons.

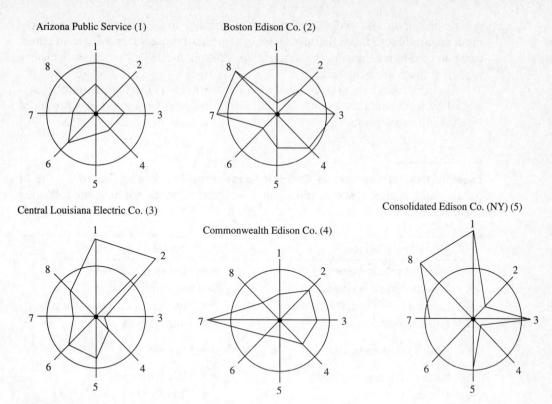

Figure 1.16 Stars for the first five public utilities.

The observations on all variables were standardized. Among the first five utilities, the smallest standardized observation for any variable was −1.6. Treating this value as zero, the variables are plotted on identical scales along eight equiangular rays originating from the center of the circle. The variables are ordered in a clockwise direction, beginning in the 12 o'clock position.

At first glance, none of these utilities appears to be similar to any other. However, because of the way the stars are constructed, each variable gets equal weight in the visual impression. If we concentrate on the variables 6 (sales in kilowatt-hour [kWh] use per year) and 8 (total fuel costs in cents per kWh), then Boston Edison and Consolidated Edison are similar (small variable 6, large variable 8), and Arizona Public Service, Central Louisiana Electric, and Commonwealth Edison are similar (moderate variable 6, moderate variable 8). ∎

Chernoff Faces

People react to faces. Chernoff [4] suggested representing p-dimensional observations as a two-dimensional face whose characteristics (face shape, mouth curvature, nose length, eye size, pupil position, and so forth) are determined by the measurements on the p variables.

As originally designed, Chernoff faces can handle up to 18 variables. The assignment of variables to facial features is done by the experimenter, and different choices produce different results. Some iteration is usually necessary before satisfactory representations are achieved.

Chernoff faces appear to be most useful for verifying (1) an initial grouping suggested by subject-matter knowledge and intuition or (2) final groupings produced by clustering algorithms.

Example 1.12 (Utility data as Chernoff faces) From the data in Table 12.4, the 22 public utility companies were represented as Chernoff faces. We have the following correspondences:

Variable	Facial characteristic
X_1: Fixed-charge coverage	↔ Half-height of face
X_2: Rate of return on capital	↔ Face width
X_3: Cost per kW capacity in place	↔ Position of center of mouth
X_4: Annual load factor	↔ Slant of eyes
X_5: Peak kWh demand growth from 1974	↔ Eccentricity $\left(\dfrac{\text{height}}{\text{width}}\right)$ of eyes
X_6: Sales (kWh use per year)	↔ Half-length of eye
X_7: Percent nuclear	↔ Curvature of mouth
X_8: Total fuel costs (cents per kWh)	↔ Length of nose

The Chernoff faces are shown in Figure 1.17. We have subjectively grouped "similar" faces into seven clusters. If a smaller number of clusters is desired, we might combine clusters 5, 6, and 7 and, perhaps, clusters 2 and 3 to obtain four or five clusters. For our assignment of variables to facial features, the firms group largely according to geographical location. ∎

Constructing Chernoff faces is a task that must be done with the aid of a computer. The data are ordinarily standardized within the computer program as part of the process for determining the locations, sizes, and orientations of the facial characteristics. With some training, we can use Chernoff faces to communicate similarities or dissimilarities, as the next example indicates.

Example 1.13 (Using Chernoff faces to show changes over time) Figure 1.18 illustrates an additional use of Chernoff faces. (See [24].) In the figure, the faces are used to track the financial well-being of a company over time. As indicated, each facial feature represents a single financial indicator, and the longitudinal changes in these indicators are thus evident at a glance. ∎

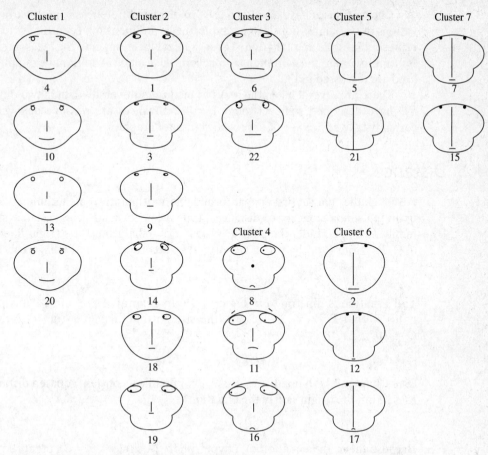

Figure 1.17 Chernoff faces for 22 public utilities.

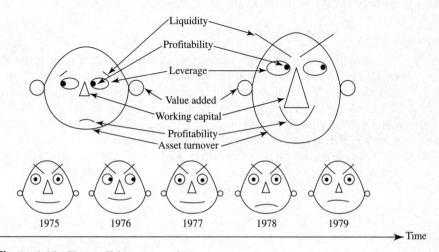

Figure 1.18 Chernoff faces over time.

Chernoff faces have also been used to display differences in multivariate observations in two dimensions. For example, the two-dimensional coordinate axes might represent latitude and longitude (geographical location), and the faces might represent multivariate measurements on several U.S. cities. Additional examples of this kind are discussed in [30].

There are several ingenious ways to picture multivariate data in two dimensions. We have described some of them. Further advances are possible and will almost certainly take advantage of improved computer graphics.

1.5 Distance

Although they may at first appear formidable, most multivariate techniques are based upon the simple concept of distance. Straight-line, or Euclidean, distance should be familiar. If we consider the point $P = (x_1, x_2)$ in the plane, the straight-line distance, $d(O, P)$, from P to the origin $O = (0, 0)$ is, according to the Pythagorean theorem,

$$d(O, P) = \sqrt{x_1^2 + x_2^2} \tag{1-9}$$

The situation is illustrated in Figure 1.19. In general, if the point P has p coordinates so that $P = (x_1, x_2, \ldots, x_p)$, the straight-line distance from P to the origin $O = (0, 0, \ldots, 0)$ is

$$d(O, P) = \sqrt{x_1^2 + x_2^2 + \cdots + x_p^2} \tag{1-10}$$

(See Chapter 2.) All points $(x_1, x_2, \ldots, x_p)$ that lie a constant squared distance, such as c^2, from the origin satisfy the equation

$$d^2(O, P) = x_1^2 + x_2^2 + \cdots + x_p^2 = c^2 \tag{1-11}$$

Because this is the equation of a hypersphere (a circle if $p = 2$), points equidistant from the origin lie on a hypersphere.

The straight-line distance between two arbitrary points P and Q with coordinates $P = (x_1, x_2, \ldots, x_p)$ and $Q = (y_1, y_2, \ldots, y_p)$ is given by

$$d(P, Q) = \sqrt{(x_1 - y_1)^2 + (x_2 - y_2)^2 + \cdots + (x_p - y_p)^2} \tag{1-12}$$

Straight-line, or Euclidean, distance is unsatisfactory for most statistical purposes. This is because each coordinate contributes equally to the calculation of Euclidean distance. When the coordinates represent measurements that are subject to random fluctuations of differing magnitudes, it is often desirable to weight coordinates subject to a great deal of variability less heavily than those that are not highly variable. This suggests a different measure of distance.

Our purpose now is to develop a "statistical" distance that accounts for differences in variation and, in due course, the presence of correlation. Because our

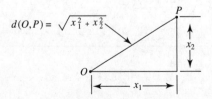

$d(O,P) = \sqrt{x_1^2 + x_2^2}$

Figure 1.19 Distance given by the Pythagorean theorem.

choice will depend upon the sample variances and covariances, at this point we use the term *statistical distance* to distinguish it from ordinary Euclidean distance. It is statistical distance that is fundamental to multivariate analysis.

To begin, we take as *fixed* the set of observations graphed as the p-dimensional scatter plot. From these, we shall construct a measure of distance from the origin to a point $P = (x_1, x_2, \ldots, x_p)$. In our arguments, the coordinates $(x_1, x_2, \ldots, x_p)$ of P can vary to produce different locations for the point. The data that determine distance will, however, remain fixed.

To illustrate, suppose we have n pairs of measurements on two variables each having mean zero. Call the variables x_1 and x_2, and assume that the x_1 measurements vary independently of the x_2 measurements.[1] In addition, assume that the variability in the x_1 measurements is larger than the variability in the x_2 measurements. A scatter plot of the data would look something like the one pictured in Figure 1.20.

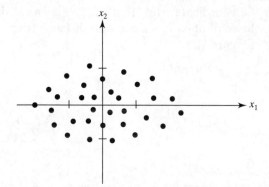

Figure 1.20 A scatter plot with greater variability in the x_1 direction than in the x_2 direction.

Glancing at Figure 1.20, we see that values which are a given deviation from the origin in the x_1 direction are not as "surprising" or "unusual" as are values equidistant from the origin in the x_2 direction. This is because the inherent variability in the x_1 direction is greater than the variability in the x_2 direction. Consequently, large x_1 coordinates (in absolute value) are not as unexpected as large x_2 coordinates. It seems reasonable, then, to weight an x_2 coordinate more heavily than an x_1 coordinate of the same value when computing the "distance" to the origin.

One way to proceed is to divide each coordinate by the sample standard deviation. Therefore, upon division by the standard deviations, we have the "standardized" coordinates $x_1^* = x_1/\sqrt{s_{11}}$ and $x_2^* = x_2/\sqrt{s_{22}}$. The standardized coordinates are now on an equal footing with one another. After taking the differences in variability into account, we determine distance using the standard Euclidean formula.

Thus, a statistical distance of the point $P = (x_1, x_2)$ from the origin $O = (0, 0)$ can be computed from its standardized coordinates $x_1^* = x_1/\sqrt{s_{11}}$ and $x_2^* = x_2/\sqrt{s_{22}}$ as

$$d(O, P) = \sqrt{(x_1^*)^2 + (x_2^*)^2}$$

$$= \sqrt{\left(\frac{x_1}{\sqrt{s_{11}}}\right)^2 + \left(\frac{x_2}{\sqrt{s_{22}}}\right)^2} = \sqrt{\frac{x_1^2}{s_{11}} + \frac{x_2^2}{s_{22}}} \qquad (1\text{-}13)$$

---

[1]At this point, "independently" means that the x_2 measurements cannot be predicted with any accuracy from the x_1 measurements, and vice versa.

Comparing (1-13) with (1-9), we see that the difference between the two expressions is due to the weights $k_1 = 1/s_{11}$ and $k_2 = 1/s_{22}$ attached to x_1^2 and x_2^2 in (1-13). Note that if the sample variances are the same, $k_1 = k_2$, then x_1^2 and x_2^2 will receive the same weight. In cases where the weights are the same, it is convenient to ignore the common divisor and use the usual Euclidean distance formula. In other words, if the variability in the x_1 direction is the same as the variability in the x_2 direction, and the x_1 values vary independently of the x_2 values, Euclidean distance is appropriate.

Using (1-13), we see that all points which have coordinates (x_1, x_2) and are a constant squared distance c^2 from the origin must satisfy

$$\frac{x_1^2}{s_{11}} + \frac{x_2^2}{s_{22}} = c^2 \tag{1-14}$$

Equation (1-14) is the equation of an ellipse centered at the origin whose major and minor axes coincide with the coordinate axes. That is, the statistical distance in (1-13) has an ellipse as the locus of all points a constant distance from the origin. This general case is shown in Figure 1.21.

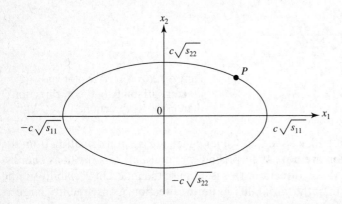

Figure 1.21 The ellipse of constant statistical distance $d^2(O, P) = x_1^2/s_{11} + x_2^2/s_{22} = c^2$.

Example 1.14 (Calculating a statistical distance) A set of paired measurements (x_1, x_2) on two variables yields $\bar{x}_1 = \bar{x}_2 = 0$, $s_{11} = 4$, and $s_{22} = 1$. Suppose the x_1 measurements are unrelated to the x_2 measurements; that is, measurements within a pair vary independently of one another. Since the sample variances are unequal, we measure the square of the distance of an arbitrary point $P = (x_1, x_2)$ to the origin $O = (0, 0)$ by

$$d^2(O, P) = \frac{x_1^2}{4} + \frac{x_2^2}{1}$$

All points (x_1, x_2) that are a constant distance 1 from the origin satisfy the equation

$$\frac{x_1^2}{4} + \frac{x_2^2}{1} = 1$$

The coordinates of some points a unit distance from the origin are presented in the following table:

Coordinates: (x_1, x_2)	Distance: $\dfrac{x_1^2}{4} + \dfrac{x_2^2}{1} = 1$
$(0, 1)$	$\dfrac{0^2}{4} + \dfrac{1^2}{1} = 1$
$(0, -1)$	$\dfrac{0^2}{4} + \dfrac{(-1)^2}{1} = 1$
$(2, 0)$	$\dfrac{2^2}{4} + \dfrac{0^2}{1} = 1$
$(1, \sqrt{3}/2)$	$\dfrac{1^2}{4} + \dfrac{(\sqrt{3}/2)^2}{1} = 1$

A plot of the equation $x_1^2/4 + x_2^2/1 = 1$ is an ellipse centered at $(0, 0)$ whose major axis lies along the x_1 coordinate axis and whose minor axis lies along the x_2 coordinate axis. The half-lengths of these major and minor axes are $\sqrt{4} = 2$ and $\sqrt{1} = 1$, respectively. The ellipse of unit distance is plotted in Figure 1.22. All points on the ellipse are regarded as being the same statistical distance from the origin—in this case, a distance of 1. ∎

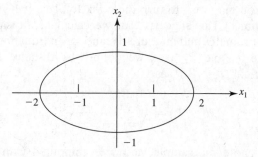

Figure 1.22 Ellipse of unit distance, $\dfrac{x_1^2}{4} + \dfrac{x_2^2}{1} = 1$.

The expression in (1-13) can be generalized to accommodate the calculation of statistical distance from an arbitrary point $P = (x_1, x_2)$ to any *fixed* point $Q = (y_1, y_2)$. If we assume that the coordinate variables vary independently of one another, the distance from P to Q is given by

$$d(P, Q) = \sqrt{\frac{(x_1 - y_1)^2}{s_{11}} + \frac{(x_2 - y_2)^2}{s_{22}}} \tag{1-15}$$

The extension of this statistical distance to more than two dimensions is straightforward. Let the points P and Q have p coordinates such that $P = (x_1, x_2, \ldots, x_p)$ and $Q = (y_1, y_2, \ldots, y_p)$. Suppose Q is a fixed point [it may be the origin $O = (0, 0, \ldots, 0)$] and the coordinate variables vary independently of one another. Let $s_{11}, s_{22}, \ldots, s_{pp}$ be sample variances constructed from n measurements on $x_1, x_2, \ldots, x_p$, respectively. Then the statistical distance from P to Q is

$$d(P, Q) = \sqrt{\frac{(x_1 - y_1)^2}{s_{11}} + \frac{(x_2 - y_2)^2}{s_{22}} + \cdots + \frac{(x_p - y_p)^2}{s_{pp}}} \tag{1-16}$$

All points P that are a constant squared distance from Q lie on a hyperellipsoid centered at Q whose major and minor axes are parallel to the coordinate axes. We note the following:

1. The distance of P to the origin O is obtained by setting $y_1 = y_2 = \cdots = y_p = 0$ in (1-16).

2. If $s_{11} = s_{22} = \cdots = s_{pp}$, the Euclidean distance formula in (1-12) is appropriate.

The distance in (1-16) still does not include most of the important cases we shall encounter, because of the assumption of independent coordinates. The scatter plot in Figure 1.23 depicts a two-dimensional situation in which the x_1 measurements do not vary independently of the x_2 measurements. In fact, the coordinates of the pairs (x_1, x_2) exhibit a tendency to be large or small together, and the sample correlation coefficient is positive. Moreover, the variability in the x_2 direction is larger than the variability in the x_1 direction.

What is a meaningful measure of distance when the variability in the x_1 direction is different from the variability in the x_2 direction and the variables x_1 and x_2 are correlated? Actually, we can use what we have already introduced, provided that we look at things in the right way. From Figure 1.23, we see that if we rotate the original coordinate system through the angle θ while keeping the scatter fixed and label the rotated axes $\tilde{x}_1$ and $\tilde{x}_2$, the scatter in terms of the new axes looks very much like that in Figure 1.20. (You may wish to turn the book to place the $\tilde{x}_1$ and $\tilde{x}_2$ axes in their customary positions.) This suggests that we calculate the sample variances using the $\tilde{x}_1$ and $\tilde{x}_2$ coordinates and measure distance as in Equation (1-13). That is, with reference to the $\tilde{x}_1$ and $\tilde{x}_2$ axes, we define the distance from the point $P = (\tilde{x}_1, \tilde{x}_2)$ to the origin $O = (0, 0)$ as

$$d(O, P) = \sqrt{\frac{\tilde{x}_1^2}{\tilde{s}_{11}} + \frac{\tilde{x}_2^2}{\tilde{s}_{22}}} \qquad (1\text{-}17)$$

where $\tilde{s}_{11}$ and $\tilde{s}_{22}$ denote the sample variances computed with the $\tilde{x}_1$ and $\tilde{x}_2$ measurements.

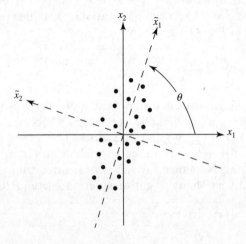

Figure 1.23 A scatter plot for positively correlated measurements and a rotated coordinate system.

The relation between the original coordinates (x_1, x_2) and the rotated coordinates $(\tilde{x}_1, \tilde{x}_2)$ is provided by

$$\tilde{x}_1 = x_1 \cos(\theta) + x_2 \sin(\theta)$$
$$\tilde{x}_2 = -x_1 \sin(\theta) + x_2 \cos(\theta)$$

(1-18)

Given the relations in (1-18), we can formally substitute for $\tilde{x}_1$ and $\tilde{x}_2$ in (1-17) and express the distance in terms of the original coordinates.

After some straightforward algebraic manipulations, the distance from $P = (\tilde{x}_1, \tilde{x}_2)$ to the origin $O = (0, 0)$ can be written in terms of the original coordinates x_1 and x_2 of P as

$$d(O, P) = \sqrt{a_{11}x_1^2 + 2a_{12}x_1x_2 + a_{22}x_2^2}$$

(1-19)

where the a's are numbers such that the distance is nonnegative for all possible values of x_1 and x_2. Here a_{11}, a_{12}, and a_{22} are determined by the angle θ, and s_{11}, s_{12}, and s_{22} calculated from the original data.[2] The particular forms for a_{11}, a_{12}, and a_{22} are not important at this point. What is important is the appearance of the cross-product term $2a_{12}x_1x_2$ necessitated by the nonzero correlation r_{12}.

Equation (1-19) can be compared with (1-13). The expression in (1-13) can be regarded as a special case of (1-19) with $a_{11} = 1/s_{11}$, $a_{22} = 1/s_{22}$, and $a_{12} = 0$.

In general, the statistical distance of the point $P = (x_1, x_2)$ from the *fixed* point $Q = (y_1, y_2)$ for situations in which the variables are correlated has the general form

$$d(P, Q) = \sqrt{a_{11}(x_1 - y_1)^2 + 2a_{12}(x_1 - y_1)(x_2 - y_2) + a_{22}(x_2 - y_2)^2}$$

(1-20)

and can always be computed once a_{11}, a_{12}, and a_{22} are known. In addition, the coordinates of all points $P = (x_1, x_2)$ that are a constant squared distance c^2 from Q satisfy

$$a_{11}(x_1 - y_1)^2 + 2a_{12}(x_1 - y_1)(x_2 - y_2) + a_{22}(x_2 - y_2)^2 = c^2$$

(1-21)

By definition, this is the equation of an ellipse centered at Q. The graph of such an equation is displayed in Figure 1.24. The major (long) and minor (short) axes are indicated. They are parallel to the $\tilde{x}_1$ and $\tilde{x}_2$ axes. For the choice of a_{11}, a_{12}, and a_{22} in footnote 2, the $\tilde{x}_1$ and $\tilde{x}_2$ axes are at an angle θ with respect to the x_1 and x_2 axes.

The generalization of the distance formulas of (1-19) and (1-20) to p dimensions is straightforward. Let $P = (x_1, x_2, \ldots, x_p)$ be a point whose coordinates represent variables that are correlated and subject to inherent variability. Let

---

[2]Specifically,

$$a_{11} = \frac{\cos^2(\theta)}{\cos^2(\theta)s_{11} + 2\sin(\theta)\cos(\theta)s_{12} + \sin^2(\theta)s_{22}} + \frac{\sin^2(\theta)}{\cos^2(\theta)s_{22} - 2\sin(\theta)\cos(\theta)s_{12} + \sin^2(\theta)s_{11}}$$

$$a_{22} = \frac{\sin^2(\theta)}{\cos^2(\theta)s_{11} + 2\sin(\theta)\cos(\theta)s_{12} + \sin^2(\theta)s_{22}} + \frac{\cos^2(\theta)}{\cos^2(\theta)s_{22} - 2\sin(\theta)\cos(\theta)s_{12} + \sin^2(\theta)s_{11}}$$

and

$$a_{12} = \frac{\cos(\theta)\sin(\theta)}{\cos^2(\theta)s_{11} + 2\sin(\theta)\cos(\theta)s_{12} + \sin^2(\theta)s_{22}} - \frac{\sin(\theta)\cos(\theta)}{\cos^2(\theta)s_{22} - 2\sin(\theta)\cos(\theta)s_{12} + \sin^2(\theta)s_{11}}$$

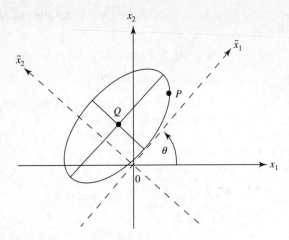

Figure 1.24 Ellipse of points a constant distance from the point Q.

$O = (0, 0, \ldots, 0)$ denote the origin, and let $Q = (y_1, y_2, \ldots, y_p)$ be a specified *fixed* point. Then the distances from P to O and from P to Q have the general forms

$$d(O, P) = \sqrt{a_{11}x_1^2 + a_{22}x_2^2 + \cdots + a_{pp}x_p^2 + 2a_{12}x_1x_2 + 2a_{13}x_1x_3 + \cdots + 2a_{p-1,p}x_{p-1}x_p}$$

(1-22)

and

$$d(P, Q) = \sqrt{\begin{aligned}[a_{11}(x_1 - y_1)^2 + a_{22}(x_2 - y_2)^2 + \cdots + a_{pp}(x_p - y_p)^2 + 2a_{12}(x_1 - y_1)(x_2 - y_2) \\ + 2a_{13}(x_1 - y_1)(x_3 - y_3) + \cdots + 2a_{p-1,p}(x_{p-1} - y_{p-1})(x_p - y_p)]\end{aligned}}$$

(1-23)

where the a's are numbers such that the distances are always nonnegative.[3]

We note that the distances in (1-22) and (1-23) are completely determined by the coefficients (weights) a_{ik}, $i = 1, 2, \ldots, p$, $k = 1, 2, \ldots, p$. These coefficients can be set out in the rectangular array

$$\begin{bmatrix} a_{11} & a_{12} & \cdots & a_{1p} \\ a_{12} & a_{22} & \cdots & a_{2p} \\ \vdots & \vdots & \ddots & \vdots \\ a_{1p} & a_{2p} & \cdots & a_{pp} \end{bmatrix}$$

(1-24)

where the a_{ik}'s with $i \neq k$ are displayed twice, since they are multiplied by 2 in the distance formulas. Consequently, the entries in this array specify the distance functions. The a_{ik}'s cannot be arbitrary numbers; they must be such that the computed distance is nonnegative for every pair of points. (See Exercise 1.10.)

Contours of constant distances computed from (1-22) and (1-23) are hyperellipsoids. A hyperellipsoid resembles a football when $p = 3$; it is impossible to visualize in more than three dimensions.

---

[3]The algebraic expressions for the *squares* of the distances in (1-22) and (1-23) are known as *quadratic forms* and, in particular, *positive definite quadratic forms*. It is possible to display these quadratic forms in a simpler manner using matrix algebra; we shall do so in Section 2.3 of Chapter 2.

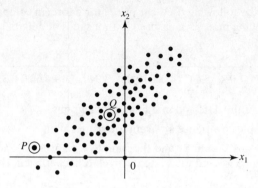

Figure 1.25 A cluster of points relative to a point P and the origin.

The need to consider statistical rather than Euclidean distance is illustrated heuristically in Figure 1.25. Figure 1.25 depicts a cluster of points whose center of gravity (sample mean) is indicated by the point Q. Consider the Euclidean distances from the point Q to the point P and the origin O. The Euclidean distance from Q to P is larger than the Euclidean distance from Q to O. However, P appears to be more like the points in the cluster than does the origin. If we take into account the variability of the points in the cluster and measure distance by the statistical distance in (1-20), then Q will be closer to P than to O. This result seems reasonable, given the nature of the scatter.

Other measures of distance can be advanced. (See Exercise 1.12.) At times, it is useful to consider distances that are not related to circles or ellipses. Any distance measure $d(P, Q)$ between two points P and Q is valid provided that it satisfies the following properties, where R is any other intermediate point:

$$d(P, Q) = d(Q, P)$$
$$d(P, Q) > 0 \text{ if } P \neq Q$$
$$d(P, Q) = 0 \text{ if } P = Q \tag{1-25}$$
$$d(P, Q) \leq d(P, R) + d(R, Q) \qquad \text{(triangle inequality)}$$

1.6 Final Comments

We have attempted to motivate the study of multivariate analysis and to provide you with some rudimentary, but important, methods for organizing, summarizing, and displaying data. In addition, a general concept of distance has been introduced that will be used repeatedly in later chapters.

Exercises

1.1. Consider the seven pairs of measurements (x_1, x_2) plotted in Figure 1.1:

x_1	3	4	2	6	8	2	5
x_2	5	5.5	4	7	10	5	7.5

Calculate the sample means $\bar{x}_1$ and $\bar{x}_2$, the sample variances s_{11} and s_{22}, and the sample covariance s_{12}.

1.2. A morning newspaper lists the following used-car prices for a foreign compact with age x_1 measured in years and selling price x_2 measured in thousands of dollars:

x_1	1	2	3	3	4	5	6	8	9	11
x_2	18.95	19.00	17.95	15.54	14.00	12.95	8.94	7.49	6.00	3.99

(a) Construct a scatter plot of the data and marginal dot diagrams.

(b) Infer the sign of the sample covariance s_{12} from the scatter plot.

(c) Compute the sample means $\bar{x}_1$ and $\bar{x}_2$ and the sample variances s_{11} and s_{22}. Compute the sample covariance s_{12} and the sample correlation coefficient r_{12}. Interpret these quantities.

(d) Display the sample mean array $\bar{x}$, the sample variance-covariance array S_n, and the sample correlation array R using (1-8).

1.3. The following are five measurements on the variables x_1, x_2, and x_3:

x_1	9	2	6	5	8
x_2	12	8	6	4	10
x_3	3	4	0	2	1

Find the arrays $\bar{x}$, S_n, and R.

1.4. The world's 10 largest companies yield the following data:

The World's 10 Largest Companies[1]

Company	x_1 = sales (billions)	x_2 = profits (billions)	x_3 = assets (billions)
Citigroup	108.28	17.05	1,484.10
General Electric	152.36	16.59	750.33
American Intl Group	95.04	10.91	766.42
Bank of America	65.45	14.14	1,110.46
HSBC Group	62.97	9.52	1,031.29
ExxonMobil	263.99	25.33	195.26
Royal Dutch/Shell	265.19	18.54	193.83
BP	285.06	15.73	191.11
ING Group	92.01	8.10	1,175.16
Toyota Motor	165.68	11.13	211.15

[1]From www.Forbes.com partially based on *Forbes* The Forbes Global 2000, April 18, 2005.

(a) Plot the scatter diagram and marginal dot diagrams for variables x_1 and x_2. Comment on the appearance of the diagrams.

(b) Compute $\bar{x}_1$, $\bar{x}_2$, s_{11}, s_{22}, s_{12}, and r_{12}. Interpret r_{12}.

1.5. Use the data in Exercise 1.4.

(a) Plot the scatter diagrams and dot diagrams for (x_2, x_3) and (x_1, x_3). Comment on the patterns.

(b) Compute the $\bar{x}$, S_n, and R arrays for (x_1, x_2, x_3).

1.6. The data in Table 1.5 are 42 measurements on air-pollution variables recorded at 12:00 noon in the Los Angeles area on different days. (See also the air-pollution data on the web at www.prenhall.com/statistics.)

(a) Plot the marginal dot diagrams for all the variables.

(b) Construct the $\bar{x}$, S_n, and R arrays, and interpret the entries in R.

Table 1.5 Air-Pollution Data

Wind (x_1)	Solar radiation (x_2)	CO (x_3)	NO (x_4)	NO$_2$ (x_5)	O$_3$ (x_6)	HC (x_7)
8	98	7	2	12	8	2
7	107	4	3	9	5	3
7	103	4	3	5	6	3
10	88	5	2	8	15	4
6	91	4	2	8	10	3
8	90	5	2	12	12	4
9	84	7	4	12	15	5
5	72	6	4	21	14	4
7	82	5	1	11	11	3
8	64	5	2	13	9	4
6	71	5	4	10	3	3
6	91	4	2	12	7	3
7	72	7	4	18	10	3
10	70	4	2	11	7	3
10	72	4	1	8	10	3
9	77	4	1	9	10	3
8	76	4	1	7	7	3
8	71	5	3	16	4	4
9	67	4	2	13	2	3
9	69	3	3	9	5	3
10	62	5	3	14	4	4
9	88	4	2	7	6	3
8	80	4	2	13	11	4
5	30	3	3	5	2	3
6	83	5	1	10	23	4
8	84	3	2	7	6	3
6	78	4	2	11	11	3
8	79	2	1	7	10	3
6	62	4	3	9	8	3
10	37	3	1	7	2	3
8	71	4	1	10	7	3
7	52	4	1	12	8	4
5	48	6	5	8	4	3
6	75	4	1	10	24	3
10	35	4	1	6	9	2
8	85	4	1	9	10	2
5	86	3	1	6	12	2
5	86	7	2	13	18	2
7	79	7	4	9	25	3
7	79	5	2	8	6	2
6	68	6	2	11	14	3
8	40	4	3	6	5	2

Source: Data courtesy of Professor G. C. Tiao.

1.7. You are given the following $n = 3$ observations on $p = 2$ variables:

$$\text{Variable 1:} \quad x_{11} = 2 \quad x_{21} = 3 \quad x_{31} = 4$$

$$\text{Variable 2:} \quad x_{12} = 1 \quad x_{22} = 2 \quad x_{32} = 4$$

(a) Plot the pairs of observations in the two-dimensional "variable space." That is, construct a two-dimensional scatter plot of the data.

(b) Plot the data as two points in the three-dimensional "item space."

1.8. Evaluate the distance of the point $P = (-1, -1)$ to the point $Q = (1, 0)$ using the Euclidean distance formula in (1-12) with $p = 2$ and using the statistical distance in (1-20) with $a_{11} = 1/3$, $a_{22} = 4/27$, and $a_{12} = 1/9$. Sketch the locus of points that are a constant squared statistical distance 1 from the point Q.

1.9. Consider the following eight pairs of measurements on two variables x_1 and x_2:

x_1	-6	-3	-2	1	2	5	6	8
x_2	-2	-3	1	-1	2	1	5	3

(a) Plot the data as a scatter diagram, and compute s_{11}, s_{22}, and s_{12}.

(b) Using (1-18), calculate the corresponding measurements on variables $\tilde{x}_1$ and $\tilde{x}_2$, assuming that the original coordinate axes are rotated through an angle of $\theta = 26°$ [given $\cos(26°) = .899$ and $\sin(26°) = .438$].

(c) Using the $\tilde{x}_1$ and $\tilde{x}_2$ measurements from (b), compute the sample variances $\tilde{s}_{11}$ and $\tilde{s}_{22}$.

(d) Consider the *new* pair of measurements $(x_1, x_2) = (4, -2)$. Transform these to measurements on $\tilde{x}_1$ and $\tilde{x}_2$ using (1-18), and calculate the distance $d(O, P)$ of the new point $P = (\tilde{x}_1, \tilde{x}_2)$ from the origin $O = (0, 0)$ using (1-17).
Note: You will need $\tilde{s}_{11}$ and $\tilde{s}_{22}$ from (c).

(e) Calculate the distance from $P = (4, -2)$ to the origin $O = (0, 0)$ using (1-19) and the expressions for a_{11}, a_{22}, and a_{12} in footnote 2.
Note: You will need s_{11}, s_{22}, and s_{12} from (a).
Compare the distance calculated here with the distance calculated using the $\tilde{x}_1$ and $\tilde{x}_2$ values in (d). (Within rounding error, the numbers should be the same.)

1.10. Are the following distance functions valid for distance from the origin? Explain.

(a) $x_1^2 + 4x_2^2 + x_1 x_2 = (\text{distance})^2$

(b) $x_1^2 - 2x_2^2 = (\text{distance})^2$

1.11. Verify that distance defined by (1-20) with $a_{11} = 4$, $a_{22} = 1$, and $a_{12} = -1$ satisfies the first three conditions in (1-25). (The triangle inequality is more difficult to verify.)

1.12. Define the distance from the point $P = (x_1, x_2)$ to the origin $O = (0, 0)$ as

$$d(O, P) = \max(|x_1|, |x_2|)$$

(a) Compute the distance from $P = (-3, 4)$ to the origin.

(b) Plot the locus of points whose squared distance from the origin is 1.

(c) Generalize the foregoing distance expression to points in p dimensions.

1.13. A large city has major roads laid out in a grid pattern, as indicated in the following diagram. Streets 1 through 5 run north–south (NS), and streets A through E run east–west (EW). Suppose there are retail stores located at intersections $(A, 2)$, $(E, 3)$, and $(C, 5)$.

Assume the distance along a street between two intersections in either the NS or EW direction is 1 unit. Define the distance between any two intersections (points) on the grid to be the "city block" distance. [For example, the distance between intersections $(D, 1)$ and $(C, 2)$, which we might call $d((D, 1), (C, 2))$, is given by $d((D, 1), (C, 2)) = d((D, 1), (D, 2)) + d((D, 2), (C, 2)) = 1 + 1 = 2.$ Also, $d((D, 1), (C, 2)) = d((D, 1), (C, 1)) + d((C, 1), (C, 2)) = 1 + 1 = 2.$]

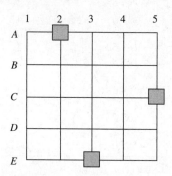

Locate a supply facility (warehouse) at an intersection such that the sum of the distances from the warehouse to the three retail stores is minimized.

The following exercises contain fairly extensive data sets. A computer may be necessary for the required calculations.

1.14. Table 1.6 contains some of the raw data discussed in Section 1.2. (See also the multiple-sclerosis data on the web at www.prenhall.com/statistics.) Two different visual stimuli ($S1$ and $S2$) produced responses in both the left eye (L) and the right eye (R) of subjects in the study groups. The values recorded in the table include x_1 (subject's age); x_2 (total response of both eyes to stimulus $S1$, that is, $S1L + S1R$); x_3 (difference between responses of eyes to stimulus $S1$, $|S1L - S1R|$); and so forth.

(a) Plot the two-dimensional scatter diagram for the variables x_2 and x_4 for the multiple-sclerosis group. Comment on the appearance of the diagram.

(b) Compute the $\bar{x}, \mathbf{S}_n$, and $\mathbf{R}$ arrays for the non-multiple-sclerosis and multiple-sclerosis groups separately.

1.15. Some of the 98 measurements described in Section 1.2 are listed in Table 1.7 (See also the radiotherapy data on the web at www.prenhall.com/statistics.) The data consist of average ratings over the course of treatment for patients undergoing radiotherapy. Variables measured include x_1 (number of symptoms, such as sore throat or nausea); x_2 (amount of activity, on a 1–5 scale); x_3 (amount of sleep, on a 1–5 scale); x_4 (amount of food consumed, on a 1–3 scale); x_5 (appetite, on a 1–5 scale); and x_6 (skin reaction, on a 0–3 scale).

(a) Construct the two-dimensional scatter plot for variables x_2 and x_3 and the marginal dot diagrams (or histograms). Do there appear to be any errors in the x_3 data?

(b) Compute the $\bar{x}, \mathbf{S}_n$, and $\mathbf{R}$ arrays. Interpret the pairwise correlations.

1.16. At the start of a study to determine whether exercise or dietary supplements would slow bone loss in older women, an investigator measured the mineral content of bones by photon absorptiometry. Measurements were recorded for three bones on the dominant and nondominant sides and are shown in Table 1.8. (See also the mineral-content data on the web at www.prenhall.com/statistics.)

Compute the $\bar{x}, \mathbf{S}_n$, and $\mathbf{R}$ arrays. Interpret the pairwise correlations.

Table 1.6 Multiple-Sclerosis Data

Non-Multiple-Sclerosis Group Data

Subject number	x_1 (Age)	x_2 $(S1L + S1R)$	x_3 $|S1L - S1R|$	x_4 $(S2L + S2R)$	x_5 $|S2L - S2R|$
1	18	152.0	1.6	198.4	.0
2	19	138.0	.4	180.8	1.6
3	20	144.0	.0	186.4	.8
4	20	143.6	3.2	194.8	.0
5	20	148.8	.0	217.6	.0
⋮	⋮	⋮	⋮	⋮	⋮
65	67	154.4	2.4	205.2	6.0
66	69	171.2	1.6	210.4	.8
67	73	157.2	.4	204.8	.0
68	74	175.2	5.6	235.6	.4
69	79	155.0	1.4	204.4	.0

Multiple-Sclerosis Group Data

Subject number	x_1	x_2	x_3	x_4	x_5
1	23	148.0	.8	205.4	.6
2	25	195.2	3.2	262.8	.4
3	25	158.0	8.0	209.8	12.2
4	28	134.4	.0	198.4	3.2
5	29	190.2	14.2	243.8	10.6
⋮	⋮	⋮	⋮	⋮	⋮
25	57	165.6	16.8	229.2	15.6
26	58	238.4	8.0	304.4	6.0
27	58	164.0	.8	216.8	.8
28	58	169.8	.0	219.2	1.6
29	59	199.8	4.6	250.2	1.0

Source: Data courtesy of Dr. G. G. Celesia.

Table 1.7 Radiotherapy Data

x_1 Symptoms	x_2 Activity	x_3 Sleep	x_4 Eat	x_5 Appetite	x_6 Skin reaction
.889	1.389	1.555	2.222	1.945	1.000
2.813	1.437	.999	2.312	2.312	2.000
1.454	1.091	2.364	2.455	2.909	3.000
.294	.941	1.059	2.000	1.000	1.000
2.727	2.545	2.819	2.727	4.091	.000
⋮	⋮	⋮	⋮	⋮	⋮
4.100	1.900	2.800	2.000	2.600	2.000
.125	1.062	1.437	1.875	1.563	.000
6.231	2.769	1.462	2.385	4.000	2.000
3.000	1.455	2.090	2.273	3.272	2.000
.889	1.000	1.000	2.000	1.000	2.000

Source: Data courtesy of Mrs. Annette Tealey, R.N. Values of x_2 and x_3 less than 1.0 are due to errors in the data-collection process. Rows containing values of x_2 and x_3 less than 1.0 may be omitted.

Table 1.8 Mineral Content in Bones

Subject number	Dominant radius	Radius	Dominant humerus	Humerus	Dominant ulna	Ulna
1	1.103	1.052	2.139	2.238	.873	.872
2	.842	.859	1.873	1.741	.590	.744
3	.925	.873	1.887	1.809	.767	.713
4	.857	.744	1.739	1.547	.706	.674
5	.795	.809	1.734	1.715	.549	.654
6	.787	.779	1.509	1.474	.782	.571
7	.933	.880	1.695	1.656	.737	.803
8	.799	.851	1.740	1.777	.618	.682
9	.945	.876	1.811	1.759	.853	.777
10	.921	.906	1.954	2.009	.823	.765
11	.792	.825	1.624	1.657	.686	.668
12	.815	.751	2.204	1.846	.678	.546
13	.755	.724	1.508	1.458	.662	.595
14	.880	.866	1.786	1.811	.810	.819
15	.900	.838	1.902	1.606	.723	.677
16	.764	.757	1.743	1.794	.586	.541
17	.733	.748	1.863	1.869	.672	.752
18	.932	.898	2.028	2.032	.836	.805
19	.856	.786	1.390	1.324	.578	.610
20	.890	.950	2.187	2.087	.758	.718
21	.688	.532	1.650	1.378	.533	.482
22	.940	.850	2.334	2.225	.757	.731
23	.493	.616	1.037	1.268	.546	.615
24	.835	.752	1.509	1.422	.618	.664
25	.915	.936	1.971	1.869	.869	.868

Source: Data courtesy of Everett Smith.

1.17. Some of the data described in Section 1.2 are listed in Table 1.9. (See also the national-track-records data on the web at www.prenhall.com/statistics.) The national track records for women in 54 countries can be examined for the relationships among the running events. Compute the $\bar{x}$, S_n, and R arrays. Notice the magnitudes of the correlation coefficients as you go from the shorter (100-meter) to the longer (marathon) running distances. Interpret these pairwise correlations.

1.18. Convert the national track records for women in Table 1.9 to speeds measured in meters per second. For example, the record speed for the 100-m dash for Argentinian women is 100 m/11.57 sec = 8.643 m/sec. Notice that the records for the 800-m, 1500-m, 3000-m and marathon runs are measured in minutes. The marathon is 26.2 miles, or 42,195 meters, long. Compute the $\bar{x}$, S_n, and R arrays. Notice the magnitudes of the correlation coefficients as you go from the shorter (100 m) to the longer (marathon) running distances. Interpret these pairwise correlations. Compare your results with the results you obtained in Exercise 1.17.

1.19. Create the scatter plot and boxplot displays of Figure 1.5 for (a) the mineral-content data in Table 1.8 and (b) the national-track-records data in Table 1.9.

Table 1.9 National Track Records for Women

Country	100 m (s)	200 m (s)	400 m (s)	800 m (min)	1500 m (min)	3000 m (min)	Marathon (min)
Argentina	11.57	22.94	52.50	2.05	4.25	9.19	150.32
Australia	11.12	22.23	48.63	1.98	4.02	8.63	143.51
Austria	11.15	22.70	50.62	1.94	4.05	8.78	154.35
Belgium	11.14	22.48	51.45	1.97	4.08	8.82	143.05
Bermuda	11.46	23.05	53.30	2.07	4.29	9.81	174.18
Brazil	11.17	22.60	50.62	1.97	4.17	9.04	147.41
Canada	10.98	22.62	49.91	1.97	4.00	8.54	148.36
Chile	11.65	23.84	53.68	2.00	4.22	9.26	152.23
China	10.79	22.01	49.81	1.93	3.84	8.10	139.39
Columbia	11.31	22.92	49.64	2.04	4.34	9.37	155.19
Cook Islands	12.52	25.91	61.65	2.28	4.82	11.10	212.33
Costa Rica	11.72	23.92	52.57	2.10	4.52	9.84	164.33
Czech Republic	11.09	21.97	47.99	1.89	4.03	8.87	145.19
Denmark	11.42	23.36	52.92	2.02	4.12	8.71	149.34
Dominican Republic	11.63	23.91	53.02	2.09	4.54	9.89	166.46
Finland	11.13	22.39	50.14	2.01	4.10	8.69	148.00
France	10.73	21.99	48.25	1.94	4.03	8.64	148.27
Germany	10.81	21.71	47.60	1.92	3.96	8.51	141.45
Great Britain	11.10	22.10	49.43	1.94	3.97	8.37	135.25
Greece	10.83	22.67	50.56	2.00	4.09	8.96	153.40
Guatemala	11.92	24.50	55.64	2.15	4.48	9.71	171.33
Hungary	11.41	23.06	51.50	1.99	4.02	8.55	148.50
India	11.56	23.86	55.08	2.10	4.36	9.50	154.29
Indonesia	11.38	22.82	51.05	2.00	4.10	9.11	158.10
Ireland	11.43	23.02	51.07	2.01	3.98	8.36	142.23
Israel	11.45	23.15	52.06	2.07	4.24	9.33	156.36
Italy	11.14	22.60	51.31	1.96	3.98	8.59	143.47
Japan	11.36	23.33	51.93	2.01	4.16	8.74	139.41
Kenya	11.62	23.37	51.56	1.97	3.96	8.39	138.47
Korea, South	11.49	23.80	53.67	2.09	4.24	9.01	146.12
Korea, North	11.80	25.10	56.23	1.97	4.25	8.96	145.31
Luxembourg	11.76	23.96	56.07	2.07	4.35	9.21	149.23
Malaysia	11.50	23.37	52.56	2.12	4.39	9.31	169.28
Mauritius	11.72	23.83	54.62	2.06	4.33	9.24	167.09
Mexico	11.09	23.13	48.89	2.02	4.19	8.89	144.06
Myanmar(Burma)	11.66	23.69	52.96	2.03	4.20	9.08	158.42
Netherlands	11.08	22.81	51.35	1.93	4.06	8.57	143.43
New Zealand	11.32	23.13	51.60	1.97	4.10	8.76	146.46
Norway	11.41	23.31	52.45	2.03	4.01	8.53	141.06
Papua New Guinea	11.96	24.68	55.18	2.24	4.62	10.21	221.14
Philippines	11.28	23.35	54.75	2.12	4.41	9.81	165.48
Poland	10.93	22.13	49.28	1.95	3.99	8.53	144.18
Portugal	11.30	22.88	51.92	1.98	3.96	8.50	143.29
Romania	11.30	22.35	49.88	1.92	3.90	8.36	142.50
Russia	10.77	21.87	49.11	1.91	3.87	8.38	141.31
Samoa	12.38	25.45	56.32	2.29	5.42	13.12	191.58

(continues)

Country	100 m (s)	200 m (s)	400 m (s)	800 m (min)	1500 m (min)	3000 m (min)	Marathon (min)
Singapore	12.13	24.54	55.08	2.12	4.52	9.94	154.41
Spain	11.06	22.38	49.67	1.96	4.01	8.48	146.51
Sweden	11.16	22.82	51.69	1.99	4.09	8.81	150.39
Switzerland	11.34	22.88	51.32	1.98	3.97	8.60	145.51
Taiwan	11.22	22.56	52.74	2.08	4.38	9.63	159.53
Thailand	11.33	23.30	52.60	2.06	4.38	10.07	162.39
Turkey	11.25	22.71	53.15	2.01	3.92	8.53	151.43
U.S.A.	10.49	21.34	48.83	1.94	3.95	8.43	141.16

Source: *IAAF/ATFS Track and Field Handbook for Helsinki* 2005 (courtesy of Ottavio Castellini).

1.20. Refer to the bankruptcy data in Table 11.4, page 657, and on the following website www.prenhall.com/statistics. Using appropriate computer software,

(a) View the entire data set in x_1, x_2, x_3 space. Rotate the coordinate axes in various directions. Check for unusual observations.

(b) Highlight the set of points corresponding to the bankrupt firms. Examine various three-dimensional perspectives. Are there some orientations of three-dimensional space for which the bankrupt firms can be distinguished from the nonbankrupt firms? Are there observations in each of the two groups that are likely to have a significant impact on any rule developed to classify firms based on the sample means, variances, and covariances calculated from these data? (See Exercise 11.24.)

1.21. Refer to the milk transportation-cost data in Table 6.10, page 345, and on the web at www.prenhall.com/statistics. Using appropriate computer software,

(a) View the entire data set in three dimensions. Rotate the coordinate axes in various directions. Check for unusual observations.

(b) Highlight the set of points corresponding to gasoline trucks. Do any of the gasoline-truck points appear to be multivariate outliers? (See Exercise 6.17.) Are there some orientations of x_1, x_2, x_3 space for which the set of points representing gasoline trucks can be readily distinguished from the set of points representing diesel trucks?

1.22. Refer to the oxygen-consumption data in Table 6.12, page 348, and on the web at www.prenhall.com/statistics. Using appropriate computer software,

(a) View the entire data set in three dimensions employing various combinations of three variables to represent the coordinate axes. Begin with the x_1, x_2, x_3 space.

(b) Check this data set for outliers.

1.23. Using the data in Table 11.9, page 666, and on the web at www.prenhall.com/statistics, represent the cereals in each of the following ways.

(a) Stars.

(b) Chernoff faces. (Experiment with the assignment of variables to facial characteristics.)

1.24. Using the utility data in Table 12.4, page 688, and on the web at www.prenhall.com/statistics, represent the public utility companies as Chernoff faces with assignments of variables to facial characteristics different from those considered in Example 1.12. Compare your faces with the faces in Figure 1.17. Are different groupings indicated?

1.25. Using the data in Table 12.4 and on the web at www.prenhall.com/statistics, represent the 22 public utility companies as stars. Visually group the companies into four or five clusters.

1.26. The data in Table 1.10 (see the bull data on the web at www.prenhall.com/statistics) are the measured characteristics of 76 young (less than two years old) bulls sold at auction. Also included in the table are the selling prices (SalePr) of these bulls. The column headings (variables) are defined as follows:

$$\text{Breed} = \begin{cases} 1 & \text{Angus} \\ 5 & \text{Hereford} \\ 8 & \text{Simental} \end{cases} \qquad \begin{aligned} \text{YrHgt} &= \text{Yearling height at} \\ &\quad \text{shoulder (inches)} \end{aligned}$$

FtFrBody = Fat free body   PrctFFB = Percent fat-free
(pounds)   body

Frame = Scale from 1 (small)   BkFat = Back fat
to 8 (large)   (inches)

SaleHt = Sale height at   SaleWt = Sale weight
shoulder (inches)   (pounds)

(a) Compute the $\bar{x}$, S_n, and **R** arrays. Interpret the pairwise correlations. Do some of these variables appear to distinguish one breed from another?

(b) View the data in three dimensions using the variables Breed, Frame, and BkFat. Rotate the coordinate axes in various directions. Check for outliers. Are the breeds well separated in this coordinate system?

(c) Repeat part b using Breed, FtFrBody, and SaleHt. Which three-dimensional display appears to result in the best separation of the three breeds of bulls?

Table 1.10 Data on Bulls

Breed	SalePr	YrHgt	FtFrBody	PrctFFB	Frame	BkFat	SaleHt	SaleWt
1	2200	51.0	1128	70.9	7	.25	54.8	1720
1	2250	51.9	1108	72.1	7	.25	55.3	1575
1	1625	49.9	1011	71.6	6	.15	53.1	1410
1	4600	53.1	993	68.9	8	.35	56.4	1595
1	2150	51.2	996	68.6	7	.25	55.0	1488
⋮	⋮	⋮	⋮	⋮	⋮	⋮	⋮	⋮
8	1450	51.4	997	73.4	7	.10	55.2	1454
8	1200	49.8	991	70.8	6	.15	54.6	1475
8	1425	50.0	928	70.8	6	.10	53.9	1375
8	1250	50.1	990	71.0	6	.10	54.9	1564
8	1500	51.7	992	70.6	7	.15	55.1	1458

Source: Data courtesy of Mark Ellersieck.

1.27. Table 1.11 presents the 2005 attendance (millions) at the fifteen most visited national parks and their size (acres).

(a) Create a scatter plot and calculate the correlation coefficient.

(b) Identify the park that is unusual. Drop this point and recalculate the correlation coefficient. Comment on the effect of this one point on correlation.

(c) Would the correlation in Part b change if you measure size in square miles instead of acres? Explain.

Table 1.11 Attendance and Size of National Parks

National Park	Size (acres)	Visitors (millions)
Arcadia	47.4	2.05
Bryce Canyon	35.8	1.02
Cuyahoga Valley	32.9	2.53
Everglades	1508.5	1.23
Grand Canyon	1217.4	4.40
Grand Teton	310.0	2.46
Great Smoky	521.8	9.19
Hot Springs	5.6	1.34
Olympic	922.7	3.14
Mount Rainier	235.6	1.17
Rocky Mountain	265.8	2.80
Shenandoah	199.0	1.09
Yellowstone	2219.8	2.84
Yosemite	761.3	3.30
Zion	146.6	2.59

References

1. Becker, R. A., W. S. Cleveland, and A. R. Wilks. "Dynamic Graphics for Data Analysis." *Statistical Science*, **2**, no. 4 (1987), 355–395.

2. Benjamin, Y., and M. Igbaria. "Clustering Categories for Better Prediction of Computer Resources Utilization." *Applied Statistics*, **40**, no. 2 (1991), 295–307.

3. Capon, N., J. Farley, D. Lehman, and J. Hulbert. "Profiles of Product Innovators among Large U. S. Manufacturers." *Management Science*, **38**, no. 2 (1992), 157–169.

4. Chernoff, H. "Using Faces to Represent Points in *K*-Dimensional Space Graphically." *Journal of the American Statistical Association*, **68**, no. 342 (1973), 361–368.

5. Cochran, W. G. *Sampling Techniques* (3rd ed.). New York: John Wiley, 1977.

6. Cochran, W. G., and G. M. Cox. *Experimental Designs* (2nd ed., paperback). New York: John Wiley, 1992.

7. Davis, J. C. "Information Contained in Sediment Size Analysis." *Mathematical Geology*, **2**, no. 2 (1970), 105–112.

8. Dawkins, B. "Multivariate Analysis of National Track Records." *The American Statistician*, **43**, no. 2 (1989), 110–115.

9. Dudoit, S., J. Fridlyand, and T. P. Speed. "Comparison of Discrimination Methods for the Classification of Tumors Using Gene Expression Data." *Journal of the American Statistical Association*, **97**, no. 457 (2002), 77-87.

10. Dunham, R. B., and D. J. Kravetz. "Canonical Correlation Analysis in a Predictive System." *Journal of Experimental Education*, **43**, no. 4 (1975), 35–42.

11. Everitt, B. *Graphical Techniques for Multivariate Data*. New York: North-Holland, 1978.

12. Gable, G. G. "A Multidimensional Model of Client Success when Engaging External Consultants." *Management Science*, **42**, no. 8 (1996) 1175–1198.

13. Halinar, J. C. "Principal Component Analysis in Plant Breeding." Unpublished report based on data collected by Dr. F. A. Bliss, University of Wisconsin, 1979.

14. Johnson, R. A., and G. K. Bhattacharyya. *Statistics: Principles and Methods* (5th ed.). New York: John Wiley, 2005.

15. Kim, L., and Y. Kim. "Innovation in a Newly Industrializing Country: A Multiple Discriminant Analysis." *Management Science*, **31**, no. 3 (1985) 312–322.

16. Klatzky, S. R., and R. W. Hodge. "A Canonical Correlation Analysis of Occupational Mobility." *Journal of the American Statistical Association*, **66**, no. 333 (1971), 16–22.

17. Lee, J., "Relationships Between Properties of Pulp-Fibre and Paper." Unpublished doctoral thesis, University of Toronto. Faculty of Forestry (1992).

18. MacCrimmon, K., and D. Wehrung. "Characteristics of Risk Taking Executives." *Management Science*, **36**, no. 4 (1990), 422–435.

19. Marriott, F. H. C. *The Interpretation of Multiple Observations*. London: Academic Press, 1974.

20. Mather, P. M. "Study of Factors Influencing Variation in Size Characteristics in Fluvioglacial Sediments." *Mathematical Geology*, **4**, no. 3 (1972), 219–234.

21. McLaughlin, M., et al. "Professional Mediators' Judgments of Mediation Tactics: Multidimensional Scaling and Cluster Analysis." *Journal of Applied Psychology*, **76**, no. 3 (1991), 465–473.

22. Naik, D. N., and R. Khattree. "Revisiting Olympic Track Records: Some Practical Considerations in the Principal Component Analysis." *The American Statistician*, **50**, no. 2 (1996), 140–144.

23. Nason, G. "Three-dimensional Projection Pursuit." *Applied Statistics*, **44**, no. 4 (1995), 411–430.

24. Smith, M., and R. Taffler. "Improving the Communication Function of Published Accounting Statements." *Accounting and Business Research*, **14**, no. 54 (1984), 139–146.

25. Spenner, K. I. "From Generation to Generation: The Transmission of Occupation." Ph.D. dissertation, University of Wisconsin, 1977.

26. Tabakoff, B., et al. "Differences in Platelet Enzyme Activity between Alcoholics and Nonalcoholics." *New England Journal of Medicine*, **318**, no. 3 (1988), 134–139.

27. Timm, N. H. *Multivariate Analysis with Applications in Education and Psychology*. Monterey, CA: Brooks/Cole, 1975.

28. Trieschmann, J. S., and G. E. Pinches. "A Multivariate Model for Predicting Financially Distressed P-L Insurers." *Journal of Risk and Insurance*, **40**, no. 3 (1973), 327–338.

29. Tukey, J. W. *Exploratory Data Analysis*. Reading, MA: Addison-Wesley, 1977.

30. Wainer, H., and D. Thissen. "Graphical Data Analysis." *Annual Review of Psychology*, **32**, (1981), 191–241.

31. Wartzman, R. "Don't Wave a Red Flag at the IRS." *The Wall Street Journal* (February 24, 1993), C1, C15.

32. Weihs, C., and H. Schmidli. "OMEGA (On Line Multivariate Exploratory Graphical Analysis): Routine Searching for Structure." *Statistical Science*, **5**, no. 2 (1990), 175–226.

Matrix Algebra and Random Vectors

2.1 Introduction

We saw in Chapter 1 that multivariate data can be conveniently displayed as an array of numbers. In general, a rectangular array of numbers with, for instance, n rows and p columns is called a *matrix* of dimension $n \times p$. The study of multivariate methods is greatly facilitated by the use of matrix algebra.

The matrix algebra results presented in this chapter will enable us to concisely state statistical models. Moreover, the formal relations expressed in matrix terms are easily programmed on computers to allow the routine calculation of important statistical quantities.

We begin by introducing some very basic concepts that are essential to both our geometrical interpretations and algebraic explanations of subsequent statistical techniques. If you have not been previously exposed to the rudiments of matrix algebra, you may prefer to follow the brief refresher in the next section by the more detailed review provided in Supplement 2A.

2.2 Some Basics of Matrix and Vector Algebra

Vectors

An array $\mathbf{x}$ of n real numbers $x_1, x_2, \ldots, x_n$ is called a *vector*, and it is written as

$$\mathbf{x} = \begin{bmatrix} x_1 \\ x_2 \\ \vdots \\ x_n \end{bmatrix} \quad \text{or} \quad \mathbf{x}' = [x_1, x_2, \ldots, x_n]$$

where the prime denotes the operation of *transposing* a column to a row.

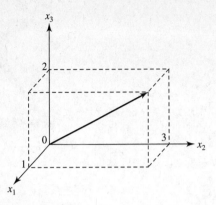

Figure 2.1 The vector $x' = [1, 3, 2]$.

A vector **x** can be represented geometrically as a directed line in n dimensions with component x_1 along the first axis, x_2 along the second axis, ..., and x_n along the nth axis. This is illustrated in Figure 2.1 for $n = 3$.

A vector can be *expanded* or *contracted* by multiplying it by a constant c. In particular, we define the vector $c\mathbf{x}$ as

$$c\mathbf{x} = \begin{bmatrix} cx_1 \\ cx_2 \\ \vdots \\ cx_n \end{bmatrix}$$

That is, $c\mathbf{x}$ is the vector obtained by multiplying each element of **x** by c. [See Figure 2.2(a).]

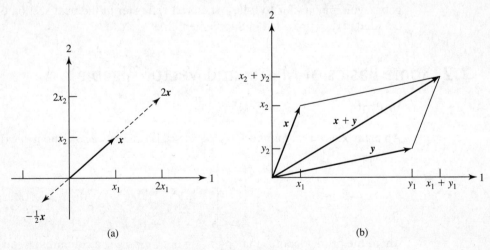

(a)   (b)

Figure 2.2 Scalar multiplication and vector addition.

Two vectors may be added. *Addition* of **x** and **y** is defined as

$$\mathbf{x} + \mathbf{y} = \begin{bmatrix} x_1 \\ x_2 \\ \vdots \\ x_n \end{bmatrix} + \begin{bmatrix} y_1 \\ y_2 \\ \vdots \\ y_n \end{bmatrix} = \begin{bmatrix} x_1 + y_1 \\ x_2 + y_2 \\ \vdots \\ x_n + y_n \end{bmatrix}$$

so that **x** + **y** is the vector with ith element $x_i + y_i$.

The sum of two vectors emanating from the origin is the diagonal of the parallelogram formed with the two original vectors as adjacent sides. This geometrical interpretation is illustrated in Figure 2.2(b).

A vector has both direction and length. In $n = 2$ dimensions, we consider the vector

$$\mathbf{x} = \begin{bmatrix} x_1 \\ x_2 \end{bmatrix}$$

The length of **x**, written $L_{\mathbf{x}}$, is defined to be

$$L_{\mathbf{x}} = \sqrt{x_1^2 + x_2^2}$$

Geometrically, the length of a vector in two dimensions can be viewed as the hypotenuse of a right triangle. This is demonstrated schematically in Figure 2.3.

The *length* of a vector $\mathbf{x}' = [x_1, x_2, \ldots, x_n]$, with n components, is defined by

$$L_{\mathbf{x}} = \sqrt{x_1^2 + x_2^2 + \cdots + x_n^2} \tag{2-1}$$

Multiplication of a vector **x** by a scalar c changes the length. From Equation (2-1),

$$L_{c\mathbf{x}} = \sqrt{c^2 x_1^2 + c^2 x_2^2 + \cdots + c^2 x_n^2}$$
$$= |c| \sqrt{x_1^2 + x_2^2 + \cdots + x_n^2} = |c| L_{\mathbf{x}}$$

Multiplication by c does not change the direction of the vector **x** if $c > 0$. However, a negative value of c creates a vector with a direction opposite that of **x**. From

$$L_{c\mathbf{x}} = |c| L_{\mathbf{x}} \tag{2-2}$$

it is clear that **x** is expanded if $|c| > 1$ and contracted if $0 < |c| < 1$. [Recall Figure 2.2(a).] Choosing $c = L_{\mathbf{x}}^{-1}$, we obtain the *unit vector* $L_{\mathbf{x}}^{-1}\mathbf{x}$, which has length 1 and lies in the direction of **x**.

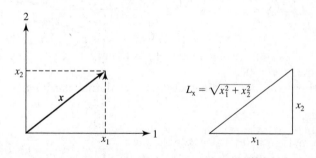

Figure 2.3
Length of $\mathbf{x} = \sqrt{x_1^2 + x_2^2}$.

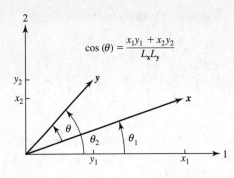

Figure 2.4 The angle θ between $\mathbf{x}' = [x_1, x_2]$ and $\mathbf{y}' = [y_1, y_2]$.

A second geometrical concept is *angle*. Consider two vectors in a plane and the angle θ between them, as in Figure 2.4. From the figure, θ can be represented as the difference between the angles θ_1 and θ_2 formed by the two vectors and the first coordinate axis. Since, by definition,

$$\cos(\theta_1) = \frac{x_1}{L_\mathbf{x}} \qquad \cos(\theta_2) = \frac{y_1}{L_\mathbf{y}}$$

$$\sin(\theta_1) = \frac{x_2}{L_\mathbf{x}} \qquad \sin(\theta_2) = \frac{y_2}{L_\mathbf{y}}$$

and

$$\cos(\theta) = \cos(\theta_2 - \theta_1) = \cos(\theta_2)\cos(\theta_1) + \sin(\theta_2)\sin(\theta_1)$$

the angle θ between the two vectors $\mathbf{x}' = [x_1, x_2]$ and $\mathbf{y}' = [y_1, y_2]$ is specified by

$$\cos(\theta) = \cos(\theta_2 - \theta_1) = \left(\frac{y_1}{L_\mathbf{y}}\right)\left(\frac{x_1}{L_\mathbf{x}}\right) + \left(\frac{y_2}{L_\mathbf{y}}\right)\left(\frac{x_2}{L_\mathbf{x}}\right) = \frac{x_1 y_1 + x_2 y_2}{L_\mathbf{x} L_\mathbf{y}} \qquad (2\text{-}3)$$

We find it convenient to introduce the *inner product* of two vectors. For $n = 2$ dimensions, the inner product of $\mathbf{x}$ and $\mathbf{y}$ is

$$\mathbf{x}'\mathbf{y} = x_1 y_1 + x_2 y_2$$

With this definition and Equation (2-3),

$$L_\mathbf{x} = \sqrt{\mathbf{x}'\mathbf{x}} \qquad \cos(\theta) = \frac{\mathbf{x}'\mathbf{y}}{L_\mathbf{x} L_\mathbf{y}} = \frac{\mathbf{x}'\mathbf{y}}{\sqrt{\mathbf{x}'\mathbf{x}}\sqrt{\mathbf{y}'\mathbf{y}}}$$

Since $\cos(90°) = \cos(270°) = 0$ and $\cos(\theta) = 0$ only if $\mathbf{x}'\mathbf{y} = 0$, $\mathbf{x}$ and $\mathbf{y}$ are perpendicular when $\mathbf{x}'\mathbf{y} = 0$.

For an arbitrary number of dimensions n, we define the inner product of $\mathbf{x}$ and $\mathbf{y}$ as

$$\mathbf{x}'\mathbf{y} = x_1 y_1 + x_2 y_2 + \cdots + x_n y_n \qquad (2\text{-}4)$$

The inner product is denoted by either $\mathbf{x}'\mathbf{y}$ or $\mathbf{y}'\mathbf{x}$.

Using the inner product, we have the natural extension of length and angle to vectors of n components:

$$L_x = \text{length of } \mathbf{x} = \sqrt{\mathbf{x'x}} \tag{2-5}$$

$$\cos(\theta) = \frac{\mathbf{x'y}}{L_x L_y} = \frac{\mathbf{x'y}}{\sqrt{\mathbf{x'x}}\sqrt{\mathbf{y'y}}} \tag{2-6}$$

Since, again, $\cos(\theta) = 0$ only if $\mathbf{x'y} = 0$, we say that $\mathbf{x}$ and $\mathbf{y}$ are *perpendicular* when $\mathbf{x'y} = 0$.

Example 2.1 (Calculating lengths of vectors and the angle between them) Given the vectors $\mathbf{x'} = [1, 3, 2]$ and $\mathbf{y'} = [-2, 1, -1]$, find $3\mathbf{x}$ and $\mathbf{x} + \mathbf{y}$. Next, determine the length of $\mathbf{x}$, the length of $\mathbf{y}$, and the angle between $\mathbf{x}$ and $\mathbf{y}$. Also, check that the length of $3\mathbf{x}$ is three times the length of $\mathbf{x}$.

First,

$$3\mathbf{x} = 3 \begin{bmatrix} 1 \\ 3 \\ 2 \end{bmatrix} = \begin{bmatrix} 3 \\ 9 \\ 6 \end{bmatrix}$$

$$\mathbf{x} + \mathbf{y} = \begin{bmatrix} 1 \\ 3 \\ 2 \end{bmatrix} + \begin{bmatrix} -2 \\ 1 \\ -1 \end{bmatrix} = \begin{bmatrix} 1 - 2 \\ 3 + 1 \\ 2 - 1 \end{bmatrix} = \begin{bmatrix} -1 \\ 4 \\ 1 \end{bmatrix}$$

Next, $\mathbf{x'x} = 1^2 + 3^2 + 2^2 = 14$, $\mathbf{y'y} = (-2)^2 + 1^2 + (-1)^2 = 6$, and $\mathbf{x'y} = 1(-2) + 3(1) + 2(-1) = -1$. Therefore,

$$L_x = \sqrt{\mathbf{x'x}} = \sqrt{14} = 3.742 \qquad L_y = \sqrt{\mathbf{y'y}} = \sqrt{6} = 2.449$$

and

$$\cos(\theta) = \frac{\mathbf{x'y}}{L_x L_y} = \frac{-1}{3.742 \times 2.449} = -.109$$

so $\theta = 96.3°$. Finally,

$$L_{3x} = \sqrt{3^2 + 9^2 + 6^2} = \sqrt{126} \quad \text{and} \quad 3L_x = 3\sqrt{14} = \sqrt{126}$$

showing $L_{3x} = 3L_x$. ∎

A pair of vectors $\mathbf{x}$ and $\mathbf{y}$ of the same dimension is said to be *linearly dependent* if there exist constants c_1 and c_2, both not zero, such that

$$c_1 \mathbf{x} + c_2 \mathbf{y} = \mathbf{0}$$

A set of vectors $\mathbf{x}_1, \mathbf{x}_2, \ldots, \mathbf{x}_k$ is said to be *linearly dependent* if there exist constants $c_1, c_2, \ldots, c_k$, not all zero, such that

$$c_1 \mathbf{x}_1 + c_2 \mathbf{x}_2 + \cdots + c_k \mathbf{x}_k = \mathbf{0} \tag{2-7}$$

Linear dependence implies that at least one vector in the set can be written as a linear combination of the other vectors. Vectors of the same dimension that are not linearly dependent are said to be *linearly independent*.

Example 2.2 (Identifying linearly independent vectors) Consider the set of vectors

$$\mathbf{x}_1 = \begin{bmatrix} 1 \\ 2 \\ 1 \end{bmatrix} \quad \mathbf{x}_2 = \begin{bmatrix} 1 \\ 0 \\ -1 \end{bmatrix} \quad \mathbf{x}_3 = \begin{bmatrix} 1 \\ -2 \\ 1 \end{bmatrix}$$

Setting

$$c_1 \mathbf{x}_1 + c_2 \mathbf{x}_2 + c_3 \mathbf{x}_3 = \mathbf{0}$$

implies that

$$\begin{aligned} c_1 + c_2 + \ c_3 &= 0 \\ 2c_1 \quad\quad - 2c_3 &= 0 \\ c_1 - c_2 + \ c_3 &= 0 \end{aligned}$$

with the unique solution $c_1 = c_2 = c_3 = 0$. As we cannot find three constants c_1, c_2, and c_3, *not all zero*, such that $c_1 \mathbf{x}_1 + c_2 \mathbf{x}_2 + c_3 \mathbf{x}_3 = \mathbf{0}$, the vectors $\mathbf{x}_1$, $\mathbf{x}_2$, and $\mathbf{x}_3$ are *linearly independent*. ∎

The *projection* (or shadow) of a vector $\mathbf{x}$ on a vector $\mathbf{y}$ is

$$\text{Projection of } \mathbf{x} \text{ on } \mathbf{y} = \frac{(\mathbf{x}'\mathbf{y})}{\mathbf{y}'\mathbf{y}} \mathbf{y} = \frac{(\mathbf{x}'\mathbf{y})}{L_y} \frac{1}{L_y} \mathbf{y} \tag{2-8}$$

where the vector $L_y^{-1}\mathbf{y}$ has unit length. The *length* of the projection is

$$\text{Length of projection} = \frac{|\mathbf{x}'\mathbf{y}|}{L_y} = L_x \left| \frac{\mathbf{x}'\mathbf{y}}{L_x L_y} \right| = L_x |\cos(\theta)| \tag{2-9}$$

where θ is the angle between $\mathbf{x}$ and $\mathbf{y}$. (See Figure 2.5.)

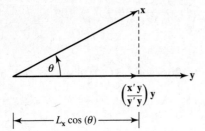

$$\left(\frac{\mathbf{x}'\mathbf{y}}{\mathbf{y}'\mathbf{y}}\right)\mathbf{y}$$

$\longmapsto L_x \cos(\theta) \longrightarrow$

Figure 2.5 The projection of $\mathbf{x}$ on $\mathbf{y}$.

Matrices

A *matrix* is any rectangular array of real numbers. We denote an arbitrary array of n rows and p columns by

$$\underset{(n \times p)}{\mathbf{A}} = \begin{bmatrix} a_{11} & a_{12} & \cdots & a_{1p} \\ a_{21} & a_{22} & \cdots & a_{2p} \\ \vdots & \vdots & \ddots & \vdots \\ a_{n1} & a_{n2} & \cdots & a_{np} \end{bmatrix}$$

Many of the vector concepts just introduced have direct generalizations to matrices.

The *transpose* operation $\mathbf{A}'$ of a matrix changes the columns into rows, so that the first column of $\mathbf{A}$ becomes the first row of $\mathbf{A}'$, the second column becomes the second row, and so forth.

Example 2.3 (The transpose of a matrix) If

$$\underset{(2\times3)}{\mathbf{A}} = \begin{bmatrix} 3 & -1 & 2 \\ 1 & 5 & 4 \end{bmatrix}$$

then

$$\underset{(3\times2)}{\mathbf{A}'} = \begin{bmatrix} 3 & 1 \\ -1 & 5 \\ 2 & 4 \end{bmatrix}$$

∎

A matrix may also be multiplied by a constant c. The product $c\mathbf{A}$ is the matrix that results from multiplying each element of $\mathbf{A}$ by c. Thus

$$\underset{(n\times p)}{c\mathbf{A}} = \begin{bmatrix} ca_{11} & ca_{12} & \cdots & ca_{1p} \\ ca_{21} & ca_{22} & \cdots & ca_{2p} \\ \vdots & \vdots & \ddots & \vdots \\ ca_{n1} & ca_{n2} & \cdots & ca_{np} \end{bmatrix}$$

Two matrices $\mathbf{A}$ and $\mathbf{B}$ of the same dimensions can be added. The sum $\mathbf{A} + \mathbf{B}$ has (i, j)th entry $a_{ij} + b_{ij}$.

Example 2.4 (The sum of two matrices and multiplication of a matrix by a constant) If

$$\underset{(2\times3)}{\mathbf{A}} = \begin{bmatrix} 0 & 3 & 1 \\ 1 & -1 & 1 \end{bmatrix} \quad \text{and} \quad \underset{(2\times3)}{\mathbf{B}} = \begin{bmatrix} 1 & -2 & -3 \\ 2 & 5 & 1 \end{bmatrix}$$

then

$$\underset{(2\times3)}{4\mathbf{A}} = \begin{bmatrix} 0 & 12 & 4 \\ 4 & -4 & 4 \end{bmatrix} \quad \text{and}$$

$$\underset{(2\times3)}{\mathbf{A}} + \underset{(2\times3)}{\mathbf{B}} = \begin{bmatrix} 0 + 1 & 3 - 2 & 1 - 3 \\ 1 + 2 & -1 + 5 & 1 + 1 \end{bmatrix} = \begin{bmatrix} 1 & 1 & -2 \\ 3 & 4 & 2 \end{bmatrix}$$

∎

It is also possible to define the multiplication of two matrices if the dimensions of the matrices conform in the following manner: When $\mathbf{A}$ is $(n \times k)$ and $\mathbf{B}$ is $(k \times p)$, so that the number of elements in a row of $\mathbf{A}$ is the same as the number of elements in a column of $\mathbf{B}$, we can form the matrix product $\mathbf{AB}$. An element of the new matrix $\mathbf{AB}$ is formed by taking the inner product of each row of $\mathbf{A}$ with each column of $\mathbf{B}$.

The *matrix product* **AB** is

$$\underset{(n\times k)(k\times p)}{\mathbf{A}\quad\mathbf{B}} = \begin{array}{l} \text{the } (n \times p) \text{ matrix whose entry in the } i\text{th row} \\ \text{and } j\text{th column is the inner product of the } i\text{th row} \\ \text{of } \mathbf{A} \text{ and the } j\text{th column of } \mathbf{B} \end{array}$$

or

$$(i, j) \text{ entry of } \mathbf{AB} = a_{i1}b_{1j} + a_{i2}b_{2j} + \cdots + a_{ik}b_{kj} = \sum_{\ell=1}^{k} a_{i\ell}b_{\ell j} \qquad (2\text{-}10)$$

When $k = 4$, we have four products to add for each entry in the matrix **AB**. Thus,

$$\underset{(n\times 4)(4\times p)}{\mathbf{A}\quad\mathbf{B}} = \begin{bmatrix} a_{11} & a_{12} & a_{13} & a_{14} \\ \vdots & \vdots & \vdots & \vdots \\ a_{i1} & a_{i2} & a_{i3} & a_{i4} \\ \vdots & \vdots & \vdots & \vdots \\ a_{n1} & a_{n2} & a_{n3} & a_{n4} \end{bmatrix} \begin{bmatrix} b_{11} & \cdots & b_{1j} & \cdots & b_{1p} \\ b_{21} & \cdots & b_{2j} & \cdots & b_{2p} \\ b_{31} & \cdots & b_{3j} & \cdots & b_{3p} \\ b_{41} & \cdots & b_{4j} & \cdots & b_{4p} \end{bmatrix}$$

$$\text{Column}$$
$$j$$

$$= \text{Row } i \begin{bmatrix} & \vdots & \\ \cdots & (a_{i1}b_{1j} + a_{i2}b_{2j} + a_{i3}b_{3j} + a_{i4}b_{4j}) & \cdots \\ & \vdots & \end{bmatrix}$$

Example 2.5 (Matrix multiplication) If

$$\mathbf{A} = \begin{bmatrix} 3 & -1 & 2 \\ 1 & 5 & 4 \end{bmatrix}, \quad \mathbf{B} = \begin{bmatrix} -2 \\ 7 \\ 9 \end{bmatrix}, \quad \text{and} \quad \mathbf{C} = \begin{bmatrix} 2 & 0 \\ 1 & -1 \end{bmatrix}$$

then

$$\underset{(2\times 3)(3\times 1)}{\mathbf{A}\quad\mathbf{B}} = \begin{bmatrix} 3 & -1 & 2 \\ 1 & 5 & 4 \end{bmatrix} \begin{bmatrix} -2 \\ 7 \\ 9 \end{bmatrix} = \begin{bmatrix} 3(-2) + (-1)(7) + 2(9) \\ 1(-2) + 5(7) \quad + 4(9) \end{bmatrix}$$

$$= \underset{(2\times 1)}{\begin{bmatrix} 5 \\ 69 \end{bmatrix}}$$

and

$$\underset{(2\times 2)(2\times 3)}{\mathbf{C}\quad\mathbf{A}} = \begin{bmatrix} 2 & 0 \\ 1 & -1 \end{bmatrix} \begin{bmatrix} 3 & -1 & 2 \\ 1 & 5 & 4 \end{bmatrix}$$

$$= \begin{bmatrix} 2(3) + 0(1) & 2(-1) + 0(5) & 2(2) + 0(4) \\ 1(3) - 1(1) & 1(-1) - 1(5) & 1(2) - 1(4) \end{bmatrix}$$

$$= \underset{(2\times 3)}{\begin{bmatrix} 6 & -2 & 4 \\ 2 & -6 & -2 \end{bmatrix}}$$

∎

When a matrix **B** consists of a single column, it is customary to use the lower-case **b** vector notation.

_____

Example 2.6 (Some typical products and their dimensions) Let

$$\mathbf{A} = \begin{bmatrix} 1 & -2 & 3 \\ 2 & 4 & -1 \end{bmatrix} \qquad \mathbf{b} = \begin{bmatrix} 7 \\ -3 \\ 6 \end{bmatrix} \qquad \mathbf{c} = \begin{bmatrix} 5 \\ 8 \\ -4 \end{bmatrix} \qquad \mathbf{d} = \begin{bmatrix} 2 \\ 9 \end{bmatrix}$$

Then $\mathbf{A}\mathbf{b}, \mathbf{b}\mathbf{c}', \mathbf{b}'\mathbf{c}$, and $\mathbf{d}'\mathbf{A}\mathbf{b}$ are typical products.

$$\mathbf{A}\mathbf{b} = \begin{bmatrix} 1 & -2 & 3 \\ 2 & 4 & -1 \end{bmatrix} \begin{bmatrix} 7 \\ -3 \\ 6 \end{bmatrix} = \begin{bmatrix} 31 \\ -4 \end{bmatrix}$$

The product $\mathbf{A}\mathbf{b}$ is a vector with dimension equal to the number of rows of $\mathbf{A}$.

$$\mathbf{b}'\mathbf{c} = \begin{bmatrix} 7 & -3 & 6 \end{bmatrix} \begin{bmatrix} 5 \\ 8 \\ -4 \end{bmatrix} = \begin{bmatrix} -13 \end{bmatrix}$$

The product $\mathbf{b}'\mathbf{c}$ is a 1×1 vector or a single number, here -13.

$$\mathbf{b}\mathbf{c}' = \begin{bmatrix} 7 \\ -3 \\ 6 \end{bmatrix} \begin{bmatrix} 5 & 8 & -4 \end{bmatrix} = \begin{bmatrix} 35 & 56 & -28 \\ -15 & -24 & 12 \\ 30 & 48 & -24 \end{bmatrix}$$

The product $\mathbf{b}\mathbf{c}'$ is a matrix whose row dimension equals the dimension of $\mathbf{b}$ and whose column dimension equals that of $\mathbf{c}$. This product is unlike $\mathbf{b}'\mathbf{c}$, which is a single number.

$$\mathbf{d}'\mathbf{A}\mathbf{b} = \begin{bmatrix} 2 & 9 \end{bmatrix} \begin{bmatrix} 1 & -2 & 3 \\ 2 & 4 & -1 \end{bmatrix} \begin{bmatrix} 7 \\ -3 \\ 6 \end{bmatrix} = \begin{bmatrix} 26 \end{bmatrix}$$

The product $\mathbf{d}'\mathbf{A}\mathbf{b}$ is a 1×1 vector or a single number, here 26. ∎

Square matrices will be of special importance in our development of statistical methods. A square matrix is said to be *symmetric* if $\mathbf{A} = \mathbf{A}'$ or $a_{ij} = a_{ji}$ for all i and j.

Example 2.7 (A symmetric matrix) The matrix

$$\begin{bmatrix} 3 & 5 \\ 5 & -2 \end{bmatrix}$$

is symmetric; the matrix

$$\begin{bmatrix} 3 & 6 \\ 4 & -2 \end{bmatrix}$$

is not symmetric. ∎

When two square matrices **A** and **B** are of the same dimension, both products **AB** and **BA** are defined, although they need not be equal. (See Supplement 2A.) If we let **I** denote the square matrix with ones on the diagonal and zeros elsewhere, it follows from the definition of matrix multiplication that the (i, j)th entry of **AI** is $a_{i1} \times 0 + \cdots + a_{i,j-1} \times 0 + a_{ij} \times 1 + a_{i,j+1} \times 0 + \cdots + a_{ik} \times 0 = a_{ij}$, so **AI** = **A**. Similarly, **IA** = **A**, so

$$\underset{(k \times k)(k \times k)}{\mathbf{I} \ \mathbf{A}} = \underset{(k \times k)(k \times k)}{\mathbf{A} \ \mathbf{I}} = \underset{(k \times k)}{\mathbf{A}} \quad \text{for any} \quad \underset{(k \times k)}{\mathbf{A}} \qquad (2\text{-}11)$$

The matrix **I** acts like 1 in ordinary multiplication $(1 \cdot a = a \cdot 1 = a)$, so it is called the *identity* matrix.

The fundamental scalar relation about the existence of an inverse number a^{-1} such that $a^{-1}a = aa^{-1} = 1$ if $a \neq 0$ has the following matrix algebra extension: If there exists a matrix **B** such that

$$\underset{(k \times k)(k \times k)}{\mathbf{B} \ \mathbf{A}} = \underset{(k \times k)(k \times k)}{\mathbf{A} \ \mathbf{B}} = \underset{(k \times k)}{\mathbf{I}}$$

then **B** is called the *inverse* of **A** and is denoted by $\mathbf{A}^{-1}$.

The technical condition that an inverse exists is that the k columns $\mathbf{a}_1, \mathbf{a}_2, \ldots, \mathbf{a}_k$ of **A** are linearly independent. That is, the existence of $\mathbf{A}^{-1}$ is equivalent to

$$c_1 \mathbf{a}_1 + c_2 \mathbf{a}_2 + \cdots + c_k \mathbf{a}_k = \mathbf{0} \quad \text{only if} \quad c_1 = \cdots = c_k = 0 \qquad (2\text{-}12)$$

(See Result 2A.9 in Supplement 2A.)

Example 2.8 (The existence of a matrix inverse) For

$$\mathbf{A} = \begin{bmatrix} 3 & 2 \\ 4 & 1 \end{bmatrix}$$

you may verify that

$$\begin{bmatrix} -.2 & .4 \\ .8 & -.6 \end{bmatrix} \begin{bmatrix} 3 & 2 \\ 4 & 1 \end{bmatrix} = \begin{bmatrix} (-.2)3 + (.4)4 & (-.2)2 + (.4)1 \\ (.8)3 + (-.6)4 & (.8)2 + (-.6)1 \end{bmatrix}$$

$$= \begin{bmatrix} 1 & 0 \\ 0 & 1 \end{bmatrix}$$

so

$$\begin{bmatrix} -.2 & .4 \\ .8 & -.6 \end{bmatrix}$$

is $\mathbf{A}^{-1}$. We note that

$$c_1 \begin{bmatrix} 3 \\ 4 \end{bmatrix} + c_2 \begin{bmatrix} 2 \\ 1 \end{bmatrix} = \begin{bmatrix} 0 \\ 0 \end{bmatrix}$$

implies that $c_1 = c_2 = 0$, so the columns of $\mathbf{A}$ are linearly independent. This confirms the condition stated in (2-12). ∎

A method for computing an inverse, when one exists, is given in Supplement 2A. The routine, but lengthy, calculations are usually relegated to a computer, especially when the dimension is greater than three. Even so, you must be forewarned that if the column sum in (2-12) is *nearly* $\mathbf{0}$ for some constants $c_1, \ldots, c_k$, then the computer may produce incorrect inverses due to extreme errors in rounding. It is always good to check the products $\mathbf{A}\mathbf{A}^{-1}$ and $\mathbf{A}^{-1}\mathbf{A}$ for equality with $\mathbf{I}$ when $\mathbf{A}^{-1}$ is produced by a computer package. (See Exercise 2.10.)

Diagonal matrices have inverses that are easy to compute. For example,

$$\begin{bmatrix} a_{11} & 0 & 0 & 0 & 0 \\ 0 & a_{22} & 0 & 0 & 0 \\ 0 & 0 & a_{33} & 0 & 0 \\ 0 & 0 & 0 & a_{44} & 0 \\ 0 & 0 & 0 & 0 & a_{55} \end{bmatrix} \quad \text{has inverse} \quad \begin{bmatrix} \dfrac{1}{a_{11}} & 0 & 0 & 0 & 0 \\ 0 & \dfrac{1}{a_{22}} & 0 & 0 & 0 \\ 0 & 0 & \dfrac{1}{a_{33}} & 0 & 0 \\ 0 & 0 & 0 & \dfrac{1}{a_{44}} & 0 \\ 0 & 0 & 0 & 0 & \dfrac{1}{a_{55}} \end{bmatrix}$$

if all the $a_{ii} \neq 0$.

Another special class of square matrices with which we shall become familiar are the *orthogonal* matrices, characterized by

$$\mathbf{Q}\mathbf{Q}' = \mathbf{Q}'\mathbf{Q} = \mathbf{I} \quad \text{or} \quad \mathbf{Q}' = \mathbf{Q}^{-1} \tag{2-13}$$

The name derives from the property that if $\mathbf{Q}$ has ith row $\mathbf{q}_i'$, then $\mathbf{Q}\mathbf{Q}' = \mathbf{I}$ implies that $\mathbf{q}_i'\mathbf{q}_i = 1$ and $\mathbf{q}_i'\mathbf{q}_j = 0$ for $i \neq j$, so the rows have unit length and are mutually perpendicular (orthogonal). According to the condition $\mathbf{Q}'\mathbf{Q} = \mathbf{I}$, the columns have the same property.

We conclude our brief introduction to the elements of matrix algebra by introducing a concept fundamental to multivariate statistical analysis. A square matrix $\mathbf{A}$ is said to have an *eigenvalue* λ, with corresponding *eigenvector* $\mathbf{x} \neq \mathbf{0}$, if

$$\mathbf{A}\mathbf{x} = \lambda\mathbf{x} \tag{2-14}$$

Ordinarily, we normalize $\mathbf{x}$ so that it has length unity; that is, $1 = \mathbf{x}'\mathbf{x}$. It is convenient to denote normalized eigenvectors by $\mathbf{e}$, and we do so in what follows. Sparing you the details of the derivation (see [1]), we state the following basic result:

Let $\mathbf{A}$ be a $k \times k$ square symmetric matrix. Then $\mathbf{A}$ has k pairs of eigenvalues and eigenvectors namely,

$$\lambda_1, \mathbf{e}_1 \qquad \lambda_2, \mathbf{e}_2 \qquad \cdots \qquad \lambda_k, \mathbf{e}_k \qquad\qquad (2\text{-}15)$$

The eigenvectors can be chosen to satisfy $1 = \mathbf{e}_1'\mathbf{e}_1 = \cdots = \mathbf{e}_k'\mathbf{e}_k$ and be mutually perpendicular. The eigenvectors are unique unless two or more eigenvalues are equal.

Example 2.9 (Verifying eigenvalues and eigenvectors) Let

$$\mathbf{A} = \begin{bmatrix} 1 & -5 \\ -5 & 1 \end{bmatrix}$$

Then, since

$$\begin{bmatrix} 1 & -5 \\ -5 & 1 \end{bmatrix} \begin{bmatrix} \dfrac{1}{\sqrt{2}} \\ -\dfrac{1}{\sqrt{2}} \end{bmatrix} = 6 \begin{bmatrix} \dfrac{1}{\sqrt{2}} \\ -\dfrac{1}{\sqrt{2}} \end{bmatrix}$$

$\lambda_1 = 6$ is an eigenvalue, and

$$\mathbf{e}_1 = \begin{bmatrix} \dfrac{1}{\sqrt{2}} \\ -\dfrac{1}{\sqrt{2}} \end{bmatrix}$$

is its corresponding normalized eigenvector. You may wish to show that a second eigenvalue–eigenvector pair is $\lambda_2 = -4$, $\mathbf{e}_2' = [1/\sqrt{2}, 1/\sqrt{2}]$. ∎

A method for calculating the λ's and $\mathbf{e}$'s is described in Supplement 2A. It is instructive to do a few sample calculations to understand the technique. We usually rely on a computer when the dimension of the square matrix is greater than two or three.

2.3 Positive Definite Matrices

The study of the variation and interrelationships in multivariate data is often based upon distances and the assumption that the data are multivariate normally distributed. Squared distances (see Chapter 1) and the multivariate normal density can be expressed in terms of matrix products called *quadratic forms* (see Chapter 4). Consequently, it should not be surprising that quadratic forms play a central role in multivariate analysis. In this section, we consider quadratic forms that are always nonnegative and the associated *positive definite* matrices.

Results involving quadratic forms and symmetric matrices are, in many cases, a direct consequence of an expansion for symmetric matrices known as the *spectral decomposition*. The spectral decomposition of a $k \times k$ symmetric matrix **A** is given by[1]

$$\underset{(k \times k)}{\mathbf{A}} = \lambda_1 \underset{(k \times 1)}{\mathbf{e}_1} \underset{(1 \times k)}{\mathbf{e}_1'} + \lambda_2 \underset{(k \times 1)}{\mathbf{e}_2} \underset{(1 \times k)}{\mathbf{e}_2'} + \cdots + \lambda_k \underset{(k \times 1)}{\mathbf{e}_k} \underset{(1 \times k)}{\mathbf{e}_k'} \qquad (2\text{-}16)$$

where $\lambda_1, \lambda_2, \ldots, \lambda_k$ are the eigenvalues of **A** and $\mathbf{e}_1, \mathbf{e}_2, \ldots, \mathbf{e}_k$ are the associated normalized eigenvectors. (See also Result 2A.14 in Supplement 2A). Thus, $\mathbf{e}_i'\mathbf{e}_i = 1$ for $i = 1, 2, \ldots, k$, and $\mathbf{e}_i'\mathbf{e}_j = 0$ for $i \neq j$.

Example 2.10 (The spectral decomposition of a matrix) Consider the symmetric matrix

$$\mathbf{A} = \begin{bmatrix} 13 & -4 & 2 \\ -4 & 13 & -2 \\ 2 & -2 & 10 \end{bmatrix}$$

The eigenvalues obtained from the characteristic equation $|\mathbf{A} - \lambda\mathbf{I}| = 0$ are $\lambda_1 = 9$, $\lambda_2 = 9$, and $\lambda_3 = 18$ (Definition 2A.30). The corresponding eigenvectors $\mathbf{e}_1, \mathbf{e}_2$, and $\mathbf{e}_3$ are the (normalized) solutions of the equations $\mathbf{A}\mathbf{e}_i = \lambda_i \mathbf{e}_i$ for $i = 1, 2, 3$. Thus, $\mathbf{A}\mathbf{e}_1 = \lambda\mathbf{e}_1$ gives

$$\begin{bmatrix} 13 & -4 & 2 \\ -4 & 13 & -2 \\ 2 & -2 & 10 \end{bmatrix} \begin{bmatrix} e_{11} \\ e_{21} \\ e_{31} \end{bmatrix} = 9 \begin{bmatrix} e_{11} \\ e_{21} \\ e_{31} \end{bmatrix}$$

or

$$13e_{11} - 4e_{21} + 2e_{31} = 9e_{11}$$
$$-4e_{11} + 13e_{21} - 2e_{31} = 9e_{21}$$
$$2e_{11} - 2e_{21} + 10e_{31} = 9e_{31}$$

Moving the terms on the right of the equals sign to the left yields three homogeneous equations in three unknowns, but two of the equations are redundant. Selecting one of the equations and arbitrarily setting $e_{11} = 1$ and $e_{21} = 1$, we find that $e_{31} = 0$. Consequently, the normalized eigenvector is $\mathbf{e}_1' = [1/\sqrt{1^2 + 1^2 + 0^2}, 1/\sqrt{1^2 + 1^2 + 0^2}, 0/\sqrt{1^2 + 1^2 + 0^2}] = [1/\sqrt{2}, 1/\sqrt{2}, 0]$, since the sum of the squares of its elements is unity. You may verify that $\mathbf{e}_2' = [1/\sqrt{18}, -1/\sqrt{18}, -4/\sqrt{18}]$ is also an eigenvector for $9 = \lambda_2$, and $\mathbf{e}_3' = [2/3, -2/3, 1/3]$ is the normalized eigenvector corresponding to the eigenvalue $\lambda_3 = 18$. Moreover, $\mathbf{e}_i'\mathbf{e}_j = 0$ for $i \neq j$.

---

[1]A proof of Equation (2-16) is beyond the scope of this book. The interested reader will find a proof in [6], Chapter 8.

The spectral decomposition of $\mathbf{A}$ is then

$$\mathbf{A} = \lambda_1 \mathbf{e}_1 \mathbf{e}_1' + \lambda_2 \mathbf{e}_2 \mathbf{e}_2' + \lambda_3 \mathbf{e}_3 \mathbf{e}_3'$$

or

$$\begin{bmatrix} 13 & -4 & 2 \\ -4 & 13 & -2 \\ 2 & -2 & 10 \end{bmatrix} = 9 \begin{bmatrix} \dfrac{1}{\sqrt{2}} \\[6pt] \dfrac{1}{\sqrt{2}} \\[6pt] 0 \end{bmatrix} \begin{bmatrix} \dfrac{1}{\sqrt{2}} & \dfrac{1}{\sqrt{2}} & 0 \end{bmatrix}$$

$$+ 9 \begin{bmatrix} \dfrac{1}{\sqrt{18}} \\[6pt] \dfrac{-1}{\sqrt{18}} \\[6pt] \dfrac{-4}{\sqrt{18}} \end{bmatrix} \begin{bmatrix} \dfrac{1}{\sqrt{18}} & \dfrac{-1}{\sqrt{18}} & \dfrac{-4}{\sqrt{18}} \end{bmatrix} + 18 \begin{bmatrix} \dfrac{2}{3} \\[6pt] -\dfrac{2}{3} \\[6pt] \dfrac{1}{3} \end{bmatrix} \begin{bmatrix} \dfrac{2}{3} & -\dfrac{2}{3} & \dfrac{1}{3} \end{bmatrix}$$

$$= 9 \begin{bmatrix} \dfrac{1}{2} & \dfrac{1}{2} & 0 \\[6pt] \dfrac{1}{2} & \dfrac{1}{2} & 0 \\[6pt] 0 & 0 & 0 \end{bmatrix} + 9 \begin{bmatrix} \dfrac{1}{18} & -\dfrac{1}{18} & -\dfrac{4}{18} \\[6pt] -\dfrac{1}{18} & \dfrac{1}{18} & \dfrac{4}{18} \\[6pt] -\dfrac{4}{18} & \dfrac{4}{18} & \dfrac{16}{18} \end{bmatrix}$$

$$+ 18 \begin{bmatrix} \dfrac{4}{9} & -\dfrac{4}{9} & \dfrac{2}{9} \\[6pt] -\dfrac{4}{9} & \dfrac{4}{9} & -\dfrac{2}{9} \\[6pt] \dfrac{2}{9} & -\dfrac{2}{9} & \dfrac{1}{9} \end{bmatrix}$$

as you may readily verify. ∎

The spectral decomposition is an important analytical tool. With it, we are very easily able to demonstrate certain statistical results. The first of these is a matrix explanation of distance, which we now develop.

Because $\mathbf{x}'\mathbf{A}\mathbf{x}$ has only squared terms x_i^2 and product terms $x_i x_k$, it is called a *quadratic form*. When a $k \times k$ symmetric matrix $\mathbf{A}$ is such that

$$0 \le \mathbf{x}'\mathbf{A}\mathbf{x} \tag{2-17}$$

for all $\mathbf{x}' = [x_1, x_2, \ldots, x_k]$, both the matrix $\mathbf{A}$ and the quadratic form are said to be *nonnegative definite*. If equality holds in (2-17) only for the vector $\mathbf{x}' = [0, 0, \ldots, 0]$, then $\mathbf{A}$ or the quadratic form is said to be *positive definite*. In other words, $\mathbf{A}$ is positive definite if

$$0 < \mathbf{x}'\mathbf{A}\mathbf{x} \tag{2-18}$$

for all vectors $\mathbf{x} \ne \mathbf{0}$.

Example 2.11 (A positive definite matrix and quadratic form) Show that the matrix for the following quadratic form is positive definite:

$$3x_1^2 + 2x_2^2 - 2\sqrt{2}\, x_1 x_2$$

To illustrate the general approach, we first write the quadratic form in matrix notation as

$$[x_1 \quad x_2] \begin{bmatrix} 3 & -\sqrt{2} \\ -\sqrt{2} & 2 \end{bmatrix} \begin{bmatrix} x_1 \\ x_2 \end{bmatrix} = \mathbf{x}'\mathbf{A}\mathbf{x}$$

By Definition 2A.30, the eigenvalues of $\mathbf{A}$ are the solutions of the equation $|\mathbf{A} - \lambda\mathbf{I}| = 0$, or $(3 - \lambda)(2 - \lambda) - 2 = 0$. The solutions are $\lambda_1 = 4$ and $\lambda_2 = 1$. Using the spectral decomposition in (2-16), we can write

$$\underset{(2\times2)}{\mathbf{A}} = \underset{(2\times1)(1\times2)}{\lambda_1 \mathbf{e}_1 \ \mathbf{e}_1'} + \underset{(2\times1)(1\times2)}{\lambda_2 \mathbf{e}_2 \ \mathbf{e}_2'}$$

$$= \underset{(2\times1)(1\times2)}{4\mathbf{e}_1 \ \mathbf{e}_1'} + \underset{(2\times1)(1\times2)}{\mathbf{e}_2 \ \mathbf{e}_2'}$$

where $\mathbf{e}_1$ and $\mathbf{e}_2$ are the normalized and orthogonal eigenvectors associated with the eigenvalues $\lambda_1 = 4$ and $\lambda_2 = 1$, respectively. Because 4 and 1 are scalars, premultiplication and postmultiplication of $\mathbf{A}$ by $\mathbf{x}'$ and $\mathbf{x}$, respectively, where $\mathbf{x}' = [x_1, x_2]$ is any *nonzero* vector, give

$$\underset{(1\times2)(2\times2)(2\times1)}{\mathbf{x}' \ \mathbf{A} \ \mathbf{x}} = \underset{(1\times2)(2\times1)(1\times2)(2\times1)}{4\mathbf{x}' \ \mathbf{e}_1 \ \mathbf{e}_1' \ \mathbf{x}} + \underset{(1\times2)(2\times1)(1\times2)(2\times1)}{\mathbf{x}' \ \mathbf{e}_2 \ \mathbf{e}_2' \ \mathbf{x}}$$

$$= 4y_1^2 + y_2^2 \geq 0$$

with

$$y_1 = \mathbf{x}'\mathbf{e}_1 = \mathbf{e}_1'\mathbf{x} \quad \text{and} \quad y_2 = \mathbf{x}'\mathbf{e}_2 = \mathbf{e}_2'\mathbf{x}$$

We now show that y_1 and y_2 are not both zero and, consequently, that $\mathbf{x}'\mathbf{A}\mathbf{x} = 4y_1^2 + y_2^2 > 0$, or $\mathbf{A}$ is *positive definite*.

From the definitions of y_1 and y_2, we have

$$\begin{bmatrix} y_1 \\ y_2 \end{bmatrix} = \begin{bmatrix} \mathbf{e}_1' \\ \mathbf{e}_2' \end{bmatrix} \begin{bmatrix} x_1 \\ x_2 \end{bmatrix}$$

or

$$\underset{(2\times1)}{\mathbf{y}} = \underset{(2\times2)(2\times1)}{\mathbf{E} \ \mathbf{x}}$$

Now $\mathbf{E}$ is an orthogonal matrix and hence has inverse $\mathbf{E}'$. Thus, $\mathbf{x} = \mathbf{E}'\mathbf{y}$. But $\mathbf{x}$ is a nonzero vector, and $\mathbf{0} \neq \mathbf{x} = \mathbf{E}'\mathbf{y}$ implies that $\mathbf{y} \neq \mathbf{0}$. ∎

Using the spectral decomposition, we can easily show that a $k \times k$ symmetric matrix $\mathbf{A}$ is a positive definite matrix if and only if every eigenvalue of $\mathbf{A}$ is positive. (See Exercise 2.17.) $\mathbf{A}$ is a nonnegative definite matrix if and only if all of its eigenvalues are greater than or equal to zero.

Assume for the moment that the p elements $x_1, x_2, \ldots, x_p$ of a vector $\mathbf{x}$ are realizations of p random variables $X_1, X_2, \ldots, X_p$. As we pointed out in Chapter 1, we can regard these elements as the coordinates of a point in p-dimensional space, and the "distance" of the point $[x_1, x_2, \ldots, x_p]'$ to the origin can, and in this case should, be interpreted in terms of standard deviation units. In this way, we can account for the inherent uncertainty (variability) in the observations. Points with the same associated "uncertainty" are regarded as being at the same distance from the origin.

If we use the distance formula introduced in Chapter 1 [see Equation (1-22)], the distance from the origin satisfies the general formula

$$(\text{distance})^2 = a_{11}x_1^2 + a_{22}x_2^2 + \cdots + a_{pp}x_p^2$$
$$+ 2(a_{12}x_1x_2 + a_{13}x_1x_3 + \cdots + a_{p-1,p}x_{p-1}x_p)$$

provided that $(\text{distance})^2 > 0$ for all $[x_1, x_2, \ldots, x_p] \neq [0, 0, \ldots, 0]$. Setting $a_{ij} = a_{ji}$, $i \neq j$, $i = 1, 2, \ldots, p$, $j = 1, 2, \ldots, p$, we have

$$0 < (\text{distance})^2 = [x_1, x_2, \ldots, x_p] \begin{bmatrix} a_{11} & a_{12} & \cdots & a_{1p} \\ a_{21} & a_{22} & \cdots & a_{2p} \\ \vdots & \vdots & \ddots & \vdots \\ a_{p1} & a_{p2} & \cdots & a_{pp} \end{bmatrix} \begin{bmatrix} x_1 \\ x_2 \\ \vdots \\ x_p \end{bmatrix}$$

or

$$0 < (\text{distance})^2 = \mathbf{x}'\mathbf{A}\mathbf{x} \qquad \text{for } \mathbf{x} \neq \mathbf{0} \qquad (2\text{-}19)$$

From (2-19), we see that the $p \times p$ symmetric matrix $\mathbf{A}$ is positive definite. In sum, distance is determined from a positive definite quadratic form $\mathbf{x}'\mathbf{A}\mathbf{x}$. Conversely, a positive definite quadratic form can be interpreted as a squared distance.

Comment. Let the square of the distance from the point $\mathbf{x}' = [x_1, x_2, \ldots, x_p]$ to the origin be given by $\mathbf{x}'\mathbf{A}\mathbf{x}$, where $\mathbf{A}$ is a $p \times p$ symmetric positive definite matrix. Then the square of the distance from $\mathbf{x}$ to an arbitrary fixed point $\boldsymbol{\mu}' = [\mu_1, \mu_2, \ldots, \mu_p]$ is given by the general expression $(\mathbf{x} - \boldsymbol{\mu})'\mathbf{A}(\mathbf{x} - \boldsymbol{\mu})$.

Expressing distance as the square root of a positive definite quadratic form allows us to give a geometrical interpretation based on the eigenvalues and eigenvectors of the matrix $\mathbf{A}$. For example, suppose $p = 2$. Then the points $\mathbf{x}' = [x_1, x_2]$ of constant distance c from the origin satisfy

$$\mathbf{x}'\mathbf{A}\mathbf{x} = a_{11}x_1^2 + a_{22}x_2^2 + 2a_{12}x_1x_2 = c^2$$

By the spectral decomposition, as in Example 2.11,

$$\mathbf{A} = \lambda_1\mathbf{e}_1\mathbf{e}_1' + \lambda_2\mathbf{e}_2\mathbf{e}_2' \quad \text{so} \quad \mathbf{x}'\mathbf{A}\mathbf{x} = \lambda_1(\mathbf{x}'\mathbf{e}_1)^2 + \lambda_2(\mathbf{x}'\mathbf{e}_2)^2$$

Now, $c^2 = \lambda_1 y_1^2 + \lambda_2 y_2^2$ is an ellipse in $y_1 = \mathbf{x}'\mathbf{e}_1$ and $y_2 = \mathbf{x}'\mathbf{e}_2$ because $\lambda_1, \lambda_2 > 0$ when $\mathbf{A}$ is positive definite. (See Exercise 2.17.) We easily verify that $\mathbf{x} = c\lambda_1^{-1/2}\mathbf{e}_1$ satisfies $\mathbf{x}'\mathbf{A}\mathbf{x} = \lambda_1(c\lambda_1^{-1/2}\mathbf{e}_1'\mathbf{e}_1)^2 = c^2$. Similarly, $\mathbf{x} = c\lambda_2^{-1/2}\mathbf{e}_2$ gives the appropriate distance in the $\mathbf{e}_2$ direction. Thus, the points at distance c lie on an ellipse whose axes are given by the eigenvectors of $\mathbf{A}$ with lengths proportional to the reciprocals of the square roots of the eigenvalues. The constant of proportionality is c. The situation is illustrated in Figure 2.6.

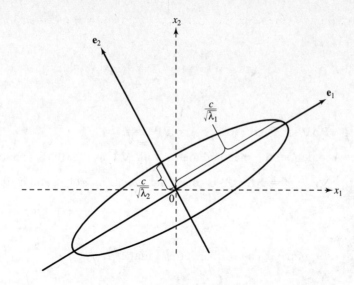

Figure 2.6 Points a constant distance c from the origin $(p = 2, 1 \le \lambda_1 < \lambda_2)$.

If $p > 2$, the points $\mathbf{x}' = [x_1, x_2, \ldots, x_p]$ a constant distance $c = \sqrt{\mathbf{x}'\mathbf{A}\mathbf{x}}$ from the origin lie on hyperellipsoids $c^2 = \lambda_1(\mathbf{x}'\mathbf{e}_1)^2 + \cdots + \lambda_p(\mathbf{x}'\mathbf{e}_p)^2$, whose axes are given by the eigenvectors of $\mathbf{A}$. The half-length in the direction $\mathbf{e}_i$ is equal to $c/\sqrt{\lambda_i}$, $i = 1, 2, \ldots, p$, where $\lambda_1, \lambda_2, \ldots, \lambda_p$ are the eigenvalues of $\mathbf{A}$.

2.4 A Square-Root Matrix

The spectral decomposition allows us to express the inverse of a square matrix in terms of its eigenvalues and eigenvectors, and this leads to a useful *square-root matrix*.

Let $\mathbf{A}$ be a $k \times k$ positive definite matrix with the spectral decomposition $\mathbf{A} = \sum_{i=1}^{k} \lambda_i \mathbf{e}_i \mathbf{e}_i'$. Let the normalized eigenvectors be the columns of another matrix $\mathbf{P} = [\mathbf{e}_1, \mathbf{e}_2, \ldots, \mathbf{e}_k]$. Then

$$\underset{(k \times k)}{\mathbf{A}} = \sum_{i=1}^{k} \lambda_i \underset{(k \times 1)}{\mathbf{e}_i} \underset{(1 \times k)}{\mathbf{e}_i'} = \underset{(k \times k)}{\mathbf{P}} \underset{(k \times k)}{\mathbf{\Lambda}} \underset{(k \times k)}{\mathbf{P}'} \tag{2-20}$$

where $\mathbf{PP}' = \mathbf{P}'\mathbf{P} = \mathbf{I}$ and $\mathbf{\Lambda}$ is the diagonal matrix

$$\underset{(k \times k)}{\mathbf{\Lambda}} = \begin{bmatrix} \lambda_1 & 0 & \cdots & 0 \\ 0 & \lambda_2 & \cdots & 0 \\ \vdots & \vdots & \ddots & \vdots \\ 0 & 0 & \cdots & \lambda_k \end{bmatrix} \quad \text{with } \lambda_i > 0$$

Thus,

$$\mathbf{A}^{-1} = \mathbf{P}\Lambda^{-1}\mathbf{P}' = \sum_{i=1}^{k} \frac{1}{\lambda_i} \mathbf{e}_i \mathbf{e}_i' \tag{2-21}$$

since $(\mathbf{P}\Lambda^{-1}\mathbf{P}')\mathbf{P}\Lambda\mathbf{P}' = \mathbf{P}\Lambda\mathbf{P}'(\mathbf{P}\Lambda^{-1}\mathbf{P}') = \mathbf{P}\mathbf{P}' = \mathbf{I}$.

Next, let $\Lambda^{1/2}$ denote the diagonal matrix with $\sqrt{\lambda_i}$ as the ith diagonal element. The matrix $\sum_{i=1}^{k} \sqrt{\lambda_i}\ \mathbf{e}_i \mathbf{e}_i' = \mathbf{P}\Lambda^{1/2}\mathbf{P}'$ is called the *square root* of $\mathbf{A}$ and is denoted by $\mathbf{A}^{1/2}$.

The square-root matrix, of a positive definite matrix $\mathbf{A}$,

$$\mathbf{A}^{1/2} = \sum_{i=1}^{k} \sqrt{\lambda_i}\ \mathbf{e}_i \mathbf{e}_i' = \mathbf{P}\Lambda^{1/2}\mathbf{P}' \tag{2-22}$$

has the following properties:

1. $(\mathbf{A}^{1/2})' = \mathbf{A}^{1/2}$ (that is, $\mathbf{A}^{1/2}$ is symmetric).

2. $\mathbf{A}^{1/2}\mathbf{A}^{1/2} = \mathbf{A}$.

3. $(\mathbf{A}^{1/2})^{-1} = \sum_{i=1}^{k} \frac{1}{\sqrt{\lambda_i}} \mathbf{e}_i \mathbf{e}_i' = \mathbf{P}\Lambda^{-1/2}\mathbf{P}'$, where $\Lambda^{-1/2}$ is a diagonal matrix with $1/\sqrt{\lambda_i}$ as the ith diagonal element.

4. $\mathbf{A}^{1/2}\mathbf{A}^{-1/2} = \mathbf{A}^{-1/2}\mathbf{A}^{1/2} = \mathbf{I}$, and $\mathbf{A}^{-1/2}\mathbf{A}^{-1/2} = \mathbf{A}^{-1}$, where $\mathbf{A}^{-1/2} = (\mathbf{A}^{1/2})^{-1}$.

2.5 Random Vectors and Matrices

A *random vector* is a vector whose elements are random variables. Similarly, a *random matrix* is a matrix whose elements are random variables. The expected value of a random matrix (or vector) is the matrix (vector) consisting of the expected values of each of its elements. Specifically, let $\mathbf{X} = \{X_{ij}\}$ be an $n \times p$ random matrix. Then the expected value of $\mathbf{X}$, denoted by $E(\mathbf{X})$, is the $n \times p$ matrix of numbers (if they exist)

$$E(\mathbf{X}) = \begin{bmatrix} E(X_{11}) & E(X_{12}) & \cdots & E(X_{1p}) \\ E(X_{21}) & E(X_{22}) & \cdots & E(X_{2p}) \\ \vdots & \vdots & \ddots & \vdots \\ E(X_{n1}) & E(X_{n2}) & \cdots & E(X_{np}) \end{bmatrix} \tag{2-23}$$

where, for each element of the matrix,[2]

$$
E(X_{ij}) = \begin{cases} \displaystyle\int_{-\infty}^{\infty} x_{ij}\, f_{ij}(x_{ij})\, dx_{ij} & \text{if } X_{ij} \text{ is a continuous random variable with} \\ & \text{probability density function } f_{ij}(x_{ij}) \\[2ex] \displaystyle\sum_{\text{all } x_{ij}} x_{ij} p_{ij}(x_{ij}) & \text{if } X_{ij} \text{ is a discrete random variable with} \\ & \text{probability function } p_{ij}(x_{ij}) \end{cases}
$$

Example 2.12 (Computing expected values for discrete random variables) Suppose $p = 2$ and $n = 1$, and consider the random vector $\mathbf{X}' = [X_1, X_2]$. Let the discrete random variable X_1 have the following probability function:

x_1	-1	0	1
$p_1(x_1)$	.3	.3	.4

Then $E(X_1) = \sum_{\text{all } x_1} x_1 p_1(x_1) = (-1)(.3) + (0)(.3) + (1)(.4) = .1.$

Similarly, let the discrete random variable X_2 have the probability function

x_2	0	1
$p_2(x_2)$	.8	.2

Then $E(X_2) = \sum_{\text{all } x_2} x_2 p_2(x_2) = (0)(.8) + (1)(.2) = .2.$

Thus,

$$
E(\mathbf{X}) = \begin{bmatrix} E(X_1) \\ E(X_2) \end{bmatrix} = \begin{bmatrix} .1 \\ .2 \end{bmatrix}
$$

∎

Two results involving the expectation of sums and products of matrices follow directly from the definition of the expected value of a random matrix and the univariate properties of expectation, $E(X_1 + Y_1) = E(X_1) + E(Y_1)$ and $E(cX_1) = cE(X_1)$. Let $\mathbf{X}$ and $\mathbf{Y}$ be random matrices of the same dimension, and let $\mathbf{A}$ and $\mathbf{B}$ be conformable matrices of constants. Then (see Exercise 2.40)

$$
E(\mathbf{X} + \mathbf{Y}) = E(\mathbf{X}) + E(\mathbf{Y}) \tag{2-24}
$$

$$
E(\mathbf{A}\mathbf{X}\mathbf{B}) = \mathbf{A}E(\mathbf{X})\mathbf{B}
$$

---

[2]If you are unfamiliar with calculus, you should concentrate on the interpretation of the expected value and, eventually, variance. Our development is based primarily on the properties of expectation rather than its particular evaluation for continuous or discrete random variables.

2.6 Mean Vectors and Covariance Matrices

Suppose $\mathbf{X}' = [X_1, X_2, \ldots, X_p]$ is a $p \times 1$ random vector. Then each element of $\mathbf{X}$ is a random variable with its own marginal probability distribution. (See Example 2.12.) The marginal means μ_i and variances σ_i^2 are defined as $\mu_i = E(X_i)$ and $\sigma_i^2 = E(X_i - \mu_i)^2$, $i = 1, 2, \ldots, p$, respectively. Specifically,

$$\mu_i = \begin{cases} \displaystyle\int_{-\infty}^{\infty} x_i f_i(x_i)\, dx_i & \text{if } X_i \text{ is a continuous random variable with probability} \\ & \text{density function } f_i(x_i) \\[2em] \displaystyle\sum_{\text{all } x_i} x_i p_i(x_i) & \text{if } X_i \text{ is a discrete random variable with probability} \\ & \text{function } p_i(x_i) \end{cases}$$

$$\sigma_i^2 = \begin{cases} \displaystyle\int_{-\infty}^{\infty} (x_i - \mu_i)^2 f_i(x_i)\, dx_i & \text{if } X_i \text{ is a continuous random variable} \\ & \text{with probability density function } f_i(x_i) \\[2em] \displaystyle\sum_{\text{all } x_i} (x_i - \mu_i)^2 p_i(x_i) & \text{if } X_i \text{ is a discrete random variable} \\ & \text{with probability function } p_i(x_i) \end{cases} \qquad (2\text{-}25)$$

It will be convenient in later sections to denote the marginal variances by σ_{ii} rather than the more traditional σ_i^2, and consequently, we shall adopt this notation.

The behavior of any pair of random variables, such as X_i and X_k, is described by their joint probability function, and a measure of the linear association between them is provided by the covariance

$$\sigma_{ik} = E(X_i - \mu_i)(X_k - \mu_k)$$

$$= \begin{cases} \displaystyle\int_{-\infty}^{\infty}\int_{-\infty}^{\infty} (x_i - \mu_i)(x_k - \mu_k)f_{ik}(x_i, x_k)dx_i\, dx_k & \text{if } X_i, X_k \text{ are continuous} \\ & \text{random variables with} \\ & \text{the joint density} \\ & \text{function } f_{ik}(x_i, x_k) \\[2em] \displaystyle\sum_{\text{all } x_i}\sum_{\text{all } x_k} (x_i - \mu_i)(x_k - \mu_k)p_{ik}(x_i, x_k) & \text{if } X_i, X_k \text{ are discrete} \\ & \text{random variables with} \\ & \text{joint probability} \\ & \text{function } p_{ik}(x_i, x_k) \end{cases}$$

$$(2\text{-}26)$$

and μ_i and μ_k, $i, k = 1, 2, \ldots, p$, are the marginal means. When $i = k$, the covariance becomes the marginal variance.

More generally, the collective behavior of the p random variables $X_1, X_2, \ldots, X_p$ or, equivalently, the random vector $\mathbf{X}' = [X_1, X_2, \ldots, X_p]$, is described by a joint probability density function $f(x_1, x_2, \ldots, x_p) = f(\mathbf{x})$. As we have already noted in this book, $f(\mathbf{x})$ will often be the multivariate normal density function. (See Chapter 4.)

If the joint probability $P[X_i \le x_i \text{ and } X_k \le x_k]$ can be written as the product of the corresponding marginal probabilities, so that

$$P[X_i \le x_i \text{ and } X_k \le x_k] = P[X_i \le x_i]P[X_k \le x_k] \qquad (2\text{-}27)$$

for all pairs of values x_i, x_k, then X_i and X_k are said to be *statistically independent*. When X_i and X_k are continuous random variables with joint density $f_{ik}(x_i, x_k)$ and marginal densities $f_i(x_i)$ and $f_k(x_k)$, the independence condition becomes

$$f_{ik}(x_i, x_k) = f_i(x_i)f_k(x_k)$$

for all pairs (x_i, x_k).

The p continuous random variables $X_1, X_2, \ldots, X_p$ are *mutually statistically independent* if their joint density can be factored as

$$f_{12\cdots p}(x_1, x_2, \ldots, x_p) = f_1(x_1)f_2(x_2) \cdots f_p(x_p) \qquad (2\text{-}28)$$

for all p-tuples $(x_1, x_2, \ldots, x_p)$.

Statistical independence has an important implication for covariance. The factorization in (2-28) implies that $\text{Cov}(X_i, X_k) = 0$. Thus,

$$\text{Cov}(X_i, X_k) = 0 \qquad \text{if } X_i \text{ and } X_k \text{ are independent} \qquad (2\text{-}29)$$

The converse of (2-29) is not true in general; there are situations where $\text{Cov}(X_i, X_k) = 0$, but X_i and X_k are not independent. (See [5].)

The means and covariances of the $p \times 1$ random vector $\mathbf{X}$ can be set out as matrices. The expected value of each element is contained in the vector of means $\boldsymbol{\mu} = E(\mathbf{X})$, and the p variances σ_{ii} and the $p(p-1)/2$ distinct covariances $\sigma_{ik}(i < k)$ are contained in the symmetric variance-covariance matrix $\boldsymbol{\Sigma} = E(\mathbf{X} - \boldsymbol{\mu})(\mathbf{X} - \boldsymbol{\mu})'$. Specifically,

$$E(\mathbf{X}) = \begin{bmatrix} E(X_1) \\ E(X_2) \\ \vdots \\ E(X_p) \end{bmatrix} = \begin{bmatrix} \mu_1 \\ \mu_2 \\ \vdots \\ \mu_p \end{bmatrix} = \boldsymbol{\mu} \qquad (2\text{-}30)$$

and

$$\boldsymbol{\Sigma} = E(\mathbf{X} - \boldsymbol{\mu})(\mathbf{X} - \boldsymbol{\mu})'$$

$$= E\left(\begin{bmatrix} X_1 - \mu_1 \\ X_2 - \mu_2 \\ \vdots \\ X_p - \mu_p \end{bmatrix} [X_1 - \mu_1, X_2 - \mu_2, \ldots, X_p - \mu_p] \right)$$

$$= E \begin{bmatrix} (X_1 - \mu_1)^2 & (X_1 - \mu_1)(X_2 - \mu_2) & \cdots & (X_1 - \mu_1)(X_p - \mu_p) \\ (X_2 - \mu_2)(X_1 - \mu_1) & (X_2 - \mu_2)^2 & \cdots & (X_2 - \mu_2)(X_p - \mu_p) \\ \vdots & \vdots & \ddots & \vdots \\ (X_p - \mu_p)(X_1 - \mu_1) & (X_p - \mu_p)(X_2 - \mu_2) & \cdots & (X_p - \mu_p)^2 \end{bmatrix}$$

$$= \begin{bmatrix} E(X_1 - \mu_1)^2 & E(X_1 - \mu_1)(X_2 - \mu_2) & \cdots & E(X_1 - \mu_1)(X_p - \mu_p) \\ E(X_2 - \mu_2)(X_1 - \mu_1) & E(X_2 - \mu_2)^2 & \cdots & E(X_2 - \mu_2)(X_p - \mu_p) \\ \vdots & \vdots & \ddots & \vdots \\ E(X_p - \mu_p)(X_1 - \mu_1) & E(X_p - \mu_p)(X_2 - \mu_2) & \cdots & E(X_p - \mu_p)^2 \end{bmatrix}$$

or

$$\Sigma = \text{Cov}(\mathbf{X}) = \begin{bmatrix} \sigma_{11} & \sigma_{12} & \cdots & \sigma_{1p} \\ \sigma_{21} & \sigma_{22} & \cdots & \sigma_{2p} \\ \vdots & \vdots & \ddots & \vdots \\ \sigma_{p1} & \sigma_{p2} & \cdots & \sigma_{pp} \end{bmatrix} \tag{2-31}$$

Example 2.13 (Computing the covariance matrix) Find the covariance matrix for the two random variables X_1 and X_2 introduced in Example 2.12 when their joint probability function $p_{12}(x_1, x_2)$ is represented by the entries in the body of the following table:

x_1 \ x_2	0	1	$p_1(x_1)$
-1	.24	.06	.3
0	.16	.14	.3
1	.40	.00	.4
$p_2(x_2)$	.8	.2	1

We have already shown that $\mu_1 = E(X_1) = .1$ and $\mu_2 = E(X_2) = .2$. (See Example 2.12.) In addition,

$$\sigma_{11} = E(X_1 - \mu_1)^2 = \sum_{\text{all } x_1} (x_1 - .1)^2 p_1(x_1)$$

$$= (-1 - .1)^2(.3) + (0 - .1)^2(.3) + (1 - .1)^2(.4) = .69$$

$$\sigma_{22} = E(X_2 - \mu_2)^2 = \sum_{\text{all } x_2} (x_2 - .2)^2 p_2(x_2)$$

$$= (0 - .2)^2(.8) + (1 - .2)^2(.2)$$

$$= .16$$

$$\sigma_{12} = E(X_1 - \mu_1)(X_2 - \mu_2) = \sum_{\text{all pairs } (x_1, x_2)} (x_1 - .1)(x_2 - .2)p_{12}(x_1, x_2)$$

$$= (-1 - .1)(0 - .2)(.24) + (-1 - .1)(1 - .2)(.06)$$

$$+ \cdots + (1 - .1)(1 - .2)(.00) = -.08$$

$$\sigma_{21} = E(X_2 - \mu_2)(X_1 - \mu_1) = E(X_1 - \mu_1)(X_2 - \mu_2) = \sigma_{12} = -.08$$

Consequently, with $\mathbf{X}' = [X_1, X_2]$,

$$\boldsymbol{\mu} = E(\mathbf{X}) = \begin{bmatrix} E(X_1) \\ E(X_2) \end{bmatrix} = \begin{bmatrix} \mu_1 \\ \mu_2 \end{bmatrix} = \begin{bmatrix} .1 \\ .2 \end{bmatrix}$$

and

$$\boldsymbol{\Sigma} = E(\mathbf{X} - \boldsymbol{\mu})(\mathbf{X} - \boldsymbol{\mu})'$$

$$= E \begin{bmatrix} (X_1 - \mu_1)^2 & (X_1 - \mu_1)(X_2 - \mu_2) \\ (X_2 - \mu_2)(X_1 - \mu_1) & (X_2 - \mu_2)^2 \end{bmatrix}$$

$$= \begin{bmatrix} E(X_1 - \mu_1)^2 & E(X_1 - \mu_1)(X_2 - \mu_2) \\ E(X_2 - \mu_2)(X_1 - \mu_1) & E(X_2 - \mu_2)^2 \end{bmatrix}$$

$$= \begin{bmatrix} \sigma_{11} & \sigma_{12} \\ \sigma_{21} & \sigma_{22} \end{bmatrix} = \begin{bmatrix} .69 & -.08 \\ -.08 & .16 \end{bmatrix} \qquad \blacksquare$$

We note that the computation of means, variances, and covariances for *discrete* random variables involves summation (as in Examples 2.12 and 2.13), while analogous computations for *continuous* random variables involve integration.

Because $\sigma_{ik} = E(X_i - \mu_i)(X_k - \mu_k) = \sigma_{ki}$, it is convenient to write the matrix appearing in (2-31) as

$$\boldsymbol{\Sigma} = E(\mathbf{X} - \boldsymbol{\mu})(\mathbf{X} - \boldsymbol{\mu})' = \begin{bmatrix} \sigma_{11} & \sigma_{12} & \cdots & \sigma_{1p} \\ \sigma_{12} & \sigma_{22} & \cdots & \sigma_{2p} \\ \vdots & \vdots & \ddots & \vdots \\ \sigma_{1p} & \sigma_{2p} & \cdots & \sigma_{pp} \end{bmatrix} \qquad (2\text{-}32)$$

We shall refer to $\boldsymbol{\mu}$ and $\boldsymbol{\Sigma}$ as the *population mean* (vector) and *population variance–covariance* (matrix), respectively.

The multivariate normal distribution is completely specified once the mean vector $\boldsymbol{\mu}$ and variance–covariance matrix $\boldsymbol{\Sigma}$ are given (see Chapter 4), so it is not surprising that these quantities play an important role in many multivariate procedures.

It is frequently informative to separate the information contained in variances σ_{ii} from that contained in measures of association and, in particular, the measure of association known as the *population correlation coefficient* ρ_{ik}. The correlation coefficient ρ_{ik} is defined in terms of the covariance σ_{ik} and variances σ_{ii} and σ_{kk} as

$$\rho_{ik} = \frac{\sigma_{ik}}{\sqrt{\sigma_{ii}}\,\sqrt{\sigma_{kk}}} \qquad (2\text{-}33)$$

The correlation coefficient measures the amount of *linear* association between the random variables X_i and X_k. (See, for example, [5].)

Let the population correlation matrix be the $p \times p$ symmetric matrix

$$
\boldsymbol{\rho} = \begin{bmatrix}
\dfrac{\sigma_{11}}{\sqrt{\sigma_{11}}\,\sqrt{\sigma_{11}}} & \dfrac{\sigma_{12}}{\sqrt{\sigma_{11}}\,\sqrt{\sigma_{22}}} & \cdots & \dfrac{\sigma_{1p}}{\sqrt{\sigma_{11}}\,\sqrt{\sigma_{pp}}} \\[2mm]
\dfrac{\sigma_{12}}{\sqrt{\sigma_{11}}\,\sqrt{\sigma_{22}}} & \dfrac{\sigma_{22}}{\sqrt{\sigma_{22}}\,\sqrt{\sigma_{22}}} & \cdots & \dfrac{\sigma_{2p}}{\sqrt{\sigma_{22}}\,\sqrt{\sigma_{pp}}} \\[2mm]
\vdots & \vdots & \ddots & \vdots \\[2mm]
\dfrac{\sigma_{1p}}{\sqrt{\sigma_{11}}\,\sqrt{\sigma_{pp}}} & \dfrac{\sigma_{2p}}{\sqrt{\sigma_{22}}\,\sqrt{\sigma_{pp}}} & \cdots & \dfrac{\sigma_{pp}}{\sqrt{\sigma_{pp}}\,\sqrt{\sigma_{pp}}}
\end{bmatrix}
$$

$$
= \begin{bmatrix}
1 & \rho_{12} & \cdots & \rho_{1p} \\
\rho_{12} & 1 & \cdots & \rho_{2p} \\
\vdots & \vdots & \ddots & \vdots \\
\rho_{1p} & \rho_{2p} & \cdots & 1
\end{bmatrix}
\tag{2-34}
$$

and let the $p \times p$ *standard deviation* matrix be

$$
\mathbf{V}^{1/2} = \begin{bmatrix}
\sqrt{\sigma_{11}} & 0 & \cdots & 0 \\
0 & \sqrt{\sigma_{22}} & \cdots & 0 \\
\vdots & \vdots & \ddots & \vdots \\
0 & 0 & \cdots & \sqrt{\sigma_{pp}}
\end{bmatrix}
\tag{2-35}
$$

Then it is easily verified (see Exercise 2.23) that

$$
\mathbf{V}^{1/2}\boldsymbol{\rho}\mathbf{V}^{1/2} = \boldsymbol{\Sigma}
\tag{2-36}
$$

and

$$
\boldsymbol{\rho} = (\mathbf{V}^{1/2})^{-1}\boldsymbol{\Sigma}(\mathbf{V}^{1/2})^{-1}
\tag{2-37}
$$

That is, $\boldsymbol{\Sigma}$ can be obtained from $\mathbf{V}^{1/2}$ and $\boldsymbol{\rho}$, whereas $\boldsymbol{\rho}$ can be obtained from $\boldsymbol{\Sigma}$. Moreover, the expression of these relationships in terms of matrix operations allows the calculations to be conveniently implemented on a computer.

Example 2.14 (Computing the correlation matrix from the covariance matrix)
Suppose

$$
\boldsymbol{\Sigma} = \begin{bmatrix}
4 & 1 & 2 \\
1 & 9 & -3 \\
2 & -3 & 25
\end{bmatrix} = \begin{bmatrix}
\sigma_{11} & \sigma_{12} & \sigma_{13} \\
\sigma_{12} & \sigma_{22} & \sigma_{23} \\
\sigma_{13} & \sigma_{23} & \sigma_{33}
\end{bmatrix}
$$

Obtain $\mathbf{V}^{1/2}$ and $\boldsymbol{\rho}$.

Here

$$\mathbf{V}^{1/2} = \begin{bmatrix} \sqrt{\sigma_{11}} & 0 & 0 \\ 0 & \sqrt{\sigma_{22}} & 0 \\ 0 & 0 & \sqrt{\sigma_{33}} \end{bmatrix} = \begin{bmatrix} 2 & 0 & 0 \\ 0 & 3 & 0 \\ 0 & 0 & 5 \end{bmatrix}$$

and

$$(\mathbf{V}^{1/2})^{-1} = \begin{bmatrix} \frac{1}{2} & 0 & 0 \\ 0 & \frac{1}{3} & 0 \\ 0 & 0 & \frac{1}{5} \end{bmatrix}$$

Consequently, from (2-37), the correlation matrix $\boldsymbol{\rho}$ is given by

$$(\mathbf{V}^{1/2})^{-1}\boldsymbol{\Sigma}(\mathbf{V}^{1/2})^{-1} = \begin{bmatrix} \frac{1}{2} & 0 & 0 \\ 0 & \frac{1}{3} & 0 \\ 0 & 0 & \frac{1}{5} \end{bmatrix} \begin{bmatrix} 4 & 1 & 2 \\ 1 & 9 & -3 \\ 2 & -3 & 25 \end{bmatrix} \begin{bmatrix} \frac{1}{2} & 0 & 0 \\ 0 & \frac{1}{3} & 0 \\ 0 & 0 & \frac{1}{5} \end{bmatrix}$$

$$= \begin{bmatrix} 1 & \frac{1}{6} & \frac{1}{5} \\ \frac{1}{6} & 1 & -\frac{1}{5} \\ \frac{1}{5} & -\frac{1}{5} & 1 \end{bmatrix}$$

∎

Partitioning the Covariance Matrix

Often, the characteristics measured on individual trials will fall naturally into two or more groups. As examples, consider measurements of variables representing consumption and income or variables representing personality traits and physical characteristics. One approach to handling these situations is to let the characteristics defining the distinct groups be subsets of the *total* collection of characteristics. If the total collection is represented by a $(p \times 1)$-dimensional random vector $\mathbf{X}$, the subsets can be regarded as components of $\mathbf{X}$ and can be sorted by partitioning $\mathbf{X}$.

In general, we can partition the p characteristics contained in the $p \times 1$ random vector $\mathbf{X}$ into, for instance, two groups of size q and $p - q$, respectively. For example, we can write

$$\mathbf{X} = \begin{bmatrix} X_1 \\ \vdots \\ X_q \\ \hdashline X_{q+1} \\ \vdots \\ X_p \end{bmatrix} \begin{matrix} \left.\vphantom{\begin{matrix}X_1\\ \vdots \\ X_q\end{matrix}}\right\} q \\ \\ \left.\vphantom{\begin{matrix}X_{q+1}\\ \vdots \\ X_p\end{matrix}}\right\} p-q \end{matrix} = \begin{bmatrix} \mathbf{X}^{(1)} \\ \hdashline \mathbf{X}^{(2)} \end{bmatrix} \quad \text{and} \quad \boldsymbol{\mu} = E(\mathbf{X}) = \begin{bmatrix} \mu_1 \\ \vdots \\ \mu_q \\ \hdashline \mu_{q+1} \\ \vdots \\ \mu_p \end{bmatrix} = \begin{bmatrix} \boldsymbol{\mu}^{(1)} \\ \hdashline \boldsymbol{\mu}^{(2)} \end{bmatrix}$$

$$(2\text{-}38)$$

From the definitions of the transpose and matrix multiplication,

$$(\mathbf{X}^{(1)} - \boldsymbol{\mu}^{(1)})(\mathbf{X}^{(2)} - \boldsymbol{\mu}^{(2)})'$$

$$= \begin{bmatrix} X_1 - \mu_1 \\ X_2 - \mu_2 \\ \vdots \\ X_q - \mu_q \end{bmatrix} [X_{q+1} - \mu_{q+1}, X_{q+2} - \mu_{q+2}, \ldots, X_p - \mu_p]$$

$$= \begin{bmatrix} (X_1 - \mu_1)(X_{q+1} - \mu_{q+1}) & (X_1 - \mu_1)(X_{q+2} - \mu_{q+2}) & \cdots & (X_1 - \mu_1)(X_p - \mu_p) \\ (X_2 - \mu_2)(X_{q+1} - \mu_{q+1}) & (X_2 - \mu_2)(X_{q+2} - \mu_{q+2}) & \cdots & (X_2 - \mu_2)(X_p - \mu_p) \\ \vdots & \vdots & \ddots & \vdots \\ (X_q - \mu_q)(X_{q+1} - \mu_{q+1}) & (X_q - \mu_q)(X_{q+2} - \mu_{q+2}) & \cdots & (X_q - \mu_q)(X_p - \mu_p) \end{bmatrix}$$

Upon taking the expectation of the matrix $(\mathbf{X}^{(1)} - \boldsymbol{\mu}^{(1)})(\mathbf{X}^{(2)} - \boldsymbol{\mu}^{(2)})'$, we get

$$E(\mathbf{X}^{(1)} - \boldsymbol{\mu}^{(1)})(\mathbf{X}^{(2)} - \boldsymbol{\mu}^{(2)})' = \begin{bmatrix} \sigma_{1,q+1} & \sigma_{1,q+2} & \cdots & \sigma_{1p} \\ \sigma_{2,q+1} & \sigma_{2,q+2} & \cdots & \sigma_{2p} \\ \vdots & \vdots & \ddots & \vdots \\ \sigma_{q,q+1} & \sigma_{q,q+2} & \cdots & \sigma_{qp} \end{bmatrix} = \boldsymbol{\Sigma}_{12} \quad (2\text{-}39)$$

which gives all the covariances, σ_{ij}, $i = 1, 2, \ldots, q$, $j = q + 1, q + 2, \ldots, p$, between a component of $\mathbf{X}^{(1)}$ and a component of $\mathbf{X}^{(2)}$. Note that the matrix $\boldsymbol{\Sigma}_{12}$ is not necessarily symmetric or even square.

Making use of the partitioning in Equation (2–38), we can easily demonstrate that

$$(\mathbf{X} - \boldsymbol{\mu})(\mathbf{X} - \boldsymbol{\mu})'$$

$$= \begin{bmatrix} \underset{(q\times1)}{(\mathbf{X}^{(1)} - \boldsymbol{\mu}^{(1)})}\underset{(1\times q)}{(\mathbf{X}^{(1)} - \boldsymbol{\mu}^{(1)})'} & \underset{(q\times1)}{(\mathbf{X}^{(1)} - \boldsymbol{\mu}^{(1)})}\underset{(1\times(p-q))}{(\mathbf{X}^{(2)} - \boldsymbol{\mu}^{(2)})'} \\ \underset{((p-q)\times1)}{(\mathbf{X}^{(2)} - \boldsymbol{\mu}^{(2)})}\underset{(1\times q)}{(\mathbf{X}^{(1)} - \boldsymbol{\mu}^{(1)})'} & \underset{((p-q)\times1)}{(\mathbf{X}^{(2)} - \boldsymbol{\mu}^{(2)})}\underset{(1\times(p-q))}{(\mathbf{X}^{(2)} - \boldsymbol{\mu}^{(2)})'} \end{bmatrix}$$

and consequently,

$$\underset{(p\times p)}{\boldsymbol{\Sigma}} = E(\mathbf{X} - \boldsymbol{\mu})(\mathbf{X} - \boldsymbol{\mu})' = \begin{array}{c} q \\ p-q \end{array}\overset{\displaystyle \overset{q}{} \quad \overset{p-q}{}}{\left[\begin{array}{c|c} \boldsymbol{\Sigma}_{11} & \boldsymbol{\Sigma}_{12} \\ \hline \boldsymbol{\Sigma}_{21} & \boldsymbol{\Sigma}_{22} \end{array}\right]}$$
$$\underset{(p\times p)}{}$$

$$= \left[\begin{array}{ccc|ccc} \sigma_{11} & \cdots & \sigma_{1q} & \sigma_{1,q+1} & \cdots & \sigma_{1p} \\ \vdots & \ddots & \vdots & \vdots & \ddots & \vdots \\ \sigma_{q1} & \cdots & \sigma_{qq} & \sigma_{q,q+1} & \cdots & \sigma_{qp} \\ \hline \sigma_{q+1,1} & \cdots & \sigma_{q+1,q} & \sigma_{q+1,q+1} & \cdots & \sigma_{q+1,p} \\ \vdots & \ddots & \vdots & \vdots & \ddots & \vdots \\ \sigma_{p1} & \cdots & \sigma_{pq} & \sigma_{p,q+1} & \cdots & \sigma_{pp} \end{array}\right] \quad (2\text{-}40)$$

Note that $\boldsymbol{\Sigma}_{12} = \boldsymbol{\Sigma}'_{21}$. The covariance matrix of $\mathbf{X}^{(1)}$ is $\boldsymbol{\Sigma}_{11}$, that of $\mathbf{X}^{(2)}$ is $\boldsymbol{\Sigma}_{22}$, and that of elements from $\mathbf{X}^{(1)}$ and $\mathbf{X}^{(2)}$ is $\boldsymbol{\Sigma}_{12}$ (or $\boldsymbol{\Sigma}_{21}$).

It is sometimes convenient to use the Cov $(\mathbf{X}^{(1)}, \mathbf{X}^{(2)})$ notation where

$$\text{Cov} (\mathbf{X}^{(1)}, \mathbf{X}^{(2)}) = \boldsymbol{\Sigma}_{12}$$

is a matrix containing all of the covariances between a component of $\mathbf{X}^{(1)}$ and a component of $\mathbf{X}^{(2)}$.

The Mean Vector and Covariance Matrix for Linear Combinations of Random Variables

Recall that if a single random variable, such as X_1, is multiplied by a constant c, then

$$E(cX_1) = cE(X_1) = c\mu_1$$

and

$$\text{Var}\,(cX_1) = E(cX_1 - c\mu_1)^2 = c^2 \text{Var}\,(X_1) = c^2 \sigma_{11}$$

If X_2 is a second random variable and a and b are constants, then, using additional properties of expectation, we get

$$
\begin{aligned}
\text{Cov}\,(aX_1, bX_2) &= E(aX_1 - a\mu_1)(bX_2 - b\mu_2)\\
&= abE(X_1 - \mu_1)(X_2 - \mu_2)\\
&= ab\,\text{Cov}\,(X_1, X_2) = ab\,\sigma_{12}
\end{aligned}
$$

Finally, for the linear combination $aX_1 + bX_2$, we have

$$
E(aX_1 + bX_2) = aE(X_1) + bE(X_2) = a\mu_1 + b\mu_2
$$
$$
\begin{aligned}
\text{Var}\,(aX_1 + bX_2) &= E[(aX_1 + bX_2) - (a\mu_1 + b\mu_2)]^2\\
&= E[a(X_1 - \mu_1) + b(X_2 - \mu_2)]^2\\
&= E[a^2(X_1 - \mu_1)^2 + b^2(X_2 - \mu_2)^2 + 2ab(X_1 - \mu_1)(X_2 - \mu_2)]\\
&= a^2 \text{Var}\,(X_1) + b^2 \text{Var}\,(X_2) + 2ab\,\text{Cov}\,(X_1, X_2)\\
&= a^2 \sigma_{11} + b^2 \sigma_{22} + 2ab\,\sigma_{12}
\end{aligned}
\qquad (2\text{-}41)
$$

With $\mathbf{c}' = [a, b]$, $aX_1 + bX_2$ can be written as

$$[a \quad b] \begin{bmatrix} X_1 \\ X_2 \end{bmatrix} = \mathbf{c}'\mathbf{X}$$

Similarly, $E(aX_1 + bX_2) = a\mu_1 + b\mu_2$ can be expressed as

$$[a \quad b] \begin{bmatrix} \mu_1 \\ \mu_2 \end{bmatrix} = \mathbf{c}'\boldsymbol{\mu}$$

If we let

$$\boldsymbol{\Sigma} = \begin{bmatrix} \sigma_{11} & \sigma_{12} \\ \sigma_{12} & \sigma_{22} \end{bmatrix}$$

be the variance–covariance matrix of $\mathbf{X}$, Equation (2–41) becomes

$$\text{Var}(aX_1 + bX_2) = \text{Var}(\mathbf{c}'\mathbf{X}) = \mathbf{c}'\boldsymbol{\Sigma}\mathbf{c} \tag{2-42}$$

since

$$\mathbf{c}'\boldsymbol{\Sigma}\mathbf{c} = [a \quad b]\begin{bmatrix} \sigma_{11} & \sigma_{12} \\ \sigma_{12} & \sigma_{22} \end{bmatrix}\begin{bmatrix} a \\ b \end{bmatrix} = a^2\sigma_{11} + 2ab\,\sigma_{12} + b^2\sigma_{22}$$

The preceding results can be extended to a linear combination of p random variables:

The linear combination $\mathbf{c}'\mathbf{X} = c_1X_1 + \cdots + c_pX_p$ has

$$\text{mean} = E(\mathbf{c}'\mathbf{X}) = \mathbf{c}'\boldsymbol{\mu}$$

$$\text{variance} = \text{Var}(\mathbf{c}'\mathbf{X}) = \mathbf{c}'\boldsymbol{\Sigma}\mathbf{c} \tag{2-43}$$

where $\boldsymbol{\mu} = E(\mathbf{X})$ and $\boldsymbol{\Sigma} = \text{Cov}(\mathbf{X})$.

In general, consider the q linear combinations of the p random variables $X_1, \ldots, X_p$:

$$Z_1 = c_{11}X_1 + c_{12}X_2 + \cdots + c_{1p}X_p$$
$$Z_2 = c_{21}X_1 + c_{22}X_2 + \cdots + c_{2p}X_p$$
$$\vdots \qquad\qquad\qquad \vdots$$
$$Z_q = c_{q1}X_1 + c_{q2}X_2 + \cdots + c_{qp}X_p$$

or

$$\mathbf{Z} = \underset{(q\times 1)}{\begin{bmatrix} Z_1 \\ Z_2 \\ \vdots \\ Z_q \end{bmatrix}} = \underset{(q\times p)}{\begin{bmatrix} c_{11} & c_{12} & \cdots & c_{1p} \\ c_{21} & c_{22} & \cdots & c_{2p} \\ \vdots & \vdots & \ddots & \vdots \\ c_{q1} & c_{q2} & \cdots & c_{qp} \end{bmatrix}} \underset{(p\times 1)}{\begin{bmatrix} X_1 \\ X_2 \\ \vdots \\ X_p \end{bmatrix}} = \mathbf{CX} \tag{2-44}$$

The linear combinations $\mathbf{Z} = \mathbf{CX}$ have

$$\boldsymbol{\mu}_\mathbf{Z} = E(\mathbf{Z}) = E(\mathbf{CX}) = \mathbf{C}\boldsymbol{\mu}_\mathbf{X}$$

$$\boldsymbol{\Sigma}_\mathbf{Z} = \text{Cov}(\mathbf{Z}) = \text{Cov}(\mathbf{CX}) = \mathbf{C}\boldsymbol{\Sigma}_\mathbf{X}\mathbf{C}' \tag{2-45}$$

where $\boldsymbol{\mu}_\mathbf{X}$ and $\boldsymbol{\Sigma}_\mathbf{X}$ are the mean vector and variance-covariance matrix of $\mathbf{X}$, respectively. (See Exercise 2.28 for the computation of the off-diagonal terms in $\mathbf{C}\boldsymbol{\Sigma}_\mathbf{X}\mathbf{C}'$.)

We shall rely heavily on the result in (2-45) in our discussions of principal components and factor analysis in Chapters 8 and 9.

Example 2.15 (Means and covariances of linear combinations) Let $\mathbf{X}' = [X_1, X_2]$ be a random vector with mean vector $\boldsymbol{\mu}'_\mathbf{X} = [\mu_1, \mu_2]$ and variance–covariance matrix

$$\boldsymbol{\Sigma}_\mathbf{X} = \begin{bmatrix} \sigma_{11} & \sigma_{12} \\ \sigma_{12} & \sigma_{22} \end{bmatrix}$$

Find the mean vector and covariance matrix for the linear combinations

$$Z_1 = X_1 - X_2$$
$$Z_2 = X_1 + X_2$$

or

$$\mathbf{Z} = \begin{bmatrix} Z_1 \\ Z_2 \end{bmatrix} = \begin{bmatrix} 1 & -1 \\ 1 & 1 \end{bmatrix} \begin{bmatrix} X_1 \\ X_2 \end{bmatrix} = \mathbf{CX}$$

in terms of $\boldsymbol{\mu}_\mathbf{X}$ and $\boldsymbol{\Sigma}_\mathbf{X}$.

Here

$$\boldsymbol{\mu}_\mathbf{Z} = E(\mathbf{Z}) = \mathbf{C}\boldsymbol{\mu}_\mathbf{X} = \begin{bmatrix} 1 & -1 \\ 1 & 1 \end{bmatrix} \begin{bmatrix} \mu_1 \\ \mu_2 \end{bmatrix} = \begin{bmatrix} \mu_1 - \mu_2 \\ \mu_1 + \mu_2 \end{bmatrix}$$

and

$$\boldsymbol{\Sigma}_\mathbf{Z} = \text{Cov}(\mathbf{Z}) = \mathbf{C}\boldsymbol{\Sigma}_\mathbf{X}\mathbf{C}' = \begin{bmatrix} 1 & -1 \\ 1 & 1 \end{bmatrix} \begin{bmatrix} \sigma_{11} & \sigma_{12} \\ \sigma_{12} & \sigma_{22} \end{bmatrix} \begin{bmatrix} 1 & 1 \\ -1 & 1 \end{bmatrix}$$

$$= \begin{bmatrix} \sigma_{11} - 2\sigma_{12} + \sigma_{22} & \sigma_{11} - \sigma_{22} \\ \sigma_{11} - \sigma_{22} & \sigma_{11} + 2\sigma_{12} + \sigma_{22} \end{bmatrix}$$

Note that if $\sigma_{11} = \sigma_{22}$—that is, if X_1 and X_2 have equal variances—the off-diagonal terms in $\boldsymbol{\Sigma}_\mathbf{Z}$ vanish. This demonstrates the well-known result that the sum and difference of two random variables with identical variances are uncorrelated. ∎

Partitioning the Sample Mean Vector and Covariance Matrix

Many of the matrix results in this section have been expressed in terms of population means and variances (covariances). The results in (2-36), (2-37), (2-38), and (2-40) also hold if the population quantities are replaced by their appropriately defined sample counterparts.

Let $\bar{\mathbf{x}}' = [\bar{x}_1, \bar{x}_2, \ldots, \bar{x}_p]$ be the vector of sample averages constructed from n observations on p variables $X_1, X_2, \ldots, X_p$, and let

$$\mathbf{S}_n = \begin{bmatrix} s_{11} & \cdots & s_{1p} \\ \vdots & \ddots & \vdots \\ s_{1p} & \cdots & s_{pp} \end{bmatrix}$$

$$= \begin{bmatrix} \dfrac{1}{n} \sum_{j=1}^{n} (x_{j1} - \bar{x}_1)^2 & \cdots & \dfrac{1}{n} \sum_{j=1}^{n} (x_{j1} - \bar{x}_1)(x_{jp} - \bar{x}_p) \\ \vdots & \ddots & \vdots \\ \dfrac{1}{n} \sum_{j=1}^{n} (x_{j1} - \bar{x}_1)(x_{jp} - \bar{x}_p) & \cdots & \dfrac{1}{n} \sum_{j=1}^{n} (x_{jp} - \bar{x}_p)^2 \end{bmatrix}$$

be the corresponding sample variance–covariance matrix.

The sample mean vector and the covariance matrix can be partitioned in order to distinguish quantities corresponding to groups of variables. Thus,

$$
\underset{(p\times 1)}{\bar{\mathbf{x}}} =
\begin{bmatrix}
\bar{x}_1 \\
\vdots \\
\bar{x}_q \\
\hline
\bar{x}_{q+1} \\
\vdots \\
\bar{x}_p
\end{bmatrix}
=
\begin{bmatrix}
\bar{\mathbf{x}}^{(1)} \\
\hline
\bar{\mathbf{x}}^{(2)}
\end{bmatrix}
\tag{2-46}
$$

and

$$
\underset{(p\times p)}{\mathbf{S}_n} =
\begin{bmatrix}
s_{11} & \cdots & s_{1q} & s_{1,q+1} & \cdots & s_{1p} \\
\vdots & \ddots & \vdots & \vdots & \ddots & \vdots \\
s_{q1} & \cdots & s_{qq} & s_{q,q+1} & \cdots & s_{qp} \\
\hline
s_{q+1,1} & \cdots & s_{q+1,q} & s_{q+1,q+1} & \cdots & s_{q+1,p} \\
\vdots & \ddots & \vdots & \vdots & \ddots & \vdots \\
s_{p1} & \cdots & s_{pq} & s_{p,q+1} & \cdots & s_{pp}
\end{bmatrix}
$$

$$
= \begin{array}{c} q \\ p-q \end{array}
\overset{\begin{array}{cc} q & p-q \end{array}}{\begin{bmatrix}
\mathbf{S}_{11} & \mathbf{S}_{12} \\
\mathbf{S}_{21} & \mathbf{S}_{22}
\end{bmatrix}}
\tag{2-47}
$$

where $\bar{\mathbf{x}}^{(1)}$ and $\bar{\mathbf{x}}^{(2)}$ are the sample mean vectors constructed from observations $\mathbf{x}^{(1)} = [x_1, \ldots, x_q]'$ and $\mathbf{x}^{(2)} = [x_{q+1}, \ldots, x_p]'$, respectively; $\mathbf{S}_{11}$ is the sample covariance matrix computed from observations $\mathbf{x}^{(1)}$; $\mathbf{S}_{22}$ is the sample covariance matrix computed from observations $\mathbf{x}^{(2)}$; and $\mathbf{S}_{12} = \mathbf{S}_{21}'$ is the sample covariance matrix for elements of $\mathbf{x}^{(1)}$ and elements of $\mathbf{x}^{(2)}$.

2.7 Matrix Inequalities and Maximization

Maximization principles play an important role in several multivariate techniques. Linear discriminant analysis, for example, is concerned with allocating observations to predetermined groups. The allocation rule is often a linear function of measurements that *maximizes* the separation between groups relative to their within-group variability. As another example, principal components are linear combinations of measurements with *maximum* variability.

The matrix inequalities presented in this section will easily allow us to derive certain maximization results, which will be referenced in later chapters.

Cauchy–Schwarz Inequality. Let $\mathbf{b}$ and $\mathbf{d}$ be *any* two $p \times 1$ vectors. Then

$$
(\mathbf{b}'\mathbf{d})^2 \leq (\mathbf{b}'\mathbf{b})(\mathbf{d}'\mathbf{d})
\tag{2-48}
$$

with equality if and only if $\mathbf{b} = c\,\mathbf{d}$ (or $\mathbf{d} = c\,\mathbf{b}$) for some constant c.

Proof. The inequality is obvious if either $\mathbf{b} = \mathbf{0}$ or $\mathbf{d} = \mathbf{0}$. Excluding this possibility, consider the vector $\mathbf{b} - x\mathbf{d}$, where x is an arbitrary scalar. Since the length of $\mathbf{b} - x\mathbf{d}$ is positive for $\mathbf{b} - x\mathbf{d} \neq \mathbf{0}$, in this case

$$0 < (\mathbf{b} - x\mathbf{d})'(\mathbf{b} - x\mathbf{d}) = \mathbf{b}'\mathbf{b} - x\mathbf{d}'\mathbf{b} - \mathbf{b}'(x\mathbf{d}) + x^2\mathbf{d}'\mathbf{d}$$
$$= \mathbf{b}'\mathbf{b} - 2x(\mathbf{b}'\mathbf{d}) + x^2(\mathbf{d}'\mathbf{d})$$

The last expression is quadratic in x. If we complete the square by adding and subtracting the scalar $(\mathbf{b}'\mathbf{d})^2/\mathbf{d}'\mathbf{d}$, we get

$$0 < \mathbf{b}'\mathbf{b} - \frac{(\mathbf{b}'\mathbf{d})^2}{\mathbf{d}'\mathbf{d}} + \frac{(\mathbf{b}'\mathbf{d})^2}{\mathbf{d}'\mathbf{d}} - 2x(\mathbf{b}'\mathbf{d}) + x^2(\mathbf{d}'\mathbf{d})$$

$$= \mathbf{b}'\mathbf{b} - \frac{(\mathbf{b}'\mathbf{d})^2}{\mathbf{d}'\mathbf{d}} + (\mathbf{d}'\mathbf{d})\left(x - \frac{\mathbf{b}'\mathbf{d}}{\mathbf{d}'\mathbf{d}}\right)^2$$

The term in brackets is zero if we choose $x = \mathbf{b}'\mathbf{d}/\mathbf{d}'\mathbf{d}$, so we conclude that

$$0 < \mathbf{b}'\mathbf{b} - \frac{(\mathbf{b}'\mathbf{d})^2}{\mathbf{d}'\mathbf{d}}$$

or $(\mathbf{b}'\mathbf{d})^2 < (\mathbf{b}'\mathbf{b})(\mathbf{d}'\mathbf{d})$ if $\mathbf{b} \neq x\mathbf{d}$ for some x.

Note that if $\mathbf{b} = c\mathbf{d}, 0 = (\mathbf{b} - c\mathbf{d})'(\mathbf{b} - c\mathbf{d})$, and the same argument produces $(\mathbf{b}'\mathbf{d})^2 = (\mathbf{b}'\mathbf{b})(\mathbf{d}'\mathbf{d})$. ∎

A simple, but important, extension of the Cauchy–Schwarz inequality follows directly.

Extended Cauchy–Schwarz Inequality. Let $\underset{(p\times 1)}{\mathbf{b}}$ and $\underset{(p\times 1)}{\mathbf{d}}$ be any two vectors, and let $\underset{(p\times p)}{\mathbf{B}}$ be a positive definite matrix. Then

$$(\mathbf{b}'\mathbf{d})^2 \leq (\mathbf{b}'\mathbf{B}\mathbf{b})(\mathbf{d}'\mathbf{B}^{-1}\mathbf{d}) \tag{2-49}$$

with equality if and only if $\mathbf{b} = c\mathbf{B}^{-1}\mathbf{d}$ (or $\mathbf{d} = c\mathbf{B}\mathbf{b}$) for some constant c.

Proof. The inequality is obvious when $\mathbf{b} = \mathbf{0}$ or $\mathbf{d} = \mathbf{0}$. For cases other than these, consider the square-root matrix $\mathbf{B}^{1/2}$ defined in terms of its eigenvalues λ_i and the normalized eigenvectors $\mathbf{e}_i$ as $\mathbf{B}^{1/2} = \sum_{i=1}^{p} \sqrt{\lambda_i}\, \mathbf{e}_i\mathbf{e}_i'$. If we set [see also (2-22)]

$$\mathbf{B}^{-1/2} = \sum_{i=1}^{p} \frac{1}{\sqrt{\lambda_i}}\, \mathbf{e}_i\mathbf{e}_i'$$

it follows that

$$\mathbf{b}'\mathbf{d} = \mathbf{b}'\mathbf{I}\mathbf{d} = \mathbf{b}'\mathbf{B}^{1/2}\mathbf{B}^{-1/2}\mathbf{d} = (\mathbf{B}^{1/2}\mathbf{b})'(\mathbf{B}^{-1/2}\mathbf{d})$$

and the proof is completed by applying the Cauchy–Schwarz inequality to the vectors $(\mathbf{B}^{1/2}\mathbf{b})$ and $(\mathbf{B}^{-1/2}\mathbf{d})$. ∎

The extended Cauchy–Schwarz inequality gives rise to the following maximization result.

Maximization Lemma. Let $\underset{(p\times p)}{\mathbf{B}}$ be positive definite and $\underset{(p\times 1)}{\mathbf{d}}$ be a given vector. Then, for an arbitrary nonzero vector $\underset{(p\times 1)}{\mathbf{x}}$,

$$\max_{\mathbf{x}\neq 0} \frac{\left(\mathbf{x}'\mathbf{d}\right)^2}{\mathbf{x}'\mathbf{B}\mathbf{x}} = \mathbf{d}'\mathbf{B}^{-1}\mathbf{d} \tag{2-50}$$

with the maximum attained when $\underset{(p\times 1)}{\mathbf{x}} = c\underset{(p\times p)(p\times 1)}{\mathbf{B}^{-1}\ \mathbf{d}}$ for any constant $c \neq 0$.

Proof. By the extended Cauchy–Schwarz inequality, $(\mathbf{x}'\mathbf{d})^2 \leq (\mathbf{x}'\mathbf{B}\mathbf{x})(\mathbf{d}'\mathbf{B}^{-1}\mathbf{d})$. Because $\mathbf{x} \neq \mathbf{0}$ and $\mathbf{B}$ is positive definite, $\mathbf{x}'\mathbf{B}\mathbf{x} > 0$. Dividing both sides of the inequality by the positive scalar $\mathbf{x}'\mathbf{B}\mathbf{x}$ yields the upper bound

$$\frac{\left(\mathbf{x}'\mathbf{d}\right)^2}{\mathbf{x}'\mathbf{B}\mathbf{x}} \leq \mathbf{d}'\mathbf{B}^{-1}\mathbf{d}$$

Taking the maximum over $\mathbf{x}$ gives Equation (2-50) because the bound is attained for $\mathbf{x} = c\mathbf{B}^{-1}\mathbf{d}$. ∎

A final maximization result will provide us with an interpretation of eigenvalues.

Maximization of Quadratic Forms for Points on the Unit Sphere. Let $\underset{(p\times p)}{\mathbf{B}}$ be a positive definite matrix with eigenvalues $\lambda_1 \geq \lambda_2 \geq \cdots \geq \lambda_p \geq 0$ and associated normalized eigenvectors $\mathbf{e}_1, \mathbf{e}_2, \ldots, \mathbf{e}_p$. Then

$$\begin{aligned}
\max_{\mathbf{x}\neq 0} \frac{\mathbf{x}'\mathbf{B}\mathbf{x}}{\mathbf{x}'\mathbf{x}} &= \lambda_1 \quad (\text{attained when } \mathbf{x} = \mathbf{e}_1) \\
\min_{\mathbf{x}\neq 0} \frac{\mathbf{x}'\mathbf{B}\mathbf{x}}{\mathbf{x}'\mathbf{x}} &= \lambda_p \quad (\text{attained when } \mathbf{x} = \mathbf{e}_p)
\end{aligned} \tag{2-51}$$

Moreover,

$$\max_{\mathbf{x}\perp \mathbf{e}_1,\ldots,\mathbf{e}_k} \frac{\mathbf{x}'\mathbf{B}\mathbf{x}}{\mathbf{x}'\mathbf{x}} = \lambda_{k+1} \quad (\text{attained when } \mathbf{x} = \mathbf{e}_{k+1}, k = 1, 2, \ldots, p-1) \tag{2-52}$$

where the symbol $\perp$ is read "is perpendicular to."

Proof. Let $\underset{(p\times p)}{\mathbf{P}}$ be the orthogonal matrix whose columns are the eigenvectors $\mathbf{e}_1, \mathbf{e}_2, \ldots, \mathbf{e}_p$ and Λ be the diagonal matrix with eigenvalues $\lambda_1, \lambda_2, \ldots, \lambda_p$ along the main diagonal. Let $\mathbf{B}^{1/2} = \mathbf{P}\Lambda^{1/2}\mathbf{P}'$ [see (2-22)] and $\underset{(p\times 1)}{\mathbf{y}} = \underset{(p\times p)(p\times 1)}{\mathbf{P}'\ \mathbf{x}}$.

Consequently, $\mathbf{x} \neq \mathbf{0}$ implies $\mathbf{y} \neq \mathbf{0}$. Thus,

$$\frac{\mathbf{x}'\mathbf{B}\mathbf{x}}{\mathbf{x}'\mathbf{x}} = \frac{\mathbf{x}'\mathbf{B}^{1/2}\mathbf{B}^{1/2}\mathbf{x}}{\mathbf{x}'\underbrace{\mathbf{P}\mathbf{P}'}_{\substack{\mathbf{I}\\(p\times p)}}\mathbf{x}} = \frac{\mathbf{x}'\mathbf{P}\Lambda^{1/2}\mathbf{P}'\mathbf{P}\Lambda^{1/2}\mathbf{P}'\mathbf{x}}{\mathbf{y}'\mathbf{y}} = \frac{\mathbf{y}'\Lambda\mathbf{y}}{\mathbf{y}'\mathbf{y}}$$

$$= \frac{\sum_{i=1}^{p}\lambda_i y_i^2}{\sum_{i=1}^{p} y_i^2} \leq \lambda_1 \frac{\sum_{i=1}^{p} y_i^2}{\sum_{i=1}^{p} y_i^2} = \lambda_1 \tag{2-53}$$

Setting $\mathbf{x} = \mathbf{e}_1$ gives

$$
\mathbf{y} = \mathbf{P}'\mathbf{e}_1 = \begin{bmatrix} 1 \\ 0 \\ \vdots \\ 0 \end{bmatrix}
$$

since

$$
\mathbf{e}_k'\mathbf{e}_1 = \begin{cases} 1, & k = 1 \\ 0, & k \neq 1 \end{cases}
$$

For this choice of $\mathbf{x}$, we have $\mathbf{y}'\boldsymbol{\Lambda}\mathbf{y}/\mathbf{y}'\mathbf{y} = \lambda_1/1 = \lambda_1$, or

$$
\frac{\mathbf{e}_1'\mathbf{B}\mathbf{e}_1}{\mathbf{e}_1'\mathbf{e}_1} = \mathbf{e}_1'\mathbf{B}\mathbf{e}_1 = \lambda_1 \tag{2-54}
$$

A similar argument produces the second part of (2-51).

Now, $\mathbf{x} = \mathbf{P}\mathbf{y} = y_1\mathbf{e}_1 + y_2\mathbf{e}_2 + \cdots + y_p\mathbf{e}_p$, so $\mathbf{x} \perp \mathbf{e}_1, \ldots, \mathbf{e}_k$ implies

$$
0 = \mathbf{e}_i'\mathbf{x} = y_1\mathbf{e}_i'\mathbf{e}_1 + y_2\mathbf{e}_i'\mathbf{e}_2 + \cdots + y_p\mathbf{e}_i'\mathbf{e}_p = y_i, \quad i \leq k
$$

Therefore, for $\mathbf{x}$ perpendicular to the first k eigenvectors $\mathbf{e}_i$, the left-hand side of the inequality in (2-53) becomes

$$
\frac{\mathbf{x}'\mathbf{B}\mathbf{x}}{\mathbf{x}'\mathbf{x}} = \frac{\displaystyle\sum_{i=k+1}^{p} \lambda_i y_i^2}{\displaystyle\sum_{i=k+1}^{p} y_i^2}
$$

Taking $y_{k+1} = 1$, $y_{k+2} = \cdots = y_p = 0$ gives the asserted maximum. ∎

For a fixed $\mathbf{x}_0 \neq \mathbf{0}$, $\mathbf{x}_0'\mathbf{B}\mathbf{x}_0/\mathbf{x}_0'\mathbf{x}_0$ has the same value as $\mathbf{x}'\mathbf{B}\mathbf{x}$, where $\mathbf{x}' = \mathbf{x}_0/\sqrt{\mathbf{x}_0'\mathbf{x}_0}$ is of unit length. Consequently, Equation (2-51) says that the largest eigenvalue, λ_1, is the maximum value of the quadratic form $\mathbf{x}'\mathbf{B}\mathbf{x}$ for all points $\mathbf{x}$ whose distance from the origin is unity. Similarly, λ_p is the smallest value of the quadratic form for all points $\mathbf{x}$ one unit from the origin. The largest and smallest eigenvalues thus represent extreme values of $\mathbf{x}'\mathbf{B}\mathbf{x}$ for points on the unit sphere. The "intermediate" eigenvalues of the $p \times p$ positive definite matrix $\mathbf{B}$ also have an interpretation as extreme values when $\mathbf{x}$ is further restricted to be perpendicular to the earlier choices.

Supplement

2A

Vectors and Matrices: Basic Concepts

Vectors

Many concepts, such as a person's health, intellectual abilities, or personality, cannot be adequately quantified as a single number. Rather, several different measurements $x_1, x_2, \ldots, x_m$ are required.

Definition 2A.1. An m-tuple of real numbers $(x_1, x_2, \ldots, x_i, \ldots, x_m)$ arranged in a column is called a *vector* and is denoted by a boldfaced, lowercase letter.

Examples of vectors are

$$\mathbf{x} = \begin{bmatrix} x_1 \\ x_2 \\ \vdots \\ x_m \end{bmatrix}, \quad \mathbf{a} = \begin{bmatrix} 1 \\ 0 \\ 0 \end{bmatrix}, \quad \mathbf{b} = \begin{bmatrix} 1 \\ -1 \\ 1 \\ -1 \end{bmatrix}, \quad \mathbf{y} = \begin{bmatrix} 1 \\ 2 \\ -2 \end{bmatrix}$$

Vectors are said to be equal if their corresponding entries are the same.

Definition 2A.2 (Scalar multiplication). Let c be an arbitrary scalar. Then the *product* $c\mathbf{x}$ is a vector with ith entry cx_i.

To illustrate scalar multiplication, take $c_1 = 5$ and $c_2 = -1.2$. Then

$$c_1\mathbf{y} = 5 \begin{bmatrix} 1 \\ 2 \\ -2 \end{bmatrix} = \begin{bmatrix} 5 \\ 10 \\ -10 \end{bmatrix} \quad \text{and} \quad c_2\mathbf{y} = (-1.2) \begin{bmatrix} 1 \\ 2 \\ -2 \end{bmatrix} = \begin{bmatrix} -1.2 \\ -2.4 \\ 2.4 \end{bmatrix}$$

Definition 2A.3 (Vector addition). The sum of two vectors $\mathbf{x}$ and $\mathbf{y}$, each having the same number of entries, is that vector

$$\mathbf{z} = \mathbf{x} + \mathbf{y} \quad \text{with } i\text{th entry} \quad z_i = x_i + y_i$$

Thus,

$$\begin{bmatrix} 3 \\ -1 \\ 4 \end{bmatrix} + \begin{bmatrix} 1 \\ 2 \\ -2 \end{bmatrix} = \begin{bmatrix} 4 \\ 1 \\ 2 \end{bmatrix}$$

$$\mathbf{x} \quad + \quad \mathbf{y} \quad = \quad \mathbf{z}$$

Taking the zero vector, $\mathbf{0}$, to be the m-tuple $(0, 0, \ldots, 0)$ and the vector $-\mathbf{x}$ to be the m-tuple $(-x_1, -x_2, \ldots, -x_m)$, the two operations of scalar multiplication and vector addition can be combined in a useful manner.

Definition 2A.4. The space of all real m-tuples, with scalar multiplication and vector addition as just defined, is called a *vector space*.

Definition 2A.5. The vector $\mathbf{y} = a_1\mathbf{x}_1 + a_2\mathbf{x}_2 + \cdots + a_k\mathbf{x}_k$ is a *linear combination* of the vectors $\mathbf{x}_1, \mathbf{x}_2, \ldots, \mathbf{x}_k$. The set of all linear combinations of $\mathbf{x}_1, \mathbf{x}_2, \ldots, \mathbf{x}_k$, is called their *linear span*.

Definition 2A.6. A set of vectors $\mathbf{x}_1, \mathbf{x}_2, \ldots, \mathbf{x}_k$ is said to be *linearly dependent* if there exist k numbers $(a_1, a_2, \ldots, a_k)$, not all zero, such that

$$a_1\mathbf{x}_1 + a_2\mathbf{x}_2 + \cdots + a_k\mathbf{x}_k = \mathbf{0}$$

Otherwise the set of vectors is said to be *linearly independent*.

If one of the vectors, for example, $\mathbf{x}_i$, is $\mathbf{0}$, the set is linearly dependent. (Let a_i be the only nonzero coefficient in Definition 2A.6.)

The familiar vectors with a one as an entry and zeros elsewhere are linearly independent. For $m = 4$,

$$\mathbf{x}_1 = \begin{bmatrix} 1 \\ 0 \\ 0 \\ 0 \end{bmatrix}, \quad \mathbf{x}_2 = \begin{bmatrix} 0 \\ 1 \\ 0 \\ 0 \end{bmatrix}, \quad \mathbf{x}_3 = \begin{bmatrix} 0 \\ 0 \\ 1 \\ 0 \end{bmatrix}, \quad \mathbf{x}_4 = \begin{bmatrix} 0 \\ 0 \\ 0 \\ 1 \end{bmatrix}$$

so

$$\mathbf{0} = a_1\mathbf{x}_1 + a_2\mathbf{x}_2 + a_3\mathbf{x}_3 + a_4\mathbf{x}_4 = \begin{bmatrix} a_1 \cdot 1 + a_2 \cdot 0 + a_3 \cdot 0 + a_4 \cdot 0 \\ a_1 \cdot 0 + a_2 \cdot 1 + a_3 \cdot 0 + a_4 \cdot 0 \\ a_1 \cdot 0 + a_2 \cdot 0 + a_3 \cdot 1 + a_4 \cdot 0 \\ a_1 \cdot 0 + a_2 \cdot 0 + a_3 \cdot 0 + a_4 \cdot 1 \end{bmatrix} = \begin{bmatrix} a_1 \\ a_2 \\ a_3 \\ a_4 \end{bmatrix}$$

implies that $a_1 = a_2 = a_3 = a_4 = 0$.

As another example, let $k = 3$ and $m = 3$, and let

$$\mathbf{x}_1 = \begin{bmatrix} 1 \\ 1 \\ 1 \end{bmatrix}, \qquad \mathbf{x}_2 = \begin{bmatrix} 2 \\ 5 \\ -1 \end{bmatrix}, \qquad \mathbf{x}_3 = \begin{bmatrix} 0 \\ 1 \\ -1 \end{bmatrix}$$

Then

$$2\mathbf{x}_1 - \mathbf{x}_2 + 3\mathbf{x}_3 = \mathbf{0}$$

Thus, $\mathbf{x}_1, \mathbf{x}_2, \mathbf{x}_3$ are a linearly dependent set of vectors, since any one can be written as a linear combination of the others (for example, $\mathbf{x}_2 = 2\mathbf{x}_1 + 3\mathbf{x}_3$).

Definition 2A.7. Any set of m linearly independent vectors is called a *basis* for the vector space of all m-tuples of real numbers.

Result 2A.1. Every vector can be expressed as a unique linear combination of a fixed basis. ∎

With $m = 4$, the usual choice of a basis is

$$\begin{bmatrix} 1 \\ 0 \\ 0 \\ 0 \end{bmatrix}, \quad \begin{bmatrix} 0 \\ 1 \\ 0 \\ 0 \end{bmatrix}, \quad \begin{bmatrix} 0 \\ 0 \\ 1 \\ 0 \end{bmatrix}, \quad \begin{bmatrix} 0 \\ 0 \\ 0 \\ 1 \end{bmatrix}$$

These four vectors were shown to be linearly independent. Any vector $\mathbf{x}$ can be uniquely expressed as

$$x_1 \begin{bmatrix} 1 \\ 0 \\ 0 \\ 0 \end{bmatrix} + x_2 \begin{bmatrix} 0 \\ 1 \\ 0 \\ 0 \end{bmatrix} + x_3 \begin{bmatrix} 0 \\ 0 \\ 1 \\ 0 \end{bmatrix} + x_4 \begin{bmatrix} 0 \\ 0 \\ 0 \\ 1 \end{bmatrix} = \begin{bmatrix} x_1 \\ x_2 \\ x_3 \\ x_4 \end{bmatrix} = \mathbf{x}$$

A vector consisting of m elements may be regarded geometrically as a point in m-dimensional space. For example, with $m = 2$, the vector $\mathbf{x}$ may be regarded as representing the point in the plane with coordinates x_1 and x_2.

Vectors have the geometrical properties of length and direction.

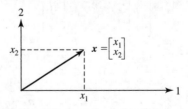

Definition 2A.8. The *length* of a vector of m elements emanating from the origin is given by the Pythagorean formula:

$$\text{length of } \mathbf{x} = L_{\mathbf{x}} = \sqrt{x_1^2 + x_2^2 + \cdots + x_m^2}$$

Definition 2A.9. The *angle* θ between two vectors **x** and **y**, both having m entries, is defined from

$$\cos(\theta) = \frac{(x_1 y_1 + x_2 y_2 + \cdots + x_m y_m)}{L_x L_y}$$

where L_x = length of **x** and L_y = length of **y**, $x_1, x_2, \ldots, x_m$ are the elements of **x**, and $y_1, y_2, \ldots, y_m$ are the elements of **y**.

Let

$$\mathbf{x} = \begin{bmatrix} -1 \\ 5 \\ 2 \\ -2 \end{bmatrix} \quad \text{and} \quad \mathbf{y} = \begin{bmatrix} 4 \\ -3 \\ 0 \\ 1 \end{bmatrix}$$

Then the length of **x**, the length of **y**, and the cosine of the angle between the two vectors are

$$\text{length of } \mathbf{x} = \sqrt{(-1)^2 + 5^2 + 2^2 + (-2)^2} = \sqrt{34} = 5.83$$

$$\text{length of } \mathbf{y} = \sqrt{4^2 + (-3)^2 + 0^2 + 1^2} = \sqrt{26} = 5.10$$

and

$$\cos(\theta) = \frac{1}{L_x}\frac{1}{L_y} [x_1 y_1 + x_2 y_2 + x_3 y_3 + x_4 y_4]$$

$$= \frac{1}{\sqrt{34}}\frac{1}{\sqrt{26}} [(-1)4 + 5(-3) + 2(0) + (-2)1]$$

$$= \frac{1}{5.83 \times 5.10} [-21] = -.706$$

Consequently, $\theta = 135°$.

Definition 2A.10. The *inner* (or *dot*) *product* of two vectors **x** and **y** with the same number of entries is defined as the sum of component products:

$$x_1 y_1 + x_2 y_2 + \cdots + x_m y_m$$

We use the notation $\mathbf{x}'\mathbf{y}$ or $\mathbf{y}'\mathbf{x}$ to denote this inner product.

With the $\mathbf{x}'\mathbf{y}$ notation, we may express the length of a vector and the cosine of the angle between two vectors as

$$L_\mathbf{x} = \text{length of } \mathbf{x} = \sqrt{x_1^2 + x_2^2 + \cdots + x_m^2} = \sqrt{\mathbf{x}'\mathbf{x}}$$

$$\cos(\theta) = \frac{\mathbf{x}'\mathbf{y}}{\sqrt{\mathbf{x}'\mathbf{x}}\sqrt{\mathbf{y}'\mathbf{y}}}$$

Definition 2A.11. When the angle between two vectors $\mathbf{x}$, $\mathbf{y}$ is $\theta = 90°$ or $270°$, we say that $\mathbf{x}$ and $\mathbf{y}$ are perpendicular. Since $\cos(\theta) = 0$ only if $\theta = 90°$ or $270°$, the condition becomes

$$\mathbf{x} \text{ and } \mathbf{y} \text{ are } \textit{perpendicular} \text{ if } \mathbf{x}'\mathbf{y} = 0$$

We write $\mathbf{x} \perp \mathbf{y}$.

The basis vectors

$$\begin{bmatrix} 1 \\ 0 \\ 0 \\ 0 \end{bmatrix}, \quad \begin{bmatrix} 0 \\ 1 \\ 0 \\ 0 \end{bmatrix}, \quad \begin{bmatrix} 0 \\ 0 \\ 1 \\ 0 \end{bmatrix}, \quad \begin{bmatrix} 0 \\ 0 \\ 0 \\ 1 \end{bmatrix}$$

are mutually perpendicular. Also, each has length unity. The same construction holds for any number of entries m.

Result 2A.2.

(a) $\mathbf{z}$ is perpendicular to every vector if and only if $\mathbf{z} = \mathbf{0}$.

(b) If $\mathbf{z}$ is perpendicular to each vector $\mathbf{x}_1, \mathbf{x}_2, \ldots, \mathbf{x}_k$, then $\mathbf{z}$ is perpendicular to their linear span.

(c) Mutually perpendicular vectors are linearly independent.   ∎

Definition 2A.12. The *projection* (or *shadow*) of a vector $\mathbf{x}$ on a vector $\mathbf{y}$ is

$$\textit{projection of } \mathbf{x} \textit{ on } \mathbf{y} = \frac{(\mathbf{x}'\mathbf{y})}{L_\mathbf{y}^2} \mathbf{y}$$

If $\mathbf{y}$ has unit length so that $L_y = 1$,

$$\textit{projection of } \mathbf{x} \textit{ on } \mathbf{y} = (\mathbf{x}'\mathbf{y})\mathbf{y}$$

If $\mathbf{y}_1, \mathbf{y}_2, \ldots, \mathbf{y}_r$ are mutually perpendicular, the *projection* (or *shadow*) of a vector $\mathbf{x}$ *on the linear span of* $\mathbf{y}_1, \mathbf{y}_2, \ldots, \mathbf{y}_r$ is

$$\frac{(\mathbf{x}'\mathbf{y}_1)}{\mathbf{y}_1'\mathbf{y}_1} \mathbf{y}_1 + \frac{(\mathbf{x}'\mathbf{y}_2)}{\mathbf{y}_2'\mathbf{y}_2} \mathbf{y}_2 + \cdots + \frac{(\mathbf{x}'\mathbf{y}_r)}{\mathbf{y}_r'\mathbf{y}_r} \mathbf{y}_r$$

Result 2A.3 (Gram–Schmidt Process). Given linearly independent vectors $\mathbf{x}_1$, $\mathbf{x}_2, \ldots, \mathbf{x}_k$, there exist mutually perpendicular vectors $\mathbf{u}_1, \mathbf{u}_2, \ldots, \mathbf{u}_k$ with the same linear span. These may be constructed sequentially by setting

$$\mathbf{u}_1 = \mathbf{x}_1$$

$$\mathbf{u}_2 = \mathbf{x}_2 - \frac{(\mathbf{x}_2'\mathbf{u}_1)}{\mathbf{u}_1'\mathbf{u}_1} \mathbf{u}_1$$

$$\vdots \qquad \vdots$$

$$\mathbf{u}_k = \mathbf{x}_k - \frac{(\mathbf{x}_k'\mathbf{u}_1)}{\mathbf{u}_1'\mathbf{u}_1} \mathbf{u}_1 - \cdots - \frac{(\mathbf{x}_k'\mathbf{u}_{k-1})}{\mathbf{u}_{k-1}'\mathbf{u}_{k-1}} \mathbf{u}_{k-1}$$

We can also convert the $\mathbf{u}$'s to unit length by setting $\mathbf{z}_j = \mathbf{u}_j/\sqrt{\mathbf{u}_j'\mathbf{u}_j}$. In this construction, $(\mathbf{x}_k'\mathbf{z}_j)\,\mathbf{z}_j$ is the projection of $\mathbf{x}_k$ on $\mathbf{z}_j$ and $\sum_{j=1}^{k-1} (\mathbf{x}_k'\mathbf{z}_j)\mathbf{z}_j$ is the *projection of $\mathbf{x}_k$ on the linear span of* $\mathbf{x}_1, \mathbf{x}_2, \dots, \mathbf{x}_{k-1}$. ∎

For example, to construct perpendicular vectors from

$$\mathbf{x}_1 = \begin{bmatrix} 4 \\ 0 \\ 0 \\ 2 \end{bmatrix} \quad \text{and} \quad \mathbf{x}_2 = \begin{bmatrix} 3 \\ 1 \\ 0 \\ -1 \end{bmatrix}$$

we take

$$\mathbf{u}_1 = \mathbf{x}_1 = \begin{bmatrix} 4 \\ 0 \\ 0 \\ 2 \end{bmatrix}$$

so

$$\mathbf{u}_1'\mathbf{u}_1 = 4^2 + 0^2 + 0^2 + 2^2 = 20$$

and

$$\mathbf{x}_2'\mathbf{u}_1 = 3(4) + 1(0) + 0(0) - 1(2) = 10$$

Thus,

$$\mathbf{u}_2 = \begin{bmatrix} 3 \\ 1 \\ 0 \\ -1 \end{bmatrix} - \frac{10}{20}\begin{bmatrix} 4 \\ 0 \\ 0 \\ 2 \end{bmatrix} = \begin{bmatrix} 1 \\ 1 \\ 0 \\ -2 \end{bmatrix} \quad \text{and} \quad \mathbf{z}_1 = \frac{1}{\sqrt{20}}\begin{bmatrix} 4 \\ 0 \\ 0 \\ 2 \end{bmatrix}, \quad \mathbf{z}_2 = \frac{1}{\sqrt{6}}\begin{bmatrix} 1 \\ 1 \\ 0 \\ -2 \end{bmatrix}$$

Matrices

Definition 2A.13. An $m \times k$ *matrix*, generally denoted by a boldface uppercase letter such as $\mathbf{A}, \mathbf{R}, \boldsymbol{\Sigma}$, and so forth, is a rectangular array of elements having m rows and k columns.

Examples of matrices are

$$\mathbf{A} = \begin{bmatrix} -7 & 2 \\ 0 & 1 \\ 3 & 4 \end{bmatrix}, \quad \mathbf{B} = \begin{bmatrix} x & 3 & 0 \\ 4 & -2 & 1/x \end{bmatrix}, \quad \mathbf{I} = \begin{bmatrix} 1 & 0 & 0 \\ 0 & 1 & 0 \\ 0 & 0 & 1 \end{bmatrix}$$

$$\boldsymbol{\Sigma} = \begin{bmatrix} 1 & .7 & -.3 \\ .7 & 2 & 1 \\ -.3 & 1 & 8 \end{bmatrix}, \quad \mathbf{E} = [e_1]$$

In our work, the matrix elements will be real numbers or functions taking on values in the real numbers.

Definition 2A.14. The *dimension* (abbreviated *dim*) of an $m \times k$ matrix is the ordered pair (m, k); m is the row dimension and k is the column dimension. The dimension of a matrix is frequently indicated in parentheses below the letter representing the matrix. Thus, the $m \times k$ matrix $\mathbf{A}$ is denoted by $\underset{(m \times k)}{\mathbf{A}}$.

In the preceding examples, the dimension of the matrix $\mathbf{\Sigma}$ is 3×3, and this information can be conveyed by writing $\underset{(3 \times 3)}{\mathbf{\Sigma}}$.

An $m \times k$ matrix, say, $\mathbf{A}$, of arbitrary constants can be written

$$\underset{(m \times k)}{\mathbf{A}} = \begin{bmatrix} a_{11} & a_{12} & \cdots & a_{1k} \\ a_{21} & a_{22} & \cdots & a_{2k} \\ \vdots & \vdots & \ddots & \vdots \\ a_{m1} & a_{m2} & \cdots & a_{mk} \end{bmatrix}$$

or more compactly as $\underset{(m \times k)}{\mathbf{A}} = \{a_{ij}\}$, where the index i refers to the row and the index j refers to the column.

An $m \times 1$ matrix is referred to as a column *vector*. A $1 \times k$ matrix is referred to as a row *vector*. Since matrices can be considered as vectors side by side, it is natural to define multiplication by a scalar and the addition of two matrices with the same dimensions.

Definition 2A.15. Two matrices $\underset{(m \times k)}{\mathbf{A}} = \{a_{ij}\}$ and $\underset{(m \times k)}{\mathbf{B}} = \{b_{ij}\}$ are said to be *equal*, written $\mathbf{A} = \mathbf{B}$, if $a_{ij} = b_{ij}$, $i = 1, 2, \ldots, m$, $j = 1, 2, \ldots, k$. That is, two matrices are equal if

(a) Their dimensionality is the same.

(b) Every corresponding element is the same.

Definition 2A.16 (Matrix addition). Let the matrices $\mathbf{A}$ and $\mathbf{B}$ both be of dimension $m \times k$ with arbitrary elements a_{ij} and b_{ij}, $i = 1, 2, \ldots, m$, $j = 1, 2, \ldots, k$, respectively. The sum of the matrices $\mathbf{A}$ and $\mathbf{B}$ is an $m \times k$ matrix $\mathbf{C}$, written $\mathbf{C} = \mathbf{A} + \mathbf{B}$, such that the arbitrary element of $\mathbf{C}$ is given by

$$c_{ij} = a_{ij} + b_{ij} \qquad i = 1, 2, \ldots, m, \quad j = 1, 2, \ldots, k$$

Note that the addition of matrices is defined only for matrices of the same dimension.

For example,

$$\underset{\mathbf{A}}{\begin{bmatrix} 3 & 2 & 3 \\ 4 & 1 & 1 \end{bmatrix}} + \underset{\mathbf{B}}{\begin{bmatrix} 3 & 6 & 7 \\ 2 & -1 & 0 \end{bmatrix}} = \underset{\mathbf{C}}{\begin{bmatrix} 6 & 8 & 10 \\ 6 & 0 & 1 \end{bmatrix}}$$

Definition 2A.17 (Scalar multiplication). Let c be an arbitrary scalar and $\underset{(m \times k)}{\mathbf{A}} = \{a_{ij}\}$. Then $\underset{(m \times k)}{c\mathbf{A}} = \underset{(m \times k)}{\mathbf{A}c} = \underset{(m \times k)}{\mathbf{B}} = \{b_{ij}\}$, where $b_{ij} = ca_{ij} = a_{ij}c$, $i = 1, 2, \ldots, m$, $j = 1, 2, \ldots, k$.

Multiplication of a matrix by a scalar produces a new matrix whose elements are the elements of the original matrix, *each* multiplied by the scalar.

For example, if $c = 2$,

$$2 \begin{bmatrix} 3 & -4 \\ 2 & 6 \\ 0 & 5 \end{bmatrix} = \begin{bmatrix} 3 & -4 \\ 2 & 6 \\ 0 & 5 \end{bmatrix} 2 = \begin{bmatrix} 6 & -8 \\ 4 & 12 \\ 0 & 10 \end{bmatrix}$$

$$c\mathbf{A} \qquad = \qquad \mathbf{A}c \qquad = \qquad \mathbf{B}$$

Definition 2A.18 (Matrix subtraction). Let $\underset{(m \times k)}{\mathbf{A}} = \{a_{ij}\}$ and $\underset{(m \times k)}{\mathbf{B}} = \{b_{ij}\}$ be two matrices of equal dimension. Then the difference between $\mathbf{A}$ and $\mathbf{B}$, written $\mathbf{A} - \mathbf{B}$, is an $m \times k$ matrix $\mathbf{C} = \{c_{ij}\}$ given by

$$\mathbf{C} = \mathbf{A} - \mathbf{B} = \mathbf{A} + (-1)\mathbf{B}$$

That is, $c_{ij} = a_{ij} + (-1)b_{ij} = a_{ij} - b_{ij}$, $i = 1, 2, \ldots, m$, $j = 1, 2, \ldots, k$.

Definition 2A.19. Consider the $m \times k$ matrix $\mathbf{A}$ with arbitrary elements a_{ij}, $i = 1, 2, \ldots, m$, $j = 1, 2, \ldots, k$. The *transpose* of the matrix $\mathbf{A}$, denoted by $\mathbf{A}'$, is the $k \times m$ matrix with elements a_{ji}, $j = 1, 2, \ldots, k$, $i = 1, 2, \ldots, m$. That is, the transpose of the matrix $\mathbf{A}$ is obtained from $\mathbf{A}$ by interchanging the rows and columns.

As an example, if

$$\underset{(2 \times 3)}{\mathbf{A}} = \begin{bmatrix} 2 & 1 & 3 \\ 7 & -4 & 6 \end{bmatrix}, \quad \text{then} \quad \underset{(3 \times 2)}{\mathbf{A}'} = \begin{bmatrix} 2 & 7 \\ 1 & -4 \\ 3 & 6 \end{bmatrix}$$

Result 2A.4. For all matrices $\mathbf{A}$, $\mathbf{B}$, and $\mathbf{C}$ (of equal dimension) and scalars c and d, the following hold:

(a) $(\mathbf{A} + \mathbf{B}) + \mathbf{C} = \mathbf{A} + (\mathbf{B} + \mathbf{C})$

(b) $\mathbf{A} + \mathbf{B} = \mathbf{B} + \mathbf{A}$

(c) $c(\mathbf{A} + \mathbf{B}) = c\mathbf{A} + c\mathbf{B}$

(d) $(c + d)\mathbf{A} = c\mathbf{A} + d\mathbf{A}$

(e) $(\mathbf{A} + \mathbf{B})' = \mathbf{A}' + \mathbf{B}'$ (That is, the transpose of the sum is equal to the sum of the transposes.)

(f) $(cd)\mathbf{A} = c(d\mathbf{A})$

(g) $(c\mathbf{A})' = c\mathbf{A}'$

Definition 2A.20. If an arbitrary matrix $\mathbf{A}$ has the *same* number of rows and columns, then $\mathbf{A}$ is called a *square* matrix. The matrices $\boldsymbol{\Sigma}$, $\mathbf{I}$, and $\mathbf{E}$ given after Definition 2A.13 are square matrices.

Definition 2A.21. Let $\mathbf{A}$ be a $k \times k$ (square) matrix. Then $\mathbf{A}$ is said to be *symmetric* if $\mathbf{A} = \mathbf{A}'$. That is, $\mathbf{A}$ is symmetric if $a_{ij} = a_{ji}$, $i = 1, 2, \ldots, k$, $j = 1, 2, \ldots, k$.

Examples of symmetric matrices are

$$
\underset{(3 \times 3)}{\mathbf{I}} = \begin{bmatrix} 1 & 0 & 0 \\ 0 & 1 & 0 \\ 0 & 0 & 1 \end{bmatrix}, \qquad \underset{(2 \times 2)}{\mathbf{A}} = \begin{bmatrix} 2 & 4 \\ 4 & 1 \end{bmatrix}, \qquad \underset{(4 \times 4)}{\mathbf{B}} = \begin{bmatrix} a & c & e & f \\ c & b & g & d \\ e & g & c & a \\ f & d & a & d \end{bmatrix}
$$

Definition 2A.22. The $k \times k$ *identity* matrix, denoted by $\underset{(k \times k)}{\mathbf{I}}$, is the square matrix with ones on the main (NW–SE) diagonal and zeros elsewhere. The 3×3 identity matrix is shown before this definition.

Definition 2A.23 (Matrix multiplication). The product $\mathbf{AB}$ of an $m \times n$ matrix $\mathbf{A} = \{a_{ij}\}$ and an $n \times k$ matrix $\mathbf{B} = \{b_{ij}\}$ is the $m \times k$ matrix $\mathbf{C}$ whose elements are

$$
c_{ij} = \sum_{\ell=1}^{n} a_{i\ell} b_{\ell j} \qquad i = 1, 2, \ldots, m \quad j = 1, 2, \ldots, k
$$

Note that for the product $\mathbf{AB}$ to be defined, the column dimension of $\mathbf{A}$ must equal the row dimension of $\mathbf{B}$. If that is so, then the row dimension of $\mathbf{AB}$ equals the row dimension of $\mathbf{A}$, and the column dimension of $\mathbf{AB}$ equals the column dimension of $\mathbf{B}$.

For example, let

$$
\underset{(2 \times 3)}{\mathbf{A}} = \begin{bmatrix} 3 & -1 & 2 \\ 4 & 0 & 5 \end{bmatrix} \quad \text{and} \quad \underset{(3 \times 2)}{\mathbf{B}} = \begin{bmatrix} 3 & 4 \\ 6 & -2 \\ 4 & 3 \end{bmatrix}
$$

Then

$$
\underset{(2 \times 3)}{\begin{bmatrix} 3 & -1 & 2 \\ 4 & 0 & 5 \end{bmatrix}} \underset{(3 \times 2)}{\begin{bmatrix} 3 & 4 \\ 6 & -2 \\ 4 & 3 \end{bmatrix}} = \underset{(2 \times 2)}{\begin{bmatrix} 11 & 20 \\ 32 & 31 \end{bmatrix}} = \begin{bmatrix} c_{11} & c_{12} \\ c_{21} & c_{22} \end{bmatrix}
$$

where

$$c_{11} = (3)(3) + (-1)(6) + (2)(4) = 11$$

$$c_{12} = (3)(4) + (-1)(-2) + (2)(3) = 20$$

$$c_{21} = (4)(3) + (0)(6) + (5)(4) = 32$$

$$c_{22} = (4)(4) + (0)(-2) + (5)(3) = 31$$

As an additional example, consider the product of two vectors. Let

$$\mathbf{x} = \begin{bmatrix} 1 \\ 0 \\ -2 \\ 3 \end{bmatrix} \quad \text{and} \quad \mathbf{y} = \begin{bmatrix} 2 \\ -3 \\ -1 \\ -8 \end{bmatrix}$$

Then $\mathbf{x}' = \begin{bmatrix} 1 & 0 & -2 & 3 \end{bmatrix}$ and

$$\mathbf{x}'\mathbf{y} = \begin{bmatrix} 1 & 0 & -2 & 3 \end{bmatrix} \begin{bmatrix} 2 \\ -3 \\ -1 \\ -8 \end{bmatrix} = [-20] = \begin{bmatrix} 2 & -3 & -1 & -8 \end{bmatrix} \begin{bmatrix} 1 \\ 0 \\ -2 \\ 3 \end{bmatrix} = \mathbf{y}'\mathbf{x}$$

Note that the product $\mathbf{xy}$ is undefined, since $\mathbf{x}$ is a 4×1 matrix and $\mathbf{y}$ is a 4×1 matrix, so the column dim of $\mathbf{x}$, 1, is unequal to the row dim of $\mathbf{y}$, 4. If $\mathbf{x}$ and $\mathbf{y}$ are vectors of the same dimension, such as $n \times 1$, both of the products $\mathbf{x}'\mathbf{y}$ and $\mathbf{xy}'$ are defined. In particular, $\mathbf{y}'\mathbf{x} = \mathbf{x}'\mathbf{y} = x_1 y_1 + x_2 y_2 + \cdots + x_n y_n$, and $\mathbf{xy}'$ is an $n \times n$ matrix with i,jth element $x_i y_j$.

Result 2A.5. For all matrices $\mathbf{A}$, $\mathbf{B}$, and $\mathbf{C}$ (of dimensions such that the indicated products are defined) and a scalar c,

(a) $c(\mathbf{AB}) = (c\mathbf{A})\mathbf{B}$

(b) $\mathbf{A}(\mathbf{BC}) = (\mathbf{AB})\mathbf{C}$

(c) $\mathbf{A}(\mathbf{B} + \mathbf{C}) = \mathbf{AB} + \mathbf{AC}$

(d) $(\mathbf{B} + \mathbf{C})\mathbf{A} = \mathbf{BA} + \mathbf{CA}$

(e) $(\mathbf{AB})' = \mathbf{B}'\mathbf{A}'$

More generally, for any $\mathbf{x}_j$ such that $\mathbf{A}\mathbf{x}_j$ is defined,

(f) $\displaystyle\sum_{j=1}^{n} \mathbf{A}\mathbf{x}_j = \mathbf{A} \sum_{j=1}^{n} \mathbf{x}_j$

(g) $\displaystyle\sum_{j=1}^{n} (\mathbf{A}\mathbf{x}_j)(\mathbf{A}\mathbf{x}_j)' = \mathbf{A} \left(\sum_{j=1}^{n} \mathbf{x}_j \mathbf{x}_j' \right) \mathbf{A}'$

■

There are several important differences between the algebra of matrices and the algebra of real numbers. Two of these differences are as follows:

1. Matrix multiplication is, in general, not commutative. That is, in general, $\mathbf{AB} \neq \mathbf{BA}$. Several examples will illustrate the failure of the commutative law (for matrices).

$$\begin{bmatrix} 3 & -1 \\ 4 & 7 \end{bmatrix} \begin{bmatrix} 0 \\ 2 \end{bmatrix} = \begin{bmatrix} -2 \\ 14 \end{bmatrix}$$

but

$$\begin{bmatrix} 0 \\ 2 \end{bmatrix} \begin{bmatrix} 3 & -1 \\ 4 & 7 \end{bmatrix}$$

is not defined.

$$\begin{bmatrix} 1 & 0 & 1 \\ 2 & -3 & 6 \end{bmatrix} \begin{bmatrix} 7 & 6 \\ -3 & 1 \\ 2 & 4 \end{bmatrix} = \begin{bmatrix} 9 & 10 \\ 35 & 33 \end{bmatrix}$$

but

$$\begin{bmatrix} 7 & 6 \\ -3 & 1 \\ 2 & 4 \end{bmatrix} \begin{bmatrix} 1 & 0 & 1 \\ 2 & -3 & 6 \end{bmatrix} = \begin{bmatrix} 19 & -18 & 43 \\ -1 & -3 & 3 \\ 10 & -12 & 26 \end{bmatrix}$$

Also,

$$\begin{bmatrix} 4 & -1 \\ 0 & 1 \end{bmatrix} \begin{bmatrix} 2 & 1 \\ -3 & 4 \end{bmatrix} = \begin{bmatrix} 11 & 0 \\ -3 & 4 \end{bmatrix}$$

but

$$\begin{bmatrix} 2 & 1 \\ -3 & 4 \end{bmatrix} \begin{bmatrix} 4 & -1 \\ 0 & 1 \end{bmatrix} = \begin{bmatrix} 8 & -1 \\ -12 & 7 \end{bmatrix}$$

2. Let $\mathbf{0}$ denote the zero matrix, that is, the matrix with zero for every element. In the algebra of real numbers, if the product of two numbers, ab, is zero, then $a = 0$ or $b = 0$. In matrix algebra, however, the product of two *nonzero* matrices may be the zero matrix. Hence,

$$\underset{(m \times n)(n \times k)}{\mathbf{AB}} = \underset{(m \times k)}{\mathbf{0}}$$

does not imply that $\mathbf{A} = \mathbf{0}$ or $\mathbf{B} = \mathbf{0}$. For example,

$$\begin{bmatrix} 3 & 1 & 3 \\ 1 & 2 & 2 \end{bmatrix} \begin{bmatrix} 4 \\ 3 \\ -5 \end{bmatrix} = \begin{bmatrix} 0 \\ 0 \end{bmatrix}$$

It is true, however, that if either $\underset{(m \times n)}{\mathbf{A}} = \underset{(m \times n)}{\mathbf{0}}$ or $\underset{(n \times k)}{\mathbf{B}} = \underset{(n \times k)}{\mathbf{0}}$, then $\underset{(m \times n)(n \times k)}{\mathbf{A} \ \mathbf{B}} = \underset{(m \times k)}{\mathbf{0}}$.

Definition 2A.24. The *determinant* of the square $k \times k$ matrix $\mathbf{A} = \{a_{ij}\}$, denoted by $|\mathbf{A}|$, is the scalar

$$|\mathbf{A}| = a_{11} \qquad\qquad \text{if } k = 1$$

$$|\mathbf{A}| = \sum_{j=1}^{k} a_{1j} |\mathbf{A}_{1j}| (-1)^{1+j} \quad \text{if } k > 1$$

where $\mathbf{A}_{1j}$ is the $(k-1) \times (k-1)$ matrix obtained by deleting the first row and jth column of $\mathbf{A}$. Also, $|\mathbf{A}| = \sum_{j=1}^{k} a_{ij} |\mathbf{A}_{ij}| (-1)^{i+j}$, with the ith row in place of the first row.

Examples of determinants (evaluated using Definition 2A.24) are

$$\begin{vmatrix} 1 & 3 \\ 6 & 4 \end{vmatrix} = 1|4|(-1)^2 + 3|6|(-1)^3 = 1(4) + 3(6)(-1) = -14$$

In general,

$$\begin{vmatrix} a_{11} & a_{12} \\ a_{21} & a_{22} \end{vmatrix} = a_{11}a_{22}(-1)^2 + a_{12}a_{21}(-1)^3 = a_{11}a_{22} - a_{12}a_{21}$$

$$\begin{vmatrix} 3 & 1 & 6 \\ 7 & 4 & 5 \\ 2 & -7 & 1 \end{vmatrix} = 3 \begin{vmatrix} 4 & 5 \\ -7 & 1 \end{vmatrix}(-1)^2 + 1 \begin{vmatrix} 7 & 5 \\ 2 & 1 \end{vmatrix}(-1)^3 + 6 \begin{vmatrix} 7 & 4 \\ 2 & -7 \end{vmatrix}(-1)^4$$

$$= 3(39) - 1(-3) + 6(-57) = -222$$

$$\begin{vmatrix} 1 & 0 & 0 \\ 0 & 1 & 0 \\ 0 & 0 & 1 \end{vmatrix} = 1 \begin{vmatrix} 1 & 0 \\ 0 & 1 \end{vmatrix}(-1)^2 + 0 \begin{vmatrix} 0 & 0 \\ 0 & 1 \end{vmatrix}(-1)^3 + 0 \begin{vmatrix} 0 & 1 \\ 0 & 0 \end{vmatrix}(-1)^4 = 1(1) = 1$$

If $\mathbf{I}$ is the $k \times k$ identity matrix, $|\mathbf{I}| = 1$.

$$\begin{vmatrix} a_{11} & a_{12} & a_{13} \\ a_{21} & a_{22} & a_{23} \\ a_{31} & a_{32} & a_{33} \end{vmatrix}$$

$$= a_{11} \begin{vmatrix} a_{22} & a_{23} \\ a_{32} & a_{33} \end{vmatrix}(-1)^2 + a_{12} \begin{vmatrix} a_{21} & a_{23} \\ a_{31} & a_{33} \end{vmatrix}(-1)^3 + a_{13} \begin{vmatrix} a_{21} & a_{22} \\ a_{31} & a_{32} \end{vmatrix}(-1)^4$$

$$= a_{11}a_{22}a_{33} + a_{12}a_{23}a_{31} + a_{21}a_{32}a_{13} - a_{31}a_{22}a_{13} - a_{21}a_{12}a_{33} - a_{32}a_{23}a_{11}$$

The determinant of any 3×3 matrix can be computed by summing the products of elements along the solid lines and subtracting the products along the dashed lines in the following diagram. This procedure is *not* valid for matrices of higher dimension, but in general, Definition 2A.24 can be employed to evaluate these determinants.

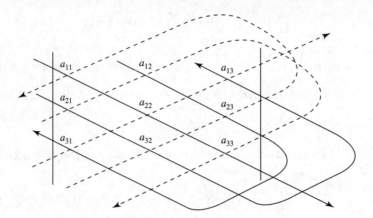

We next want to state a result that describes some properties of the determinant. However, we must first introduce some notions related to matrix inverses.

Definition 2A.25. The *row rank* of a matrix is the maximum number of linearly independent rows, considered as vectors (that is, row vectors). The *column rank* of a matrix is the rank of its set of columns, considered as vectors.

For example, let the matrix

$$\mathbf{A} = \begin{bmatrix} 1 & 1 & 1 \\ 2 & 5 & -1 \\ 0 & 1 & -1 \end{bmatrix}$$

The rows of $\mathbf{A}$, written as vectors, were shown to be linearly dependent after Definition 2A.6. Note that the column rank of $\mathbf{A}$ is also 2, since

$$-2\begin{bmatrix} 1 \\ 2 \\ 0 \end{bmatrix} + \begin{bmatrix} 1 \\ 5 \\ 1 \end{bmatrix} + \begin{bmatrix} 1 \\ -1 \\ -1 \end{bmatrix} = \begin{bmatrix} 0 \\ 0 \\ 0 \end{bmatrix}$$

but columns 1 and 2 are linearly independent. This is no coincidence, as the following result indicates.

Result 2A.6. The row rank and the column rank of a matrix are equal.   ∎

Thus, the *rank of a matrix* is either the row rank or the column rank.

Definition 2A.26. A square matrix $\underset{(k \times k)}{\mathbf{A}}$ is *nonsingular* if $\underset{(k \times k)}{\mathbf{A}} \underset{(k \times 1)}{\mathbf{x}} = \underset{(k \times 1)}{\mathbf{0}}$ implies that $\underset{(k \times 1)}{\mathbf{x}} = \underset{(k \times 1)}{\mathbf{0}}$. If a matrix fails to be nonsingular, it is called *singular*. Equivalently, a *square* matrix is nonsingular if its rank is equal to the number of rows (or columns) it has.

Note that $\mathbf{A}\mathbf{x} = x_1\mathbf{a}_1 + x_2\mathbf{a}_2 + \cdots + x_k\mathbf{a}_k$, where $\mathbf{a}_i$ is the ith column of $\mathbf{A}$, so that the condition of nonsingularity is just the statement that the columns of $\mathbf{A}$ are linearly independent.

Result 2A.7. Let $\mathbf{A}$ be a nonsingular square matrix of dimension $k \times k$. Then there is a unique $k \times k$ matrix $\mathbf{B}$ such that

$$\mathbf{A}\mathbf{B} = \mathbf{B}\mathbf{A} = \mathbf{I}$$

where $\mathbf{I}$ is the $k \times k$ identity matrix. ∎

Definition 2A.27. The $\mathbf{B}$ such that $\mathbf{A}\mathbf{B} = \mathbf{B}\mathbf{A} = \mathbf{I}$ is called the *inverse* of $\mathbf{A}$ and is denoted by $\mathbf{A}^{-1}$. In fact, if $\mathbf{B}\mathbf{A} = \mathbf{I}$ *or* $\mathbf{A}\mathbf{B} = \mathbf{I}$, then $\mathbf{B} = \mathbf{A}^{-1}$, and both products must equal $\mathbf{I}$.

For example,

$$\mathbf{A} = \begin{bmatrix} 2 & 3 \\ 1 & 5 \end{bmatrix} \quad \text{has} \quad \mathbf{A}^{-1} = \begin{bmatrix} \frac{5}{7} & -\frac{3}{7} \\ -\frac{1}{7} & \frac{2}{7} \end{bmatrix}$$

since

$$\begin{bmatrix} 2 & 3 \\ 1 & 5 \end{bmatrix} \begin{bmatrix} \frac{5}{7} & -\frac{3}{7} \\ -\frac{1}{7} & \frac{2}{7} \end{bmatrix} = \begin{bmatrix} \frac{5}{7} & -\frac{3}{7} \\ -\frac{1}{7} & \frac{2}{7} \end{bmatrix} \begin{bmatrix} 2 & 3 \\ 1 & 5 \end{bmatrix} = \begin{bmatrix} 1 & 0 \\ 0 & 1 \end{bmatrix}$$

Result 2A.8.

(a) The inverse of any 2×2 matrix

$$\mathbf{A} = \begin{bmatrix} a_{11} & a_{12} \\ a_{21} & a_{22} \end{bmatrix}$$

is given by

$$\mathbf{A}^{-1} = \frac{1}{|\mathbf{A}|} \begin{bmatrix} a_{22} & -a_{12} \\ -a_{21} & a_{11} \end{bmatrix}$$

(b) The inverse of any 3×3 matrix

$$\mathbf{A} = \begin{bmatrix} a_{11} & a_{12} & a_{13} \\ a_{21} & a_{22} & a_{23} \\ a_{31} & a_{32} & a_{33} \end{bmatrix}$$

is given by

$$
\mathbf{A}^{-1} = \frac{1}{|\mathbf{A}|}
\begin{bmatrix}
\begin{vmatrix} a_{22} & a_{23} \\ a_{32} & a_{33} \end{vmatrix} &
-\begin{vmatrix} a_{12} & a_{13} \\ a_{32} & a_{33} \end{vmatrix} &
\begin{vmatrix} a_{12} & a_{13} \\ a_{22} & a_{23} \end{vmatrix} \\[2ex]
-\begin{vmatrix} a_{21} & a_{23} \\ a_{31} & a_{33} \end{vmatrix} &
\begin{vmatrix} a_{11} & a_{13} \\ a_{31} & a_{33} \end{vmatrix} &
-\begin{vmatrix} a_{11} & a_{13} \\ a_{21} & a_{23} \end{vmatrix} \\[2ex]
\begin{vmatrix} a_{21} & a_{22} \\ a_{31} & a_{32} \end{vmatrix} &
-\begin{vmatrix} a_{11} & a_{12} \\ a_{31} & a_{32} \end{vmatrix} &
\begin{vmatrix} a_{11} & a_{12} \\ a_{21} & a_{22} \end{vmatrix}
\end{bmatrix}
$$

In both (a) and (b), it is clear that $|\mathbf{A}| \neq 0$ if the inverse is to exist.

(c) In general, $\mathbf{A}^{-1}$ has j, ith entry $[|\mathbf{A}_{ij}|/|\mathbf{A}|](-1)^{i+j}$, where $\mathbf{A}_{ij}$ is the matrix obtained from $\mathbf{A}$ by deleting the ith row and jth column. ∎

Result 2A.9. For a square matrix $\mathbf{A}$ of dimension $k \times k$, the following are equivalent:

(a) $\underset{(k \times k)(k \times 1)}{\mathbf{A} \quad \mathbf{x}} = \underset{(k \times 1)}{\mathbf{0}}$ implies $\underset{(k \times 1)}{\mathbf{x}} = \underset{(k \times 1)}{\mathbf{0}}$ ($\mathbf{A}$ is nonsingular).

(b) $|\mathbf{A}| \neq 0$.

(c) There exists a matrix $\mathbf{A}^{-1}$ such that $\mathbf{A}\mathbf{A}^{-1} = \mathbf{A}^{-1}\mathbf{A} = \underset{(k \times k)}{\mathbf{I}}$. ∎

Result 2A.10. Let $\mathbf{A}$ and $\mathbf{B}$ be square matrices of the same dimension, and let the indicated inverses exist. Then the following hold:

(a) $(\mathbf{A}^{-1})' = (\mathbf{A}')^{-1}$

(b) $(\mathbf{AB})^{-1} = \mathbf{B}^{-1}\mathbf{A}^{-1}$ ∎

The determinant has the following properties.

Result 2A.11. Let $\mathbf{A}$ and $\mathbf{B}$ be $k \times k$ square matrices.

(a) $|\mathbf{A}| = |\mathbf{A}'|$

(b) If each element of a row (column) of $\mathbf{A}$ is zero, then $|\mathbf{A}| = 0$

(c) If any two rows (columns) of $\mathbf{A}$ are identical, then $|\mathbf{A}| = 0$

(d) If $\mathbf{A}$ is nonsingular, then $|\mathbf{A}| = 1/|\mathbf{A}^{-1}|$; that is, $|\mathbf{A}||\mathbf{A}^{-1}| = 1$.

(e) $|\mathbf{AB}| = |\mathbf{A}||\mathbf{B}|$

(f) $|c\mathbf{A}| = c^k|\mathbf{A}|$, where c is a scalar.

You are referred to [6] for proofs of parts of Results 2A.9 and 2A.11. Some of these proofs are rather complex and beyond the scope of this book. ∎

Definition 2A.28. Let $\mathbf{A} = \{a_{ij}\}$ be a $k \times k$ square matrix. The *trace* of the matrix $\mathbf{A}$, written $\mathrm{tr}\,(\mathbf{A})$, is the sum of the diagonal elements; that is, $\mathrm{tr}\,(\mathbf{A}) = \sum\limits_{i=1}^{k} a_{ii}$.

Result 2A.12. Let **A** and **B** be $k \times k$ matrices and c be a scalar.

(a) $\text{tr}(c\mathbf{A}) = c\,\text{tr}(\mathbf{A})$

(b) $\text{tr}(\mathbf{A} \pm \mathbf{B}) = \text{tr}(\mathbf{A}) \pm \text{tr}(\mathbf{B})$

(c) $\text{tr}(\mathbf{AB}) = \text{tr}(\mathbf{BA})$

(d) $\text{tr}(\mathbf{B}^{-1}\mathbf{AB}) = \text{tr}(\mathbf{A})$

(e) $\text{tr}(\mathbf{AA}') = \displaystyle\sum_{i=1}^{k}\sum_{j=1}^{k} a_{ij}^2$ ∎

Definition 2A.29. A square matrix **A** is said to be *orthogonal* if its rows, considered as vectors, are mutually perpendicular and have unit lengths; that is, $\mathbf{AA}' = \mathbf{I}$.

Result 2A.13. A matrix **A** is orthogonal if and only if $\mathbf{A}^{-1} = \mathbf{A}'$. For an orthogonal matrix, $\mathbf{AA}' = \mathbf{A}'\mathbf{A} = \mathbf{I}$, so the columns are also mutually perpendicular and have unit lengths. ∎

An example of an orthogonal matrix is

$$
\mathbf{A} = \begin{bmatrix}
-\frac{1}{2} & \frac{1}{2} & \frac{1}{2} & \frac{1}{2} \\
\frac{1}{2} & -\frac{1}{2} & \frac{1}{2} & \frac{1}{2} \\
\frac{1}{2} & \frac{1}{2} & -\frac{1}{2} & \frac{1}{2} \\
\frac{1}{2} & \frac{1}{2} & \frac{1}{2} & -\frac{1}{2}
\end{bmatrix}
$$

Clearly, $\mathbf{A} = \mathbf{A}'$, so $\mathbf{AA}' = \mathbf{A}'\mathbf{A} = \mathbf{AA}$. We verify that $\mathbf{AA} = \mathbf{I} = \mathbf{AA}' = \mathbf{A}'\mathbf{A}$, or

$$
\underbrace{\begin{bmatrix}
-\frac{1}{2} & \frac{1}{2} & \frac{1}{2} & \frac{1}{2} \\
\frac{1}{2} & -\frac{1}{2} & \frac{1}{2} & \frac{1}{2} \\
\frac{1}{2} & \frac{1}{2} & -\frac{1}{2} & \frac{1}{2} \\
\frac{1}{2} & \frac{1}{2} & \frac{1}{2} & -\frac{1}{2}
\end{bmatrix}}_{\mathbf{A}}
\underbrace{\begin{bmatrix}
-\frac{1}{2} & \frac{1}{2} & \frac{1}{2} & \frac{1}{2} \\
\frac{1}{2} & -\frac{1}{2} & \frac{1}{2} & \frac{1}{2} \\
\frac{1}{2} & \frac{1}{2} & -\frac{1}{2} & \frac{1}{2} \\
\frac{1}{2} & \frac{1}{2} & \frac{1}{2} & -\frac{1}{2}
\end{bmatrix}}_{\mathbf{A}}
=
\underbrace{\begin{bmatrix}
1 & 0 & 0 & 0 \\
0 & 1 & 0 & 0 \\
0 & 0 & 1 & 0 \\
0 & 0 & 0 & 1
\end{bmatrix}}_{\mathbf{I}}
$$

so $\mathbf{A}' = \mathbf{A}^{-1}$, and **A** must be an orthogonal matrix.

Square matrices are best understood in terms of quantities called eigenvalues and eigenvectors.

Definition 2A.30. Let **A** be a $k \times k$ square matrix and **I** be the $k \times k$ identity matrix. Then the scalars $\lambda_1, \lambda_2, \ldots, \lambda_k$ satisfying the polynomial equation $|\mathbf{A} - \lambda\mathbf{I}| = 0$ are called the *eigenvalues* (or *characteristic roots*) of a matrix **A**. The equation $|\mathbf{A} - \lambda\mathbf{I}| = 0$ (as a function of λ) is called the *characteristic equation*.

For example, let

$$
\mathbf{A} = \begin{bmatrix} 1 & 0 \\ 1 & 3 \end{bmatrix}
$$

Then

$$|\mathbf{A} - \lambda\mathbf{I}| = \left| \begin{bmatrix} 1 & 0 \\ 1 & 3 \end{bmatrix} - \lambda \begin{bmatrix} 1 & 0 \\ 0 & 1 \end{bmatrix} \right|$$

$$= \begin{vmatrix} 1 - \lambda & 0 \\ 1 & 3 - \lambda \end{vmatrix} = (1 - \lambda)(3 - \lambda) = 0$$

implies that there are two roots, $\lambda_1 = 1$ and $\lambda_2 = 3$. The eigenvalues of $\mathbf{A}$ are 3 and 1. Let

$$\mathbf{A} = \begin{bmatrix} 13 & -4 & 2 \\ -4 & 13 & -2 \\ 2 & -2 & 10 \end{bmatrix}$$

Then the equation

$$|\mathbf{A} - \lambda\mathbf{I}| = \begin{vmatrix} 13 - \lambda & -4 & 2 \\ -4 & 13 - \lambda & -2 \\ 2 & -2 & 10 - \lambda \end{vmatrix} = -\lambda^3 + 36\lambda^2 - 405\lambda + 1458 = 0$$

has three roots: $\lambda_1 = 9$, $\lambda_2 = 9$, and $\lambda_3 = 18$; that is, 9, 9, and 18 are the eigenvalues of $\mathbf{A}$.

Definition 2A.31. Let $\mathbf{A}$ be a square matrix of dimension $k \times k$ and let λ be an eigenvalue of $\mathbf{A}$. If $\underset{(k\times1)}{\mathbf{x}}$ is a *nonzero vector* ($\underset{(k\times1)}{\mathbf{x}} \neq \underset{(k\times1)}{\mathbf{0}}$) such that

$$\mathbf{A}\mathbf{x} = \lambda\mathbf{x}$$

then $\mathbf{x}$ is said to be an *eigenvector (characteristic vector)* of the matrix $\mathbf{A}$ associated with the *eigenvalue* λ.

An equivalent condition for λ to be a solution of the eigenvalue–eigenvector equation is $|\mathbf{A} - \lambda\mathbf{I}| = 0$. This follows because the statement that $\mathbf{A}\mathbf{x} = \lambda\mathbf{x}$ for some λ and $\mathbf{x} \neq \mathbf{0}$ implies that

$$\mathbf{0} = (\mathbf{A} - \lambda\mathbf{I})\mathbf{x} = x_1 \operatorname{col}_1(\mathbf{A} - \lambda\mathbf{I}) + \cdots + x_k \operatorname{col}_k(\mathbf{A} - \lambda\mathbf{I})$$

That is, the columns of $\mathbf{A} - \lambda\mathbf{I}$ are linearly dependent so, by Result 2A.9(b), $|\mathbf{A} - \lambda\mathbf{I}| = 0$, as asserted. Following Definition 2A.30, we have shown that the eigenvalues of

$$\mathbf{A} = \begin{bmatrix} 1 & 0 \\ 1 & 3 \end{bmatrix}$$

are $\lambda_1 = 1$ and $\lambda_2 = 3$. The eigenvectors associated with these eigenvalues can be determined by solving the following equations:

$$\begin{bmatrix} 1 & 0 \\ 1 & 3 \end{bmatrix} \begin{bmatrix} x_1 \\ x_2 \end{bmatrix} = 1 \begin{bmatrix} x_1 \\ x_2 \end{bmatrix}$$

$$\mathbf{A}\mathbf{x} = \lambda_1\mathbf{x}$$

$$\begin{bmatrix} 1 & 0 \\ 1 & 3 \end{bmatrix} \begin{bmatrix} x_1 \\ x_2 \end{bmatrix} = 3 \begin{bmatrix} x_1 \\ x_2 \end{bmatrix}$$

$$\mathbf{A}\mathbf{x} = \lambda_2 \mathbf{x}$$

From the first expression,

$$x_1 = x_1$$
$$x_1 + 3x_2 = x_2$$

or

$$x_1 = -2x_2$$

There are many solutions for x_1 and x_2.

Setting $x_2 = 1$ (arbitrarily) gives $x_1 = -2$, and hence,

$$\mathbf{x} = \begin{bmatrix} -2 \\ 1 \end{bmatrix}$$

is an eigenvector corresponding to the eigenvalue 1. From the second expression,

$$x_1 = 3x_1$$
$$x_1 + 3x_2 = 3x_2$$

implies that $x_1 = 0$ and $x_2 = 1$ (arbitrarily), and hence,

$$\mathbf{x} = \begin{bmatrix} 0 \\ 1 \end{bmatrix}$$

is an eigenvector corresponding to the eigenvalue 3. It is usual practice to determine an eigenvector so that it has length unity. That is, if $\mathbf{A}\mathbf{x} = \lambda\mathbf{x}$, we take $\mathbf{e} = \mathbf{x}/\sqrt{\mathbf{x}'\mathbf{x}}$ as the eigenvector corresponding to λ. For example, the eigenvector for $\lambda_1 = 1$ is $\mathbf{e}'_1 = [-2/\sqrt{5}, \quad 1/\sqrt{5}]$.

Definition 2A.32. A *quadratic form* $Q(\mathbf{x})$ in the k variables $x_1, x_2, \ldots, x_k$ is $Q(\mathbf{x}) = \mathbf{x}'\mathbf{A}\mathbf{x}$, where $\mathbf{x}' = [x_1, x_2, \ldots, x_k]$ and $\mathbf{A}$ is a $k \times k$ symmetric matrix.

Note that a quadratic form can be written as $Q(\mathbf{x}) = \sum\limits_{i=1}^{k} \sum\limits_{j=1}^{k} a_{ij} x_i x_j$. For example,

$$Q(\mathbf{x}) = [x_1 \quad x_2] \begin{bmatrix} 1 & 1 \\ 1 & 1 \end{bmatrix} \begin{bmatrix} x_1 \\ x_2 \end{bmatrix} = x_1^2 + 2x_1 x_2 + x_2^2$$

$$Q(\mathbf{x}) = [x_1 \quad x_2 \quad x_3] \begin{bmatrix} 1 & 3 & 0 \\ 3 & -1 & -2 \\ 0 & -2 & 2 \end{bmatrix} \begin{bmatrix} x_1 \\ x_2 \\ x_3 \end{bmatrix} = x_1^2 + 6x_1 x_2 - x_2^2 - 4x_2 x_3 + 2x_3^2$$

Any symmetric square matrix can be reconstructured from its eigenvalues and eigenvectors. The particular expression reveals the relative importance of each pair according to the relative size of the eigenvalue and the direction of the eigenvector.

Result 2A.14. *The Spectral Decomposition.* Let $\mathbf{A}$ be a $k \times k$ symmetric matrix. Then $\mathbf{A}$ can be expressed in terms of its k eigenvalue–eigenvector pairs $(\lambda_i, \mathbf{e}_i)$ as

$$\mathbf{A} = \sum_{i=1}^{k} \lambda_i \, \mathbf{e}_i \, \mathbf{e}_i' \qquad\qquad \blacksquare$$

For example, let

$$\mathbf{A} = \begin{bmatrix} 2.2 & .4 \\ .4 & 2.8 \end{bmatrix}$$

Then

$$|\mathbf{A} - \lambda\mathbf{I}| = \lambda^2 - 5\lambda + 6.16 - .16 = (\lambda - 3)(\lambda - 2)$$

so $\mathbf{A}$ has eigenvalues $\lambda_1 = 3$ and $\lambda_2 = 2$. The corresponding eigenvectors are $\mathbf{e}_1' = \left[1/\sqrt{5}, 2/\sqrt{5} \right]$ and $\mathbf{e}_2' = \left[2/\sqrt{5}, -1/\sqrt{5} \right]$, respectively. Consequently,

$$\mathbf{A} = \begin{bmatrix} 2.2 & .4 \\ .4 & 2.8 \end{bmatrix} = 3 \begin{bmatrix} \dfrac{1}{\sqrt{5}} \\[2ex] \dfrac{2}{\sqrt{5}} \end{bmatrix} \begin{bmatrix} \dfrac{1}{\sqrt{5}} & \dfrac{2}{\sqrt{5}} \end{bmatrix} + 2 \begin{bmatrix} \dfrac{2}{\sqrt{5}} \\[2ex] \dfrac{-1}{\sqrt{5}} \end{bmatrix} \begin{bmatrix} \dfrac{2}{\sqrt{5}} & \dfrac{-1}{\sqrt{5}} \end{bmatrix}$$

$$= \begin{bmatrix} .6 & 1.2 \\ 1.2 & 2.4 \end{bmatrix} + \begin{bmatrix} 1.6 & -.8 \\ -.8 & .4 \end{bmatrix}$$

The ideas that lead to the spectral decomposition can be extended to provide a decomposition for a rectangular, rather than a square, matrix. If $\mathbf{A}$ is a rectangular matrix, then the vectors in the expansion of $\mathbf{A}$ are the eigenvectors of the square matrices $\mathbf{A}\mathbf{A}'$ and $\mathbf{A}'\mathbf{A}$.

Result 2A.15. *Singular-Value Decomposition.* Let $\mathbf{A}$ be an $m \times k$ matrix of real numbers. Then there exist an $m \times m$ orthogonal matrix $\mathbf{U}$ and a $k \times k$ orthogonal matrix $\mathbf{V}$ such that

$$\mathbf{A} = \mathbf{U}\mathbf{\Lambda}\mathbf{V}'$$

where the $m \times k$ matrix $\mathbf{\Lambda}$ has (i, i) entry $\lambda_i \geq 0$ for $i = 1, 2, \ldots, \min(m, k)$ and the other entries are zero. The positive constants λ_i are called the *singular values* of $\mathbf{A}$. $\blacksquare$

The singular-value decomposition can also be expressed as a matrix expansion that depends on the rank r of $\mathbf{A}$. Specifically, there exist r positive constants $\lambda_1, \lambda_2, \ldots, \lambda_r,\, r$ orthogonal $m \times 1$ unit vectors $\mathbf{u}_1, \mathbf{u}_2, \ldots, \mathbf{u}_r$, and r orthogonal $k \times 1$ unit vectors $\mathbf{v}_1, \mathbf{v}_2, \ldots, \mathbf{v}_r$, such that

$$\mathbf{A} = \sum_{i=1}^{r} \lambda_i \, \mathbf{u}_i \, \mathbf{v}_i' = \mathbf{U}_r \mathbf{\Lambda}_r \mathbf{V}_r'$$

where $\mathbf{U}_r = [\mathbf{u}_1, \mathbf{u}_2, \ldots, \mathbf{u}_r]$, $\mathbf{V}_r = [\mathbf{v}_1, \mathbf{v}_2, \ldots, \mathbf{v}_r]$, and $\mathbf{\Lambda}_r$ is an $r \times r$ diagonal matrix with diagonal entries λ_i.

Here $\mathbf{AA}'$ has eigenvalue–eigenvector pairs $(\lambda_i^2, \mathbf{u}_i)$, so

$$\mathbf{AA}'\mathbf{u}_i = \lambda_i^2 \mathbf{u}_i$$

with $\lambda_1^2, \lambda_2^2, \ldots, \lambda_r^2 > 0 = \lambda_{r+1}^2, \lambda_{r+2}^2, \ldots, \lambda_m^2$ (for $m > k$). Then $\mathbf{v}_i = \lambda_i^{-1}\mathbf{A}'\mathbf{u}_i$. Alternatively, the $\mathbf{v}_i$ are the eigenvectors of $\mathbf{A}'\mathbf{A}$ with the same nonzero eigenvalues λ_i^2.

The matrix expansion for the singular-value decomposition written in terms of the full dimensional matrices $\mathbf{U}, \mathbf{V}, \Lambda$ is

$$\underset{(m \times k)}{\mathbf{A}} = \underset{(m \times m)}{\mathbf{U}} \; \underset{(m \times k)}{\Lambda} \; \underset{(k \times k)}{\mathbf{V}'}$$

where $\mathbf{U}$ has m orthogonal eigenvectors of $\mathbf{AA}'$ as its columns, $\mathbf{V}$ has k orthogonal eigenvectors of $\mathbf{A}'\mathbf{A}$ as its columns, and Λ is specified in Result 2A.15.

For example, let

$$\mathbf{A} = \begin{bmatrix} 3 & 1 & 1 \\ -1 & 3 & 1 \end{bmatrix}$$

Then

$$\mathbf{AA}' = \begin{bmatrix} 3 & 1 & 1 \\ -1 & 3 & 1 \end{bmatrix} \begin{bmatrix} 3 & -1 \\ 1 & 3 \\ 1 & 1 \end{bmatrix} = \begin{bmatrix} 11 & 1 \\ 1 & 11 \end{bmatrix}$$

You may verify that the eigenvalues $\gamma = \lambda^2$ of $\mathbf{AA}'$ satisfy the equation $\gamma^2 - 22\gamma + 120 = (\gamma - 12)(\gamma - 10)$, and consequently, the eigenvalues are $\gamma_1 = \lambda_1^2 = 12$ and $\gamma_2 = \lambda_2^2 = 10$. The corresponding eigenvectors are

$$\mathbf{u}_1' = \begin{bmatrix} \dfrac{1}{\sqrt{2}} & \dfrac{1}{\sqrt{2}} \end{bmatrix} \text{ and } \mathbf{u}_2' = \begin{bmatrix} \dfrac{1}{\sqrt{2}} & \dfrac{-1}{\sqrt{2}} \end{bmatrix}, \text{ respectively.}$$

Also,

$$\mathbf{A}'\mathbf{A} = \begin{bmatrix} 3 & -1 \\ 1 & 3 \\ 1 & 1 \end{bmatrix} \begin{bmatrix} 3 & 1 & 1 \\ -1 & 3 & 1 \end{bmatrix} = \begin{bmatrix} 10 & 0 & 2 \\ 0 & 10 & 4 \\ 2 & 4 & 2 \end{bmatrix}$$

so $|\mathbf{A}'\mathbf{A} - \gamma\mathbf{I}| = -\gamma^3 - 22\gamma^2 - 120\gamma = -\gamma(\gamma - 12)(\gamma - 10)$, and the eigenvalues are $\gamma_1 = \lambda_1^2 = 12, \gamma_2 = \lambda_2^2 = 10$, and $\gamma_3 = \lambda_3^2 = 0$. The nonzero eigenvalues are the same as those of $\mathbf{AA}'$. A computer calculation gives the eigenvectors

$$\mathbf{v}_1' = \begin{bmatrix} \dfrac{1}{\sqrt{6}} & \dfrac{2}{\sqrt{6}} & \dfrac{1}{\sqrt{6}} \end{bmatrix}, \mathbf{v}_2' = \begin{bmatrix} \dfrac{2}{\sqrt{5}} & \dfrac{-1}{\sqrt{5}} & 0 \end{bmatrix}, \text{ and } \mathbf{v}_3' = \begin{bmatrix} \dfrac{1}{\sqrt{30}} & \dfrac{2}{\sqrt{30}} & \dfrac{-5}{\sqrt{30}} \end{bmatrix}.$$

Eigenvectors $\mathbf{v}_1$ and $\mathbf{v}_2$ can be verified by checking:

$$\mathbf{A}'\mathbf{A}\mathbf{v}_1 = \begin{bmatrix} 10 & 0 & 2 \\ 0 & 10 & 4 \\ 2 & 4 & 2 \end{bmatrix} \frac{1}{\sqrt{6}} \begin{bmatrix} 1 \\ 2 \\ 1 \end{bmatrix} = 12 \frac{1}{\sqrt{6}} \begin{bmatrix} 1 \\ 2 \\ 1 \end{bmatrix} = \lambda_1^2 \mathbf{v}_1$$

$$\mathbf{A}'\mathbf{A}\mathbf{v}_2 = \begin{bmatrix} 10 & 0 & 2 \\ 0 & 10 & 4 \\ 2 & 4 & 2 \end{bmatrix} \frac{1}{\sqrt{5}} \begin{bmatrix} 2 \\ -1 \\ 0 \end{bmatrix} = 10 \frac{1}{\sqrt{5}} \begin{bmatrix} 2 \\ -1 \\ 0 \end{bmatrix} = \lambda_2^2 \mathbf{v}_2$$

Taking $\lambda_1 = \sqrt{12}$ and $\lambda_2 = \sqrt{10}$, we find that the singular-value decomposition of **A** is

$$\mathbf{A} = \begin{bmatrix} 3 & 1 & 1 \\ -1 & 3 & 1 \end{bmatrix}$$

$$= \sqrt{12} \begin{bmatrix} \dfrac{1}{\sqrt{2}} \\ \dfrac{1}{\sqrt{2}} \end{bmatrix} \begin{bmatrix} \dfrac{1}{\sqrt{6}} & \dfrac{2}{\sqrt{6}} & \dfrac{1}{\sqrt{6}} \end{bmatrix} + \sqrt{10} \begin{bmatrix} \dfrac{1}{\sqrt{2}} \\ \dfrac{-1}{\sqrt{2}} \end{bmatrix} \begin{bmatrix} \dfrac{2}{\sqrt{5}} & \dfrac{-1}{\sqrt{5}} & 0 \end{bmatrix}$$

The equality may be checked by carrying out the operations on the right-hand side.

The singular-value decomposition is closely connected to a result concerning the approximation of a rectangular matrix by a lower-dimensional matrix, due to Eckart and Young ([2]). If a $m \times k$ matrix **A** is approximated by **B**, having the same dimension but lower rank, the sum of squared differences

$$\sum_{i=1}^{m} \sum_{j=1}^{k} (a_{ij} - b_{ij})^2 = \mathrm{tr}[(\mathbf{A} - \mathbf{B})(\mathbf{A} - \mathbf{B})']$$

Result 2A.16. Let **A** be an $m \times k$ matrix of real numbers with $m \geq k$ and singular value decomposition $\mathbf{U}\mathbf{\Lambda}\mathbf{V}'$. Let $s < k = \mathrm{rank}(\mathbf{A})$. Then

$$\mathbf{B} = \sum_{i=1}^{s} \lambda_i \mathbf{u}_i \mathbf{v}_i'$$

is the rank-s least squares approximation to **A**. It minimizes

$$\mathrm{tr}[(\mathbf{A} - \mathbf{B})(\mathbf{A} - \mathbf{B})']$$

over all $m \times k$ matrices **B** having rank no greater than s. The minimum value, or error of approximation, is $\displaystyle\sum_{i=s+1}^{k} \lambda_i^2$. ∎

To establish this result, we use $\mathbf{U}\mathbf{U}' = \mathbf{I}_m$ and $\mathbf{V}\mathbf{V}' = \mathbf{I}_k$ to write the sum of squares as

$$\mathrm{tr}[(\mathbf{A} - \mathbf{B})(\mathbf{A} - \mathbf{B})'] = \mathrm{tr}[\mathbf{U}\mathbf{U}'(\mathbf{A} - \mathbf{B})\mathbf{V}\mathbf{V}'(\mathbf{A} - \mathbf{B})']$$

$$= \mathrm{tr}[\mathbf{U}'(\mathbf{A} - \mathbf{B})\mathbf{V}\mathbf{V}'(\mathbf{A} - \mathbf{B})'\mathbf{U}]$$

$$= \mathrm{tr}[(\mathbf{\Lambda} - \mathbf{C})(\mathbf{\Lambda} - \mathbf{C})'] = \sum_{i=1}^{m} \sum_{j=1}^{k} (\lambda_{ij} - c_{ij})^2 = \sum_{i=1}^{m} (\lambda_i - c_{ii})^2 + \sum_{i \neq j} \sum c_{ij}^2$$

where $\mathbf{C} = \mathbf{U}'\mathbf{B}\mathbf{V}$. Clearly, the minimum occurs when $c_{ij} = 0$ for $i \neq j$ and $c_{ii} = \lambda_i$ for the s largest singular values. The other $c_{ii} = 0$. That is, $\mathbf{U}\mathbf{B}\mathbf{V}' = \mathbf{\Lambda}_s$ or $\mathbf{B} = \displaystyle\sum_{i=1}^{s} \lambda_i \mathbf{u}_i \mathbf{v}_i'$.

Exercises

2.1. Let $\mathbf{x}' = [5, \quad 1, \quad 3]$ and $\mathbf{y}' = [-1, \quad 3, \quad 1]$.

(a) Graph the two vectors.

(b) Find (i) the length of $\mathbf{x}$, (ii) the angle between $\mathbf{x}$ and $\mathbf{y}$, and (iii) the projection of $\mathbf{y}$ on $\mathbf{x}$.

(c) Since $\bar{x} = 3$ and $\bar{y} = 1$, graph $[5 - 3, 1 - 3, 3 - 3] = [2, -2, 0]$ and $[-1 - 1, 3 - 1, 1 - 1] = [-2, 2, 0]$.

2.2. Given the matrices

$$\mathbf{A} = \begin{bmatrix} -1 & 3 \\ 4 & 2 \end{bmatrix}, \qquad \mathbf{B} = \begin{bmatrix} 4 & -3 \\ 1 & -2 \\ -2 & 0 \end{bmatrix}, \quad \text{and} \quad \mathbf{C} = \begin{bmatrix} 5 \\ -4 \\ 2 \end{bmatrix}$$

perform the indicated multiplications.

(a) $5\mathbf{A}$

(b) $\mathbf{BA}$

(c) $\mathbf{A}'\mathbf{B}'$

(d) $\mathbf{C}'\mathbf{B}$

(e) Is $\mathbf{AB}$ defined?

2.3. Verify the following properties of the transpose when

$$\mathbf{A} = \begin{bmatrix} 2 & 1 \\ 1 & 3 \end{bmatrix}, \qquad \mathbf{B} = \begin{bmatrix} 1 & 4 & 2 \\ 5 & 0 & 3 \end{bmatrix}, \quad \text{and} \quad \mathbf{C} = \begin{bmatrix} 1 & 4 \\ 3 & 2 \end{bmatrix}$$

(a) $(\mathbf{A}')' = \mathbf{A}$

(b) $(\mathbf{C}')^{-1} = (\mathbf{C}^{-1})'$

(c) $(\mathbf{AB})' = \mathbf{B}'\mathbf{A}'$

(d) For general $\underset{(m \times k)}{\mathbf{A}}$ and $\underset{(k \times \ell)}{\mathbf{B}}$, $(\mathbf{AB})' = \mathbf{B}'\mathbf{A}'$.

2.4. When $\mathbf{A}^{-1}$ and $\mathbf{B}^{-1}$ exist, prove each of the following.

(a) $(\mathbf{A}')^{-1} = (\mathbf{A}^{-1})'$

(b) $(\mathbf{AB})^{-1} = \mathbf{B}^{-1}\mathbf{A}^{-1}$

Hint: Part a can be proved by noting that $\mathbf{AA}^{-1} = \mathbf{I}$, $\mathbf{I} = \mathbf{I}'$, and $(\mathbf{AA}^{-1})' = (\mathbf{A}^{-1})'\mathbf{A}'$. Part b follows from $(\mathbf{B}^{-1}\mathbf{A}^{-1})\mathbf{AB} = \mathbf{B}^{-1}(\mathbf{A}^{-1}\mathbf{A})\mathbf{B} = \mathbf{B}^{-1}\mathbf{B} = \mathbf{I}$.

2.5. Check that

$$\mathbf{Q} = \begin{bmatrix} \frac{5}{13} & \frac{12}{13} \\ -\frac{12}{13} & \frac{5}{13} \end{bmatrix}$$

is an orthogonal matrix.

2.6. Let

$$\mathbf{A} = \begin{bmatrix} 9 & -2 \\ -2 & 6 \end{bmatrix}$$

(a) Is $\mathbf{A}$ symmetric?

(b) Show that $\mathbf{A}$ is positive definite.

2.7. Let **A** be as given in Exercise 2.6.

   (a) Determine the eigenvalues and eigenvectors of **A**.

   (b) Write the spectral decomposition of **A**.

   (c) Find $\mathbf{A}^{-1}$.

   (d) Find the eigenvalues and eigenvectors of $\mathbf{A}^{-1}$.

2.8. Given the matrix

$$\mathbf{A} = \begin{bmatrix} 1 & 2 \\ 2 & -2 \end{bmatrix}$$

find the eigenvalues λ_1 and λ_2 and the associated normalized eigenvectors $\mathbf{e}_1$ and $\mathbf{e}_2$. Determine the spectral decomposition (2-16) of **A**.

2.9. Let **A** be as in Exercise 2.8.

   (a) Find $\mathbf{A}^{-1}$.

   (b) Compute the eigenvalues and eigenvectors of $\mathbf{A}^{-1}$.

   (c) Write the spectral decomposition of $\mathbf{A}^{-1}$, and compare it with that of **A** from Exercise 2.8.

2.10. Consider the matrices

$$\mathbf{A} = \begin{bmatrix} 4 & 4.001 \\ 4.001 & 4.002 \end{bmatrix} \quad \text{and} \quad \mathbf{B} = \begin{bmatrix} 4 & 4.001 \\ 4.001 & 4.002001 \end{bmatrix}$$

These matrices are identical except for a small difference in the $(2, 2)$ position. Moreover, the columns of **A** (and **B**) are nearly linearly dependent. Show that $\mathbf{A}^{-1} \doteq (-3)\mathbf{B}^{-1}$. Consequently, small changes—perhaps caused by rounding—can give substantially different inverses.

2.11. Show that the determinant of the $p \times p$ diagonal matrix $\mathbf{A} = \{a_{ij}\}$ with $a_{ij} = 0, i \neq j$, is given by the product of the diagonal elements; thus, $|\mathbf{A}| = a_{11}a_{22}\cdots a_{pp}$.
Hint: By Definition 2A.24, $|\mathbf{A}| = a_{11}\mathbf{A}_{11} + 0 + \cdots + 0$. Repeat for the submatrix $\mathbf{A}_{11}$ obtained by deleting the first row and first column of **A**.

2.12. Show that the determinant of a square symmetric $p \times p$ matrix **A** can be expressed as the product of its eigenvalues $\lambda_1, \lambda_2, \ldots, \lambda_p$; that is, $|\mathbf{A}| = \prod_{i=1}^{p} \lambda_i$.
Hint: From (2-16) and (2-20), $\mathbf{A} = \mathbf{P}\boldsymbol{\Lambda}\mathbf{P}'$ with $\mathbf{P}'\mathbf{P} = \mathbf{I}$. From Result 2A.11(e), $|\mathbf{A}| = |\mathbf{P}\boldsymbol{\Lambda}\mathbf{P}'| = |\mathbf{P}||\boldsymbol{\Lambda}\mathbf{P}'| = |\mathbf{P}||\boldsymbol{\Lambda}||\mathbf{P}'| = |\boldsymbol{\Lambda}||\mathbf{I}|$, since $|\mathbf{I}| = |\mathbf{P}'\mathbf{P}| = |\mathbf{P}'||\mathbf{P}|$. Apply Exercise 2.11.

2.13. Show that $|\mathbf{Q}| = +1$ or -1 if **Q** is a $p \times p$ orthogonal matrix.
Hint: $|\mathbf{Q}\mathbf{Q}'| = |\mathbf{I}|$. Also, from Result 2A.11, $|\mathbf{Q}||\mathbf{Q}'| = |\mathbf{Q}|^2$. Thus, $|\mathbf{Q}|^2 = |\mathbf{I}|$. Now use Exercise 2.11.

2.14. Show that $\underset{(p \times p)(p \times p)(p \times p)}{\mathbf{Q}' \; \mathbf{A} \; \mathbf{Q}}$ and $\underset{(p \times p)}{\mathbf{A}}$ have the same eigenvalues if **Q** is orthogonal.
Hint: Let λ be an eigenvalue of **A**. Then $0 = |\mathbf{A} - \lambda\mathbf{I}|$. By Exercise 2.13 and Result 2A.11(e), we can write $0 = |\mathbf{Q}'||\mathbf{A} - \lambda\mathbf{I}||\mathbf{Q}| = |\mathbf{Q}'\mathbf{A}\mathbf{Q} - \lambda\mathbf{I}|$, since $\mathbf{Q}'\mathbf{Q} = \mathbf{I}$.

2.15. A quadratic form $\mathbf{x}'\mathbf{A}\mathbf{x}$ is said to be positive definite if the matrix **A** is positive definite. Is the quadratic form $3x_1^2 + 3x_2^2 - 2x_1x_2$ positive definite?

2.16. Consider an arbitrary $n \times p$ matrix **A**. Then $\mathbf{A}'\mathbf{A}$ is a symmetric $p \times p$ matrix. Show that $\mathbf{A}'\mathbf{A}$ is necessarily nonnegative definite.
Hint: Set $\mathbf{y} = \mathbf{A}\mathbf{x}$ so that $\mathbf{y}'\mathbf{y} = \mathbf{x}'\mathbf{A}'\mathbf{A}\mathbf{x}$.

2.17. Prove that every eigenvalue of a $k \times k$ positive definite matrix $\mathbf{A}$ is positive.
Hint: Consider the definition of an eigenvalue, where $\mathbf{Ae} = \lambda\mathbf{e}$. Multiply on the left by $\mathbf{e}'$ so that $\mathbf{e}'\mathbf{Ae} = \lambda\mathbf{e}'\mathbf{e}$.

2.18. Consider the sets of points (x_1, x_2) whose "distances" from the origin are given by

$$c^2 = 4x_1^2 + 3x_2^2 - 2\sqrt{2}x_1x_2$$

for $c^2 = 1$ and for $c^2 = 4$. Determine the major and minor axes of the ellipses of constant distances and their associated lengths. Sketch the ellipses of constant distances and comment on their positions. What will happen as c^2 increases?

2.19. Let $\underset{(m \times m)}{\mathbf{A}^{1/2}} = \sum_{i=1}^{m} \sqrt{\lambda_i}\, \mathbf{e}_i\mathbf{e}_i' = \mathbf{P}\mathbf{\Lambda}^{1/2}\mathbf{P}'$, where $\mathbf{PP}' = \mathbf{P}'\mathbf{P} = \mathbf{I}$. (The λ_i's and the $\mathbf{e}_i$'s are the eigenvalues and associated normalized eigenvectors of the matrix $\mathbf{A}$.) Show Properties (1)–(4) of the square-root matrix in (2-22).

2.20. Determine the square-root matrix $\mathbf{A}^{1/2}$, using the matrix $\mathbf{A}$ in Exercise 2.3. Also, determine $\mathbf{A}^{-1/2}$, and show that $\mathbf{A}^{1/2}\mathbf{A}^{-1/2} = \mathbf{A}^{-1/2}\mathbf{A}^{1/2} = \mathbf{I}$.

2.21. (See Result 2A.15) Using the matrix

$$\mathbf{A} = \begin{bmatrix} 1 & 1 \\ 2 & -2 \\ 2 & 2 \end{bmatrix}$$

(a) Calculate $\mathbf{A}'\mathbf{A}$ and obtain its eigenvalues and eigenvectors.
(b) Calculate $\mathbf{AA}'$ and obtain its eigenvalues and eigenvectors. Check that the nonzero eigenvalues are the same as those in part a.
(c) Obtain the singular-value decomposition of $\mathbf{A}$.

2.22. (See Result 2A.15) Using the matrix

$$\mathbf{A} = \begin{bmatrix} 4 & 8 & 8 \\ 3 & 6 & -9 \end{bmatrix}$$

(a) Calculate $\mathbf{AA}'$ and obtain its eigenvalues and eigenvectors.
(b) Calculate $\mathbf{A}'\mathbf{A}$ and obtain its eigenvalues and eigenvectors. Check that the nonzero eigenvalues are the same as those in part a.
(c) Obtain the singular-value decomposition of $\mathbf{A}$.

2.23. Verify the relationships $\mathbf{V}^{1/2}\boldsymbol{\rho}\mathbf{V}^{1/2} = \mathbf{\Sigma}$ and $\boldsymbol{\rho} = (\mathbf{V}^{1/2})^{-1}\mathbf{\Sigma}(\mathbf{V}^{1/2})^{-1}$, where $\mathbf{\Sigma}$ is the $p \times p$ population covariance matrix [Equation (2-32)], $\boldsymbol{\rho}$ is the $p \times p$ population correlation matrix [Equation (2-34)], and $\mathbf{V}^{1/2}$ is the population standard deviation matrix [Equation (2-35)].

2.24. Let $\mathbf{X}$ have covariance matrix

$$\mathbf{\Sigma} = \begin{bmatrix} 4 & 0 & 0 \\ 0 & 9 & 0 \\ 0 & 0 & 1 \end{bmatrix}$$

Find
(a) $\mathbf{\Sigma}^{-1}$
(b) The eigenvalues and eigenvectors of $\mathbf{\Sigma}$.
(c) The eigenvalues and eigenvectors of $\mathbf{\Sigma}^{-1}$.

2.25. Let $\mathbf{X}$ have covariance matrix

$$\boldsymbol{\Sigma} = \begin{bmatrix} 25 & -2 & 4 \\ -2 & 4 & 1 \\ 4 & 1 & 9 \end{bmatrix}$$

(a) Determine $\boldsymbol{\rho}$ and $\mathbf{V}^{1/2}$.

(b) Multiply your matrices to check the relation $\mathbf{V}^{1/2} \boldsymbol{\rho} \mathbf{V}^{1/2} = \boldsymbol{\Sigma}$.

2.26. Use $\boldsymbol{\Sigma}$ as given in Exercise 2.25.

(a) Find ρ_{13}.

(b) Find the correlation between X_1 and $\frac{1}{2}X_2 + \frac{1}{2}X_3$.

2.27. Derive expressions for the mean and variances of the following linear combinations in terms of the means and covariances of the random variables X_1, X_2, and X_3.

(a) $X_1 - 2X_2$

(b) $-X_1 + 3X_2$

(c) $X_1 + X_2 + X_3$

(e) $X_1 + 2X_2 - X_3$

(f) $3X_1 - 4X_2$ if X_1 and X_2 are independent random variables.

2.28. Show that

$$\mathrm{Cov}\,(c_{11}X_1 + c_{12}X_2 + \cdots + c_{1p}X_p, c_{21}X_1 + c_{22}X_2 + \cdots + c_{2p}X_p) = \mathbf{c}_1'\boldsymbol{\Sigma}_{\mathbf{X}}\mathbf{c}_2$$

where $\mathbf{c}_1' = [c_{11}, c_{12}, \ldots, c_{1p}]$ and $\mathbf{c}_2' = [c_{21}, c_{22}, \ldots, c_{2p}]$. This verifies the off-diagonal elements $\mathbf{C}\boldsymbol{\Sigma}_{\mathbf{X}}\mathbf{C}'$ in (2–45) or diagonal elements if $\mathbf{c}_1 = \mathbf{c}_2$.

Hint: By (2–43), $Z_1 - E(Z_1) = c_{11}(X_1 - \mu_1) + \cdots + c_{1p}(X_p - \mu_p)$ and $Z_2 - E(Z_2) = c_{21}(X_1 - \mu_1) + \cdots + c_{2p}(X_p - \mu_p)$. So $\mathrm{Cov}\,(Z_1, Z_2) = E[(Z_1 - E(Z_1))(Z_2 - E(Z_2))] = E[(c_{11}(X_1 - \mu_1) + \cdots + c_{1p}(X_p - \mu_p))(c_{21}(X_1 - \mu_1) + c_{22}(X_2 - \mu_2) + \cdots + c_{2p}(X_p - \mu_p))]$.

The product

$$(c_{11}(X_1 - \mu_1) + c_{12}(X_2 - \mu_2) + \cdots$$
$$+ c_{1p}(X_p - \mu_p))(c_{21}(X_1 - \mu_1) + c_{22}(X_2 - \mu_2) + \cdots + c_{2p}(X_p - \mu_p))$$

$$= \left(\sum_{\ell=1}^{p} c_{1\ell}(X_\ell - \mu_\ell) \right) \left(\sum_{m=1}^{p} c_{2m}(X_m - \mu_m) \right)$$

$$= \sum_{\ell=1}^{p} \sum_{m=1}^{p} c_{1\ell}c_{2m}(X_\ell - \mu_\ell)(X_m - \mu_m)$$

has expected value

$$\sum_{\ell=1}^{p} \sum_{m=1}^{p} c_{1\ell}c_{2m}\sigma_{\ell m} = [c_{11}, \ldots, c_{1p}]\boldsymbol{\Sigma}[c_{21}, \ldots, c_{2p}]'.$$

Verify the last step by the definition of matrix multiplication. The same steps hold for all elements.

2.29. Consider the arbitrary random vector $\mathbf{X}' = [X_1, X_2, X_3, X_4, X_5]$ with mean vector $\boldsymbol{\mu}' = [\mu_1, \mu_2, \mu_3, \mu_4, \mu_5]$. Partition $\mathbf{X}$ into

$$\mathbf{X} = \left[\frac{\mathbf{X}^{(1)}}{\mathbf{X}^{(2)}} \right]$$

where

$$\mathbf{X}^{(1)} = \begin{bmatrix} X_1 \\ X_2 \end{bmatrix} \quad \text{and} \quad \mathbf{X}^{(2)} = \begin{bmatrix} X_3 \\ X_4 \\ X_5 \end{bmatrix}$$

Let $\boldsymbol{\Sigma}$ be the covariance matrix of $\mathbf{X}$ with general element σ_{ik}. Partition $\boldsymbol{\Sigma}$ into the covariance matrices of $\mathbf{X}^{(1)}$ and $\mathbf{X}^{(2)}$ and the covariance matrix of an element of $\mathbf{X}^{(1)}$ and an element of $\mathbf{X}^{(2)}$.

2.30. You are given the random vector $\mathbf{X}' = [X_1, X_2, X_3, X_4]$ with mean vector $\boldsymbol{\mu}_{\mathbf{X}}' = [4, 3, 2, 1]$ and variance–covariance matrix

$$\boldsymbol{\Sigma}_{\mathbf{X}} = \begin{bmatrix} 3 & 0 & 2 & 2 \\ 0 & 1 & 1 & 0 \\ 2 & 1 & 9 & -2 \\ 2 & 0 & -2 & 4 \end{bmatrix}$$

Partition $\mathbf{X}$ as

$$\mathbf{X} = \begin{bmatrix} X_1 \\ X_2 \\ \hline X_3 \\ X_4 \end{bmatrix} = \left[\frac{\mathbf{X}^{(1)}}{\mathbf{X}^{(2)}} \right]$$

Let

$$\mathbf{A} = [1 \quad 2] \quad \text{and} \quad \mathbf{B} = \begin{bmatrix} 1 & -2 \\ 2 & -1 \end{bmatrix}$$

and consider the linear combinations $\mathbf{A}\mathbf{X}^{(1)}$ and $\mathbf{B}\mathbf{X}^{(2)}$. Find

(a) $E(\mathbf{X}^{(1)})$

(b) $E(\mathbf{A}\mathbf{X}^{(1)})$

(c) $\mathrm{Cov}(\mathbf{X}^{(1)})$

(d) $\mathrm{Cov}(\mathbf{A}\mathbf{X}^{(1)})$

(e) $E(\mathbf{X}^{(2)})$

(f) $E(\mathbf{B}\mathbf{X}^{(2)})$

(g) $\mathrm{Cov}(\mathbf{X}^{(2)})$

(h) $\mathrm{Cov}(\mathbf{B}\mathbf{X}^{(2)})$

(i) $\mathrm{Cov}(\mathbf{X}^{(1)}, \mathbf{X}^{(2)})$

(j) $\mathrm{Cov}(\mathbf{A}\mathbf{X}^{(1)}, \mathbf{B}\mathbf{X}^{(2)})$

2.31. Repeat Exercise 2.30, but with $\mathbf{A}$ and $\mathbf{B}$ replaced by

$$\mathbf{A} = [1 \quad -1] \quad \text{and} \quad \mathbf{B} = \begin{bmatrix} 2 & -1 \\ 0 & 1 \end{bmatrix}$$

2.32. You are given the random vector $\mathbf{X}' = [X_1, X_2, \ldots, X_5]$ with mean vector $\boldsymbol{\mu}'_{\mathbf{X}} = [2, 4, -1, 3, 0]$ and variance–covariance matrix

$$
\boldsymbol{\Sigma}_{\mathbf{X}} = \begin{bmatrix}
4 & -1 & \frac{1}{2} & -\frac{1}{2} & 0 \\
-1 & 3 & 1 & -1 & 0 \\
\frac{1}{2} & 1 & 6 & 1 & -1 \\
-\frac{1}{2} & -1 & 1 & 4 & 0 \\
0 & 0 & -1 & 0 & 2
\end{bmatrix}
$$

Partition $\mathbf{X}$ as

$$
\mathbf{X} = \begin{bmatrix}
X_1 \\
X_2 \\
\hline
X_3 \\
X_4 \\
X_5
\end{bmatrix} = \begin{bmatrix}
\mathbf{X}^{(1)} \\
\hline
\mathbf{X}^{(2)}
\end{bmatrix}
$$

Let

$$
\mathbf{A} = \begin{bmatrix} 1 & -1 \\ 1 & 1 \end{bmatrix} \quad \text{and} \quad \mathbf{B} = \begin{bmatrix} 1 & 1 & 1 \\ 1 & 1 & -2 \end{bmatrix}
$$

and consider the linear combinations $\mathbf{A}\mathbf{X}^{(1)}$ and $\mathbf{B}\mathbf{X}^{(2)}$. Find

(a) $E(\mathbf{X}^{(1)})$

(b) $E(\mathbf{A}\mathbf{X}^{(1)})$

(c) $\text{Cov}(\mathbf{X}^{(1)})$

(d) $\text{Cov}(\mathbf{A}\mathbf{X}^{(1)})$

(e) $E(\mathbf{X}^{(2)})$

(f) $E(\mathbf{B}\mathbf{X}^{(2)})$

(g) $\text{Cov}(\mathbf{X}^{(2)})$

(h) $\text{Cov}(\mathbf{B}\mathbf{X}^{(2)})$

(i) $\text{Cov}(\mathbf{X}^{(1)}, \mathbf{X}^{(2)})$

(j) $\text{Cov}(\mathbf{A}\mathbf{X}^{(1)}, \mathbf{B}\mathbf{X}^{(2)})$

2.33. Repeat Exercise 2.32, but with $\mathbf{X}$ partitioned as

$$
\mathbf{X} = \begin{bmatrix}
X_1 \\
X_2 \\
X_3 \\
\hline
X_4 \\
X_5
\end{bmatrix} = \begin{bmatrix}
\mathbf{X}^{(1)} \\
\hline
\mathbf{X}^{(2)}
\end{bmatrix}
$$

and with $\mathbf{A}$ and $\mathbf{B}$ replaced by

$$
\mathbf{A} = \begin{bmatrix} 2 & -1 & 0 \\ 1 & 1 & 3 \end{bmatrix} \quad \text{and} \quad \mathbf{B} = \begin{bmatrix} 1 & 2 \\ 1 & -1 \end{bmatrix}
$$

2.34. Consider the vectors $\mathbf{b}' = [2, -1, 4, 0]$ and $\mathbf{d}' = [-1, 3, -2, 1]$. Verify the Cauchy–Schwarz inequality $(\mathbf{b}'\mathbf{d})^2 \leq (\mathbf{b}'\mathbf{b})(\mathbf{d}'\mathbf{d})$.

2.35. Using the vectors $\mathbf{b}' = [-4, 3]$ and $\mathbf{d}' = [1, 1]$, verify the extended Cauchy–Schwarz inequality $(\mathbf{b}'\mathbf{d})^2 \leq (\mathbf{b}'\mathbf{B}\mathbf{b})(\mathbf{d}'\mathbf{B}^{-1}\mathbf{d})$ if

$$\mathbf{B} = \begin{bmatrix} 2 & -2 \\ -2 & 5 \end{bmatrix}$$

2.36. Find the maximum and minimum values of the quadratic form $4x_1^2 + 4x_2^2 + 6x_1x_2$ for all points $\mathbf{x}' = [x_1, x_2]$ such that $\mathbf{x}'\mathbf{x} = 1$.

2.37. With $\mathbf{A}$ as given in Exercise 2.6, find the maximum value of $\mathbf{x}'\mathbf{A}\mathbf{x}$ for $\mathbf{x}'\mathbf{x} = 1$.

2.38. Find the maximum and minimum values of the ratio $\mathbf{x}'\mathbf{A}\mathbf{x}/\mathbf{x}'\mathbf{x}$ for any nonzero vectors $\mathbf{x}' = [x_1, x_2, x_3]$ if

$$\mathbf{A} = \begin{bmatrix} 13 & -4 & 2 \\ -4 & 13 & -2 \\ 2 & -2 & 10 \end{bmatrix}$$

2.39. Show that

$$\underset{(r\times s)(s\times t)(t\times v)}{\mathbf{A} \quad \mathbf{B} \quad \mathbf{C}} \text{ has } (i, j)\text{th entry } \sum_{\ell=1}^{s} \sum_{k=1}^{t} a_{i\ell} b_{\ell k} c_{kj}$$

Hint: $\mathbf{BC}$ has (ℓ, j)th entry $\sum_{k=1}^{t} b_{\ell k} c_{kj} = d_{\ell j}$. So $\mathbf{A}(\mathbf{BC})$ has (i, j)th element

$$a_{i1}d_{1j} + a_{i2}d_{2j} + \cdots + a_{is}d_{sj} = \sum_{\ell=1}^{s} a_{i\ell}\left(\sum_{k=1}^{t} b_{\ell k} c_{kj}\right) = \sum_{\ell=1}^{s} \sum_{k=1}^{t} a_{i\ell} b_{\ell k} c_{kj}$$

2.40. Verify (2-24): $E(\mathbf{X} + \mathbf{Y}) = E(\mathbf{X}) + E(\mathbf{Y})$ and $E(\mathbf{A}\mathbf{X}\mathbf{B}) = \mathbf{A}E(\mathbf{X})\mathbf{B}$.
Hint: $\mathbf{X} + \mathbf{Y}$ has $X_{ij} + Y_{ij}$ as its (i, j)th element. Now, $E(X_{ij} + Y_{ij}) = E(X_{ij}) + E(Y_{ij})$ by a univariate property of expectation, and this last quantity is the (i, j)th element of $E(\mathbf{X}) + E(\mathbf{Y})$. Next (see Exercise 2.39), $\mathbf{A}\mathbf{X}\mathbf{B}$ has (i, j)th entry $\sum_{\ell} \sum_{k} a_{i\ell} X_{\ell k} b_{kj}$, and by the additive property of expectation,

$$E\left(\sum_{\ell} \sum_{k} a_{i\ell} X_{\ell k} b_{kj}\right) = \sum_{\ell} \sum_{k} a_{i\ell} E(X_{\ell k}) b_{kj}$$

which is the (i, j)th element of $\mathbf{A}E(\mathbf{X})\mathbf{B}$.

2.41. You are given the random vector $\mathbf{X}' = [X_1, X_2, X_3, X_4]$ with mean vector $\boldsymbol{\mu}_\mathbf{X}' = [3, 2, -2, 0]$ and variance–covariance matrix

$$\Sigma_\mathbf{X} = \begin{bmatrix} 3 & 0 & 0 & 0 \\ 0 & 3 & 0 & 0 \\ 0 & 0 & 3 & 0 \\ 0 & 0 & 0 & 3 \end{bmatrix}$$

Let

$$\mathbf{A} = \begin{bmatrix} 1 & -1 & 0 & 0 \\ 1 & 1 & -2 & 0 \\ 1 & 1 & 1 & -3 \end{bmatrix}$$

(a) Find $E(\mathbf{A}\mathbf{X})$, the mean of $\mathbf{A}\mathbf{X}$.

(b) Find Cov $(\mathbf{A}\mathbf{X})$, the variances and covariances of $\mathbf{A}\mathbf{X}$.

(c) Which pairs of linear combinations have zero covariances?

2.42. Repeat Exercise 2.41, but with

$$\Sigma_{\mathbf{X}} = \begin{bmatrix} 3 & 1 & 1 & 1 \\ 1 & 3 & 1 & 1 \\ 1 & 1 & 3 & 1 \\ 1 & 1 & 1 & 3 \end{bmatrix}$$

References

1. Bellman, R. *Introduction to Matrix Analysis* (2nd ed.) Philadelphia: Soc for Industrial & Applied Math (SIAM), 1997.

2. Eckart, C., and G. Young. "The Approximation of One Matrix by Another of Lower Rank." *Psychometrika*, **1** (1936), 211–218.

3. Graybill, F. A. *Introduction to Matrices with Applications in Statistics*. Belmont, CA: Wadsworth, 1969.

4. Halmos, P. R. *Finite-Dimensional Vector Spaces*. New York: Springer-Verlag, 1993.

5. Johnson, R. A., and G. K. Bhattacharyya. *Statistics: Principles and Methods* (5th ed.) New York: John Wiley, 2005.

6. Noble, B., and J. W. Daniel. *Applied Linear Algebra* (3rd ed.). Englewood Cliffs, NJ: Prentice Hall, 1988.

Sample Geometry and Random Sampling

3.1 Introduction

With the vector concepts introduced in the previous chapter, we can now delve deeper into the geometrical interpretations of the descriptive statistics $\bar{\mathbf{x}}$, $\mathbf{S}_n$, and $\mathbf{R}$; we do so in Section 3.2. Many of our explanations use the representation of the columns of $\mathbf{X}$ as p vectors in n dimensions. In Section 3.3 we introduce the assumption that the observations constitute a random sample. Simply stated, random sampling implies that (1) measurements taken on different items (or trials) are unrelated to one another and (2) the joint distribution of all p variables remains the same for all items. Ultimately, it is this structure of the random sample that justifies a particular choice of distance and dictates the geometry for the n-dimensional representation of the data. Furthermore, when data can be treated as a random sample, statistical inferences are based on a solid foundation.

Returning to geometric interpretations in Section 3.4, we introduce a single number, called *generalized variance*, to describe variability. This generalization of variance is an integral part of the comparison of multivariate means. In later sections we use matrix algebra to provide concise expressions for the matrix products and sums that allow us to calculate $\bar{\mathbf{x}}$ and $\mathbf{S}_n$ directly from the data matrix $\mathbf{X}$. The connection between $\bar{\mathbf{x}}$, $\mathbf{S}_n$, and the means and covariances for linear combinations of variables is also clearly delineated, using the notion of matrix products.

3.2 The Geometry of the Sample

A single multivariate observation is the collection of measurements on p different variables taken on the same item or trial. As in Chapter 1, if n observations have been obtained, the entire data set can be placed in an $n \times p$ array (matrix):

$$\underset{(n \times p)}{\mathbf{X}} = \begin{bmatrix} x_{11} & x_{12} & \cdots & x_{1p} \\ x_{21} & x_{22} & \cdots & x_{2p} \\ \vdots & \vdots & \ddots & \vdots \\ x_{n1} & x_{n2} & \cdots & x_{np} \end{bmatrix}$$

Each row of $\mathbf{X}$ represents a multivariate observation. Since the entire set of measurements is often one particular realization of what might have been observed, we say that the data are a *sample* of size n from a p-variate "population." The sample then consists of n measurements, each of which has p components.

As we have seen, the data can be plotted in two different ways. For the p-dimensional scatter plot, the *rows* of $\mathbf{X}$ represent n points in p-dimensional space. We can write

$$\underset{(n \times p)}{\mathbf{X}} = \begin{bmatrix} x_{11} & x_{12} & \cdots & x_{1p} \\ x_{21} & x_{22} & \cdots & x_{2p} \\ \vdots & \vdots & \ddots & \vdots \\ x_{n1} & x_{n2} & \cdots & x_{np} \end{bmatrix} = \begin{bmatrix} \mathbf{x}_1' \\ \mathbf{x}_2' \\ \vdots \\ \mathbf{x}_n' \end{bmatrix} \begin{matrix} \leftarrow \text{1st (multivariate) observation} \\ \\ \\ \leftarrow n\text{th (multivariate) observation} \end{matrix} \qquad (3\text{-}1)$$

The row vector $\mathbf{x}_j'$, representing the jth observation, contains the coordinates of a point.

The scatter plot of n points in p-dimensional space provides information on the locations and variability of the points. If the points are regarded as solid spheres, the sample mean vector $\bar{\mathbf{x}}$, given by (1-8), is the center of balance. Variability occurs in more than one direction, and it is quantified by the sample variance–covariance matrix $\mathbf{S}_n$. A *single* numerical measure of variability is provided by the determinant of the sample variance–covariance matrix. When p is greater than 3, this scatter plot representation cannot actually be graphed. Yet the consideration of the data as n points in p dimensions provides insights that are not readily available from algebraic expressions. Moreover, the concepts illustrated for $p = 2$ or $p = 3$ remain valid for the other cases.

Example 3.1 (Computing the mean vector) Compute the mean vector $\bar{\mathbf{x}}$ from the data matrix.

$$\mathbf{X} = \begin{bmatrix} 4 & 1 \\ -1 & 3 \\ 3 & 5 \end{bmatrix}$$

Plot the $n = 3$ data points in $p = 2$ space, and locate $\bar{\mathbf{x}}$ on the resulting diagram.

The first point, $\mathbf{x}_1$, has coordinates $\mathbf{x}_1' = [4, 1]$. Similarly, the remaining two points are $\mathbf{x}_2' = [-1, 3]$ and $\mathbf{x}_3' = [3, 5]$. Finally,

$$\bar{\mathbf{x}} = \begin{bmatrix} \dfrac{4 - 1 + 3}{3} \\ \dfrac{1 + 3 + 5}{3} \end{bmatrix} = \begin{bmatrix} 2 \\ 3 \end{bmatrix}$$

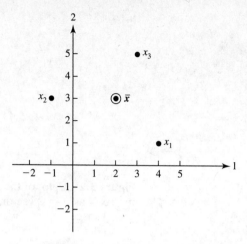

Figure 3.1 A plot of the data matrix **X** as $n = 3$ points in $p = 2$ space.

Figure 3.1 shows that $\bar{\mathbf{x}}$ is the balance point (center of gravity) of the scatter plot. ∎

The alternative geometrical representation is constructed by considering the data as p vectors in n-dimensional space. Here we take the elements of the *columns* of the data matrix to be the coordinates of the vectors. Let

$$
\underset{(n\times p)}{\mathbf{X}} = \begin{bmatrix} x_{11} & x_{12} & \cdots & x_{1p} \\ x_{21} & x_{22} & \cdots & x_{2p} \\ \vdots & \vdots & \ddots & \vdots \\ x_{n1} & x_{n2} & \cdots & x_{np} \end{bmatrix} = [\mathbf{y}_1 \mid \mathbf{y}_2 \mid \cdots \mid \mathbf{y}_p] \tag{3-2}
$$

Then the coordinates of the first point $\mathbf{y}_1' = [x_{11}, x_{21}, \ldots, x_{n1}]$ are the n measurements on the first variable. In general, the ith point $\mathbf{y}_i' = [x_{1i}, x_{2i}, \ldots, x_{ni}]$ is determined by the n-tuple of all measurements on the ith variable. In this geometrical representation, we depict $\mathbf{y}_1, \ldots, \mathbf{y}_p$ as vectors rather than points, as in the p-dimensional scatter plot. We shall be manipulating these quantities shortly using the algebra of vectors discussed in Chapter 2.

Example 3.2 (Data as p vectors in n dimensions) Plot the following data as $p = 2$ vectors in $n = 3$ space:

$$
\mathbf{X} = \begin{bmatrix} 4 & 1 \\ -1 & 3 \\ 3 & 5 \end{bmatrix}
$$

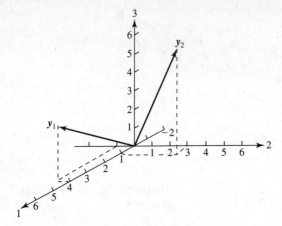

Figure 3.2 A plot of the data matrix $\mathbf{X}$ as $p = 2$ vectors in $n = 3$ space.

Here $\mathbf{y}_1' = [4, -1, 3]$ and $\mathbf{y}_2' = [1, 3, 5]$. These vectors are shown in Figure 3.2. ∎

Many of the algebraic expressions we shall encounter in multivariate analysis can be related to the geometrical notions of length, angle, and volume. This is important because geometrical representations ordinarily facilitate understanding and lead to further insights.

Unfortunately, we are limited to visualizing objects in three dimensions, and consequently, the n-dimensional representation of the data matrix $\mathbf{X}$ may not seem like a particularly useful device for $n > 3$. It turns out, however, that geometrical relationships and the associated statistical concepts depicted for any three vectors remain valid regardless of their dimension. This follows because three vectors, even if n dimensional, can span no more than a three-dimensional space, just as two vectors with any number of components must lie in a plane. By selecting an appropriate three-dimensional perspective—that is, a portion of the n-dimensional space containing the three vectors of interest—a view is obtained that preserves both lengths and angles. Thus, it is possible, with the right choice of axes, to illustrate certain algebraic statistical concepts in terms of only two or three vectors of any dimension n. Since the specific choice of axes is not relevant to the geometry, we shall always label the coordinate axes 1, 2, and 3.

It is possible to give a geometrical interpretation of the process of finding a sample mean. We start by defining the $n \times 1$ vector $\mathbf{1}_n' = [1, 1, \ldots, 1]$. (To simplify the notation, the subscript n will be dropped when the dimension of the vector $\mathbf{1}_n$ is clear from the context.) The vector $\mathbf{1}$ forms equal angles with each of the n coordinate axes, so the vector $(1/\sqrt{n})\mathbf{1}$ has unit length in the equal-angle direction. Consider the vector $\mathbf{y}_i' = [x_{1i}, x_{2i}, \ldots, x_{ni}]$. The projection of $\mathbf{y}_i$ on the unit vector $(1/\sqrt{n})\mathbf{1}$ is, by (2-8),

$$\mathbf{y}_i'\left(\frac{1}{\sqrt{n}}\mathbf{1}\right)\frac{1}{\sqrt{n}}\mathbf{1} = \frac{x_{1i} + x_{2i} + \cdots + x_{ni}}{n}\mathbf{1} = \bar{x}_i\mathbf{1} \qquad (3\text{-}3)$$

That is, the sample mean $\bar{x}_i = (x_{1i} + x_{2i} + \cdots + x_{ni})/n = \mathbf{y}_i'\mathbf{1}/n$ corresponds to the multiple of $\mathbf{1}$ required to give the projection of $\mathbf{y}_i$ onto the line determined by $\mathbf{1}$.

Further, for each $\mathbf{y}_i$, we have the decomposition

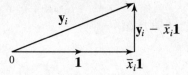

where $\bar{x}_i\mathbf{1}$ is perpendicular to $\mathbf{y}_i - \bar{x}_i\mathbf{1}$. The deviation, or mean corrected, vector is

$$\mathbf{d}_i = \mathbf{y}_i - \bar{x}_i\mathbf{1} = \begin{bmatrix} x_{1i} - \bar{x}_i \\ x_{2i} - \bar{x}_i \\ \vdots \\ x_{ni} - \bar{x}_i \end{bmatrix} \qquad (3\text{-}4)$$

The elements of $\mathbf{d}_i$ are the deviations of the measurements on the ith variable from their sample mean. Decomposition of the $\mathbf{y}_i$ vectors into mean components and deviation from the mean components is shown in Figure 3.3 for $p = 3$ and $n = 3$.

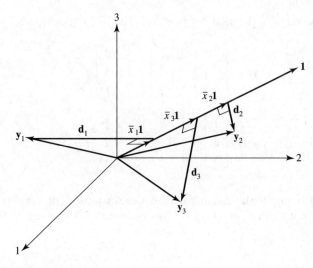

Figure 3.3 The decomposition of $\mathbf{y}_i$ into a mean component $\bar{x}_i\mathbf{1}$ and a deviation component $\mathbf{d}_i = \mathbf{y}_i - \bar{x}_i\mathbf{1}$, $i = 1, 2, 3$.

Example 3.3 (Decomposing a vector into its mean and deviation components) Let us carry out the decomposition of $\mathbf{y}_i$ into $\bar{x}_i\mathbf{1}$ and $\mathbf{d}_i = \mathbf{y}_i - \bar{x}_i\mathbf{1}$, $i = 1, 2$, for the data given in Example 3.2:

$$\mathbf{X} = \begin{bmatrix} 4 & 1 \\ -1 & 3 \\ 3 & 5 \end{bmatrix}$$

Here, $\bar{x}_1 = (4 - 1 + 3)/3 = 2$ and $\bar{x}_2 = (1 + 3 + 5)/3 = 3$, so

$$\bar{x}_1\mathbf{1} = 2\begin{bmatrix} 1 \\ 1 \\ 1 \end{bmatrix} = \begin{bmatrix} 2 \\ 2 \\ 2 \end{bmatrix} \qquad \bar{x}_2\mathbf{1} = 3\begin{bmatrix} 1 \\ 1 \\ 1 \end{bmatrix} = \begin{bmatrix} 3 \\ 3 \\ 3 \end{bmatrix}$$

Consequently,

$$\mathbf{d}_1 = \mathbf{y}_1 - \bar{x}_1 \mathbf{1} = \begin{bmatrix} 4 \\ -1 \\ 3 \end{bmatrix} - \begin{bmatrix} 2 \\ 2 \\ 2 \end{bmatrix} = \begin{bmatrix} 2 \\ -3 \\ 1 \end{bmatrix}$$

and

$$\mathbf{d}_2 = \mathbf{y}_2 - \bar{x}_2 \mathbf{1} = \begin{bmatrix} 1 \\ 3 \\ 5 \end{bmatrix} - \begin{bmatrix} 3 \\ 3 \\ 3 \end{bmatrix} = \begin{bmatrix} -2 \\ 0 \\ 2 \end{bmatrix}$$

We note that $\bar{x}_1 \mathbf{1}$ and $\mathbf{d}_1 = \mathbf{y}_1 - \bar{x}_1 \mathbf{1}$ are perpendicular, because

$$(\bar{x}_1 \mathbf{1})'(\mathbf{y}_1 - \bar{x}_1 \mathbf{1}) = \begin{bmatrix} 2 & 2 & 2 \end{bmatrix} \begin{bmatrix} 2 \\ -3 \\ 1 \end{bmatrix} = 4 - 6 + 2 = 0$$

A similar result holds for $\bar{x}_2 \mathbf{1}$ and $\mathbf{d}_2 = \mathbf{y}_2 - \bar{x}_2 \mathbf{1}$. The decomposition is

$$\mathbf{y}_1 = \begin{bmatrix} 4 \\ -1 \\ 3 \end{bmatrix} = \begin{bmatrix} 2 \\ 2 \\ 2 \end{bmatrix} + \begin{bmatrix} 2 \\ -3 \\ 1 \end{bmatrix}$$

$$\mathbf{y}_2 = \begin{bmatrix} 1 \\ 3 \\ 5 \end{bmatrix} = \begin{bmatrix} 3 \\ 3 \\ 3 \end{bmatrix} + \begin{bmatrix} -2 \\ 0 \\ 2 \end{bmatrix}$$ ∎

For the time being, we are interested in the deviation (or residual) vectors $\mathbf{d}_i = \mathbf{y}_i - \bar{x}_i \mathbf{1}$. A plot of the deviation vectors of Figure 3.3 is given in Figure 3.4.

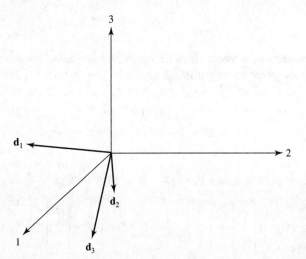

Figure 3.4 The deviation vectors $\mathbf{d}_i$ from Figure 3.3.

We have translated the deviation vectors to the origin without changing their lengths or orientations.

Now consider the squared lengths of the deviation vectors. Using (2-5) and (3-4), we obtain

$$L_{\mathbf{d}_i}^2 = \mathbf{d}_i'\mathbf{d}_i = \sum_{j=1}^{n}(x_{ji} - \bar{x}_i)^2 \tag{3-5}$$

(Length of deviation vector)^2 = sum of squared deviations

From (1-3), we see that the squared length is proportional to the variance of the measurements on the ith variable. Equivalently, the *length* is proportional to the *standard deviation*. Longer vectors represent more variability than shorter vectors.

For any two deviation vectors $\mathbf{d}_i$ and $\mathbf{d}_k$,

$$\mathbf{d}_i'\mathbf{d}_k = \sum_{j=1}^{n}(x_{ji} - \bar{x}_i)(x_{jk} - \bar{x}_k) \tag{3-6}$$

Let θ_{ik} denote the angle formed by the vectors $\mathbf{d}_i$ and $\mathbf{d}_k$. From (2-6), we get

$$\mathbf{d}_i'\mathbf{d}_k = L_{\mathbf{d}_i}L_{\mathbf{d}_k}\cos(\theta_{ik})$$

or, using (3-5) and (3-6), we obtain

$$\sum_{j=1}^{n}(x_{ji} - \bar{x}_i)(x_{jk} - \bar{x}_k) = \sqrt{\sum_{j=1}^{n}(x_{ji} - \bar{x}_i)^2}\sqrt{\sum_{j=1}^{n}(x_{jk} - \bar{x}_k)^2}\cos(\theta_{ik})$$

so that [see (1-5)]

$$r_{ik} = \frac{s_{ik}}{\sqrt{s_{ii}}\sqrt{s_{kk}}} = \cos(\theta_{ik}) \tag{3-7}$$

The *cosine* of the angle is the sample *correlation coefficient*. Thus, if the two deviation vectors have nearly the same orientation, the sample correlation will be close to 1. If the two vectors are nearly perpendicular, the sample correlation will be approximately zero. If the two vectors are oriented in nearly opposite directions, the sample correlation will be close to -1.

Example 3.4 (Calculating S$_n$ and R from deviation vectors) Given the deviation vectors in Example 3.3, let us compute the sample variance–covariance matrix $\mathbf{S}_n$ and sample correlation matrix $\mathbf{R}$ using the geometrical concepts just introduced.

From Example 3.3,

$$\mathbf{d}_1 = \begin{bmatrix} 2 \\ -3 \\ 1 \end{bmatrix} \quad \text{and} \quad \mathbf{d}_2 = \begin{bmatrix} -2 \\ 0 \\ 2 \end{bmatrix}$$

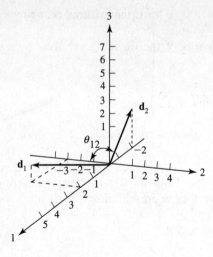

Figure 3.5 The deviation vectors $\mathbf{d}_1$ and $\mathbf{d}_2$.

These vectors, translated to the origin, are shown in Figure 3.5. Now,

$$\mathbf{d}_1'\mathbf{d}_1 = \begin{bmatrix} 2 & -3 & 1 \end{bmatrix} \begin{bmatrix} 2 \\ -3 \\ 1 \end{bmatrix} = 14 = 3s_{11}$$

or $s_{11} = \frac{14}{3}$. Also,

$$\mathbf{d}_2'\mathbf{d}_2 = \begin{bmatrix} -2 & 0 & 2 \end{bmatrix} \begin{bmatrix} -2 \\ 0 \\ 2 \end{bmatrix} = 8 = 3s_{22}$$

or $s_{22} = \frac{8}{3}$. Finally,

$$\mathbf{d}_1'\mathbf{d}_2 = \begin{bmatrix} 2 & -3 & 1 \end{bmatrix} \begin{bmatrix} -2 \\ 0 \\ 2 \end{bmatrix} = -2 = 3s_{12}$$

or $s_{12} = -\frac{2}{3}$. Consequently,

$$r_{12} = \frac{s_{12}}{\sqrt{s_{11}}\,\sqrt{s_{22}}} = \frac{-\frac{2}{3}}{\sqrt{\frac{14}{3}}\,\sqrt{\frac{8}{3}}} = -.189$$

and

$$\mathbf{S}_n = \begin{bmatrix} \frac{14}{3} & -\frac{2}{3} \\ -\frac{2}{3} & \frac{8}{3} \end{bmatrix}, \qquad \mathbf{R} = \begin{bmatrix} 1 & -.189 \\ -.189 & 1 \end{bmatrix}$$

∎

The concepts of length, angle, and projection have provided us with a geometrical interpretation of the sample. We summarize as follows:

Geometrical Interpretation of the Sample

1. The projection of a column y_i of the data matrix X onto the equal angular vector 1 is the vector $\bar{x}_i 1$. The vector $\bar{x}_i 1$ has length $\sqrt{n} |\bar{x}_i|$. Therefore, the ith sample mean, $\bar{x}_i$, is related to the length of the projection of y_i on 1.

2. The information comprising S_n is obtained from the deviation vectors $d_i = y_i - \bar{x}_i 1 = [x_{1i} - \bar{x}_i, \bar{x}_{2i} - \bar{x}_i, \ldots, x_{ni} - \bar{x}_i]'$. The square of the length of d_i is ns_{ii}, and the (inner) product between d_i and d_k is ns_{ik}.[1]

3. The sample correlation r_{ik} is the cosine of the angle between d_i and d_k.

3.3 Random Samples and the Expected Values of the Sample Mean and Covariance Matrix

In order to study the sampling variability of statistics such as $\bar{x}$ and S_n with the ultimate aim of making inferences, we need to make assumptions about the variables whose observed values constitute the data set X.

Suppose, then, that the data have not yet been observed, but we *intend* to collect n sets of measurements on p variables. Before the measurements are made, their values cannot, in general, be predicted exactly. Consequently, we treat them as random variables. In this context, let the (j, k)-th entry in the data matrix be the random variable X_{jk}. Each set of measurements X_j on p variables is a random vector, and we have the random matrix

$$\underset{(n \times p)}{X} = \begin{bmatrix} X_{11} & X_{12} & \cdots & X_{1p} \\ X_{21} & X_{22} & \cdots & X_{2p} \\ \vdots & \vdots & \ddots & \vdots \\ X_{n1} & X_{n2} & \cdots & X_{np} \end{bmatrix} = \begin{bmatrix} X_1' \\ X_2' \\ \vdots \\ X_n' \end{bmatrix} \tag{3-8}$$

A *random sample* can now be defined.

If the row vectors $X_1', X_2', \ldots, X_n'$ in (3-8) represent *independent* observations from a *common* joint distribution with density function $f(x) = f(x_1, x_2, \ldots, x_p)$, then $X_1, X_2, \ldots, X_n$ are said to form a *random sample* from $f(x)$. Mathematically, $X_1, X_2, \ldots, X_n$ form a random sample if their joint density function is given by the product $f(x_1)f(x_2) \cdots f(x_n)$, where $f(x_j) = f(x_{j1}, x_{j2}, \ldots, x_{jp})$ is the density function for the jth row vector.

Two points connected with the definition of random sample merit special attention:

1. The measurements of the p variables in a *single* trial, such as $X_j' = [X_{j1}, X_{j2}, \ldots, X_{jp}]$, will usually be correlated. Indeed, we expect this to be the case. The measurements from *different* trials must, however, be independent.

---

[1] The square of the length and the inner product are $(n-1)s_{ii}$ and $(n-1)s_{ik}$, respectively, when the divisor $n - 1$ is used in the definitions of the sample variance and covariance.

2. The independence of measurements from trial to trial may not hold when the variables are likely to drift over time, as with sets of p stock prices or p economic indicators. Violations of the tentative assumption of independence can have a serious impact on the quality of statistical inferences.

The following examples illustrate these remarks.

Example 3.5 (Selecting a random sample) As a preliminary step in designing a permit system for utilizing a wilderness canoe area without overcrowding, a natural-resource manager took a survey of users. The total wilderness area was divided into subregions, and respondents were asked to give information on the regions visited, lengths of stay, and other variables.

The method followed was to select persons randomly (perhaps using a random number table) from all those who entered the wilderness area during a particular week. All persons were equally likely to be in the sample, so the more popular entrances were represented by larger proportions of canoeists.

Here one would expect the sample observations to conform closely to the criterion for a random sample from the population of users or potential users. On the other hand, if one of the samplers had waited at a campsite far in the interior of the area and interviewed only canoeists who reached that spot, successive measurements would not be independent. For instance, lengths of stay in the wilderness area for different canoeists from this group would all tend to be large. ■

Example 3.6 (A nonrandom sample) Because of concerns with future solid-waste disposal, an ongoing study concerns the gross weight of municipal solid waste generated per year in the United States (Environmental Protection Agency). Estimated amounts attributed to x_1 = paper and paperboard waste and x_2 = plastic waste, in millions of tons, are given for selected years in Table 3.1. Should these measurements on $\mathbf{X}' = [X_1, X_2]$ be treated as a random sample of size $n = 7$? No! In fact, except for a slight but fortunate downturn in paper and paperboard waste in 2003, *both* variables are increasing over time.

Table 3.1 Solid Waste							
Year	1960	1970	1980	1990	1995	2000	2003
x_1 (paper)	29.2	44.3	55.2	72.7	81.7	87.7	83.1
x_2 (plastics)	.4	2.9	6.8	17.1	18.9	24.7	26.7

■

As we have argued heuristically in Chapter 1, the notion of statistical independence has important implications for measuring distance. Euclidean distance appears appropriate if the components of a vector are independent and have the same variances. Suppose we consider the location of the kth column $\mathbf{Y}'_k = [X_{1k}, X_{2k}, \ldots, X_{nk}]$ of $\mathbf{X}$, regarded as a point in n dimensions. The location of this point is determined by the joint probability distribution $f(\mathbf{y}_k) = f(x_{1k}, x_{2k}, \ldots, x_{nk})$. When the measurements $X_{1k}, X_{2k}, \ldots, X_{nk}$ are a random sample, $f(\mathbf{y}_k) = f(x_{1k}, x_{2k}, \ldots, x_{nk}) = f_k(x_{1k})f_k(x_{2k}) \cdots f_k(x_{nk})$ and, consequently, each coordinate x_{jk} contributes equally to the location through the identical marginal distributions $f_k(x_{jk})$.

If the n components are not independent or the marginal distributions are not identical, the influence of individual measurements (coordinates) on location is asymmetrical. We would then be led to consider a distance function in which the coordinates were weighted unequally, as in the "statistical" distances or quadratic forms introduced in Chapters 1 and 2.

Certain conclusions can be reached concerning the sampling distributions of $\overline{\mathbf{X}}$ and $\mathbf{S}_n$ without making further assumptions regarding the form of the underlying joint distribution of the variables. In particular, we can see how $\overline{\mathbf{X}}$ and $\mathbf{S}_n$ fare as point estimators of the corresponding population mean vector $\boldsymbol{\mu}$ and covariance matrix $\boldsymbol{\Sigma}$.

Result 3.1. Let $\mathbf{X}_1, \mathbf{X}_2, \ldots, \mathbf{X}_n$ be a random sample from a joint distribution that has mean vector $\boldsymbol{\mu}$ and covariance matrix $\boldsymbol{\Sigma}$. Then $\overline{\mathbf{X}}$ is an *unbiased* estimator of $\boldsymbol{\mu}$, and its covariance matrix is

$$\frac{1}{n}\boldsymbol{\Sigma}$$

That is,

$$E(\overline{\mathbf{X}}) = \boldsymbol{\mu} \qquad \text{(population mean vector)}$$

$$\text{Cov}(\overline{\mathbf{X}}) = \frac{1}{n}\boldsymbol{\Sigma} \qquad \left(\begin{array}{c}\text{population variance–covariance matrix} \\ \text{divided by sample size}\end{array}\right) \qquad (3\text{-}9)$$

For the covariance matrix $\mathbf{S}_n$,

$$E(\mathbf{S}_n) = \frac{n-1}{n}\boldsymbol{\Sigma} = \boldsymbol{\Sigma} - \frac{1}{n}\boldsymbol{\Sigma}$$

Thus,

$$E\left(\frac{n}{n-1}\mathbf{S}_n\right) = \boldsymbol{\Sigma} \qquad (3\text{-}10)$$

so $[n/(n-1)]\mathbf{S}_n$ is an *unbiased* estimator of $\boldsymbol{\Sigma}$, while $\mathbf{S}_n$ is a *biased* estimator with (bias) $= E(\mathbf{S}_n) - \boldsymbol{\Sigma} = -(1/n)\boldsymbol{\Sigma}$.

Proof. Now, $\overline{\mathbf{X}} = (\mathbf{X}_1 + \mathbf{X}_2 + \cdots + \mathbf{X}_n)/n$. The repeated use of the properties of expectation in (2-24) for two vectors gives

$$E(\overline{\mathbf{X}}) = E\left(\frac{1}{n}\mathbf{X}_1 + \frac{1}{n}\mathbf{X}_2 + \cdots + \frac{1}{n}\mathbf{X}_n\right)$$

$$= E\left(\frac{1}{n}\mathbf{X}_1\right) + E\left(\frac{1}{n}\mathbf{X}_2\right) + \cdots + E\left(\frac{1}{n}\mathbf{X}_n\right)$$

$$= \frac{1}{n}E(\mathbf{X}_1) + \frac{1}{n}E(\mathbf{X}_2) + \cdots + \frac{1}{n}E(\mathbf{X}_n) = \frac{1}{n}\boldsymbol{\mu} + \frac{1}{n}\boldsymbol{\mu} + \cdots + \frac{1}{n}\boldsymbol{\mu}$$

$$= \boldsymbol{\mu}$$

Next,

$$(\overline{\mathbf{X}} - \boldsymbol{\mu})(\overline{\mathbf{X}} - \boldsymbol{\mu})' = \left(\frac{1}{n}\sum_{j=1}^{n}(\mathbf{X}_j - \boldsymbol{\mu})\right)\left(\frac{1}{n}\sum_{\ell=1}^{n}(\mathbf{X}_\ell - \boldsymbol{\mu})\right)'$$

$$= \frac{1}{n^2}\sum_{j=1}^{n}\sum_{\ell=1}^{n}(\mathbf{X}_j - \boldsymbol{\mu})(\mathbf{X}_\ell - \boldsymbol{\mu})'$$

so

$$\text{Cov}(\overline{\mathbf{X}}) = E(\overline{\mathbf{X}} - \boldsymbol{\mu})(\overline{\mathbf{X}} - \boldsymbol{\mu})' = \frac{1}{n^2}\left(\sum_{j=1}^{n}\sum_{\ell=1}^{n} E(\mathbf{X}_j - \boldsymbol{\mu})(\mathbf{X}_\ell - \boldsymbol{\mu})'\right)$$

For $j \neq \ell$, each entry in $E(\mathbf{X}_j - \boldsymbol{\mu})(\mathbf{X}_\ell - \boldsymbol{\mu})'$ is zero because the entry is the covariance between a component of $\mathbf{X}_j$ and a component of $\mathbf{X}_\ell$, and these are independent. [See Exercise 3.17 and (2-29).]

Therefore,

$$\text{Cov}(\overline{\mathbf{X}}) = \frac{1}{n^2}\left(\sum_{j=1}^{n} E(\mathbf{X}_j - \boldsymbol{\mu})(\mathbf{X}_j - \boldsymbol{\mu})'\right)$$

Since $\boldsymbol{\Sigma} = E(\mathbf{X}_j - \boldsymbol{\mu})(\mathbf{X}_j - \boldsymbol{\mu})'$ is the common population covariance matrix for each $\mathbf{X}_j$, we have

$$\text{Cov}(\overline{\mathbf{X}}) = \frac{1}{n^2}\left(\sum_{j=1}^{n} E(\mathbf{X}_j - \boldsymbol{\mu})(\mathbf{X}_j - \boldsymbol{\mu})'\right) = \frac{1}{n^2}\underbrace{(\boldsymbol{\Sigma} + \boldsymbol{\Sigma} + \cdots + \boldsymbol{\Sigma})}_{n \text{ terms}}$$

$$= \frac{1}{n^2}(n\boldsymbol{\Sigma}) = \left(\frac{1}{n}\right)\boldsymbol{\Sigma}$$

To obtain the expected value of $\mathbf{S}_n$, we first note that $(X_{ji} - \overline{X}_i)(X_{jk} - \overline{X}_k)$ is the (i, k)th element of $(\mathbf{X}_j - \overline{\mathbf{X}})(\mathbf{X}_j - \overline{\mathbf{X}})'$. The matrix representing sums of squares and cross products can then be written as

$$\sum_{j=1}^{n}(\mathbf{X}_j - \overline{\mathbf{X}})(\mathbf{X}_j - \overline{\mathbf{X}})' = \sum_{j=1}^{n}(\mathbf{X}_j - \overline{\mathbf{X}})\mathbf{X}_j' + \left(\sum_{j=1}^{n}(\mathbf{X}_j - \overline{\mathbf{X}})\right)(-\overline{\mathbf{X}})'$$

$$= \sum_{j=1}^{n}\mathbf{X}_j\mathbf{X}_j' - n\overline{\mathbf{X}}\,\overline{\mathbf{X}}'$$

since $\sum_{j=1}^{n}(\mathbf{X}_j - \overline{\mathbf{X}}) = \mathbf{0}$ and $n\overline{\mathbf{X}}' = \sum_{i=1}^{n}\mathbf{X}_j'$. Therefore, its expected value is

$$E\left(\sum_{j=1}^{n}\mathbf{X}_j\mathbf{X}_j' - n\overline{\mathbf{X}}\,\overline{\mathbf{X}}'\right) = \sum_{j=1}^{n} E(\mathbf{X}_j\mathbf{X}_j') - nE(\overline{\mathbf{X}}\,\overline{\mathbf{X}}')$$

For any random vector $\mathbf{V}$ with $E(\mathbf{V}) = \boldsymbol{\mu}_V$ and $\text{Cov}(\mathbf{V}) = \boldsymbol{\Sigma}_V$, we have $E(\mathbf{V}\mathbf{V}') = \boldsymbol{\Sigma}_V + \boldsymbol{\mu}_V\boldsymbol{\mu}_V'$. (See Exercise 3.16.) Consequently,

$$E(\mathbf{X}_j\mathbf{X}_j') = \boldsymbol{\Sigma} + \boldsymbol{\mu}\boldsymbol{\mu}' \quad \text{and} \quad E(\overline{\mathbf{X}}\,\overline{\mathbf{X}}') = \frac{1}{n}\boldsymbol{\Sigma} + \boldsymbol{\mu}\boldsymbol{\mu}'$$

Using these results, we obtain

$$\sum_{j=1}^{n} E(\mathbf{X}_j\mathbf{X}_j') - nE(\overline{\mathbf{X}}\,\overline{\mathbf{X}}') = n\boldsymbol{\Sigma} + n\boldsymbol{\mu}\boldsymbol{\mu}' - n\left(\frac{1}{n}\boldsymbol{\Sigma} + \boldsymbol{\mu}\boldsymbol{\mu}'\right) = (n - 1)\boldsymbol{\Sigma}$$

and thus, since $\mathbf{S}_n = (1/n)\left(\sum_{j=1}^{n}\mathbf{X}_j\mathbf{X}_j' - n\overline{\mathbf{X}}\,\overline{\mathbf{X}}'\right)$, it follows immediately that

$$E(\mathbf{S}_n) = \frac{(n - 1)}{n}\boldsymbol{\Sigma}$$

∎

Result 3.1 shows that the (i, k)th entry, $(n - 1)^{-1} \sum_{j=1}^{n} (X_{ji} - \bar{X}_i)(X_{jk} - \bar{X}_k)$, of $[n/(n - 1)]\mathbf{S}_n$ is an unbiased estimator of σ_{ik}. However, the individual sample standard deviations $\sqrt{s_{ii}}$, calculated with either n or $n - 1$ as a divisor, are not unbiased estimators of the corresponding population quantities $\sqrt{\sigma_{ii}}$. Moreover, the correlation coefficients r_{ik} are *not* unbiased estimators of the population quantities ρ_{ik}. However, the bias $E\left(\sqrt{s_{ii}}\right) - \sqrt{\sigma_{ii}}$, or $E(r_{ik}) - \rho_{ik}$, can usually be ignored if the sample size n is moderately large.

Consideration of bias motivates a slightly modified definition of the sample variance–covariance matrix. Result 3.1 provides us with an unbiased estimator $\mathbf{S}$ of $\boldsymbol{\Sigma}$:

(Unbiased) Sample Variance–Covariance Matrix

$$\mathbf{S} = \left(\frac{n}{n - 1}\right)\mathbf{S}_n = \frac{1}{n - 1} \sum_{j=1}^{n} (\mathbf{X}_j - \bar{\mathbf{X}})(\mathbf{X}_j - \bar{\mathbf{X}})' \qquad (3\text{-}11)$$

Here S, without a subscript, has (i, k)th entry $(n - 1)^{-1} \sum_{j=1}^{n} (X_{ji} - \bar{X}_i)(X_{jk} - \bar{X}_k)$. This definition of sample covariance is commonly used in many multivariate test statistics. Therefore, it will replace $\mathbf{S}_n$ as the sample covariance matrix in most of the material throughout the rest of this book.

3.4 Generalized Variance

With a single variable, the sample variance is often used to describe the amount of variation in the measurements on that variable. When p variables are observed on each unit, the variation is described by the sample variance–covariance matrix

$$\mathbf{S} = \begin{bmatrix} s_{11} & s_{12} & \cdots & s_{1p} \\ s_{12} & s_{22} & \cdots & s_{2p} \\ \vdots & \vdots & \ddots & \vdots \\ s_{1p} & s_{2p} & \cdots & s_{pp} \end{bmatrix} = \left\{ s_{ik} = \frac{1}{n - 1} \sum_{j=1}^{n} (x_{ji} - \bar{x}_i)(x_{jk} - \bar{x}_k) \right\}$$

The sample covariance matrix contains p variances and $\frac{1}{2}p(p - 1)$ potentially different covariances. Sometimes it is desirable to assign a *single* numerical value for the variation expressed by $\mathbf{S}$. One choice for a value is the determinant of $\mathbf{S}$, which reduces to the usual sample variance of a single characteristic when $p = 1$. This determinant[2] is called the *generalized sample variance:*

Generalized sample variance $= |\mathbf{S}|$ \qquad (3-12)

---

[2] Definition 2A.24 defines "determinant" and indicates one method for calculating the value of a determinant.

Example 3.7 (Calculating a generalized variance) Employees (x_1) and profits per employee (x_2) for the 16 largest publishing firms in the United States are shown in Figure 1.3. The sample covariance matrix, obtained from the data in the April 30, 1990, *Forbes* magazine article, is

$$S = \begin{bmatrix} 252.04 & -68.43 \\ -68.43 & 123.67 \end{bmatrix}$$

Evaluate the generalized variance.

In this case, we compute

$$|S| = (252.04)(123.67) - (-68.43)(-68.43) = 26{,}487 \qquad \blacksquare$$

The generalized sample variance provides one way of writing the information on all variances and covariances as a single number. Of course, when $p > 1$, some information about the sample is lost in the process. A geometrical interpretation of $|S|$ will help us appreciate its strengths and weaknesses as a descriptive summary.

Consider the area generated within the plane by two deviation vectors $d_1 = y_1 - \bar{x}_1 1$ and $d_2 = y_2 - \bar{x}_2 1$. Let L_{d_1} be the length of d_1 and L_{d_2} the length of d_2. By elementary geometry, we have the diagram

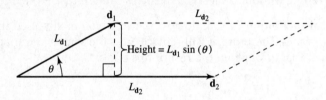

and the area of the trapezoid is $|L_{d_1} \sin(\theta)| L_{d_2}$. Since $\cos^2(\theta) + \sin^2(\theta) = 1$, we can express this area as

$$\text{Area} = L_{d_1} L_{d_2} \sqrt{1 - \cos^2(\theta)}$$

From (3-5) and (3-7),

$$L_{d_1} = \sqrt{\sum_{j=1}^{n} (x_{j1} - \bar{x}_1)^2} = \sqrt{(n-1)s_{11}}$$

$$L_{d_2} = \sqrt{\sum_{j=1}^{n} (x_{j2} - \bar{x}_2)^2} = \sqrt{(n-1)s_{22}}$$

and

$$\cos(\theta) = r_{12}$$

Therefore,

$$\text{Area} = (n-1)\sqrt{s_{11}}\sqrt{s_{22}}\sqrt{1 - r_{12}^2} = (n-1)\sqrt{s_{11}s_{22}(1 - r_{12}^2)} \qquad (3\text{-}13)$$

Also,

$$|S| = \left| \begin{bmatrix} s_{11} & s_{12} \\ s_{12} & s_{22} \end{bmatrix} \right| = \left| \begin{bmatrix} s_{11} & \sqrt{s_{11}}\sqrt{s_{22}}\, r_{12} \\ \sqrt{s_{11}}\sqrt{s_{22}}\, r_{12} & s_{22} \end{bmatrix} \right|$$

$$= s_{11}s_{22} - s_{11}s_{22}r_{12}^2 = s_{11}s_{22}(1 - r_{12}^2) \qquad (3\text{-}14)$$

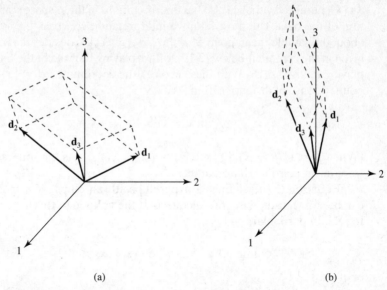

Figure 3.6 (a) "Large" generalized sample variance for $p = 3$.
(b) "Small" generalized sample variance for $p = 3$.

If we compare (3-14) with (3-13), we see that

$$|\mathbf{S}| = (\text{area})^2/(n - 1)^2$$

Assuming now that $|\mathbf{S}| = (n - 1)^{-(p-1)}(\text{volume})^2$ holds for the volume generated in n space by the $p - 1$ deviation vectors $\mathbf{d}_1, \mathbf{d}_2, \ldots, \mathbf{d}_{p-1}$, we can establish the following general result for p deviation vectors by induction (see [1], p. 266):

$$\text{Generalized sample variance} = |\mathbf{S}| = (n - 1)^{-p}(\text{volume})^2 \qquad (3\text{-}15)$$

Equation (3-15) says that the generalized sample variance, for a fixed set of data, is proportional to the square of the volume generated by the p deviation vectors[3] $\mathbf{d}_1 = \mathbf{y}_1 - \bar{x}_1\mathbf{1}$, $\mathbf{d}_2 = \mathbf{y}_2 - \bar{x}_2\mathbf{1}, \ldots, \mathbf{d}_p = \mathbf{y}_p - \bar{x}_p\mathbf{1}$. Figures 3.6(a) and (b) show trapezoidal regions, generated by $p = 3$ residual vectors, corresponding to "large" and "small" generalized variances.

For a fixed sample size, it is clear from the geometry that volume, or $|\mathbf{S}|$, will increase when the length of any $\mathbf{d}_i = \mathbf{y}_i - \bar{x}_i\mathbf{1}$ (or $\sqrt{s_{ii}}$) is increased. In addition, volume will increase if the residual vectors of fixed length are moved until they are at right angles to one another, as in Figure 3.6(a). On the other hand, the volume, or $|\mathbf{S}|$, will be small if just one of the s_{ii} is small or one of the deviation vectors lies nearly in the (hyper) plane formed by the others, or both. In the second case, the trapezoid has very little height above the plane. This is the situation in Figure 3.6(b), where $\mathbf{d}_3$ lies nearly in the plane formed by $\mathbf{d}_1$ and $\mathbf{d}_2$.

---

[3] If generalized variance is defined in terms of the sample covariance matrix $\mathbf{S}_n = [(n - 1)/n]\mathbf{S}$, then, using Result 2A.11, $|\mathbf{S}_n| = |[(n - 1)/n]\mathbf{I}_p\mathbf{S}| = |[(n - 1)/n]\mathbf{I}_p||\mathbf{S}| = [(n - 1)/n]^p|\mathbf{S}|$. Consequently, using (3-15), we can also write the following: Generalized sample variance $= |\mathbf{S}_n| = n^{-p}(\text{volume})^2$.

Generalized variance also has interpretations in the p-space scatter plot representation of the data. The most intuitive interpretation concerns the spread of the scatter about the sample mean point $\bar{\mathbf{x}}' = [\bar{x}_1, \bar{x}_2, \ldots, \bar{x}_p]$. Consider the measure of distance given in the comment below (2-19), with $\bar{\mathbf{x}}$ playing the role of the fixed point $\boldsymbol{\mu}$ and $\mathbf{S}^{-1}$ playing the role of $\mathbf{A}$. With these choices, the coordinates $\mathbf{x}' = [x_1, x_2, \ldots, x_p]$ of the points a constant distance c from $\bar{\mathbf{x}}$ satisfy

$$(\mathbf{x} - \bar{\mathbf{x}})'\mathbf{S}^{-1}(\mathbf{x} - \bar{\mathbf{x}}) = c^2 \tag{3-16}$$

[When $p = 1$, $(\mathbf{x} - \bar{\mathbf{x}})'\mathbf{S}^{-1}(\mathbf{x} - \bar{\mathbf{x}}) = (x_1 - \bar{x}_1)^2/s_{11}$ is the squared distance from x_1 to $\bar{x}_1$ in standard deviation units.]

Equation (3-16) defines a hyperellipsoid (an ellipse if $p = 2$) centered at $\bar{\mathbf{x}}$. It can be shown using integral calculus that the volume of this hyperellipsoid is related to $|\mathbf{S}|$. In particular,

$$\text{Volume of } \{\mathbf{x}: (\mathbf{x} - \bar{\mathbf{x}})'\mathbf{S}^{-1}(\mathbf{x} - \bar{\mathbf{x}}) \leq c^2\} = k_p|\mathbf{S}|^{1/2}c^p \tag{3-17}$$

or

$$(\text{Volume of ellipsoid})^2 = (\text{constant})\,(\text{generalized sample variance})$$

where the constant k_p is rather formidable.[4] A large volume corresponds to a large generalized variance.

Although the generalized variance has some intuitively pleasing geometrical interpretations, it suffers from a basic weakness as a descriptive summary of the sample covariance matrix $\mathbf{S}$, as the following example shows.

Example 3.8 (Interpreting the generalized variance) Figure 3.7 gives three scatter plots with very different patterns of correlation.

All three data sets have $\bar{\mathbf{x}}' = [2, 1]$, and the covariance matrices are

$$\mathbf{S} = \begin{bmatrix} 5 & 4 \\ 4 & 5 \end{bmatrix}, r = .8 \quad \mathbf{S} = \begin{bmatrix} 3 & 0 \\ 0 & 3 \end{bmatrix}, r = 0 \quad \mathbf{S} = \begin{bmatrix} 5 & -4 \\ -4 & 5 \end{bmatrix}, r = -.8$$

Each covariance matrix $\mathbf{S}$ contains the information on the variability of the component variables and also the information required to calculate the correlation coefficient. In this sense, $\mathbf{S}$ captures the orientation and size of the pattern of scatter.

The eigenvalues and eigenvectors extracted from $\mathbf{S}$ further describe the pattern in the scatter plot. For

$$\mathbf{S} = \begin{bmatrix} 5 & 4 \\ 4 & 5 \end{bmatrix}, \quad \text{the eigenvalues satisfy} \quad \begin{aligned} 0 &= (\lambda - 5)^2 - 4^2 \\ &= (\lambda - 9)(\lambda - 1) \end{aligned}$$

---

[4] For those who are curious, $k_p = 2\pi^{p/2}/p\,\Gamma(p/2)$, where $\Gamma(z)$ denotes the gamma function evaluated at z.

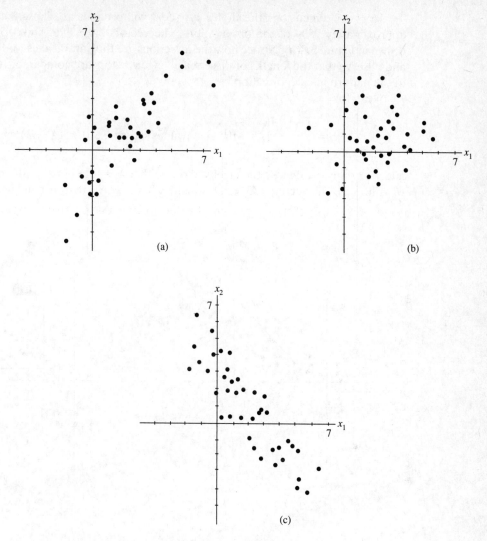

Figure 3.7 Scatter plots with three different orientations.

and we determine the eigenvalue–eigenvector pairs $\lambda_1 = 9$, $\mathbf{e}_1' = \left[1/\sqrt{2}, 1/\sqrt{2}\right]$ and $\lambda_2 = 1$, $\mathbf{e}_2' = \left[1/\sqrt{2}, -1/\sqrt{2}\right]$.

The mean-centered ellipse, with center $\bar{\mathbf{x}}' = [2, 1]$ for all three cases, is

$$(\mathbf{x} - \bar{\mathbf{x}})'\mathbf{S}^{-1}(\mathbf{x} - \bar{\mathbf{x}}) \leq c^2$$

To describe this ellipse, as in Section 2.3, with $\mathbf{A} = \mathbf{S}^{-1}$, we notice that if $(\lambda, \mathbf{e})$ is an eigenvalue–eigenvector pair for $\mathbf{S}$, then $(\lambda^{-1}, \mathbf{e})$ is an eigenvalue–eigenvector pair for $\mathbf{S}^{-1}$. That is, if $\mathbf{S}\mathbf{e} = \lambda\mathbf{e}$, then multiplying on the left by $\mathbf{S}^{-1}$ gives $\mathbf{S}^{-1}\mathbf{S}\mathbf{e} = \lambda\mathbf{S}^{-1}\mathbf{e}$, or $\mathbf{S}^{-1}\mathbf{e} = \lambda^{-1}\mathbf{e}$. Therefore, using the eigenvalues from $\mathbf{S}$, we know that the ellipse extends $c\sqrt{\lambda_i}$ in the direction of $\mathbf{e}_i$ from $\bar{\mathbf{x}}$.

In $p = 2$ dimensions, the choice $c^2 = 5.99$ will produce an ellipse that contains approximately 95% of the observations. The vectors $3\sqrt{5.99}\ \mathbf{e}_1$ and $\sqrt{5.99}\ \mathbf{e}_2$ are drawn in Figure 3.8(a). Notice how the directions are the natural axes for the ellipse, and observe that the lengths of these scaled eigenvectors are comparable to the size of the pattern in each direction.

Next, for

$$\mathbf{S} = \begin{bmatrix} 3 & 0 \\ 0 & 3 \end{bmatrix}, \qquad \text{the eigenvalues satisfy} \qquad 0 = (\lambda - 3)^2$$

and we arbitrarily choose the eigenvectors so that $\lambda_1 = 3$, $\mathbf{e}_1' = [1,\quad 0]$ and $\lambda_2 = 3$, $\mathbf{e}_2' = [0,\quad 1]$. The vectors $\sqrt{3}\ \sqrt{5.99}\ \mathbf{e}_1$ and $\sqrt{3}\ \sqrt{5.99}\ \mathbf{e}_2$ are drawn in Figure 3.8(b).

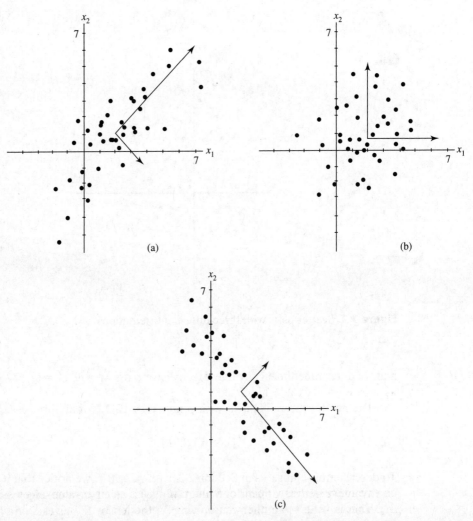

Figure 3.8 Axes of the mean-centered 95% ellipses for the scatter plots in Figure 3.7.

Finally, for

$$\mathbf{S} = \begin{bmatrix} 5 & -4 \\ -4 & 5 \end{bmatrix}, \qquad \text{the eigenvalues satisfy} \qquad \begin{aligned} 0 &= (\lambda - 5)^2 - (-4)^2 \\ &= (\lambda - 9)(\lambda - 1) \end{aligned}$$

and we determine the eigenvalue–eigenvector pairs $\lambda_1 = 9$, $\mathbf{e}_1' = [1/\sqrt{2}, \ -1/\sqrt{2}]$ and $\lambda_2 = 1$, $\mathbf{e}_2' = [1/\sqrt{2}, \ 1/\sqrt{2}]$. The scaled eigenvectors $3\sqrt{5.99}\,\mathbf{e}_1$ and $\sqrt{5.99}\,\mathbf{e}_2$ are drawn in Figure 3.8(c).

In two dimensions, we can often sketch the axes of the mean-centered ellipse by eye. However, the eigenvector approach also works for high dimensions where the data cannot be examined visually.

Note: Here the generalized variance $|\mathbf{S}|$ gives the same value, $|\mathbf{S}| = 9$, for all three patterns. But generalized variance does not contain any information on the orientation of the patterns. Generalized variance is easier to interpret when the two or more samples (patterns) being compared have nearly the same orientations.

Notice that our three patterns of scatter appear to cover approximately the same area. The ellipses that summarize the variability

$$(\mathbf{x} - \bar{\mathbf{x}})'\mathbf{S}^{-1}(\mathbf{x} - \bar{\mathbf{x}}) \le c^2$$

do have exactly the same area [see (3-17)], since all have $|\mathbf{S}| = 9$. ∎

As Example 3.8 demonstrates, different correlation structures are not detected by $|\mathbf{S}|$. The situation for $p > 2$ can be even more obscure.

Consequently, it is often desirable to provide more than the single number $|\mathbf{S}|$ as a summary of $\mathbf{S}$. From Exercise 2.12, $|\mathbf{S}|$ can be expressed as the product $\lambda_1 \lambda_2 \cdots \lambda_p$ of the eigenvalues of $\mathbf{S}$. Moreover, the mean-centered ellipsoid based on $\mathbf{S}^{-1}$ [see (3-16)] has axes whose lengths are proportional to the square roots of the λ_i's (see Section 2.3). These eigenvalues then provide information on the variability in all directions in the p-space representation of the data. It is useful, therefore, to report their individual values, as well as their product. We shall pursue this topic later when we discuss principal components.

Situations in which the Generalized Sample Variance Is Zero

The generalized sample variance will be zero in certain situations. A generalized variance of zero is indicative of extreme degeneracy, in the sense that at least one column of the matrix of deviations,

$$\begin{bmatrix} \mathbf{x}_1' - \bar{\mathbf{x}}' \\ \mathbf{x}_2' - \bar{\mathbf{x}}' \\ \vdots \\ \mathbf{x}_n' - \bar{\mathbf{x}}' \end{bmatrix} = \begin{bmatrix} x_{11} - \bar{x}_1 & x_{12} - \bar{x}_2 & \cdots & x_{1p} - \bar{x}_p \\ x_{21} - \bar{x}_1 & x_{22} - \bar{x}_2 & \cdots & x_{2p} - \bar{x}_p \\ \vdots & \vdots & \ddots & \vdots \\ x_{n1} - \bar{x}_1 & x_{n2} - \bar{x}_2 & \cdots & x_{np} - \bar{x}_p \end{bmatrix}$$

$$= \underset{(n \times p)}{\mathbf{X}} - \underset{(n \times 1)(1 \times p)}{\mathbf{1}\,\bar{\mathbf{x}}'} \tag{3-18}$$

can be expressed as a linear combination of the other columns. As we have shown geometrically, this is a case where one of the deviation vectors—for instance, $\mathbf{d}_i' = [x_{1i} - \bar{x}_i, \ldots, x_{ni} - \bar{x}_i]$—lies in the (hyper) plane generated by $\mathbf{d}_1, \ldots, \mathbf{d}_{i-1}$, $\mathbf{d}_{i+1}, \ldots, \mathbf{d}_p$.

Result 3.2. The generalized variance is zero when, and only when, at least one deviation vector lies in the (hyper) plane formed by all linear combinations of the others—that is, when the columns of the matrix of deviations in (3-18) are linearly dependent.

Proof. If the columns of the deviation matrix $(\mathbf{X} - \mathbf{1}\bar{\mathbf{x}}')$ are linearly dependent, there is a linear combination of the columns such that

$$\mathbf{0} = a_1 \, \text{col}_1(\mathbf{X} - \mathbf{1}\bar{\mathbf{x}}') + \cdots + a_p \, \text{col}_p(\mathbf{X} - \mathbf{1}\bar{\mathbf{x}}')$$
$$= (\mathbf{X} - \mathbf{1}\bar{\mathbf{x}}')\mathbf{a} \quad \text{for some } \mathbf{a} \neq \mathbf{0}$$

But then, as you may verify, $(n-1)\mathbf{S} = (\mathbf{X} - \mathbf{1}\bar{\mathbf{x}}')'(\mathbf{X} - \mathbf{1}\bar{\mathbf{x}}')$ and

$$(n-1)\mathbf{S}\mathbf{a} = (\mathbf{X} - \mathbf{1}\bar{\mathbf{x}}')'(\mathbf{X} - \mathbf{1}\bar{\mathbf{x}}')\mathbf{a} = \mathbf{0}$$

so the same $\mathbf{a}$ corresponds to a linear dependency, $a_1 \, \text{col}_1(\mathbf{S}) + \cdots + a_p \, \text{col}_p(\mathbf{S}) = \mathbf{S}\mathbf{a} = \mathbf{0}$, in the columns of $\mathbf{S}$. So, by Result 2A.9, $|\mathbf{S}| = 0$.

In the other direction, if $|\mathbf{S}| = 0$, then there is some linear combination $\mathbf{S}\mathbf{a}$ of the columns of $\mathbf{S}$ such that $\mathbf{S}\mathbf{a} = \mathbf{0}$. That is, $\mathbf{0} = (n-1)\mathbf{S}\mathbf{a} = (\mathbf{X} - \mathbf{1}\bar{\mathbf{x}}')'(\mathbf{X} - \mathbf{1}\bar{\mathbf{x}}')\mathbf{a}$. Premultiplying by $\mathbf{a}'$ yields

$$\mathbf{0} = \mathbf{a}'(\mathbf{X} - \mathbf{1}\bar{\mathbf{x}}')'(\mathbf{X} - \mathbf{1}\bar{\mathbf{x}}')\mathbf{a} = L^2_{(\mathbf{X}-\mathbf{1}\bar{\mathbf{x}}')\mathbf{a}}$$

and, for the length to equal zero, we must have $(\mathbf{X} - \mathbf{1}\bar{\mathbf{x}}')\mathbf{a} = \mathbf{0}$. Thus, the columns of $(\mathbf{X} - \mathbf{1}\bar{\mathbf{x}}')$ are linearly dependent. ∎

Example 3.9 (A case where the generalized variance is zero) Show that $|\mathbf{S}| = 0$ for

$$\mathop{\mathbf{X}}_{(3\times 3)} = \begin{bmatrix} 1 & 2 & 5 \\ 4 & 1 & 6 \\ 4 & 0 & 4 \end{bmatrix}$$

and determine the degeneracy.

Here $\bar{\mathbf{x}}' = [3, 1, 5]$, so

$$\mathbf{X} - \mathbf{1}\bar{\mathbf{x}}' = \begin{bmatrix} 1-3 & 2-1 & 5-5 \\ 4-3 & 1-1 & 6-5 \\ 4-3 & 0-1 & 4-5 \end{bmatrix} = \begin{bmatrix} -2 & 1 & 0 \\ 1 & 0 & 1 \\ 1 & -1 & -1 \end{bmatrix}$$

The deviation (column) vectors are $\mathbf{d}_1' = [-2, 1, 1]$, $\mathbf{d}_2' = [1, 0, -1]$, and $\mathbf{d}_3' = [0, 1, -1]$. Since $\mathbf{d}_3 = \mathbf{d}_1 + 2\mathbf{d}_2$, there is column degeneracy. (Note that there is row degeneracy also.) This means that one of the deviation vectors—for example, $\mathbf{d}_3$—lies in the plane generated by the other two residual vectors. Consequently, the *three*-dimensional volume is zero. This case is illustrated in Figure 3.9 and may be verified algebraically by showing that $|\mathbf{S}| = 0$. We have

$$\mathop{\mathbf{S}}_{(3\times 3)} = \begin{bmatrix} 3 & -\frac{3}{2} & 0 \\ -\frac{3}{2} & 1 & \frac{1}{2} \\ 0 & \frac{1}{2} & 1 \end{bmatrix}$$

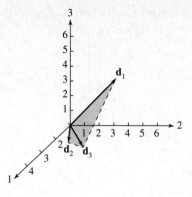

Figure 3.9 A case where the three-dimensional volume is zero ($|\mathbf{S}| = 0$).

and from Definition 2A.24,

$$|\mathbf{S}| = 3 \begin{vmatrix} 1 & \frac{1}{2} \\ \frac{1}{2} & 1 \end{vmatrix} (-1)^2 + \left(-\frac{3}{2}\right) \begin{vmatrix} -\frac{3}{2} & \frac{1}{2} \\ 0 & 1 \end{vmatrix} (-1)^3 + (0) \begin{vmatrix} -\frac{3}{2} & 1 \\ 0 & \frac{1}{2} \end{vmatrix} (-1)^4$$

$$= 3\left(1 - \frac{1}{4}\right) + \left(\frac{3}{2}\right)\left(-\frac{3}{2} - 0\right) + 0 = \frac{9}{4} - \frac{9}{4} = 0 \qquad \blacksquare$$

When large data sets are sent and received electronically, investigators are sometimes unpleasantly surprised to find a case of zero generalized variance, so that **S** does not have an inverse. We have encountered several such cases, with their associated difficulties, before the situation was unmasked. A singular covariance matrix occurs when, for instance, the data are test scores and the investigator has included variables that are sums of the others. For example, an algebra score and a geometry score could be combined to give a total math score, or class midterm and final exam scores summed to give total points. Once, the total weight of a number of chemicals was included along with that of each component.

This common practice of creating new variables that are sums of the original variables and then including them in the data set has caused enough lost time that we emphasize the necessity of being alert to avoid these consequences.

Example 3.10 (Creating new variables that lead to a zero generalized variance) Consider the data matrix

$$\mathbf{X} = \begin{bmatrix} 1 & 9 & 10 \\ 4 & 12 & 16 \\ 2 & 10 & 12 \\ 5 & 8 & 13 \\ 3 & 11 & 14 \end{bmatrix}$$

where the third column is the sum of first two columns. These data could be the number of successful phone solicitations per day by a part-time and a full-time employee, respectively, so the third column is the total number of successful solicitations per day.

Show that the generalized variance $|\mathbf{S}| = 0$, and determine the nature of the dependency in the data.

We find that the mean corrected data matrix, with entries $x_{jk} - \bar{x}_k$, is

$$\mathbf{X} - \mathbf{1}\bar{\mathbf{x}}' = \begin{bmatrix} -2 & -1 & -3 \\ 1 & 2 & 3 \\ -1 & 0 & -1 \\ 2 & -2 & 0 \\ 0 & 1 & 1 \end{bmatrix}$$

The resulting covariance matrix is

$$\mathbf{S} = \begin{bmatrix} 2.5 & 0 & 2.5 \\ 0 & 2.5 & 2.5 \\ 2.5 & 2.5 & 5.0 \end{bmatrix}$$

We verify that, in this case, the generalized variance

$$|\mathbf{S}| = 2.5^2 \times 5 + 0 + 0 - 2.5^3 - 2.5^3 - 0 = 0$$

In general, if the three columns of the data matrix $\mathbf{X}$ satisfy a linear constraint $a_1 x_{j1} + a_2 x_{j2} + a_3 x_{j3} = c$, a constant for all j, then $a_1\bar{x}_1 + a_2\bar{x}_2 + a_3\bar{x}_3 = c$, so that

$$a_1(x_{j1} - \bar{x}_1) + a_2(x_{j2} - \bar{x}_2) + a_3(x_{j3} - \bar{x}_3) = 0$$

for all j. That is,

$$(\mathbf{X} - \mathbf{1}\bar{\mathbf{x}}')\mathbf{a} = \mathbf{0}$$

and the columns of the mean corrected data matrix are linearly dependent. Thus, the inclusion of the third variable, which is linearly related to the first two, has led to the case of a zero generalized variance.

Whenever the columns of the mean corrected data matrix are linearly dependent,

$$(n - 1)\mathbf{S}\mathbf{a} = (\mathbf{X} - \mathbf{1}\bar{\mathbf{x}}')'(\mathbf{X} - \mathbf{1}\bar{\mathbf{x}}')\mathbf{a} = (\mathbf{X} - \mathbf{1}\bar{\mathbf{x}}')\mathbf{0} = \mathbf{0}$$

and $\mathbf{S}\mathbf{a} = \mathbf{0}$ establishes the linear dependency of the columns of $\mathbf{S}$. Hence, $|\mathbf{S}| = 0$.

Since $\mathbf{S}\mathbf{a} = \mathbf{0} = 0\mathbf{a}$, we see that $\mathbf{a}$ is a scaled eigenvector of $\mathbf{S}$ associated with an eigenvalue of zero. This gives rise to an important diagnostic: If we are unaware of any extra variables that are linear combinations of the others, we can find them by calculating the eigenvectors of $\mathbf{S}$ and identifying the one associated with a zero eigenvalue. That is, if we were unaware of the dependency in this example, a computer calculation would find an eigenvalue proportional to $\mathbf{a}' = [1, 1, -1]$, since

$$\mathbf{S}\mathbf{a} = \begin{bmatrix} 2.5 & 0 & 2.5 \\ 0 & 2.5 & 2.5 \\ 2.5 & 2.5 & 5.0 \end{bmatrix}\begin{bmatrix} 1 \\ 1 \\ -1 \end{bmatrix} = \begin{bmatrix} 0 \\ 0 \\ 0 \end{bmatrix} = 0\begin{bmatrix} 1 \\ 1 \\ -1 \end{bmatrix}$$

The coefficients reveal that

$$1(x_{j1} - \bar{x}_1) + 1(x_{j2} - \bar{x}_2) + (-1)(x_{j3} - \bar{x}_3) = 0 \quad \text{for all } j$$

In addition, the sum of the first two variables minus the third is a constant c for all n units. Here the third variable is actually the sum of the first two variables, so the columns of the original data matrix satisfy a linear constraint with $c = 0$. Because we have the special case $c = 0$, the constraint establishes the fact that the columns of the data matrix are linearly dependent. ∎

Let us summarize the important equivalent conditions for a generalized variance to be zero that we discussed in the preceding example. Whenever a nonzero vector $\mathbf{a}$ satisfies one of the following three conditions, it satisfies all of them:

(1) $\mathbf{Sa} = \mathbf{0}$	(2) $\mathbf{a}'(\mathbf{x}_j - \bar{\mathbf{x}}) = 0$ for all j	(3) $\mathbf{a}'\mathbf{x}_j = c$ for all j $(c = \mathbf{a}'\bar{\mathbf{x}})$
$\mathbf{a}$ is a scaled eigenvector of $\mathbf{S}$ with eigenvalue 0.	The linear combination of the mean corrected data, using $\mathbf{a}$, is zero.	The linear combination of the original data, using $\mathbf{a}$, is a constant.

We showed that if condition (3) is satisfied—that is, if the values for one variable can be expressed in terms of the others—then the generalized variance is zero because $\mathbf{S}$ has a zero eigenvalue. In the other direction, if condition (1) holds, then the eigenvector $\mathbf{a}$ gives coefficients for the linear dependency of the mean corrected data.

In any statistical analysis, $|\mathbf{S}| = 0$ means that the measurements on some variables should be removed from the study as far as the mathematical computations are concerned. The corresponding reduced data matrix will then lead to a covariance matrix of full rank and a nonzero generalized variance. The question of which measurements to remove in degenerate cases is not easy to answer. When there is a choice, one should retain measurements on a (presumed) causal variable instead of those on a secondary characteristic. We shall return to this subject in our discussion of principal components.

At this point, we settle for delineating some simple conditions for $\mathbf{S}$ to be of full rank or of reduced rank.

Result 3.3. If $n \le p$, that is, (sample size) $\le$ (number of variables), then $|\mathbf{S}| = 0$ for all samples.

Proof. We must show that the rank of $\mathbf{S}$ is less than or equal to p and then apply Result 2A.9.

For any fixed sample, the n row vectors in (3-18) sum to the zero vector. The existence of this linear combination means that the rank of $\mathbf{X} - \mathbf{1}\bar{\mathbf{x}}'$ is less than or equal to $n - 1$, which, in turn, is less than or equal to $p - 1$ because $n \le p$. Since

$$(n-1)\underset{(p \times p)}{\mathbf{S}} = (\mathbf{X} - \mathbf{1}\bar{\mathbf{x}})'_{(p \times n)}(\mathbf{X} - \mathbf{1}\bar{\mathbf{x}}')_{(n \times p)}$$

the kth column of $\mathbf{S}$, $\text{col}_k(\mathbf{S})$, can be written as a linear combination of the columns of $(\mathbf{X} - \mathbf{1}\bar{\mathbf{x}}')'$. In particular,

$$(n-1)\,\text{col}_k(\mathbf{S}) = (\mathbf{X} - \mathbf{1}\bar{\mathbf{x}}')'\,\text{col}_k(\mathbf{X} - \mathbf{1}\bar{\mathbf{x}}')$$
$$= (x_{1k} - \bar{x}_k)\,\text{col}_1(\mathbf{X} - \mathbf{1}\bar{\mathbf{x}}')' + \cdots + (x_{nk} - \bar{x}_k)\,\text{col}_n(\mathbf{X} - \mathbf{1}\bar{\mathbf{x}}')'$$

Since the column vectors of $(\mathbf{X} - \mathbf{1}\bar{\mathbf{x}}')'$ sum to the zero vector, we can write, for example, $\text{col}_1(\mathbf{X} - \mathbf{1}\bar{\mathbf{x}}')'$ as the negative of the sum of the remaining column vectors. After substituting for $\text{row}_1(\mathbf{X} - \mathbf{1}\bar{\mathbf{x}}')'$ in the preceding equation, we can express $\text{col}_k(\mathbf{S})$ as a linear combination of the at most $n - 1$ linearly independent row vectors $\text{col}_2(\mathbf{X} - \mathbf{1}\bar{\mathbf{x}}')', \ldots, \text{col}_n(\mathbf{X} - \mathbf{1}\bar{\mathbf{x}}')'$. The rank of $\mathbf{S}$ is therefore less than or equal to $n - 1$, which—as noted at the beginning of the proof—is less than or equal to $p - 1$, and $\mathbf{S}$ is singular. This implies, from Result 2A.9, that $|\mathbf{S}| = 0$. ∎

Result 3.4. Let the $p \times 1$ vectors $\mathbf{x}_1, \mathbf{x}_2, \ldots, \mathbf{x}_n$, where $\mathbf{x}'_j$ is the jth row of the data matrix $\mathbf{X}$, be realizations of the independent random vectors $\mathbf{X}_1, \mathbf{X}_2, \ldots, \mathbf{X}_n$. Then

1. If the linear combination $\mathbf{a}'\mathbf{X}_j$ has positive variance for each constant vector $\mathbf{a} \neq \mathbf{0}$, then, provided that $p < n$, $\mathbf{S}$ has full rank with probability 1 and $|\mathbf{S}| > 0$.

2. If, with probability 1, $\mathbf{a}'\mathbf{X}_j$ is a constant (for example, c) *for all* j, then $|\mathbf{S}| = 0$.

Proof. (Part 2). If $\mathbf{a}'\mathbf{X}_j = a_1 X_{j1} + a_2 X_{j2} + \cdots + a_p X_{jp} = c$ with probability 1, $\mathbf{a}'\mathbf{x}_j = c$ for all j, and the sample mean of this linear combination is $c = \sum_{j=1}^{n} (a_1 x_{j1} + a_2 x_{j2} + \cdots + a_p x_{jp})/n = a_1 \bar{x}_1 + a_2 \bar{x}_2 + \cdots + a_p \bar{x}_p = \mathbf{a}'\bar{\mathbf{x}}$. Then

$$(\mathbf{X} - \mathbf{1}\bar{\mathbf{x}}')\mathbf{a} = a_1 \begin{bmatrix} x_{11} - \bar{x}_1 \\ \vdots \\ x_{n1} - \bar{x}_1 \end{bmatrix} + \cdots + a_p \begin{bmatrix} x_{1p} - \bar{x}_p \\ \vdots \\ x_{np} - \bar{x}_p \end{bmatrix}$$

$$= \begin{bmatrix} \mathbf{a}'\mathbf{x}_1 - \mathbf{a}'\bar{\mathbf{x}} \\ \vdots \\ \mathbf{a}'\mathbf{x}_n - \mathbf{a}'\bar{\mathbf{x}} \end{bmatrix} = \begin{bmatrix} c - c \\ \vdots \\ c - c \end{bmatrix} = \mathbf{0}$$

indicating linear dependence; the conclusion follows from Result 3.2.

The proof of Part (1) is difficult and can be found in [2]. ∎

Generalized Variance Determined by $|\mathbf{R}|$ and Its Geometrical Interpretation

The generalized sample variance is unduly affected by the variability of measurements on a single variable. For example, suppose some s_{ii} is either large or quite small. Then, geometrically, the corresponding deviation vector $\mathbf{d}_i = (\mathbf{y}_i - \bar{x}_i\mathbf{1})$ will be very long or very short and will therefore clearly be an important factor in determining volume. Consequently, it is sometimes useful to scale all the deviation vectors so that they have the same length.

Scaling the residual vectors is equivalent to replacing each original observation x_{jk} by its standardized value $(x_{jk} - \bar{x}_k)/\sqrt{s_{kk}}$. The sample covariance matrix of the standardized variables is then $\mathbf{R}$, the sample correlation matrix of the original variables. (See Exercise 3.13.) We define

$$\begin{pmatrix} \text{Generalized sample variance} \\ \text{of the standardized variables} \end{pmatrix} = |\mathbf{R}| \qquad (3\text{-}19)$$

Since the resulting vectors

$$[(x_{1k} - \bar{x}_k)/\sqrt{s_{kk}}, (x_{2k} - \bar{x}_k)/\sqrt{s_{kk}}, \ldots, (x_{nk} - \bar{x}_k)/\sqrt{s_{kk}}] = (\mathbf{y}_k - \bar{x}_k\mathbf{1})'/\sqrt{s_{kk}}$$

all have length $\sqrt{n-1}$, the generalized sample variance of the standardized variables will be large when these vectors are nearly perpendicular and will be small when two or more of these vectors are in almost the same direction. Employing the argument leading to (3-7), we readily find that the cosine of the angle θ_{ik} between $(\mathbf{y}_i - \bar{x}_i\mathbf{1})/\sqrt{s_{ii}}$ and $(\mathbf{y}_k - \bar{x}_k\mathbf{1})/\sqrt{s_{kk}}$ is the sample correlation coefficient r_{ik}. Therefore, we can make the statement that $|\mathbf{R}|$ is large when all the r_{ik} are nearly zero and it is small when one or more of the r_{ik} are nearly $+1$ or -1.

In sum, we have the following result: Let

$$\frac{(\mathbf{y}_i - \bar{x}_i\mathbf{1})}{\sqrt{s_{ii}}} = \begin{bmatrix} \dfrac{x_{1i} - \bar{x}_i}{\sqrt{s_{ii}}} \\[2mm] \dfrac{x_{2i} - \bar{x}_i}{\sqrt{s_{ii}}} \\[2mm] \vdots \\[2mm] \dfrac{x_{ni} - \bar{x}_i}{\sqrt{s_{ii}}} \end{bmatrix}, \qquad i = 1, 2, \ldots, p$$

be the deviation vectors of the standardized variables. The ith deviation vectors lie in the direction of $\mathbf{d}_i$, but all have a squared length of $n - 1$. The volume generated in p-space by the deviation vectors can be related to the generalized sample variance. The same steps that lead to (3-15) produce

$$\left(\begin{array}{l} \text{Generalized sample variance} \\ \text{of the standardized variables} \end{array}\right) = |\mathbf{R}| = (n - 1)^{-p}(\text{volume})^2 \qquad (3\text{-}20)$$

The volume generated by deviation vectors of the standardized variables is illustrated in Figure 3.10 for the two sets of deviation vectors graphed in Figure 3.6. A comparison of Figures 3.10 and 3.6 reveals that the influence of the $\mathbf{d}_2$ vector (large variability in x_2) on the squared volume $|\mathbf{S}|$ is much greater than its influence on the squared volume $|\mathbf{R}|$.

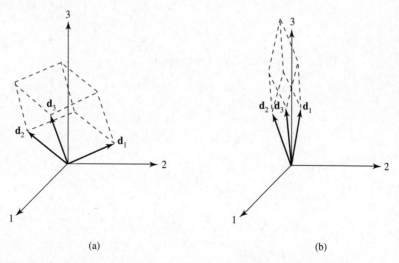

Figure 3.10 The volume generated by equal-length deviation vectors of the standardized variables.

The quantities $|\mathbf{S}|$ and $|\mathbf{R}|$ are connected by the relationship

$$|\mathbf{S}| = (s_{11}s_{22}\cdots s_{pp})|\mathbf{R}| \qquad (3\text{-}21)$$

so

$$(n-1)^p|\mathbf{S}| = (n-1)^p(s_{11}s_{22}\cdots s_{pp})|\mathbf{R}| \qquad (3\text{-}22)$$

[The proof of (3-21) is left to the reader as Exercise 3.12.]

Interpreting (3-22) in terms of volumes, we see from (3-15) and (3-20) that the squared volume $(n-1)^p|\mathbf{S}|$ is proportional to the squared volume $(n-1)^p|\mathbf{R}|$. The constant of proportionality is the product of the variances, which, in turn, is proportional to the product of the squares of the lengths $(n-1)s_{ii}$ of the $\mathbf{d}_i$. Equation (3-21) shows, algebraically, how a change in the measurement scale of X_1, for example, will alter the relationship between the generalized variances. Since $|\mathbf{R}|$ is based on standardized measurements, it is unaffected by the change in scale. However, the relative value of $|\mathbf{S}|$ will be changed whenever the multiplicative factor s_{11} changes.

Example 3.11 (Illustrating the relation between $|\mathbf{S}|$ and $|\mathbf{R}|$) Let us illustrate the relationship in (3-21) for the generalized variances $|\mathbf{S}|$ and $|\mathbf{R}|$ when $p = 3$. Suppose

$$\underset{(3\times3)}{\mathbf{S}} = \begin{bmatrix} 4 & 3 & 1 \\ 3 & 9 & 2 \\ 1 & 2 & 1 \end{bmatrix}$$

Then $s_{11} = 4$, $s_{22} = 9$, and $s_{33} = 1$. Moreover,

$$\mathbf{R} = \begin{bmatrix} 1 & \frac{1}{2} & \frac{1}{2} \\ \frac{1}{2} & 1 & \frac{2}{3} \\ \frac{1}{2} & \frac{2}{3} & 1 \end{bmatrix}$$

Using Definition 2A.24, we obtain

$$|\mathbf{S}| = 4\begin{vmatrix} 9 & 2 \\ 2 & 1 \end{vmatrix}(-1)^2 + 3\begin{vmatrix} 3 & 2 \\ 1 & 1 \end{vmatrix}(-1)^3 + 1\begin{vmatrix} 3 & 9 \\ 1 & 2 \end{vmatrix}(-1)^4$$

$$= 4(9-4) - 3(3-2) + 1(6-9) = 14$$

$$|\mathbf{R}| = 1\begin{vmatrix} 1 & \frac{2}{3} \\ \frac{2}{3} & 1 \end{vmatrix}(-1)^2 + \frac{1}{2}\begin{vmatrix} \frac{1}{2} & \frac{2}{3} \\ \frac{1}{2} & 1 \end{vmatrix}(-1)^3 + \frac{1}{2}\begin{vmatrix} \frac{1}{2} & 1 \\ \frac{1}{2} & \frac{2}{3} \end{vmatrix}(-1)^4$$

$$= \left(1 - \frac{4}{9}\right) - \left(\frac{1}{2}\right)\left(\frac{1}{2} - \frac{1}{3}\right) + \left(\frac{1}{2}\right)\left(\frac{1}{3} - \frac{1}{2}\right) = \frac{7}{18}$$

It then follows that

$$14 = |\mathbf{S}| = s_{11}s_{22}s_{33}|\mathbf{R}| = (4)(9)(1)\left(\tfrac{7}{18}\right) = 14 \qquad \text{(check)} \quad \blacksquare$$

Another Generalization of Variance

We conclude this discussion by mentioning another generalization of variance. Specifically, we define the *total sample variance* as the sum of the diagonal elements of the sample variance–covariance matrix **S**. Thus,

$$\text{Total sample variance} = s_{11} + s_{22} + \cdots + s_{pp} \qquad (3\text{-}23)$$

Example 3.12 (Calculating the total sample variance) Calculate the total sample variance for the variance–covariance matrices **S** in Examples 3.7 and 3.9.
   From Example 3.7.

$$\mathbf{S} = \begin{bmatrix} 252.04 & -68.43 \\ -68.43 & 123.67 \end{bmatrix}$$

and

$$\text{Total sample variance} = s_{11} + s_{22} = 252.04 + 123.67 = 375.71$$

From Example 3.9,

$$\mathbf{S} = \begin{bmatrix} 3 & -\frac{3}{2} & 0 \\ -\frac{3}{2} & 1 & \frac{1}{2} \\ 0 & \frac{1}{2} & 1 \end{bmatrix}$$

and

$$\text{Total sample variance} = s_{11} + s_{22} + s_{33} = 3 + 1 + 1 = 5 \qquad \blacksquare$$

Geometrically, the total sample variance is the sum of the squared lengths of the p deviation vectors $\mathbf{d}_1 = (\mathbf{y}_1 - \bar{x}_1 \mathbf{1}), \ldots, \mathbf{d}_p = (\mathbf{y}_p - \bar{x}_p \mathbf{1})$, divided by $n - 1$. The total sample variance criterion pays no attention to the orientation (correlation structure) of the residual vectors. For instance, it assigns the same values to both sets of residual vectors (a) and (b) in Figure 3.6.

3.5 Sample Mean, Covariance, and Correlation as Matrix Operations

We have developed geometrical representations of the data matrix **X** and the derived descriptive statistics $\bar{\mathbf{x}}$ and **S**. In addition, it is possible to link algebraically the calculation of $\bar{\mathbf{x}}$ and **S** directly to **X** using matrix operations. The resulting expressions, which depict the relation between $\bar{\mathbf{x}}$, **S**, and the full data set **X** concisely, are easily programmed on electronic computers.

We have it that $\bar{x}_i = (x_{1i} \cdot 1 + x_{2i} \cdot 1 + \cdots + x_{ni} \cdot 1)/n = \mathbf{y}_i' \mathbf{1}/n$. Therefore,

$$\bar{\mathbf{x}} = \begin{bmatrix} \bar{x}_1 \\ \bar{x}_2 \\ \vdots \\ \bar{x}_p \end{bmatrix} = \begin{bmatrix} \dfrac{\mathbf{y}_1'\mathbf{1}}{n} \\ \dfrac{\mathbf{y}_2'\mathbf{1}}{n} \\ \vdots \\ \dfrac{\mathbf{y}_p'\mathbf{1}}{n} \end{bmatrix} = \frac{1}{n} \begin{bmatrix} x_{11} & x_{12} & \cdots & x_{1n} \\ x_{21} & x_{22} & \cdots & x_{2n} \\ \vdots & \vdots & \ddots & \vdots \\ x_{p1} & x_{p2} & \cdots & x_{pn} \end{bmatrix} \begin{bmatrix} 1 \\ 1 \\ \vdots \\ 1 \end{bmatrix}$$

or

$$\bar{\mathbf{x}} = \frac{1}{n} \mathbf{X}' \mathbf{1} \tag{3-24}$$

That is, $\bar{\mathbf{x}}$ is calculated from the transposed data matrix by postmultiplying by the vector $\mathbf{1}$ and then multiplying the result by the constant $1/n$.

Next, we create an $n \times p$ *matrix of means* by transposing both sides of (3-24) and premultiplying by $\mathbf{1}$; that is,

$$\mathbf{1}\bar{\mathbf{x}}' = \frac{1}{n} \mathbf{1}\mathbf{1}'\mathbf{X} = \begin{bmatrix} \bar{x}_1 & \bar{x}_2 & \cdots & \bar{x}_p \\ \bar{x}_1 & \bar{x}_2 & \cdots & \bar{x}_p \\ \vdots & \vdots & \ddots & \vdots \\ \bar{x}_1 & \bar{x}_2 & \cdots & \bar{x}_p \end{bmatrix} \tag{3-25}$$

Subtracting this result from $\mathbf{X}$ produces the $n \times p$ *matrix of deviations* (residuals)

$$\mathbf{X} - \frac{1}{n}\mathbf{1}\mathbf{1}'\mathbf{X} = \begin{bmatrix} x_{11} - \bar{x}_1 & x_{12} - \bar{x}_2 & \cdots & x_{1p} - \bar{x}_p \\ x_{21} - \bar{x}_1 & x_{22} - \bar{x}_2 & \cdots & x_{2p} - \bar{x}_p \\ \vdots & \vdots & \ddots & \vdots \\ x_{n1} - \bar{x}_1 & x_{n2} - \bar{x}_2 & \cdots & x_{np} - \bar{x}_p \end{bmatrix} \tag{3-26}$$

Now, the matrix $(n-1)\mathbf{S}$ representing sums of squares and cross products is just the transpose of the matrix (3-26) times the matrix itself, or

$$(n-1)\mathbf{S} = \begin{bmatrix} x_{11} - \bar{x}_1 & x_{21} - \bar{x}_1 & \cdots & x_{n1} - \bar{x}_1 \\ x_{12} - \bar{x}_2 & x_{22} - \bar{x}_2 & \cdots & x_{n2} - \bar{x}_2 \\ \vdots & \vdots & \ddots & \vdots \\ x_{1p} - \bar{x}_p & x_{2p} - \bar{x}_p & \cdots & x_{np} - \bar{x}_p \end{bmatrix}$$

$$\times \begin{bmatrix} x_{11} - \bar{x}_1 & x_{12} - \bar{x}_2 & \cdots & x_{1p} - \bar{x}_p \\ x_{21} - \bar{x}_1 & x_{22} - \bar{x}_2 & \cdots & x_{2p} - \bar{x}_p \\ \vdots & \vdots & \ddots & \vdots \\ x_{n1} - \bar{x}_1 & x_{n2} - \bar{x}_2 & \cdots & x_{np} - \bar{x}_p \end{bmatrix}$$

$$= \left(\mathbf{X} - \frac{1}{n}\mathbf{1}\mathbf{1}'\mathbf{X} \right)' \left(\mathbf{X} - \frac{1}{n}\mathbf{1}\mathbf{1}'\mathbf{X} \right) = \mathbf{X}' \left(\mathbf{I} - \frac{1}{n}\mathbf{1}\mathbf{1}' \right) \mathbf{X}$$

since

$$\left(\mathbf{I} - \frac{1}{n}\mathbf{1}\mathbf{1}'\right)'\left(\mathbf{I} - \frac{1}{n}\mathbf{1}\mathbf{1}'\right) = \mathbf{I} - \frac{1}{n}\mathbf{1}\mathbf{1}' - \frac{1}{n}\mathbf{1}\mathbf{1}' + \frac{1}{n^2}\mathbf{1}\mathbf{1}'\mathbf{1}\mathbf{1}' = \mathbf{I} - \frac{1}{n}\mathbf{1}\mathbf{1}'$$

To summarize, the matrix expressions relating $\bar{\mathbf{x}}$ and $\mathbf{S}$ to the data set $\mathbf{X}$ are

$$\bar{\mathbf{x}} = \frac{1}{n}\mathbf{X}'\mathbf{1}$$

$$\mathbf{S} = \frac{1}{n-1}\mathbf{X}'\left(\mathbf{I} - \frac{1}{n}\mathbf{1}\mathbf{1}'\right)\mathbf{X} \tag{3-27}$$

The result for $\mathbf{S}_n$ is similar, except that $1/n$ replaces $1/(n-1)$ as the first factor.

The relations in (3-27) show clearly how matrix operations on the data matrix $\mathbf{X}$ lead to $\bar{\mathbf{x}}$ and $\mathbf{S}$.

Once $\mathbf{S}$ is computed, it can be related to the sample correlation matrix $\mathbf{R}$. The resulting expression can also be "inverted" to relate $\mathbf{R}$ to $\mathbf{S}$. We first define the $p \times p$ *sample standard deviation matrix* $\mathbf{D}^{1/2}$ and compute its inverse, $(\mathbf{D}^{1/2})^{-1} = \mathbf{D}^{-1/2}$. Let

$$\underset{(p\times p)}{\mathbf{D}^{1/2}} = \begin{bmatrix} \sqrt{s_{11}} & 0 & \cdots & 0 \\ 0 & \sqrt{s_{22}} & \cdots & 0 \\ \vdots & \vdots & \ddots & \vdots \\ 0 & 0 & \cdots & \sqrt{s_{pp}} \end{bmatrix} \tag{3-28}$$

Then

$$\underset{(p\times p)}{\mathbf{D}^{-1/2}} = \begin{bmatrix} \dfrac{1}{\sqrt{s_{11}}} & 0 & \cdots & 0 \\ 0 & \dfrac{1}{\sqrt{s_{22}}} & \cdots & 0 \\ \vdots & \vdots & \ddots & \vdots \\ 0 & 0 & \cdots & \dfrac{1}{\sqrt{s_{pp}}} \end{bmatrix}$$

Since

$$\mathbf{S} = \begin{bmatrix} s_{11} & s_{12} & \cdots & s_{1p} \\ \vdots & \vdots & \ddots & \vdots \\ s_{1p} & s_{2p} & \cdots & s_{pp} \end{bmatrix}$$

and

$$\mathbf{R} = \begin{bmatrix} \dfrac{s_{11}}{\sqrt{s_{11}}\sqrt{s_{11}}} & \dfrac{s_{12}}{\sqrt{s_{11}}\sqrt{s_{22}}} & \cdots & \dfrac{s_{1p}}{\sqrt{s_{11}}\sqrt{s_{pp}}} \\ \vdots & \vdots & \ddots & \vdots \\ \dfrac{s_{1p}}{\sqrt{s_{11}}\sqrt{s_{pp}}} & \dfrac{s_{2p}}{\sqrt{s_{22}}\sqrt{s_{pp}}} & \cdots & \dfrac{s_{pp}}{\sqrt{s_{pp}}\sqrt{s_{pp}}} \end{bmatrix} = \begin{bmatrix} 1 & r_{12} & \cdots & r_{1p} \\ \vdots & \vdots & \ddots & \vdots \\ r_{1p} & r_{2p} & \cdots & 1 \end{bmatrix}$$

we have

$$\mathbf{R} = \mathbf{D}^{-1/2}\mathbf{S}\mathbf{D}^{-1/2} \tag{3-29}$$

Postmultiplying and premultiplying both sides of (3-29) by $\mathbf{D}^{1/2}$ and noting that $\mathbf{D}^{-1/2}\mathbf{D}^{1/2} = \mathbf{D}^{1/2}\mathbf{D}^{-1/2} = \mathbf{I}$ gives

$$\mathbf{S} = \mathbf{D}^{1/2}\,\mathbf{R}\mathbf{D}^{1/2} \qquad (3\text{-}30)$$

That is, $\mathbf{R}$ can be obtained from the information in $\mathbf{S}$, whereas $\mathbf{S}$ can be obtained from $\mathbf{D}^{1/2}$ and $\mathbf{R}$. Equations (3-29) and (3-30) are sample analogs of (2-36) and (2-37).

3.6 Sample Values of Linear Combinations of Variables

We have introduced linear combinations of p variables in Section 2.6. In many multi-variate procedures, we are led naturally to consider a linear combination of the form

$$\mathbf{c}'\mathbf{X} = c_1 X_1 + c_2 X_2 + \cdots + c_p X_p$$

whose observed value on the jth trial is

$$\mathbf{c}'\mathbf{x}_j = c_1 x_{j1} + c_2 x_{j2} + \cdots + c_p x_{jp}, \qquad j = 1, 2, \ldots, n \qquad (3\text{-}31)$$

The n derived observations in (3-31) have

$$\text{Sample mean} = \frac{(\mathbf{c}'\mathbf{x}_1 + \mathbf{c}'\mathbf{x}_2 + \cdots + \mathbf{c}'\mathbf{x}_n)}{n}$$

$$= \mathbf{c}'(\mathbf{x}_1 + \mathbf{x}_2 + \cdots + \mathbf{x}_n)\frac{1}{n} = \mathbf{c}'\bar{\mathbf{x}} \qquad (3\text{-}32)$$

Since $(\mathbf{c}'\mathbf{x}_j - \mathbf{c}'\bar{\mathbf{x}})^2 = (\mathbf{c}'(\mathbf{x}_j - \bar{\mathbf{x}}))^2 = \mathbf{c}'(\mathbf{x}_j - \bar{\mathbf{x}})(\mathbf{x}_j - \bar{\mathbf{x}})'\mathbf{c}$, we have

$$\text{Sample variance} = \frac{(\mathbf{c}'\mathbf{x}_1 - \mathbf{c}'\bar{\mathbf{x}})^2 + (\mathbf{c}'\mathbf{x}_2 - \mathbf{c}'\bar{\mathbf{x}})^2 + \cdots + (\mathbf{c}'\mathbf{x}_n - \mathbf{c}'\bar{\mathbf{x}})^2}{n - 1}$$

$$= \frac{\mathbf{c}'(\mathbf{x}_1 - \bar{\mathbf{x}})(\mathbf{x}_1 - \bar{\mathbf{x}})'\mathbf{c} + \mathbf{c}'(\mathbf{x}_2 - \bar{\mathbf{x}})(\mathbf{x}_2 - \bar{\mathbf{x}})'\mathbf{c} + \cdots + \mathbf{c}'(\mathbf{x}_n - \bar{\mathbf{x}})(\mathbf{x}_n - \bar{\mathbf{x}})'\mathbf{c}}{n - 1}$$

$$= \mathbf{c}'\left[\frac{(\mathbf{x}_1 - \bar{\mathbf{x}})(\mathbf{x}_1 - \bar{\mathbf{x}})' + (\mathbf{x}_2 - \bar{\mathbf{x}})(\mathbf{x}_2 - \bar{\mathbf{x}})' + \cdots + (\mathbf{x}_n - \bar{\mathbf{x}})(\mathbf{x}_n - \bar{\mathbf{x}})'}{n - 1}\right]\mathbf{c}$$

or

$$\text{Sample variance of } \mathbf{c}'\mathbf{X} = \mathbf{c}'\mathbf{S}\mathbf{c} \qquad (3\text{-}33)$$

Equations (3-32) and (3-33) are sample analogs of (2-43). They correspond to substituting the sample quantities $\bar{\mathbf{x}}$ and $\mathbf{S}$ for the "population" quantities $\boldsymbol{\mu}$ and $\boldsymbol{\Sigma}$, respectively, in (2-43).

Now consider a second linear combination

$$\mathbf{b}'\mathbf{X} = b_1 X_1 + b_2 X_2 + \cdots + b_p X_p$$

whose observed value on the jth trial is

$$\mathbf{b}'\mathbf{x}_j = b_1 x_{j1} + b_2 x_{j2} + \cdots + b_p x_{jp}, \qquad j = 1, 2, \ldots, n \qquad (3\text{-}34)$$

It follows from (3-32) and (3-33) that the sample mean and variance of these derived observations are

$$\text{Sample mean of } \mathbf{b}'\mathbf{X} = \mathbf{b}'\bar{\mathbf{x}}$$

$$\text{Sample variance of } \mathbf{b}'\mathbf{X} = \mathbf{b}'\mathbf{S}\mathbf{b}$$

Moreover, the sample covariance computed from pairs of observations on $\mathbf{b}'\mathbf{X}$ and $\mathbf{c}'\mathbf{X}$ is

Sample covariance

$$= \frac{(\mathbf{b}'\mathbf{x}_1 - \mathbf{b}'\bar{\mathbf{x}})(\mathbf{c}'\mathbf{x}_1 - \mathbf{c}'\bar{\mathbf{x}}) + (\mathbf{b}'\mathbf{x}_2 - \mathbf{b}'\bar{\mathbf{x}})(\mathbf{c}'\mathbf{x}_2 - \mathbf{c}'\bar{\mathbf{x}}) + \cdots + (\mathbf{b}'\mathbf{x}_n - \mathbf{b}'\bar{\mathbf{x}})(\mathbf{c}'\mathbf{x}_n - \mathbf{c}'\bar{\mathbf{x}})}{n-1}$$

$$= \frac{\mathbf{b}'(\mathbf{x}_1 - \bar{\mathbf{x}})(\mathbf{x}_1 - \bar{\mathbf{x}})'\mathbf{c} + \mathbf{b}'(\mathbf{x}_2 - \bar{\mathbf{x}})(\mathbf{x}_2 - \bar{\mathbf{x}})'\mathbf{c} + \cdots + \mathbf{b}'(\mathbf{x}_n - \bar{\mathbf{x}})(\mathbf{x}_n - \bar{\mathbf{x}})'\mathbf{c}}{n-1}$$

$$= \mathbf{b}'\left[\frac{(\mathbf{x}_1 - \bar{\mathbf{x}})(\mathbf{x}_1 - \bar{\mathbf{x}})' + (\mathbf{x}_2 - \bar{\mathbf{x}})(\mathbf{x}_2 - \bar{\mathbf{x}})' + \cdots + (\mathbf{x}_n - \bar{\mathbf{x}})(\mathbf{x}_n - \bar{\mathbf{x}})'}{n-1}\right]\mathbf{c}$$

or

$$\text{Sample covariance of } \mathbf{b}'\mathbf{X} \text{ and } \mathbf{c}'\mathbf{X} = \mathbf{b}'\mathbf{S}\mathbf{c} \qquad (3\text{-}35)$$

In sum, we have the following result.

Result 3.5. The linear combinations

$$\mathbf{b}'\mathbf{X} = b_1 X_1 + b_2 X_2 + \cdots + b_p X_p$$

$$\mathbf{c}'\mathbf{X} = c_1 X_1 + c_2 X_2 + \cdots + c_p X_p$$

have sample means, variances, and covariances that are related to $\bar{\mathbf{x}}$ and $\mathbf{S}$ by

$$\text{Sample mean of } \mathbf{b}'\mathbf{X} = \mathbf{b}'\bar{\mathbf{x}}$$

$$\text{Sample mean of } \mathbf{c}'\mathbf{X} = \mathbf{c}'\bar{\mathbf{x}}$$

$$\text{Sample variance of } \mathbf{b}'\mathbf{X} = \mathbf{b}'\mathbf{S}\mathbf{b} \qquad (3\text{-}36)$$

$$\text{Sample variance of } \mathbf{c}'\mathbf{X} = \mathbf{c}'\mathbf{S}\mathbf{c}$$

$$\text{Sample covariance of } \mathbf{b}'\mathbf{X} \text{ and } \mathbf{c}'\mathbf{X} = \mathbf{b}'\mathbf{S}\mathbf{c}$$

∎

Example 3.13 (Means and covariances for linear combinations) We shall consider two linear combinations and their derived values for the $n = 3$ observations given in Example 3.9 as

$$\mathbf{X} = \begin{bmatrix} x_{11} & x_{12} & x_{13} \\ x_{21} & x_{22} & x_{23} \\ x_{31} & x_{32} & x_{33} \end{bmatrix} = \begin{bmatrix} 1 & 2 & 5 \\ 4 & 1 & 6 \\ 4 & 0 & 4 \end{bmatrix}$$

Consider the two linear combinations

$$\mathbf{b}'\mathbf{X} = \begin{bmatrix} 2 & 2 & -1 \end{bmatrix} \begin{bmatrix} X_1 \\ X_2 \\ X_3 \end{bmatrix} = 2X_1 + 2X_2 - X_3$$

and

$$\mathbf{c}'\mathbf{X} = \begin{bmatrix} 1 & -1 & 3 \end{bmatrix} \begin{bmatrix} X_1 \\ X_2 \\ X_3 \end{bmatrix} = X_1 - X_2 + 3X_3$$

The means, variances, and covariance will first be evaluated directly and then be evaluated by (3-36).

Observations on these linear combinations are obtained by replacing X_1, X_2, and X_3 with their observed values. For example, the $n = 3$ observations on $\mathbf{b}'\mathbf{X}$ are

$$\mathbf{b}'\mathbf{x}_1 = 2x_{11} + 2x_{12} - x_{13} = 2(1) + 2(2) - (5) = 1$$
$$\mathbf{b}'\mathbf{x}_2 = 2x_{21} + 2x_{22} - x_{23} = 2(4) + 2(1) - (6) = 4$$
$$\mathbf{b}'\mathbf{x}_3 = 2x_{31} + 2x_{32} - x_{33} = 2(4) + 2(0) - (4) = 4$$

The sample mean and variance of these values are, respectively,

$$\text{Sample mean} = \frac{(1 + 4 + 4)}{3} = 3$$

$$\text{Sample variance} = \frac{(1 - 3)^2 + (4 - 3)^2 + (4 - 3)^2}{3 - 1} = 3$$

In a similar manner, the $n = 3$ observations on $\mathbf{c}'\mathbf{X}$ are

$$\mathbf{c}'\mathbf{x}_1 = 1x_{11} - 1x_{12} + 3x_{13} = 1(1) - 1(2) + 3(5) = 14$$
$$\mathbf{c}'\mathbf{x}_2 = 1(4) - 1(1) + 3(6) = 21$$
$$\mathbf{c}'\mathbf{x}_3 = 1(4) - 1(0) + 3(4) = 16$$

and

$$\text{Sample mean} = \frac{(14 + 21 + 16)}{3} = 17$$

$$\text{Sample variance} = \frac{(14 - 17)^2 + (21 - 17)^2 + (16 - 17)^2}{3 - 1} = 13$$

Moreover, the sample covariance, computed from the pairs of observations $(\mathbf{b}'\mathbf{x}_1, \mathbf{c}'\mathbf{x}_1)$, $(\mathbf{b}'\mathbf{x}_2, \mathbf{c}'\mathbf{x}_2)$, and $(\mathbf{b}'\mathbf{x}_3, \mathbf{c}'\mathbf{x}_3)$, is

Sample covariance

$$= \frac{(1 - 3)(14 - 17) + (4 - 3)(21 - 17) + (4 - 3)(16 - 17)}{3 - 1} = \frac{9}{2}$$

Alternatively, we use the sample mean vector $\bar{\mathbf{x}}$ and sample covariance matrix $\mathbf{S}$ derived from the original data matrix $\mathbf{X}$ to calculate the sample means, variances, and covariances for the linear combinations. Thus, if only the descriptive statistics are of interest, we do not even need to calculate the observations $\mathbf{b}'\mathbf{x}_j$ and $\mathbf{c}'\mathbf{x}_j$.

From Example 3.9,

$$\bar{\mathbf{x}} = \begin{bmatrix} 3 \\ 1 \\ 5 \end{bmatrix} \quad \text{and} \quad \mathbf{S} = \begin{bmatrix} 3 & -\frac{3}{2} & 0 \\ -\frac{3}{2} & 1 & \frac{1}{2} \\ 0 & \frac{1}{2} & 1 \end{bmatrix}$$

Consequently, using (3-36), we find that the two sample means for the derived observations are

$$\text{Sample mean of } \mathbf{b'X} = \mathbf{b'\bar{x}} = [2 \quad 2 \quad -1]\begin{bmatrix} 3 \\ 1 \\ 5 \end{bmatrix} = 3 \quad \text{(check)}$$

$$\text{Sample mean of } \mathbf{c'X} = \mathbf{c'\bar{x}} = [1 \quad -1 \quad 3]\begin{bmatrix} 3 \\ 1 \\ 5 \end{bmatrix} = 17 \quad \text{(check)}$$

Using (3-36), we also have

$$\text{Sample variance of } \mathbf{b'X} = \mathbf{b'Sb}$$

$$= [2 \quad 2 \quad -1]\begin{bmatrix} 3 & -\frac{3}{2} & 0 \\ -\frac{3}{2} & 1 & \frac{1}{2} \\ 0 & \frac{1}{2} & 1 \end{bmatrix}\begin{bmatrix} 2 \\ 2 \\ -1 \end{bmatrix}$$

$$= [2 \quad 2 \quad -1]\begin{bmatrix} 3 \\ -\frac{3}{2} \\ 0 \end{bmatrix} = 3 \quad \text{(check)}$$

$$\text{Sample variance of } \mathbf{c'X} = \mathbf{c'Sc}$$

$$= [1 \quad -1 \quad 3]\begin{bmatrix} 3 & -\frac{3}{2} & 0 \\ -\frac{3}{2} & 1 & \frac{1}{2} \\ 0 & \frac{1}{2} & 1 \end{bmatrix}\begin{bmatrix} 1 \\ -1 \\ 3 \end{bmatrix}$$

$$= [1 \quad -1 \quad 3]\begin{bmatrix} \frac{9}{2} \\ -1 \\ \frac{5}{2} \end{bmatrix} = 13 \quad \text{(check)}$$

$$\text{Sample covariance of } \mathbf{b'X} \text{ and } \mathbf{c'X} = \mathbf{b'Sc}$$

$$= [2 \quad 2 \quad -1]\begin{bmatrix} 3 & -\frac{3}{2} & 0 \\ -\frac{3}{2} & 1 & \frac{1}{2} \\ 0 & \frac{1}{2} & 1 \end{bmatrix}\begin{bmatrix} 1 \\ -1 \\ 3 \end{bmatrix}$$

$$= [2 \quad 2 \quad -1]\begin{bmatrix} \frac{9}{2} \\ -1 \\ \frac{5}{2} \end{bmatrix} = \frac{9}{2} \quad \text{(check)}$$

As indicated, these last results check with the corresponding sample quantities computed directly from the observations on the linear combinations. ∎

The sample mean and covariance relations in Result 3.5 pertain to any number of linear combinations. Consider the q linear combinations

$$a_{i1}X_1 + a_{i2}X_2 + \cdots + a_{ip}X_p, \quad i = 1, 2, \ldots, q \quad (3\text{-}37)$$

These can be expressed in matrix notation as

$$
\begin{bmatrix}
a_{11}X_1 & + & a_{12}X_2 & + \cdots + & a_{1p}X_p \\
a_{21}X_1 & + & a_{22}X_2 & + \cdots + & a_{2p}X_p \\
\vdots & & \vdots & \vdots & \vdots \\
a_{q1}X_1 & + & a_{q2}X_2 & + \cdots + & a_{qp}X_p
\end{bmatrix}
=
\begin{bmatrix}
a_{11} & a_{12} & \cdots & a_{1p} \\
a_{21} & a_{22} & \cdots & a_{2p} \\
\vdots & \vdots & \ddots & \vdots \\
a_{q1} & a_{q2} & \cdots & a_{qp}
\end{bmatrix}
\begin{bmatrix}
X_1 \\
X_2 \\
\vdots \\
X_p
\end{bmatrix}
= \mathbf{AX}
$$

$$(3\text{-}38)$$

Taking the ith row of $\mathbf{A}$, $\mathbf{a}_i'$, to be $\mathbf{b}'$ and the kth row of $\mathbf{A}$, $\mathbf{a}_k'$, to be $\mathbf{c}'$, we see that Equations (3-36) imply that the ith row of $\mathbf{AX}$ has sample mean $\mathbf{a}_i'\bar{\mathbf{x}}$ and the ith and kth rows of $\mathbf{AX}$ have sample covariance $\mathbf{a}_i'\mathbf{S}\,\mathbf{a}_k$. Note that $\mathbf{a}_i'\mathbf{S}\,\mathbf{a}_k$ is the (i, k)th element of $\mathbf{ASA}'$.

Result 3.6. The q linear combinations $\mathbf{AX}$ in (3-38) have sample mean vector $\mathbf{A}\bar{\mathbf{x}}$ and sample covariance matrix $\mathbf{ASA}'$.   ∎

Exercises

3.1. Given the data matrix

$$
\mathbf{X} = \begin{bmatrix}
9 & 1 \\
5 & 3 \\
1 & 2
\end{bmatrix}
$$

(a) Graph the scatter plot in $p = 2$ dimensions. Locate the sample mean on your diagram.

(b) Sketch the $n = 3$-dimensional representation of the data, and plot the deviation vectors $\mathbf{y}_1 - \bar{x}_1\mathbf{1}$ and $\mathbf{y}_2 - \bar{x}_2\mathbf{1}$.

(c) Sketch the deviation vectors in (b) emanating from the origin. Calculate the lengths of these vectors and the cosine of the angle between them. Relate these quantities to $\mathbf{S}_n$ and $\mathbf{R}$.

3.2. Given the data matrix

$$
\mathbf{X} = \begin{bmatrix}
3 & 4 \\
6 & -2 \\
3 & 1
\end{bmatrix}
$$

(a) Graph the scatter plot in $p = 2$ dimensions, and locate the sample mean on your diagram.

(b) Sketch the $n = 3$-space representation of the data, and plot the deviation vectors $\mathbf{y}_1 - \bar{x}_1\mathbf{1}$ and $\mathbf{y}_2 - \bar{x}_2\mathbf{1}$.

(c) Sketch the deviation vectors in (b) emanating from the origin. Calculate their lengths and the cosine of the angle between them. Relate these quantities to $\mathbf{S}_n$ and $\mathbf{R}$.

3.3. Perform the decomposition of $\mathbf{y}_1$ into $\bar{x}_1\mathbf{1}$ and $\mathbf{y}_1 - \bar{x}_1\mathbf{1}$ using the first column of the data matrix in Example 3.9.

3.4. Use the six observations on the variable X_1, in units of millions, from Table 1.1.

(a) Find the projection on $\mathbf{1}' = [1, 1, 1, 1, 1, 1]$.

(b) Calculate the deviation vector $\mathbf{y}_1 - \bar{x}_1\mathbf{1}$. Relate its length to the sample standard deviation.

(c) Graph (to scale) the triangle formed by y_1, $\bar{x}_1\mathbf{1}$, and $y_1 - \bar{x}_1\mathbf{1}$. Identify the length of each component in your graph.

(d) Repeat Parts a–c for the variable X_2 in Table 1.1.

(e) Graph (to scale) the two deviation vectors $y_1 - \bar{x}_1\mathbf{1}$ and $y_2 - \bar{x}_2\mathbf{1}$. Calculate the value of the angle between them.

3.5. Calculate the generalized sample variance $|\mathbf{S}|$ for (a) the data matrix $\mathbf{X}$ in Exercise 3.1 and (b) the data matrix $\mathbf{X}$ in Exercise 3.2.

3.6. Consider the data matrix

$$\mathbf{X} = \begin{bmatrix} -1 & 3 & -2 \\ 2 & 4 & 2 \\ 5 & 2 & 3 \end{bmatrix}$$

(a) Calculate the matrix of deviations (residuals), $\mathbf{X} - \mathbf{1}\bar{\mathbf{x}}'$. Is this matrix of full rank? Explain.

(b) Determine $\mathbf{S}$ and calculate the generalized sample variance $|\mathbf{S}|$. Interpret the latter geometrically.

(c) Using the results in (b), calculate the total sample variance. [See (3-23).]

3.7. Sketch the solid ellipsoids $(\mathbf{x} - \bar{\mathbf{x}})'\mathbf{S}^{-1}(\mathbf{x} - \bar{\mathbf{x}}) \leq 1$ [see (3-16)] for the three matrices

$$\mathbf{S} = \begin{bmatrix} 5 & 4 \\ 4 & 5 \end{bmatrix}, \quad \mathbf{S} = \begin{bmatrix} 5 & -4 \\ -4 & 5 \end{bmatrix}, \quad \mathbf{S} = \begin{bmatrix} 3 & 0 \\ 0 & 3 \end{bmatrix}$$

(Note that these matrices have the *same* generalized variance $|\mathbf{S}|$.)

3.8. Given

$$\mathbf{S} = \begin{bmatrix} 1 & 0 & 0 \\ 0 & 1 & 0 \\ 0 & 0 & 1 \end{bmatrix} \quad \text{and} \quad \mathbf{S} = \begin{bmatrix} 1 & -\frac{1}{2} & -\frac{1}{2} \\ -\frac{1}{2} & 1 & -\frac{1}{2} \\ -\frac{1}{2} & -\frac{1}{2} & 1 \end{bmatrix}$$

(a) Calculate the total sample variance for each $\mathbf{S}$. Compare the results.

(b) Calculate the generalized sample variance for each $\mathbf{S}$, and compare the results. Comment on the discrepancies, if any, found between Parts a and b.

3.9. The following data matrix contains data on test scores, with x_1 = score on first test, x_2 = score on second test, and x_3 = total score on the two tests:

$$\mathbf{X} = \begin{bmatrix} 12 & 17 & 29 \\ 18 & 20 & 38 \\ 14 & 16 & 30 \\ 20 & 18 & 38 \\ 16 & 19 & 35 \end{bmatrix}$$

(a) Obtain the mean corrected data matrix, and verify that the columns are linearly dependent. Specify an $\mathbf{a}' = [a_1, a_2, a_3]$ vector that establishes the linear dependence.

(b) Obtain the sample covariance matrix $\mathbf{S}$, and verify that the generalized variance is zero. Also, show that $\mathbf{Sa} = \mathbf{0}$, so $\mathbf{a}$ can be rescaled to be an eigenvector corresponding to eigenvalue zero.

(c) Verify that the third column of the data matrix is the sum of the first two columns. That is, show that there is linear dependence, with $a_1 = 1$, $a_2 = 1$, and $a_3 = -1$.

3.10. When the generalized variance is zero, it is the columns of the mean corrected data matrix $\mathbf{X}_c = \mathbf{X} - \mathbf{1}\bar{\mathbf{x}}'$ that are linearly dependent, not necessarily those of the data matrix itself. Given the data

$$
\begin{bmatrix}
3 & 1 & 0 \\
6 & 4 & 6 \\
4 & 2 & 2 \\
7 & 0 & 3 \\
5 & 3 & 4
\end{bmatrix}
$$

(a) Obtain the mean corrected data matrix, and verify that the columns are linearly dependent. Specify an $\mathbf{a}' = [a_1, a_2, a_3]$ vector that establishes the dependence.

(b) Obtain the sample covariance matrix $\mathbf{S}$, and verify that the generalized variance is zero.

(c) Show that the columns of the data matrix are linearly independent in this case.

3.11. Use the sample covariance obtained in Example 3.7 to verify (3-29) and (3-30), which state that $\mathbf{R} = \mathbf{D}^{-1/2}\mathbf{S}\mathbf{D}^{-1/2}$ and $\mathbf{D}^{1/2}\mathbf{R}\mathbf{D}^{1/2} = \mathbf{S}$.

3.12. Show that $|\mathbf{S}| = (s_{11}s_{22}\cdots s_{pp})|\mathbf{R}|$.

Hint: From Equation (3-30), $\mathbf{S} = \mathbf{D}^{1/2}\mathbf{R}\mathbf{D}^{1/2}$. Taking determinants gives $|\mathbf{S}| = |\mathbf{D}^{1/2}||\mathbf{R}||\mathbf{D}^{1/2}|$. (See Result 2A.11.) Now examine $|\mathbf{D}^{1/2}|$.

3.13. Given a data matrix $\mathbf{X}$ and the resulting sample correlation matrix $\mathbf{R}$, consider the standardized observations $(x_{jk} - \bar{x}_k)/\sqrt{s_{kk}}$, $k = 1, 2, \ldots, p$, $j = 1, 2, \ldots, n$. Show that these standardized quantities have sample covariance matrix $\mathbf{R}$.

3.14. Consider the data matrix $\mathbf{X}$ in Exercise 3.1. We have $n = 3$ observations on $p = 2$ variables X_1 and X_2. Form the linear combinations

$$
\mathbf{c}'\mathbf{X} = \begin{bmatrix} -1 & 2 \end{bmatrix}\begin{bmatrix} X_1 \\ X_2 \end{bmatrix} = -X_1 + 2X_2
$$

$$
\mathbf{b}'\mathbf{X} = \begin{bmatrix} 2 & 3 \end{bmatrix}\begin{bmatrix} X_1 \\ X_2 \end{bmatrix} = 2X_1 + 3X_2
$$

(a) Evaluate the sample means, variances, and covariance of $\mathbf{b}'\mathbf{X}$ and $\mathbf{c}'\mathbf{X}$ from first principles. That is, calculate the observed values of $\mathbf{b}'\mathbf{X}$ and $\mathbf{c}'\mathbf{X}$, and then use the sample mean, variance, and covariance formulas.

(b) Calculate the sample means, variances, and covariance of $\mathbf{b}'\mathbf{X}$ and $\mathbf{c}'\mathbf{X}$ using (3-36). Compare the results in (a) and (b).

3.15. Repeat Exercise 3.14 using the data matrix

$$
\mathbf{X} = \begin{bmatrix}
1 & 4 & 3 \\
6 & 2 & 6 \\
8 & 3 & 3
\end{bmatrix}
$$

and the linear combinations

$$\mathbf{b'X} = [1 \quad 1 \quad 1] \begin{bmatrix} X_1 \\ X_2 \\ X_3 \end{bmatrix}$$

and

$$\mathbf{c'X} = [1 \quad 2 \quad -3] \begin{bmatrix} X_1 \\ X_2 \\ X_3 \end{bmatrix}$$

3.16. Let $\mathbf{V}$ be a vector random variable with mean vector $E(\mathbf{V}) = \boldsymbol{\mu}_\mathbf{V}$ and covariance matrix $E(\mathbf{V} - \boldsymbol{\mu}_\mathbf{V})(\mathbf{V} - \boldsymbol{\mu}_\mathbf{V})' = \boldsymbol{\Sigma}_\mathbf{V}$. Show that $E(\mathbf{VV'}) = \boldsymbol{\Sigma}_\mathbf{V} + \boldsymbol{\mu}_\mathbf{V}\boldsymbol{\mu}_\mathbf{V}'$.

3.17. Show that, if $\underset{(p\times1)}{\mathbf{X}}$ and $\underset{(q\times1)}{\mathbf{Z}}$ are independent, then each component of $\mathbf{X}$ is independent of each component of $\mathbf{Z}$.

Hint: $P[X_1 \leq x_1, X_2 \leq x_2, \dots, X_p \leq x_p \text{ and } Z_1 \leq z_1, \dots, Z_q \leq z_q]$

$$= P[X_1 \leq x_1, X_2 \leq x_2, \dots, X_p \leq x_p] \cdot P[Z_1 \leq z_1, \dots, Z_q \leq z_q]$$

by independence. Let $x_2, \dots, x_p$ and $z_2, \dots, z_q$ tend to infinity, to obtain

$$P[X_1 \leq x_1 \text{ and } Z_1 \leq z_1] = P[X_1 \leq x_1] \cdot P[Z_1 \leq z_1]$$

for all x_1, z_1. So X_1 and Z_1 are independent. Repeat for other pairs.

3.18. Energy consumption in 2001, by state, from the major sources

$$x_1 = \text{petroleum} \qquad\qquad x_2 = \text{natural gas}$$
$$x_3 = \text{hydroelectric power} \qquad x_4 = \text{nuclear electric power}$$

is recorded in quadrillions (10^{15}) of BTUs (Source: *Statistical Abstract of the United States 2006*).

The resulting mean and covariance matrix are

$$\bar{\mathbf{x}} = \begin{bmatrix} 0.766 \\ 0.508 \\ 0.438 \\ 0.161 \end{bmatrix} \qquad \mathbf{S} = \begin{bmatrix} 0.856 & 0.635 & 0.173 & 0.096 \\ 0.635 & 0.568 & 0.128 & 0.067 \\ 0.173 & 0.127 & 0.171 & 0.039 \\ 0.096 & 0.067 & 0.039 & 0.043 \end{bmatrix}$$

(a) Using the summary statistics, determine the sample mean and variance of a state's total energy consumption for these major sources.

(b) Determine the sample mean and variance of the excess of petroleum consumption over natural gas consumption. Also find the sample covariance of this variable with the total variable in part a.

3.19. Using the summary statistics for the first three variables in Exercise 3.18, verify the relation

$$|\mathbf{S}| = (s_{11}\, s_{22}\, s_{33})\, |\mathbf{R}|$$

3.20. In northern climates, roads must be cleared of snow quickly following a storm. One measure of storm severity is x_1 = its duration in hours, while the effectiveness of snow removal can be quantified by x_2 = the number of hours crews, men, and machine, spend to clear snow. Here are the results for 25 incidents in Wisconsin.

Table 3.2 Snow Data

x_1	x_2	x_1	x_2	x_1	x_2
12.5	13.7	9.0	24.4	3.5	26.1
14.5	16.5	6.5	18.2	8.0	14.5
8.0	17.4	10.5	22.0	17.5	42.3
9.0	11.0	10.0	32.5	10.5	17.5
19.5	23.6	4.5	18.7	12.0	21.8
8.0	13.2	7.0	15.8	6.0	10.4
9.0	32.1	8.5	15.6	13.0	25.6
7.0	12.3	6.5	12.0		
7.0	11.8	8.0	12.8		

(a) Find the sample mean and variance of the difference $x_2 - x_1$ by first obtaining the summary statistics.

(b) Obtain the mean and variance by first obtaining the individual values $x_{j2} - x_{j1}$, for $j = 1, 2, \ldots, 25$ and then calculating the mean and variance. Compare these values with those obtained in part a.

References

1. Anderson, T. W. *An Introduction to Multivariate Statistical Analysis* (3rd ed.). New York: John Wiley, 2003.

2. Eaton, M., and M. Perlman. "The Non-Singularity of Generalized Sample Covariance Matrices." *Annals of Statistics*, **1** (1973), 710–717.

4

THE MULTIVARIATE NORMAL DISTRIBUTION

4.1 Introduction

A generalization of the familiar bell-shaped normal density to several dimensions plays a fundamental role in multivariate analysis. In fact, most of the techniques encountered in this book are based on the assumption that the data were generated from a *multivariate* normal distribution. While real data are never *exactly* multivariate normal, the normal density is often a useful approximation to the "true" population distribution.

One advantage of the multivariate normal distribution stems from the fact that it is mathematically tractable and "nice" results can be obtained. This is frequently not the case for other data-generating distributions. Of course, mathematical attractiveness per se is of little use to the practitioner. It turns out, however, that normal distributions are useful in practice for two reasons: First, the normal distribution serves as a bona fide population model in some instances; second, the sampling distributions of many multivariate statistics are approximately normal, regardless of the form of the parent population, because of a *central limit* effect.

To summarize, many real-world problems fall naturally within the framework of normal theory. The importance of the normal distribution rests on its dual role as both population model for certain natural phenomena and approximate sampling distribution for many statistics.

4.2 The Multivariate Normal Density and Its Properties

The multivariate normal density is a generalization of the univariate normal density to $p \geq 2$ dimensions. Recall that the univariate normal distribution, with mean μ and variance σ^2, has the probability density function

$$f(x) = \frac{1}{\sqrt{2\pi\sigma^2}} e^{-[(x-\mu)/\sigma]^2/2} \qquad -\infty < x < \infty \qquad (4\text{-}1)$$

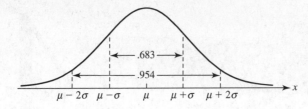

Figure 4.1 A normal density with mean μ and variance σ^2 and selected areas under the curve.

A plot of this function yields the familiar bell-shaped curve shown in Figure 4.1. Also shown in the figure are approximate areas under the curve within ± 1 standard deviations and ± 2 standard deviations of the mean. These areas represent probabilities, and thus, for the normal random variable X,

$$P(\mu - \sigma \le X \le \mu + \sigma) \doteq .68$$

$$P(\mu - 2\sigma \le X \le \mu + 2\sigma) \doteq .95$$

It is convenient to denote the normal density function with mean μ and variance σ^2 by $N(\mu, \sigma^2)$. Therefore, $N(10, 4)$ refers to the function in (4-1) with $\mu = 10$ and $\sigma = 2$. This notation will be extended to the multivariate case later.

The term

$$\left(\frac{x - \mu}{\sigma}\right)^2 = (x - \mu)(\sigma^2)^{-1}(x - \mu) \tag{4-2}$$

in the exponent of the univariate normal density function measures the square of the distance from x to μ in standard deviation units. This can be generalized for a $p \times 1$ vector $\mathbf{x}$ of observations on several variables as

$$(\mathbf{x} - \boldsymbol{\mu})'\boldsymbol{\Sigma}^{-1}(\mathbf{x} - \boldsymbol{\mu}) \tag{4-3}$$

The $p \times 1$ vector $\boldsymbol{\mu}$ represents the expected value of the random vector $\mathbf{X}$, and the $p \times p$ matrix $\boldsymbol{\Sigma}$ is the variance–covariance matrix of $\mathbf{X}$. [See (2–30) and (2–31).] We shall assume that the symmetric matrix $\boldsymbol{\Sigma}$ is positive definite, so the expression in (4-3) is the square of the generalized distance from $\mathbf{x}$ to $\boldsymbol{\mu}$.

The multivariate normal density is obtained by replacing the univariate distance in (4-2) by the multivariate generalized distance of (4-3) in the density function of (4-1). When this replacement is made, the univariate normalizing constant $(2\pi)^{-1/2}(\sigma^2)^{-1/2}$ must be changed to a more general constant that makes the *volume* under the surface of the multivariate density function unity for any p. This is necessary because, in the multivariate case, probabilities are represented by volumes under the surface over regions defined by intervals of the x_i values. It can be shown (see [1]) that this constant is $(2\pi)^{-p/2}|\boldsymbol{\Sigma}|^{-1/2}$, and consequently, a p-dimensional normal density for the random vector $\mathbf{X}' = [X_1, X_2, \dots, X_p]$ has the form

MND
$$f(\mathbf{x}) = \frac{1}{(2\pi)^{p/2}|\boldsymbol{\Sigma}|^{1/2}} e^{-(\mathbf{x}-\boldsymbol{\mu})'\boldsymbol{\Sigma}^{-1}(\mathbf{x}-\boldsymbol{\mu})/2} \tag{4-4}$$

where $-\infty < x_i < \infty$, $i = 1, 2, \dots, p$. We shall denote this p-dimensional normal density by $N_p(\boldsymbol{\mu}, \boldsymbol{\Sigma})$, which is analogous to the normal density in the univariate case.

Example 4.1 (Bivariate normal density) Let us evaluate the $p = 2$-variate normal density in terms of the individual parameters $\mu_1 = E(X_1)$, $\mu_2 = E(X_2)$, $\sigma_{11} = \text{Var}(X_1)$, $\sigma_{22} = \text{Var}(X_2)$, and $\rho_{12} = \sigma_{12}/(\sqrt{\sigma_{11}}\sqrt{\sigma_{22}}) = \text{Corr}(X_1, X_2)$.

Using Result 2A.8, we find that the inverse of the covariance matrix

$$\Sigma = \begin{bmatrix} \sigma_{11} & \sigma_{12} \\ \sigma_{12} & \sigma_{22} \end{bmatrix}$$

is

$$\Sigma^{-1} = \frac{1}{\sigma_{11}\sigma_{22} - \sigma_{12}^2} \begin{bmatrix} \sigma_{22} & -\sigma_{12} \\ -\sigma_{12} & \sigma_{11} \end{bmatrix}$$

Introducing the correlation coefficient ρ_{12} by writing $\sigma_{12} = \rho_{12}\sqrt{\sigma_{11}}\sqrt{\sigma_{22}}$, we obtain $\sigma_{11}\sigma_{22} - \sigma_{12}^2 = \sigma_{11}\sigma_{22}(1 - \rho_{12}^2)$, and the squared distance becomes

$$(\mathbf{x} - \boldsymbol{\mu})'\Sigma^{-1}(\mathbf{x} - \boldsymbol{\mu})$$

$$= [x_1 - \mu_1, x_2 - \mu_2] \frac{1}{\sigma_{11}\sigma_{22}(1 - \rho_{12}^2)}$$

$$\begin{bmatrix} \sigma_{22} & -\rho_{12}\sqrt{\sigma_{11}}\sqrt{\sigma_{22}} \\ -\rho_{12}\sqrt{\sigma_{11}}\sqrt{\sigma_{22}} & \sigma_{11} \end{bmatrix} \begin{bmatrix} x_1 - \mu_1 \\ x_2 - \mu_2 \end{bmatrix}$$

$$= \frac{\sigma_{22}(x_1 - \mu_1)^2 + \sigma_{11}(x_2 - \mu_2)^2 - 2\rho_{12}\sqrt{\sigma_{11}}\sqrt{\sigma_{22}}(x_1 - \mu_1)(x_2 - \mu_2)}{\sigma_{11}\sigma_{22}(1 - \rho_{12}^2)}$$

$$= \frac{1}{1 - \rho_{12}^2}\left[\left(\frac{x_1 - \mu_1}{\sqrt{\sigma_{11}}} \right)^2 + \left(\frac{x_2 - \mu_2}{\sqrt{\sigma_{22}}} \right)^2 - 2\rho_{12}\left(\frac{x_1 - \mu_1}{\sqrt{\sigma_{11}}} \right)\left(\frac{x_2 - \mu_2}{\sqrt{\sigma_{22}}} \right) \right] \qquad (4\text{-}5)$$

The last expression is written in terms of the standardized values $(x_1 - \mu_1)/\sqrt{\sigma_{11}}$ and $(x_2 - \mu_2)/\sqrt{\sigma_{22}}$.

Next, since $|\Sigma| = \sigma_{11}\sigma_{22} - \sigma_{12}^2 = \sigma_{11}\sigma_{22}(1 - \rho_{12}^2)$, we can substitute for Σ^{-1} and $|\Sigma|$ in (4-4) to get the expression for the bivariate $(p = 2)$ normal density involving the individual parameters $\mu_1, \mu_2, \sigma_{11}, \sigma_{22}$, and ρ_{12}:

$$f(x_1, x_2) = \frac{1}{2\pi\sqrt{\sigma_{11}\sigma_{22}(1 - \rho_{12}^2)}} \qquad (4\text{-}6)$$

$$\times \exp\left\{ -\frac{1}{2(1 - \rho_{12}^2)}\left[\left(\frac{x_1 - \mu_1}{\sqrt{\sigma_{11}}} \right)^2 + \left(\frac{x_2 - \mu_2}{\sqrt{\sigma_{22}}} \right)^2 \right.\right.$$

$$\left.\left. - 2\rho_{12}\left(\frac{x_1 - \mu_1}{\sqrt{\sigma_{11}}} \right)\left(\frac{x_2 - \mu_2}{\sqrt{\sigma_{22}}} \right) \right] \right\}$$

The expression in (4-6) is somewhat unwieldy, and the compact general form in (4-4) is more informative in many ways. On the other hand, the expression in (4-6) is useful for discussing certain properties of the normal distribution. For example, if the random variables X_1 and X_2 are uncorrelated, so that $\rho_{12} = 0$, the joint density can be written as the product of two univariate normal densities each of the form of (4-1).

That is, $f(x_1, x_2) = f(x_1)f(x_2)$ and X_1 and X_2 are independent. [See (2-28).] This result is true in general. (See Result 4.5.)

Two bivariate distributions with $\sigma_{11} = \sigma_{22}$ are shown in Figure 4.2. In Figure 4.2(a), X_1 and X_2 are independent ($\rho_{12} = 0$). In Figure 4.2(b), $\rho_{12} = .75$. Notice how the presence of correlation causes the probability to concentrate along a line. ∎

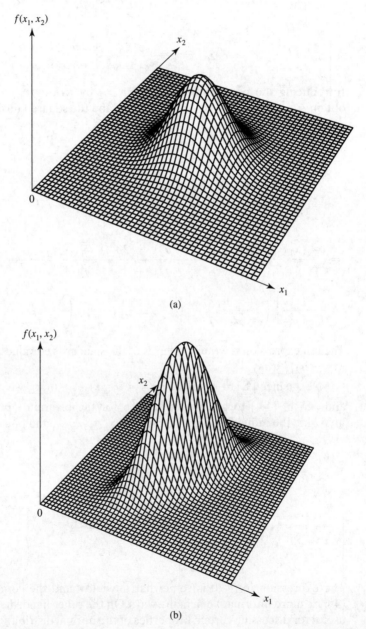

(a)

(b)

Figure 4.2 Two bivariate normal distributions. (a) $\sigma_{11} = \sigma_{22}$ and $\rho_{12} = 0$. (b) $\sigma_{11} = \sigma_{22}$ and $\rho_{12} = .75$.

From the expression in (4-4) for the density of a p-dimensional normal variable, it should be clear that the paths of $\mathbf{x}$ values yielding a constant height for the density are ellipsoids. That is, the multivariate normal density is constant on surfaces where the square of the distance $(\mathbf{x} - \boldsymbol{\mu})'\boldsymbol{\Sigma}^{-1}(\mathbf{x} - \boldsymbol{\mu})$ is constant. These paths are called *contours*:

$$Constant\ probability\ density\ contour = \{all\ \mathbf{x}\ such\ that\ (\mathbf{x} - \boldsymbol{\mu})'\boldsymbol{\Sigma}^{-1}(\mathbf{x} - \boldsymbol{\mu}) = c^2\}$$

$$= surface\ of\ an\ ellipsoid\ centered\ at\ \boldsymbol{\mu}$$

The axes of each ellipsoid of constant density are in the direction of the eigenvectors of $\boldsymbol{\Sigma}^{-1}$, and their lengths are proportional to the reciprocals of the square roots of the eigenvalues of $\boldsymbol{\Sigma}^{-1}$. Fortunately, we can avoid the calculation of $\boldsymbol{\Sigma}^{-1}$ when determining the axes, since these ellipsoids are also determined by the eigenvalues and eigenvectors of $\boldsymbol{\Sigma}$. We state the correspondence formally for later reference.

Result 4.1. If $\boldsymbol{\Sigma}$ is positive definite, so that $\boldsymbol{\Sigma}^{-1}$ exists, then

$$\boldsymbol{\Sigma}\mathbf{e} = \lambda\mathbf{e} \quad implies \quad \boldsymbol{\Sigma}^{-1}\mathbf{e} = \left(\frac{1}{\lambda}\right)\mathbf{e}$$

so $(\lambda, \mathbf{e})$ is an eigenvalue–eigenvector pair for $\boldsymbol{\Sigma}$ corresponding to the pair $(1/\lambda, \mathbf{e})$ for $\boldsymbol{\Sigma}^{-1}$. Also, $\boldsymbol{\Sigma}^{-1}$ is positive definite.

Proof. For $\boldsymbol{\Sigma}$ positive definite and $\mathbf{e} \neq \mathbf{0}$ an eigenvector, we have $0 < \mathbf{e}'\boldsymbol{\Sigma}\mathbf{e} = \mathbf{e}'(\boldsymbol{\Sigma}\mathbf{e})$ $= \mathbf{e}'(\lambda\mathbf{e}) = \lambda\mathbf{e}'\mathbf{e} = \lambda$. Moreover, $\mathbf{e} = \boldsymbol{\Sigma}^{-1}(\boldsymbol{\Sigma}\mathbf{e}) = \boldsymbol{\Sigma}^{-1}(\lambda\mathbf{e})$, or $\mathbf{e} = \lambda\boldsymbol{\Sigma}^{-1}\mathbf{e}$, and division by $\lambda > 0$ gives $\boldsymbol{\Sigma}^{-1}\mathbf{e} = (1/\lambda)\mathbf{e}$. Thus, $(1/\lambda, \mathbf{e})$ is an eigenvalue–eigenvector pair for $\boldsymbol{\Sigma}^{-1}$. Also, for any $p \times 1$ $\mathbf{x}$, by (2-21)

$$\mathbf{x}'\boldsymbol{\Sigma}^{-1}\mathbf{x} = \mathbf{x}'\left(\sum_{i=1}^{p}\left(\frac{1}{\lambda_i}\right)\mathbf{e}_i\mathbf{e}_i'\right)\mathbf{x}$$

$$= \sum_{i=1}^{p}\left(\frac{1}{\lambda_i}\right)(\mathbf{x}'\mathbf{e}_i)^2 \geq 0$$

since each term $\lambda_i^{-1}(\mathbf{x}'\mathbf{e}_i)^2$ is nonnegative. In addition, $\mathbf{x}'\mathbf{e}_i = 0$ for all i only if $\mathbf{x} = \mathbf{0}$. So $\mathbf{x} \neq \mathbf{0}$ implies that $\sum_{i=1}^{p}(1/\lambda_i)(\mathbf{x}'\mathbf{e}_i)^2 > 0$, and it follows that $\boldsymbol{\Sigma}^{-1}$ is positive definite. ∎

The following summarizes these concepts:

Contours of constant density for the p-dimensional normal distribution are ellipsoids defined by $\mathbf{x}$ such the that

$$(\mathbf{x} - \boldsymbol{\mu}.)'\boldsymbol{\Sigma}^{-1}(\mathbf{x} - \boldsymbol{\mu}) = c^2 \tag{4-7}$$

These ellipsoids are centered at $\boldsymbol{\mu}$ and have axes $\pm c\sqrt{\lambda_i}\,\mathbf{e}_i$, where $\boldsymbol{\Sigma}\mathbf{e}_i = \lambda_i\mathbf{e}_i$ for $i = 1, 2, \ldots, p$.

A contour of constant density for a bivariate normal distribution with $\sigma_{11} = \sigma_{22}$ is obtained in the following example.

Example 4.2 (Contours of the bivariate normal density) We shall obtain the axes of constant probability density contours for a bivariate normal distribution when $\sigma_{11} = \sigma_{22}$. From (4-7), these axes are given by the eigenvalues and eigenvectors of Σ. Here $|\Sigma - \lambda \mathbf{I}| = 0$ becomes

$$0 = \begin{vmatrix} \sigma_{11} - \lambda & \sigma_{12} \\ \sigma_{12} & \sigma_{11} - \lambda \end{vmatrix} = (\sigma_{11} - \lambda)^2 - \sigma_{12}^2$$

$$= (\lambda - \sigma_{11} - \sigma_{12})(\lambda - \sigma_{11} + \sigma_{12})$$

Consequently, the eigenvalues are $\lambda_1 = \sigma_{11} + \sigma_{12}$ and $\lambda_2 = \sigma_{11} - \sigma_{12}$. The eigenvector $\mathbf{e}_1$ is determined from

$$\begin{bmatrix} \sigma_{11} & \sigma_{12} \\ \sigma_{12} & \sigma_{11} \end{bmatrix} \begin{bmatrix} e_1 \\ e_2 \end{bmatrix} = (\sigma_{11} + \sigma_{12}) \begin{bmatrix} e_1 \\ e_2 \end{bmatrix}$$

or

$$\sigma_{11} e_1 + \sigma_{12} e_2 = (\sigma_{11} + \sigma_{12}) e_1$$

$$\sigma_{12} e_1 + \sigma_{11} e_2 = (\sigma_{11} + \sigma_{12}) e_2$$

These equations imply that $e_1 = e_2$, and after normalization, the first eigenvalue–eigenvector pair is

$$\lambda_1 = \sigma_{11} + \sigma_{12}, \qquad \mathbf{e}_1 = \begin{bmatrix} \dfrac{1}{\sqrt{2}} \\ \dfrac{1}{\sqrt{2}} \end{bmatrix}$$

Similarly, $\lambda_2 = \sigma_{11} - \sigma_{12}$ yields the eigenvector $\mathbf{e}_2' = [1/\sqrt{2}, -1/\sqrt{2}]$.

When the covariance σ_{12} (or correlation ρ_{12}) is positive, $\lambda_1 = \sigma_{11} + \sigma_{12}$ is the *largest* eigenvalue, and its associated eigenvector $\mathbf{e}_1' = [1/\sqrt{2}, 1/\sqrt{2}]$ lies along the 45° line through the point $\boldsymbol{\mu}' = [\mu_1, \mu_2]$. This is true for any positive value of the covariance (correlation). Since the axes of the constant-density ellipses are given by $\pm c\sqrt{\lambda_1}\, \mathbf{e}_1$ and $\pm c\sqrt{\lambda_2}\, \mathbf{e}_2$ [see (4–7)], and the eigenvectors each have length unity, the major axis will be associated with the largest eigenvalue. For positively correlated normal random variables, then, the *major* axis of the constant-density ellipses will be along the 45° line through $\boldsymbol{\mu}$. (See Figure 4.3.)

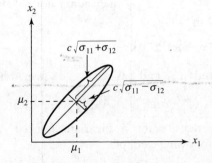

Figure 4.3 A constant-density contour for a bivariate normal distribution with $\sigma_{11} = \sigma_{22}$ and $\sigma_{12} > 0$ (or $\rho_{12} > 0$).

When the covariance (correlation) is negative, $\lambda_2 = \sigma_{11} - \sigma_{12}$ will be the largest eigenvalue, and the major axes of the constant-density ellipses will lie along a line at right angles to the $45°$ line through $\boldsymbol{\mu}$. (These results are true only for $\sigma_{11} = \sigma_{22}$.)

To summarize, the axes of the ellipses of constant density for a bivariate normal distribution with $\sigma_{11} = \sigma_{22}$ are determined by

$$\pm c\sqrt{\sigma_{11} + \sigma_{12}}\begin{bmatrix} \dfrac{1}{\sqrt{2}} \\ \dfrac{1}{\sqrt{2}} \end{bmatrix} \quad \text{and} \quad \pm c\sqrt{\sigma_{11} - \sigma_{12}}\begin{bmatrix} \dfrac{1}{\sqrt{2}} \\ \dfrac{-1}{\sqrt{2}} \end{bmatrix}$$ ∎

We show in Result 4.7 that the choice $c^2 = \chi_p^2(\alpha)$, where $\chi_p^2(\alpha)$ is the upper (100α)th percentile of a chi-square distribution with p degrees of freedom, leads to contours that contain $(1 - \alpha) \times 100\%$ of the probability. Specifically, the following is true for a p-dimensional normal distribution:

The solid ellipsoid of $\mathbf{x}$ values satisfying

$$(\mathbf{x} - \boldsymbol{\mu})'\boldsymbol{\Sigma}^{-1}(\mathbf{x} - \boldsymbol{\mu}) \leq \chi_p^2(\alpha) \tag{4-8}$$

has probability $1 - \alpha$.   *is the volume of the ellipsoid*

The constant-density contours containing 50% and 90% of the probability under the bivariate normal surfaces in Figure 4.2 are pictured in Figure 4.4.

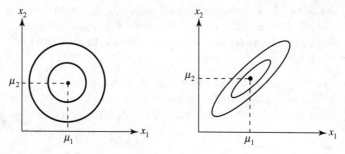

Figure 4.4 The 50% and 90% contours for the bivariate normal distributions in Figure 4.2.

The p-variate normal density in (4-4) has a maximum value when the squared distance in (4-3) is zero—that is, when $\mathbf{x} = \boldsymbol{\mu}$. Thus, $\boldsymbol{\mu}$ is the point of maximum density, or *mode*, as well as the expected value of $\mathbf{X}$, or *mean*. The fact that $\boldsymbol{\mu}$ is the mean of the multivariate normal distribution follows from the symmetry exhibited by the constant-density contours: These contours are centered, or balanced, at $\boldsymbol{\mu}$.

Additional Properties of the Multivariate Normal Distribution

Certain properties of the normal distribution will be needed repeatedly in our explanations of statistical models and methods. These properties make it possible to manipulate normal distributions easily and, as we suggested in Section 4.1, are partly responsible for the popularity of the normal distribution. The key properties, which we shall soon discuss in some mathematical detail, can be stated rather simply.

The following are true for a random vector $\mathbf{X}$ having a multivariate normal distribution:

$\rightarrow y = \beta_0 + \beta_1 X_1 + \cdots \beta_n X_n$

1. Linear combinations of the components of $\mathbf{X}$ are normally distributed. $\rightarrow X_i = N(\ )$

2. All subsets of the components of $\mathbf{X}$ have a (multivariate) normal distribution.

3. Zero covariance implies that the corresponding components are independently distributed. $\leftarrow p 160 \rightarrow$ may have to group sections of a matrix

   iff, not a requirement

4. The conditional distributions of the components are (multivariate) normal.

These statements are reproduced mathematically in the results that follow. Many of these results are illustrated with examples. The proofs that are included should help improve your understanding of matrix manipulations and also lead you to an appreciation for the manner in which the results successively build on themselves.

Result 4.2 can be taken as a working definition of the normal distribution. With this in hand, the subsequent properties are almost immediate. Our partial proof of Result 4.2 indicates how the linear combination definition of a normal density relates to the multivariate density in (4-4).

Result 4.2. If $\mathbf{X}$ is distributed as $N_p(\boldsymbol{\mu}, \boldsymbol{\Sigma})$, then any linear combination of variables $\mathbf{a}'\mathbf{X} = a_1 X_1 + a_2 X_2 + \cdots + a_p X_p$ is distributed as $N(\mathbf{a}'\boldsymbol{\mu}, \mathbf{a}'\boldsymbol{\Sigma}\mathbf{a})$. Also, if $\mathbf{a}'\mathbf{X}$ is distributed as $N(\mathbf{a}'\boldsymbol{\mu}, \mathbf{a}'\boldsymbol{\Sigma}\mathbf{a})$ for every $\mathbf{a}$, then $\mathbf{X}$ must be $N_p(\boldsymbol{\mu}, \boldsymbol{\Sigma})$.

Proof. The expected value and variance of $\mathbf{a}'\mathbf{X}$ follow from (2-43). Proving that $\mathbf{a}'\mathbf{X}$ is normally distributed if $\mathbf{X}$ is multivariate normal is more difficult. You can find a proof in [1]. The second part of result 4.2 is also demonstrated in [1]. ∎

Example 4.3 (The distribution of a linear combination of the components of a normal random vector) Consider the linear combination $\mathbf{a}'\mathbf{X}$ of a multivariate normal random vector determined by the choice $\mathbf{a}' = [1, 0, \ldots, 0]$. Since

$$\mathbf{a}'\mathbf{X} = [1, 0, \ldots, 0] \begin{bmatrix} X_1 \\ X_2 \\ \vdots \\ X_p \end{bmatrix} = X_1$$

and

$$
\mathbf{a}'\boldsymbol{\mu} = [1,0,\ldots,0] \begin{bmatrix} \mu_1 \\ \mu_2 \\ \vdots \\ \mu_p \end{bmatrix} = \mu_1
$$

we have

$$
\mathbf{a}'\boldsymbol{\Sigma}\mathbf{a} = [1,0,\ldots,0] \begin{bmatrix} \sigma_{11} & \sigma_{12} & \cdots & \sigma_{1p} \\ \sigma_{12} & \sigma_{22} & \cdots & \sigma_{2p} \\ \vdots & \vdots & \ddots & \vdots \\ \sigma_{1p} & \sigma_{2p} & \cdots & \sigma_{pp} \end{bmatrix} \begin{bmatrix} 1 \\ 0 \\ \vdots \\ 0 \end{bmatrix} = \sigma_{11}
$$

and it follows from Result 4.2 that X_1 is distributed as $N(\mu_1, \sigma_{11})$. More generally, the marginal distribution of any component X_i of $\mathbf{X}$ is $N(\mu_i, \sigma_{ii})$.   ∎

The next result considers several linear combinations of a multivariate normal vector $\mathbf{X}$.

Result 4.3. If $\mathbf{X}$ is distributed as $N_p(\boldsymbol{\mu}, \boldsymbol{\Sigma})$, the q linear combinations

$$
\underset{(q \times p)(p \times 1)}{\mathbf{A} \quad \mathbf{X}} = \begin{bmatrix} a_{11}X_1 + \cdots + a_{1p}X_p \\ a_{21}X_1 + \cdots + a_{2p}X_p \\ \vdots \\ a_{q1}X_1 + \cdots + a_{qp}X_p \end{bmatrix}
$$

are distributed as $N_q(\mathbf{A}\boldsymbol{\mu}, \mathbf{A}\boldsymbol{\Sigma}\mathbf{A}')$. Also, $\underset{(p\times 1)}{\mathbf{X}} + \underset{(p\times 1)}{\mathbf{d}}$, where $\mathbf{d}$ is a vector of constants, is distributed as $N_p(\boldsymbol{\mu} + \mathbf{d}, \boldsymbol{\Sigma})$.

Proof. The expected value $E(\mathbf{AX})$ and the covariance matrix of $\mathbf{AX}$ follow from (2–45). Any linear combination $\mathbf{b}'(\mathbf{AX})$ is a linear combination of $\mathbf{X}$, of the form $\mathbf{a}'\mathbf{X}$ with $\mathbf{a} = \mathbf{A}'\mathbf{b}$. Thus, the conclusion concerning $\mathbf{AX}$ follows directly from Result 4.2.

The second part of the result can be obtained by considering $\mathbf{a}'(\mathbf{X} + \mathbf{d}) = \mathbf{a}'\mathbf{X} + (\mathbf{a}'\mathbf{d})$, where $\mathbf{a}'\mathbf{X}$ is distributed as $N(\mathbf{a}'\boldsymbol{\mu}, \mathbf{a}'\boldsymbol{\Sigma}\mathbf{a})$. It is known from the univariate case that adding a constant $\mathbf{a}'\mathbf{d}$ to the random variable $\mathbf{a}'\mathbf{X}$ leaves the variance unchanged and translates the mean to $\mathbf{a}'\boldsymbol{\mu} + \mathbf{a}'\mathbf{d} = \mathbf{a}'(\boldsymbol{\mu} + \mathbf{d})$. Since $\mathbf{a}$ was arbitrary, $\mathbf{X} + \mathbf{d}$ is distributed as $N_p(\boldsymbol{\mu} + \mathbf{d}, \boldsymbol{\Sigma})$.   ∎

Example 4.4 (The distribution of two linear combinations of the components of a normal random vector) For $\mathbf{X}$ distributed as $N_3(\boldsymbol{\mu}, \boldsymbol{\Sigma})$, find the distribution of

$$
\begin{bmatrix} X_1 - X_2 \\ X_2 - X_3 \end{bmatrix} = \begin{bmatrix} 1 & -1 & 0 \\ 0 & 1 & -1 \end{bmatrix} \begin{bmatrix} X_1 \\ X_2 \\ X_3 \end{bmatrix} = \mathbf{AX}
$$

By Result 4.3, the distribution of $\mathbf{AX}$ is multivariate normal with mean

$$\mathbf{A\boldsymbol{\mu}} = \begin{bmatrix} 1 & -1 & 0 \\ 0 & 1 & -1 \end{bmatrix} \begin{bmatrix} \mu_1 \\ \mu_2 \\ \mu_3 \end{bmatrix} = \begin{bmatrix} \mu_1 - \mu_2 \\ \mu_2 - \mu_3 \end{bmatrix}$$

and covariance matrix

$$\mathbf{A\boldsymbol{\Sigma}A'} = \begin{bmatrix} 1 & -1 & 0 \\ 0 & 1 & -1 \end{bmatrix} \begin{bmatrix} \sigma_{11} & \sigma_{12} & \sigma_{13} \\ \sigma_{12} & \sigma_{22} & \sigma_{23} \\ \sigma_{13} & \sigma_{23} & \sigma_{33} \end{bmatrix} \begin{bmatrix} 1 & 0 \\ -1 & 1 \\ 0 & -1 \end{bmatrix}$$

$$= \begin{bmatrix} \sigma_{11} - \sigma_{12} & \sigma_{12} - \sigma_{22} & \sigma_{13} - \sigma_{23} \\ \sigma_{12} - \sigma_{13} & \sigma_{22} - \sigma_{23} & \sigma_{23} - \sigma_{33} \end{bmatrix} \begin{bmatrix} 1 & 0 \\ -1 & 1 \\ 0 & -1 \end{bmatrix}$$

$$= \begin{bmatrix} \sigma_{11} - 2\sigma_{12} + \sigma_{22} & \sigma_{12} + \sigma_{23} - \sigma_{22} - \sigma_{13} \\ \sigma_{12} + \sigma_{23} - \sigma_{22} - \sigma_{13} & \sigma_{22} - 2\sigma_{23} + \sigma_{33} \end{bmatrix}$$

Alternatively, the mean vector $\mathbf{A\boldsymbol{\mu}}$ and covariance matrix $\mathbf{A\boldsymbol{\Sigma}A'}$ may be verified by direct calculation of the means and covariances of the two random variables $Y_1 = X_1 - X_2$ and $Y_2 = X_2 - X_3$. ∎

We have mentioned that all subsets of a multivariate normal random vector $\mathbf{X}$ are themselves normally distributed. We state this property formally as Result 4.4.

Result 4.4. All subsets of $\mathbf{X}$ are normally distributed. If we respectively partition $\mathbf{X}$, its mean vector $\boldsymbol{\mu}$, and its covariance matrix $\boldsymbol{\Sigma}$ as

$$\underset{(p\times 1)}{\mathbf{X}} = \left[\begin{array}{c} \underset{(q\times 1)}{\mathbf{X}_1} \\ \hline \underset{((p-q)\times 1)}{\mathbf{X}_2} \end{array} \right] \qquad \underset{(p\times 1)}{\boldsymbol{\mu}} = \left[\begin{array}{c} \underset{(q\times 1)}{\boldsymbol{\mu}_1} \\ \hline \underset{((p-q)\times 1)}{\boldsymbol{\mu}_2} \end{array} \right]$$

and

$$\underset{(p\times p)}{\boldsymbol{\Sigma}} = \left[\begin{array}{c|c} \underset{(q\times q)}{\boldsymbol{\Sigma}_{11}} & \underset{(q\times(p-q))}{\boldsymbol{\Sigma}_{12}} \\ \hline \underset{((p-q)\times q)}{\boldsymbol{\Sigma}_{21}} & \underset{((p-q)\times(p-q))}{\boldsymbol{\Sigma}_{22}} \end{array} \right]$$

then $\mathbf{X}_1$ is distributed as $N_q(\boldsymbol{\mu}_1, \boldsymbol{\Sigma}_{11})$.

Proof. Set $\underset{(q\times p)}{\mathbf{A}} = \left[\underset{(q\times q)}{\mathbf{I}} \,\vdots\, \underset{(q\times(p-q))}{\mathbf{0}} \right]$ in Result 4.3, and the conclusion follows. To apply Result 4.4 to an *arbitrary* subset of the components of $\mathbf{X}$, we simply relabel the subset of interest as $\mathbf{X}_1$ and select the corresponding component means and covariances as $\boldsymbol{\mu}_1$ and $\boldsymbol{\Sigma}_{11}$, respectively. ∎

Example 4.5 (The distribution of a subset of a normal random vector)

If $\mathbf{X}$ is distributed as $N_5(\boldsymbol{\mu}, \boldsymbol{\Sigma})$, find the distribution of $\begin{bmatrix} X_2 \\ X_4 \end{bmatrix}$. We set

$$\mathbf{X}_1 = \begin{bmatrix} X_2 \\ X_4 \end{bmatrix}, \qquad \boldsymbol{\mu}_1 = \begin{bmatrix} \mu_2 \\ \mu_4 \end{bmatrix}, \qquad \boldsymbol{\Sigma}_{11} = \begin{bmatrix} \sigma_{22} & \sigma_{24} \\ \sigma_{24} & \sigma_{44} \end{bmatrix}$$

and note that with this assignment, $\mathbf{X}$, $\boldsymbol{\mu}$, and $\boldsymbol{\Sigma}$ can respectively be rearranged and partitioned as

$$\mathbf{X} = \begin{bmatrix} X_2 \\ X_4 \\ \hline X_1 \\ X_3 \\ X_5 \end{bmatrix}, \qquad \boldsymbol{\mu} = \begin{bmatrix} \mu_2 \\ \mu_4 \\ \hline \mu_1 \\ \mu_3 \\ \mu_5 \end{bmatrix}, \qquad \boldsymbol{\Sigma} = \begin{bmatrix} \sigma_{22} & \sigma_{24} & \sigma_{12} & \sigma_{23} & \sigma_{25} \\ \sigma_{24} & \sigma_{44} & \sigma_{14} & \sigma_{34} & \sigma_{45} \\ \hline \sigma_{12} & \sigma_{14} & \sigma_{11} & \sigma_{13} & \sigma_{15} \\ \sigma_{23} & \sigma_{34} & \sigma_{13} & \sigma_{33} & \sigma_{35} \\ \sigma_{25} & \sigma_{45} & \sigma_{15} & \sigma_{35} & \sigma_{55} \end{bmatrix}$$

or

$$\mathbf{X} = \begin{bmatrix} \mathbf{X}_1 \\ {\scriptstyle(2\times1)} \\ \hline \mathbf{X}_2 \\ {\scriptstyle(3\times1)} \end{bmatrix}, \qquad \boldsymbol{\mu} = \begin{bmatrix} \boldsymbol{\mu}_1 \\ {\scriptstyle(2\times1)} \\ \hline \boldsymbol{\mu}_2 \\ {\scriptstyle(3\times1)} \end{bmatrix}, \qquad \boldsymbol{\Sigma} = \begin{bmatrix} \boldsymbol{\Sigma}_{11} & \boldsymbol{\Sigma}_{12} \\ {\scriptstyle(2\times2)} & {\scriptstyle(2\times3)} \\ \hline \boldsymbol{\Sigma}_{21} & \boldsymbol{\Sigma}_{22} \\ {\scriptstyle(3\times2)} & {\scriptstyle(3\times3)} \end{bmatrix}$$

Thus, from Result 4.4, for

$$\mathbf{X}_1 = \begin{bmatrix} X_2 \\ X_4 \end{bmatrix}$$

we have the distribution

$$N_2(\boldsymbol{\mu}_1, \boldsymbol{\Sigma}_{11}) = N_2\left(\begin{bmatrix} \mu_2 \\ \mu_4 \end{bmatrix}, \begin{bmatrix} \sigma_{22} & \sigma_{24} \\ \sigma_{24} & \sigma_{44} \end{bmatrix} \right)$$

It is clear from this example that the normal distribution for any subset can be expressed by simply selecting the appropriate means and covariances from the original $\boldsymbol{\mu}$ and $\boldsymbol{\Sigma}$. The formal process of relabeling and partitioning is unnecessary. ∎

We are now in a position to state that zero correlation between normal random variables or sets of normal random variables is equivalent to statistical independence.

Result 4.5.

(a) If $\underset{(q_1\times1)}{\mathbf{X}_1}$ and $\underset{(q_2\times1)}{\mathbf{X}_2}$ are independent, then $\text{Cov}(\mathbf{X}_1, \mathbf{X}_2) = \mathbf{0}$, a $q_1 \times q_2$ matrix of zeros.

(b) If $\begin{bmatrix} \mathbf{X}_1 \\ \hline \mathbf{X}_2 \end{bmatrix}$ is $N_{q_1+q_2}\left(\begin{bmatrix} \boldsymbol{\mu}_1 \\ \hline \boldsymbol{\mu}_2 \end{bmatrix}, \begin{bmatrix} \boldsymbol{\Sigma}_{11} & \boldsymbol{\Sigma}_{12} \\ \hline \boldsymbol{\Sigma}_{21} & \boldsymbol{\Sigma}_{22} \end{bmatrix} \right)$, then $\mathbf{X}_1$ and $\mathbf{X}_2$ are independent if and only if $\boldsymbol{\Sigma}_{12} = \mathbf{0}$.

(c) If $\mathbf{X}_1$ and $\mathbf{X}_2$ are independent and are distributed as $N_{q_1}(\boldsymbol{\mu}_1, \boldsymbol{\Sigma}_{11})$ and $N_{q_2}(\boldsymbol{\mu}_2, \boldsymbol{\Sigma}_{22})$, respectively, then $\begin{bmatrix} \mathbf{X}_1 \\ \hline \mathbf{X}_2 \end{bmatrix}$ has the multivariate normal distribution

$$N_{q_1+q_2}\left(\begin{bmatrix} \boldsymbol{\mu}_1 \\ \hline \boldsymbol{\mu}_2 \end{bmatrix}, \begin{bmatrix} \boldsymbol{\Sigma}_{11} & \mathbf{0} \\ \hline \mathbf{0}' & \boldsymbol{\Sigma}_{22} \end{bmatrix} \right)$$

Proof. (See Exercise 4.14 for partial proofs based upon factoring the density function when $\boldsymbol{\Sigma}_{12} = \mathbf{0}$.) ∎

Example 4.6 (The equivalence of zero covariance and independence for normal variables) Let $\underset{(3\times1)}{\mathbf{X}}$ be $N_3(\boldsymbol{\mu}, \boldsymbol{\Sigma})$ with

$$\boldsymbol{\Sigma} = \begin{bmatrix} 4 & 1 & 0 \\ 1 & 3 & 0 \\ 0 & 0 & 2 \end{bmatrix}$$

Are X_1 and X_2 independent? What about (X_1, X_2) and X_3?

Since X_1 and X_2 have covariance $\sigma_{12} = 1$, they are not independent. However, partitioning $\mathbf{X}$ and $\boldsymbol{\Sigma}$ as

$$\mathbf{X} = \begin{bmatrix} X_1 \\ X_2 \\ \hline X_3 \end{bmatrix}, \quad \boldsymbol{\Sigma} = \begin{bmatrix} 4 & 1 & 0 \\ 1 & 3 & 0 \\ \hline 0 & 0 & 2 \end{bmatrix} = \begin{bmatrix} \boldsymbol{\Sigma}_{11} & \boldsymbol{\Sigma}_{12} \\ (2\times2) & (2\times1) \\ \hline \boldsymbol{\Sigma}_{21} & \boldsymbol{\Sigma}_{22} \\ (1\times2) & (1\times1) \end{bmatrix}$$

we see that $\mathbf{X}_1 = \begin{bmatrix} X_1 \\ X_2 \end{bmatrix}$ and X_3 have covariance matrix $\boldsymbol{\Sigma}_{12} = \begin{bmatrix} 0 \\ 0 \end{bmatrix}$. Therefore, (X_1, X_2) and X_3 are independent by Result 4.5. This implies X_3 is independent of X_1 and also of X_2. ∎

We pointed out in our discussion of the bivariate normal distribution that $\rho_{12} = 0$ (zero correlation) implied independence because the joint density function [see (4-6)] could then be written as the product of the marginal (normal) densities of X_1 and X_2. This fact, which we encouraged you to verify directly, is simply a special case of Result 4.5 with $q_1 = q_2 = 1$.

Result 4.6. Let $\mathbf{X} = \begin{bmatrix} \mathbf{X}_1 \\ \hline \mathbf{X}_2 \end{bmatrix}$ be distributed as $N_p(\boldsymbol{\mu}, \boldsymbol{\Sigma})$ with $\boldsymbol{\mu} = \begin{bmatrix} \boldsymbol{\mu}_1 \\ \hline \boldsymbol{\mu}_2 \end{bmatrix}$, $\boldsymbol{\Sigma} = \begin{bmatrix} \boldsymbol{\Sigma}_{11} & \boldsymbol{\Sigma}_{12} \\ \hline \boldsymbol{\Sigma}_{21} & \boldsymbol{\Sigma}_{22} \end{bmatrix}$, and $|\boldsymbol{\Sigma}_{22}| > 0$. Then the conditional distribution of $\mathbf{X}_1$, given that $\mathbf{X}_2 = \mathbf{x}_2$, is normal and has

$$\text{Mean} = \boldsymbol{\mu}_1 + \boldsymbol{\Sigma}_{12}\boldsymbol{\Sigma}_{22}^{-1}(\mathbf{x}_2 - \boldsymbol{\mu}_2)$$

and

$$\text{Covariance} = \mathbf{\Sigma}_{11} - \mathbf{\Sigma}_{12}\mathbf{\Sigma}_{22}^{-1}\mathbf{\Sigma}_{21}$$

Note that the covariance does not depend on the value $\mathbf{x}_2$ of the conditioning variable.

Proof. We shall give an indirect proof. (See Exercise 4.13, which uses the densities directly.) Take

$$\underset{(p \times p)}{\mathbf{A}} = \begin{bmatrix} \underset{(q \times q)}{\mathbf{I}} & -\mathbf{\Sigma}_{12}\mathbf{\Sigma}_{22}^{-1} \\ & q \times (p-q) \\ \hline \underset{(p-q) \times q}{\mathbf{0}} & \underset{(p-q) \times (p-q)}{\mathbf{I}} \end{bmatrix}$$

so

$$\mathbf{A}(\mathbf{X} - \boldsymbol{\mu}) = \mathbf{A}\begin{bmatrix} \mathbf{X}_1 - \boldsymbol{\mu}_1 \\ \hline \mathbf{X}_2 - \boldsymbol{\mu}_2 \end{bmatrix} = \begin{bmatrix} \mathbf{X}_1 - \boldsymbol{\mu}_1 - \mathbf{\Sigma}_{12}\mathbf{\Sigma}_{22}^{-1}(\mathbf{X}_2 - \boldsymbol{\mu}_2) \\ \hline \mathbf{X}_2 - \boldsymbol{\mu}_2 \end{bmatrix}$$

is jointly normal with covariance matrix $\mathbf{A}\mathbf{\Sigma}\mathbf{A}'$ given by

$$\begin{bmatrix} \mathbf{I} & -\mathbf{\Sigma}_{12}\mathbf{\Sigma}_{22}^{-1} \\ \hline \mathbf{0} & \mathbf{I} \end{bmatrix}\begin{bmatrix} \mathbf{\Sigma}_{11} & \mathbf{\Sigma}_{12} \\ \hline \mathbf{\Sigma}_{21} & \mathbf{\Sigma}_{22} \end{bmatrix}\begin{bmatrix} \mathbf{I} & \mathbf{0}' \\ \hline (-\mathbf{\Sigma}_{12}\mathbf{\Sigma}_{22}^{-1})' & \mathbf{I} \end{bmatrix} = \begin{bmatrix} \mathbf{\Sigma}_{11} - \mathbf{\Sigma}_{12}\mathbf{\Sigma}_{22}^{-1}\mathbf{\Sigma}_{21} & \mathbf{0}' \\ \hline \mathbf{0} & \mathbf{\Sigma}_{22} \end{bmatrix}$$

Since $\mathbf{X}_1 - \boldsymbol{\mu}_1 - \mathbf{\Sigma}_{12}\mathbf{\Sigma}_{22}^{-1}(\mathbf{X}_2 - \boldsymbol{\mu}_2)$ and $\mathbf{X}_2 - \boldsymbol{\mu}_2$ have zero covariance, they are independent. Moreover, the quantity $\mathbf{X}_1 - \boldsymbol{\mu}_1 - \mathbf{\Sigma}_{12}\mathbf{\Sigma}_{22}^{-1}(\mathbf{X}_2 - \boldsymbol{\mu}_2)$ has distribution $N_q(\mathbf{0}, \mathbf{\Sigma}_{11} - \mathbf{\Sigma}_{12}\mathbf{\Sigma}_{22}^{-1}\mathbf{\Sigma}_{21})$. Given that $\mathbf{X}_2 = \mathbf{x}_2, \boldsymbol{\mu}_1 + \mathbf{\Sigma}_{12}\mathbf{\Sigma}_{22}^{-1}(\mathbf{x}_2 - \boldsymbol{\mu}_2)$ is a constant. Because $\mathbf{X}_1 - \boldsymbol{\mu}_1 - \mathbf{\Sigma}_{12}\mathbf{\Sigma}_{22}^{-1}(\mathbf{X}_2 - \boldsymbol{\mu}_2)$ and $\mathbf{X}_2 - \boldsymbol{\mu}_2$ are independent, the conditional distribution of $\mathbf{X}_1 - \boldsymbol{\mu}_1 - \mathbf{\Sigma}_{12}\mathbf{\Sigma}_{22}^{-1}(\mathbf{x}_2 - \boldsymbol{\mu}_2)$ is the same as the unconditional distribution of $\mathbf{X}_1 - \boldsymbol{\mu}_1 - \mathbf{\Sigma}_{12}\mathbf{\Sigma}_{22}^{-1}(\mathbf{X}_2 - \boldsymbol{\mu}_2)$. Since $\mathbf{X}_1 - \boldsymbol{\mu}_1 - \mathbf{\Sigma}_{12}\mathbf{\Sigma}_{22}^{-1}(\mathbf{X}_2 - \boldsymbol{\mu}_2)$ is $N_q(\mathbf{0}, \mathbf{\Sigma}_{11} - \mathbf{\Sigma}_{12}\mathbf{\Sigma}_{22}^{-1}\mathbf{\Sigma}_{21})$, so is the random vector $\mathbf{X}_1 - \boldsymbol{\mu}_1 - \mathbf{\Sigma}_{12}\mathbf{\Sigma}_{22}^{-1}(\mathbf{x}_2 - \boldsymbol{\mu}_2)$ when $\mathbf{X}_2$ has the particular value $\mathbf{x}_2$. Equivalently, given that $\mathbf{X}_2 = \mathbf{x}_2, \mathbf{X}_1$ is distributed as $N_q(\boldsymbol{\mu}_1 + \mathbf{\Sigma}_{12}\mathbf{\Sigma}_{22}^{-1}(\mathbf{x}_2 - \boldsymbol{\mu}_2), \mathbf{\Sigma}_{11} - \mathbf{\Sigma}_{12}\mathbf{\Sigma}_{22}^{-1}\mathbf{\Sigma}_{21})$. ∎

Example 4.7 (The conditional density of a bivariate normal distribution) The conditional density of X_1, given that $X_2 = x_2$ for any bivariate distribution, is defined by

$$f(x_1 | x_2) = \{\text{conditional density of } X_1 \text{ given that } X_2 = x_2\} = \frac{f(x_1, x_2)}{f(x_2)}$$

where $f(x_2)$ is the marginal distribution of X_2. If $f(x_1, x_2)$ is the bivariate normal density, show that $f(x_1 | x_2)$ is

$$N\left(\mu_1 + \frac{\sigma_{12}}{\sigma_{22}}(x_2 - \mu_2), \sigma_{11} - \frac{\sigma_{12}^2}{\sigma_{22}}\right)$$

Here $\sigma_{11} - \sigma_{12}^2/\sigma_{22} = \sigma_{11}(1 - \rho_{12}^2)$. The two terms involving $x_1 - \mu_1$ in the exponent of the bivariate normal density [see Equation (4-6)] become, apart from the multiplicative constant $-1/2(1 - \rho_{12}^2)$,

$$\frac{(x_1 - \mu_1)^2}{\sigma_{11}} - 2\rho_{12}\frac{(x_1 - \mu_1)(x_2 - \mu_2)}{\sqrt{\sigma_{11}}\sqrt{\sigma_{22}}}$$

$$= \frac{1}{\sigma_{11}}\left[x_1 - \mu_1 - \rho_{12}\frac{\sqrt{\sigma_{11}}}{\sqrt{\sigma_{22}}}(x_2 - \mu_2)\right]^2 - \frac{\rho_{12}^2}{\sigma_{22}}(x_2 - \mu_2)^2$$

Because $\rho_{12} = \sigma_{12}/\sqrt{\sigma_{11}}\sqrt{\sigma_{22}}$, or $\rho_{12}\sqrt{\sigma_{11}}/\sqrt{\sigma_{22}} = \sigma_{12}/\sigma_{22}$, the complete exponent is

$$\frac{-1}{2(1 - \rho_{12}^2)}\left(\frac{(x_1 - \mu_1)^2}{\sigma_{11}} - 2\rho_{12}\frac{(x_1 - \mu_1)(x_2 - \mu_2)}{\sqrt{\sigma_{11}}\sqrt{\sigma_{22}}} + \frac{(x_2 - \mu_2)^2}{\sigma_{22}}\right)$$

$$= \frac{-1}{2\sigma_{11}(1 - \rho_{12}^2)}\left(x_1 - \mu_1 - \rho_{12}\frac{\sqrt{\sigma_{11}}}{\sqrt{\sigma_{22}}}(x_2 - \mu_2)\right)^2$$

$$- \frac{1}{2(1 - \rho_{12}^2)}\left(\frac{1}{\sigma_{22}} - \frac{\rho_{12}^2}{\sigma_{22}}\right)(x_2 - \mu_2)^2$$

$$= \frac{-1}{2\sigma_{11}(1 - \rho_{12}^2)}\left(x_1 - \mu_1 - \frac{\sigma_{12}}{\sigma_{22}}(x_2 - \mu_2)\right)^2 - \frac{1}{2}\frac{(x_2 - \mu_2)^2}{\sigma_{22}}$$

The constant term $2\pi\sqrt{\sigma_{11}\sigma_{22}(1 - \rho_{12}^2)}$ also factors as

$$\sqrt{2\pi}\sqrt{\sigma_{22}} \times \sqrt{2\pi}\sqrt{\sigma_{11}(1 - \rho_{12}^2)}$$

Dividing the joint density of X_1 and X_2 by the marginal density

$$f(x_2) = \frac{1}{\sqrt{2\pi}\sqrt{\sigma_{22}}}e^{-(x_2-\mu_2)^2/2\sigma_{22}}$$

and canceling terms yields the conditional density

$$f(x_1 | x_2) = \frac{f(x_1, x_2)}{f(x_2)}$$

$$= \frac{1}{\sqrt{2\pi}\sqrt{\sigma_{11}(1 - \rho_{12}^2)}}e^{-[x_1-\mu_1-(\sigma_{12}/\sigma_{22})(x_2-\mu_2)]^2/2\sigma_{11}(1-\rho_{12}^2)},$$

$$-\infty < x_1 < \infty$$

Thus, with our customary notation, the conditional distribution of X_1 given that $X_2 = x_2$ is $N(\mu_1 + (\sigma_{12}/\sigma_{22})(x_2 - \mu_2), \sigma_{11}(1 - \rho_{12}^2))$. Now, $\Sigma_{11} - \Sigma_{12}\Sigma_{22}^{-1}\Sigma_{21} = \sigma_{11} - \sigma_{12}^2/\sigma_{22} = \sigma_{11}(1 - \rho_{12}^2)$ and $\Sigma_{12}\Sigma_{22}^{-1} = \sigma_{12}/\sigma_{22}$, agreeing with Result 4.6, which we obtained by an indirect method.   ■

For the multivariate normal situation, it is worth emphasizing the following:

1. All conditional distributions are (multivariate) normal.
2. The conditional mean is of the form

$$\mu_1 + \beta_{1,q+1}(x_{q+1} - \mu_{q+1}) + \cdots + \beta_{1,p}(x_p - \mu_p)$$

$$\vdots$$

$$\mu_q + \beta_{q,q+1}(x_{q+1} - \mu_{q+1}) + \cdots + \beta_{q,p}(x_p - \mu_p)$$

(4-9)

where the β's are defined by

$$\Sigma_{12}\Sigma_{22}^{-1} = \begin{bmatrix} \beta_{1,q+1} & \beta_{1,q+2} & \cdots & \beta_{1,p} \\ \beta_{2,q+1} & \beta_{2,q+2} & \cdots & \beta_{2,p} \\ \vdots & \vdots & \ddots & \vdots \\ \beta_{q,q+1} & \beta_{q,q+2} & \cdots & \beta_{q,p} \end{bmatrix}$$

3. The conditional covariance, $\Sigma_{11} - \Sigma_{12}\Sigma_{22}^{-1}\Sigma_{21}$, does not depend upon the value(s) of the conditioning variable(s).

We conclude this section by presenting two final properties of multivariate normal random vectors. One has to do with the probability content of the ellipsoids of constant density. The other discusses the distribution of another form of linear combinations.

The chi-square distribution determines the variability of the sample variance $s^2 = s_{11}$ for samples from a univariate normal population. It also plays a basic role in the multivariate case.

Result 4.7. Let $\mathbf{X}$ be distributed as $N_p(\boldsymbol{\mu}, \boldsymbol{\Sigma})$ with $|\boldsymbol{\Sigma}| > 0$. Then

(a) $(\mathbf{X} - \boldsymbol{\mu})'\boldsymbol{\Sigma}^{-1}(\mathbf{X} - \boldsymbol{\mu})$ is distributed as χ_p^2, where χ_p^2 denotes the chi-square distribution with p degrees of freedom.

(b) The $N_p(\boldsymbol{\mu}, \boldsymbol{\Sigma})$ distribution assigns probability $1 - \alpha$ to the solid ellipsoid $\{\mathbf{x}: (\mathbf{x} - \boldsymbol{\mu})'\boldsymbol{\Sigma}^{-1}(\mathbf{x} - \boldsymbol{\mu}) \leq \chi_p^2(\alpha)\}$, where $\chi_p^2(\alpha)$ denotes the upper (100α)th percentile of the χ_p^2 distribution.

Proof. We know that χ_p^2 is defined as the distribution of the sum $Z_1^2 + Z_2^2 + \cdots + Z_p^2$, where $Z_1, Z_2, \ldots, Z_p$ are independent $N(0, 1)$ random variables. Next, by the spectral decomposition [see Equations (2-16) and (2-21) with $\mathbf{A} = \boldsymbol{\Sigma}$, and see Result 4.1], $\boldsymbol{\Sigma}^{-1} = \sum_{i=1}^{p} \frac{1}{\lambda_i} \mathbf{e}_i\mathbf{e}_i'$, where $\boldsymbol{\Sigma}\mathbf{e}_i = \lambda_i\mathbf{e}_i$, so $\boldsymbol{\Sigma}^{-1}\mathbf{e}_i = (1/\lambda_i)\mathbf{e}_i$. Consequently,

$$(\mathbf{X} - \boldsymbol{\mu})'\boldsymbol{\Sigma}^{-1}(\mathbf{X} - \boldsymbol{\mu}) = \sum_{i=1}^{p} (1/\lambda_i)(\mathbf{X} - \boldsymbol{\mu})'\mathbf{e}_i\mathbf{e}_i'(\mathbf{X} - \boldsymbol{\mu}) = \sum_{i=1}^{p} (1/\lambda_i)(\mathbf{e}_i'(\mathbf{X} - \boldsymbol{\mu}))^2 = $$

$$\sum_{i=1}^{p} [(1/\sqrt{\lambda_i})\,\mathbf{e}_i'(\mathbf{X} - \boldsymbol{\mu})]^2 = \sum_{i=1}^{p} Z_i^2, \text{ for instance. Now, we can write } \mathbf{Z} = \mathbf{A}(\mathbf{X} - \boldsymbol{\mu}),$$

where

$$
\mathbf{Z}_{(p\times 1)} = \begin{bmatrix} Z_1 \\ Z_2 \\ \vdots \\ Z_p \end{bmatrix}, \qquad \mathbf{A}_{(p\times p)} = \begin{bmatrix} \dfrac{1}{\sqrt{\lambda_1}} \mathbf{e}_1' \\[6pt] \dfrac{1}{\sqrt{\lambda_2}} \mathbf{e}_2' \\ \vdots \\ \dfrac{1}{\sqrt{\lambda_p}} \mathbf{e}_p' \end{bmatrix}
$$

and $\mathbf{X} - \boldsymbol{\mu}$ is distributed as $N_p(\mathbf{0}, \boldsymbol{\Sigma})$. Therefore, by Result 4.3, $\mathbf{Z} = \mathbf{A}(\mathbf{X} - \boldsymbol{\mu})$ is distributed as $N_p(\mathbf{0}, \mathbf{A}\boldsymbol{\Sigma}\mathbf{A}')$, where

$$
\underset{(p\times p)(p\times p)(p\times p)}{\mathbf{A}\ \ \boldsymbol{\Sigma}\ \ \mathbf{A}'} = \begin{bmatrix} \dfrac{1}{\sqrt{\lambda_1}} \mathbf{e}_1' \\[6pt] \dfrac{1}{\sqrt{\lambda_2}} \mathbf{e}_2' \\ \vdots \\ \dfrac{1}{\sqrt{\lambda_p}} \mathbf{e}_p' \end{bmatrix} \left[\sum_{i=1}^{p} \lambda_i \mathbf{e}_i \mathbf{e}_i' \right] \left[\dfrac{1}{\sqrt{\lambda_1}} \mathbf{e}_1 \ \bigg| \ \dfrac{1}{\sqrt{\lambda_2}} \mathbf{e}_2 \ \bigg| \cdots \bigg| \ \dfrac{1}{\sqrt{\lambda_p}} \mathbf{e}_p \right]
$$

$$
= \begin{bmatrix} \sqrt{\lambda_1}\, \mathbf{e}_1' \\ \sqrt{\lambda_2}\, \mathbf{e}_2' \\ \vdots \\ \sqrt{\lambda_p}\, \mathbf{e}_p' \end{bmatrix} \left[\dfrac{1}{\sqrt{\lambda_1}} \mathbf{e}_1 \ \bigg| \ \dfrac{1}{\sqrt{\lambda_2}} \mathbf{e}_2 \ \bigg| \cdots \bigg| \ \dfrac{1}{\sqrt{\lambda_p}} \mathbf{e}_p \right] = \mathbf{I}
$$

By Result 4.5, $Z_1, Z_2, \ldots, Z_p$ are *independent* standard normal variables, and we conclude that $(\mathbf{X} - \boldsymbol{\mu})'\boldsymbol{\Sigma}^{-1}(\mathbf{X} - \boldsymbol{\mu})$ has a χ_p^2-distribution.

For Part b, we note that $P[(\mathbf{X} - \boldsymbol{\mu})'\boldsymbol{\Sigma}^{-1}(\mathbf{X} - \boldsymbol{\mu}) \leq c^2]$ is the probability assigned to the ellipsoid $(\mathbf{X} - \boldsymbol{\mu})'\boldsymbol{\Sigma}^{-1}(\mathbf{X} - \boldsymbol{\mu}) \leq c^2$ by the density $N_p(\boldsymbol{\mu}, \boldsymbol{\Sigma})$. But from Part a, $P[(\mathbf{X} - \boldsymbol{\mu})'\boldsymbol{\Sigma}^{-1}(\mathbf{X} - \boldsymbol{\mu}) \leq \chi_p^2(\alpha)] = 1 - \alpha$, and Part b holds.   ∎

Remark: (Interpretation of statistical distance) Result 4.7 provides an interpretation of a squared statistical distance. When $\mathbf{X}$ is distributed as $N_p(\boldsymbol{\mu}, \boldsymbol{\Sigma})$,

$$
(\mathbf{X} - \boldsymbol{\mu})'\boldsymbol{\Sigma}^{-1}(\mathbf{X} - \boldsymbol{\mu})
$$

is the squared statistical distance from $\mathbf{X}$ to the population mean vector $\boldsymbol{\mu}$. If one component has a much larger variance than another, it will contribute less to the squared distance. Moreover, two highly correlated random variables will contribute less than two variables that are nearly uncorrelated. Essentially, the use of the inverse of the covariance matrix, (1) standardizes all of the variables and (2) eliminates the effects of correlation. From the proof of Result 4.7,

$$
(\mathbf{X} - \boldsymbol{\mu})'\boldsymbol{\Sigma}^{-1}(\mathbf{X} - \boldsymbol{\mu}) = Z_1^2 + Z_2^2 + \cdots + Z_p^2
$$

In terms of $\boldsymbol{\Sigma}^{-\frac{1}{2}}$ (see (2-22)), $\mathbf{Z} = \boldsymbol{\Sigma}^{-\frac{1}{2}}(\mathbf{X} - \boldsymbol{\mu})$ has a $N_p(\mathbf{0}, \mathbf{I}_p)$ distribution, and

$$(\mathbf{X} - \boldsymbol{\mu})'\boldsymbol{\Sigma}^{-1}(\mathbf{X} - \boldsymbol{\mu}) = (\mathbf{X} - \boldsymbol{\mu})'\boldsymbol{\Sigma}^{-\frac{1}{2}}\boldsymbol{\Sigma}^{-\frac{1}{2}}(\mathbf{X} - \boldsymbol{\mu})$$

$$= \mathbf{Z}'\mathbf{Z} = Z_1^2 + Z_2^2 + \cdots + Z_p^2$$

The squared statistical distance is calculated as if, first, the random vector $\mathbf{X}$ were transformed to p independent standard normal random variables and then the usual squared distance, the sum of the squares of the variables, were applied.

Next, consider the linear combination of vector random variables

$$c_1\mathbf{X}_1 + c_2\mathbf{X}_2 + \cdots + c_n\mathbf{X}_n = \underset{(p \times n)}{[\mathbf{X}_1 \mid \mathbf{X}_2 \mid \cdots \mid \mathbf{X}_n]} \underset{(n \times 1)}{\mathbf{c}} \qquad (4\text{-}10)$$

This linear combination differs from the linear combinations considered earlier in that it defines a $p \times 1$ *vector* random variable that is a linear combination of vectors. Previously, we discussed a *single* random variable that could be written as a linear combination of other univariate random variables.

Result 4.8. Let $\mathbf{X}_1, \mathbf{X}_2, \ldots, \mathbf{X}_n$ be mutually independent with $\mathbf{X}_j$ distributed as $N_p(\boldsymbol{\mu}_j, \boldsymbol{\Sigma})$. (Note that each $\mathbf{X}_j$ has the *same* covariance matrix $\boldsymbol{\Sigma}$.) Then

$$\mathbf{V}_1 = c_1\mathbf{X}_1 + c_2\mathbf{X}_2 + \cdots + c_n\mathbf{X}_n$$

is distributed as $N_p\left(\sum_{j=1}^{n} c_j\boldsymbol{\mu}_j, \left(\sum_{j=1}^{n} c_j^2\right)\boldsymbol{\Sigma}\right)$. Moreover, $\mathbf{V}_1$ and $\mathbf{V}_2 = b_1\mathbf{X}_1 + b_2\mathbf{X}_2 + \cdots + b_n\mathbf{X}_n$ are jointly multivariate normal with covariance matrix

$$\begin{bmatrix} \left(\sum_{j=1}^{n} c_j^2\right)\boldsymbol{\Sigma} & (\mathbf{b}'\mathbf{c})\,\boldsymbol{\Sigma} \\ (\mathbf{b}'\mathbf{c})\,\boldsymbol{\Sigma} & \left(\sum_{j=1}^{n} b_j^2\right)\boldsymbol{\Sigma} \end{bmatrix}$$

Consequently, $\mathbf{V}_1$ and $\mathbf{V}_2$ are independent if $\mathbf{b}'\mathbf{c} = \sum_{j=1}^{n} c_j b_j = 0$.

Proof. By Result 4.5(c), the np component vector

$$[X_{11}, \ldots, X_{1p}, X_{21}, \ldots, X_{2p}, \ldots, X_{np}] = [\mathbf{X}_1', \mathbf{X}_2', \ldots, \mathbf{X}_n'] = \underset{(1 \times np)}{\mathbf{X}'}$$

is multivariate normal. In particular, $\underset{(np \times 1)}{\mathbf{X}}$ is distributed as $N_{np}(\boldsymbol{\mu}, \boldsymbol{\Sigma}_{\mathbf{x}})$, where

$$\underset{(np \times 1)}{\boldsymbol{\mu}} = \begin{bmatrix} \boldsymbol{\mu}_1 \\ \boldsymbol{\mu}_2 \\ \vdots \\ \boldsymbol{\mu}_n \end{bmatrix} \quad \text{and} \quad \underset{(np \times np)}{\boldsymbol{\Sigma}_{\mathbf{x}}} = \begin{bmatrix} \boldsymbol{\Sigma} & \mathbf{0} & \cdots & \mathbf{0} \\ \mathbf{0} & \boldsymbol{\Sigma} & \cdots & \mathbf{0} \\ \vdots & \vdots & \ddots & \vdots \\ \mathbf{0} & \mathbf{0} & \cdots & \boldsymbol{\Sigma} \end{bmatrix}$$

The choice

$$\underset{(2p \times np)}{\mathbf{A}} = \begin{bmatrix} c_1\mathbf{I} & c_2\mathbf{I} & \cdots & c_n\mathbf{I} \\ b_1\mathbf{I} & b_2\mathbf{I} & \cdots & b_n\mathbf{I} \end{bmatrix}$$

where $\mathbf{I}$ is the $p \times p$ identity matrix, gives

$$\mathbf{AX} = \begin{bmatrix} \sum_{j=1}^{n} c_j\mathbf{X}_j \\ \sum_{j=1}^{n} b_j\mathbf{X}_j \end{bmatrix} = \begin{bmatrix} \mathbf{V}_1 \\ \mathbf{V}_2 \end{bmatrix}$$

and $\mathbf{AX}$ is normal $N_{2p}(\mathbf{A}\boldsymbol{\mu}, \mathbf{A}\boldsymbol{\Sigma}_\mathbf{x}\mathbf{A}')$ by Result 4.3. Straightforward block multiplication shows that $\mathbf{A}\boldsymbol{\Sigma}_\mathbf{x}\mathbf{A}'$ has the first block diagonal term

$$[c_1\boldsymbol{\Sigma}, c_2\boldsymbol{\Sigma}, \ldots, c_n\boldsymbol{\Sigma}][c_1\mathbf{I}, c_2\mathbf{I}, \ldots, c_n\mathbf{I}]' = \left(\sum_{j=1}^{n} c_j^2 \right) \boldsymbol{\Sigma}$$

The off-diagonal term is

$$[c_1\boldsymbol{\Sigma}, c_2\boldsymbol{\Sigma}, \ldots, c_n\boldsymbol{\Sigma}][b_1\mathbf{I}, b_2\mathbf{I}, \ldots, b_n\mathbf{I}]' = \left(\sum_{j=1}^{n} c_j b_j \right) \boldsymbol{\Sigma}$$

This term is the covariance matrix for $\mathbf{V}_1, \mathbf{V}_2$. Consequently, when $\sum_{j=1}^{n} c_j b_j = \mathbf{b}'\mathbf{c} = 0$, so that $\left(\sum_{j=1}^{n} c_j b_j \right) \boldsymbol{\Sigma} = \underset{(p \times p)}{\mathbf{0}}$, $\mathbf{V}_1$ and $\mathbf{V}_2$ are independent by Result 4.5(b). ∎

For sums of the type in (4-10), the property of zero correlation is equivalent to requiring the coefficient vectors $\mathbf{b}$ and $\mathbf{c}$ to be perpendicular.

Example 4.8 (Linear combinations of random vectors) Let $\mathbf{X}_1, \mathbf{X}_2, \mathbf{X}_3$, and $\mathbf{X}_4$ be independent and identically distributed 3×1 random vectors with

$$\boldsymbol{\mu} = \begin{bmatrix} 3 \\ -1 \\ 1 \end{bmatrix} \quad \text{and} \quad \boldsymbol{\Sigma} = \begin{bmatrix} 3 & -1 & 1 \\ -1 & 1 & 0 \\ 1 & 0 & 2 \end{bmatrix}$$

We first consider a linear combination $\mathbf{a}'\mathbf{X}_1$ of the three components of $\mathbf{X}_1$. This is a random variable with mean

$$\mathbf{a}'\boldsymbol{\mu} = 3a_1 - a_2 + a_3$$

and variance

$$\mathbf{a}'\boldsymbol{\Sigma}\mathbf{a} = 3a_1^2 + a_2^2 + 2a_3^2 - 2a_1 a_2 + 2a_1 a_3$$

That is, a linear combination $\mathbf{a}'\mathbf{X}_1$ of the components of a random vector is a single random variable consisting of a sum of terms that are each a constant times a variable. This is very different from a linear combination of random vectors, say,

$$c_1\mathbf{X}_1 + c_2\mathbf{X}_2 + c_3\mathbf{X}_3 + c_4\mathbf{X}_4$$

which is itself a random vector. Here each term in the sum is a constant times a random vector.

Now consider two linear combinations of random vectors

$$\frac{1}{2}\mathbf{X}_1 + \frac{1}{2}\mathbf{X}_2 + \frac{1}{2}\mathbf{X}_3 + \frac{1}{2}\mathbf{X}_4$$

and

$$\mathbf{X}_1 + \mathbf{X}_2 + \mathbf{X}_3 - 3\mathbf{X}_4$$

Find the mean vector and covariance matrix for each linear combination of vectors and also the covariance between them.

By Result 4.8 with $c_1 = c_2 = c_3 = c_4 = 1/2$, the first linear combination has mean vector

$$(c_1 + c_2 + c_3 + c_4)\boldsymbol{\mu} = 2\boldsymbol{\mu} = \begin{bmatrix} 6 \\ -2 \\ 2 \end{bmatrix}$$

and covariance matrix

$$(c_1^2 + c_2^2 + c_3^2 + c_4^2)\boldsymbol{\Sigma} = 1 \times \boldsymbol{\Sigma} = \begin{bmatrix} 3 & -1 & 1 \\ -1 & 1 & 0 \\ 1 & 0 & 2 \end{bmatrix}$$

For the second linear combination of random vectors, we apply Result 4.8 with $b_1 = b_2 = b_3 = 1$ and $b_4 = -3$ to get mean vector

$$(b_1 + b_2 + b_3 + b_4)\boldsymbol{\mu} = 0\boldsymbol{\mu} = \begin{bmatrix} 0 \\ 0 \\ 0 \end{bmatrix}$$

and covariance matrix

$$(b_1^2 + b_2^2 + b_3^2 + b_4^2)\boldsymbol{\Sigma} = 12 \times \boldsymbol{\Sigma} = \begin{bmatrix} 36 & -12 & 12 \\ -12 & 12 & 0 \\ 12 & 0 & 24 \end{bmatrix}$$

Finally, the covariance matrix for the two linear combinations of random vectors is

$$(c_1 b_1 + c_2 b_2 + c_3 b_3 + c_4 b_4)\boldsymbol{\Sigma} = 0\boldsymbol{\Sigma} = \begin{bmatrix} 0 & 0 & 0 \\ 0 & 0 & 0 \\ 0 & 0 & 0 \end{bmatrix}$$

Every component of the first linear combination of random vectors has zero covariance with every component of the second linear combination of random vectors.

If, in addition, each $\mathbf{X}$ has a trivariate normal distribution, then the two linear combinations have a joint six-variate normal distribution, and the two linear combinations of vectors are independent. ∎

4.3 Sampling from a Multivariate Normal Distribution and Maximum Likelihood Estimation

We discussed sampling and selecting random samples briefly in Chapter 3. In this section, we shall be concerned with samples from a multivariate normal population—in particular, with the sampling distribution of $\overline{\mathbf{X}}$ and $\mathbf{S}$.

The Multivariate Normal Likelihood

Let us assume that the $p \times 1$ vectors $\mathbf{X}_1, \mathbf{X}_2, \ldots, \mathbf{X}_n$ represent a random sample from a multivariate normal population with mean vector $\boldsymbol{\mu}$ and covariance matrix $\boldsymbol{\Sigma}$. Since $\mathbf{X}_1, \mathbf{X}_2, \ldots, \mathbf{X}_n$ are mutually independent and each has distribution $N_p(\boldsymbol{\mu}, \boldsymbol{\Sigma})$, the joint density function of all the observations is the product of the marginal normal densities:

$$\left\{ \begin{array}{c} \text{Joint density} \\ \text{of } \mathbf{X}_1, \mathbf{X}_2, \ldots, \mathbf{X}_n \end{array} \right\} = \prod_{j=1}^{n} \left\{ \frac{1}{(2\pi)^{p/2} |\boldsymbol{\Sigma}|^{1/2}} e^{-(\mathbf{x}_j - \boldsymbol{\mu})' \boldsymbol{\Sigma}^{-1} (\mathbf{x}_j - \boldsymbol{\mu})/2} \right\}$$

$$= \frac{1}{(2\pi)^{np/2} |\boldsymbol{\Sigma}|^{n/2}} e^{-\sum\limits_{j=1}^{n} (\mathbf{x}_j - \boldsymbol{\mu})' \boldsymbol{\Sigma}^{-1} (\mathbf{x}_j - \boldsymbol{\mu})/2} \qquad (4\text{-}11)$$

When the numerical values of the observations become available, they may be substituted for the $\mathbf{x}_j$ in Equation (4-11). The resulting expression, now considered as a function of $\boldsymbol{\mu}$ and $\boldsymbol{\Sigma}$ for the fixed set of observations $\mathbf{x}_1, \mathbf{x}_2, \ldots, \mathbf{x}_n$, is called the *likelihood*.

Many good statistical procedures employ values for the population parameters that "best" explain the observed data. One meaning of *best* is to select the parameter values that *maximize* the joint density evaluated at the observations. This technique is called *maximum likelihood estimation*, and the maximizing parameter values are called *maximum likelihood estimates*.

At this point, we shall consider maximum likelihood estimation of the parameters $\boldsymbol{\mu}$ and $\boldsymbol{\Sigma}$ for a multivariate normal population. To do so, we take the observations $\mathbf{x}_1, \mathbf{x}_2, \ldots, \mathbf{x}_n$ as fixed and consider the joint density of Equation (4-11) evaluated at these values. The result is the likelihood function. In order to simplify matters, we rewrite the likelihood function in another form. We shall need some additional properties for the trace of a square matrix. (The trace of a matrix is the sum of its diagonal elements, and the properties of the trace are discussed in Definition 2A.28 and Result 2A.12.)

Result 4.9. Let $\mathbf{A}$ be a $k \times k$ symmetric matrix and $\mathbf{x}$ be a $k \times 1$ vector. Then

(a) $\mathbf{x}' \mathbf{A} \mathbf{x} = \text{tr}(\mathbf{x}' \mathbf{A} \mathbf{x}) = \text{tr}(\mathbf{A} \mathbf{x} \mathbf{x}')$

(b) $\text{tr}(\mathbf{A}) = \sum\limits_{i=1}^{k} \lambda_i$, where the λ_i are the eigenvalues of $\mathbf{A}$.

Proof. For Part a, we note that $\mathbf{x}' \mathbf{A} \mathbf{x}$ is a scalar, so $\mathbf{x}' \mathbf{A} \mathbf{x} = \text{tr}(\mathbf{x}' \mathbf{A} \mathbf{x})$. We pointed out in Result 2A.12 that $\text{tr}(\mathbf{B}\mathbf{C}) = \text{tr}(\mathbf{C}\mathbf{B})$ for any two matrices $\mathbf{B}$ and $\mathbf{C}$ of dimensions $m \times k$ and $k \times m$, respectively. This follows because $\mathbf{B}\mathbf{C}$ has $\sum\limits_{j=1}^{k} b_{ij} c_{ji}$ as its ith diagonal element, so $\mathrm{tr}\,(\mathbf{BC}) = \sum\limits_{i=1}^{m} \left(\sum\limits_{j=1}^{k} b_{ij}c_{ji} \right)$. Similarly, the jth diagonal

element of $\mathbf{CB}$ is $\sum\limits_{i=1}^{m} c_{ji}b_{ij}$, so $\mathrm{tr}\,(\mathbf{CB}) = \sum\limits_{j=1}^{k} \left(\sum\limits_{i=1}^{m} c_{ji}b_{ij} \right) = \sum\limits_{i=1}^{m} \left(\sum\limits_{j=1}^{k} b_{ij}c_{ji} \right) = \mathrm{tr}\,(\mathbf{BC})$.

Let $\mathbf{x}'$ be the matrix $\mathbf{B}$ with $m = 1$, and let $\mathbf{Ax}$ play the role of the matrix $\mathbf{C}$. Then $\mathrm{tr}\,(\mathbf{x}'(\mathbf{Ax})) = \mathrm{tr}\,((\mathbf{Ax})\mathbf{x}')$, and the result follows.

Part b is proved by using the spectral decomposition of (2-20) to write $\mathbf{A} = \mathbf{P}'\mathbf{\Lambda}\mathbf{P}$, where $\mathbf{PP}' = \mathbf{I}$ and $\mathbf{\Lambda}$ is a diagonal matrix with entries $\lambda_1, \lambda_2, \ldots, \lambda_k$. Therefore, $\mathrm{tr}\,(\mathbf{A}) = \mathrm{tr}\,(\mathbf{P}'\mathbf{\Lambda}\mathbf{P}) = \mathrm{tr}\,(\mathbf{\Lambda}\mathbf{PP}') = \mathrm{tr}\,(\mathbf{\Lambda}) = \lambda_1 + \lambda_2 + \cdots + \lambda_k$.   ∎

Now the exponent in the joint density in (4–11) can be simplified. By Result 4.9(a),

$$(\mathbf{x}_j - \boldsymbol{\mu})'\boldsymbol{\Sigma}^{-1}(\mathbf{x}_j - \boldsymbol{\mu}) = \mathrm{tr}\,[(\mathbf{x}_j - \boldsymbol{\mu})'\boldsymbol{\Sigma}^{-1}(\mathbf{x}_j - \boldsymbol{\mu})]$$

$$= \mathrm{tr}\,[\boldsymbol{\Sigma}^{-1}(\mathbf{x}_j - \boldsymbol{\mu})(\mathbf{x}_j - \boldsymbol{\mu})'] \qquad (4\text{-}12)$$

Next,

$$\sum_{j=1}^{n} (\mathbf{x}_j - \boldsymbol{\mu})'\boldsymbol{\Sigma}^{-1}(\mathbf{x}_j - \boldsymbol{\mu}) = \sum_{j=1}^{n} \mathrm{tr}\,[(\mathbf{x}_j - \boldsymbol{\mu})'\boldsymbol{\Sigma}^{-1}(\mathbf{x}_j - \boldsymbol{\mu})]$$

$$= \sum_{j=1}^{n} \mathrm{tr}\,[\boldsymbol{\Sigma}^{-1}(\mathbf{x}_j - \boldsymbol{\mu})(\mathbf{x}_j - \boldsymbol{\mu})']$$

$$= \mathrm{tr}\,\left[\boldsymbol{\Sigma}^{-1}\left(\sum_{j=1}^{n} (\mathbf{x}_j - \boldsymbol{\mu})(\mathbf{x}_j - \boldsymbol{\mu})'\right)\right] \qquad (4\text{-}13)$$

since the trace of a sum of matrices is equal to the sum of the traces of the matrices, according to Result 2A.12(b). We can add and subtract $\bar{\mathbf{x}} = (1/n)\sum\limits_{j=1}^{n} \mathbf{x}_j$ in each term $(\mathbf{x}_j - \boldsymbol{\mu})$ in $\sum\limits_{j=1}^{n} (\mathbf{x}_j - \boldsymbol{\mu})(\mathbf{x}_j - \boldsymbol{\mu})'$ to give

$$\sum_{j=1}^{n} (\mathbf{x}_j - \bar{\mathbf{x}} + \bar{\mathbf{x}} - \boldsymbol{\mu})(\mathbf{x}_j - \bar{\mathbf{x}} + \bar{\mathbf{x}} - \boldsymbol{\mu})'$$

$$= \sum_{j=1}^{n} (\mathbf{x}_j - \bar{\mathbf{x}})(\mathbf{x}_j - \bar{\mathbf{x}})' + \sum_{j=1}^{n} (\bar{\mathbf{x}} - \boldsymbol{\mu})(\bar{\mathbf{x}} - \boldsymbol{\mu})'$$

$$= \sum_{j=1}^{n} (\mathbf{x}_j - \bar{\mathbf{x}})(\mathbf{x}_j - \bar{\mathbf{x}})' + n(\bar{\mathbf{x}} - \boldsymbol{\mu})(\bar{\mathbf{x}} - \boldsymbol{\mu})' \qquad (4\text{-}14)$$

because the cross-product terms, $\sum\limits_{j=1}^{n} (\mathbf{x}_j - \bar{\mathbf{x}})(\bar{\mathbf{x}} - \boldsymbol{\mu})'$ and $\sum\limits_{j=1}^{n} (\bar{\mathbf{x}} - \boldsymbol{\mu})(\mathbf{x}_j - \bar{\mathbf{x}})'$, are both matrices of zeros. (See Exercise 4.15.) Consequently, using Equations (4-13) and (4-14), we can write the joint density of a random sample from a multivariate normal population as

$$\left\{ \begin{array}{l} \text{Joint density of} \\ \mathbf{X}_1, \mathbf{X}_2, \ldots, \mathbf{X}_n \end{array} \right\} = (2\pi)^{-np/2} |\boldsymbol{\Sigma}|^{-n/2}$$

$$\times \exp\left\{ -\mathrm{tr}\left[\boldsymbol{\Sigma}^{-1}\left(\sum_{j=1}^{n} (\mathbf{x}_j - \bar{\mathbf{x}})(\mathbf{x}_j - \bar{\mathbf{x}})' + n(\bar{\mathbf{x}} - \boldsymbol{\mu})(\bar{\mathbf{x}} - \boldsymbol{\mu})' \right) \right] \Big/ 2 \right\} \qquad (4\text{-}15)$$

Substituting the observed values $\mathbf{x}_1, \mathbf{x}_2, \ldots, \mathbf{x}_n$ into the joint density yields the likelihood function. We shall denote this function by $L(\boldsymbol{\mu}, \boldsymbol{\Sigma})$, to stress the fact that it is a function of the (unknown) population parameters $\boldsymbol{\mu}$ and $\boldsymbol{\Sigma}$. Thus, when the vectors $\mathbf{x}_j$ contain the specific numbers actually observed, we have

$$L(\boldsymbol{\mu}, \boldsymbol{\Sigma}) = \frac{1}{(2\pi)^{np/2}|\boldsymbol{\Sigma}|^{n/2}} e^{-\mathrm{tr}\left[\boldsymbol{\Sigma}^{-1}\left(\sum\limits_{j=1}^{n}(\mathbf{x}_j-\bar{\mathbf{x}})(\mathbf{x}_j-\bar{\mathbf{x}})'+n(\bar{\mathbf{x}}-\boldsymbol{\mu})(\bar{\mathbf{x}}-\boldsymbol{\mu})'\right)\right]/2} \tag{4-16}$$

It will be convenient in later sections of this book to express the exponent in the likelihood function (4-16) in different ways. In particular, we shall make use of the identity

$$\mathrm{tr}\left[\boldsymbol{\Sigma}^{-1}\left(\sum_{j=1}^{n}(\mathbf{x}_j - \bar{\mathbf{x}})(\mathbf{x}_j - \bar{\mathbf{x}})' + n(\bar{\mathbf{x}} - \boldsymbol{\mu})(\bar{\mathbf{x}} - \boldsymbol{\mu})'\right)\right]$$

$$= \mathrm{tr}\left[\boldsymbol{\Sigma}^{-1}\left(\sum_{j=1}^{n}(\mathbf{x}_j - \bar{\mathbf{x}})(\mathbf{x}_j - \bar{\mathbf{x}})'\right)\right] + n\,\mathrm{tr}[\boldsymbol{\Sigma}^{-1}(\bar{\mathbf{x}} - \boldsymbol{\mu})(\bar{\mathbf{x}} - \boldsymbol{\mu})']$$

$$= \mathrm{tr}\left[\boldsymbol{\Sigma}^{-1}\left(\sum_{j=1}^{n}(\mathbf{x}_j - \bar{\mathbf{x}})(\mathbf{x}_j - \bar{\mathbf{x}})'\right)\right] + n(\bar{\mathbf{x}} - \boldsymbol{\mu})'\boldsymbol{\Sigma}^{-1}(\bar{\mathbf{x}} - \boldsymbol{\mu}) \tag{4-17}$$

Maximum Likelihood Estimation of μ and Σ

The next result will eventually allow us to obtain the maximum likelihood estimators of $\boldsymbol{\mu}$ and $\boldsymbol{\Sigma}$.

Result 4.10. Given a $p \times p$ symmetric positive definite matrix $\mathbf{B}$ and a scalar $b > 0$, it follows that

$$\frac{1}{|\boldsymbol{\Sigma}|^b} e^{-\mathrm{tr}(\boldsymbol{\Sigma}^{-1}\mathbf{B})/2} \leq \frac{1}{|\mathbf{B}|^b}(2b)^{pb}e^{-bp}$$

for all positive definite $\underset{(p\times p)}{\boldsymbol{\Sigma}}$, with equality holding only for $\boldsymbol{\Sigma} = (1/2b)\mathbf{B}$.

Proof. Let $\mathbf{B}^{1/2}$ be the symmetric square root of $\mathbf{B}$ [see Equation (2-22)], so $\mathbf{B}^{1/2}\mathbf{B}^{1/2} = \mathbf{B}$, $\mathbf{B}^{1/2}\mathbf{B}^{-1/2} = \mathbf{I}$, and $\mathbf{B}^{-1/2}\mathbf{B}^{-1/2} = \mathbf{B}^{-1}$. Then $\mathrm{tr}(\boldsymbol{\Sigma}^{-1}\mathbf{B}) = \mathrm{tr}[(\boldsymbol{\Sigma}^{-1}\mathbf{B}^{1/2})\mathbf{B}^{1/2}] = \mathrm{tr}[\mathbf{B}^{1/2}(\boldsymbol{\Sigma}^{-1}\mathbf{B}^{1/2})]$. Let η be an eigenvalue of $\mathbf{B}^{1/2}\boldsymbol{\Sigma}^{-1}\mathbf{B}^{1/2}$. This matrix is positive definite because $\mathbf{y}'\mathbf{B}^{1/2}\boldsymbol{\Sigma}^{-1}\mathbf{B}^{1/2}\mathbf{y} = (\mathbf{B}^{1/2}\mathbf{y})'\boldsymbol{\Sigma}^{-1}(\mathbf{B}^{1/2}\mathbf{y}) > 0$ if $\mathbf{B}^{1/2}\mathbf{y} \neq \mathbf{0}$ or, equivalently, $\mathbf{y} \neq \mathbf{0}$. Thus, the eigenvalues η_i of $\mathbf{B}^{1/2}\boldsymbol{\Sigma}^{-1}\mathbf{B}^{1/2}$ are positive by Exercise 2.17. Result 4.9(b) then gives

$$\mathrm{tr}(\boldsymbol{\Sigma}^{-1}\mathbf{B}) = \mathrm{tr}(\mathbf{B}^{1/2}\boldsymbol{\Sigma}^{-1}\mathbf{B}^{1/2}) = \sum_{i=1}^{p}\eta_i$$

and $|\mathbf{B}^{1/2}\boldsymbol{\Sigma}^{-1}\mathbf{B}^{1/2}| = \prod\limits_{i=1}^{p}\eta_i$ by Exercise 2.12. From the properties of determinants in Result 2A.11, we can write

$$|\mathbf{B}^{1/2}\boldsymbol{\Sigma}^{-1}\mathbf{B}^{1/2}| = |\mathbf{B}^{1/2}||\boldsymbol{\Sigma}^{-1}||\mathbf{B}^{1/2}| = |\boldsymbol{\Sigma}^{-1}||\mathbf{B}^{1/2}||\mathbf{B}^{1/2}|$$

$$= |\boldsymbol{\Sigma}^{-1}||\mathbf{B}| = \frac{1}{|\boldsymbol{\Sigma}|}|\mathbf{B}|$$

or

$$\frac{1}{|\mathbf{\Sigma}|} = \frac{|\mathbf{B}^{1/2}\mathbf{\Sigma}^{-1}\mathbf{B}^{1/2}|}{|\mathbf{B}|} = \frac{\prod\limits_{i=1}^{p} \eta_i}{|\mathbf{B}|}$$

Combining the results for the trace and the determinant yields

$$\frac{1}{|\mathbf{\Sigma}|^b} e^{-\mathrm{tr}[\mathbf{\Sigma}^{-1}\mathbf{B}]/2} = \frac{\left(\prod\limits_{i=1}^{p} \eta_i\right)^b}{|\mathbf{B}|^b} e^{-\sum\limits_{i=1}^{p} \eta_i/2} = \frac{1}{|\mathbf{B}|^b} \prod_{i=1}^{p} \eta_i^b e^{-\eta_i/2}$$

But the function $\eta^b e^{-\eta/2}$ has a maximum, with respect to η, of $(2b)^b e^{-b}$, occurring at $\eta = 2b$. The choice $\eta_i = 2b$, for each i, therefore gives

$$\frac{1}{|\mathbf{\Sigma}|^b} e^{-\mathrm{tr}(\mathbf{\Sigma}^{-1}\mathbf{B})/2} \le \frac{1}{|\mathbf{B}|^b} (2b)^{pb} e^{-bp}$$

The upper bound is uniquely attained when $\mathbf{\Sigma} = (1/2b)\mathbf{B}$, since, for this choice,

$$\mathbf{B}^{1/2}\mathbf{\Sigma}^{-1}\mathbf{B}^{1/2} = \mathbf{B}^{1/2}(2b)\mathbf{B}^{-1}\mathbf{B}^{1/2} = (2b) \underset{(p\times p)}{\mathbf{I}}$$

and

$$\mathrm{tr}[\mathbf{\Sigma}^{-1}\mathbf{B}] = \mathrm{tr}[\mathbf{B}^{1/2}\mathbf{\Sigma}^{-1}\mathbf{B}^{1/2}] = \mathrm{tr}[(2b)\mathbf{I}] = 2bp$$

Moreover,

$$\frac{1}{|\mathbf{\Sigma}|} = \frac{|\mathbf{B}^{1/2}\mathbf{\Sigma}^{-1}\mathbf{B}^{1/2}|}{|\mathbf{B}|} = \frac{|(2b)\mathbf{I}|}{|\mathbf{B}|} = \frac{(2b)^p}{|\mathbf{B}|}$$

Straightforward substitution for $\mathrm{tr}[\mathbf{\Sigma}^{-1}\mathbf{B}]$ and $1/|\mathbf{\Sigma}|^b$ yields the bound asserted. ∎

The maximum likelihood estimates of $\boldsymbol{\mu}$ and $\mathbf{\Sigma}$ are those values—denoted by $\hat{\boldsymbol{\mu}}$ and $\hat{\mathbf{\Sigma}}$—that maximize the function $L(\boldsymbol{\mu}, \mathbf{\Sigma})$ in (4-16). The estimates $\hat{\boldsymbol{\mu}}$ and $\hat{\mathbf{\Sigma}}$ will depend on the observed values $\mathbf{x}_1, \mathbf{x}_2, \dots, \mathbf{x}_n$ through the summary statistics $\bar{\mathbf{x}}$ and $\mathbf{S}$.

Result 4.11. Let $\mathbf{X}_1, \mathbf{X}_2, \dots, \mathbf{X}_n$ be a random sample from a normal population with mean $\boldsymbol{\mu}$ and covariance $\mathbf{\Sigma}$. Then

$$\hat{\boldsymbol{\mu}} = \overline{\mathbf{X}} \qquad \text{and} \qquad \hat{\mathbf{\Sigma}} = \frac{1}{n}\sum_{j=1}^{n}(\mathbf{X}_j - \overline{\mathbf{X}})(\mathbf{X}_j - \overline{\mathbf{X}})' = \frac{(n-1)}{n}\mathbf{S}$$

are the *maximum likelihood estimators* of $\boldsymbol{\mu}$ and $\mathbf{\Sigma}$, respectively. Their observed values, $\bar{\mathbf{x}}$ and $(1/n)\sum\limits_{j=1}^{n}(\mathbf{x}_j - \bar{\mathbf{x}})(\mathbf{x}_j - \bar{\mathbf{x}})'$, are called the *maximum likelihood estimates* of $\boldsymbol{\mu}$ and $\mathbf{\Sigma}$.

Proof. The exponent in the likelihood function [see Equation (4-16)], apart from the multiplicative factor $-\frac{1}{2}$, is [see (4-17)]

$$\mathrm{tr}\left[\mathbf{\Sigma}^{-1}\left(\sum_{j=1}^{n}(\mathbf{x}_j - \bar{\mathbf{x}})(\mathbf{x}_j - \bar{\mathbf{x}})'\right)\right] + n(\bar{\mathbf{x}} - \boldsymbol{\mu})'\mathbf{\Sigma}^{-1}(\bar{\mathbf{x}} - \boldsymbol{\mu})$$

By Result 4.1, $\boldsymbol{\Sigma}^{-1}$ is positive definite, so the distance $(\bar{\mathbf{x}} - \boldsymbol{\mu})'\boldsymbol{\Sigma}^{-1}(\bar{\mathbf{x}} - \boldsymbol{\mu}) > 0$ unless $\boldsymbol{\mu} = \bar{\mathbf{x}}$. Thus, the likelihood is maximized with respect to $\boldsymbol{\mu}$ at $\hat{\boldsymbol{\mu}} = \bar{\mathbf{x}}$. It remains to maximize

$$L(\boldsymbol{\mu}, \boldsymbol{\Sigma}) = \frac{1}{(2\pi)^{np/2}|\boldsymbol{\Sigma}|^{n/2}} e^{-\operatorname{tr}\left[\boldsymbol{\Sigma}^{-1}\left(\sum_{j=1}^{n}(\mathbf{x}_j - \bar{\mathbf{x}})(\mathbf{x}_j - \bar{\mathbf{x}})'\right)\right]/2}$$

over $\boldsymbol{\Sigma}$. By Result 4.10 with $b = n/2$ and $\mathbf{B} = \sum_{j=1}^{n}(\mathbf{x}_j - \bar{\mathbf{x}})(\mathbf{x}_j - \bar{\mathbf{x}})'$, the maximum occurs at $\hat{\boldsymbol{\Sigma}} = (1/n)\sum_{j=1}^{n}(\mathbf{x}_j - \bar{\mathbf{x}})(\mathbf{x}_j - \bar{\mathbf{x}})'$, as stated.

The maximum likelihood estimators are random quantities. They are obtained by replacing the observations $\mathbf{x}_1, \mathbf{x}_2, \ldots, \mathbf{x}_n$ in the expressions for $\hat{\boldsymbol{\mu}}$ and $\hat{\boldsymbol{\Sigma}}$ with the corresponding random vectors, $\mathbf{X}_1, \mathbf{X}_2, \ldots, \mathbf{X}_n$.   ∎

We note that the maximum likelihood estimator $\overline{\mathbf{X}}$ is a random vector and the maximum likelihood estimator $\hat{\boldsymbol{\Sigma}}$ is a random matrix. The maximum likelihood estimates are their particular values for the given data set. In addition, the maximum of the likelihood is

$$L(\hat{\boldsymbol{\mu}}, \hat{\boldsymbol{\Sigma}}) = \frac{1}{(2\pi)^{np/2}} e^{-np/2} \frac{1}{|\hat{\boldsymbol{\Sigma}}|^{n/2}} \tag{4-18}$$

or, since $|\hat{\boldsymbol{\Sigma}}| = [(n-1)/n]^p |\mathbf{S}|$,

$$L(\hat{\boldsymbol{\mu}}, \hat{\boldsymbol{\Sigma}}) = \text{constant} \times (\text{generalized variance})^{-n/2} \tag{4-19}$$

The generalized variance determines the "peakedness" of the likelihood function and, consequently, is a natural measure of variability when the parent population is multivariate normal.

Maximum likelihood estimators possess an *invariance property*. Let $\hat{\boldsymbol{\theta}}$ be the maximum likelihood estimator of $\boldsymbol{\theta}$, and consider estimating the parameter $h(\boldsymbol{\theta})$, which is a function of $\boldsymbol{\theta}$. Then the *maximum likelihood estimate* of

$$\underset{\text{(a function of } \boldsymbol{\theta})}{h(\boldsymbol{\theta})} \qquad \text{is given by} \qquad \underset{\text{(same function of } \hat{\boldsymbol{\theta}})}{h(\hat{\boldsymbol{\theta}})} \tag{4-20}$$

(See [1] and [15].) For example,

1. The maximum likelihood estimator of $\boldsymbol{\mu}'\boldsymbol{\Sigma}^{-1}\boldsymbol{\mu}$ is $\hat{\boldsymbol{\mu}}'\hat{\boldsymbol{\Sigma}}^{-1}\hat{\boldsymbol{\mu}}$, where $\hat{\boldsymbol{\mu}} = \overline{\mathbf{X}}$ and $\hat{\boldsymbol{\Sigma}} = ((n-1)/n)\mathbf{S}$ are the maximum likelihood estimators of $\boldsymbol{\mu}$ and $\boldsymbol{\Sigma}$, respectively.

2. The maximum likelihood estimator of $\sqrt{\sigma_{ii}}$ is $\sqrt{\hat{\sigma}_{ii}}$, where

$$\hat{\sigma}_{ii} = \frac{1}{n}\sum_{j=1}^{n}(X_{ij} - \overline{X}_i)^2$$

is the maximum likelihood estimator of $\sigma_{ii} = \text{Var}(X_i)$.

Sufficient Statistics

From expression (4-15), the joint density depends on the whole set of observations $\mathbf{x}_1, \mathbf{x}_2, \ldots, \mathbf{x}_n$ only through the sample mean $\overline{\mathbf{x}}$ and the sum-of-squares-and-cross-products matrix $\sum_{j=1}^{n} (\mathbf{x}_j - \overline{\mathbf{x}})(\mathbf{x}_j - \overline{\mathbf{x}})' = (n - 1)\mathbf{S}$. We express this fact by saying that $\overline{\mathbf{x}}$ and $(n - 1)\mathbf{S}$ (or $\mathbf{S}$) are *sufficient statistics*:

> Let $\mathbf{X}_1, \mathbf{X}_2, \ldots, \mathbf{X}_n$ be a random sample from a multivariate normal population with mean $\boldsymbol{\mu}$ and covariance $\boldsymbol{\Sigma}$. Then
>
> $$\overline{\mathbf{X}} \text{ and } \mathbf{S} \text{ are } sufficient\ statistics \qquad (4\text{-}21)$$

The importance of sufficient statistics for normal populations is that all of the information about $\boldsymbol{\mu}$ and $\boldsymbol{\Sigma}$ in the data matrix $\mathbf{X}$ is contained in $\overline{\mathbf{x}}$ and $\mathbf{S}$, regardless of the sample size n. This generally is not true for nonnormal populations. Since many multivariate techniques begin with sample means and covariances, it is prudent to check on the *adequacy* of the multivariate normal assumption. (See Section 4.6.) If the data cannot be regarded as multivariate normal, techniques that depend solely on $\overline{\mathbf{x}}$ and $\mathbf{S}$ may be ignoring other useful sample information.

4.4 The Sampling Distribution of $\overline{X}$ and S

The tentative assumption that $\mathbf{X}_1, \mathbf{X}_2, \ldots, \mathbf{X}_n$ constitute a random sample from a normal population with mean $\boldsymbol{\mu}$ and covariance $\boldsymbol{\Sigma}$ completely determines the sampling distributions of $\overline{\mathbf{X}}$ and $\mathbf{S}$. Here we present the results on the sampling distributions of $\overline{\mathbf{X}}$ and $\mathbf{S}$ by drawing a parallel with the familiar univariate conclusions.

In the univariate case ($p = 1$), we know that $\overline{X}$ is normal with mean $\mu =$ (population mean) and variance

$$\frac{1}{n}\sigma^2 = \frac{\text{population variance}}{\text{sample size}}$$

The result for the multivariate case ($p \geq 2$) is analogous in that $\overline{\mathbf{X}}$ has a normal distribution with mean $\boldsymbol{\mu}$ and covariance matrix $(1/n)\boldsymbol{\Sigma}$.

For the sample variance, recall that $(n - 1)s^2 = \sum_{j=1}^{n} (X_j - \overline{X})^2$ is distributed as σ^2 times a chi-square variable having $n - 1$ degrees of freedom (d.f.). In turn, this chi-square is the distribution of a sum of squares of independent standard normal random variables. That is, $(n - 1)s^2$ is distributed as $\sigma^2(Z_1^2 + \cdots + Z_{n-1}^2) = (\sigma Z_1)^2 + \cdots + (\sigma Z_{n-1})^2$. The individual terms σZ_i are independently distributed as $N(0, \sigma^2)$. It is this latter form that is suitably generalized to the basic sampling distribution for the sample covariance matrix.

The sampling distribution of the sample covariance matrix is called the *Wishart distribution*, after its discoverer; it is defined as the sum of independent products of multivariate normal random vectors. Specifically,

$$W_m(\cdot \mid \Sigma) = \text{Wishart distribution with } m \text{ d.f.} \qquad (4\text{-}22)$$

$$= \text{distribution of } \sum_{j=1}^{m} \mathbf{Z}_j \mathbf{Z}_j'$$

where the $\mathbf{Z}_j$ are each independently distributed as $N_p(\mathbf{0}, \Sigma)$.

We summarize the sampling distribution results as follows:

Let $\mathbf{X}_1, \mathbf{X}_2, \dots, \mathbf{X}_n$ be a random sample of size n from a p-variate *normal* distribution with mean $\boldsymbol{\mu}$ and covariance matrix Σ. Then

1. $\overline{\mathbf{X}}$ is distributed as $N_p(\boldsymbol{\mu}, (1/n)\Sigma)$.
2. $(n-1)\mathbf{S}$ is distributed as a Wishart random matrix with $n-1$ d.f.   (4-23)
3. $\overline{\mathbf{X}}$ and $\mathbf{S}$ are independent.

Because Σ is unknown, the distribution of $\overline{\mathbf{X}}$ cannot be used directly to make inferences about $\boldsymbol{\mu}$. However, $\mathbf{S}$ provides independent information about Σ, and the distribution of $\mathbf{S}$ does not depend on $\boldsymbol{\mu}$. This allows us to construct a statistic for making inferences about $\boldsymbol{\mu}$, as we shall see in Chapter 5.

For the present, we record some further results from multivariable distribution theory. The following properties of the Wishart distribution are derived directly from its definition as a sum of the independent products, $\mathbf{Z}_j \mathbf{Z}_j'$. Proofs can be found in [1].

Properties of the Wishart Distribution

1. If $\mathbf{A}_1$ is distributed as $W_{m_1}(\mathbf{A}_1 \mid \Sigma)$ independently of $\mathbf{A}_2$, which is distributed as $W_{m_2}(\mathbf{A}_2 \mid \Sigma)$, then $\mathbf{A}_1 + \mathbf{A}_2$ is distributed as $W_{m_1+m_2}(\mathbf{A}_1 + \mathbf{A}_2 \mid \Sigma)$. That is, the degrees of freedom add.   (4-24)
2. If $\mathbf{A}$ is distributed as $W_m(\mathbf{A} \mid \Sigma)$, then $\mathbf{CAC}'$ is distributed as $W_m(\mathbf{CAC}' \mid \mathbf{C\Sigma C}')$.

Although we do not have any particular need for the probability density function of the Wishart distribution, it may be of some interest to see its rather complicated form. The density does not exist unless the sample size n is greater than the number of variables p. When it does exist, its value at the positive definite matrix $\mathbf{A}$ is

$$w_{n-1}(\mathbf{A} \mid \Sigma) = \frac{|\mathbf{A}|^{(n-p-2)/2} e^{-\text{tr}[\mathbf{A}\Sigma^{-1}]/2}}{2^{p(n-1)/2} \pi^{p(p-1)/4} |\Sigma|^{(n-1)/2} \prod_{i=1}^{p} \Gamma\left(\frac{1}{2}(n-i)\right)}, \qquad \mathbf{A} \text{ positive definite}$$

$$(4\text{-}25)$$

where $\Gamma(\cdot)$ is the gamma function. (See [1] and [11].)

4.5 Large-Sample Behavior of $\overline{X}$ and S

Suppose the quantity X is determined by a large number of independent causes $V_1, V_2, \ldots, V_n$, where the random variables V_i representing the causes have approximately the same variability. If X is the sum

$$X = V_1 + V_2 + \cdots + V_n$$

then the central limit theorem applies, and we conclude that X has a distribution that is nearly normal. This is true for virtually any parent distribution of the V_i's, provided that n is large enough.

The univariate central limit theorem also tells us that the sampling distribution of the sample mean, $\overline{X}$ for a large sample size is nearly normal, whatever the form of the underlying population distribution. A similar result holds for many other important univariate statistics.

It turns out that certain multivariate statistics, like $\overline{\mathbf{X}}$ and S, have large-sample properties analogous to their univariate counterparts. As the sample size is increased without bound, certain regularities govern the sampling variation in $\overline{\mathbf{X}}$ and S, irrespective of the form of the parent population. Therefore, the conclusions presented in this section do not require multivariate normal populations. The only requirements are that the parent population, whatever its form, have a mean $\boldsymbol{\mu}$ and a finite covariance $\boldsymbol{\Sigma}$.

Result 4.12 (Law of large numbers). Let $Y_1, Y_2, \ldots, Y_n$ be independent observations from a population with mean $E(Y_i) = \mu$. Then

$$\overline{Y} = \frac{Y_1 + Y_2 + \cdots + Y_n}{n}$$

converges in probability to μ as n increases without bound. That is, for any prescribed accuracy $\varepsilon > 0$, $P[-\varepsilon < \overline{Y} - \mu < \varepsilon]$ approaches unity as $n \to \infty$.

Proof. See [9]. ∎

As a direct consequence of the law of large numbers, which says that each $\overline{X}_i$ converges in probability to $\mu_i, i = 1, 2, \ldots, p,$

$$\overline{\mathbf{X}} \text{ converges in probability to } \boldsymbol{\mu} \tag{4-26}$$

Also, each sample covariance s_{ik} converges in probability to $\sigma_{ik}, i, k = 1, 2, \ldots, p,$ and

$$\mathbf{S} \text{ (or } \hat{\boldsymbol{\Sigma}} = \mathbf{S}_n) \text{ converges in probability to } \boldsymbol{\Sigma} \tag{4-27}$$

Statement (4-27) follows from writing

$$(n-1)s_{ik} = \sum_{j=1}^{n} (X_{ji} - \overline{X}_i)(X_{jk} - \overline{X}_k)$$

$$= \sum_{j=1}^{n} (X_{ji} - \mu_i + \mu_i - \overline{X}_i)(X_{jk} - \mu_k + \mu_k - \overline{X}_k)$$

$$= \sum_{j=1}^{n} (X_{ji} - \mu_i)(X_{jk} - \mu_k) + n(\overline{X}_i - \mu_i)(\overline{X}_k - \mu_k)$$

Letting $Y_j = (X_{ji} - \mu_i)(X_{jk} - \mu_k)$, with $E(Y_j) = \sigma_{ik}$, we see that the first term in s_{ik} converges to σ_{ik} and the second term converges to zero, by applying the law of large numbers.

The practical interpretation of statements (4-26) and (4-27) is that, with high probability, $\overline{\mathbf{X}}$ will be close to $\boldsymbol{\mu}$ and $\mathbf{S}$ will be close to $\boldsymbol{\Sigma}$ whenever the sample size is large. The statement concerning $\overline{\mathbf{X}}$ is made even more precise by a multivariate version of the central limit theorem.

Result 4.13 (The central limit theorem). Let $\mathbf{X}_1, \mathbf{X}_2, \ldots, \mathbf{X}_n$ be independent observations from any population with mean $\boldsymbol{\mu}$ and finite covariance $\boldsymbol{\Sigma}$. Then

$$\sqrt{n}\,(\overline{\mathbf{X}} - \boldsymbol{\mu}) \text{ has an approximate } N_p(\mathbf{0}, \boldsymbol{\Sigma}) \text{ distribution}$$

for large sample sizes. Here n should also be large relative to p.

Proof. See [1].   ■

The approximation provided by the central limit theorem applies to discrete, as well as continuous, multivariate populations. Mathematically, the limit is exact, and the approach to normality is often fairly rapid. Moreover, from the results in Section 4.4, we know that $\overline{\mathbf{X}}$ is exactly normally distributed when the underlying population is normal. Thus, we would expect the central limit theorem approximation to be quite good for moderate n when the parent population is nearly normal.

As we have seen, when n is large, $\mathbf{S}$ is close to $\boldsymbol{\Sigma}$ with high probability. Consequently, replacing $\boldsymbol{\Sigma}$ by $\mathbf{S}$ in the approximating normal distribution for $\overline{\mathbf{X}}$ will have a negligible effect on subsequent probability calculations.

Result 4.7 can be used to show that $n(\overline{\mathbf{X}} - \boldsymbol{\mu})'\boldsymbol{\Sigma}^{-1}(\overline{\mathbf{X}} - \boldsymbol{\mu})$ has a χ_p^2 distribution when $\overline{\mathbf{X}}$ is distributed as $N_p\!\left(\boldsymbol{\mu}, \dfrac{1}{n}\boldsymbol{\Sigma}\right)$ or, equivalently, when $\sqrt{n}\,(\overline{\mathbf{X}} - \boldsymbol{\mu})$ has an $N_p(\mathbf{0}, \boldsymbol{\Sigma})$ distribution. The χ_p^2 distribution is *approximately* the sampling distribution of $n(\overline{\mathbf{X}} - \boldsymbol{\mu})'\boldsymbol{\Sigma}^{-1}(\overline{\mathbf{X}} - \boldsymbol{\mu})$ when $\overline{\mathbf{X}}$ is approximately normally distributed. Replacing $\boldsymbol{\Sigma}^{-1}$ by $\mathbf{S}^{-1}$ does not seriously affect this approximation for n large and much greater than p.

We summarize the major conclusions of this section as follows:

Let $\mathbf{X}_1, \mathbf{X}_2, \ldots, \mathbf{X}_n$ be independent observations from a population with mean $\boldsymbol{\mu}$ and finite (nonsingular) covariance $\boldsymbol{\Sigma}$. Then

$$\sqrt{n}\,(\overline{\mathbf{X}} - \boldsymbol{\mu}) \text{ is approximately } N_p(\mathbf{0}, \boldsymbol{\Sigma})$$

and   (4-28)

$$n(\overline{\mathbf{X}} - \boldsymbol{\mu})'\mathbf{S}^{-1}(\overline{\mathbf{X}} - \boldsymbol{\mu}) \text{ is approximately } \chi_p^2$$

for $n - p$ large.

In the next three sections, we consider ways of verifying the assumption of normality and methods for transforming nonnormal observations into observations that are approximately normal.

4.6 Assessing the Assumption of Normality

As we have pointed out, most of the statistical techniques discussed in subsequent chapters assume that each vector observation $\mathbf{X}_j$ comes from a multivariate normal distribution. On the other hand, in situations where the sample size is large and the techniques depend solely on the behavior of $\overline{\mathbf{X}}$, or distances involving $\overline{\mathbf{X}}$ of the form $n(\overline{\mathbf{X}} - \boldsymbol{\mu})'\mathbf{S}^{-1}(\overline{\mathbf{X}} - \boldsymbol{\mu})$, the assumption of normality for the individual observations is less crucial. But to some degree, the *quality* of inferences made by these methods depends on how closely the true parent population resembles the multivariate normal form. It is imperative, then, that procedures exist for detecting cases where the data exhibit moderate to extreme departures from what is expected under multivariate normality.

We want to answer this question: Do the observations $\mathbf{X}_j$ appear to violate the assumption that they came from a normal population? Based on the properties of normal distributions, we know that all linear combinations of normal variables are normal and the contours of the multivariate normal density are ellipsoids. Therefore, we address these questions:

1. Do the marginal distributions of the elements of $\mathbf{X}$ appear to be normal? What about a few linear combinations of the components X_i?

2. Do the scatter plots of pairs of observations on different characteristics give the elliptical appearance expected from normal populations?

3. Are there any "wild" observations that should be checked for accuracy?

It will become clear that our investigations of normality will concentrate on the behavior of the observations in one or two dimensions (for example, marginal distributions and scatter plots). As might be expected, it has proved difficult to construct a "good" overall test of joint normality in more than two dimensions because of the large number of things that can go wrong. To some extent, we must pay a price for concentrating on univariate and bivariate examinations of normality: We can never be sure that we have not missed some feature that is revealed only in higher dimensions. (It is possible, for example, to construct a nonnormal bivariate distribution with normal marginals. [See Exercise 4.8.]) Yet many types of nonnormality are often reflected in the marginal distributions and scatter plots. Moreover, for most practical work, one-dimensional and two-dimensional investigations are ordinarily sufficient. Fortunately, pathological data sets that are normal in lower dimensional representations, but nonnormal in higher dimensions, are not frequently encountered in practice.

Evaluating the Normality of the Univariate Marginal Distributions

Dot diagrams for smaller n and histograms for $n > 25$ or so help reveal situations where one tail of a univariate distribution is much longer than the other. If the histogram for a variable X_i appears reasonably symmetric, we can check further by counting the number of observations in certain intervals. A univariate normal distribution assigns probability .683 to the interval $(\mu_i - \sqrt{\sigma_{ii}}, \mu_i + \sqrt{\sigma_{ii}})$ and probability .954 to the interval $(\mu_i - 2\sqrt{\sigma_{ii}}, \mu_i + 2\sqrt{\sigma_{ii}})$. Consequently, with a large sample size n, we expect the observed proportion $\hat{p}_{i1}$ of the observations lying in the interval $(\bar{x}_i - \sqrt{s_{ii}}, \bar{x}_i + \sqrt{s_{ii}})$ to be about .683. Similarly, the observed proportion $\hat{p}_{i2}$ of the observations in $(\bar{x}_i - 2\sqrt{s_{ii}}, \bar{x}_i + 2\sqrt{s_{ii}})$ should be about .954. Using the normal approximation to the sampling distribution of $\hat{p}_i$ (see [9]), we observe that either

$$| \hat{p}_{i1} - .683 | > 3 \sqrt{\frac{(.683)(.317)}{n}} = \frac{1.396}{\sqrt{n}}$$

or

$$| \hat{p}_{i2} - .954 | > 3 \sqrt{\frac{(.954)(.046)}{n}} = \frac{.628}{\sqrt{n}} \qquad (4\text{-}29)$$

would indicate departures from an assumed normal distribution for the ith characteristic. When the observed proportions are too small, parent distributions with thicker tails than the normal are suggested.

Plots are always useful devices in any data analysis. Special plots called *Q–Q plots* can be used to assess the assumption of normality. These plots can be made for the marginal distributions of the sample observations on each variable. They are, in effect, plots of the sample quantile versus the quantile one would expect to observe if the observations actually were normally distributed. When the points lie very nearly along a straight line, the normality assumption remains tenable. Normality is suspect if the points deviate from a straight line. Moreover, the pattern of the deviations can provide clues about the nature of the nonnormality. Once the reasons for the nonnormality are identified, corrective action is often possible. (See Section 4.8.)

To simplify notation, let $x_1, x_2, \ldots, x_n$ represent n observations on any single characteristic X_i. Let $x_{(1)} \leq x_{(2)} \leq \cdots \leq x_{(n)}$ represent these observations after they are ordered according to magnitude. For example, $x_{(2)}$ is the second smallest observation and $x_{(n)}$ is the largest observation. The $x_{(j)}$'s are the sample quantiles. When the $x_{(j)}$ are distinct, exactly j observations are less than or equal to $x_{(j)}$. (This is theoretically always true when the observations are of the continuous type, which we usually assume.) The proportion j/n of the sample at or to the left of $x_{(j)}$ is often approximated by $\left(j - \tfrac{1}{2}\right)/n$ for analytical convenience.[1]

For a standard normal distribution, the quantiles $q_{(j)}$ are defined by the relation

$$P[Z \leq q_{(j)}] = \int_{-\infty}^{q(j)} \frac{1}{\sqrt{2\pi}} e^{-z^2/2} \, dz = p_{(j)} = \frac{j - \tfrac{1}{2}}{n} \qquad (4\text{-}30)$$

(See Table 1 in the appendix). Here $p_{(j)}$ is the probability of getting a value less than or equal to $q_{(j)}$ in a single drawing from a standard normal population.

The idea is to look at the pairs of quantiles $(q_{(j)}, x_{(j)})$ with the same associated cumulative probability $\left(j - \tfrac{1}{2}\right)/n$. If the data arise from a normal population, the pairs $(q_{(j)}, x_{(j)})$ will be approximately linearly related, since $\sigma q_{(j)} + \mu$ is nearly the expected sample quantile.[2]

---

[1]The $\tfrac{1}{2}$ in the numerator of $\left(j - \tfrac{1}{2}\right)/n$ is a "continuity" correction. Some authors (see [5] and [10]) have suggested replacing $\left(j - \tfrac{1}{2}\right)/n$ by $\left(j - \tfrac{3}{8}\right)/\left(n + \tfrac{1}{4}\right)$.

[2]A better procedure is to plot $(m_{(j)}, x_{(j)})$, where $m_{(j)} = E(z_{(j)})$ is the expected value of the jth-order statistic in a sample of size n from a standard normal distribution. (See [13] for further discussion.)

Example 4.9 (Constructing a Q–Q plot) A sample of $n = 10$ observations gives the values in the following table:

Ordered observations $x_{(j)}$	Probability levels $(j - \frac{1}{2})/n$	Standard normal quantiles $q_{(j)}$
−1.00	.05	−1.645
−.10	.15	−1.036
.16	.25	−.674
.41	.35	−.385
.62	.45	−.125
.80	.55	.125
1.26	.65	.385
1.54	.75	.674
1.71	.85	1.036
2.30	.95	1.645

Here, for example, $P[Z \le .385] = \int_{-\infty}^{.385} \frac{1}{\sqrt{2\pi}} e^{-z^2/2} \, dz = .65.$ [See (4-30).]

Let us now construct the Q–Q plot and comment on its appearance. The Q–Q plot for the foregoing data, which is a plot of the ordered data $x_{(j)}$ against the normal quantiles $q_{(j)}$, is shown in Figure 4.5. The pairs of points $(q_{(j)}, x_{(j)})$ lie very nearly along a straight line, and we would not reject the notion that these data are normally distributed—particularly with a sample size as small as $n = 10$.

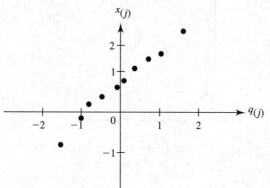

Figure 4.5 A Q–Q plot for the data in Example 4.9.   ∎

The calculations required for Q–Q plots are easily programmed for electronic computers. Many statistical programs available commercially are capable of producing such plots.

The steps leading to a Q–Q plot are as follows:

1. Order the original observations to get $x_{(1)}, x_{(2)}, \dots, x_{(n)}$ and their corresponding probability values $(1 - \frac{1}{2})/n, (2 - \frac{1}{2})/n, \dots, (n - \frac{1}{2})/n$;
2. Calculate the standard normal quantiles $q_{(1)}, q_{(2)}, \dots, q_{(n)}$; and
3. Plot the pairs of observations $(q_{(1)}, x_{(1)}), (q_{(2)}, x_{(2)}), \dots, (q_{(n)}, x_{(n)})$, and examine the "straightness" of the outcome.

Q–Q plots are not particularly informative unless the sample size is moderate to large—for instance, $n \geq 20$. There can be quite a bit of variability in the straightness of the Q–Q plot for small samples, even when the observations are known to come from a normal population.

Example 4.10 (A Q–Q plot for radiation data) The quality-control department of a manufacturer of microwave ovens is required by the federal government to monitor the amount of radiation emitted when the doors of the ovens are closed. Observations of the radiation emitted through closed doors of $n = 42$ randomly selected ovens were made. The data are listed in Table 4.1.

Table 4.1 Radiation Data (Door Closed)

Oven no.	Radiation	Oven no.	Radiation	Oven no.	Radiation
1	.15	16	.10	31	.10
2	.09	17	.02	32	.20
3	.18	18	.10	33	.11
4	.10	19	.01	34	.30
5	.05	20	.40	35	.02
6	.12	21	.10	36	.20
7	.08	22	.05	37	.20
8	.05	23	.03	38	.30
9	.08	24	.05	39	.30
10	.10	25	.15	40	.40
11	.07	26	.10	41	.30
12	.02	27	.15	42	.05
13	.01	28	.09		
14	.10	29	.08		
15	.10	30	.18		

Source: Data courtesy of J. D. Cryer.

In order to determine the probability of exceeding a prespecified tolerance level, a probability distribution for the radiation emitted was needed. Can we regard the observations here as being normally distributed?

A computer was used to assemble the pairs $(q_{(j)}, x_{(j)})$ and construct the Q–Q plot, pictured in Figure 4.6 on page 181. It appears from the plot that the data as a whole are not normally distributed. The points indicated by the circled locations in the figure are outliers—values that are too large relative to the rest of the observations.

For the radiation data, several observations are equal. When this occurs, those observations with like values are associated with the same normal quantile. This quantile is calculated using the average of the quantiles the tied observations would have if they all differed slightly. ∎

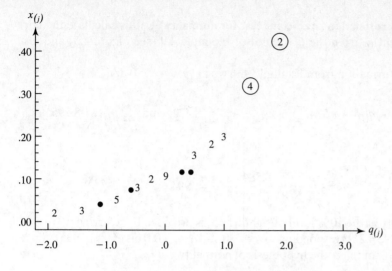

Figure 4.6 A Q–Q plot of the radiation data (door closed) from Example 4.10. (The integers in the plot indicate the number of points occupying the same location.)

The straightness of the Q–Q plot can be measured by calculating the correlation coefficient of the points in the plot. The correlation coefficient for the Q–Q plot is defined by

$$r_Q = \frac{\sum_{j=1}^{n} (x_{(j)} - \bar{x})(q_{(j)} - \bar{q})}{\sqrt{\sum_{j=1}^{n} (x_{(j)} - \bar{x})^2} \sqrt{\sum_{j=1}^{n} (q_{(j)} - \bar{q})^2}} \quad (4\text{-}31)$$

and a powerful test of normality can be based on it. (See [5], [10], and [12].) Formally, we reject the hypothesis of normality at level of significance α if r_Q falls *below* the appropriate value in Table 4.2.

Table 4.2 Critical Points for the Q–Q Plot Correlation Coefficient Test for Normality

Sample size	Significance levels α		
n	.01	.05	.10
5	.8299	.8788	.9032
10	.8801	.9198	.9351
15	.9126	.9389	.9503
20	.9269	.9508	.9604
25	.9410	.9591	.9665
30	.9479	.9652	.9715
35	.9538	.9682	.9740
40	.9599	.9726	.9771
45	.9632	.9749	.9792
50	.9671	.9768	.9809
55	.9695	.9787	.9822
60	.9720	.9801	.9836
75	.9771	.9838	.9866
100	.9822	.9873	.9895
150	.9879	.9913	.9928
200	.9905	.9931	.9942
300	.9935	.9953	.9960

Example 4.11 (A correlation coefficient test for normality) Let us calculate the correlation coefficient r_Q from the Q–Q plot of Example 4.9 (see Figure 4.5) and test for normality.

Using the information from Example 4.9, we have $\bar{x} = .770$ and

$$\sum_{j=1}^{10} (x_{(j)} - \bar{x})q_{(j)} = 8.584, \quad \sum_{j=1}^{10} (x_{(j)} - \bar{x})^2 = 8.472, \quad \text{and} \quad \sum_{j=1}^{10} q_{(j)}^2 = 8.795$$

Since always, $\bar{q} = 0$,

$$r_Q = \frac{8.584}{\sqrt{8.472}\ \sqrt{8.795}} = .994$$

A test of normality at the 10% level of significance is provided by referring $r_Q = .994$ to the entry in Table 4.2 corresponding to $n = 10$ and $\alpha = .10$. This entry is .9351. Since $r_Q > .9351$, we do not reject the hypothesis of normality. ∎

Instead of r_Q, some software packages evaluate the original statistic proposed by Shapiro and Wilk [12]. Its correlation form corresponds to replacing $q_{(j)}$ by a function of the expected value of standard normal-order statistics and their covariances. We prefer r_Q because it corresponds directly to the points in the normal-scores plot. For large sample sizes, the two statistics are nearly the same (see [13]), so either can be used to judge lack of fit.

Linear combinations of more than one characteristic can be investigated. Many statisticians suggest plotting

$$\hat{\mathbf{e}}_1'\mathbf{x}_j \quad \text{where} \quad \mathbf{S}\hat{\mathbf{e}}_1 = \hat{\lambda}_1\hat{\mathbf{e}}_1$$

in which $\hat{\lambda}_1$ is the largest eigenvalue of $\mathbf{S}$. Here $\mathbf{x}_j' = [x_{j1}, x_{j2}, \ldots, x_{jp}]$ is the jth observation on the p variables $X_1, X_2, \ldots, X_p$. The linear combination $\hat{\mathbf{e}}_p'\mathbf{x}_j$ corresponding to the smallest eigenvalue is also frequently singled out for inspection. (See Chapter 8 and [6] for further details.)

Evaluating Bivariate Normality

We would like to check on the assumption of normality for all distributions of $2, 3, \ldots, p$ dimensions. However, as we have pointed out, for practical work it is usually sufficient to investigate the univariate and bivariate distributions. We considered univariate marginal distributions earlier. It is now of interest to examine the bivariate case.

In Chapter 1, we described scatter plots for pairs of characteristics. If the observations were generated from a multivariate normal distribution, each bivariate distribution would be normal, and the contours of constant density would be ellipses. The scatter plot should conform to this structure by exhibiting an overall pattern that is nearly elliptical.

Moreover, by Result 4.7, the set of bivariate outcomes $\mathbf{x}$ such that

$$(\mathbf{x} - \boldsymbol{\mu})'\boldsymbol{\Sigma}^{-1}(\mathbf{x} - \boldsymbol{\mu}) \leq \chi_2^2(.5)$$

has probability .5. Thus, we should expect *roughly* the same percentage, 50%, of sample observations to lie in the ellipse given by

$$\{\text{all } \mathbf{x} \text{ such that } (\mathbf{x} - \bar{\mathbf{x}})'\mathbf{S}^{-1}(\mathbf{x} - \bar{\mathbf{x}}) \leq \chi_2^2(.5)\}$$

where we have replaced $\boldsymbol{\mu}$ by its estimate $\bar{\mathbf{x}}$ and $\boldsymbol{\Sigma}^{-1}$ by its estimate $\mathbf{S}^{-1}$. If not, the normality assumption is suspect.

Example 4.12 (Checking bivariate normality) Although not a random sample, data consisting of the pairs of observations (x_1 = sales, x_2 = profits) for the 10 largest companies in the world are listed in Exercise 1.4. These data give

$$\bar{\mathbf{x}} = \begin{bmatrix} 155.60 \\ 14.70 \end{bmatrix}, \quad \mathbf{S} = \begin{bmatrix} 7476.45 & 303.62 \\ 303.62 & 26.19 \end{bmatrix}$$

so

$$\mathbf{S}^{-1} = \frac{1}{103,623.12} \begin{bmatrix} 26.19 & -303.62 \\ -303.62 & 7476.45 \end{bmatrix}$$

$$= \begin{bmatrix} .000253 & -.002930 \\ -.002930 & .072148 \end{bmatrix}$$

From Table 3 in the appendix, $\chi_2^2(.5) = 1.39$. Thus, any observation $\mathbf{x}' = [x_1, x_2]$ satisfying

$$\begin{bmatrix} x_1 - 155.60 \\ x_2 - 14.70 \end{bmatrix}' \begin{bmatrix} .000253 & -.002930 \\ -.002930 & .072148 \end{bmatrix} \begin{bmatrix} x_1 - 155.60 \\ x_2 - 14.70 \end{bmatrix} \leq 1.39$$

is on or inside the estimated 50% contour. Otherwise the observation is outside this contour. The first pair of observations in Exercise 1.4 is $[x_1, x_2]' = [108.28, 17.05]$. In this case

$$\begin{bmatrix} 108.28 - 155.60 \\ 17.05 - 14.70 \end{bmatrix}' \begin{bmatrix} .000253 & -.002930 \\ -.002930 & .072148 \end{bmatrix} \begin{bmatrix} 108.28 - 155.60 \\ 17.05 - 14.70 \end{bmatrix}$$

$$= 1.61 > 1.39$$

and this point falls outside the 50% contour. The remaining nine points have generalized distances from $\bar{\mathbf{x}}$ of .30, .62, 1.79, 1.30, 4.38, 1.64, 3.53, 1.71, and 1.16, respectively. Since four of these distances are less than 1.39, a proportion, .40, of the data falls within the 50% contour. If the observations were normally distributed, we would expect about half, or 5, of them to be within this contour. This difference in proportions might ordinarily provide evidence for rejecting the notion of bivariate normality; however, our sample size of 10 is too small to reach this conclusion. (See also Example 4.13.) ■

Computing the fraction of the points within a contour and subjectively comparing it with the theoretical probability is a useful, but rather rough, procedure.

A somewhat more formal method for judging the joint normality of a data set is based on the squared generalized distances

$$d_j^2 = (\mathbf{x}_j - \bar{\mathbf{x}})'\mathbf{S}^{-1}(\mathbf{x}_j - \bar{\mathbf{x}}), \qquad j = 1, 2, \ldots, n \tag{4-32}$$

where $\mathbf{x}_1, \mathbf{x}_2, \ldots, \mathbf{x}_n$ are the sample observations. The procedure we are about to describe is not limited to the bivariate case; it can be used for all $p \geq 2$.

When the parent population is multivariate normal and both n and $n - p$ are greater than 25 or 30, each of the squared distances $d_1^2, d_2^2, \ldots, d_n^2$ should behave like a chi-square random variable. [See Result 4.7 and Equations (4-26) and (4-27).] Although these distances are *not* independent or exactly chi-square distributed, it is helpful to plot them as if they were. The resulting plot is called a *chi-square plot* or *gamma plot*, because the chi-square distribution is a special case of the more general gamma distribution. (See [6].)

To construct the chi-square plot,

1. Order the squared distances in (4-32) from smallest to largest as $d_{(1)}^2 \leq d_{(2)}^2 \leq \cdots \leq d_{(n)}^2$.
2. Graph the pairs $\left(q_{c,p}\left((j - \frac{1}{2})/n\right), d_{(j)}^2\right)$, where $q_{c,p}\left((j - \frac{1}{2})/n\right)$ is the $100(j - \frac{1}{2})/n$ quantile of the chi-square distribution with p degrees of freedom.

Quantiles are specified in terms of proportions, whereas percentiles are specified in terms of percentages.

The quantiles $q_{c,p}\left((j - \frac{1}{2})/n\right)$ are related to the upper percentiles of a chi-squared distribution. In particular, $q_{c,p}\left((j - \frac{1}{2})/n\right) = \chi_p^2\left((n - j + \frac{1}{2})/n\right)$.

The plot should resemble a straight line through the origin having slope 1. A systematic curved pattern suggests lack of normality. One or two points far above the line indicate large distances, or outlying observations, that merit further attention.

Example 4.13 (Constructing a chi-square plot) Let us construct a chi-square plot of the generalized distances given in Example 4.12. The ordered distances and the corresponding chi-square percentiles for $p = 2$ and $n = 10$ are listed in the following table:

j	$d_{(j)}^2$	$q_{c,2}\left(\dfrac{j - \frac{1}{2}}{10}\right)$
1	.30	.10
2	.62	.33
3	1.16	.58
4	1.30	.86
5	1.61	1.20
6	1.64	1.60
7	1.71	2.10
8	1.79	2.77
9	3.53	3.79
10	4.38	5.99

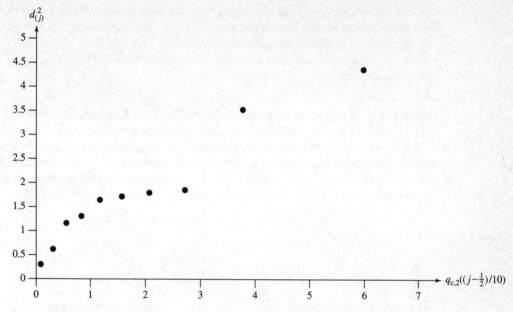

Figure 4.7 A chi-square plot of the ordered distances in Example 4.13.

A graph of the pairs $\left(q_{c,2}\left((j - \tfrac{1}{2})/10\right), d_{(j)}^2\right)$ is shown in Figure 4.7. The points in Figure 4.7 are reasonably straight. Given the small sample size it is difficult to reject bivariate normality on the evidence in this graph. If further analysis of the data were required, it might be reasonable to transform them to observations more nearly bivariate normal. Appropriate transformations are discussed in Section 4.8. ∎

In addition to inspecting univariate plots and scatter plots, we should check multivariate normality by constructing a chi-squared or d^2 plot. Figure 4.8 contains d^2

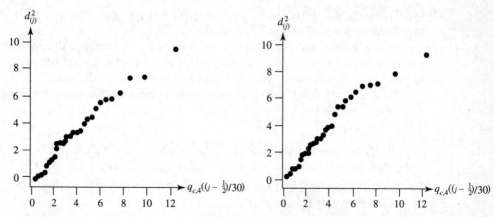

Figure 4.8 Chi-square plots for two simulated four-variate normal data sets with $n = 30$.

plots based on two computer-generated samples of 30 four-variate normal random vectors. As expected, the plots have a straight-line pattern, but the top two or three ordered squared distances are quite variable.

The next example contains a real data set comparable to the simulated data set that produced the plots in Figure 4.8.

Example 4.14 (Evaluating multivariate normality for a four-variable data set) The data in Table 4.3 were obtained by taking four different measures of stiffness, x_1, x_2, x_3, and x_4, of each of $n = 30$ boards. The first measurement involves sending a shock wave down the board, the second measurement is determined while vibrating the board, and the last two measurements are obtained from static tests. The squared distances $d_j^2 = (\mathbf{x}_j - \bar{\mathbf{x}})' \mathbf{S}^{-1} (\mathbf{x}_j - \bar{\mathbf{x}})$ are also presented in the table.

Table 4.3 Four Measurements of Stiffness

Observation no.	x_1	x_2	x_3	x_4	d^2	Observation no.	x_1	x_2	x_3	x_4	d^2
1	1889	1651	1561	1778	.60	16	1954	2149	1180	1281	16.85
2	2403	2048	2087	2197	5.48	17	1325	1170	1002	1176	3.50
3	2119	1700	1815	2222	7.62	18	1419	1371	1252	1308	3.99
4	1645	1627	1110	1533	5.21	19	1828	1634	1602	1755	1.36
5	1976	1916	1614	1883	1.40	20	1725	1594	1313	1646	1.46
6	1712	1712	1439	1546	2.22	21	2276	2189	1547	2111	9.90
7	1943	1685	1271	1671	4.99	22	1899	1614	1422	1477	5.06
8	2104	1820	1717	1874	1.49	23	1633	1513	1290	1516	.80
9	2983	2794	2412	2581	12.26	24	2061	1867	1646	2037	2.54
10	1745	1600	1384	1508	.77	25	1856	1493	1356	1533	4.58
11	1710	1591	1518	1667	1.93	26	1727	1412	1238	1469	3.40
12	2046	1907	1627	1898	.46	27	2168	1896	1701	1834	2.38
13	1840	1841	1595	1741	2.70	28	1655	1675	1414	1597	3.00
14	1867	1685	1493	1678	.13	29	2326	2301	2065	2234	6.28
15	1859	1649	1389	1714	1.08	30	1490	1382	1214	1284	2.58

Source: Data courtesy of William Galligan.

The marginal distributions appear quite normal (see Exercise 4.33), with the possible exception of specimen (board) 9.

To further evaluate multivariate normality, we constructed the chi-square plot shown in Figure 4.9. The two specimens with the largest squared distances are clearly removed from the straight-line pattern. Together, with the next largest point or two, they make the plot appear curved at the upper end. We will return to a discussion of this plot in Example 4.15. ∎

We have discussed some rather simple techniques for checking the multivariate normality assumption. Specifically, we advocate calculating the d_j^2, $j = 1, 2, \ldots, n$ [see Equation (4-32)] and comparing the results with χ^2 quantiles. For example, p-variate normality is indicated if

1. Roughly half of the d_j^2 are less than or equal to $q_{c,p}(.50)$.

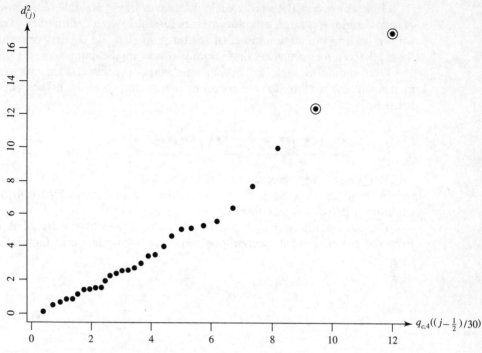

Figure 4.9 A chi-square plot for the data in Example 4.14.

2. A plot of the ordered squared distances $d^2_{(1)} \le d^2_{(2)} \le \cdots \le d^2_{(n)}$ versus $q_{c,p}\left(\dfrac{1 - \frac{1}{2}}{n}\right)$, $q_{c,p}\left(\dfrac{2 - \frac{1}{2}}{n}\right), \ldots, q_{c,p}\left(\dfrac{n - \frac{1}{2}}{n}\right)$, respectively, is nearly a straight line having slope 1 and that passes through the origin.

(See [6] for a more complete exposition of methods for assessing normality.)

We close this section by noting that all measures of goodness of fit suffer the same serious drawback. When the sample size is small, only the most aberrant behavior will be identified as lack of fit. On the other hand, very large samples invariably produce statistically significant lack of fit. Yet the departure from the specified distribution may be very small and technically unimportant to the inferential conclusions.

4.7 Detecting Outliers and Cleaning Data

Most data sets contain one or a few unusual observations that do not seem to belong to the pattern of variability produced by the other observations. With data on a single characteristic, unusual observations are those that are either very large or very small relative to the others. The situation can be more complicated with multivariate data. Before we address the issue of identifying these *outliers*, we must emphasize that not all outliers are wrong numbers. They may, justifiably, be part of the group and may lead to a better understanding of the phenomena being studied.

Outliers are best detected visually whenever this is possible. When the number of observations n is large, dot plots are not feasible. When the number of characteristics p is large, the large number of scatter plots $p(p-1)/2$ may prevent viewing them all. Even so, we suggest first visually inspecting the data whenever possible.

What should we look for? For a single random variable, the problem is one dimensional, and we look for observations that are far from the others. For instance, the dot diagram

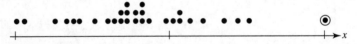

reveals a single large observation which is circled.

In the bivariate case, the situation is more complicated. Figure 4.10 shows a situation with two unusual observations.

The data point circled in the upper right corner of the figure is detached from the pattern, and its second coordinate is large relative to the rest of the x_2

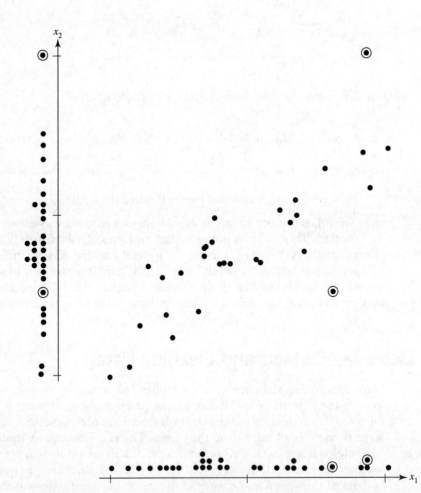

Figure 4.10 Two outliers; one univariate and one bivariate.

measurements, as shown by the vertical dot diagram. The second outlier, also circled, is far from the elliptical pattern of the rest of the points, but, separately, each of its components has a typical value. This outlier cannot be detected by inspecting the marginal dot diagrams.

In higher dimensions, there can be outliers that cannot be detected from the univariate plots or even the bivariate scatter plots. Here a large value of $(\mathbf{x}_j - \bar{\mathbf{x}})'\mathbf{S}^{-1}(\mathbf{x}_j - \bar{\mathbf{x}})$ will suggest an unusual observation, even though it cannot be seen visually.

Steps for Detecting Outliers

1. Make a dot plot for each variable.
2. Make a scatter plot for each pair of variables.
3. Calculate the standardized values $z_{jk} = (x_{jk} - \bar{x}_k)/\sqrt{s_{kk}}$ for $j = 1, 2, \ldots, n$ and each column $k = 1, 2, \ldots, p$. Examine these standardized values for large or small values.
4. Calculate the generalized squared distances $(\mathbf{x}_j - \bar{\mathbf{x}})'\mathbf{S}^{-1}(\mathbf{x}_j - \bar{\mathbf{x}})$. Examine these distances for unusually large values. In a chi-square plot, these would be the points farthest from the origin.

In step 3, "large" must be interpreted relative to the sample size and number of variables. There are $n \times p$ standardized values. When $n = 100$ and $p = 5$, there are 500 values. You expect 1 or 2 of these to exceed 3 or be less than -3, even if the data came from a multivariate distribution that is exactly normal. As a guideline, 3.5 might be considered large for moderate sample sizes.

In step 4, "large" is measured by an appropriate percentile of the chi-square distribution with p degrees of freedom. If the sample size is $n = 100$, we would expect 5 observations to have values of d_j^2 that exceed the upper fifth percentile of the chi-square distribution. A more extreme percentile must serve to determine observations that do not fit the pattern of the remaining data.

The data we presented in Table 4.3 concerning lumber have already been cleaned up somewhat. Similar data sets from the same study also contained data on x_5 = tensile strength. Nine observation vectors, out of the total of 112, are given as rows in the following table, along with their standardized values.

x_1	x_2	x_3	x_4	x_5	z_1	z_2	z_3	z_4	z_5
⋮	⋮	⋮	⋮	⋮⋮	⋮	⋮	⋮	⋮	
1631	1528	1452	1559	1602	.06	−.15	.05	.28	−.12
1770	1677	1707	1738	1785	.64	.43	1.07	.94	.60
1376	1190	723	1285	2791	−1.01	−1.47	−2.87	−.73	⟨4.57⟩
1705	1577	1332	1703	1664	.37	.04	−.43	.81	.13
1643	1535	1510	1494	1582	.11	−.12	.28	.04	−.20
1567	1510	1301	1405	1553	−.21	−.22	−.56	−.28	−.31
1528	1591	1714	1685	1698	−.38	.10	1.10	.75	.26
1803	1826	1748	2746	1764	.78	1.01	1.23	⟨4.65⟩	.52
1587	1554	1352	1554	1551	−.13	−.05	−.35	.26	−.32
⋮	⋮	⋮	⋮	⋮⋮	⋮	⋮	⋮	⋮	

The standardized values are based on the sample mean and variance, calculated from all 112 observations. There are two extreme standardized values. Both are too large with standardized values over 4.5. During their investigation, the researchers recorded measurements by hand in a logbook and then performed calculations that produced the values given in the table. When they checked their records regarding the values pinpointed by this analysis, errors were discovered. The value $x_5 = 2791$ was corrected to 1241, and $x_4 = 2746$ was corrected to 1670. Incorrect readings on an individual variable are quickly detected by locating a large leading digit for the standardized value.

The next example returns to the data on lumber discussed in Example 4.14.

Example 4.15 (Detecting outliers in the data on lumber) Table 4.4 contains the data in Table 4.3, along with the standardized observations. These data consist of four different measures of stiffness x_1, x_2, x_3, and x_4, on each of $n = 30$ boards. Recall that the first measurement involves sending a shock wave down the board, the second measurement is determined while vibrating the board, and the last two measurements are obtained from static tests. The standardized measurements are

Table 4.4 Four Measurements of Stiffness with Standardized Values

x_1	x_2	x_3	x_4	Observation no.	z_1	z_2	z_3	z_4	d^2
1889	1651	1561	1778	1	−.1	−.3	.2	.2	.60
2403	2048	2087	2197	2	1.5	.9	1.9	1.5	5.48
2119	1700	1815	2222	3	.7	−.2	1.0	1.5	7.62
1645	1627	1110	1533	4	−.8	−.4	−1.3	−.6	5.21
1976	1916	1614	1883	5	.2	.5	.3	.5	1.40
1712	1712	1439	1546	6	−.6	−.1	−.2	−.6	2.22
1943	1685	1271	1671	7	.1	−.2	−.8	−.2	4.99
2104	1820	1717	1874	8	.6	.2	.7	.5	1.49
2983	2794	2412	2581	9	3.3	3.3	3.0	2.7	12.26
1745	1600	1384	1508	10	−.5	−.5	−.4	−.7	.77
1710	1591	1518	1667	11	−.6	−.5	.0	−.2	1.93
2046	1907	1627	1898	12	.4	.5	.4	.5	.46
1840	1841	1595	1741	13	−.2	.3	.3	.0	2.70
1867	1685	1493	1678	14	−.1	−.2	−.1	−.1	.13
1859	1649	1389	1714	15	−.1	−.3	−.4	−.0	1.08
1954	2149	1180	1281	16	.1	1.3	−1.1	−1.4	16.85
1325	1170	1002	1176	17	−1.8	−1.8	−1.7	−1.7	3.50
1419	1371	1252	1308	18	−1.5	−1.2	−.8	−1.3	3.99
1828	1634	1602	1755	19	−.2	−.4	.3	.1	1.36
1725	1594	1313	1646	20	−.6	−.5	−.6	−.2	1.46
2276	2189	1547	2111	21	1.1	1.4	.1	1.2	9.90
1899	1614	1422	1477	22	−.0	−.4	−.3	−.8	5.06
1633	1513	1290	1516	23	−.8	−.7	−.7	−.6	.80
2061	1867	1646	2037	24	.5	.4	.5	1.0	2.54
1856	1493	1356	1533	25	−.2	−.8	−.5	−.6	4.58
1727	1412	1238	1469	26	−.6	−1.1	−.9	−.8	3.40
2168	1896	1701	1834	27	.8	.5	.6	.3	2.38
1655	1675	1414	1597	28	−.8	−.2	−.3	−.4	3.00
2326	2301	2065	2234	29	1.3	1.7	1.8	1.6	6.28
1490	1382	1214	1284	30	−1.3	−1.2	−1.0	−1.4	2.58

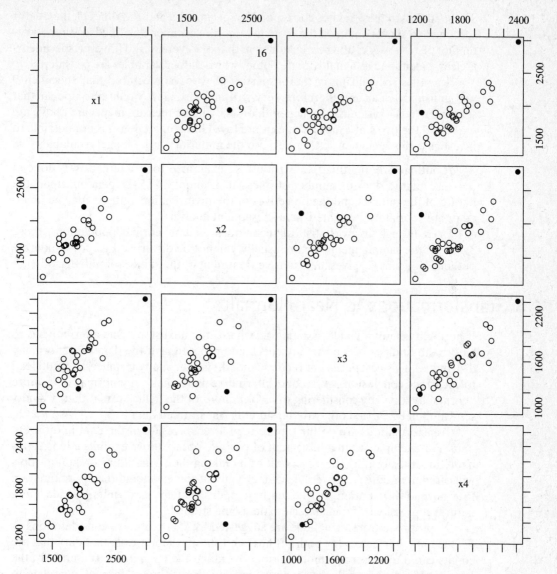

Figure 4.11 Scatter plots for the lumber stiffness data with specimens 9 and 16 plotted as solid dots.

$$z_{jk} = \frac{x_{jk} - \bar{x}_k}{\sqrt{s_{kk}}}, \qquad k = 1, 2, 3, 4; \qquad j = 1, 2, \dots, 30$$

and the squares of the distances are $d_j^2 = (\mathbf{x}_j - \bar{\mathbf{x}})'\mathbf{S}^{-1}(\mathbf{x}_j - \bar{\mathbf{x}})$.

The last column in Table 4.4 reveals that specimen 16 is a multivariate outlier, since $\chi_4^2(.005) = 14.86$; yet all of the individual measurements are well within their respective univariate scatters. Specimen 9 also has a large d^2 value.

The two specimens (9 and 16) with large squared distances stand out as clearly different from the rest of the pattern in Figure 4.9. Once these two points are removed, the remaining pattern conforms to the expected straight-line relation. Scatter plots for the lumber stiffness measurements are given in Figure 4.11 above.

The solid dots in these figures correspond to specimens 9 and 16. Although the dot for specimen 16 stands out in all the plots, the dot for specimen 9 is "hidden" in the scatter plot of x_3 versus x_4 and nearly hidden in that of x_1 versus x_3. However, specimen 9 is clearly identified as a multivariate outlier when all four variables are considered.

Scientists specializing in the properties of wood conjectured that specimen 9 was unusually clear and therefore very stiff and strong. It would also appear that specimen 16 is a bit unusual, since both of its dynamic measurements are above average and the two static measurements are low. Unfortunately, it was not possible to investigate this specimen further because the material was no longer available.   ∎

If outliers are identified, they should be examined for content, as was done in the case of the data on lumber stiffness in Example 4.15. Depending upon the nature of the outliers and the objectives of the investigation, outliers may be deleted or appropriately "weighted" in a subsequent analysis.

Even though many statistical techniques assume normal populations, those based on the sample mean vectors usually will not be disturbed by a few moderate outliers. Hawkins [7] gives an extensive treatment of the subject of outliers.

4.8 Transformations to Near Normality

If normality is not a viable assumption, what is the next step? One alternative is to ignore the findings of a normality check and proceed as if the data were normally distributed. This practice is not recommended, since, in many instances, it could lead to incorrect conclusions. A second alternative is to make nonnormal data more "normal looking" by considering *transformations* of the data. Normal-theory analyses can then be carried out with the suitably transformed data.

Transformations are nothing more than a reexpression of the data in different units. For example, when a histogram of positive observations exhibits a long right-hand tail, transforming the observations by taking their logarithms or square roots will often markedly improve the symmetry about the mean and the approximation to a normal distribution. It frequently happens that the new units provide more natural expressions of the characteristics being studied.

Appropriate transformations are suggested by (1) theoretical considerations or (2) the data themselves (or both). It has been shown theoretically that data that are counts can often be made more normal by taking their *square roots*. Similarly, the *logit transformation* applied to proportions and *Fisher's z-transformation* applied to correlation coefficients yield quantities that are approximately normally distributed.

Helpful Transformations To Near Normality

Original Scale	*Transformed Scale*	
1. Counts, y	$\sqrt{y}$	
2. Proportions, $\hat{p}$	$\mathrm{logit}(\hat{p}) = \dfrac{1}{2}\log\left(\dfrac{\hat{p}}{1-\hat{p}}\right)$	(4-33)
3. Correlations, r	Fisher's $\quad z(r) = \dfrac{1}{2}\log\left(\dfrac{1+r}{1-r}\right)$	

In many instances, the choice of a transformation to improve the approximation to normality is not obvious. For such cases, it is convenient to let the data suggest a transformation. A useful family of transformations for this purpose is the family of *power transformations*.

Power transformations are defined only for positive variables. However, this is not as restrictive as it seems, because a single constant can be added to each observation in the data set if some of the values are negative.

Let x represent an arbitrary observation. The power family of transformations is indexed by a parameter λ. A given value for λ implies a particular transformation. For example, consider x^{λ} with $\lambda = -1$. Since $x^{-1} = 1/x$, this choice of λ corresponds to the reciprocal transformation. We can trace the family of transformations as λ ranges from negative to positive powers of x. For $\lambda = 0$, we define $x^0 = \ln x$. A sequence of possible transformations is

$$\ldots, x^{-1} = \frac{1}{x}, x^0 = \ln x, x^{1/4} = \sqrt[4]{x}, x^{1/2} = \sqrt{x}, \qquad x^2, x^3, \ldots$$

shrinks large values of x	increases large values of x

To select a power transformation, an investigator looks at the marginal dot diagram or histogram and decides whether large values have to be "pulled in" or "pushed out" to improve the symmetry about the mean. Trial-and-error calculations with a few of the foregoing transformations should produce an improvement. The final choice should always be examined by a Q–Q plot or other checks to see whether the tentative normal assumption is satisfactory.

The transformations we have been discussing are data based in the sense that it is only the appearance of the data themselves that influences the choice of an appropriate transformation. There are no external considerations involved, although the transformation actually used is often determined by some mix of information supplied by the data and extra-data factors, such as simplicity or ease of interpretation.

A convenient analytical method is available for choosing a power transformation. We begin by focusing our attention on the univariate case.

Box and Cox [3] consider the slightly modified family of power transformations

$$x^{(\lambda)} = \begin{cases} \dfrac{x^{\lambda} - 1}{\lambda} & \lambda \neq 0 \\[2mm] \ln x & \lambda = 0 \end{cases} \tag{4-34}$$

which is continuous in λ for $x > 0$. (See [8].) Given the observations $x_1, x_2, \ldots, x_n$, the Box–Cox solution for the choice of an appropriate power λ is the solution that *maximizes* the expression

$$\ell(\lambda) = -\frac{n}{2} \ln \left[\frac{1}{n} \sum_{j=1}^{n} (x_j^{(\lambda)} - \overline{x^{(\lambda)}})^2 \right] + (\lambda - 1) \sum_{j=1}^{n} \ln x_j \tag{4-35}$$

We note that $x_j^{(\lambda)}$ is defined in (4-34) and

$$\overline{x^{(\lambda)}} = \frac{1}{n} \sum_{j=1}^{n} x_j^{(\lambda)} = \frac{1}{n} \sum_{j=1}^{n} \left(\frac{x_j^{\lambda} - 1}{\lambda} \right) \tag{4-36}$$

is the arithmetic average of the transformed observations. The first term in (4-35) is, apart from a constant, the logarithm of a normal likelihood function, after maximizing it with respect to the population mean and variance parameters.

The calculation of $\ell(\lambda)$ for many values of λ is an easy task for a computer. It is helpful to have a graph of $\ell(\lambda)$ versus λ, as well as a tabular display of the pairs $(\lambda, \ell(\lambda))$, in order to study the behavior near the maximizing value $\hat{\lambda}$. For instance, if either $\lambda = 0$ (logarithm) or $\lambda = \frac{1}{2}$ (square root) is near $\hat{\lambda}$, one of these may be preferred because of its simplicity.

Rather than program the calculation of (4-35), some statisticians recommend the equivalent procedure of fixing λ, creating the new variable

$$y_j^{(\lambda)} = \frac{x_j^\lambda - 1}{\lambda\left[\left(\prod_{i=1}^{n} x_i\right)^{1/n}\right]^{\lambda-1}} \qquad j = 1,\ldots,n \qquad (4\text{-}37)$$

and then calculating the sample variance. The minimum of the variance occurs at the same λ that maximizes (4-35).

Comment. It is now understood that the transformation obtained by maximizing $\ell(\lambda)$ usually improves the approximation to normality. However, there is no guarantee that even the best choice of λ will produce a transformed set of values that adequately conform to a normal distribution. The outcomes produced by a transformation selected according to (4-35) should always be carefully examined for possible violations of the tentative assumption of normality. This warning applies with equal force to transformations selected by any other technique.

Example 4.16 (Determining a power transformation for univariate data) We gave readings of the microwave radiation emitted through the closed doors of $n = 42$ ovens in Example 4.10. The Q–Q plot of these data in Figure 4.6 indicates that the observations deviate from what would be expected if they were normally distributed. Since all the observations are positive, let us perform a power transformation of the data which, we hope, will produce results that are more nearly normal. Restricting our attention to the family of transformations in (4-34), we must find that value of λ maximizing the function $\ell(\lambda)$ in (4-35).

The pairs $(\lambda, \ell(\lambda))$ are listed in the following table for several values of λ:

λ	$\ell(\lambda)$	λ	$\ell(\lambda)$
−1.00	70.52		
−.90	75.65	.40	106.20
−.80	80.46	.50	105.50
−.70	84.94	.60	104.43
−.60	89.06	.70	103.03
−.50	92.79	.80	101.33
−.40	96.10	.90	99.34
−.30	98.97	1.00	97.10
−.20	101.39	1.10	94.64
−.10	103.35	1.20	91.96
.00	104.83	1.30	89.10
.10	105.84	1.40	86.07
.20	106.39	1.50	82.88
.30	106.51		

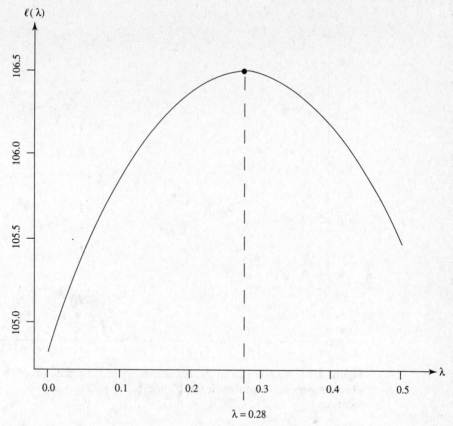

Figure 4.12 Plot of $\ell(\lambda)$ versus λ for radiation data (door closed).

The curve of $\ell(\lambda)$ versus λ that allows the more exact determination $\hat\lambda = .28$ is shown in Figure 4.12 .

It is evident from both the table and the plot that a value of $\hat\lambda$ around .30 maximizes $\ell(\lambda)$. For convenience, we choose $\hat\lambda = .25$. The data x_j were reexpressed as

$$x_j^{(1/4)} = \frac{x_j^{1/4} - 1}{\frac{1}{4}} \qquad j = 1, 2, \ldots, 42$$

and a Q–Q plot was constructed from the transformed quantities. This plot is shown in Figure 4.13 on page 196. The quantile pairs fall very close to a straight line, and we would conclude from this evidence that the $x_j^{(1/4)}$ are approximately normal. ∎

Transforming Multivariate Observations

With multivariate observations, a power transformation must be selected for each of the variables. Let $\lambda_1, \lambda_2, \ldots, \lambda_p$ be the power transformations for the p measured characteristics. Each λ_k can be selected by *maximizing*

$$\ell_k(\lambda) = -\frac{n}{2} \ln\left[\frac{1}{n} \sum_{j=1}^{n} (x_{jk}^{(\lambda_k)} - \overline{x_k^{(\lambda_k)}})^2\right] + (\lambda_k - 1) \sum_{j=1}^{n} \ln x_{jk} \qquad (4\text{-}38)$$

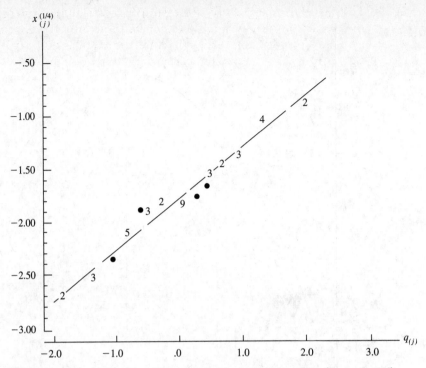

Figure 4.13 A Q–Q plot of the transformed radiation data (door closed). (The integers in the plot indicate the number of points occupying the same location.)

where $x_{1k}, x_{2k}, \ldots, x_{nk}$ are the n observations on the kth variable, $k = 1, 2, \ldots, p$. Here

$$\overline{x_k^{(\lambda_k)}} = \frac{1}{n} \sum_{j=1}^n x_{jk}^{(\lambda_k)} = \frac{1}{n} \sum_{j=1}^n \left(\frac{x_{jk}^{\lambda_k} - 1}{\lambda_k} \right) \tag{4-39}$$

is the arithmetic average of the transformed observations. The jth transformed multivariate observation is

$$\mathbf{x}_j^{(\lambda)} = \begin{bmatrix} \dfrac{x_{j1}^{\hat{\lambda}_1} - 1}{\hat{\lambda}_1} \\[2ex] \dfrac{x_{j2}^{\hat{\lambda}_2} - 1}{\hat{\lambda}_2} \\[2ex] \vdots \\[2ex] \dfrac{x_{jp}^{\hat{\lambda}_p} - 1}{\hat{\lambda}_p} \end{bmatrix}$$

where $\hat{\lambda}_1, \hat{\lambda}_2, \ldots, \hat{\lambda}_p$ are the values that individually maximize (4-38).

The procedure just described is equivalent to making each marginal distribution approximately normal. Although normal marginals are not sufficient to ensure that the joint distribution is normal, in practical applications this may be good enough. If not, we could start with the values $\hat{\lambda}_1, \hat{\lambda}_2, \ldots, \hat{\lambda}_p$ obtained from the preceding transformations and iterate toward the set of values $\boldsymbol{\lambda}' = [\lambda_1, \lambda_2, \ldots, \lambda_p]$, which collectively maximizes

$$\ell(\lambda_1, \lambda_2, \ldots, \lambda_p)$$

$$= -\frac{n}{2} \ln |\mathbf{S}(\boldsymbol{\lambda})| + (\lambda_1 - 1) \sum_{j=1}^{n} \ln x_{j1} + (\lambda_2 - 1) \sum_{j=1}^{n} \ln x_{j2}$$

$$+ \cdots + (\lambda_p - 1) \sum_{j=1}^{n} \ln x_{jp} \qquad (4\text{-}40)$$

where $\mathbf{S}(\boldsymbol{\lambda})$ is the sample covariance matrix computed from

$$\mathbf{x}_j^{(\boldsymbol{\lambda})} = \begin{bmatrix} \dfrac{x_{j1}^{\lambda_1} - 1}{\lambda_1} \\[2mm] \dfrac{x_{j2}^{\lambda_2} - 1}{\lambda_2} \\[1mm] \vdots \\[1mm] \dfrac{x_{jp}^{\lambda_p} - 1}{\lambda_p} \end{bmatrix} \qquad j = 1, 2, \ldots, n$$

Maximizing (4-40) not only is substantially more difficult than maximizing the individual expressions in (4-38), but also is unlikely to yield remarkably better results. The selection method based on Equation (4-40) is equivalent to maximizing a multivariate likelihood over $\boldsymbol{\mu}$, $\boldsymbol{\Sigma}$ and $\boldsymbol{\lambda}$, whereas the method based on (4-38) corresponds to maximizing the kth univariate likelihood over μ_k, σ_{kk}, and λ_k. The latter likelihood is generated by pretending there is some λ_k for which the observations $(x_{jk}^{\lambda_k} - 1)/\lambda_k$, $j = 1, 2, \ldots, n$ have a normal distribution. See [3] and [2] for detailed discussions of the univariate and multivariate cases, respectively. (Also, see [8].)

Example 4.17 (Determining power transformations for bivariate data) Radiation measurements were also recorded through the open doors of the $n = 42$ microwave ovens introduced in Example 4.10. The amount of radiation emitted through the open doors of these ovens is listed in Table 4.5.

In accordance with the procedure outlined in Example 4.16, a power transformation for these data was selected by maximizing $\ell(\lambda)$ in (4-35). The approximate maximizing value was $\hat{\lambda} = .30$. Figure 4.14 on page 199 shows Q–Q plots of the untransformed and transformed door-open radiation data. (These data were actually

Table 4.5 Radiation Data (Door Open)

Oven no.	Radiation	Oven no.	Radiation	Oven no.	Radiation
1	.30	16	.20	31	.10
2	.09	17	.04	32	.10
3	.30	18	.10	33	.10
4	.10	19	.01	34	.30
5	.10	20	.60	35	.12
6	.12	21	.12	36	.25
7	.09	22	.10	37	.20
8	.10	23	.05	38	.40
9	.09	24	.05	39	.33
10	.10	25	.15	40	.32
11	.07	26	.30	41	.12
12	.05	27	.15	42	.12
13	.01	28	.09		
14	.45	29	.09		
15	.12	30	.28		

Source: Data courtesy of J. D. Cryer.

transformed by taking the fourth root, as in Example 4.16.) It is clear from the figure that the transformed data are more nearly normal, although the normal approximation is not as good as it was for the door-closed data.

Let us denote the door-closed data by $x_{11}, x_{21}, \ldots, x_{42,1}$ and the door-open data by $x_{12}, x_{22}, \ldots, x_{42,2}$. Choosing a power transformation for each set by maximizing the expression in (4-35) is equivalent to maximizing $\ell_k(\lambda)$ in (4-38) with $k = 1, 2$. Thus, using the outcomes from Example 4.16 and the foregoing results, we have $\hat{\lambda}_1 = .30$ and $\hat{\lambda}_2 = .30$. These powers were determined for the *marginal* distributions of x_1 and x_2.

We can consider the *joint* distribution of x_1 and x_2 and simultaneously determine the pair of powers (λ_1, λ_2) that makes this joint distribution approximately bivariate normal. To do this, we must maximize $\ell(\lambda_1, \lambda_2)$ in (4-40) with respect to both λ_1 and λ_2.

We computed $\ell(\lambda_1, \lambda_2)$ for a grid of λ_1, λ_2 values covering $0 \leq \lambda_1 \leq .50$ and $0 \leq \lambda_2 \leq .50$, and we constructed the contour plot shown in Figure 4.15 on page 200. We see that the maximum occurs at about $(\hat{\lambda}_1, \hat{\lambda}_2) = (.16, .16)$.

The "best" power transformations for this bivariate case do not differ substantially from those obtained by considering each marginal distribution. ∎

As we saw in Example 4.17, making each marginal distribution approximately normal is roughly equivalent to addressing the bivariate distribution directly and making it approximately normal. It is generally easier to select appropriate transformations for the marginal distributions than for the joint distributions.

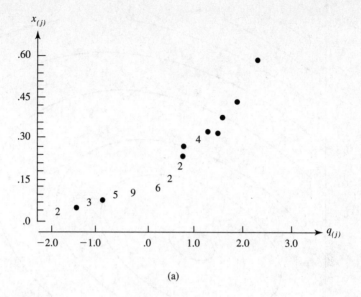

(a)

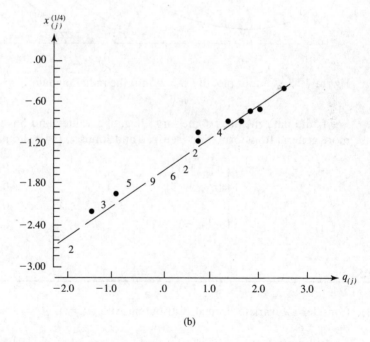

(b)

Figure 4.14 *Q–Q* plots of (a) the original and (b) the transformed radiation data (with door open). (The integers in the plot indicate the number of points occupying the same location.)

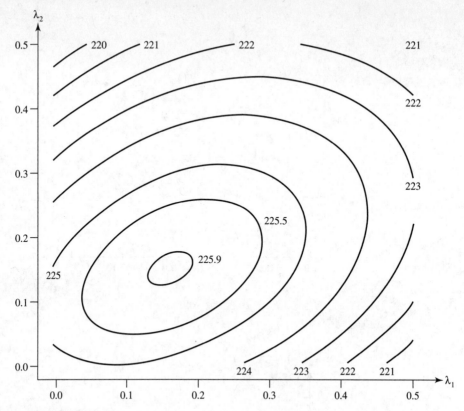

Figure 4.15 Contour plot of $\ell(\lambda_1, \lambda_2)$ for the radiation data.

If the data includes some large negative values and have a single long tail, a more general transformation (see Yeo and Johnson [14]) should be applied.

$$x^{(\lambda)} = \begin{cases} \{(x+1)^\lambda - 1\}/\lambda & x \geq 0, \lambda \neq 0 \\ \ln(x+1) & x \geq 0, \lambda = 0 \\ -\{(-x+1)^{2-\lambda} - 1\}/(2-\lambda) & x < 0, \lambda \neq 2 \\ -\ln(-x+1) & x < 0, \lambda = 2 \end{cases}$$

Exercises

4.1. Consider a bivariate normal distribution with $\mu_1 = 1$, $\mu_2 = 3$, $\sigma_{11} = 2$, $\sigma_{22} = 1$ and $\rho_{12} = -.8$.

(a) Write out the bivariate normal density.

(b) Write out the squared statistical distance expression $(\mathbf{x} - \boldsymbol{\mu})'\boldsymbol{\Sigma}^{-1}(\mathbf{x} - \boldsymbol{\mu})$ as a quadratic function of x_1 and x_2.

4.2. Consider a bivariate normal population with $\mu_1 = 0$, $\mu_2 = 2$, $\sigma_{11} = 2$, $\sigma_{22} = 1$, and $\rho_{12} = .5$.

(a) Write out the bivariate normal density.

(b) Write out the squared generalized distance expression $(\mathbf{x} - \boldsymbol{\mu})' \boldsymbol{\Sigma}^{-1} (\mathbf{x} - \boldsymbol{\mu})$ as a function of x_1 and x_2.

(c) Determine (and sketch) the constant-density contour that contains 50% of the probability.

4.3. Let $\mathbf{X}$ be $N_3(\boldsymbol{\mu}, \boldsymbol{\Sigma})$ with $\boldsymbol{\mu}' = [-3, 1, 4]$ and

$$\boldsymbol{\Sigma} = \begin{bmatrix} 1 & -2 & 0 \\ -2 & 5 & 0 \\ 0 & 0 & 2 \end{bmatrix}$$

Which of the following random variables are independent? Explain.

(a) X_1 and X_2

(b) X_2 and X_3

(c) (X_1, X_2) and X_3

(d) $\dfrac{X_1 + X_2}{2}$ and X_3

(e) X_2 and $X_2 - \frac{5}{2}X_1 - X_3$

4.4. Let $\mathbf{X}$ be $N_3(\boldsymbol{\mu}, \boldsymbol{\Sigma})$ with $\boldsymbol{\mu}' = [2, -3, 1]$ and

$$\boldsymbol{\Sigma} = \begin{bmatrix} 1 & 1 & 1 \\ 1 & 3 & 2 \\ 1 & 2 & 2 \end{bmatrix}$$

(a) Find the distribution of $3X_1 - 2X_2 + X_3$.

(b) Relabel the variables if necessary, and find a 2×1 vector $\mathbf{a}$ such that X_2 and

$X_2 - \mathbf{a}' \begin{bmatrix} X_1 \\ X_3 \end{bmatrix}$ are independent.

4.5. Specify each of the following.

(a) The conditional distribution of X_1, given that $X_2 = x_2$ for the joint distribution in Exercise 4.2.

(b) The conditional distribution of X_2, given that $X_1 = x_1$ and $X_3 = x_3$ for the joint distribution in Exercise 4.3.

(c) The conditional distribution of X_3, given that $X_1 = x_1$ and $X_2 = x_2$ for the joint distribution in Exercise 4.4.

4.6. Let $\mathbf{X}$ be distributed as $N_3(\boldsymbol{\mu}, \boldsymbol{\Sigma})$, where $\boldsymbol{\mu}' = [1, -1, 2]$ and

$$\boldsymbol{\Sigma} = \begin{bmatrix} 4 & 0 & -1 \\ 0 & 5 & 0 \\ -1 & 0 & 2 \end{bmatrix}$$

Which of the following random variables are independent? Explain.

(a) X_1 and X_2

(b) X_1 and X_3

(c) X_2 and X_3

(d) (X_1, X_3) and X_2

(e) X_1 and $X_1 + 3X_2 - 2X_3$

4.7. Refer to Exercise 4.6 and specify each of the following.

(a) The conditional distribution of X_1, given that $X_3 = x_3$.

(b) The conditional distribution of X_1, given that $X_2 = x_2$ and $X_3 = x_3$.

4.8. (Example of a nonnormal bivariate distribution with normal marginals.) Let X_1 be $N(0, 1)$, and let

$$X_2 = \begin{cases} -X_1 & \text{if } -1 \le X_1 \le 1 \\ X_1 & \text{otherwise} \end{cases}$$

Show each of the following.

(a) X_2 also has an $N(0, 1)$ distribution.

(b) X_1 and X_2 do *not* have a bivariate normal distribution.

Hint:

(a) Since X_1 is $N(0,1)$, $P[-1 < X_1 \le x] = P[-x \le X_1 < 1]$ for any x. When $-1 < x_2 < 1$, $P[X_2 \le x_2] = P[X_2 \le -1] + P[-1 < X_2 \le x_2] = P[X_1 \le -1] + P[-1 < -X_1 \le x_2] = P[X_1 \le -1] + P[-x_2 \le X_1 < 1]$. But $P[-x_2 \le X_1 < 1] = P[-1 < X_1 \le x_2]$ from the symmetry argument in the first line of this hint. Thus, $P[X_2 \le x_2] = P[X_1 \le -1] + P[-1 < X_1 \le x_2] = P[X_1 \le x_2]$, which is a standard normal probability.

(b) Consider the linear combination $X_1 - X_2$, which equals zero with probability $P[|X_1| > 1] = .3174$.

4.9. Refer to Exercise 4.8, but modify the construction by replacing the break point 1 by c so that

$$X_2 = \begin{cases} -X_1 & \text{if } -c \le X_1 \le c \\ X_1 & \text{elsewhere} \end{cases}$$

Show that c can be chosen so that $\text{Cov}(X_1, X_2) = 0$, but that the two random variables are not independent.

Hint:

For $c = 0$, evaluate $\text{Cov}(X_1, X_2) = E[X_1(X_1)]$

For c very large, evaluate $\text{Cov}(X_1, X_2) \doteq E[X_1(-X_1)]$.

4.10. Show each of the following.

(a)

$$\begin{vmatrix} \mathbf{A} & \mathbf{0} \\ \mathbf{0}' & \mathbf{B} \end{vmatrix} = |\mathbf{A}||\mathbf{B}|$$

(b)

$$\begin{vmatrix} \mathbf{A} & \mathbf{C} \\ \mathbf{0}' & \mathbf{B} \end{vmatrix} = |\mathbf{A}||\mathbf{B}| \quad \text{for} \quad |\mathbf{A}| \ne 0$$

Hint:

(a) $\begin{vmatrix} \mathbf{A} & \mathbf{0} \\ \mathbf{0}' & \mathbf{B} \end{vmatrix} = \begin{vmatrix} \mathbf{A} & \mathbf{0} \\ \mathbf{0}' & \mathbf{I} \end{vmatrix}\begin{vmatrix} \mathbf{I} & \mathbf{0} \\ \mathbf{0}' & \mathbf{B} \end{vmatrix}$. Expanding the determinant $\begin{vmatrix} \mathbf{I} & \mathbf{0} \\ \mathbf{0}' & \mathbf{B} \end{vmatrix}$ by the first row (see Definition 2A.24) gives 1 times a determinant of the same form, with the order of $\mathbf{I}$ reduced by one. This procedure is repeated until $1 \times |\mathbf{B}|$ is obtained. Similarly, expanding the determinant $\begin{vmatrix} \mathbf{A} & \mathbf{0} \\ \mathbf{0}' & \mathbf{I} \end{vmatrix}$ by the last row gives $\begin{vmatrix} \mathbf{A} & \mathbf{0} \\ \mathbf{0}' & \mathbf{I} \end{vmatrix} = |\mathbf{A}|$.

(b) $\begin{vmatrix} \mathbf{A} & \mathbf{C} \\ \mathbf{0}' & \mathbf{B} \end{vmatrix} = \begin{vmatrix} \mathbf{A} & \mathbf{0} \\ \mathbf{0}' & \mathbf{B} \end{vmatrix} \begin{vmatrix} \mathbf{I} & \mathbf{A}^{-1}\mathbf{C} \\ \mathbf{0}' & \mathbf{I} \end{vmatrix}$. But expanding the determinant $\begin{vmatrix} \mathbf{I} & \mathbf{A}^{-1}\mathbf{C} \\ \mathbf{0}' & \mathbf{I} \end{vmatrix}$

by the last row gives $\begin{vmatrix} \mathbf{I} & \mathbf{A}^{-1}\mathbf{C} \\ \mathbf{0}' & \mathbf{I} \end{vmatrix} = 1$. Now use the result in Part a.

4.11. Show that, if $\mathbf{A}$ is square,

$$|\mathbf{A}| = |\mathbf{A}_{22}||\mathbf{A}_{11} - \mathbf{A}_{12}\mathbf{A}_{22}^{-1}\mathbf{A}_{21}| \quad \text{for } |\mathbf{A}_{22}| \neq 0$$

$$= |\mathbf{A}_{11}||\mathbf{A}_{22} - \mathbf{A}_{21}\mathbf{A}_{11}^{-1}\mathbf{A}_{12}| \quad \text{for } |\mathbf{A}_{11}| \neq 0$$

Hint: Partition $\mathbf{A}$ and verify that

$$\begin{bmatrix} \mathbf{I} & -\mathbf{A}_{12}\mathbf{A}_{22}^{-1} \\ \mathbf{0}' & \mathbf{I} \end{bmatrix} \begin{bmatrix} \mathbf{A}_{11} & \mathbf{A}_{12} \\ \mathbf{A}_{21} & \mathbf{A}_{22} \end{bmatrix} \begin{bmatrix} \mathbf{I} & \mathbf{0} \\ -\mathbf{A}_{22}^{-1}\mathbf{A}_{21} & \mathbf{I} \end{bmatrix} = \begin{bmatrix} \mathbf{A}_{11} - \mathbf{A}_{12}\mathbf{A}_{22}^{-1}\mathbf{A}_{21} & \mathbf{0} \\ \mathbf{0}' & \mathbf{A}_{22} \end{bmatrix}$$

Take determinants on both sides of this equality. Use Exercise 4.10 for the first and third determinants on the left and for the determinant on the right. The second equality for $|\mathbf{A}|$ follows by considering

$$\begin{bmatrix} \mathbf{I} & \mathbf{0} \\ -\mathbf{A}_{21}\mathbf{A}_{11}^{-1} & \mathbf{I} \end{bmatrix} \begin{bmatrix} \mathbf{A}_{11} & \mathbf{A}_{12} \\ \mathbf{A}_{21} & \mathbf{A}_{22} \end{bmatrix} \begin{bmatrix} \mathbf{I} & -\mathbf{A}_{11}^{-1}\mathbf{A}_{12} \\ \mathbf{0}' & \mathbf{I} \end{bmatrix} = \begin{bmatrix} \mathbf{A}_{11} & \mathbf{0} \\ \mathbf{0}' & \mathbf{A}_{22} - \mathbf{A}_{21}\mathbf{A}_{11}^{-1}\mathbf{A}_{12} \end{bmatrix}$$

4.12. Show that, for $\mathbf{A}$ symmetric,

$$\mathbf{A}^{-1} = \begin{bmatrix} \mathbf{I} & \mathbf{0} \\ -\mathbf{A}_{22}^{-1}\mathbf{A}_{21} & \mathbf{I} \end{bmatrix} \begin{bmatrix} (\mathbf{A}_{11} - \mathbf{A}_{12}\mathbf{A}_{22}^{-1}\mathbf{A}_{21})^{-1} & \mathbf{0} \\ \mathbf{0}' & \mathbf{A}_{22}^{-1} \end{bmatrix} \begin{bmatrix} \mathbf{I} & -\mathbf{A}_{12}\mathbf{A}_{22}^{-1} \\ \mathbf{0}' & \mathbf{I} \end{bmatrix}$$

Thus, $(\mathbf{A}_{11} - \mathbf{A}_{12}\mathbf{A}_{22}^{-1}\mathbf{A}_{21})^{-1}$ is the upper left-hand block of $\mathbf{A}^{-1}$.

Hint: Premultiply the expression in the hint to Exercise 4.11 by $\begin{bmatrix} \mathbf{I} & -\mathbf{A}_{12}\mathbf{A}_{22}^{-1} \\ \mathbf{0}' & \mathbf{I} \end{bmatrix}^{-1}$ and

postmultiply by $\begin{bmatrix} \mathbf{I} & \mathbf{0} \\ -\mathbf{A}_{22}^{-1}\mathbf{A}_{21} & \mathbf{I} \end{bmatrix}^{-1}$. Take inverses of the resulting expression.

4.13. Show the following if $\mathbf{X}$ is $N_p(\boldsymbol{\mu}, \boldsymbol{\Sigma})$ with $|\boldsymbol{\Sigma}| \neq 0$.

(a) Check that $|\boldsymbol{\Sigma}| = |\boldsymbol{\Sigma}_{22}||\boldsymbol{\Sigma}_{11} - \boldsymbol{\Sigma}_{12}\boldsymbol{\Sigma}_{22}^{-1}\boldsymbol{\Sigma}_{21}|$. (Note that $|\boldsymbol{\Sigma}|$ can be factored into the product of contributions from the marginal and conditional distributions.)

(b) Check that

$$(\mathbf{x} - \boldsymbol{\mu})'\boldsymbol{\Sigma}^{-1}(\mathbf{x} - \boldsymbol{\mu}) = [\mathbf{x}_1 - \boldsymbol{\mu}_1 - \boldsymbol{\Sigma}_{12}\boldsymbol{\Sigma}_{22}^{-1}(\mathbf{x}_2 - \boldsymbol{\mu}_2)]'$$

$$\times (\boldsymbol{\Sigma}_{11} - \boldsymbol{\Sigma}_{12}\boldsymbol{\Sigma}_{22}^{-1}\boldsymbol{\Sigma}_{21})^{-1}[\mathbf{x}_1 - \boldsymbol{\mu}_1 - \boldsymbol{\Sigma}_{12}\boldsymbol{\Sigma}_{22}^{-1}(\mathbf{x}_2 - \boldsymbol{\mu}_2)]$$

$$+ (\mathbf{x}_2 - \boldsymbol{\mu}_2)'\boldsymbol{\Sigma}_{22}^{-1}(\mathbf{x}_2 - \boldsymbol{\mu}_2)$$

(Thus, the joint density exponent can be written as the sum of two terms corresponding to contributions from the conditional and marginal distributions.)

(c) Given the results in Parts a and b, identify the marginal distribution of $\mathbf{X}_2$ and the conditional distribution of $\mathbf{X}_1 | \mathbf{X}_2 = \mathbf{x}_2$.

Hint:

(a) Apply Exercise 4.11.

(b) Note from Exercise 4.12 that we can write $(\mathbf{x} - \boldsymbol{\mu})'\boldsymbol{\Sigma}^{-1}(\mathbf{x} - \boldsymbol{\mu})$ as

$$\begin{bmatrix} \mathbf{x}_1 - \boldsymbol{\mu}_1 \\ \mathbf{x}_2 - \boldsymbol{\mu}_2 \end{bmatrix}' \begin{bmatrix} \mathbf{I} & \mathbf{0} \\ -\boldsymbol{\Sigma}_{22}^{-1}\boldsymbol{\Sigma}_{21} & \mathbf{I} \end{bmatrix} \begin{bmatrix} (\boldsymbol{\Sigma}_{11} - \boldsymbol{\Sigma}_{12}\boldsymbol{\Sigma}_{22}^{-1}\boldsymbol{\Sigma}_{21})^{-1} & \mathbf{0} \\ \mathbf{0}' & \boldsymbol{\Sigma}_{22}^{-1} \end{bmatrix}$$

$$\times \begin{bmatrix} \mathbf{I} & -\boldsymbol{\Sigma}_{12}\boldsymbol{\Sigma}_{22}^{-1} \\ \mathbf{0}' & \mathbf{I} \end{bmatrix} \begin{bmatrix} \mathbf{x}_1 - \boldsymbol{\mu}_1 \\ \mathbf{x}_2 - \boldsymbol{\mu}_2 \end{bmatrix}$$

If we group the product so that

$$\begin{bmatrix} \mathbf{I} & -\boldsymbol{\Sigma}_{12}\boldsymbol{\Sigma}_{22}^{-1} \\ \mathbf{0}' & \mathbf{I} \end{bmatrix} \begin{bmatrix} \mathbf{x}_1 - \boldsymbol{\mu}_1 \\ \mathbf{x}_2 - \boldsymbol{\mu}_2 \end{bmatrix} = \begin{bmatrix} \mathbf{x}_1 - \boldsymbol{\mu}_1 - \boldsymbol{\Sigma}_{12}\boldsymbol{\Sigma}_{22}^{-1}(\mathbf{x}_2 - \boldsymbol{\mu}_2) \\ \mathbf{x}_2 - \boldsymbol{\mu}_2 \end{bmatrix}$$

the result follows.

4.14. If $\mathbf{X}$ is distributed as $N_p(\boldsymbol{\mu}, \boldsymbol{\Sigma})$ with $|\boldsymbol{\Sigma}| \neq 0$, show that the joint density can be written as the product of marginal densities for

$$\underset{(q \times 1)}{\mathbf{X}_1} \quad \text{and} \quad \underset{((p-q) \times 1)}{\mathbf{X}_2} \quad \text{if} \quad \underset{(q \times (p-q))}{\boldsymbol{\Sigma}_{12}} = \mathbf{0}$$

Hint: Show by block multiplication that

$$\begin{bmatrix} \boldsymbol{\Sigma}_{11}^{-1} & \mathbf{0} \\ \mathbf{0}' & \boldsymbol{\Sigma}_{22}^{-1} \end{bmatrix} \text{ is the inverse of } \boldsymbol{\Sigma} = \begin{bmatrix} \boldsymbol{\Sigma}_{11} & \mathbf{0} \\ \mathbf{0}' & \boldsymbol{\Sigma}_{22} \end{bmatrix}$$

Then write

$$(\mathbf{x} - \boldsymbol{\mu})'\boldsymbol{\Sigma}^{-1}(\mathbf{x} - \boldsymbol{\mu}) = [(\mathbf{x}_1 - \boldsymbol{\mu}_1)', (\mathbf{x}_2 - \boldsymbol{\mu}_2)'] \begin{bmatrix} \boldsymbol{\Sigma}_{11}^{-1} & \mathbf{0} \\ \mathbf{0}' & \boldsymbol{\Sigma}_{22}^{-1} \end{bmatrix} \begin{bmatrix} \mathbf{x}_1 - \boldsymbol{\mu}_1 \\ \mathbf{x}_2 - \boldsymbol{\mu}_2 \end{bmatrix}$$

$$= (\mathbf{x}_1 - \boldsymbol{\mu}_1)'\boldsymbol{\Sigma}_{11}^{-1}(\mathbf{x}_1 - \boldsymbol{\mu}_1) + (\mathbf{x}_2 - \boldsymbol{\mu}_2)'\boldsymbol{\Sigma}_{22}^{-1}(\mathbf{x}_2 - \boldsymbol{\mu}_2)$$

Note that $|\boldsymbol{\Sigma}| = |\boldsymbol{\Sigma}_{11}||\boldsymbol{\Sigma}_{22}|$ from Exercise 4.10(a). Now factor the joint density.

4.15. Show that $\sum_{j=1}^{n} (\mathbf{x}_j - \bar{\mathbf{x}})(\bar{\mathbf{x}} - \boldsymbol{\mu})'$ and $\sum_{j=1}^{n} (\bar{\mathbf{x}} - \boldsymbol{\mu})(\mathbf{x}_j - \bar{\mathbf{x}})'$ are both $p \times p$ matrices of zeros. Here $\mathbf{x}_j' = [x_{j1}, x_{j2}, \dots, x_{jp}], j = 1, 2, \dots, n$, and

$$\bar{\mathbf{x}} = \frac{1}{n} \sum_{j=1}^{n} \mathbf{x}_j$$

4.16. Let $\mathbf{X}_1, \mathbf{X}_2, \mathbf{X}_3,$ and $\mathbf{X}_4$ be independent $N_p(\boldsymbol{\mu}, \boldsymbol{\Sigma})$ random vectors.

(a) Find the marginal distributions for each of the random vectors

$$\mathbf{V}_1 = \tfrac{1}{4}\mathbf{X}_1 - \tfrac{1}{4}\mathbf{X}_2 + \tfrac{1}{4}\mathbf{X}_3 - \tfrac{1}{4}\mathbf{X}_4$$

and

$$\mathbf{V}_2 = \tfrac{1}{4}\mathbf{X}_1 + \tfrac{1}{4}\mathbf{X}_2 - \tfrac{1}{4}\mathbf{X}_3 - \tfrac{1}{4}\mathbf{X}_4$$

(b) Find the joint density of the random vectors $\mathbf{V}_1$ and $\mathbf{V}_2$ defined in (a).

4.17. Let $\mathbf{X}_1, \mathbf{X}_2, \mathbf{X}_3, \mathbf{X}_4,$ and $\mathbf{X}_5$ be independent and identically distributed random vectors with mean vector $\boldsymbol{\mu}$ and covariance matrix $\boldsymbol{\Sigma}$. Find the mean vector and covariance matrices for each of the two linear combinations of random vectors

$$\tfrac{1}{5}\mathbf{X}_1 + \tfrac{1}{5}\mathbf{X}_2 + \tfrac{1}{5}\mathbf{X}_3 + \tfrac{1}{5}\mathbf{X}_4 + \tfrac{1}{5}\mathbf{X}_5$$

and

$$\mathbf{X}_1 - \mathbf{X}_2 + \mathbf{X}_3 - \mathbf{X}_4 + \mathbf{X}_5$$

in terms of $\boldsymbol{\mu}$ and $\boldsymbol{\Sigma}$. Also, obtain the covariance between the two linear combinations of random vectors.

4.18. Find the maximum likelihood estimates of the 2×1 mean vector $\boldsymbol{\mu}$ and the 2×2 covariance matrix $\boldsymbol{\Sigma}$ based on the random sample

$$\mathbf{X} = \begin{bmatrix} 3 & 6 \\ 4 & 4 \\ 5 & 7 \\ 4 & 7 \end{bmatrix}.$$

from a bivariate normal population.

4.19. Let $\mathbf{X}_1, \mathbf{X}_2, \ldots, \mathbf{X}_{20}$ be a random sample of size $n = 20$ from an $N_6(\boldsymbol{\mu}, \boldsymbol{\Sigma})$ population. Specify each of the following completely.

(a) The distribution of $(\mathbf{X}_1 - \boldsymbol{\mu})' \boldsymbol{\Sigma}^{-1} (\mathbf{X}_1 - \boldsymbol{\mu})$

(b) The distributions of $\overline{\mathbf{X}}$ and $\sqrt{n}(\overline{\mathbf{X}} - \boldsymbol{\mu})$

(c) The distribution of $(n - 1)\mathbf{S}$

4.20. For the random variables $\mathbf{X}_1, \mathbf{X}_2, \ldots, \mathbf{X}_{20}$ in Exercise 4.19, specify the distribution of $\mathbf{B}(19\mathbf{S})\mathbf{B}'$ in each case.

(a) $\mathbf{B} = \begin{bmatrix} 1 & -\frac{1}{2} & -\frac{1}{2} & 0 & 0 & 0 \\ 0 & 0 & 0 & -\frac{1}{2} & -\frac{1}{2} & 1 \end{bmatrix}$

(b) $\mathbf{B} = \begin{bmatrix} 1 & 0 & 0 & 0 & 0 & 0 \\ 0 & 0 & 1 & 0 & 0 & 0 \end{bmatrix}$

4.21. Let $\mathbf{X}_1, \ldots, \mathbf{X}_{60}$ be a random sample of size 60 from a four-variate normal distribution having mean $\boldsymbol{\mu}$ and covariance $\boldsymbol{\Sigma}$. Specify each of the following completely.

(a) The distribution of $\overline{\mathbf{X}}$

(b) The distribution of $(\mathbf{X}_1 - \boldsymbol{\mu})' \boldsymbol{\Sigma}^{-1} (\mathbf{X}_1 - \boldsymbol{\mu})$

(c) The distribution of $n(\overline{\mathbf{X}} - \boldsymbol{\mu})' \boldsymbol{\Sigma}^{-1} (\overline{\mathbf{X}} - \boldsymbol{\mu})$

(d) The approximate distribution of $n(\overline{\mathbf{X}} - \boldsymbol{\mu})' \mathbf{S}^{-1} (\overline{\mathbf{X}} - \boldsymbol{\mu})$

4.22. Let $\mathbf{X}_1, \mathbf{X}_2, \ldots, \mathbf{X}_{75}$ be a random sample from a population distribution with mean $\boldsymbol{\mu}$ and covariance matrix $\boldsymbol{\Sigma}$. What is the approximate distribution of each of the following?

(a) $\overline{\mathbf{X}}$

(b) $n(\overline{\mathbf{X}} - \boldsymbol{\mu})' \mathbf{S}^{-1} (\overline{\mathbf{X}} - \boldsymbol{\mu})$

4.23. Consider the annual rates of return (including dividends) on the Dow-Jones industrial average for the years 1996–2005. These data, multiplied by 100, are

$-0.6 \quad 3.1 \quad 25.3 \quad -16.8 \quad -7.1 \quad -6.2 \quad 16.1 \quad 25.2 \quad 22.6 \quad 26.0.$

Use these 10 observations to complete the following.

(a) Construct a Q–Q plot. Do the data seem to be normally distributed? Explain.

(b) Carry out a test of normality based on the correlation coefficient r_Q. [See (4–31).] Let the significance level be $\alpha = .10$.

4.24. Exercise 1.4 contains data on three variables for the world's 10 largest companies as of April 2005. For the sales (x_1) and profits (x_2) data:

(a) Construct Q–Q plots. Do these data appear to be normally distributed? Explain.

(b) Carry out a test of normality based on the correlation coefficient r_Q. [See (4–31).] Set the significance level at $\alpha = .10$. Do the results of these tests corroborate the results in Part a?

4.25. Refer to the data for the world's 10 largest companies in Exercise 1.4. Construct a chi-square plot using all *three* variables. The chi-square quantiles are

0.3518 0.7978 1.2125 1.6416 2.1095 2.6430 3.2831 4.1083 5.3170 7.8147

4.26. Exercise 1.2 gives the age x_1, measured in years, as well as the selling price x_2, measured in thousands of dollars, for $n = 10$ used cars. These data are reproduced as follows:

x_1	1	2	3	3	4	5	6	8	9	11
x_2	18.95	19.00	17.95	15.54	14.00	12.95	8.94	7.49	6.00	3.99

(a) Use the results of Exercise 1.2 to calculate the squared statistical distances $(\mathbf{x}_j - \bar{\mathbf{x}})'\mathbf{S}^{-1}(\mathbf{x}_j - \bar{\mathbf{x}})$, $j = 1, 2, \ldots, 10$, where $\mathbf{x}_j' = [x_{j1}, x_{j2}]$.

(b) Using the distances in Part a, determine the proportion of the observations falling within the estimated 50% probability contour of a bivariate normal distribution.

(c) Order the distances in Part a and construct a chi-square plot.

(d) Given the results in Parts b and c, are these data approximately bivariate normal? Explain.

4.27. Consider the radiation data (with door closed) in Example 4.10. Construct a Q–Q plot for the natural logarithms of these data. [Note that the natural logarithm transformation corresponds to the value $\lambda = 0$ in (4-34).] Do the natural logarithms appear to be normally distributed? Compare your results with Figure 4.13. Does the choice $\lambda = \frac{1}{4}$ or $\lambda = 0$ make much difference in this case?

The following exercises may require a computer.

4.28. Consider the air-pollution data given in Table 1.5. Construct a Q–Q plot for the solar radiation measurements and carry out a test for normality based on the correlation coefficient r_Q [see (4-31)]. Let $\alpha = .05$ and use the entry corresponding to $n = 40$ in Table 4.2.

4.29. Given the air-pollution data in Table 1.5, examine the pairs $X_5 = NO_2$ and $X_6 = O_3$ for bivariate normality.

(a) Calculate statistical distances $(\mathbf{x}_j - \bar{\mathbf{x}})'\mathbf{S}^{-1}(\mathbf{x}_j - \bar{\mathbf{x}})$, $j = 1, 2, \ldots, 42$, where $\mathbf{x}_j' = [x_{j5}, x_{j6}]$.

(b) Determine the proportion of observations $\mathbf{x}_j' = [x_{j5}, x_{j6}]$, $j = 1, 2, \ldots, 42$, falling within the approximate 50% probability contour of a bivariate normal distribution.

(c) Construct a chi-square plot of the ordered distances in Part a.

4.30. Consider the used-car data in Exercise 4.26.

(a) Determine the power transformation $\hat{\lambda}_1$ that makes the x_1 values approximately normal. Construct a Q–Q plot for the transformed data.

(b) Determine the power transformations $\hat{\lambda}_2$ that makes the x_2 values approximately normal. Construct a Q–Q plot for the transformed data.

(c) Determine the power transformations $\hat{\mathbf{\lambda}}' = [\hat{\lambda}_1, \hat{\lambda}_2]$ that make the $[x_1, x_2]$ values jointly normal using (4-40). Compare the results with those obtained in Parts a and b.

4.31. Examine the marginal normality of the observations on variables $X_1, X_2, \ldots, X_5$ for the multiple-sclerosis data in Table 1.6. Treat the non-multiple-sclerosis and multiple-sclerosis groups separately. Use whatever methodology, including transformations, you feel is appropriate.

4.32. Examine the marginal normality of the observations on variables $X_1, X_2, \ldots, X_6$ for the radiotherapy data in Table 1.7. Use whatever methodology, including transformations, you feel is appropriate.

4.33. Examine the marginal and bivariate normality of the observations on variables X_1, X_2, X_3, and X_4 for the data in Table 4.3.

4.34. Examine the data on bone mineral content in Table 1.8 for marginal and bivariate normality.

4.35. Examine the data on paper-quality measurements in Table 1.2 for marginal and multivariate normality.

4.36. Examine the data on women's national track records in Table 1.9 for marginal and multivariate normality.

4.37. Refer to Exercise 1.18. Convert the women's track records in Table 1.9 to speeds measured in meters per second. Examine the data on speeds for marginal and multivariate normality.

4.38. Examine the data on bulls in Table 1.10 for marginal and multivariate normality. Consider only the variables YrHgt, FtFrBody, PrctFFB, BkFat, SaleHt, and SaleWt.

4.39. The data in Table 4.6 (see the psychological profile data: www.prenhall.com/statistics) consist of 130 observations generated by scores on a psychological test administered to Peruvian teenagers (ages 15, 16, and 17). For each of these teenagers the gender (male = 1, female = 2) and socioeconomic status (low = 1, medium = 2) were also recorded. The scores were accumulated into five subscale scores labeled *independence* (indep), *support* (supp), *benevolence* (benev), *conformity* (conform), and *leadership* (leader).

Table 4.6 Psychological Profile Data

Indep	Supp	Benev	Conform	Leader	Gender	Socio
27	13	14	20	11	2	1
12	13	24	25	6	2	1
14	20	15	16	7	2	1
18	20	17	12	6	2	1
9	22	22	21	6	2	1
⋮	⋮	⋮	⋮	⋮	⋮	⋮
10	11	26	17	10	1	2
14	12	14	11	29	1	2
19	11	23	18	13	2	2
27	19	22	7	9	2	2
10	17	22	22	8	2	2

Source: Data courtesy of C. Soto.

(a) Examine each of the variables independence, support, benevolence, conformity and leadership for marginal normality.

(b) Using all five variables, check for multivariate normality.

(c) Refer to part (a). For those variables that are nonnormal, determine the transformation that makes them more nearly normal.

4.40. Consider the data on national parks in Exercise 1.27.

(a) Comment on any possible outliers in a scatter plot of the original variables.

(b) Determine the power transformation $\hat{\lambda}_1$ the makes the x_1 values approximately normal. Construct a $Q-Q$ plot of the transformed observations.

(c) Determine the power transformation $\hat{\lambda}_2$ the makes the x_2 values approximately normal. Construct a $Q-Q$ plot of the transformed observations.

(d) Determine the power transformation for approximate bivariate normality using (4-40).

4.41. Consider the data on snow removal in Exercise 3.20.

(a) Comment on any possible outliers in a scatter plot of the original variables.

(b) Determine the power transformation $\hat{\lambda}_1$ the makes the x_1 values approximately normal. Construct a $Q-Q$ plot of the transformed observations.

(c) Determine the power transformation $\hat{\lambda}_2$ the makes the x_2 values approximately normal. Construct a $Q-Q$ plot of the transformed observations.

(d) Determine the power transformation for approximate bivariate normality using (4-40).

References

1. Anderson, T. W. *An Introduction to Multivariate Statistical Analysis* (3rd ed.). New York: John Wiley, 2003.

2. Andrews, D. F., R. Gnanadesikan, and J. L. Warner. "Transformations of Multivariate Data." *Biometrics*, **27**, no. 4 (1971), 825–840.

3. Box, G. E. P., and D. R. Cox. "An Analysis of Transformations" (with discussion). *Journal of the Royal Statistical Society (B)*, **26**, no. 2 (1964), 211–252.

4. Daniel, C. and F. S. Wood, *Fitting Equations to Data: Computer Analysis of Multifactor Data*. New York: John Wiley, 1980.

5. Filliben, J. J. "The Probability Plot Correlation Coefficient Test for Normality." *Technometrics*, **17**, no. 1 (1975), 111–117.

6. Gnanadesikan, R. *Methods for Statistical Data Analysis of Multivariate Observations* (2nd ed.). New York: Wiley-Interscience, 1977.

7. Hawkins, D. M. *Identification of Outliers*. London, UK: Chapman and Hall, 1980.

8. Hernandez, F., and R. A. Johnson. "The Large-Sample Behavior of Transformations to Normality." *Journal of the American Statistical Association*, **75**, no. 372 (1980), 855–861.

9. Hogg, R. V., Craig. A. T. and J. W. Mckean *Introduction to Mathematical Statistics* (6th ed.). Upper Saddle River, N.J.: Prentice Hall, 2004.

10. Looney, S. W., and T. R. Gulledge, Jr. "Use of the Correlation Coefficient with Normal Probability Plots." *The American Statistician*, **39**, no. 1 (1985), 75–79.

11. Mardia, K. V., Kent, J. T. and J. M. Bibby. *Multivariate Analysis* (Paperback). London: Academic Press, 2003.

12. Shapiro, S. S., and M. B. Wilk. "An Analysis of Variance Test for Normality (Complete Samples)." *Biometrika*, **52**, no. 4 (1965), 591–611.

13. Verrill, S., and R. A. Johnson. "Tables and Large-Sample Distribution Theory for Censored-Data Correlation Statistics for Testing Normality." *Journal of the American Statistical Association*, **83**, no. 404 (1988), 1192–1197.

14. Yeo, I. and R. A. Johnson "A New Family of Power Transformations to Improve Normality or Symmetry." *Biometrika*, **87**, no. 4 (2000), 954–959.

15. Zehna, P. "Invariance of Maximum Likelihood Estimators." *Annals of Mathematical Statistics*, **37**, no. 3 (1966), 744.

Chapter

5

INFERENCES ABOUT A MEAN VECTOR

5.1 Introduction

This chapter is the first of the methodological sections of the book. We shall now use the concepts and results set forth in Chapters 1 through 4 to develop techniques for analyzing data. A large part of any analysis is concerned with *inference*—that is, reaching valid conclusions concerning a population on the basis of information from a sample.

At this point, we shall concentrate on inferences about a population mean vector and its component parts. Although we introduce statistical inference through initial discussions of tests of hypotheses, our ultimate aim is to present a full statistical analysis of the component means based on simultaneous confidence statements.

One of the central messages of multivariate analysis is that p correlated variables must be analyzed jointly. This principle is exemplified by the methods presented in this chapter.

5.2 The Plausibility of μ_0 as a Value for a Normal Population Mean

Let us start by recalling the univariate theory for determining whether a specific value μ_0 is a plausible value for the population mean μ. From the point of view of hypothesis testing, this problem can be formulated as a *test* of the competing *hypotheses*

$$H_0: \mu = \mu_0 \quad \text{and} \quad H_1: \mu \neq \mu_0$$

Here H_0 is the null hypothesis and H_1 is the (two-sided) alternative hypothesis. If $X_1, X_2, \ldots, X_n$ denote a random sample from a normal population, the appropriate test statistic is

$$t = \frac{(\overline{X} - \mu_0)}{s/\sqrt{n}}, \quad \text{where} \quad \overline{X} = \frac{1}{n} \sum_{j=1}^{n} X_j \quad \text{and} \quad s^2 = \frac{1}{n-1} \sum_{j=1}^{n} (X_j - \overline{X})^2$$

This test statistic has a student's t-distribution with $n - 1$ degrees of freedom (d.f.). We reject H_0, that μ_0 is a plausible value of μ, if the observed $|t|$ exceeds a specified percentage point of a t-distribution with $n - 1$ d.f.

Rejecting H_0 when $|t|$ is large is equivalent to rejecting H_0 if its square,

$$t^2 = \frac{(\overline{X} - \mu_0)^2}{s^2/n} = n(\overline{X} - \mu_0)(s^2)^{-1}(\overline{X} - \mu_0) \qquad (5\text{-}1)$$

is large. The variable t^2 in (5-1) is the square of the distance from the sample mean $\overline{X}$ to the test value μ_0. The units of distance are expressed in terms of $s/\sqrt{n}$, or estimated standard deviations of $\overline{X}$. Once $\overline{X}$ and s^2 are observed, the test becomes: Reject H_0 in favor of H_1, at significance level α, if

$$n(\overline{x} - \mu_0)(s^2)^{-1}(\overline{x} - \mu_0) > t_{n-1}^2(\alpha/2) \qquad (5\text{-}2)$$

where $t_{n-1}(\alpha/2)$ denotes the upper $100(\alpha/2)$th percentile of the t-distribution with $n - 1$ d.f.

If H_0 is not rejected, we conclude that μ_0 is a plausible value for the normal population mean. Are there other values of μ which are also consistent with the data? The answer is yes! In fact, there is always a *set* of plausible values for a normal population mean. From the well-known correspondence between acceptance regions for tests of $H_0\colon \mu = \mu_0$ versus $H_1\colon \mu \neq \mu_0$ and confidence intervals for μ, we have

$$\{\text{Do not reject } H_0\colon \mu = \mu_0 \text{ at level } \alpha\} \quad \text{or} \quad \left| \frac{\overline{x} - \mu_0}{s/\sqrt{n}} \right| \leq t_{n-1}(\alpha/2)$$

is equivalent to

$$\left\{ \mu_0 \text{ lies in the } 100(1 - \alpha)\% \text{ confidence interval } \overline{x} \pm t_{n-1}(\alpha/2)\frac{s}{\sqrt{n}} \right\}$$

or

$$\overline{x} - t_{n-1}(\alpha/2)\frac{s}{\sqrt{n}} \leq \mu_0 \leq \overline{x} + t_{n-1}(\alpha/2)\frac{s}{\sqrt{n}} \qquad (5\text{-}3)$$

The confidence interval consists of all those values μ_0 that would not be rejected by the level α test of $H_0\colon \mu = \mu_0$.

Before the sample is selected, the $100(1 - \alpha)\%$ confidence interval in (5-3) is a *random interval* because the endpoints depend upon the random variables $\overline{X}$ and s. The probability that the interval contains μ is $1 - \alpha$; among large numbers of such independent intervals, approximately $100(1 - \alpha)\%$ of them will contain μ.

Consider now the problem of determining whether a given $p \times 1$ vector $\boldsymbol{\mu}_0$ is a plausible value for the mean of a multivariate normal distribution. We shall proceed by analogy to the univariate development just presented.

A natural generalization of the squared distance in (5-1) is its multivariate analog

$$T^2 = (\overline{\mathbf{X}} - \boldsymbol{\mu}_0)'\left(\frac{1}{n}\mathbf{S}\right)^{-1}(\overline{\mathbf{X}} - \boldsymbol{\mu}_0) = n(\overline{\mathbf{X}} - \boldsymbol{\mu}_0)'\mathbf{S}^{-1}(\overline{\mathbf{X}} - \boldsymbol{\mu}_0) \qquad (5\text{-}4)$$

where

$$\underset{(p \times 1)}{\overline{\mathbf{X}}} = \frac{1}{n} \sum_{j=1}^{n} \mathbf{X}_j, \qquad \underset{(p \times p)}{\mathbf{S}} = \frac{1}{n-1} \sum_{j=1}^{n} (\mathbf{X}_j - \overline{\mathbf{X}})(\mathbf{X}_j - \overline{\mathbf{X}})', \text{ and } \underset{(p \times 1)}{\boldsymbol{\mu}_0} = \begin{bmatrix} \mu_{10} \\ \mu_{20} \\ \vdots \\ \mu_{p0} \end{bmatrix}$$

The statistic T^2 is called *Hotelling's T^2* in honor of Harold Hotelling, a pioneer in multivariate analysis, who first obtained its sampling distribution. Here $(1/n)\mathbf{S}$ is the estimated covariance matrix of $\overline{\mathbf{X}}$. (See Result 3.1.)

If the observed statistical distance T^2 is too large—that is, if $\overline{\mathbf{x}}$ is "too far" from $\boldsymbol{\mu}_0$—the hypothesis $H_0: \boldsymbol{\mu} = \boldsymbol{\mu}_0$ is rejected. It turns out that special tables of T^2 percentage points are not required for formal tests of hypotheses. This is true because

$$T^2 \text{ is distributed as } \frac{(n-1)p}{(n-p)} F_{p, n-p} \tag{5-5}$$

where $F_{p,n-p}$ denotes a random variable with an F-distribution with p and $n - p$ d.f.

To summarize, we have the following:

Let $\mathbf{X}_1, \mathbf{X}_2, \ldots, \mathbf{X}_n$ be a random sample from an $N_p(\boldsymbol{\mu}, \boldsymbol{\Sigma})$ population. Then with $\overline{\mathbf{X}} = \dfrac{1}{n} \sum_{j=1}^{n} \mathbf{X}_j$ and $\mathbf{S} = \dfrac{1}{(n-1)} \sum_{j=1}^{n} (\mathbf{X}_j - \overline{\mathbf{X}})(\mathbf{X}_j - \overline{\mathbf{X}})'$,

$$\alpha = P\left[T^2 > \frac{(n-1)p}{(n-p)} F_{p,n-p}(\alpha) \right]$$

$$= P\left[n(\overline{\mathbf{X}} - \boldsymbol{\mu})' \mathbf{S}^{-1}(\overline{\mathbf{X}} - \boldsymbol{\mu}) > \frac{(n-1)p}{(n-p)} F_{p,n-p}(\alpha) \right] \tag{5-6}$$

whatever the true $\boldsymbol{\mu}$ and $\boldsymbol{\Sigma}$. Here $F_{p,n-p}(\alpha)$ is the upper (100α)th percentile of the $F_{p,n-p}$ distribution.

Statement (5-6) leads immediately to a test of the hypothesis $H_0: \boldsymbol{\mu} = \boldsymbol{\mu}_0$ versus $H_1: \boldsymbol{\mu} \neq \boldsymbol{\mu}_0$. At the α level of significance, we reject H_0 in favor of H_1 if the observed

$$T^2 = n(\overline{\mathbf{x}} - \boldsymbol{\mu}_0)' \mathbf{S}^{-1}(\overline{\mathbf{x}} - \boldsymbol{\mu}_0) > \frac{(n-1)p}{(n-p)} F_{p,n-p}(\alpha) \tag{5-7}$$

It is informative to discuss the nature of the T^2-distribution briefly and its correspondence with the univariate test statistic. In Section 4.4, we described the manner in which the Wishart distribution generalizes the chi-square distribution. We can write

$$T^2 = \sqrt{n}\,(\overline{\mathbf{X}} - \boldsymbol{\mu}_0)' \left(\frac{\displaystyle\sum_{j=1}^{n} (\mathbf{X}_j - \overline{\mathbf{X}})(\mathbf{X}_j - \overline{\mathbf{X}})'}{n-1} \right)^{-1} \sqrt{n}\,(\overline{\mathbf{X}} - \boldsymbol{\mu}_0)$$

which combines a normal, $N_p(\mathbf{0}, \boldsymbol{\Sigma})$, random vector and a Wishart, $\mathbf{W}_{p,n-1}(\boldsymbol{\Sigma})$, random matrix in the form

$$
T^2_{p,n-1} = \left(\begin{array}{c} \text{multivariate normal} \\ \text{random vector} \end{array} \right)' \left(\frac{\begin{array}{c} \text{Wishart random} \\ \text{matrix} \end{array}}{\text{d.f.}} \right)^{-1} \left(\begin{array}{c} \text{multivariate normal} \\ \text{random vector} \end{array} \right)
$$

$$
= N_p(\mathbf{0}, \boldsymbol{\Sigma})' \left[\frac{1}{n-1} \mathbf{W}_{p,n-1}(\boldsymbol{\Sigma}) \right]^{-1} N_p(\mathbf{0}, \boldsymbol{\Sigma}) \tag{5-8}
$$

This is analogous to

$$
t^2 = \sqrt{n}\,(\bar{X} - \mu_0)(s^2)^{-1}\sqrt{n}\,(\bar{X} - \mu_0)
$$

or

$$
t^2_{n-1} = \left(\begin{array}{c} \text{normal} \\ \text{random variable} \end{array} \right) \left(\frac{\begin{array}{c} \text{(scaled) chi-square} \\ \text{random variable} \end{array}}{\text{d.f.}} \right)^{-1} \left(\begin{array}{c} \text{normal} \\ \text{random variable} \end{array} \right)
$$

for the univariate case. Since the multivariate normal and Wishart random variables are independently distributed [see (4-23)], their joint density function is the product of the marginal normal and Wishart distributions. Using calculus, the distribution (5-5) of T^2 as given previously can be derived from this joint distribution and the representation (5-8).

It is rare, in multivariate situations, to be content with a test of $H_0: \boldsymbol{\mu} = \boldsymbol{\mu}_0$, where all of the mean vector components are specified under the null hypothesis. Ordinarily, it is preferable to find regions of $\boldsymbol{\mu}$ values that are plausible in light of the observed data. We shall return to this issue in Section 5.4.

Example 5.1 (Evaluating T^2) Let the data matrix for a random sample of size $n = 3$ from a bivariate normal population be

$$
\mathbf{X} = \begin{bmatrix} 6 & 9 \\ 10 & 6 \\ 8 & 3 \end{bmatrix}
$$

Evaluate the observed T^2 for $\boldsymbol{\mu}'_0 = [9, 5]$. What is the sampling distribution of T^2 in this case? We find

$$
\bar{\mathbf{x}} = \begin{bmatrix} \bar{x}_1 \\ \bar{x}_2 \end{bmatrix} = \begin{bmatrix} \dfrac{6 + 10 + 8}{3} \\ \dfrac{9 + 6 + 3}{3} \end{bmatrix} = \begin{bmatrix} 8 \\ 6 \end{bmatrix}
$$

and

$$
s_{11} = \frac{(6-8)^2 + (10-8)^2 + (8-8)^2}{2} = 4
$$

$$
s_{12} = \frac{(6-8)(9-6) + (10-8)(6-6) + (8-8)(3-6)}{2} = -3
$$

$$
s_{22} = \frac{(9-6)^2 + (6-6)^2 + (3-6)^2}{2} = 9
$$

so

$$\mathbf{S} = \begin{bmatrix} 4 & -3 \\ -3 & 9 \end{bmatrix}$$

Thus,

$$\mathbf{S}^{-1} = \frac{1}{(4)(9) - (-3)(-3)}\begin{bmatrix} 9 & 3 \\ 3 & 4 \end{bmatrix} = \begin{bmatrix} \frac{1}{3} & \frac{1}{9} \\ \frac{1}{9} & \frac{4}{27} \end{bmatrix}$$

and, from (5-4),

$$T^2 = 3[8 - 9, \quad 6 - 5]\begin{bmatrix} \frac{1}{3} & \frac{1}{9} \\ \frac{1}{9} & \frac{4}{27} \end{bmatrix}\begin{bmatrix} 8 - 9 \\ 6 - 5 \end{bmatrix} = 3[-1, \quad 1]\begin{bmatrix} -\frac{2}{9} \\ \frac{1}{27} \end{bmatrix} = \frac{7}{9}$$

Before the sample is selected, T^2 has the distribution of a

$$\frac{(3 - 1)2}{(3 - 2)}F_{2,3-2} = 4F_{2,1}$$

random variable.   ∎

The next example illustrates a test of the hypothesis $H_0: \boldsymbol{\mu} = \boldsymbol{\mu}_0$ using data collected as part of a search for new diagnostic techniques at the University of Wisconsin Medical School.

Example 5.2 (Testing a multivariate mean vector with T^2) Perspiration from 20 healthy females was analyzed. Three components, X_1 = sweat rate, X_2 = sodium content, and X_3 = potassium content, were measured, and the results, which we call the *sweat data*, are presented in Table 5.1.

Test the hypothesis $H_0: \boldsymbol{\mu}' = [4, 50, 10]$ against $H_1: \boldsymbol{\mu}' \neq [4, 50, 10]$ at level of significance $\alpha = .10$.

Computer calculations provide

$$\bar{\mathbf{x}} = \begin{bmatrix} 4.640 \\ 45.400 \\ 9.965 \end{bmatrix}, \quad \mathbf{S} = \begin{bmatrix} 2.879 & 10.010 & -1.810 \\ 10.010 & 199.788 & -5.640 \\ -1.810 & -5.640 & 3.628 \end{bmatrix}$$

and

$$\mathbf{S}^{-1} = \begin{bmatrix} .586 & -.022 & .258 \\ -.022 & .006 & -.002 \\ .258 & -.002 & .402 \end{bmatrix}$$

We evaluate

$$T^2 =$$

$$20[4.640 - 4, \quad 45.400 - 50, \quad 9.965 - 10]\begin{bmatrix} .586 & -.022 & .258 \\ -.022 & .006 & -.002 \\ .258 & -.002 & .402 \end{bmatrix}\begin{bmatrix} 4.640 - 4 \\ 45.400 - 50 \\ 9.965 - 10 \end{bmatrix}$$

$$= 20[.640, \quad -4.600, \quad -.035]\begin{bmatrix} .467 \\ -.042 \\ .160 \end{bmatrix} = 9.74$$

Table 5.1 Sweat Data

Individual	X_1 (Sweat rate)	X_2 (Sodium)	X_3 (Potassium)
1	3.7	48.5	9.3
2	5.7	65.1	8.0
3	3.8	47.2	10.9
4	3.2	53.2	12.0
5	3.1	55.5	9.7
6	4.6	36.1	7.9
7	2.4	24.8	14.0
8	7.2	33.1	7.6
9	6.7	47.4	8.5
10	5.4	54.1	11.3
11	3.9	36.9	12.7
12	4.5	58.8	12.3
13	3.5	27.8	9.8
14	4.5	40.2	8.4
15	1.5	13.5	10.1
16	8.5	56.4	7.1
17	4.5	71.6	8.2
18	6.5	52.8	10.9
19	4.1	44.1	11.2
20	5.5	40.9	9.4

Source: Courtesy of Dr. Gerald Bargman.

Comparing the observed $T^2 = 9.74$ with the critical value

$$\frac{(n-1)p}{(n-p)} F_{p,n-p}(.10) = \frac{19(3)}{17} F_{3,17}(.10) = 3.353(2.44) = 8.18$$

we see that $T^2 = 9.74 > 8.18$, and consequently, we reject H_0 at the 10% level of significance.

We note that H_0 will be rejected if one or more of the component means, or some combination of means, differs too much from the hypothesized values [4, 50, 10]. At this point, we have no idea which of these hypothesized values may not be supported by the data.

We have assumed that the sweat data are multivariate normal. The Q–Q plots constructed from the marginal distributions of X_1, X_2, and X_3 all approximate straight lines. Moreover, scatter plots for pairs of observations have approximate elliptical shapes, and we conclude that the normality assumption was reasonable in this case. (See Exercise 5.4.) ∎

One feature of the T^2-statistic is that it is invariant (unchanged) under changes in the units of measurements for **X** of the form

$$\underset{(p\times1)}{\mathbf{Y}} = \underset{(p\times p)(p\times1)}{\mathbf{C}\ \mathbf{X}} + \underset{(p\times1)}{\mathbf{d}}, \quad \mathbf{C}\ \text{nonsingular} \tag{5-9}$$

A transformation of the observations of this kind arises when a constant b_i is subtracted from the ith variable to form $X_i - b_i$ and the result is multiplied by a constant $a_i > 0$ to get $a_i(X_i - b_i)$. Premultiplication of the *centered* and *scaled* quantities $a_i(X_i - b_i)$ by any nonsingular matrix will yield Equation (5-9). As an example, the operations involved in changing X_i to $a_i(X_i - b_i)$ correspond exactly to the process of converting temperature from a Fahrenheit to a Celsius reading.

Given observations $\mathbf{x}_1, \mathbf{x}_2, \dots, \mathbf{x}_n$ and the transformation in (5-9), it immediately follows from Result 3.6 that

$$\bar{\mathbf{y}} = \mathbf{C}\bar{\mathbf{x}} + \mathbf{d} \quad \text{and} \quad \mathbf{S_y} = \frac{1}{n-1} \sum_{j=1}^{n} (\mathbf{y}_j - \bar{\mathbf{y}})(\mathbf{y}_j - \bar{\mathbf{y}})' = \mathbf{CSC}'$$

Moreover, by (2-24) and (2-45),

$$\boldsymbol{\mu_Y} = E(\mathbf{Y}) = E(\mathbf{CX} + \mathbf{d}) = E(\mathbf{CX}) + E(\mathbf{d}) = \mathbf{C}\boldsymbol{\mu} + \mathbf{d}$$

Therefore, T^2 computed with the $\mathbf{y}$'s and a hypothesized value $\boldsymbol{\mu}_{\mathbf{Y},0} = \mathbf{C}\boldsymbol{\mu}_0 + \mathbf{d}$ is

$$
\begin{aligned}
T^2 &= n(\bar{\mathbf{y}} - \boldsymbol{\mu}_{\mathbf{Y},0})'\mathbf{S_y}^{-1}(\bar{\mathbf{y}} - \boldsymbol{\mu}_{\mathbf{Y},0}) \\
&= n(\mathbf{C}(\bar{\mathbf{x}} - \boldsymbol{\mu}_0))'(\mathbf{CSC}')^{-1}(\mathbf{C}(\bar{\mathbf{x}} - \boldsymbol{\mu}_0)) \\
&= n(\bar{\mathbf{x}} - \boldsymbol{\mu}_0)'\mathbf{C}'(\mathbf{CSC}')^{-1}\mathbf{C}(\bar{\mathbf{x}} - \boldsymbol{\mu}_0) \\
&= n(\bar{\mathbf{x}} - \boldsymbol{\mu}_0)'\mathbf{C}'(\mathbf{C}')^{-1}\mathbf{S}^{-1}\mathbf{C}^{-1}\mathbf{C}(\bar{\mathbf{x}} - \boldsymbol{\mu}_0) = n(\bar{\mathbf{x}} - \boldsymbol{\mu}_0)'\mathbf{S}^{-1}(\bar{\mathbf{x}} - \boldsymbol{\mu}_0)
\end{aligned}
$$

The last expression is recognized as the value of T^2 computed with the $\mathbf{x}$'s.

5.3 Hotelling's T^2 and Likelihood Ratio Tests

We introduced the T^2-statistic by analogy with the univariate squared distance t^2. There is a general principle for constructing test procedures called the *likelihood ratio method*, and the T^2-statistic can be derived as the likelihood ratio test of H_0: $\boldsymbol{\mu} = \boldsymbol{\mu}_0$. The general theory of likelihood ratio tests is beyond the scope of this book. (See [3] for a treatment of the topic.) Likelihood ratio tests have several optimal properties for reasonably large samples, and they are particularly convenient for hypotheses formulated in terms of multivariate normal parameters.

We know from (4-18) that the maximum of the multivariate normal likelihood as $\boldsymbol{\mu}$ and $\boldsymbol{\Sigma}$ are varied over their possible values is given by

$$\max_{\boldsymbol{\mu}, \boldsymbol{\Sigma}} L(\boldsymbol{\mu}, \boldsymbol{\Sigma}) = \frac{1}{(2\pi)^{np/2}|\hat{\boldsymbol{\Sigma}}|^{n/2}} e^{-np/2} \tag{5-10}$$

where

$$\hat{\boldsymbol{\Sigma}} = \frac{1}{n} \sum_{j=1}^{n} (\mathbf{x}_j - \bar{\mathbf{x}})(\mathbf{x}_j - \bar{\mathbf{x}})' \quad \text{and} \quad \hat{\boldsymbol{\mu}} = \bar{\mathbf{x}} = \frac{1}{n} \sum_{j=1}^{n} \mathbf{x}_j$$

are the maximum likelihood estimates. Recall that $\hat{\boldsymbol{\mu}}$ and $\hat{\boldsymbol{\Sigma}}$ are those choices for $\boldsymbol{\mu}$ and $\boldsymbol{\Sigma}$ that best explain the observed values of the random sample.

Under the hypothesis H_0: $\boldsymbol{\mu} = \boldsymbol{\mu}_0$, the normal likelihood specializes to

$$L(\boldsymbol{\mu}_0, \boldsymbol{\Sigma}) = \frac{1}{(2\pi)^{np/2}|\boldsymbol{\Sigma}|^{n/2}} \exp\left(-\frac{1}{2} \sum_{j=1}^{n} (\mathbf{x}_j - \boldsymbol{\mu}_0)'\boldsymbol{\Sigma}^{-1}(\mathbf{x}_j - \boldsymbol{\mu}_0) \right)$$

The mean $\boldsymbol{\mu}_0$ is now fixed, but $\boldsymbol{\Sigma}$ can be varied to find the value that is "most likely" to have led, with $\boldsymbol{\mu}_0$ fixed, to the observed sample. This value is obtained by maximizing $L(\boldsymbol{\mu}_0, \boldsymbol{\Sigma})$ with respect to $\boldsymbol{\Sigma}$.

Following the steps in (4-13), the exponent in $L(\boldsymbol{\mu}_0, \boldsymbol{\Sigma})$ may be written as

$$-\frac{1}{2} \sum_{j=1}^{n} (\mathbf{x}_j - \boldsymbol{\mu}_0)'\boldsymbol{\Sigma}^{-1}(\mathbf{x}_j - \boldsymbol{\mu}_0) = -\frac{1}{2} \sum_{j=1}^{n} \mathrm{tr}\left[\boldsymbol{\Sigma}^{-1}(\mathbf{x}_j - \boldsymbol{\mu}_0)(\mathbf{x}_j - \boldsymbol{\mu}_0)' \right]$$

$$= -\frac{1}{2} \mathrm{tr}\left[\boldsymbol{\Sigma}^{-1}\left(\sum_{j=1}^{n} (\mathbf{x}_j - \boldsymbol{\mu}_0)(\mathbf{x}_j - \boldsymbol{\mu}_0)' \right) \right]$$

Applying Result 4.10 with $\mathbf{B} = \sum_{j=1}^{n} (\mathbf{x}_j - \boldsymbol{\mu}_0)(\mathbf{x}_j - \boldsymbol{\mu}_0)'$ and $b = n/2$, we have

$$\max_{\boldsymbol{\Sigma}} L(\boldsymbol{\mu}_0, \boldsymbol{\Sigma}) = \frac{1}{(2\pi)^{np/2}|\hat{\boldsymbol{\Sigma}}_0|^{n/2}} e^{-np/2} \tag{5-11}$$

with

$$\hat{\boldsymbol{\Sigma}}_0 = \frac{1}{n} \sum_{j=1}^{n} (\mathbf{x}_j - \boldsymbol{\mu}_0)(\mathbf{x}_j - \boldsymbol{\mu}_0)'$$

To determine whether $\boldsymbol{\mu}_0$ is a plausible value of $\boldsymbol{\mu}$, the maximum of $L(\boldsymbol{\mu}_0, \boldsymbol{\Sigma})$ is compared with the unrestricted maximum of $L(\boldsymbol{\mu}, \boldsymbol{\Sigma})$. The resulting ratio is called the *likelihood ratio statistic*.

Using Equations (5-10) and (5-11), we get

$$\text{Likelihood ratio} = \Lambda = \frac{\max_{\boldsymbol{\Sigma}} L(\boldsymbol{\mu}_0, \boldsymbol{\Sigma})}{\max_{\boldsymbol{\mu}, \boldsymbol{\Sigma}} L(\boldsymbol{\mu}, \boldsymbol{\Sigma})} = \left(\frac{|\hat{\boldsymbol{\Sigma}}|}{|\hat{\boldsymbol{\Sigma}}_0|} \right)^{n/2} \tag{5-12}$$

The equivalent statistic $\Lambda^{2/n} = |\hat{\boldsymbol{\Sigma}}|/|\hat{\boldsymbol{\Sigma}}_0|$ is called *Wilks' lambda*. If the observed value of this likelihood ratio is too small, the hypothesis H_0: $\boldsymbol{\mu} = \boldsymbol{\mu}_0$ is unlikely to be true and is, therefore, rejected. Specifically, the likelihood ratio test of H_0: $\boldsymbol{\mu} = \boldsymbol{\mu}_0$ against H_1: $\boldsymbol{\mu} \neq \boldsymbol{\mu}_0$ rejects H_0 if

$$\Lambda = \left(\frac{|\hat{\boldsymbol{\Sigma}}|}{|\hat{\boldsymbol{\Sigma}}_0|} \right)^{n/2} = \left(\frac{\left| \sum_{j=1}^{n} (\mathbf{x}_j - \bar{\mathbf{x}})(\mathbf{x}_j - \bar{\mathbf{x}})' \right|}{\left| \sum_{j=1}^{n} (\mathbf{x}_j - \boldsymbol{\mu}_0)(\mathbf{x}_j - \boldsymbol{\mu}_0)' \right|} \right)^{n/2} < c_\alpha \tag{5-13}$$

where c_α is the lower (100α)th percentile of the distribution of Λ. (Note that the likelihood ratio test statistic is a power of the ratio of generalized variances.) Fortunately, because of the following relation between T^2 and Λ, we do not need the distribution of the latter to carry out the test.

Result 5.1. Let $\mathbf{X}_1, \mathbf{X}_2, \ldots, \mathbf{X}_n$ be a random sample from an $N_p(\boldsymbol{\mu}, \boldsymbol{\Sigma})$ population. Then the test in (5-7) based on T^2 is equivalent to the likelihood ratio test of $H_0: \boldsymbol{\mu} = \boldsymbol{\mu}_0$ versus $H_1: \boldsymbol{\mu} \neq \boldsymbol{\mu}_0$ because

$$\Lambda^{2/n} = \left(1 + \frac{T^2}{(n-1)}\right)^{-1}$$

Proof. Let the $(p+1) \times (p+1)$ matrix

$$\mathbf{A} = \left[\begin{array}{c|c} \sum\limits_{j=1}^{n} (\mathbf{x}_j - \bar{\mathbf{x}})(\mathbf{x}_j - \bar{\mathbf{x}})' & \sqrt{n}\,(\bar{\mathbf{x}} - \boldsymbol{\mu}_0) \\ \hline \sqrt{n}\,(\bar{\mathbf{x}} - \boldsymbol{\mu}_0)' & -1 \end{array}\right] = \left[\begin{array}{c|c} \mathbf{A}_{11} & \mathbf{A}_{12} \\ \hline \mathbf{A}_{21} & \mathbf{A}_{22} \end{array}\right]$$

By Exercise 4.11, $|\mathbf{A}| = |\mathbf{A}_{22}||\mathbf{A}_{11} - \mathbf{A}_{12}\mathbf{A}_{22}^{-1}\mathbf{A}_{21}| = |\mathbf{A}_{11}||\mathbf{A}_{22} - \mathbf{A}_{21}\mathbf{A}_{11}^{-1}\mathbf{A}_{12}|$, from which we obtain

$$(-1)\left|\sum_{j=1}^{n} (\mathbf{x}_j - \bar{\mathbf{x}})(\mathbf{x}_j - \bar{\mathbf{x}})' + n(\bar{\mathbf{x}} - \boldsymbol{\mu}_0)(\bar{\mathbf{x}} - \boldsymbol{\mu}_0)'\right|$$

$$= \left|\sum_{j=1}^{n} (\mathbf{x}_j - \bar{\mathbf{x}})(\mathbf{x}_j - \bar{\mathbf{x}})'\right|\left|-1 - n(\bar{\mathbf{x}} - \boldsymbol{\mu}_0)'\left(\sum_{j=1}^{n} (\mathbf{x}_j - \bar{\mathbf{x}})(\mathbf{x}_j - \bar{\mathbf{x}})'\right)^{-1}(\bar{\mathbf{x}} - \boldsymbol{\mu}_0)\right|$$

Since, by (4-14),

$$\sum_{j=1}^{n} (\mathbf{x}_j - \boldsymbol{\mu}_0)(\mathbf{x}_j - \boldsymbol{\mu}_0)' = \sum_{j=1}^{n} (\mathbf{x}_j - \bar{\mathbf{x}} + \bar{\mathbf{x}} - \boldsymbol{\mu}_0)(\mathbf{x}_j - \bar{\mathbf{x}} + \bar{\mathbf{x}} - \boldsymbol{\mu}_0)'$$

$$= \sum_{j=1}^{n} (\mathbf{x}_j - \bar{\mathbf{x}})(\mathbf{x}_j - \bar{\mathbf{x}})' + n(\bar{\mathbf{x}} - \boldsymbol{\mu}_0)(\bar{\mathbf{x}} - \boldsymbol{\mu}_0)'$$

the foregoing equality involving determinants can be written

$$(-1)\left|\sum_{j=1}^{n} (\mathbf{x}_j - \boldsymbol{\mu}_0)(\mathbf{x}_j - \boldsymbol{\mu}_0)'\right| = \left|\sum_{j=1}^{n} (\mathbf{x}_j - \bar{\mathbf{x}})(\mathbf{x}_j - \bar{\mathbf{x}})'\right|(-1)\left(1 + \frac{T^2}{(n-1)}\right)$$

or

$$|n\hat{\boldsymbol{\Sigma}}_0| = |n\hat{\boldsymbol{\Sigma}}|\left(1 + \frac{T^2}{(n-1)}\right)$$

Thus,

$$\Lambda^{2/n} = \frac{|\hat{\boldsymbol{\Sigma}}|}{|\hat{\boldsymbol{\Sigma}}_0|} = \left(1 + \frac{T^2}{(n-1)}\right)^{-1} \tag{5-14}$$

Here H_0 is rejected for small values of $\Lambda^{2/n}$ or, equivalently, large values of T^2. The critical values of T^2 are determined by (5-6). ∎

Incidentally, relation (5-14) shows that T^2 may be calculated from two determinants, thus avoiding the computation of $\mathbf{S}^{-1}$. Solving (5-14) for T^2, we have

$$T^2 = \frac{(n-1)\,|\,\hat{\boldsymbol{\Sigma}}_0\,|}{|\,\hat{\boldsymbol{\Sigma}}\,|} - (n-1)$$

$$= \frac{(n-1)\left|\sum_{j=1}^{n}(\mathbf{x}_j - \boldsymbol{\mu}_0)(\mathbf{x}_j - \boldsymbol{\mu}_0)'\right|}{\left|\sum_{j=1}^{n}(\mathbf{x}_j - \bar{\mathbf{x}})(\mathbf{x}_j - \bar{\mathbf{x}})'\right|} - (n-1) \tag{5-15}$$

Likelihood ratio tests are common in multivariate analysis. Their optimal large sample properties hold in very general contexts, as we shall indicate shortly. They are well suited for the testing situations considered in this book. Likelihood ratio methods yield test statistics that reduce to the familiar F- and t-statistics in univariate situations.

General Likelihood Ratio Method

We shall now consider the general likelihood ratio method. Let $\boldsymbol{\theta}$ be a vector consisting of all the *unknown* population parameters, and let $L(\boldsymbol{\theta})$ be the likelihood function obtained by evaluating the joint density of $\mathbf{X}_1, \mathbf{X}_2, \ldots, \mathbf{X}_n$ at their observed values $\mathbf{x}_1, \mathbf{x}_2, \ldots, \mathbf{x}_n$. The parameter vector $\boldsymbol{\theta}$ takes its value in the parameter set $\boldsymbol{\Theta}$. For example, in the p-dimensional multivariate normal case, $\boldsymbol{\theta}' = [\mu_1, \ldots, \mu_p, \sigma_{11}, \ldots, \sigma_{1p}, \sigma_{22}, \ldots, \sigma_{2p}, \ldots, \sigma_{p-1,p}, \sigma_{pp}]$ and $\boldsymbol{\Theta}$ consists of the p-dimensional space, where $-\infty < \mu_1 < \infty, \ldots, -\infty < \mu_p < \infty$ combined with the $[p(p+1)/2]$-dimensional space of variances and covariances such that $\boldsymbol{\Sigma}$ is positive definite. Therefore, $\boldsymbol{\Theta}$ has dimension $\nu = p + p(p+1)/2$. Under the null hypothesis $H_0: \boldsymbol{\theta} = \boldsymbol{\theta}_0$, $\boldsymbol{\theta}$ is restricted to lie in a subset $\boldsymbol{\Theta}_0$ of $\boldsymbol{\Theta}$. For the multivariate normal situation with $\boldsymbol{\mu} = \boldsymbol{\mu}_0$ and $\boldsymbol{\Sigma}$ unspecified, $\boldsymbol{\Theta}_0 = \{\mu_1 = \mu_{10}, \mu_2 = \mu_{20}, \ldots, \mu_p = \mu_{p0}; \sigma_{11}, \ldots, \sigma_{1p}, \sigma_{22}, \ldots, \sigma_{2p}, \ldots, \sigma_{p-1,p}, \sigma_{pp}$ with $\boldsymbol{\Sigma}$ positive definite$\}$, so $\boldsymbol{\Theta}_0$ has dimension $\nu_0 = 0 + p(p+1)/2 = p(p+1)/2$.

A likelihood ratio test of $H_0: \boldsymbol{\theta} \in \boldsymbol{\Theta}_0$ rejects H_0 in favor of $H_1: \boldsymbol{\theta} \notin \boldsymbol{\Theta}_0$ if

$$\Lambda = \frac{\max_{\boldsymbol{\theta} \in \boldsymbol{\Theta}_0} L(\boldsymbol{\theta})}{\max_{\boldsymbol{\theta} \in \boldsymbol{\Theta}} L(\boldsymbol{\theta})} < c \tag{5-16}$$

where c is a suitably chosen constant. Intuitively, we reject H_0 if the maximum of the likelihood obtained by allowing $\boldsymbol{\theta}$ to vary over the set $\boldsymbol{\Theta}_0$ is much smaller than the maximum of the likelihood obtained by varying $\boldsymbol{\theta}$ over all values in $\boldsymbol{\Theta}$. When the maximum in the numerator of expression (5-16) is much smaller than the maximum in the denominator, $\boldsymbol{\Theta}_0$ does not contain plausible values for $\boldsymbol{\theta}$.

In each application of the likelihood ratio method, we must obtain the sampling distribution of the likelihood-ratio test statistic Λ. Then c can be selected to produce a test with a specified significance level α. However, when the sample size is large and certain regularity conditions are satisfied, the sampling distribution of $-2 \ln \Lambda$ is well approximated by a chi-square distribution. This attractive feature accounts, in part, for the popularity of likelihood ratio procedures.

Result 5.2. When the sample size n is large, under the null hypothesis H_0,

$$-2 \ln \Lambda = -2 \ln \left(\frac{\max\limits_{\boldsymbol{\theta} \in \Theta_0} L(\boldsymbol{\theta})}{\max\limits_{\boldsymbol{\theta} \in \Theta} L(\boldsymbol{\theta})} \right)$$

is, approximately, a $\chi^2_{\nu-\nu_0}$ random variable. Here the degrees of freedom are $\nu - \nu_0$ = (dimension of Θ) $-$ (dimension of Θ_0). ∎

Statistical tests are compared on the basis of their *power*, which is defined as the curve or surface whose height is $P[\text{test rejects } H_0 | \boldsymbol{\theta}]$, evaluated at each parameter vector $\boldsymbol{\theta}$. Power measures the ability of a test to reject H_0 when it is not true. In the rare situation where $\boldsymbol{\theta} = \boldsymbol{\theta}_0$ is completely specified under H_0 and the alternative H_1 consists of the single specified value $\boldsymbol{\theta} = \boldsymbol{\theta}_1$, the likelihood ratio test has the highest power among all tests with the same significance level $\alpha = P[\text{test rejects } H_0 | \boldsymbol{\theta} = \boldsymbol{\theta}_0]$. In many single-parameter cases ($\boldsymbol{\theta}$ has one component), the likelihood ratio test is uniformly most powerful against all alternatives to one side of $H_0: \theta = \theta_0$. In other cases, this property holds approximately for large samples.

We shall not give the technical details required for discussing the optimal properties of likelihood ratio tests in the multivariate situation. The general import of these properties, for our purposes, is that they have the highest possible (average) power when the sample size is large.

5.4 Confidence Regions and Simultaneous Comparisons of Component Means

To obtain our primary method for making inferences from a sample, we need to extend the concept of a univariate confidence interval to a multivariate *confidence region*. Let $\boldsymbol{\theta}$ be a vector of unknown population parameters and Θ be the set of all possible values of $\boldsymbol{\theta}$. A confidence region is a region of likely $\boldsymbol{\theta}$ values. This region is determined by the data, and for the moment, we shall denote it by $R(\mathbf{X})$, where $\mathbf{X} = [\mathbf{X}_1, \mathbf{X}_2, \ldots, \mathbf{X}_n]'$ is the data matrix.

The region $R(\mathbf{X})$ is said to be a $100(1 - \alpha)\%$ *confidence region* if, before the sample is selected,

$$P[R(\mathbf{X}) \text{ will cover the true } \boldsymbol{\theta}] = 1 - \alpha \qquad (5\text{-}17)$$

This probability is calculated under the true, but unknown, value of $\boldsymbol{\theta}$.

The confidence region for the mean $\boldsymbol{\mu}$ of a p-dimensional normal population is available from (5-6). Before the sample is selected,

$$P\left[n(\overline{\mathbf{X}} - \boldsymbol{\mu})' \mathbf{S}^{-1} (\overline{\mathbf{X}} - \boldsymbol{\mu}) \leq \frac{(n-1)p}{(n-p)} F_{p,n-p}(\alpha) \right] = 1 - \alpha$$

whatever the values of the unknown $\boldsymbol{\mu}$ and $\boldsymbol{\Sigma}$. In words, $\overline{\mathbf{X}}$ will be within

$$[(n-1)p F_{p,n-p}(\alpha)/(n-p)]^{1/2}$$

of $\boldsymbol{\mu}$, with probability $1 - \alpha$, provided that distance is defined in terms of $n\mathbf{S}^{-1}$. For a particular sample, $\overline{\mathbf{x}}$ and $\mathbf{S}$ can be computed, and the inequality

$n(\bar{\mathbf{x}} - \boldsymbol{\mu})'\mathbf{S}^{-1}(\bar{\mathbf{x}} - \boldsymbol{\mu}) \leq (n - 1)pF_{p,n-p}(\alpha)/(n - p)$ will define a region $R(\mathbf{X})$ within the space of all possible parameter values. In this case, the region will be an ellipsoid centered at $\bar{\mathbf{x}}$. This ellipsoid is the $100(1 - \alpha)\%$ confidence region for $\boldsymbol{\mu}$.

A $100(1 - \alpha)\%$ *confidence region* for the mean of a p-dimensional normal distribution is the ellipsoid determined by all $\boldsymbol{\mu}$ such that

$$n(\bar{\mathbf{x}} - \boldsymbol{\mu})'\mathbf{S}^{-1}(\bar{\mathbf{x}} - \boldsymbol{\mu}) \leq \frac{p(n - 1)}{(n - p)} F_{p,n-p}(\alpha) \qquad (5\text{-}18)$$

where $\bar{\mathbf{x}} = \dfrac{1}{n}\sum_{j=1}^{n} \mathbf{x}_j$, $\mathbf{S} = \dfrac{1}{(n - 1)}\sum_{j=1}^{n} (\mathbf{x}_j - \bar{\mathbf{x}})(\mathbf{x}_j - \bar{\mathbf{x}})'$ and $\mathbf{x}_1, \mathbf{x}_2, \ldots, \mathbf{x}_n$ are the sample observations.

To determine whether any $\boldsymbol{\mu}_0$ lies within the confidence region (is a plausible value for $\boldsymbol{\mu}$), we need to compute the generalized squared distance $n(\bar{\mathbf{x}} - \boldsymbol{\mu}_0)'\mathbf{S}^{-1}(\bar{\mathbf{x}} - \boldsymbol{\mu}_0)$ and compare it with $[p(n - 1)/(n - p)]F_{p,n-p}(\alpha)$. If the squared distance is larger than $[p(n - 1)/(n - p)]F_{p,n-p}(\alpha)$, $\boldsymbol{\mu}_0$ is not in the confidence region. Since this is analogous to testing $H_0\colon \boldsymbol{\mu} = \boldsymbol{\mu}_0$ versus $H_1\colon \boldsymbol{\mu} \neq \boldsymbol{\mu}_0$ [see (5-7)], we see that the confidence region of (5-18) consists of all $\boldsymbol{\mu}_0$ vectors for which the T^2-test would *not* reject H_0 in favor of H_1 at significance level α.

For $p \geq 4$, we cannot graph the joint confidence region for $\boldsymbol{\mu}$. However, we can calculate the axes of the confidence ellipsoid and their relative lengths. These are determined from the eigenvalues λ_i and eigenvectors $\mathbf{e}_i$ of $\mathbf{S}$. As in (4-7), the directions and lengths of the axes of

$$n(\bar{\mathbf{x}} - \boldsymbol{\mu})'\mathbf{S}^{-1}(\bar{\mathbf{x}} - \boldsymbol{\mu}) \leq c^2 = \frac{p(n - 1)}{(n - p)} F_{p,n-p}(\alpha)$$

are determined by going

$$\sqrt{\lambda_i}\, c/\sqrt{n} = \sqrt{\lambda_i}\, \sqrt{p(n - 1)F_{p,n-p}(\alpha)/n(n - p)}$$

units along the eigenvectors $\mathbf{e}_i$. Beginning at the center $\bar{\mathbf{x}}$, the axes of the confidence ellipsoid are

$$\pm\sqrt{\lambda_i}\, \sqrt{\frac{p(n - 1)}{n(n - p)} F_{p,n-p}(\alpha)}\, \mathbf{e}_i \qquad \text{where } \mathbf{S}\mathbf{e}_i = \lambda_i \mathbf{e}_i, \quad i = 1, 2, \ldots, p \qquad (5\text{-}19)$$

The ratios of the λ_i's will help identify relative amounts of elongation along pairs of axes.

Example 5.3 (Constructing a confidence ellipse for $\boldsymbol{\mu}$) Data for radiation from microwave ovens were introduced in Examples 4.10 and 4.17. Let

$$x_1 = \sqrt[4]{\text{measured radiation with door closed}}$$

and

$$x_2 = \sqrt[4]{\text{measured radiation with door open}}$$

For the $n = 42$ pairs of transformed observations, we find that

$$\bar{x} = \begin{bmatrix} .564 \\ .603 \end{bmatrix}, \qquad S = \begin{bmatrix} .0144 & .0117 \\ .0117 & .0146 \end{bmatrix},$$

$$S^{-1} = \begin{bmatrix} 203.018 & -163.391 \\ -163.391 & 200.228 \end{bmatrix}$$

The eigenvalue and eigenvector pairs for S are

$$\lambda_1 = .026, \qquad e_1' = [.704, \ .710]$$
$$\lambda_2 = .002, \qquad e_2' = [-.710, \ .704]$$

The 95% confidence ellipse for μ consists of all values (μ_1, μ_2) satisfying

$$42[.564 - \mu_1, \ \ .603 - \mu_2] \begin{bmatrix} 203.018 & -163.391 \\ -163.391 & 200.228 \end{bmatrix} \begin{bmatrix} .564 - \mu_1 \\ .603 - \mu_2 \end{bmatrix}$$

$$\leq \frac{2(41)}{40} F_{2,40}(.05)$$

or, since $F_{2,40}(.05) = 3.23$,

$$42(203.018)(.564 - \mu_1)^2 + 42(200.228)(.603 - \mu_2)^2$$
$$- 84(163.391)(.564 - \mu_1)(.603 - \mu_2) \leq 6.62$$

To see whether $\mu' = [.562, .589]$ is in the confidence region, we compute

$$42(203.018)(.564 - .562)^2 + 42(200.228)(.603 - .589)^2$$
$$- 84(163.391)(.564 - .562)(.603 - .589) = 1.30 \leq 6.62$$

We conclude that $\mu' = [.562, .589]$ is in the region. Equivalently, a test of H_0: $\mu = \begin{bmatrix} .562 \\ .589 \end{bmatrix}$ would not be rejected in favor of H_1: $\mu \neq \begin{bmatrix} .562 \\ .589 \end{bmatrix}$ at the $\alpha = .05$ level of significance.

The joint confidence ellipsoid is plotted in Figure 5.1. The center is at $\bar{x}' = [.564, .603]$, and the half-lengths of the major and minor axes are given by

$$\sqrt{\lambda_1} \sqrt{\frac{p(n-1)}{n(n-p)} F_{p,n-p}(\alpha)} = \sqrt{.026} \sqrt{\frac{2(41)}{42(40)}(3.23)} = .064$$

and

$$\sqrt{\lambda_2} \sqrt{\frac{p(n-1)}{n(n-p)} F_{p,n-p}(\alpha)} = \sqrt{.002} \sqrt{\frac{2(41)}{42(40)}(3.23)} = .018$$

respectively. The axes lie along $e_1' = [.704, .710]$ and $e_2' = [-.710, .704]$ when these vectors are plotted with $\bar{x}$ as the origin. An indication of the elongation of the confidence ellipse is provided by the ratio of the lengths of the major and minor axes. This ratio is

$$\frac{2\sqrt{\lambda_1} \sqrt{\dfrac{p(n-1)}{n(n-p)} F_{p,n-p}(\alpha)}}{2\sqrt{\lambda_2} \sqrt{\dfrac{p(n-1)}{n(n-p)} F_{p,n-p}(\alpha)}} = \frac{\sqrt{\lambda_1}}{\sqrt{\lambda_2}} = \frac{.161}{.045} = 3.6$$

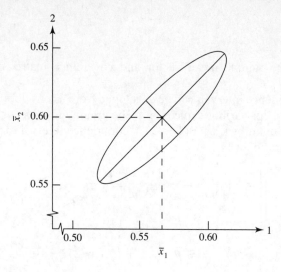

Figure 5.1 A 95% confidence ellipse for μ based on microwave-radiation data.

The length of the major axis is 3.6 times the length of the minor axis. ∎

Simultaneous Confidence Statements

While the confidence region $n(\bar{\mathbf{x}} - \boldsymbol{\mu})'\mathbf{S}^{-1}(\bar{\mathbf{x}} - \boldsymbol{\mu}) \leq c^2$, for c a constant, correctly assesses the joint knowledge concerning plausible values of $\boldsymbol{\mu}$, any summary of conclusions ordinarily includes confidence statements about the individual component means. In so doing, we adopt the attitude that all of the separate confidence statements should hold *simultaneously* with a specified high probability. It is the guarantee of a specified probability against *any* statement being incorrect that motivates the term *simultaneous confidence intervals*. We begin by considering simultaneous confidence statements which are intimately related to the joint confidence region based on the T^2-statistic.

Let $\mathbf{X}$ have an $N_p(\boldsymbol{\mu}, \boldsymbol{\Sigma})$ distribution and form the linear combination

$$Z = a_1 X_1 + a_2 X_2 + \cdots + a_p X_p = \mathbf{a}'\mathbf{X}$$

From (2-43),

$$\mu_Z = E(Z) = \mathbf{a}'\boldsymbol{\mu}$$

and

$$\sigma_Z^2 = \mathrm{Var}\,(Z) = \mathbf{a}'\boldsymbol{\Sigma}\mathbf{a}$$

Moreover, by Result 4.2, Z has an $N(\mathbf{a}'\boldsymbol{\mu}, \mathbf{a}'\boldsymbol{\Sigma}\mathbf{a})$ distribution. If a random sample $\mathbf{X}_1, \mathbf{X}_2, \ldots, \mathbf{X}_n$ from the $N_p(\boldsymbol{\mu}, \boldsymbol{\Sigma})$ population is available, a corresponding sample of Z's can be created by taking linear combinations. Thus,

$$Z_j = a_1 X_{j1} + a_2 X_{j2} + \cdots + a_p X_{jp} = \mathbf{a}'\mathbf{X}_j \qquad j = 1, 2, \ldots, n$$

The sample mean and variance of the observed values $z_1, z_2, \ldots, z_n$ are, by (3-36),

$$\bar{z} = \mathbf{a}'\bar{\mathbf{x}}$$

and

$$s_z^2 = \mathbf{a}'\mathbf{S}\mathbf{a}$$

where $\bar{\mathbf{x}}$ and $\mathbf{S}$ are the sample mean vector and covariance matrix of the $\mathbf{x}_j$'s, respectively.

Simultaneous confidence intervals can be developed from a consideration of confidence intervals for $\mathbf{a}'\boldsymbol{\mu}$ for various choices of $\mathbf{a}$. The argument proceeds as follows.

For $\mathbf{a}$ *fixed* and σ_z^2 unknown, a $100(1-\alpha)\%$ confidence interval for $\mu_Z = \mathbf{a}'\boldsymbol{\mu}$ is based on student's t-ratio

$$t = \frac{\bar{z} - \mu_Z}{s_z/\sqrt{n}} = \frac{\sqrt{n}\,(\mathbf{a}'\bar{\mathbf{x}} - \mathbf{a}'\boldsymbol{\mu})}{\sqrt{\mathbf{a}'\mathbf{S}\mathbf{a}}} \tag{5-20}$$

and leads to the statement

$$\bar{z} - t_{n-1}(\alpha/2)\frac{s_z}{\sqrt{n}} \le \mu_Z \le \bar{z} + t_{n-1}(\alpha/2)\frac{s_z}{\sqrt{n}}$$

or

$$\mathbf{a}'\bar{\mathbf{x}} - t_{n-1}(\alpha/2)\frac{\sqrt{\mathbf{a}'\mathbf{S}\mathbf{a}}}{\sqrt{n}} \le \mathbf{a}'\boldsymbol{\mu} \le \mathbf{a}'\bar{\mathbf{x}} + t_{n-1}(\alpha/2)\frac{\sqrt{\mathbf{a}'\mathbf{S}\mathbf{a}}}{\sqrt{n}} \tag{5-21}$$

where $t_{n-1}(\alpha/2)$ is the upper $100(\alpha/2)$th percentile of a t-distribution with $n-1$ d.f.

Inequality (5-21) can be interpreted as a statement about the components of the mean vector $\boldsymbol{\mu}$. For example, with $\mathbf{a}' = [1, 0, \ldots, 0]$, $\mathbf{a}'\boldsymbol{\mu} = \mu_1$, and (5-21) becomes the usual confidence interval for a normal population mean. (Note, in this case, that $\mathbf{a}'\mathbf{S}\mathbf{a} = s_{11}$.) Clearly, we could make several confidence statements about the components of $\boldsymbol{\mu}$, each with associated confidence coefficient $1 - \alpha$, by choosing different coefficient vectors $\mathbf{a}$. However, the confidence associated with all of the statements taken together is *not* $1 - \alpha$.

Intuitively, it would be desirable to associate a "collective" confidence coefficient of $1 - \alpha$ with the confidence intervals that can be generated by all choices of $\mathbf{a}$. However, a price must be paid for the convenience of a large simultaneous confidence coefficient: intervals that are wider (less precise) than the interval of (5-21) for a specific choice of $\mathbf{a}$.

Given a data set $\mathbf{x}_1, \mathbf{x}_2, \ldots, \mathbf{x}_n$ and a particular $\mathbf{a}$, the confidence interval in (5-21) is that set of $\mathbf{a}'\boldsymbol{\mu}$ values for which

$$|t| = \left| \frac{\sqrt{n}\,(\mathbf{a}'\bar{\mathbf{x}} - \mathbf{a}'\boldsymbol{\mu})}{\sqrt{\mathbf{a}'\mathbf{S}\mathbf{a}}} \right| \le t_{n-1}(\alpha/2)$$

or, equivalently,

$$t^2 = \frac{n(\mathbf{a}'\bar{\mathbf{x}} - \mathbf{a}'\boldsymbol{\mu})^2}{\mathbf{a}'\mathbf{S}\mathbf{a}} = \frac{n(\mathbf{a}'(\bar{\mathbf{x}} - \boldsymbol{\mu}))^2}{\mathbf{a}'\mathbf{S}\mathbf{a}} \le t_{n-1}^2(\alpha/2) \tag{5-22}$$

A simultaneous confidence region is given by the set of $\mathbf{a}'\boldsymbol{\mu}$ values such that t^2 is relatively small for *all* choices of $\mathbf{a}$. It seems reasonable to expect that the constant $t_{n-1}^2(\alpha/2)$ in (5-22) will be replaced by a larger value, c^2, when statements are developed for many choices of $\mathbf{a}$.

Considering the values of $\mathbf{a}$ for which $t^2 \leq c^2$, we are naturally led to the determination of

$$\max_{\mathbf{a}} t^2 = \max_{\mathbf{a}} \frac{n(\mathbf{a}'(\bar{\mathbf{x}} - \boldsymbol{\mu}))^2}{\mathbf{a}'\mathbf{S}\mathbf{a}}$$

Using the maximization lemma (2-50) with $\mathbf{x} = \mathbf{a}$, $\mathbf{d} = (\bar{\mathbf{x}} - \boldsymbol{\mu})$, and $\mathbf{B} = \mathbf{S}$, we get

$$\max_{\mathbf{a}} \frac{n(\mathbf{a}'(\bar{\mathbf{x}} - \boldsymbol{\mu}))^2}{\mathbf{a}'\mathbf{S}\mathbf{a}} = n\left[\max_{\mathbf{a}} \frac{(\mathbf{a}'(\bar{\mathbf{x}} - \boldsymbol{\mu}))^2}{\mathbf{a}'\mathbf{S}\mathbf{a}}\right] = n(\bar{\mathbf{x}} - \boldsymbol{\mu})'\mathbf{S}^{-1}(\bar{\mathbf{x}} - \boldsymbol{\mu}) = T^2 \quad (5\text{-}23)$$

with the maximum occurring for $\mathbf{a}$ proportional to $\mathbf{S}^{-1}(\bar{\mathbf{x}} - \boldsymbol{\mu})$.

Result 5.3. Let $\mathbf{X}_1, \mathbf{X}_2, \ldots, \mathbf{X}_n$ be a random sample from an $N_p(\boldsymbol{\mu}, \boldsymbol{\Sigma})$ population with $\boldsymbol{\Sigma}$ positive definite. Then, simultaneously for all $\mathbf{a}$, the interval

$$\left(\mathbf{a}'\bar{\mathbf{X}} - \sqrt{\frac{p(n-1)}{n(n-p)}F_{p,n-p}(\alpha)\,\mathbf{a}'\mathbf{S}\mathbf{a}}, \quad \mathbf{a}'\bar{\mathbf{X}} + \sqrt{\frac{p(n-1)}{n(n-p)}F_{p,n-p}(\alpha)\,\mathbf{a}'\mathbf{S}\mathbf{a}}\right)$$

will contain $\mathbf{a}'\boldsymbol{\mu}$ with probability $1 - \alpha$.

Proof. From (5-23),

$$T^2 = n(\bar{\mathbf{x}} - \boldsymbol{\mu})'\mathbf{S}^{-1}(\bar{\mathbf{x}} - \boldsymbol{\mu}) \leq c^2 \quad \text{implies} \quad \frac{n(\mathbf{a}'\bar{\mathbf{x}} - \mathbf{a}'\boldsymbol{\mu})^2}{\mathbf{a}'\mathbf{S}\mathbf{a}} \leq c^2$$

for every $\mathbf{a}$, or

$$\mathbf{a}'\bar{\mathbf{x}} - c\sqrt{\frac{\mathbf{a}'\mathbf{S}\mathbf{a}}{n}} \leq \mathbf{a}'\boldsymbol{\mu} \leq \mathbf{a}'\bar{\mathbf{x}} + c\sqrt{\frac{\mathbf{a}'\mathbf{S}\mathbf{a}}{n}}$$

for every $\mathbf{a}$. Choosing $c^2 = p(n-1)F_{p,n-p}(\alpha)/(n-p)$ [see (5-6)] gives intervals that will contain $\mathbf{a}'\boldsymbol{\mu}$ for all $\mathbf{a}$, with probability $1 - \alpha = P[T^2 \leq c^2]$. ∎

It is convenient to refer to the simultaneous intervals of Result 5.3 as T^2-*intervals*, since the coverage probability is determined by the distribution of T^2. The successive choices $\mathbf{a}' = [1, 0, \ldots, 0]$, $\mathbf{a}' = [0, 1, \ldots, 0]$, and so on through $\mathbf{a}' = [0, 0, \ldots, 1]$ for the T^2-intervals allow us to conclude that

$$\bar{x}_1 - \sqrt{\frac{p(n-1)}{(n-p)}F_{p,n-p}(\alpha)}\sqrt{\frac{s_{11}}{n}} \leq \mu_1 \leq \bar{x}_1 + \sqrt{\frac{p(n-1)}{(n-p)}F_{p,n-p}(\alpha)}\sqrt{\frac{s_{11}}{n}}$$

$$\bar{x}_2 - \sqrt{\frac{p(n-1)}{(n-p)}F_{p,n-p}(\alpha)}\sqrt{\frac{s_{22}}{n}} \leq \mu_2 \leq \bar{x}_2 + \sqrt{\frac{p(n-1)}{(n-p)}F_{p,n-p}(\alpha)}\sqrt{\frac{s_{22}}{n}}$$

$$\vdots \qquad\qquad \vdots \qquad\qquad \vdots \qquad (5\text{-}24)$$

$$\bar{x}_p - \sqrt{\frac{p(n-1)}{(n-p)}F_{p,n-p}(\alpha)}\sqrt{\frac{s_{pp}}{n}} \leq \mu_p \leq \bar{x}_p + \sqrt{\frac{p(n-1)}{(n-p)}F_{p,n-p}(\alpha)}\sqrt{\frac{s_{pp}}{n}}$$

all hold simultaneously with confidence coefficient $1 - \alpha$. Note that, without modifying the coefficient $1 - \alpha$, we can make statements about the differences $\mu_i - \mu_k$ corresponding to $\mathbf{a}' = [0, \ldots, 0, \ a_i, 0, \ldots, 0, \ a_k, 0, \ldots, 0]$, where $a_i = 1$ and

$a_k = -1$. In this case $\mathbf{a}'\mathbf{S}\mathbf{a} = s_{ii} - 2s_{ik} + s_{kk}$, and we have the statement

$$\bar{x}_i - \bar{x}_k - \sqrt{\frac{p(n-1)}{(n-p)}F_{p,n-p}(\alpha)}\sqrt{\frac{s_{ii} - 2s_{ik} + s_{kk}}{n}} \leq \mu_i - \mu_k$$

$$\leq \bar{x}_i - \bar{x}_k + \sqrt{\frac{p(n-1)}{(n-p)}F_{p,n-p}(\alpha)}\sqrt{\frac{s_{ii} - 2s_{ik} + s_{kk}}{n}} \quad (5\text{-}25)$$

The simultaneous T^2 confidence intervals are ideal for "data snooping." The confidence coefficient $1 - \alpha$ remains unchanged for any choice of $\mathbf{a}$, so linear combinations of the components μ_i that merit inspection *based upon an examination of the data* can be estimated.

In addition, according to the results in Supplement 5A, we can include the statements about (μ_i, μ_k) belonging to the sample mean-centered ellipses

$$n[\bar{x}_i - \mu_i, \ \bar{x}_k - \mu_k]\begin{bmatrix} s_{ii} & s_{ik} \\ s_{ik} & s_{kk} \end{bmatrix}^{-1}\begin{bmatrix} \bar{x}_i - \mu_i \\ \bar{x}_k - \mu_k \end{bmatrix} \leq \frac{p(n-1)}{n-p}F_{p,n-p}(\alpha) \quad (5\text{-}26)$$

and still maintain the confidence coefficient $(1 - \alpha)$ for the whole set of statements.

The simultaneous T^2 confidence intervals for the individual components of a mean vector are just the shadows, or projections, of the confidence ellipsoid on the component axes. This connection between the shadows of the ellipsoid and the simultaneous confidence intervals given by (5-24) is illustrated in the next example.

Example 5.4 (Simultaneous confidence intervals as shadows of the confidence ellipsoid)
In Example 5.3, we obtained the 95% confidence ellipse for the means of the fourth roots of the door-closed and door-open microwave radiation measurements. The 95% simultaneous T^2 intervals for the two component means are, from (5-24),

$$\left(\bar{x}_1 - \sqrt{\frac{p(n-1)}{(n-p)}F_{p,n-p}(.05)}\sqrt{\frac{s_{11}}{n}}, \ \bar{x}_1 + \sqrt{\frac{p(n-1)}{(n-p)}F_{p,n-p}(.05)}\sqrt{\frac{s_{11}}{n}}\right)$$

$$= \left(.564 - \sqrt{\frac{2(41)}{40}3.23}\sqrt{\frac{.0144}{42}}, \ .564 + \sqrt{\frac{2(41)}{40}3.23}\sqrt{\frac{.0144}{42}}\right) \quad \text{or} \quad (.516, \ .612)$$

$$\left(\bar{x}_2 - \sqrt{\frac{p(n-1)}{(n-p)}F_{p,n-p}(.05)}\sqrt{\frac{s_{22}}{n}}, \ \bar{x}_2 + \sqrt{\frac{p(n-1)}{(n-p)}F_{p,n-p}(.05)}\sqrt{\frac{s_{22}}{n}}\right)$$

$$= \left(.603 - \sqrt{\frac{2(41)}{40}3.23}\sqrt{\frac{.0146}{42}}, \ .603 + \sqrt{\frac{2(41)}{40}3.23}\sqrt{\frac{.0146}{42}}\right) \quad \text{or} \quad (.555, \ .651)$$

In Figure 5.2, we have redrawn the 95% confidence ellipse from Example 5.3. The 95% simultaneous intervals are shown as shadows, or projections, of this ellipse on the axes of the component means. ∎

Example 5.5 (Constructing simultaneous confidence intervals and ellipses) The scores obtained by $n = 87$ college students on the College Level Examination Program (CLEP) subtest X_1 and the College Qualification Test (CQT) subtests X_2 and X_3 are given in Table 5.2 on page 228 for $X_1 =$ social science and history, $X_2 =$ verbal, and $X_3 =$ science. These data give

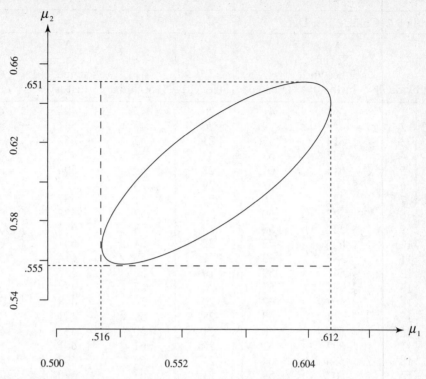

Figure 5.2 Simultaneous T^2-intervals for the component means as shadows of the confidence ellipse on the axes—microwave radiation data.

$$\bar{\mathbf{x}} = \begin{bmatrix} 526.29 \\ 54.69 \\ 25.13 \end{bmatrix} \quad \text{and} \quad \mathbf{S} = \begin{bmatrix} 5808.06 & 597.84 & 222.03 \\ 597.84 & 126.05 & 23.39 \\ 222.03 & 23.39 & 23.11 \end{bmatrix}$$

Let us compute the 95% simultaneous confidence intervals for μ_1, μ_2, and μ_3. We have

$$\frac{p(n-1)}{n-p} F_{p,n-p}(\alpha) = \frac{3(87-1)}{(87-3)} F_{3,84}(.05) = \frac{3(86)}{84}(2.7) = 8.29$$

and we obtain the simultaneous confidence statements [see (5-24)]

$$526.29 - \sqrt{8.29}\sqrt{\frac{5808.06}{87}} \le \mu_1 \le 526.29 + \sqrt{8.29}\sqrt{\frac{5808.06}{87}}$$

or

$$503.06 \le \mu_1 \le 550.12$$

$$54.69 - \sqrt{8.29}\sqrt{\frac{126.05}{87}} \le \mu_2 \le 54.69 + \sqrt{8.29}\sqrt{\frac{126.05}{87}}$$

or

$$51.22 \le \mu_2 \le 58.16$$

$$25.13 - \sqrt{8.29}\sqrt{\frac{23.11}{87}} \le \mu_3 \le 25.13 + \sqrt{8.29}\sqrt{\frac{23.11}{87}}$$

Table 5.2 College Test Data

Individual	X_1 (Social science and history)	X_2 (Verbal)	X_3 (Science)	Individual	X_1 (Social science and history)	X_2 (Verbal)	X_3 (Science)
1	468	41	26	45	494	41	24
2	428	39	26	46	541	47	25
3	514	53	21	47	362	36	17
4	547	67	33	48	408	28	17
5	614	61	27	49	594	68	23
6	501	67	29	50	501	25	26
7	421	46	22	51	687	75	33
8	527	50	23	52	633	52	31
9	527	55	19	53	647	67	29
10	620	72	32	54	647	65	34
11	587	63	31	55	614	59	25
12	541	59	19	56	633	65	28
13	561	53	26	57	448	55	24
14	468	62	20	58	408	51	19
15	614	65	28	59	441	35	22
16	527	48	21	60	435	60	20
17	507	32	27	61	501	54	21
18	580	64	21	62	507	42	24
19	507	59	21	63	620	71	36
20	521	54	23	64	415	52	20
21	574	52	25	65	554	69	30
22	587	64	31	66	348	28	18
23	488	51	27	67	468	49	25
24	488	62	18	68	507	54	26
25	587	56	26	69	527	47	31
26	421	38	16	70	527	47	26
27	481	52	26	71	435	50	28
28	428	40	19	72	660	70	25
29	640	65	25	73	733	73	33
30	574	61	28	74	507	45	28
31	547	64	27	75	527	62	29
32	580	64	28	76	428	37	19
33	494	53	26	77	481	48	23
34	554	51	21	78	507	61	19
35	647	58	23	79	527	66	23
36	507	65	23	80	488	41	28
37	454	52	28	81	607	69	28
38	427	57	21	82	561	59	34
39	521	66	26	83	614	70	23
40	468	57	14	84	527	49	30
41	587	55	30	85	474	41	16
42	507	61	31	86	441	47	26
43	574	54	31	87	607	67	32
44	507	53	23				

Source: Data courtesy of Richard W. Johnson.

or

$$23.65 \le \mu_3 \le 26.61$$

With the possible exception of the verbal scores, the marginal Q–Q plots and two-dimensional scatter plots do not reveal any serious departures from normality for the college qualification test data. (See Exercise 5.18.) Moreover, the sample size is large enough to justify the methodology, even though the data are not quite normally distributed. (See Section 5.5.)

The simultaneous T^2-intervals above are wider than univariate intervals because all three must hold with 95% confidence. They may also be wider than necessary, because, with the same confidence, we can make statements about differences.

For instance, with $\mathbf{a}' = [0, 1, -1]$, the interval for $\mu_2 - \mu_3$ has endpoints

$$(\bar{x}_2 - \bar{x}_3) \pm \sqrt{\frac{p(n-1)}{(n-p)} F_{p,n-p}(.05)} \sqrt{\frac{s_{22} + s_{33} - 2s_{23}}{n}}$$

$$= (54.69 - 25.13) \pm \sqrt{8.29} \sqrt{\frac{126.05 + 23.11 - 2(23.39)}{87}} = 29.56 \pm 3.12$$

so (26.44, 32.68) is a 95% confidence interval for $\mu_2 - \mu_3$. Simultaneous intervals can also be constructed for the other differences.

Finally, we can construct confidence ellipses for pairs of means, and the same 95% confidence holds. For example, for the pair (μ_2, μ_3), we have

$$87[54.69 - \mu_2, \quad 25.13 - \mu_3] \begin{bmatrix} 126.05 & 23.39 \\ 23.39 & 23.11 \end{bmatrix}^{-1} \begin{bmatrix} 54.69 - \mu_2 \\ 25.13 - \mu_3 \end{bmatrix}$$

$$= 0.849(54.69 - \mu_2)^2 + 4.633(25.13 - \mu_3)^2$$

$$- 2 \times 0.859(54.69 - \mu_2)(25.13 - \mu_3) \le 8.29$$

This ellipse is shown in Figure 5.3 on page 230, along with the 95% confidence ellipses for the other two pairs of means. The projections or shadows of these ellipses on the axes are also indicated, and these projections are the T^2-intervals. ■

A Comparison of Simultaneous Confidence Intervals with One-at-a-Time Intervals

An alternative approach to the construction of confidence intervals is to consider the components μ_i one at a time, as suggested by (5-21) with $\mathbf{a}' = [0, \ldots, 0, a_i, 0, \ldots, 0]$ where $a_i = 1$. This approach ignores the covariance structure of the p variables and leads to the intervals

$$\bar{x}_1 - t_{n-1}(\alpha/2) \sqrt{\frac{s_{11}}{n}} \le \mu_1 \le \bar{x}_1 + t_{n-1}(\alpha/2) \sqrt{\frac{s_{11}}{n}}$$

$$\bar{x}_2 - t_{n-1}(\alpha/2) \sqrt{\frac{s_{22}}{n}} \le \mu_2 \le \bar{x}_2 + t_{n-1}(\alpha/2) \sqrt{\frac{s_{22}}{n}} \qquad (5\text{-}27)$$

$$\vdots \qquad \qquad \vdots \qquad \qquad \vdots$$

$$\bar{x}_p - t_{n-1}(\alpha/2) \sqrt{\frac{s_{pp}}{n}} \le \mu_p \le \bar{x}_p + t_{n-1}(\alpha/2) \sqrt{\frac{s_{pp}}{n}}$$

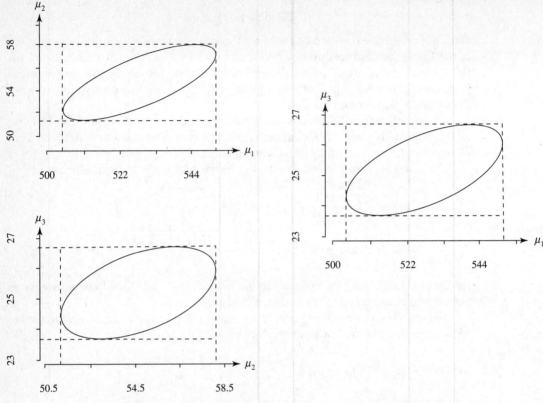

Figure 5.3 95% confidence ellipses for pairs of means and the simultaneous T^2-intervals—college test data.

Although prior to sampling, the ith interval has probability $1 - \alpha$ of covering μ_i, we do not know what to assert, in general, about the probability of *all* intervals containing their respective μ_i's. As we have pointed out, this probability is not $1 - \alpha$.

To shed some light on the problem, consider the special case where the observations have a joint normal distribution and

$$\mathbf{\Sigma} = \begin{bmatrix} \sigma_{11} & 0 & \cdots & 0 \\ 0 & \sigma_{22} & \cdots & 0 \\ \vdots & \vdots & \ddots & \vdots \\ 0 & 0 & \cdots & \sigma_{pp} \end{bmatrix}$$

Since the observations on the first variable are independent of those on the second variable, and so on, the product rule for independent events can be applied. Before the sample is selected,

$$P[\text{all } t\text{-intervals in (5-27) contain the } \mu_i\text{'s}] = (1 - \alpha)(1 - \alpha) \cdots (1 - \alpha)$$
$$= (1 - \alpha)^p$$

If $1 - \alpha = .95$ and $p = 6$, this probability is $(.95)^6 = .74$.

To guarantee a probability of $1 - \alpha$ that all of the statements about the component means hold simultaneously, the individual intervals must be wider than the separate t-intervals; just how much wider depends on both p and n, as well as on $1 - \alpha$.

For $1 - \alpha = .95$, $n = 15$, and $p = 4$, the multipliers of $\sqrt{s_{ii}/n}$ in (5-24) and (5-27) are

$$\sqrt{\frac{p(n-1)}{(n-p)}} \, F_{p,n-p}(.05) = \sqrt{\frac{4(14)}{11}} \, (3.36) = 4.14$$

and $t_{n-1}(.025) = 2.145$, respectively. Consequently, in this case the simultaneous intervals are $100(4.14 - 2.145)/2.145 = 93\%$ wider than those derived from the one-at-a-time t method.

Table 5.3 gives some critical distance multipliers for one-at-a-time t-intervals computed according to (5-21), as well as the corresponding simultaneous T^2-intervals. In general, the width of the T^2-intervals, relative to the t-intervals, increases for fixed n as p increases and decreases for fixed p as n increases.

Table 5.3 Critical Distance Multipliers for One-at-a-Time t- Intervals and T^2-Intervals for Selected n and p $(1 - \alpha = .95)$

n	$t_{n-1}(.025)$	$\sqrt{\dfrac{(n-1)p}{(n-p)}} \, F_{p,n-p}(.05)$	
		$p = 4$	$p = 10$
15	2.145	4.14	11.52
25	2.064	3.60	6.39
50	2.010	3.31	5.05
100	1.970	3.19	4.61
∞	1.960	3.08	4.28

The comparison implied by Table 5.3 is a bit unfair, since the confidence level associated with any collection of T^2-intervals, for fixed n and p, is .95, and the overall confidence associated with a collection of individual t intervals, for the same n, can, as we have seen, be much less than .95. The one-at-a-time t intervals are too short to maintain an overall confidence level for separate statements about, say, all p means. Nevertheless, we sometimes look at them as the best possible information concerning a mean, if this is the only inference to be made. Moreover, if the one-at-a-time intervals are calculated only when the T^2-test rejects the null hypothesis, some researchers think they may more accurately represent the information about the means than the T^2-intervals do.

The T^2-intervals are too wide if they are applied only to the p component means. To see why, consider the confidence ellipse and the simultaneous intervals shown in Figure 5.2. If μ_1 lies in its T^2-interval and μ_2 lies in its T^2-interval, then (μ_1, μ_2) lies in the rectangle formed by these two intervals. This rectangle contains the confidence ellipse and more. The confidence ellipse is smaller but has probability .95 of covering the mean vector $\boldsymbol{\mu}$ with its component means μ_1 and μ_2. Consequently, the probability of covering the two individual means μ_1 and μ_2 will be larger than .95 for the rectangle formed by the T^2-intervals. This result leads us to consider a second approach to making multiple comparisons known as the Bonferroni method.

The Bonferroni Method of Multiple Comparisons

Often, attention is restricted to a small number of individual confidence statements. In these situations it is possible to do better than the simultaneous intervals of Result 5.3. If the number m of specified component means μ_i or linear combinations $\mathbf{a}'\boldsymbol{\mu} = a_1\mu_1 + a_2\mu_2 + \cdots + a_p\mu_p$ is small, simultaneous confidence intervals can be developed that are shorter (more precise) than the simultaneous T^2-intervals. The alternative method for multiple comparisons is called the *Bonferroni method*, because it is developed from a probability inequality carrying that name.

Suppose that, prior to the collection of data, confidence statements about m linear combinations $\mathbf{a}_1'\boldsymbol{\mu}, \mathbf{a}_2'\boldsymbol{\mu}, \ldots, \mathbf{a}_m'\boldsymbol{\mu}$ are required. Let C_i denote a confidence statement about the value of $\mathbf{a}_i'\boldsymbol{\mu}$ with $P[C_i \text{ true}] = 1 - \alpha_i$, $i = 1, 2, \ldots, m$. Now (see Exercise 5.6),

$$P[\text{all } C_i \text{ true}] = 1 - P[\text{at least one } C_i \text{ false}]$$

$$\geq 1 - \sum_{i=1}^{m} P(C_i \text{ false}) = 1 - \sum_{i=1}^{m} (1 - P(C_i \text{ true}))$$

$$= 1 - (\alpha_1 + \alpha_2 + \cdots + \alpha_m) \tag{5-28}$$

Inequality (5-28), a special case of the Bonferroni inequality, allows an investigator to control the overall error rate $\alpha_1 + \alpha_2 + \cdots + \alpha_m$, regardless of the correlation structure behind the confidence statements. There is also the flexibility of controlling the error rate for a group of important statements and balancing it by another choice for the less important statements.

Let us develop simultaneous interval estimates for the restricted set consisting of the components μ_i of $\boldsymbol{\mu}$. Lacking information on the relative importance of these components, we consider the individual t-intervals

$$\bar{x}_i \pm t_{n-1}\left(\frac{\alpha_i}{2}\right)\sqrt{\frac{s_{ii}}{n}} \qquad i = 1, 2, \ldots, m$$

with $\alpha_i = \alpha/m$. Since $P[\bar{X}_i \pm t_{n-1}(\alpha/2m)\sqrt{s_{ii}/n} \text{ contains } \mu_i] = 1 - \alpha/m$, $i = 1, 2, \ldots, m$, we have, from (5-28),

$$P\left[\bar{X}_i \pm t_{n-1}\left(\frac{\alpha}{2m}\right)\sqrt{\frac{s_{ii}}{n}} \text{ contains } \mu_i, \text{ all } i\right] \geq 1 - \underbrace{\left(\frac{\alpha}{m} + \frac{\alpha}{m} + \cdots + \frac{\alpha}{m}\right)}_{m \text{ terms}}$$

$$= 1 - \alpha$$

Therefore, with an overall confidence level greater than or equal to $1 - \alpha$, we can make the following $m = p$ statements:

$$\bar{x}_1 - t_{n-1}\left(\frac{\alpha}{2p}\right)\sqrt{\frac{s_{11}}{n}} \leq \mu_1 \leq \bar{x}_1 + t_{n-1}\left(\frac{\alpha}{2p}\right)\sqrt{\frac{s_{11}}{n}}$$

$$\bar{x}_2 - t_{n-1}\left(\frac{\alpha}{2p}\right)\sqrt{\frac{s_{22}}{n}} \leq \mu_2 \leq \bar{x}_2 + t_{n-1}\left(\frac{\alpha}{2p}\right)\sqrt{\frac{s_{22}}{n}} \tag{5-29}$$

$$\vdots \qquad\qquad \vdots \qquad\qquad \vdots$$

$$\bar{x}_p - t_{n-1}\left(\frac{\alpha}{2p}\right)\sqrt{\frac{s_{pp}}{n}} \leq \mu_p \leq \bar{x}_p + t_{n-1}\left(\frac{\alpha}{2p}\right)\sqrt{\frac{s_{pp}}{n}}$$

The statements in (5-29) can be compared with those in (5-24). The percentage point $t_{n-1}(\alpha/2p)$ replaces $\sqrt{(n-1)pF_{p,n-p}(\alpha)/(n-p)}$, but otherwise the intervals are of the same structure.

Example 5.6 (Constructing Bonferroni simultaneous confidence intervals and comparing them with T^2-intervals) Let us return to the microwave oven radiation data in Examples 5.3 and 5.4. We shall obtain the simultaneous 95% Bonferroni confidence intervals for the means, μ_1 and μ_2, of the fourth roots of the door-closed and door-open measurements with $\alpha_i = .05/2$, $i = 1, 2$. We make use of the results in Example 5.3, noting that $n = 42$ and $t_{41}(.05/2(2)) = t_{41}(.0125) = 2.327$, to get

$$\bar{x}_1 \pm t_{41}(.0125)\sqrt{\frac{s_{11}}{n}} = .564 \pm 2.327\sqrt{\frac{.0144}{42}} \quad \text{or} \quad .521 \le \mu_1 \le .607$$

$$\bar{x}_2 \pm t_{41}(.0125)\sqrt{\frac{s_{22}}{n}} = .603 \pm 2.327\sqrt{\frac{.0146}{42}} \quad \text{or} \quad .560 \le \mu_2 \le .646$$

Figure 5.4 shows the 95% T^2 simultaneous confidence intervals for μ_1, μ_2 from Figure 5.2, along with the corresponding 95% Bonferroni intervals. For each component mean, the Bonferroni interval falls within the T^2-interval. Consequently, the rectangular (joint) region formed by the two Bonferroni intervals is contained in the rectangular region formed by the two T^2-intervals. If we are interested only in the component means, the Bonferroni intervals provide more precise estimates than

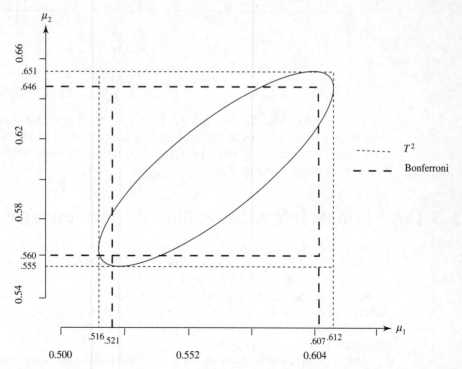

Figure 5.4 The 95% T^2 and 95% Bonferroni simultaneous confidence intervals for the component means—microwave radiation data.

the T^2-intervals. On the other hand, the 95% confidence region for μ gives the plausible values for the pairs (μ_1, μ_2) when the correlation between the measured variables is taken into account. ■

The Bonferroni intervals for linear combinations $a'\mu$ and the analogous T^2-intervals (recall Result 5.3) have the same general form:

$$a'\overline{X} \pm (\text{critical value})\sqrt{\frac{a'Sa}{n}}$$

Consequently, in every instance where $\alpha_i = \alpha/m$,

$$\frac{\text{Length of Bonferroni interval}}{\text{Length of } T^2\text{-interval}} = \frac{t_{n-1}(\alpha/2m)}{\sqrt{\dfrac{p(n-1)}{n-p} F_{p,n-p}(\alpha)}} \tag{5-30}$$

which does not depend on the random quantities $\overline{X}$ and S. As we have pointed out, for a small number m of specified parametric functions $a'\mu$, the Bonferroni intervals will always be shorter. How much shorter is indicated in Table 5.4 for selected n and p.

Table 5.4 (Length of Bonferroni Interval)/(Length of T^2-Interval) for $1 - \alpha = .95$ and $\alpha_i = .05/m$

n	$m = p$		
	2	4	10
15	.88	.69	.29
25	.90	.75	.48
50	.91	.78	.58
100	.91	.80	.62
∞	.91	.81	.66

We see from Table 5.4 that the Bonferroni method provides shorter intervals when $m = p$. Because they are easy to apply and provide the relatively short confidence intervals needed for inference, we will often apply simultaneous t-intervals based on the Bonferroni method.

5.5 Large Sample Inferences about a Population Mean Vector

When the sample size is large, tests of hypotheses and confidence regions for μ can be constructed without the assumption of a normal population. As illustrated in Exercises 5.15, 5.16, and 5.17, for large n, we are able to make inferences about the population mean even though the parent distribution is discrete. In fact, serious departures from a normal population can be overcome by large sample sizes. Both tests of hypotheses and simultaneous confidence statements will then possess (approximately) their nominal levels.

The advantages associated with large samples may be partially offset by a loss in sample information caused by using only the summary statistics $\bar{x}$, and S. On the other hand, since $(\bar{x}, S)$ is a sufficient summary for *normal* populations [see (4-21)], the closer the underlying population is to multivariate normal, the more efficiently the sample information will be utilized in making inferences.

All large-sample inferences about μ are based on a χ^2-distribution. From (4-28), we know that $(\overline{\mathbf{X}} - \boldsymbol{\mu})'(n^{-1}\mathbf{S})^{-1}(\overline{\mathbf{X}} - \boldsymbol{\mu}) = n(\overline{\mathbf{X}} - \boldsymbol{\mu})'\mathbf{S}^{-1}(\overline{\mathbf{X}} - \boldsymbol{\mu})$ is approximately χ^2 with p d.f., and thus,

$$P[n(\overline{\mathbf{X}} - \boldsymbol{\mu})'\mathbf{S}^{-1}(\overline{\mathbf{X}} - \boldsymbol{\mu}) \leq \chi_p^2(\alpha)] \doteq 1 - \alpha \qquad (5\text{-}31)$$

where $\chi_p^2(\alpha)$ is the upper (100α)th percentile of the χ_p^2-distribution.

Equation (5-31) immediately leads to large sample tests of hypotheses and simultaneous confidence regions. These procedures are summarized in Results 5.4 and 5.5.

Result 5.4. Let $\mathbf{X}_1, \mathbf{X}_2, \ldots, \mathbf{X}_n$ be a random sample from a population with mean $\boldsymbol{\mu}$ and positive definite covariance matrix $\boldsymbol{\Sigma}$. When $n - p$ is large, the hypothesis $H_0: \boldsymbol{\mu} = \boldsymbol{\mu}_0$ is rejected in favor of $H_1: \boldsymbol{\mu} \neq \boldsymbol{\mu}_0$, at a level of significance approximately α, if the observed

$$n(\overline{\mathbf{x}} - \boldsymbol{\mu}_0)'\mathbf{S}^{-1}(\overline{\mathbf{x}} - \boldsymbol{\mu}_0) > \chi_p^2(\alpha)$$

Here $\chi_p^2(\alpha)$ is the upper (100α)th percentile of a chi-square distribution with p d.f. ∎

Comparing the test in Result 5.4 with the corresponding *normal theory* test in (5-7), we see that the test statistics have the same structure, but the critical values are different. A closer examination, however, reveals that both tests yield essentially the same result in situations where the χ^2-test of Result 5.4 is appropriate. This follows directly from the fact that $(n - 1)pF_{p,n-p}(\alpha)/(n - p)$ and $\chi_p^2(\alpha)$ are approximately equal for n large relative to p. (See Tables 3 and 4 in the appendix.)

Result 5.5. Let $\mathbf{X}_1, \mathbf{X}_2, \ldots, \mathbf{X}_n$ be a random sample from a population with mean $\boldsymbol{\mu}$ and positive definite covariance $\boldsymbol{\Sigma}$. If $n - p$ is large,

$$\mathbf{a}'\overline{\mathbf{X}} \pm \sqrt{\chi_p^2(\alpha)} \sqrt{\frac{\mathbf{a}'\mathbf{Sa}}{n}}$$

will contain $\mathbf{a}'\boldsymbol{\mu}$, for every $\mathbf{a}$, with probability approximately $1 - \alpha$. Consequently, we can make the $100(1 - \alpha)\%$ simultaneous confidence statements

$$\overline{x}_1 \pm \sqrt{\chi_p^2(\alpha)} \sqrt{\frac{s_{11}}{n}} \qquad \text{contains } \mu_1$$

$$\overline{x}_2 \pm \sqrt{\chi_p^2(\alpha)} \sqrt{\frac{s_{22}}{n}} \qquad \text{contains } \mu_2$$

$$\vdots \qquad\qquad\qquad \vdots$$

$$\overline{x}_p \pm \sqrt{\chi_p^2(\alpha)} \sqrt{\frac{s_{pp}}{n}} \qquad \text{contains } \mu_p$$

and, in addition, for all pairs (μ_i, μ_k), $i, k = 1, 2, \ldots, p$, the sample mean-centered ellipses

$$n[\overline{x}_i - \mu_i, \;\; \overline{x}_k - \mu_k] \begin{bmatrix} s_{ii} & s_{ik} \\ s_{ik} & s_{kk} \end{bmatrix}^{-1} \begin{bmatrix} \overline{x}_i - \mu_i \\ \overline{x}_k - \mu_k \end{bmatrix} \leq \chi_p^2(\alpha) \quad \text{contain} \quad (\mu_i, \mu_k)$$

Proof. The first part follows from Result 5A.1, with $c^2 = \chi_p^2(\alpha)$. The probability level is a consequence of (5-31). The statements for the μ_i are obtained by the special choices $\mathbf{a}' = [0, \ldots, 0, a_i, 0, \ldots, 0]$, where $a_i = 1$, $i = 1, 2, \ldots, p$. The ellipsoids for pairs of means follow from Result 5A.2 with $c^2 = \chi_p^2(\alpha)$. The overall confidence level of approximately $1 - \alpha$ for all statements is, once again, a result of the large sample distribution theory summarized in (5-31).   ■

The question of what is a large sample size is not easy to answer. In one or two dimensions, sample sizes in the range 30 to 50 can usually be considered large. As the number characteristics becomes large, certainly larger sample sizes are required for the asymptotic distributions to provide good approximations to the true distributions of various test statistics. Lacking definitive studies, we simply state that $n - p$ must be large and realize that the true case is more complicated. An application with $p = 2$ and sample size 50 is much different than an application with $p = 52$ and sample size 100 although both have $n - p = 48$.

It is good statistical practice to subject these large sample inference procedures to the same checks required of the normal-theory methods. Although small to moderate departures from normality do not cause any difficulties for n large, *extreme* deviations could cause problems. Specifically, the true error rate may be far removed from the nominal level α. If, on the basis of Q–Q plots and other investigative devices, outliers and other forms of extreme departures are indicated (see, for example, [2]), appropriate corrective actions, including transformations, are desirable. Methods for testing mean vectors of symmetric multivariate distributions that are relatively insensitive to departures from normality are discussed in [11]. In some instances, Results 5.4 and 5.5 are useful only for *very* large samples.

The next example allows us to illustrate the construction of large sample simultaneous statements for all single mean components.

Example 5.7 (Constructing large sample simultaneous confidence intervals) A music educator tested thousands of Finnish students on their native musical ability in order to set national norms in Finland. Summary statistics for part of the data set are given in Table 5.5. These statistics are based on a sample of $n = 96$ Finnish 12th graders.

Table 5.5 Musical Aptitude Profile Means and Standard Deviations for 96 12th-Grade Finnish Students Participating in a Standardization Program

Variable	Raw score	
	Mean ($\bar{x}_i$)	Standard deviation ($\sqrt{s_{ii}}$)
X_1 = melody	28.1	5.76
X_2 = harmony	26.6	5.85
X_3 = tempo	35.4	3.82
X_4 = meter	34.2	5.12
X_5 = phrasing	23.6	3.76
X_6 = balance	22.0	3.93
X_7 = style	22.7	4.03

Source: Data courtesy of V. Sell.

Let us construct 90% simultaneous confidence intervals for the individual mean components μ_i, $i = 1, 2, \ldots, 7$.

From Result 5.5, simultaneous 90% confidence limits are given by $\bar{x}_i \pm \sqrt{\chi_7^2(.10)} \sqrt{\dfrac{s_{ii}}{n}}$, $i = 1, 2, \ldots, 7$, where $\chi_7^2(.10) = 12.02$. Thus, with approximately 90% confidence,

$$28.1 \pm \sqrt{12.02}\,\frac{5.76}{\sqrt{96}} \quad \text{contains } \mu_1 \quad \text{or} \quad 26.06 \le \mu_1 \le 30.14$$

$$26.6 \pm \sqrt{12.02}\,\frac{5.85}{\sqrt{96}} \quad \text{contains } \mu_2 \quad \text{or} \quad 24.53 \le \mu_2 \le 28.67$$

$$35.4 \pm \sqrt{12.02}\,\frac{3.82}{\sqrt{96}} \quad \text{contains } \mu_3 \quad \text{or} \quad 34.05 \le \mu_3 \le 36.75$$

$$34.2 \pm \sqrt{12.02}\,\frac{5.12}{\sqrt{96}} \quad \text{contains } \mu_4 \quad \text{or} \quad 32.39 \le \mu_4 \le 36.01$$

$$23.6 \pm \sqrt{12.02}\,\frac{3.76}{\sqrt{96}} \quad \text{contains } \mu_5 \quad \text{or} \quad 22.27 \le \mu_5 \le 24.93$$

$$22.0 \pm \sqrt{12.02}\,\frac{3.93}{\sqrt{96}} \quad \text{contains } \mu_6 \quad \text{or} \quad 20.61 \le \mu_6 \le 23.39$$

$$22.7 \pm \sqrt{12.02}\,\frac{4.03}{\sqrt{96}} \quad \text{contains } \mu_7 \quad \text{or} \quad 21.27 \le \mu_7 \le 24.13$$

Based, perhaps, upon thousands of American students, the investigator could hypothesize the musical aptitude profile to be

$$\boldsymbol{\mu}_0' = [31, 27, 34, 31, 23, 22, 22]$$

We see from the simultaneous statements above that the melody, tempo, and meter components of $\boldsymbol{\mu}_0$ do not appear to be plausible values for the corresponding means of Finnish scores. ∎

When the sample size is large, the one-at-a-time confidence intervals for individual means are

$$\bar{x}_i - z\left(\frac{\alpha}{2}\right)\sqrt{\frac{s_{ii}}{n}} \le \mu_i \le \bar{x}_i + z\left(\frac{\alpha}{2}\right)\sqrt{\frac{s_{ii}}{n}} \qquad i = 1, 2, \ldots, p$$

where $z(\alpha/2)$ is the upper $100(\alpha/2)$th percentile of the standard normal distribution. The Bonferroni simultaneous confidence intervals for the $m = p$ statements about the individual means take the same form, but use the modified percentile $z(\alpha/2p)$ to give

$$\bar{x}_i - z\left(\frac{\alpha}{2p}\right)\sqrt{\frac{s_{ii}}{n}} \le \mu_i \le \bar{x}_i + z\left(\frac{\alpha}{2p}\right)\sqrt{\frac{s_{ii}}{n}} \qquad i = 1, 2, \ldots, p$$

Table 5.6 gives the individual, Bonferroni, and chi-square-based (or shadow of the confidence ellipsoid) intervals for the musical aptitude data in Example 5.7.

Table 5.6 The Large Sample 95% Individual, Bonferroni, and T^2-Intervals for the Musical Aptitude Data

The one-at-a-time confidence intervals use $z(.025) = 1.96$.
The simultaneous Bonferroni intervals use $z(.025/7) = 2.69$.
The simultaneous T^2, or shadows of the ellipsoid, use $\chi_7^2(.05) = 14.07$.

Variable	One-at-a-time Lower	Upper	Bonferroni Intervals Lower	Upper	Shadow of Ellipsoid Lower	Upper
X_1 = melody	26.95	29.25	26.52	29.68	25.90	30.30
X_2 = harmony	25.43	27.77	24.99	28.21	24.36	28.84
X_3 = tempo	34.64	36.16	34.35	36.45	33.94	36.86
X_4 = meter	33.18	35.22	32.79	35.61	32.24	36.16
X_5 = phrasing	22.85	24.35	22.57	24.63	22.16	25.04
X_6 = balance	21.21	22.79	20.92	23.08	20.50	23.50
X_7 = style	21.89	23.51	21.59	23.81	21.16	24.24

Although the sample size may be large, some statisticians prefer to retain the F- and t-based percentiles rather than use the chi-square or standard normal-based percentiles. The latter constants are the infinite sample size limits of the former constants. The F and t percentiles produce larger intervals and, hence, are more conservative. Table 5.7 gives the individual, Bonferroni, and F-based, or shadow of the confidence ellipsoid, intervals for the musical aptitude data. Comparing Table 5.7 with Table 5.6, we see that all of the intervals in Table 5.7 are larger. However, with the relatively large sample size $n = 96$, the differences are typically in the third, or tenths, digit.

Table 5.7 The 95% Individual, Bonferroni, and T^2-Intervals for the Musical Aptitude Data

The one-at-a-time confidence intervals use $t_{95}(.025) = 1.99$.
The simultaneous Bonferroni intervals use $t_{95}(.025/7) = 2.75$.
The simultaneous T^2, or shadows of the ellipsoid, use $F_{7,89}(.05) = 2.11$.

Variable	One-at-a-time Lower	Upper	Bonferroni Intervals Lower	Upper	Shadow of Ellipsoid Lower	Upper
X_1 = melody	26.93	29.27	26.48	29.72	25.76	30.44
X_2 = harmony	25.41	27.79	24.96	28.24	24.23	28.97
X_3 = tempo	34.63	36.17	34.33	36.47	33.85	36.95
X_4 = meter	33.16	35.24	32.76	35.64	32.12	36.28
X_5 = phrasing	22.84	24.36	22.54	24.66	22.07	25.13
X_6 = balance	21.20	22.80	20.90	23.10	20.41	23.59
X_7 = style	21.88	23.52	21.57	23.83	21.07	24.33

5.6 Multivariate Quality Control Charts

To improve the quality of goods and services, data need to be examined for causes of variation. When a manufacturing process is continuously producing items or when we are monitoring activities of a service, data should be collected to evaluate the capabilities and stability of the process. When a process is stable, the variation is produced by common causes that are always present, and no one cause is a major source of variation.

The purpose of any control chart is to identify occurrences of *special causes* of variation that come from outside of the usual process. These causes of variation often indicate a need for a timely repair, but they can also suggest improvements to the process. Control charts make the variation visible and allow one to distinguish common from special causes of variation.

A control chart typically consists of data plotted in time order and horizontal lines, called control limits, that indicate the amount of variation due to common causes. One useful control chart is the $\overline{X}$-chart (read X-bar chart). To create an $\overline{X}$-chart,

1. Plot the individual observations or sample means in time order.
2. Create and plot the centerline $\bar{\bar{x}}$, the sample mean of all of the observations.
3. Calculate and plot the control limits given by

$$\text{Upper control limit (UCL)} = \bar{\bar{x}} + 3(\text{standard deviation})$$

$$\text{Lower control limit (LCL)} = \bar{\bar{x}} - 3(\text{standard deviation})$$

The standard deviation in the control limits is the estimated standard deviation of the observations being plotted. For single observations, it is often the sample standard deviation. If the means of subsamples of size m are plotted, then the standard deviation is the sample standard deviation divided by $\sqrt{m}$. The control limits of plus and minus three standard deviations are chosen so that there is a very small chance, assuming normally distributed data, of falsely signaling an out-of-control observation—that is, an observation suggesting a special cause of variation.

Example 5.8 (Creating a univariate control chart) The Madison, Wisconsin, police department regularly monitors many of its activities as part of an ongoing quality improvement program. Table 5.8 gives the data on five different kinds of overtime hours. Each observation represents a total for 12 pay periods, or about half a year.

We examine the stability of the legal appearances overtime hours. A computer calculation gives $\bar{x}_1 = 3558$. Since individual values will be plotted, $\bar{x}_1$ is the same as $\bar{\bar{x}}_1$. Also, the sample standard deviation is $\sqrt{s_{11}} = 607$, and the control limits are

$$\text{UCL} = \bar{\bar{x}}_1 + 3(\sqrt{s_{11}}) = 3558 + 3(607) = 5379$$

$$\text{LCL} = \bar{\bar{x}}_1 - 3(\sqrt{s_{11}}) = 3558 - 3(607) = 1737$$

Table 5.8 Five Types of Overtime Hours for the Madison, Wisconsin, Police Department

x_1 Legal Appearances Hours	x_2 Extraordinary Event Hours	x_3 Holdover Hours	x_4 COA[1] Hours	x_5 Meeting Hours
3387	2200	1181	14,861	236
3109	875	3532	11,367	310
2670	957	2502	13,329	1182
3125	1758	4510	12,328	1208
3469	868	3032	12,847	1385
3120	398	2130	13,979	1053
3671	1603	1982	13,528	1046
4531	523	4675	12,699	1100
3678	2034	2354	13,534	1349
3238	1136	4606	11,609	1150
3135	5326	3044	14,189	1216
5217	1658	3340	15,052	660
3728	1945	2111	12,236	299
3506	344	1291	15,482	206
3824	807	1365	14,900	239
3516	1223	1175	15,078	161

[1] Compensatory overtime allowed.

The data, along with the centerline and control limits, are plotted as an $\overline{X}$-chart in Figure 5.5.

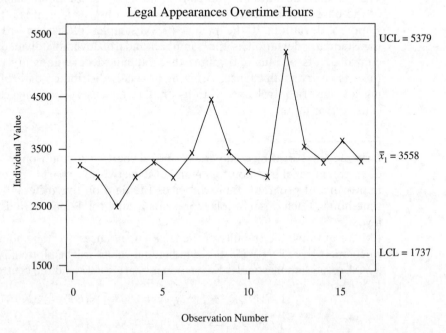

Figure 5.5 The $\overline{X}$-chart for x_1 = legal appearances overtime hours.

The legal appearances overtime hours are stable over the period in which the data were collected. The variation in overtime hours appears to be due to common causes, so no special-cause variation is indicated. ∎

With more than one important characteristic, a multivariate approach should be used to monitor process stability. Such an approach can account for correlations between characteristics and will control the overall probability of falsely signaling a special cause of variation when one is not present. High correlations among the variables can make it impossible to assess the overall error rate that is implied by a large number of univariate charts.

The two most common multivariate charts are (i) the ellipse format chart and (ii) the T^2-chart.

Two cases that arise in practice need to be treated differently:

1. Monitoring the stability of a given sample of multivariate observations
2. Setting a control region for future observations

Initially, we consider the use of multivariate control procedures for a sample of multivariate observations $\mathbf{x}_1, \mathbf{x}_2, \ldots, \mathbf{x}_n$. Later, we discuss these procedures when the observations are subgroup means.

Charts for Monitoring a Sample of Individual Multivariate Observations for Stability

We assume that $\mathbf{X}_1, \mathbf{X}_2, \ldots, \mathbf{X}_n$ are independently distributed as $N_p(\boldsymbol{\mu}, \boldsymbol{\Sigma})$. By Result 4.8,

$$\mathbf{X}_j - \overline{\mathbf{X}} = \left(1 - \frac{1}{n}\right)\mathbf{X}_j - \frac{1}{n}\mathbf{X}_1 - \cdots - \frac{1}{n}\mathbf{X}_{j-1} - \frac{1}{n}\mathbf{X}_{j+1} - \cdots - \frac{1}{n}\mathbf{X}_n$$

has

$$E(\mathbf{X}_j - \overline{\mathbf{X}}) = \mathbf{0} = (1 - n^{-1})\boldsymbol{\mu} - (n-1)n^{-1}\boldsymbol{\mu}$$

and

$$\text{Cov}(\mathbf{X}_j - \overline{\mathbf{X}}) = \left(1 - \frac{1}{n}\right)^2 \boldsymbol{\Sigma} + (n-1)n^{-2}\boldsymbol{\Sigma} = \frac{(n-1)}{n}\boldsymbol{\Sigma}$$

Each $\mathbf{X}_j - \overline{\mathbf{X}}$ has a normal distribution but, $\mathbf{X}_j - \overline{\mathbf{X}}$ is not independent of the sample covariance matrix $\mathbf{S}$. However to set control limits, we approximate that $(\mathbf{X}_j - \overline{\mathbf{X}})'\mathbf{S}^{-1}(\mathbf{X}_j - \overline{\mathbf{X}})$ has a chi-square distribution.

Ellipse Format Chart. The ellipse format chart for a bivariate control region is the more intuitive of the charts, but its approach is limited to two variables. The two characteristics on the jth unit are plotted as a pair (x_{j1}, x_{j2}). The 95% quality ellipse consists of all $\mathbf{x}$ that satisfy

$$(\mathbf{x} - \overline{\mathbf{x}})'\mathbf{S}^{-1}(\mathbf{x} - \overline{\mathbf{x}}) \leq \chi_2^2(.05) \tag{5-32}$$

Example 5.9 (An ellipse format chart for overtime hours) Let us refer to Example 5.8 and create a quality ellipse for the pair of overtime characteristics (legal appearances, extraordinary event) hours. A computer calculation gives

$$\bar{\mathbf{x}} = \begin{bmatrix} 3558 \\ 1478 \end{bmatrix} \quad \text{and} \quad \mathbf{S} = \begin{bmatrix} 367{,}884.7 & -72{,}093.8 \\ -72{,}093.8 & 1{,}399{,}053.1 \end{bmatrix}$$

We illustrate the quality ellipse format chart using the 99% ellipse, which consists of all $\mathbf{x}$ that satisfy

$$(\mathbf{x} - \bar{\mathbf{x}})' \mathbf{S}^{-1} (\mathbf{x} - \bar{\mathbf{x}}) \le \chi_2^2(.01)$$

Here $p = 2$, so $\chi_2^2(.01) = 9.21$, and the ellipse becomes

$$\frac{s_{11}s_{22}}{s_{11}s_{22} - s_{12}^2} \left(\frac{(x_1 - \bar{x}_1)^2}{s_{11}} - 2s_{12}\frac{(x_1 - \bar{x}_1)(x_2 - \bar{x}_2)}{s_{11}s_{22}} + \frac{(x_2 - \bar{x}_2)^2}{s_{22}} \right)$$

$$= \frac{(367844.7 \times 1399053.1)}{367844.7 \times 1399053.1 - (-72093.8)^2}$$

$$\times \left(\frac{(x_1 - 3558)^2}{367844.7} - 2(-72093.8)\frac{(x_1 - 3558)(x_2 - 1478)}{367844.7 \times 1399053.1} + \frac{(x_2 - 1478)^2}{1399053.1} \right) \le 9.21$$

This ellipse format chart is graphed, along with the pairs of data, in Figure 5.6.

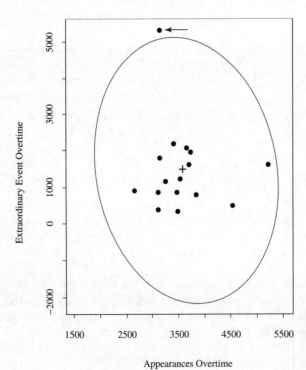

Figure 5.6 The quality control 99% ellipse for legal appearances and extraordinary event overtime.

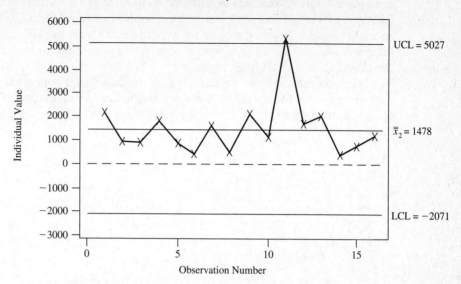

Figure 5.7 The $\overline{X}$ -chart for x_2 = extraordinary event hours.

Notice that one point, indicated with an arrow, is definitely outside of the el-lipse. When a point is out of the control region, individual $\overline{X}$ charts are constructed. The $\overline{X}$ -chart for x_1 was given in Figure 5.5; that for x_2 is given in Figure 5.7.

When the lower control limit is less than zero for data that must be non-negative, it is generally set to zero. The LCL = 0 limit is shown by the dashed line in Figure 5.7.

Was there a special cause of the single point for extraordinary event overtime that is outside the upper control limit in Figure 5.7? During this period, the United States bombed a foreign capital, and students at Madison were protesting. A major-ity of the extraordinary overtime was used in that four-week period. Although, by its very definition, extraordinary overtime occurs only when special events occur and is therefore unpredictable, it still has a certain stability. ∎

T^2-Chart. A T^2-chart can be applied to a large number of characteristics. Unlike the ellipse format, it is not limited to two variables. Moreover, the points are displayed in time order rather than as a scatter plot, and this makes patterns and trends visible.

For the jth point, we calculate the T^2-statistic

$$T_j^2 = (\mathbf{x}_j - \bar{\mathbf{x}})'\mathbf{S}^{-1}(\mathbf{x}_j - \bar{\mathbf{x}}) \qquad (5\text{-}33)$$

We then plot the T^2-values on a time axis. The lower control limit is zero, and we use the upper control limit

$$\text{UCL} = \chi_p^2(.05)$$

or, sometimes, $\chi_p^2(.01)$.

There is no centerline in the T^2-chart. Notice that the T^2-statistic is the same as the quantity d_j^2 used to test normality in Section 4.6.

Example 5.10 (A T^2-chart for overtime hours) Using the police department data in Example 5.8, we construct a T^2-plot based on the two variables X_1 = legal appearances hours and X_2 = extraordinary event hours. T^2-charts with more than two variables are considered in Exercise 5.26. We take $\alpha = .01$ to be consistent with the ellipse format chart in Example 5.9.

The T^2-chart in Figure 5.8 reveals that the pair (legal appearances, extraordinary event) hours for period 11 is out of control. Further investigation, as in Example 5.9, confirms that this is due to the large value of extraordinary event overtime during that period. ∎

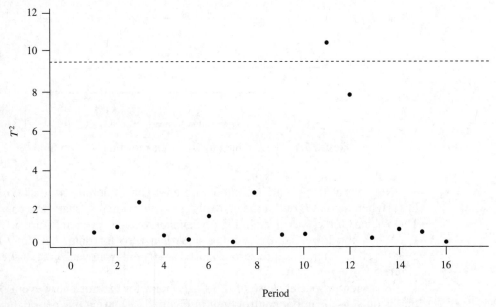

Figure 5.8 The T^2-chart for legal appearances hours and extraordinary event hours, $\alpha = .01$.

When the multivariate T^2-chart signals that the jth unit is out of control, it should be determined which variables are responsible. A modified region based on Bonferroni intervals is frequently chosen for this purpose. The kth variable is out of control if x_{jk} does not lie in the interval

$$(\bar{x}_k - t_{n-1}(.005/p)\sqrt{s_{kk}},\quad \bar{x}_k + t_{n-1}(.005/p)\sqrt{s_{kk}})$$

where p is the total number of measured variables.

Example 5.11 (Control of robotic welders—more than T^2 needed) The assembly of a driveshaft for an automobile requires the circle welding of tube yokes to a tube. The inputs to the automated welding machines must be controlled to be within certain operating limits where a machine produces welds of good quality. In order to control the process, one process engineer measured four critical variables:

$$X_1 = \text{Voltage (volts)}$$
$$X_2 = \text{Current (amps)}$$
$$X_3 = \text{Feed speed(in/min)}$$
$$X_4 = \text{(inert) Gas flow (cfm)}$$

Table 5.9 gives the values of these variables at five-second intervals.

Table 5.9 Welder Data

Case	Voltage (X_1)	Current (X_2)	Feed speed (X_3)	Gas flow (X_4)
1	23.0	276	289.6	51.0
2	22.0	281	289.0	51.7
3	22.8	270	288.2	51.3
4	22.1	278	288.0	52.3
5	22.5	275	288.0	53.0
6	22.2	273	288.0	51.0
7	22.0	275	290.0	53.0
8	22.1	268	289.0	54.0
9	22.5	277	289.0	52.0
10	22.5	278	289.0	52.0
11	22.3	269	287.0	54.0
12	21.8	274	287.6	52.0
13	22.3	270	288.4	51.0
14	22.2	273	290.2	51.3
15	22.1	274	286.0	51.0
16	22.1	277	287.0	52.0
17	21.8	277	287.0	51.0
18	22.6	276	290.0	51.0
19	22.3	278	287.0	51.7
20	23.0	266	289.1	51.0
21	22.9	271	288.3	51.0
22	21.3	274	289.0	52.0
23	21.8	280	290.0	52.0
24	22.0	268	288.3	51.0
25	22.8	269	288.7	52.0
26	22.0	264	290.0	51.0
27	22.5	273	288.6	52.0
28	22.2	269	288.2	52.0
29	22.6	273	286.0	52.0
30	21.7	283	290.0	52.7
31	21.9	273	288.7	55.3
32	22.3	264	287.0	52.0
33	22.2	263	288.0	52.0
34	22.3	266	288.6	51.7
35	22.0	263	288.0	51.7
36	22.8	272	289.0	52.3
37	22.0	277	287.7	53.3
38	22.7	272	289.0	52.0
39	22.6	274	287.2	52.7
40	22.7	270	290.0	51.0

Source: Data courtesy of Mark Abbotoy.

The normal assumption is reasonable for most variables, but we take the natural logarithm of gas flow. In addition, there is no appreciable serial correlation for successive observations on each variable.

A T^2-chart for the four welding variables is given in Figure 5.9. The dotted line is the 95% limit and the solid line is the 99% limit. Using the 99% limit, no points are out of control, but case 31 is outside the 95% limit.

What do the quality control ellipses (ellipse format charts) show for two variables? Most of the variables are in control. However, the 99% quality ellipse for gas flow and voltage, shown in Figure 5.10, reveals that case 31 is out of control and this is due to an unusually large volume of gas flow. The univariate $\overline{X}$ chart for ln(gas flow), in Figure 5.11, shows that this point is outside the three sigma limits. It appears that gas flow was reset at the target for case 32. All the other univariate $\overline{X}$-charts have all points within their three sigma control limits.

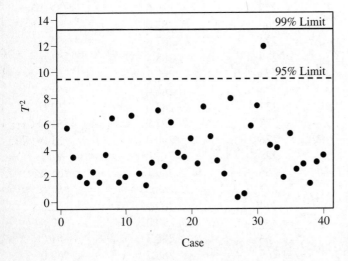

Figure 5.9 The T^2-chart for the welding data with 95% and 99% limits.

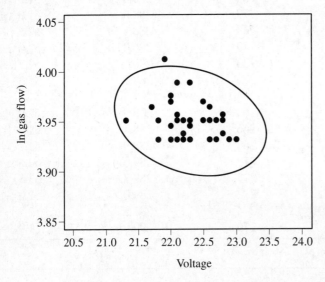

Figure 5.10 The 99% quality control ellipse for ln(gas flow) and voltage.

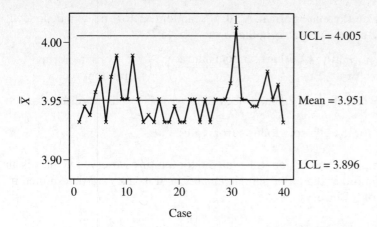

Figure 5.11 The univariate $\overline{X}$ -chart for ln(gas flow).

In this example, a shift in a single variable was masked with 99% limits, or almost masked (with 95% limits), by being combined into a single T^2-value. ∎

Control Regions for Future Individual Observations

The goal now is to use data $\mathbf{x}_1, \mathbf{x}_2, \ldots, \mathbf{x}_n$, collected when a process is stable, to set a control region for a future observation $\mathbf{x}$ or future observations. The region in which a future observation is expected to lie is called a *forecast*, or *prediction, region*. If the process is stable, we take the observations to be independently distributed as $N_p(\boldsymbol{\mu}, \boldsymbol{\Sigma})$. Because these regions are of more general importance than just for monitoring quality, we give the basic distribution theory as Result 5.6.

Result 5.6. Let $\mathbf{X}_1, \mathbf{X}_2, \ldots, \mathbf{X}_n$ be independently distributed as $N_p(\boldsymbol{\mu}, \boldsymbol{\Sigma})$, and let $\mathbf{X}$ be a future observation from the same distribution. Then

$$T^2 = \frac{n}{n + 1}(\mathbf{X} - \overline{\mathbf{X}})'\mathbf{S}^{-1}(\mathbf{X} - \overline{\mathbf{X}}) \text{ is distributed as } \frac{(n - 1)p}{n - p}F_{p,n-p}$$

and a $100(1 - \alpha)\%$ p-dimensional prediction ellipsoid is given by all $\mathbf{x}$ satisfying

$$(\mathbf{x} - \overline{\mathbf{x}})'\mathbf{S}^{-1}(\mathbf{x} - \overline{\mathbf{x}}) \leq \frac{(n^2 - 1)p}{n(n - p)}F_{p,n-p}(\alpha)$$

Proof. We first note that $\mathbf{X} - \overline{\mathbf{X}}$ has mean $\mathbf{0}$. Since $\mathbf{X}$ is a future observation, $\mathbf{X}$ and $\overline{\mathbf{X}}$ are independent, so

$$\text{Cov}(\mathbf{X} - \overline{\mathbf{X}}) = \text{Cov}(\mathbf{X}) + \text{Cov}(\overline{\mathbf{X}}) = \boldsymbol{\Sigma} + \frac{1}{n}\boldsymbol{\Sigma} = \frac{(n + 1)}{n}\boldsymbol{\Sigma}$$

and, by Result 4.8, $\sqrt{n/(n + 1)}\,(\mathbf{X} - \overline{\mathbf{X}})$ is distributed as $N_p(\mathbf{0}, \boldsymbol{\Sigma})$. Now,

$$\sqrt{\frac{n}{n + 1}}(\mathbf{X} - \overline{\mathbf{X}})'\mathbf{S}^{-1}\sqrt{\frac{n}{n + 1}}(\mathbf{X} - \overline{\mathbf{X}})$$

which combines a multivariate normal, $N_p(\mathbf{0}, \boldsymbol{\Sigma})$, random vector and an independent Wishart, $W_{p,n-1}(\boldsymbol{\Sigma})$, random matrix in the form

$$\left(\begin{array}{c} \text{multivariate normal} \\ \text{random vector} \end{array} \right)' \left(\frac{\text{Wishart random matrix}}{\text{d.f.}} \right)^{-1} \left(\begin{array}{c} \text{multivariate normal} \\ \text{random vector} \end{array} \right)$$

has the scaled F distribution claimed according to (5-8) and the discussion on page 213.

The constant for the ellipsoid follows from (5-6). ∎

Note that the prediction region in Result 5.6 for a future observed value $\mathbf{x}$ is an ellipsoid. It is centered at the initial sample mean $\bar{\mathbf{x}}$, and its axes are determined by the eigenvectors of $\mathbf{S}$. Since

$$P\left[(\mathbf{X} - \bar{\mathbf{X}})'\mathbf{S}^{-1}(\mathbf{X} - \bar{\mathbf{X}}) \leq \frac{(n^2 - 1)p}{n(n - p)} F_{p,n-p}(\alpha) \right] = 1 - \alpha$$

before any new observations are taken, the probability that $\mathbf{X}$ will fall in the prediction ellipse is $1 - \alpha$.

Keep in mind that the current observations must be stable before they can be used to determine control regions for future observations.

Based on Result 5.6, we obtain the two charts for future observations.

Control Ellipse for Future Observations

With $p = 2$, the 95% prediction ellipse in Result 5.6 specializes to

$$(\mathbf{x} - \bar{\mathbf{x}})'\mathbf{S}^{-1}(\mathbf{x} - \bar{\mathbf{x}}) \leq \frac{(n^2 - 1)2}{n(n - 2)} F_{2,n-2}(.05) \qquad (5\text{-}34)$$

Any future observation $\mathbf{x}$ is declared to be out of control if it falls out of the control ellipse.

Example 5.12 (A control ellipse for future overtime hours) In Example 5.9, we checked the stability of legal appearances and extraordinary event overtime hours. Let's use these data to determine a control region for future pairs of values.

From Example 5.9 and Figure 5.6, we find that the pair of values for period 11 were out of control. We removed this point and determined the new 99% ellipse. All of the points are then in control, so they can serve to determine the 95% prediction region just defined for $p = 2$. This control ellipse is shown in Figure 5.12 along with the initial 15 stable observations.

Any future observation falling in the ellipse is regarded as stable or in control. An observation outside of the ellipse represents a potential out-of-control observation or special-cause variation. ∎

T^2-Chart for Future Observations

For each new observation $\mathbf{x}$, plot

$$T^2 = \frac{n}{n + 1} (\mathbf{x} - \bar{\mathbf{x}})'\mathbf{S}^{-1}(\mathbf{x} - \bar{\mathbf{x}})$$

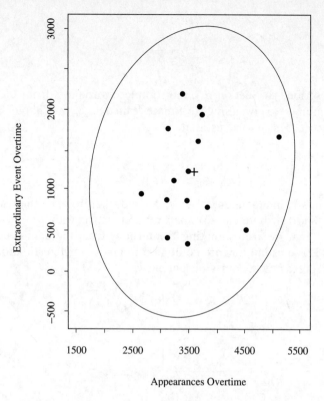

Figure 5.12 The 95% control ellipse for future legal appearances and extraordinary event overtime.

in time order. Set LCL = 0, and take

$$\text{UCL} = \frac{(n-1)p}{(n-p)} F_{p,n-p}(.05)$$

Points above the upper control limit represent potential special cause variation and suggest that the process in question should be examined to determine whether immediate corrective action is warranted. See [9] for discussion of other procedures.

Control Charts Based on Subsample Means

It is assumed that each random vector of observations from the process is independently distributed as $N_p(\mathbf{0}, \boldsymbol{\Sigma})$. We proceed differently when the sampling procedure specifies that $m > 1$ units be selected, at the same time, from the process. From the first sample, we determine its sample mean $\overline{\mathbf{X}}_1$ and covariance matrix $\mathbf{S}_1$. When the population is normal, these two random quantities are independent.

For a general subsample mean $\overline{\mathbf{X}}_j$, $\overline{\mathbf{X}}_j - \overline{\overline{\mathbf{X}}}$ has a normal distribution with mean $\mathbf{0}$ and

$$\text{Cov}(\overline{\mathbf{X}}_j - \overline{\overline{\mathbf{X}}}) = \left(1 - \frac{1}{n}\right)^2 \text{Cov}(\overline{\mathbf{X}}_j) + \frac{n-1}{n^2} \text{Cov}(\overline{\mathbf{X}}_1) = \frac{(n-1)}{nm} \boldsymbol{\Sigma}$$

where

$$\overline{\overline{\mathbf{X}}} = \frac{1}{n} \sum_{j=1}^{n} \overline{\mathbf{X}}_j$$

As will be described in Section 6.4, the sample covariances from the n sub-samples can be combined to give a single estimate (called $\mathbf{S}_{\text{pooled}}$ in Chapter 6) of the common covariance $\boldsymbol{\Sigma}$. This pooled estimate is

$$\mathbf{S} = \frac{1}{n} (\mathbf{S}_1 + \mathbf{S}_2 + \cdots + \mathbf{S}_n)$$

Here $(nm - n)\mathbf{S}$ is independent of each $\overline{\mathbf{X}}_j$ and, therefore, of their mean $\overline{\overline{\mathbf{X}}}$. Further, $(nm - n)\mathbf{S}$ is distributed as a Wishart random matrix with $nm - n$ degrees of freedom. Notice that we are estimating $\boldsymbol{\Sigma}$ internally from the data collected in any given period. These estimators are combined to give a single estimator with a large number of degrees of freedom. Consequently,

$$T^2 = \frac{nm}{n-1} (\overline{\mathbf{X}}_j - \overline{\overline{\mathbf{X}}})' \mathbf{S}^{-1} (\overline{\mathbf{X}}_j - \overline{\overline{\mathbf{X}}}) \tag{5-35}$$

is distributed as

$$\frac{(nm - n)p}{(nm - n - p + 1)} F_{p, nm-n-p+1}$$

Ellipse Format Chart.   In an analogous fashion to our discussion on individual multivariate observations, the ellipse format chart for pairs of subsample means is

$$(\overline{\mathbf{x}} - \overline{\overline{\mathbf{x}}})' \mathbf{S}^{-1} (\overline{\mathbf{x}} - \overline{\overline{\mathbf{x}}}) \leq \frac{(n-1)(m-1)2}{m(nm - n - 1)} F_{2, nm-n-1}(.05) \tag{5-36}$$

although the right-hand side is usually approximated as $\chi_2^2(.05)/m$.

Subsamples corresponding to points outside of the control ellipse should be carefully checked for changes in the behavior of the quality characteristics being measured. The interested reader is referred to [10] for additional discussion.

T^2-Chart.   To construct a T^2-chart with subsample data and p characteristics, we plot the quantity

$$T_j^2 = m(\overline{\mathbf{X}}_j - \overline{\overline{\mathbf{X}}})' \mathbf{S}^{-1} (\overline{\mathbf{X}}_j - \overline{\overline{\mathbf{X}}})$$

for $j = 1, 2, \ldots, n$, where the

$$\text{UCL} = \frac{(n-1)(m-1)p}{(nm - n - p + 1)} F_{p, nm-n-p+1}(.05)$$

The UCL is often approximated as $\chi_p^2(.05)$ when n is large.

Values of T_j^2 that exceed the UCL correspond to potentially out-of-control or special cause variation, which should be checked. (See [10].)

Control Regions for Future Subsample Observations

Once data are collected from the stable operation of a process, they can be used to set control limits for future observed subsample means.

If $\overline{\mathbf{X}}$ is a future subsample mean, then $\overline{\mathbf{X}} - \overline{\overline{\mathbf{X}}}$ has a multivariate normal distribution with mean $\mathbf{0}$ and

$$\mathrm{Cov}(\overline{\mathbf{X}} - \overline{\overline{\mathbf{X}}}) = \mathrm{Cov}(\overline{\mathbf{X}}) + \frac{1}{n}\mathrm{Cov}(\overline{\mathbf{X}}_1) = \frac{(n+1)}{nm}\boldsymbol{\Sigma}$$

Consequently,

$$\frac{nm}{n+1}(\overline{\mathbf{X}} - \overline{\overline{\mathbf{X}}})'\mathbf{S}^{-1}(\overline{\mathbf{X}} - \overline{\overline{\mathbf{X}}})$$

is distributed as

$$\frac{(nm - n)p}{(nm - n - p + 1)}F_{p,nm-n-p+1}$$

Control Ellipse for Future Subsample Means. The prediction ellipse for a future subsample mean for $p = 2$ characteristics is defined by the set of all $\overline{\mathbf{x}}$ such that

$$(\overline{\mathbf{x}} - \overline{\overline{\mathbf{x}}})'\mathbf{S}^{-1}(\overline{\mathbf{x}} - \overline{\overline{\mathbf{x}}}) \le \frac{(n+1)(m-1)2}{m(nm - n - 1)}F_{2,nm-n-1}(.05) \qquad (5\text{-}37)$$

where, again, the right-hand side is usually approximated as $\chi_2^2(.05)/m$.

T^2-Chart for Future Subsample Means. As before, we bring $n/(n+1)$ into the control limit and plot the quantity

$$T^2 = m(\overline{\mathbf{X}} - \overline{\overline{\mathbf{X}}})'\mathbf{S}^{-1}(\overline{\mathbf{X}} - \overline{\overline{\mathbf{X}}})$$

for future sample means in chronological order. The upper control limit is then

$$\mathrm{UCL} = \frac{(n+1)(m-1)p}{(nm - n - p + 1)}F_{p,nm-n-p+1}(.05)$$

The UCL is often approximated as $\chi_p^2(.05)$ when n is large.

Points outside of the prediction ellipse or above the UCL suggest that the current values of the quality characteristics are different in some way from those of the previous stable process. This may be good or bad, but almost certainly warrants a careful search for the reasons for the change.

5.7 Inferences about Mean Vectors When Some Observations Are Missing

Often, some components of a vector observation are unavailable. This may occur because of a breakdown in the recording equipment or because of the unwillingness of a respondent to answer a particular item on a survey questionnaire. The best way to handle incomplete observations, or missing values, depends, to a large extent, on the experimental context. If the pattern of missing values is closely tied to the value of the response, such as people with extremely high incomes who refuse to respond in a survey on salaries, subsequent inferences may be seriously biased. To date, no statistical techniques have been developed for these cases. However, we are able to treat situations where data are missing at random—that is, cases in which the chance mechanism responsible for the missing values is *not* influenced by the value of the variables.

A general approach for computing maximum likelihood estimates from incomplete data is given by Dempster, Laird, and Rubin [5]. Their technique, called the *EM algorithm*, consists of an iterative calculation involving two steps. We call them the *prediction* and *estimation* steps:

1. *Prediction step.* Given some estimate $\widetilde{\boldsymbol{\theta}}$ of the unknown parameters, predict the contribution of any missing observation to the (complete-data) sufficient statistics.

2. *Estimation step.* Use the predicted sufficient statistics to compute a revised estimate of the parameters.

The calculation cycles from one step to the other, until the revised estimates do not differ appreciably from the estimate obtained in the previous iteration.

When the observations $\mathbf{X}_1, \mathbf{X}_2, \ldots, \mathbf{X}_n$ are a random sample from a p-variate normal population, the prediction–estimation algorithm is based on the complete-data sufficient statistics [see (4-21)]

$$\mathbf{T}_1 = \sum_{j=1}^{n} \mathbf{X}_j = n\overline{\mathbf{X}}$$

and

$$\mathbf{T}_2 = \sum_{j=1}^{n} \mathbf{X}_j \mathbf{X}_j' = (n - 1)\mathbf{S} + n\overline{\mathbf{X}}\,\overline{\mathbf{X}}'$$

In this case, the algorithm proceeds as follows: We assume that the population mean and variance—$\boldsymbol{\mu}$ and $\boldsymbol{\Sigma}$, respectively—are unknown and must be estimated.

Prediction step. For each vector $\mathbf{x}_j$ with missing values, let $\mathbf{x}_j^{(1)}$ denote the missing components and $\mathbf{x}_j^{(2)}$ denote those components which are available. Thus, $\mathbf{x}_j' = [\mathbf{x}_j^{(1)\prime}, \mathbf{x}_j^{(2)\prime}]$.

Given estimates $\widetilde{\boldsymbol{\mu}}$ and $\widetilde{\boldsymbol{\Sigma}}$ from the estimation step, use the mean of the conditional normal distribution of $\mathbf{x}^{(1)}$, given $\mathbf{x}^{(2)}$, to estimate the missing values. That is,[1]

$$\widetilde{\mathbf{x}}_j^{(1)} = E(\mathbf{X}_j^{(1)} \mid \mathbf{x}_j^{(2)}; \widetilde{\boldsymbol{\mu}}, \widetilde{\boldsymbol{\Sigma}}) = \widetilde{\boldsymbol{\mu}}^{(1)} + \widetilde{\boldsymbol{\Sigma}}_{12} \widetilde{\boldsymbol{\Sigma}}_{22}^{-1} (\mathbf{x}_j^{(2)} - \widetilde{\boldsymbol{\mu}}^{(2)}) \tag{5-38}$$

estimates the contribution of $\mathbf{x}_j^{(1)}$ to $\mathbf{T}_1$.

Next, the predicted contribution of $\mathbf{x}_j^{(1)}$ to $\mathbf{T}_2$ is

$$\widetilde{\mathbf{x}_j^{(1)} \mathbf{x}_j^{(1)\prime}} = E(\mathbf{X}_j^{(1)} \mathbf{X}_j^{(1)\prime} \mid \mathbf{x}_j^{(2)}; \widetilde{\boldsymbol{\mu}}, \widetilde{\boldsymbol{\Sigma}}) = \widetilde{\boldsymbol{\Sigma}}_{11} - \widetilde{\boldsymbol{\Sigma}}_{12} \widetilde{\boldsymbol{\Sigma}}_{22}^{-1} \widetilde{\boldsymbol{\Sigma}}_{21} + \widetilde{\mathbf{x}}_j^{(1)} \widetilde{\mathbf{x}}_j^{(1)\prime} \tag{5-39}$$

---

[1] If all the components $\mathbf{x}_j$ are missing, set $\widetilde{\mathbf{x}}_j = \widetilde{\boldsymbol{\mu}}$ and $\widetilde{\mathbf{x}_j \mathbf{x}_j'} = \widetilde{\boldsymbol{\Sigma}} + \widetilde{\boldsymbol{\mu}} \widetilde{\boldsymbol{\mu}}'$.

and

$$\widehat{x_j^{(1)}x_j^{(2)\prime}} = E(X_j^{(1)}X_j^{(2)\prime} \mid x_j^{(2)}; \ \tilde{\mu}, \tilde{\Sigma}) = \tilde{x}_j^{(1)}x_j^{(2)\prime}$$

The contributions in (5-38) and (5-39) are summed over all x_j with missing components. The results are combined with the sample data to yield $\tilde{T}_1$ and $\tilde{T}_2$.

Estimation step. Compute the revised maximum likelihood estimates (see Result 4.11):

$$\tilde{\mu} = \frac{\tilde{T}_1}{n}, \qquad \tilde{\Sigma} = \frac{1}{n}\tilde{T}_2 - \tilde{\mu}\tilde{\mu}' \tag{5-40}$$

We illustrate the computational aspects of the prediction–estimation algorithm in Example 5.13.

Example 5.13 (Illustrating the *EM* algorithm) Estimate the normal population mean μ and covariance Σ using the incomplete data set

$$\mathbf{X} = \begin{bmatrix} - & 0 & 3 \\ 7 & 2 & 6 \\ 5 & 1 & 2 \\ - & - & 5 \end{bmatrix}$$

Here $n = 4$, $p = 3$, and parts of observation vectors x_1 and x_4 are missing. We obtain the initial sample averages

$$\tilde{\mu}_1 = \frac{7+5}{2} = 6, \qquad \tilde{\mu}_2 = \frac{0+2+1}{3} = 1, \qquad \tilde{\mu}_3 = \frac{3+6+2+5}{4} = 4$$

from the available observations. Substituting these averages for any missing values, so that $\tilde{x}_{11} = 6$, for example, we can obtain initial covariance estimates. We shall construct these estimates using the divisor n because the algorithm eventually produces the maximum likelihood estimate $\hat{\Sigma}$ Thus,

$$\tilde{\sigma}_{11} = \frac{(6-6)^2 + (7-6)^2 + (5-6)^2 + (6-6)^2}{4} = \frac{1}{2}$$

$$\tilde{\sigma}_{22} = \frac{1}{2}, \qquad \tilde{\sigma}_{33} = \frac{5}{2}$$

$$\tilde{\sigma}_{12} = \frac{(6-6)(0-1) + (7-6)(2-1) + (5-6)(1-1) + (6-6)(1-1)}{4}$$

$$= \frac{1}{4}$$

$$\tilde{\sigma}_{23} = \frac{3}{4}, \qquad \tilde{\sigma}_{13} = 1$$

The prediction step consists of using the initial estimates $\tilde{\mu}$ and $\tilde{\Sigma}$ to predict the contributions of the missing values to the sufficient statistics T_1 and T_2. [See (5-38) and (5-39).]

The first component of $\mathbf{x}_1$ is missing, so we partition $\widetilde{\boldsymbol{\mu}}$ and $\widetilde{\boldsymbol{\Sigma}}$ as

$$
\widetilde{\boldsymbol{\mu}} = \begin{bmatrix} \widetilde{\mu}_1 \\ \widetilde{\mu}_2 \\ \widetilde{\mu}_3 \end{bmatrix} = \begin{bmatrix} \widetilde{\mu}^{(1)} \\ \widetilde{\mu}^{(2)} \end{bmatrix}, \qquad
\widetilde{\boldsymbol{\Sigma}} = \begin{bmatrix} \widetilde{\sigma}_{11} & \widetilde{\sigma}_{12} & \widetilde{\sigma}_{13} \\ \widetilde{\sigma}_{12} & \widetilde{\sigma}_{22} & \widetilde{\sigma}_{23} \\ \widetilde{\sigma}_{13} & \widetilde{\sigma}_{23} & \widetilde{\sigma}_{33} \end{bmatrix} = \begin{bmatrix} \widetilde{\boldsymbol{\Sigma}}_{11} & \widetilde{\boldsymbol{\Sigma}}_{12} \\ \widetilde{\boldsymbol{\Sigma}}_{21} & \widetilde{\boldsymbol{\Sigma}}_{22} \end{bmatrix}
$$

and predict

$$
\widetilde{x}_{11} = \widetilde{\mu}_1 + \widetilde{\boldsymbol{\Sigma}}_{12}\widetilde{\boldsymbol{\Sigma}}_{22}^{-1} \begin{bmatrix} x_{12} - \widetilde{\mu}_2 \\ x_{13} - \widetilde{\mu}_3 \end{bmatrix} = 6 + \begin{bmatrix} \frac{1}{4}, & 1 \end{bmatrix} \begin{bmatrix} \frac{1}{2} & \frac{3}{4} \\ \frac{3}{4} & \frac{5}{2} \end{bmatrix}^{-1} \begin{bmatrix} 0 - 1 \\ 3 - 4 \end{bmatrix} = 5.73
$$

$$
\widetilde{x^2_{11}} = \widetilde{\sigma}_{11} - \widetilde{\boldsymbol{\Sigma}}_{12}\widetilde{\boldsymbol{\Sigma}}_{22}^{-1}\widetilde{\boldsymbol{\Sigma}}_{21} + \widetilde{x}^2_{11} = \frac{1}{2} - \begin{bmatrix} \frac{1}{4}, & 1 \end{bmatrix} \begin{bmatrix} \frac{1}{2} & \frac{3}{4} \\ \frac{3}{4} & \frac{5}{2} \end{bmatrix}^{-1} \begin{bmatrix} \frac{1}{4} \\ 1 \end{bmatrix} + (5.73)^2 = 32.99
$$

$$
\widetilde{x_{11}[x_{12}, \quad x_{13}]} = \widetilde{x}_{11}[x_{12}, \quad x_{13}] = 5.73[0, \quad 3] = [0, \quad 17.18]
$$

For the two missing components of $\mathbf{x}_4$, we partition $\widetilde{\boldsymbol{\mu}}$ and $\widetilde{\boldsymbol{\Sigma}}$ as

$$
\widetilde{\boldsymbol{\mu}} = \begin{bmatrix} \widetilde{\mu}_1 \\ \widetilde{\mu}_2 \\ \widetilde{\mu}_3 \end{bmatrix} = \begin{bmatrix} \widetilde{\mu}^{(1)} \\ \widetilde{\mu}^{(2)} \end{bmatrix}, \qquad
\widetilde{\boldsymbol{\Sigma}} = \begin{bmatrix} \widetilde{\sigma}_{11} & \widetilde{\sigma}_{12} & \widetilde{\sigma}_{13} \\ \widetilde{\sigma}_{12} & \widetilde{\sigma}_{22} & \widetilde{\sigma}_{23} \\ \widetilde{\sigma}_{13} & \widetilde{\sigma}_{23} & \widetilde{\sigma}_{33} \end{bmatrix} = \begin{bmatrix} \widetilde{\boldsymbol{\Sigma}}_{11} & \widetilde{\boldsymbol{\Sigma}}_{12} \\ \widetilde{\boldsymbol{\Sigma}}_{21} & \widetilde{\boldsymbol{\Sigma}}_{22} \end{bmatrix}
$$

and predict

$$
\begin{bmatrix} \widetilde{x_{41}} \\ x_{42} \end{bmatrix} = E\left(\begin{bmatrix} X_{41} \\ X_{42} \end{bmatrix} \middle| x_{43} = 5; \widetilde{\boldsymbol{\mu}}, \widetilde{\boldsymbol{\Sigma}} \right) = \begin{bmatrix} \widetilde{\mu}_1 \\ \widetilde{\mu}_2 \end{bmatrix} + \widetilde{\boldsymbol{\Sigma}}_{12}\widetilde{\boldsymbol{\Sigma}}_{22}^{-1}(x_{43} - \widetilde{\mu}_3)
$$

$$
= \begin{bmatrix} 6 \\ 1 \end{bmatrix} + \begin{bmatrix} 1 \\ \frac{3}{4} \end{bmatrix} \left(\frac{5}{2}\right)^{-1} (5 - 4) = \begin{bmatrix} 6.4 \\ 1.3 \end{bmatrix}
$$

for the contribution to $\mathbf{T}_1$. Also, from (5-39),

$$
\begin{bmatrix} \widetilde{x^2_{41}} & \widetilde{x_{41}x_{42}} \\ \widetilde{x_{41}x_{42}} & \widetilde{x^2_{42}} \end{bmatrix} = E\left(\begin{bmatrix} X^2_{41} & X_{41}X_{42} \\ X_{41}X_{42} & X^2_{42} \end{bmatrix} \middle| x_{43} = 5; \widetilde{\boldsymbol{\mu}}, \widetilde{\boldsymbol{\Sigma}} \right)
$$

$$
= \begin{bmatrix} \frac{1}{2} & \frac{1}{4} \\ \frac{1}{4} & \frac{1}{2} \end{bmatrix} - \begin{bmatrix} 1 \\ \frac{3}{4} \end{bmatrix} \left(\frac{5}{2}\right)^{-1} \begin{bmatrix} 1 & \frac{3}{4} \end{bmatrix} + \begin{bmatrix} 6.4 \\ 1.3 \end{bmatrix} \begin{bmatrix} 6.4 & 1.3 \end{bmatrix}
$$

$$
= \begin{bmatrix} 41.06 & 8.27 \\ 8.27 & 1.97 \end{bmatrix}
$$

and

$$
\widetilde{\begin{bmatrix} x_{41} \\ x_{42} \end{bmatrix}(x_{43})} = E\left(\begin{bmatrix} X_{41}X_{43} \\ X_{42}X_{43} \end{bmatrix} \middle| x_{43} = 5; \widetilde{\boldsymbol{\mu}}, \widetilde{\boldsymbol{\Sigma}} \right) = \begin{bmatrix} \widetilde{x}_{41} \\ \widetilde{x}_{42} \end{bmatrix}(x_{43})
$$

$$
= \begin{bmatrix} 6.4 \\ 1.3 \end{bmatrix}(5) = \begin{bmatrix} 32.0 \\ 6.5 \end{bmatrix}
$$

are the contributions to $\mathbf{T}_2$. Thus, the predicted complete-data sufficient statistics
are

$$\widetilde{\mathbf{T}}_1 = \begin{bmatrix} \tilde{x}_{11} + x_{21} + x_{31} + \tilde{x}_{41} \\ x_{12} + x_{22} + x_{32} + \tilde{x}_{42} \\ x_{13} + x_{23} + x_{33} + x_{43} \end{bmatrix} = \begin{bmatrix} 5.73 + 7 + 5 + 6.4 \\ 0 + 2 + 1 + 1.3 \\ 3 + 6 + 2 + 5 \end{bmatrix} = \begin{bmatrix} 24.13 \\ 4.30 \\ 16.00 \end{bmatrix}$$

$$\widetilde{\mathbf{T}}_2 = \begin{bmatrix} \widetilde{x_{11}^2} + \widetilde{x_{21}^2} + \widetilde{x_{31}^2} + \widetilde{x_{41}^2} & & \\ \widetilde{x_{11}x_{12}} + x_{21}x_{22} + x_{31}x_{32} + \widetilde{x_{41}x_{42}} & x_{12}^2 + x_{22}^2 + x_{32}^2 + \widetilde{x_{42}^2} & \\ \widetilde{x_{11}x_{13}} + x_{21}x_{23} + x_{31}x_{33} + \widetilde{x_{41}x_{43}} & x_{12}x_{13} + x_{22}x_{23} + x_{32}x_{33} + \widetilde{x_{42}x_{43}} & x_{13}^2 + x_{23}^2 + x_{33}^2 + x_{43}^2 \end{bmatrix}$$

$$= \begin{bmatrix} 32.99 + 7^2 + 5^2 + 41.06 & & \\ 0 + 7(2) + 5(1) + 8.27 & 0^2 + 2^2 + 1^2 + 1.97 & \\ 17.18 + 7(6) + 5(2) + 32 & 0(3) + 2(6) + 1(2) + 6.5 & 3^2 + 6^2 + 2^2 + 5^2 \end{bmatrix}$$

$$= \begin{bmatrix} 148.05 & 27.27 & 101.18 \\ 27.27 & 6.97 & 20.50 \\ 101.18 & 20.50 & 74.00 \end{bmatrix}$$

This completes one prediction step.

The next estimation step, using (5-40), provides the revised estimates[2]

$$\tilde{\boldsymbol{\mu}} = \frac{1}{n}\widetilde{\mathbf{T}}_1 = \frac{1}{4}\begin{bmatrix} 24.13 \\ 4.30 \\ 16.00 \end{bmatrix} = \begin{bmatrix} 6.03 \\ 1.08 \\ 4.00 \end{bmatrix}$$

$$\widetilde{\boldsymbol{\Sigma}} = \frac{1}{n}\widetilde{\mathbf{T}}_2 - \tilde{\boldsymbol{\mu}}\tilde{\boldsymbol{\mu}}'$$

$$= \frac{1}{4}\begin{bmatrix} 148.05 & 27.27 & 101.18 \\ 27.27 & 6.97 & 20.50 \\ 101.18 & 20.50 & 74.00 \end{bmatrix} - \begin{bmatrix} 6.03 \\ 1.08 \\ 4.00 \end{bmatrix}\begin{bmatrix} 6.03 & 1.08 & 4.00 \end{bmatrix}$$

$$= \begin{bmatrix} .61 & .33 & 1.17 \\ .33 & .59 & .83 \\ 1.17 & .83 & 2.50 \end{bmatrix}$$

Note that $\tilde{\sigma}_{11} = .61$ and $\tilde{\sigma}_{22} = .59$ are larger than the corresponding initial estimates obtained by replacing the missing observations on the first and second variables by the sample means of the remaining values. The third variance estimate $\tilde{\sigma}_{33}$ remains unchanged, because it is not affected by the missing components.

The iteration between the prediction and estimation steps continues until the elements of $\tilde{\boldsymbol{\mu}}$ and $\widetilde{\boldsymbol{\Sigma}}$ remain essentially unchanged. Calculations of this sort are easily handled with a computer. ∎

---

[2]The final entries in $\widetilde{\boldsymbol{\Sigma}}$ are exact to two decimal places.

Once final estimates $\hat{\boldsymbol{\mu}}$ and $\hat{\boldsymbol{\Sigma}}$ are obtained and relatively few missing components occur in $\mathbf{X}$, it seems reasonable to treat

$$\text{all } \boldsymbol{\mu} \text{ such that } n(\hat{\boldsymbol{\mu}} - \boldsymbol{\mu})'\hat{\boldsymbol{\Sigma}}^{-1}(\hat{\boldsymbol{\mu}} - \boldsymbol{\mu}) \le \chi_p^2(\alpha) \qquad (5\text{-}41)$$

as an approximate $100(1 - \alpha)\%$ confidence ellipsoid. The simultaneous confidence statements would then follow as in Section 5.5, but with $\bar{\mathbf{x}}$ replaced by $\hat{\boldsymbol{\mu}}$ and $\mathbf{S}$ replaced by $\hat{\boldsymbol{\Sigma}}$.

Caution. The prediction–estimation algorithm we discussed is developed on the basis that component observations are missing at random. If missing values are related to the response levels, then handling the missing values as suggested may introduce serious biases into the estimation procedures. Typically, missing values *are* related to the responses being measured. Consequently, we must be dubious of any computational scheme that fills in values as if they were lost at random. When more than a few values are missing, it is imperative that the investigator search for the systematic causes that created them.

5.8 Difficulties Due to Time Dependence in Multivariate Observations

For the methods described in this chapter, we have assumed that the multivariate observations $\mathbf{X}_1, \mathbf{X}_2, \dots, \mathbf{X}_n$ constitute a random sample; that is, they are independent of one another. If the observations are collected over time, this assumption may not be valid. The presence of even a moderate amount of time dependence among the observations can cause serious difficulties for tests, confidence regions, and simultaneous confidence intervals, which are all constructed assuming that independence holds.

We will illustrate the nature of the difficulty when the time dependence can be represented as a multivariate first order autoregressive [AR(1)] model. Let the $p \times 1$ random vector $\mathbf{X}_t$ follow the multivariate AR(1) model

$$\mathbf{X}_t - \boldsymbol{\mu} = \boldsymbol{\Phi}(\mathbf{X}_{t-1} - \boldsymbol{\mu}) + \boldsymbol{\varepsilon}_t \qquad (5\text{-}42)$$

where the $\boldsymbol{\varepsilon}_t$ are independent and identically distributed with $E[\boldsymbol{\varepsilon}_t] = \mathbf{0}$ and $\text{Cov}(\boldsymbol{\varepsilon}_t) = \boldsymbol{\Sigma}_\varepsilon$ and all of the eigenvalues of the coefficient matrix $\boldsymbol{\Phi}$ are between -1 and 1. Under this model $\text{Cov}(\mathbf{X}_t, \mathbf{X}_{t-r}) = \boldsymbol{\Phi}^r \boldsymbol{\Sigma}_\mathbf{x}$ where

$$\boldsymbol{\Sigma}_\mathbf{x} = \sum_{j=0}^{\infty} \boldsymbol{\Phi}^j \boldsymbol{\Sigma}_\varepsilon \boldsymbol{\Phi}'^j$$

The AR(1) model (5-42) relates the observation at time t, to the observation at time $t - 1$, through the coefficient matrix $\boldsymbol{\Phi}$. Further, the autoregressive model says the observations are independent, under multivariate normality, if all the entries in the coefficient matrix $\boldsymbol{\Phi}$ are 0. The name autoregressive model comes from the fact that (5-42) looks like a multivariate version of a regression with $\mathbf{X}_t$ as the dependent variable and the previous value $\mathbf{X}_{t-1}$ as the independent variable.

As shown in Johnson and Langeland [8],·

$$\overline{\mathbf{X}} \rightarrow \boldsymbol{\mu}, \quad \mathbf{S} = \frac{1}{n-1} \sum_{t=1}^{n} (\mathbf{X}_t - \overline{\mathbf{X}})(\mathbf{X}_t - \overline{\mathbf{X}})' \rightarrow \boldsymbol{\Sigma}_{\mathbf{X}}$$

where the arrow above indicates convergence in probability, and

$$\text{Cov}\left(n^{-1/2} \sum_{t=1}^{n} \mathbf{X}_t \right) \rightarrow (\mathbf{I} - \boldsymbol{\Phi})^{-1} \boldsymbol{\Sigma}_{\mathbf{X}} + \boldsymbol{\Sigma}_{\mathbf{X}}(\mathbf{I} - \boldsymbol{\Phi}')^{-1} - \boldsymbol{\Sigma}_{\mathbf{X}} \qquad (5\text{-}43)$$

Moreover, for large n, $\sqrt{n}\,(\overline{\mathbf{X}} - \boldsymbol{\mu})$ is approximately normal with mean $\mathbf{0}$ and covariance matrix given by (5-43).

To make the calculations easy, suppose the underlying process has $\boldsymbol{\Phi} = \phi \mathbf{I}$ where $|\phi| < 1$. Now consider the large sample nominal 95% confidence ellipsoid for $\boldsymbol{\mu}$.

$$\{\text{all } \boldsymbol{\mu} \text{ such that } n(\overline{\mathbf{X}} - \boldsymbol{\mu})' \mathbf{S}^{-1}(\overline{\mathbf{X}} - \boldsymbol{\mu}) \leq \chi_p^2(.05)\}$$

This ellipsoid has large sample coverage probability .95 if the observations are independent. If the observations are related by our autoregressive model, however, this ellipsoid has large sample coverage probability

$$P[\chi_p^2 \leq (1 - \phi)(1 + \phi)^{-1}\chi_p^2(.05)]$$

Table 5.10 shows how the coverage probability is related to the coefficient ϕ and the number of variables p.

According to Table 5.10, the coverage probability can drop very low, to .632, even for the bivariate case.

The independence assumption is crucial, and the results based on this assumption can be very misleading if the observations are, in fact, dependent.

Table 5.10 Coverage Probability of the Nominal 95% Confidence Ellipsoid

		$-.25$	ϕ 0	.25	.5
	1	.989	.950	.871	.742
	2	.993	.950	.834	.632
p	5	.998	.950	.751	.405
	10	.999	.950	.641	.193
	15	1.000	.950	.548	.090

5A

Simultaneous Confidence Intervals and Ellipses as Shadows of the p-Dimensional Ellipsoids

We begin this supplementary section by establishing the general result concerning the projection (shadow) of an ellipsoid onto a line.

Result 5A.1. Let the constant $c > 0$ and positive definite $p \times p$ matrix $\mathbf{A}$ determine the ellipsoid $\{\mathbf{z}: \mathbf{z}'\mathbf{A}^{-1}\mathbf{z} \leq c^2\}$. For a given vector $\mathbf{u} \neq \mathbf{0}$, and $\mathbf{z}$ belonging to the ellipsoid, the

$$\begin{pmatrix} \text{Projection (shadow) of} \\ \{\mathbf{z}'\mathbf{A}^{-1}\mathbf{z} \leq c^2\} \text{ on } \mathbf{u} \end{pmatrix} = c \, \frac{\sqrt{\mathbf{u}'\mathbf{A}\mathbf{u}}}{\mathbf{u}'\mathbf{u}} \, \mathbf{u}$$

which extends from $\mathbf{0}$ along $\mathbf{u}$ with length $c\sqrt{\mathbf{u}'\mathbf{A}\mathbf{u}/\mathbf{u}'\mathbf{u}}$. When $\mathbf{u}$ is a unit vector, the shadow extends $c\sqrt{\mathbf{u}'\mathbf{A}\mathbf{u}}$ units, so $|\mathbf{z}'\mathbf{u}| \leq c\sqrt{\mathbf{u}'\mathbf{A}\mathbf{u}}$. The shadow also extends $c\sqrt{\mathbf{u}'\mathbf{A}\mathbf{u}}$ units in the $-\mathbf{u}$ direction.

Proof. By Definition 2A.12, the projection of any $\mathbf{z}$ on $\mathbf{u}$ is given by $(\mathbf{z}'\mathbf{u})\mathbf{u}/\mathbf{u}'\mathbf{u}$. Its squared length is $(\mathbf{z}'\mathbf{u})^2/\mathbf{u}'\mathbf{u}$. We want to maximize this shadow over all $\mathbf{z}$ with $\mathbf{z}'\mathbf{A}^{-1}\mathbf{z} \leq c^2$. The extended Cauchy–Schwarz inequality in (2-49) states that $(\mathbf{b}'\mathbf{d})^2 \leq (\mathbf{b}'\mathbf{B}\mathbf{d})(\mathbf{d}'\mathbf{B}^{-1}\mathbf{d})$, with equality when $\mathbf{b} = k\mathbf{B}^{-1}\mathbf{d}$. Setting $\mathbf{b} = \mathbf{z}$, $\mathbf{d} = \mathbf{u}$, and $\mathbf{B} = \mathbf{A}^{-1}$, we obtain

$$(\mathbf{u}'\mathbf{u})\,(\text{length of projection})^2 = (\mathbf{z}'\mathbf{u})^2 \leq (\mathbf{z}'\mathbf{A}^{-1}\mathbf{z})\,(\mathbf{u}'\mathbf{A}\mathbf{u})$$

$$\leq c^2\mathbf{u}'\mathbf{A}\mathbf{u} \qquad \text{for all } \mathbf{z}: \mathbf{z}'\mathbf{A}^{-1}\mathbf{z} \leq c^2$$

The choice $\mathbf{z} = c\mathbf{A}\mathbf{u}/\sqrt{\mathbf{u}'\mathbf{A}\mathbf{u}}$ yields equalities and thus gives the maximum shadow, besides belonging to the boundary of the ellipsoid. That is, $\mathbf{z}'\mathbf{A}^{-1}\mathbf{z} = c^2\mathbf{u}'\mathbf{A}\mathbf{u}/\mathbf{u}'\mathbf{A}\mathbf{u} = c^2$ for this $\mathbf{z}$ that provides the longest shadow. Consequently, the projection of the ellipsoid on $\mathbf{u}$ is $c\sqrt{\mathbf{u'Au}}\ \mathbf{u}/\mathbf{u'u}$, and its length is $c\sqrt{\mathbf{u'Au}}/\mathbf{u'u}$. With the unit vector $\mathbf{e_u} = \mathbf{u}/\sqrt{\mathbf{u'u}}$, the projection extends

$$\sqrt{c^2 \mathbf{e_u'Ae_u}} = \frac{c}{\sqrt{\mathbf{u'u}}}\sqrt{\mathbf{u'Au}} \quad \text{units along } \mathbf{u}$$

The projection of the ellipsoid also extends the same length in the direction $-\mathbf{u}$. ∎

Result 5A.2. Suppose that the ellipsoid $\{\mathbf{z}: \mathbf{z'A^{-1}z} \le c^2\}$ is given and that $\mathbf{U} = [\mathbf{u}_1 \ \vdots \ \mathbf{u}_2]$ is arbitrary but of rank two. Then

$$\left\{\begin{array}{c} \mathbf{z} \text{ in the ellipsoid} \\ \text{based on } \mathbf{A}^{-1} \text{ and } c^2 \end{array}\right\} \quad \text{implies that} \quad \left\{\begin{array}{c} \text{for all } \mathbf{U}, \mathbf{U'z} \text{ is in the ellipsoid} \\ \text{based on } (\mathbf{U'AU})^{-1} \text{ and } c^2 \end{array}\right\}$$

or

$$\mathbf{z'A^{-1}z} \le c^2 \quad \text{implies that} \quad (\mathbf{U'z})'(\mathbf{U'AU})^{-1}(\mathbf{U'z}) \le c^2 \quad \text{for all } \mathbf{U}$$

Proof. We first establish a basic inequality. Set $\mathbf{P} = \mathbf{A}^{1/2}\mathbf{U}(\mathbf{U'AU})^{-1}\mathbf{U'A}^{1/2}$, where $\mathbf{A} = \mathbf{A}^{1/2}\mathbf{A}^{1/2}$. Note that $\mathbf{P} = \mathbf{P'}$ and $\mathbf{P}^2 = \mathbf{P}$, so $(\mathbf{I} - \mathbf{P})\mathbf{P'} = \mathbf{P} - \mathbf{P}^2 = \mathbf{0}$. Next, using $\mathbf{A}^{-1} = \mathbf{A}^{-1/2}\mathbf{A}^{-1/2}$, we write $\mathbf{z'A^{-1}z} = (\mathbf{A}^{-1/2}\mathbf{z})'(\mathbf{A}^{-1/2}\mathbf{z})$ and $\mathbf{A}^{-1/2}\mathbf{z} = \mathbf{PA}^{-1/2}\mathbf{z} + (\mathbf{I} - \mathbf{P})\mathbf{A}^{-1/2}\mathbf{z}$. Then

$$\begin{aligned} \mathbf{z'A^{-1}z} &= (\mathbf{A}^{-1/2}\mathbf{z})'(\mathbf{A}^{-1/2}\mathbf{z}) \\ &= (\mathbf{PA}^{-1/2}\mathbf{z} + (\mathbf{I} - \mathbf{P})\mathbf{A}^{-1/2}\mathbf{z})'(\mathbf{PA}^{-1/2}\mathbf{z} + (\mathbf{I} - \mathbf{P})\mathbf{A}^{-1/2}\mathbf{z}) \\ &= (\mathbf{PA}^{-1/2}\mathbf{z})'(\mathbf{PA}^{-1/2}\mathbf{z}) + ((\mathbf{I} - \mathbf{P})\mathbf{A}^{-1/2}\mathbf{z})'((\mathbf{I} - \mathbf{P})\mathbf{A}^{-1/2}\mathbf{z}) \\ &\ge \mathbf{z'A}^{-1/2}\mathbf{P'PA}^{-1/2}\mathbf{z} = \mathbf{z'A}^{-1/2}\mathbf{PA}^{-1/2}\mathbf{z} = \mathbf{z'U}(\mathbf{U'AU})^{-1}\mathbf{U'z} \end{aligned} \quad (5\text{A-}1)$$

Since $\mathbf{z'A^{-1}z} \le c^2$ and $\mathbf{U}$ was arbitrary, the result follows. ∎

Our next result establishes the two-dimensional confidence ellipse as a projection of the p-dimensional ellipsoid. (See Figure 5.13.)

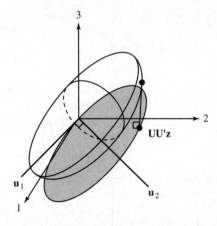

Figure 5.13 The shadow of the ellipsoid $\mathbf{z'A^{-1}z} \le c^2$ on the $\mathbf{u}_1, \mathbf{u}_2$ plane is an ellipse.

Projection on a plane is simplest when the two vectors $\mathbf{u}_1$ and $\mathbf{u}_2$ determining the plane are first converted to perpendicular vectors of unit length. (See Result 2A.3.)

Result 5A.3. Given the ellipsoid $\{\mathbf{z}: \mathbf{z}'\mathbf{A}^{-1}\mathbf{z} \leq c^2\}$ and two perpendicular unit vectors $\mathbf{u}_1$ and $\mathbf{u}_2$, the projection (or shadow) of $\{\mathbf{z}'\mathbf{A}^{-1}\mathbf{z} \leq c^2\}$ on the $\mathbf{u}_1, \mathbf{u}_2$ plane results in the two-dimensional ellipse $\{(\mathbf{U}'\mathbf{z})'(\mathbf{U}'\mathbf{A}\mathbf{U})^{-1}(\mathbf{U}'\mathbf{z}) \leq c^2\}$, where $\mathbf{U} = [\mathbf{u}_1 \vdots \mathbf{u}_2]$.

Proof. By Result 2A.3, the projection of a vector $\mathbf{z}$ on the $\mathbf{u}_1, \mathbf{u}_2$ plane is

$$(\mathbf{u}_1'\mathbf{z})\,\mathbf{u}_1 + (\mathbf{u}_2'\mathbf{z})\,\mathbf{u}_2 = [\mathbf{u}_1 \vdots \mathbf{u}_2] \begin{bmatrix} \mathbf{u}_1'\mathbf{z} \\ \mathbf{u}_2'\mathbf{z} \end{bmatrix} = \mathbf{U}\mathbf{U}'\mathbf{z}$$

The projection of the ellipsoid $\{\mathbf{z}: \mathbf{z}'\mathbf{A}^{-1}\mathbf{z} \leq c^2\}$ consists of all $\mathbf{U}\mathbf{U}'\mathbf{z}$ with $\mathbf{z}'\mathbf{A}^{-1}\mathbf{z} \leq c^2$. Consider the two coordinates $\mathbf{U}'\mathbf{z}$ of the projection $\mathbf{U}(\mathbf{U}'\mathbf{z})$. Let $\mathbf{z}$ belong to the set $\{\mathbf{z}: \mathbf{z}'\mathbf{A}^{-1}\mathbf{z} \leq c^2\}$ so that $\mathbf{U}\mathbf{U}'\mathbf{z}$ belongs to the shadow of the ellipsoid. By Result 5A.2,

$$(\mathbf{U}'\mathbf{z})'(\mathbf{U}'\mathbf{A}\mathbf{U})^{-1}(\mathbf{U}'\mathbf{z}) \leq c^2$$

so the ellipse $\{(\mathbf{U}'\mathbf{z})'(\mathbf{U}'\mathbf{A}\mathbf{U})^{-1}(\mathbf{U}'\mathbf{z}) \leq c^2\}$ contains the coefficient vectors for the shadow of the ellipsoid.

Let $\mathbf{U}\mathbf{a}$ be a vector in the $\mathbf{u}_1, \mathbf{u}_2$ plane whose coefficients $\mathbf{a}$ belong to the ellipse $\{\mathbf{a}'(\mathbf{U}'\mathbf{A}\mathbf{U})^{-1}\mathbf{a} \leq c^2\}$. If we set $\mathbf{z} = \mathbf{A}\mathbf{U}(\mathbf{U}'\mathbf{A}\mathbf{U})^{-1}\mathbf{a}$, it follows that

$$\mathbf{U}'\mathbf{z} = \mathbf{U}'\mathbf{A}\mathbf{U}(\mathbf{U}'\mathbf{A}\mathbf{U})^{-1}\mathbf{a} = \mathbf{a}$$

and

$$\mathbf{z}'\mathbf{A}^{-1}\mathbf{z} = \mathbf{a}'(\mathbf{U}'\mathbf{A}\mathbf{U})^{-1}\mathbf{U}'\mathbf{A}\mathbf{A}^{-1}\mathbf{A}\mathbf{U}(\mathbf{U}'\mathbf{A}\mathbf{U})^{-1}\mathbf{a} = \mathbf{a}'(\mathbf{U}'\mathbf{A}\mathbf{U})^{-1}\mathbf{a} \leq c^2$$

Thus, $\mathbf{U}'\mathbf{z}$ belongs to the coefficient vector ellipse, and $\mathbf{z}$ belongs to the ellipsoid $\mathbf{z}'\mathbf{A}^{-1}\mathbf{z} \leq c^2$. Consequently, the ellipse contains only coefficient vectors from the projection of $\{\mathbf{z}: \mathbf{z}'\mathbf{A}^{-1}\mathbf{z} \leq c^2\}$ onto the $\mathbf{u}_1, \mathbf{u}_2$ plane. ∎

Remark. Projecting the ellipsoid $\mathbf{z}'\mathbf{A}^{-1}\mathbf{z} \leq c^2$ first to the $\mathbf{u}_1, \mathbf{u}_2$ plane and then to the line $\mathbf{u}_1$ is the same as projecting it directly to the line determined by $\mathbf{u}_1$. In the context of confidence ellipsoids, the shadows of the two-dimensional ellipses give the single component intervals.

Remark. Results 5A.2 and 5A.3 remain valid if $\mathbf{U} = [\mathbf{u}_1, \ldots, \mathbf{u}_q]$ consists of $2 < q \leq p$ linearly independent columns.

Exercises

5.1. (a) Evaluate T^2, for testing H_0: $\boldsymbol{\mu}' = [7, \quad 11]$, using the data

$$\mathbf{X} = \begin{bmatrix} 2 & 12 \\ 8 & 9 \\ 6 & 9 \\ 8 & 10 \end{bmatrix}$$

(b) Specify the distribution of T^2 for the situation in (a).

(c) Using (a) and (b), test H_0 at the $\alpha = .05$ level. What conclusion do you reach?

5.2. Using the data in Example 5.1, verify that T^2 remains unchanged if each observation $\mathbf{x}_j, j = 1, 2, 3$, is replaced by $\mathbf{C}\mathbf{x}_j$, where

$$\mathbf{C} = \begin{bmatrix} 1 & -1 \\ 1 & 1 \end{bmatrix}$$

Note that the observations

$$\mathbf{C}\mathbf{x}_j = \begin{bmatrix} x_{j1} - x_{j2} \\ x_{j1} + x_{j2} \end{bmatrix}$$

yield the data matrix

$$\begin{bmatrix} (6 - 9) & (10 - 6) & (8 - 3) \\ (6 + 9) & (10 + 6) & (8 + 3) \end{bmatrix}'$$

5.3. (a) Use expression (5-15) to evaluate T^2 for the data in Exercise 5.1.

(b) Use the data in Exercise 5.1 to evaluate Λ in (5-13). Also, evaluate Wilks' lambda.

5.4. Use the sweat data in Table 5.1. (See Example 5.2.)

(a) Determine the axes of the 90% confidence ellipsoid for $\boldsymbol{\mu}$. Determine the lengths of these axes.

(b) Construct Q–Q plots for the observations on sweat rate, sodium content, and potassium content, respectively. Construct the three possible scatter plots for pairs of observations. Does the multivariate normal assumption seem justified in this case? Comment.

5.5. The quantities $\bar{\mathbf{x}}$, $\mathbf{S}$, and $\mathbf{S}^{-1}$ are given in Example 5.3 for the transformed microwave-radiation data. Conduct a test of the null hypothesis H_0: $\boldsymbol{\mu}' = [.55, .60]$ at the $\alpha = .05$ level of significance. Is your result consistent with the 95% confidence ellipse for $\boldsymbol{\mu}$ pictured in Figure 5.1? Explain.

5.6. Verify the Bonferroni inequality in (5-28) for $m = 3$.

Hint: A Venn diagram for the three events C_1, C_2, and C_3 may help.

5.7. Use the sweat data in Table 5.1 (See Example 5.2.) Find simultaneous 95% T^2 confidence intervals for μ_1, μ_2, and μ_3 using Result 5.3. Construct the 95% Bonferroni intervals using (5-29). Compare the two sets of intervals.

5.8. From (5-23), we know that T^2 is equal to the largest squared univariate t-value constructed from the linear combination $\mathbf{a}'\mathbf{x}_j$ with $\mathbf{a} = \mathbf{S}^{-1}(\bar{\mathbf{x}} - \boldsymbol{\mu}_0)$. Using the results in Example 5.3 and the H_0 in Exercise 5.5, evaluate $\mathbf{a}$ for the transformed microwave-radiation data. Verify that the t^2-value computed with this $\mathbf{a}$ is equal to T^2 in Exercise 5.5.

5.9. Harry Roberts, a naturalist for the Alaska Fish and Game department, studies grizzly bears with the goal of maintaining a healthy population. Measurements on $n = 61$ bears provided the following summary statistics (see also Exercise 8.23):

Variable	Weight (kg)	Body length (cm)	Neck (cm)	Girth (cm)	Head length (cm)	Head width (cm)
Sample mean $\bar{x}$	95.52	164.38	55.69	93.39	17.98	31.13

Covariance matrix

$$\mathbf{S} = \begin{bmatrix} 3266.46 & 1343.97 & 731.54 & 1175.50 & 162.68 & 238.37 \\ 1343.97 & 721.91 & 324.25 & 537.35 & 80.17 & 117.73 \\ 731.54 & 324.25 & 179.28 & 281.17 & 39.15 & 56.80 \\ 1175.50 & 537.35 & 281.17 & 474.98 & 63.73 & 94.85 \\ 162.68 & 80.17 & 39.15 & 63.73 & 9.95 & 13.88 \\ 238.37 & 117.73 & 56.80 & 94.85 & 13.88 & 21.26 \end{bmatrix}$$

(a) Obtain the large sample 95% simultaneous confidence intervals for the six population mean body measurements.

(b) Obtain the large sample 95% simultaneous confidence ellipse for mean weight and mean girth.

(c) Obtain the 95% Bonferroni confidence intervals for the six means in Part a.

(d) Refer to Part b. Construct the 95% Bonferroni confidence rectangle for the mean weight and mean girth using $m = 6$. Compare this rectangle with the confidence ellipse in Part b.

(e) Obtain the 95% Bonferroni confidence interval for

$$\text{mean head width} - \text{mean head length}$$

using $m = 6 + 1 = 7$ to allow for this statement as well as statements about each individual mean.

5.10. Refer to the bear growth data in Example 1.10 (see Table 1.4). Restrict your attention to the measurements of length.

(a) Obtain the 95% T^2 simultaneous confidence intervals for the four population means for length.

(b) Refer to Part a. Obtain the 95% T^2 simultaneous confidence intervals for the three successive yearly increases in mean length.

(c) Obtain the 95% T^2 confidence ellipse for the mean increase in length from 2 to 3 years and the mean increase in length from 4 to 5 years.

(d) Refer to Parts a and b. Construct the 95% Bonferroni confidence intervals for the set consisting of four mean lengths and three successive yearly increases in mean length.

(e) Refer to Parts c and d. Compare the 95% Bonferroni confidence rectangle for the mean increase in length from 2 to 3 years and the mean increase in length from 4 to 5 years with the confidence ellipse produced by the T^2-procedure.

5.11. A physical anthropologist performed a mineral analysis of nine ancient Peruvian hairs. The results for the chromium (x_1) and strontium (x_2) levels, in parts per million (ppm), were as follows:

x_1(Cr)	.48	40.53	2.19	.55	.74	.66	.93	.37	.22
x_2(St)	12.57	73.68	11.13	20.03	20.29	.78	4.64	.43	1.08

Source: Benfer and others, "Mineral Analysis of Ancient Peruvian Hair," *American Journal of Physical Anthropology*, **48**, no. 3 (1978), 277–282.

It is known that low levels (less than or equal to .100 ppm) of chromium suggest the presence of diabetes, while strontium is an indication of animal protein intake.

(a) Construct and plot a 90% joint confidence ellipse for the population mean vector $\mu' = [\mu_1, \mu_2]$, assuming that these nine Peruvian hairs represent a random sample from individuals belonging to a particular ancient Peruvian culture.

(b) Obtain the individual simultaneous 90% confidence intervals for μ_1 and μ_2 by "projecting" the ellipse constructed in Part a on each coordinate axis. (Alternatively, we could use Result 5.3.) Does it appear as if this Peruvian culture has a mean strontium level of 10? That is, are any of the points (μ_1 arbitrary, 10) in the confidence regions? Is $[.30, 10]'$ a plausible value for μ? Discuss.

(c) Do these data appear to be bivariate normal? Discuss their status with reference to Q–Q plots and a scatter diagram. If the data are not bivariate normal, what implications does this have for the results in Parts a and b?

(d) Repeat the analysis with the obvious "outlying" observation removed. Do the inferences change? Comment.

5.12. Given the data

$$\mathbf{X} = \begin{bmatrix} 3 & 6 & 0 \\ 4 & 4 & 3 \\ - & 8 & 3 \\ 5 & - & - \end{bmatrix}$$

with missing components, use the prediction–estimation algorithm of Section 5.7 to estimate μ and Σ. Determine the initial estimates, and iterate to find the *first* revised estimates.

5.13. Determine the approximate distribution of $-n \ln(|\hat{\Sigma}|/|\hat{\Sigma}_0|)$ for the sweat data in Table 5.1. (See Result 5.2.)

5.14. Create a table similar to Table 5.4 using the entries (length of one-at-a-time t-interval)/ (length of Bonferroni t-interval).

Exercises 5.15, 5.16, and 5.17 refer to the following information:

Frequently, some or all of the population characteristics of interest are in the form of *attributes*. Each individual in the population may then be described in terms of the attributes it possesses. For convenience, attributes are usually numerically coded with respect to their presence or absence. If we let the variable X pertain to a specific attribute, then we can distinguish between the presence or absence of this attribute by defining

$$X = \begin{cases} 1 & \text{if attribute present} \\ 0 & \text{if attribute absent} \end{cases}$$

In this way, we can assign numerical values to qualitative characteristics.

When attributes are numerically coded as 0–1 variables, a random sample from the population of interest results in statistics that consist of the *counts* of the number of sample items that have each distinct set of characteristics. If the sample counts are large, methods for producing simultaneous confidence statements can be easily adapted to situations involving proportions.

We consider the situation where an individual with a particular combination of attributes can be classified into one of $q + 1$ mutually exclusive and exhaustive categories. The corresponding probabilities are denoted by $p_1, p_2, \ldots, p_q, p_{q+1}$. Since the categories include all possibilities, we take $p_{q+1} = 1 - (p_1 + p_2 + \cdots + p_q)$. An individual from category k will be assigned the $((q + 1) \times 1)$ vector value $[0, \ldots, 0, 1, 0, \ldots, 0]'$ with 1 in the kth position.

The probability distribution for an observation from the population of individuals in $q + 1$ mutually exclusive and exhaustive categories is known as the *multinomial distribution*. It has the following structure:

Category	1	2	$\cdots$	k	$\cdots$	q	$q + 1$
Outcome (value)	$\begin{bmatrix} 1 \\ 0 \\ 0 \\ \vdots \\ \vdots \\ \vdots \\ 0 \end{bmatrix}$	$\begin{bmatrix} 0 \\ 1 \\ 0 \\ \vdots \\ \vdots \\ \vdots \\ 0 \end{bmatrix}$	$\cdots$	$\begin{bmatrix} 0 \\ \vdots \\ 0 \\ 1 \\ 0 \\ \vdots \\ 0 \end{bmatrix}$	$\cdots$	$\begin{bmatrix} 0 \\ 0 \\ 0 \\ \vdots \\ 0 \\ 1 \\ 0 \end{bmatrix}$	$\begin{bmatrix} 0 \\ 0 \\ 0 \\ \vdots \\ \vdots \\ 0 \\ 1 \end{bmatrix}$
Probability (proportion)	p_1	p_2	$\cdots$	p_k	$\cdots$	p_q	$p_{q+1} = 1 - \sum_{i=1}^{q} p_i$

Let $\mathbf{X}_j, j = 1, 2, \ldots, n$, be a random sample of size n from the multinomial distribution.

The kth component, X_{jk}, of $\mathbf{X}_j$ is 1 if the observation (individual) is from category k and is 0 otherwise. The random sample $\mathbf{X}_1, \mathbf{X}_2, \ldots, \mathbf{X}_n$ can be converted to a sample proportion vector, which, given the nature of the preceding observations, is a sample mean vector. Thus,

$$\hat{\mathbf{p}} = \begin{bmatrix} \hat{p}_1 \\ \hat{p}_2 \\ \vdots \\ \hat{p}_{q+1} \end{bmatrix} = \frac{1}{n} \sum_{j=1}^{n} \mathbf{X}_j \quad \text{with} \quad E(\hat{\mathbf{p}}) = \mathbf{p} = \begin{bmatrix} p_1 \\ p_2 \\ \vdots \\ p_{q+1} \end{bmatrix}$$

and

$$\mathrm{Cov}\,(\hat{\mathbf{p}}) = \frac{1}{n}\,\mathrm{Cov}\,(\mathbf{X}_j) = \frac{1}{n}\boldsymbol{\Sigma} = \frac{1}{n}\begin{bmatrix} \sigma_{11} & \sigma_{12} & \cdots & \sigma_{1,q+1} \\ \sigma_{21} & \sigma_{22} & \cdots & \sigma_{2,q+1} \\ \vdots & \vdots & \ddots & \vdots \\ \sigma_{1,q+1} & \sigma_{2,q+1} & \cdots & \sigma_{q+1,q+1} \end{bmatrix}$$

For large n, the approximate sampling distribution of $\hat{\mathbf{p}}$ is provided by the central limit theorem. We have

$$\sqrt{n}\,(\hat{\mathbf{p}} - \mathbf{p}) \quad \text{is approximately} \quad N(\mathbf{0}, \boldsymbol{\Sigma})$$

where the elements of $\boldsymbol{\Sigma}$ are $\sigma_{kk} = p_k(1 - p_k)$ and $\sigma_{ik} = -p_i p_k$. The normal approximation remains valid when σ_{kk} is estimated by $\hat{\sigma}_{kk} = \hat{p}_k(1 - \hat{p}_k)$ and σ_{ik} is estimated by $\hat{\sigma}_{ik} = -\hat{p}_i\hat{p}_k$, $i \neq k$.

Since each individual must belong to exactly one category, $X_{q+1,j} = 1 - (X_{1j} + X_{2j} + \cdots + X_{qj})$, so $\hat{p}_{q+1} = 1 - (\hat{p}_1 + \hat{p}_2 + \cdots + \hat{p}_q)$, and as a result, $\hat{\boldsymbol{\Sigma}}$ has rank q. The usual inverse of $\hat{\boldsymbol{\Sigma}}$ does not exist, but it is still possible to develop simultaneous $100(1 - \alpha)\%$ confidence intervals for all linear combinations $\mathbf{a}'\mathbf{p}$.

Result. Let $\mathbf{X}_1, \mathbf{X}_2, \ldots, \mathbf{X}_n$ be a random sample from a $q + 1$ category multinomial distribution with $P[X_{jk} = 1] = p_k$, $k = 1, 2, \ldots, q + 1$, $j = 1, 2, \ldots, n$. Approximate simultaneous $100(1 - \alpha)\%$ confidence regions for all linear combinations $\mathbf{a}'\mathbf{p} = a_1 p_1 + a_2 p_2 + \cdots + a_{q+1}p_{q+1}$ are given by the observed values of

$$\mathbf{a}'\hat{\mathbf{p}} \pm \sqrt{\chi_q^2(\alpha)}\,\sqrt{\frac{\mathbf{a}'\hat{\boldsymbol{\Sigma}}\mathbf{a}}{n}}$$

provided that $n - q$ is large. Here $\hat{\mathbf{p}} = (1/n)\sum_{j=1}^{n}\mathbf{X}_j$, and $\hat{\boldsymbol{\Sigma}} = \{\hat{\sigma}_{ik}\}$ is a $(q + 1) \times (q + 1)$ matrix with $\hat{\sigma}_{kk} = \hat{p}_k(1 - \hat{p}_k)$ and $\hat{\sigma}_{ik} = -\hat{p}_i\hat{p}_k$, $i \neq k$. Also, $\chi_q^2(\alpha)$ is the upper (100α)th percentile of the chi-square distribution with q d.f. ∎

In this result, the requirement that $n - q$ is large is interpreted to mean $n\hat{p}_k$ is about 20 or more for each category.

We have only touched on the possibilities for the analysis of categorical data. Complete discussions of categorical data analysis are available in [1] and [4].

5.15. Let X_{ji} and X_{jk} be the ith and kth components, respectively, of $\mathbf{X}_j$.

(a) Show that $\mu_i = E(X_{ji}) = p_i$ and $\sigma_{ii} = \mathrm{Var}(X_{ji}) = p_i(1 - p_i)$, $i = 1, 2, \ldots, p$.

(b) Show that $\sigma_{ik} = \mathrm{Cov}(X_{ji}, X_{jk}) = -p_i p_k$, $i \neq k$. Why must this covariance necessarily be negative?

5.16. As part of a larger marketing research project, a consultant for the Bank of Shorewood wants to know the proportion of savers that uses the bank's facilities as their primary vehicle for saving. The consultant would also like to know the proportions of savers who use the three major competitors: Bank B, Bank C, and Bank D. Each individual contacted in a survey responded to the following question:

Which bank is your primary savings bank?

Response:	Bank of Shorewood	Bank B	Bank C	Bank D	Another Bank	No Savings

A sample of $n = 355$ people with savings accounts produced the following counts when asked to indicate their primary savings banks (the people with no savings will be ignored in the comparison of savers, so there are five categories):

Bank (category)	Bank of Shorewood	Bank B	Bank C	Bank D	Another bank	
Observed number	105	119	56	25	50	Total $n = 355$
Population proportion	p_1	p_2	p_3	p_4	$p_5 = 1 - (p_1 + p_2 + p_3 + p_4)$	
Observed sample proportion	$\hat{p}_1 = \dfrac{105}{355} = .30$	$\hat{p}_2 = .33$	$\hat{p}_3 = .16$	$\hat{p}_4 = .07$	$\hat{p}_5 = .14$	

Let the population proportions be

$$p_1 = \text{proportion of savers at Bank of Shorewood}$$

$$p_2 = \text{proportion of savers at Bank B}$$

$$p_3 = \text{proportion of savers at Bank C}$$

$$p_4 = \text{proportion of savers at Bank D}$$

$$1 - (p_1 + p_2 + p_3 + p_4) = \text{proportion of savers at other banks}$$

(a) Construct simultaneous 95% confidence intervals for $p_1, p_2, \ldots, p_5$.

(b) Construct a simultaneous 95% confidence interval that allows a comparison of the Bank of Shorewood with its major competitor, Bank B. Interpret this interval.

5.17. In order to assess the prevalence of a drug problem among high school students in a particular city, a random sample of 200 students from the city's five high schools were surveyed. One of the survey questions and the corresponding responses are as follows:

What is your typical weekly marijuana usage?

	Category		
	None	Moderate (1–3 joints)	Heavy (4 or more joints)
Number of responses	117	62	21

Construct 95% simultaneous confidence intervals for the three proportions p_1, p_2, and $p_3 = 1 - (p_1 + p_2)$.

The following exercises may require a computer.

5.18. Use the college test data in Table 5.2. (See Example 5.5.)

(a) Test the null hypothesis $H_0: \boldsymbol{\mu}' = [500, 50, 30]$ versus $H_1: \boldsymbol{\mu}' \neq [500, 50, 30]$ at the $\alpha = .05$ level of significance. Suppose $[500, 50, 30]'$ represent average scores for thousands of college students over the last 10 years. Is there reason to believe that the group of students represented by the scores in Table 5.2 is scoring differently? Explain.

(b) Determine the lengths and directions for the axes of the 95% confidence ellipsoid for $\boldsymbol{\mu}$.

(c) Construct Q–Q plots from the marginal distributions of social science and history, verbal, and science scores. Also, construct the three possible scatter diagrams from the pairs of observations on different variables. Do these data appear to be normally distributed? Discuss.

5.19. Measurements of x_1 = stiffness and x_2 = bending strength for a sample of $n = 30$ pieces of a particular grade of lumber are given in Table 5.11. The units are pounds/(inches)^2. Using the data in the table,

Table 5.11 Lumber Data

x_1 (Stiffness: modulus of elasticity)	x_2 (Bending strength)	x_1 (Stiffness: modulus of elasticity)	x_2 (Bending strength)
1232	4175	1712	7749
1115	6652	1932	6818
2205	7612	1820	9307
1897	10,914	1900	6457
1932	10,850	2426	10,102
1612	7627	1558	7414
1598	6954	1470	7556
1804	8365	1858	7833
1752	9469	1587	8309
2067	6410	2208	9559
2365	10,327	1487	6255
1646	7320	2206	10,723
1579	8196	2332	5430
1880	9709	2540	12,090
1773	10,370	2322	10,072

Source: Data courtesy of U.S. Forest Products Laboratory.

(a) Construct and sketch a 95% confidence ellipse for the pair $[\mu_1, \mu_2]'$, where $\mu_1 = E(X_1)$ and $\mu_2 = E(X_2)$.

(b) Suppose $\mu_{10} = 2000$ and $\mu_{20} = 10{,}000$ represent "typical" values for stiffness and bending strength, respectively. Given the result in (a), are the data in Table 5.11 consistent with these values? Explain.

(c) Is the bivariate normal distribution a viable population model? Explain with reference to Q–Q plots and a scatter diagram.

5.20. A wildlife ecologist measured x_1 = tail length (in millimeters) and x_2 = wing length (in millimeters) for a sample of n = 45 female hook-billed kites. These data are displayed in Table 5.12. Using the data in the table,

Table 5.12 Bird Data

x_1 (Tail length)	x_2 (Wing length)	x_1 (Tail length)	x_2 (Wing length)	x_1 (Tail length)	x_2 (Wing length)
191	284	186	266	173	271
197	285	197	285	194	280
208	288	201	295	198	300
180	273	190	282	180	272
180	275	209	305	190	292
188	280	187	285	191	286
210	283	207	297	196	285
196	288	178	268	207	286
191	271	202	271	209	303
179	257	205	285	179	261
208	289	190	280	186	262
202	285	189	277	174	245
200	272	211	310	181	250
192	282	216	305	189	262
199	280	189	274	188	258

Source: Data courtesy of S. Temple.

(a) Find and sketch the 95% confidence ellipse for the population means μ_1 and μ_2. Suppose it is known that μ_1 = 190 mm and μ_2 = 275 mm for *male* hook-billed kites. Are these plausible values for the mean tail length and mean wing length for the female birds? Explain.

(b) Construct the simultaneous 95% T^2-intervals for μ_1 and μ_2 and the 95% Bonferroni intervals for μ_1 and μ_2. Compare the two sets of intervals. What advantage, if any, do the T^2-intervals have over the Bonferroni intervals?

(c) Is the bivariate normal distribution a viable population model? Explain with reference to Q–Q plots and a scatter diagram.

5.21. Using the data on bone mineral content in Table 1.8, construct the 95% Bonferroni intervals for the individual means. Also, find the 95% simultaneous T^2-intervals. Compare the two sets of intervals.

5.22. A portion of the data contained in Table 6.10 in Chapter 6 is reproduced in Table 5.13. These data represent various costs associated with transporting milk from farms to dairy plants for gasoline trucks. Only the first 25 multivariate observations for gasoline trucks are given. Observations 9 and 21 have been identified as outliers from the full data set of 36 observations. (See [2].)

Table 5.13 Milk Transportation-Cost Data

Fuel (x_1)	Repair (x_2)	Capital (x_3)
16.44	12.43	11.23
7.19	2.70	3.92
9.92	1.35	9.75
4.24	5.78	7.78
11.20	5.05	10.67
14.25	5.78	9.88
13.50	10.98	10.60
13.32	14.27	9.45
29.11	15.09	3.28
12.68	7.61	10.23
7.51	5.80	8.13
9.90	3.63	9.13
10.25	5.07	10.17
11.11	6.15	7.61
12.17	14.26	14.39
10.24	2.59	6.09
10.18	6.05	12.14
8.88	2.70	12.23
12.34	7.73	11.68
8.51	14.02	12.01
26.16	17.44	16.89
12.95	8.24	7.18
16.93	13.37	17.59
14.70	10.78	14.58
10.32	5.16	17.00

(a) Construct Q–Q plots of the marginal distributions of fuel, repair, and capital costs. Also, construct the three possible scatter diagrams from the pairs of observations on different variables. Are the outliers evident? Repeat the Q–Q plots and the scatter diagrams with the apparent outliers removed. Do the data now appear to be normally distributed? Discuss.

(b) Construct 95% Bonferroni intervals for the individual cost means. Also, find the 95% T^2-intervals. Compare the two sets of intervals.

5.23. Consider the 30 observations on male Egyptian skulls for the first time period given in Table 6.13 on page 349.

(a) Construct Q–Q plots of the marginal distributions of the maxbreath, basheight, baslength and nasheight variables. Also, construct a chi-square plot of the multivariate observations. Do these data appear to be normally distributed? Explain.

(b) Construct 95% Bonferroni intervals for the individual skull dimension variables. Also, find the 95% T^2-intervals. Compare the two sets of intervals.

5.24. Using the Madison, Wisconsin, Police Department data in Table 5.8, construct individual $\overline{X}$ charts for x_3 = holdover hours and x_4 = COA hours. Do these individual process characteristics seem to be in control? (That is, are they stable?) Comment.

5.25. Refer to Exercise 5.24. Using the data on the holdover and COA overtime hours, construct a quality ellipse and a T^2-chart. Does the process represented by the bivariate observations appear to be in control? (That is, is it stable?) Comment. Do you learn something from the multivariate control charts that was not apparent in the individual $\overline{X}$-charts?

5.26. Construct a T^2-chart using the data on x_1 = legal appearances overtime hours, x_2 = extraordinary event overtime hours, and x_3 = holdover overtime hours from Table 5.8. Compare this chart with the chart in Figure 5.8 of Example 5.10. Does plotting T^2 with an additional characteristic change your conclusion about process stability? Explain.

5.27. Using the data on x_3 = holdover hours and x_4 = COA hours from Table 5.8, construct a prediction ellipse for a future observation $\mathbf{x}' = (x_3, x_4)$. Remember, a prediction ellipse should be calculated from a stable process. Interpret the result.

5.28 As part of a study of its sheet metal assembly process, a major automobile manufacturer uses sensors that record the deviation from the nominal thickness (millimeters) at six locations on a car. The first four are measured when the car body is complete and the last two are measured on the underbody at an earlier stage of assembly. Data on 50 cars are given in Table 5.14.

(a) The process seems stable for the first 30 cases. Use these cases to estimate $\mathbf{S}$ and $\bar{\mathbf{x}}$. Then construct a T^2 chart using all of the variables. Include all 50 cases.

(b) Which individual locations seem to show a cause for concern?

5.29 Refer to the car body data in Exercise 5.28. These are all measured as deviations from target value so it is appropriate to test the null hypothesis that the mean vector is zero. Using the first 30 cases, test $H_0: \boldsymbol{\mu} = \mathbf{0}$ at $\alpha = .05$

5.30 Refer to the data on energy consumption in Exercise 3.18.

(a) Obtain the large sample 95% Bonferroni confidence intervals for the mean consumption of each of the four types, the total of the four, and the difference, petroleum minus natural gas.

(b) Obtain the large sample 95% simultaneous T^2 intervals for the mean consumption of each of the four types, the total of the four, and the difference, petroleum minus natural gas. Compare with your results for Part a.

5.31 Refer to the data on snow storms in Exercise 3.20.

(a) Find a 95% confidence region for the mean vector after taking an appropriate transformation.

(b) On the same scale, find the 95% Bonferroni confidence intervals for the two component means.

TABLE 5.14 Car Body Assembly Data

Index	x_1	x_2	x_3	x_4	x_5	x_6
1	−0.12	0.36	0.40	0.25	1.37	−0.13
2	−0.60	−0.35	0.04	−0.28	−0.25	−0.15
3	−0.13	0.05	0.84	0.61	1.45	0.25
4	−0.46	−0.37	0.30	0.00	−0.12	−0.25
5	−0.46	−0.24	0.37	0.13	0.78	−0.15
6	−0.46	−0.16	0.07	0.10	1.15	−0.18
7	−0.46	−0.24	0.13	0.02	0.26	−0.20
8	−0.13	0.05	−0.01	0.09	−0.15	−0.18
9	−0.31	−0.16	−0.20	0.23	0.65	0.15
10	−0.37	−0.24	0.37	0.21	1.15	0.05
11	−1.08	−0.83	−0.81	0.05	0.21	0.00
12	−0.42	−0.30	0.37	−0.58	0.00	−0.45
13	−0.31	0.10	−0.24	0.24	0.65	0.35
14	−0.14	0.06	0.18	−0.50	1.25	0.05
15	−0.61	−0.35	−0.24	0.75	0.15	−0.20
16	−0.61	−0.30	−0.20	−0.21	−0.50	−0.25
17	−0.84	−0.35	−0.14	−0.22	1.65	−0.05
18	−0.96	−0.85	0.19	−0.18	1.00	−0.08
19	−0.90	−0.34	−0.78	−0.15	0.25	0.25
20	−0.46	0.36	0.24	−0.58	0.15	0.25
21	−0.90	−0.59	0.13	0.13	0.60	−0.08
22	−0.61	−0.50	−0.34	−0.58	0.95	−0.08
23	−0.61	−0.20	−0.58	−0.20	1.10	0.00
24	−0.46	−0.30	−0.10	−0.10	0.75	−0.10
25	−0.60	−0.35	−0.45	0.37	1.18	−0.30
26	−0.60	−0.36	−0.34	−0.11	1.68	−0.32
27	−0.31	0.35	−0.45	−0.10	1.00	−0.25
28	−0.60	−0.25	−0.42	0.28	0.75	0.10
29	−0.31	0.25	−0.34	−0.24	0.65	0.10
30	−0.36	−0.16	0.15	−0.38	1.18	−0.10
31	−0.40	−0.12	−0.48	−0.34	0.30	−0.20
32	−0.60	−0.40	−0.20	0.32	0.50	0.10
33	−0.47	−0.16	−0.34	−0.31	0.85	0.60
34	−0.46	−0.18	0.16	0.01	0.60	0.35
35	−0.44	−0.12	−0.20	−0.48	1.40	0.10
36	−0.90	−0.40	0.75	−0.31	0.60	−0.10
37	−0.50	−0.35	0.84	−0.52	0.35	−0.75
38	−0.38	0.08	0.55	−0.15	0.80	−0.10
39	−0.60	−0.35	−0.35	−0.34	0.60	0.85
40	0.11	0.24	0.15	0.40	0.00	−0.10
41	0.05	0.12	0.85	0.55	1.65	−0.10
42	−0.85	−0.65	0.50	0.35	0.80	−0.21
43	−0.37	−0.10	−0.10	−0.58	1.85	−0.11
44	−0.11	0.24	0.75	−0.10	0.65	−0.10
45	−0.60	−0.24	0.13	0.84	0.85	0.15
46	−0.84	−0.59	0.05	0.61	1.00	0.20
47	−0.46	−0.16	0.37	−0.15	0.68	0.25
48	−0.56	−0.35	−0.10	0.75	0.45	0.20
49	−0.56	−0.16	0.37	−0.25	1.05	0.15
50	−0.25	−0.12	−0.05	−0.20	1.21	0.10

Source: Data Courtesy of Darek Ceglarek.

References

1. Agresti, A. *Categorical Data Analysis* (2nd ed.), New York: John Wiley, 2002.

2. Bacon-Sone, J., and W. K. Fung. "A New Graphical Method for Detecting Single and Multiple Outliers in Univariate and Multivariate Data." *Applied Statistics,* **36**, no. 2 (1987), 153–162.

3. Bickel, P. J., and K. A. Doksum. *Mathematical Statistics: Basic Ideas and Selected Topics*, Vol. I (2nd ed.), Upper Saddle River, NJ: Prentice Hall, 2000.

4. Bishop, Y. M. M., S. E. Feinberg, and P. W. Holland. *Discrete Multivariate Analysis: Theory and Practice* (Paperback). Cambridge, MA: The MIT Press, 1977.

5. Dempster, A. P., N. M. Laird, and D. B. Rubin. "Maximum Likelihood from Incomplete Data via the EM Algorithm (with Discussion)." *Journal of the Royal Statistical Society* (B), **39**, no. 1 (1977), 1–38.

6. Hartley, H. O. "Maximum Likelihood Estimation from Incomplete Data." *Biometrics*, **14** (1958), 174–194.

7. Hartley, H. O., and R. R. Hocking. "The Analysis of Incomplete Data." *Biometrics*, **27** (1971), 783–808.

8. Johnson, R. A. and T. Langeland "A Linear Combinations Test for Detecting Serial Correlation in Multivariate Samples." *Topics in Statistical Dependence.* (1991) Institute of Mathematical Statistics Monograph, Eds. Block, H. et al., 299–313.

9. Johnson, R. A. and R. Li "Multivariate Statistical Process Control Schemes for Controlling a Mean." *Springer Handbook of Engineering Statistics* (2006), H. Pham, Ed. Springer, Berlin.

10. Ryan, T. P. *Statistical Methods for Quality Improvement* (2nd ed.). New York: John Wiley, 2000.

11. Tiku, M. L., and M. Singh. "Robust Statistics for Testing Mean Vectors of Multivariate Distributions." *Communications in Statistics—Theory and Methods*, **11**, no. 9 (1982), 985–1001.

6

Comparisons of Several Multivariate Means

6.1 Introduction

The ideas developed in Chapter 5 can be extended to handle problems involving the comparison of several mean vectors. The theory is a little more complicated and rests on an assumption of multivariate normal distributions or large sample sizes. Similarly, the notation becomes a bit cumbersome. To circumvent these problems, we shall often review univariate procedures for comparing several means and then generalize to the corresponding multivariate cases by analogy. The numerical examples we present will help cement the concepts.

Because comparisons of means frequently (and should) emanate from designed experiments, we take the opportunity to discuss some of the tenets of good experimental practice. A *repeated measures* design, useful in behavioral studies, is explicitly considered, along with modifications required to analyze *growth curves*.

We begin by considering pairs of mean vectors. In later sections, we discuss several comparisons among mean vectors arranged according to treatment levels. The corresponding test statistics depend upon a partitioning of the total variation into pieces of variation attributable to the treatment sources and error. This partitioning is known as the *multivariate analysis of variance* (MANOVA).

6.2 Paired Comparisons and a Repeated Measures Design

Paired Comparisons

Measurements are often recorded under different sets of experimental conditions to see whether the responses differ significantly over these sets. For example, the efficacy of a new drug or of a saturation advertising campaign may be determined by comparing measurements before the "treatment" (drug or advertising) with those after the treatment. In other situations, *two or more* treatments can be administered to the same or similar experimental units, and responses can be compared to assess the effects of the treatments.

One rational approach to comparing two treatments, or the presence and absence of a single treatment, is to assign both treatments to the *same* or *identical* units (individuals, stores, plots of land, and so forth). The paired responses may then be analyzed by computing their differences, thereby eliminating much of the influence of extraneous unit-to-unit variation.

In the single response (univariate) case, let X_{j1} denote the response to treatment 1 (or the response before treatment), and let X_{j2} denote the response to treatment 2 (or the response after treatment) for the jth trial. That is, (X_{j1}, X_{j2}) are measurements recorded on the jth unit or jth pair of like units. By design, the n differences

$$D_j = X_{j1} - X_{j2}, \qquad j = 1, 2, \ldots, n \tag{6-1}$$

should reflect only the differential effects of the treatments.

Given that the differences D_j in (6-1) represent independent observations from an $N(\delta, \sigma_d^2)$ distribution, the variable

$$t = \frac{\bar{D} - \delta}{s_d / \sqrt{n}} \tag{6-2}$$

where

$$\bar{D} = \frac{1}{n} \sum_{j=1}^{n} D_j \quad \text{and} \quad s_d^2 = \frac{1}{n-1} \sum_{j=1}^{n} (D_j - \bar{D})^2 \tag{6-3}$$

has a t-distribution with $n - 1$ d.f. Consequently, an α-level test of

$$H_0: \delta = 0 \quad \text{(zero mean difference for treatments)}$$

versus

$$H_1: \delta \neq 0$$

may be conducted by comparing $|t|$ with $t_{n-1}(\alpha/2)$—the upper $100(\alpha/2)$th percentile of a t-distribution with $n - 1$ d.f. A $100(1 - \alpha)\%$ confidence interval for the mean difference $\delta = E(X_{j1} - X_{j2})$ is provided the statement

$$\bar{d} - t_{n-1}(\alpha/2) \frac{s_d}{\sqrt{n}} \leq \delta \leq \bar{d} + t_{n-1}(\alpha/2) \frac{s_d}{\sqrt{n}} \tag{6-4}$$

(For example, see [11].)

Additional notation is required for the multivariate extension of the paired-comparison procedure. It is necessary to distinguish between p responses, two treatments, and n experimental units. We label the p responses within the *jth unit* as

$$X_{1j1} = \text{variable 1 under treatment 1}$$
$$X_{1j2} = \text{variable 2 under treatment 1}$$
$$\vdots \qquad \qquad \vdots$$
$$X_{1jp} = \text{variable } p \text{ under treatment 1}$$
$$X_{2j1} = \text{variable 1 under treatment 2}$$
$$X_{2j2} = \text{variable 2 under treatment 2}$$
$$\vdots \qquad \qquad \vdots$$
$$X_{2jp} = \text{variable } p \text{ under treatment 2}$$

and the p paired-difference random variables become

$$
\begin{aligned}
D_{j1} &= X_{1j1} - X_{2j1} \\
D_{j2} &= X_{1j2} - X_{2j2} \\
&\vdots \qquad\qquad \vdots \\
D_{jp} &= X_{1jp} - X_{2jp}
\end{aligned}
\tag{6-5}
$$

Let $\mathbf{D}_j' = [D_{j1}, D_{j2}, \ldots, D_{jp}]$, and assume, for $j = 1, 2, \ldots, n$, that

$$
E(\mathbf{D}_j) = \boldsymbol{\delta} = \begin{bmatrix} \delta_1 \\ \delta_2 \\ \vdots \\ \delta_p \end{bmatrix} \quad \text{and} \quad \text{Cov}(\mathbf{D}_j) = \boldsymbol{\Sigma}_d
\tag{6-6}
$$

If, in addition, $\mathbf{D}_1, \mathbf{D}_2, \ldots, \mathbf{D}_n$ are independent $N_p(\boldsymbol{\delta}, \boldsymbol{\Sigma}_d)$ random vectors, inferences about the vector of mean differences $\boldsymbol{\delta}$ can be based upon a T^2-statistic. Specifically,

$$
T^2 = n(\overline{\mathbf{D}} - \boldsymbol{\delta})' \mathbf{S}_d^{-1} (\overline{\mathbf{D}} - \boldsymbol{\delta})
\tag{6-7}
$$

where

$$
\overline{\mathbf{D}} = \frac{1}{n} \sum_{j=1}^{n} \mathbf{D}_j \quad \text{and} \quad \mathbf{S}_d = \frac{1}{n-1} \sum_{j=1}^{n} (\mathbf{D}_j - \overline{\mathbf{D}})(\mathbf{D}_j - \overline{\mathbf{D}})'
\tag{6-8}
$$

Result 6.1. Let the differences $\mathbf{D}_1, \mathbf{D}_2, \ldots, \mathbf{D}_n$ be a random sample from an $N_p(\boldsymbol{\delta}, \boldsymbol{\Sigma}_d)$ population. Then

$$
T^2 = n(\overline{\mathbf{D}} - \boldsymbol{\delta})' \mathbf{S}_d^{-1} (\overline{\mathbf{D}} - \boldsymbol{\delta})
$$

is distributed as an $[(n-1)p/(n-p)]F_{p,n-p}$ random variable, whatever the true $\boldsymbol{\delta}$ and $\boldsymbol{\Sigma}_d$.

If n and $n - p$ are both large, T^2 is approximately distributed as a χ_p^2 random variable, regardless of the form of the underlying population of differences.

Proof. The exact distribution of T^2 is a restatement of the summary in (5-6), with vectors of differences for the observation vectors. The approximate distribution of T^2, for n and $n - p$ large, follows from (4-28). ∎

The condition $\boldsymbol{\delta} = \mathbf{0}$ is equivalent to "no average difference between the two treatments." For the ith variable, $\delta_i > 0$ implies that treatment 1 is larger, on average, than treatment 2. In general, inferences about $\boldsymbol{\delta}$ can be made using Result 6.1.

Given the observed differences $\mathbf{d}_j' = [d_{j1}, d_{j2}, \ldots, d_{jp}]$, $j = 1, 2, \ldots, n$, corresponding to the random variables in (6-5), an α-*level test* of $H_0: \boldsymbol{\delta} = \mathbf{0}$ *versus* $H_1: \boldsymbol{\delta} \neq \mathbf{0}$ for an $N_p(\boldsymbol{\delta}, \boldsymbol{\Sigma}_d)$ population rejects H_0 if the observed

$$
T^2 = n\overline{\mathbf{d}}' \mathbf{S}_d^{-1} \overline{\mathbf{d}} > \frac{(n-1)p}{(n-p)} F_{p,n-p}(\alpha)
$$

where $F_{p,n-p}(\alpha)$ is the upper (100α)th percentile of an F-distribution with p and $n - p$ d.f. Here $\overline{\mathbf{d}}$ and $\mathbf{S}_d$ are given by (6-8).

A $100(1 - \alpha)\%$ *confidence region for* $\boldsymbol{\delta}$ consists of all $\boldsymbol{\delta}$ such that

$$(\bar{\mathbf{d}} - \boldsymbol{\delta})'\mathbf{S}_d^{-1}(\bar{\mathbf{d}} - \boldsymbol{\delta}) \leq \frac{(n - 1)p}{n(n - p)} F_{p,n-p}(\alpha) \qquad (6\text{-}9)$$

Also, $100(1 - \alpha)\%$ *simultaneous confidence intervals for the individual mean differences* δ_i are given by

$$\delta_i: \quad \bar{d}_i \pm \sqrt{\frac{(n - 1)p}{(n - p)} F_{p,n-p}(\alpha)} \sqrt{\frac{s_{d_i}^2}{n}} \qquad (6\text{-}10)$$

where $\bar{d}_i$ is the ith element of $\bar{\mathbf{d}}$ and $s_{d_i}^2$ is the ith diagonal element of $\mathbf{S}_d$.

For $n - p$ large, $[(n - 1)p/(n - p)]F_{p,n-p}(\alpha) \doteq \chi_p^2(\alpha)$ and normality need not be assumed.

The *Bonferroni* $100(1 - \alpha)\%$ *simultaneous confidence intervals* for the individual mean differences are

$$\delta_i: \quad \bar{d}_i \pm t_{n-1}\left(\frac{\alpha}{2p}\right)\sqrt{\frac{s_{d_i}^2}{n}} \qquad (6\text{-}10a)$$

where $t_{n-1}(\alpha/2p)$ is the upper $100(\alpha/2p)$th percentile of a t-distribution with $n - 1$ d.f.

Example 6.1 (Checking for a mean difference with paired observations) Municipal wastewater treatment plants are required by law to monitor their discharges into rivers and streams on a regular basis. Concern about the reliability of data from one of these self-monitoring programs led to a study in which samples of effluent were divided and sent to two laboratories for testing. One-half of each sample was sent to the Wisconsin State Laboratory of Hygiene, and one-half was sent to a private commercial laboratory routinely used in the monitoring program. Measurements of biochemical oxygen demand (BOD) and suspended solids (SS) were obtained, for $n = 11$ sample splits, from the two laboratories. The data are displayed in Table 6.1.

Table 6.1 Effluent Data

Sample j	Commercial lab		State lab of hygiene	
	x_{1j1} (BOD)	x_{1j2} (SS)	x_{2j1} (BOD)	x_{2j2} (SS)
1	6	27	25	15
2	6	23	28	13
3	18	64	36	22
4	8	44	35	29
5	11	30	15	31
6	34	75	44	64
7	28	26	42	30
8	71	124	54	64
9	43	54	34	56
10	33	30	29	20
11	20	14	39	21

Source: Data courtesy of S. Weber.

Do the two laboratories' chemical analyses agree? If differences exist, what is their nature?

The T^2-statistic for testing $H_0: \boldsymbol{\delta}' = [\delta_1, \delta_2] = [0, 0]$ is constructed from the differences of paired observations:

$d_{j1} = x_{1j1} - x_{2j1}$	-19	-22	-18	-27	-4	-10	-14	17	9	4	-19
$d_{j2} = x_{1j2} - x_{2j2}$	12	10	42	15	-1	11	-4	60	-2	10	-7

Here

$$\bar{\mathbf{d}} = \begin{bmatrix} \bar{d}_1 \\ \bar{d}_2 \end{bmatrix} = \begin{bmatrix} -9.36 \\ 13.27 \end{bmatrix}, \qquad \mathbf{S}_d = \begin{bmatrix} 199.26 & 88.38 \\ 88.38 & 418.61 \end{bmatrix}$$

and

$$T^2 = 11[-9.36, \quad 13.27] \begin{bmatrix} .0055 & -.0012 \\ -.0012 & .0026 \end{bmatrix} \begin{bmatrix} -9.36 \\ 13.27 \end{bmatrix} = 13.6$$

Taking $\alpha = .05$, we find that $[p(n-1)/(n-p)]F_{p,n-p}(.05) = [2(10)/9]F_{2,9}(.05) = 9.47$. Since $T^2 = 13.6 > 9.47$, we reject H_0 and conclude that there is a nonzero mean difference between the measurements of the two laboratories. It appears, from inspection of the data, that the commercial lab tends to produce lower BOD measurements and higher SS measurements than the State Lab of Hygiene. The 95% simultaneous confidence intervals for the mean differences δ_1 and δ_2 can be computed using (6-10). These intervals are

$$\delta_1: \bar{d}_1 \pm \sqrt{\frac{(n-1)p}{(n-p)}\, F_{p,n-p}(\alpha)} \sqrt{\frac{s_{d_1}^2}{n}} = -9.36 \pm \sqrt{9.47} \sqrt{\frac{199.26}{11}}$$

$$\text{or} \quad (-22.46, 3.74)$$

$$\delta_2: 13.27 \pm \sqrt{9.47} \sqrt{\frac{418.61}{11}} \quad \text{or} \quad (-5.71, 32.25)$$

The 95% *simultaneous confidence intervals* include zero, yet the hypothesis $H_0: \boldsymbol{\delta} = \mathbf{0}$ was rejected at the 5% level. What are we to conclude?

The evidence points toward real differences. The point $\boldsymbol{\delta} = \mathbf{0}$ falls outside the 95% *confidence region* for $\boldsymbol{\delta}$ (see Exercise 6.1), and this result is consistent with the T^2-test. The 95% simultaneous confidence coefficient applies to the *entire* set of intervals that could be constructed for all possible linear combinations of the form $a_1\delta_1 + a_2\delta_2$. The particular intervals corresponding to the choices $(a_1 = 1, a_2 = 0)$ and $(a_1 = 0, a_2 = 1)$ contain zero. Other choices of a_1 and a_2 will produce simultaneous intervals that do *not* contain zero. (If the hypothesis $H_0: \boldsymbol{\delta} = \mathbf{0}$ were not rejected, then *all* simultaneous intervals would include zero.)

The Bonferroni simultaneous intervals also cover zero. (See Exercise 6.2.)

Our analysis assumed a normal distribution for the $\mathbf{D}_j$. In fact, the situation is further complicated by the presence of one or, possibly, two outliers. (See Exercise 6.3.) These data can be transformed to data more nearly normal, but with such a small sample, it is difficult to remove the effects of the outlier(s). (See Exercise 6.4.)

The numerical results of this example illustrate an unusual circumstance that can occur when making inferences. ∎

The experimenter in Example 6.1 actually divided a sample by first shaking it and then pouring it rapidly back and forth into two bottles for chemical analysis. This was prudent because a simple division of the sample into two pieces obtained by pouring the top half into one bottle and the remainder into another bottle might result in more suspended solids in the lower half due to setting. The two laboratories would then not be working with the same, or even like, experimental units, and the conclusions would not pertain to laboratory competence, measuring techniques, and so forth.

Whenever an investigator can control the assignment of treatments to experimental units, an appropriate pairing of units and a randomized assignment of treatments can enhance the statistical analysis. Differences, if any, between supposedly identical units must be identified and most-alike units paired. Further, a random assignment of treatment 1 to one unit and treatment 2 to the other unit will help eliminate the systematic effects of uncontrolled sources of variation. Randomization can be implemented by flipping a coin to determine whether the first unit in a pair receives treatment 1 (heads) or treatment 2 (tails). The remaining treatment is then assigned to the other unit. A separate independent randomization is conducted for each pair. One can conceive of the process as follows:

Experimental Design for Paired Comparisons

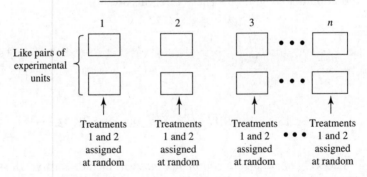

We conclude our discussion of paired comparisons by noting that $\bar{\mathbf{d}}$ and $\mathbf{S}_d$, and hence T^2, may be calculated from the full-sample quantities $\bar{\mathbf{x}}$ and $\mathbf{S}$. Here $\bar{\mathbf{x}}$ is the $2p \times 1$ vector of sample averages for the p variables on the two treatments given by

$$\bar{\mathbf{x}}' = [\bar{x}_{11}, \bar{x}_{12}, \ldots, \bar{x}_{1p}, \bar{x}_{21}, \bar{x}_{22}, \ldots, \bar{x}_{2p}] \tag{6-11}$$

and $\mathbf{S}$ is the $2p \times 2p$ matrix of sample variances and covariances arranged as

$$\mathbf{S} = \begin{bmatrix} \underset{(p \times p)}{\mathbf{S}_{11}} & \underset{(p \times p)}{\mathbf{S}_{12}} \\ \underset{(p \times p)}{\mathbf{S}_{21}} & \underset{(p \times p)}{\mathbf{S}_{22}} \end{bmatrix} \tag{6-12}$$

The matrix $\mathbf{S}_{11}$ contains the sample variances and covariances for the p variables on treatment 1. Similarly, $\mathbf{S}_{22}$ contains the sample variances and covariances computed for the p variables on treatment 2. Finally, $\mathbf{S}_{12} = \mathbf{S}'_{21}$ are the matrices of sample covariances computed from observations on pairs of treatment 1 and treatment 2 variables.

Defining the matrix

$$\underset{(p\times 2p)}{\mathbf{C}} = \begin{bmatrix} 1 & 0 & \cdots & 0 & -1 & 0 & \cdots & 0 \\ 0 & 1 & \cdots & 0 & 0 & -1 & \cdots & 0 \\ \vdots & \vdots & \ddots & \vdots & \vdots & \vdots & \ddots & \vdots \\ 0 & 0 & \cdots & 1 & 0 & 0 & \cdots & -1 \end{bmatrix} \qquad (6\text{-}13)$$

$$\uparrow$$
$$(p + 1)\text{st column}$$

we can verify (see Exercise 6.9) that

$$\mathbf{d}_j = \mathbf{C}\mathbf{x}_j, \qquad j = 1, 2, \ldots, n$$

$$\bar{\mathbf{d}} = \mathbf{C}\bar{\mathbf{x}} \quad \text{and} \quad \mathbf{S}_d = \mathbf{C}\mathbf{S}\mathbf{C}' \qquad (6\text{-}14)$$

Thus,

$$T^2 = n\bar{\mathbf{x}}'\mathbf{C}'(\mathbf{C}\mathbf{S}\mathbf{C}')^{-1}\mathbf{C}\bar{\mathbf{x}} \qquad (6\text{-}15)$$

and it is not necessary first to calculate the differences $\mathbf{d}_1, \mathbf{d}_2, \ldots, \mathbf{d}_n$. On the other hand, it is wise to calculate these differences in order to check normality and the assumption of a random sample.

Each row $\mathbf{c}'_i$ of the matrix $\mathbf{C}$ in (6-13) is a *contrast vector*, because its elements sum to zero. Attention is usually centered on contrasts when comparing treatments. Each contrast is perpendicular to the vector $\mathbf{1}' = [1, 1, \ldots, 1]$ since $\mathbf{c}'_i \mathbf{1} = 0$. The component $\mathbf{1}'\mathbf{x}_j$, representing the overall treatment sum, is ignored by the test statistic T^2 presented in this section.

A Repeated Measures Design for Comparing Treatments

Another generalization of the univariate paired t-statistic arises in situations where q treatments are compared with respect to a *single* response variable. Each subject or experimental unit receives each treatment once over successive periods of time. The jth observation is

$$\mathbf{X}_j = \begin{bmatrix} X_{j1} \\ X_{j2} \\ \vdots \\ X_{jq} \end{bmatrix}, \qquad j = 1, 2, \ldots, n$$

where X_{ji} is the response to the ith treatment on the jth unit. The name *repeated measures* stems from the fact that all treatments are administered to each unit.

For comparative purposes, we consider contrasts of the components of $\boldsymbol{\mu} = E(\mathbf{X}_j)$. These could be

$$\begin{bmatrix} \mu_1 - \mu_2 \\ \mu_1 - \mu_3 \\ \vdots \\ \mu_1 - \mu_q \end{bmatrix} = \begin{bmatrix} 1 & -1 & 0 & \cdots & 0 \\ 1 & 0 & -1 & \cdots & 0 \\ \vdots & \vdots & \vdots & \ddots & \vdots \\ 1 & 0 & 0 & \cdots & -1 \end{bmatrix} \begin{bmatrix} \mu_1 \\ \mu_2 \\ \vdots \\ \mu_q \end{bmatrix} = \mathbf{C}_1 \boldsymbol{\mu}$$

or

$$\begin{bmatrix} \mu_2 - \mu_1 \\ \mu_3 - \mu_2 \\ \vdots \\ \mu_q - \mu_{q-1} \end{bmatrix} = \begin{bmatrix} -1 & 1 & 0 & \cdots & 0 & 0 \\ 0 & -1 & 1 & \cdots & 0 & 0 \\ \vdots & \vdots & \vdots & \ddots & \vdots & \vdots \\ 0 & 0 & 0 & \cdots & -1 & 1 \end{bmatrix} \begin{bmatrix} \mu_1 \\ \mu_2 \\ \vdots \\ \mu_q \end{bmatrix} = \mathbf{C}_2 \boldsymbol{\mu}$$

Both $\mathbf{C}_1$ and $\mathbf{C}_2$ are called *contrast matrices*, because their $q - 1$ rows are linearly independent and each is a contrast vector. The nature of the design eliminates much of the influence of unit-to-unit variation on treatment comparisons. Of course, the experimenter should randomize the order in which the treatments are presented to each subject.

When the treatment means are equal, $\mathbf{C}_1\boldsymbol{\mu} = \mathbf{C}_2\boldsymbol{\mu} = \mathbf{0}$. In general, the hypothesis that there are no differences in treatments (equal treatment means) becomes $\mathbf{C}\boldsymbol{\mu} = \mathbf{0}$ for any choice of the contrast matrix $\mathbf{C}$.

Consequently, based on the contrasts $\mathbf{C}\mathbf{x}_j$ in the observations, we have means $\mathbf{C}\bar{\mathbf{x}}$ and covariance matrix $\mathbf{C}\mathbf{S}\mathbf{C}'$, and we test $\mathbf{C}\boldsymbol{\mu} = \mathbf{0}$ using the T^2-statistic

$$T^2 = n(\mathbf{C}\bar{\mathbf{x}})'(\mathbf{C}\mathbf{S}\mathbf{C}')^{-1}\mathbf{C}\bar{\mathbf{x}}$$

Test for Equality of Treatments in a Repeated Measures Design

Consider an $N_q(\boldsymbol{\mu}, \boldsymbol{\Sigma})$ population, and let $\mathbf{C}$ be a contrast matrix. An α-level test of H_0: $\mathbf{C}\boldsymbol{\mu} = \mathbf{0}$ (equal treatment means) versus H_1: $\mathbf{C}\boldsymbol{\mu} \neq \mathbf{0}$ is as follows: Reject H_0 if

$$T^2 = n(\mathbf{C}\bar{\mathbf{x}})'(\mathbf{C}\mathbf{S}\mathbf{C}')^{-1}\mathbf{C}\bar{\mathbf{x}} > \frac{(n-1)(q-1)}{(n-q+1)} F_{q-1,n-q+1}(\alpha) \qquad (6\text{-}16)$$

where $F_{q-1,n-q+1}(\alpha)$ is the upper (100α)th percentile of an F-distribution with $q - 1$ and $n - q + 1$ d.f. Here $\bar{\mathbf{x}}$ and $\mathbf{S}$ are the sample mean vector and covariance matrix defined, respectively, by

$$\bar{\mathbf{x}} = \frac{1}{n}\sum_{j=1}^{n}\mathbf{x}_j \quad \text{and} \quad \mathbf{S} = \frac{1}{n-1}\sum_{j=1}^{n}(\mathbf{x}_j - \bar{\mathbf{x}})(\mathbf{x}_j - \bar{\mathbf{x}})'$$

It can be shown that T^2 does not depend on the particular choice of $\mathbf{C}$.[1]

---

[1] Any pair of contrast matrices $\mathbf{C}_1$ and $\mathbf{C}_2$ must be related by $\mathbf{C}_1 = \mathbf{B}\mathbf{C}_2$, with $\mathbf{B}$ nonsingular. This follows because each $\mathbf{C}$ has the largest possible number, $q - 1$, of linearly independent rows, all perpendicular to the vector $\mathbf{1}$. Then $(\mathbf{B}\mathbf{C}_2)'(\mathbf{B}\mathbf{C}_2\mathbf{S}\mathbf{C}_2'\mathbf{B}')^{-1}(\mathbf{B}\mathbf{C}_2) = \mathbf{C}_2'\mathbf{B}'(\mathbf{B}')^{-1}(\mathbf{C}_2\mathbf{S}\mathbf{C}_2')^{-1}\mathbf{B}^{-1}\mathbf{B}\mathbf{C}_2 = \mathbf{C}_2'(\mathbf{C}_2\mathbf{S}\mathbf{C}_2')^{-1}\mathbf{C}_2$, so T^2 computed with $\mathbf{C}_2$ or $\mathbf{C}_1 = \mathbf{B}\mathbf{C}_2$ gives the same result.

A confidence region for contrasts $C\mu$, with μ the mean of a normal population, is determined by the set of all $C\mu$ such that

$$n(C\bar{x} - C\mu)'(CSC')^{-1}(C\bar{x} - C\mu) \le \frac{(n-1)(q-1)}{(n-q+1)} F_{q-1,n-q+1}(\alpha) \qquad (6\text{-}17)$$

where $\bar{x}$ and S are as defined in (6-16). Consequently, simultaneous $100(1 - \alpha)\%$ confidence intervals for single contrasts $c'\mu$ for any contrast vectors of interest are given by (see Result 5A.1)

$$c'\mu: \quad c'\bar{x} \pm \sqrt{\frac{(n-1)(q-1)}{(n-q+1)} F_{q-1,n-q+1}(\alpha)} \sqrt{\frac{c'Sc}{n}} \qquad (6\text{-}18)$$

Example 6.2 (Testing for equal treatments in a repeated measures design) Improved anesthetics are often developed by first studying their effects on animals. In one study, 19 dogs were initially given the drug pentobarbitol. Each dog was then administered carbon dioxide CO_2 at each of two pressure levels. Next, halothane (H) was added, and the administration of CO_2 was repeated. The response, milliseconds between heartbeats, was measured for the four treatment combinations:

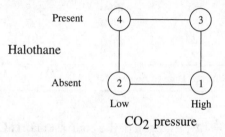

Table 6.2 contains the four measurements for each of the 19 dogs, where

Treatment 1 = high CO_2 pressure without H

Treatment 2 = low CO_2 pressure without H

Treatment 3 = high CO_2 pressure with H

Treatment 4 = low CO_2 pressure with H

We shall analyze the anesthetizing effects of CO_2 pressure and halothane from this repeated-measures design.

There are three treatment contrasts that might be of interest in the experiment. Let μ_1, μ_2, μ_3, and μ_4 correspond to the mean responses for treatments 1, 2, 3, and 4, respectively. Then

$$(\mu_3 + \mu_4) - (\mu_1 + \mu_2) = \left(\begin{array}{c} \text{Halothane contrast representing the} \\ \text{difference between the presence and} \\ \text{absence of halothane} \end{array} \right)$$

$$(\mu_1 + \mu_3) - (\mu_2 + \mu_4) = \left(\begin{array}{c} CO_2 \text{ contrast representing the difference} \\ \text{between high and low } CO_2 \text{ pressure} \end{array} \right)$$

$$(\mu_1 + \mu_4) - (\mu_2 + \mu_3) = \left(\begin{array}{c} \text{Contrast representing the influence} \\ \text{of halothane on } CO_2 \text{ pressure differences} \\ (\text{H}-CO_2 \text{ pressure "interaction"}) \end{array} \right)$$

Table 6.2 Sleeping-Dog Data

Dog	Treatment 1	Treatment 2	Treatment 3	Treatment 4
1	426	609	556	600
2	253	236	392	395
3	359	433	349	357
4	432	431	522	600
5	405	426	513	513
6	324	438	507	539
7	310	312	410	456
8	326	326	350	504
9	375	447	547	548
10	286	286	403	422
11	349	382	473	497
12	429	410	488	547
13	348	377	447	514
14	412	473	472	446
15	347	326	455	468
16	434	458	637	524
17	364	367	432	469
18	420	395	508	531
19	397	556	645	625

Source: Data courtesy of Dr. J. Atlee.

With $\boldsymbol{\mu}' = [\mu_1, \mu_2, \mu_3, \mu_4]$, the contrast matrix $\mathbf{C}$ is

$$\mathbf{C} = \begin{bmatrix} -1 & -1 & 1 & 1 \\ 1 & -1 & 1 & -1 \\ 1 & -1 & -1 & 1 \end{bmatrix}$$

The data (see Table 6.2) give

$$\bar{\mathbf{x}} = \begin{bmatrix} 368.21 \\ 404.63 \\ 479.26 \\ 502.89 \end{bmatrix} \quad \text{and} \quad \mathbf{S} = \begin{bmatrix} 2819.29 & & & \\ 3568.42 & 7963.14 & & \\ 2943.49 & 5303.98 & 6851.32 & \\ 2295.35 & 4065.44 & 4499.63 & 4878.99 \end{bmatrix}$$

It can be verified that

$$\mathbf{C}\bar{\mathbf{x}} = \begin{bmatrix} 209.31 \\ -60.05 \\ -12.79 \end{bmatrix}; \quad \mathbf{C}\mathbf{S}\mathbf{C}' = \begin{bmatrix} 9432.32 & 1098.92 & 927.62 \\ 1098.92 & 5195.84 & 914.54 \\ 927.62 & 914.54 & 7557.44 \end{bmatrix}$$

and

$$T^2 = n(\mathbf{C}\bar{\mathbf{x}})'(\mathbf{C}\mathbf{S}\mathbf{C}')^{-1}(\mathbf{C}\bar{\mathbf{x}}) = 19(6.11) = 116$$

With $\alpha = .05$,

$$\frac{(n-1)(q-1)}{(n-q+1)} F_{q-1,n-q+1}(\alpha) = \frac{18(3)}{16} F_{3,16}(.05) = \frac{18(3)}{16}(3.24) = 10.94$$

From (6-16), $T^2 = 116 > 10.94$, and we reject H_0: $\mathbf{C}\boldsymbol{\mu} = \mathbf{0}$ (no treatment effects). To see which of the contrasts are responsible for the rejection of H_0, we construct 95% simultaneous confidence intervals for these contrasts. From (6-18), the contrast

$$\mathbf{c}_1'\boldsymbol{\mu} = (\mu_3 + \mu_4) - (\mu_1 + \mu_2) = \text{halothane influence}$$

is estimated by the interval

$$(\bar{x}_3 + \bar{x}_4) - (\bar{x}_1 + \bar{x}_2) \pm \sqrt{\frac{18(3)}{16} F_{3,16}(.05)} \sqrt{\frac{\mathbf{c}_1'\mathbf{Sc}_1}{19}} = 209.31 \pm \sqrt{10.94} \sqrt{\frac{9432.32}{19}}$$

$$= 209.31 \pm 73.70$$

where $\mathbf{c}_1'$ is the first row of $\mathbf{C}$. Similarly, the remaining contrasts are estimated by

CO_2 pressure influence $= (\mu_1 + \mu_3) - (\mu_2 + \mu_4)$:

$$-60.05 \pm \sqrt{10.94} \sqrt{\frac{5195.84}{19}} = -60.05 \pm 54.70$$

H–CO_2 pressure "interaction" $= (\mu_1 + \mu_4) - (\mu_2 + \mu_3)$:

$$-12.79 \pm \sqrt{10.94} \sqrt{\frac{7557.44}{19}} = -12.79 \pm 65.97$$

The first confidence interval implies that there is a halothane effect. The presence of halothane produces longer times between heartbeats. This occurs at both levels of CO_2 pressure, since the H–CO_2 pressure interaction contrast, $(\mu_1 + \mu_4) - (\mu_2 - \mu_3)$, is not significantly different from zero. (See the third confidence interval.) The second confidence interval indicates that there is an effect due to CO_2 pressure: The *lower* CO_2 pressure produces longer times between heartbeats.

Some caution must be exercised in our interpretation of the results because the trials with halothane must follow those without. The apparent H-effect may be due to a time trend. (Ideally, the time order of *all* treatments should be determined at random.) ■

The test in (6-16) is appropriate when the covariance matrix, $\text{Cov}(\mathbf{X}) = \boldsymbol{\Sigma}$, cannot be assumed to have any special structure. If it is reasonable to assume that $\boldsymbol{\Sigma}$ has a particular structure, tests designed with this structure in mind have higher power than the one in (6-16). (For $\boldsymbol{\Sigma}$ with the equal correlation structure (8-14), see a discussion of the "randomized block" design in [17] or [22].)

6.3 Comparing Mean Vectors from Two Populations

A T^2-statistic for testing the equality of vector means from two multivariate populations can be developed by analogy with the univariate procedure. (See [11] for a discussion of the univariate case.) This T^2-statistic is appropriate for comparing responses from one set of experimental settings (population 1) with independent responses from another set of experimental settings (population 2). The comparison can be made without explicitly controlling for unit-to-unit variability, as in the paired-comparison case.

If possible, the experimental units should be randomly assigned to the sets of experimental conditions. Randomization will, to some extent, mitigate the effect of unit-to-unit variability in a subsequent comparison of treatments. Although some precision is lost relative to paired comparisons, the inferences in the two-population case are, ordinarily, applicable to a more general collection of experimental units simply because unit homogeneity is not required.

Consider a random sample of size n_1 from population 1 and a sample of size n_2 from population 2. The observations on p variables can be arranged as follows:

Sample	Summary statistics	
(Population 1) $\mathbf{x}_{11}, \mathbf{x}_{12}, \ldots, \mathbf{x}_{1n_1}$	$\bar{\mathbf{x}}_1 = \dfrac{1}{n_1} \sum\limits_{j=1}^{n_1} \mathbf{x}_{1j}$	$\mathbf{S}_1 = \dfrac{1}{n_1 - 1} \sum\limits_{j=1}^{n_1} (\mathbf{x}_{1j} - \bar{\mathbf{x}}_1)(\mathbf{x}_{1j} - \bar{\mathbf{x}}_1)'$
(Population 2) $\mathbf{x}_{21}, \mathbf{x}_{22}, \ldots, \mathbf{x}_{2n_2}$	$\bar{\mathbf{x}}_2 = \dfrac{1}{n_2} \sum\limits_{j=1}^{n_2} \mathbf{x}_{2j}$	$\mathbf{S}_2 = \dfrac{1}{n_2 - 1} \sum\limits_{j=1}^{n_2} (\mathbf{x}_{2j} - \bar{\mathbf{x}}_2)(\mathbf{x}_{2j} - \bar{\mathbf{x}}_2)'$

In this notation, the first subscript—1 or 2—denotes the population.

We want to make inferences about

(mean vector of population 1) − (mean vector of population 2) = $\boldsymbol{\mu}_1 - \boldsymbol{\mu}_2$.

For instance, we shall want to answer the question, Is $\boldsymbol{\mu}_1 = \boldsymbol{\mu}_2$ (or, equivalently, is $\boldsymbol{\mu}_1 - \boldsymbol{\mu}_2 = \mathbf{0}$)? Also, if $\boldsymbol{\mu}_1 - \boldsymbol{\mu}_2 \neq \mathbf{0}$, which component means are different?

With a few tentative assumptions, we are able to provide answers to these questions.

Assumptions Concerning the Structure of the Data

1. The sample $\mathbf{X}_{11}, \mathbf{X}_{12}, \ldots, \mathbf{X}_{1n_1}$, is a random sample of size n_1 from a p-variate population with mean vector $\boldsymbol{\mu}_1$ and covariance matrix $\boldsymbol{\Sigma}_1$.

2. The sample $\mathbf{X}_{21}, \mathbf{X}_{22}, \ldots, \mathbf{X}_{2n_2}$, is a random sample of size n_2 from a p-variate population with mean vector $\boldsymbol{\mu}_2$ and covariance matrix $\boldsymbol{\Sigma}_2$.

3. Also, $\mathbf{X}_{11}, \mathbf{X}_{12}, \ldots, \mathbf{X}_{1n_1}$, are independent of $\mathbf{X}_{21}, \mathbf{X}_{22}, \ldots, \mathbf{X}_{2n_2}$.   (6-19)

We shall see later that, for large samples, this structure is sufficient for making inferences about the $p \times 1$ vector $\boldsymbol{\mu}_1 - \boldsymbol{\mu}_2$. However, when the sample sizes n_1 and n_2 are small, more assumptions are needed.

Further Assumptions When n_1 and n_2 Are Small

1. Both populations are multivariate normal.

2. Also, $\Sigma_1 = \Sigma_2$ (same covariance matrix). $\qquad$ (6-20)

The second assumption, that $\Sigma_1 = \Sigma_2$, is much stronger than its univariate counterpart. Here we are assuming that several pairs of variances and covariances are nearly equal.

When $\Sigma_1 = \Sigma_2 = \Sigma$, $\sum_{j=1}^{n_1} (\mathbf{x}_{1j} - \bar{\mathbf{x}}_1)(\mathbf{x}_{1j} - \bar{\mathbf{x}}_1)'$ is an estimate of $(n_1 - 1)\Sigma$ and $\sum_{j=1}^{n_2} (\mathbf{x}_{2j} - \bar{\mathbf{x}}_2)(\mathbf{x}_{2j} - \bar{\mathbf{x}}_2)'$ is an estimate of $(n_2 - 1)\Sigma$. Consequently, we can pool the information in both samples in order to estimate the common covariance Σ.

We set

$$
\begin{aligned}
\mathbf{S}_{\text{pooled}} &= \frac{\sum_{j=1}^{n_1} (\mathbf{x}_{1j} - \bar{\mathbf{x}}_1)(\mathbf{x}_{1j} - \bar{\mathbf{x}}_1)' + \sum_{j=1}^{n_2} (\mathbf{x}_{2j} - \bar{\mathbf{x}}_2)(\mathbf{x}_{2j} - \bar{\mathbf{x}}_2)'}{n_1 + n_2 - 2} \\
&= \frac{n_1 - 1}{n_1 + n_2 - 2} \mathbf{S}_1 + \frac{n_2 - 1}{n_1 + n_2 - 2} \mathbf{S}_2
\end{aligned}
\qquad (6\text{-}21)
$$

Since $\sum_{j=1}^{n_1} (\mathbf{x}_{1j} - \bar{\mathbf{x}}_1)(\mathbf{x}_{1j} - \bar{\mathbf{x}}_1)'$ has $n_1 - 1$ d.f. and $\sum_{j=1}^{n_2} (\mathbf{x}_{2j} - \bar{\mathbf{x}}_2)(\mathbf{x}_{2j} - \bar{\mathbf{x}}_2)'$ has $n_2 - 1$ d.f., the divisor $(n_1 - 1) + (n_2 - 1)$ in (6-21) is obtained by combining the two component degrees of freedom. [See (4-24).] Additional support for the pooling procedure comes from consideration of the multivariate normal likelihood. (See Exercise 6.11.)

To test the hypothesis that $\boldsymbol{\mu}_1 - \boldsymbol{\mu}_2 = \boldsymbol{\delta}_0$, a specified vector, we consider the squared statistical distance from $\bar{\mathbf{x}}_1 - \bar{\mathbf{x}}_2$ to $\boldsymbol{\delta}_0$. Now,

$$
E(\bar{\mathbf{X}}_1 - \bar{\mathbf{X}}_2) = E(\bar{\mathbf{X}}_1) - E(\bar{\mathbf{X}}_2) = \boldsymbol{\mu}_1 - \boldsymbol{\mu}_2
$$

Since the independence assumption in (6-19) implies that $\bar{\mathbf{X}}_1$ and $\bar{\mathbf{X}}_2$ are independent and thus $\text{Cov}(\bar{\mathbf{X}}_1, \bar{\mathbf{X}}_2) = \mathbf{0}$ (see Result 4.5), by (3-9), it follows that

$$
\text{Cov}(\bar{\mathbf{X}}_1 - \bar{\mathbf{X}}_2) = \text{Cov}(\bar{\mathbf{X}}_1) + \text{Cov}(\bar{\mathbf{X}}_2) = \frac{1}{n_1}\Sigma + \frac{1}{n_2}\Sigma = \left(\frac{1}{n_1} + \frac{1}{n_2}\right)\Sigma \qquad (6\text{-}22)
$$

Because $\mathbf{S}_{\text{pooled}}$ estimates Σ, we see that

$$
\left(\frac{1}{n_1} + \frac{1}{n_2}\right)\mathbf{S}_{\text{pooled}}
$$

is an estimator of $\text{Cov}(\bar{\mathbf{X}}_1 - \bar{\mathbf{X}}_2)$.

The likelihood ratio test of

$$
H_0: \boldsymbol{\mu}_1 - \boldsymbol{\mu}_2 = \boldsymbol{\delta}_0
$$

is based on the square of the statistical distance, T^2, and is given by (see [1]). Reject H_0 if

$$
T^2 = (\bar{\mathbf{x}}_1 - \bar{\mathbf{x}}_2 - \boldsymbol{\delta}_0)' \left[\left(\frac{1}{n_1} + \frac{1}{n_2}\right)\mathbf{S}_{\text{pooled}}\right]^{-1} (\bar{\mathbf{x}}_1 - \bar{\mathbf{x}}_2 - \boldsymbol{\delta}_0) > c^2 \qquad (6\text{-}23)
$$

where the critical distance c^2 is determined from the distribution of the two-sample T^2-statistic.

Result 6.2. If $X_{11}, X_{12}, \ldots, X_{1n_1}$ is a random sample of size n_1 from $N_p(\mu_1, \Sigma)$ and $X_{21}, X_{22}, \ldots, X_{2n_2}$ is an independent random sample of size n_2 from $N_p(\mu_2, \Sigma)$, then

$$T^2 = [\overline{X}_1 - \overline{X}_2 - (\mu_1 - \mu_2)]' \left[\left(\frac{1}{n_1} + \frac{1}{n_2} \right) S_{pooled} \right]^{-1} [\overline{X}_1 - \overline{X}_2 - (\mu_1 - \mu_2)]$$

is distributed as

$$\frac{(n_1 + n_2 - 2)p}{(n_1 + n_2 - p - 1)} F_{p, n_1 + n_2 - p - 1}$$

Consequently,

$$P\left[(\overline{X}_1 - \overline{X}_2 - (\mu_1 - \mu_2))' \left[\left(\frac{1}{n_1} + \frac{1}{n_2} \right) S_{pooled} \right]^{-1} (\overline{X}_1 - \overline{X}_2 - (\mu_1 - \mu_2)) \le c^2 \right] = 1 - \alpha$$

$$(6\text{-}24)$$

where

$$c^2 = \frac{(n_1 + n_2 - 2)p}{(n_1 + n_2 - p - 1)} F_{p, n_1 + n_2 - p - 1}(\alpha)$$

Proof. We first note that

$$\overline{X}_1 - \overline{X}_2 = \frac{1}{n_1} X_{11} + \frac{1}{n_1} X_{12} + \cdots + \frac{1}{n_1} X_{1n_1} - \frac{1}{n_2} X_{21} - \frac{1}{n_2} X_{22} - \cdots - \frac{1}{n_2} X_{2n_2}$$

is distributed as

$$N_p\left(\mu_1 - \mu_2, \left(\frac{1}{n_1} + \frac{1}{n_2} \right) \Sigma \right)$$

by Result 4.8, with $c_1 = c_2 = \cdots = c_{n_1} = 1/n_1$ and $c_{n_1+1} = c_{n_1+2} = \cdots = c_{n_1+n_2} = -1/n_2$. According to (4-23),

$$(n_1 - 1)S_1 \text{ is distributed as } W_{n_1-1}(\Sigma) \text{ and } (n_2 - 1)S_2 \text{ as } W_{n_2-1}(\Sigma)$$

By assumption, the X_{1j}'s and the X_{2j}'s are independent, so $(n_1 - 1)S_1$ and $(n_2 - 1)S_2$ are also independent. From (4-24), $(n_1 - 1)S_1 + (n_2 - 1)S_2$ is then distributed as $W_{n_1+n_2-2}(\Sigma)$. Therefore,

$$T^2 = \left(\frac{1}{n_1} + \frac{1}{n_2} \right)^{-1/2} (\overline{X}_1 - \overline{X}_2 - (\mu_1 - \mu_2))' S_{pooled}^{-1} \left(\frac{1}{n_1} + \frac{1}{n_2} \right)^{-1/2} (\overline{X}_1 - \overline{X}_2 - (\mu_1 - \mu_2))$$

$$= \left(\begin{array}{c} \text{multivariate normal} \\ \text{random vector} \end{array} \right)' \left(\frac{\text{Wishart random matrix}}{\text{d.f.}} \right)^{-1} \left(\begin{array}{c} \text{multivariate normal} \\ \text{random vector} \end{array} \right)$$

$$= N_p(0, \Sigma)' \left[\frac{W_{n_1+n_2-2}(\Sigma)}{n_1 + n_2 - 2} \right]^{-1} N_p(0, \Sigma)$$

which is the T^2-distribution specified in (5-8), with n replaced by $n_1 + n_2 - 1$. [See (5-5) for the relation to F.]　∎

We are primarily interested in confidence regions for $\mu_1 - \mu_2$. From (6-24), we conclude that all $\mu_1 - \mu_2$ within squared statistical distance c^2 of $\bar{x}_1 - \bar{x}_2$ constitute the confidence region. This region is an ellipsoid centered at the observed difference $\bar{x}_1 - \bar{x}_2$ and whose axes are determined by the eigenvalues and eigenvectors of S_{pooled} (or S_{pooled}^{-1}).

Example 6.3 (Constructing a confidence region for the difference of two mean vectors)
Fifty bars of soap are manufactured in each of two ways. Two characteristics, X_1 = lather and X_2 = mildness, are measured. The summary statistics for bars produced by methods 1 and 2 are

$$\bar{x}_1 = \begin{bmatrix} 8.3 \\ 4.1 \end{bmatrix}, \qquad S_1 = \begin{bmatrix} 2 & 1 \\ 1 & 6 \end{bmatrix}$$

$$\bar{x}_2 = \begin{bmatrix} 10.2 \\ 3.9 \end{bmatrix}, \qquad S_2 = \begin{bmatrix} 2 & 1 \\ 1 & 4 \end{bmatrix}$$

Obtain a 95% confidence region for $\mu_1 - \mu_2$.

We first note that S_1 and S_2 are approximately equal, so that it is reasonable to pool them. Hence, from (6-21),

$$S_{pooled} = \frac{49}{98} S_1 + \frac{49}{98} S_2 = \begin{bmatrix} 2 & 1 \\ 1 & 5 \end{bmatrix}$$

Also,

$$\bar{x}_1 - \bar{x}_2 = \begin{bmatrix} -1.9 \\ .2 \end{bmatrix}$$

so the confidence ellipse is centered at $[-1.9, .2]'$. The eigenvalues and eigenvectors of S_{pooled} are obtained from the equation

$$0 = |S_{pooled} - \lambda I| = \begin{vmatrix} 2-\lambda & 1 \\ 1 & 5-\lambda \end{vmatrix} = \lambda^2 - 7\lambda + 9$$

so $\lambda = (7 \pm \sqrt{49-36})/2$. Consequently, $\lambda_1 = 5.303$ and $\lambda_2 = 1.697$, and the corresponding eigenvectors, e_1 and e_2, determined from

$$S_{pooled} e_i = \lambda_i e_i, \qquad i = 1,2$$

are

$$e_1 = \begin{bmatrix} .290 \\ .957 \end{bmatrix} \quad \text{and} \quad e_2 = \begin{bmatrix} .957 \\ -.290 \end{bmatrix}$$

By Result 6.2,

$$\left(\frac{1}{n_1} + \frac{1}{n_2}\right)c^2 = \left(\frac{1}{50} + \frac{1}{50}\right)\frac{(98)(2)}{(97)} F_{2,97}(.05) = .25$$

since $F_{2,97}(.05) = 3.1$. The confidence ellipse extends

$$\sqrt{\lambda_i}\sqrt{\left(\frac{1}{n_1} + \frac{1}{n_2}\right)c^2} = \sqrt{\lambda_i}\sqrt{.25}$$

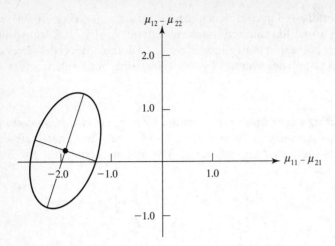

Figure 6.1 95% confidence ellipse for $\boldsymbol{\mu}_1 - \boldsymbol{\mu}_2$.

units along the eigenvector $\mathbf{e}_i$, or 1.15 units in the $\mathbf{e}_1$ direction and .65 units in the $\mathbf{e}_2$ direction. The 95% confidence ellipse is shown in Figure 6.1. Clearly, $\boldsymbol{\mu}_1 - \boldsymbol{\mu}_2 = \mathbf{0}$ is not in the ellipse, and we conclude that the two methods of manufacturing soap produce different results. It appears as if the two processes produce bars of soap with about the same mildness (X_2), but those from the second process have more lather (X_1). ∎

Simultaneous Confidence Intervals

It is possible to derive simultaneous confidence intervals for the components of the vector $\boldsymbol{\mu}_1 - \boldsymbol{\mu}_2$. These confidence intervals are developed from a consideration of all possible linear combinations of the differences in the mean vectors. It is assumed that the parent multivariate populations are normal with a common covariance $\boldsymbol{\Sigma}$.

Result 6.3. Let $c^2 = [(n_1 + n_2 - 2)p/(n_1 + n_2 - p - 1)]F_{p,n_1+n_2-p-1}(\alpha)$. With probability $1 - \alpha$.

$$\mathbf{a}'(\overline{\mathbf{X}}_1 - \overline{\mathbf{X}}_2) \pm c \sqrt{\mathbf{a}'\left(\frac{1}{n_1} + \frac{1}{n_2}\right)\mathbf{S}_{\text{pooled}}\mathbf{a}}$$

will cover $\mathbf{a}'(\boldsymbol{\mu}_1 - \boldsymbol{\mu}_2)$ for all $\mathbf{a}$. In particular $\mu_{1i} - \mu_{2i}$ will be covered by

$$(\overline{X}_{1i} - \overline{X}_{2i}) \pm c \sqrt{\left(\frac{1}{n_1} + \frac{1}{n_2}\right)s_{ii,\text{pooled}}} \qquad \text{for } i = 1, 2, \ldots, p$$

Proof. Consider univariate linear combinations of the observations

$$\mathbf{X}_{11}, \mathbf{X}_{12}, \ldots, \mathbf{X}_{1n_1} \quad \text{and} \quad \mathbf{X}_{21}, \mathbf{X}_{22}, \ldots, \mathbf{X}_{2n_2}$$

given by $\mathbf{a}'\mathbf{X}_{1j} = a_1 X_{1j1} + a_2 X_{1j2} + \cdots + a_p X_{1jp}$ and $\mathbf{a}'\mathbf{X}_{2j} = a_1 X_{2j1} + a_2 X_{2j2} + \cdots + a_p X_{2jp}$. These linear combinations have sample means and covariances $\mathbf{a}'\overline{\mathbf{X}}_1$, $\mathbf{a}'\mathbf{S}_1\mathbf{a}$ and $\mathbf{a}'\overline{\mathbf{X}}_2$, $\mathbf{a}'\mathbf{S}_2\mathbf{a}$, respectively, where $\overline{\mathbf{X}}_1$, $\mathbf{S}_1$, and $\overline{\mathbf{X}}_2$, $\mathbf{S}_2$ are the mean and covariance statistics for the two original samples. (See Result 3.5.) When both parent populations have the same covariance matrix, $s_{1,\mathbf{a}}^2 = \mathbf{a}'\mathbf{S}_1\mathbf{a}$ and $s_{2,\mathbf{a}}^2 = \mathbf{a}'\mathbf{S}_2\mathbf{a}$

are both estimators of $\mathbf{a}'\boldsymbol{\Sigma}\mathbf{a}$, the common population variance of the linear combinations $\mathbf{a}'\mathbf{X}_1$ and $\mathbf{a}'\mathbf{X}_2$. Pooling these estimators, we obtain

$$
\begin{aligned}
s^2_{\mathbf{a},\text{pooled}} &= \frac{(n_1 - 1)s^2_{1,\mathbf{a}} + (n_2 - 1)s^2_{2,\mathbf{a}}}{(n_1 + n_2 - 2)} \\[2mm]
&= \mathbf{a}'\left[\frac{n_1 - 1}{n_1 + n_2 - 2}\mathbf{S}_1 + \frac{n_2 - 1}{n_1 + n_2 - 2}\mathbf{S}_2\right]\mathbf{a} \quad\quad (6\text{-}25)\\[2mm]
&= \mathbf{a}'\mathbf{S}_{\text{pooled}}\mathbf{a}
\end{aligned}
$$

To test $H_0: \mathbf{a}'(\boldsymbol{\mu}_1 - \boldsymbol{\mu}_2) = \mathbf{a}'\boldsymbol{\delta}_0$, on the basis of the $\mathbf{a}'\mathbf{X}_{1j}$ and $\mathbf{a}'\mathbf{X}_{2j}$, we can form the square of the univariate two-sample t-statistic

$$
t^2_{\mathbf{a}} = \frac{[\mathbf{a}'(\overline{\mathbf{X}}_1 - \overline{\mathbf{X}}_2) - \mathbf{a}'(\boldsymbol{\mu}_1 - \boldsymbol{\mu}_2)]^2}{\left(\dfrac{1}{n_1} + \dfrac{1}{n_2}\right)s^2_{\mathbf{a},\text{pooled}}} = \frac{[\mathbf{a}'(\overline{\mathbf{X}}_1 - \overline{\mathbf{X}}_2 - (\boldsymbol{\mu}_1 - \boldsymbol{\mu}_2))]^2}{\mathbf{a}'\left(\dfrac{1}{n_1} + \dfrac{1}{n_2}\right)\mathbf{S}_{\text{pooled}}\mathbf{a}} \quad (6\text{-}26)
$$

According to the maximization lemma with $\mathbf{d} = (\overline{\mathbf{X}}_1 - \overline{\mathbf{X}}_2 - (\boldsymbol{\mu}_1 - \boldsymbol{\mu}_2))$ and $\mathbf{B} = (1/n_1 + 1/n_2)\mathbf{S}_{\text{pooled}}$ in (2-50),

$$
\begin{aligned}
t^2_{\mathbf{a}} &\le (\overline{\mathbf{X}}_1 - \overline{\mathbf{X}}_2 - (\boldsymbol{\mu}_1 - \boldsymbol{\mu}_2))'\left[\left(\frac{1}{n_1} + \frac{1}{n_2}\right)\mathbf{S}_{\text{pooled}}\right]^{-1}(\overline{\mathbf{X}}_1 - \overline{\mathbf{X}}_2 - (\boldsymbol{\mu}_1 - \boldsymbol{\mu}_2)) \\[2mm]
&= T^2
\end{aligned}
$$

for all $\mathbf{a} \ne \mathbf{0}$. Thus,

$$
\begin{aligned}
(1 - \alpha) &= P[T^2 \le c^2] = P[t^2_{\mathbf{a}} \le c^2, \quad \text{for all } \mathbf{a}] \\[2mm]
&= P\left[\left|\mathbf{a}'(\overline{\mathbf{X}}_1 - \overline{\mathbf{X}}_2) - \mathbf{a}'(\boldsymbol{\mu}_1 - \boldsymbol{\mu}_2)\right| \le c\sqrt{\mathbf{a}'\left(\frac{1}{n_1} + \frac{1}{n_2}\right)\mathbf{S}_{\text{pooled}}\mathbf{a}} \quad \text{for all } \mathbf{a}\right]
\end{aligned}
$$

where c^2 is selected according to Result 6.2. ∎

Remark. For testing $H_0: \boldsymbol{\mu}_1 - \boldsymbol{\mu}_2 = \mathbf{0}$, the linear combination $\hat{\mathbf{a}}'(\overline{\mathbf{x}}_1 - \overline{\mathbf{x}}_2)$, with coefficient vector $\hat{\mathbf{a}} \propto \mathbf{S}^{-1}_{\text{pooled}}(\overline{\mathbf{x}}_1 - \overline{\mathbf{x}}_2)$, quantifies the largest population difference. That is, if T^2 rejects H_0, then $\hat{\mathbf{a}}'(\overline{\mathbf{x}}_1 - \overline{\mathbf{x}}_2)$ will have a nonzero mean. Frequently, we try to interpret the components of this linear combination for both subject matter and statistical importance.

***

Example 6.4 (Calculating simultaneous confidence intervals for the differences in mean components) Samples of sizes $n_1 = 45$ and $n_2 = 55$ were taken of Wisconsin homeowners with and without air conditioning, respectively. (Data courtesy of Statistical Laboratory, University of Wisconsin.) Two measurements of electrical usage (in kilowatt hours) were considered. The first is a measure of total *on*-peak consumption (X_1) during July, and the second is a measure of total *off*-peak consumption (X_2) during July. The resulting summary statistics are

$$
\overline{\mathbf{x}}_1 = \begin{bmatrix} 204.4 \\ 556.6 \end{bmatrix}, \quad \mathbf{S}_1 = \begin{bmatrix} 13825.3 & 23823.4 \\ 23823.4 & 73107.4 \end{bmatrix}, \quad n_1 = 45
$$

$$
\overline{\mathbf{x}}_2 = \begin{bmatrix} 130.0 \\ 355.0 \end{bmatrix}, \quad \mathbf{S}_2 = \begin{bmatrix} 8632.0 & 19616.7 \\ 19616.7 & 55964.5 \end{bmatrix}, \quad n_2 = 55
$$

(The off-peak consumption is higher than the on-peak consumption because there are more off-peak hours in a month.)

Let us find 95% simultaneous confidence intervals for the differences in the mean components.

Although there appears to be somewhat of a discrepancy in the sample variances, for illustrative purposes we proceed to a calculation of the pooled sample covariance matrix. Here

$$\mathbf{S}_{\text{pooled}} = \frac{n_1 - 1}{n_1 + n_2 - 2}\mathbf{S}_1 + \frac{n_2 - 1}{n_1 + n_2 - 2}\mathbf{S}_2 = \begin{bmatrix} 10963.7 & 21505.5 \\ 21505.5 & 63661.3 \end{bmatrix}$$

and

$$c^2 = \frac{(n_1 + n_2 - 2)p}{n_1 + n_2 - p - 1}F_{p,n_1+n_2-p-1}(\alpha) = \frac{98(2)}{97}F_{2,97}(.05)$$

$$= (2.02)(3.1) = 6.26$$

With $\boldsymbol{\mu}_1' - \boldsymbol{\mu}_2' = [\mu_{11} - \mu_{21}, \mu_{12} - \mu_{22}]$, the 95% simultaneous confidence intervals for the population differences are

$$\mu_{11} - \mu_{21}: (204.4 - 130.0) \pm \sqrt{6.26}\sqrt{\left(\frac{1}{45} + \frac{1}{55}\right)10963.7}$$

or

$$21.7 \le \mu_{11} - \mu_{21} \le 127.1 \qquad (\textit{on}\text{-peak})$$

$$\mu_{12} - \mu_{22}: (556.6 - 355.0) \pm \sqrt{6.26}\sqrt{\left(\frac{1}{45} + \frac{1}{55}\right)63661.3}$$

or

$$74.7 \le \mu_{12} - \mu_{22} \le 328.5 \qquad (\textit{off}\text{-peak})$$

We conclude that there is a difference in electrical consumption between those with air-conditioning and those without. This difference is evident in both on-peak and off-peak consumption.

The 95% confidence ellipse for $\boldsymbol{\mu}_1 - \boldsymbol{\mu}_2$ is determined from the eigenvalue-eigenvector pairs $\lambda_1 = 71323.5$, $\mathbf{e}_1' = [.336, .942]$ and $\lambda_2 = 3301.5$, $\mathbf{e}_2' = [.942, -.336]$.

Since

$$\sqrt{\lambda_1}\sqrt{\left(\frac{1}{n_1} + \frac{1}{n_2}\right)c^2} = \sqrt{71323.5}\sqrt{\left(\frac{1}{45} + \frac{1}{55}\right)6.26} = 134.3$$

and

$$\sqrt{\lambda_2}\sqrt{\left(\frac{1}{n_1} + \frac{1}{n_2}\right)c^2} = \sqrt{3301.5}\sqrt{\left(\frac{1}{45} + \frac{1}{55}\right)6.26} = 28.9$$

we obtain the 95% confidence ellipse for $\boldsymbol{\mu}_1 - \boldsymbol{\mu}_2$ sketched in Figure 6.2 on page 291. Because the confidence ellipse for the difference in means does not cover $\mathbf{0}' = [0, 0]$, the T^2-statistic will reject $H_0: \boldsymbol{\mu}_1 - \boldsymbol{\mu}_2 = \mathbf{0}$ at the 5% level.

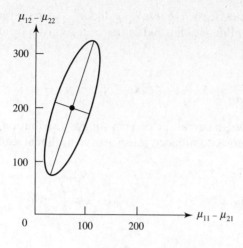

Figure 6.2 95% confidence ellipse for $\boldsymbol{\mu}'_1 - \boldsymbol{\mu}'_2 = (\mu_{11} - \mu_{21}, \mu_{12} - \mu_{22})$.

The coefficient vector for the linear combination most responsible for rejection is proportional to $\mathbf{S}^{-1}_{\text{pooled}}(\bar{\mathbf{x}}_1 - \bar{\mathbf{x}}_2)$. (See Exercise 6.7.)   ∎

The *Bonferroni* $100(1 - \alpha)\%$ *simultaneous confidence intervals* for the p population mean differences are

$$\mu_{1i} - \mu_{2i}: \quad (\bar{x}_{1i} - \bar{x}_{2i}) \pm t_{n_1+n_2-2}\left(\frac{\alpha}{2p}\right)\sqrt{\left(\frac{1}{n_1} + \frac{1}{n_2}\right)s_{ii,\text{pooled}}}$$

where $t_{n_1+n_2-2}(\alpha/2p)$ is the upper $100(\alpha/2p)$th percentile of a t-distribution with $n_1 + n_2 - 2$ d.f.

The Two-Sample Situation When $\boldsymbol{\Sigma}_1 \neq \boldsymbol{\Sigma}_2$

When $\boldsymbol{\Sigma}_1 \neq \boldsymbol{\Sigma}_2$. we are unable to find a "distance" measure like T^2, whose distribution does not depend on the unknowns $\boldsymbol{\Sigma}_1$ and $\boldsymbol{\Sigma}_2$. Bartlett's test [3] is used to test the equality of $\boldsymbol{\Sigma}_1$ and $\boldsymbol{\Sigma}_2$ in terms of generalized variances. Unfortunately, the conclusions can be seriously misleading when the populations are nonnormal. Nonnormality and unequal covariances cannot be separated with Bartlett's test. (See also Section 6.6.) A method of testing the equality of two covariance matrices that is less sensitive to the assumption of multivariate normality has been proposed by Tiku and Balakrishnan [23]. However, more practical experience is needed with this test before we can recommend it unconditionally.

We suggest, without much factual support, that any discrepancy of the order $\sigma_{1,ii} = 4\sigma_{2,ii}$, or vice versa, is probably serious. This is true in the univariate case. The size of the discrepancies that are critical in the multivariate situation probably depends, to a large extent, on the number of variables p.

A transformation may improve things when the marginal variances are quite different. However, for n_1 and n_2 large, we can avoid the complexities due to unequal covariance matrices.

Result 6.4. Let the sample sizes be such that $n_1 - p$ and $n_2 - p$ are large. Then, an approximate $100(1 - \alpha)\%$ confidence ellipsoid for $\boldsymbol{\mu}_1 - \boldsymbol{\mu}_2$ is given by all $\boldsymbol{\mu}_1 - \boldsymbol{\mu}_2$ satisfying

$$[\bar{\mathbf{x}}_1 - \bar{\mathbf{x}}_2 - (\boldsymbol{\mu}_1 - \boldsymbol{\mu}_2)]' \left[\frac{1}{n_1} \mathbf{S}_1 + \frac{1}{n_2} \mathbf{S}_2 \right]^{-1} [\bar{\mathbf{x}}_1 - \bar{\mathbf{x}}_2 - (\boldsymbol{\mu}_1 - \boldsymbol{\mu}_2)] \le \chi_p^2(\alpha)$$

where $\chi_p^2(\alpha)$ is the upper (100α)th percentile of a chi-square distribution with p d.f. Also, $100(1 - \alpha)\%$ simultaneous confidence intervals for all linear combinations $\mathbf{a}'(\boldsymbol{\mu}_1 - \boldsymbol{\mu}_2)$ are provided by

$$\mathbf{a}'(\boldsymbol{\mu}_1 - \boldsymbol{\mu}_2) \quad \text{belongs to} \quad \mathbf{a}'(\bar{\mathbf{x}}_1 - \bar{\mathbf{x}}_2) \pm \sqrt{\chi_p^2(\alpha)} \sqrt{\mathbf{a}'\left(\frac{1}{n_1} \mathbf{S}_1 + \frac{1}{n_2} \mathbf{S}_2 \right) \mathbf{a}}$$

Proof. From (6-22) and (3-9),

$$E(\bar{\mathbf{X}}_1 - \bar{\mathbf{X}}_2) = \boldsymbol{\mu}_1 - \boldsymbol{\mu}_2$$

and

$$\text{Cov}(\bar{\mathbf{X}}_1 - \bar{\mathbf{X}}_2) = \text{Cov}(\bar{\mathbf{X}}_1) + \text{Cov}(\bar{\mathbf{X}}_2) = \frac{1}{n_1} \boldsymbol{\Sigma}_1 + \frac{1}{n_2} \boldsymbol{\Sigma}_2$$

By the central limit theorem, $\bar{\mathbf{X}}_1 - \bar{\mathbf{X}}_2$ is nearly $N_p[\boldsymbol{\mu}_1 - \boldsymbol{\mu}_2, n_1^{-1}\boldsymbol{\Sigma}_1 + n_2^{-1}\boldsymbol{\Sigma}_2]$. If $\boldsymbol{\Sigma}_1$ and $\boldsymbol{\Sigma}_2$ were known, the square of the statistical distance from $\bar{\mathbf{X}}_1 - \bar{\mathbf{X}}_2$ to $\boldsymbol{\mu}_1 - \boldsymbol{\mu}_2$ would be

$$[\bar{\mathbf{X}}_1 - \bar{\mathbf{X}}_2 - (\boldsymbol{\mu}_1 - \boldsymbol{\mu}_2)]' \left(\frac{1}{n_1} \boldsymbol{\Sigma}_1 + \frac{1}{n_2} \boldsymbol{\Sigma}_2 \right)^{-1} [\bar{\mathbf{X}}_1 - \bar{\mathbf{X}}_2 - (\boldsymbol{\mu}_1 - \boldsymbol{\mu}_2)]$$

This squared distance has an approximate χ_p^2-distribution, by Result 4.7. When n_1 and n_2 are large, with high probability, $\mathbf{S}_1$ will be close to $\boldsymbol{\Sigma}_1$ and $\mathbf{S}_2$ will be close to $\boldsymbol{\Sigma}_2$. Consequently, the approximation holds with $\mathbf{S}_1$ and $\mathbf{S}_2$ in place of $\boldsymbol{\Sigma}_1$ and $\boldsymbol{\Sigma}_2$, respectively.

The results concerning the simultaneous confidence intervals follow from Result 5A.1. ∎

Remark. If $n_1 = n_2 = n$, then $(n - 1)/(n + n - 2) = 1/2$, so

$$\frac{1}{n_1} \mathbf{S}_1 + \frac{1}{n_2} \mathbf{S}_2 = \frac{1}{n}(\mathbf{S}_1 + \mathbf{S}_2) = \frac{(n - 1)\mathbf{S}_1 + (n - 1)\mathbf{S}_2}{n + n - 2} \left(\frac{1}{n} + \frac{1}{n} \right)$$

$$= \mathbf{S}_{\text{pooled}} \left(\frac{1}{n} + \frac{1}{n} \right)$$

With equal sample sizes, the large sample procedure is essentially the same as the procedure based on the pooled covariance matrix. (See Result 6.2.) In one dimension, it is well known that the effect of unequal variances is least when $n_1 = n_2$ and greatest when n_1 is much less than n_2 or vice versa.

Example 6.5 (Large sample procedures for inferences about the difference in means)
We shall analyze the electrical-consumption data discussed in Example 6.4 using the
large sample approach. We first calculate

$$\frac{1}{n_1}\mathbf{S}_1 + \frac{1}{n_2}\mathbf{S}_2 = \frac{1}{45}\begin{bmatrix} 13825.3 & 23823.4 \\ 23823.4 & 73107.4 \end{bmatrix} + \frac{1}{55}\begin{bmatrix} 8632.0 & 19616.7 \\ 19616.7 & 55964.5 \end{bmatrix}$$

$$= \begin{bmatrix} 464.17 & 886.08 \\ 886.08 & 2642.15 \end{bmatrix}$$

The 95% simultaneous confidence intervals for the linear combinations

$$\mathbf{a}'(\boldsymbol{\mu}_1 - \boldsymbol{\mu}_2) = [1, 0]\begin{bmatrix} \mu_{11} - \mu_{21} \\ \mu_{12} - \mu_{22} \end{bmatrix} = \mu_{11} - \mu_{21}$$

and

$$\mathbf{a}'(\boldsymbol{\mu}_1 - \boldsymbol{\mu}_2) = [0, 1]\begin{bmatrix} \mu_{11} - \mu_{21} \\ \mu_{12} - \mu_{22} \end{bmatrix} = \mu_{12} - \mu_{22}$$

are (see Result 6.4)

$$\mu_{11} - \mu_{21}: \quad 74.4 \pm \sqrt{5.99}\sqrt{464.17} \quad \text{or} \quad (21.7, 127.1)$$
$$\mu_{12} - \mu_{22}: \quad 201.6 \pm \sqrt{5.99}\sqrt{2642.15} \quad \text{or} \quad (75.8, 327.4)$$

Notice that these intervals differ negligibly from the intervals in Example 6.4, where
the pooling procedure was employed. The T^2-statistic for testing $H_0: \boldsymbol{\mu}_1 - \boldsymbol{\mu}_2 = \mathbf{0}$ is

$$T^2 = [\bar{\mathbf{x}}_1 - \bar{\mathbf{x}}_2]'\left[\frac{1}{n_1}\mathbf{S}_1 + \frac{1}{n_2}\mathbf{S}_2\right]^{-1}[\bar{\mathbf{x}}_1 - \bar{\mathbf{x}}_2]$$

$$= \begin{bmatrix} 204.4 - 130.0 \\ 556.6 - 355.0 \end{bmatrix}'\begin{bmatrix} 464.17 & 886.08 \\ 886.08 & 2642.15 \end{bmatrix}^{-1}\begin{bmatrix} 204.4 - 130.0 \\ 556.6 - 355.0 \end{bmatrix}$$

$$= [74.4 \quad 201.6](10^{-4})\begin{bmatrix} 59.874 & -20.080 \\ -20.080 & 10.519 \end{bmatrix}\begin{bmatrix} 74.4 \\ 201.6 \end{bmatrix} = 15.66$$

For $\alpha = .05$, the critical value is $\chi_2^2(.05) = 5.99$ and, since $T^2 = 15.66 > \chi_2^2(.05)$
$= 5.99$, we reject H_0.

The most critical linear combination leading to the rejection of H_0 has coefficient vector

$$\hat{\mathbf{a}} \propto \left(\frac{1}{n_1}\mathbf{S}_1 + \frac{1}{n_2}\mathbf{S}_2\right)^{-1}(\bar{\mathbf{x}}_1 - \bar{\mathbf{x}}_2) = (10^{-4})\begin{bmatrix} 59.874 & -20.080 \\ -20.080 & 10.519 \end{bmatrix}\begin{bmatrix} 74.4 \\ 201.6 \end{bmatrix}$$

$$= \begin{bmatrix} .041 \\ .063 \end{bmatrix}$$

The difference in *off*-peak electrical consumption between those with air conditioning and those without contributes more than the corresponding difference in
on-peak consumption to the rejection of $H_0: \boldsymbol{\mu}_1 - \boldsymbol{\mu}_2 = \mathbf{0}$. ∎

A statistic similar to T^2 that is less sensitive to outlying observations for small and moderately sized samples has been developed by Tiku and Singh [24]. However, if the sample size is moderate to large, Hotelling's T^2 is remarkably unaffected by slight departures from normality and/or the presence of a few outliers.

An Approximation to the Distribution of T^2 for Normal Populations When Sample Sizes Are Not Large

One can test $H_0: \mu_1 - \mu_2 = 0$ when the population covariance matrices are unequal even if the two sample sizes are not large, provided the two populations are multivariate normal. This situation is often called the multivariate Behrens-Fisher problem. The result requires that both sample sizes n_1 and n_2 are greater than p, the number of variables. The approach depends on an approximation to the distribution of the statistic

$$T^2 = (\bar{\mathbf{X}}_1 - \bar{\mathbf{X}}_2 - (\mu_1 - \mu_2))' \left[\frac{1}{n_1} \mathbf{S}_1 + \frac{1}{n_2} \mathbf{S}_2 \right]^{-1} (\bar{\mathbf{X}}_1 - \bar{\mathbf{X}}_2 - (\mu_1 - \mu_2)) \quad (6\text{-}27)$$

which is identical to the large sample statistic in Result 6.4. However, instead of using the chi-square approximation to obtain the critical value for testing H_0 the recommended approximation for smaller samples (see [15] and [19]) is given by

$$T^2 = \frac{vp}{v - p + 1} F_{p,v-p+1} \quad (6\text{-}28)$$

where the degrees of freedom v are estimated from the sample covariance matrices using the relation

$$v = \frac{p + p^2}{\sum_{i=1}^{2} \frac{1}{n_i} \left\{ \text{tr} \left[\left(\frac{1}{n_i} \mathbf{S}_i \left(\frac{1}{n_1} \mathbf{S}_1 + \frac{1}{n_2} \mathbf{S}_2 \right)^{-1} \right)^2 \right] + \left(\text{tr} \left[\frac{1}{n_i} \mathbf{S}_i \left(\frac{1}{n_1} \mathbf{S}_1 + \frac{1}{n_2} \mathbf{S}_2 \right)^{-1} \right] \right)^2 \right\}} \quad (6\text{-}29)$$

where $\min(n_1, n_2) \le v \le n_1 + n_2$. This approximation reduces to the usual Welch solution to the Behrens-Fisher problem in the univariate ($p = 1$) case.

With moderate sample sizes and two normal populations, the approximate level α test for equality of means rejects $H_0: \mu_1 - \mu_2 = 0$ if

$$(\bar{\mathbf{x}}_1 - \bar{\mathbf{x}}_2 - (\mu_1 - \mu_2))' \left[\frac{1}{n_1} \mathbf{S}_1 + \frac{1}{n_2} \mathbf{S}_2 \right]^{-1} (\bar{\mathbf{x}}_1 - \bar{\mathbf{x}}_2 - (\mu_1 - \mu_2)) > \frac{vp}{v - p + 1} F_{p,v-p+1}(\alpha)$$

where the degrees of freedom v are given by (6-29). This procedure is consistent with the large samples procedure in Result 6.4 except that the critical value $\chi_p^2(\alpha)$ is replaced by the larger constant $\dfrac{vp}{v - p + 1} F_{p,v-p+1}(\alpha)$.

Similarly, the approximate $100(1 - \alpha)\%$ confidence region is given by all $\mu_1 - \mu_2$ such that

$$(\bar{\mathbf{x}}_1 - \bar{\mathbf{x}}_2 - (\mu_1 - \mu_2))' \left[\frac{1}{n_1} \mathbf{S}_1 + \frac{1}{n_2} \mathbf{S}_2 \right]^{-1} (\bar{\mathbf{x}}_1 - \bar{\mathbf{x}}_2 - (\mu_1 - \mu_2)) \le \frac{vp}{v - p + 1} F_{p, v-p+1}(\alpha)$$

$$(6\text{-}30)$$

For normal populations, the approximation to the distribution of T^2 given by (6-28) and (6-29) usually gives reasonable results.

Example 6.6 (The approximate T^2 distribution when $\Sigma_1 \neq \Sigma_2$) Although the sample sizes are rather large for the electrical consumption data in Example 6.4, we use these data and the calculations in Example 6.5 to illustrate the computations leading to the approximate distribution of T^2 when the population covariance matrices are unequal.

We first calculate

$$\frac{1}{n_1}\mathbf{S}_1 = \frac{1}{45}\begin{bmatrix} 13825.2 & 23823.4 \\ 23823.4 & 73107.4 \end{bmatrix} = \begin{bmatrix} 307.227 & 529.409 \\ 529.409 & 1624.609 \end{bmatrix}$$

$$\frac{1}{n_2}\mathbf{S}_2 = \frac{1}{55}\begin{bmatrix} 8632.0 & 19616.7 \\ 19616.7 & 55964.5 \end{bmatrix} = \begin{bmatrix} 156.945 & 356.667 \\ 356.667 & 1017.536 \end{bmatrix}$$

and using a result from Example 6.5,

$$\left[\frac{1}{n_1}\mathbf{S}_1 + \frac{1}{n_2}\mathbf{S}_2\right]^{-1} = (10^{-4})\begin{bmatrix} 59.874 & -20.080 \\ -20.080 & 10.519 \end{bmatrix}$$

Consequently,

$$\frac{1}{n_1}\mathbf{S}_1\left[\frac{1}{n_1}\mathbf{S}_1 + \frac{1}{n_2}\mathbf{S}_2\right]^{-1} =$$

$$\begin{bmatrix} 307.227 & 529.409 \\ 529.409 & 1624.609 \end{bmatrix}(10^{-4})\begin{bmatrix} 59.874 & -20.080 \\ -20.080 & 10.519 \end{bmatrix} = \begin{bmatrix} .776 & -.060 \\ -.092 & .646 \end{bmatrix}$$

and

$$\left(\frac{1}{n_1}\mathbf{S}_1\left[\frac{1}{n_1}\mathbf{S}_1 + \frac{1}{n_2}\mathbf{S}_2\right]^{-1}\right)^2 = \begin{bmatrix} .776 & -.060 \\ -.092 & .646 \end{bmatrix}\begin{bmatrix} .776 & -.060 \\ -.092 & .646 \end{bmatrix} = \begin{bmatrix} .608 & -.085 \\ -.131 & .423 \end{bmatrix}$$

Further,

$$\frac{1}{n_2}\mathbf{S}_2\left[\frac{1}{n_1}\mathbf{S}_1 + \frac{1}{n_2}\mathbf{S}_2\right]^{-1} =$$

$$\begin{bmatrix} 156.945 & 356.667 \\ 356.667 & 1017.536 \end{bmatrix}(10^{-4})\begin{bmatrix} 59.874 & -20.080 \\ -20.080 & 10.519 \end{bmatrix} = \begin{bmatrix} .224 & -.060 \\ .092 & .354 \end{bmatrix}$$

and

$$\left(\frac{1}{n_2}\mathbf{S}_2\left[\frac{1}{n_1}\mathbf{S}_1 + \frac{1}{n_2}\mathbf{S}_2\right]^{-1}\right)^2 = \begin{bmatrix} .224 & .060 \\ -.092 & .354 \end{bmatrix}\begin{bmatrix} .224 & .060 \\ -.092 & .354 \end{bmatrix} = \begin{bmatrix} .055 & .035 \\ .053 & .131 \end{bmatrix}$$

Then

$$\frac{1}{n_1}\left\{\text{tr}\left[\left(\frac{1}{n_1}\mathbf{S}_1\left(\frac{1}{n_1}\mathbf{S}_1 + \frac{1}{n_2}\mathbf{S}_2\right)^{-1}\right)^2\right] + \left(\text{tr}\left[\frac{1}{n_1}\mathbf{S}_1\left(\frac{1}{n_1}\mathbf{S}_1 + \frac{1}{n_2}\mathbf{S}_2\right)^{-1}\right]\right)^2\right\}$$

$$= \frac{1}{45}\left\{(.608 + .423) + (.776 + .646)^2\right\} = .0678$$

$$\frac{1}{n_2}\left\{\text{tr}\left[\left(\frac{1}{n_2}\mathbf{S}_2\left(\frac{1}{n_1}\mathbf{S}_1 + \frac{1}{n_2}\mathbf{S}_2\right)^{-1}\right)^2\right] + \left(\text{tr}\left[\frac{1}{n_2}\mathbf{S}_2\left(\frac{1}{n_1}\mathbf{S}_1 + \frac{1}{n_2}\mathbf{S}_2\right)^{-1}\right]\right)^2\right\}$$

$$= \frac{1}{55}\left\{(.055 + .131) + (.224 + .354)^2\right\} = .0095$$

Using (6-29), the estimated degrees of freedom v is

$$v = \frac{2 + 2^2}{.0678 + .0095} = 77.6$$

and the $\alpha = .05$ critical value is

$$\frac{vp}{v - p + 1}F_{p, v-p+1}(.05) = \frac{77.6 \times 2}{77.6 - 2 + 1}F_{2,77.6-2+1}(.05) = \frac{155.2}{76.6}3.12 = 6.32$$

From Example 6.5, the observed value of the test statistic is $T^2 = 15.66$ so the hypothesis $H_0: \boldsymbol{\mu}_1 - \boldsymbol{\mu}_2 = \mathbf{0}$ is rejected at the 5% level. This is the same conclusion reached with the large sample procedure described in Example 6.5. ∎

As was the case in Example 6.6, the $F_{p, v-p+1}$ distribution can be defined with noninteger degrees of freedom. A slightly more conservative approach is to use the integer part of v.

6.4 Comparing Several Multivariate Population Means (One-Way MANOVA)

Often, more than two populations need to be compared. Random samples, collected from each of g populations, are arranged as

Population 1: $\mathbf{X}_{11}, \mathbf{X}_{12}, \ldots, \mathbf{X}_{1n_1}$

Population 2: $\mathbf{X}_{21}, \mathbf{X}_{22}, \ldots, \mathbf{X}_{2n_2}$ (6-31)

⋮   ⋮

Population g: $\mathbf{X}_{g1}, \mathbf{X}_{g2}, \ldots, \mathbf{X}_{gn_g}$

MANOVA is used first to investigate whether the population mean vectors are the same and, if not, which mean components differ significantly.

Assumptions about the Structure of the Data for One-Way MANOVA

1. $\mathbf{X}_{\ell1}, \mathbf{X}_{\ell2}, \ldots, \mathbf{X}_{\ell n_\ell}$, is a random sample of size n_ℓ from a population with mean $\boldsymbol{\mu}_\ell$, $\ell = 1, 2, \ldots, g$. The random samples from different populations are independent.

2. All populations have a common covariance matrix Σ.

3. Each population is multivariate normal.

Condition 3 can be relaxed by appealing to the central limit theorem (Result 4.13) when the sample sizes n_ℓ are large.

A review of the univariate analysis of variance (ANOVA) will facilitate our discussion of the multivariate assumptions and solution methods.

A Summary of Univariate ANOVA

In the univariate situation, the assumptions are that $X_{\ell 1}, X_{\ell 2}, \ldots, X_{\ell n_\ell}$ is a random sample from an $N(\mu_\ell, \sigma^2)$ population, $\ell = 1, 2, \ldots, g$, and that the random samples are independent. Although the null hypothesis of equality of means could be formulated as $\mu_1 = \mu_2 = \cdots = \mu_g$, it is customary to regard μ_ℓ as the sum of an overall mean component, such as μ, and a component due to the specific population. For instance, we can write $\mu_\ell = \mu + (\mu_\ell - \mu)$ or $\mu_\ell = \mu + \tau_\ell$ where $\tau_\ell = \mu_\ell - \mu$.

Populations usually correspond to different sets of experimental conditions, and therefore, it is convenient to investigate the deviations τ_ℓ associated with the ℓth population (treatment).

The *reparameterization*

$$
\underset{\left(\substack{\ell\text{th population} \\ \text{mean}}\right)}{\mu_\ell} \quad = \quad \underset{\left(\substack{\text{overall} \\ \text{mean}}\right)}{\mu} \quad + \quad \underset{\left(\substack{\ell\text{th population} \\ \text{(treatment) effect}}\right)}{\tau_\ell} \tag{6-32}
$$

leads to a restatement of the hypothesis of equality of means. The null hypothesis becomes

$$
H_0: \tau_1 = \tau_2 = \cdots = \tau_g = 0
$$

The response $X_{\ell j}$, distributed as $N(\mu + \tau_\ell, \sigma^2)$, can be expressed in the suggestive form

$$
\underset{(\text{overall mean})}{X_{\ell j} =} \quad \mu \quad + \quad \underset{\left(\substack{\text{treatment} \\ \text{effect}}\right)}{\tau_\ell} \quad + \quad \underset{\left(\substack{\text{random} \\ \text{error}}\right)}{e_{\ell j}} \tag{6-33}
$$

where the $e_{\ell j}$ are independent $N(0, \sigma^2)$ random variables. To define uniquely the model parameters and their least squares estimates, it is customary to impose the constraint $\sum_{\ell=1}^{g} n_\ell \tau_\ell = 0$.

Motivated by the decomposition in (6-33), the analysis of variance is based upon an analogous decomposition of the observations,

$$
\underset{(\text{observation})}{x_{\ell j}} \quad = \quad \underset{\left(\substack{\text{overall} \\ \text{sample mean}}\right)}{\bar{x}} \quad + \quad \underset{\left(\substack{\text{estimated} \\ \text{treatment effect}}\right)}{(\bar{x}_\ell - \bar{x})} \quad + \quad \underset{(\text{residual})}{(x_{\ell j} - \bar{x}_\ell)} \tag{6-34}
$$

where $\bar{x}$ is an estimate of μ, $\hat{\tau}_\ell = (\bar{x}_\ell - \bar{x})$ is an estimate of τ_ℓ, and $(x_{\ell j} - \bar{x}_\ell)$ is an estimate of the error $e_{\ell j}$.

Example 6.7 (The sum of squares decomposition for univariate ANOVA) Consider the following independent samples.

$$\text{Population 1:}\quad 9, 6, 9$$
$$\text{Population 2:}\quad 0, 2$$
$$\text{Population 3:}\quad 3, 1, 2$$

Since, for example, $\bar{x}_3 = (3 + 1 + 2)/3 = 2$ and $\bar{x} = (9 + 6 + 9 + 0 + 2 + 3 + 1 + 2)/8 = 4$, we find that

$$
\begin{aligned}
3 = x_{31} &= \bar{x} + (\bar{x}_3 - \bar{x}) + (x_{31} - \bar{x}_3) \\
&= 4 + (2 - 4) + (3 - 2) \\
&= 4 + (-2) + 1
\end{aligned}
$$

Repeating this operation for each observation, we obtain the arrays

$$
\underset{\substack{\text{observation} \\ (x_{\ell j})}}{\begin{pmatrix} 9 & 6 & 9 \\ 0 & 2 & \\ 3 & 1 & 2 \end{pmatrix}} = \underset{\substack{\text{mean} \\ (\bar{x})}}{\begin{pmatrix} 4 & 4 & 4 \\ 4 & 4 & \\ 4 & 4 & 4 \end{pmatrix}} + \underset{\substack{\text{treatment effect} \\ (\bar{x}_\ell - \bar{x})}}{\begin{pmatrix} 4 & 4 & 4 \\ -3 & -3 & \\ -2 & -2 & -2 \end{pmatrix}} + \underset{\substack{\text{residual} \\ (x_{\ell j} - \bar{x}_\ell)}}{\begin{pmatrix} 1 & -2 & 1 \\ -1 & 1 & \\ 1 & -1 & 0 \end{pmatrix}}
$$

The question of equality of means is answered by assessing whether the contribution of the treatment array is large relative to the residuals. (Our estimates $\hat{\tau}_\ell = \bar{x}_\ell - \bar{x}$ of τ_ℓ always satisfy $\sum_{\ell=1}^{g} n_\ell \hat{\tau}_\ell = 0$. Under H_0, each $\hat{\tau}_\ell$ is an estimate of zero.) If the treatment contribution is large, H_0 should be rejected. The size of an array is quantified by stringing the rows of the array out into a vector and calculating its squared length. This quantity is called the *sum of squares* (SS). For the observations, we construct the vector $\mathbf{y}' = [9, 6, 9, 0, 2, 3, 1, 2]$. Its squared length is

$$SS_{obs} = 9^2 + 6^2 + 9^2 + 0^2 + 2^2 + 3^2 + 1^2 + 2^2 = 216$$

Similarly,

$$SS_{mean} = 4^2 + 4^2 + 4^2 + 4^2 + 4^2 + 4^2 + 4^2 + 4^2 = 8(4^2) = 128$$
$$
\begin{aligned}
SS_{tr} &= 4^2 + 4^2 + 4^2 + (-3)^2 + (-3)^2 + (-2)^2 + (-2)^2 + (-2)^2 \\
&= 3(4^2) + 2(-3)^2 + 3(-2)^2 = 78
\end{aligned}
$$

and the residual sum of squares is

$$SS_{res} = 1^2 + (-2)^2 + 1^2 + (-1)^2 + 1^2 + 1^2 + (-1)^2 + 0^2 = 10$$

The sums of squares satisfy the same decomposition, (6-34), as the observations. Consequently,

$$SS_{obs} = SS_{mean} + SS_{tr} + SS_{res}$$

or $216 = 128 + 78 + 10$. The breakup into sums of squares apportions variability in the combined samples into mean, treatment, and residual (error) components. An analysis of variance proceeds by comparing the relative sizes of SS_{tr} and SS_{res}. If H_0 is true, variances computed from SS_{tr} and SS_{res} should be approximately equal. ∎

The sum of squares decomposition illustrated numerically in Example 6.7 is so basic that the algebraic equivalent will now be developed.

Subtracting $\bar{x}$ from both sides of (6-34) and squaring gives

$$(x_{\ell j} - \bar{x})^2 = (\bar{x}_\ell - \bar{x})^2 + (x_{\ell j} - \bar{x}_\ell)^2 + 2(\bar{x}_\ell - \bar{x})(x_{\ell j} - \bar{x}_\ell)$$

We can sum both sides over j, note that $\displaystyle\sum_{j=1}^{n_\ell} (x_{\ell j} - \bar{x}_\ell) = 0$, and obtain

$$\sum_{j=1}^{n_\ell} (x_{\ell j} - \bar{x})^2 = n_\ell(\bar{x}_\ell - \bar{x})^2 + \sum_{j=1}^{n_\ell} (x_{\ell j} - \bar{x}_\ell)^2$$

Next, summing both sides over ℓ we get

$$\sum_{\ell=1}^{g} \sum_{j=1}^{n_\ell} (x_{\ell j} - \bar{x})^2 = \sum_{\ell=1}^{g} n_\ell(\bar{x}_\ell - \bar{x})^2 + \sum_{\ell=1}^{g} \sum_{j=1}^{n_\ell} (x_{\ell j} - \bar{x}_\ell)^2 \qquad (6\text{-}35)$$

$$\left(\begin{array}{c} \text{SS}_{\text{cor}} \\ \text{total (corrected) SS} \end{array}\right) = \left(\begin{array}{c} \text{SS}_{\text{tr}} \\ \text{between (samples) SS} \end{array}\right) + \left(\begin{array}{c} \text{SS}_{\text{res}} \\ \text{within (samples) SS} \end{array}\right)$$

or

$$\sum_{\ell=1}^{g} \sum_{j=1}^{n_\ell} x_{\ell j}^2 = (n_1 + n_2 + \cdots + n_g)\bar{x}^2 + \sum_{\ell=1}^{g} n_\ell(\bar{x}_\ell - \bar{x})^2 + \sum_{\ell=1}^{g} \sum_{j=1}^{n_\ell} (x_{\ell j} - \bar{x}_\ell)^2$$

$$(\text{SS}_{\text{obs}}) \quad = \quad (\text{SS}_{\text{mean}}) \quad + \quad (\text{SS}_{\text{tr}}) \quad + \quad (\text{SS}_{\text{res}}) \quad (6\text{-}36)$$

In the course of establishing (6-36), we have verified that the arrays representing the mean, treatment effects, and residuals are *orthogonal*. That is, these arrays, considered as vectors, are perpendicular whatever the observation vector $\mathbf{y}' = [x_{11}, \ldots, x_{1n_1}, x_{21}, \ldots, x_{2n_2}, \ldots, x_{gn_g}]$. Consequently, we could obtain SS_{res} by subtraction, without having to calculate the individual residuals, because $\text{SS}_{\text{res}} = \text{SS}_{\text{obs}} - \text{SS}_{\text{mean}} - \text{SS}_{\text{tr}}$. However, this is false economy because plots of the residuals provide checks on the assumptions of the model.

The vector representations of the arrays involved in the decomposition (6-34) also have geometric interpretations that provide the degrees of freedom. For an arbitrary set of observations, let $[x_{11}, \ldots, x_{1n_1}, x_{21}, \ldots, x_{2n_2}, \ldots, x_{gn_g}] = \mathbf{y}'$. The observation vector $\mathbf{y}$ can lie anywhere in $n = n_1 + n_2 + \cdots + n_g$ dimensions; the mean vector $\bar{x}\mathbf{1} = [\bar{x}, \ldots, \bar{x}]'$ must lie along the equiangular line of $\mathbf{1}$, and the treatment effect vector

$$(\bar{x}_1 - \bar{x})\begin{bmatrix} 1 \\ \vdots \\ 1 \\ 0 \\ \vdots \\ 0 \\ 0 \\ \vdots \\ 0 \end{bmatrix}\Big\}n_1 + (\bar{x}_2 - \bar{x})\begin{bmatrix} 0 \\ \vdots \\ 0 \\ 1 \\ \vdots \\ 1 \\ 0 \\ \vdots \\ 0 \end{bmatrix}\Big\}n_2 + \cdots + (\bar{x}_g - \bar{x})\begin{bmatrix} 0 \\ \vdots \\ 0 \\ 0 \\ \vdots \\ 0 \\ 1 \\ \vdots \\ 1 \end{bmatrix}\Big\}n_g$$

$$= (\bar{x}_1 - \bar{x})\mathbf{u}_1 + (\bar{x}_2 - \bar{x})\mathbf{u}_2 + \cdots + (\bar{x}_g - \bar{x})\mathbf{u}_g$$

lies in the hyperplane of linear combinations of the g vectors $\mathbf{u}_1, \mathbf{u}_2, \ldots, \mathbf{u}_g$. Since $\mathbf{1} = \mathbf{u}_1 + \mathbf{u}_2 + \cdots + \mathbf{u}_g$, the mean vector also lies in this hyperplane, and it is *always* perpendicular to the treatment vector. (See Exercise 6.10.) Thus, the mean vector has the freedom to lie anywhere along the one-dimensional equiangular line, and the treatment vector has the freedom to lie anywhere in the other $g - 1$ dimensions. The residual vector, $\hat{\mathbf{e}} = \mathbf{y} - (\bar{x}\mathbf{1}) - [(\bar{x}_1 - \bar{x})\mathbf{u}_1 + \cdots + (\bar{x}_g - \bar{x})\mathbf{u}_g]$ is perpendicular to both the mean vector and the treatment effect vector and has the freedom to lie anywhere in the subspace of dimension $n - (g - 1) - 1 = n - g$ that is perpendicular to their hyperplane.

To summarize, we attribute 1 d.f. to SS_{mean}, $g - 1$ d.f. to SS_{tr}, and $n - g = (n_1 + n_2 + \cdots + n_g) - g$ d.f. to SS_{res}. The total number of degrees of freedom is $n = n_1 + n_2 + \cdots + n_g$. Alternatively, by appealing to the univariate distribution theory, we find that these are the degrees of freedom for the chi-square distributions associated with the corresponding sums of squares.

The calculations of the sums of squares and the associated degrees of freedom are conveniently summarized by an ANOVA table.

ANOVA Table for Comparing Univariate Population Means

Source of variation	Sum of squares (SS)	Degrees of freedom (d.f.)
Treatments	$SS_{tr} = \sum\limits_{\ell=1}^{g} n_\ell (\bar{x}_\ell - \bar{x})^2$	$g - 1$
Residual (error)	$SS_{res} = \sum\limits_{\ell=1}^{g} \sum\limits_{j=1}^{n_\ell} (x_{\ell j} - \bar{x}_\ell)^2$	$\sum\limits_{\ell=1}^{g} n_\ell - g$
Total (corrected for the mean)	$SS_{cor} = \sum\limits_{\ell=1}^{g} \sum\limits_{j=1}^{n_\ell} (x_{\ell j} - \bar{x})^2$	$\sum\limits_{\ell=1}^{g} n_\ell - 1$

The usual F-test rejects $H_0 : \tau_1 = \tau_2 = \cdots = \tau_g = 0$ at level α if

$$F = \frac{SS_{tr}/(g - 1)}{SS_{res} / \left(\sum\limits_{\ell=1}^{g} n_\ell - g \right)} > F_{g-1, \Sigma n_\ell - g}(\alpha)$$

where $F_{g-1, \Sigma n_\ell - g}(\alpha)$ is the upper (100α)th percentile of the F-distribution with $g - 1$ and $\Sigma n_\ell - g$ degrees of freedom. This is equivalent to rejecting H_0 for large values of SS_{tr}/SS_{res} or for large values of $1 + SS_{tr}/SS_{res}$. The statistic appropriate for a multivariate generalization rejects H_0 for *small* values of the reciprocal

$$\frac{1}{1 + SS_{tr}/SS_{res}} = \frac{SS_{res}}{SS_{res} + SS_{tr}} \tag{6-37}$$

Example 6.8 (A univariate ANOVA table and F-test for treatment effects) Using the information in Example 6.7, we have the following ANOVA table:

Source of variation	Sum of squares	Degrees of freedom
Treatments	$SS_{tr} = 78$	$g - 1 = 3 - 1 = 2$
Residual	$SS_{res} = 10$	$\sum_{\ell=1}^{g} n_\ell - g = (3 + 2 + 3) - 3 = 5$
Total (corrected)	$SS_{cor} = 88$	$\sum_{\ell=1}^{g} n_\ell - 1 = 7$

Consequently,

$$F = \frac{SS_{tr}/(g - 1)}{SS_{res}/(\Sigma n_\ell - g)} = \frac{78/2}{10/5} = 19.5$$

Since $F = 19.5 > F_{2,5}(.01) = 13.27$, we reject $H_0 : \tau_1 = \tau_2 = \tau_3 = 0$ (no treatment effect) at the 1% level of significance. ∎

Multivariate Analysis of Variance (MANOVA)

Paralleling the univariate reparameterization, we specify the MANOVA model:

MANOVA Model For Comparing g Population Mean Vectors

$$\mathbf{X}_{\ell j} = \boldsymbol{\mu} + \boldsymbol{\tau}_\ell + \mathbf{e}_{\ell j}, \quad j = 1, 2, \ldots, n_\ell \quad \text{and} \quad \ell = 1, 2, \ldots, g \quad (6\text{-}38)$$

where the $\mathbf{e}_{\ell j}$ are independent $N_p(\mathbf{0}, \boldsymbol{\Sigma})$ variables. Here the parameter vector $\boldsymbol{\mu}$ is an overall mean (level), and $\boldsymbol{\tau}_\ell$ represents the ℓth treatment effect with

$$\sum_{\ell=1}^{g} n_\ell \boldsymbol{\tau}_\ell = \mathbf{0}.$$

According to the model in (6-38), *each component* of the observation vector $\mathbf{X}_{\ell j}$ satisfies the univariate model (6-33). The errors for the components of $\mathbf{X}_{\ell j}$ are correlated, but the covariance matrix $\boldsymbol{\Sigma}$ is the same for all populations.

A vector of observations may be decomposed as suggested by the model. Thus,

$$\begin{matrix} \mathbf{x}_{\ell j} & = & \bar{\mathbf{x}} & + & (\bar{\mathbf{x}}_\ell - \bar{\mathbf{x}}) & + & (\mathbf{x}_{\ell j} - \bar{\mathbf{x}}_\ell) \\ \text{(observation)} & & \begin{pmatrix} \text{overall sample} \\ \text{mean } \hat{\boldsymbol{\mu}} \end{pmatrix} & & \begin{pmatrix} \text{estimated} \\ \text{treatment} \\ \text{effect } \hat{\boldsymbol{\tau}}_\ell \end{pmatrix} & & \begin{pmatrix} \text{residual} \\ \hat{\mathbf{e}}_{\ell j} \end{pmatrix} \end{matrix} \quad (6\text{-}39)$$

The decomposition in (6-39) leads to the multivariate analog of the univariate sum of squares breakup in (6-35). First we note that the product

$$(\mathbf{x}_{\ell j} - \bar{\mathbf{x}})(\mathbf{x}_{\ell j} - \bar{\mathbf{x}})'$$

can be written as

$$(\mathbf{x}_{\ell j} - \bar{\mathbf{x}})(\mathbf{x}_{\ell j} - \bar{\mathbf{x}})' = [(\mathbf{x}_{\ell j} - \bar{\mathbf{x}}_\ell) + (\bar{\mathbf{x}}_\ell - \bar{\mathbf{x}})][(\mathbf{x}_{\ell j} - \bar{\mathbf{x}}_\ell) + (\bar{\mathbf{x}}_\ell - \bar{\mathbf{x}})]'$$

$$= (\mathbf{x}_{\ell j} - \bar{\mathbf{x}}_\ell)(\mathbf{x}_{\ell j} - \bar{\mathbf{x}}_\ell)' + (\mathbf{x}_{\ell j} - \bar{\mathbf{x}}_\ell)(\bar{\mathbf{x}}_\ell - \bar{\mathbf{x}})'$$

$$+ (\bar{\mathbf{x}}_\ell - \bar{\mathbf{x}})(\mathbf{x}_{\ell j} - \bar{\mathbf{x}}_\ell)' + (\bar{\mathbf{x}}_\ell - \bar{\mathbf{x}})(\bar{\mathbf{x}}_\ell - \bar{\mathbf{x}})'$$

The sum over j of the middle two expressions is the zero matrix, because $\sum_{j=1}^{n_\ell}(\mathbf{x}_{\ell j} - \bar{\mathbf{x}}_\ell) = \mathbf{0}$. Hence, summing the cross product over ℓ and j yields

$$\sum_{\ell=1}^{g}\sum_{j=1}^{n_\ell}(\mathbf{x}_{\ell j} - \bar{\mathbf{x}})(\mathbf{x}_{\ell j} - \bar{\mathbf{x}})' = \sum_{\ell=1}^{g} n_\ell(\bar{\mathbf{x}}_\ell - \bar{\mathbf{x}})(\bar{\mathbf{x}}_\ell - \bar{\mathbf{x}})' + \sum_{\ell=1}^{g}\sum_{j=1}^{n_\ell}(\mathbf{x}_{\ell j} - \bar{\mathbf{x}}_\ell)(\mathbf{x}_{\ell j} - \bar{\mathbf{x}}_\ell)'$$

$$\begin{pmatrix} \text{total (corrected) sum} \\ \text{of squares and cross} \\ \text{products} \end{pmatrix} \qquad \begin{pmatrix} \text{treatment (\underline{B}etween)} \\ \text{sum of squares and} \\ \text{cross products} \end{pmatrix} \qquad \begin{pmatrix} \text{residual (\underline{W}ithin) sum} \\ \text{of squares and cross} \\ \text{products} \end{pmatrix} \qquad (6\text{-}40)$$

The *within* sum of squares and cross products matrix can be expressed as

$$\mathbf{W} = \sum_{\ell=1}^{g}\sum_{j=1}^{n_\ell}(\mathbf{x}_{\ell j} - \bar{\mathbf{x}}_\ell)(\mathbf{x}_{\ell j} - \bar{\mathbf{x}}_\ell)'$$

$$= (n_1 - 1)\mathbf{S}_1 + (n_2 - 1)\mathbf{S}_2 + \cdots + (n_g - 1)\mathbf{S}_g \qquad (6\text{-}41)$$

where $\mathbf{S}_\ell$ is the sample covariance matrix for the ℓth sample. This matrix is a generalization of the $(n_1 + n_2 - 2)\mathbf{S}_{\text{pooled}}$ matrix encountered in the two-sample case. It plays a dominant role in testing for the presence of treatment effects.

Analogous to the univariate result, the hypothesis of no treatment effects,

$$H_0: \boldsymbol{\tau}_1 = \boldsymbol{\tau}_2 = \cdots = \boldsymbol{\tau}_g = \mathbf{0}$$

is tested by considering the relative sizes of the treatment and residual sums of squares and cross products. Equivalently, we may consider the relative sizes of the residual and total (corrected) sum of squares and cross products. Formally, we summarize the calculations leading to the test statistic in a MANOVA table.

MANOVA Table for Comparing Population Mean Vectors

Source of variation	Matrix of sum of squares and cross products (SSP)	Degrees of freedom (d.f.)
Treatment	$\mathbf{B} = \sum_{\ell=1}^{g} n_\ell(\bar{\mathbf{x}}_\ell - \bar{\mathbf{x}})(\bar{\mathbf{x}}_\ell - \bar{\mathbf{x}})'$	$g - 1$
Residual (Error)	$\mathbf{W} = \sum_{\ell=1}^{g}\sum_{j=1}^{n_\ell}(\mathbf{x}_{\ell j} - \bar{\mathbf{x}}_\ell)(\mathbf{x}_{\ell j} - \bar{\mathbf{x}}_\ell)'$	$\sum_{\ell=1}^{g} n_\ell - g$
Total (corrected for the mean)	$\mathbf{B} + \mathbf{W} = \sum_{\ell=1}^{g}\sum_{j=1}^{n_\ell}(\mathbf{x}_{\ell j} - \bar{\mathbf{x}})(\mathbf{x}_{\ell j} - \bar{\mathbf{x}})'$	$\sum_{\ell=1}^{g} n_\ell - 1$

This table is exactly the same form, component by component, as the ANOVA table, except that squares of scalars are replaced by their vector counterparts. For example, $(\bar{x}_\ell - \bar{x})^2$ becomes $(\bar{\mathbf{x}}_\ell - \bar{\mathbf{x}})(\bar{\mathbf{x}}_\ell - \bar{\mathbf{x}})'$. The degrees of freedom correspond to the univariate geometry and also to some multivariate distribution theory involving Wishart densities. (See [1].)

One test of $H_0: \boldsymbol{\tau}_1 = \boldsymbol{\tau}_2 = \cdots = \boldsymbol{\tau}_g = \mathbf{0}$ involves generalized variances. We reject H_0 if the ratio of generalized variances

$$\Lambda^* = \frac{|\mathbf{W}|}{|\mathbf{B} + \mathbf{W}|} = \frac{\left| \sum_{\ell=1}^{g} \sum_{j=1}^{n_\ell} (\mathbf{x}_{\ell j} - \bar{\mathbf{x}}_\ell)(\mathbf{x}_{\ell j} - \bar{\mathbf{x}}_\ell)' \right|}{\left| \sum_{\ell=1}^{g} \sum_{j=1}^{n_\ell} (\mathbf{x}_{\ell j} - \bar{\mathbf{x}})(\mathbf{x}_{\ell j} - \bar{\mathbf{x}})' \right|} \tag{6-42}$$

is too small. The quantity $\Lambda^* = |\mathbf{W}|/|\mathbf{B} + \mathbf{W}|$, proposed originally by Wilks (see [25]), corresponds to the equivalent form (6-37) of the F-test of H_0: no treatment effects in the univariate case. *Wilks' lambda* has the virtue of being convenient and related to the likelihood ratio criterion.[2] The exact distribution of Λ^* can be derived for the special cases listed in Table 6.3. For other cases and large sample sizes, a modification of Λ^* due to Bartlett (see [4]) can be used to test H_0.

Table 6.3 Distribution of Wilks' Lambda, $\Lambda^* = |\mathbf{W}|/|\mathbf{B} + \mathbf{W}|$

No. of variables	No. of groups	Sampling distribution for multivariate normal data
$p = 1$	$g \geq 2$	$\left(\dfrac{\Sigma n_\ell - g}{g - 1} \right) \left(\dfrac{1 - \Lambda^*}{\Lambda^*} \right) \sim F_{g-1, \Sigma n_\ell - g}$
$p = 2$	$g \geq 2$	$\left(\dfrac{\Sigma n_\ell - g - 1}{g - 1} \right) \left(\dfrac{1 - \sqrt{\Lambda^*}}{\sqrt{\Lambda^*}} \right) \sim F_{2(g-1), 2(\Sigma n_\ell - g - 1)}$
$p \geq 1$	$g = 2$	$\left(\dfrac{\Sigma n_\ell - p - 1}{p} \right) \left(\dfrac{1 - \Lambda^*}{\Lambda^*} \right) \sim F_{p, \Sigma n_\ell - p - 1}$
$p \geq 1$	$g = 3$	$\left(\dfrac{\Sigma n_\ell - p - 2}{p} \right) \left(\dfrac{1 - \sqrt{\Lambda^*}}{\sqrt{\Lambda^*}} \right) \sim F_{2p, 2(\Sigma n_\ell - p - 2)}$

[2] Wilks' lambda can also be expressed as a function of the eigenvalues of $\hat{\lambda}_1, \hat{\lambda}_2, \ldots, \hat{\lambda}_s$ of $\mathbf{W}^{-1}\mathbf{B}$ as

$$\Lambda^* = \prod_{i=1}^{s} \left(\frac{1}{1 + \hat{\lambda}_i} \right)$$

where $s = \min(p, g - 1)$, the rank of $\mathbf{B}$. Other statistics for checking the equality of several multivariate means, such as Pillai's statistic, the Lawley–Hotelling statistic, and Roy's largest root statistic can also be written as particular functions of the eigenvalues of $\mathbf{W}^{-1}\mathbf{B}$. For large samples, all of these statistics are, essentially equivalent. (See the additional discussion on page 336.)

Bartlett (see [4]) has shown that if H_0 is true and $\Sigma n_\ell = n$ is large,

$$-\left(n - 1 - \frac{(p + g)}{2}\right) \ln \Lambda^* = -\left(n - 1 - \frac{(p + g)}{2}\right) \ln\left(\frac{|\mathbf{W}|}{|\mathbf{B} + \mathbf{W}|}\right) \qquad (6\text{-}43)$$

has approximately a chi-square distribution with $p(g - 1)$ d.f. Consequently, for $\Sigma n_\ell = n$ large, we reject H_0 at significance level α if

$$-\left(n - 1 - \frac{(p + g)}{2}\right) \ln\left(\frac{|\mathbf{W}|}{|\mathbf{B} + \mathbf{W}|}\right) > \chi^2_{p(g-1)}(\alpha) \qquad (6\text{-}44)$$

where $\chi^2_{p(g-1)}(\alpha)$ is the upper (100α)th percentile of a chi-square distribution with $p(g - 1)$ d.f.

Example 6.9 (A MANOVA table and Wilks' lambda for testing the equality of three mean vectors) Suppose an additional variable is observed along with the variable introduced in Example 6.7. The sample sizes are $n_1 = 3, n_2 = 2$, and $n_3 = 3$. Arranging the observation pairs $\mathbf{x}_{\ell j}$ in rows, we obtain

$$\left(\begin{array}{ccc} \begin{bmatrix} 9 \\ 3 \end{bmatrix} & \begin{bmatrix} 6 \\ 2 \end{bmatrix} & \begin{bmatrix} 9 \\ 7 \end{bmatrix} \\ \begin{bmatrix} 0 \\ 4 \end{bmatrix} & \begin{bmatrix} 2 \\ 0 \end{bmatrix} & \\ \begin{bmatrix} 3 \\ 8 \end{bmatrix} & \begin{bmatrix} 1 \\ 9 \end{bmatrix} & \begin{bmatrix} 2 \\ 7 \end{bmatrix} \end{array}\right) \qquad \text{with } \bar{\mathbf{x}}_1 = \begin{bmatrix} 8 \\ 4 \end{bmatrix}, \quad \bar{\mathbf{x}}_2 = \begin{bmatrix} 1 \\ 2 \end{bmatrix}, \quad \bar{\mathbf{x}}_3 = \begin{bmatrix} 2 \\ 8 \end{bmatrix},$$

$$\text{and } \bar{\mathbf{x}} = \begin{bmatrix} 4 \\ 5 \end{bmatrix}$$

We have already expressed the observations on the first variable as the sum of an overall mean, treatment effect, and residual in our discussion of univariate ANOVA. We found that

$$\begin{pmatrix} 9 & 6 & 9 \\ 0 & 2 & \\ 3 & 1 & 2 \end{pmatrix} = \begin{pmatrix} 4 & 4 & 4 \\ 4 & 4 & \\ 4 & 4 & 4 \end{pmatrix} + \begin{pmatrix} 4 & 4 & 4 \\ -3 & -3 & \\ -2 & -2 & -2 \end{pmatrix} + \begin{pmatrix} 1 & -2 & 1 \\ -1 & 1 & \\ 1 & -1 & 0 \end{pmatrix}$$

$$\text{(observation)} \qquad \text{(mean)} \qquad \begin{pmatrix} \text{treatment} \\ \text{effect} \end{pmatrix} \qquad \text{(residual)}$$

and

$$\text{SS}_{\text{obs}} = \text{SS}_{\text{mean}} + \text{SS}_{\text{tr}} + \text{SS}_{\text{res}}$$
$$216 = 128 + 78 + 10$$
$$\text{Total SS (corrected)} = \text{SS}_{\text{obs}} - \text{SS}_{\text{mean}} = 216 - 128 = 88$$

Repeating this operation for the observations on the second variable, we have

$$\begin{pmatrix} 3 & 2 & 7 \\ 4 & 0 & \\ 8 & 9 & 7 \end{pmatrix} = \begin{pmatrix} 5 & 5 & 5 \\ 5 & 5 & \\ 5 & 5 & 5 \end{pmatrix} + \begin{pmatrix} -1 & -1 & -1 \\ -3 & -3 & \\ 3 & 3 & 3 \end{pmatrix} + \begin{pmatrix} -1 & -2 & 3 \\ 2 & -2 & \\ 0 & 1 & -1 \end{pmatrix}$$

$$\text{(observation)} \qquad \text{(mean)} \qquad \begin{pmatrix} \text{treatment} \\ \text{effect} \end{pmatrix} \qquad \text{(residual)}$$

and

$$SS_{obs} = SS_{mean} + SS_{tr} + SS_{res}$$
$$272 = 200 + 48 + 24$$
$$\text{Total SS (corrected)} = SS_{obs} - SS_{mean} = 272 - 200 = 72$$

These two single-component analyses must be augmented with the sum of entry-by-entry *cross products* in order to complete the entries in the MANOVA table. Proceeding row by row in the arrays for the two variables, we obtain the cross product contributions:

$$\text{Mean: } 4(5) + 4(5) + \cdots + 4(5) = 8(4)(5) = 160$$
$$\text{Treatment: } 3(4)(-1) + 2(-3)(-3) + 3(-2)(3) = -12$$
$$\text{Residual: } 1(-1) + (-2)(-2) + 1(3) + (-1)(2) + \cdots + 0(-1) = 1$$
$$\text{Total: } 9(3) + 6(2) + 9(7) + 0(4) + \cdots + 2(7) = 149$$

Total (corrected) cross product = total cross product − mean cross product

$$= 149 - 160 = -11$$

Thus, the MANOVA table takes the following form:

Source of variation	Matrix of sum of squares and cross products	Degrees of freedom
Treatment	$\begin{bmatrix} 78 & -12 \\ -12 & 48 \end{bmatrix}$	$3 - 1 = 2$
Residual	$\begin{bmatrix} 10 & 1 \\ 1 & 24 \end{bmatrix}$	$3 + 2 + 3 - 3 = 5$
Total (corrected)	$\begin{bmatrix} 88 & -11 \\ -11 & 72 \end{bmatrix}$	7

Equation (6-40) is verified by noting that

$$\begin{bmatrix} 88 & -11 \\ -11 & 72 \end{bmatrix} = \begin{bmatrix} 78 & -12 \\ -12 & 48 \end{bmatrix} + \begin{bmatrix} 10 & 1 \\ 1 & 24 \end{bmatrix}$$

Using (6-42), we get

$$\Lambda^* = \frac{|\mathbf{W}|}{|\mathbf{B} + \mathbf{W}|} = \frac{\begin{vmatrix} 10 & 1 \\ 1 & 24 \end{vmatrix}}{\begin{vmatrix} 88 & -11 \\ -11 & 72 \end{vmatrix}} = \frac{10(24) - (1)^2}{88(72) - (-11)^2} = \frac{239}{6215} = .0385$$

Since $p = 2$ and $g = 3$, Table 6.3 indicates that an exact test (assuming normality and equal group covariance matrices) of $H_0: \boldsymbol{\tau}_1 = \boldsymbol{\tau}_2 = \boldsymbol{\tau}_3 = \mathbf{0}$ (no treatment effects) versus H_1: at least one $\boldsymbol{\tau}_\ell \neq \mathbf{0}$ is available. To carry out the test, we compare the test statistic

$$\left(\frac{1 - \sqrt{\Lambda^*}}{\sqrt{\Lambda^*}}\right) \frac{(\Sigma n_\ell - g - 1)}{(g - 1)} = \left(\frac{1 - \sqrt{.0385}}{\sqrt{.0385}}\right) \left(\frac{8 - 3 - 1}{3 - 1}\right) = 8.19$$

with a percentage point of an F-distribution having $\nu_1 = 2(g - 1) = 4$ and $\nu_2 = 2(\Sigma n_\ell - g - 1) = 8$ d.f. Since $8.19 > F_{4,8}(.01) = 7.01$, we reject H_0 at the $\alpha = .01$ level and conclude that treatment differences exist. ∎

When the number of variables, p, is large, the MANOVA table is usually not constructed. Still, it is good practice to have the computer print the matrices $\mathbf{B}$ and $\mathbf{W}$ so that especially large entries can be located. Also, the residual vectors

$$\hat{\mathbf{e}}_{\ell j} = \mathbf{x}_{\ell j} - \bar{\mathbf{x}}_\ell$$

should be examined for normality and the presence of outliers using the techniques discussed in Sections 4.6 and 4.7 of Chapter 4.

Example 6.10 (A multivariate analysis of Wisconsin nursing home data) The Wisconsin Department of Health and Social Services reimburses nursing homes in the state for the services provided. The department develops a set of formulas for rates for each facility, based on factors such as level of care, mean wage rate, and average wage rate in the state.

Nursing homes can be classified on the basis of ownership (private party, nonprofit organization, and government) and certification (skilled nursing facility, intermediate care facility, or a combination of the two).

One purpose of a recent study was to investigate the effects of ownership or certification (or both) on costs. Four costs, computed on a per-patient-day basis and measured in hours per patient day, were selected for analysis: $X_1 = $ cost of nursing labor, $X_2 = $ cost of dietary labor, $X_3 = $ cost of plant operation and maintenance labor, and $X_4 = $ cost of housekeeping and laundry labor. A total of $n = 516$ observations on each of the $p = 4$ cost variables were initially separated according to ownership. Summary statistics for each of the $g = 3$ groups are given in the following table.

Group	Number of observations	Sample mean vectors		
$\ell = 1$ (private)	$n_1 = 271$	$\bar{\mathbf{x}}_1 = \begin{bmatrix} 2.066 \\ .480 \\ .082 \\ .360 \end{bmatrix}$	$\bar{\mathbf{x}}_2 = \begin{bmatrix} 2.167 \\ .596 \\ .124 \\ .418 \end{bmatrix}$	$\bar{\mathbf{x}}_3 = \begin{bmatrix} 2.273 \\ .521 \\ .125 \\ .383 \end{bmatrix}$
$\ell = 2$ (nonprofit)	$n_2 = 138$			
$\ell = 3$ (government)	$n_3 = 107$			
	$\sum_{\ell=1}^{3} n_\ell = 516$			

Sample covariance matrices

$$
S_1 = \begin{bmatrix} .291 \\ -.001 & .011 \\ .002 & .000 & .001 \\ .010 & .003 & .000 & .010 \end{bmatrix} ; \qquad
S_2 = \begin{bmatrix} .561 \\ .011 & .025 \\ .001 & .004 & .005 \\ .037 & .007 & .002 & .019 \end{bmatrix} ;
$$

$$
S_3 = \begin{bmatrix} .261 \\ .030 & .017 \\ .003 & -.000 & .004 \\ .018 & .006 & .001 & .013 \end{bmatrix}
$$

Source: Data courtesy of State of Wisconsin Department of Health and Social Services.

Since the S_ℓ's seem to be reasonably compatible,[3] they were pooled [see (6-41)] to obtain

$$
W = (n_1 - 1)S_1 + (n_2 - 1)S_2 + (n_3 - 1)S_3
$$

$$
= \begin{bmatrix} 182.962 \\ 4.408 & 8.200 \\ 1.695 & .633 & 1.484 \\ 9.581 & 2.428 & .394 & 6.538 \end{bmatrix}
$$

Also,

$$
\bar{x} = \frac{n_1\bar{x}_1 + n_2\bar{x}_2 + n_3\bar{x}_3}{n_1 + n_2 + n_3} = \begin{bmatrix} 2.136 \\ .519 \\ .102 \\ .380 \end{bmatrix}
$$

and

$$
B = \sum_{\ell=1}^{3} n_\ell(\bar{x}_\ell - \bar{x})(\bar{x}_\ell - \bar{x})' = \begin{bmatrix} 3.475 \\ 1.111 & 1.225 \\ .821 & .453 & .235 \\ .584 & .610 & .230 & .304 \end{bmatrix}
$$

To test $H_0: \tau_1 = \tau_2 = \tau_3$ (no ownership effects or, equivalently, no difference in average costs among the three types of owners—private, nonprofit, and government), we can use the result in Table 6.3 for $g = 3$.

Computer-based calculations give

$$
\Lambda^* = \frac{|W|}{|B + W|} = .7714
$$

---

[3]However, a normal-theory test of $H_0: \Sigma_1 = \Sigma_2 = \Sigma_3$ would reject H_0 at any reasonable significance level because of the large sample sizes (see Example 6.12).

and

$$\left(\frac{\Sigma n_\ell - p - 2}{p}\right)\left(\frac{1 - \sqrt{\Lambda^*}}{\sqrt{\Lambda^*}}\right) = \left(\frac{516 - 4 - 2}{4}\right)\left(\frac{1 - \sqrt{.7714}}{\sqrt{.7714}}\right) = 17.67$$

Let $\alpha = .01$, so that $F_{2(4),2(510)}(.01) \doteq \chi_8^2(.01)/8 = 2.51$. Since $17.67 > F_{8,1020}(.01) \doteq 2.51$, we reject H_0 at the 1% level and conclude that average costs differ, depending on type of ownership.

It is informative to compare the results based on this "exact" test with those obtained using the large-sample procedure summarized in (6-43) and (6-44). For the present example, $\Sigma n_\ell = n = 516$ is large, and H_0 can be tested at the $\alpha = .01$ level by comparing

$$-(n - 1 - (p + g)/2)\ln\left(\frac{|\mathbf{W}|}{|\mathbf{B} + \mathbf{W}|}\right) = -511.5\ln(.7714) = 132.76$$

with $\chi_{p(g-1)}^2(.01) = \chi_8^2(.01) = 20.09$. Since $132.76 > \chi_8^2(.01) = 20.09$, we reject H_0 at the 1% level. This result is consistent with the result based on the foregoing F-statistic. ∎

6.5 Simultaneous Confidence Intervals for Treatment Effects

When the hypothesis of equal treatment effects is rejected, those effects that led to the rejection of the hypothesis are of interest. For pairwise comparisons, the Bonferroni approach (see Section 5.4) can be used to construct simultaneous confidence intervals for the components of the differences $\boldsymbol{\tau}_k - \boldsymbol{\tau}_\ell$ (or $\boldsymbol{\mu}_k - \boldsymbol{\mu}_\ell$). These intervals are shorter than those obtained for all contrasts, and they require critical values only for the univariate t-statistic.

Let τ_{ki} be the ith component of $\boldsymbol{\tau}_k$. Since $\boldsymbol{\tau}_k$ is estimated by $\hat{\boldsymbol{\tau}}_k = \bar{\mathbf{x}}_k - \bar{\mathbf{x}}$

$$\hat{\tau}_{ki} = \bar{x}_{ki} - \bar{x}_i \tag{6-45}$$

and $\hat{\tau}_{ki} - \hat{\tau}_{\ell i} = \bar{x}_{ki} - \bar{x}_{\ell i}$ is the difference between two independent sample means. The two-sample t-based confidence interval is valid with an appropriately modified α. Notice that

$$\text{Var}(\hat{\tau}_{ki} - \hat{\tau}_{\ell i}) = \text{Var}(\bar{X}_{ki} - \bar{X}_{\ell i}) = \left(\frac{1}{n_k} + \frac{1}{n_\ell}\right)\sigma_{ii}$$

where σ_{ii} is the ith diagonal element of $\boldsymbol{\Sigma}$. As suggested by (6-41), $\text{Var}(\bar{X}_{ki} - \bar{X}_{\ell i})$ is estimated by dividing the corresponding element of $\mathbf{W}$ by its degrees of freedom. That is,

$$\widehat{\text{Var}}(\bar{X}_{ki} - \bar{X}_{\ell i}) = \left(\frac{1}{n_k} + \frac{1}{n_\ell}\right)\frac{w_{ii}}{n - g}$$

where w_{ii} is the ith diagonal element of $\mathbf{W}$ and $n = n_1 + \cdots + n_g$.

It remains to apportion the error rate over the numerous confidence statements. Relation (5-28) still applies. There are p variables and $g(g-1)/2$ pairwise differences, so each two-sample t-interval will employ the critical value $t_{n-g}(\alpha/2m)$, where

$$m = pg(g-1)/2 \tag{6-46}$$

is the number of simultaneous confidence statements.

Result 6.5. Let $n = \sum\limits_{k=1}^{g} n_k$. For the model in (6-38), with confidence at least $(1 - \alpha)$,

$$\tau_{ki} - \tau_{\ell i} \text{ belongs to } \bar{x}_{ki} - \bar{x}_{\ell i} \pm t_{n-g}\left(\frac{\alpha}{pg(g-1)}\right)\sqrt{\frac{w_{ii}}{n-g}\left(\frac{1}{n_k} + \frac{1}{n_\ell}\right)}$$

for all components $i = 1, \ldots, p$ and all differences $\ell < k = 1, \ldots, g$. Here w_{ii} is the ith diagonal element of $\mathbf{W}$.

We shall illustrate the construction of simultaneous interval estimates for the pairwise differences in treatment means using the nursing-home data introduced in Example 6.10.

Example 6.11 (Simultaneous intervals for treatment differences—nursing homes)
We saw in Example 6.10 that average costs for nursing homes differ, depending on the type of ownership. We can use Result 6.5 to estimate the magnitudes of the differences. A comparison of the variable X_3, costs of plant operation and maintenance labor, between privately owned nursing homes and government-owned nursing homes can be made by estimating $\tau_{13} - \tau_{33}$. Using (6-39) and the information in Example 6.10, we have

$$\hat{\boldsymbol{\tau}}_1 = (\bar{\mathbf{x}}_1 - \bar{\mathbf{x}}) = \begin{bmatrix} -.070 \\ -.039 \\ -.020 \\ -.020 \end{bmatrix}, \quad \hat{\boldsymbol{\tau}}_3 = (\bar{\mathbf{x}}_3 - \bar{\mathbf{x}}) = \begin{bmatrix} .137 \\ .002 \\ .023 \\ .003 \end{bmatrix}$$

$$\mathbf{W} = \begin{bmatrix} 182.962 & & & \\ 4.408 & 8.200 & & \\ 1.695 & .633 & 1.484 & \\ 9.581 & 2.428 & .394 & 6.538 \end{bmatrix}$$

Consequently,

$$\hat{\tau}_{13} - \hat{\tau}_{33} = -.020 - .023 = -.043$$

and $n = 271 + 138 + 107 = 516$, so that

$$\sqrt{\left(\frac{1}{n_1} + \frac{1}{n_3}\right)\frac{w_{33}}{n-g}} = \sqrt{\left(\frac{1}{271} + \frac{1}{107}\right)\frac{1.484}{516-3}} = .00614$$

Since $p = 4$ and $g = 3$, for 95% simultaneous confidence statements we require that $t_{513}(.05/4(3)2) \doteq 2.87$. (See Appendix, Table 1.) The 95% simultaneous confidence statement is

$$\tau_{13} - \tau_{33} \text{ belongs to } \quad \hat{\tau}_{13} - \hat{\tau}_{33} \pm t_{513}(.00208) \sqrt{\left(\frac{1}{n_1} + \frac{1}{n_3}\right) \frac{w_{33}}{n - g}}$$

$$= -.043 \pm 2.87(.00614)$$

$$= -.043 \pm .018, \text{ or } (-.061, -.025)$$

We conclude that the average maintenance and labor cost for government-owned nursing homes is higher by .025 to .061 hour per patient day than for privately owned nursing homes. With the same 95% confidence, we can say that

$$\tau_{13} - \tau_{23} \text{ belongs to the interval } (-.058, -.026)$$

and

$$\tau_{23} - \tau_{33} \text{ belongs to the interval } (-.021, .019)$$

Thus, a difference in this cost exists between private and nonprofit nursing homes, but no difference is observed between nonprofit and government nursing homes. ∎

6.6 Testing for Equality of Covariance Matrices

One of the assumptions made when comparing two or more multivariate mean vectors is that the covariance matrices of the potentially different populations are the same. (This assumption will appear again in Chapter 11 when we discuss discrimination and classification.) Before pooling the variation across samples to form a pooled covariance matrix when comparing mean vectors, it can be worthwhile to test the equality of the population covariance matrices. One commonly employed test for equal covariance matrices is Box's M-test ([8], [9]).

With g populations, the null hypothesis is

$$H_0 : \Sigma_1 = \Sigma_2 = \cdots = \Sigma_g = \Sigma \tag{6-47}$$

where Σ_ℓ is the covariance matrix for the ℓth population, $\ell = 1, 2, \ldots, g$, and Σ is the presumed common covariance matrix. The alternative hypothesis is that at least two of the covariance matrices are not equal.

Assuming multivariate normal populations, a likelihood ratio statistic for testing (6–47) is given by (see [1])

$$\Lambda = \prod_\ell \left(\frac{|S_\ell|}{|S_{\text{pooled}}|}\right)^{(n_\ell - 1)/2} \tag{6-48}$$

Here n_ℓ is the sample size for the ℓth group, S_ℓ is the ℓth group sample covariance matrix and S_{pooled} is the pooled sample covariance matrix given by

$$S_{\text{pooled}} = \frac{1}{\sum_\ell (n_\ell - 1)} \{(n_1 - 1)S_1 + (n_2 - 1)S_2 + \cdots + (n_g - 1)S_g\} \tag{6-49}$$

Box's test is based on his χ^2 approximation to the sampling distribution of $-2 \ln \Lambda$ (see Result 5.2). Setting $-2 \ln \Lambda = M$ (Box's M statistic) gives

$$M = \left[\sum_{\ell}(n_\ell - 1)\right]\ln|\mathbf{S}_{\text{pooled}}| - \sum_{\ell}[(n_\ell - 1)\ln|\mathbf{S}_\ell|] \qquad (6\text{-}50)$$

If the null hypothesis is true, the individual sample covariance matrices are not expected to differ too much and, consequently, do not differ too much from the pooled covariance matrix. In this case, the ratio of the determinants in (6-48) will all be close to 1, Λ will be near 1 and Box's M statistic will be small. If the null hypothesis is false, the sample covariance matrices can differ more and the differences in their determinants will be more pronounced. In this case Λ will be small and M will be relatively large. To illustrate, note that the determinant of the pooled covariance matrix, $|\mathbf{S}_{\text{pooled}}|$, will lie somewhere near the "middle" of the determinants $|\mathbf{S}_\ell|$'s of the individual group covariance matrices. As the latter quantities become more disparate, the product of the ratios in (6-44) will get closer to 0. In fact, as the $|\mathbf{S}_\ell|$'s increase in spread, $|\mathbf{S}_{(1)}|/|\mathbf{S}_{\text{pooled}}|$ reduces the product proportionally more than $|\mathbf{S}_{(g)}|/|\mathbf{S}_{\text{pooled}}|$ increases it, where $|\mathbf{S}_{(1)}|$ and $|\mathbf{S}_{(g)}|$ are the minimum and maximum determinant values, respectively.

Box's Test for Equality of Covariance Matrices

Set

$$u = \left[\sum_\ell \frac{1}{(n_\ell - 1)} - \frac{1}{\sum_\ell(n_\ell - 1)}\right]\left[\frac{2p^2 + 3p - 1}{6(p+1)(g-1)}\right] \qquad (6\text{-}51)$$

where p is the number of variables and g is the number of groups. Then

$$C = (1-u)M = (1-u)\left\{\left[\sum_\ell(n_\ell-1)\right]\ln|\mathbf{S}_{\text{pooled}}| - \sum_\ell[(n_\ell-1)\ln|\mathbf{S}_\ell|]\right\} (6\text{-}52)$$

has an approximate χ^2 distribution with

$$v = g\frac{1}{2}p(p+1) - \frac{1}{2}p(p+1) = \frac{1}{2}p(p+1)(g-1) \qquad (6\text{-}53)$$

degrees of freedom. At significance level α, reject H_0 if $C > \chi^2_{p(p+1)(g-1)/2}(\alpha)$.

Box's χ^2 approximation works well if each n_ℓ exceeds 20 and if p and g do not exceed 5. In situations where these conditions do not hold, Box ([7], [8]) has provided a more precise F approximation to the sampling distribution of M.

Example 6.12 (Testing equality of covariance matrices—nursing homes) We introduced the Wisconsin nursing home data in Example 6.10. In that example the sample covariance matrices for $p = 4$ cost variables associated with $g = 3$ groups of nursing homes are displayed. Assuming multivariate normal data, we test the hypothesis $H_0: \Sigma_1 = \Sigma_2 = \Sigma_3 = \Sigma$.

Using the information in Example 6.10, we have $n_1 = 271$, $n_2 = 138$, $n_3 = 107$ and $|S_1| = 2.783 \times 10^{-8}$, $|S_2| = 89.539 \times 10^{-8}$, $|S_3| = 14.579 \times 10^{-8}$, and $|S_{pooled}| = 17.398 \times 10^{-8}$. Taking the natural logarithms of the determinants gives $\ln|S_1| = -17.397$, $\ln|S_2| = -13.926$, $\ln|S_3| = -15.741$ and $\ln|S_{pooled}| = -15.564$. We calculate

$$u = \left[\frac{1}{270} + \frac{1}{137} + \frac{1}{106} - \frac{1}{270 + 137 + 106} \right]\left[\frac{2(4^2) + 3(4) - 1}{6(4 + 1)(3 - 1)} \right] = .0133$$

$$M = [270 + 137 + 106](-15.564) - [270(-17.397) + 137(-13.926) + 106(-15.741)]$$

$$= 289.3$$

and $C = (1 - .0133)289.3 = 285.5$. Referring C to a χ^2 table with $v = 4(4 + 1)(3 - 1)/2 = 20$ degrees of freedom, it is clear that H_0 is rejected at any reasonable level of significance. We conclude that the covariance matrices of the cost variables associated with the three populations of nursing homes are not the same. ∎

Box's M-test is routinely calculated in many statistical computer packages that do MANOVA and other procedures requiring equal covariance matrices. It is known that the M-test is sensitive to some forms of non-normality. More broadly, in the presence of non-normality, normal theory tests on covariances are influenced by the kurtosis of the parent populations (see [16]). However, with reasonably large samples, the MANOVA tests of means or treatment effects are rather robust to nonnormality. Thus the M-test may reject H_0 in some non-normal cases where it is not damaging to the MANOVA tests. Moreover, with equal sample sizes, some differences in covariance matrices have little effect on the MANOVA tests. To summarize, we may decide to continue with the usual MANOVA tests even though the M-test leads to rejection of H_0.

6.7 Two-Way Multivariate Analysis of Variance

Following our approach to the one-way MANOVA, we shall briefly review the analysis for a *univariate* two-way fixed-effects model and then simply generalize to the multivariate case by analogy.

Univariate Two-Way Fixed-Effects Model with Interaction

We assume that measurements are recorded at various levels of two factors. In some cases, these experimental conditions represent levels of a single treatment arranged within several blocks. The particular experimental design employed will not concern us in this book. (See [10] and [17] for discussions of experimental design.) We shall, however, assume that observations at different combinations of experimental conditions are independent of one another.

Let the two sets of experimental conditions be the levels of, for instance, factor 1 and factor 2, respectively.[4] Suppose there are g levels of factor 1 and b levels of factor 2, and that n independent observations can be observed at each of the gb combi-

---

[4]The use of the term "factor" to indicate an experimental condition is convenient. The factors discussed here should not be confused with the unobservable factors considered in Chapter 9 in the context of factor analysis.

nations of levels. Denoting the rth observation at level ℓ of factor 1 and level k of factor 2 by $X_{\ell k r}$, we specify the univariate two-way model as

$$X_{\ell k r} = \mu + \tau_\ell + \beta_k + \gamma_{\ell k} + e_{\ell k r}$$
$$\ell = 1, 2, \ldots, g$$
$$k = 1, 2, \ldots, b \qquad (6\text{-}54)$$
$$r = 1, 2, \ldots, n$$

where $\sum\limits_{\ell=1}^{g} \tau_\ell = \sum\limits_{k=1}^{b} \beta_k = \sum\limits_{\ell=1}^{g} \gamma_{\ell k} = \sum\limits_{k=1}^{b} \gamma_{\ell k} = 0$ and the $e_{\ell k r}$ are independent $N(0, \sigma^2)$ random variables. Here μ represents an overall level, τ_ℓ represents the fixed effect of factor 1, β_k represents the fixed effect of factor 2, and $\gamma_{\ell k}$ is the interaction between factor 1 and factor 2. The expected response at the ℓth level of factor 1 and the kth level of factor 2 is thus

$$E(X_{\ell k r}) \quad = \quad \mu \quad + \quad \tau_\ell \quad + \quad \beta_k \quad + \quad \gamma_{\ell k}$$

$$\begin{pmatrix} \text{mean} \\ \text{response} \end{pmatrix} = \begin{pmatrix} \text{overall} \\ \text{level} \end{pmatrix} + \begin{pmatrix} \text{effect of} \\ \text{factor 1} \end{pmatrix} + \begin{pmatrix} \text{effect of} \\ \text{factor 2} \end{pmatrix} + \begin{pmatrix} \text{factor 1-factor 2} \\ \text{interaction} \end{pmatrix}$$

$$\ell = 1, 2, \ldots, g, \qquad k = 1, 2, \ldots, b \qquad (6\text{-}55)$$

The presence of interaction, $\gamma_{\ell k}$, implies that the factor effects are not additive and complicates the interpretation of the results. Figures 6.3(a) and (b) show

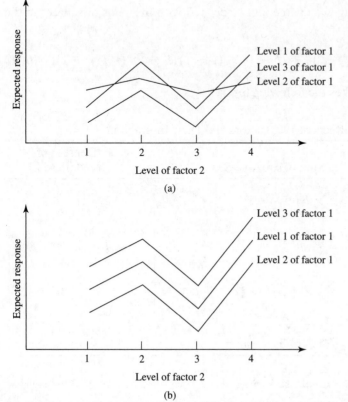

Level 1 of factor 1
Level 3 of factor 1
Level 2 of factor 1

Level of factor 2

(a)

Level 3 of factor 1
Level 1 of factor 1
Level 2 of factor 1

Level of factor 2

(b)

Figure 6.3 Curves for expected responses (a) with interaction and (b) without interaction.

expected responses as a function of the factor levels with and without interaction, respectively. The absense of interaction means $\gamma_{\ell k} = 0$ for all ℓ and k.

In a manner analogous to (6-55), each observation can be decomposed as

$$x_{\ell kr} = \bar{x} + (\bar{x}_{\ell\cdot} - \bar{x}) + (\bar{x}_{\cdot k} - \bar{x}) + (\bar{x}_{\ell k} - \bar{x}_{\ell\cdot} - \bar{x}_{\cdot k} + \bar{x}) + (x_{\ell kr} - \bar{x}_{\ell k}) \qquad (6\text{-}56)$$

where $\bar{x}$ is the overall average, $\bar{x}_{\ell\cdot}$ is the average for the ℓth level of factor 1, $\bar{x}_{\cdot k}$ is the average for the kth level of factor 2, and $\bar{x}_{\ell k}$ is the average for the ℓth level of factor 1 *and* the kth level of factor 2. Squaring and summing the deviations $(x_{\ell kr} - \bar{x})$ gives

$$\sum_{\ell=1}^{g} \sum_{k=1}^{b} \sum_{r=1}^{n} (x_{\ell kr} - \bar{x})^2 = \sum_{\ell=1}^{g} bn(\bar{x}_{\ell\cdot} - \bar{x})^2 + \sum_{k=1}^{b} gn(\bar{x}_{\cdot k} - \bar{x})^2$$

$$+ \sum_{\ell=1}^{g} \sum_{k=1}^{b} n(\bar{x}_{\ell k} - \bar{x}_{\ell\cdot} - \bar{x}_{\cdot k} + \bar{x})^2$$

$$+ \sum_{\ell=1}^{g} \sum_{k=1}^{b} \sum_{r=1}^{n} (x_{\ell kr} - \bar{x}_{\ell k})^2 \qquad (6\text{-}57)$$

or

$$SS_{cor} = SS_{fac1} + SS_{fac2} + SS_{int} + SS_{res}$$

The corresponding degrees of freedom associated with the sums of squares in the breakup in (6-57) are

$$gbn - 1 = (g - 1) + (b - 1) + (g - 1)(b - 1) + gb(n - 1) \qquad (6\text{-}58)$$

The ANOVA table takes the following form:

ANOVA Table for Comparing Effects of Two Factors and Their Interaction

Source of variation	Sum of squares (SS)	Degrees of freedom (d.f.)
Factor 1	$SS_{fac1} = \sum_{\ell=1}^{g} bn(\bar{x}_{\ell\cdot} - \bar{x})^2$	$g - 1$
Factor 2	$SS_{fac2} = \sum_{k=1}^{b} gn(\bar{x}_{\cdot k} - \bar{x})^2$	$b - 1$
Interaction	$SS_{int} = \sum_{\ell=1}^{g} \sum_{k=1}^{b} n(\bar{x}_{\ell k} - \bar{x}_{\ell\cdot} - \bar{x}_{\cdot k} + \bar{x})^2$	$(g - 1)(b - 1)$
Residual (Error)	$SS_{res} = \sum_{\ell=1}^{g} \sum_{k=1}^{b} \sum_{r=1}^{n} (x_{\ell kr} - \bar{x}_{\ell k})^2$	$gb(n - 1)$
Total (corrected)	$SS_{cor} = \sum_{\ell=1}^{g} \sum_{k=1}^{b} \sum_{r=1}^{n} (x_{\ell kr} - \bar{x})^2$	$gbn - 1$

The F-ratios of the mean squares, $SS_{fac1}/(g - 1)$, $SS_{fac2}/(b - 1)$, and $SS_{int}/(g - 1)(b - 1)$ to the mean square, $SS_{res}/(gb(n - 1))$ can be used to test for the effects of factor 1, factor 2, and factor 1–factor 2 interaction, respectively. (See [11] for a discussion of univariate two-way analysis of variance.)

Multivariate Two-Way Fixed-Effects Model with Interaction

Proceeding by analogy, we specify the two-way fixed-effects model for a *vector* response consisting of p components [see (6-54)]

$$\mathbf{X}_{\ell kr} = \boldsymbol{\mu} + \boldsymbol{\tau}_\ell + \boldsymbol{\beta}_k + \boldsymbol{\gamma}_{\ell k} + \mathbf{e}_{\ell kr}$$

$$\ell = 1, 2, \ldots, g$$

$$k = 1, 2, \ldots, b \qquad (6\text{-}59)$$

$$r = 1, 2, \ldots, n$$

where $\sum\limits_{\ell=1}^{g} \boldsymbol{\tau}_\ell = \sum\limits_{k=1}^{b} \boldsymbol{\beta}_k = \sum\limits_{\ell=1}^{g} \boldsymbol{\gamma}_{\ell k} = \sum\limits_{k=1}^{b} \boldsymbol{\gamma}_{\ell k} = \mathbf{0}$. The vectors are all of order $p \times 1$, and the $\mathbf{e}_{\ell kr}$ are independent $N_p(\mathbf{0}, \boldsymbol{\Sigma})$ random vectors. Thus, the responses consist of p measurements replicated n times at each of the possible combinations of levels of factors 1 and 2.

Following (6-56), we can decompose the observation vectors $\mathbf{x}_{\ell kr}$ as

$$\mathbf{x}_{\ell kr} = \bar{\mathbf{x}} + (\bar{\mathbf{x}}_{\ell \cdot} - \bar{\mathbf{x}}) + (\bar{\mathbf{x}}_{\cdot k} - \bar{\mathbf{x}}) + (\bar{\mathbf{x}}_{\ell k} - \bar{\mathbf{x}}_{\ell \cdot} - \bar{\mathbf{x}}_{\cdot k} + \bar{\mathbf{x}}) + (\mathbf{x}_{\ell kr} - \bar{\mathbf{x}}_{\ell k}) \quad (6\text{-}60)$$

where $\bar{\mathbf{x}}$ is the overall average of the observation vectors, $\bar{\mathbf{x}}_{\ell \cdot}$ is the average of the observation vectors at the ℓth level of factor 1, $\bar{\mathbf{x}}_{\cdot k}$ is the average of the observation vectors at the kth level of factor 2, and $\bar{\mathbf{x}}_{\ell k}$ is the average of the observation vectors at the ℓth level of factor 1 *and* the kth level of factor 2.

Straightforward generalizations of (6-57) and (6-58) give the breakups of the sum of squares and cross products and degrees of freedom:

$$\sum_{\ell=1}^{g} \sum_{k=1}^{b} \sum_{r=1}^{n} (\mathbf{x}_{\ell kr} - \bar{\mathbf{x}})(\mathbf{x}_{\ell kr} - \bar{\mathbf{x}})' = \sum_{\ell=1}^{g} bn(\bar{\mathbf{x}}_{\ell \cdot} - \bar{\mathbf{x}})(\bar{\mathbf{x}}_{\ell \cdot} - \bar{\mathbf{x}})'$$

$$+ \sum_{k=1}^{b} gn(\bar{\mathbf{x}}_{\cdot k} - \bar{\mathbf{x}})(\bar{\mathbf{x}}_{\cdot k} - \bar{\mathbf{x}})'$$

$$+ \sum_{\ell=1}^{g} \sum_{k=1}^{b} n(\bar{\mathbf{x}}_{\ell k} - \bar{\mathbf{x}}_{\ell \cdot} - \bar{\mathbf{x}}_{\cdot k} + \bar{\mathbf{x}})(\bar{\mathbf{x}}_{\ell k} - \bar{\mathbf{x}}_{\ell \cdot} - \bar{\mathbf{x}}_{\cdot k} + \bar{\mathbf{x}})'$$

$$+ \sum_{\ell=1}^{g} \sum_{k=1}^{b} \sum_{r=1}^{n} (\mathbf{x}_{\ell kr} - \bar{\mathbf{x}}_{\ell k})(\mathbf{x}_{\ell kr} - \bar{\mathbf{x}}_{\ell k})' \qquad (6\text{-}61)$$

$$gbn - 1 = (g - 1) + (b - 1) + (g - 1)(b - 1) + gb(n - 1) \qquad (6\text{-}62)$$

Again, the generalization from the univariate to the multivariate analysis consists simply of replacing a scalar such as $(\bar{x}_{\ell \cdot} - \bar{x})^2$ with the corresponding matrix $(\bar{\mathbf{x}}_{\ell \cdot} - \bar{\mathbf{x}})(\bar{\mathbf{x}}_{\ell \cdot} - \bar{\mathbf{x}})'$.

The MANOVA table is the following:

MANOVA Table for Comparing Factors and Their Interaction

Source of variation	Matrix of sum of squares and cross products (SSP)	Degrees of freedom (d.f.)
Factor 1	$\text{SSP}_{\text{fac1}} = \sum_{\ell=1}^{g} bn(\bar{\mathbf{x}}_{\ell \cdot} - \bar{\mathbf{x}})(\bar{\mathbf{x}}_{\ell \cdot} - \bar{\mathbf{x}})'$	$g - 1$
Factor 2	$\text{SSP}_{\text{fac2}} = \sum_{k=1}^{b} gn(\bar{\mathbf{x}}_{\cdot k} - \bar{\mathbf{x}})(\bar{\mathbf{x}}_{\cdot k} - \bar{\mathbf{x}})'$	$b - 1$
Interaction	$\text{SSP}_{\text{int}} = \sum_{\ell=1}^{g} \sum_{k=1}^{b} n(\bar{\mathbf{x}}_{\ell k} - \bar{\mathbf{x}}_{\ell \cdot} - \bar{\mathbf{x}}_{\cdot k} + \bar{\mathbf{x}})(\bar{\mathbf{x}}_{\ell k} - \bar{\mathbf{x}}_{\ell \cdot} - \bar{\mathbf{x}}_{\cdot k} + \bar{\mathbf{x}})'$	$(g-1)(b-1)$
Residual (Error)	$\text{SSP}_{\text{res}} = \sum_{\ell=1}^{g} \sum_{k=1}^{b} \sum_{r=1}^{n} (\mathbf{x}_{\ell k r} - \bar{\mathbf{x}}_{\ell k})(\mathbf{x}_{\ell k r} - \bar{\mathbf{x}}_{\ell k})'$	$gb(n-1)$
Total (corrected)	$\text{SSP}_{\text{cor}} = \sum_{\ell=1}^{g} \sum_{k=1}^{b} \sum_{r=1}^{n} (\mathbf{x}_{\ell k r} - \bar{\mathbf{x}})(\mathbf{x}_{\ell k r} - \bar{\mathbf{x}})'$	$gbn - 1$

A test (the likelihood ratio test)[5] of

$$H_0: \boldsymbol{\gamma}_{11} = \boldsymbol{\gamma}_{12} = \cdots = \boldsymbol{\gamma}_{gb} = \mathbf{0} \quad \text{(no interaction effects)} \quad (6\text{-}63)$$

versus

$$H_1: \text{At least one } \boldsymbol{\gamma}_{\ell k} \neq \mathbf{0}$$

is conducted by rejecting H_0 for small values of the ratio

$$\Lambda^* = \frac{|\text{SSP}_{\text{res}}|}{|\text{SSP}_{\text{int}} + \text{SSP}_{\text{res}}|} \quad (6\text{-}64)$$

For large samples, Wilks' lambda, Λ^*, can be referred to a chi-square percentile. Using Bartlett's multiplier (see [6]) to improve the chi-square approximation, we reject $H_0: \boldsymbol{\gamma}_{11} = \boldsymbol{\gamma}_{12} = \cdots = \boldsymbol{\gamma}_{gb} = \mathbf{0}$ at the α level if

$$-\left[gb(n-1) - \frac{p + 1 - (g-1)(b-1)}{2} \right] \ln \Lambda^* > \chi^2_{(g-1)(b-1)p}(\alpha) \quad (6\text{-}65)$$

where Λ^* is given by (6-64) and $\chi^2_{(g-1)(b-1)p}(\alpha)$ is the upper (100α)th percentile of a chi-square distribution with $(g-1)(b-1)p$ d.f.

Ordinarily, the test for interaction is carried out before the tests for main factor effects. If interaction effects exist, the factor effects do not have a clear interpretation. From a practical standpoint, it is not advisable to proceed with the additional multivariate tests. Instead, p univariate two-way analyses of variance (one for each variable) are often conducted to see whether the interaction appears in some responses but not

---

[5]The likelihood test procedures require that $p \leq gb(n-1)$, so that SSP_{res} will be positive definite (with probability 1).

others. Those responses without interaction may be interpreted in terms of additive factor 1 and 2 effects, provided that the latter effects exist. In any event, interaction plots similar to Figure 6.3, but with treatment sample means replacing expected values, best clarify the relative magnitudes of the main and interaction effects.

In the multivariate model, we test for factor 1 and factor 2 main effects as follows. First, consider the hypotheses $H_0: \boldsymbol{\tau}_1 = \boldsymbol{\tau}_2 = \cdots = \boldsymbol{\tau}_g = \mathbf{0}$ and H_1: at least one $\boldsymbol{\tau}_\ell \neq \mathbf{0}$. These hypotheses specify *no* factor 1 effects and *some* factor 1 effects, respectively. Let

$$\Lambda^* = \frac{|\text{SSP}_{\text{res}}|}{|\text{SSP}_{\text{fac1}} + \text{SSP}_{\text{res}}|} \tag{6-66}$$

so that small values of Λ^* are consistent with H_1. Using Bartlett's correction, the likelihood ratio test is as follows:

Reject $H_0: \boldsymbol{\tau}_1 = \boldsymbol{\tau}_2 = \cdots = \boldsymbol{\tau}_g = \mathbf{0}$ (no factor 1 effects) at level α if

$$-\left[gb(n-1) - \frac{p + 1 - (g-1)}{2} \right] \ln \Lambda^* > \chi^2_{(g-1)p}(\alpha) \tag{6-67}$$

where Λ^* is given by (6-66) and $\chi^2_{(g-1)p}(\alpha)$ is the upper (100α)th percentile of a chi-square distribution with $(g-1)p$ d.f.

In a similar manner, factor 2 effects are tested by considering $H_0: \boldsymbol{\beta}_1 = \boldsymbol{\beta}_2 = \cdots = \boldsymbol{\beta}_b = \mathbf{0}$ and H_1: at least one $\boldsymbol{\beta}_k \neq \mathbf{0}$. Small values of

$$\Lambda^* = \frac{|\text{SSP}_{\text{res}}|}{|\text{SSP}_{\text{fac2}} + \text{SSP}_{\text{res}}|} \tag{6-68}$$

are consistent with H_1. Once again, for large samples and using Bartlett's correction: Reject $H_0: \boldsymbol{\beta}_1 = \boldsymbol{\beta}_2 = \cdots = \boldsymbol{\beta}_b = \mathbf{0}$ (no factor 2 effects) at level α if

$$-\left[gb(n-1) - \frac{p + 1 - (b-1)}{2} \right] \ln \Lambda^* > \chi^2_{(b-1)p}(\alpha) \tag{6-69}$$

where Λ^* is given by (6-68) and $\chi^2_{(b-1)p}(\alpha)$ is the upper (100α)th percentile of a chi-square distribution with $(b-1)p$ degrees of freedom.

Simultaneous confidence intervals for contrasts in the model parameters can provide insights into the nature of the factor effects. Results comparable to Result 6.5 are available for the two-way model. When interaction effects are negligible, we may concentrate on contrasts in the factor 1 and factor 2 main effects. The Bonferroni approach applies to the components of the differences $\boldsymbol{\tau}_\ell - \boldsymbol{\tau}_m$ of the factor 1 effects and the components of $\boldsymbol{\beta}_k - \boldsymbol{\beta}_q$ of the factor 2 effects, respectively.

The $100(1 - \alpha)\%$ simultaneous confidence intervals for $\tau_{\ell i} - \tau_{mi}$ are

$$\tau_{\ell i} - \tau_{mi} \quad \text{belongs to} \quad (\bar{x}_{\ell \cdot i} - \bar{x}_{m \cdot i}) \pm t_\nu \left(\frac{\alpha}{pg(g-1)} \right) \sqrt{\frac{E_{ii}}{\nu} \frac{2}{bn}} \tag{6-70}$$

where $\nu = gb(n-1)$, E_{ii} is the ith diagonal element of $\mathbf{E} = \text{SSP}_{\text{res}}$, and $\bar{x}_{\ell \cdot i} - \bar{x}_{m \cdot i}$ is the ith component of $\bar{\mathbf{x}}_{\ell \cdot} - \bar{\mathbf{x}}_{m \cdot \cdot}$.

Similarly, the $100(1 - \alpha)$ percent simultaneous confidence intervals for $\beta_{ki} - \beta_{qi}$ are

$$\beta_{ki} - \beta_{qi} \text{ belongs to }\quad (\bar{x}_{\cdot ki} - \bar{x}_{\cdot qi}) \pm t_\nu \left(\frac{\alpha}{pb(b-1)}\right)\sqrt{\frac{E_{ii}}{\nu}\frac{2}{gn}} \qquad (6\text{-}71)$$

where ν and E_{ii} are as just defined and $\bar{x}_{\cdot ki} - \bar{x}_{\cdot qi}$ is the ith component of $\bar{\mathbf{x}}_{\cdot k} - \bar{\mathbf{x}}_{\cdot q}$.

Comment. We have considered the multivariate two-way model with replications. That is, the model allows for n replications of the responses at each combination of factor levels. This enables us to examine the "interaction" of the factors. If only one observation vector is available at each combination of factor levels, the two-way model does not allow for the possibility of a general interaction term $\gamma_{\ell k}$. The corresponding MANOVA table includes only factor 1, factor 2, and residual sources of variation as components of the total variation. (See Exercise 6.13.)

Example 6.13 (A two-way multivariate analysis of variance of plastic film data) The optimum conditions for extruding plastic film have been examined using a technique called Evolutionary Operation. (See [9].) In the course of the study that was done, three responses—X_1 = tear resistance, X_2 = gloss, and X_3 = opacity—were measured at two levels of the factors, *rate of extrusion* and *amount of an additive*. The measurements were repeated $n = 5$ times at each combination of the factor levels. The data are displayed in Table 6.4.

Table 6.4 Plastic Film Data						
x_1 = tear resistance, x_2 = gloss, and x_3 = opacity						
		Factor 2: Amount of additive				
		Low (1.0%)			High (1.5%)	
		x_1 $\quad$ x_2 $\quad$ x_3			x_1 $\quad$ x_2 $\quad$ x_3	
Factor 1: Change in rate of extrusion	Low (−10)%	[6.5 $\quad$ 9.5 $\quad$ 4.4] [6.2 $\quad$ 9.9 $\quad$ 6.4] [5.8 $\quad$ 9.6 $\quad$ 3.0] [6.5 $\quad$ 9.6 $\quad$ 4.1] [6.5 $\quad$ 9.2 $\quad$ 0.8]			[6.9 $\quad$ 9.1 $\quad$ 5.7] [7.2 $\quad$ 10.0 $\quad$ 2.0] [6.9 $\quad$ 9.9 $\quad$ 3.9] [6.1 $\quad$ 9.5 $\quad$ 1.9] [6.3 $\quad$ 9.4 $\quad$ 5.7]	
	High (10%)	[6.7 $\quad$ 9.1 $\quad$ 2.8] [6.6 $\quad$ 9.3 $\quad$ 4.1] [7.2 $\quad$ 8.3 $\quad$ 3.8] [7.1 $\quad$ 8.4 $\quad$ 1.6] [6.8 $\quad$ 8.5 $\quad$ 3.4]			[7.1 $\quad$ 9.2 $\quad$ 8.4] [7.0 $\quad$ 8.8 $\quad$ 5.2] [7.2 $\quad$ 9.7 $\quad$ 6.9] [7.5 $\quad$ 10.1 $\quad$ 2.7] [7.6 $\quad$ 9.2 $\quad$ 1.9]	

The matrices of the appropriate sum of squares and cross products were calculated (see the SAS statistical software output in Panel 6.1[6]), leading to the following MANOVA table:

---

[6]Additional SAS programs for MANOVA and other procedures discussed in this chapter are available in [13].

Source of variation		SSP			d.f.
Factor 1:	change in rate of extrusion	$\begin{bmatrix} 1.7405 & -1.5045 & .8555 \\ & 1.3005 & -.7395 \\ & & .4205 \end{bmatrix}$			1
Factor 2:	amount of additive	$\begin{bmatrix} .7605 & .6825 & 1.9305 \\ & .6125 & 1.7325 \\ & & 4.9005 \end{bmatrix}$			1
Interaction		$\begin{bmatrix} .0005 & .0165 & .0445 \\ & .5445 & 1.4685 \\ & & 3.9605 \end{bmatrix}$			1
Residual		$\begin{bmatrix} 1.7640 & .0200 & -3.0700 \\ & 2.6280 & -.5520 \\ & & 64.9240 \end{bmatrix}$			16
Total (corrected)		$\begin{bmatrix} 4.2655 & -.7855 & -.2395 \\ & 5.0855 & 1.9095 \\ & & 74.2055 \end{bmatrix}$			19

PANEL 6.1   SAS ANALYSIS FOR EXAMPLE 6.13 USING PROC GLM

```
title 'MANOVA';
data film;
infile 'T6-4.dat';
input x1 x2 x3 factor1 factor2;
proc glm data = film;
class factor1 factor2;
model x1 x2 x3 = factor1 factor2 factor1*factor2 /ss3;
manova h = factor1 factor2 factor1*factor2 /printe;
means factor1 factor2;
```
PROGRAM COMMANDS

General Linear Models Procedure
Class Level Information

Class	Levels	Values		OUTPUT
FACTOR1	2	0 1		
FACTOR2	2	0 1		

Number of observations in data set = 20

Dependent Variable: X1

Source	DF	Sum of Squares	Mean Square	F Value	Pr > F
Model	3	2.50150000	0.83383333	7.56	0.0023
Error	16	1.76400000	0.11025000		
Corrected Total	19	4.26550000			

R-Square	C.V.	Root MSE		X1 Mean
0.586449	4.893724	0.332039		6.78500000

Source	DF	Type III SS	Mean Square	F Value	Pr > F
FACTOR1	1	1.74050000	1.74050000	15.79	0.0011
FACTOR2	1	0.76050000	0.76050000	6.90	0.0183
FACTOR1*FACTOR2	1	0.00050000	0.00050000	0.00	0.9471

(continues on next page)

PANEL 6.1   (continued)

Dependent Variable: X2

Source	DF	Sum of Squares	Mean Square	F Value	Pr > F
Model	3	2.45750000	0.81916667	4.99	0.0125
Error	16	2.62800000	0.16425000		
Corrected Total	19	5.08550000			

	R-Square	C.V.	Root MSE		X2 Mean
	0.483237	4.350807	0.405278		9.31500000

Source	DF	Type III SS	Mean Square	F Value	Pr > F
FACTOR1	1	1.30050000	1.30050000	7.92	0.0125
FACTOR2	1	0.61250000	0.61250000	3.73	0.0714
FACTOR1*FACTOR2	1	0.54450000	0.54450000	3.32	0.0874

Dependent Variable: X3

Source	DF	Sum of Squares	Mean Square	F Value	Pr > F
Model	3	9.28150000	3.09383333	0.76	0.5315
Error	16	64.92400000	4.05775000		
Corrected Total	19	74.20550000			

	R-Square	C.V.	Root MSE		X3 Mean
	0.125078	51.19151	2.014386		3.93500000

Source	DF	Type III SS	Mean Square	F Value	Pr > F
FACTOR	1	0.42050000	0.42050000	0.10	0.7517
FACTOR2	1	4.90050000	4.90050000	1.21	0.2881
FACTOR1*FACTOR2	1	3.96050000	3.96050000	0.98	0.3379

E = Error SS&CP Matrix

	X1	X2	X3
X1	1.764	0.02	−3.07
X2	0.02	2.628	−0.552
X3	−3.07	−0.552	64.924

Manova Test Criteria and Exact F Statistics for

the   **Hypothesis of no Overall FACTOR1 Effect**

H = Type III SS&CP Matrix for FACTOR1   E = Error SS&CP Matrix
S = 1   M = 0.5   N = 6

Statistic	Value	F	Num DF	Den DF	Pr > F
Wilks' Lambda	0.38185838	7.5543	3	14	0.0030
Pillai's Trace	0.61814162	7.5543	3	14	0.0030
Hotelling–Lawley Trace	1.61877188	7.5543	3	14	0.0030
Roy's Greatest Root	1.61877188	7.5543	3	14	0.0030

(continues on next page)

PANEL 6.1   *(continued)*

Manova Test Criteria and Exact F Statistics for

the | **Hypothesis of no Overall FACTOR2 Effect**

H = Type III SS&CP Matrix for FACTOR2   E = Error SS&CP Matrix
S = 1   M = 0.5   N = 6

Statistic	Value	F	Num DF	Den DF	Pr > F
Wilks' Lambda	**0.52303490**	**4.2556**	**3**	**14**	**0.0247**
Pillai's Trace	0.47696510	4.2556	3	14	0.0247
Hotelling–Lawley Trace	0.91191832	4.2556	3	14	0.0247
Roy's Greatest Root	0.91191832	4.2556	3	14	0.0247

Manova Test Criteria and Exact F Statistics for

the | **Hypothesis of no Overall FACTOR1*FACTOR2 Effect**

H = Type III SS&CP Matrix for FACTOR1*FACTOR2   E = Error SS&CP Matrix
S = 1   M = 0.5   N = 6

Statistic	Value	F	Num DF	Den DF	Pr > F
Wilks' Lambda	**0.77710576**	**1.3385**	**3**	**14**	**0.3018**
Pillai's Trace	0.22289424	1.3385	3	14	0.3018
Hotelling–Lawley Trace	0.28682614	1.3385	3	14	0.3018
Roy's Greatest Root	0.28682614	1.3385	3	14	0.3018

Level of FACTOR1	N	――――X1―――― Mean	SD	――――X2―――― Mean	SD
0	10	6.49000000	0.42018514	9.57000000	0.29832868
1	10	7.08000000	0.32249031	9.06000000	0.57580861

Level of FACTOR1	N	――――X3―――― Mean	SD
0	10	3.79000000	1.85379491
1	10	4.08000000	2.18214981

Level of FACTOR2	N	――――X1―――― Mean	SD	――――X2―――― Mean	SD
0	10	6.59000000	0.40674863	9.14000000	0.56015871
1	10	6.98000000	0.47328638	9.49000000	0.42804465

Level of FACTOR2	N	――――X3―――― Mean	SD
0	10	3.44000000	1.55077042
1	10	4.43000000	2.30123155

To test for interaction, we compute

$$\Lambda^* = \frac{|SSP_{res}|}{|SSP_{int} + SSP_{res}|} = \frac{275.7098}{354.7906} = .7771$$

For $(g - 1)(b - 1) = 1$,

$$F = \left(\frac{1 - \Lambda^*}{\Lambda^*}\right) \frac{(gb(n - 1) - p + 1)/2}{(|(g - 1)(b - 1) - p| + 1)/2}$$

has an exact F-distribution with $\nu_1 = |(g - 1)(b - 1) - p| + 1$ and $\nu_2 = gb(n - 1) - p + 1$ d.f. (See [1].) For our example.

$$F = \left(\frac{1 - .7771}{.7771}\right) \frac{(2(2)(4) - 3 + 1)/2}{(|1(1) - 3| + 1)/2} = 1.34$$

$$\nu_1 = (|1(1) - 3| + 1) = 3$$

$$\nu_2 = (2(2)(4) - 3 + 1) = 14$$

and $F_{3,14}(.05) = 3.34$. Since $F = 1.34 < F_{3,14}(.05) = 3.34$, we do not reject the hypothesis $H_0: \boldsymbol{\gamma}_{11} = \boldsymbol{\gamma}_{12} = \boldsymbol{\gamma}_{21} = \boldsymbol{\gamma}_{22} = \mathbf{0}$ (no interaction effects).

Note that the approximate chi-square statistic for this test is $-[2(2)(4) - (3 + 1 - 1(1))/2] \ln(.7771) = 3.66$, from (6-65). Since $\chi_3^2(.05) = 7.81$, we would reach the same conclusion as provided by the exact F-test.

To test for factor 1 and factor 2 effects (see page 317), we calculate

$$\Lambda_1^* = \frac{|\text{SSP}_{\text{res}}|}{|\text{SSP}_{\text{fac1}} + \text{SSP}_{\text{res}}|} = \frac{275.7098}{722.0212} = .3819$$

and

$$\Lambda_2^* = \frac{|\text{SSP}_{\text{res}}|}{|\text{SSP}_{\text{fac2}} + \text{SSP}_{\text{res}}|} = \frac{275.7098}{527.1347} = .5230$$

For both $g - 1 = 1$ and $b - 1 = 1$,

$$F_1 = \left(\frac{1 - \Lambda_1^*}{\Lambda_1^*}\right) \frac{(gb(n - 1) - p + 1)/2}{(|(g - 1) - p| + 1)/2}$$

and

$$F_2 = \left(\frac{1 - \Lambda_2^*}{\Lambda_2^*}\right) \frac{(gb(n - 1) - p + 1)/2}{(|(b - 1) - p| + 1)/2}$$

have F-distributions with degrees of freedom $\nu_1 = |(g - 1) - p| + 1$, $\nu_2 = gb(n - 1) - p + 1$ and $\nu_1 = |(b - 1) - p| + 1$, $\nu_2 = gb(n - 1) - p + 1$, respectively. (See [1].) In our case,

$$F_1 = \left(\frac{1 - .3819}{.3819}\right) \frac{(16 - 3 + 1)/2}{(|1 - 3| + 1)/2} = 7.55$$

$$F_2 = \left(\frac{1 - .5230}{.5230}\right) \frac{(16 - 3 + 1)/2}{(|1 - 3| + 1)/2} = 4.26$$

and

$$\nu_1 = |1 - 3| + 1 = 3 \qquad \nu_2 = (16 - 3 + 1) = 14$$

From before, $F_{3,14}(.05) = 3.34$. We have $F_1 = 7.55 > F_{3,14}(.05) = 3.34$, and therefore, we reject $H_0: \tau_1 = \tau_2 = 0$ (no factor 1 effects) at the 5% level. Similarly, $F_2 = 4.26 > F_{3,14}(.05) = 3.34$, and we reject $H_0: \beta_1 = \beta_2 = 0$ (no factor 2 effects) at the 5% level. We conclude that both the *change in rate of extrusion* and the *amount of additive* affect the responses, and they do so in an additive manner.

The *nature* of the effects of factors 1 and 2 on the responses is explored in Exercise 6.15. In that exercise, simultaneous confidence intervals for contrasts in the components of τ_ℓ and β_k are considered. ∎

6.8 Profile Analysis

Profile analysis pertains to situations in which a battery of p treatments (tests, questions, and so forth) are administered to two or more groups of subjects. All responses must be expressed in similar units. Further, it is assumed that the responses for the different groups are independent of one another. Ordinarily, we might pose the question, are the population mean vectors the same? In profile analysis, the question of equality of mean vectors is divided into several specific possibilities.

Consider the population means $\mu_1' = [\mu_{11}, \mu_{12}, \mu_{13}, \mu_{14}]$ representing the average responses to four treatments for the first group. A plot of these means, connected by straight lines, is shown in Figure 6.4. This broken-line graph is the *profile* for population 1.

Profiles can be constructed for each population (group). We shall concentrate on two groups. Let $\mu_1' = [\mu_{11}, \mu_{12}, \ldots, \mu_{1p}]$ and $\mu_2' = [\mu_{21}, \mu_{22}, \ldots, \mu_{2p}]$ be the mean responses to p treatments for populations 1 and 2, respectively. The hypothesis $H_0: \mu_1 = \mu_2$ implies that the treatments have the same (average) effect on the two populations. In terms of the population profiles, we can formulate the question of equality in a stepwise fashion.

1. Are the profiles parallel?
   Equivalently: Is $H_{01}: \mu_{1i} - \mu_{1i-1} = \mu_{2i} - \mu_{2i-1},\ i = 2, 3, \ldots, p$, acceptable?
2. Assuming that the profiles *are* parallel, are the profiles coincident? [7]
   Equivalently: Is $H_{02}: \mu_{1i} = \mu_{2i},\ i = 1, 2, \ldots, p$, acceptable?

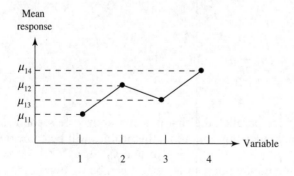

Figure 6.4 The population profile $p = 4$.

[7] The question, "Assuming that the profiles are parallel, are the profiles linear?" is considered in Exercise 6.12. The null hypothesis of parallel linear profiles can be written $H_0: (\mu_{1i} + \mu_{2i}) - (\mu_{1i-1} + \mu_{2i-1}) = (\mu_{1i-1} + \mu_{2i-1}) - (\mu_{1i-2} + \mu_{2i-2}),\ i = 3, \ldots, p$. Although this hypothesis may be of interest in a particular situation, in practice the question of whether two parallel profiles are the same (coincident), whatever their nature, is usually of greater interest.

3. Assuming that the profiles *are* coincident, are the profiles level? That is, are all the means equal to the same constant?

Equivalently: Is $H_{03}: \mu_{11} = \mu_{12} = \cdots = \mu_{1p} = \mu_{21} = \mu_{22} = \cdots = \mu_{2p}$ acceptable?

The null hypothesis in stage 1 can be written

$$H_{01}: \mathbf{C}\boldsymbol{\mu}_1 = \mathbf{C}\boldsymbol{\mu}_2$$

where $\mathbf{C}$ is the contrast matrix

$$\underset{((p-1)\times p)}{\mathbf{C}} = \begin{bmatrix} -1 & 1 & 0 & 0 & \cdots & 0 & 0 \\ 0 & -1 & 1 & 0 & \cdots & 0 & 0 \\ \vdots & \vdots & \vdots & \vdots & \ddots & \vdots & \vdots \\ 0 & 0 & 0 & 0 & \cdots & -1 & 1 \end{bmatrix} \qquad (6\text{-}72)$$

For independent samples of sizes n_1 and n_2 from the two populations, the null hypothesis can be tested by constructing the transformed observations

$$\mathbf{C}\mathbf{x}_{1j}, \qquad j = 1, 2, \ldots, n_1$$

and

$$\mathbf{C}\mathbf{x}_{2j}, \qquad j = 1, 2, \ldots, n_2$$

These have sample mean vectors $\mathbf{C}\bar{\mathbf{x}}_1$ and $\mathbf{C}\bar{\mathbf{x}}_2$, respectively, and pooled covariance matrix $\mathbf{C}\mathbf{S}_{\text{pooled}}\mathbf{C}'$.

Since the two sets of transformed observations have $N_{p-1}(\mathbf{C}\boldsymbol{\mu}_1, \mathbf{C}\boldsymbol{\Sigma}\mathbf{C}')$ and $N_{p-1}(\mathbf{C}\boldsymbol{\mu}_2, \mathbf{C}\boldsymbol{\Sigma}\mathbf{C}')$ distributions, respectively, an application of Result 6.2 provides a test for parallel profiles.

Test for Parallel Profiles for Two Normal Populations

Reject $H_{01}: \mathbf{C}\boldsymbol{\mu}_1 = \mathbf{C}\boldsymbol{\mu}_2$ (parallel profiles) at level α if

$$T^2 = (\bar{\mathbf{x}}_1 - \bar{\mathbf{x}}_2)'\mathbf{C}'\left[\left(\frac{1}{n_1} + \frac{1}{n_2}\right)\mathbf{C}\mathbf{S}_{\text{pooled}}\mathbf{C}'\right]^{-1}\mathbf{C}(\bar{\mathbf{x}}_1 - \bar{\mathbf{x}}_2) > c^2 \qquad (6\text{-}73)$$

where

$$c^2 = \frac{(n_1 + n_2 - 2)(p - 1)}{n_1 + n_2 - p}F_{p-1, n_1+n_2-p}(\alpha)$$

When the profiles are parallel, the first is either above the second ($\mu_{1i} > \mu_{2i}$, for all i), or vice versa. Under this condition, the profiles will be coincident only if the total heights $\mu_{11} + \mu_{12} + \cdots + \mu_{1p} = \mathbf{1}'\boldsymbol{\mu}_1$ and $\mu_{21} + \mu_{22} + \cdots + \mu_{2p} = \mathbf{1}'\boldsymbol{\mu}_2$ are equal. Therefore, the null hypothesis at stage 2 can be written in the equivalent form

$$H_{02}: \mathbf{1}'\boldsymbol{\mu}_1 = \mathbf{1}'\boldsymbol{\mu}_2$$

We can then test H_{02} with the usual two-sample t-statistic based on the univariate observations $\mathbf{1}'\mathbf{x}_{1j}, j = 1, 2, \ldots, n_1$, and $\mathbf{1}'\mathbf{x}_{2j}, j = 1, 2, \ldots, n_2$.

Test for Coincident Profiles, Given That Profiles Are Parallel

For two normal populations, reject $H_{02}: \mathbf{1}'\boldsymbol{\mu}_1 = \mathbf{1}'\boldsymbol{\mu}_2$ (profiles coincident) at level α if

$$T^2 = \mathbf{1}'(\bar{\mathbf{x}}_1 - \bar{\mathbf{x}}_2) \left[\left(\frac{1}{n_1} + \frac{1}{n_2} \right) \mathbf{1}' \mathbf{S}_{\text{pooled}} \mathbf{1} \right]^{-1} \mathbf{1}'(\bar{\mathbf{x}}_1 - \bar{\mathbf{x}}_2)$$

$$= \left(\frac{\mathbf{1}'(\bar{\mathbf{x}}_1 - \bar{\mathbf{x}}_2)}{\sqrt{\left(\dfrac{1}{n_1} + \dfrac{1}{n_2} \right) \mathbf{1}' \mathbf{S}_{\text{pooled}} \mathbf{1}}} \right)^2 > t^2_{n_1+n_2-2}\left(\frac{\alpha}{2} \right) = F_{1,n_1+n_2-2}(\alpha) \qquad (6\text{-}74)$$

For coincident profiles, $\mathbf{x}_{11}, \mathbf{x}_{12}, \ldots, \mathbf{x}_{1n_1}$ and $\mathbf{x}_{21}, \mathbf{x}_{22}, \ldots, \mathbf{x}_{2n_2}$ are all observations from the same normal population? The next step is to see whether all variables have the same mean, so that the common profile is level.

When H_{01} and H_{02} are tenable, the common mean vector $\boldsymbol{\mu}$ is estimated, using all $n_1 + n_2$ observations, by

$$\bar{\mathbf{x}} = \frac{1}{n_1 + n_2} \left(\sum_{j=1}^{n_1} \mathbf{x}_{1j} + \sum_{j=1}^{n_2} \mathbf{x}_{2j} \right) = \frac{n_1}{(n_1 + n_2)} \bar{\mathbf{x}}_1 + \frac{n_2}{(n_1 + n_2)} \bar{\mathbf{x}}_2$$

If the common profile is level, then $\mu_1 = \mu_2 = \cdots = \mu_p$, and the null hypothesis at stage 3 can be written as

$$H_{03}: \mathbf{C}\boldsymbol{\mu} = \mathbf{0}$$

where $\mathbf{C}$ is given by (6-72). Consequently, we have the following test.

Test for Level Profiles, Given That Profiles Are Coincident

For two normal populations: Reject $H_{03}: \mathbf{C}\boldsymbol{\mu} = \mathbf{0}$ (profiles level) at level α if

$$(n_1 + n_2)\bar{\mathbf{x}}'\mathbf{C}'[\mathbf{C}\mathbf{S}\mathbf{C}']^{-1}\mathbf{C}\bar{\mathbf{x}} > c^2 \qquad (6\text{-}75)$$

where $\mathbf{S}$ is the sample covariance matrix based on all $n_1 + n_2$ observations and

$$c^2 = \frac{(n_1 + n_2 - 1)(p - 1)}{(n_1 + n_2 - p + 1)} F_{p-1, n_1+n_2-p+1}(\alpha)$$

Example 6.14 (A profile analysis of love and marriage data) As part of a larger study of love and marriage, E. Hatfield, a sociologist, surveyed adults with respect to their marriage "contributions" and "outcomes" and their levels of "passionate" and "companionate" love. Recently married males and females were asked to respond to the following questions, using the 8-point scale in the figure below.

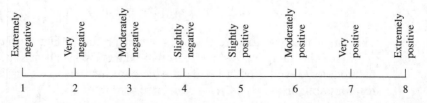

1. All things considered, how would you describe *your contributions* to the marriage?
2. All things considered, how would you describe *your outcomes* from the marriage?

Subjects were also asked to respond to the following questions, using the 5-point scale shown.

3. What is the level of *passionate* love that you feel for your partner?
4. What is the level of *companionate* love that you feel for your partner?

None at all	Very little	Some	A great deal	Tremendous amount
1	2	3	4	5

Let

$$x_1 = \text{an 8-point scale response to Question 1}$$
$$x_2 = \text{an 8-point scale response to Question 2}$$
$$x_3 = \text{a 5-point scale response to Question 3}$$
$$x_4 = \text{a 5-point scale response to Question 4}$$

and the two populations be defined as

$$\text{Population 1} = \text{married men}$$
$$\text{Population 2} = \text{married women}$$

The population means are the average responses to the $p = 4$ questions for the populations of males and females. Assuming a common covariance matrix Σ, it is of interest to see whether the profiles of males and females are the same.

A sample of $n_1 = 30$ males and $n_2 = 30$ females gave the sample mean vectors

$$\bar{\mathbf{x}}_1 = \begin{bmatrix} 6.833 \\ 7.033 \\ 3.967 \\ 4.700 \end{bmatrix}, \quad \bar{\mathbf{x}}_2 = \begin{bmatrix} 6.633 \\ 7.000 \\ 4.000 \\ 4.533 \end{bmatrix}$$
$$\text{(males)} \qquad\qquad \text{(females)}$$

and pooled covariance matrix

$$\mathbf{S}_{\text{pooled}} = \begin{bmatrix} .606 & .262 & .066 & .161 \\ .262 & .637 & .173 & .143 \\ .066 & .173 & .810 & .029 \\ .161 & .143 & .029 & .306 \end{bmatrix}$$

The sample mean vectors are plotted as sample profiles in Figure 6.5 on page 327.

Since the sample sizes are reasonably large, we shall use the normal theory methodology, even though the data, which are integers, are clearly nonnormal. To test for parallelism ($H_{01}: \mathbf{C}\boldsymbol{\mu}_1 = \mathbf{C}\boldsymbol{\mu}_2$), we compute

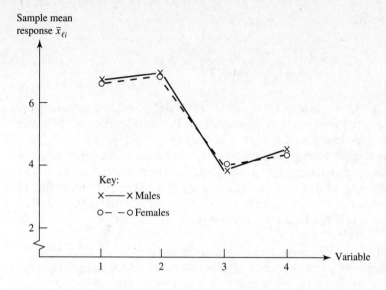

Sample mean response $\bar{x}_{\ell i}$

Key:
×——× Males
o— —o Females

Variable

Figure 6.5 Sample profiles for marriage–love responses.

$$\mathbf{CS}_{\text{pooled}}\mathbf{C'} = \begin{bmatrix} -1 & 1 & 0 & 0 \\ 0 & -1 & 1 & 0 \\ 0 & 0 & -1 & 1 \end{bmatrix} \mathbf{S}_{\text{pooled}} \begin{bmatrix} -1 & 0 & 0 \\ 1 & -1 & 0 \\ 0 & 1 & -1 \\ 0 & 0 & 1 \end{bmatrix}$$

$$= \begin{bmatrix} .719 & -.268 & -.125 \\ -.268 & 1.101 & -.751 \\ -.125 & -.751 & 1.058 \end{bmatrix}$$

and

$$\mathbf{C}(\bar{\mathbf{x}}_1 - \bar{\mathbf{x}}_2) = \begin{bmatrix} -1 & 1 & 0 & 0 \\ 0 & -1 & 1 & 0 \\ 0 & 0 & -1 & 1 \end{bmatrix} \begin{bmatrix} .200 \\ .033 \\ -.033 \\ .167 \end{bmatrix} = \begin{bmatrix} -.167 \\ -.066 \\ .200 \end{bmatrix}$$

Thus,

$$T^2 = [-.167, -.066, .200]\left(\tfrac{1}{30} + \tfrac{1}{30}\right)^{-1} \begin{bmatrix} .719 & -.268 & -.125 \\ -.268 & 1.101 & -.751 \\ -.125 & -.751 & 1.058 \end{bmatrix}^{-1} \begin{bmatrix} -.167 \\ -.066 \\ .200 \end{bmatrix}$$

$$= 15(.067) = 1.005$$

Moreover, with $\alpha = .05$, $c^2 = [(30+30-2)(4-1)/(30+30-4)]F_{3,56}(.05) = 3.11(2.8) = 8.7$. Since $T^2 = 1.005 < 8.7$, we conclude that the hypothesis of parallel profiles for men and women is tenable. Given the plot in Figure 6.5, this finding is not surprising.

Assuming that the profiles are parallel, we can test for *coincident* profiles. To test H_{02}: $\mathbf{1'}\boldsymbol{\mu}_1 = \mathbf{1'}\boldsymbol{\mu}_2$ (profiles coincident), we need

Sum of elements in $(\bar{\mathbf{x}}_1 - \bar{\mathbf{x}}_2) = \mathbf{1'}(\bar{\mathbf{x}}_1 - \bar{\mathbf{x}}_2) = .367$

Sum of elements in $\mathbf{S}_{\text{pooled}} = \mathbf{1'}\mathbf{S}_{\text{pooled}}\mathbf{1} = 4.207$

Using (6-74), we obtain

$$T^2 = \left(\frac{.367}{\sqrt{\left(\frac{1}{30} + \frac{1}{30} \right) 4.027}} \right)^2 = .501$$

With $\alpha = .05$, $F_{1,58}(.05) = 4.0$, and $T^2 = .501 < F_{1,58}(.05) = 4.0$, we cannot reject the hypothesis that the profiles are coincident. That is, the responses of men and women to the four questions posed appear to be the same.

We could now test for level profiles; however, it does not make sense to carry out this test for our example, since Questions 1 and 2 were measured on a scale of 1–8, while Questions 3 and 4 were measured on a scale of 1–5. The incompatibility of these scales makes the test for level profiles meaningless and illustrates the need for similar measurements in order to carry out a complete profile analysis. ∎

When the sample sizes are small, a profile analysis will depend on the normality assumption. This assumption can be checked, using methods discussed in Chapter 4, with the original observations $\mathbf{x}_{\ell j}$ or the contrast observations $\mathbf{C}\mathbf{x}_{\ell j}$.

The analysis of profiles for several populations proceeds in much the same fashion as that for two populations. In fact, the general measures of comparison are analogous to those just discussed. (See [13], [18].)

6.9 Repeated Measures Designs and Growth Curves

As we said earlier, the term "repeated measures" refers to situations where the same characteristic is observed, at different times or locations, on the same subject.

(a) The observations on a subject may correspond to different treatments as in Example 6.2 where the time between heartbeats was measured under the 2×2 treatment combinations applied to each dog. The treatments need to be compared when the responses on the same subject are correlated.

(b) A single treatment may be applied to each subject and a single characteristic observed over a period of time. For instance, we could measure the weight of a puppy at birth and then once a month. It is the curve traced by a typical dog that must be modeled. In this context, we refer to the curve as a *growth curve*.

When some subjects receive one treatment and others another treatment, the growth curves for the treatments need to be compared.

To illustrate the growth curve model introduced by Potthoff and Roy [21], we consider calcium measurements of the dominant ulna bone in older women. Besides an initial reading, Table 6.5 gives readings after one year, two years, and three years for the control group. Readings obtained by photon absorptiometry from the same subject are correlated but those from different subjects should be independent. The model assumes that the same covariance matrix $\mathbf{\Sigma}$ holds for each subject. Unlike univariate approaches, this model does not require the four measurements to have equal variances. A profile, constructed from the four sample means $(\bar{x}_1, \bar{x}_2, \bar{x}_3, \bar{x}_4)$, summarizes the growth which here is a loss of calcium over time. Can the growth pattern be adequately represented by a polynomial in time?

Table 6.5 Calcium Measurements on the Dominant Ulna; Control Group

Subject	Initial	1 year	2 year	3 year
1	87.3	86.9	86.7	75.5
2	59.0	60.2	60.0	53.6
3	76.7	76.5	75.7	69.5
4	70.6	76.1	72.1	65.3
5	54.9	55.1	57.2	49.0
6	78.2	75.3	69.1	67.6
7	73.7	70.8	71.8	74.6
8	61.8	68.7	68.2	57.4
9	85.3	84.4	79.2	67.0
10	82.3	86.9	79.4	77.4
11	68.6	65.4	72.3	60.8
12	67.8	69.2	66.3	57.9
13	66.2	67.0	67.0	56.2
14	81.0	82.3	86.8	73.9
15	72.3	74.6	75.3	66.1
Mean	72.38	73.29	72.47	64.79

Source: Data courtesy of Everett Smith.

When the p measurements on all subjects are taken at times $t_1, t_2, \ldots, t_p$, the Potthoff–Roy model for quadratic growth becomes

$$
E[\mathbf{X}] = E \begin{bmatrix} X_1 \\ X_2 \\ \vdots \\ X_p \end{bmatrix} = \begin{bmatrix} \beta_0 + \beta_1 t_1 + \beta_2 t_1^2 \\ \beta_0 + \beta_1 t_2 + \beta_2 t_2^2 \\ \vdots \\ \beta_0 + \beta_1 t_p + \beta_2 t_p^2 \end{bmatrix}
$$

where the ith mean μ_i is the quadratic expression evaluated at t_i.

Usually groups need to be compared. Table 6.6 gives the calcium measurements for a second set of women, the treatment group, that received special help with diet and a regular exercise program.

When a study involves several treatment groups, an extra subscript is needed as in the one-way MANOVA model. Let $\mathbf{X}_{\ell 1}, \mathbf{X}_{\ell 2}, \ldots, \mathbf{X}_{\ell n_\ell}$ be the n_ℓ vectors of measurements on the n_ℓ subjects in group ℓ, for $\ell = 1, \ldots, g$.

Assumptions. All of the $\mathbf{X}_{\ell j}$ are independent and have the same covariance matrix $\boldsymbol{\Sigma}$. Under the quadratic growth model, the mean vectors are

$$
E[\mathbf{X}_{\ell j}] = \begin{bmatrix} \beta_{\ell 0} + \beta_{\ell 1} t_1 + \beta_{\ell 2} t_1^2 \\ \beta_{\ell 0} + \beta_{\ell 1} t_2 + \beta_{\ell 2} t_2^2 \\ \vdots \\ \beta_{\ell 0} + \beta_{\ell 1} t_p + \beta_{\ell 2} t_p^2 \end{bmatrix} = \begin{bmatrix} 1 & t_1 & t_1^2 \\ 1 & t_2 & t_2^2 \\ \vdots & \vdots & \vdots \\ 1 & t_p & t_p^2 \end{bmatrix} \begin{bmatrix} \beta_{\ell 0} \\ \beta_{\ell 1} \\ \beta_{\ell 2} \end{bmatrix} = \mathbf{B}\boldsymbol{\beta}_\ell
$$

Table 6.6 Calcium Measurements on the Dominant Ulna; Treatment Group

Subject	Initial	1 year	2 year	3 year
1	83.8	85.5	86.2	81.2
2	65.3	66.9	67.0	60.6
3	81.2	79.5	84.5	75.2
4	75.4	76.7	74.3	66.7
5	55.3	58.3	59.1	54.2
6	70.3	72.3	70.6	68.6
7	76.5	79.9	80.4	71.6
8	66.0	70.9	70.3	64.1
9	76.7	79.0	76.9	70.3
10	77.2	74.0	77.8	67.9
11	67.3	70.7	68.9	65.9
12	50.3	51.4	53.6	48.0
13	57.7	57.0	57.5	51.5
14	74.3	77.7	72.6	68.0
15	74.0	74.7	74.5	65.7
16	57.3	56.0	64.7	53.0
Mean	69.29	70.66	71.18	64.53

Source: Data courtesy of Everett Smith.

where

$$\mathbf{B} = \begin{bmatrix} 1 & t_1 & t_1^2 \\ 1 & t_2 & t_2^2 \\ \vdots & \vdots & \vdots \\ 1 & t_p & t_p^2 \end{bmatrix} \quad \text{and} \quad \boldsymbol{\beta}_\ell = \begin{bmatrix} \beta_{\ell 0} \\ \beta_{\ell 1} \\ \beta_{\ell 2} \end{bmatrix} \tag{6-76}$$

If a qth-order polynomial is fit to the growth data, then

$$\mathbf{B} = \begin{bmatrix} 1 & t_1 & \cdots & t_1^q \\ 1 & t_2 & \cdots & t_2^q \\ \cdot & \cdot & & \cdot \\ \cdot & \cdot & & \cdot \\ \cdot & \cdot & & \cdot \\ 1 & t_p & \cdots & t_p^q \end{bmatrix} \quad \text{and} \quad \boldsymbol{\beta}_\ell = \begin{bmatrix} \beta_{\ell 0} \\ \beta_{\ell 1} \\ \cdot \\ \cdot \\ \cdot \\ \beta_{\ell q} \end{bmatrix} \tag{6-77}$$

Under the assumption of multivariate normality, the maximum likelihood estimators of the $\boldsymbol{\beta}_\ell$ are

$$\hat{\boldsymbol{\beta}}_\ell = (\mathbf{B}'\mathbf{S}_{\text{pooled}}^{-1}\mathbf{B})^{-1}\mathbf{B}'\mathbf{S}_{\text{pooled}}^{-1}\overline{\mathbf{X}}_\ell \quad \text{for} \quad \ell = 1, 2, \ldots, g \tag{6-78}$$

where

$$\mathbf{S}_{\text{pooled}} = \frac{1}{(N - g)}\left((n_1 - 1)\mathbf{S_1} + \cdots + (n_g - 1)\mathbf{S_g}\right) = \frac{1}{N - g}\mathbf{W}$$

with $N = \sum_{\ell=1}^{g} n_\ell$, is the pooled estimator of the common covariance matrix $\boldsymbol{\Sigma}$. The estimated covariances of the maximum likelihood estimators are

$$\widehat{\mathrm{Cov}}(\hat{\boldsymbol{\beta}}_\ell) = \frac{k}{n_\ell} (\mathbf{B}'\mathbf{S}_{\mathrm{pooled}}^{-1}\mathbf{B})^{-1} \quad \text{for} \quad \ell = 1, 2, \ldots, g \tag{6-79}$$

where $k = (N - g)(N - g - 1)/(N - g - p + q)(N - g - p + q + 1)$.

Also, $\hat{\boldsymbol{\beta}}_\ell$ and $\hat{\boldsymbol{\beta}}_h$ are independent, for $\ell \neq h$, so their covariance is $\mathbf{0}$.

We can formally test that a qth-order polynomial is adequate. The model is fit without restrictions, the error sum of squares and cross products matrix is just the within groups $\mathbf{W}$ that has $N - g$ degrees of freedom. Under a qth-order polynomial, the error sum of squares and cross products

$$\mathbf{W}_q = \sum_{\ell=1}^{g} \sum_{j=1}^{n_\ell} (\mathbf{X}_{\ell j} - \mathbf{B}\hat{\boldsymbol{\beta}}_\ell)(\mathbf{X}_{\ell j} - \mathbf{B}\hat{\boldsymbol{\beta}}_\ell)' \tag{6-80}$$

has $n_g - g + p - q - 1$ degrees of freedom. The likelihood ratio test of the null hypothesis that the q-order polynomial is adequate can be based on Wilks' lambda

$$\Lambda^* = \frac{|\mathbf{W}|}{|\mathbf{W}_q|} \tag{6-81}$$

Under the polynomial growth model, there are $q + 1$ terms instead of the p means for each of the groups. Thus there are $(p - q - 1)g$ fewer parameters. For large sample sizes, the null hypothesis that the polynomial is adequate is rejected if

$$-\left(N - \frac{1}{2}(p - q + g)\right) \ln \Lambda^* > \chi^2_{(p-q-1)g}(\alpha) \tag{6-82}$$

Example 6.15 (Fitting a quadratic growth curve to calcium loss) Refer to the data in Tables 6.5 and 6.6. Fit the model for quadratic growth.

A computer calculation gives

$$[\hat{\boldsymbol{\beta}}_1, \hat{\boldsymbol{\beta}}_2] = \begin{bmatrix} 73.0701 & 70.1387 \\ 3.6444 & 4.0900 \\ -2.0274 & -1.8534 \end{bmatrix}$$

so the estimated growth curves are

Control group: $73.07 \;+\; 3.64t \;-\; 2.03t^2$
                (2.58) $(.83)$ $(.28)$

Treatment group: $70.14 \;+\; 4.09t \;-\; 1.85t^2$
                (2.50) $(.80)$ $(.27)$

where

$$(\mathbf{B}'\mathbf{S}_{\mathrm{pooled}}^{-1}\mathbf{B})^{-1} = \begin{bmatrix} 93.1744 & -5.8368 & 0.2184 \\ -5.8368 & 9.5699 & -3.0240 \\ 0.2184 & -3.0240 & 1.1051 \end{bmatrix}$$

and, by (6-79), the standard errors given below the parameter estimates were obtained by dividing the diagonal elements by n_ℓ and taking the square root.

Examination of the estimates and the standard errors reveals that the t^2 terms are needed. Loss of calcium is predicted after 3 years for both groups. Further, there does not seem to be any substantial difference between the two groups.

Wilks' lambda for testing the null hypothesis that the quadratic growth model is adequate becomes

$$\Lambda^* = \frac{|\mathbf{W}|}{|\mathbf{W}_2|} = \frac{\begin{vmatrix} 2726.282 & 2660.749 & 2369.308 & 2335.912 \\ 2660.749 & 2756.009 & 2343.514 & 2327.961 \\ 2369.308 & 2343.514 & 2301.714 & 2098.544 \\ 2335.912 & 2327.961 & 2098.544 & 2277.452 \end{vmatrix}}{\begin{vmatrix} 2781.017 & 2698.589 & 2363.228 & 2362.253 \\ 2698.589 & 2832.430 & 2331.235 & 2381.160 \\ 2363.228 & 2331.235 & 2303.687 & 2089.996 \\ 2362.253 & 2381.160 & 2089.996 & 2314.485 \end{vmatrix}} = .7627$$

Since, with $\alpha = .01$,

$$-\left(N - \frac{1}{2}(p - q + g)\right) \ln \Lambda^* = -\left(31 - \frac{1}{2}(4 - 2 + 2)\right) \ln .7627$$

$$= 7.86 < \chi^2_{(4-2-1)2}(.01) = 9.21$$

we fail to reject the adequacy of the quadratic fit at $\alpha = .01$. Since the p-value is less than .05 there is, however, some evidence that the quadratic does not fit well.

We could, without restricting to quadratic growth, test for parallel and coincident calcium loss using profile analysis. ∎

The Potthoff and Roy growth curve model holds for more general designs than one-way MANOVA. However, the $\hat{\boldsymbol{\beta}}_\ell$ are no longer given by (6-78) and the expression for its covariance matrix becomes more complicated than (6-79). We refer the reader to [14] for more examples and further tests.

There are many other modifications to the model treated here. They include the following:

(a) Dropping the restriction to polynomial growth. Use nonlinear parametric models or even nonparametric splines.

(b) Restricting the covariance matrix to a special form such as equally correlated responses on the same individual.

(c) Observing more than one response variable, over time, on the same individual. This results in a multivariate version of the growth curve model.

6.10 Perspectives and a Strategy for Analyzing Multivariate Models

We emphasize that, with several characteristics, it is important to control the overall probability of making any incorrect decision. This is particularly important when testing for the equality of two or more treatments as the examples in this chapter indicate. A single multivariate test, with its associated single p-value, is preferable to performing a large number of univariate tests. The outcome tells us whether or not it is worthwhile to look closer on a variable by variable and group by group analysis.

A single multivariate test is recommended over, say, p univariate tests because, as the next example demonstrates, univariate tests ignore important information and can give misleading results.

Example 6.16 (Comparing multivariate and univariate tests for the differences in means) Suppose we collect measurements on two variables X_1 and X_2 for ten randomly selected experimental units from each of two groups. The hypothetical data are noted here and displayed as scatter plots and marginal dot diagrams in Figure 6.6 on page 334.

x_1	x_2	Group
5.0	3.0	1
4.5	3.2	1
6.0	3.5	1
6.0	4.6	1
6.2	5.6	1
6.9	5.2	1
6.8	6.0	1
5.3	5.5	1
6.6	7.3	1
7.3	6.5	1
4.6	4.9	2
4.9	5.9	2
4.0	4.1	2
3.8	5.4	2
6.2	6.1	2
5.0	7.0	2
5.3	4.7	2
7.1	6.6	2
5.8	7.8	2
6.8	8.0	2

It is clear from the horizontal marginal dot diagram that there is considerable overlap in the x_1 values for the two groups. Similarly, the vertical marginal dot diagram shows there is considerable overlap in the x_2 values for the two groups. The scatter plots suggest that there is fairly strong positive correlation between the two variables for each group, and that, although there is some overlap, the group 1 measurements are generally to the southeast of the group 2 measurements.

Let $\boldsymbol{\mu}_1' = [\mu_{11}, \mu_{12}]$ be the population mean vector for the first group, and let $\boldsymbol{\mu}_2' = [\mu_{21}, \mu_{22}]$ be the population mean vector for the second group. Using the x_1 observations, a univariate analysis of variance gives $F = 2.46$ with $\nu_1 = 1$ and $\nu_2 = 18$ degrees of freedom. Consequently, we cannot reject $H_0: \mu_{11} = \mu_{21}$ at any reasonable significance level ($F_{1,18}(.10) = 3.01$). Using the x_2 observations, a univariate analysis of variance gives $F = 2.68$ with $\nu_1 = 1$ and $\nu_2 = 18$ degrees of freedom. Again, we cannot reject $H_0: \mu_{12} = \mu_{22}$ at any reasonable significance level.

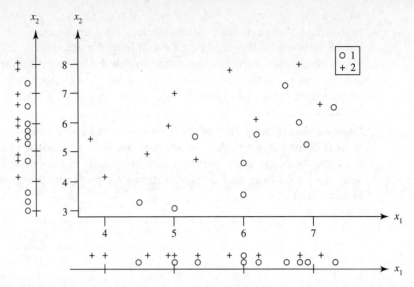

Figure 6.6 Scatter plots and marginal dot diagrams for the data from two groups.

The univariate tests suggest there is no difference between the component means for the two groups, and hence we cannot discredit $\boldsymbol{\mu}_1 = \boldsymbol{\mu}_2$.

On the other hand, if we use Hotelling's T^2 to test for the equality of the mean vectors, we find

$$T^2 = 17.29 > c^2 = \frac{(18)(2)}{17} F_{2,17}(.01) = 2.118 \times 6.11 = 12.94$$

and we reject $H_0 \colon \boldsymbol{\mu}_1 = \boldsymbol{\mu}_2$ at the 1% level. The multivariate test takes into account the positive correlation between the two measurements for each group—information that is unfortunately ignored by the univariate tests. This T^2-test is equivalent to the MANOVA test (6-42). ∎

Example 6.17 (Data on lizards that require a bivariate test to establish a difference in means) A zoologist collected lizards in the southwestern United States. Among other variables, he measured mass (in grams) and the snout-vent length (in millimeters). Because the tails sometimes break off in the wild, the snout-vent length is a more representative measure of length. The data for the lizards from two genera, Cnemidophorus (C) and Sceloporus (S), collected in 1997 and 1999 are given in Table 6.7. Notice that there are $n_1 = 20$ measurements for C lizards and $n_2 = 40$ measurements for S lizards.

After taking natural logarithms, the summary statistics are

$$C \colon n_1 = 20 \quad \bar{\mathbf{x}}_1 = \begin{bmatrix} 2.240 \\ 4.394 \end{bmatrix} \quad \mathbf{S}_1 = \begin{bmatrix} 0.35305 & 0.09417 \\ 0.09417 & 0.02595 \end{bmatrix}$$

$$S \colon n_2 = 40 \quad \bar{\mathbf{x}}_2 = \begin{bmatrix} 2.368 \\ 4.308 \end{bmatrix} \quad \mathbf{S}_2 = \begin{bmatrix} 0.50684 & 0.14539 \\ 0.14539 & 0.04255 \end{bmatrix}$$

Table 6.7 Lizard Data for Two Genera

C		S		S	
Mass	SVL	Mass	SVL	Mass	SVL
7.513	74.0	13.911	77.0	14.666	80.0
5.032	69.5	5.236	62.0	4.790	62.0
5.867	72.0	37.331	108.0	5.020	61.5
11.088	80.0	41.781	115.0	5.220	62.0
2.419	56.0	31.995	106.0	5.690	64.0
13.610	94.0	3.962	56.0	6.763	63.0
18.247	95.5	4.367	60.5	9.977	71.0
16.832	99.5	3.048	52.0	8.831	69.5
15.910	97.0	4.838	60.0	9.493	67.5
17.035	90.5	6.525	64.0	7.811	66.0
16.526	91.0	22.610	96.0	6.685	64.5
4.530	67.0	13.342	79.5	11.980	79.0
7.230	75.0	4.109	55.5	16.520	84.0
5.200	69.5	12.369	75.0	13.630	81.0
13.450	91.5	7.120	64.5	13.700	82.5
14.080	91.0	21.077	87.5	10.350	74.0
14.665	90.0	42.989	109.0	7.900	68.5
6.092	73.0	27.201	96.0	9.103	70.0
5.264	69.5	38.901	111.0	13.216	77.5
16.902	94.0	19.747	84.5	9.787	70.0

SVL = snout-vent length.
Source: Data courtesy of Kevin E. Bonine.

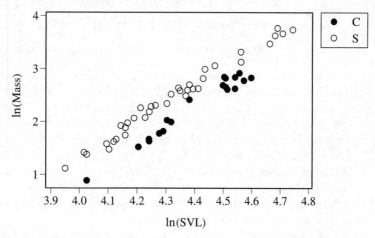

Figure 6.7   Scatter plot of ln(Mass) versus ln(SVL) for the lizard data in Table 6.7.

A plot of mass (Mass) versus snout-vent length (SVL), after taking natural logarithms, is shown in Figure 6.7. The large sample individual 95% confidence intervals for the difference in ln(Mass) means and the difference in ln(SVL) means both cover 0.

$$\ln(\text{Mass}): \quad \mu_{11} - \mu_{21}: \quad (-0.476, 0.220)$$
$$\ln(\text{SVL}): \quad \mu_{12} - \mu_{22}: \quad (-0.011, 0.183)$$

The corresponding univariate Student's t-test statistics for testing for no difference in the individual means have p-values of .46 and .08, respectively. Clearly, from a univariate perspective, we cannot detect a difference in mass means or a difference in snout-vent length means for the two genera of lizards.

However, consistent with the scatter diagram in Figure 6.7, a bivariate analysis strongly supports a difference in size between the two groups of lizards. Using Result 6.4 (also see Example 6.5), the T^2-statistic has an approximate χ_2^2 distribution. For this example, $T^2 = 225.4$ with a p-value less than .0001. A multivariate method is essential in this case.   ∎

Examples 6.16 and 6.17 demonstrate the efficacy of a multivariate test relative to its univariate counterparts. We encountered exactly this situation with the effluent data in Example 6.1.

In the context of random samples from several populations (recall the one-way MANOVA in Section 6.4), multivariate tests are based on the matrices

$$\mathbf{W} = \sum_{\ell=1}^{g} \sum_{j=1}^{n_\ell} (\mathbf{x}_{\ell j} - \bar{\mathbf{x}}_\ell)(\mathbf{x}_{\ell j} - \bar{\mathbf{x}}_\ell)' \quad \text{and} \quad \mathbf{B} = \sum_{\ell=1}^{g} n_\ell (\bar{\mathbf{x}}_\ell - \bar{\mathbf{x}})(\bar{\mathbf{x}}_\ell - \bar{\mathbf{x}})'$$

Throughout this chapter, we have used

$$\text{Wilks' lambda statistic } \Lambda^* = \frac{|\mathbf{W}|}{|\mathbf{B} + \mathbf{W}|}$$

which is equivalent to the likelihood ratio test. Three other multivariate test statistics are regularly included in the output of statistical packages.

$$\text{Lawley–Hotelling trace} = \text{tr}[\mathbf{B}\mathbf{W}^{-1}]$$
$$\text{Pillai trace} = \text{tr}[\mathbf{B}(\mathbf{B} + \mathbf{W})^{-1}]$$
$$\text{Roy's largest root} = \text{maximum eigenvalue of } \mathbf{W}(\mathbf{B} + \mathbf{W})^{-1}$$

All four of these tests appear to be nearly equivalent for extremely large samples. For moderate sample sizes, all comparisons are based on what is necessarily a limited number of cases studied by simulation. From the simulations reported to date, the first three tests have similar power, while the last, Roy's test, behaves differently. Its power is best only when there is a single nonzero eigenvalue and, at the same time, the power is large. This may approximate situations where a large difference exists in just one characteristic and it is between one group and all of the others. There is also some suggestion that Pillai's trace is slightly more robust against nonnormality. However, we suggest trying transformations on the original data when the residuals are nonnormal.

All four statistics apply in the two-way setting and in even more complicated MANOVA. More discussion is given in terms of the multivariate regression model in Chapter 7.

When, and only when, the multivariate tests signals a difference, or departure from the null hypothesis, do we probe deeper. We recommend calculating the Bonferonni intervals for all pairs of groups and all characteristics. The simultaneous confidence statements determined from the shadows of the confidence ellipse are, typically, too large. The one-at-a-time intervals may be suggestive of differences that merit further study but, with the current data, cannot be taken as conclusive evidence for the existence of differences. We summarize the procedure developed in this chapter for comparing treatments. The first step is to check the data for outliers using visual displays and other calculations.

A Strategy for the Multivariate Comparison of Treatments

1. *Try to identify outliers*. Check the data group by group for outliers. Also check the collection of residual vectors from any fitted model for outliers. Be aware of any outliers so calculations can be performed with and without them.

2. *Perform a multivariate test of hypothesis*. Our choice is the likelihood ratio test, which is equivalent to Wilks' lambda test.

3. *Calculate the Bonferroni simultaneous confidence intervals*. If the multivariate test reveals a difference, then proceed to calculate the Bonferroni confidence intervals for all pairs of groups or treatments, and all characteristics. If no differences are significant, try looking at Bonferroni intervals for the larger set of responses that includes the differences and sums of pairs of responses.

We must issue one caution concerning the proposed strategy. It may be the case that differences would appear in only one of the many characteristics and, further, the differences hold for only a few treatment combinations. Then, these few active differences may become lost among all the inactive ones. That is, the overall test may not show significance whereas a univariate test restricted to the specific active variable would detect the difference. The best preventative is a good experimental design. To design an effective experiment when one specific variable is expected to produce differences, do not include too many other variables that are not expected to show differences among the treatments.

Exercises

6.1. Construct and sketch a joint 95% confidence region for the mean difference vector δ using the effluent data and results in Example 6.1. Note that the point $\delta = 0$ falls outside the 95% contour. Is this result consistent with the test of H_0: $\delta = 0$ considered in Example 6.1? Explain.

6.2. Using the information in Example 6.1. construct the 95% Bonferroni simultaneous intervals for the components of the mean difference vector δ. Compare the lengths of these intervals with those of the simultaneous intervals constructed in the example.

6.3. The data corresponding to sample 8 in Table 6.1 seem unusually large. Remove sample 8. Construct a joint 95% confidence region for the mean difference vector δ and the 95% Bonferroni simultaneous intervals for the components of the mean difference vector. Are the results consistent with a test of H_0: $\delta = 0$? Discuss. Does the "outlier" make a difference in the analysis of these data?

6.4. Refer to Example 6.1.

(a) Redo the analysis in Example 6.1 after transforming the pairs of observations to ln(BOD) and ln(SS).

(b) Construct the 95% Bonferroni simultaneous intervals for the components of the mean vector $\boldsymbol{\delta}$ of transformed variables.

(c) Discuss any possible violation of the assumption of a bivariate normal distribution for the difference vectors of transformed observations.

6.5. A researcher considered three indices measuring the severity of heart attacks. The values of these indices for $n = 40$ heart-attack patients arriving at a hospital emergency room produced the summary statistics

$$\bar{\mathbf{x}} = \begin{bmatrix} 46.1 \\ 57.3 \\ 50.4 \end{bmatrix} \quad \text{and} \quad \mathbf{S} = \begin{bmatrix} 101.3 & 63.0 & 71.0 \\ 63.0 & 80.2 & 55.6 \\ 71.0 & 55.6 & 97.4 \end{bmatrix}$$

(a) All three indices are evaluated for each patient. Test for the equality of mean indices using (6-16) with $\alpha = .05$.

(b) Judge the differences in pairs of mean indices using 95% simultaneous confidence intervals. [See (6-18).]

6.6. Use the data for treatments 2 and 3 in Exercise 6.8.

(a) Calculate $\mathbf{S}_{\text{pooled}}$.

(b) Test $H_0: \boldsymbol{\mu}_2 - \boldsymbol{\mu}_3 = \mathbf{0}$ employing a two-sample approach with $\alpha = .01$.

(c) Construct 99% simultaneous confidence intervals for the differences $\mu_{2i} - \mu_{3i}$, $i = 1, 2$.

6.7. Using the summary statistics for the electricity-demand data given in Example 6.4, compute T^2 and test the hypothesis $H_0: \boldsymbol{\mu}_1 - \boldsymbol{\mu}_2 = \mathbf{0}$, assuming that $\boldsymbol{\Sigma}_1 = \boldsymbol{\Sigma}_2$. Set $\alpha = .05$. Also, determine the linear combination of mean components most responsible for the rejection of H_0.

6.8. Observations on two responses are collected for three treatments. The observation vectors $\begin{bmatrix} x_1 \\ x_2 \end{bmatrix}$ are

$$\text{Treatment 1:} \quad \begin{bmatrix} 6 \\ 7 \end{bmatrix}, \begin{bmatrix} 5 \\ 9 \end{bmatrix}, \begin{bmatrix} 8 \\ 6 \end{bmatrix}, \begin{bmatrix} 4 \\ 9 \end{bmatrix}, \begin{bmatrix} 7 \\ 9 \end{bmatrix}$$

$$\text{Treatment 2:} \quad \begin{bmatrix} 3 \\ 3 \end{bmatrix}, \begin{bmatrix} 1 \\ 6 \end{bmatrix}, \begin{bmatrix} 2 \\ 3 \end{bmatrix}$$

$$\text{Treatment 3:} \quad \begin{bmatrix} 2 \\ 3 \end{bmatrix}, \begin{bmatrix} 5 \\ 1 \end{bmatrix}, \begin{bmatrix} 3 \\ 1 \end{bmatrix}, \begin{bmatrix} 2 \\ 3 \end{bmatrix}$$

(a) Break up the observations into mean, treatment, and residual components, as in (6-39). Construct the corresponding arrays for each variable. (See Example 6.9.)

(b) Using the information in Part a, construct the one-way MANOVA table.

(c) Evaluate Wilks' lambda, Λ^*, and use Table 6.3 to test for treatment effects. Set $\alpha = .01$. Repeat the test using the chi-square approximation with Bartlett's correction. [See (6-43).] Compare the conclusions.

6.9. Using the contrast matrix $\mathbf{C}$ in (6-13), verify the relationships $\mathbf{d}_j = \mathbf{C}\mathbf{x}_j$, $\bar{\mathbf{d}} = \mathbf{C}\bar{\mathbf{x}}$, and $\mathbf{S}_d = \mathbf{C}\mathbf{S}\mathbf{C}'$ in (6-14).

6.10. Consider the univariate one-way decomposition of the observation $x_{\ell j}$ given by (6-34). Show that the mean vector $\bar{x}\mathbf{1}$ is always perpendicular to the treatment effect vector $(\bar{x}_1 - \bar{x})\mathbf{u}_1 + (\bar{x}_2 - \bar{x})\mathbf{u}_2 + \cdots + (\bar{x}_g - \bar{x})\mathbf{u}_g$ where

$$
\mathbf{u}_1 = \begin{bmatrix} 1 \\ \vdots \\ 1 \\ 0 \\ \vdots \\ 0 \\ 0 \\ \vdots \\ 0 \end{bmatrix} \Big\} n_1 \;,\; \mathbf{u}_2 = \begin{bmatrix} 0 \\ \vdots \\ 0 \\ 1 \\ \vdots \\ 1 \\ 0 \\ \vdots \\ 0 \end{bmatrix} \Big\} n_2 \;,\dots,\; \mathbf{u}_g = \begin{bmatrix} 0 \\ \vdots \\ 0 \\ 0 \\ \vdots \\ 0 \\ 1 \\ \vdots \\ 1 \end{bmatrix} \Big\} n_g
$$

6.11. A likelihood argument provides additional support for pooling the two independent sample covariance matrices to estimate a common covariance matrix in the case of two normal populations. Give the likelihood function, $L(\boldsymbol{\mu}_1, \boldsymbol{\mu}_2, \boldsymbol{\Sigma})$, for two independent samples of sizes n_1 and n_2 from $N_p(\boldsymbol{\mu}_1, \boldsymbol{\Sigma})$ and $N_p(\boldsymbol{\mu}_2, \boldsymbol{\Sigma})$ populations, respectively. Show that this likelihood is maximized by the choices $\hat{\boldsymbol{\mu}}_1 = \bar{\mathbf{x}}_1$, $\hat{\boldsymbol{\mu}}_2 = \bar{\mathbf{x}}_2$ and

$$
\hat{\boldsymbol{\Sigma}} = \frac{1}{n_1 + n_2}[(n_1 - 1)\mathbf{S}_1 + (n_2 - 1)\mathbf{S}_2] = \left(\frac{n_1 + n_2 - 2}{n_1 + n_2}\right)\mathbf{S}_{\text{pooled}}
$$

Hint: Use (4-16) and the maximization Result 4.10.

6.12. *(Test for linear profiles, given that the profiles are parallel.)* Let $\boldsymbol{\mu}_1' = [\mu_{11}, \mu_{12}, \dots, \mu_{1p}]$ and $\boldsymbol{\mu}_2' = [\mu_{21}, \mu_{22}, \dots, \mu_{2p}]$ be the mean responses to p treatments for populations 1 and 2, respectively. Assume that the profiles given by the two mean vectors are parallel.

(a) Show that the hypothesis that the profiles are linear can be written as $H_0: (\mu_{1i} + \mu_{2i}) - (\mu_{1i-1} + \mu_{2i-1}) = (\mu_{1i-1} + \mu_{2i-1}) - (\mu_{1i-2} + \mu_{2i-2})$, $i = 3, \dots, p$ or as $H_0: \mathbf{C}(\boldsymbol{\mu}_1 + \boldsymbol{\mu}_2) = \mathbf{0}$, where the $(p - 2) \times p$ matrix

$$
\mathbf{C} = \begin{bmatrix} 1 & -2 & 1 & 0 & \cdots & 0 & 0 & 0 \\ 0 & 1 & -2 & 1 & \cdots & 0 & 0 & 0 \\ \vdots & \vdots & \vdots & \vdots & & \vdots & \vdots & \vdots \\ 0 & 0 & 0 & 0 & \cdots & 1 & -2 & 1 \end{bmatrix}
$$

(b) Following an argument similar to the one leading to (6-73), we reject $H_0: \mathbf{C}(\boldsymbol{\mu}_1 + \boldsymbol{\mu}_2) = \mathbf{0}$ at level α if

$$
T^2 = (\bar{\mathbf{x}}_1 + \bar{\mathbf{x}}_2)'\mathbf{C}'\left[\left(\frac{1}{n_1} + \frac{1}{n_2}\right)\mathbf{C}\mathbf{S}_{\text{pooled}}\mathbf{C}'\right]^{-1}\mathbf{C}(\bar{\mathbf{x}}_1 + \bar{\mathbf{x}}_2) > c^2
$$

where

$$
c^2 = \frac{(n_1 + n_2 - 2)(p - 2)}{n_1 + n_2 - p + 1}F_{p-2, n_1+n_2-p+1}(\alpha)
$$

Let $n_1 = 30$, $n_2 = 30$, $\bar{\mathbf{x}}_1' = [6.4, 6.8, 7.3, 7.0]$, $\bar{\mathbf{x}}_2' = [4.3, 4.9, 5.3, 5.1]$, and

$$
\mathbf{S}_{\text{pooled}} = \begin{bmatrix}
.61 & .26 & .07 & .16 \\
.26 & .64 & .17 & .14 \\
.07 & .17 & .81 & .03 \\
.16 & .14 & .03 & .31
\end{bmatrix}
$$

Test for linear profiles, assuming that the profiles are parallel. Use $\alpha = .05$.

6.13. *(Two-way MANOVA without replications.)* Consider the observations on two responses, x_1 and x_2, displayed in the form of the following two-way table (note that there is a *single* observation vector at each combination of factor levels):

		Factor 2			
		Level 1	Level 2	Level 3	Level 4
	Level 1	$\begin{bmatrix} 6 \\ 8 \end{bmatrix}$	$\begin{bmatrix} 4 \\ 6 \end{bmatrix}$	$\begin{bmatrix} 8 \\ 12 \end{bmatrix}$	$\begin{bmatrix} 2 \\ 6 \end{bmatrix}$
Factor 1	Level 2	$\begin{bmatrix} 3 \\ 8 \end{bmatrix}$	$\begin{bmatrix} -3 \\ 2 \end{bmatrix}$	$\begin{bmatrix} 4 \\ 3 \end{bmatrix}$	$\begin{bmatrix} -4 \\ 3 \end{bmatrix}$
	Level 3	$\begin{bmatrix} -3 \\ 2 \end{bmatrix}$	$\begin{bmatrix} -4 \\ -5 \end{bmatrix}$	$\begin{bmatrix} 3 \\ -3 \end{bmatrix}$	$\begin{bmatrix} -4 \\ -6 \end{bmatrix}$

With no replications, the two-way MANOVA model is

$$
\mathbf{X}_{\ell k} = \boldsymbol{\mu} + \boldsymbol{\tau}_\ell + \boldsymbol{\beta}_k + \mathbf{e}_{\ell k}; \qquad \sum_{\ell=1}^{g} \boldsymbol{\tau}_\ell = \sum_{k=1}^{b} \boldsymbol{\beta}_k = \mathbf{0}
$$

where the $\mathbf{e}_{\ell k}$ are independent $N_p(\mathbf{0}, \boldsymbol{\Sigma})$ random vectors.

(a) Decompose the observations for each of the two variables as

$$
x_{\ell k} = \bar{x} + (\bar{x}_{\ell \cdot} - \bar{x}) + (\bar{x}_{\cdot k} - \bar{x}) + (x_{\ell k} - \bar{x}_{\ell \cdot} - \bar{x}_{\cdot k} + \bar{x})
$$

similar to the arrays in Example 6.9. *For each response*, this decomposition will result in several 3×4 matrices. Here $\bar{x}$ is the overall average, $\bar{x}_\ell$ is the average for the ℓth level of factor 1, and $\bar{x}_{\cdot k}$ is the average for the kth level of factor 2.

(b) Regard the rows of the matrices in Part a as strung out in a single "long" vector, and compute the sums of squares

$$
SS_{\text{tot}} = SS_{\text{mean}} + SS_{\text{fac1}} + SS_{\text{fac2}} + SS_{\text{res}}
$$

and sums of cross products

$$
SCP_{\text{tot}} = SCP_{\text{mean}} + SCP_{\text{fac1}} + SCP_{\text{fac2}} + SCP_{\text{res}}
$$

Consequently, obtain the matrices $\mathbf{SSP}_{\text{cor}}$, $\mathbf{SSP}_{\text{fac1}}$, $\mathbf{SSP}_{\text{fac2}}$, and $\mathbf{SSP}_{\text{res}}$ with degrees of freedom $gb - 1$, $g - 1$, $b - 1$, and $(g - 1)(b - 1)$, respectively.

(c) Summarize the calculations in Part b in a MANOVA table.

Hint: This MANOVA table is consistent with the two-way MANOVA table for comparing factors and their interactions where $n = 1$. Note that, with $n = 1$, $\mathbf{SSP}_{res}$ in the general two-way MANOVA table is a zero matrix with zero degrees of freedom. The matrix of interaction sum of squares and cross products now becomes the *residual* sum of squares and cross products matrix.

(d) Given the summary in Part c, test for factor 1 and factor 2 main effects at the $\alpha = .05$ level.

Hint: Use the results in (6-67) and (6-69) with $gb(n - 1)$ replaced by $(g - 1)(b - 1)$.

Note: The tests require that $p \le (g - 1)(b - 1)$ so that $\mathbf{SSP}_{res}$ will be positive definite (with probability 1).

6.14. A *replicate* of the experiment in Exercise 6.13 yields the following data:

		Factor 2			
		Level 1	Level 2	Level 3	Level 4
	Level 1	$\begin{bmatrix} 14 \\ 8 \end{bmatrix}$	$\begin{bmatrix} 6 \\ 2 \end{bmatrix}$	$\begin{bmatrix} 8 \\ 2 \end{bmatrix}$	$\begin{bmatrix} 16 \\ -4 \end{bmatrix}$
Factor 1	Level 2	$\begin{bmatrix} 1 \\ 6 \end{bmatrix}$	$\begin{bmatrix} 5 \\ 12 \end{bmatrix}$	$\begin{bmatrix} 0 \\ 15 \end{bmatrix}$	$\begin{bmatrix} 2 \\ 7 \end{bmatrix}$
	Level 3	$\begin{bmatrix} 3 \\ -2 \end{bmatrix}$	$\begin{bmatrix} -2 \\ 7 \end{bmatrix}$	$\begin{bmatrix} -11 \\ 1 \end{bmatrix}$	$\begin{bmatrix} -6 \\ 6 \end{bmatrix}$

(a) Use these data to decompose each of the two measurements in the observation vector as

$$x_{\ell k} = \bar{x} + (\bar{x}_{\ell \cdot} - \bar{x}) + (\bar{x}_{\cdot k} - \bar{x}) + (x_{\ell k} - \bar{x}_{\ell \cdot} - \bar{x}_{\cdot k} + \bar{x})$$

where $\bar{x}$ is the overall average, $\bar{x}_{\ell \cdot}$ is the average for the ℓth level of factor 1, and $\bar{x}_{\cdot k}$ is the average for the kth level of factor 2. Form the corresponding arrays for each of the two responses.

(b) Combine the preceding data with the data in Exercise 6.13 and carry out the necessary calculations to complete the general two-way MANOVA table.

(c) Given the results in Part b, test for interactions, and if the interactions do not exist, test for factor 1 and factor 2 main effects. Use the likelihood ratio test with $\alpha = .05$.

(d) If main effects, but no interactions, exist, examine the nature of the main effects by constructing Bonferroni simultaneous 95% confidence intervals for differences of the components of the factor effect parameters.

6.15. Refer to Example 6.13.

(a) Carry out approximate chi-square (likelihood ratio) tests for the factor 1 and factor 2 effects. Set $\alpha = .05$. Compare these results with the results for the exact F-tests given in the example. Explain any differences.

(b) Using (6-70), construct simultaneous 95% confidence intervals for differences in the factor 1 effect parameters for *pairs* of the three responses. Interpret these intervals. Repeat these calculations for factor 2 effect parameters.

The following exercises may require the use of a computer.

6.16. Four measures of the response *stiffness* on each of 30 boards are listed in Table 4.3 (see Example 4.14). The measures, on a given board, are repeated in the sense that they were made one after another. Assuming that the measures of stiffness arise from four treatments, test for the equality of treatments in a *repeated measures design* context. Set $\alpha = .05$. Construct a 95% (simultaneous) confidence interval for a contrast in the mean levels representing a comparison of the dynamic measurements with the static measurements.

6.17. The data in Table 6.8 were collected to test two psychological models of numerical cognition. Does the processing of numbers depend on the way the numbers are presented (words, Arabic digits)? Thirty-two subjects were required to make a series of

Table 6.8 Number Parity Data (Median Times in Milliseconds)

WordDiff (x_1)	WordSame (x_2)	ArabicDiff (x_3)	ArabicSame (x_4)
869.0	860.5	691.0	601.0
995.0	875.0	678.0	659.0
1056.0	930.5	833.0	826.0
1126.0	954.0	888.0	728.0
1044.0	909.0	865.0	839.0
925.0	856.5	1059.5	797.0
1172.5	896.5	926.0	766.0
1408.5	1311.0	854.0	986.0
1028.0	887.0	915.0	735.0
1011.0	863.0	761.0	657.0
726.0	674.0	663.0	583.0
982.0	894.0	831.0	640.0
1225.0	1179.0	1037.0	905.5
731.0	662.0	662.5	624.0
975.5	872.5	814.0	735.0
1130.5	811.0	843.0	657.0
945.0	909.0	867.5	754.0
747.0	752.5	777.0	687.5
656.5	659.5	572.0	539.0
919.0	833.0	752.0	611.0
751.0	744.0	683.0	553.0
774.0	735.0	671.0	612.0
941.0	931.0	901.5	700.0
751.0	785.0	789.0	735.0
767.0	737.5	724.0	639.0
813.5	750.5	711.0	625.0
1289.5	1140.0	904.5	784.5
1096.5	1009.0	1076.0	983.0
1083.0	958.0	918.0	746.5
1114.0	1046.0	1081.0	796.0
708.0	669.0	657.0	572.5
1201.0	925.0	1004.5	673.5

Source: Data courtesy of J. Carr.

quick numerical judgments about two numbers presented as either two number words ("two," "four") or two single Arabic digits ("2," "4"). The subjects were asked to respond "same" if the two numbers had the same numerical parity (both even or both odd) and "different" if the two numbers had a different parity (one even, one odd). Half of the subjects were assigned a block of Arabic digit trials, followed by a block of number word trials, and half of the subjects received the blocks of trials in the reverse order. Within each block, the order of "same" and "different" parity trials was randomized for each subject. For each of the four combinations of parity and format, the median reaction times for correct responses were recorded for each subject. Here

X_1 = median reaction time for word format–different parity combination

X_2 = median reaction time for word format–same parity combination

X_3 = median reaction time for Arabic format–different parity combination

X_4 = median reaction time for Arabic format–same parity combination

(a) Test for treatment effects using a *repeated measures design*. Set α = .05.

(b) Construct 95% (simultaneous) confidence intervals for the contrasts representing the number format effect, the parity type effect and the interaction effect. Interpret the resulting intervals.

(c) The absence of interaction supports the M model of numerical cognition, while the presence of interaction supports the C and C model of numerical cognition. Which model is supported in this experiment?

(d) For each subject, construct three difference scores corresponding to the number format contrast, the parity type contrast, and the interaction contrast. Is a multivariate normal distribution a reasonable population model for these data? Explain.

6.18. Jolicoeur and Mosimann [12] studied the relationship of size and shape for painted turtles. Table 6.9 contains their measurements on the carapaces of 24 female and 24 male turtles.

(a) Test for equality of the two population mean vectors using α = .05.

(b) If the hypothesis in Part a is rejected, find the linear combination of mean components most responsible for rejecting H_0.

(c) Find simultaneous confidence intervals for the component mean differences. Compare with the Bonferroni intervals.

Hint: You may wish to consider logarithmic transformations of the observations.

6.19. In the first phase of a study of the cost of transporting milk from farms to dairy plants, a survey was taken of firms engaged in milk transportation. Cost data on X_1 = fuel, X_2 = repair, and X_3 = capital, all measured on a per-mile basis, are presented in Table 6.10 on page 345 for n_1 = 36 gasoline and n_2 = 23 diesel trucks.

(a) Test for differences in the mean cost vectors. Set α = .01.

(b) If the hypothesis of equal cost vectors is rejected in Part a, find the linear combination of mean components most responsible for the rejection.

(c) Construct 99% simultaneous confidence intervals for the pairs of mean components. Which costs, if any, appear to be quite different?

(d) Comment on the validity of the assumptions used in your analysis. Note in particular that observations 9 and 21 for gasoline trucks have been identified as multivariate outliers. (See Exercise 5.22 and [2].) Repeat Part a with these observations deleted. Comment on the results.

Table 6.9 Carapace Measurements (in Millimeters) for Painted Turtles

Female			Male		
Length (x_1)	Width (x_2)	Height (x_3)	Length (x_1)	Width (x_2)	Height (x_3)
98	81	38	93	74	37
103	84	38	94	78	35
103	86	42	96	80	35
105	86	42	101	84	39
109	88	44	102	85	38
123	92	50	103	81	37
123	95	46	104	83	39
133	99	51	106	83	39
133	102	51	107	82	38
133	102	51	112	89	40
134	100	48	113	88	40
136	102	49	114	86	40
138	98	51	116	90	43
138	99	51	117	90	41
141	105	53	117	91	41
147	108	57	119	93	41
149	107	55	120	89	40
153	107	56	120	93	44
155	115	63	121	95	42
155	117	60	125	93	45
158	115	62	127	96	45
159	118	63	128	95	45
162	124	61	131	95	46
177	132	67	135	106	47

6.20. The tail lengths in millimeters (x_1) and wing lengths in millimeters (x_2) for 45 *male* hook-billed kites are given in Table 6.11 on page 346. Similar measurements for female hook-billed kites were given in Table 5.12.

(a) Plot the male hook-billed kite data as a scatter diagram, and (visually) check for outliers. (Note, in particular, observation 31 with $x_1 = 284$.)

(b) Test for equality of mean vectors for the populations of male and female hook-billed kites. Set $\alpha = .05$. If $H_0: \mu_1 - \mu_2 = 0$ is rejected, find the linear combination most responsible for the rejection of H_0. (You may want to eliminate any outliers found in Part a for the male hook-billed kite data before conducting this test. Alternatively, you may want to interpret $x_1 = 284$ for observation 31 as a misprint and conduct the test with $x_1 = 184$ for this observation. Does it make any difference in this case how observation 31 for the male hook-billed kite data is treated?)

(c) Determine the 95% confidence region for $\mu_1 - \mu_2$ and 95% simultaneous confidence intervals for the components of $\mu_1 - \mu_2$.

(d) Are male or female birds generally larger?

Table 6.10 Milk Transportation-Cost Data

Gasoline trucks			Diesel trucks		
x_1	x_2	x_3	x_1	x_2	x_3
16.44	12.43	11.23	8.50	12.26	9.11
7.19	2.70	3.92	7.42	5.13	17.15
9.92	1.35	9.75	10.28	3.32	11.23
4.24	5.78	7.78	10.16	14.72	5.99
11.20	5.05	10.67	12.79	4.17	29.28
14.25	5.78	9.88	9.60	12.72	11.00
13.50	10.98	10.60	6.47	8.89	19.00
13.32	14.27	9.45	11.35	9.95	14.53
29.11	15.09	3.28	9.15	2.94	13.68
12.68	7.61	10.23	9.70	5.06	20.84
7.51	5.80	8.13	9.77	17.86	35.18
9.90	3.63	9.13	11.61	11.75	17.00
10.25	5.07	10.17	9.09	13.25	20.66
11.11	6.15	7.61	8.53	10.14	17.45
12.17	14.26	14.39	8.29	6.22	16.38
10.24	2.59	6.09	15.90	12.90	19.09
10.18	6.05	12.14	11.94	5.69	14.77
8.88	2.70	12.23	9.54	16.77	22.66
12.34	7.73	11.68	10.43	17.65	10.66
8.51	14.02	12.01	10.87	21.52	28.47
26.16	17.44	16.89	7.13	13.22	19.44
12.95	8.24	7.18	11.88	12.18	21.20
16.93	13.37	17.59	12.03	9.22	23.09
14.70	10.78	14.58			
10.32	5.16	17.00			
8.98	4.49	4.26			
9.70	11.59	6.83			
12.72	8.63	5.59			
9.49	2.16	6.23			
8.22	7.95	6.72			
13.70	11.22	4.91			
8.21	9.85	8.17			
15.86	11.42	13.06			
9.18	9.18	9.49			
12.49	4.67	11.94			
17.32	6.86	4.44			

Source: Data courtesy of M. Keaton.

6.21. Using Moody's bond ratings, samples of 20 Aa (middle-high quality) corporate bonds and 20 Baa (top-medium quality) corporate bonds were selected. For each of the corresponding companies, the ratios

X_1 = current ratio (a measure of short-term liquidity)

X_2 = long-term interest rate (a measure of interest coverage)

X_3 = debt-to-equity ratio (a measure of financial risk or leverage)

X_4 = rate of return on equity (a measure of profitability)

Table 6.11 Male Hook-Billed Kite Data					
x_1 (Tail length)	x_2 (Wing length)	x_1 (Tail length)	x_2 (Wing length)	x_1 (Tail length)	x_2 (Wing length)
180	278	185	282	284	277
186	277	195	285	176	281
206	308	183	276	185	287
184	290	202	308	191	295
177	273	177	254	177	267
177	284	177	268	197	310
176	267	170	260	199	299
200	281	186	274	190	273
191	287	177	272	180	278
193	271	178	266	189	280
212	302	192	281	194	290
181	254	204	276	186	287
195	297	191	290	191	286
187	281	178	265	187	288
190	284	177	275	186	275

Source: Data courtesy of S. Temple.

were recorded. The summary statistics are as follows:

Aa bond companies: $n_1 = 20, \bar{\mathbf{x}}_1' = [2.287, 12.600, .347, 14.830]$, and

$$
\mathbf{S}_1 = \begin{bmatrix}
.459 & .254 & -.026 & -.244 \\
.254 & 27.465 & -.589 & -.267 \\
-.026 & -.589 & .030 & .102 \\
-.244 & -.267 & .102 & 6.854
\end{bmatrix}
$$

Baa bond companies: $n_2 = 20, \bar{\mathbf{x}}_2' = [2.404, 7.155, .524, 12.840]$,

$$
\mathbf{S}_2 = \begin{bmatrix}
.944 & -.089 & .002 & -.719 \\
-.089 & 16.432 & -.400 & 19.044 \\
.002 & -.400 & .024 & -.094 \\
-.719 & 19.044 & -.094 & 61.854
\end{bmatrix}
$$

and

$$
\mathbf{S}_{\text{pooled}} = \begin{bmatrix}
.701 & .083 & -.012 & -.481 \\
.083 & 21.949 & -.494 & 9.388 \\
-.012 & -.494 & .027 & .004 \\
-.481 & 9.388 & .004 & 34.354
\end{bmatrix}
$$

(a) Does pooling appear reasonable here? Comment on the pooling procedure in this case.

(b) Are the financial characteristics of firms with Aa bonds different from those with Baa bonds? Using the pooled covariance matrix, test for the equality of mean vectors. Set $\alpha = .05$.

(c) Calculate the linear combinations of mean components most responsible for rejecting $H_0: \boldsymbol{\mu}_1 - \boldsymbol{\mu}_2 = \mathbf{0}$ in Part b.

(d) Bond rating companies are interested in a company's ability to satisfy its outstanding debt obligations as they mature. Does it appear as if one or more of the foregoing financial ratios might be useful in helping to classify a bond as "high" or "medium" quality? Explain.

(e) Repeat part (b) assuming normal populations with unequal covariance matices (see (6-27), (6-28) and (6-29)). Does your conclusion change?

6.22. Researchers interested in assessing pulmonary function in nonpathological populations asked subjects to run on a treadmill until exhaustion. Samples of air were collected at definite intervals and the gas contents analyzed. The results on 4 measures of oxygen consumption for 25 males and 25 females are given in Table 6.12 on page 348. The variables were

$$X_1 = \text{resting volume } O_2 \text{ (L/min)}$$
$$X_2 = \text{resting volume } O_2 \text{ (mL/kg/min)}$$
$$X_3 = \text{maximum volume } O_2 \text{ (L/min)}$$
$$X_4 = \text{maximum volume } O_2 \text{ (mL/kg/min)}$$

(a) Look for gender differences by testing for equality of group means. Use $\alpha = .05$. If you reject $H_0: \boldsymbol{\mu}_1 - \boldsymbol{\mu}_2 = \mathbf{0}$, find the linear combination most responsible.

(b) Construct the 95% simultaneous confidence intervals for each $\mu_{1i} - \mu_{2i}, i = 1, 2, 3, 4$. Compare with the corresponding Bonferroni intervals.

(c) The data in Table 6.12 were collected from graduate-student volunteers, and thus they do not represent a random sample. Comment on the possible implications of this information.

6.23. Construct a one-way MANOVA using the width measurements from the iris data in Table 11.5. Construct 95% simultaneous confidence intervals for differences in mean components for the two responses for each pair of populations. Comment on the validity of the assumption that $\boldsymbol{\Sigma}_1 = \boldsymbol{\Sigma}_2 = \boldsymbol{\Sigma}_3$.

6.24. Researchers have suggested that a change in skull size over time is evidence of the inter-breeding of a resident population with immigrant populations. Four measurements were made of male Egyptian skulls for three different time periods: period 1 is 4000 B.C., period 2 is 3300 B.C., and period 3 is 1850 B.C. The data are shown in Table 6.13 on page 349 (see the skull data on the website www.prenhall.com/statistics). The measured variables are

$$X_1 = \text{maximum breadth of skull (mm)}$$
$$X_2 = \text{basibregmatic height of skull (mm)}$$
$$X_3 = \text{basialveolar length of skull (mm)}$$
$$X_4 = \text{nasal height of skull (mm)}$$

Construct a one-way MANOVA of the Egyptian skull data. Use $\alpha = .05$. Construct 95% simultaneous confidence intervals to determine which mean components differ among the populations represented by the three time periods. Are the usual MANOVA assumptions realistic for these data? Explain.

6.25. Construct a one-way MANOVA of the crude-oil data listed in Table 11.7 on page 662. Construct 95% simultaneous confidence intervals to determine which mean components differ among the populations. (You may want to consider transformations of the data to make them more closely conform to the usual MANOVA assumptions.)

Table 6.12 Oxygen-Consumption Data

	Males				Females		
x_1 Resting O_2 (L/min)	x_2 Resting O_2 (mL/kg/min)	x_3 Maximum O_2 (L/min)	x_4 Maximum O_2 (mL/kg/min)	x_1 Resting O_2 (L/min)	x_2 Resting O_2 (mL/kg/min)	x_3 Maximum O_2 (L/min)	x_4 Maximum O_2 (mL/kg/min)
0.34	3.71	2.87	30.87	0.29	5.04	1.93	33.85
0.39	5.08	3.38	43.85	0.28	3.95	2.51	35.82
0.48	5.13	4.13	44.51	0.31	4.88	2.31	36.40
0.31	3.95	3.60	46.00	0.30	5.97	1.90	37.87
0.36	5.51	3.11	47.02	0.28	4.57	2.32	38.30
0.33	4.07	3.95	48.50	0.11	1.74	2.49	39.19
0.43	4.77	4.39	48.75	0.25	4.66	2.12	39.21
0.48	6.69	3.50	48.86	0.26	5.28	1.98	39.94
0.21	3.71	2.82	48.92	0.39	7.32	2.25	42.41
0.32	4.35	3.59	48.38	0.37	6.22	1.71	28.97
0.54	7.89	3.47	50.56	0.31	4.20	2.76	37.80
0.32	5.37	3.07	51.15	0.35	5.10	2.10	31.10
0.40	4.95	4.43	55.34	0.29	4.46	2.50	38.30
0.31	4.97	3.56	56.67	0.33	5.60	3.06	51.80
0.44	6.68	3.86	58.49	0.18	2.80	2.40	37.60
0.32	4.80	3.31	49.99	0.28	4.01	2.58	36.78
0.50	6.43	3.29	42.25	0.44	6.69	3.05	46.16
0.36	5.99	3.10	51.70	0.22	4.55	1.85	38.95
0.48	6.30	4.80	63.30	0.34	5.73	2.43	40.60
0.40	6.00	3.06	46.23	0.30	5.12	2.58	43.69
0.42	6.04	3.85	55.08	0.31	4.77	1.97	30.40
0.55	6.45	5.00	58.80	0.27	5.16	2.03	39.46
0.50	5.55	5.23	57.46	0.66	11.05	2.32	39.34
0.34	4.27	4.00	50.35	0.37	5.23	2.48	34.86
0.40	4.58	2.82	32.48	0.35	5.37	2.25	35.07

Source: Data courtesy of S. Rokicki.

Table 6.13 Egyptian Skull Data

MaxBreath (x_1)	BasHeight (x_2)	BasLength (x_3)	NasHeight (x_4)	Time Period
131	138	89	49	1
125	131	92	48	1
131	132	99	50	1
119	132	96	44	1
136	143	100	54	1
138	137	89	56	1
139	130	108	48	1
125	136	93	48	1
131	134	102	51	1
134	134	99	51	1
⋮	⋮	⋮	⋮	⋮
124	138	101	48	2
133	134	97	48	2
138	134	98	45	2
148	129	104	51	2
126	124	95	45	2
135	136	98	52	2
132	145	100	54	2
133	130	102	48	2
131	134	96	50	2
133	125	94	46	2
⋮	⋮	⋮	⋮	⋮
132	130	91	52	3
133	131	100	50	3
138	137	94	51	3
130	127	99	45	3
136	133	91	49	3
134	123	95	52	3
136	137	101	54	3
133	131	96	49	3
138	133	100	55	3
138	133	91	46	3

Source: Data courtesy of J. Jackson.

6.26. A project was designed to investigate how consumers in Green Bay, Wisconsin, would react to an electrical time-of-use pricing scheme. The cost of electricity during peak periods for some customers was set at eight times the cost of electricity during off-peak hours. Hourly consumption (in kilowatt-hours) was measured on a hot summer day in July and compared, for both the test group and the control group, with baseline consumption measured on a similar day before the experimental rates began. The responses,

$$\log(\text{current consumption}) - \log(\text{baseline consumption})$$

for the hours ending 9 A.M. 11 A.M. (a peak hour), 1 P.M., and 3 P.M. (a peak hour) produced the following summary statistics:

Test group:	$n_1 = 28, \bar{x}_1' = [.153, -.231, -.322, -.339]$
Control group:	$n_2 = 58, \bar{x}_2' = [.151, .180, .256, .257]$
and	

$$\mathbf{S}_{\text{pooled}} = \begin{bmatrix} .804 & .355 & .228 & .232 \\ .355 & .722 & .233 & .199 \\ .228 & .233 & .592 & .239 \\ .232 & .199 & .239 & .479 \end{bmatrix}$$

Source: Data courtesy of Statistical Laboratory, University of Wisconsin.

Perform a profile analysis. Does time-of-use pricing seem to make a difference in electrical consumption? What is the nature of this difference, if any? Comment. (Use a significance level of $\alpha = .05$ for any statistical tests.)

6.27. As part of the study of love and marriage in Example 6.14, a sample of husbands and wives were asked to respond to these questions:

1. What is the level of passionate love you feel for your partner?
2. What is the level of passionate love that your partner feels for you?
3. What is the level of companionate love that you feel for your partner?
4. What is the level of companionate love that your partner feels for you?

The responses were recorded on the following 5-point scale.

None at all	Very little	Some	A great deal	Tremendous amount
1	2	3	4	5

Thirty husbands and 30 wives gave the responses in Table 6.14, where X_1 = a 5-point-scale response to Question 1, X_2 = a 5-point-scale response to Question 2, X_3 = a 5-point-scale response to Question 3, and X_4 = a 5-point-scale response to Question 4.

(a) Plot the mean vectors for husbands and wives as sample profiles.

(b) Is the husband rating wife profile parallel to the wife rating husband profile? Test for parallel profiles with $\alpha = .05$. If the profiles appear to be parallel, test for coincident profiles at the same level of significance. Finally, if the profiles are coincident, test for level profiles with $\alpha = .05$. What conclusion(s) can be drawn from this analysis?

6.28. Two species of biting flies (genus *Leptoconops*) are so similar morphologically, that for many years they were thought to be the same. Biological differences such as sex ratios of emerging flies and biting habits were found to exist. Do the taxonomic data listed in part in Table 6.15 on page 352 and on the website www.prenhall.com/statistics indicate any difference in the two species *L. carteri* and *L. torrens*? Test for the equality of the two population mean vectors using $\alpha = .05$. If the hypotheses of equal mean vectors is rejected, determine the mean components (or linear combinations of mean components) most responsible for rejecting H_0. Justify your use of normal-theory methods for these data.

6.29. Using the data on bone mineral content in Table 1.8, investigate equality between the dominant and nondominant bones.

Table 6.14 Spouse Data

Husband rating wife				Wife rating husband			
x_1	x_2	x_3	x_4	x_1	x_2	x_3	x_4
2	3	5	5	4	4	5	5
5	5	4	4	4	5	5	5
4	5	5	5	4	4	5	5
4	3	4	4	4	5	5	5
3	3	5	5	4	4	5	5
3	3	4	5	3	3	4	4
3	4	4	4	4	3	5	4
4	4	5	5	3	4	5	5
4	5	5	5	4	4	5	4
4	4	3	3	3	4	4	4
4	4	5	5	4	5	5	5
5	5	4	4	5	5	5	5
4	4	4	4	4	4	5	5
4	3	5	5	4	4	4	4
4	4	5	5	4	4	5	5
3	3	4	5	3	4	4	4
4	5	4	4	5	5	5	5
5	5	5	5	4	5	4	4
5	5	4	4	3	4	4	4
4	4	4	4	5	3	4	4
4	4	4	4	5	3	4	4
4	4	4	4	4	5	4	4
3	4	5	5	2	5	5	5
5	3	5	5	3	4	5	5
5	5	3	3	4	3	5	5
3	3	4	4	4	4	4	4
4	4	4	4	4	4	5	5
3	3	5	5	3	4	4	4
4	4	3	3	4	4	5	4
4	4	5	5	4	4	5	5

Source: Data courtesy of E. Hatfield.

(a) Test using $\alpha = .05$.

(b) Construct 95% simultaneous confidence intervals for the mean differences.

(c) Construct the Bonferroni 95% simultaneous intervals, and compare these with the intervals in Part b.

6.30. Table 6.16 on page 353 contains the bone mineral contents, for the first 24 subjects in Table 1.8, 1 year after their participation in an experimental program. Compare the data from both tables to determine whether there has been bone loss.

(a) Test using $\alpha = .05$.

(b) Construct 95% simultaneous confidence intervals for the mean differences.

(c) Construct the Bonferroni 95% simultaneous intervals, and compare these with the intervals in Part b.

Table 6.15   Biting-Fly Data

	x_1 (Wing length)	x_2 (Wing width)	x_3 (Third palp length)	x_4 (Third palp width)	x_5 (Fourth palp length)	x_6 (Length of antennal segment 12)	x_7 (Length of antennal segment 13)
	85	41	31	13	25	9	8
	87	38	32	14	22	13	13
	94	44	36	15	27	8	9
	92	43	32	17	28	9	9
	96	43	35	14	26	10	10
	91	44	36	12	24	9	9
	90	42	36	16	26	9	9
	92	43	36	17	26	9	9
	91	41	36	14	23	9	9
	87	38	35	11	24	9	10
L. torrens	⋮	⋮	⋮	⋮	⋮	⋮	⋮
	106	47	38	15	26	10	10
	105	46	34	14	31	10	11
	103	44	34	15	23	10	10
	100	41	35	14	24	10	10
	109	44	36	13	27	11	10
	104	45	36	15	30	10	10
	95	40	35	14	23	9	10
	104	44	34	15	29	9	10
	90	40	37	12	22	9	10
	104	46	37	14	30	10	10
	86	19	37	11	25	9	9
	94	40	38	14	31	6	7
	103	48	39	14	33	10	10
	82	41	35	12	25	9	8
	103	43	42	15	32	9	9
	101	43	40	15	25	9	9
	103	45	44	14	29	11	11
	100	43	40	18	31	11	10
	99	41	42	15	31	10	10
	100	44	43	16	34	10	10
L. carteri	⋮	⋮	⋮	⋮	⋮	⋮	⋮
	99	42	38	14	33	9	9
	110	45	41	17	36	9	10
	99	44	35	16	31	10	10
	103	43	38	14	32	10	10
	95	46	36	15	31	8	8
	101	47	38	14	37	11	11
	103	47	40	15	32	11	11
	99	43	37	14	23	11	10
	105	50	40	16	33	12	11
	99	47	39	14	34	7	7

Source: Data courtesy of William Atchley.

Table 6.16 Mineral Content in Bones (After 1 Year)

Subject number	Dominant radius	Radius	Dominant humerus	Humerus	Dominant ulna	Ulna
1	1.027	1.051	2.268	2.246	.869	.964
2	.857	.817	1.718	1.710	.602	.689
3	.875	.880	1.953	1.756	.765	.738
4	.873	.698	1.668	1.443	.761	.698
5	.811	.813	1.643	1.661	.551	.619
6	.640	.734	1.396	1.378	.753	.515
7	.947	.865	1.851	1.686	.708	.787
8	.886	.806	1.742	1.815	.687	.715
9	.991	.923	1.931	1.776	.844	.656
10	.977	.925	1.933	2.106	.869	.789
11	.825	.826	1.609	1.651	.654	.726
12	.851	.765	2.352	1.980	.692	.526
13	.770	.730	1.470	1.420	.670	.580
14	.912	.875	1.846	1.809	.823	.773
15	.905	.826	1.842	1.579	.746	.729
16	.756	.727	1.747	1.860	.656	.506
17	.765	.764	1.923	1.941	.693	.740
18	.932	.914	2.190	1.997	.883	.785
19	.843	.782	1.242	1.228	.577	.627
20	.879	.906	2.164	1.999	.802	.769
21	.673	.537	1.573	1.330	.540	.498
22	.949	.900	2.130	2.159	.804	.779
23	.463	.637	1.041	1.265	.570	.634
24	.776	.743	1.442	1.411	.585	.640

Source: Data courtesy of Everett Smith.

6.31. Peanuts are an important crop in parts of the southern United States. In an effort to develop improved plants, crop scientists routinely compare varieties with respect to several variables. The data for one two-factor experiment are given in Table 6.17 on page 354. Three varieties (5, 6, and 8) were grown at two geographical locations (1, 2) and, in this case, the three variables representing yield and the two important grade–grain characteristics were measured. The three variables are

$$X_1 = \text{Yield (plot weight)}$$
$$X_2 = \text{Sound mature kernels (weight in grams—maximum of 250 grams)}$$
$$X_3 = \text{Seed size (weight, in grams, of 100 seeds)}$$

There were two replications of the experiment.

(a) Perform a two-factor MANOVA using the data in Table 6.17. Test for a location effect, a variety effect, and a location–variety interaction. Use $\alpha = .05$.

(b) Analyze the residuals from Part a. Do the usual MANOVA assumptions appear to be satisfied? Discuss.

(c) Using the results in Part a, can we conclude that the location and/or variety effects are additive? If not, does the interaction effect show up for some variables, but not for others? Check by running three separate univariate two-factor ANOVAs.

Table 6.17 Peanut Data

Factor 1 Location	Factor 2 Variety	x_1 Yield	x_2 SdMatKer	x_3 SeedSize
1	5	195.3	153.1	51.4
1	5	194.3	167.7	53.7
2	5	189.7	139.5	55.5
2	5	180.4	121.1	44.4
1	6	203.0	156.8	49.8
1	6	195.9	166.0	45.8
2	6	202.7	166.1	60.4
2	6	197.6	161.8	54.1
1	8	193.5	164.5	57.8
1	8	187.0	165.1	58.6
2	8	201.5	166.8	65.0
2	8	200.0	173.8	67.2

Source: Data courtesy of Yolanda Lopez.

(d) Larger numbers correspond to better yield and grade–grain characteristics. Using location 2, can we conclude that one variety is better than the other two for each characteristic? Discuss your answer, using 95% Bonferroni simultaneous intervals for pairs of varieties.

6.32. In one experiment involving remote sensing, the spectral reflectance of three species of 1-year-old seedlings was measured at various wavelengths during the growing season. The seedlings were grown with two different levels of nutrient: the optimal level, coded +, and a suboptimal level, coded −. The species of seedlings used were sitka spruce (SS), Japanese larch (JL), and lodgepole pine (LP). Two of the variables measured were

$$X_1 = \text{percent spectral reflectance at wavelength 560 nm (green)}$$
$$X_2 = \text{percent spectral reflectance at wavelength 720 nm (near infrared)}$$

The cell means (CM) for Julian day 235 for each combination of species and nutrient level are as follows. These averages are based on four replications.

560CM	720CM	Species	Nutrient
10.35	25.93	SS	+
13.41	38.63	JL	+
7.78	25.15	LP	+
10.40	24.25	SS	−
17.78	41.45	JL	−
10.40	29.20	LP	−

(a) Treating the cell means as individual observations, perform a two-way MANOVA to test for a species effect and a nutrient effect. Use $\alpha = .05$.

(b) Construct a two-way ANOVA for the 560CM observations and another two-way ANOVA for the 720CM observations. Are these results consistent with the MANOVA results in Part a? If not, can you explain any differences?

6.33. Refer to Exercise 6.32. The data in Table 6.18 are measurements on the variables

X_1 = percent spectral reflectance at wavelength 560 nm (green)

X_2 = percent spectral reflectance at wavelength 720 nm (near infrared)

for three species (sitka spruce [SS], Japanese larch [JL], and lodgepole pine [LP]) of 1-year-old seedlings taken at three different times (Julian day 150 [1], Julian day 235 [2], and Julian day 320 [3]) during the growing season. The seedlings were all grown with the optimal level of nutrient.

(a) Perform a two-factor MANOVA using the data in Table 6.18. Test for a species effect, a time effect and species–time interaction. Use α = .05.

Table 6.18 Spectral Reflectance Data

560 nm	720 nm	Species	Time	Replication
9.33	19.14	SS	1	1
8.74	19.55	SS	1	2
9.31	19.24	SS	1	3
8.27	16.37	SS	1	4
10.22	25.00	SS	2	1
10.13	25.32	SS	2	2
10.42	27.12	SS	2	3
10.62	26.28	SS	2	4
15.25	38.89	SS	3	1
16.22	36.67	SS	3	2
17.24	40.74	SS	3	3
12.77	67.50	SS	3	4
12.07	33.03	JL	1	1
11.03	32.37	JL	1	2
12.48	31.31	JL	1	3
12.12	33.33	JL	1	4
15.38	40.00	JL	2	1
14.21	40.48	JL	2	2
9.69	33.90	JL	2	3
14.35	40.15	JL	2	4
38.71	77.14	JL	3	1
44.74	78.57	JL	3	2
36.67	71.43	JL	3	3
37.21	45.00	JL	3	4
8.73	23.27	LP	1	1
7.94	20.87	LP	1	2
8.37	22.16	LP	1	3
7.86	21.78	LP	1	4
8.45	26.32	LP	2	1
6.79	22.73	LP	2	2
8.34	26.67	LP	2	3
7.54	24.87	LP	2	4
14.04	44.44	LP	3	1
13.51	37.93	LP	3	2
13.33	37.93	LP	3	3
12.77	60.87	LP	3	4

Source: Data courtesy of Mairtin Mac Siurtain.

(b) Do you think the usual MANOVA assumptions are satisfied for the these data? Discuss with reference to a residual analysis, and the possibility of correlated observations over time.

(c) Foresters are particularly interested in the interaction of species and time. Does interaction show up for one variable but not for the other? Check by running a univariate two-factor ANOVA for each of the two responses.

(d) Can you think of another method of analyzing these data (or a different experimental design) that would allow for a potential time trend in the spectral reflectance numbers?

6.34. Refer to Example 6.15.

(a) Plot the profiles, the components of $\bar{x}_1$ versus time and those of $\bar{x}_2$ versus time, on the same graph. Comment on the comparison.

(b) Test that linear growth is adequate. Take $\alpha = .01$.

6.35. Refer to Example 6.15 but treat all 31 subjects as a single group. The maximum likelihood estimate of the $(q + 1) \times 1$ $\boldsymbol{\beta}$ is

$$\hat{\boldsymbol{\beta}} = (\mathbf{B}'\mathbf{S}^{-1}\mathbf{B})^{-1}\mathbf{B}'\mathbf{S}^{-1}\bar{\mathbf{x}}$$

where $\mathbf{S}$ is the sample covariance matrix.

The estimated covariances of the maximum likelihood estimators are

$$\widehat{\text{Cov}}(\hat{\boldsymbol{\beta}}) = \frac{(n-1)(n-2)}{(n-1-p+q)(n-p+q)n}(\mathbf{B}'\mathbf{S}^{-1}\mathbf{B})^{-1}$$

Fit a quadratic growth curve to this single group and comment on the fit.

6.36. Refer to Example 6.4. Given the summary information on electrical usage in this example, use Box's M-test to test the hypothesis $H_0: \boldsymbol{\Sigma}_1 = \boldsymbol{\Sigma}_2 = \boldsymbol{\Sigma}$. Here $\boldsymbol{\Sigma}_1$ is the covariance matrix for the two measures of usage for the population of Wisconsin homeowners *with* air conditioning, and $\boldsymbol{\Sigma}_2$ is the electrical usage covariance matrix for the population of Wisconsin homeowners *without* air conditioning. Set $\alpha = .05$.

6.37. Table 6.9 page 344 contains the carapace measurements for 24 female and 24 male turtles. Use Box's M-test to test $H_0: \boldsymbol{\Sigma}_1 = \boldsymbol{\Sigma}_2 = \boldsymbol{\Sigma}$. where $\boldsymbol{\Sigma}_1$ is the population covariance matrix for carapace measurements for female turtles, and $\boldsymbol{\Sigma}_2$ is the population covariance matrix for carapace measurements for male turtles. Set $\alpha = .05$.

6.38. Table 11.7 page 662 contains the values of three trace elements and two measures of hydrocarbons for crude oil samples taken from three groups (zones) of sandstone. Use Box's M-test to test equality of population covariance matrices for the three sandstone groups. Set $\alpha = .05$. Here there are $p = 5$ variables and you may wish to consider transformations of the measurements on these variables to make them more nearly normal.

6.39. Anacondas are some of the largest snakes in the world. Jesus Ravis and his fellow researchers capture a snake and measure its (i) snout vent length (cm) or the length from the snout of the snake to its vent where it evacuates waste and (ii) weight (kilograms). A sample of these measurements in shown in Table 6.19.

(a) Test for equality of means between males and females using $\alpha = .05$. Apply the large sample statistic.

(b) Is it reasonable to pool variances in this case? Explain.

(c) Find the 95% Boneferroni confidence intervals for the mean differences between males and females on both length and weight.

Table 6.19 Anaconda Data

Snout vent Length	Weight	Gender	Snout vent length	Weight	Gender
271.0	18.50	F	176.7	3.00	M
477.0	82.50	F	259.5	9.75	M
306.3	23.40	F	258.0	10.07	M
365.3	33.50	F	229.8	7.50	M
466.0	69.00	F	233.0	6.25	M
440.7	54.00	F	237.5	9.85	M
315.0	24.97	F	268.3	10.00	M
417.5	56.75	F	222.5	9.00	M
307.3	23.15	F	186.5	3.75	M
319.0	29.51	F	238.8	9.75	M
303.9	19.98	F	257.6	9.75	M
331.7	24.00	F	172.0	3.00	M
435.0	70.37	F	244.7	10.00	M
261.3	15.50	F	224.7	7.25	M
384.8	63.00	F	231.7	9.25	M
360.3	39.00	F	235.9	7.50	M
441.4	53.00	F	236.5	5.75	M
246.7	15.75	F	247.4	7.75	M
365.3	44.00	F	223.0	5.75	M
336.8	30.00	F	223.7	5.75	M
326.7	34.00	F	212.5	7.65	M
312.0	25.00	F	223.2	7.75	M
226.7	9.25	F	225.0	5.84	M
347.4	30.00	F	228.0	7.53	M
280.2	15.25	F	215.6	5.75	M
290.7	21.50	F	221.0	6.45	M
438.6	57.00	F	236.7	6.49	M
377.1	61.50	F	235.3	6.00	M

Source: Data Courtesy of Jesus Ravis.

6.40. Compare the male national track records in Table 8.6 with the female national track records in Table 1.9 using the results for the 100m, 200m, 400m, 800m and 1500m races. Treat the data as a random sample of size 64 of the twelve record values.

(a) Test for equality of means between males and females using $\alpha = .05$. Explain why it may be appropriate to analyze differences.

(b) Find the 95% Bonferroni confidence intervals for the mean differences between male and females on all of the races.

6.41. When cell phone relay towers are not working properly, wireless providers can lose great amounts of money so it is important to be able to fix problems expeditiously. A first step toward understanding the problems involved is to collect data from a designed experiment involving three factors. A problem was initially classified as low or high severity, simple or complex, and the engineer assigned was rated as relatively new (novice) or expert (guru).

Two times were observed. The time to assess the problem and plan an attack and the time to implement the solution were each measured in hours. The data are given in Table 6.20.

Perform a MANOVA including appropriate confidence intervals for important effects.

Table 6.20 Fixing Breakdowns

Problem Severity Level	Problem Complexity Level	Engineer Experience Level	Problem Assessment Time	Problem Implementation Time	Total Resolution Time
Low	Simple	Novice	3.0	6.3	9.3
Low	Simple	Novice	2.3	5.3	7.6
Low	Simple	Guru	1.7	2.1	3.8
Low	Simple	Guru	1.2	1.6	2.8
Low	Complex	Novice	6.7	12.6	19.3
Low	Complex	Novice	7.1	12.8	19.9
Low	Complex	Guru	5.6	8.8	14.4
Low	Complex	Guru	4.5	9.2	13.7
High	Simple	Novice	4.5	9.5	14.0
High	Simple	Novice	4.7	10.7	15.4
High	Simple	Guru	3.1	6.3	9.4
High	Simple	Guru	3.0	5.6	8.6
High	Complex	Novice	7.9	15.6	23.5
High	Complex	Novice	6.9	14.9	21.8
High	Complex	Guru	5.0	10.4	15.4
High	Complex	Guru	5.3	10.4	15.7

Source: Data courtesy of Dan Porter.

References

1. Anderson, T. W. *An Introduction to Multivariate Statistical Analysis* (3rd ed.). New York: John Wiley, 2003.

2. Bacon-Shone, J., and W. K. Fung. "A New Graphical Method for Detecting Single and Multiple Outliers in Univariate and Multivariate Data." *Applied Statistics*, **36**, no. 2 (1987), 153–162.

3. Bartlett, M. S. "Properties of Sufficiency and Statistical Tests." *Proceedings of the Royal Society of London (A)*, **160** (1937), 268–282.

4. Bartlett, M. S. "Further Aspects of the Theory of Multiple Regression." *Proceedings of the Cambridge Philosophical Society*, **34** (1938), 33–40.

5. Bartlett, M. S. "Multivariate Analysis." *Journal of the Royal Statistical Society Supplement (B)*, **9** (1947), 176–197.

6. Bartlett, M. S. "A Note on the Multiplying Factors for Various χ^2 Approximations." *Journal of the Royal Statistical Society (B)*, **16** (1954), 296–298.

7. Box, G. E. P., "A General Distribution Theory for a Class of Likelihood Criteria." *Biometrika,* **36** (1949), 317–346.

8. Box, G. E. P., "Problems in the Analysis of Growth and Wear Curves." *Biometrics,* **6** (1950), 362–389.

9. Box, G. E. P., and N. R. Draper. *Evolutionary Operation: A Statistical Method for Process Improvement*. New York: John Wiley, 1969.

10. Box, G. E. P., W. G. Hunter, and J. S. Hunter. *Statistics for Experimenters* (2nd ed.). New York: John Wiley, 2005.

11. Johnson, R. A. and G. K. Bhattacharyya. *Statistics: Principles and Methods* (5th ed.). New York: John Wiley, 2005.

12. Jolicoeur, P., and J. E. Mosimann. "Size and Shape Variation in the Painted Turtle: A Principal Component Analysis." *Growth*, **24** (1960), 339–354.

13. Khattree, R. and D. N. Naik, *Applied Multivariate Statistics with SAS® Software* (2nd ed.). Cary, NC: SAS Institute Inc., 1999.

14. Kshirsagar, A. M., and W. B. Smith, *Growth Curves*. New York: Marcel Dekker, 1995.

15. Krishnamoorthy, K., and J. Yu. "Modified Nel and Van der Merwe Test for the Multivariate Behrens-Fisher Problem." *Statistics & Probability Letters*, **66** (2004), 161–169.

16. Mardia, K. V., "The Effect of Nonnormality on some Multivariate Tests and Robustnes to Nonnormality in the Linear Model." *Biometrika*, **58** (1971), 105-121.

17. Montgomery, D. C. *Design and Analysis of Experiments* (6th ed.). New York: John Wiley, 2005.

18. Morrison, D. F. *Multivariate Statistical Methods* (4th ed.). Belmont, CA: Brooks/Cole Thomson Learning, 2005.

19. Nel, D. G., and C. A. Van der Merwe. "A Solution to the Multivariate Behrens-Fisher Problem." *Communications in Statistics—Theory and Methods*, **15** (1986), 3719–3735.

20. Pearson, E. S., and H. O. Hartley, eds. *Biometrika Tables for Statisticians*. vol. II. Cambridge, England: Cambridge University Press, 1972.

21. Potthoff, R. F. and S. N. Roy. "A Generalized Multivariate Analysis of Variance Model Useful Especially for Growth Curve Problems." *Biometrika*, **51** (1964), 313–326.

22. Scheffé, H. *The Analysis of Variance*. New York: John Wiley, 1959.

23. Tiku, M. L., and N. Balakrishnan. "Testing the Equality of Variance–Covariance Matrices the Robust Way." *Communications in Statistics—Theory and Methods*, **14**, no. 12 (1985), 3033–3051.

24. Tiku, M. L., and M. Singh. "Robust Statistics for Testing Mean Vectors of Multivariate Distributions." *Communications in Statistics—Theory and Methods*, **11**, no. 9 (1982), 985–1001.

25. Wilks, S. S. "Certain Generalizations in the Analysis of Variance." *Biometrika*, **24** (1932), 471–494.

Chapter
7

MULTIVARIATE LINEAR REGRESSION MODELS

7.1 Introduction

→ y axis (handwritten)

Regression analysis is the statistical methodology for predicting values of one or more *response* (dependent) variables from a collection of *predictor* (independent) variable values. It can also be used for assessing the effects of the predictor variables on the responses. Unfortunately, the name *regression*, culled from the title of the first paper on the subject by F. Galton [15], in no way reflects either the importance or breadth of application of this methodology.

In this chapter, we first discuss the multiple regression model for the prediction of a *single* response. This model is then generalized to handle the prediction of *several* dependent variables. Our treatment must be somewhat terse, as a vast literature exists on the subject. (If you are interested in pursuing regression analysis, see the following books, in ascending order of difficulty: Abraham and Ledolter [1], Bowerman and O'Connell [6], Neter, Wasserman, Kutner, and Nachtsheim [20], Draper and Smith [13], Cook and Weisberg [11], Seber [23], and Goldberger [16].) Our abbreviated treatment highlights the regression assumptions and their consequences, alternative formulations of the regression model, and the general applicability of regression techniques to seemingly different situations.

x-axis (handwritten)

7.2 The Classical Linear Regression Model

Let $z_1, z_2, \ldots, z_r$ be r predictor variables thought to be related to a response variable Y. For example, with $r = 4$, we might have

$$Y = \text{current market value of home}$$

→ i.e. $y = mx + b$ (handwritten)

value m · (people's naivity) + b (handwritten)
← phtete (handwritten)

and

These are like the x's in y=mx+b (handwritten)

z_1 = square feet of living area

z_2 = location (indicator for zone of city)

z_3 = appraised value last year

z_4 = quality of construction (price per square foot)

The classical linear regression model states that Y is composed of a mean, which depends in a continuous manner on the z_i's, and a random error ε, which accounts for measurement error and the effects of other variables not explicitly considered in the model. The values of the predictor variables recorded from the experiment or set by the investigator are treated as *fixed*. The error (and hence the response) is viewed as a random variable whose behavior is characterized by a set of distributional assumptions.

Specifically, the linear regression model with a single response takes the form

will be linear in multidim space (handwritten)

$$Y = \beta_0 + \beta_1 z_1 + \cdots + \beta_r z_r + \varepsilon$$

$$[\text{Response}] = [\text{mean (depending on } z_1, z_2, \ldots, z_r)] + [\text{error}]$$

The term "linear" refers to the fact that the mean is a linear function of the unknown parameters $\beta_0, \beta_1, \ldots, \beta_r$. The predictor variables may or may not enter the model as first-order terms.

With n independent observations on Y and the associated values of z_i, the complete model becomes

Can't plot multidimensional (handwritten)

$$
\begin{aligned}
Y_1 &= \beta_0 + \beta_1 z_{11} + \beta_2 z_{12} + \cdots + \beta_r z_{1r} + \varepsilon_1 \\
Y_2 &= \beta_0 + \beta_1 z_{21} + \beta_2 z_{22} + \cdots + \beta_r z_{2r} + \varepsilon_2 \\
&\;\;\vdots \qquad\qquad\qquad \vdots \\
Y_n &= \beta_0 + \beta_1 z_{n1} + \beta_2 z_{n2} + \cdots + \beta_r z_{nr} + \varepsilon_n
\end{aligned}
\tag{7-1}
$$

where the error terms are assumed to have the following properties:

1. $E(\varepsilon_j) = 0$;
2. $\text{Var}(\varepsilon_j) = \sigma^2$ (constant); and
3. $\text{Cov}(\varepsilon_j, \varepsilon_k) = 0, j \neq k.$ → *no relation between each subsequent point* (handwritten)

$$\tag{7-2}$$

In matrix notation, (7-1) becomes

$$
\begin{bmatrix} Y_1 \\ Y_2 \\ \vdots \\ Y_n \end{bmatrix}
=
\begin{bmatrix}
1 & z_{11} & z_{12} & \cdots & z_{1r} \\
1 & z_{21} & z_{22} & \cdots & z_{2r} \\
\vdots & \vdots & \vdots & \ddots & \vdots \\
1 & z_{n1} & z_{n2} & \cdots & z_{nr}
\end{bmatrix}
\begin{bmatrix} \beta_0 \\ \beta_1 \\ \vdots \\ \beta_r \end{bmatrix}
+
\begin{bmatrix} \varepsilon_1 \\ \varepsilon_2 \\ \vdots \\ \varepsilon_n \end{bmatrix}
$$

or

$$
\underset{(n\times 1)}{\mathbf{Y}} = \underset{(n\times(r+1))}{\mathbf{Z}} \underset{((r+1)\times 1)}{\boldsymbol{\beta}} + \underset{(n\times 1)}{\boldsymbol{\varepsilon}}
$$

and the specifications in (7-2) become

1. $E(\boldsymbol{\varepsilon}) = \mathbf{0}$; and
2. $\text{Cov}(\boldsymbol{\varepsilon}) = E(\boldsymbol{\varepsilon}\boldsymbol{\varepsilon}') = \sigma^2 \mathbf{I}.$

Note that a one in the first column of the *design matrix* $\mathbf{Z}$ is the multiplier of the constant term β_0. It is customary to introduce the artificial variable $z_{j0} = 1$, so that

$$\beta_0 + \beta_1 z_{j1} + \cdots + \beta_r z_{jr} = \beta_0 z_{j0} + \beta_1 z_{j1} + \cdots + \beta_r z_{jr}$$

Each column of $\mathbf{Z}$ consists of the n values of the corresponding predictor variable, while the jth row of $\mathbf{Z}$ contains the values for all predictor variables on the jth trial.

Classical Linear Regression Model

$$\underset{(n \times 1)}{\mathbf{Y}} = \underset{(n \times (r+1))}{\mathbf{Z}} \underset{((r+1) \times 1)}{\boldsymbol{\beta}} + \underset{(n \times 1)}{\boldsymbol{\varepsilon}},$$

$$E(\boldsymbol{\varepsilon}) = \underset{(n \times 1)}{\mathbf{0}} \text{ and } \operatorname{Cov}(\boldsymbol{\varepsilon}) = \underset{(n \times n)}{\sigma^2 \mathbf{I}}, \qquad (7\text{-}3)$$

where $\boldsymbol{\beta}$ and σ^2 are unknown parameters and the design matrix $\mathbf{Z}$ has jth row $[z_{j0}, z_{j1}, \ldots, z_{jr}]$.

Although the error-term assumptions in (7-2) are very modest, we shall later need to add the assumption of joint normality for making confidence statements and testing hypotheses.

We now provide some examples of the linear regression model.

Example 7.1 (Fitting a straight-line regression model) Determine the linear regression model for fitting a straight line

$$\text{Mean response} = E(Y) = \beta_0 + \beta_1 z_1$$

to the data

z_1	0	1	2	3	4
y	1	4	3	8	9

Before the responses $\mathbf{Y}' = [Y_1, Y_2, \ldots, Y_5]$ are observed, the errors $\boldsymbol{\varepsilon}' = [\varepsilon_1, \varepsilon_2, \ldots, \varepsilon_5]$ are random, and we can write

$$\mathbf{Y} = \mathbf{Z}\boldsymbol{\beta} + \boldsymbol{\varepsilon}$$

where

$$\mathbf{Y} = \begin{bmatrix} Y_1 \\ Y_2 \\ \vdots \\ Y_5 \end{bmatrix}, \quad \mathbf{Z} = \begin{bmatrix} 1 & z_{11} \\ 1 & z_{21} \\ \vdots & \vdots \\ 1 & z_{51} \end{bmatrix}, \quad \boldsymbol{\beta} = \begin{bmatrix} \beta_0 \\ \beta_1 \end{bmatrix}, \quad \boldsymbol{\varepsilon} = \begin{bmatrix} \varepsilon_1 \\ \varepsilon_2 \\ \vdots \\ \varepsilon_5 \end{bmatrix}$$

The data for this model are contained in the observed response vector $\mathbf{y}$ and the design matrix $\mathbf{Z}$, where

$$\mathbf{y} = \begin{bmatrix} 1 \\ 4 \\ 3 \\ 8 \\ 9 \end{bmatrix}, \quad \mathbf{Z} = \begin{bmatrix} 1 & 0 \\ 1 & 1 \\ 1 & 2 \\ 1 & 3 \\ 1 & 4 \end{bmatrix}$$

Note that we can handle a quadratic expression for the mean response by introducing the term $\beta_2 z_2$, with $z_2 = z_1^2$. The linear regression model for the jth trial in this latter case is

$$Y_j = \beta_0 + \beta_1 z_{j1} + \beta_2 z_{j2} + \varepsilon_j$$

or

$$Y_j = \beta_0 + \beta_1 z_{j1} + \beta_2 z_{j1}^2 + \varepsilon_j$$

∎

Example 7.2 (The design matrix for one-way ANOVA as a regression model)
Determine the design matrix if the linear regression model is applied to the one-way ANOVA situation in Example 6.6.

We create so-called *dummy* variables to handle the three population means: $\mu_1 = \mu + \tau_1$, $\mu_2 = \mu + \tau_2$, and $\mu_3 = \mu + \tau_3$. We set

$$z_1 = \begin{cases} 1 & \text{if the observation is} \\ & \text{from population 1} \\ 0 & \text{otherwise} \end{cases} \qquad z_2 = \begin{cases} 1 & \text{if the observation is} \\ & \text{from population 2} \\ 0 & \text{otherwise} \end{cases}$$

$$z_3 = \begin{cases} 1 & \text{if the observation is} \\ & \text{from population 3} \\ 0 & \text{otherwise} \end{cases}$$

and $\beta_0 = \mu$, $\beta_1 = \tau_1$, $\beta_2 = \tau_2$, $\beta_3 = \tau_3$. Then

$$Y_j = \beta_0 + \beta_1 z_{j1} + \beta_2 z_{j2} + \beta_3 z_{j3} + \varepsilon_j, \qquad j = 1, 2, \ldots, 8$$

where we arrange the observations from the three populations in sequence. Thus, we obtain the observed response vector and design matrix

$$\mathbf{Y}_{(8 \times 1)} = \begin{bmatrix} 9 \\ 6 \\ 9 \\ 0 \\ 2 \\ 3 \\ 1 \\ 2 \end{bmatrix}; \qquad \mathbf{Z}_{(8 \times 4)} = \begin{bmatrix} 1 & 1 & 0 & 0 \\ 1 & 1 & 0 & 0 \\ 1 & 1 & 0 & 0 \\ 1 & 0 & 1 & 0 \\ 1 & 0 & 1 & 0 \\ 1 & 0 & 0 & 1 \\ 1 & 0 & 0 & 1 \\ 1 & 0 & 0 & 1 \end{bmatrix}$$

∎

The construction of dummy variables, as in Example 7.2, allows the whole of analysis of variance to be treated within the multiple linear regression framework.

7.3 Least Squares Estimation

One of the objectives of regression analysis is to develop an equation that will allow the investigator to predict the response for given values of the predictor variables. Thus, it is necessary to "fit" the model in (7-3) to the observed y_j corresponding to the known values $1, z_{j1}, \ldots, z_{jr}$. That is, we must determine the values for the *regression coefficients* $\boldsymbol{\beta}$ and the *error variance* σ^2 consistent with the available data.

Let $\mathbf{b}$ be trial values for $\boldsymbol{\beta}$. Consider the difference $y_j - b_0 - b_1 z_{j1} - \cdots - b_r z_{jr}$ between the observed response y_j and the value $b_0 + b_1 z_{j1} + \cdots + b_r z_{jr}$ that would be expected if $\mathbf{b}$ were the "true" parameter vector. Typically, the differences $y_j - b_0 - b_1 z_{j1} - \cdots - b_r z_{jr}$ will not be zero, because the response fluctuates (in a manner characterized by the error term assumptions) about its expected value. The *method of least squares* selects $\mathbf{b}$ so as to minimize the sum of the squares of the differences:

R codes: lm()

matlab: ...get β

or regress regstat

$$S(\mathbf{b}) = \sum_{j=1}^{n} (y_j - b_0 - b_1 z_{j1} - \cdots - b_r z_{jr})^2$$

$$= (\mathbf{y} - \mathbf{Zb})'(\mathbf{y} - \mathbf{Zb})$$

(7-4)

The coefficients $\mathbf{b}$ chosen by the least squares criterion are called *least squares estimates* of the regression parameters $\boldsymbol{\beta}$. They will henceforth be denoted by $\hat{\boldsymbol{\beta}}$ to emphasize their role as estimates of $\boldsymbol{\beta}$.

The coefficients $\hat{\boldsymbol{\beta}}$ are consistent with the data in the sense that they produce estimated (fitted) mean responses, $\hat{\beta}_0 + \hat{\beta}_1 z_{j1} + \cdots + \hat{\beta}_r z_{jr}$, the sum of whose squares of the differences from the observed y_j is as small as possible. The deviations

$$\hat{\varepsilon}_j = y_j - \hat{\beta}_0 - \hat{\beta}_1 z_{j1} - \cdots - \hat{\beta}_r z_{jr}, \qquad j = 1, 2, \ldots, n$$

(7-5)

are called *residuals*. The vector of residuals $\hat{\boldsymbol{\varepsilon}} = \mathbf{y} - \mathbf{Z}\hat{\boldsymbol{\beta}}$ contains the information about the remaining unknown parameter σ^2. (See Result 7.2.)

Result 7.1. Let $\mathbf{Z}$ have full rank $r + 1 \leq n$.[1] The least squares estimate of $\boldsymbol{\beta}$ in (7-3) is given by

$$\hat{\boldsymbol{\beta}} = (\mathbf{Z}'\mathbf{Z})^{-1}\mathbf{Z}'\mathbf{y}$$

Let $\hat{\mathbf{y}} = \mathbf{Z}\hat{\boldsymbol{\beta}} = \mathbf{Hy}$ denote the *fitted values* of $\mathbf{y}$, where $\mathbf{H} = \mathbf{Z}(\mathbf{Z}'\mathbf{Z})^{-1}\mathbf{Z}'$ is called "hat" matrix. Then the *residuals*

$$\hat{\boldsymbol{\varepsilon}} = \mathbf{y} - \hat{\mathbf{y}} = [\mathbf{I} - \mathbf{Z}(\mathbf{Z}'\mathbf{Z})^{-1}\mathbf{Z}']\mathbf{y} = (\mathbf{I} - \mathbf{H})\mathbf{y}$$

satisfy $\mathbf{Z}'\hat{\boldsymbol{\varepsilon}} = \mathbf{0}$ and $\hat{\mathbf{y}}'\hat{\boldsymbol{\varepsilon}} = 0$. Also, the

$$residual\ sum\ of\ squares = \sum_{j=1}^{n} (y_j - \hat{\beta}_0 - \hat{\beta}_1 z_{j1} - \cdots - \hat{\beta}_r z_{jr})^2 = \hat{\boldsymbol{\varepsilon}}'\hat{\boldsymbol{\varepsilon}}$$

$$= \mathbf{y}'[\mathbf{I} - \mathbf{Z}(\mathbf{Z}'\mathbf{Z})^{-1}\mathbf{Z}']\mathbf{y} = \mathbf{y}'\mathbf{y} - \mathbf{y}'\mathbf{Z}\hat{\boldsymbol{\beta}}$$

---

[1] If $\mathbf{Z}$ is not full rank, $(\mathbf{Z}'\mathbf{Z})^{-1}$ is replaced by $(\mathbf{Z}'\mathbf{Z})^-$, a *generalized inverse* of $\mathbf{Z}'\mathbf{Z}$. (See Exercise 7.6.)

Proof. Let $\hat{\boldsymbol{\beta}} = (\mathbf{Z}'\mathbf{Z})^{-1}\mathbf{Z}'\mathbf{y}$ as asserted. Then $\hat{\boldsymbol{\varepsilon}} = \mathbf{y} - \hat{\mathbf{y}} = \mathbf{y} - \mathbf{Z}\hat{\boldsymbol{\beta}} = [\mathbf{I} - \mathbf{Z}(\mathbf{Z}'\mathbf{Z})^{-1}\mathbf{Z}']\mathbf{y}$. The matrix $[\mathbf{I} - \mathbf{Z}(\mathbf{Z}'\mathbf{Z})^{-1}\mathbf{Z}']$ satisfies

1. $[\mathbf{I} - \mathbf{Z}(\mathbf{Z}'\mathbf{Z})^{-1}\mathbf{Z}']' = [\mathbf{I} - \mathbf{Z}(\mathbf{Z}'\mathbf{Z})^{-1}\mathbf{Z}']$ (symmetric);

2. $[\mathbf{I} - \mathbf{Z}(\mathbf{Z}'\mathbf{Z})^{-1}\mathbf{Z}'][\mathbf{I} - \mathbf{Z}(\mathbf{Z}'\mathbf{Z})^{-1}\mathbf{Z}']$

$$= \mathbf{I} - 2\mathbf{Z}(\mathbf{Z}'\mathbf{Z})^{-1}\mathbf{Z}' + \mathbf{Z}(\mathbf{Z}'\mathbf{Z})^{-1}\mathbf{Z}'\mathbf{Z}(\mathbf{Z}'\mathbf{Z})^{-1}\mathbf{Z}' \qquad (7\text{-}6)$$

$$= [\mathbf{I} - \mathbf{Z}(\mathbf{Z}'\mathbf{Z})^{-1}\mathbf{Z}'] \quad \text{(idempotent)};$$

3. $\mathbf{Z}'[\mathbf{I} - \mathbf{Z}(\mathbf{Z}'\mathbf{Z})^{-1}\mathbf{Z}'] = \mathbf{Z}' - \mathbf{Z}' = \mathbf{0}$.

Consequently, $\mathbf{Z}'\hat{\boldsymbol{\varepsilon}} = \mathbf{Z}'(\mathbf{y} - \hat{\mathbf{y}}) = \mathbf{Z}'[\mathbf{I} - \mathbf{Z}(\mathbf{Z}'\mathbf{Z})^{-1}\mathbf{Z}']\mathbf{y} = \mathbf{0}$, so $\hat{\mathbf{y}}'\hat{\boldsymbol{\varepsilon}} = \hat{\boldsymbol{\beta}}'\mathbf{Z}'\hat{\boldsymbol{\varepsilon}} = 0$. Additionally, $\hat{\boldsymbol{\varepsilon}}'\hat{\boldsymbol{\varepsilon}} = \mathbf{y}'[\mathbf{I} - \mathbf{Z}(\mathbf{Z}'\mathbf{Z})^{-1}\mathbf{Z}'][\mathbf{I} - \mathbf{Z}(\mathbf{Z}'\mathbf{Z})^{-1}\mathbf{Z}']\mathbf{y} = \mathbf{y}'[\mathbf{I} - \mathbf{Z}(\mathbf{Z}'\mathbf{Z})^{-1}\mathbf{Z}']\mathbf{y} = \mathbf{y}'\mathbf{y} - \mathbf{y}'\mathbf{Z}\hat{\boldsymbol{\beta}}$. To verify the expression for $\hat{\boldsymbol{\beta}}$, we write

$$\mathbf{y} - \mathbf{Z}\mathbf{b} = \mathbf{y} - \mathbf{Z}\hat{\boldsymbol{\beta}} + \mathbf{Z}\hat{\boldsymbol{\beta}} - \mathbf{Z}\mathbf{b} = \mathbf{y} - \mathbf{Z}\hat{\boldsymbol{\beta}} + \mathbf{Z}(\hat{\boldsymbol{\beta}} - \mathbf{b})$$

so

$$\begin{aligned} S(\mathbf{b}) &= (\mathbf{y} - \mathbf{Z}\mathbf{b})'(\mathbf{y} - \mathbf{Z}\mathbf{b}) \\ &= (\mathbf{y} - \mathbf{Z}\hat{\boldsymbol{\beta}})'(\mathbf{y} - \mathbf{Z}\hat{\boldsymbol{\beta}}) + (\hat{\boldsymbol{\beta}} - \mathbf{b})'\mathbf{Z}'\mathbf{Z}(\hat{\boldsymbol{\beta}} - \mathbf{b}) \\ &\quad + 2(\mathbf{y} - \mathbf{Z}\hat{\boldsymbol{\beta}})'\mathbf{Z}(\hat{\boldsymbol{\beta}} - \mathbf{b}) \\ &= (\mathbf{y} - \mathbf{Z}\hat{\boldsymbol{\beta}})'(\mathbf{y} - \mathbf{Z}\hat{\boldsymbol{\beta}}) + (\hat{\boldsymbol{\beta}} - \mathbf{b})'\mathbf{Z}'\mathbf{Z}(\hat{\boldsymbol{\beta}} - \mathbf{b}) \end{aligned}$$

since $(\mathbf{y} - \mathbf{Z}\hat{\boldsymbol{\beta}})'\mathbf{Z} = \hat{\boldsymbol{\varepsilon}}'\mathbf{Z} = \mathbf{0}'$. The first term in $S(\mathbf{b})$ does not depend on $\mathbf{b}$ and the second is the squared length of $\mathbf{Z}(\hat{\boldsymbol{\beta}} - \mathbf{b})$. Because $\mathbf{Z}$ has full rank, $\mathbf{Z}(\hat{\boldsymbol{\beta}} - \mathbf{b}) \neq \mathbf{0}$ if $\hat{\boldsymbol{\beta}} \neq \mathbf{b}$, so the minimum sum of squares is unique and occurs for $\mathbf{b} = \hat{\boldsymbol{\beta}} = (\mathbf{Z}'\mathbf{Z})^{-1}\mathbf{Z}'\mathbf{y}$. Note that $(\mathbf{Z}'\mathbf{Z})^{-1}$ exists since $\mathbf{Z}'\mathbf{Z}$ has rank $r + 1 \leq n$. (If $\mathbf{Z}'\mathbf{Z}$ is not of full rank, $\mathbf{Z}'\mathbf{Z}\mathbf{a} = \mathbf{0}$ for some $\mathbf{a} \neq \mathbf{0}$, but then $\mathbf{a}'\mathbf{Z}'\mathbf{Z}\mathbf{a} = 0$ or $\mathbf{Z}\mathbf{a} = \mathbf{0}$, which contradicts $\mathbf{Z}$ having full rank $r + 1$.) ∎

Result 7.1 shows how the least squares estimates $\hat{\boldsymbol{\beta}}$ and the residuals $\hat{\boldsymbol{\varepsilon}}$ can be obtained from the design matrix $\mathbf{Z}$ and responses $\mathbf{y}$ by simple matrix operations.

Example 7.3 (Calculating the least squares estimates, the residuals, and the residual sum of squares) Calculate the least square estimates $\hat{\boldsymbol{\beta}}$, the residuals $\hat{\boldsymbol{\varepsilon}}$, and the residual sum of squares for a straight-line model

$$Y_j = \beta_0 + \beta_1 z_{j1} + \varepsilon_j$$

fit to the data

z_1	0	1	2	3	4
y	1	4	3	8	9

We have

$$\frac{1}{det}\begin{bmatrix} d & -b \\ -c & a \end{bmatrix}$$

$\mathbf{Z}'$	$\mathbf{y}$	$\mathbf{Z}'\mathbf{Z}$	$(\mathbf{Z}'\mathbf{Z})^{-1}$	$\mathbf{Z}'\mathbf{y}$

introduce artificial variable

$$\begin{bmatrix} 1 & 1 & 1 & 1 & 1 \\ 0 & 1 & 2 & 3 & 4 \end{bmatrix} \quad \begin{bmatrix} 1 \\ 4 \\ 3 \\ 8 \\ 9 \end{bmatrix} \quad \begin{bmatrix} 5 & 10 \\ 10 & 30 \end{bmatrix} \quad \begin{bmatrix} .6 & -.2 \\ -.2 & .1 \end{bmatrix} \quad \begin{bmatrix} 25 \\ 70 \end{bmatrix}$$

Consequently,

$$\hat{\boldsymbol{\beta}} = \begin{bmatrix} \hat{\beta}_0 \\ \hat{\beta}_1 \end{bmatrix} = (\mathbf{Z}'\mathbf{Z})^{-1}\mathbf{Z}'\mathbf{y} = \begin{bmatrix} .6 & -.2 \\ -.2 & .1 \end{bmatrix}\begin{bmatrix} 25 \\ 70 \end{bmatrix} = \begin{bmatrix} 1 \\ 2 \end{bmatrix} \begin{matrix} \hat{\beta}_0 \\ \hat{\beta}_1 \end{matrix}$$

and the fitted equation is

$$\hat{y} = 1 + 2z$$

The vector of fitted (predicted) values is

$$\hat{\mathbf{y}} = \mathbf{Z}\hat{\boldsymbol{\beta}} = \begin{bmatrix} 1 & 0 \\ 1 & 1 \\ 1 & 2 \\ 1 & 3 \\ 1 & 4 \end{bmatrix}\begin{bmatrix} 1 \\ 2 \end{bmatrix} = \begin{bmatrix} 1 \\ 3 \\ 5 \\ 7 \\ 9 \end{bmatrix}$$

→ linear

so

$$\hat{\boldsymbol{\varepsilon}} = \mathbf{y} - \hat{\mathbf{y}} = \begin{bmatrix} 1 \\ 4 \\ 3 \\ 8 \\ 9 \end{bmatrix} - \begin{bmatrix} 1 \\ 3 \\ 5 \\ 7 \\ 9 \end{bmatrix} = \begin{bmatrix} 0 \\ 1 \\ -2 \\ 1 \\ 0 \end{bmatrix}$$

The residual sum of squares is

$$\hat{\boldsymbol{\varepsilon}}'\hat{\boldsymbol{\varepsilon}} = \begin{bmatrix} 0 & 1 & -2 & 1 & 0 \end{bmatrix}\begin{bmatrix} 0 \\ 1 \\ -2 \\ 1 \\ 0 \end{bmatrix} = 0^2 + 1^2 + (-2)^2 + 1^2 + 0^2 = 6 \quad \blacksquare$$

Sum-of-Squares Decomposition

According to Result 7.1, $\hat{\mathbf{y}}'\hat{\boldsymbol{\varepsilon}} = 0$, so the total response sum of squares $\mathbf{y}'\mathbf{y} = \sum_{j=1}^{n} y_j^2$ satisfies

$$\mathbf{y}'\mathbf{y} = (\hat{\mathbf{y}} + \mathbf{y} - \hat{\mathbf{y}})'(\hat{\mathbf{y}} + \mathbf{y} - \hat{\mathbf{y}}) = (\hat{\mathbf{y}} + \hat{\boldsymbol{\varepsilon}})'(\hat{\mathbf{y}} + \hat{\boldsymbol{\varepsilon}}) = \hat{\mathbf{y}}'\hat{\mathbf{y}} + \hat{\boldsymbol{\varepsilon}}'\hat{\boldsymbol{\varepsilon}} \quad (7\text{-}7)$$

Since the first column of $\mathbf{Z}$ is $\mathbf{1}$, the condition $\mathbf{Z}'\hat{\boldsymbol{\varepsilon}} = \mathbf{0}$ includes the requirement

$$0 = \mathbf{1}'\hat{\boldsymbol{\varepsilon}} = \sum_{j=1}^{n} \hat{\varepsilon}_j = \sum_{j=1}^{n} y_j - \sum_{j=1}^{n} \hat{y}_j, \text{ or } \bar{y} = \bar{\hat{y}}.$$

Subtracting $n\bar{y}^2 = n(\bar{\hat{y}})^2$ from both sides of the decomposition in (7-7), we obtain the basic decomposition of the sum of squares about the mean:

$$\mathbf{y}'\mathbf{y} - n\bar{y}^2 = \hat{\mathbf{y}}'\hat{\mathbf{y}} - n(\bar{\hat{y}})^2 + \hat{\boldsymbol{\varepsilon}}'\hat{\boldsymbol{\varepsilon}}$$

or

$$\sum_{j=1}^{n} (y_j - \bar{y})^2 = \sum_{j=1}^{n} (\hat{y}_j - \bar{y})^2 + \sum_{j=1}^{n} \hat{\varepsilon}_j^2 \tag{7-8}$$

$$\begin{pmatrix} \text{total sum} \\ \text{of squares} \\ \text{about mean} \end{pmatrix} = \begin{pmatrix} \text{regression} \\ \text{sum of} \\ \text{squares} \end{pmatrix} + \begin{pmatrix} \text{residual (error)} \\ \text{sum of squares} \end{pmatrix}$$

The preceding sum of squares decomposition suggests that the quality of the models fit can be measured by the *coefficient of determination*

$$R^2 = 1 - \frac{\displaystyle\sum_{j=1}^{n} \hat{\varepsilon}_j^2}{\displaystyle\sum_{j=1}^{n} (y_j - \bar{y})^2} = \frac{\displaystyle\sum_{j=1}^{n} (\hat{y}_j - \bar{y})^2}{\displaystyle\sum_{j=1}^{n} (y_j - \bar{y})^2} \tag{7-9}$$

The quantity R^2 gives the proportion of the total variation in the y_j's "explained" by, or attributable to, the predictor variables $z_1, z_2, \ldots, z_r$. Here R^2 (or the *multiple correlation coefficient* $R = +\sqrt{R^2}$) equals 1 if the fitted equation passes through all the data points, so that $\hat{\varepsilon}_j = 0$ for all j. At the other extreme, R^2 is 0 if $\hat{\beta}_0 = \bar{y}$ and $\hat{\beta}_1 = \hat{\beta}_2 = \cdots = \hat{\beta}_r = 0$. In this case, the predictor variables $z_1, z_2, \ldots, z_r$ have no influence on the response.

Geometry of Least Squares

A geometrical interpretation of the least squares technique highlights the nature of the concept. According to the classical linear regression model,

$$\text{Mean response vector} = E(\mathbf{Y}) = \mathbf{Z}\boldsymbol{\beta} = \beta_0 \begin{bmatrix} 1 \\ 1 \\ \vdots \\ 1 \end{bmatrix} + \beta_1 \begin{bmatrix} z_{11} \\ z_{21} \\ \vdots \\ z_{n1} \end{bmatrix} + \cdots + \beta_r \begin{bmatrix} z_{1r} \\ z_{2r} \\ \vdots \\ z_{nr} \end{bmatrix}$$

Thus, $E(\mathbf{Y})$ is a linear combination of the columns of $\mathbf{Z}$. As $\boldsymbol{\beta}$ varies, $\mathbf{Z}\boldsymbol{\beta}$ spans the model plane of all linear combinations. Usually, the observation vector $\mathbf{y}$ will not lie in the model plane, because of the random error $\boldsymbol{\varepsilon}$; that is, $\mathbf{y}$ is not (exactly) a linear combination of the columns of $\mathbf{Z}$. Recall that

$$\begin{array}{ccccc} \mathbf{Y} & = & \mathbf{Z}\boldsymbol{\beta} & + & \boldsymbol{\varepsilon} \\ \begin{pmatrix} \text{response} \\ \text{vector} \end{pmatrix} & & \begin{pmatrix} \text{vector} \\ \text{in model} \\ \text{plane} \end{pmatrix} & & \begin{pmatrix} \text{error} \\ \text{vector} \end{pmatrix} \end{array}$$

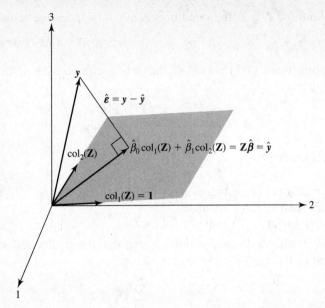

Figure 7.1 Least squares as a projection for $n = 3, r = 1$.

Once the observations become available, the least squares solution is derived from the deviation vector

$$\mathbf{y} - \mathbf{Z}\mathbf{b} = (\text{observation vector}) - (\text{vector in model plane})$$

The squared length $(\mathbf{y} - \mathbf{Z}\mathbf{b})'(\mathbf{y} - \mathbf{Z}\mathbf{b})$ is the sum of squares $S(\mathbf{b})$. As illustrated in Figure 7.1, $S(\mathbf{b})$ is as small as possible when $\mathbf{b}$ is selected such that $\mathbf{Z}\mathbf{b}$ is the point in the model plane closest to $\mathbf{y}$. This point occurs at the tip of the perpendicular projection of $\mathbf{y}$ on the plane. That is, for the choice $\mathbf{b} = \hat{\boldsymbol{\beta}}$, $\hat{\mathbf{y}} = \mathbf{Z}\hat{\boldsymbol{\beta}}$ is the projection of $\mathbf{y}$ on the plane consisting of all linear combinations of the columns of $\mathbf{Z}$. The residual vector $\hat{\boldsymbol{\varepsilon}} = \mathbf{y} - \hat{\mathbf{y}}$ is perpendicular to that plane. This geometry holds even when $\mathbf{Z}$ is not of full rank.

When $\mathbf{Z}$ has full rank, the projection operation is expressed analytically as multiplication by the matrix $\mathbf{Z}(\mathbf{Z}'\mathbf{Z})^{-1}\mathbf{Z}'$. To see this, we use the spectral decomposition (2-16) to write

$$\mathbf{Z}'\mathbf{Z} = \lambda_1\mathbf{e}_1\mathbf{e}_1' + \lambda_2\mathbf{e}_2\mathbf{e}_2' + \cdots + \lambda_{r+1}\mathbf{e}_{r+1}\mathbf{e}_{r+1}'$$

where $\lambda_1 \geq \lambda_2 \geq \cdots \geq \lambda_{r+1} > 0$ are the eigenvalues of $\mathbf{Z}'\mathbf{Z}$ and $\mathbf{e}_1, \mathbf{e}_2, \ldots, \mathbf{e}_{r+1}$ are the corresponding eigenvectors. If $\mathbf{Z}$ is of full rank,

$$(\mathbf{Z}'\mathbf{Z})^{-1} = \frac{1}{\lambda_1}\mathbf{e}_1\mathbf{e}_1' + \frac{1}{\lambda_2}\mathbf{e}_2\mathbf{e}_2' + \cdots + \frac{1}{\lambda_{r+1}}\mathbf{e}_{r+1}\mathbf{e}_{r+1}'$$

Consider $\mathbf{q}_i = \lambda_i^{-1/2}\mathbf{Z}\mathbf{e}_i$, which is a linear combination of the columns of $\mathbf{Z}$. Then $\mathbf{q}_i'\mathbf{q}_k = \lambda_i^{-1/2}\lambda_k^{-1/2}\mathbf{e}_i'\mathbf{Z}'\mathbf{Z}\mathbf{e}_k = \lambda_i^{-1/2}\lambda_k^{-1/2}\mathbf{e}_i'\lambda_k\mathbf{e}_k = 0$ if $i \neq k$ or 1 if $i = k$. That is, the $r + 1$ vectors $\mathbf{q}_i$ are mutually perpendicular and have unit length. Their linear combinations span the space of all linear combinations of the columns of $\mathbf{Z}$. Moreover,

$$\mathbf{Z}(\mathbf{Z}'\mathbf{Z})^{-1}\mathbf{Z}' = \sum_{i=1}^{r+1}\lambda_i^{-1}\mathbf{Z}\mathbf{e}_i\mathbf{e}_i'\mathbf{Z}' = \sum_{i=1}^{r+1}\mathbf{q}_i\mathbf{q}_i'$$

According to Result 2A.2 and Definition 2A.12, the projection of $\mathbf{y}$ on a linear combination of $\{\mathbf{q}_1, \mathbf{q}_2, \ldots, \mathbf{q}_{r+1}\}$ is $\sum_{i=1}^{r+1} (\mathbf{q}_i'\mathbf{y})\,\mathbf{q}_i = \left(\sum_{i=1}^{r+1} \mathbf{q}_i\mathbf{q}_i'\right)\mathbf{y} = \mathbf{Z}(\mathbf{Z}'\mathbf{Z})^{-1}\mathbf{Z}'\mathbf{y} = \mathbf{Z}\hat{\boldsymbol{\beta}}$.

Thus, multiplication by $\mathbf{Z}(\mathbf{Z}'\mathbf{Z})^{-1}\mathbf{Z}'$ projects a vector onto the space spanned by the columns of $\mathbf{Z}$.[2]

Similarly, $[\mathbf{I} - \mathbf{Z}(\mathbf{Z}'\mathbf{Z})^{-1}\mathbf{Z}']$ is the matrix for the projection of $\mathbf{y}$ on the plane perpendicular to the plane spanned by the columns of $\mathbf{Z}$.

Sampling Properties of Classical Least Squares Estimators

The least squares estimator $\hat{\boldsymbol{\beta}}$ and the residuals $\hat{\boldsymbol{\varepsilon}}$ have the sampling properties detailed in the next result.

Result 7.2. Under the general linear regression model in (7-3), the least squares estimator $\hat{\boldsymbol{\beta}} = (\mathbf{Z}'\mathbf{Z})^{-1}\mathbf{Z}'\mathbf{Y}$ has

$$E(\hat{\boldsymbol{\beta}}) = \boldsymbol{\beta} \quad \text{and} \quad \text{Cov}(\hat{\boldsymbol{\beta}}) = \sigma^2(\mathbf{Z}'\mathbf{Z})^{-1}$$

The residuals $\hat{\boldsymbol{\varepsilon}}$ have the properties

$$E(\hat{\boldsymbol{\varepsilon}}) = \mathbf{0} \quad \text{and} \quad \text{Cov}(\hat{\boldsymbol{\varepsilon}}) = \sigma^2[\mathbf{I} - \mathbf{Z}(\mathbf{Z}'\mathbf{Z})^{-1}\mathbf{Z}'] = \sigma^2[\mathbf{I} - \mathbf{H}]$$

Also, $E(\hat{\boldsymbol{\varepsilon}}'\hat{\boldsymbol{\varepsilon}}) = (n - r - 1)\sigma^2$, so defining

$$s^2 = \frac{\hat{\boldsymbol{\varepsilon}}'\hat{\boldsymbol{\varepsilon}}}{n - (r+1)} = \frac{\mathbf{Y}'[\mathbf{I} - \mathbf{Z}(\mathbf{Z}'\mathbf{Z})^{-1}\mathbf{Z}']\mathbf{Y}}{n - r - 1} = \frac{\mathbf{Y}'[\mathbf{I} - \mathbf{H}]\mathbf{Y}}{n - r - 1}$$

we have

$$E(s^2) = \sigma^2$$

Moreover, $\hat{\boldsymbol{\beta}}$ and $\hat{\boldsymbol{\varepsilon}}$ are uncorrelated.

Proof. (See webpage: www.prenhall.com/statistics)   ∎

The least squares estimator $\hat{\boldsymbol{\beta}}$ possesses a minimum variance property that was first established by Gauss. The following result concerns "best" estimators of linear parametric functions of the form $\mathbf{c}'\boldsymbol{\beta} = c_0\beta_0 + c_1\beta_1 + \cdots + c_r\beta_r$ for any $\mathbf{c}$.

Result 7.3 (Gauss'[3] least squares theorem). Let $\mathbf{Y} = \mathbf{Z}\boldsymbol{\beta} + \boldsymbol{\varepsilon}$, where $E(\boldsymbol{\varepsilon}) = \mathbf{0}$, $\text{Cov}(\boldsymbol{\varepsilon}) = \sigma^2\mathbf{I}$, and $\mathbf{Z}$ has full rank $r + 1$. For any $\mathbf{c}$, the estimator

$$\mathbf{c}'\hat{\boldsymbol{\beta}} = c_0\hat{\beta}_0 + c_1\hat{\beta}_1 + \cdots + c_r\hat{\beta}_r$$

---

[2] If $\mathbf{Z}$ is not of full rank, we can use the *generalized inverse* $(\mathbf{Z}'\mathbf{Z})^- = \sum_{i=1}^{r_1+1} \lambda_i^{-1}\mathbf{e}_i\mathbf{e}_i'$, where $\lambda_1 \geq \lambda_2 \geq \cdots \geq \lambda_{r_1+1} > 0 = \lambda_{r_1+2} = \cdots = \lambda_{r+1}$, as described in Exercise 7.6. Then $\mathbf{Z}(\mathbf{Z}'\mathbf{Z})^-\mathbf{Z}' = \sum_{i=1}^{r_1+1} \mathbf{q}_i\mathbf{q}_i'$ has rank $r_1 + 1$ and generates the unique projection of $\mathbf{y}$ on the space spanned by the linearly independent columns of $\mathbf{Z}$. This is true for any choice of the generalized inverse. (See [23].)

[3] Much later, Markov proved a less general result, which misled many writers into attaching his name to this theorem.

of $\mathbf{c}'\boldsymbol{\beta}$ has the smallest possible variance among all linear estimators of the form

$$\mathbf{a}'\mathbf{Y} = a_1 Y_1 + a_2 Y_2 + \cdots + a_n Y_n$$

that are unbiased for $\mathbf{c}'\boldsymbol{\beta}$.

Proof. For any fixed $\mathbf{c}$, let $\mathbf{a}'\mathbf{Y}$ be any unbiased estimator of $\mathbf{c}'\boldsymbol{\beta}$. Then $E(\mathbf{a}'\mathbf{Y}) = \mathbf{c}'\boldsymbol{\beta}$, whatever the value of $\boldsymbol{\beta}$. Also, by assumption, $E(\mathbf{a}'\mathbf{Y}) = E(\mathbf{a}'\mathbf{Z}\boldsymbol{\beta} + \mathbf{a}'\boldsymbol{\varepsilon}) = \mathbf{a}'\mathbf{Z}\boldsymbol{\beta}$. Equating the two expected value expressions yields $\mathbf{a}'\mathbf{Z}\boldsymbol{\beta} = \mathbf{c}'\boldsymbol{\beta}$ or $(\mathbf{c}' - \mathbf{a}'\mathbf{Z})\boldsymbol{\beta} = 0$ for all $\boldsymbol{\beta}$, including the choice $\boldsymbol{\beta} = (\mathbf{c}' - \mathbf{a}'\mathbf{Z})'$. This implies that $\mathbf{c}' = \mathbf{a}'\mathbf{Z}$ for any unbiased estimator.

Now, $\mathbf{c}'\hat{\boldsymbol{\beta}} = \mathbf{c}'(\mathbf{Z}'\mathbf{Z})^{-1}\mathbf{Z}'\mathbf{Y} = \mathbf{a}^{*\prime}\mathbf{Y}$ with $\mathbf{a}^* = \mathbf{Z}(\mathbf{Z}'\mathbf{Z})^{-1}\mathbf{c}$. Moreover, from Result 7.2 $E(\hat{\boldsymbol{\beta}}) = \boldsymbol{\beta}$, so $\mathbf{c}'\hat{\boldsymbol{\beta}} = \mathbf{a}^{*\prime}\mathbf{Y}$ is an unbiased estimator of $\mathbf{c}'\boldsymbol{\beta}$. Thus, for any $\mathbf{a}$ satisfying the unbiased requirement $\mathbf{c}' = \mathbf{a}'\mathbf{Z}$,

$$\begin{aligned}
\mathrm{Var}\,(\mathbf{a}'\mathbf{Y}) &= \mathrm{Var}\,(\mathbf{a}'\mathbf{Z}\boldsymbol{\beta} + \mathbf{a}'\boldsymbol{\varepsilon}) = \mathrm{Var}\,(\mathbf{a}'\boldsymbol{\varepsilon}) = \mathbf{a}'\mathbf{I}\sigma^2\mathbf{a} \\
&= \sigma^2(\mathbf{a} - \mathbf{a}^* + \mathbf{a}^*)'(\mathbf{a} - \mathbf{a}^* + \mathbf{a}^*) \\
&= \sigma^2[(\mathbf{a} - \mathbf{a}^*)'(\mathbf{a} - \mathbf{a}^*) + \mathbf{a}^{*\prime}\mathbf{a}^*]
\end{aligned}$$

since $(\mathbf{a} - \mathbf{a}^*)'\mathbf{a}^* = (\mathbf{a} - \mathbf{a}^*)'\mathbf{Z}(\mathbf{Z}'\mathbf{Z})^{-1}\mathbf{c} = 0$ from the condition $(\mathbf{a} - \mathbf{a}^*)'\mathbf{Z} = \mathbf{a}'\mathbf{Z} - \mathbf{a}^{*\prime}\mathbf{Z} = \mathbf{c}' - \mathbf{c}' = \mathbf{0}'$. Because $\mathbf{a}^*$ is fixed and $(\mathbf{a} - \mathbf{a}^*)'(\mathbf{a} - \mathbf{a}^*)$ is positive unless $\mathbf{a} = \mathbf{a}^*$, $\mathrm{Var}\,(\mathbf{a}'\mathbf{Y})$ is minimized by the choice $\mathbf{a}^{*\prime}\mathbf{Y} = \mathbf{c}'(\mathbf{Z}'\mathbf{Z})^{-1}\mathbf{Z}'\mathbf{Y} = \mathbf{c}'\hat{\boldsymbol{\beta}}$. ∎

This powerful result states that substitution of $\hat{\boldsymbol{\beta}}$ for $\boldsymbol{\beta}$ leads to the best estimator of $\mathbf{c}'\boldsymbol{\beta}$ for any $\mathbf{c}$ of interest. In statistical terminology, the estimator $\mathbf{c}'\hat{\boldsymbol{\beta}}$ is called the *best (minimum-variance) linear unbiased estimator* (BLUE) of $\mathbf{c}'\boldsymbol{\beta}$.

7.4 Inferences About the Regression Model

We describe inferential procedures based on the classical linear regression model in (7-3) with the additional (tentative) assumption that the errors $\boldsymbol{\varepsilon}$ have a normal distribution. Methods for checking the general adequacy of the model are considered in Section 7.6.

Inferences Concerning the Regression Parameters

Before we can assess the importance of particular variables in the *regression function*

$$E(Y) = \beta_0 + \beta_1 z_1 + \cdots + \beta_r z_r \tag{7-10}$$

we must determine the sampling distributions of $\hat{\boldsymbol{\beta}}$ and the residual sum of squares, $\hat{\boldsymbol{\varepsilon}}'\hat{\boldsymbol{\varepsilon}}$. To do so, we shall assume that the errors $\boldsymbol{\varepsilon}$ have a normal distribution.

Result 7.4. Let $\mathbf{Y} = \mathbf{Z}\boldsymbol{\beta} + \boldsymbol{\varepsilon}$, where $\mathbf{Z}$ has full rank $r + 1$ and $\boldsymbol{\varepsilon}$ is distributed as $N_n(\mathbf{0}, \sigma^2\mathbf{I})$. Then the maximum likelihood estimator of $\boldsymbol{\beta}$ is the same as the least squares estimator $\hat{\boldsymbol{\beta}}$. Moreover,

$$\hat{\boldsymbol{\beta}} = (\mathbf{Z}'\mathbf{Z})^{-1}\mathbf{Z}'\mathbf{Y} \quad \text{is distributed as} \quad N_{r+1}(\boldsymbol{\beta}, \sigma^2(\mathbf{Z}'\mathbf{Z})^{-1})$$

and is distributed independently of the residuals $\hat{\boldsymbol{\varepsilon}} = \mathbf{Y} - \mathbf{Z}\hat{\boldsymbol{\beta}}$. Further,

$$n\hat{\sigma}^2 = \hat{\boldsymbol{\varepsilon}}'\hat{\boldsymbol{\varepsilon}} \quad \text{is distributed as} \quad \sigma^2\chi^2_{n-r-1}$$

where $\hat{\sigma}^2$ is the maximum likelihood estimator of σ^2.

Proof. (See webpage: www.prenhall.com/statistics) ∎

A confidence ellipsoid for $\boldsymbol{\beta}$ is easily constructed. It is expressed in terms of the estimated covariance matrix $s^2(\mathbf{Z}'\mathbf{Z})^{-1}$, where $s^2 = \hat{\boldsymbol{\varepsilon}}'\hat{\boldsymbol{\varepsilon}}/(n-r-1)$.

Result 7.5. Let $\mathbf{Y} = \mathbf{Z}\boldsymbol{\beta} + \boldsymbol{\varepsilon}$, where $\mathbf{Z}$ has full rank $r+1$ and $\boldsymbol{\varepsilon}$ is $N_n(\mathbf{0}, \sigma^2\mathbf{I})$. Then a $100(1-\alpha)$ percent confidence region for $\boldsymbol{\beta}$ is given by

$$(\boldsymbol{\beta}-\hat{\boldsymbol{\beta}})'\mathbf{Z}'\mathbf{Z}(\boldsymbol{\beta}-\hat{\boldsymbol{\beta}}) \le (r+1)\,s^2 F_{r+1,n-r-1}(\alpha)$$

where $F_{r+1,n-r-1}(\alpha)$ is the upper (100α)th percentile of an F-distribution with $r+1$ and $n-r-1$ d.f.

Also, *simultaneous* $100(1-\alpha)$ percent confidence intervals for the β_i are given by

$$\hat{\beta}_i \pm \sqrt{\widehat{\mathrm{Var}}(\hat{\beta}_i)}\,\sqrt{(r+1)F_{r+1,n-r-1}(\alpha)}, \quad i = 0, 1, \ldots, r$$

where $\widehat{\mathrm{Var}}(\hat{\beta}_i)$ is the diagonal element of $s^2(\mathbf{Z}'\mathbf{Z})^{-1}$ corresponding to $\hat{\beta}_i$.

Proof. Consider the symmetric square-root matrix $(\mathbf{Z}'\mathbf{Z})^{1/2}$. [See (2-22).] Set $\mathbf{V} = (\mathbf{Z}'\mathbf{Z})^{1/2}(\hat{\boldsymbol{\beta}} - \boldsymbol{\beta})$ and note that $E(\mathbf{V}) = \mathbf{0}$,

$$\mathrm{Cov}(\mathbf{V}) = (\mathbf{Z}'\mathbf{Z})^{1/2}\,\mathrm{Cov}(\hat{\boldsymbol{\beta}})(\mathbf{Z}'\mathbf{Z})^{1/2} = \sigma^2(\mathbf{Z}'\mathbf{Z})^{1/2}(\mathbf{Z}'\mathbf{Z})^{-1}(\mathbf{Z}'\mathbf{Z})^{1/2} = \sigma^2\mathbf{I}$$

and $\mathbf{V}$ is normally distributed, since it consists of linear combinations of the $\hat{\beta}_i$'s. Therefore, $\mathbf{V}'\mathbf{V} = (\hat{\boldsymbol{\beta}} - \boldsymbol{\beta})'(\mathbf{Z}'\mathbf{Z})^{1/2}(\mathbf{Z}'\mathbf{Z})^{1/2}(\hat{\boldsymbol{\beta}} - \boldsymbol{\beta}) = (\hat{\boldsymbol{\beta}} - \boldsymbol{\beta})'(\mathbf{Z}'\mathbf{Z})(\hat{\boldsymbol{\beta}} - \boldsymbol{\beta})$ is distributed as $\sigma^2\chi^2_{r+1}$. By Result 7.4 $(n-r-1)s^2 = \hat{\boldsymbol{\varepsilon}}'\hat{\boldsymbol{\varepsilon}}$ is distributed as $\sigma^2\chi^2_{n-r-1}$, independently of $\hat{\boldsymbol{\beta}}$ and, hence, independently of $\mathbf{V}$. Consequently, $[\chi^2_{r+1}/(r+1)]/[\chi^2_{n-r-1}/(n-r-1)] = [\mathbf{V}'\mathbf{V}/(r+1)]/s^2$ has an $F_{r+1,n-r-1}$ distribution, and the confidence ellipsoid for $\boldsymbol{\beta}$ follows. Projecting this ellipsoid for $(\hat{\boldsymbol{\beta}} - \boldsymbol{\beta})$ using Result 5A.1 with $\mathbf{A}^{-1} = \mathbf{Z}'\mathbf{Z}/s^2$, $c^2 = (r+1)F_{r+1,n-r-1}(\alpha)$, and $\mathbf{u}' = [0, \ldots, 0, 1, 0, \ldots, 0]$ yields $|\beta_i - \hat{\beta}_i| \le \sqrt{(r+1)F_{r+1,n-r-1}(\alpha)}\,\sqrt{\widehat{\mathrm{Var}}(\hat{\beta}_i)}$, where $\widehat{\mathrm{Var}}(\hat{\beta}_i)$ is the diagonal element of $s^2(\mathbf{Z}'\mathbf{Z})^{-1}$ corresponding to $\hat{\beta}_i$. ∎

The confidence ellipsoid is centered at the maximum likelihood estimate $\hat{\boldsymbol{\beta}}$, and its orientation and size are determined by the eigenvalues and eigenvectors of $\mathbf{Z}'\mathbf{Z}$. If an eigenvalue is nearly zero, the confidence ellipsoid will be very long in the direction of the corresponding eigenvector.

Practitioners often ignore the "simultaneous" confidence property of the interval estimates in Result 7.5. Instead, they replace $(r + 1)F_{r+1,n-r-1}(\alpha)$ with the one-at-a-time t value $t_{n-r-1}(\alpha/2)$ and use the intervals

$$\hat{\beta} \pm t_{n-r-1}\left(\frac{\alpha}{2}\right)\sqrt{\widehat{Var}(\hat{\beta}_i)} \qquad (7\text{-}11)$$

when searching for important predictor variables.

Example 7.4 (Fitting a regression model to real-estate data) The assessment data in Table 7.1 were gathered from 20 homes in a Milwaukee, Wisconsin, neighborhood. Fit the regression model

$$Y_j = \beta_0 + \beta_1 z_{j1} + \beta_2 z_{j2} + \varepsilon_j$$

where z_1 = total dwelling size (in hundreds of square feet), z_2 = assessed value (in thousands of dollars), and Y = selling price (in thousands of dollars), to these data using the method of least squares. A computer calculation yields

$$(\mathbf{Z'Z})^{-1} = \begin{bmatrix} 5.1523 & & \\ .2544 & .0512 & \\ -.1463 & -.0172 & .0067 \end{bmatrix}$$

Table 7.1 Real-Estate Data

z_1 Total dwelling size $(100\ \text{ft}^2)$	z_2 Assessed value ($1000)	Y Selling price ($1000)
15.31	57.3	74.8
15.20	63.8	74.0
16.25	65.4	72.9
14.33	57.0	70.0
14.57	63.8	74.9
17.33	63.2	76.0
14.48	60.2	72.0
14.91	57.7	73.5
15.25	56.4	74.5
13.89	55.6	73.5
15.18	62.6	71.5
14.44	63.4	71.0
14.87	60.2	78.9
18.63	67.2	86.5
15.20	57.1	68.0
25.76	89.6	102.0
19.05	68.6	84.0
15.37	60.1	69.0
18.06	66.3	88.0
16.35	65.8	76.0

and

$$\hat{\boldsymbol{\beta}} = (\mathbf{Z'Z})^{-1}\mathbf{Z'y} = \begin{bmatrix} 30.967 \\ 2.634 \\ .045 \end{bmatrix}$$

Thus, the fitted equation is

$$\hat{y} = 30.967 + 2.634z_1 + .045z_2$$
$$\quad\quad\;\;(7.88)\quad\;\;(.785)\quad\;(.285)$$

with $s = 3.473$. The numbers in parentheses are the estimated standard deviations of the least squares coefficients. Also, $R^2 = .834$, indicating that the data exhibit a strong regression relationship. (See Panel 7.1, which contains the regression analysis of these data using the SAS statistical software package.) If the residuals $\hat{\varepsilon}$ pass the diagnostic checks described in Section 7.6, the fitted equation could be used to predict the selling price of another house in the neighborhood from its size

PANEL 7.1   SAS ANALYSIS FOR EXAMPLE 7.4 USING PROC REG.

```
title 'Regression Analysis';
data estate;                          ⎫
infile 'T7-1.dat';                    ⎬  PROGRAM COMMANDS
input z1 z2 y;                        ⎭
proc reg data = estate;
model y = z1 z2;
```

Model: MODEL 1   OUTPUT
Dependent Variable:

Analysis of Variance

Source	DF	Sum of Squares	Mean Square	F value	Prob > F
Model	2	1032.87506	516.43753	42.828	0.0001
Error	17	204.99494	12.05853		
C Total	19	1237.87000			

Root MSE	3.47254		R-square	0.8344	
Deep Mean	76.55000		Adj R-sq	0.8149	
C.V.	4.53630				

Parameter Estimates

Variable	DF	Parameter Estimate	Standard Error	T for H0: Parameter = 0	Prob > |T|
INTERCEP	1	30.966566	7.88220844	3.929	0.0011
z1	1	2.634400	0.78559872	3.353	0.0038
z2	1	0.045184	0.28518271	0.158	0.8760

and assessed value. We note that a 95% confidence interval for β_2 [see (7-14)] is given by

$$\hat{\beta}_2 \pm t_{17}(.025) \sqrt{\widehat{\text{Var}}(\hat{\beta}_2)} = .045 \pm 2.110(.285)$$

or

$$(-.556, .647)$$

Since the confidence interval includes $\beta_2 = 0$, the variable z_2 might be dropped from the regression model and the analysis repeated with the single predictor variable z_1. Given dwelling size, assessed value seems to add little to the prediction of selling price. ∎

Likelihood Ratio Tests for the Regression Parameters

Part of regression analysis is concerned with assessing the effects of particular predictor variables on the response variable. One null hypothesis of interest states that certain of the z_i's do not influence the response Y. These predictors will be labeled $z_{q+1}, z_{q+2}, \ldots, z_r$. The statement that $z_{q+1}, z_{q+2}, \ldots, z_r$ do not influence Y translates into the statistical hypothesis

$$H_0\!: \beta_{q+1} = \beta_{q+2} = \cdots = \beta_r = 0 \quad \text{or} \quad H_0\!: \boldsymbol{\beta}_{(2)} = \mathbf{0} \qquad (7\text{-}12)$$

where $\boldsymbol{\beta}'_{(2)} = [\beta_{q+1}, \beta_{q+2}, \ldots, \beta_r]$.

Setting

$$\mathbf{Z} = \left[\underset{n\times(q+1)}{\mathbf{Z}_1} \ \vdots \ \underset{n\times(r-q)}{\mathbf{Z}_2} \right], \qquad \boldsymbol{\beta} = \left[\begin{array}{c} \boldsymbol{\beta}_{(1)} \\ {\scriptstyle((q+1)\times1)} \\ \hline \boldsymbol{\beta}_{(2)} \\ {\scriptstyle((r-q)\times1)} \end{array} \right]$$

we can express the general linear model as

$$\mathbf{Y} = \mathbf{Z}\boldsymbol{\beta} + \boldsymbol{\varepsilon} = [\mathbf{Z}_1 \ \vdots \ \mathbf{Z}_2] \left[\begin{array}{c} \boldsymbol{\beta}_{(1)} \\ \hline \boldsymbol{\beta}_{(2)} \end{array} \right] + \boldsymbol{\varepsilon} = \mathbf{Z}_1\boldsymbol{\beta}_{(1)} + \mathbf{Z}_2\boldsymbol{\beta}_{(2)} + \boldsymbol{\varepsilon}$$

Under the null hypothesis $H_0\!: \boldsymbol{\beta}_{(2)} = \mathbf{0}$, $\mathbf{Y} = \mathbf{Z}_1\boldsymbol{\beta}_{(1)} + \boldsymbol{\varepsilon}$. The likelihood ratio test of H_0 is based on the

$$\text{Extra sum of squares} = \text{SS}_{\text{res}}(\mathbf{Z}_1) - \text{SS}_{\text{res}}(\mathbf{Z}) \qquad (7\text{-}13)$$

$$= (\mathbf{y} - \mathbf{Z}_1\hat{\boldsymbol{\beta}}_{(1)})'(\mathbf{y} - \mathbf{Z}_1\hat{\boldsymbol{\beta}}_{(1)}) - (\mathbf{y} - \mathbf{Z}\hat{\boldsymbol{\beta}})'(\mathbf{y} - \mathbf{Z}\hat{\boldsymbol{\beta}})$$

where $\hat{\boldsymbol{\beta}}_{(1)} = (\mathbf{Z}_1'\mathbf{Z}_1)^{-1}\mathbf{Z}_1'\mathbf{y}$.

Result 7.6. Let $\mathbf{Z}$ have full rank $r + 1$ and $\boldsymbol{\varepsilon}$ be distributed as $N_n(\mathbf{0}, \sigma^2\mathbf{I})$. The likelihood ratio test of $H_0\!: \boldsymbol{\beta}_{(2)} = \mathbf{0}$ is equivalent to a test of H_0 based on the extra sum of squares in (7-13) and $s^2 = (\mathbf{y} - \mathbf{Z}\hat{\boldsymbol{\beta}})'(\mathbf{y} - \mathbf{Z}\hat{\boldsymbol{\beta}})/(n - r - 1)$. In particular, the likelihood ratio test rejects H_0 if

$$\frac{(\text{SS}_{\text{res}}(\mathbf{Z}_1) - \text{SS}_{\text{res}}(\mathbf{Z}))/(r - q)}{s^2} > F_{r-q, n-r-1}(\alpha)$$

where $F_{r-q, n-r-1}(\alpha)$ is the upper (100α)th percentile of an F-distribution with $r - q$ and $n - r - 1$ d.f.

Proof. Given the data and the normal assumption, the likelihood associated with the parameters $\boldsymbol{\beta}$ and σ^2 is

$$L(\boldsymbol{\beta}, \sigma^2) = \frac{1}{(2\pi)^{n/2}\sigma^n} e^{-(\mathbf{y}-\mathbf{Z}\boldsymbol{\beta})'(\mathbf{y}-\mathbf{Z}\boldsymbol{\beta})/2\sigma^2} \leq \frac{1}{(2\pi)^{n/2}\hat{\sigma}^n} e^{-n/2}$$

with the maximum occurring at $\hat{\boldsymbol{\beta}} = (\mathbf{Z}'\mathbf{Z})^{-1}\mathbf{Z}'\mathbf{y}$ and $\hat{\sigma}^2 = (\mathbf{y} - \mathbf{Z}\hat{\boldsymbol{\beta}})'(\mathbf{y} - \mathbf{Z}\hat{\boldsymbol{\beta}})/n$. Under the restriction of the null hypothesis, $\mathbf{Y} = \mathbf{Z}_1\boldsymbol{\beta}_{(1)} + \boldsymbol{\varepsilon}$ and

$$\max_{\boldsymbol{\beta}_{(1)},\sigma^2} L(\boldsymbol{\beta}_{(1)}, \sigma^2) = \frac{1}{(2\pi)^{n/2}\hat{\sigma}_1^n} e^{-n/2}$$

where the maximum occurs at $\hat{\boldsymbol{\beta}}_{(1)} = (\mathbf{Z}_1'\mathbf{Z}_1)^{-1}\mathbf{Z}_1'\mathbf{y}$. Moreover,

$$\hat{\sigma}_1^2 = (\mathbf{y} - \mathbf{Z}_1\hat{\boldsymbol{\beta}}_{(1)})'(\mathbf{y} - \mathbf{Z}_1\hat{\boldsymbol{\beta}}_{(1)})/n$$

Rejecting $H_0: \boldsymbol{\beta}_{(2)} = \mathbf{0}$ for small values of the likelihood ratio

$$\frac{\max_{\boldsymbol{\beta}_{(1)},\sigma^2} L(\boldsymbol{\beta}_{(1)}, \sigma^2)}{\max_{\boldsymbol{\beta},\sigma^2} L(\boldsymbol{\beta}, \sigma^2)} = \left(\frac{\hat{\sigma}_1^2}{\hat{\sigma}^2}\right)^{-n/2} = \left(\frac{\hat{\sigma}^2 + \hat{\sigma}_1^2 - \hat{\sigma}^2}{\hat{\sigma}^2}\right)^{-n/2} = \left(1 + \frac{\hat{\sigma}_1^2 - \hat{\sigma}^2}{\hat{\sigma}^2}\right)^{-n/2}$$

is equivalent to rejecting H_0 for large values of $(\hat{\sigma}_1^2 - \hat{\sigma}^2)/\hat{\sigma}^2$ or its scaled version,

$$\frac{n(\hat{\sigma}_1^2 - \hat{\sigma}^2)/(r - q)}{n\hat{\sigma}^2/(n - r - 1)} = \frac{(\mathrm{SS}_{\mathrm{res}}(\mathbf{Z}_1) - \mathrm{SS}_{\mathrm{res}}(\mathbf{Z}))/(r - q)}{s^2} = F$$

The preceding F-ratio has an F-distribution with $r - q$ and $n - r - 1$ d.f. (See [22] or Result 7.11 with $m = 1$.) ∎

Comment. The likelihood ratio test is implemented as follows. To test whether all coefficients in a subset are zero, fit the model with and without the terms corresponding to these coefficients. The improvement in the residual sum of squares (the extra sum of squares) is compared to the residual sum of squares for the full model via the F-ratio. The same procedure applies even in analysis of variance situations where $\mathbf{Z}$ is not of full rank.[4]

More generally, it is possible to formulate null hypotheses concerning $r - q$ linear combinations of $\boldsymbol{\beta}$ of the form $H_0: \mathbf{C}\boldsymbol{\beta} = \mathbf{A}_0$. Let the $(r - q) \times (r + 1)$ matrix $\mathbf{C}$ have full rank, let $\mathbf{A}_0 = \mathbf{0}$, and consider

$$H_0: \mathbf{C}\boldsymbol{\beta} = \mathbf{0}$$

$\left(\text{This null hypothesis reduces to the previous choice when } \mathbf{C} = \left[\mathbf{0} \,\vdots\, \underset{(r-q)\times(r-q)}{\mathbf{I}}\right].\right)$

---

[4]In situations where $\mathbf{Z}$ is not of full rank, rank($\mathbf{Z}$) replaces $r + 1$ and rank($\mathbf{Z}_1$) replaces $q + 1$ in Result 7.6.

Under the full model, $\mathbf{C}\hat{\boldsymbol{\beta}}$ is distributed as $N_{r-q}(\mathbf{C}\boldsymbol{\beta}, \sigma^2 \mathbf{C}(\mathbf{Z}'\mathbf{Z})^{-1}\mathbf{C}')$. We reject $H_0\colon \mathbf{C}\boldsymbol{\beta} = \mathbf{0}$ at level α if $\mathbf{0}$ does *not* lie in the $100(1 - \alpha)\%$ confidence ellipsoid for $\mathbf{C}\boldsymbol{\beta}$. Equivalently, we reject $H_0\colon \mathbf{C}\boldsymbol{\beta} = \mathbf{0}$ if

$$\frac{(\mathbf{C}\hat{\boldsymbol{\beta}})'(\mathbf{C}(\mathbf{Z}'\mathbf{Z})^{-1}\mathbf{C}')^{-1}(\mathbf{C}\hat{\boldsymbol{\beta}})}{s^2} > (r - q)F_{r-q,n-r-1}(\alpha) \qquad (7\text{-}14)$$

where $s^2 = (\mathbf{y} - \mathbf{Z}\hat{\boldsymbol{\beta}})'(\mathbf{y} - \mathbf{Z}\hat{\boldsymbol{\beta}})/(n - r - 1)$ and $F_{r-q,n-r-1}(\alpha)$ is the upper (100α)th percentile of an F-distribution with $r - q$ and $n - r - 1$ d.f. The test in (7-14) is the likelihood ratio test, and the numerator in the F-ratio is the extra residual sum of squares incurred by fitting the model, subject to the restriction that $\mathbf{C}\boldsymbol{\beta} = \mathbf{0}$. (See [23]).

The next example illustrates how unbalanced experimental designs are easily handled by the general theory just described.

Example 7.5 (Testing the importance of additional predictors using the extra sum-of-squares approach) Male and female patrons rated the service in three establishments (locations) of a large restaurant chain. The service ratings were converted into an index. Table 7.2 contains the data for $n = 18$ customers. Each data point in the table is categorized according to location (1, 2, or 3) and gender (male = 0 and female = 1). This categorization has the format of a two-way table with unequal numbers of observations per cell. For instance, the combination of location 1 and male has 5 responses, while the combination of location 2 and female has 2 responses. Introducing three dummy variables to account for location and two dummy variables to account for gender, we can develop a regression model linking the service index Y to location, gender, and their "interaction" using the design matrix

Table 7.2 Restaurant-Service Data

Location	Gender	Service (Y)
1	0	15.2
1	0	21.2
1	0	27.3
1	0	21.2
1	0	21.2
1	1	36.4
1	1	92.4
2	0	27.3
2	0	15.2
2	0	9.1
2	0	18.2
2	0	50.0
2	1	44.0
2	1	63.6
3	0	15.2
3	0	30.3
3	1	36.4
3	1	40.9

$$
\mathbf{Z} =
\begin{array}{c}
\overbrace{\text{constant}}\quad \overbrace{\text{location}}\quad \overbrace{\text{gender}}\quad\;\; \overbrace{\text{interaction}} \\
\left[
\begin{array}{c|ccc|cc|cccccc}
1 & 1 & 0 & 0 & 1 & 0 & 1 & 0 & 0 & 0 & 0 & 0 \\
1 & 1 & 0 & 0 & 1 & 0 & 1 & 0 & 0 & 0 & 0 & 0 \\
1 & 1 & 0 & 0 & 1 & 0 & 1 & 0 & 0 & 0 & 0 & 0 \\
1 & 1 & 0 & 0 & 1 & 0 & 1 & 0 & 0 & 0 & 0 & 0 \\
1 & 1 & 0 & 0 & 1 & 0 & 1 & 0 & 0 & 0 & 0 & 0 \\
1 & 1 & 0 & 0 & 0 & 1 & 0 & 1 & 0 & 0 & 0 & 0 \\
1 & 1 & 0 & 0 & 0 & 1 & 0 & 1 & 0 & 0 & 0 & 0 \\
1 & 0 & 1 & 0 & 1 & 0 & 0 & 0 & 1 & 0 & 0 & 0 \\
1 & 0 & 1 & 0 & 1 & 0 & 0 & 0 & 1 & 0 & 0 & 0 \\
1 & 0 & 1 & 0 & 1 & 0 & 0 & 0 & 1 & 0 & 0 & 0 \\
1 & 0 & 1 & 0 & 1 & 0 & 0 & 0 & 1 & 0 & 0 & 0 \\
1 & 0 & 1 & 0 & 1 & 0 & 0 & 0 & 1 & 0 & 0 & 0 \\
1 & 0 & 1 & 0 & 0 & 1 & 0 & 0 & 0 & 1 & 0 & 0 \\
1 & 0 & 1 & 0 & 0 & 1 & 0 & 0 & 0 & 1 & 0 & 0 \\
1 & 0 & 0 & 1 & 1 & 0 & 0 & 0 & 0 & 0 & 1 & 0 \\
1 & 0 & 0 & 1 & 1 & 0 & 0 & 0 & 0 & 0 & 1 & 0 \\
1 & 0 & 0 & 1 & 0 & 1 & 0 & 0 & 0 & 0 & 0 & 1 \\
1 & 0 & 0 & 1 & 0 & 1 & 0 & 0 & 0 & 0 & 0 & 1 \\
\end{array}
\right]
\begin{array}{l}
\left.\begin{array}{c}\\\\\\\\\end{array}\right\}\text{5 responses} \\
\left.\begin{array}{c}\\\end{array}\right\}\text{2 responses} \\
\left.\begin{array}{c}\\\\\\\\\end{array}\right\}\text{5 responses} \\
\left.\begin{array}{c}\\\end{array}\right\}\text{2 responses} \\
\left.\begin{array}{c}\\\end{array}\right\}\text{2 responses} \\
\left.\begin{array}{c}\\\end{array}\right\}\text{2 responses}
\end{array}
\end{array}
$$

The coefficient vector can be set out as

$$\boldsymbol{\beta}' = [\beta_0, \beta_1, \beta_2, \beta_3, \tau_1, \tau_2, \gamma_{11}, \gamma_{12}, \gamma_{21}, \gamma_{22}, \gamma_{31}, \gamma_{32}]$$

where the β_i's $(i > 0)$ represent the effects of the locations on the determination of service, the τ_i's represent the effects of gender on the service index, and the γ_{ik}'s represent the location-gender interaction effects.

The design matrix $\mathbf{Z}$ is not of full rank. (For instance, column 1 equals the sum of columns 2–4 or columns 5–6.) In fact, $\text{rank}(\mathbf{Z}) = 6$.

For the complete model, results from a computer program give

$$\text{SS}_{\text{res}}(\mathbf{Z}) = 2977.4$$

and $n - \text{rank}(\mathbf{Z}) = 18 - 6 = 12$.

The model without the interaction terms has the design matrix $\mathbf{Z}_1$ consisting of the first six columns of $\mathbf{Z}$. We find that

$$\text{SS}_{\text{res}}(\mathbf{Z}_1) = 3419.1$$

with $n - \text{rank}(\mathbf{Z}_1) = 18 - 4 = 14$. To test $H_0 : \gamma_{11} = \gamma_{12} = \gamma_{21} = \gamma_{22} = \gamma_{31} = \gamma_{32} = 0$ (no location–gender interaction), we compute

$$F = \frac{(\text{SS}_{\text{res}}(\mathbf{Z}_1) - \text{SS}_{\text{res}}(\mathbf{Z}))/(6 - 4)}{s^2} = \frac{(\text{SS}_{\text{res}}(\mathbf{Z}_1) - \text{SS}_{\text{res}}(\mathbf{Z}))/2}{\text{SS}_{\text{res}}(\mathbf{Z})/12}$$

$$= \frac{(3419.1 - 2977.4)/2}{2977.4/12} = .89$$

The F-ratio may be compared with an appropriate percentage point of an F-distribution with 2 and 12 d.f. This F-ratio is not significant for any reasonable significance level α. Consequently, we conclude that the service index does not depend upon any location–gender interaction, and these terms can be dropped from the model.

Using the extra sum-of-squares approach, we may verify that there is no difference between locations (no location effect), but that gender is significant; that is, males and females do not give the same ratings to service.

In analysis-of-variance situations where the cell counts are unequal, the variation in the response attributable to different predictor variables and their interactions cannot usually be separated into independent amounts. To evaluate the relative influences of the predictors on the response in this case, it is necessary to fit the model with and without the terms in question and compute the appropriate F-test statistics. ∎

7.5 Inferences from the Estimated Regression Function

Once an investigator is satisfied with the fitted regression model, it can be used to solve two prediction problems. Let $\mathbf{z}_0' = [1, z_{01}, \ldots, z_{0r}]$ be selected values for the predictor variables. Then $\mathbf{z}_0$ and $\hat{\boldsymbol{\beta}}$ can be used (1) to estimate the regression function $\beta_0 + \beta_1 z_{01} + \cdots + \beta_r z_{0r}$ at $\mathbf{z}_0$ and (2) to estimate the value of the response Y at $\mathbf{z}_0$.

Estimating the Regression Function at z_0

Let Y_0 denote the value of the response when the predictor variables have values $\mathbf{z}_0' = [1, z_{01}, \ldots, z_{0r}]$. According to the model in (7-3), the expected value of Y_0 is

$$E(Y_0 | \mathbf{z}_0) = \beta_0 + \beta_1 z_{01} + \cdots + \beta_r z_{0r} = \mathbf{z}_0' \boldsymbol{\beta} \tag{7-15}$$

Its least squares estimate is $\mathbf{z}_0' \hat{\boldsymbol{\beta}}$.

Result 7.7. For the linear regression model in (7-3), $\mathbf{z}_0' \hat{\boldsymbol{\beta}}$ is the unbiased linear estimator of $E(Y_0 | \mathbf{z}_0)$ with minimum variance, $\mathrm{Var}(\mathbf{z}_0' \hat{\boldsymbol{\beta}}) = \mathbf{z}_0' (\mathbf{Z}'\mathbf{Z})^{-1} \mathbf{z}_0 \sigma^2$. If the errors $\boldsymbol{\varepsilon}$ are normally distributed, then a $100(1 - \alpha)\%$ confidence interval for $E(Y_0 | \mathbf{z}_0) = \mathbf{z}_0' \boldsymbol{\beta}$ is provided by

$$\mathbf{z}_0' \hat{\boldsymbol{\beta}} \pm t_{n-r-1}\left(\frac{\alpha}{2}\right) \sqrt{(\mathbf{z}_0'(\mathbf{Z}'\mathbf{Z})^{-1}\mathbf{z}_0) s^2}$$

where $t_{n-r-1}(\alpha/2)$ is the upper $100(\alpha/2)$th percentile of a t-distribution with $n - r - 1$ d.f.

Proof. For a fixed $\mathbf{z}_0$, $\mathbf{z}_0' \hat{\boldsymbol{\beta}}$ is just a linear combination of the β_i's, so Result 7.3 applies. Also, $\mathrm{Var}(\mathbf{z}_0' \hat{\boldsymbol{\beta}}) = \mathbf{z}_0' \mathrm{Cov}(\hat{\boldsymbol{\beta}}) \mathbf{z}_0 = \mathbf{z}_0'(\mathbf{Z}'\mathbf{Z})^{-1} \mathbf{z}_0 \sigma^2$ since $\mathrm{Cov}(\hat{\boldsymbol{\beta}}) = \sigma^2 (\mathbf{Z}'\mathbf{Z})^{-1}$ by Result 7.2. Under the further assumption that $\boldsymbol{\varepsilon}$ is normally distributed, Result 7.4 asserts that $\hat{\boldsymbol{\beta}}$ is $N_{r+1}(\boldsymbol{\beta}, \sigma^2(\mathbf{Z}'\mathbf{Z})^{-1})$ independently of s^2/σ^2, which is distributed as $\chi^2_{n-r-1}/(n - r - 1)$. Consequently, the linear combination $\mathbf{z}_0'\hat{\boldsymbol{\beta}}$ is $N(\mathbf{z}_0'\boldsymbol{\beta}, \sigma^2 \mathbf{z}_0'(\mathbf{Z}'\mathbf{Z})^{-1}\mathbf{z}_0)$ and

$$\frac{(\mathbf{z}_0'\hat{\boldsymbol{\beta}} - \mathbf{z}_0'\boldsymbol{\beta})/\sqrt{\sigma^2 \mathbf{z}_0'(\mathbf{Z}'\mathbf{Z})^{-1}\mathbf{z}_0}}{\sqrt{s^2/\sigma^2}} = \frac{(\mathbf{z}_0'\hat{\boldsymbol{\beta}} - \mathbf{z}_0'\boldsymbol{\beta})}{\sqrt{s^2(\mathbf{z}_0'(\mathbf{Z}'\mathbf{Z})^{-1}\mathbf{z}_0)}}$$

is distributed as t_{n-r-1}. The confidence interval follows. ∎

Forecasting a New Observation at $\mathbf{z}_0$

Prediction of a new observation, such as Y_0, at $\mathbf{z}_0' = [1, z_{01}, \ldots, z_{0r}]$ is more uncertain than estimating the *expected value* of Y_0. According to the regression model of (7-3),

$$Y_0 = \mathbf{z}_0'\boldsymbol{\beta} + \varepsilon_0$$

or

$$(\text{new response } Y_0) = (\text{expected value of } Y_0 \text{ at } \mathbf{z}_0) + (\text{new error})$$

where ε_0 is distributed as $N(0, \sigma^2)$ and is independent of $\boldsymbol{\varepsilon}$ and, hence, of $\hat{\boldsymbol{\beta}}$ and s^2. The errors $\boldsymbol{\varepsilon}$ influence the estimators $\hat{\boldsymbol{\beta}}$ and s^2 through the responses $\mathbf{Y}$, but ε_0 does not.

Result 7.8. Given the linear regression model of (7-3), a new observation Y_0 has the *unbiased predictor*

$$\mathbf{z}_0'\hat{\boldsymbol{\beta}} = \hat{\beta}_0 + \hat{\beta}_1 z_{01} + \cdots + \hat{\beta}_r z_{0r}$$

The variance of the *forecast error* $Y_0 - \mathbf{z}_0'\hat{\boldsymbol{\beta}}$ is

$$\text{Var}(Y_0 - \mathbf{z}_0'\hat{\boldsymbol{\beta}}) = \sigma^2(1 + \mathbf{z}_0'(\mathbf{Z}'\mathbf{Z})^{-1}\mathbf{z}_0)$$

When the errors $\boldsymbol{\varepsilon}$ have a normal distribution, a $100(1 - \alpha)\%$ *prediction interval* for Y_0 is given by

$$\mathbf{z}_0'\hat{\boldsymbol{\beta}} \pm t_{n-r-1}\left(\frac{\alpha}{2}\right)\sqrt{s^2(1 + \mathbf{z}_0'(\mathbf{Z}'\mathbf{Z})^{-1}\mathbf{z}_0)}$$

where $t_{n-r-1}(\alpha/2)$ is the upper $100(\alpha/2)$th percentile of a t-distribution with $n - r - 1$ degrees of freedom.

Proof. We forecast Y_0 by $\mathbf{z}_0'\hat{\boldsymbol{\beta}}$, which estimates $E(Y_0 \mid \mathbf{z}_0)$. By Result 7.7, $\mathbf{z}_0'\hat{\boldsymbol{\beta}}$ has $E(\mathbf{z}_0'\hat{\boldsymbol{\beta}}) = \mathbf{z}_0'\boldsymbol{\beta}$ and $\text{Var}(\mathbf{z}_0'\hat{\boldsymbol{\beta}}) = \mathbf{z}_0'(\mathbf{Z}'\mathbf{Z})^{-1}\mathbf{z}_0\sigma^2$. The forecast error is then $Y_0 - \mathbf{z}_0'\hat{\boldsymbol{\beta}} = \mathbf{z}_0'\boldsymbol{\beta} + \varepsilon_0 - \mathbf{z}_0'\hat{\boldsymbol{\beta}} = \varepsilon_0 + \mathbf{z}_0'(\boldsymbol{\beta} - \hat{\boldsymbol{\beta}})$. Thus, $E(Y_0 - \mathbf{z}_0'\hat{\boldsymbol{\beta}}) = E(\varepsilon_0) + E(\mathbf{z}_0'(\boldsymbol{\beta} - \hat{\boldsymbol{\beta}})) = 0$ so the predictor is unbiased. Since ε_0 and $\hat{\boldsymbol{\beta}}$ are independent, $\text{Var}(Y_0 - \mathbf{z}_0'\hat{\boldsymbol{\beta}}) = \text{Var}(\varepsilon_0) + \text{Var}(\mathbf{z}_0'\hat{\boldsymbol{\beta}}) = \sigma^2 + \mathbf{z}_0'(\mathbf{Z}'\mathbf{Z})^{-1}\mathbf{z}_0\sigma^2 = \sigma^2(1 + \mathbf{z}_0'(\mathbf{Z}'\mathbf{Z})^{-1}\mathbf{z}_0)$. If it is further assumed that $\boldsymbol{\varepsilon}$ has a normal distribution, then $\hat{\boldsymbol{\beta}}$ is normally distributed, and so is the linear combination $Y_0 - \mathbf{z}_0'\hat{\boldsymbol{\beta}}$. Consequently, $(Y_0 - \mathbf{z}_0'\hat{\boldsymbol{\beta}})/\sqrt{\sigma^2(1 + \mathbf{z}_0'(\mathbf{Z}'\mathbf{Z})^{-1}\mathbf{z}_0)}$ is distributed as $N(0, 1)$. Dividing this ratio by $\sqrt{s^2/\sigma^2}$, which is distributed as $\sqrt{\chi^2_{n-r-1}/(n - r - 1)}$, we obtain

$$\frac{(Y_0 - \mathbf{z}_0'\hat{\boldsymbol{\beta}})}{\sqrt{s^2(1 + \mathbf{z}_0'(\mathbf{Z}'\mathbf{Z})^{-1}\mathbf{z}_0)}}$$

which is distributed as t_{n-r-1}. The prediction interval follows immediately. ∎

The prediction interval for Y_0 is wider than the confidence interval for estimating the value of the regression function $E(Y_0|\mathbf{z}_0) = \mathbf{z}_0'\boldsymbol{\beta}$. The additional uncertainty in forecasting Y_0, which is represented by the extra term s^2 in the expression $s^2(1 + \mathbf{z}_0'(\mathbf{Z}'\mathbf{Z})^{-1}\mathbf{z}_0)$, comes from the presence of the unknown error term ε_0.

Example 7.6 (Interval estimates for a mean response and a future response) Companies considering the purchase of a computer must first assess their future needs in order to determine the proper equipment. A computer scientist collected data from seven similar company sites so that a forecast equation of computer-hardware requirements for inventory management could be developed. The data are given in Table 7.3 for

$$z_1 = \text{customer orders (in thousands)}$$
$$z_2 = \text{add-delete item count (in thousands)}$$
$$Y = \text{CPU (central processing unit) time (in hours)}$$

Construct a 95% confidence interval for the mean CPU time, $E(Y_0|\mathbf{z}_0) = \beta_0 + \beta_1 z_{01} + \beta_2 z_{02}$ at $\mathbf{z}_0' = [1, 130, 7.5]$. Also, find a 95% prediction interval for a new facility's CPU requirement corresponding to the same $\mathbf{z}_0$.

A computer program provides the estimated regression function

$$\hat{y} = 8.42 + 1.08z_1 + .42z_2$$

$$(\mathbf{Z}'\mathbf{Z})^{-1} = \begin{bmatrix} 8.17969 & & \\ -.06411 & .00052 & \\ .08831 & -.00107 & .01440 \end{bmatrix}$$

and $s = 1.204$. Consequently,

$$\mathbf{z}_0'\hat{\boldsymbol{\beta}} = 8.42 + 1.08(130) + .42(7.5) = 151.97$$

and $s\sqrt{\mathbf{z}_0'(\mathbf{Z}'\mathbf{Z})^{-1}\mathbf{z}_0} = 1.204(.58928) = .71$. We have $t_4(.025) = 2.776$, so the 95% confidence interval for the mean CPU time at $\mathbf{z}_0$ is

$$\mathbf{z}_0'\hat{\boldsymbol{\beta}} \pm t_4(.025)s\sqrt{\mathbf{z}_0'(\mathbf{Z}'\mathbf{Z})^{-1}\mathbf{z}_0} = 151.97 \pm 2.776(.71)$$

or $(150.00, 153.94)$.

Table 7.3 Computer Data

z_1 (Orders)	z_2 (Add–delete items)	Y (CPU time)
123.5	2.108	141.5
146.1	9.213	168.9
133.9	1.905	154.8
128.5	.815	146.5
151.5	1.061	172.8
136.2	8.603	160.1
92.0	1.125	108.5

Source: Data taken from H. P. Artis, *Forecasting Computer Requirements: A Forecaster's Dilemma* (Piscataway, NJ: Bell Laboratories, 1979).

Since $s\sqrt{1 + \mathbf{z}_0'(\mathbf{Z}'\mathbf{Z})^{-1}\mathbf{z}_0} = (1.204)(1.16071) = 1.40$, a 95% prediction interval for the CPU time at a new facility with conditions $\mathbf{z}_0$ is

$$\mathbf{z}_0'\hat{\boldsymbol{\beta}} \pm t_4(.025)s\sqrt{1 + \mathbf{z}_0'(\mathbf{Z}'\mathbf{Z})^{-1}\mathbf{z}_0} = 151.97 \pm 2.776(1.40)$$

or $(148.08, 155.86)$. ∎

7.6 Model Checking and Other Aspects of Regression

Does the Model Fit?

Assuming that the model is "correct," we have used the estimated regression function to make inferences. Of course, it is imperative to examine the adequacy of the model *before* the estimated function becomes a permanent part of the decision-making apparatus.

All the sample information on lack of fit is contained in the residuals

$$\hat{\varepsilon}_1 = y_1 - \hat{\beta}_0 - \hat{\beta}_1 z_{11} - \cdots - \hat{\beta}_r z_{1r}$$
$$\hat{\varepsilon}_2 = y_2 - \hat{\beta}_0 - \hat{\beta}_1 z_{21} - \cdots - \hat{\beta}_r z_{2r}$$
$$\vdots \qquad \qquad \vdots$$
$$\hat{\varepsilon}_n = y_n - \hat{\beta}_0 - \hat{\beta}_1 z_{n1} - \cdots - \hat{\beta}_r z_{nr}$$

or

$$\hat{\boldsymbol{\varepsilon}} = [\mathbf{I} - \mathbf{Z}(\mathbf{Z}'\mathbf{Z})^{-1}\mathbf{Z}']\mathbf{y} = [\mathbf{I} - \mathbf{H}]\mathbf{y} \tag{7-16}$$

If the model is valid, each residual $\hat{\varepsilon}_j$ is an estimate of the error ε_j, which is assumed to be a normal random variable with mean zero and variance σ^2. Although the residuals $\hat{\boldsymbol{\varepsilon}}$ have expected value $\mathbf{0}$, their covariance matrix $\sigma^2[\mathbf{I} - \mathbf{Z}(\mathbf{Z}'\mathbf{Z})^{-1}\mathbf{Z}'] = \sigma^2[\mathbf{I} - \mathbf{H}]$ is not diagonal. Residuals have unequal variances and nonzero correlations. Fortunately, the correlations are often small and the variances are nearly equal.

Because the residuals $\hat{\boldsymbol{\varepsilon}}$ have covariance matrix $\sigma^2[\mathbf{I} - \mathbf{H}]$, the variances of the ε_j can vary greatly if the diagonal elements of $\mathbf{H}$, the *leverages* h_{jj}, are substantially different. Consequently, many statisticians prefer graphical diagnostics based on studentized residuals. Using the residual mean square s^2 as an estimate of σ^2, we have

$$\widehat{\text{Var}}(\hat{\varepsilon}_j) = s^2(1 - h_{jj}), \qquad j = 1, 2, \ldots, n \tag{7-17}$$

and the *studentized residuals* are

$$\hat{\varepsilon}_j^* = \frac{\hat{\varepsilon}_j}{\sqrt{s^2(1 - h_{jj})}}, \qquad j = 1, 2, \ldots, n \tag{7-18}$$

We expect the studentized residuals to look, approximately, like independent drawings from an $N(0, 1)$ distribution. Some software packages go one step further and studentize $\hat{\varepsilon}_j$ using the delete-one estimated variance $s^2(j)$, which is the residual mean square when the jth observation is dropped from the analysis.

Residuals should be plotted in various ways to detect possible anomalies. For general diagnostic purposes, the following are useful graphs:

1. *Plot the residuals* $\hat{\varepsilon}_j$ *against the predicted values* $\hat{y}_j = \hat{\beta}_0 + \hat{\beta}_1 z_{j1} + \cdots + \hat{\beta}_r z_{jr}$. Departures from the assumptions of the model are typically indicated by two types of phenomena:

    (a) *A dependence of the residuals on the predicted value.* This is illustrated in Figure 7.2(a). The numerical calculations are incorrect, or a β_0 term has been omitted from the model.

    (b) *The variance is not constant.* The pattern of residuals may be funnel shaped, as in Figure 7.2(b), so that there is large variability for large $\hat{y}$ and small variability for small $\hat{y}$. If this is the case, the variance of the error is not constant, and transformations or a weighted least squares approach (or both) are required. (See Exercise 7.3.) In Figure 7.2(d), the residuals form a horizontal band. This is ideal and indicates equal variances and no dependence on $\hat{y}$.

2. *Plot the residuals* $\hat{\varepsilon}_j$ *against a predictor variable, such as* z_1, *or products of predictor variables, such as* z_1^2 *or* $z_1 z_2$. A systematic pattern in these plots suggests the need for more terms in the model. This situation is illustrated in Figure 7.2(c).

3. *Q–Q plots and histograms.* Do the errors appear to be normally distributed? To answer this question, the residuals $\hat{\varepsilon}_j$ or $\hat{\varepsilon}_j^*$ can be examined using the techniques discussed in Section 4.6. The Q–Q plots, histograms, and dot diagrams help to detect the presence of unusual observations or severe departures from normality that may require special attention in the analysis. If n is large, minor departures from normality will not greatly affect inferences about $\boldsymbol{\beta}$.

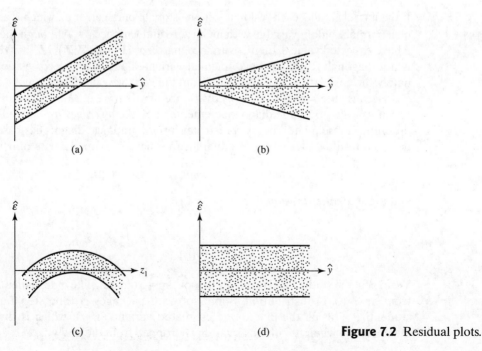

Figure 7.2 Residual plots.

4. *Plot the residuals versus time.* The assumption of independence is crucial, but hard to check. If the data are naturally chronological, a plot of the residuals versus time may reveal a systematic pattern. (A plot of the positions of the residuals in space may also reveal associations among the errors.) For instance, residuals that increase over time indicate a strong positive dependence. A statistical test of independence can be constructed from the first autocorrelation,

$$r_1 = \frac{\sum_{j=2}^{n} \hat{\varepsilon}_j \hat{\varepsilon}_{j-1}}{\sum_{j=1}^{n} \hat{\varepsilon}_j^2} \tag{7-19}$$

of residuals from adjacent periods. A popular test based on the statistic $\sum_{j=2}^{n} (\hat{\varepsilon}_j - \hat{\varepsilon}_{j-1})^2 \Big/ \sum_{j=1}^{n} \hat{\varepsilon}_j^2 \doteq 2(1 - r_1)$ is called the *Durbin–Watson test.* (See [14] for a description of this test and tables of critical values.)

Example 7.7 (Residual plots) Three residual plots for the computer data discussed in Example 7.6 are shown in Figure 7.3. The sample size $n = 7$ is really too small to allow definitive judgments; however, it appears as if the regression assumptions are tenable. ∎

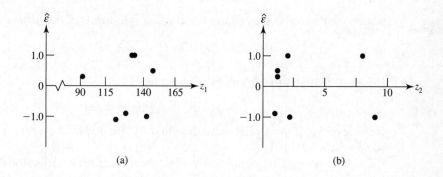

(a)   (b)

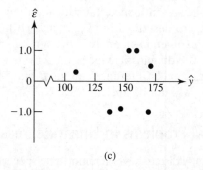

(c)

Figure 7.3 Residual plots for the computer data of Example 7.6.

If several observations of the response are available for the *same* values of the predictor variables, then a formal test for lack of fit can be carried out. (See [13] for a discussion of the pure-error lack-of-fit test.)

Leverage and Influence

Although a residual analysis is useful in assessing the fit of a model, departures from the regression model are often hidden by the fitting process. For example, there may be "outliers" in either the response or explanatory variables that can have a considerable effect on the analysis yet are not easily detected from an examination of residual plots. In fact, these outliers may *determine* the fit.

The leverage h_{jj} the (j, j) diagonal element of $\mathbf{H} = \mathbf{Z}(\mathbf{Z}'\mathbf{Z})^{-1}\mathbf{Z}$, can be interpreted in two related ways. First, the leverage is associated with the jth data point measures, in the space of the explanatory variables, how far the jth observation is from the other $n - 1$ observations. For simple linear regression with one explanatory variable z,

$$h_{jj} = \frac{1}{n} + \frac{(z_j - \bar{z})^2}{\sum\limits_{j=1}^{n} (z_j - \bar{z})^2}$$

The average leverage is $(r + 1)/n$. (See Exercise 7.8.)

Second, the leverage h_{jj}, is a measure of pull that a single case exerts on the fit. The vector of predicted values is

$$\hat{\mathbf{y}} = \mathbf{Z}\hat{\boldsymbol{\beta}} = \mathbf{Z}(\mathbf{Z}'\mathbf{Z})^{-1}\mathbf{Z}\mathbf{y} = \mathbf{H}\mathbf{y}$$

where the jth row expresses the fitted value $\hat{y}_j$ in terms of the observations as

$$\hat{y}_j = h_{jj}y_j + \sum_{k \neq j} h_{jk} y_k$$

Provided that all other y values are held fixed

$$(\text{change in } \hat{y}_j) = h_{jj} (\text{change in } y_j)$$

If the leverage is large relative to the other h_{jk}, then y_j will be a major contributor to the predicted value $\hat{y}_j$.

Observations that significantly affect inferences drawn from the data are said to be *influential*. Methods for assessing influence are typically based on the change in the vector of parameter estimates, $\hat{\boldsymbol{\beta}}$, when observations are deleted. Plots based upon leverage and influence statistics and their use in diagnostic checking of regression models are described in [3], [5], and [10]. These references are recommended for anyone involved in an analysis of regression models.

If, after the diagnostic checks, no serious violations of the assumptions are detected, we can make inferences about $\boldsymbol{\beta}$ and the future Y values with some assurance that we will not be misled.

Additional Problems in Linear Regression

We shall briefly discuss several important aspects of regression that deserve and receive extensive treatments in texts devoted to regression analysis. (See [10], [11], [13], and [23].)

Selecting predictor variables from a large set. In practice, it is often difficult to formulate an appropriate regression function immediately. Which predictor variables should be included? What form should the regression function take?

When the list of possible predictor variables is very large, not all of the variables can be included in the regression function. Techniques and computer programs designed to select the "best" subset of predictors are now readily available. The good ones try all subsets: z_1 alone, z_2 alone, ..., z_1 and z_2, ... The best choice is decided by examining some criterion quantity like R^2. [See (7-9).] However, R^2 always increases with the inclusion of additional predictor variables. Although this problem can be circumvented by using the adjusted R^2, $\bar{R}^2 = 1 - (1 - R^2)(n - 1)/(n - r - 1)$, a better statistic for selecting variables seems to be Mallow's C_p statistic (see [12]),

$$C_p = \left(\frac{\begin{array}{c} \text{residual sum of squares for subset model} \\ \text{with } p \text{ parameters, including an intercept} \end{array}}{\text{(residual variance for full model)}} \right) - (n - 2p)$$

A plot of the pairs (p, C_p), one for each subset of predictors, will indicate models that forecast the observed responses well. Good models typically have (p, C_p) coordinates near the 45° line. In Figure 7.4, we have circled the point corresponding to the "best" subset of predictor variables.

If the list of predictor variables is very long, cost considerations limit the number of models that can be examined. Another approach, called *stepwise regression* (see [13]), attempts to select important predictors without considering all the possibilities.

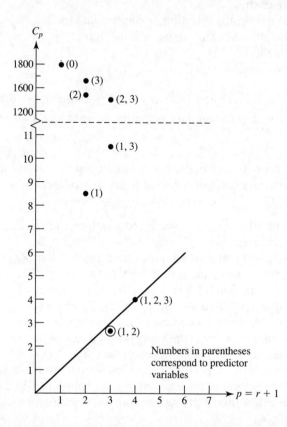

Figure 7.4 C_p plot for computer data from Example 7.6 with three predictor variables (z_1 = orders, z_2 = add–delete count, z_3 = number of items; see the example and original source).

The procedure can be described by listing the basic steps (algorithm) involved in the computations:

Step 1. All possible *simple* linear regressions are considered. The predictor variable that explains the largest significant proportion of the variation in Y (the variable that has the largest correlation with the response) is the first variable to enter the regression function.

Step 2. The next variable to enter is the one (out of those not yet included) that makes the largest significant contribution to the regression sum of squares. The significance of the contribution is determined by an F-test. (See Result 7.6.) The value of the F-statistic that must be exceeded before the contribution of a variable is deemed significant is often called the *F to enter*.

Step 3. Once an additional variable has been included in the equation, the individual contributions to the regression sum of squares of the other variables already in the equation are checked for significance using F-tests. If the F-statistic is less than the one (called the *F to remove*) corresponding to a prescribed significance level, the variable is deleted from the regression function.

Step 4. Steps 2 and 3 are repeated until all possible additions are nonsignificant and all possible deletions are significant. At this point the selection stops.

Because of the step-by-step procedure, there is no guarantee that this approach will select, for example, the best three variables for prediction. A second drawback is that the (automatic) selection methods are not capable of indicating when transformations of variables are useful.

Another popular criterion for selecting an appropriate model, called an information criterion, also balances the size of the residual sum of squares with the number of parameters in the model.

Akaike's information criterion (AIC) is

$$
\text{AIC} = n \ln \left(\frac{\substack{\text{residual sum of squares for subset model} \\ \text{with } p \text{ parameters, including an intercept}}}{n} \right) + 2p
$$

It is desirable that residual sum of squares be small, but the second term penalizes for too many parameters. Overall, we want to select models from those having the smaller values of AIC.

Colinearity. If $\mathbf{Z}$ is not of full rank, some linear combination, such as $\mathbf{Za}$, must equal $\mathbf{0}$. In this situation, the columns are said to be *colinear*. This implies that $\mathbf{Z'Z}$ does not have an inverse. For most regression analyses, it is unlikely that $\mathbf{Za} = \mathbf{0}$ exactly. Yet, if linear combinations of the columns of $\mathbf{Z}$ exist that are nearly $\mathbf{0}$, the calculation of $(\mathbf{Z'Z})^{-1}$ is numerically unstable. Typically, the diagonal entries of $(\mathbf{Z'Z})^{-1}$ will be large. This yields large estimated variances for the $\hat{\beta}_i$'s and it is then difficult to detect the "significant" regression coefficients $\hat{\beta}_i$. The problems caused by colinearity can be overcome somewhat by (1) deleting one of a pair of predictor variables that are *strongly* correlated or (2) relating the response Y to the *principal components* of the predictor variables—that is, the rows $\mathbf{z}'_j$ of $\mathbf{Z}$ are treated as a sample, and the first few principal components are calculated as is subsequently described in Section 8.3. The response Y is then regressed on these new predictor variables.

Bias caused by a misspecified model. Suppose some important predictor variables are omitted from the proposed regression model. That is, suppose the true model has $\mathbf{Z} = [\mathbf{Z}_1 \mathbin{\vdots} \mathbf{Z}_2]$ with rank $r + 1$ and

$$
\underset{(n \times 1)}{\mathbf{Y}} = \left[\underset{(n \times (q+1))}{\mathbf{Z}_1} \mathbin{\vdots} \underset{(n \times (r-q))}{\mathbf{Z}_2} \right] \begin{bmatrix} \underset{((q+1) \times 1)}{\boldsymbol{\beta}_{(1)}} \\ \text{-----} \\ \underset{((r-q) \times 1)}{\boldsymbol{\beta}_{(2)}} \end{bmatrix} + \underset{(n \times 1)}{\boldsymbol{\varepsilon}} \tag{7-20}
$$

$$
= \mathbf{Z}_1 \boldsymbol{\beta}_{(1)} + \mathbf{Z}_2 \boldsymbol{\beta}_{(2)} + \boldsymbol{\varepsilon}
$$

where $E(\boldsymbol{\varepsilon}) = \mathbf{0}$ and $\text{Var}(\boldsymbol{\varepsilon}) = \sigma^2 \mathbf{I}$. However, the investigator unknowingly fits a model using only the first q predictors by minimizing the error sum of squares $(\mathbf{Y} - \mathbf{Z}_1 \boldsymbol{\beta}_{(1)})'(\mathbf{Y} - \mathbf{Z}_1 \boldsymbol{\beta}_{(1)})$. The least squares estimator of $\boldsymbol{\beta}_{(1)}$ is $\hat{\boldsymbol{\beta}}_{(1)} = (\mathbf{Z}_1' \mathbf{Z}_1)^{-1} \mathbf{Z}_1' \mathbf{Y}$. Then, unlike the situation when the model is correct,

$$
E(\hat{\boldsymbol{\beta}}_{(1)}) = (\mathbf{Z}_1' \mathbf{Z}_1)^{-1} \mathbf{Z}_1' E(\mathbf{Y}) = (\mathbf{Z}_1' \mathbf{Z}_1)^{-1} \mathbf{Z}_1' (\mathbf{Z}_1 \boldsymbol{\beta}_{(1)} + \mathbf{Z}_2 \boldsymbol{\beta}_{(2)} + E(\boldsymbol{\varepsilon}))
$$

$$
= \boldsymbol{\beta}_{(1)} + (\mathbf{Z}_1' \mathbf{Z}_1)^{-1} \mathbf{Z}_1' \mathbf{Z}_2 \boldsymbol{\beta}_{(2)} \tag{7-21}
$$

That is, $\hat{\boldsymbol{\beta}}_{(1)}$ is a biased estimator of $\boldsymbol{\beta}_{(1)}$ unless the columns of $\mathbf{Z}_1$ are perpendicular to those of $\mathbf{Z}_2$ (that is, $\mathbf{Z}_1' \mathbf{Z}_2 = \mathbf{0}$). If important variables are missing from the model, the least squares estimates $\hat{\boldsymbol{\beta}}_{(1)}$ may be misleading.

7.7 Multivariate Multiple Regression

In this section, we consider the problem of modeling the relationship between m responses $Y_1, Y_2, \ldots, Y_m$ and a single set of predictor variables $z_1, z_2, \ldots, z_r$. Each response is assumed to follow its own regression model, so that

$$
\begin{aligned}
Y_1 &= \beta_{01} + \beta_{11} z_1 + \cdots + \beta_{r1} z_r + \varepsilon_1 \\
Y_2 &= \beta_{02} + \beta_{12} z_1 + \cdots + \beta_{r2} z_r + \varepsilon_2 \\
&\vdots \qquad\qquad \vdots \\
Y_m &= \beta_{0m} + \beta_{1m} z_1 + \cdots + \beta_{rm} z_r + \varepsilon_m
\end{aligned} \tag{7-22}
$$

The error term $\boldsymbol{\varepsilon}' = [\varepsilon_1, \varepsilon_2, \ldots, \varepsilon_m]$ has $E(\boldsymbol{\varepsilon}) = \mathbf{0}$ and $\text{Var}(\boldsymbol{\varepsilon}) = \boldsymbol{\Sigma}$. Thus, the error terms associated with different responses may be correlated.

To establish notation conforming to the classical linear regression model, let $[z_{j0}, z_{j1}, \ldots, z_{jr}]$ denote the values of the predictor variables for the jth trial, let $\mathbf{Y}_j' = [Y_{j1}, Y_{j2}, \ldots, Y_{jm}]$ be the responses, and let $\boldsymbol{\varepsilon}_j' = [\varepsilon_{j1}, \varepsilon_{j2}, \ldots, \varepsilon_{jm}]$ be the errors. In matrix notation, the design matrix

$$
\underset{(n \times (r+1))}{\mathbf{Z}} = \begin{bmatrix} z_{10} & z_{11} & \cdots & z_{1r} \\ z_{20} & z_{21} & \cdots & z_{2r} \\ \vdots & \vdots & \ddots & \vdots \\ z_{n0} & z_{n1} & \cdots & z_{nr} \end{bmatrix}
$$

is the same as that for the single-response regression model. [See (7-3).] The other matrix quantities have multivariate counterparts. Set

$$
\mathbf{Y}_{(n \times m)} =
\begin{bmatrix}
Y_{11} & Y_{12} & \cdots & Y_{1m} \\
Y_{21} & Y_{22} & \cdots & Y_{2m} \\
\vdots & \vdots & \ddots & \vdots \\
Y_{n1} & Y_{n2} & \cdots & Y_{nm}
\end{bmatrix}
= [\mathbf{Y}_{(1)} \;\vdots\; \mathbf{Y}_{(2)} \;\vdots\; \cdots \;\vdots\; \mathbf{Y}_{(m)}]
$$

$$
\boldsymbol{\beta}_{((r+1) \times m)} =
\begin{bmatrix}
\beta_{01} & \beta_{02} & \cdots & \beta_{0m} \\
\beta_{11} & \beta_{12} & \cdots & \beta_{1m} \\
\vdots & \vdots & \ddots & \vdots \\
\beta_{r1} & \beta_{r2} & \cdots & \beta_{rm}
\end{bmatrix}
= [\boldsymbol{\beta}_{(1)} \;\vdots\; \boldsymbol{\beta}_{(2)} \;\vdots\; \cdots \;\vdots\; \boldsymbol{\beta}_{(m)}]
$$

$$
\boldsymbol{\varepsilon}_{(n \times m)} =
\begin{bmatrix}
\varepsilon_{11} & \varepsilon_{12} & \cdots & \varepsilon_{1m} \\
\varepsilon_{21} & \varepsilon_{22} & \cdots & \varepsilon_{2m} \\
\vdots & \vdots & \ddots & \vdots \\
\varepsilon_{n1} & \varepsilon_{n2} & \cdots & \varepsilon_{nm}
\end{bmatrix}
= [\boldsymbol{\varepsilon}_{(1)} \;\vdots\; \boldsymbol{\varepsilon}_{(2)} \;\vdots\; \cdots \;\vdots\; \boldsymbol{\varepsilon}_{(m)}]
$$

$$
=
\begin{bmatrix}
\boldsymbol{\varepsilon}_1' \\
\boldsymbol{\varepsilon}_2' \\
\vdots \\
\boldsymbol{\varepsilon}_n'
\end{bmatrix}
$$

The *multivariate linear regression model* is

$$
\mathbf{Y}_{(n \times m)} = \mathbf{Z}_{(n \times (r+1))} \boldsymbol{\beta}_{((r+1) \times m)} + \boldsymbol{\varepsilon}_{(n \times m)}
$$

with   (7-23)

$$
E(\boldsymbol{\varepsilon}_{(i)}) = \mathbf{0} \quad \text{and} \quad \text{Cov}(\boldsymbol{\varepsilon}_{(i)}, \boldsymbol{\varepsilon}_{(k)}) = \sigma_{ik}\mathbf{I} \qquad i, k = 1, 2, \ldots, m
$$

The m observations on the jth trial have covariance matrix $\boldsymbol{\Sigma} = \{\sigma_{ik}\}$, but observations from different trials are uncorrelated. Here $\boldsymbol{\beta}$ and σ_{ik} are unknown parameters; the design matrix $\mathbf{Z}$ has jth row $[z_{j0}, z_{j1}, \ldots, z_{jr}]$.

Simply stated, the ith response $\mathbf{Y}_{(i)}$ follows the linear regression model

$$
\mathbf{Y}_{(i)} = \mathbf{Z}\boldsymbol{\beta}_{(i)} + \boldsymbol{\varepsilon}_{(i)}, \qquad i = 1, 2, \ldots, m \tag{7-24}
$$

with $\text{Cov}(\boldsymbol{\varepsilon}_{(i)}) = \sigma_{ii}\mathbf{I}$. However, the errors for *different* responses on the *same* trial can be correlated.

Given the outcomes $\mathbf{Y}$ and the values of the predictor variables $\mathbf{Z}$ with full column rank, we determine the least squares estimates $\hat{\boldsymbol{\beta}}_{(i)}$ exclusively from the observations $\mathbf{Y}_{(i)}$ on the ith response. In conformity with the single-response solution, we take

$$
\hat{\boldsymbol{\beta}}_{(i)} = (\mathbf{Z}'\mathbf{Z})^{-1}\mathbf{Z}'\mathbf{Y}_{(i)} \tag{7-25}
$$

Collecting these univariate least squares estimates, we obtain

$$\hat{\boldsymbol{\beta}} = [\hat{\boldsymbol{\beta}}_{(1)} \mid \hat{\boldsymbol{\beta}}_{(2)} \mid \cdots \mid \hat{\boldsymbol{\beta}}_{(m)}] = (\mathbf{Z}'\mathbf{Z})^{-1}\mathbf{Z}'[\mathbf{Y}_{(1)} \mid \mathbf{Y}_{(2)} \mid \cdots \mid \mathbf{Y}_{(m)}]$$

or

$$\hat{\boldsymbol{\beta}} = (\mathbf{Z}'\mathbf{Z})^{-1}\mathbf{Z}'\mathbf{Y} \qquad (7\text{-}26)$$

For any choice of parameters $\mathbf{B} = [\mathbf{b}_{(1)} \mid \mathbf{b}_{(2)} \mid \cdots \mid \mathbf{b}_{(m)}]$, the matrix of errors is $\mathbf{Y} - \mathbf{ZB}$. The error sum of squares and cross products matrix is

$$(\mathbf{Y} - \mathbf{ZB})'(\mathbf{Y} - \mathbf{ZB})$$

$$= \begin{bmatrix} (\mathbf{Y}_{(1)} - \mathbf{Zb}_{(1)})'(\mathbf{Y}_{(1)} - \mathbf{Zb}_{(1)}) & \cdots & (\mathbf{Y}_{(1)} - \mathbf{Zb}_{(1)})'(\mathbf{Y}_{(m)} - \mathbf{Zb}_{(m)}) \\ \vdots & & \vdots \\ (\mathbf{Y}_{(m)} - \mathbf{Zb}_{(m)})'(\mathbf{Y}_{(1)} - \mathbf{Zb}_{(1)}) & \cdots & (\mathbf{Y}_{(m)} - \mathbf{Zb}_{(m)})'(\mathbf{Y}_{(m)} - \mathbf{Zb}_{(m)}) \end{bmatrix}$$

$$(7\text{-}27)$$

The selection $\mathbf{b}_{(i)} = \hat{\boldsymbol{\beta}}_{(i)}$ minimizes the ith diagonal sum of squares $(\mathbf{Y}_{(i)} - \mathbf{Zb}_{(i)})'(\mathbf{Y}_{(i)} - \mathbf{Zb}_{(i)})$. Consequently, $\text{tr}[(\mathbf{Y} - \mathbf{ZB})'(\mathbf{Y} - \mathbf{ZB})]$ is minimized by the choice $\mathbf{B} = \hat{\boldsymbol{\beta}}$. Also, the generalized variance $|(\mathbf{Y} - \mathbf{ZB})'(\mathbf{Y} - \mathbf{ZB})|$ is minimized by the least squares estimates $\hat{\boldsymbol{\beta}}$. (See Exercise 7.11 for an additional generalized sum of squares property.)

Using the least squares estimates $\hat{\boldsymbol{\beta}}$, we can form the matrices of

$$\text{Predicted values:} \quad \hat{\mathbf{Y}} = \mathbf{Z}\hat{\boldsymbol{\beta}} = \mathbf{Z}(\mathbf{Z}'\mathbf{Z})^{-1}\mathbf{Z}'\mathbf{Y}$$

$$\text{Residuals:} \qquad \hat{\boldsymbol{\varepsilon}} = \mathbf{Y} - \hat{\mathbf{Y}} = [\mathbf{I} - \mathbf{Z}(\mathbf{Z}'\mathbf{Z})^{-1}\mathbf{Z}']\mathbf{Y} \qquad (7\text{-}28)$$

The orthogonality conditions among the residuals, predicted values, and columns of $\mathbf{Z}$, which hold in classical linear regression, hold in multivariate multiple regression. They follow from $\mathbf{Z}'[\mathbf{I} - \mathbf{Z}(\mathbf{Z}'\mathbf{Z})^{-1}\mathbf{Z}'] = \mathbf{Z}' - \mathbf{Z}' = \mathbf{0}$. Specifically,

$$\mathbf{Z}'\hat{\boldsymbol{\varepsilon}} = \mathbf{Z}'[\mathbf{I} - \mathbf{Z}(\mathbf{Z}'\mathbf{Z})^{-1}\mathbf{Z}']\mathbf{Y} = \mathbf{0} \qquad (7\text{-}29)$$

so the residuals $\hat{\boldsymbol{\varepsilon}}_{(i)}$ are perpendicular to the columns of $\mathbf{Z}$. Also,

$$\hat{\mathbf{Y}}'\hat{\boldsymbol{\varepsilon}} = \hat{\boldsymbol{\beta}}'\mathbf{Z}'[\mathbf{I} - \mathbf{Z}(\mathbf{Z}'\mathbf{Z})^{-1}\mathbf{Z}']\mathbf{Y} = \mathbf{0} \qquad (7\text{-}30)$$

confirming that the predicted values $\hat{\mathbf{Y}}_{(i)}$ are perpendicular to all residual vectors $\hat{\boldsymbol{\varepsilon}}_{(k)}$. Because $\mathbf{Y} = \hat{\mathbf{Y}} + \hat{\boldsymbol{\varepsilon}}$,

$$\mathbf{Y}'\mathbf{Y} = (\hat{\mathbf{Y}} + \hat{\boldsymbol{\varepsilon}})'(\hat{\mathbf{Y}} + \hat{\boldsymbol{\varepsilon}}) = \hat{\mathbf{Y}}'\hat{\mathbf{Y}} + \hat{\boldsymbol{\varepsilon}}'\hat{\boldsymbol{\varepsilon}} + \mathbf{0} + \mathbf{0}'$$

or

$$\underset{\begin{pmatrix} \text{total sum of squares} \\ \text{and cross products} \end{pmatrix}}{\mathbf{Y}'\mathbf{Y}} = \underset{\begin{pmatrix} \text{predicted sum of squares} \\ \text{and cross products} \end{pmatrix}}{\hat{\mathbf{Y}}'\hat{\mathbf{Y}}} + \underset{\begin{pmatrix} \text{residual (error) sum} \\ \text{of squares and} \\ \text{cross products} \end{pmatrix}}{\hat{\boldsymbol{\varepsilon}}'\hat{\boldsymbol{\varepsilon}}}$$

$$(7\text{-}31)$$

The residual sum of squares and cross products can also be written as

$$\hat{\boldsymbol{\varepsilon}}'\hat{\boldsymbol{\varepsilon}} = \mathbf{Y}'\mathbf{Y} - \hat{\mathbf{Y}}'\hat{\mathbf{Y}} = \mathbf{Y}'\mathbf{Y} - \hat{\boldsymbol{\beta}}'\mathbf{Z}'\mathbf{Z}\hat{\boldsymbol{\beta}} \qquad (7\text{-}32)$$

Example 7.8 (Fitting a multivariate straight-line regression model) To illustrate the calculations of $\hat{\boldsymbol{\beta}}, \hat{\mathbf{Y}}$, and $\hat{\boldsymbol{\varepsilon}}$, we fit a straight-line regression model (see Panel 7.2),

$$Y_{j1} = \beta_{01} + \beta_{11}z_{j1} + \varepsilon_{j1}$$
$$Y_{j2} = \beta_{02} + \beta_{12}z_{j1} + \varepsilon_{j2}, \qquad j = 1, 2, \ldots, 5$$

to two responses Y_1 and Y_2 using the data in Example 7.3. These data, augmented by observations on an additional response, are as follows:

z_1	0	1	2	3	4
y_1	1	4	3	8	9
y_2	−1	−1	2	3	2

The design matrix $\mathbf{Z}$ remains unchanged from the single-response problem. We find that

$$\mathbf{Z}' = \begin{bmatrix} 1 & 1 & 1 & 1 & 1 \\ 0 & 1 & 2 & 3 & 4 \end{bmatrix} \qquad (\mathbf{Z}'\mathbf{Z})^{-1} = \begin{bmatrix} .6 & -.2 \\ -.2 & .1 \end{bmatrix}$$

PANEL 7.2   SAS ANALYSIS FOR EXAMPLE 7.8 USING PROC. GLM.

```
title 'Multivariate Regression Analysis';
data mra;
infile 'Example 7-8 data;            PROGRAM COMMANDS
input y1 y2 z1;
proc glm data = mra;
model y1 y2 = z1/ss3;
manova h = z1/printe;
```

General Linear Models Procedure

Dependent Variable: Y1   OUTPUT

Source	DF	Sum of Squares	Mean Square	F Value	Pr > F
Model	1	40.00000000	40.00000000	20.00	0.0208
Error	3	6.00000000	2.00000000		
Corrected Total	4	46.00000000			

R-Square	C.V.	Root MSE	Y1 Mean
0.869565	28.28427	1.414214	5.00000000

(continues on next page)

PANEL 7.2   *(continued)*

Source	DF	Type III SS	Mean Square	F Value	Pr > F
Z1	1	40.00000000	40.00000000	20.00	0.0208

Parameter	Estimate	T for H0: Parameter = 0	Pr > ITI	Std Error of Estimate
INTERCEPT	1.000000000	0.91	0.4286	1.09544512
Z1	2.000000000	4.47	0.0208	0.44721360

Dependent Variable: Y2

Source	DF	Sum of Squares	Mean Square	F Value	Pr > F
Model	1	10.00000000	10.00000000	7.50	0.0714
Error	3	4.00000000	1.33333333		
Corrected Total	4	14.00000000			

R-Square	C.V.	Root MSE	Y2 Mean
0.714286	115.4701	1.154701	1.00000000

Source	DF	Type III SS	Mean Square	F Value	Pr > F
Z1	1	10.00000000	10.00000000	7.50	0.0714

Parameter	Estimate	T for H0: Parameter = 0	Pr > ITI	Std Error of Estimate
INTERCEPT	-1.000000000	-1.12	0.3450	0.89442719
Z1	1.000000000	2.74	0.0714	0.36514837

E = Error SS & CP Matrix

	Y1	Y2
Y1	6	-2
Y2	-2	4

Manova Test Criteria and Exact F Statistics for
the Hypothesis of no Overall Z1 Effect
H = Type III SS&CP Matrix for Z1   E = Error SS&CP Matrix
S = 1   M = 0   N = 0

Statistic	Value	F	Num DF	Den DF	Pr > F
Wilks' Lambda	0.06250000	15.0000	2	2	0.0625
Pillai's Trace	0.93750000	15.0000	2	2	0.0625
Hotelling-Lawley Trace	15.00000000	15.0000	2	2	0.0625
Roy's Greatest Root	15.00000000	15.0000	2	2	0.0625

and

$$\mathbf{Z}'\mathbf{y}_{(2)} = \begin{bmatrix} 1 & 1 & 1 & 1 & 1 \\ 0 & 1 & 2 & 3 & 4 \end{bmatrix} \begin{bmatrix} -1 \\ -1 \\ 2 \\ 3 \\ 2 \end{bmatrix} = \begin{bmatrix} 5 \\ 20 \end{bmatrix}$$

so

$$\hat{\boldsymbol{\beta}}_{(2)} = (\mathbf{Z}'\mathbf{Z})^{-1}\mathbf{Z}'\mathbf{y}_{(2)} = \begin{bmatrix} .6 & -.2 \\ -.2 & .1 \end{bmatrix} \begin{bmatrix} 5 \\ 20 \end{bmatrix} = \begin{bmatrix} -1 \\ 1 \end{bmatrix}$$

From Example 7.3,

$$\hat{\boldsymbol{\beta}}_{(1)} = (\mathbf{Z}'\mathbf{Z})^{-1}\mathbf{Z}'\mathbf{y}_{(1)} = \begin{bmatrix} 1 \\ 2 \end{bmatrix}$$

Hence,

$$\hat{\boldsymbol{\beta}} = [\hat{\boldsymbol{\beta}}_{(1)} \mid \hat{\boldsymbol{\beta}}_{(2)}] = \begin{bmatrix} 1 & -1 \\ 2 & 1 \end{bmatrix} = (\mathbf{Z}'\mathbf{Z})^{-1}\mathbf{Z}'[\mathbf{y}_{(1)} \mid \mathbf{y}_{(2)}]$$

The fitted values are generated from $\hat{y}_1 = 1 + 2z_1$ and $\hat{y}_2 = -1 + z_2$. Collectively,

$$\hat{\mathbf{Y}} = \mathbf{z}\hat{\boldsymbol{\beta}} = \begin{bmatrix} 1 & 0 \\ 1 & 1 \\ 1 & 2 \\ 1 & 3 \\ 1 & 4 \end{bmatrix} \begin{bmatrix} 1 & -1 \\ 2 & 1 \end{bmatrix} = \begin{bmatrix} 1 & -1 \\ 3 & 0 \\ 5 & 1 \\ 7 & 2 \\ 9 & 3 \end{bmatrix}$$

and

$$\hat{\boldsymbol{\varepsilon}} = \mathbf{Y} - \hat{\mathbf{Y}} = \begin{bmatrix} 0 & 1 & -2 & 1 & 0 \\ 0 & -1 & 1 & 1 & -1 \end{bmatrix}'$$

Note that

$$\hat{\boldsymbol{\varepsilon}}'\hat{\mathbf{Y}} = \begin{bmatrix} 0 & 1 & -2 & 1 & 0 \\ 0 & -1 & 1 & 1 & -1 \end{bmatrix} \begin{bmatrix} 1 & -1 \\ 3 & 0 \\ 5 & 1 \\ 7 & 2 \\ 9 & 3 \end{bmatrix} = \begin{bmatrix} 0 & 0 \\ 0 & 0 \end{bmatrix}$$

Since

$$\mathbf{Y}'\mathbf{Y} = \begin{bmatrix} 1 & 4 & 3 & 8 & 9 \\ -1 & -1 & 2 & 3 & 2 \end{bmatrix} \begin{bmatrix} 1 & -1 \\ 4 & -1 \\ 3 & 2 \\ 8 & 3 \\ 9 & 2 \end{bmatrix} = \begin{bmatrix} 171 & 43 \\ 43 & 19 \end{bmatrix}$$

$$\hat{\mathbf{Y}}'\hat{\mathbf{Y}} = \begin{bmatrix} 165 & 45 \\ 45 & 15 \end{bmatrix} \quad \text{and} \quad \hat{\boldsymbol{\varepsilon}}'\hat{\boldsymbol{\varepsilon}} = \begin{bmatrix} 6 & -2 \\ -2 & 4 \end{bmatrix}$$

the sum of squares and cross-products decomposition

$$\mathbf{Y}'\mathbf{Y} = \hat{\mathbf{Y}}'\hat{\mathbf{Y}} + \hat{\boldsymbol{\varepsilon}}'\hat{\boldsymbol{\varepsilon}}$$

is easily verified.   ∎

Result 7.9. For the least squares estimator $\hat{\boldsymbol{\beta}} = [\hat{\boldsymbol{\beta}}_{(1)} \vdots \hat{\boldsymbol{\beta}}_{(2)} \vdots \cdots \vdots \hat{\boldsymbol{\beta}}_{(m)}]$ determined under the multivariate multiple regression model (7-23) with full rank $(\mathbf{Z}) = r + 1 < n$,

$$E(\hat{\boldsymbol{\beta}}_{(i)}) = \boldsymbol{\beta}_{(i)} \quad \text{or} \quad E(\hat{\boldsymbol{\beta}}) = \boldsymbol{\beta}$$

and

$$\text{Cov}(\hat{\boldsymbol{\beta}}_{(i)}, \hat{\boldsymbol{\beta}}_{(k)}) = \sigma_{ik}(\mathbf{Z}'\mathbf{Z})^{-1}, \quad i, k = 1, 2, \ldots, m$$

The residuals $\hat{\boldsymbol{\varepsilon}} = [\hat{\boldsymbol{\varepsilon}}_{(1)} \vdots \hat{\boldsymbol{\varepsilon}}_{(2)} \vdots \cdots \vdots \hat{\boldsymbol{\varepsilon}}_{(m)}] = \mathbf{Y} - \mathbf{Z}\hat{\boldsymbol{\beta}}$ satisfy $E(\hat{\boldsymbol{\varepsilon}}_{(i)}) = \mathbf{0}$ and $E(\hat{\boldsymbol{\varepsilon}}'_{(i)}\hat{\boldsymbol{\varepsilon}}_{(k)}) = (n - r - 1)\sigma_{ik}$, so

$$E(\hat{\boldsymbol{\varepsilon}}) = \mathbf{0} \quad \text{and} \quad E\left(\frac{1}{n - r - 1}\hat{\boldsymbol{\varepsilon}}'\hat{\boldsymbol{\varepsilon}}\right) = \boldsymbol{\Sigma}$$

Also, $\hat{\boldsymbol{\varepsilon}}$ and $\hat{\boldsymbol{\beta}}$ are uncorrelated.

Proof. The ith response follows the multiple regression model

$$\mathbf{Y}_{(i)} = \mathbf{Z}\boldsymbol{\beta}_{(i)} + \boldsymbol{\varepsilon}_{(i)}, \quad E(\boldsymbol{\varepsilon}_{(i)}) = \mathbf{0}, \quad \text{and} \quad E(\boldsymbol{\varepsilon}_{(i)}\boldsymbol{\varepsilon}'_{(i)}) = \sigma_{ii}\mathbf{I}$$

Also, from (7-24),

$$\hat{\boldsymbol{\beta}}_{(i)} - \boldsymbol{\beta}_{(i)} = (\mathbf{Z}'\mathbf{Z})^{-1}\mathbf{Z}'\mathbf{Y}_{(i)} - \boldsymbol{\beta}_{(i)} = (\mathbf{Z}'\mathbf{Z})^{-1}\mathbf{Z}'\boldsymbol{\varepsilon}_{(i)} \qquad (7\text{-}33)$$

and

$$\hat{\boldsymbol{\varepsilon}}_{(i)} = \mathbf{Y}_{(i)} - \hat{\mathbf{Y}}_{(i)} = [\mathbf{I} - \mathbf{Z}(\mathbf{Z}'\mathbf{Z})^{-1}\mathbf{Z}']\mathbf{Y}_{(i)} = [\mathbf{I} - \mathbf{Z}(\mathbf{Z}'\mathbf{Z})^{-1}\mathbf{Z}']\boldsymbol{\varepsilon}_{(i)}$$

so $E(\hat{\boldsymbol{\beta}}_{(i)}) = \boldsymbol{\beta}_{(i)}$ and $E(\hat{\boldsymbol{\varepsilon}}_{(i)}) = \mathbf{0}$.

Next,

$$\text{Cov}(\hat{\boldsymbol{\beta}}_{(i)}, \hat{\boldsymbol{\beta}}_{(k)}) = E(\hat{\boldsymbol{\beta}}_{(i)} - \boldsymbol{\beta}_{(i)})(\hat{\boldsymbol{\beta}}_{(k)} - \boldsymbol{\beta}_{(k)})'$$

$$= (\mathbf{Z}'\mathbf{Z})^{-1}\mathbf{Z}'E(\boldsymbol{\varepsilon}_{(i)}\boldsymbol{\varepsilon}'_{(k)})\mathbf{Z}(\mathbf{Z}'\mathbf{Z})^{-1} = \sigma_{ik}(\mathbf{Z}'\mathbf{Z})^{-1}$$

Using Result 4.9, with $\mathbf{U}$ any random vector and $\mathbf{A}$ a fixed matrix, we have that $E[\mathbf{U}'\mathbf{A}\mathbf{U}] = E[\text{tr}(\mathbf{A}\mathbf{U}\mathbf{U}')] = \text{tr}[\mathbf{A}E(\mathbf{U}\mathbf{U}')]$. Consequently, from the proof of Result 7.1 and using Result 2A.12

$$E(\hat{\boldsymbol{\varepsilon}}'_{(i)}\hat{\boldsymbol{\varepsilon}}_{(k)}) = E(\boldsymbol{\varepsilon}'_{(i)}(\mathbf{I} - \mathbf{Z}(\mathbf{Z}'\mathbf{Z})^{-1}\mathbf{Z}')\boldsymbol{\varepsilon}_{(k)}) = \text{tr}[(\mathbf{I} - \mathbf{Z}(\mathbf{Z}'\mathbf{Z})^{-1}\mathbf{Z}')\sigma_{ik}\mathbf{I}]$$

$$= \sigma_{ik}\,\text{tr}[(\mathbf{I} - \mathbf{Z}(\mathbf{Z}'\mathbf{Z})^{-1}\mathbf{Z}')] = \sigma_{ik}(n - r - 1)$$

Dividing each entry $\hat{\boldsymbol{\varepsilon}}'_{(i)}\hat{\boldsymbol{\varepsilon}}_{(k)}$ of $\hat{\boldsymbol{\varepsilon}}'\hat{\boldsymbol{\varepsilon}}$ by $n - r - 1$, we obtain the unbiased estimator of $\boldsymbol{\Sigma}$. Finally,

$$
\begin{aligned}
\text{Cov}\,(\hat{\boldsymbol{\beta}}_{(i)}, \hat{\boldsymbol{\varepsilon}}_{(k)}) &= E[(\mathbf{Z}'\mathbf{Z})^{-1}\mathbf{Z}'\boldsymbol{\varepsilon}_{(i)}\boldsymbol{\varepsilon}'_{(k)}(\mathbf{I} - \mathbf{Z}(\mathbf{Z}'\mathbf{Z})^{-1}\mathbf{Z}')] \\
&= (\mathbf{Z}'\mathbf{Z})^{-1}\mathbf{Z}'E(\boldsymbol{\varepsilon}_{(i)}\boldsymbol{\varepsilon}'_{(k)})(\mathbf{I} - \mathbf{Z}(\mathbf{Z}'\mathbf{Z})^{-1}\mathbf{Z}') \\
&= (\mathbf{Z}'\mathbf{Z})^{-1}\mathbf{Z}'\sigma_{ik}\mathbf{I}(\mathbf{I} - \mathbf{Z}(\mathbf{Z}'\mathbf{Z})^{-1}\mathbf{Z}') \\
&= \sigma_{ik}((\mathbf{Z}'\mathbf{Z})^{-1}\mathbf{Z}' - (\mathbf{Z}'\mathbf{Z})^{-1}\mathbf{Z}') = \mathbf{0}
\end{aligned}
$$

so each element of $\hat{\boldsymbol{\beta}}$ is uncorrelated with each element of $\hat{\boldsymbol{\varepsilon}}$. ∎

The mean vectors and covariance matrices determined in Result 7.9 enable us to obtain the sampling properties of the least squares predictors.

We first consider the problem of estimating the mean vector when the predictor variables have the values $\mathbf{z}'_0 = [1, z_{01}, \ldots, z_{0r}]$. The mean of the ith response variable is $\mathbf{z}'_0\boldsymbol{\beta}_{(i)}$, and this is estimated by $\mathbf{z}'_0\hat{\boldsymbol{\beta}}_{(i)}$, the ith component of the fitted regression relationship. Collectively,

$$
\mathbf{z}'_0\hat{\boldsymbol{\beta}} = [\mathbf{z}'_0\hat{\boldsymbol{\beta}}_{(1)} \;\vdots\; \mathbf{z}'_0\hat{\boldsymbol{\beta}}_{(2)} \;\vdots\; \cdots \;\vdots\; \mathbf{z}'_0\hat{\boldsymbol{\beta}}_{(m)}] \tag{7-34}
$$

is an unbiased estimator $\mathbf{z}'_0\boldsymbol{\beta}$ since $E(\mathbf{z}'_0\hat{\boldsymbol{\beta}}_{(i)}) = \mathbf{z}'_0 E(\hat{\boldsymbol{\beta}}_{(i)}) = \mathbf{z}'_0\boldsymbol{\beta}_{(i)}$ for each component. From the covariance matrix for $\hat{\boldsymbol{\beta}}_{(i)}$ and $\hat{\boldsymbol{\beta}}_{(k)}$, the estimation errors $\mathbf{z}'_0\boldsymbol{\beta}_{(i)} - \mathbf{z}'_0\hat{\boldsymbol{\beta}}_{(i)}$ have covariances

$$
\begin{aligned}
E[\mathbf{z}'_0(\boldsymbol{\beta}_{(i)} - \hat{\boldsymbol{\beta}}_{(i)})(\boldsymbol{\beta}_{(k)} - \hat{\boldsymbol{\beta}}_{(k)})'\mathbf{z}_0] &= \mathbf{z}'_0(E(\boldsymbol{\beta}_{(i)} - \hat{\boldsymbol{\beta}}_{(i)})(\boldsymbol{\beta}_{(k)} - \hat{\boldsymbol{\beta}}_{(k)})')\mathbf{z}_0 \\
&= \sigma_{ik}\mathbf{z}'_0(\mathbf{Z}'\mathbf{Z})^{-1}\mathbf{z}_0
\end{aligned} \tag{7-35}
$$

The related problem is that of forecasting a new observation vector $\mathbf{Y}'_0 = [Y_{01}, Y_{02}, \ldots, Y_{0m}]$ at $\mathbf{z}_0$. According to the regression model, $Y_{0i} = \mathbf{z}'_0\boldsymbol{\beta}_{(i)} + \varepsilon_{0i}$ where the "new" error $\boldsymbol{\varepsilon}'_0 = [\varepsilon_{01}, \varepsilon_{02}, \ldots, \varepsilon_{0m}]$ is independent of the errors $\boldsymbol{\varepsilon}$ and satisfies $E(\varepsilon_{0i}) = 0$ and $E(\varepsilon_{0i}\varepsilon_{0k}) = \sigma_{ik}$. The *forecast error* for the ith component of $\mathbf{Y}_0$ is

$$
\begin{aligned}
Y_{0i} - \mathbf{z}'_0\hat{\boldsymbol{\beta}}_{(i)} &= Y_{0i} - \mathbf{z}'_0\boldsymbol{\beta}_{(i)} + \mathbf{z}'_0\boldsymbol{\beta}_{(i)} - \mathbf{z}'_0\hat{\boldsymbol{\beta}}_{(i)} \\
&= \varepsilon_{0i} - \mathbf{z}'_0(\hat{\boldsymbol{\beta}}_{(i)} - \boldsymbol{\beta}_{(i)})
\end{aligned}
$$

so $E(Y_{0i} - \mathbf{z}'_0\hat{\boldsymbol{\beta}}_{(i)}) = E(\varepsilon_{0i}) - \mathbf{z}'_0 E(\hat{\boldsymbol{\beta}}_{(i)} - \boldsymbol{\beta}_{(i)}) = 0$, indicating that $\mathbf{z}'_0\hat{\boldsymbol{\beta}}_{(i)}$ is an *unbiased predictor* of Y_{0i}. The forecast errors have covariances

$$
\begin{aligned}
E(Y_{0i} &- \mathbf{z}'_0\hat{\boldsymbol{\beta}}_{(i)})(Y_{0k} - \mathbf{z}'_0\hat{\boldsymbol{\beta}}_{(k)}) \\
&= E(\varepsilon_{0i} - \mathbf{z}'_0(\hat{\boldsymbol{\beta}}_{(i)} - \boldsymbol{\beta}_{(i)}))(\varepsilon_{0k} - \mathbf{z}'_0(\hat{\boldsymbol{\beta}}_{(k)} - \boldsymbol{\beta}_{(k)})) \\
&= E(\varepsilon_{0i}\varepsilon_{0k}) + \mathbf{z}'_0 E(\hat{\boldsymbol{\beta}}_{(i)} - \boldsymbol{\beta}_{(i)})(\hat{\boldsymbol{\beta}}_{(k)} - \boldsymbol{\beta}_{(k)})'\mathbf{z}_0 \\
&\quad - \mathbf{z}'_0 E((\hat{\boldsymbol{\beta}}_{(i)} - \boldsymbol{\beta}_{(i)})\varepsilon_{0k}) - E(\varepsilon_{0i}(\hat{\boldsymbol{\beta}}_{(k)} - \boldsymbol{\beta}_{(k)})')\mathbf{z}_0 \\
&= \sigma_{ik}(1 + \mathbf{z}'_0(\mathbf{Z}'\mathbf{Z})^{-1}\mathbf{z}_0)
\end{aligned} \tag{7-36}
$$

Note that $E((\hat{\boldsymbol{\beta}}_{(i)} - \boldsymbol{\beta}_{(i)})\varepsilon_{0k}) = \mathbf{0}$ since $\hat{\boldsymbol{\beta}}_{(i)} = (\mathbf{Z}'\mathbf{Z})^{-1}\mathbf{Z}'\boldsymbol{\varepsilon}_{(i)} + \boldsymbol{\beta}_{(i)}$ is independent of $\boldsymbol{\varepsilon}_0$. A similar result holds for $E(\varepsilon_{0i}(\hat{\boldsymbol{\beta}}_{(k)} - \boldsymbol{\beta}_{(k)})')$.

Maximum likelihood estimators and their distributions can be obtained when the errors $\boldsymbol{\varepsilon}$ have a normal distribution.

Result 7.10. Let the multivariate multiple regression model in (7-23) hold with full rank $(\mathbf{Z}) = r + 1$, $n \geq (r + 1) + m$, and let the errors $\boldsymbol{\varepsilon}$ have a normal distribution. Then

$$\hat{\boldsymbol{\beta}} = (\mathbf{Z}'\mathbf{Z})^{-1}\mathbf{Z}'\mathbf{Y}$$

is the maximum likelihood estimator of $\boldsymbol{\beta}$ and $\hat{\boldsymbol{\beta}}$ has a normal distribution with $E(\hat{\boldsymbol{\beta}}) = \boldsymbol{\beta}$ and $\text{Cov}(\hat{\boldsymbol{\beta}}_{(i)}, \hat{\boldsymbol{\beta}}_{(k)}) = \sigma_{ik}(\mathbf{Z}'\mathbf{Z})^{-1}$. Also, $\hat{\boldsymbol{\beta}}$ is independent of the maximum likelihood estimator of the positive definite $\boldsymbol{\Sigma}$ given by

$$\hat{\boldsymbol{\Sigma}} = \frac{1}{n}\hat{\boldsymbol{\varepsilon}}'\hat{\boldsymbol{\varepsilon}} = \frac{1}{n}(\mathbf{Y} - \mathbf{z}\hat{\boldsymbol{\beta}})'(\mathbf{Y} - \mathbf{z}\hat{\boldsymbol{\beta}})$$

and

$$n\hat{\boldsymbol{\Sigma}} \quad \text{is distributed as} \quad W_{p,n-r-1}(\boldsymbol{\Sigma})$$

The maximized likelihood $L(\hat{\boldsymbol{\mu}}, \hat{\boldsymbol{\Sigma}}) = (2\pi)^{-mn/2}|\hat{\boldsymbol{\Sigma}}|^{-n/2}e^{-mn/2}$.

Proof. (See website: www.prenhall.com/statistics) ∎

Result 7.10 provides additional support for using least squares estimates. When the errors are normally distributed, $\hat{\boldsymbol{\beta}}$ and $n^{-1}\hat{\boldsymbol{\varepsilon}}'\hat{\boldsymbol{\varepsilon}}$ are the maximum likelihood estimators of $\boldsymbol{\beta}$ and $\boldsymbol{\Sigma}$, respectively. Therefore, for large samples, they have nearly the smallest possible variances.

Comment. The multivariate multiple regression model poses no new computational problems. Least squares (maximum likelihood) estimates, $\hat{\boldsymbol{\beta}}_{(i)} = (\mathbf{Z}'\mathbf{Z})^{-1}\mathbf{Z}'\mathbf{y}_{(i)}$, are computed individually for each response variable. Note, however, that the model requires that the *same* predictor variables be used for all responses.

Once a multivariate multiple regression model has been fit to the data, it should be subjected to the diagnostic checks described in Section 7.6 for the single-response model. The residual vectors $[\hat{\varepsilon}_{j1}, \hat{\varepsilon}_{j2}, \ldots, \hat{\varepsilon}_{jm}]$ can be examined for normality or outliers using the techniques in Section 4.6.

The remainder of this section is devoted to brief discussions of inference for the normal theory multivariate multiple regression model. Extended accounts of these procedures appear in [2] and [18].

Likelihood Ratio Tests for Regression Parameters

The multiresponse analog of (7-12), the hypothesis that the responses do not depend on $z_{q+1}, z_{q+2}, \ldots, z_r$, becomes

$$H_0: \boldsymbol{\beta}_{(2)} = \mathbf{0} \quad \text{where} \quad \boldsymbol{\beta} = \begin{bmatrix} \boldsymbol{\beta}_{(1)} \\ {\scriptstyle ((q+1)\times m)} \\ \hline \boldsymbol{\beta}_{(2)} \\ {\scriptstyle ((r-q)\times m)} \end{bmatrix} \tag{7-37}$$

Setting $\mathbf{Z} = \begin{bmatrix} \mathbf{Z}_1 & \vdots & \mathbf{Z}_2 \\ {\scriptstyle (n\times(q+1))} & \vdots & {\scriptstyle (n\times(r-q))} \end{bmatrix}$, we can write the general model as

$$E(\mathbf{Y}) = \mathbf{Z}\boldsymbol{\beta} = [\mathbf{Z}_1 \;\vdots\; \mathbf{Z}_2] \begin{bmatrix} \boldsymbol{\beta}_{(1)} \\ \hline \boldsymbol{\beta}_{(2)} \end{bmatrix} = \mathbf{Z}_1\boldsymbol{\beta}_{(1)} + \mathbf{Z}_2\boldsymbol{\beta}_{(2)}$$

Under H_0: $\boldsymbol{\beta}_{(2)} = \mathbf{0}$, $\mathbf{Y} = \mathbf{Z}_1 \boldsymbol{\beta}_{(1)} + \boldsymbol{\varepsilon}$ and the likelihood ratio test of H_0 is based on the quantities involved in the

extra sum of squares and cross products

$$= (\mathbf{Y} - \mathbf{Z}_1 \hat{\boldsymbol{\beta}}_{(1)})'(\mathbf{Y} - \mathbf{Z}_1 \hat{\boldsymbol{\beta}}_{(1)}) - (\mathbf{Y} - \mathbf{z}\hat{\boldsymbol{\beta}})'(\mathbf{Y} - \mathbf{z}\hat{\boldsymbol{\beta}})$$

$$= n(\hat{\boldsymbol{\Sigma}}_1 - \hat{\boldsymbol{\Sigma}})$$

where $\hat{\boldsymbol{\beta}}_{(1)} = (\mathbf{Z}_1'\mathbf{Z}_1)^{-1}\mathbf{Z}_1'\mathbf{Y}$ and $\hat{\boldsymbol{\Sigma}}_1 = n^{-1}(\mathbf{Y} - \mathbf{Z}_1\hat{\boldsymbol{\beta}}_{(1)})'(\mathbf{Y} - \mathbf{Z}_1\hat{\boldsymbol{\beta}}_{(1)})$.

From Result 7.10, the likelihood ratio, Λ, can be expressed in terms of generalized variances:

$$\Lambda = \frac{\max\limits_{\boldsymbol{\beta}_{(1)}, \boldsymbol{\Sigma}} L(\boldsymbol{\beta}_{(1)}, \boldsymbol{\Sigma})}{\max\limits_{\boldsymbol{\beta}, \boldsymbol{\Sigma}} L(\boldsymbol{\beta}, \boldsymbol{\Sigma})} = \frac{L(\hat{\boldsymbol{\beta}}_{(1)}, \hat{\boldsymbol{\Sigma}}_1)}{L(\hat{\boldsymbol{\beta}}, \hat{\boldsymbol{\Sigma}})} = \left(\frac{|\hat{\boldsymbol{\Sigma}}|}{|\hat{\boldsymbol{\Sigma}}_1|}\right)^{n/2} \tag{7-38}$$

Equivalently, *Wilks' lambda statistic*

$$\Lambda^{2/n} = \frac{|\hat{\boldsymbol{\Sigma}}|}{|\hat{\boldsymbol{\Sigma}}_1|}$$

can be used.

Result 7.11. Let the multivariate multiple regression model of (7-23) hold with $\mathbf{Z}$ of full rank $r + 1$ and $(r + 1) + m \le n$. Let the errors $\boldsymbol{\varepsilon}$ be normally distributed. Under H_0: $\boldsymbol{\beta}_{(2)} = \mathbf{0}$, $n\hat{\boldsymbol{\Sigma}}$ is distributed as $W_{p, n-r-1}(\boldsymbol{\Sigma})$ independently of $n(\hat{\boldsymbol{\Sigma}}_1 - \hat{\boldsymbol{\Sigma}})$ which, in turn, is distributed as $W_{p, r-q}(\boldsymbol{\Sigma})$. The likelihood ratio test of H_0 is equivalent to rejecting H_0 for large values of

$$-2\ln\Lambda = -n\ln\left(\frac{|\hat{\boldsymbol{\Sigma}}|}{|\hat{\boldsymbol{\Sigma}}_1|}\right) = -n\ln\frac{|n\hat{\boldsymbol{\Sigma}}|}{|n\hat{\boldsymbol{\Sigma}} + n(\hat{\boldsymbol{\Sigma}}_1 - \hat{\boldsymbol{\Sigma}})|}$$

For n large,[5] the modified statistic

$$-\left[n - r - 1 - \frac{1}{2}(m - r + q + 1)\right]\ln\left(\frac{|\hat{\boldsymbol{\Sigma}}|}{|\hat{\boldsymbol{\Sigma}}_1|}\right)$$

has, to a close approximation, a chi-square distribution with $m(r - q)$ d.f.

Proof. (See Supplement 7A.)   ∎

If $\mathbf{Z}$ is not of full rank, but has rank $r_1 + 1$, then $\hat{\boldsymbol{\beta}} = (\mathbf{Z}'\mathbf{Z})^-\mathbf{Z}'\mathbf{Y}$, where $(\mathbf{Z}'\mathbf{Z})^-$ is the *generalized inverse* discussed in [22]. (See also Exercise 7.6.) The distributional conclusions stated in Result 7.11 remain the same, provided that r is replaced by r_1 and $q + 1$ by rank $(\mathbf{Z}_1)$. However, not all hypotheses concerning $\boldsymbol{\beta}$ can be tested due to the lack of uniqueness in the identification of $\boldsymbol{\beta}$ caused by the linear dependencies among the columns of $\mathbf{Z}$. Nevertheless, the generalized inverse allows all of the important MANOVA models to be analyzed as special cases of the multivariate multiple regression model.

---

[5] Technically, both $n - r$ and $n - m$ should also be large to obtain a good chi-square approximation.

Example 7.9 (Testing the importance of additional predictors with a multivariate response) The service in three locations of a large restaurant chain was rated according to two measures of quality by male and female patrons. The first service-quality index was introduced in Example 7.5. Suppose we consider a regression model that allows for the effects of location, gender, and the location–gender interaction on both service-quality indices. The design matrix (see Example 7.5) remains the same for the two-response situation. We shall illustrate the test of no location-gender interaction in either response using Result 7.11. A computer program provides

$$\begin{pmatrix} \text{residual sum of squares} \\ \text{and cross products} \end{pmatrix} = n\hat{\boldsymbol{\Sigma}} = \begin{bmatrix} 2977.39 & 1021.72 \\ 1021.72 & 2050.95 \end{bmatrix}$$

$$\begin{pmatrix} \text{extra sum of squares} \\ \text{and cross products} \end{pmatrix} = n(\hat{\boldsymbol{\Sigma}}_1 - \hat{\boldsymbol{\Sigma}}) = \begin{bmatrix} 441.76 & 246.16 \\ 246.16 & 366.12 \end{bmatrix}$$

Let $\boldsymbol{\beta}_{(2)}$ be the matrix of interaction parameters for the two responses. Although the sample size $n = 18$ is not large, we shall illustrate the calculations involved in the test of $H_0\colon \boldsymbol{\beta}_{(2)} = \mathbf{0}$ given in Result 7.11. Setting $\alpha = .05$, we test H_0 by referring

$$-\left[n - r_1 - 1 - \frac{1}{2}(m - r_1 + q_1 + 1) \right] \ln\left(\frac{|n\hat{\boldsymbol{\Sigma}}|}{|n\hat{\boldsymbol{\Sigma}} + n(\hat{\boldsymbol{\Sigma}}_1 - \hat{\boldsymbol{\Sigma}})|} \right)$$

$$= -\left[18 - 5 - 1 - \frac{1}{2}(2 - 5 + 3 + 1) \right] \ln(.7605) = 3.28$$

to a chi-square percentage point with $m(r_1 - q_1) = 2(2) = 4$ d.f. Since $3.28 < \chi_4^2(.05) = 9.49$, we do not reject H_0 at the 5% level. The interaction terms are not needed. ∎

Information criterion are also available to aid in the selection of a simple but adequate multivariate multiple regresson model. For a model that includes d predictor variables counting the intercept, let

$$\hat{\boldsymbol{\Sigma}}_d = \frac{1}{n}(\text{residual sum of squares and cross products matrix})$$

Then, the multivariate multiple regression version of the Akaike's information criterion is

$$\text{AIC} = n \ln(|\hat{\boldsymbol{\Sigma}}_d|) - 2p \times d$$

This criterion attempts to balance the generalized variance with the number of parameters. Models with smaller AIC's are preferable.

In the context of Example 7.9, under the null hypothesis of no interaction terms, we have $n = 18$, $p = 2$ response variables, and $d = 4$ terms, so

$$\text{AIC} = n \ln(|\boldsymbol{\Sigma}|) - 2p \times d = 18 \ln\left(\left| \frac{1}{18} \begin{bmatrix} 3419.15 & 1267.88 \\ 1267.88 & 2417.07 \end{bmatrix} \right| \right) - 2 \times 2 \times 4$$

$$= 18 \times \ln(20545.7) - 16 = 162.75$$

More generally, we could consider a null hypothesis of the form $H_0\colon \mathbf{C}\boldsymbol{\beta} = \boldsymbol{\Gamma}_0$, where $\mathbf{C}$ is $(r - q) \times (r + 1)$ and is of full rank $(r - q)$. For the choices

$\mathbf{C} = \begin{bmatrix} \mathbf{0} & \vdots & \mathbf{I} \\ & & {\scriptstyle (r-q)\times(r-q)} \end{bmatrix}$ and $\boldsymbol{\Gamma}_0 = \mathbf{0}$, this null hypothesis becomes $H_0: \mathbf{C}\boldsymbol{\beta} = \boldsymbol{\beta}_{(2)} = \mathbf{0}$,

the case considered earlier. It can be shown that the extra sum of squares and cross products generated by the hypothesis H_0 is

$$n(\hat{\boldsymbol{\Sigma}}_1 - \hat{\boldsymbol{\Sigma}}) = (\mathbf{C}\hat{\boldsymbol{\beta}} - \boldsymbol{\Gamma}_0)'(\mathbf{C}(\mathbf{Z}'\mathbf{Z})^{-1}\mathbf{C}')^{-1}(\mathbf{C}\hat{\boldsymbol{\beta}} - \boldsymbol{\Gamma}_0)$$

Under the null hypothesis, the statistic $n(\hat{\boldsymbol{\Sigma}}_1 - \hat{\boldsymbol{\Sigma}})$ is distributed as $W_{r-q}(\boldsymbol{\Sigma})$ independently of $\hat{\boldsymbol{\Sigma}}$. This distribution theory can be employed to develop a test of $H_0: \mathbf{C}\boldsymbol{\beta} = \boldsymbol{\Gamma}_0$ similar to the test discussed in Result 7.11. (See, for example, [18].)

Other Multivariate Test Statistics

Tests other than the likelihood ratio test have been proposed for testing $H_0: \boldsymbol{\beta}_{(2)} = \mathbf{0}$ in the multivariate multiple regression model.

Popular computer-package programs routinely calculate four multivariate test statistics. To connect with their output, we introduce some alternative notation. Let $\mathbf{E}$ be the $p \times p$ *error*, or residual, sum of squares and cross products matrix

$$\mathbf{E} = n\hat{\boldsymbol{\Sigma}}$$

that results from fitting the full model. The $p \times p$ *hypothesis*, or extra, sum of squares and cross-products matrix

$$\mathbf{H} = n(\hat{\boldsymbol{\Sigma}}_1 - \hat{\boldsymbol{\Sigma}})$$

The statistics can be defined in terms of $\mathbf{E}$ and $\mathbf{H}$ directly, or in terms of the nonzero eigenvalues $\eta_1 \geq \eta_2 \geq \ldots \geq \eta_s$ of $\mathbf{HE}^{-1}$, where $s = \min(p, r - q)$. Equivalently, they are the roots of $|(\hat{\boldsymbol{\Sigma}}_1 - \hat{\boldsymbol{\Sigma}}) - \eta\hat{\boldsymbol{\Sigma}}| = 0$. The definitions are

$$\text{Wilks' lambda} = \prod_{i=1}^{s} \frac{1}{1 + \eta_i} = \frac{|\mathbf{E}|}{|\mathbf{E} + \mathbf{H}|}$$

$$\text{Pillai's trace} = \sum_{i=1}^{s} \frac{\eta_i}{1 + \eta_i} = \text{tr}[\mathbf{H}(\mathbf{H} + \mathbf{E})^{-1}]$$

$$\text{Hotelling–Lawley trace} = \sum_{i=1}^{s} \eta_i = \text{tr}[\mathbf{HE}^{-1}]$$

$$\text{Roy's greatest root} = \frac{\eta_1}{1 + \eta_1}$$

Roy's test selects the coefficient vector $\mathbf{a}$ so that the univariate F-statistic based on a $\mathbf{a}'\mathbf{Y}_j$ has its maximum possible value. When several of the eigenvalues η_i are moderately large, Roy's test will perform poorly relative to the other three. Simulation studies suggest that its power will be best when there is only one large eigenvalue.

Charts and tables of critical values are available for Roy's test. (See [21] and [17].) Wilks' lambda, Roy's greatest root, and the Hotelling–Lawley trace test are nearly equivalent for large sample sizes.

If there is a large discrepancy in the reported P-values for the four tests, the eigenvalues and vectors may lead to an interpretation. In this text, we report Wilks' lambda, which is the likelihood ratio test.

Predictions from Multivariate Multiple Regressions

Suppose the model $\mathbf{Y} = \mathbf{Z}\boldsymbol{\beta} + \boldsymbol{\varepsilon}$, with normal errors $\boldsymbol{\varepsilon}$, has been fit and checked for any inadequacies. If the model is adequate, it can be employed for predictive purposes.

One problem is to predict the mean responses corresponding to fixed values $\mathbf{z}_0$ of the predictor variables. Inferences about the mean responses can be made using the distribution theory in Result 7.10. From this result, we determine that

$$\hat{\boldsymbol{\beta}}'\mathbf{z}_0 \quad \text{is distributed as} \quad N_m(\boldsymbol{\beta}'\mathbf{z}_0, \mathbf{z}_0'(\mathbf{Z}'\mathbf{Z})^{-1}\mathbf{z}_0\,\boldsymbol{\Sigma})$$

and

$$n\hat{\boldsymbol{\Sigma}} \quad \text{is independently distributed as} \quad W_{n-r-1}(\boldsymbol{\Sigma})$$

The unknown value of the regression function at $\mathbf{z}_0$ is $\boldsymbol{\beta}'\mathbf{z}_0$. So, from the discussion of the T^2-statistic in Section 5.2, we can write

$$T^2 = \left(\frac{\hat{\boldsymbol{\beta}}'\mathbf{z}_0 - \boldsymbol{\beta}'\mathbf{z}_0}{\sqrt{\mathbf{z}_0'(\mathbf{Z}'\mathbf{Z})^{-1}\mathbf{z}_0}}\right)'\left(\frac{n}{n-r-1}\hat{\boldsymbol{\Sigma}}\right)^{-1}\left(\frac{\hat{\boldsymbol{\beta}}'\mathbf{z}_0 - \boldsymbol{\beta}'\mathbf{z}_0}{\sqrt{\mathbf{z}_0'(\mathbf{Z}'\mathbf{Z})^{-1}\mathbf{z}_0}}\right) \qquad (7\text{-}39)$$

and the $100(1-\alpha)\%$ confidence ellipsoid for $\boldsymbol{\beta}'\mathbf{z}_0$ is provided by the inequality

$$(\boldsymbol{\beta}'\mathbf{z}_0 - \hat{\boldsymbol{\beta}}'\mathbf{z}_0)'\left(\frac{n}{n-r-1}\hat{\boldsymbol{\Sigma}}\right)^{-1}(\boldsymbol{\beta}'\mathbf{z}_0 - \hat{\boldsymbol{\beta}}'\mathbf{z}_0)$$

$$\leq \mathbf{z}_0'(\mathbf{Z}'\mathbf{Z})^{-1}\mathbf{z}_0\left[\left(\frac{m(n-r-1)}{n-r-m}\right)F_{m,n-r-m}(\alpha)\right] \qquad (7\text{-}40)$$

where $F_{m,n-r-m}(\alpha)$ is the upper (100α)th percentile of an F-distribution with m and $n - r - m$ d.f.

The $100(1-\alpha)\%$ *simultaneous* confidence intervals for $E(Y_i) = \mathbf{z}_0'\boldsymbol{\beta}_{(i)}$ are

$$\mathbf{z}_0'\hat{\boldsymbol{\beta}}_{(i)} \pm \sqrt{\left(\frac{m(n-r-1)}{n-r-m}\right)F_{m,n-r-m}(\alpha)}\sqrt{\mathbf{z}_0'(\mathbf{Z}'\mathbf{Z})^{-1}\mathbf{z}_0\left(\frac{n}{n-r-1}\hat{\sigma}_{ii}\right)},$$

$$i = 1, 2, \ldots, m \qquad (7\text{-}41)$$

where $\hat{\boldsymbol{\beta}}_{(i)}$ is the ith column of $\hat{\boldsymbol{\beta}}$ and $\hat{\sigma}_{ii}$ is the ith diagonal element of $\hat{\boldsymbol{\Sigma}}$.

The second prediction problem is concerned with forecasting new responses $\mathbf{Y}_0 = \boldsymbol{\beta}'\mathbf{z}_0 + \boldsymbol{\varepsilon}_0$ at $\mathbf{z}_0$. Here $\boldsymbol{\varepsilon}_0$ is independent of $\boldsymbol{\varepsilon}$. Now,

$$\mathbf{Y}_0 - \hat{\boldsymbol{\beta}}'\mathbf{z}_0 = (\boldsymbol{\beta} - \hat{\boldsymbol{\beta}})'\mathbf{z}_0 + \boldsymbol{\varepsilon}_0 \quad \text{is distributed as} \quad N_m(\mathbf{0}, (1 + \mathbf{z}_0'(\mathbf{Z}'\mathbf{Z})^{-1}\mathbf{z}_0)\boldsymbol{\Sigma})$$

independently of $n\hat{\boldsymbol{\Sigma}}$, so the $100(1-\alpha)\%$ *prediction ellipsoid* for $\mathbf{Y}_0$ becomes

$$(\mathbf{Y}_0 - \hat{\boldsymbol{\beta}}'\mathbf{z}_0)'\left(\frac{n}{n-r-1}\hat{\boldsymbol{\Sigma}}\right)^{-1}(\mathbf{Y}_0 - \hat{\boldsymbol{\beta}}'\mathbf{z}_0)$$

$$\leq (1 + \mathbf{z}_0'(\mathbf{Z}'\mathbf{Z})^{-1}\mathbf{z}_0)\left[\left(\frac{m(n-r-1)}{n-r-m}\right)F_{m,n-r-m}(\alpha)\right] \qquad (7\text{-}42)$$

The $100(1-\alpha)\%$ *simultaneous prediction intervals* for the individual responses Y_{0i} are

$$\mathbf{z}_0'\hat{\boldsymbol{\beta}}_{(i)} \pm \sqrt{\left(\frac{m(n-r-1)}{n-r-m}\right)F_{m,n-r-m}(\alpha)}\sqrt{(1 + \mathbf{z}_0'(\mathbf{Z}'\mathbf{Z})^{-1}\mathbf{z}_0)\left(\frac{n}{n-r-1}\hat{\sigma}_{ii}\right)},$$

$$i = 1, 2, \ldots, m \qquad (7\text{-}43)$$

where $\hat{\boldsymbol{\beta}}_{(i)}$, $\hat{\sigma}_{ii}$, and $F_{m,n-r-m}(\alpha)$ are the same quantities appearing in (7-41). Comparing (7-41) and (7-43), we see that the prediction intervals for the *actual* values of the response variables are wider than the corresponding intervals for the *expected* values. The extra width reflects the presence of the random error ε_{0i}.

Example 7.10 (Constructing a confidence ellipse and a prediction ellipse for bivariate responses) A second response variable was measured for the computer-requirement problem discussed in Example 7.6. Measurements on the response Y_2, disk input/output capacity, corresponding to the z_1 and z_2 values in that example were

$$\mathbf{y}_2' = [301.8, 396.1, 328.2, 307.4, 362.4, 369.5, 229.1]$$

Obtain the 95% confidence ellipse for $\boldsymbol{\beta}'\mathbf{z}_0$ and the 95% prediction ellipse for $\mathbf{Y}_0' = [Y_{01}, Y_{02}]$ for a site with the configuration $\mathbf{z}_0' = [1, 130, 7.5]$.

Computer calculations provide the fitted equation

$$\hat{y}_2 = 14.14 + 2.25z_1 + 5.67z_2$$

with $s = 1.812$. Thus, $\hat{\boldsymbol{\beta}}_{(2)}' = [14.14, 2.25, 5.67]$. From Example 7.6,

$$\hat{\boldsymbol{\beta}}_{(1)}' = [8.42, 1.08, 42], \quad \mathbf{z}_0'\hat{\boldsymbol{\beta}}_{(1)} = 151.97, \quad \text{and} \quad \mathbf{z}_0'(\mathbf{Z}'\mathbf{Z})^{-1}\mathbf{z}_0 = .34725$$

We find that

$$\mathbf{z}_0'\hat{\boldsymbol{\beta}}_{(2)} = 14.14 + 2.25(130) + 5.67(7.5) = 349.17$$

and

$$n\hat{\boldsymbol{\Sigma}} = \begin{bmatrix} (\mathbf{y}_{(1)} - \mathbf{Z}\hat{\boldsymbol{\beta}}_{(1)})'(\mathbf{y}_{(1)} - \mathbf{Z}\hat{\boldsymbol{\beta}}_{(1)}) & (\mathbf{y}_{(1)} - \mathbf{Z}\hat{\boldsymbol{\beta}}_{(1)})'(\mathbf{y}_{(2)} - \mathbf{Z}\hat{\boldsymbol{\beta}}_{(2)}) \\ (\mathbf{y}_{(2)} - \mathbf{Z}\hat{\boldsymbol{\beta}}_{(2)})'(\mathbf{y}_{(1)} - \mathbf{Z}\hat{\boldsymbol{\beta}}_{(1)}) & (\mathbf{y}_{(2)} - \mathbf{Z}\hat{\boldsymbol{\beta}}_{(2)})'(\mathbf{y}_{(2)} - \mathbf{Z}\hat{\boldsymbol{\beta}}_{(2)}) \end{bmatrix}$$

$$= \begin{bmatrix} 5.80 & 5.30 \\ 5.30 & 13.13 \end{bmatrix}$$

Since

$$\hat{\boldsymbol{\beta}}'\mathbf{z}_0 = \begin{bmatrix} \hat{\boldsymbol{\beta}}_{(1)}' \\ \hline \hat{\boldsymbol{\beta}}_{(2)}' \end{bmatrix} \mathbf{z}_0 = \begin{bmatrix} \mathbf{z}_0'\hat{\boldsymbol{\beta}}_{(1)} \\ \hline \mathbf{z}_0'\hat{\boldsymbol{\beta}}_{(2)} \end{bmatrix} = \begin{bmatrix} 151.97 \\ 349.17 \end{bmatrix}$$

$n = 7$, $r = 2$, and $m = 2$, a 95% confidence ellipse for $\boldsymbol{\beta}'\mathbf{z}_0 = \begin{bmatrix} \mathbf{z}_0'\boldsymbol{\beta}_{(1)} \\ \hline \mathbf{z}_0'\boldsymbol{\beta}_{(2)} \end{bmatrix}$ is, from (7-40), the set

$$[\mathbf{z}_0'\boldsymbol{\beta}_{(1)} - 151.97, \mathbf{z}_0'\boldsymbol{\beta}_{(2)} - 349.17](4)\begin{bmatrix} 5.80 & 5.30 \\ 5.30 & 13.13 \end{bmatrix}^{-1}\begin{bmatrix} \mathbf{z}_0'\boldsymbol{\beta}_{(1)} - 151.97 \\ \mathbf{z}_0'\boldsymbol{\beta}_{(2)} - 349.17 \end{bmatrix}$$

$$\leq (.34725)\left[\left(\frac{2(4)}{3}\right)F_{2,3}(.05)\right]$$

with $F_{2,3}(.05) = 9.55$. This ellipse is centered at $(151.97, 349.17)$. Its orientation and the lengths of the major and minor axes can be determined from the eigenvalues and eigenvectors of $n\hat{\boldsymbol{\Sigma}}$.

Comparing (7-40) and (7-42), we see that the only change required for the calculation of the 95% prediction ellipse is to replace $\mathbf{z}_0'(\mathbf{Z}'\mathbf{Z})^{-1}\mathbf{z}_0 = .34725$ with

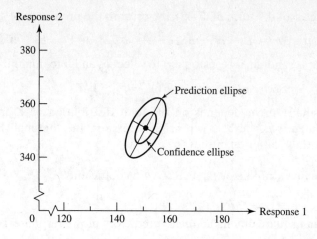

Figure 7.5 95% confidence and prediction ellipses for the computer data with two responses.

$1 + \mathbf{z}_0'(\mathbf{Z}'\mathbf{Z})^{-1}\mathbf{z}_0 = 1.34725$. Thus, the 95% prediction ellipse for $\mathbf{Y}_0' = [Y_{01}, Y_{02}]$ is also centered at $(151.97, 349.17)$, but is larger than the confidence ellipse. Both ellipses are sketched in Figure 7.5.

It is the *prediction* ellipse that is relevant to the determination of computer requirements for a particular site with the given $\mathbf{z}_0$. ∎

7.8 The Concept of Linear Regression

The classical linear regression model is concerned with the association between a single dependent variable Y and a collection of predictor variables $z_1, z_2, \ldots, z_r$. The regression model that we have considered treats Y as a random variable whose mean depends upon *fixed* values of the z_i's. This mean is assumed to be a linear function of the regression *coefficients* $\beta_0, \beta_1, \ldots, \beta_r$.

The linear regression model also arises in a different setting. Suppose all the variables $Y, Z_1, Z_2, \ldots, Z_r$ are random and have a joint distribution, not necessarily normal, with mean vector $\underset{(r+1)\times1}{\boldsymbol{\mu}}$ and covariance matrix $\underset{(r+1)\times(r+1)}{\boldsymbol{\Sigma}}$. Partitioning $\boldsymbol{\mu}$ and $\boldsymbol{\Sigma}$ in an obvious fashion, we write

$$\boldsymbol{\mu} = \begin{bmatrix} \mu_Y \\ {\scriptstyle(1\times1)} \\ \text{-------} \\ \boldsymbol{\mu_Z} \\ {\scriptstyle(r\times1)} \end{bmatrix} \quad \text{and} \quad \boldsymbol{\Sigma} = \begin{bmatrix} \sigma_{YY} & \boldsymbol{\sigma}_{ZY}' \\ {\scriptstyle(1\times1)} & {\scriptstyle(1\times r)} \\ \text{-------} & \text{-------} \\ \boldsymbol{\sigma_{ZY}} & \boldsymbol{\Sigma_{ZZ}} \\ {\scriptstyle(r\times1)} & {\scriptstyle(r\times r)} \end{bmatrix}$$

with

$$\boldsymbol{\sigma}_{ZY}' = [\sigma_{YZ_1}, \sigma_{YZ_2}, \ldots, \sigma_{YZ_r}] \tag{7-44}$$

$\boldsymbol{\Sigma_{ZZ}}$ can be taken to have full rank.[6] Consider the problem of predicting Y using the

$$linear\ predictor = b_0 + b_1 Z_1 + \cdots + b_r Z_r = b_0 + \mathbf{b}'\mathbf{Z} \tag{7-45}$$

---

[6] If $\boldsymbol{\Sigma_{ZZ}}$ is not of full rank, one variable—for example, Z_k—can be written as a linear combination of the other Z_i's and thus is redundant in forming the linear regression function $\mathbf{Z}'\boldsymbol{\beta}$. That is, $\mathbf{Z}$ may be replaced by any subset of components whose nonsingular covariance matrix has the same rank as $\boldsymbol{\Sigma_{ZZ}}$.

For a given predictor of the form of (7-45), the error in the prediction of Y is

$$\text{prediction error} = Y - b_0 - b_1 Z_1 - \cdots - b_r Z_r = Y - b_0 - \mathbf{b}'\mathbf{Z} \quad (7\text{-}46)$$

Because this error is random, it is customary to select b_0 and $\mathbf{b}$ to minimize the

$$\text{mean square error} = E(Y - b_0 - \mathbf{b}'\mathbf{Z})^2 \quad (7\text{-}47)$$

Now the mean square error depends on the joint distribution of Y and $\mathbf{Z}$ only through the parameters $\boldsymbol{\mu}$ and $\boldsymbol{\Sigma}$. It is possible to express the "optimal" linear predictor in terms of these latter quantities.

Result 7.12. The linear predictor $\beta_0 + \boldsymbol{\beta}'\mathbf{Z}$ with coefficients

$$\boldsymbol{\beta} = \boldsymbol{\Sigma}_{ZZ}^{-1}\boldsymbol{\sigma}_{ZY}, \qquad \beta_0 = \mu_Y - \boldsymbol{\beta}'\boldsymbol{\mu}_Z$$

has minimum mean square among all *linear* predictors of the response Y. Its mean square error is

$$E(Y - \beta_0 - \boldsymbol{\beta}'\mathbf{Z})^2 = E(Y - \mu_Y - \boldsymbol{\sigma}_{ZY}'\boldsymbol{\Sigma}_{ZZ}^{-1}(\mathbf{Z} - \boldsymbol{\mu}_Z))^2 = \sigma_{YY} - \boldsymbol{\sigma}_{ZY}'\boldsymbol{\Sigma}_{ZZ}^{-1}\boldsymbol{\sigma}_{ZY}$$

Also, $\beta_0 + \boldsymbol{\beta}'\mathbf{Z} = \mu_Y + \boldsymbol{\sigma}_{ZY}'\boldsymbol{\Sigma}_{ZZ}^{-1}(\mathbf{Z} - \boldsymbol{\mu}_Z)$ is the linear predictor having maximum correlation with Y; that is,

$$\text{Corr}(Y, \beta_0 + \boldsymbol{\beta}'\mathbf{Z}) = \max_{b_0, \mathbf{b}} \text{Corr}(Y, b_0 + \mathbf{b}'\mathbf{Z})$$

$$= \sqrt{\frac{\boldsymbol{\beta}'\boldsymbol{\Sigma}_{ZZ}\boldsymbol{\beta}}{\sigma_{YY}}} = \sqrt{\frac{\boldsymbol{\sigma}_{ZY}'\boldsymbol{\Sigma}_{ZZ}^{-1}\boldsymbol{\sigma}_{ZY}}{\sigma_{YY}}}$$

Proof. Writing $b_0 + \mathbf{b}'\mathbf{Z} = b_0 + \mathbf{b}'\mathbf{Z} + (\mu_Y - \mathbf{b}'\boldsymbol{\mu}_Z) - (\mu_Y - \mathbf{b}'\boldsymbol{\mu}_Z)$, we get

$$E(Y - b_0 - \mathbf{b}'\mathbf{Z})^2 = E[Y - \mu_Y - (\mathbf{b}'\mathbf{Z} - \mathbf{b}'\boldsymbol{\mu}_Z) + (\mu_Y - b_0 - \mathbf{b}'\boldsymbol{\mu}_Z)]^2$$

$$= E(Y - \mu_Y)^2 + E(\mathbf{b}'(\mathbf{Z} - \boldsymbol{\mu}_Z))^2 + (\mu_Y - b_0 - \mathbf{b}'\boldsymbol{\mu}_Z)^2$$

$$- 2E[\mathbf{b}'(\mathbf{Z} - \boldsymbol{\mu}_Z)(Y - \mu_Y)]$$

$$= \sigma_{YY} + \mathbf{b}'\boldsymbol{\Sigma}_{ZZ}\mathbf{b} + (\mu_Y - b_0 - \mathbf{b}'\boldsymbol{\mu}_Z)^2 - 2\mathbf{b}'\boldsymbol{\sigma}_{ZY}$$

Adding and subtracting $\boldsymbol{\sigma}_{ZY}'\boldsymbol{\Sigma}_{ZZ}^{-1}\boldsymbol{\sigma}_{ZY}$, we obtain

$$E(Y - b_0 - \mathbf{b}'\mathbf{Z})^2 = \sigma_{YY} - \boldsymbol{\sigma}_{ZY}'\boldsymbol{\Sigma}_{ZZ}^{-1}\boldsymbol{\sigma}_{ZY} + (\mu_Y - b_0 - \mathbf{b}'\boldsymbol{\mu}_Z)^2$$

$$+ (\mathbf{b} - \boldsymbol{\Sigma}_{ZZ}^{-1}\boldsymbol{\sigma}_{ZY})'\boldsymbol{\Sigma}_{ZZ}(\mathbf{b} - \boldsymbol{\Sigma}_{ZZ}^{-1}\boldsymbol{\sigma}_{ZY})$$

The mean square error is minimized by taking $\mathbf{b} = \boldsymbol{\Sigma}_{ZZ}^{-1}\boldsymbol{\sigma}_{ZY} = \boldsymbol{\beta}$, making the last term zero, and then choosing $b_0 = \mu_Y - (\boldsymbol{\Sigma}_{ZZ}^{-1}\boldsymbol{\sigma}_{ZY})'\boldsymbol{\mu}_Z = \beta_0$ to make the third term zero. The minimum mean square error is thus $\sigma_{YY} - \boldsymbol{\sigma}_{ZY}'\boldsymbol{\Sigma}_{ZZ}^{-1}\boldsymbol{\sigma}_{ZY}$.

Next, we note that $\text{Cov}(b_0 + \mathbf{b}'\mathbf{Z}, Y) = \text{Cov}(\mathbf{b}'\mathbf{Z}, Y) = \mathbf{b}'\boldsymbol{\sigma}_{ZY}$ so

$$[\text{Corr}(b_0 + \mathbf{b}'\mathbf{Z}, Y)]^2 = \frac{[\mathbf{b}'\boldsymbol{\sigma}_{ZY}]^2}{\sigma_{YY}(\mathbf{b}'\boldsymbol{\Sigma}_{ZZ}\mathbf{b})}, \qquad \text{for all } b_0, \mathbf{b}$$

Employing the extended Cauchy–Schwartz inequality of (2-49) with $\mathbf{B} = \boldsymbol{\Sigma}_{ZZ}$, we obtain

$$(\mathbf{b}'\boldsymbol{\sigma}_{ZY})^2 \leq \mathbf{b}'\boldsymbol{\Sigma}_{ZZ}\mathbf{b}\,\boldsymbol{\sigma}_{ZY}'\boldsymbol{\Sigma}_{ZZ}^{-1}\boldsymbol{\sigma}_{ZY}$$

or

$$[\text{Corr}(b_0 + \mathbf{b}'\mathbf{Z}, Y)]^2 \le \frac{\boldsymbol{\sigma}'_{\mathbf{Z}Y}\boldsymbol{\Sigma}_{\mathbf{Z}\mathbf{Z}}^{-1}\boldsymbol{\sigma}_{\mathbf{Z}Y}}{\sigma_{YY}}$$

with equality for $\mathbf{b} = \boldsymbol{\Sigma}_{\mathbf{Z}\mathbf{Z}}^{-1}\boldsymbol{\sigma}_{\mathbf{Z}Y} = \boldsymbol{\beta}$. The alternative expression for the maximum correlation follows from the equation $\boldsymbol{\sigma}'_{\mathbf{Z}Y}\boldsymbol{\Sigma}_{\mathbf{Z}\mathbf{Z}}^{-1}\boldsymbol{\sigma}_{\mathbf{Z}Y} = \boldsymbol{\sigma}'_{\mathbf{Z}Y}\boldsymbol{\beta} = \boldsymbol{\sigma}'_{\mathbf{Z}Y}\boldsymbol{\Sigma}_{\mathbf{Z}\mathbf{Z}}^{-1}\boldsymbol{\Sigma}_{\mathbf{Z}\mathbf{Z}}\boldsymbol{\beta} = \boldsymbol{\beta}'\boldsymbol{\Sigma}_{\mathbf{Z}\mathbf{Z}}\boldsymbol{\beta}$. ∎

The correlation between Y and its best linear predictor is called the *population multiple correlation coefficient*

$$\rho_{Y(\mathbf{Z})} = +\sqrt{\frac{\boldsymbol{\sigma}'_{\mathbf{Z}Y}\boldsymbol{\Sigma}_{\mathbf{Z}\mathbf{Z}}^{-1}\boldsymbol{\sigma}_{\mathbf{Z}Y}}{\sigma_{YY}}} \qquad (7\text{-}48)$$

The square of the population multiple correlation coefficient, $\rho_{Y(\mathbf{Z})}^2$, is called the *population coefficient of determination*. Note that, unlike other correlation coefficients, the multiple correlation coefficient is a *positive* square root, so $0 \le \rho_{Y(\mathbf{Z})} \le 1$.

The population coefficient of determination has an important interpretation. From Result 7.12, the mean square error in using $\beta_0 + \boldsymbol{\beta}'\mathbf{Z}$ to forecast Y is

$$\sigma_{YY} - \boldsymbol{\sigma}'_{\mathbf{Z}Y}\boldsymbol{\Sigma}_{\mathbf{Z}\mathbf{Z}}^{-1}\boldsymbol{\sigma}_{\mathbf{Z}Y} = \sigma_{YY} - \sigma_{YY}\left(\frac{\boldsymbol{\sigma}'_{\mathbf{Z}Y}\boldsymbol{\Sigma}_{\mathbf{Z}\mathbf{Z}}^{-1}\boldsymbol{\sigma}_{\mathbf{Z}Y}}{\sigma_{YY}}\right) = \sigma_{YY}(1 - \rho_{Y(\mathbf{Z})}^2) \qquad (7\text{-}49)$$

If $\rho_{Y(\mathbf{Z})}^2 = 0$, there is no predictive power in $\mathbf{Z}$. At the other extreme, $\rho_{Y(\mathbf{Z})}^2 = 1$ implies that Y can be predicted with no error.

Example 7.11 (Determining the best linear predictor, its mean square error, and the multiple correlation coefficient) Given the mean vector and covariance matrix of Y, Z_1, Z_2,

$$\boldsymbol{\mu} = \begin{bmatrix} \mu_Y \\ \hline \boldsymbol{\mu}_{\mathbf{Z}} \end{bmatrix} = \begin{bmatrix} 5 \\ \hline 2 \\ 0 \end{bmatrix} \quad \text{and} \quad \boldsymbol{\Sigma} = \begin{bmatrix} \sigma_{YY} & \boldsymbol{\sigma}'_{\mathbf{Z}Y} \\ \hline \boldsymbol{\sigma}_{\mathbf{Z}Y} & \boldsymbol{\Sigma}_{\mathbf{Z}\mathbf{Z}} \end{bmatrix} = \begin{bmatrix} 10 & 1 & -1 \\ \hline 1 & 7 & 3 \\ -1 & 3 & 2 \end{bmatrix}$$

determine (a) the best linear predictor $\beta_0 + \beta_1 Z_1 + \beta_2 Z_2$, (b) its mean square error, and (c) the multiple correlation coefficient. Also, verify that the mean square error equals $\sigma_{YY}(1 - \rho_{Y(\mathbf{Z})}^2)$.

First,

$$\boldsymbol{\beta} = \boldsymbol{\Sigma}_{\mathbf{Z}\mathbf{Z}}^{-1}\boldsymbol{\sigma}_{\mathbf{Z}Y} = \begin{bmatrix} 7 & 3 \\ 3 & 2 \end{bmatrix}^{-1}\begin{bmatrix} 1 \\ -1 \end{bmatrix} = \begin{bmatrix} .4 & -.6 \\ -.6 & 1.4 \end{bmatrix}\begin{bmatrix} 1 \\ -1 \end{bmatrix} = \begin{bmatrix} 1 \\ -2 \end{bmatrix}$$

$$\beta_0 = \mu_Y - \boldsymbol{\beta}'\boldsymbol{\mu}_{\mathbf{Z}} = 5 - [1, -2]\begin{bmatrix} 2 \\ 0 \end{bmatrix} = 3$$

so the best linear predictor is $\beta_0 + \boldsymbol{\beta}'\mathbf{Z} = 3 + Z_1 - 2Z_2$. The mean square error is

$$\sigma_{YY} - \boldsymbol{\sigma}'_{\mathbf{Z}Y}\boldsymbol{\Sigma}_{\mathbf{Z}\mathbf{Z}}^{-1}\boldsymbol{\sigma}_{\mathbf{Z}Y} = 10 - [1, -1]\begin{bmatrix} .4 & -.6 \\ -.6 & 1.4 \end{bmatrix}\begin{bmatrix} 1 \\ -1 \end{bmatrix} = 10 - 3 = 7$$

and the multiple correlation coefficient is

$$\rho_{Y(\mathbf{Z})} = \sqrt{\frac{\boldsymbol{\sigma}'_{\mathbf{Z}Y}\boldsymbol{\Sigma}_{\mathbf{ZZ}}^{-1}\boldsymbol{\sigma}_{\mathbf{Z}Y}}{\sigma_{YY}}} = \sqrt{\frac{3}{10}} = .548$$

Note that $\sigma_{YY}(1 - \rho_{Y(\mathbf{Z})}^2) = 10\left(1 - \frac{3}{10}\right) = 7$ is the mean square error.   ■

It is possible to show (see Exercise 7.5) that

$$1 - \rho_{Y(\mathbf{Z})}^2 = \frac{1}{\rho^{YY}} \tag{7-50}$$

where ρ^{YY} is the upper-left-hand corner of the inverse of the correlation matrix determined from $\boldsymbol{\Sigma}$.

The restriction to linear predictors is closely connected to the assumption of normality. Specifically, if we take

$$\begin{bmatrix} Y \\ Z_1 \\ Z_2 \\ \vdots \\ Z_r \end{bmatrix} \quad \text{to be distributed as} \quad N_{r+1}(\boldsymbol{\mu}, \boldsymbol{\Sigma})$$

then the conditional distribution of Y with $z_1, z_2, \ldots, z_r$ fixed (see Result 4.6) is

$$N(\mu_Y + \boldsymbol{\sigma}'_{\mathbf{Z}Y}\boldsymbol{\Sigma}_{\mathbf{ZZ}}^{-1}(\mathbf{Z} - \boldsymbol{\mu}_{\mathbf{Z}}), \sigma_{YY} - \boldsymbol{\sigma}'_{\mathbf{Z}Y}\boldsymbol{\Sigma}_{\mathbf{ZZ}}^{-1}\boldsymbol{\sigma}_{\mathbf{Z}Y})$$

The mean of this conditional distribution is the linear predictor in Result 7.12. That is,

$$E(Y|z_1, z_2, \ldots, z_r) = \mu_Y + \boldsymbol{\sigma}'_{\mathbf{Z}Y}\boldsymbol{\Sigma}_{\mathbf{ZZ}}^{-1}(\mathbf{z} - \boldsymbol{\mu}_{\mathbf{Z}}) \tag{7-51}$$
$$= \beta_0 + \boldsymbol{\beta}'\mathbf{z}$$

and we conclude that $E(Y|z_1, z_2, \ldots, z_r)$ is the best linear predictor of Y when the population is $N_{r+1}(\boldsymbol{\mu}, \boldsymbol{\Sigma})$. The conditional expectation of Y in (7-51) is called the *regression function*. For normal populations, it is linear.

When the population is *not* normal, the regression function $E(Y|z_1, z_2, \ldots, z_r)$ need not be of the form $\beta_0 + \boldsymbol{\beta}'\mathbf{z}$. Nevertheless, it can be shown (see [22]) that $E(Y|z_1, z_2, \ldots, z_r)$, whatever its form, predicts Y with the smallest mean square error. Fortunately, this wider optimality among all estimators is possessed by the *linear* predictor when the population is normal.

Result 7.13. Suppose the joint distribution of Y and $\mathbf{Z}$ is $N_{r+1}(\boldsymbol{\mu}, \boldsymbol{\Sigma})$. Let

$$\hat{\boldsymbol{\mu}} = \begin{bmatrix} \overline{Y} \\ \hline \overline{\mathbf{Z}} \end{bmatrix} \quad \text{and} \quad \mathbf{S} = \begin{bmatrix} s_{YY} & s'_{\mathbf{Z}Y} \\ \hline s_{\mathbf{Z}Y} & \mathbf{S}_{\mathbf{ZZ}} \end{bmatrix}$$

be the sample mean vector and sample covariance matrix, respectively, for a random sample of size n from this population. Then the maximum likelihood estimators of the coefficients in the linear predictor are

$$\hat{\boldsymbol{\beta}} = \mathbf{S}_{\mathbf{ZZ}}^{-1}s_{\mathbf{Z}Y}, \qquad \hat{\beta}_0 = \overline{Y} - s'_{\mathbf{Z}Y}\mathbf{S}_{\mathbf{ZZ}}^{-1}\overline{\mathbf{Z}} = \overline{Y} - \hat{\boldsymbol{\beta}}'\overline{\mathbf{Z}}$$

Consequently, the maximum likelihood estimator of the linear regression function is

$$\hat{\beta}_0 + \hat{\boldsymbol{\beta}}'\mathbf{z} = \overline{Y} + \mathbf{s}'_{ZY}\mathbf{S}^{-1}_{ZZ}(\mathbf{z} - \overline{\mathbf{Z}})$$

and the maximum likelihood estimator of the mean square error $E[Y - \beta_0 - \boldsymbol{\beta}'\mathbf{Z}]^2$ is

$$\hat{\sigma}_{YY\cdot\mathbf{Z}} = \frac{n-1}{n}(s_{YY} - \mathbf{s}'_{ZY}\mathbf{S}^{-1}_{ZZ}\mathbf{s}_{ZY})$$

Proof. We use Result 4.11 and the invariance property of maximum likelihood estimators. [See (4-20).] Since, from Result 7.12,

$$\beta_0 = \mu_Y - (\boldsymbol{\Sigma}^{-1}_{ZZ}\boldsymbol{\sigma}_{ZY})'\boldsymbol{\mu}_\mathbf{Z},$$

$$\boldsymbol{\beta} = \boldsymbol{\Sigma}^{-1}_{ZZ}\boldsymbol{\sigma}_{ZY}, \qquad \beta_0 + \boldsymbol{\beta}'\mathbf{z} = \mu_Y + \boldsymbol{\sigma}'_{ZY}\boldsymbol{\Sigma}^{-1}_{ZZ}(\mathbf{z} - \boldsymbol{\mu}_\mathbf{Z})$$

and

$$\text{mean square error} = \sigma_{YY\cdot\mathbf{Z}} = \sigma_{YY} - \boldsymbol{\sigma}'_{ZY}\boldsymbol{\Sigma}^{-1}_{ZZ}\boldsymbol{\sigma}_{ZY}$$

the conclusions follow upon substitution of the maximum likelihood estimators

$$\hat{\boldsymbol{\mu}} = \begin{bmatrix} \overline{Y} \\ \hline \overline{\mathbf{Z}} \end{bmatrix} \quad \text{and} \quad \hat{\boldsymbol{\Sigma}} = \begin{bmatrix} \hat{\sigma}_{YY} & \hat{\boldsymbol{\sigma}}'_{ZY} \\ \hline \hat{\boldsymbol{\sigma}}_{ZY} & \hat{\boldsymbol{\Sigma}}_{ZZ} \end{bmatrix} = \left(\frac{n-1}{n}\right)\mathbf{S}$$

for

$$\boldsymbol{\mu} = \begin{bmatrix} \mu_Y \\ \hline \boldsymbol{\mu}_\mathbf{Z} \end{bmatrix} \quad \text{and} \quad \boldsymbol{\Sigma} = \begin{bmatrix} \sigma_{YY} & \boldsymbol{\sigma}'_{ZY} \\ \hline \boldsymbol{\sigma}_{ZY} & \boldsymbol{\Sigma}_{ZZ} \end{bmatrix} \qquad \blacksquare$$

It is customary to change the divisor from n to $n - (r + 1)$ in the estimator of the mean square error, $\sigma_{YY\cdot\mathbf{Z}} = E(Y - \beta_0 - \boldsymbol{\beta}'\mathbf{Z})^2$, in order to obtain the *unbiased* estimator

$$\left(\frac{n-1}{n-r-1}\right)(s_{YY} - \mathbf{s}'_{ZY}\mathbf{S}^{-1}_{ZZ}\mathbf{s}_{ZY}) = \frac{\sum_{j=1}^{n}(Y_j - \hat{\beta}_0 - \hat{\boldsymbol{\beta}}'\mathbf{Z}_j)^2}{n-r-1} \qquad (7\text{-}52)$$

Example 7.12 (Maximum likelihood estimate of the regression function—single response) For the computer data of Example 7.6, the $n = 7$ observations on Y (CPU time), Z_1 (orders), and Z_2 (add–delete items) give the sample mean vector and sample covariance matrix:

$$\hat{\boldsymbol{\mu}} = \begin{bmatrix} \overline{y} \\ \hline \overline{\mathbf{Z}} \end{bmatrix} = \begin{bmatrix} 150.44 \\ \hline 130.24 \\ 3.547 \end{bmatrix}$$

$$\mathbf{S} = \begin{bmatrix} s_{YY} & \mathbf{s}'_{ZY} \\ \hline \mathbf{s}_{ZY} & \mathbf{S}_{ZZ} \end{bmatrix} = \begin{bmatrix} 467.913 & 418.763 & 35.983 \\ \hline 418.763 & 377.200 & 28.034 \\ 35.983 & 28.034 & 13.657 \end{bmatrix}$$

Assuming that Y, Z_1, and Z_2 are jointly normal, obtain the estimated regression function and the estimated mean square error.

Result 7.13 gives the maximum likelihood estimates

$$\hat{\boldsymbol{\beta}} = \mathbf{S}_{ZZ}^{-1}\mathbf{s}_{ZY} = \begin{bmatrix} .003128 & -.006422 \\ -.006422 & .086404 \end{bmatrix} \begin{bmatrix} 418.763 \\ 35.983 \end{bmatrix} = \begin{bmatrix} 1.079 \\ .420 \end{bmatrix}$$

$$\hat{\beta}_0 = \bar{y} - \hat{\boldsymbol{\beta}}'\bar{\mathbf{z}} = 150.44 - [1.079, .420]\begin{bmatrix} 130.24 \\ 3.547 \end{bmatrix} = 150.44 - 142.019$$

$$= 8.421$$

and the estimated regression function

$$\hat{\beta}_0 + \hat{\boldsymbol{\beta}}'\mathbf{z} = 8.42 - 1.08z_1 + .42z_2$$

The maximum likelihood estimate of the mean square error arising from the prediction of Y with this regression function is

$$\left(\frac{n-1}{n}\right)(s_{YY} - \mathbf{s}_{ZY}'\mathbf{S}_{ZZ}^{-1}\mathbf{s}_{ZY})$$

$$= \left(\frac{6}{7}\right)\left(467.913 - [418.763, 35.983]\begin{bmatrix} .003128 & -.006422 \\ -.006422 & .086404 \end{bmatrix}\begin{bmatrix} 418.763 \\ 35.983 \end{bmatrix}\right)$$

$$= .894 \qquad\blacksquare$$

Prediction of Several Variables

The extension of the previous results to the prediction of several responses Y_1, $Y_2, \ldots, Y_m$ is almost immediate. We present this extension for normal populations. Suppose

$$\begin{bmatrix} \mathbf{Y} \\ {\scriptstyle (m\times 1)} \\ \hline \mathbf{Z} \\ {\scriptstyle (r\times 1)} \end{bmatrix} \quad \text{is distributed as} \quad N_{m+r}(\boldsymbol{\mu}, \boldsymbol{\Sigma})$$

with

$$\boldsymbol{\mu} = \begin{bmatrix} \boldsymbol{\mu}_{\mathbf{Y}} \\ {\scriptstyle (m\times 1)} \\ \hline \boldsymbol{\mu}_{\mathbf{Z}} \\ {\scriptstyle (r\times 1)} \end{bmatrix} \quad \text{and} \quad \boldsymbol{\Sigma} = \begin{bmatrix} \boldsymbol{\Sigma}_{\mathbf{YY}} & \boldsymbol{\Sigma}_{\mathbf{YZ}} \\ {\scriptstyle (m\times m)} & {\scriptstyle (m\times r)} \\ \hline \boldsymbol{\Sigma}_{\mathbf{ZY}} & \boldsymbol{\Sigma}_{\mathbf{ZZ}} \\ {\scriptstyle (r\times m)} & {\scriptstyle (r\times r)} \end{bmatrix}$$

By Result 4.6, the conditional expectation of $[Y_1, Y_2, \ldots, Y_m]'$, given the fixed values $z_1, z_2, \ldots, z_r$ of the predictor variables, is

$$E[\mathbf{Y} \mid z_1, z_2, \ldots, z_r] = \boldsymbol{\mu}_{\mathbf{Y}} + \boldsymbol{\Sigma}_{\mathbf{YZ}}\boldsymbol{\Sigma}_{\mathbf{ZZ}}^{-1}(\mathbf{z} - \boldsymbol{\mu}_{\mathbf{Z}}) \qquad (7\text{-}53)$$

This conditional expected value, considered as a function of $z_1, z_2, \ldots, z_r$, is called the *multivariate regression* of the vector $\mathbf{Y}$ on $\mathbf{Z}$. It is composed of m univariate regressions. For instance, the first component of the conditional mean vector is $\mu_{Y_1} + \boldsymbol{\Sigma}_{Y_1\mathbf{Z}}\boldsymbol{\Sigma}_{\mathbf{ZZ}}^{-1}(\mathbf{z} - \boldsymbol{\mu}_{\mathbf{Z}}) = E(Y_1 \mid z_1, z_2, \ldots, z_r)$, which minimizes the mean square error for the prediction of Y_1. The $m \times r$ matrix $\boldsymbol{\beta} = \boldsymbol{\Sigma}_{\mathbf{YZ}}\boldsymbol{\Sigma}_{\mathbf{ZZ}}^{-1}$ is called the matrix of *regression coefficients*.

The error of prediction vector

$$\mathbf{Y} - \boldsymbol{\mu_Y} - \boldsymbol{\Sigma_{YZ}}\boldsymbol{\Sigma_{ZZ}^{-1}}(\mathbf{Z} - \boldsymbol{\mu_Z})$$

has the expected squares and cross-products matrix

$$
\begin{aligned}
\boldsymbol{\Sigma_{YY\cdot Z}} &= E[\mathbf{Y} - \boldsymbol{\mu_Y} - \boldsymbol{\Sigma_{YZ}}\boldsymbol{\Sigma_{ZZ}^{-1}}(\mathbf{Z} - \boldsymbol{\mu_Z})][\mathbf{Y} - \boldsymbol{\mu_Y} - \boldsymbol{\Sigma_{YZ}}\boldsymbol{\Sigma_{ZZ}^{-1}}(\mathbf{Z} - \boldsymbol{\mu_Z})]' \\
&= \boldsymbol{\Sigma_{YY}} - \boldsymbol{\Sigma_{YZ}}\boldsymbol{\Sigma_{ZZ}^{-1}}(\boldsymbol{\Sigma_{YZ}})' - \boldsymbol{\Sigma_{YZ}}\boldsymbol{\Sigma_{ZZ}^{-1}}\boldsymbol{\Sigma_{ZY}} + \boldsymbol{\Sigma_{YZ}}\boldsymbol{\Sigma_{ZZ}^{-1}}\boldsymbol{\Sigma_{ZZ}}\boldsymbol{\Sigma_{ZZ}^{-1}}(\boldsymbol{\Sigma_{YZ}})' \quad (7\text{-}54) \\
&= \boldsymbol{\Sigma_{YY}} - \boldsymbol{\Sigma_{YZ}}\boldsymbol{\Sigma_{ZZ}^{-1}}\boldsymbol{\Sigma_{ZY}}
\end{aligned}
$$

Because $\boldsymbol{\mu}$ and $\boldsymbol{\Sigma}$ are typically unknown, they must be estimated from a random sample in order to construct the multivariate linear predictor and determine expected prediction errors.

Result 7.14. Suppose $\mathbf{Y}$ and $\mathbf{Z}$ are jointly distributed as $N_{m+r}(\boldsymbol{\mu}, \boldsymbol{\Sigma})$. Then the regression of the vector $\mathbf{Y}$ on $\mathbf{Z}$ is

$$\boldsymbol{\beta_0} + \boldsymbol{\beta}\mathbf{z} = \boldsymbol{\mu_Y} - \boldsymbol{\Sigma_{YZ}}\boldsymbol{\Sigma_{ZZ}^{-1}}\boldsymbol{\mu_Z} + \boldsymbol{\Sigma_{YZ}}\boldsymbol{\Sigma_{ZZ}^{-1}}\mathbf{z} = \boldsymbol{\mu_Y} + \boldsymbol{\Sigma_{YZ}}\boldsymbol{\Sigma_{ZZ}^{-1}}(\mathbf{z} - \boldsymbol{\mu_Z})$$

The expected squares and cross-products matrix for the errors is

$$E(\mathbf{Y} - \boldsymbol{\beta_0} - \boldsymbol{\beta}\mathbf{Z})(\mathbf{Y} - \boldsymbol{\beta_0} - \boldsymbol{\beta}\mathbf{Z})' = \boldsymbol{\Sigma_{YY\cdot Z}} = \boldsymbol{\Sigma_{YY}} - \boldsymbol{\Sigma_{YZ}}\boldsymbol{\Sigma_{ZZ}^{-1}}\boldsymbol{\Sigma_{ZY}}$$

Based on a random sample of size n, the maximum likelihood estimator of the regression function is

$$\hat{\boldsymbol{\beta_0}} + \hat{\boldsymbol{\beta}}\mathbf{z} = \overline{\mathbf{Y}} + \mathbf{S_{YZ}}\mathbf{S_{ZZ}^{-1}}(\mathbf{z} - \overline{\mathbf{Z}})$$

and the maximum likelihood estimator of $\boldsymbol{\Sigma_{YY\cdot Z}}$ is

$$\hat{\boldsymbol{\Sigma}}_{\mathbf{YY\cdot Z}} = \left(\frac{n-1}{n}\right)(\mathbf{S_{YY}} - \mathbf{S_{YZ}}\mathbf{S_{ZZ}^{-1}}\mathbf{S_{ZY}})$$

Proof. The regression function and the covariance matrix for the prediction errors follow from Result 4.6. Using the relationships

$$\boldsymbol{\beta_0} = \boldsymbol{\mu_Y} - \boldsymbol{\Sigma_{YZ}}\boldsymbol{\Sigma_{ZZ}^{-1}}\boldsymbol{\mu_Z}, \qquad \boldsymbol{\beta} = \boldsymbol{\Sigma_{YZ}}\boldsymbol{\Sigma_{ZZ}^{-1}}$$

$$\boldsymbol{\beta_0} + \boldsymbol{\beta}\mathbf{z} = \boldsymbol{\mu_Y} + \boldsymbol{\Sigma_{YZ}}\boldsymbol{\Sigma_{ZZ}^{-1}}(\mathbf{z} - \boldsymbol{\mu_Z})$$

$$\boldsymbol{\Sigma_{YY\cdot Z}} = \boldsymbol{\Sigma_{YY}} - \boldsymbol{\Sigma_{YZ}}\boldsymbol{\Sigma_{ZZ}^{-1}}\boldsymbol{\Sigma_{ZY}} = \boldsymbol{\Sigma_{YY}} - \boldsymbol{\beta}\boldsymbol{\Sigma_{ZZ}}\boldsymbol{\beta}'$$

we deduce the maximum likelihood statements from the invariance property [see (4-20)] of maximum likelihood estimators upon substitution of

$$\hat{\boldsymbol{\mu}} = \begin{bmatrix} \overline{\mathbf{Y}} \\ \hline \overline{\mathbf{Z}} \end{bmatrix}; \quad \hat{\boldsymbol{\Sigma}} = \begin{bmatrix} \hat{\boldsymbol{\Sigma}}_{\mathbf{YY}} & \hat{\boldsymbol{\Sigma}}_{\mathbf{YZ}} \\ \hline \hat{\boldsymbol{\Sigma}}_{\mathbf{ZY}} & \hat{\boldsymbol{\Sigma}}_{\mathbf{ZZ}} \end{bmatrix} = \left(\frac{n-1}{n}\right)\mathbf{S} = \left(\frac{n-1}{n}\right)\begin{bmatrix} \mathbf{S_{YY}} & \mathbf{S_{YZ}} \\ \hline \mathbf{S_{ZY}} & \mathbf{S_{ZZ}} \end{bmatrix} \quad \blacksquare$$

It can be shown that an unbiased estimator of $\boldsymbol{\Sigma_{YY\cdot Z}}$ is

$$\left(\frac{n-1}{n-r-1}\right)(\mathbf{S_{YY}} - \mathbf{S_{YZ}}\mathbf{S_{ZZ}^{-1}}\mathbf{S_{ZY}})$$

$$= \frac{1}{n-r-1}\sum_{j=1}^{n}(\mathbf{Y}_j - \hat{\boldsymbol{\beta_0}} - \hat{\boldsymbol{\beta}}\mathbf{Z}_j)(\mathbf{Y}_j - \hat{\boldsymbol{\beta_0}} - \hat{\boldsymbol{\beta}}\mathbf{Z}_j)' \quad (7\text{-}55)$$

Example 7.13 (Maximum likelihood estimates of the regression functions—two responses) We return to the computer data given in Examples 7.6 and 7.10. For Y_1 = CPU time, Y_2 = disk I/O capacity, Z_1 = orders, and Z_2 = add–delete items, we have

$$\hat{\boldsymbol{\mu}} = \begin{bmatrix} \bar{\mathbf{y}} \\ \hline \bar{\mathbf{z}} \end{bmatrix} = \begin{bmatrix} 150.44 \\ 327.79 \\ \hline 130.24 \\ 3.547 \end{bmatrix}$$

and

$$\mathbf{S} = \begin{bmatrix} \mathbf{S_{YY}} & \mathbf{S_{YZ}} \\ \hline \mathbf{S_{ZY}} & \mathbf{S_{ZZ}} \end{bmatrix} = \begin{bmatrix} 467.913 & 1148.556 & 418.763 & 35.983 \\ 1148.556 & 3072.491 & 1008.976 & 140.558 \\ \hline 418.763 & 1008.976 & 377.200 & 28.034 \\ 35.983 & 140.558 & 28.034 & 13.657 \end{bmatrix}$$

Assuming normality, we find that the estimated regression function is

$$\hat{\boldsymbol{\beta}}_0 + \hat{\boldsymbol{\beta}}\mathbf{z} = \bar{\mathbf{y}} + \mathbf{S_{YZ}S_{ZZ}^{-1}}(\mathbf{z} - \bar{\mathbf{z}})$$

$$= \begin{bmatrix} 150.44 \\ 327.79 \end{bmatrix} + \begin{bmatrix} 418.763 & 35.983 \\ 1008.976 & 140.558 \end{bmatrix}$$

$$\times \begin{bmatrix} .003128 & -.006422 \\ -.006422 & .086404 \end{bmatrix} \begin{bmatrix} z_1 - 130.24 \\ z_2 - 3.547 \end{bmatrix}$$

$$= \begin{bmatrix} 150.44 \\ 327.79 \end{bmatrix} + \begin{bmatrix} 1.079\,(z_1 - 130.24) + .420\,(z_2 - 3.547) \\ 2.254\,(z_1 - 130.24) + 5.665\,(z_2 - 3.547) \end{bmatrix}$$

Thus, the minimum mean square error predictor of Y_1 is

$$150.44 + 1.079(z_1 - 130.24) + .420(z_2 - 3.547) = 8.42 + 1.08z_1 + .42z_2$$

Similarly, the best predictor of Y_2 is

$$14.14 + 2.25z_1 + 5.67z_2$$

The maximum likelihood estimate of the expected squared errors and cross-products matrix $\boldsymbol{\Sigma_{YY \cdot Z}}$ is given by

$$\left(\frac{n-1}{n}\right)(\mathbf{S_{YY}} - \mathbf{S_{YZ}S_{ZZ}^{-1}S_{ZY}})$$

$$= \left(\frac{6}{7}\right)\left(\begin{bmatrix} 467.913 & 1148.536 \\ 1148.536 & 3072.491 \end{bmatrix}\right.$$

$$\left. - \begin{bmatrix} 418.763 & 35.983 \\ 1008.976 & 140.558 \end{bmatrix} \begin{bmatrix} .003128 & -.006422 \\ -.006422 & .086404 \end{bmatrix} \begin{bmatrix} 418.763 & 1008.976 \\ 35.983 & 140.558 \end{bmatrix}\right)$$

$$= \left(\frac{6}{7}\right)\begin{bmatrix} 1.043 & 1.042 \\ 1.042 & 2.572 \end{bmatrix} = \begin{bmatrix} .894 & .893 \\ .893 & 2.205 \end{bmatrix}$$

The first estimated regression function, $8.42 + 1.08z_1 + .42z_2$, and the associated mean square error, $.894$, are the same as those in Example 7.12 for the single-response case. Similarly, the second estimated regression function, $14.14 + 2.25z_1 + 5.67z_2$, is the same as that given in Example 7.10.

We see that the data enable us to predict the first response, Y_1, with smaller error than the second response, Y_2. The positive covariance $.893$ indicates that over-prediction (underprediction) of CPU time tends to be accompanied by overprediction (underprediction) of disk capacity. ∎

Comment. Result 7.14 states that the assumption of a joint normal distribution for the whole collection $Y_1, Y_2, \ldots, Y_m, Z_1, Z_2, \ldots, Z_r$ leads to the prediction equations

$$\hat{y}_1 = \hat{\beta}_{01} + \hat{\beta}_{11}z_1 + \cdots + \hat{\beta}_{r1}z_r$$
$$\hat{y}_2 = \hat{\beta}_{02} + \hat{\beta}_{12}z_1 + \cdots + \hat{\beta}_{r2}z_r$$
$$\vdots \qquad\qquad \vdots$$
$$\hat{y}_m = \hat{\beta}_{0m} + \hat{\beta}_{1m}z_1 + \cdots + \hat{\beta}_{rm}z_r$$

We note the following:

1. The same values, $z_1, z_2, \ldots, z_r$ are used to predict each Y_i.
2. The $\hat{\beta}_{ik}$ are estimates of the (i, k)th entry of the regression coefficient matrix $\boldsymbol{\beta} = \boldsymbol{\Sigma_{YZ}\Sigma_{ZZ}^{-1}}$ for $i, k \geq 1$.

We conclude this discussion of the regression problem by introducing one further correlation coefficient.

Partial Correlation Coefficient

Consider the pair of errors

$$Y_1 - \mu_{Y_1} - \boldsymbol{\Sigma_{Y_1Z}\Sigma_{ZZ}^{-1}}(\mathbf{Z} - \boldsymbol{\mu_Z})$$
$$Y_2 - \mu_{Y_2} - \boldsymbol{\Sigma_{Y_2Z}\Sigma_{ZZ}^{-1}}(\mathbf{Z} - \boldsymbol{\mu_Z})$$

obtained from using the best linear predictors to predict Y_1 and Y_2. Their correlation, determined from the error covariance matrix $\boldsymbol{\Sigma_{YY\cdot Z}} = \boldsymbol{\Sigma_{YY}} - \boldsymbol{\Sigma_{YZ}\Sigma_{ZZ}^{-1}\Sigma_{ZY}}$, measures the association between Y_1 and Y_2 after eliminating the effects of Z_1, $Z_2, \ldots, Z_r$.

We define the *partial correlation coefficient* between Y_1 and Y_2, eliminating Z_1, $Z_2, \ldots, Z_r$, by

$$\rho_{Y_1Y_2\cdot\mathbf{Z}} = \frac{\sigma_{Y_1Y_2\cdot\mathbf{Z}}}{\sqrt{\sigma_{Y_1Y_1\cdot\mathbf{Z}}}\sqrt{\sigma_{Y_2Y_2\cdot\mathbf{Z}}}} \tag{7-56}$$

where $\sigma_{Y_iY_k\cdot\mathbf{Z}}$ is the (i, k)th entry in the matrix $\boldsymbol{\Sigma_{YY\cdot Z}} = \boldsymbol{\Sigma_{YY}} - \boldsymbol{\Sigma_{YZ}\Sigma_{ZZ}^{-1}\Sigma_{ZY}}$. The corresponding *sample partial correlation coefficient* is

$$r_{Y_1Y_2\cdot\mathbf{Z}} = \frac{s_{Y_1Y_2\cdot\mathbf{Z}}}{\sqrt{s_{Y_1Y_1\cdot\mathbf{Z}}}\sqrt{s_{Y_2Y_2\cdot\mathbf{Z}}}} \tag{7-57}$$

with $s_{Y_i Y_k \cdot \mathbf{Z}}$ the (i, k)th element of $\mathbf{S_{YY}} - \mathbf{S_{YZ}S_{ZZ}^{-1}S_{ZY}}$. Assuming that $\mathbf{Y}$ and $\mathbf{Z}$ have a joint multivariate normal distribution, we find that the sample partial correlation coefficient in (7-57) is the maximum likelihood estimator of the partial correlation coefficient in (7-56).

Example 7.14 (Calculating a partial correlation) From the computer data in Example 7.13,

$$\mathbf{S_{YY}} - \mathbf{S_{YZ}S_{ZZ}^{-1}S_{ZY}} = \begin{bmatrix} 1.043 & 1.042 \\ 1.042 & 2.572 \end{bmatrix}$$

Therefore,

$$r_{Y_1 Y_2 \cdot \mathbf{Z}} = \frac{s_{Y_1 Y_2 \cdot \mathbf{Z}}}{\sqrt{s_{Y_1 Y_1 \cdot \mathbf{Z}}} \sqrt{s_{Y_2 Y_2 \cdot \mathbf{Z}}}} = \frac{1.042}{\sqrt{1.043} \sqrt{2.572}} = .64 \qquad (7\text{-}58)$$

Calculating the ordinary correlation coefficient, we obtain $r_{Y_1 Y_2} = .96$. Comparing the two correlation coefficients, we see that the association between Y_1 and Y_2 has been sharply reduced after eliminating the effects of the variables $\mathbf{Z}$ on both responses. ∎

7.9 Comparing the Two Formulations of the Regression Model

In Sections 7.2 and 7.7, we presented the multiple regression models for one and several response variables, respectively. In these treatments, the predictor variables had *fixed* values $\mathbf{z}_j$ at the jth trial. Alternatively, we can start—as in Section 7.8—with a set of variables that have a joint normal distribution. The process of conditioning on one subset of variables in order to predict values of the other set leads to a conditional expectation that is a multiple regression model. The two approaches to multiple regression are related. To show this relationship explicitly, we introduce two minor variants of the regression model formulation.

Mean Corrected Form of the Regression Model

For any response variable Y, the multiple regression model asserts that

$$Y_j = \beta_0 + \beta_1 z_{1j} + \cdots + \beta_r z_{rj} + \varepsilon_j$$

The predictor variables can be "centered" by subtracting their means. For instance, $\beta_1 z_{1j} = \beta_1(z_{1j} - \bar{z}_1) + \beta_1 \bar{z}_1$ and we can write

$$Y_j = (\beta_0 + \beta_1 \bar{z}_1 + \cdots + \beta_r \bar{z}_r) + \beta_1(z_{1j} - \bar{z}_1) + \cdots + \beta_r(z_{rj} - \bar{z}_r) + \varepsilon_j$$

$$= \beta_* + \beta_1(z_{1j} - \bar{z}_1) + \cdots + \beta_r(z_{rj} - \bar{z}_r) + \varepsilon_j \qquad (7\text{-}59)$$

with $\beta_* = \beta_0 + \beta_1 \bar{z}_1 + \cdots + \beta_r \bar{z}_r$. The *mean corrected* design matrix corresponding to the reparameterization in (7-59) is

$$
\mathbf{Z}_c = \begin{bmatrix} 1 & z_{11} - \bar{z}_1 & \cdots & z_{1r} - \bar{z}_r \\ 1 & z_{21} - \bar{z}_1 & \cdots & z_{2r} - \bar{z}_r \\ \vdots & \vdots & \ddots & \vdots \\ 1 & z_{n1} - \bar{z}_1 & \cdots & z_{nr} - \bar{z}_r \end{bmatrix}
$$

where the last r columns are each perpendicular to the first column, since

$$
\sum_{j=1}^{n} 1(z_{ji} - \bar{z}_i) = 0, \qquad i = 1, 2, \ldots, r
$$

Further, setting $\mathbf{Z}_c = [\mathbf{1} \,|\, \mathbf{Z}_{c2}]$ with $\mathbf{Z}'_{c2}\mathbf{1} = \mathbf{0}$, we obtain

$$
\mathbf{Z}'_c \mathbf{Z}_c = \begin{bmatrix} \mathbf{1}'\mathbf{1} & \mathbf{1}'\mathbf{Z}_{c2} \\ \mathbf{Z}'_{c2}\mathbf{1} & \mathbf{Z}'_{c2}\mathbf{Z}_{c2} \end{bmatrix} = \begin{bmatrix} n & \mathbf{0}' \\ \mathbf{0} & \mathbf{Z}'_{c2}\mathbf{Z}_{c2} \end{bmatrix}
$$

so

$$
\begin{bmatrix} \hat{\beta}_* \\ \hline \hat{\beta}_1 \\ \vdots \\ \hat{\beta}_r \end{bmatrix} = (\mathbf{Z}'_c \mathbf{Z}_c)^{-1} \mathbf{Z}'_c \mathbf{y}
$$

$$
= \begin{bmatrix} \dfrac{1}{n} & \mathbf{0}' \\ \mathbf{0} & (\mathbf{Z}'_{c2}\mathbf{Z}_{c2})^{-1} \end{bmatrix} \begin{bmatrix} \mathbf{1}'\mathbf{y} \\ \mathbf{Z}'_{c2}\mathbf{y} \end{bmatrix} = \begin{bmatrix} \bar{y} \\ \hline (\mathbf{Z}'_{c2}\mathbf{Z}_{c2})^{-1}\mathbf{Z}'_{c2}\mathbf{y} \end{bmatrix} \tag{7-60}
$$

That is, the regression coefficients $[\beta_1, \beta_2, \ldots, \beta_r]'$ are unbiasedly estimated by $(\mathbf{Z}'_{c2}\mathbf{Z}_{c2})^{-1}\mathbf{Z}'_{c2}\mathbf{y}$ and β_* is estimated by $\bar{y}$. Because the definitions $\beta_1, \beta_2, \ldots, \beta_r$ remain unchanged by the reparameterization in (7-59), their best estimates computed from the design matrix $\mathbf{Z}_c$ are exactly the same as the best estimates computed from the design matrix $\mathbf{Z}$. Thus, setting $\hat{\boldsymbol{\beta}}'_c = [\hat{\beta}_1, \hat{\beta}_2, \ldots, \hat{\beta}_r]$, the linear predictor of Y can be written as

$$
\hat{y} = \hat{\beta}_* + \hat{\boldsymbol{\beta}}'_c(\mathbf{z} - \bar{\mathbf{z}}) = \bar{y} + \mathbf{y}'\mathbf{Z}_{c2}(\mathbf{Z}'_{c2}\mathbf{Z}_{c2})^{-1}(\mathbf{z} - \bar{\mathbf{z}}) \tag{7-61}
$$

with $(\mathbf{z} - \bar{\mathbf{z}}) = [z_1 - \bar{z}_1, z_2 - \bar{z}_2, \ldots, z_r - \bar{z}_r]'$. Finally,

$$
\begin{bmatrix} \mathrm{Var}(\hat{\beta}_*) & \mathrm{Cov}(\hat{\beta}_*, \hat{\boldsymbol{\beta}}_c) \\ \mathrm{Cov}(\hat{\boldsymbol{\beta}}_c, \hat{\beta}_*) & \mathrm{Cov}(\hat{\boldsymbol{\beta}}_c) \end{bmatrix} = (\mathbf{Z}'_c \mathbf{Z}_c)^{-1}\sigma^2 = \begin{bmatrix} \dfrac{\sigma^2}{n} & \mathbf{0}' \\ \mathbf{0} & (\mathbf{Z}'_{c2}\mathbf{Z}_{c2})^{-1}\sigma^2 \end{bmatrix} \tag{7-62}
$$

Comment. The *multivariate* multiple regression model yields the same mean corrected design matrix for each response. The least squares estimates of the coefficient vectors for the ith response are given by

$$\hat{\boldsymbol{\beta}}_{(i)} = \left[\begin{array}{c} \bar{y}_{(i)} \\ \hline (\mathbf{Z}'_{c2}\mathbf{Z}_{c2})^{-1}\mathbf{Z}'_{c2}\,\mathbf{y}_{(i)} \end{array} \right], \qquad i = 1, 2, \ldots, m \qquad (7\text{-}63)$$

Sometimes, for even further numerical stability, "standardized" input variables $(z_{ji} - \bar{z}_i) \big/ \sqrt{\sum_{j=1}^{n} (z_{ji} - \bar{z}_i)^2} = (z_{ji} - \bar{z}_i)/\sqrt{(n-1)s_{z_iz_i}}$ are used. In this case, the slope coefficients β_i in the regression model are replaced by $\tilde{\beta}_i = \beta_i \sqrt{(n-1)s_{z_iz_i}}$. The least squares estimates of the *beta coefficients* $\tilde{\beta}_i$ become $\hat{\tilde{\beta}}_i = \hat{\beta}_i \sqrt{(n-1)s_{z_iz_i}}$, $i = 1, 2, \ldots, r$. These relationships hold for each response in the multivariate multiple regression situation as well.

Relating the Formulations

When the variables $Y, Z_1, Z_2, \ldots, Z_r$ are jointly normal, the estimated predictor of Y (see Result 7.13) is

$$\hat{\beta}_0 + \hat{\boldsymbol{\beta}}'\mathbf{z} = \bar{y} + \mathbf{s}'_{\mathbf{Z}Y}\mathbf{S}_{\mathbf{ZZ}}^{-1}(\mathbf{z} - \bar{\mathbf{z}}) = \hat{\mu}_Y + \hat{\boldsymbol{\sigma}}'_{\mathbf{Z}Y}\hat{\boldsymbol{\Sigma}}_{\mathbf{ZZ}}^{-1}(\mathbf{z} - \hat{\boldsymbol{\mu}}_{\mathbf{Z}}) \qquad (7\text{-}64)$$

where the estimation procedure leads naturally to the introduction of centered z_i's.

Recall from the mean corrected form of the regression model that the best linear predictor of Y [see (7-61)] is

$$\hat{y} = \hat{\beta}_* + \hat{\boldsymbol{\beta}}'_c(\mathbf{z} - \bar{\mathbf{z}})$$

with $\hat{\beta}_* = \bar{y}$ and $\hat{\boldsymbol{\beta}}'_c = \mathbf{y}'\mathbf{Z}_{c2}(\mathbf{Z}'_{c2}\mathbf{Z}_{c2})^{-1}$. Comparing (7-61) and (7-64), we see that $\hat{\beta}_* = \bar{y} = \hat{\beta}_0$ and $\hat{\boldsymbol{\beta}}_c = \hat{\boldsymbol{\beta}}$ since[7]

$$\mathbf{s}'_{\mathbf{Z}Y}\mathbf{S}_{\mathbf{ZZ}}^{-1} = \mathbf{y}'\mathbf{Z}_{c2}(\mathbf{Z}'_{c2}\mathbf{Z}_{c2})^{-1} \qquad (7\text{-}65)$$

Therefore, both the normal theory conditional mean and the classical regression model approaches yield exactly the *same* linear predictors.

A similar argument indicates that the best linear predictors of the responses in the two multivariate multiple regression setups are also exactly the same.

Example 7.15 (Two approaches yield the same linear predictor) The computer data with the single response Y_1 = CPU time were analyzed in Example 7.6 using the classical linear regression model. The same data were analyzed again in Example 7.12, assuming that the variables Y_1, Z_1, and Z_2 were jointly normal so that the best predictor of Y_1 is the conditional mean of Y_1 given z_1 and z_2. Both approaches yielded the same predictor,

$$\hat{y} = 8.42 + 1.08z_1 + .42z_2$$ ∎

---

[7]The identify in (7-65) is established by writing $\mathbf{y} = (\mathbf{y} - \bar{y}\mathbf{1}) + \bar{y}\mathbf{1}$ so that

$$\mathbf{y}'\mathbf{Z}_{c2} = (\mathbf{y} - \bar{y}\mathbf{1})'\mathbf{Z}_{c2} + \bar{y}\mathbf{1}'\mathbf{Z}_{c2} = (\mathbf{y} - \bar{y}\mathbf{1})'\mathbf{Z}_{c2} + \mathbf{0}' = (\mathbf{y} - \bar{y}\mathbf{1})'\mathbf{Z}_{c2}$$

Consequently,

$$\mathbf{y}\mathbf{Z}_{c2}(\mathbf{Z}'_{c2}\mathbf{Z}_{c2})^{-1} = (\mathbf{y} - \bar{y}\mathbf{1})'\mathbf{Z}_{c2}(\mathbf{Z}'_{c2}\mathbf{Z}_{c2})^{-1} = (n-1)\mathbf{s}'_{\mathbf{Z}Y}[(n-1)\mathbf{S}_{\mathbf{ZZ}}]^{-1} = \mathbf{s}'_{\mathbf{Z}Y}\mathbf{S}_{\mathbf{ZZ}}^{-1}$$

Although the two formulations of the linear prediction problem yield the same predictor equations, conceptually they are quite different. For the model in (7-3) or (7-23), the values of the input variables are assumed to be set by the experimenter. In the conditional mean model of (7-51) or (7-53), the values of the predictor variables are random variables that are observed along with the values of the response variable(s). The assumptions underlying the second approach are more stringent, but they yield an *optimal* predictor among *all* choices, rather than merely among linear predictors.

We close by noting that the multivariate regression calculations in either case can be couched in terms of the sample mean vectors $\bar{\mathbf{y}}$ and $\bar{\mathbf{z}}$ and the sample sums of squares and cross-products:

$$
\left[
\begin{array}{c|c}
\sum\limits_{j=1}^{n} (\mathbf{y}_j - \bar{\mathbf{y}})(\mathbf{y}_j - \bar{\mathbf{y}})' & \sum\limits_{j=1}^{n} (\mathbf{y}_j - \bar{\mathbf{y}})(\mathbf{z}_j - \bar{\mathbf{z}})' \\
\hline
\sum\limits_{j=1}^{n} (\mathbf{z}_j - \bar{\mathbf{z}})(\mathbf{y}_j - \bar{\mathbf{y}})' & \sum\limits_{j=1}^{n} (\mathbf{z}_j - \bar{\mathbf{z}})(\mathbf{z}_j - \bar{\mathbf{z}})'
\end{array}
\right]
=
\left[
\begin{array}{c|c}
\mathbf{Y}_c'\mathbf{Y}_c & \mathbf{Y}_c'\mathbf{Z}_{c2} \\
\hline
\mathbf{Z}_{c2}'\mathbf{Y}_c & \mathbf{Z}_{c2}'\mathbf{Z}_{c2}
\end{array}
\right]
$$

$$
= n
\left[
\begin{array}{c|c}
\hat{\boldsymbol{\Sigma}}_{\mathbf{YY}} & \hat{\boldsymbol{\Sigma}}_{\mathbf{YZ}} \\
\hline
\hat{\boldsymbol{\Sigma}}_{\mathbf{ZY}} & \hat{\boldsymbol{\Sigma}}_{\mathbf{ZZ}}
\end{array}
\right]
$$

This is the only information necessary to compute the estimated regression coefficients and their estimated covariances. Of course, an important part of regression analysis is model checking. This requires the residuals (errors), which must be calculated using all the original data.

7.10 Multiple Regression Models with Time Dependent Errors

For data collected over time, observations in different time periods are often related, or autocorrelated. Consequently, in a regression context, the observations on the dependent variable or, equivalently, the errors, cannot be independent. As indicated in our discussion of dependence in Section 5.8, time dependence in the observations can invalidate inferences made using the usual independence assumption. Similarly, inferences in regression can be misleading when regression models are fit to time ordered data and the standard regression assumptions are used. This issue is important so, in the example that follows, we not only show how to detect the presence of time dependence, but also how to incorporate this dependence into the multiple regression model.

Example 7.16 (Incorporating time dependent errors in a regression model) Power companies must have enough natural gas to heat all of their customers' homes and businesses, particularly during the coldest days of the year. A major component of the planning process is a forecasting exercise based on a model relating the send-outs of natural gas to factors, like temperature, that clearly have some relationship to the amount of gas consumed. More gas is required on cold days. Rather than use the daily average temperature, it is customary to use degree heating days

(DHD) = 65 deg − daily average temperature. A large number for DHD indicates a cold day. Wind speed, again a 24-hour average, can also be a factor in the sendout amount. Because many businesses close for the weekend, the demand for natural gas is typically less on a weekend day. Data on these variables for one winter in a major northern city are shown, in part, in Table 7.4. (See website: www.prenhall.com/statistics for the complete data set. There are $n = 63$ observations.)

Table 7.4 Natural Gas Data

Y Sendout	Z_1 DHD	Z_2 DHDLag	Z_3 Windspeed	Z_4 Weekend
227	32	30	12	1
236	31	32	8	1
228	30	31	8	0
252	34	30	8	0
238	28	34	12	0
$\vdots$	$\vdots$	$\vdots$	$\vdots$	$\vdots$
333	46	41	8	0
266	33	46	8	0
280	38	33	18	0
386	52	38	22	0
415	57	52	18	0

Initially, we developed a regression model relating gas sendout to degree heating days, wind speed and a weekend dummy variable. Other variables likely to have some affect on natural gas consumption, like percent cloud cover, are subsumed in the error term. After several attempted fits, we decided to include not only the current DHD but also that of the previous day. (The degree heating day lagged one time period is denoted by DHDLag in Table 7.4.) The fitted model is

$$\text{Sendout} = 1.858 + 5.874\,\text{DHD} + 1.405\,\text{DHDLag}$$
$$+ 1.315\,\text{Windspeed} - 15.857\,\text{Weekend}$$

with $R^2 = .952$. All the coefficients, with the exception of the intercept, are significant and it looks like we have a very good fit. (The intercept term could be dropped. When this is done, the results do not change substantially.) However, if we calculate the correlation of the residuals that are adjacent in time, the lag 1 autocorrelation, we get

$$\text{lag 1 autocorrelation} = r_1(\hat{\varepsilon}) = \frac{\sum_{j=2}^{n} \hat{\varepsilon}_j \hat{\varepsilon}_{j-1}}{\sum_{j=1}^{n} \hat{\varepsilon}_j^2} = .52$$

The value, .52, of the lag 1 autocorrelation is too large to be ignored. A plot of the residual autocorrelations for the first 15 lags shows that there might also be some dependence among the errors 7 time periods, or one week, apart. This amount of dependence invalidates the t-tests and P-values associated with the coefficients in the model.

The first step toward correcting the model is to replace the presumed independent errors in the regression model for sendout with a possibly dependent series of noise terms N_j. That is, we formulate a regression model for the N_j where we relate each N_j to its previous value N_{j-1}, its value one week ago, N_{j-7}, and an independent error ε_j. Thus, we consider

$$N_j = \phi_1 N_{j-1} + \phi_7 N_{j-7} + \varepsilon_j$$

where the ε_j are independent normal random variables with mean 0 and variance σ^2. The form of the equation for N_j is known as an *autoregressive* model. (See [8].) The SAS commands and part of the output from fitting this combined regression model for sendout with an autoregressive model for the noise are shown in Panel 7.3 on page 416.

The fitted model is

Sendout $= 2.130 + 5.810$ DHD $+ 1.426$ DHDLag

$+ 1.207$ Windspeed $- 10.109$ Weekend

and the time dependence in the noise terms is estimated by

$$N_j = .470 N_{j-1} + .240 N_{j-7} + \varepsilon_j$$

The variance of ε is estimated to be $\hat{\sigma}^2 = 228.89$.

From Panel 7.3, we see that the autocorrelations of the residuals from the enriched model are all negligible. Each is within two estimated standard errors of 0. Also, a weighted sum of squares of residual autocorrelations for a group of consecutive lags is not large as judged by the P-value for this statistic. That is, there is no reason to reject the hypothesis that a group of consecutive autocorrelations are simultaneously equal to 0. The groups examined in Panel 7.3 are those for lags 1–6, 1–12, 1–18, and 1–24.

The noise is now adequately modeled. The tests concerning the coefficient of each predictor variable, the significance of the regression, and so forth, are now valid.[8] The intercept term in the final model can be dropped. When this is done, there is very little change in the resulting model. The enriched model has better forecasting potential and can now be used to forecast sendout of natural gas for given values of the predictor variables. We will not pursue prediction here, since it involves ideas beyond the scope of this book. (See [8].) ∎

---

[8]These tests are obtained by the extra sum of squares procedure but applied to the regression plus autoregressive noise model. The tests are those described in the computer output.

When modeling relationships using time ordered data, regression models with noise structures that allow for the time dependence are often useful. Modern software packages, like SAS, allow the analyst to easily fit these expanded models.

PANEL 7.3   SAS ANALYSIS FOR EXAMPLE 7.16 USING PROC ARIMA

```
data a;
  infile 'T7-4.dat';
  time = _n_;
  input obsend dhd dhdlag wind xweekend;

proc arima data = a;
  identify var = obsend crosscor = (
  dhd dhdlag wind xweekend );
  estimate p = (1 7) method = ml input = (
  dhd dhdlag wind xweekend ) plot;
  estimate p = (1 7) noconstant method = ml input = (
  dhd dhdlag wind xweekend ) plot;
```
PROGRAM COMMANDS

ARIMA Procedure

Maximum Likelihood Estimation   **OUTPUT**

Parameter	Estimate	Approx. Std Error	T Ratio	Lag	Variable	Shift
MU	2.12957	13.12340	0.16	0	OBSEND	0
AR1, 1	0.47008	0.11779	3.99	1	OBSEND	0
AR1, 2	0.23986	0.11528	2.08	7	OBSEND	0
NUM1	5.80976	0.24047	24.16	0	DHD	0
NUM2	1.42632	0.24932	5.72	0	DHDLAG	0
NUM3	1.20740	0.44681	2.70	0	WIND	0
NUM4	−10.10890	6.03445	−1.68	0	XWEEKEND	0

Constant Estimate   = 0.61770069

Variance Estimate   = 228.894028

Std Error Estimate   = 15.1292441
AIC   = 528.490321
SBC   = 543.492264
Number of Residuals =   63

Autocorrelation Check of Residuals

To Lag	Chi Square	DF	Prob	Autocorrelations					
6	6.04	4	0.196	0.079	0.012	0.022	0.192	−0.127	0.161
12	10.27	10	0.417	0.144	−0.067	−0.111	−0.056	−0.056	−0.108
18	15.92	16	0.458	0.013	0.106	−0.137	−0.170	−0.079	0.018
24	23.44	22	0.377	0.018	0.004	0.250	−0.080	−0.069	−0.051

(continues on next page)

PANEL 7.3   *(continued)*

Autocorrelation Plot of Residuals

Lag	Covariance	Correlation	−1 9 8 7 6 5 4 3 2 1 0 1 2 3 4 5 6 7 8 9 1
0	228.894	1.00000	| |********************|
1	18.194945	0.07949	| . |** . |
2	2.763255	0.01207	| . | . |
3	5.038727	0.02201	| . | . |
4	44.059835	0.19249	| . |**** . |
5	−29.118892	−0.12722	| . *** | . |
6	36.904291	0.16123	| . |*** . |
7	33.008858	0.14421	| . |*** . |
8	−15.424015	−0.06738	| . *| . |
9	−25.379057	−0.11088	| . **| . |
10	−12.890888	−0.05632	| . *| . |
11	−12.777280	−0.05582	| . *| . |
12	−24.825623	−0.10846	| . **| . |
13	2.970197	0.01298	| . | . |
14	24.150168	0.10551	| . |** . |
15	−31.407314	−0.13721	| . *** | . |

"." marks two standard errors

7A

THE DISTRIBUTION OF THE LIKELIHOOD RATIO FOR THE MULTIVARIATE MULTIPLE REGRESSION MODEL

The development in this supplement establishes Result 7.11.

We know that $n\hat{\boldsymbol{\Sigma}} = \mathbf{Y}'(\mathbf{I} - \mathbf{Z}(\mathbf{Z}'\mathbf{Z})^{-1}\mathbf{Z}')\mathbf{Y}$ and under H_0, $n\hat{\boldsymbol{\Sigma}}_1 = \mathbf{Y}'[\mathbf{I} - \mathbf{Z}_1(\mathbf{Z}_1'\mathbf{Z}_1)^{-1}\mathbf{Z}_1']\mathbf{Y}$ with $\mathbf{Y} = \mathbf{Z}_1\boldsymbol{\beta}_{(1)} + \boldsymbol{\varepsilon}$. Set $\mathbf{P} = [\mathbf{I} - \mathbf{Z}(\mathbf{Z}'\mathbf{Z})^{-1}\mathbf{Z}']$. Since $\mathbf{0} = [\mathbf{I} - \mathbf{Z}(\mathbf{Z}'\mathbf{Z})^{-1}\mathbf{Z}']\mathbf{Z} = [\mathbf{I} - \mathbf{Z}(\mathbf{Z}'\mathbf{Z})^{-1}\mathbf{Z}'][\mathbf{Z}_1 \,\vdots\, \mathbf{Z}_2] = [\mathbf{PZ}_1 \,\vdots\, \mathbf{PZ}_2]$ the columns of $\mathbf{Z}$ are perpendicular to $\mathbf{P}$. Thus, we can write

$$n\hat{\boldsymbol{\Sigma}} = (\mathbf{Z}\boldsymbol{\beta} + \boldsymbol{\varepsilon})'\mathbf{P}(\mathbf{Z}\boldsymbol{\beta} + \boldsymbol{\varepsilon}) = \boldsymbol{\varepsilon}'\mathbf{P}\boldsymbol{\varepsilon}$$

$$n\hat{\boldsymbol{\Sigma}}_1 = (\mathbf{Z}_1\boldsymbol{\beta}_{(1)} + \boldsymbol{\varepsilon})'\mathbf{P}_1(\mathbf{Z}_1\boldsymbol{\beta}_{(1)} + \boldsymbol{\varepsilon}) = \boldsymbol{\varepsilon}'\mathbf{P}_1\boldsymbol{\varepsilon}$$

where $\mathbf{P}_1 = \mathbf{I} - \mathbf{Z}_1(\mathbf{Z}_1'\mathbf{Z}_1)^{-1}\mathbf{Z}_1'$. We then use the Gram–Schmidt process (see Result 2A.3) to construct the orthonormal vectors $[\mathbf{g}_1, \mathbf{g}_2, \ldots, \mathbf{g}_{q+1}] = \mathbf{G}$ from the columns of $\mathbf{Z}_1$. Then we continue, obtaining the orthonormal set from $[\mathbf{G}, \mathbf{Z}_2]$, and finally complete the set to n dimensions by constructing an arbitrary orthonormal set of $n - r - 1$ vectors orthogonal to the previous vectors. Consequently, we have

$$\underbrace{\mathbf{g}_1, \mathbf{g}_2, \ldots, \mathbf{g}_{q+1},}_{\substack{\text{from columns} \\ \text{of } \mathbf{Z}_1}} \quad \underbrace{\mathbf{g}_{q+2}, \mathbf{g}_{q+3}, \ldots, \mathbf{g}_{r+1},}_{\substack{\text{from columns of } \mathbf{Z}_2 \\ \text{but perpendicular} \\ \text{to columns of } \mathbf{Z}_1}} \quad \underbrace{\mathbf{g}_{r+2}, \mathbf{g}_{r+3}, \ldots, \mathbf{g}_n}_{\substack{\text{arbitrary set of} \\ \text{orthonormal} \\ \text{vectors orthogonal} \\ \text{to columns of } \mathbf{Z}}}$$

Let $(\lambda, \mathbf{e})$ be an eigenvalue-eigenvector pair of $\mathbf{Z}_1(\mathbf{Z}_1'\mathbf{Z}_1)^{-1}\mathbf{Z}_1'$. Then, since $[\mathbf{Z}_1(\mathbf{Z}_1'\mathbf{Z}_1)^{-1}\mathbf{Z}_1'][\mathbf{Z}_1(\mathbf{Z}_1'\mathbf{Z}_1)^{-1}\mathbf{Z}_1'] = \mathbf{Z}_1(\mathbf{Z}_1'\mathbf{Z}_1)^{-1}\mathbf{Z}_1$, it follows that

$$\lambda\mathbf{e} = \mathbf{Z}_1(\mathbf{Z}_1'\mathbf{Z}_1)^{-1}\mathbf{Z}_1'\mathbf{e} = (\mathbf{Z}_1(\mathbf{Z}_1'\mathbf{Z}_1)^{-1}\mathbf{Z}_1')^2\mathbf{e} = \lambda(\mathbf{Z}_1(\mathbf{Z}_1'\mathbf{Z}_1)^{-1}\mathbf{Z}_1')\mathbf{e} = \lambda^2\mathbf{e}$$

and the eigenvalues of $\mathbf{Z}_1(\mathbf{Z}_1'\mathbf{Z}_1)^{-1}\mathbf{Z}_1'$ are 0 or 1. Moreover, $\text{tr}(\mathbf{Z}_1(\mathbf{Z}_1'\mathbf{Z}_1)^{-1}\mathbf{Z}_1')$
$= \text{tr}((\mathbf{Z}_1'\mathbf{Z}_1)^{-1}\mathbf{Z}_1'\mathbf{Z}_1) = \text{tr}\left(\mathbf{I}_{(q+1)\times(q+1)}\right) = q+1 = \lambda_1 + \lambda_2 + \cdots + \lambda_{q+1}$, where
$\lambda_1 \geq \lambda_2 \geq \cdots \geq \lambda_{q+1} > 0$ are the eigenvalues of $\mathbf{Z}_1(\mathbf{Z}_1'\mathbf{Z}_1)^{-1}\mathbf{Z}_1'$. This shows that
$\mathbf{Z}_1(\mathbf{Z}_1'\mathbf{Z}_1)^{-1}\mathbf{Z}_1'$ has $q+1$ eigenvalues equal to 1. Now, $(\mathbf{Z}_1(\mathbf{Z}_1'\mathbf{Z}_1)^{-1}\mathbf{Z}_1')\mathbf{Z}_1 = \mathbf{Z}_1$, so
any linear combination $\mathbf{Z}_1\mathbf{b}_\ell$ of unit length is an eigenvector corresponding to the
eigenvalue 1. The orthonormal vectors $\mathbf{g}_\ell$, $\ell = 1, 2, \ldots, q+1$, are therefore eigenvectors of $\mathbf{Z}_1(\mathbf{Z}_1'\mathbf{Z}_1)^{-1}\mathbf{Z}_1'$, since they are formed by taking particular linear combinations of the columns of $\mathbf{Z}_1$. By the spectral decomposition (2-16), we have
$\mathbf{Z}_1(\mathbf{Z}_1'\mathbf{Z}_1)^{-1}\mathbf{Z}_1' = \sum_{\ell=1}^{q+1} \mathbf{g}_\ell\mathbf{g}_\ell'$. Similarly, by writing $(\mathbf{Z}(\mathbf{Z}'\mathbf{Z})^{-1}\mathbf{Z}')\mathbf{Z} = \mathbf{Z}$, we readily see
that the linear combination $\mathbf{Z}\mathbf{b}_\ell = \mathbf{g}_\ell$, for example, is an eigenvector of $\mathbf{Z}(\mathbf{Z}'\mathbf{Z})^{-1}\mathbf{Z}'$
with eigenvalue $\lambda = 1$, so that $\mathbf{Z}(\mathbf{Z}'\mathbf{Z})^{-1}\mathbf{Z}' = \sum_{\ell=1}^{r+1} \mathbf{g}_\ell\mathbf{g}_\ell'$.

Continuing, we have $\mathbf{PZ} = [\mathbf{I} - \mathbf{Z}(\mathbf{Z}'\mathbf{Z})^{-1}\mathbf{Z}']\mathbf{Z} = \mathbf{Z} - \mathbf{Z} = \mathbf{0}$ so $\mathbf{g}_\ell = \mathbf{Z}\mathbf{b}_\ell$,
$\ell \leq r+1$, are eigenvectors of $\mathbf{P}$ with eigenvalues $\lambda = 0$. Also, from the way the $\mathbf{g}_\ell$,
$\ell > r+1$, were constructed, $\mathbf{Z}'\mathbf{g}_\ell = \mathbf{0}$, so that $\mathbf{Pg}_\ell = \mathbf{g}_\ell$. Consequently, these $\mathbf{g}_\ell$'s
are eigenvectors of $\mathbf{P}$ corresponding to the $n - r - 1$ unit eigenvalues. By the spec-
tral decomposition (2-16), $\mathbf{P} = \sum_{\ell=r+2}^{n} \mathbf{g}_\ell\mathbf{g}_\ell'$ and

$$n\hat{\mathbf{\Sigma}} = \boldsymbol{\varepsilon}'\mathbf{P}\boldsymbol{\varepsilon} = \sum_{\ell=r+2}^{n} (\boldsymbol{\varepsilon}'\mathbf{g}_\ell)(\boldsymbol{\varepsilon}'\mathbf{g}_\ell)' = \sum_{\ell=r+2}^{n} \mathbf{V}_\ell\mathbf{V}_\ell'$$

where, because $\text{Cov}(V_{\ell i}, V_{jk}) = E(\mathbf{g}_\ell'\boldsymbol{\varepsilon}_{(i)}\boldsymbol{\varepsilon}_{(k)}'\mathbf{g}_j) = \sigma_{ik}\mathbf{g}_\ell'\mathbf{g}_j = 0$, $\ell \neq j$, the $\boldsymbol{\varepsilon}'\mathbf{g}_\ell = \mathbf{V}_\ell = [V_{\ell 1}, \ldots, V_{\ell i}, \ldots, V_{\ell m}]'$ are independently distributed as $N_m(\mathbf{0}, \mathbf{\Sigma})$. Consequently, by (4-22), $n\hat{\mathbf{\Sigma}}$ is distributed as $W_{p, n-r-1}(\mathbf{\Sigma})$. In the same manner,

$$\mathbf{P}_1\mathbf{g}_\ell = \begin{cases} \mathbf{g}_\ell & \ell > q+1 \\ \mathbf{0} & \ell \leq q+1 \end{cases}$$

so $\mathbf{P}_1 = \sum_{\ell=q+2}^{n} \mathbf{g}_\ell\mathbf{g}_\ell'$. We can write the extra sum of squares and cross products as

$$n(\hat{\mathbf{\Sigma}}_1 - \hat{\mathbf{\Sigma}}) = \boldsymbol{\varepsilon}'(\mathbf{P}_1 - \mathbf{P})\boldsymbol{\varepsilon} = \sum_{\ell=q+2}^{r+1} (\boldsymbol{\varepsilon}'\mathbf{g}_\ell)(\boldsymbol{\varepsilon}'\mathbf{g}_\ell)' = \sum_{\ell=q+2}^{r+1} \mathbf{V}_\ell\mathbf{V}_\ell'$$

where the $\mathbf{V}_\ell$ are independently distributed as $N_m(\mathbf{0}, \mathbf{\Sigma})$. By (4-22), $n(\hat{\mathbf{\Sigma}}_1 - \hat{\mathbf{\Sigma}})$ is
distributed as $W_{p, r-q}(\mathbf{\Sigma})$ independently of $n\hat{\mathbf{\Sigma}}$, since $n(\hat{\mathbf{\Sigma}}_1 - \hat{\mathbf{\Sigma}})$ involves a different
set of independent $\mathbf{V}_\ell$'s.

The large sample distribution for $-[n - r - 1 - \frac{1}{2}(m - r + q + 1)]\ln(|\hat{\mathbf{\Sigma}}|/|\hat{\mathbf{\Sigma}}_1|)$
follows from Result 5.2, with $\nu - \nu_0 = m(m+1)/2 + m(r+1) - m(m+1)/2 - m(q+1) = m(r-q)$ d.f. The use of $\left(n - r - 1 - \frac{1}{2}(m - r + q + 1)\right)$ instead
of n in the statistic is due to Bartlett [4] following Box [7], and it improves the
chi-square approximation.

Exercises

7.1. Given the data

z_1	10	5	7	19	11	8
y	15	9	3	25	7	13

fit the linear regression model $Y_j = \beta_0 + \beta_1 z_{j1} + \varepsilon_j$, $j = 1, 2, \ldots, 6$. Specifically, calculate the least squares estimates $\hat{\boldsymbol{\beta}}$, the fitted values $\hat{\mathbf{y}}$, the residuals $\hat{\boldsymbol{\varepsilon}}$, and the residual sum of squares, $\hat{\boldsymbol{\varepsilon}}'\hat{\boldsymbol{\varepsilon}}$.

7.2. Given the data

z_1	10	5	7	19	11	18
z_2	2	3	3	6	7	9
y	15	9	3	25	7	13

fit the regression model

$$Y_j = \beta_1 z_{j1} + \beta_2 z_{j2} + \varepsilon_j, \qquad j = 1, 2, \ldots, 6.$$

to the *standardized* form (see page 412) of the variables y, z_1, and z_2. From this fit, deduce the corresponding fitted regression equation for the original (not standardized) variables.

7.3. (*Weighted least squares estimators.*) Let

$$\underset{(n \times 1)}{\mathbf{Y}} = \underset{(n \times (r+1))}{\mathbf{Z}} \; \underset{((r+1) \times 1)}{\boldsymbol{\beta}} + \underset{(n \times 1)}{\boldsymbol{\varepsilon}}$$

where $E(\boldsymbol{\varepsilon}) = \mathbf{0}$ but $E(\boldsymbol{\varepsilon}\boldsymbol{\varepsilon}') = \sigma^2 \mathbf{V}$, with $\mathbf{V}(n \times n)$ known and positive definite. For $\mathbf{V}$ of full rank, show that the *weighted least squares* estimator is

$$\hat{\boldsymbol{\beta}}_W = (\mathbf{Z}'\mathbf{V}^{-1}\mathbf{Z})^{-1}\mathbf{Z}'\mathbf{V}^{-1}\mathbf{Y}$$

If σ^2 is unknown, it may be estimated, unbiasedly, by

$$(n - r - 1)^{-1} \times (\mathbf{Y} - \mathbf{Z}\hat{\boldsymbol{\beta}}_W)'\mathbf{V}^{-1}(\mathbf{Y} - \mathbf{Z}\hat{\boldsymbol{\beta}}_W).$$

Hint: $\mathbf{V}^{-1/2}\mathbf{Y} = (\mathbf{V}^{-1/2}\mathbf{Z})\boldsymbol{\beta} + \mathbf{V}^{-1/2}\boldsymbol{\varepsilon}$ is of the classical linear regression form $\mathbf{Y}^* = \mathbf{Z}^*\boldsymbol{\beta} + \boldsymbol{\varepsilon}^*$, with $E(\boldsymbol{\varepsilon}^*) = \mathbf{0}$ and $E(\boldsymbol{\varepsilon}^*\boldsymbol{\varepsilon}^{*\prime}) = \sigma^2\mathbf{I}$. Thus, $\hat{\boldsymbol{\beta}}_W = \hat{\boldsymbol{\beta}}^* = (\mathbf{Z}^*\mathbf{Z}^*)^{-1}\mathbf{Z}^{*\prime}\mathbf{Y}^*$.

7.4. Use the weighted least squares estimator in Exercise 7.3 to derive an expression for the estimate of the slope β in the model $Y_j = \beta z_j + \varepsilon_j, j = 1, 2, \ldots, n$, when (a) $\mathrm{Var}\,(\varepsilon_j) = \sigma^2$, (b) $\mathrm{Var}\,(\varepsilon_j) = \sigma^2 z_j$, and (c) $\mathrm{Var}\,(\varepsilon_j) = \sigma^2 z_j^2$. Comment on the manner in which the unequal variances for the errors influence the optimal choice of $\hat{\beta}_W$.

7.5. Establish (7-50): $\rho^2_{Y(\mathbf{Z})} = 1 - 1/\rho^{YY}$.
Hint: From (7-49) and Exercise 4.11

$$1 - \rho^2_{Y(\mathbf{Z})} = \frac{\sigma_{YY} - \boldsymbol{\sigma}'_{\mathbf{Z}Y}\boldsymbol{\Sigma}^{-1}_{\mathbf{Z}\mathbf{Z}}\boldsymbol{\sigma}_{\mathbf{Z}Y}}{\sigma_{YY}} = \frac{|\boldsymbol{\Sigma}_{\mathbf{Z}\mathbf{Z}}|\,(\sigma_{YY} - \boldsymbol{\sigma}'_{\mathbf{Z}Y}\boldsymbol{\Sigma}^{-1}_{\mathbf{Z}\mathbf{Z}}\boldsymbol{\sigma}_{\mathbf{Z}Y})}{|\boldsymbol{\Sigma}_{\mathbf{Z}\mathbf{Z}}|\,\sigma_{YY}} = \frac{|\boldsymbol{\Sigma}|}{|\boldsymbol{\Sigma}_{\mathbf{Z}\mathbf{Z}}|\,\sigma_{YY}}$$

From Result 2A.8(c), $\sigma^{YY} = |\boldsymbol{\Sigma}_{\mathbf{Z}\mathbf{Z}}|/|\boldsymbol{\Sigma}|$, where σ^{YY} is the entry of $\boldsymbol{\Sigma}^{-1}$ in the first row and first column. Since (see Exercise 2.23) $\boldsymbol{\rho} = \mathbf{V}^{-1/2}\,\boldsymbol{\Sigma}\,\mathbf{V}^{-1/2}$ and $\boldsymbol{\rho}^{-1} = (\mathbf{V}^{-1/2}\,\boldsymbol{\Sigma}\,\mathbf{V}^{-1/2})^{-1} = \mathbf{V}^{1/2}\boldsymbol{\Sigma}^{-1}\mathbf{V}^{1/2}$, the entry in the $(1, 1)$ position of $\boldsymbol{\rho}^{-1}$ is $\rho^{YY} = \sigma^{YY}\sigma_{YY}$.

7.6. *(Generalized inverse of $\mathbf{Z}'\mathbf{Z}$)* A matrix $(\mathbf{Z}'\mathbf{Z})^-$ is called a generalized inverse of $\mathbf{Z}'\mathbf{Z}$ if $\mathbf{Z}'\mathbf{Z}(\mathbf{Z}'\mathbf{Z})^-\mathbf{Z}'\mathbf{Z} = \mathbf{Z}'\mathbf{Z}$. Let $r_1 + 1 = \text{rank}(\mathbf{Z})$ and suppose $\lambda_1 \geq \lambda_2 \geq \cdots \geq \lambda_{r_1+1} > 0$ are the nonzero eigenvalues of $\mathbf{Z}'\mathbf{Z}$ with corresponding eigenvectors $\mathbf{e}_1, \mathbf{e}_2, \ldots, \mathbf{e}_{r_1+1}$.

(a) Show that

$$(\mathbf{Z}'\mathbf{Z})^- = \sum_{i=1}^{r_1+1} \lambda_i^{-1}\mathbf{e}_i\mathbf{e}_i'$$

is a generalized inverse of $\mathbf{Z}'\mathbf{Z}$.

(b) The coefficients $\hat{\boldsymbol{\beta}}$ that minimize the sum of squared errors $(\mathbf{y} - \mathbf{Z}\boldsymbol{\beta})'(\mathbf{y} - \mathbf{Z}\boldsymbol{\beta})$ satisfy the *normal equations* $(\mathbf{Z}'\mathbf{Z})\hat{\boldsymbol{\beta}} = \mathbf{Z}'\mathbf{y}$. Show that these equations are satisfied for any $\hat{\boldsymbol{\beta}}$ such that $\mathbf{Z}\hat{\boldsymbol{\beta}}$ is the projection of $\mathbf{y}$ on the columns of $\mathbf{Z}$.

(c) Show that $\mathbf{Z}\hat{\boldsymbol{\beta}} = \mathbf{Z}(\mathbf{Z}'\mathbf{Z})^-\mathbf{Z}'\mathbf{y}$ is the projection of $\mathbf{y}$ on the columns of $\mathbf{Z}$. (See Footnote 2 in this chapter.)

(d) Show directly that $\hat{\boldsymbol{\beta}} = (\mathbf{Z}'\mathbf{Z})^-\mathbf{Z}'\mathbf{y}$ is a solution to the normal equations $(\mathbf{Z}'\mathbf{Z})[(\mathbf{Z}'\mathbf{Z})^-\mathbf{Z}'\mathbf{y}] = \mathbf{Z}'\mathbf{y}$.

Hint: (b) If $\mathbf{Z}\hat{\boldsymbol{\beta}}$ is the projection, then $\mathbf{y} - \mathbf{Z}\hat{\boldsymbol{\beta}}$ is perpendicular to the columns of $\mathbf{Z}$.

(d) The eigenvalue–eigenvector requirement implies that $(\mathbf{Z}'\mathbf{Z})(\lambda_i^{-1}\mathbf{e}_i) = \mathbf{e}_i$ for $i \leq r_1 + 1$ and $0 = \mathbf{e}_i'(\mathbf{Z}'\mathbf{Z})\mathbf{e}_i$ for $i > r_1 + 1$. Therefore, $(\mathbf{Z}'\mathbf{Z})(\lambda_i^{-1}\mathbf{e}_i)\mathbf{e}_i'\mathbf{Z}' = \mathbf{e}_i\mathbf{e}_i'\mathbf{Z}'$. Summing over i gives

$$(\mathbf{Z}'\mathbf{Z})(\mathbf{Z}'\mathbf{Z})^-\mathbf{Z}' = \mathbf{Z}'\mathbf{Z}\left(\sum_{i=1}^{r_1+1} \lambda_i^{-1}\mathbf{e}_i\mathbf{e}_i'\right)\mathbf{Z}'$$

$$= \left(\sum_{i=1}^{r_1+1} \mathbf{e}_i\mathbf{e}_i'\right)\mathbf{Z}' = \left(\sum_{i=1}^{r+1} \mathbf{e}_i\mathbf{e}_i'\right)\mathbf{Z}' = \mathbf{I}\mathbf{Z}' = \mathbf{Z}'$$

since $\mathbf{e}_i'\mathbf{Z}' = \mathbf{0}$ for $i > r_1 + 1$.

7.7. Suppose the classical regression model is, with rank $(\mathbf{Z}) = r + 1$, written as

$$\underset{(n\times1)}{\mathbf{Y}} = \underset{(n\times(q+1))}{\mathbf{Z}_1}\underset{((q+1)\times1)}{\boldsymbol{\beta}_{(1)}} + \underset{(n\times(r-q))}{\mathbf{Z}_2}\underset{((r-q)\times1)}{\boldsymbol{\beta}_{(2)}} + \underset{(n\times1)}{\boldsymbol{\varepsilon}}$$

where $\text{rank}(\mathbf{Z}_1) = q + 1$ and $\text{rank}(\mathbf{Z}_2) = r - q$. If the parameters $\boldsymbol{\beta}_{(2)}$ are identified beforehand as being of primary interest, show that a $100(1 - \alpha)\%$ confidence region for $\boldsymbol{\beta}_{(2)}$ is given by

$$(\hat{\boldsymbol{\beta}}_{(2)} - \boldsymbol{\beta}_{(2)})'[\mathbf{Z}_2'\mathbf{Z}_2 - \mathbf{Z}_2'\mathbf{Z}_1(\mathbf{Z}_1'\mathbf{Z}_1)^{-1}\mathbf{Z}_1'\mathbf{Z}_2](\hat{\boldsymbol{\beta}}_{(2)} - \boldsymbol{\beta}_{(2)}) \leq s^2(r - q)F_{r-q,n-r-1}(\alpha)$$

Hint: By Exercise 4.12, with 1's and 2's interchanged,

$$\mathbf{C}^{22} = [\mathbf{Z}_2'\mathbf{Z}_2 - \mathbf{Z}_2'\mathbf{Z}_1(\mathbf{Z}_1'\mathbf{Z}_1)^{-1}\mathbf{Z}_1'\mathbf{Z}_2]^{-1}, \quad \text{where } (\mathbf{Z}'\mathbf{Z})^{-1} = \begin{bmatrix} \mathbf{C}^{11} & \mathbf{C}^{12} \\ \mathbf{C}^{21} & \mathbf{C}^{22} \end{bmatrix}$$

Multiply by the square-root matrix $(\mathbf{C}^{22})^{-1/2}$, and conclude that $(\mathbf{C}^{22})^{-1/2}(\hat{\boldsymbol{\beta}}_{(2)} - \boldsymbol{\beta}_{(2)})/\sigma^2$ is $N(\mathbf{0}, \mathbf{I})$, so that

$$(\hat{\boldsymbol{\beta}}_{(2)} - \boldsymbol{\beta}_{(2)})'(\mathbf{C}^{22})^{-1}(\hat{\boldsymbol{\beta}}_{(2)} - \boldsymbol{\beta}_{(2)}) \text{ is } \sigma^2\chi_{r-q}^2.$$

7.8. Recall that the hat matrix is defined by $\mathbf{H} = \mathbf{Z}(\mathbf{Z}'\mathbf{Z})^{-1}\mathbf{Z}'$ with diagonal elements h_{jj}.

(a) Show that $\mathbf{H}$ is an idempotent matrix. [See Result 7.1 and (7-6).]

(b) Show that $0 < h_{jj} < 1$, $j = 1, 2, \ldots, n$, and that $\sum_{j=1}^{n} h_{jj} = r + 1$, where r is the number of independent variables in the regression model. (In fact, $(1/n) \leq h_{jj} < 1$.)

(c) Verify, for the simple linear regression model with one independent variable z, that the leverage, h_{jj}, is given by

$$h_{jj} = \frac{1}{n} + \frac{(z_j - \bar{z})^2}{\sum_{j=1}^{n} (z_j - \bar{z})^2}$$

7.9. Consider the following data on one predictor variable z_1 and two responses Y_1 and Y_2:

z_1	-2	-1	0	1	2
y_1	5	3	4	2	1
y_2	-3	-1	-1	2	3

Determine the least squares estimates of the parameters in the bivariate straight-line regression model

$$Y_{j1} = \beta_{01} + \beta_{11}z_{j1} + \varepsilon_{j1}$$
$$Y_{j2} = \beta_{02} + \beta_{12}z_{j1} + \varepsilon_{j2}, \qquad j = 1,2,3,4,5$$

Also, calculate the matrices of fitted values $\hat{\mathbf{Y}}$ and residuals $\hat{\boldsymbol{\varepsilon}}$ with $\mathbf{Y} = [\mathbf{y}_1 \mid \mathbf{y}_2]$. Verify the sum of squares and cross-products decomposition

$$\mathbf{Y}'\mathbf{Y} = \hat{\mathbf{Y}}'\hat{\mathbf{Y}} + \hat{\boldsymbol{\varepsilon}}'\hat{\boldsymbol{\varepsilon}}$$

7.10. Using the results from Exercise 7.9, calculate each of the following.

(a) A 95% confidence interval for the mean response $E(Y_{01}) = \beta_{01} + \beta_{11}z_{01}$ corresponding to $z_{01} = 0.5$

(b) A 95% prediction interval for the response Y_{01} corresponding to $z_{01} = 0.5$

(c) A 95% prediction region for the responses Y_{01} and Y_{02} corresponding to $z_{01} = 0.5$

7.11. *(Generalized least squares for multivariate multiple regression.)* Let $\mathbf{A}$ be a positive definite matrix, so that $d_j^2(\mathbf{B}) = (\mathbf{y}_j - \mathbf{B}'\mathbf{z}_j)'\mathbf{A}(\mathbf{y}_j - \mathbf{B}'\mathbf{z}_j)$ is a squared statistical distance from the jth observation $\mathbf{y}_j$ to its regression $\mathbf{B}'\mathbf{z}_j$. Show that the choice $\mathbf{B} = \hat{\boldsymbol{\beta}} = (\mathbf{Z}'\mathbf{Z})^{-1}\mathbf{Z}'\mathbf{Y}$ minimizes the sum of squared statistical distances, $\sum_{j=1}^{n} d_j^2(\mathbf{B})$, for any choice of positive definite $\mathbf{A}$. Choices for $\mathbf{A}$ include $\boldsymbol{\Sigma}^{-1}$ and $\mathbf{I}$.
Hint: Repeat the steps in the proof of Result 7.10 with $\boldsymbol{\Sigma}^{-1}$ replaced by $\mathbf{A}$.

7.12. Given the mean vector and covariance matrix of Y, Z_1, and Z_2,

$$\boldsymbol{\mu} = \begin{bmatrix} \mu_Y \\ \hline \boldsymbol{\mu}_Z \end{bmatrix} = \begin{bmatrix} 4 \\ \hline 3 \\ -2 \end{bmatrix} \quad \text{and} \quad \boldsymbol{\Sigma} = \begin{bmatrix} \sigma_{YY} & \boldsymbol{\sigma}'_{ZY} \\ \hline \boldsymbol{\sigma}_{ZY} & \boldsymbol{\Sigma}_{ZZ} \end{bmatrix} = \begin{bmatrix} 9 & 3 & 1 \\ \hline 3 & 2 & 1 \\ 1 & 1 & 1 \end{bmatrix}$$

determine each of the following.

(a) The best linear predictor $\beta_0 + \beta_1 Z_1 + \beta_2 Z_2$ of Y

(b) The mean square error of the best linear predictor

(c) The population multiple correlation coefficient

(d) The partial correlation coefficient $\rho_{YZ_1 \cdot Z_2}$

7.13. The test scores for college students described in Example 5.5 have

$$\bar{z} = \begin{bmatrix} \bar{z}_1 \\ \bar{z}_2 \\ \bar{z}_3 \end{bmatrix} = \begin{bmatrix} 527.74 \\ 54.69 \\ 25.13 \end{bmatrix}, \quad S = \begin{bmatrix} 5691.34 & & \\ 600.51 & 126.05 & \\ 217.25 & 23.37 & 23.11 \end{bmatrix}$$

Assume joint normality.

(a) Obtain the maximum likelihood estimates of the parameters for predicting Z_1 from Z_2 and Z_3.

(b) Evaluate the estimated multiple correlation coefficient $R_{Z_1(Z_2, Z_3)}$.

(c) Determine the estimated partial correlation coefficient $R_{Z_1, Z_2 \cdot Z_3}$.

7.14. Twenty-five portfolio managers were evaluated in terms of their performance. Suppose Y represents the rate of return achieved over a period of time, Z_1 is the manager's attitude toward risk measured on a five-point scale from "very conservative" to "very risky," and Z_2 is years of experience in the investment business. The observed correlation coefficients between pairs of variables are

$$R = \begin{matrix} & Y & Z_1 & Z_2 \\ & \begin{bmatrix} 1.0 & -.35 & .82 \\ -.35 & 1.0 & -.60 \\ .82 & -.60 & 1.0 \end{bmatrix} \end{matrix}$$

(a) Interpret the sample correlation coefficients $r_{YZ_1} = -.35$ and $r_{YZ_2} = -.82$.

(b) Calculate the partial correlation coefficient $r_{YZ_1 \cdot Z_2}$ and interpret this quantity with respect to the interpretation provided for r_{YZ_1} in Part a.

The following exercises may require the use of a computer.

7.15. Use the real-estate data in Table 7.1 and the linear regression model in Example 7.4.

(a) Verify the results in Example 7.4.

(b) Analyze the residuals to check the adequacy of the model. (See Section 7.6.)

(c) Generate a 95% prediction interval for the selling price (Y_0) corresponding to total dwelling size $z_1 = 17$ and assessed value $z_2 = 46$.

(d) Carry out a likelihood ratio test of $H_0: \beta_2 = 0$ with a significance level of $\alpha = .05$. Should the original model be modified? Discuss.

7.16. Calculate a C_p plot corresponding to the possible linear regressions involving the real-estate data in Table 7.1.

7.17. Consider the *Forbes* data in Exercise 1.4.

(a) Fit a linear regression model to these data using profits as the dependent variable and sales and assets as the independent variables.

(b) Analyze the residuals to check the adequacy of the model. Compute the leverages associated with the data points. Does one (or more) of these companies stand out as an outlier in the set of independent variable data points?

(c) Generate a 95% prediction interval for profits corresponding to sales of 100 (billions of dollars) and assets of 500 (billions of dollars).

(d) Carry out a likelihood ratio test of $H_0: \beta_2 = 0$ with a significance level of $\alpha = .05$. Should the original model be modified? Discuss.

7.18. Calculate

(a) a C_p plot corresponding to the possible regressions involving the *Forbes* data in Exercise 1.4.

(b) the AIC for each possible regression.

7.19. Satellite applications motivated the development of a silver-zinc battery. Table 7.5 contains failure data collected to characterize the performance of the battery during its life cycle. Use these data.

(a) Find the estimated linear regression of $\ln(Y)$ on an appropriate ("best") subset of predictor variables.

(b) Plot the residuals from the fitted model chosen in Part a to check the normal assumption.

Table 7.5 Battery-Failure Data

Z_1 Charge rate (amps)	Z_2 Discharge rate (amps)	Z_3 Depth of discharge (% of rated ampere-hours)	Z_4 Temperature (°C)	Z_5 End of charge voltage (volts)	Y Cycles to failure
.375	3.13	60.0	40	2.00	101
1.000	3.13	76.8	30	1.99	141
1.000	3.13	60.0	20	2.00	96
1.000	3.13	60.0	20	1.98	125
1.625	3.13	43.2	10	2.01	43
1.625	3.13	60.0	20	2.00	16
1.625	3.13	60.0	20	2.02	188
.375	5.00	76.8	10	2.01	10
1.000	5.00	43.2	10	1.99	3
1.000	5.00	43.2	30	2.01	386
1.000	5.00	100.0	20	2.00	45
1.625	5.00	76.8	10	1.99	2
.375	1.25	76.8	10	2.01	76
1.000	1.25	43.2	10	1.99	78
1.000	1.25	76.8	30	2.00	160
1.000	1.25	60.0	0	2.00	3
1.625	1.25	43.2	30	1.99	216
1.625	1.25	60.0	20	2.00	73
.375	3.13	76.8	30	1.99	314
.375	3.13	60.0	20	2.00	170

Source: Selected from S. Sidik, H. Leibecki, and J. Bozek, *Failure of Silver–Zinc Cells with Competing Failure Modes—Preliminary Data Analysis*, NASA Technical Memorandum 81556 (Cleveland: Lewis Research Center, 1980).

7.20. Using the battery-failure data in Table 7.5, regress $\ln(Y)$ on the first principal component of the predictor variables $z_1, z_2, \ldots, z_5$. (See Section 8.3.) Compare the result with the fitted model obtained in Exercise 7.19(a).

7.21. Consider the air-pollution data in Table 1.5. Let $Y_1 = NO_2$ and $Y_2 = O_3$ be the two responses (pollutants) corresponding to the predictor variables $Z_1 =$ wind and $Z_2 =$ solar radiation.

(a) Perform a regression analysis using only the first response Y_1.

(i)   Suggest and fit appropriate linear regression models.

(ii)   Analyze the residuals.

(iii)   Construct a 95% prediction interval for NO_2 corresponding to $z_1 = 10$ and $z_2 = 80$.

(b) Perform a multivariate multiple regression analysis using both responses Y_1 and Y_2.

(i)   Suggest and fit appropriate linear regression models.

(ii)   Analyze the residuals.

(iii)   Construct a 95% prediction ellipse for both NO_2 and O_3 for $z_1 = 10$ and $z_2 = 80$. Compare this ellipse with the prediction interval in Part a (iii). Comment.

7.22. Using the data on bone mineral content in Table 1.8:

(a) Perform a regression analysis by fitting the response for the dominant radius bone to the measurements on the last four bones.

(i)   Suggest and fit appropriate linear regression models.

(ii)   Analyze the residuals.

(b) Perform a multivariate multiple regression analysis by fitting the responses from both radius bones.

(c) Calculate the AIC for the model you chose in (b) and for the full model.

7.23. Using the data on the characteristics of bulls sold at auction in Table 1.10:

(a) Perform a regression analysis using the response $Y_1 =$ SalePr and the predictor variables Breed, YrHgt, FtFrBody, PrctFFB, Frame, BkFat, SaleHt, and SaleWt.

(i)   Determine the "best" regression equation by retaining only those predictor variables that are individually significant.

(ii)   Using the best fitting model, construct a 95% prediction interval for selling price for the set of predictor variable values (in the order listed above) 5, 48.7, 990, 74.0, 7, .18, 54.2 and 1450.

(iii)   Examine the residuals from the best fitting model.

(b) Repeat the analysis in Part a, using the natural logarithm of the sales price as the response. That is, set $Y_1 = \text{Ln}(\text{SalePr})$. Which analysis do you prefer? Why?

7.24. Using the data on the characteristics of bulls sold at auction in Table 1.10:

(a) Perform a regression analysis, using only the response $Y_1 =$ SaleHt and the predictor variables $Z_1 =$ YrHgt and $Z_2 =$ FtFrBody.

(i)   Fit an appropriate model and analyze the residuals.

(ii)   Construct a 95% prediction interval for SaleHt corresponding to $z_1 = 50.5$ and $z_2 = 970$.

(b) Perform a multivariate regression analysis with the responses $Y_1 =$ SaleHt and $Y_2 =$ SaleWt and the predictors $Z_1 =$ YrHgt and $Z_2 =$ FtFrBody.

(i)   Fit an appropriate multivariate model and analyze the residuals.

(ii)   Construct a 95% prediction ellipse for both SaleHt and SaleWt for $z_1 = 50.5$ and $z_2 = 970$. Compare this ellipse with the prediction interval in Part a (ii). Comment.

7.25. Amitriptyline is prescribed by some physicians as an antidepressant. However, there are also conjectured side effects that seem to be related to the use of the drug: irregular heartbeat, abnormal blood pressures, and irregular waves on the electrocardiogram, among other things. Data gathered on 17 patients who were admitted to the hospital after an amitriptyline overdose are given in Table 7.6. The two response variables are

$$Y_1 = \text{Total TCAD plasma level (TOT)}$$
$$Y_2 = \text{Amount of amitriptyline present in TCAD plasma level (AMI)}$$

The five predictor variables are

$$Z_1 = \text{Gender: 1 if female, 0 if male (GEN)}$$
$$Z_2 = \text{Amount of antidepressants taken at time of overdose (AMT)}$$
$$Z_3 = \text{PR wave measurement (PR)}$$
$$Z_4 = \text{Diastolic blood pressure (DIAP)}$$
$$Z_5 = \text{QRS wave measurement (QRS)}$$

Table 7.6 Amitriptyline Data

y_1 TOT	y_2 AMI	z_1 GEN	z_2 AMT	z_3 PR	z_4 DIAP	z_5 QRS
3389	3149	1	7500	220	0	140
1101	653	1	1975	200	0	100
1131	810	0	3600	205	60	111
596	448	1	675	160	60	120
896	844	1	750	185	70	83
1767	1450	1	2500	180	60	80
807	493	1	350	154	80	98
1111	941	0	1500	200	70	93
645	547	1	375	137	60	105
628	392	1	1050	167	60	74
1360	1283	1	3000	180	60	80
652	458	1	450	160	64	60
860	722	1	1750	135	90	79
500	384	0	2000	160	60	80
781	501	0	4500	180	0	100
1070	405	0	1500	170	90	120
1754	1520	1	3000	180	0	129

Source: See [24].

(a) Perform a regression analysis using only the first response Y_1.

    (i)   Suggest and fit appropriate linear regression models.

    (ii) Analyze the residuals.

    (iii) Construct a 95% prediction interval for Total TCAD for $z_1 = 1$, $z_2 = 1200$, $z_3 = 140$, $z_4 = 70$, and $z_5 = 85$.

(b) Repeat Part a using the second response Y_2.

(c) Perform a multivariate multiple regression analysis using both responses Y_1 and Y_2.

  (i) Suggest and fit appropriate linear regression models.

  (ii) Analyze the residuals.

  (iii) Construct a 95% prediction ellipse for both Total TCAD and Amount of amitriptyline for $z_1 = 1$, $z_2 = 1200$, $z_3 = 140$, $z_4 = 70$, and $z_5 = 85$. Compare this ellipse with the prediction intervals in Parts a and b. Comment.

7.26. Measurements of properties of pulp fibers and the paper made from them are contained in Table 7.7 (see also [19] and website: www.prenhall.com/statistics). There are $n = 62$ observations of the pulp fiber characteristics, z_1 = arithmetic fiber length, z_2 = long fiber fraction, z_3 = fine fiber fraction, z_4 = zero span tensile, and the paper properties, y_1 = breaking length, y_2 = elastic modulus, y_3 = stress at failure, y_4 = burst strength.

Table 7.7 Pulp and Paper Properites Data

y_1 BL	y_2 EM	y_3 SF	y_4 BS	z_1 AFL	z_2 LFF	z_3 FFF	z_4 ZST
21.312	7.039	5.326	.932	−.030	35.239	36.991	1.057
21.206	6.979	5.237	.871	.015	35.713	36.851	1.064
20.709	6.779	5.060	.742	.025	39.220	30.586	1.053
19.542	6.601	4.479	.513	.030	39.756	21.072	1.050
20.449	6.795	4.912	.577	−.070	32.991	36.570	1.049
⋮	⋮	⋮	⋮	⋮	⋮	⋮	⋮
16.441	6.315	2.997	−.400	−.605	2.845	84.554	1.008
16.294	6.572	3.017	−.478	−.694	1.515	81.988	.998
20.289	7.719	4.866	.239	−.559	2.054	8.786	1.081
17.163	7.086	3.396	−.236	−.415	3.018	5.855	1.033
20.289	7.437	4.859	.470	−.324	17.639	28.934	1.070

Source: See Lee [19].

(a) Perform a regression analysis using each of the response variables Y_1, Y_2, Y_3 and Y_4.

  (i) Suggest and fit appropriate linear regression models.

  (ii) Analyze the residuals. Check for outliers or observations with high leverage.

  (iii) Construct a 95% prediction interval for SF (Y_3) for $z_1 = .330$, $z_2 = 45.500$, $z_3 = 20.375$, $z_4 = 1.010$.

(b) Perform a multivariate multiple regression analysis using all four response variables, Y_1, Y_2, Y_3 and Y_4, and the four independent variables, Z_1, Z_2, Z_3 and Z_4.

  (i) Suggest and fit an appropriate linear regression model. Specify the matrix of estimated coefficients $\hat{\boldsymbol{\beta}}$ and estimated error covariance matrix $\hat{\boldsymbol{\Sigma}}$.

  (ii) Analyze the residuals. Check for outliers.

  (iii) Construct simultaneous 95% prediction intervals for the individual responses $Y_{0i}, i = 1, 2, 3, 4$, for the same settings of the independent variables given in part a (iii) above. Compare the simultaneous prediction interval for Y_{03} with the prediction interval in part a (iii). Comment.

7.27. Refer to the data on fixing breakdowns in cell phone relay towers in Table 6.20. In the initial design, experience level was coded as Novice or Guru. Now consider three levels of experience: Novice, Guru and Experienced. Some additional runs for an experienced engineer are given below. Also, in the original data set, reclassify Guru in run 3 as

Experienced and Novice in run 14 as Experienced. Keep all the other numbers for these two engineers the same. With these changes and the new data below, perform a multivariate multiple regression analysis with assessment and implementation times as the responses, and problem severity, problem complexity and experience level as the predictor variables. Consider regression models with the predictor variables and two factor interaction terms as inputs. (Note: The two changes in the original data set and the additional data below unbalances the design, so the analysis is best handled with regression methods.)

Problem severity level	Problem complexity level	Engineer experience level	Problem assessment time	Problem implementation time	Total resolution time
Low	Complex	Experienced	5.3	9.2	14.5
Low	Complex	Experienced	5.0	10.9	15.9
High	Simple	Experienced	4.0	8.6	12.6
High	Simple	Experienced	4.5	8.7	13.2
High	Complex	Experienced	6.9	14.9	21.8

References

1. Abraham, B. and J. Ledolter. *Introduction to Regression Modeling*, Belmont, CA: Thompson Brooks/Cole, 2006.

2. Anderson, T. W. *An Introduction to Multivariate Statistical Analysis* (3rd ed.). New York: John Wiley, 2003.

3. Atkinson, A. C. *Plots, Transformations and Regression: An Introduction to Graphical Methods of Diagnostic Regression Analysis*. Oxford, England: Oxford University Press, 1986.

4. Bartlett, M. S. "A Note on Multiplying Factors for Various Chi-Squared Approximations." *Journal of the Royal Statistical Society (B)*, **16** (1954), 296–298.

5. Belsley, D. A., E. Kuh, and R. E. Welsh. *Regression Diagnostics: Identifying Influential Data and Sources of Collinearity* (Paperback). New York: Wiley-Interscience, 2004.

6. Bowerman, B. L., and R. T. O'Connell. *Linear Statistical Models: An Applied Approach* (2nd ed.). Belmont, CA: Thompson Brooks/Cole, 2000.

7. Box, G. E. P. "A General Distribution Theory for a Class of Likelihood Criteria." *Biometrika*, **36** (1949), 317–346.

8. Box, G. E. P., G. M. Jenkins, and G. C. Reinsel. *Time Series Analysis: Forecasting and Control* (3rd ed.). Englewood Cliffs, NJ: Prentice Hall, 1994.

9. Chatterjee, S., A. S. Hadi, and B. Price. *Regression Analysis by Example* (4th ed.). New York: Wiley-Interscience, 2006.

10. Cook, R. D., and S. Weisberg. *Applied Regression Including Computing and Graphics*. New York: John Wiley, 1999.

11. Cook, R. D., and S. Weisberg. *Residuals and Influence in Regression*. London: Chapman and Hall, 1982.

12. Daniel, C. and F. S. Wood. *Fitting Equations to Data* (2nd ed.) (paperback). New York: Wiley-Interscience, 1999.

13. Draper, N. R., and H. Smith. *Applied Regression Analysis* (3rd ed.). New York: John Wiley, 1998.

14. Durbin, J., and G. S. Watson. "Testing for Serial Correlation in Least Squares Regression, II." *Biometrika*, **38** (1951), 159–178.

15. Galton, F. "Regression Toward Mediocrity in Heredity Stature." *Journal of the Anthropological Institute*, **15** (1885), 246–263.

16. Goldberger, A. S. *Econometric Theory*. New York: John Wiley, 1964.

17. Heck, D. L. "Charts of Some Upper Percentage Points of the Distribution of the Largest Characteristic Root." *Annals of Mathematical Statistics*, **31** (1960), 625–642.

18. Khattree, R. and D. N. Naik. *Applied Multivariate Statistics with SAS® Software* (2nd ed.) Cary, NC: SAS Institute Inc., 1999.

19. Lee, J. "Relationships Between Properties of Pulp-Fibre and Paper." Unpublished doctoral thesis, University of Toronto, Faculty of Forestry, 1992.

20. Neter, J., W. Wasserman, M. Kutner, and C. Nachtsheim. *Applied Linear Regression Models* (3rd ed.). Chicago: Richard D. Irwin, 1996.

21. Pillai, K. C. S. "Upper Percentage Points of the Largest Root of a Matrix in Multivariate Analysis." *Biometrika*, **54** (1967), 189–193.

22. Rao, C. R. *Linear Statistical Inference and Its Applications* (2nd ed.) (paperback). New York: Wiley-Interscience, 2002.

23. Seber, G. A. F. *Linear Regression Analysis*. New York: John Wiley, 1977.

24. Rudorfer, M. V. "Cardiovascular Changes and Plasma Drug Levels after Amitriptyline Overdose." *Journal of Toxicology-Clinical Toxicology*, **19** (1982), 67–71.

Chapter

8

PRINCIPAL COMPONENTS

8.1 Introduction

A principal component analysis is concerned with explaining the variance–covariance structure of a set of variables through a few *linear* combinations of these variables. Its general objectives are (1) data reduction and (2) interpretation.

Although p components are required to reproduce the total system variability, often much of this variability can be accounted for by a small number k of the principal components. If so, there is (almost) as much information in the k components as there is in the original p variables. The k principal components can then replace the initial p variables, and the original data set, consisting of n measurements on p variables, is reduced to a data set consisting of n measurements on k principal components.

An analysis of principal components often reveals relationships that were not previously suspected and thereby allows interpretations that would not ordinarily result. A good example of this is provided by the stock market data discussed in Example 8.5.

Analyses of principal components are more of a means to an end rather than an end in themselves, because they frequently serve as intermediate steps in much larger investigations. For example, principal components may be inputs to a multiple regression (see Chapter 7) or cluster analysis (see Chapter 12). Moreover, (scaled) principal components are one "factoring" of the covariance matrix for the factor analysis model considered in Chapter 9.

8.2 Population Principal Components

Algebraically, principal components are particular linear combinations of the p random variables $X_1, X_2, \ldots, X_p$. Geometrically, these linear combinations represent the selection of a new coordinate system obtained by rotating the original system

:• a couple PC's are unique (uncorrelated) due to being orthogonal

•These PC's will have unique ways to represent the data.

• PC scores tell us who has most / least score in that category.

430

with $X_1, X_2, \ldots, X_p$ as the coordinate axes. The new axes represent the directions with maximum variability and provide a simpler and more parsimonious description of the covariance structure.

As we shall see, principal components depend solely on the covariance matrix Σ (or the correlation matrix ρ) of $X_1, X_2, \ldots, X_p$. Their development does not require a multivariate normal assumption. On the other hand, principal components derived for multivariate normal populations have useful interpretations in terms of the constant density ellipsoids. Further, inferences can be made from the sample components when the population is multivariate normal. (See Section 8.5.)

Let the random vector $\mathbf{X}' = [X_1, X_2, \ldots, X_p]$ have the covariance matrix Σ with eigenvalues $\lambda_1 \geq \lambda_2 \geq \cdots \geq \lambda_p \geq 0$.

Consider the linear combinations

$$
\begin{aligned}
Y_1 &= \mathbf{a}_1'\mathbf{X} = a_{11}X_1 + a_{12}X_2 + \cdots + a_{1p}X_p \\
Y_2 &= \mathbf{a}_2'\mathbf{X} = a_{21}X_1 + a_{22}X_2 + \cdots + a_{2p}X_p \\
&\;\;\vdots \qquad\qquad\qquad\qquad\;\; \vdots \\
Y_p &= \mathbf{a}_p'\mathbf{X} = a_{p1}X_1 + a_{p2}X_2 + \cdots + a_{pp}X_p
\end{aligned}
\tag{8-1}
$$

Then, using (2-45), we obtain

$$
\text{Var}(Y_i) = \mathbf{a}_i'\Sigma\mathbf{a}_i \qquad i = 1, 2, \ldots, p \tag{8-2}
$$

$$
\text{Cov}(Y_i, Y_k) = \mathbf{a}_i'\Sigma\mathbf{a}_k \qquad i, k = 1, 2, \ldots, p \tag{8-3}
$$

The principal components are those *uncorrelated* linear combinations $Y_1, Y_2, \ldots, Y_p$ whose variances in (8-2) are as large as possible.

The first principal component is the linear combination with maximum variance. That is, it maximizes $\text{Var}(Y_1) = \mathbf{a}_1'\Sigma\mathbf{a}_1$. It is clear that $\text{Var}(Y_1) = \mathbf{a}_1'\Sigma\mathbf{a}_1$ can be increased by multiplying any $\mathbf{a}_1$ by some constant. To eliminate this indeterminacy, it is convenient to restrict attention to coefficient vectors of unit length. We therefore define

First principal component = linear combination $\mathbf{a}_1'\mathbf{X}$ that maximizes
$\qquad\qquad\qquad\qquad\qquad$ Var$(\mathbf{a}_1'\mathbf{X})$ subject to $\mathbf{a}_1'\mathbf{a}_1 = 1$

Second principal component = linear combination $\mathbf{a}_2'\mathbf{X}$ that maximizes
$\qquad\qquad\qquad\qquad\qquad$ Var$(\mathbf{a}_2'\mathbf{X})$ subject to $\mathbf{a}_2'\mathbf{a}_2 = 1$ and
$\qquad\qquad\qquad\qquad\qquad$ Cov$(\mathbf{a}_1'\mathbf{X}, \mathbf{a}_2'\mathbf{X}) = 0$

At the ith step,

ith principal component = linear combination $\mathbf{a}_i'\mathbf{X}$ that maximizes
$\qquad\qquad\qquad\qquad\qquad$ Var$(\mathbf{a}_i'\mathbf{X})$ subject to $\mathbf{a}_i'\mathbf{a}_i = 1$ and
$\qquad\qquad\qquad\qquad\qquad$ Cov$(\mathbf{a}_i'\mathbf{X}, \mathbf{a}_k'\mathbf{X}) = 0$ for $k < i$

Result 8.1. Let $\boldsymbol{\Sigma}$ be the covariance matrix associated with the random vector $\mathbf{X}' = [X_1, X_2, \ldots, X_p]$. Let $\boldsymbol{\Sigma}$ have the eigenvalue-eigenvector pairs $(\lambda_1, \mathbf{e}_1)$, $(\lambda_2, \mathbf{e}_2), \ldots, (\lambda_p, \mathbf{e}_p)$ where $\lambda_1 \geq \lambda_2 \geq \cdots \geq \lambda_p \geq 0$. Then the *ith principal component* is given by

$$Y_i = \mathbf{e}_i'\mathbf{X} = e_{i1}X_1 + e_{i2}X_2 + \cdots + e_{ip}X_p, \quad i = 1, 2, \ldots, p \qquad (8\text{-}4)$$

With these choices,

$$\text{Var}(Y_i) = \mathbf{e}_i'\boldsymbol{\Sigma}\mathbf{e}_i = \lambda_i \quad i = 1, 2, \ldots, p$$

$$\text{Cov}(Y_i, Y_k) = \mathbf{e}_i'\boldsymbol{\Sigma}\mathbf{e}_k = 0 \quad i \neq k \qquad (8\text{-}5)$$

If some λ_i are equal, the choices of the corresponding coefficient vectors, $\mathbf{e}_i$, and hence Y_i, are not unique.

Proof. We know from (2-51), with $\mathbf{B} = \boldsymbol{\Sigma}$, that

$$\max_{\mathbf{a} \neq 0} \frac{\mathbf{a}'\boldsymbol{\Sigma}\mathbf{a}}{\mathbf{a}'\mathbf{a}} = \lambda_1 \qquad (\text{attained when } \mathbf{a} = \mathbf{e}_1)$$

But $\mathbf{e}_1'\mathbf{e}_1 = 1$ since the eigenvectors are normalized. Thus,

$$\max_{\mathbf{a} \neq 0} \frac{\mathbf{a}'\boldsymbol{\Sigma}\mathbf{a}}{\mathbf{a}'\mathbf{a}} = \lambda_1 = \frac{\mathbf{e}_1'\boldsymbol{\Sigma}\mathbf{e}_1}{\mathbf{e}_1'\mathbf{e}_1} = \mathbf{e}_1'\boldsymbol{\Sigma}\mathbf{e}_1 = \text{Var}(Y_1)$$

Similarly, using (2-52), we get

$$\max_{\mathbf{a} \perp \mathbf{e}_1, \mathbf{e}_2, \ldots, \mathbf{e}_k} \frac{\mathbf{a}'\boldsymbol{\Sigma}\mathbf{a}}{\mathbf{a}'\mathbf{a}} = \lambda_{k+1} \quad k = 1, 2, \ldots, p - 1$$

For the choice $\mathbf{a} = \mathbf{e}_{k+1}$, with $\mathbf{e}_{k+1}'\mathbf{e}_i = 0$, for $i = 1, 2, \ldots, k$ and $k = 1, 2, \ldots, p - 1$,

$$\mathbf{e}_{k+1}'\boldsymbol{\Sigma}\mathbf{e}_{k+1}/\mathbf{e}_{k+1}'\mathbf{e}_{k+1} = \mathbf{e}_{k+1}'\boldsymbol{\Sigma}\mathbf{e}_{k+1} = \text{Var}(Y_{k+1})$$

But $\mathbf{e}_{k+1}'(\boldsymbol{\Sigma}\mathbf{e}_{k+1}) = \lambda_{k+1}\mathbf{e}_{k+1}'\mathbf{e}_{k+1} = \lambda_{k+1}$ so $\text{Var}(Y_{k+1}) = \lambda_{k+1}$. It remains to show that $\mathbf{e}_i$ perpendicular to $\mathbf{e}_k$ (that is, $\mathbf{e}_i'\mathbf{e}_k = 0$, $i \neq k$) gives $\text{Cov}(Y_i, Y_k) = 0$. Now, the eigenvectors of $\boldsymbol{\Sigma}$ are orthogonal if all the eigenvalues $\lambda_1, \lambda_2, \ldots, \lambda_p$ are distinct. If the eigenvalues are not all distinct, the eigenvectors corresponding to common eigenvalues may be chosen to be orthogonal. Therefore, for any two eigenvectors $\mathbf{e}_i$ and $\mathbf{e}_k$, $\mathbf{e}_i'\mathbf{e}_k = 0$, $i \neq k$. Since $\boldsymbol{\Sigma}\mathbf{e}_k = \lambda_k\mathbf{e}_k$, premultiplication by $\mathbf{e}_i'$ gives

$$\text{Cov}(Y_i, Y_k) = \mathbf{e}_i'\boldsymbol{\Sigma}\mathbf{e}_k = \mathbf{e}_i'\lambda_k\mathbf{e}_k = \lambda_k\mathbf{e}_i'\mathbf{e}_k = 0$$

for any $i \neq k$, and the proof is complete.   ∎

From Result 8.1, the principal components are uncorrelated and have variances equal to the eigenvalues of $\boldsymbol{\Sigma}$.

Result 8.2. Let $\mathbf{X}' = [X_1, X_2, \ldots, X_p]$ have covariance matrix $\boldsymbol{\Sigma}$, with eigenvalue–eigenvector pairs $(\lambda_1, \mathbf{e}_1)$, $(\lambda_2, \mathbf{e}_2), \ldots, (\lambda_p, \mathbf{e}_p)$ where $\lambda_1 \geq \lambda_2 \geq \cdots \geq \lambda_p \geq 0$. Let $Y_1 = \mathbf{e}_1'\mathbf{X}, Y_2 = \mathbf{e}_2'\mathbf{X}, \ldots, Y_p = \mathbf{e}_p'\mathbf{X}$ be the principal components. Then

$$\sigma_{11} + \sigma_{22} + \cdots + \sigma_{pp} = \sum_{i=1}^{p} \text{Var}(X_i) = \lambda_1 + \lambda_2 + \cdots + \lambda_p = \sum_{i=1}^{p} \text{Var}(Y_i)$$

Proof. From Definition 2A.28, $\sigma_{11} + \sigma_{22} + \cdots + \sigma_{pp} = \text{tr}(\boldsymbol{\Sigma})$. From (2-20) with $\mathbf{A} = \boldsymbol{\Sigma}$, we can write $\boldsymbol{\Sigma} = \mathbf{P}\boldsymbol{\Lambda}\mathbf{P}'$ where $\boldsymbol{\Lambda}$ is the diagonal matrix of eigenvalues and $\mathbf{P} = [\mathbf{e}_1, \mathbf{e}_2, \ldots, \mathbf{e}_p]$ so that $\mathbf{PP}' = \mathbf{P}'\mathbf{P} = \mathbf{I}$. Using Result 2A.12(c), we have

$$\text{tr}(\boldsymbol{\Sigma}) = \text{tr}(\mathbf{P}\boldsymbol{\Lambda}\mathbf{P}') = \text{tr}(\boldsymbol{\Lambda}\mathbf{P}'\mathbf{P}) = \text{tr}(\boldsymbol{\Lambda}) = \lambda_1 + \lambda_2 + \cdots + \lambda_p$$

Thus,
$$\text{tr}(AC) = \text{tr}(CAB) \text{ etc.}$$

$$\sum_{i=1}^{p} \text{Var}(X_i) = \text{tr}(\boldsymbol{\Sigma}) = \text{tr}(\boldsymbol{\Lambda}) = \sum_{i=1}^{p} \text{Var}(Y_i) \qquad \blacksquare$$

Result 8.2 says that
$$\therefore \text{total var} = \text{sum of eigenvalues}$$

$$\text{Total population variance} = \sigma_{11} + \sigma_{22} + \cdots + \sigma_{pp}$$
$$= \lambda_1 + \lambda_2 + \cdots + \lambda_p \qquad (8\text{-}6)$$

and consequently, the proportion of total variance due to (explained by) the kth principal component is

$$\begin{pmatrix} \text{Proportion of total} \\ \text{population variance} \\ \text{due to } k\text{th principal} \\ \text{component} \end{pmatrix} = \frac{\lambda_k}{\lambda_1 + \lambda_2 + \cdots + \lambda_p} \qquad k = 1, 2, \ldots, p \quad (8\text{-}7)$$

If most (for instance, 80 to 90%) of the total population variance, for large p, can be attributed to the first one, two, or three components, then these components can "replace" the original p variables without much loss of information.

Each component of the coefficient vector $\mathbf{e}_i' = [e_{i1}, \ldots, e_{ik}, \ldots, e_{ip}]$ also merits inspection. The magnitude of e_{ik} measures the importance of the kth variable to the ith principal component, irrespective of the other variables. In particular, e_{ik} is proportional to the correlation coefficient between Y_i and X_k.

PC ✱ **Result 8.3.** If $Y_1 = \mathbf{e}_1'\mathbf{X}, Y_2 = \mathbf{e}_2'\mathbf{X}, \ldots, Y_p = \mathbf{e}_p'\mathbf{X}$ are the principal components obtained from the covariance matrix $\boldsymbol{\Sigma}$, then

$$\rho_{Y_i, X_k} = \frac{e_{ik}\sqrt{\lambda_i}}{\sqrt{\sigma_{kk}}} \qquad i, k = 1, 2, \ldots, p \qquad (8\text{-}8)$$

are the correlation coefficients between the components Y_i and the variables X_k. Here $(\lambda_1, \mathbf{e}_1), (\lambda_2, \mathbf{e}_2), \ldots, (\lambda_p, \mathbf{e}_p)$ are the eigenvalue–eigenvector pairs for $\boldsymbol{\Sigma}$.

Proof. Set $\mathbf{a}_k' = [0, \ldots, 0, 1, 0, \ldots, 0]$ so that $X_k = \mathbf{a}_k'\mathbf{X}$ and $\text{Cov}(X_k, Y_i) = \text{Cov}(\mathbf{a}_k'\mathbf{X}, \mathbf{e}_i'\mathbf{X}) = \mathbf{a}_k'\boldsymbol{\Sigma}\mathbf{e}_i$, according to (2-45). Since $\boldsymbol{\Sigma}\mathbf{e}_i = \lambda_i\mathbf{e}_i$, $\text{Cov}(X_k, Y_i) = \mathbf{a}_k'\lambda_i\mathbf{e}_i = \lambda_i e_{ik}$. Then $\text{Var}(Y_i) = \lambda_i$ [see (8-5)] and $\text{Var}(X_k) = \sigma_{kk}$ yield

$$\rho_{Y_i, X_k} = \frac{\text{Cov}(Y_i, X_k)}{\sqrt{\text{Var}(Y_i)} \sqrt{\text{Var}(X_k)}} = \frac{\lambda_i e_{ik}}{\sqrt{\lambda_i} \sqrt{\sigma_{kk}}} = \frac{e_{ik}\sqrt{\lambda_i}}{\sqrt{\sigma_{kk}}} \qquad i, k = 1, 2, \ldots, p \quad \blacksquare$$

Although the correlations of the variables with the principal components often help to interpret the components, they measure only the univariate contribution of an individual X to a component Y. That is, they do not indicate the importance of an X to a component Y in the presence of the other X's. For this reason, some statisticians (see, for example, Rencher [16]) recommend that only the coefficients e_{ik}, and not the correlations, be used to interpret the components. Although the coefficients and the correlations can lead to different rankings as measures of the importance of the variables to a given component, it is our experience that these rankings are often not *appreciably* different. In practice, variables with relatively large coefficients (in absolute value) tend to have relatively large correlations, so the two measures of importance, the first multivariate and the second univariate, frequently give similar results. We recommend that both the coefficients and the correlations be examined to help interpret the principal components.

The following hypothetical example illustrates the contents of Results 8.1, 8.2, and 8.3.

Example 8.1 (Calculating the population principal components) Suppose the random variables X_1, X_2 and X_3 have the covariance matrix

$$\Sigma = \begin{bmatrix} 1 & -2 & 0 \\ -2 & 5 & 0 \\ 0 & 0 & 2 \end{bmatrix}$$

It may be verified that the eigenvalue–eigenvector pairs are

$$\lambda_1 = 5.83, \quad e_1' = [.383, -.924, 0]$$
$$\lambda_2 = 2.00, \quad e_2' = [0, 0, 1]$$
$$\lambda_3 = 0.17, \quad e_3' = [.924, .383, 0]$$

Therefore, the principal components become

$$Y_1 = e_1'\mathbf{X} = .383X_1 - .924X_2$$
$$Y_2 = e_2'\mathbf{X} = X_3$$
$$Y_3 = e_3'\mathbf{X} = .924X_1 + .383X_2$$

The variable X_3 is one of the principal components, because it is uncorrelated with the other two variables.

Equation (8-5) can be demonstrated from first principles. For example,

$$\begin{aligned} \text{Var}(Y_1) &= \text{Var}(.383X_1 - .924X_2) \\ &= (.383)^2 \text{Var}(X_1) + (-.924)^2 \text{Var}(X_2) \\ &\quad + 2(.383)(-.924) \text{Cov}(X_1, X_2) \\ &= .147(1) + .854(5) - .708(-2) \\ &= 5.83 = \lambda_1 \end{aligned}$$

$$\begin{aligned} \text{Cov}(Y_1, Y_2) &= \text{Cov}(.383X_1 - .924X_2, X_3) \\ &= .383 \text{Cov}(X_1, X_3) - .924 \text{Cov}(X_2, X_3) \\ &= .383(0) - .924(0) = 0 \end{aligned}$$

It is also readily apparent that

$$\sigma_{11} + \sigma_{22} + \sigma_{33} = 1 + 5 + 2 = \lambda_1 + \lambda_2 + \lambda_3 = 5.83 + 2.00 + .17$$

validating Equation (8-6) for this example. The proportion of total variance accounted for by the first principal component is $\lambda_1/(\lambda_1 + \lambda_2 + \lambda_3) = 5.83/8 = .73$. Further, the first two components account for a proportion $(5.83 + 2)/8 = .98$ of the population variance. In this case, the components Y_1 and Y_2 could replace the original three variables with little loss of information.

Next, using (8-8), we obtain

$$\rho_{Y_1, X_1} = \frac{e_{11}\sqrt{\lambda_1}}{\sqrt{\sigma_{11}}} = \frac{.383\sqrt{5.83}}{\sqrt{1}} = .925$$

$$\rho_{Y_1, X_2} = \frac{e_{12}\sqrt{\lambda_1}}{\sqrt{\sigma_{22}}} = \frac{-.924\sqrt{5.83}}{\sqrt{5}} = -.998$$

Notice here that the variable X_2, with coefficient $-.924$, receives the greatest weight in the component Y_1. It also has the largest correlation (in absolute value) with Y_1. The correlation of X_1, with Y_1, $.925$, is almost as large as that for X_2, indicating that the variables are about equally important to the first principal component. The relative sizes of the coefficients of X_1 and X_2 suggest, however, that X_2 contributes more to the determination of Y_1 than does X_1. Since, in this case, both coefficients are reasonably large and they have opposite signs, we would argue that both variables aid in the interpretation of Y_1.

Finally,

$$\rho_{Y_2, X_1} = \rho_{Y_2, X_2} = 0 \quad \text{and} \quad \rho_{Y_2, X_3} = \frac{\sqrt{\lambda_2}}{\sqrt{\sigma_{33}}} = \frac{\sqrt{2}}{\sqrt{2}} = 1 \quad \text{(as it should)}$$

The remaining correlations can be neglected, since the third component is unimportant. ∎

It is informative to consider principal components derived from multivariate normal random variables. Suppose $\mathbf{X}$ is distributed as $N_p(\boldsymbol{\mu}, \boldsymbol{\Sigma})$. We know from (4-7) that the density of $\mathbf{X}$ is constant on the $\boldsymbol{\mu}$ centered ellipsoids

$$(\mathbf{x} - \boldsymbol{\mu})'\boldsymbol{\Sigma}^{-1}(\mathbf{x} - \boldsymbol{\mu}) = c^2$$

which have axes $\pm c\sqrt{\lambda_i}\,\mathbf{e}_i$, $i = 1, 2, \ldots, p$, where the $(\lambda_i, \mathbf{e}_i)$ are the eigenvalue–eigenvector pairs of $\boldsymbol{\Sigma}$. A point lying on the ith axis of the ellipsoid will have coordinates proportional to $\mathbf{e}_i' = [e_{i1}, e_{i2}, \ldots, e_{ip}]$ in the coordinate system that has origin $\boldsymbol{\mu}$ and axes that are parallel to the original axes $x_1, x_2, \ldots, x_p$. It will be convenient to set $\boldsymbol{\mu} = \mathbf{0}$ in the argument that follows.[1]

From our discussion in Section 2.3 with $\mathbf{A} = \boldsymbol{\Sigma}^{-1}$, we can write

$$c^2 = \mathbf{x}'\boldsymbol{\Sigma}^{-1}\mathbf{x} = \frac{1}{\lambda_1}(\mathbf{e}_1'\mathbf{x})^2 + \frac{1}{\lambda_2}(\mathbf{e}_2'\mathbf{x})^2 + \cdots + \frac{1}{\lambda_p}(\mathbf{e}_p'\mathbf{x})^2$$

[1]This can be done without loss of generality because the normal random vector $\mathbf{X}$ can always be translated to the normal random vector $\mathbf{W} = \mathbf{X} - \boldsymbol{\mu}$ and $E(\mathbf{W}) = \mathbf{0}$. However, $\text{Cov}(\mathbf{X}) = \text{Cov}(\mathbf{W})$.

where $\mathbf{e}_1'\mathbf{x}, \mathbf{e}_2'\mathbf{x}, \ldots, \mathbf{e}_p'\mathbf{x}$ are recognized as the principal components of $\mathbf{x}$. Setting $y_1 = \mathbf{e}_1'\mathbf{x}, y_2 = \mathbf{e}_2'\mathbf{x}, \ldots, y_p = \mathbf{e}_p'\mathbf{x}$, we have

$$c^2 = \frac{1}{\lambda_1} y_1^2 + \frac{1}{\lambda_2} y_2^2 + \cdots + \frac{1}{\lambda_p} y_p^2$$

and this equation defines an ellipsoid (since $\lambda_1, \lambda_2, \ldots, \lambda_p$ are positive) in a coordinate system with axes $y_1, y_2, \ldots, y_p$ lying in the directions of $\mathbf{e}_1, \mathbf{e}_2, \ldots, \mathbf{e}_p$, respectively. If λ_1 is the largest eigenvalue, then the major axis lies in the direction $\mathbf{e}_1$. The remaining minor axes lie in the directions defined by $\mathbf{e}_2, \ldots, \mathbf{e}_p$.

To summarize, the principal components $y_1 = \mathbf{e}_1'\mathbf{x}, y_2 = \mathbf{e}_2'\mathbf{x}, \ldots, y_p = \mathbf{e}_p'\mathbf{x}$ lie in the directions of the axes of a constant density ellipsoid. Therefore, any point on the ith ellipsoid axis has $\mathbf{x}$ coordinates proportional to $\mathbf{e}_i' = [e_{i1}, e_{i2}, \ldots, e_{ip}]$ and, necessarily, principal component coordinates of the form $[0, \ldots, 0, y_i, 0, \ldots, 0]$.

When $\boldsymbol{\mu} \neq \mathbf{0}$, it is the mean-centered principal component $y_i = \mathbf{e}_i'(\mathbf{x} - \boldsymbol{\mu})$ that has mean 0 and lies in the direction $\mathbf{e}_i$.

A constant density ellipse and the principal components for a bivariate normal random vector with $\boldsymbol{\mu} = \mathbf{0}$ and $\rho = .75$ are shown in Figure 8.1. We see that the principal components are obtained by rotating the original coordinate axes through an angle θ until they coincide with the axes of the constant density ellipse. This result holds for $p > 2$ dimensions as well.

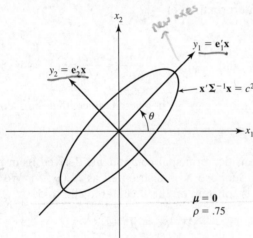

Figure 8.1 The constant density ellipse $\mathbf{x}'\boldsymbol{\Sigma}^{-1}\mathbf{x} = c^2$ and the principal components y_1, y_2 for a bivariate normal random vector $\mathbf{X}$ having mean $\mathbf{0}$.

Principal Components Obtained from Standardized Variables

Principal components may also be obtained for the standardized variables

$$Z_1 = \frac{(X_1 - \mu_1)}{\sqrt{\sigma_{11}}}$$

$$Z_2 = \frac{(X_2 - \mu_2)}{\sqrt{\sigma_{22}}}$$

$$\vdots \qquad \vdots$$

$$Z_p = \frac{(X_p - \mu_p)}{\sqrt{\sigma_{pp}}}$$

(8-9)

In matrix notation,

$$\mathbf{Z} = (\mathbf{V}^{1/2})^{-1}(\mathbf{X} - \boldsymbol{\mu}) \qquad (8\text{-}10)$$

where the diagonal standard deviation matrix $\mathbf{V}^{1/2}$ is defined in (2-35). Clearly, $E(\mathbf{Z}) = \mathbf{0}$ and

$$\text{Cov}(\mathbf{Z}) = (\mathbf{V}^{1/2})^{-1}\boldsymbol{\Sigma}(\mathbf{V}^{1/2})^{-1} = \boldsymbol{\rho}$$

by (2-37). The principal components of $\mathbf{Z}$ may be obtained from the eigenvectors of the *correlation* matrix $\boldsymbol{\rho}$ of $\mathbf{X}$. All our previous results apply, with some simplifications, since the variance of each Z_i is unity. We shall continue to use the notation Y_i to refer to the ith principal component and $(\lambda_i, \mathbf{e}_i)$ for the eigenvalue–eigenvector pair from either $\boldsymbol{\rho}$ or $\boldsymbol{\Sigma}$. *However, the $(\lambda_i, \mathbf{e}_i)$ derived from $\boldsymbol{\Sigma}$ are, in general, not the same as the ones derived from $\boldsymbol{\rho}$.*

Result 8.4. The ith principal component of the standardized variables $\mathbf{Z}' = [Z_1, Z_2, \ldots, Z_p]$ with $\text{Cov}(\mathbf{Z}) = \boldsymbol{\rho}$, is given by

$$Y_i = \mathbf{e}_i'\mathbf{Z} = \mathbf{e}_i'(\mathbf{V}^{1/2})^{-1}(\mathbf{X} - \boldsymbol{\mu}), \qquad i = 1, 2, \ldots, p$$

Moreover,

$$\sum_{i=1}^{p} \text{Var}(Y_i) = \sum_{i=1}^{p} \text{Var}(Z_i) = p \qquad (8\text{-}11)$$

and

$$\rho_{Y_i, Z_k} = e_{ik}\sqrt{\lambda_i} \qquad i, k = 1, 2, \ldots, p$$

In this case, $(\lambda_1, \mathbf{e}_1), (\lambda_2, \mathbf{e}_2), \ldots, (\lambda_p, \mathbf{e}_p)$ are the eigenvalue–eigenvector pairs for $\boldsymbol{\rho}$, with $\lambda_1 \geq \lambda_2 \geq \cdots \geq \lambda_p \geq 0$.

Proof. Result 8.4 follows from Results 8.1, 8.2, and 8.3, with $Z_1, Z_2, \ldots, Z_p$ in place of $X_1, X_2, \ldots, X_p$ and $\boldsymbol{\rho}$ in place of $\boldsymbol{\Sigma}$. ∎

We see from (8-11) that the total (standardized variables) population variance is simply p, the sum of the diagonal elements of the matrix $\boldsymbol{\rho}$. Using (8-7) with $\mathbf{Z}$ in place of $\mathbf{X}$, we find that the proportion of total variance explained by the kth principal component of $\mathbf{Z}$ is

$$\begin{pmatrix} \text{Proportion of (standardized)} \\ \text{population variance due} \\ \text{to } k\text{th principal component} \end{pmatrix} = \frac{\lambda_k}{p}, \qquad k = 1, 2, \ldots, p \qquad (8\text{-}12)$$

where the λ_k's are the eigenvalues of $\boldsymbol{\rho}$.

Example 8.2 (Principal components obtained from covariance and correlation matrices are different) Consider the covariance matrix

$$\boldsymbol{\Sigma} = \begin{bmatrix} 1 & 4 \\ 4 & 100 \end{bmatrix}$$

and the derived correlation matrix

$$\boldsymbol{\rho} = \begin{bmatrix} 1 & .4 \\ .4 & 1 \end{bmatrix}$$

The eigenvalue–eigenvector pairs from $\boldsymbol{\Sigma}$ are

$$\lambda_1 = 100.16, \quad \mathbf{e}_1' = [.040, .999]$$
$$\lambda_2 = \quad .84, \quad \mathbf{e}_2' = [.999, -.040]$$

Similarly, the eigenvalue–eigenvector pairs from $\boldsymbol{\rho}$ are

$$\lambda_1 = 1 + \rho = 1.4, \quad \mathbf{e}_1' = [.707, .707]$$
$$\lambda_2 = 1 - \rho = \quad .6, \quad \mathbf{e}_2' = [.707, -.707]$$

The respective principal components become

$$\boldsymbol{\Sigma}: \quad \begin{aligned} Y_1 &= .040X_1 + .999X_2 \\ Y_2 &= .999X_1 - .040X_2 \end{aligned}$$

and

$$Y_1 = .707Z_1 + .707Z_2 = .707\left(\frac{X_1 - \mu_1}{1}\right) + .707\left(\frac{X_2 - \mu_2}{10}\right)$$

$$\boldsymbol{\rho}: \qquad\qquad = .707(X_1 - \mu_1) + .0707(X_2 - \mu_2)$$

$$Y_2 = .707Z_1 - .707Z_2 = .707\left(\frac{X_1 - \mu_1}{1}\right) - .707\left(\frac{X_2 - \mu_2}{10}\right)$$

$$= .707(X_1 - \mu_1) - .0707(X_2 - \mu_2)$$

Because of its large variance, X_2 completely dominates the first principal component determined from $\boldsymbol{\Sigma}$. Moreover, this first principal component explains a proportion

$$\frac{\lambda_1}{\lambda_1 + \lambda_2} = \frac{100.16}{101} = .992$$

of the total population variance.

When the variables X_1 and X_2 are standardized, however, the resulting variables contribute equally to the principal components determined from $\boldsymbol{\rho}$. Using Result 8.4, we obtain

$$\rho_{Y_1, Z_1} = e_{11}\sqrt{\lambda_1} = .707\sqrt{1.4} = .837$$

and

$$\rho_{Y_1, Z_2} = e_{21}\sqrt{\lambda_1} = .707\sqrt{1.4} = .837$$

In this case, the first principal component explains a proportion

$$\frac{\lambda_1}{p} = \frac{1.4}{2} = .7$$

of the total (standardized) population variance.

Most strikingly, we see that the relative importance of the variables to, for instance, the first principal component is greatly affected by the standardization.

When the first principal component obtained from $\boldsymbol{\rho}$ is expressed in terms of X_1 and X_2, the relative magnitudes of the weights .707 and .0707 are in direct opposition to those of the weights .040 and .999 attached to these variables in the principal component obtained from $\boldsymbol{\Sigma}$. ∎

The preceding example demonstrates that the principal components derived from $\boldsymbol{\Sigma}$ are different from those derived from $\boldsymbol{\rho}$. Furthermore, one set of principal components is not a simple function of the other. This suggests that the standardization is not inconsequential.

Variables should probably be standardized if they are measured on scales with widely differing ranges or if the units of measurement are not commensurate. For example, if X_1 represents annual sales in the $10,000 to $350,000 range and X_2 is the ratio (net annual income)/(total assets) that falls in the .01 to .60 range, then the total variation will be due almost exclusively to dollar sales. In this case, we would expect a single (important) principal component with a heavy weighting of X_1. Alternatively, if both variables are standardized, their subsequent magnitudes will be of the same order, and X_2 (or Z_2) will play a larger role in the construction of the principal components. This behavior was observed in Example 8.2.

Principal Components for Covariance Matrices with Special Structures

There are certain patterned covariance and correlation matrices whose principal components can be expressed in simple forms. Suppose $\boldsymbol{\Sigma}$ is the diagonal matrix

$$\boldsymbol{\Sigma} = \begin{bmatrix} \sigma_{11} & 0 & \cdots & 0 \\ 0 & \sigma_{22} & \cdots & 0 \\ \vdots & \vdots & \ddots & \vdots \\ 0 & 0 & \cdots & \sigma_{pp} \end{bmatrix} \tag{8-13}$$

Setting $\mathbf{e}'_i = [0, \ldots, 0, 1, 0, \ldots, 0]$, with 1 in the ith position, we observe that

$$\begin{bmatrix} \sigma_{11} & 0 & \cdots & 0 \\ 0 & \sigma_{22} & \cdots & 0 \\ \vdots & \vdots & \ddots & \vdots \\ 0 & 0 & \cdots & \sigma_{pp} \end{bmatrix} \begin{bmatrix} 0 \\ \vdots \\ 0 \\ 1 \\ 0 \\ \vdots \\ 0 \end{bmatrix} = \begin{bmatrix} 0 \\ \vdots \\ 0 \\ 1\sigma_{ii} \\ 0 \\ \vdots \\ 0 \end{bmatrix} \quad \text{or} \quad \boldsymbol{\Sigma}\mathbf{e}_i = \sigma_{ii}\mathbf{e}_i$$

and we conclude that $(\sigma_{ii}, \mathbf{e}_i)$ is the ith eigenvalue–eigenvector pair. Since the linear combination $\mathbf{e}'_i \mathbf{X} = X_i$, the set of principal components is just the original set of uncorrelated random variables.

For a covariance matrix with the pattern of (8-13), nothing is gained by extracting the principal components. From another point of view, if $\mathbf{X}$ is distributed as $N_p(\boldsymbol{\mu}, \boldsymbol{\Sigma})$, the contours of constant density are ellipsoids whose axes already lie in the directions of maximum variation. Consequently, there is no need to rotate the coordinate system.

Standardization does not substantially alter the situation for the Σ in (8-13). In that case, $\boldsymbol{\rho} = \mathbf{I}$, the $p \times p$ identity matrix. Clearly, $\boldsymbol{\rho}\mathbf{e}_i = 1\mathbf{e}_i$, so the eigenvalue 1 has multiplicity p and $\mathbf{e}_i' = [0, \ldots, 0, 1, 0, \ldots, 0]$, $i = 1, 2, \ldots, p$, are convenient choices for the eigenvectors. Consequently, the principal components determined from $\boldsymbol{\rho}$ are also the original variables $Z_1, \ldots, Z_p$. Moreover, in this case of equal eigenvalues, the multivariate normal ellipsoids of constant density are spheroids.

Another patterned covariance matrix, which often describes the correspondence among certain biological variables such as the sizes of living things, has the general form

covar

$$\Sigma = \begin{bmatrix} \sigma^2 & \rho\sigma^2 & \cdots & \rho\sigma^2 \\ \rho\sigma^2 & \sigma^2 & \cdots & \rho\sigma^2 \\ \vdots & \vdots & \ddots & \vdots \\ \rho\sigma^2 & \rho\sigma^2 & \cdots & \sigma^2 \end{bmatrix} \tag{8-14}$$

The resulting correlation matrix

dit is divide by σ^{-2}

correlation

$$\boldsymbol{\rho} = \begin{bmatrix} 1 & \rho & \cdots & \rho \\ \rho & 1 & \cdots & \rho \\ \vdots & \vdots & \ddots & \vdots \\ \rho & \rho & \cdots & 1 \end{bmatrix} \tag{8-15}$$

is also the covariance matrix of the standardized variables. The matrix in (8-15) implies that the variables $X_1, X_2, \ldots, X_p$ are equally correlated.

It is not difficult to show (see Exercise 8.5) that the p eigenvalues of the correlation matrix (8-15) can be divided into two groups. When ρ is positive, the largest is

$$\lambda_1 = 1 + (p - 1)\rho \tag{8-16}$$

with associated eigenvector

$$\mathbf{e}_1' = \left[\frac{1}{\sqrt{p}}, \frac{1}{\sqrt{p}}, \ldots, \frac{1}{\sqrt{p}} \right] \tag{8-17}$$

The remaining $p - 1$ eigenvalues are

$$\lambda_2 = \lambda_3 = \cdots = \lambda_p = 1 - \rho$$

and one choice for their eigenvectors is → *any multiple ok ; want* $\frac{1}{2}$ *norm =1*

$$\mathbf{e}_2' = \left[\frac{1}{\sqrt{1 \times 2}}, \frac{-1}{\sqrt{1 \times 2}}, 0, \ldots, 0 \right]$$

$$\mathbf{e}_3' = \left[\frac{1}{\sqrt{2 \times 3}}, \frac{1}{\sqrt{2 \times 3}}, \frac{-2}{\sqrt{2 \times 3}}, 0, \ldots, 0 \right]$$

$$\vdots \qquad \vdots$$

$$\mathbf{e}_i' = \left[\frac{1}{\sqrt{(i-1)i}}, \ldots, \frac{1}{\sqrt{(i-1)i}}, \frac{-(i-1)}{\sqrt{(i-1)i}}, 0, \ldots, 0 \right]$$

$$\vdots \qquad \vdots$$

$$\mathbf{e}_p' = \left[\frac{1}{\sqrt{(p-1)p}}, \ldots, \frac{1}{\sqrt{(p-1)p}}, \frac{-(p-1)}{\sqrt{(p-1)p}} \right]$$

The first principal component

$$Y_1 = \mathbf{e}_1' \mathbf{Z} = \frac{1}{\sqrt{p}} \sum_{i=1}^{p} Z_i$$

is proportional to the sum of the p standarized variables. It might be regarded as an "index" with equal weights. This principal component explains a proportion

$$\frac{\lambda_1}{p} = \frac{1 + (p - 1)\rho}{p} = \rho + \frac{1 - \rho}{p} \qquad (8\text{-}18)$$

of the total population variation. We see that $\lambda_1/p \doteq \rho$ for ρ close to 1 or p large. For example, if $\rho = .80$ and $p = 5$, the first component explains 84% of the total variance. When ρ is near 1, the last $p - 1$ components collectively contribute very little to the total variance and can often be neglected. In this special case, retaining only the first principal component $Y_1 = (1/\sqrt{p})[1, 1, \ldots, 1]\mathbf{X}$, a measure of total size, still explains the same proportion (8-18) of total variance.

If the standardized variables $Z_1, Z_2, \ldots, Z_p$ have a multivariate normal distribution with a covariance matrix given by (8-15), then the ellipsoids of constant density are "cigar shaped," with the major axis proportional to the first principal component $Y_1 = (1/\sqrt{p})[1, 1, \ldots, 1]\mathbf{Z}$. This principal component is the projection of $\mathbf{Z}$ on the equiangular line $\mathbf{1}' = [1, 1, \ldots, 1]$. The minor axes (and remaining principal components) occur in spherically symmetric directions perpendicular to the major axis (and first principal component).

8.3 Summarizing Sample Variation by Principal Components

We now have the framework necessary to study the problem of summarizing the variation in n measurements on p variables with a few judiciously chosen linear combinations.

Suppose the data $\mathbf{x}_1, \mathbf{x}_2, \ldots, \mathbf{x}_n$ represent n independent drawings from some p-dimensional population with mean vector $\boldsymbol{\mu}$ and covariance matrix $\boldsymbol{\Sigma}$. These data yield the sample mean vector $\bar{\mathbf{x}}$, the sample covariance matrix $\mathbf{S}$, and the sample correlation matrix $\mathbf{R}$.

Our objective in this section will be to construct uncorrelated linear combinations of the measured characteristics that account for much of the variation in the sample. The uncorrelated combinations with the largest variances will be called the *sample principal components*.

Recall that the n values of any linear combination

$$\mathbf{a}_1' \mathbf{x} = a_{11} x_{j1} + a_{12} x_{j2} + \cdots + a_{1p} x_{jp}, \qquad j = 1, 2, \ldots, n$$

have sample mean $\mathbf{a}_1' \bar{\mathbf{x}}$ and sample variance $\mathbf{a}_1' \mathbf{S} \mathbf{a}_1$. Also, the pairs of values $(\mathbf{a}_1' \mathbf{x}_j, \mathbf{a}_2' \mathbf{x}_j)$, for two linear combinations, have sample covariance $\mathbf{a}_1' \mathbf{S} \mathbf{a}_2$ [see (3-36)].

The sample principal components are defined as those linear combinations which have maximum sample variance. As with the population quantities, we restrict the coefficient vectors $\mathbf{a}_i$ to satisfy $\mathbf{a}_i'\mathbf{a}_i = 1$. Specifically,

First *sample*
principal component = linear combination $\mathbf{a}_1'\mathbf{x}_j$ that maximizes the sample variance of $\mathbf{a}_1'\mathbf{x}_j$ subject to $\mathbf{a}_1'\mathbf{a}_1 = 1$

Second *sample*
principal component = linear combination $\mathbf{a}_2'\mathbf{x}_j$ that maximizes the sample variance of $\mathbf{a}_2'\mathbf{x}_j$ subject to $\mathbf{a}_2'\mathbf{a}_2 = 1$ and zero sample covariance for the pairs $(\mathbf{a}_1'\mathbf{x}_j, \mathbf{a}_2'\mathbf{x}_j)$

At the ith step, we have

ith *sample*
principal component = linear combination $\mathbf{a}_i'\mathbf{x}_j$ that maximizes the sample variance of $\mathbf{a}_i'\mathbf{x}_j$ subject to $\mathbf{a}_i'\mathbf{a}_i = 1$ and zero sample covariance for all pairs $(\mathbf{a}_i'\mathbf{x}_j, \mathbf{a}_k'\mathbf{x}_j)$, $k < i$

The first principal component maximizes $\mathbf{a}_1'\mathbf{S}\mathbf{a}_1$ or, equivalently,

$$\frac{\mathbf{a}_1'\mathbf{S}\mathbf{a}_1}{\mathbf{a}_1'\mathbf{a}_1} \tag{8-19}$$

By (2-51), the maximum is the largest eigenvalue $\hat{\lambda}_1$ attained for the choice $\mathbf{a}_1 = $ eigenvector $\hat{\mathbf{e}}_1$ of $\mathbf{S}$. Successive choices of $\mathbf{a}_i$ maximize (8-19) subject to $0 = \mathbf{a}_i'\mathbf{S}\hat{\mathbf{e}}_k = \mathbf{a}_i'\hat{\lambda}_k\hat{\mathbf{e}}_k$, or $\mathbf{a}_i$ perpendicular to $\hat{\mathbf{e}}_k$. Thus, as in the proofs of Results 8.1–8.3, we obtain the following results concerning sample principal components:

If $\mathbf{S} = \{s_{ik}\}$ is the $p \times p$ sample covariance matrix with eigenvalue-eigenvector pairs $(\hat{\lambda}_1, \hat{\mathbf{e}}_1), (\hat{\lambda}_2, \hat{\mathbf{e}}_2), \ldots, (\hat{\lambda}_p, \hat{\mathbf{e}}_p)$, the ith sample principal component is given by

$$\hat{y}_i = \hat{\mathbf{e}}_i'\mathbf{x} = \hat{e}_{i1}x_1 + \hat{e}_{i2}x_2 + \cdots + \hat{e}_{ip}x_p, \qquad i = 1, 2, \ldots, p$$

where $\hat{\lambda}_1 \geq \hat{\lambda}_2 \geq \cdots \geq \hat{\lambda}_p \geq 0$ and $\mathbf{x}$ is any observation on the variables $X_1, X_2, \ldots, X_p$. Also,

$$\text{Sample variance}(\hat{y}_k) = \hat{\lambda}_k, \quad k = 1, 2, \ldots, p$$
$$\text{Sample covariance}(\hat{y}_i, \hat{y}_k) = 0, \quad i \neq k$$

In addition, $\tag{8-20}$

$$\text{Total sample variance} = \sum_{i=1}^{p} s_{ii} = \hat{\lambda}_1 + \hat{\lambda}_2 + \cdots + \hat{\lambda}_p$$

and

$$r_{\hat{y}_i, x_k} = \frac{\hat{e}_{ik}\sqrt{\hat{\lambda}_i}}{\sqrt{s_{kk}}}, \quad i, k = 1, 2, \ldots, p$$

sqrt of var (diagelements)

We shall denote the sample principal components by $\hat{y}_1, \hat{y}_2, \ldots, \hat{y}_p$, irrespective of whether they are obtained from **S** or **R**.[2] The components constructed from **S** and **R** are *not* the same, in general, but it will be clear from the context which matrix is being used, and the single notation $\hat{y}_i$ is convenient. It is also convenient to label the component coefficient vectors $\hat{\mathbf{e}}_i$ and the component variances $\hat{\lambda}_i$ for both situations.

The observations $\mathbf{x}_j$ are often "centered" by subtracting $\bar{\mathbf{x}}$. This has no effect on the sample covariance matrix **S** and gives the ith principal component

$$\hat{y}_i = \hat{\mathbf{e}}_i'(\mathbf{x} - \bar{\mathbf{x}}), \qquad i = 1, 2, \ldots, p \qquad (8\text{-}21)$$

for any observation vector $\mathbf{x}$. If we consider the *values* of the ith component

$$\hat{y}_{ji} = \hat{\mathbf{e}}_i'(\mathbf{x}_j - \bar{\mathbf{x}}), \qquad j = 1, 2, \ldots, n \qquad (8\text{-}22)$$

generated by substituting each observation $\mathbf{x}_j$ for the arbitrary $\mathbf{x}$ in (8-21), then

$$\bar{\hat{y}}_i = \frac{1}{n} \sum_{j=1}^{n} \hat{\mathbf{e}}_i'(\mathbf{x}_j - \bar{\mathbf{x}}) = \frac{1}{n} \hat{\mathbf{e}}_i' \left(\sum_{j=1}^{n} (\mathbf{x}_j - \bar{\mathbf{x}}) \right) = \frac{1}{n} \hat{\mathbf{e}}_i' \mathbf{0} = 0 \qquad (8\text{-}23)$$

That is, the sample mean of each principal component is zero. The sample variances are still given by the $\hat{\lambda}_i$'s, as in (8-20).

Example 8.3 (Summarizing sample variability with two sample principal components)
A census provided information, by tract, on five socioeconomic variables for the Madison, Wisconsin, area. The data from 61 tracts are listed in Table 8.5 in the exercises at the end of this chapter. These data produced the following summary statistics:

$\bar{\mathbf{x}}' =$	[4.47,	3.96,	71.42,	26.91,	1.64]
	total population (thousands)	professional degree (percent)	employed age over 16 (percent)	government employment (percent)	median home value ($100,000)

sample mean vector

and

$$\mathbf{S} = \begin{bmatrix} 3.397 & -1.102 & 4.306 & -2.078 & 0.027 \\ -1.102 & 9.673 & -1.513 & 10.953 & 1.203 \\ 4.306 & -1.513 & 55.626 & -28.937 & -0.044 \\ -2.078 & 10.953 & -28.937 & 89.067 & 0.957 \\ 0.027 & 1.203 & -0.044 & 0.957 & 0.319 \end{bmatrix}$$

Can the sample variation be summarized by one or two principal components?

---

[2] Sample principal components also can be obtained from $\hat{\boldsymbol{\Sigma}} = \mathbf{S}_n$, the maximum likelihood estimate of the covariance matrix $\boldsymbol{\Sigma}$, if the $\mathbf{X}_j$ are normally distributed. (See Result 4.11.) In this case, provided that the eigenvalues of $\boldsymbol{\Sigma}$ are distinct, the sample principal components can be viewed as the maximum likelihood estimates of the corresponding population counterparts. (See [1].) We shall not consider $\hat{\boldsymbol{\Sigma}}$ because the assumption of normality is not required in this section. Also, $\hat{\boldsymbol{\Sigma}}$ has eigenvalues $[(n-1)/n]\hat{\lambda}_i$ and corresponding eigenvectors $\hat{\mathbf{e}}_i$, where $(\hat{\lambda}_i, \hat{\mathbf{e}}_i)$ are the eigenvalue–eigenvector pairs for **S**. Thus, both **S** and $\hat{\boldsymbol{\Sigma}}$ give the same sample principal components $\hat{\mathbf{e}}_i'\mathbf{x}$ [see (8-20)] and the same proportion of explained variance $\hat{\lambda}_i/(\hat{\lambda}_1 + \hat{\lambda}_2 + \cdots + \hat{\lambda}_p)$. Finally, both **S** and $\hat{\boldsymbol{\Sigma}}$ give the same sample correlation matrix **R**, so if the variables are standardized, the choice of **S** or $\hat{\boldsymbol{\Sigma}}$ is irrelevant.

We find the following:

did in class & f Feb 3 *(handwritten)*

other covariates are barely correlated (handwritten)

Coefficients for the Principal Components
(Correlation Coefficients in Parentheses)

Variable	$\hat{e}_1\,(r_{\hat{y}_1,x_k})$	$\hat{e}_2\,(r_{\hat{y}_2,x_k})$	$\hat{e}_3$	$\hat{e}_4$	$\hat{e}_5$
Total population	$-0.039(-.22)$	$0.071(.24)$	0.188	0.977	-0.058
Profession	$0.105(.35)$	$0.130(.26)$	-0.961	0.171	-0.139
Employment (%)	$-0.492(-.68)$	$0.864(.73)$	0.046	-0.091	0.005
Government employment (%)	$0.863(.95)$	$0.480(.32)$	0.153	-0.030	0.007
Medium home value	$0.009(.16)$	$0.015(.17)$	-0.125	0.082	0.989
Variance $(\hat{\lambda}_i)$:	107.02	39.67	8.37	2.87	0.15
Cumulative percentage of total variance	67.7	92.8	98.1	99.9	1.000

take either two or three components in this case (handwritten)

The first principal component explains 67.7% of the total sample variance. The first two principal components, collectively, explain 92.8% of the total sample variance. Consequently, sample variation is summarized very well by two principal components and a reduction in the data from 61 observations on 5 observations to 61 observations on 2 principal components is reasonable.

Given the foregoing component coefficients, the first principal component, appears to be essentially a weighted difference between the percent employed by government and the percent total employment. The second principal component appears to be a weighted sum of the two. ∎

As we said in our discussion of the population components, the component coefficients $\hat{e}_{ik}$ and the correlations $r_{\hat{y}_i,x_k}$ should both be examined to interpret the principal components. The correlations allow for differences in the variances of the original variables, but only measure the importance of an individual X without regard to the other X's making up the component. We notice in Example 8.3, however, that the correlation coefficients displayed in the table confirm the interpretation provided by the component coefficients.

The Number of Principal Components

There is always the question of how many components to retain. There is no definitive answer to this question. Things to consider include the amount of total sample variance explained, the relative sizes of the eigenvalues (the variances of the sample components), and the subject-matter interpretations of the components. In addition, as we discuss later, a component associated with an eigenvalue near zero and, hence, deemed unimportant, may indicate an unsuspected linear dependency in the data.

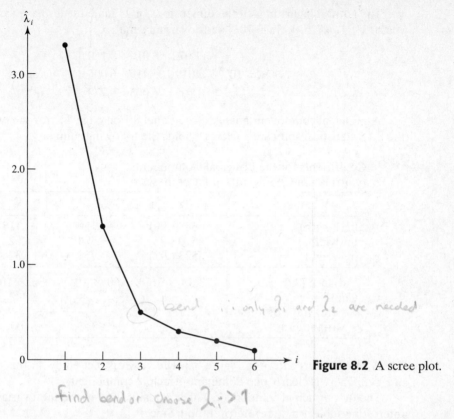

bend ... only λ_1 and λ_2 are needed

Figure 8.2 A scree plot.

find bend or choose $\lambda_i > 1$

A useful visual aid to determining an appropriate number of principal components is a *scree plot*.[3] With the eigenvalues ordered from largest to smallest, a scree plot is a plot of $\hat{\lambda}_i$ versus i—the magnitude of an eigenvalue versus its number. To determine the appropriate number of components, we look for an elbow (bend) in the scree plot. The number of components is taken to be the point at which the remaining eigenvalues are relatively small and all about the same size. Figure 8.2 shows a scree plot for a situation with six principal components.

An elbow occurs in the plot in Figure 8.2 at about $i = 3$. That is, the eigenvalues after $\hat{\lambda}_2$ are all relatively small and about the same size. In this case, it appears, without any other evidence, that two (or perhaps three) sample principal components effectively summarize the total sample variance.

Example 8.4 (Summarizing sample variability with one sample principal component) In a study of size and shape relationships for painted turtles, Jolicoeur and Mosimann [11] measured carapace length, width, and height. Their data, reproduced in Exercise 6.18, Table 6.9, suggest an analysis in terms of logarithms. (Jolicoeur [10] generally suggests a logarithmic transformation in studies of size-and-shape relationships.) Perform a principal component analysis.

---

[3]Scree is the rock debris at the bottom of a cliff.

The natural logarithms of the dimensions of 24 male turtles have sample mean vector $\bar{\mathbf{x}}' = [4.725, 4.478, 3.703]$ and covariance matrix

$$\mathbf{S} = 10^{-3} \begin{bmatrix} 11.072 & 8.019 & 8.160 \\ 8.019 & 6.417 & 6.005 \\ 8.160 & 6.005 & 6.773 \end{bmatrix}$$

A principal component analysis (see Panel 8.1 on page 447 for the output from the SAS statistical software package) yields the following summary:

Coefficients for the Principal Components
(Correlation Coefficients in Parentheses)

Variable	$\hat{\mathbf{e}}_1(r_{\hat{y}_1, x_k})$	$\hat{\mathbf{e}}_2$	$\hat{\mathbf{e}}_3$
ln (length)	.683 (.99)	−.159	−.713
ln (width)	.510 (.97)	−.594	.622
ln (height)	.523 (.97)	.788	.324
Variance $(\hat{\lambda}_i)$:	23.30×10^{-3}	$.60 \times 10^{-3}$	$.36 \times 10^{-3}$
Cumulative percentage of total variance	96.1	98.5	100

A scree plot is shown in Figure 8.3. The very distinct elbow in this plot occurs at $i = 2$. There is clearly one dominant principal component.

The first principal component, which explains 96% of the total variance, has an interesting subject-matter interpretation. Since

$$\hat{y}_1 = .683 \ln(\text{length}) + .510 \ln(\text{width}) + .523 \ln(\text{height})$$
$$= \ln[(\text{length})^{.683}(\text{width})^{.510}(\text{height})^{.523}]$$

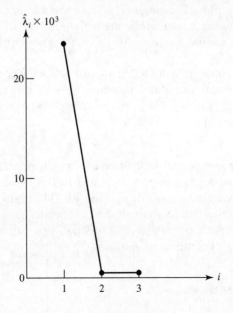

Figure 8.3 A scree plot for the turtle data.

PANEL 8.1 SAS ANALYSIS FOR EXAMPLE 8.4 USING PROC PRINCOMP.

```
title 'Principal Component Analysis';
data turtle;
infile 'E8-4.dat';                              PROGRAM COMMANDS
input length width height;
x1 = log(length); x2 =log(width); x3 =log(height);
proc princomp cov data = turtle out = result;
var x1 x2 x3;
```

Principal Components Analysis

24 Observations   OUTPUT
 3 Variables

Simple Statistics

	X1	X2	X3
Mean	4.725443647	4.477573765	3.703185794
StD	0.105223590	0.080104466	0.082296771

Covariance Matrix

	X1	X2	X3
X1	0.0110720040	0.0080191419	0.0081596480
X2	0.0080191419	0.0064167255	0.0060052707
X3	0.0081596480	0.0060052707	0.0067727585

Total Variance = 0.024261488

Eigenvalues of the Covariance Matrix

	Eigenvalue	Difference	Proportion	Cumulative
PRIN1	0.023303	0.022705	0.960508	0.96051
PRIN2	0.000598	0.000238	0.024661	0.98517
PRIN3	0.000360		0.014832	1.00000

Eigenvectors

	PRIN1	PRIN2	PRIN3
X1	0.683102	−.159479	−.712697
X2	0.510220	−.594012	0.621953
X3	0.522539	0.788490	0.324401

the first principal component may be viewed as the ln (volume) of a box with adjusted dimensions. For instance, the adjusted height is (height)$^{.523}$, which accounts, in some sense, for the rounded shape of the carapace.   ∎

Interpretation of the Sample Principal Components

The sample principal components have several interpretations. First, suppose the underlying distribution of $\mathbf{X}$ is nearly $N_p(\boldsymbol{\mu}, \boldsymbol{\Sigma})$. Then the sample principal components, $\hat{y}_i = \hat{\mathbf{e}}_i'(\mathbf{x} - \bar{\mathbf{x}})$ are realizations of population principal components $Y_i = \mathbf{e}_i'(\mathbf{X} - \boldsymbol{\mu})$, which have an $N_p(\mathbf{0}, \boldsymbol{\Lambda})$ distribution. The diagonal matrix $\boldsymbol{\Lambda}$ has entries $\lambda_1, \lambda_2, \ldots, \lambda_p$ and $(\lambda_i, \mathbf{e}_i)$ are the eigenvalue–eigenvector pairs of $\boldsymbol{\Sigma}$.

Also, from the sample values $\mathbf{x}_j$, we can approximate $\boldsymbol{\mu}$ by $\bar{\mathbf{x}}$ and $\boldsymbol{\Sigma}$ by $\mathbf{S}$. If $\mathbf{S}$ is positive definite, the contour consisting of all $p \times 1$ vectors $\mathbf{x}$ satisfying

$$(\mathbf{x} - \bar{\mathbf{x}})'\mathbf{S}^{-1}(\mathbf{x} - \bar{\mathbf{x}}) = c^2 \qquad (8\text{-}24)$$

estimates the constant density contour $(\mathbf{x} - \boldsymbol{\mu})'\boldsymbol{\Sigma}^{-1}(\mathbf{x} - \boldsymbol{\mu}) = c^2$ of the underlying normal density. The approximate contours can be drawn on the scatter plot to indicate the normal distribution that generated the data. The normality assumption is useful for the inference procedures discussed in Section 8.5, but it is not required for the development of the properties of the sample principal components summarized in (8-20).

Even when the normal assumption is suspect and the scatter plot may depart somewhat from an elliptical pattern, we can still extract the eigenvalues from $\mathbf{S}$ and obtain the sample principal components. Geometrically, the data may be plotted as n points in p-space. The data can then be expressed in the new coordinates, which coincide with the axes of the contour of (8-24). Now, (8-24) defines a hyperellipsoid that is centered at $\bar{\mathbf{x}}$ and whose axes are given by the eigenvectors of $\mathbf{S}^{-1}$ or, equivalently, of $\mathbf{S}$. (See Section 2.3 and Result 4.1, with $\mathbf{S}$ in place of $\boldsymbol{\Sigma}$.) The lengths of these hyperellipsoid axes are proportional to $\sqrt{\hat{\lambda}_i}$, $i = 1, 2, \ldots, p$, where $\hat{\lambda}_1 \geq \hat{\lambda}_2 \geq \cdots \geq \hat{\lambda}_p \geq 0$ are the eigenvalues of $\mathbf{S}$.

Because $\hat{\mathbf{e}}_i$ has length 1, the absolute value of the ith principal component, $|\hat{y}_i| = |\hat{\mathbf{e}}_i'(\mathbf{x} - \bar{\mathbf{x}})|$, gives the length of the projection of the vector $(\mathbf{x} - \bar{\mathbf{x}})$ on the unit vector $\hat{\mathbf{e}}_i$. [See (2-8) and (2-9).] Thus, the sample principal components $\hat{y}_i = \hat{\mathbf{e}}_i'(\mathbf{x} - \bar{\mathbf{x}})$, $i = 1, 2, \ldots, p$, lie along the axes of the hyperellipsoid, and their absolute values are the lengths of the projections of $\mathbf{x} - \bar{\mathbf{x}}$ in the directions of the axes $\hat{\mathbf{e}}_i$. Consequently, the sample principal components can be viewed as the result of translating the origin of the original coordinate system to $\bar{\mathbf{x}}$ and then rotating the coordinate axes until they pass through the scatter in the directions of maximum variance.

The geometrical interpretation of the sample principal components is illustrated in Figure 8.4 for $p = 2$. Figure 8.4(a) shows an ellipse of constant distance, centered at $\bar{\mathbf{x}}$, with $\hat{\lambda}_1 > \hat{\lambda}_2$. The sample principal components are well determined. They lie along the axes of the ellipse in the perpendicular directions of maximum sample variance. Figure 8.4(b) shows a constant distance ellipse, centered at $\bar{\mathbf{x}}$, with $\hat{\lambda}_1 \doteq \hat{\lambda}_2$. If $\hat{\lambda}_1 = \hat{\lambda}_2$, the axes of the ellipse (circle) of constant distance are not uniquely determined and can lie in any two perpendicular directions, including the

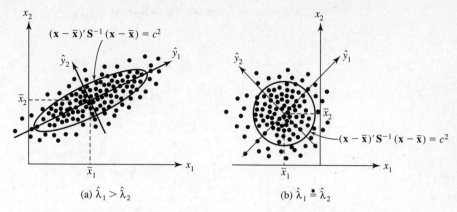

Figure 8.4 Sample principal components and ellipses of constant distance.

directions of the original coordinate axes. Similarly, the sample principal components can lie in any two perpendicular directions, including those of the original coordinate axes. When the contours of constant distance are nearly circular or, equivalently, when the eigenvalues of $\mathbf{S}$ are nearly equal, the sample variation is homogeneous in all directions. It is then not possible to represent the data well in fewer than p dimensions.

If the last few eigenvalues $\hat{\lambda}_i$ are sufficiently small such that the variation in the corresponding $\hat{\mathbf{e}}_i$ directions is negligible, the last few sample principal components can often be ignored, and the data can be adequately approximated by their representations in the space of the retained components. (See Section 8.4.)

Finally, Supplement 8A gives a further result concerning the role of the sample principal components when directly approximating the mean-centered data $\mathbf{x}_j - \bar{\mathbf{x}}$.

Standardizing the Sample Principal Components

Sample principal components are, in general, not invariant with respect to changes in scale. (See Exercises 8.6 and 8.7.) As we mentioned in the treatment of population components, variables measured on different scales or on a common scale with widely differing ranges are often standardized. For the sample, standardization is accomplished by constructing

$$\mathbf{z}_j = \mathbf{D}^{-1/2}(\mathbf{x}_j - \bar{\mathbf{x}}) = \begin{bmatrix} \dfrac{x_{j1} - \bar{x}_1}{\sqrt{s_{11}}} \\ \dfrac{x_{j2} - \bar{x}_2}{\sqrt{s_{22}}} \\ \vdots \\ \dfrac{x_{jp} - \bar{x}_p}{\sqrt{s_{pp}}} \end{bmatrix} \qquad j = 1, 2, \ldots, n \qquad (8\text{-}25)$$

The $n \times p$ data matrix of standardized observations

$$
\mathbf{Z} = \begin{bmatrix} \mathbf{z}_1' \\ \mathbf{z}_2' \\ \vdots \\ \mathbf{z}_n' \end{bmatrix} = \begin{bmatrix} z_{11} & z_{12} & \cdots & z_{1p} \\ z_{21} & z_{22} & \cdots & z_{2p} \\ \vdots & \vdots & \ddots & \vdots \\ z_{n1} & z_{n2} & \cdots & z_{np} \end{bmatrix}
$$

$$
= \begin{bmatrix} \dfrac{x_{11} - \bar{x}_1}{\sqrt{s_{11}}} & \dfrac{x_{12} - \bar{x}_2}{\sqrt{s_{22}}} & \cdots & \dfrac{x_{1p} - \bar{x}_p}{\sqrt{s_{pp}}} \\[2ex] \dfrac{x_{21} - \bar{x}_1}{\sqrt{s_{11}}} & \dfrac{x_{22} - \bar{x}_2}{\sqrt{s_{22}}} & \cdots & \dfrac{x_{2p} - \bar{x}_p}{\sqrt{s_{pp}}} \\[2ex] \vdots & \vdots & \ddots & \vdots \\[1ex] \dfrac{x_{n1} - \bar{x}_1}{\sqrt{s_{11}}} & \dfrac{x_{n2} - \bar{x}_2}{\sqrt{s_{22}}} & \cdots & \dfrac{x_{np} - \bar{x}_p}{\sqrt{s_{pp}}} \end{bmatrix} \tag{8-26}
$$

yields the sample mean vector [see (3-24)]

$$
\bar{\mathbf{z}} = \frac{1}{n}(\mathbf{1}'\mathbf{Z})' = \frac{1}{n}\mathbf{Z}'\mathbf{1} = \frac{1}{n} \begin{bmatrix} \displaystyle\sum_{j=1}^{n} \dfrac{x_{j1} - \bar{x}_1}{\sqrt{s_{11}}} \\[2ex] \displaystyle\sum_{j=1}^{n} \dfrac{x_{j2} - \bar{x}_2}{\sqrt{s_{22}}} \\[1ex] \vdots \\ \displaystyle\sum_{j=1}^{n} \dfrac{x_{jp} - \bar{x}_p}{\sqrt{s_{pp}}} \end{bmatrix} = \mathbf{0} \tag{8-27}
$$

and sample covariance matrix [see (3-27)]

$$
\mathbf{S}_z = \frac{1}{n-1}\left(\mathbf{Z} - \frac{1}{n}\mathbf{1}\mathbf{1}'\mathbf{Z}\right)'\left(\mathbf{Z} - \frac{1}{n}\mathbf{1}\mathbf{1}'\mathbf{Z}\right)
$$

$$
= \frac{1}{n-1}(\mathbf{Z} - \mathbf{1}\bar{\mathbf{z}}')'(\mathbf{Z} - \mathbf{1}\bar{\mathbf{z}}')
$$

$$
= \frac{1}{n-1}\mathbf{Z}'\mathbf{Z}
$$

$$
= \frac{1}{n-1} \begin{bmatrix} \dfrac{(n-1)s_{11}}{s_{11}} & \dfrac{(n-1)s_{12}}{\sqrt{s_{11}}\sqrt{s_{22}}} & \cdots & \dfrac{(n-1)s_{1p}}{\sqrt{s_{11}}\sqrt{s_{pp}}} \\[2ex] \dfrac{(n-1)s_{12}}{\sqrt{s_{11}}\sqrt{s_{22}}} & \dfrac{(n-1)s_{22}}{s_{22}} & \cdots & \dfrac{(n-1)s_{2p}}{\sqrt{s_{22}}\sqrt{s_{pp}}} \\[1ex] \vdots & \vdots & \ddots & \vdots \\ \dfrac{(n-1)s_{1p}}{\sqrt{s_{11}}\sqrt{s_{pp}}} & \dfrac{(n-1)s_{2p}}{\sqrt{s_{22}}\sqrt{s_{pp}}} & \cdots & \dfrac{(n-1)s_{pp}}{s_{pp}} \end{bmatrix} = \mathbf{R} \tag{8-28}
$$

The sample principal components of the standardized observations are given by (8-20), with the matrix $\mathbf{R}$ in place of $\mathbf{S}$. Since the observations are already "centered" by construction, there is no need to write the components in the form of (8-21).

If $z_1, z_2, \ldots, z_n$ are standardized observations with covariance matrix $\mathbf{R}$, the ith sample principal component is

$$\hat{y}_i = \hat{\mathbf{e}}_i' \mathbf{z} = \hat{e}_{i1} z_1 + \hat{e}_{i2} z_2 + \cdots + \hat{e}_{ip} z_p, \qquad i = 1, 2, \ldots, p$$

where $(\hat{\lambda}_i, \hat{\mathbf{e}}_i)$ is the ith eigenvalue-eigenvector pair of $\mathbf{R}$ with $\hat{\lambda}_1 \geq \hat{\lambda}_2 \geq \cdots \geq \hat{\lambda}_p \geq 0$. Also,

$$\text{Sample variance } (\hat{y}_i) = \hat{\lambda}_i \qquad i = 1, 2, \ldots, p$$
$$\text{Sample covariance } (\hat{y}_i, \hat{y}_k) = 0 \qquad i \neq k$$

In addition,   (8-29)

$$\text{Total (standardized) sample variance} = \text{tr}(\mathbf{R}) = p = \hat{\lambda}_1 + \hat{\lambda}_2 + \cdots + \hat{\lambda}_p$$

and

$$r_{\hat{y}_i, z_k} = \hat{e}_{ik} \sqrt{\hat{\lambda}_i}, \qquad i, k = 1, 2, \ldots, p$$

Using (8-29), we see that the proportion of the total sample variance explained by the ith sample principal component is

$$\begin{pmatrix} \text{Proportion of (standardized)} \\ \text{sample variance due to } i\text{th} \\ \text{sample principal component} \end{pmatrix} = \frac{\hat{\lambda}_i}{p} \qquad i = 1, 2, \ldots, p \qquad (8\text{-}30)$$

A rule of thumb suggests retaining only those components whose variances $\hat{\lambda}_i$ are greater than unity or, equivalently, only those components which, individually, explain at least a proportion $1/p$ of the total variance. This rule does not have a great deal of theoretical support, however, and it should not be applied blindly. As we have mentioned, a scree plot is also useful for selecting the appropriate number of components.

Example 8.5 (Sample principal components from standardized data) The weekly rates of return for five stocks (JP Morgan, Citibank, Wells Fargo, Royal Dutch Shell, and ExxonMobil) listed on the New York Stock Exchange were determined for the period January 2004 through December 2005. The weekly rates of return are defined as (current week closing price—previous week closing price)/(previous week closing price), adjusted for stock splits and dividends. The data are listed in Table 8.4 in the Exercises. The observations in 103 successive weeks appear to be independently distributed, but the rates of return *across* stocks are correlated, because as one expects, stocks tend to move together in response to general economic conditions.

Let $x_1, x_2, \ldots, x_5$ denote observed weekly rates of return for JP Morgan, Citibank, Wells Fargo, Royal Dutch Shell, and ExxonMobil, respectively. Then

$$\bar{\mathbf{x}}' = [.0011, .0007, .0016, .0040, .0040]$$

and

$$\mathbf{R} = \begin{bmatrix} 1.000 & .632 & .511 & .115 & .155 \\ .632 & 1.000 & .574 & .322 & .213 \\ .511 & .574 & 1.000 & .183 & .146 \\ .115 & .322 & .183 & 1.000 & .683 \\ .155 & .213 & .146 & .683 & 1.000 \end{bmatrix}$$

We note that $\mathbf{R}$ is the covariance matrix of the standardized observations

$$z_1 = \frac{x_1 - \bar{x}_1}{\sqrt{s_{11}}}, z_2 = \frac{x_2 - \bar{x}_2}{\sqrt{s_{22}}}, \ldots, z_5 = \frac{x_5 - \bar{x}_5}{\sqrt{s_{55}}}$$

The eigenvalues and corresponding normalized eigenvectors of $\mathbf{R}$, determined by a computer, are

$$\hat{\lambda}_1 = 2.437, \quad \hat{\mathbf{e}}_1' = [\ .469, \quad .532, \quad .465, \quad .387, \quad .361]$$

$$\hat{\lambda}_2 = 1.407, \quad \hat{\mathbf{e}}_2' = [-.368, -.236, -.315, \quad .585, \quad .606]$$

$$\hat{\lambda}_3 = .501, \quad \hat{\mathbf{e}}_3' = [-.604, -.136, \quad .772, \quad .093, -.109]$$

$$\hat{\lambda}_4 = .400, \quad \hat{\mathbf{e}}_4' = [\ .363, -.629, \quad .289, -.381, \quad .493]$$

$$\hat{\lambda}_5 = .255, \quad \hat{\mathbf{e}}_5' = [\ .384, -.496, \quad .071, \quad .595, -.498]$$

Using the standardized variables, we obtain the first two sample principal components:

$$\hat{y}_1 = \hat{\mathbf{e}}_1'\mathbf{z} = .469z_1 + .532z_2 + .465z_3 + .387z_4 + .361z_5$$

$$\hat{y}_2 = \hat{\mathbf{e}}_2'\mathbf{z} = -.368z_1 - .236z_2 - .315z_3 + .585z_4 + .606z_5$$

These components, which account for

$$\left(\frac{\hat{\lambda}_1 + \hat{\lambda}_2}{p}\right)100\% = \left(\frac{2.437 + 1.407}{5}\right)100\% = 77\%$$

of the total (standardized) sample variance, have interesting interpretations. The first component is a roughly equally weighted sum, or "index," of the five stocks. This component might be called a *general stock-market component*, or, simply, a *market component*.

The second component represents a contrast between the banking stocks (JP Morgan, Citibank, Wells Fargo) and the oil stocks (Royal Dutch Shell, Exxon-Mobil). It might be called an *industry component*. Thus, we see that most of the variation in these stock returns is due to market activity and uncorrelated industry activity. This interpretation of stock price behavior also has been suggested by King [12].

The remaining components are not easy to interpret and, collectively, represent variation that is probably specific to each stock. In any event, they do not explain much of the total sample variance. ∎

Example 8.6 (Components from a correlation matrix with a special structure) Geneticists are often concerned with the inheritance of characteristics that can be measured several times during an animal's lifetime. Body weight (in grams) for $n = 150$ female mice were obtained immediately after the birth of their first four litters.[4] The sample mean vector and sample correlation matrix were, respectively,

$$\bar{\mathbf{x}}' = [39.88, 45.08, 48.11, 49.95]$$

and

$$\mathbf{R} = \begin{bmatrix} 1.000 & .7501 & .6329 & .6363 \\ .7501 & 1.000 & .6925 & .7386 \\ .6329 & .6925 & 1.000 & .6625 \\ .6363 & .7386 & .6625 & 1.000 \end{bmatrix}$$

The eigenvalues of this matrix are

$$\hat{\lambda}_1 = 3.085, \quad \hat{\lambda}_2 = .382, \quad \hat{\lambda}_3 = .342, \quad \text{and} \quad \hat{\lambda}_4 = .217$$

We note that the first eigenvalue is nearly equal to $1 + (p - 1)\bar{r} = 1 + (4 - 1)(.6854) = 3.056$, where $\bar{r}$ is the arithmetic average of the off-diagonal elements of $\mathbf{R}$. The remaining eigenvalues are small and about equal, although $\hat{\lambda}_4$ is somewhat smaller than $\hat{\lambda}_2$ and $\hat{\lambda}_3$. Thus, there is some evidence that the corresponding population correlation matrix $\boldsymbol{\rho}$ may be of the "equal-correlation" form of (8-15). This notion is explored further in Example 8.9.

The first principal component

$$\hat{y}_1 = \hat{\mathbf{e}}_1' \mathbf{z} = .49z_1 + .52z_2 + .49z_3 + .50z_4$$

accounts for $100(\hat{\lambda}_1/p)\% = 100(3.058/4)\% = 76\%$ of the total variance. Although the average postbirth weights increase over time, the *variation* in weights is fairly well explained by the first principal component with (nearly) equal coefficients. ∎

Comment. An unusually small value for the *last* eigenvalue from either the sample covariance or correlation matrix can indicate an unnoticed linear dependency in the data set. If this occurs, one (or more) of the variables is redundant and should be deleted. Consider a situation where x_1, x_2, and x_3 are subtest scores and the total score x_4 is the sum $x_1 + x_2 + x_3$. Then, although the linear combination $\mathbf{e}'\mathbf{x} = [1, 1, 1, -1]\mathbf{x} = x_1 + x_2 + x_3 - x_4$ is always zero, rounding error in the computation of eigenvalues may lead to a small nonzero value. If the linear expression relating x_4 to (x_1, x_2, x_3) was initially overlooked, the smallest eigenvalue–eigenvector pair should provide a clue to its existence. (See the discussion in Section 3.4, pages 131–133.)

Thus, although "large" eigenvalues and the corresponding eigenvectors are important in a principal component analysis, eigenvalues very close to zero should not be routinely ignored. The eigenvectors associated with these latter eigenvalues may point out linear dependencies in the data set that can cause interpretive and computational problems in a subsequent analysis.

---

[4] Data courtesy of J. J. Rutledge.

8.4 Graphing the Principal Components

Plots of the principal components can reveal suspect observations, as well as provide checks on the assumption of normality. Since the principal components are linear combinations of the original variables, it is not unreasonable to expect them to be nearly normal. It is often necessary to verify that the first few principal components are approximately normally distributed when they are to be used as the input data for additional analyses.

The last principal components can help pinpoint suspect observations. Each observation can be expressed as a linear combination

$$\mathbf{x}_j = (\mathbf{x}_j'\hat{\mathbf{e}}_1)\,\hat{\mathbf{e}}_1 + (\mathbf{x}_j'\hat{\mathbf{e}}_2)\,\hat{\mathbf{e}}_2 + \cdots + (\mathbf{x}_j'\hat{\mathbf{e}}_p)\,\hat{\mathbf{e}}_p$$

$$= \hat{y}_{j1}\hat{\mathbf{e}}_1 + \hat{y}_{j2}\hat{\mathbf{e}}_2 + \cdots + \hat{y}_{jp}\hat{\mathbf{e}}_p$$

of the complete set of eigenvectors $\hat{\mathbf{e}}_1, \hat{\mathbf{e}}_2, \ldots, \hat{\mathbf{e}}_p$ of $\mathbf{S}$. Thus, the magnitudes of the last principal components determine how well the first few fit the observations. That is, $\hat{y}_{j1}\hat{\mathbf{e}}_1 + \hat{y}_{j2}\hat{\mathbf{e}}_2 + \cdots + \hat{y}_{j,q-1}\hat{\mathbf{e}}_{q-1}$ differs from $\mathbf{x}_j$ by $\hat{y}_{jq}\hat{\mathbf{e}}_q + \cdots + \hat{y}_{jp}\hat{\mathbf{e}}_p$, the square of whose length is $\hat{y}_{jq}^2 + \cdots + \hat{y}_{jp}^2$. Suspect observations will often be such that at least one of the coordinates $\hat{y}_{jq}, \ldots, \hat{y}_{jp}$ contributing to this squared length will be large. (See Supplement 8A for more general approximation results.)

The following statements summarize these ideas.

1. To help check the normal assumption, construct scatter diagrams for pairs of the first few principal components. Also, make Q–Q plots from the sample values generated by *each* principal component.

2. Construct scatter diagrams and Q–Q plots for the last few principal components. These help identify suspect observations.

Example 8.7 (Plotting the principal components for the turtle data) We illustrate the plotting of principal components for the data on male turtles discussed in Example 8.4. The three sample principal components are

$$\hat{y}_1 = \quad .683(x_1 - 4.725) + .510(x_2 - 4.478) + .523(x_3 - 3.703)$$

$$\hat{y}_2 = -.159(x_1 - 4.725) - .594(x_2 - 4.478) + .788(x_3 - 3.703)$$

$$\hat{y}_3 = -.713(x_1 - 4.725) + .622(x_2 - 4.478) + .324(x_3 - 3.703)$$

where $x_1 = \ln(\text{length})$, $x_2 = \ln(\text{width})$, and $x_3 = \ln(\text{height})$, respectively.

Figure 8.5 shows the Q–Q plot for $\hat{y}_2$ and Figure 8.6 shows the scatter plot of $(\hat{y}_1, \hat{y}_2)$. The observation for the first turtle is circled and lies in the lower right corner of the scatter plot and in the upper right corner of the Q–Q plot; it may be suspect. This point should have been checked for recording errors, or the turtle should have been examined for structural anomalies. Apart from the first turtle, the scatter plot appears to be reasonably elliptical. The plots for the other sets of principal components do not indicate any substantial departures from normality. ∎

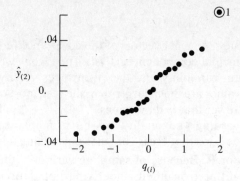

Figure 8.5 A Q–Q plot for the second principal component $\hat{y}_2$ from the data on male turtles.

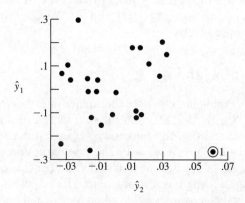

Figure 8.6 Scatter plot of the principal components $\hat{y}_1$ and $\hat{y}_2$ of the data on male turtles.

The diagnostics involving principal components apply equally well to the checking of assumptions for a multivariate multiple regression model. In fact, having fit any model by any method of estimation, it is prudent to consider the

$$\text{Residual vector} = (\text{observation vector}) - \left(\begin{matrix}\text{vector of predicted}\\ \text{(estimated) values}\end{matrix}\right)$$

or

$$\underset{(p\times 1)}{\hat{\boldsymbol{\varepsilon}}_j} = \underset{(p\times 1)}{\mathbf{y}_j} - \underset{(p\times 1)}{\hat{\boldsymbol{\beta}}'\mathbf{z}_j} \qquad j = 1, 2, \ldots, n \tag{8-31}$$

for the multivariate linear model. Principal components, derived from the covariance matrix of the residuals,

$$\frac{1}{n-p}\sum_{j=1}^{n}(\hat{\boldsymbol{\varepsilon}}_j - \bar{\hat{\boldsymbol{\varepsilon}}}_j)(\hat{\boldsymbol{\varepsilon}}_j - \bar{\hat{\boldsymbol{\varepsilon}}}_j)' \tag{8-32}$$

can be scrutinized in the same manner as those determined from a random sample. You should be aware that there *are* linear dependencies among the residuals from a linear regression analysis, so the last eigenvalues will be zero, within rounding error.

8.5 Large Sample Inferences

We have seen that the eigenvalues and eigenvectors of the covariance (correlation) matrix are the essence of a principal component analysis. The eigenvectors determine the directions of maximum variability, and the eigenvalues specify the variances. When the first few eigenvalues are much larger than the rest, most of the total variance can be "explained" in fewer than p dimensions.

In practice, decisions regarding the quality of the principal component approximation must be made on the basis of the eigenvalue–eigenvector pairs $(\hat{\lambda}_i, \hat{\mathbf{e}}_i)$ extracted from $\mathbf{S}$ or $\mathbf{R}$. Because of sampling variation, these eigenvalues and eigenvectors will differ from their underlying population counterparts. The sampling distributions of $\hat{\lambda}_i$ and $\hat{\mathbf{e}}_i$ are difficult to derive and beyond the scope of this book. If you are interested, you can find some of these derivations for multivariate normal populations in [1], [2], and [5]. We shall simply summarize the pertinent large sample results.

Large Sample Properties of $\hat{\lambda}_i$ and $\hat{\mathbf{e}}_i$

Currently available results concerning large sample confidence intervals for $\hat{\lambda}_i$ and $\hat{\mathbf{e}}_i$ assume that the observations $\mathbf{X}_1, \mathbf{X}_2, \ldots, \mathbf{X}_n$ are a random sample from a normal population. It must also be assumed that the (unknown) eigenvalues of $\boldsymbol{\Sigma}$ are distinct and positive, so that $\lambda_1 > \lambda_2 > \cdots > \lambda_p > 0$. The one exception is the case where the number of equal eigenvalues is known. Usually the conclusions for distinct eigenvalues are applied, unless there is a strong reason to believe that $\boldsymbol{\Sigma}$ has a special structure that yields equal eigenvalues. Even when the normal assumption is violated, the confidence intervals obtained in this manner still provide some indication of the uncertainty in $\hat{\lambda}_i$ and $\hat{\mathbf{e}}_i$.

Anderson [2] and Girshick [5] have established the following large sample distribution theory for the eigenvalues $\hat{\boldsymbol{\lambda}}' = [\hat{\lambda}_1, \ldots, \hat{\lambda}_p]$ and eigenvectors $\hat{\mathbf{e}}_1, \ldots, \hat{\mathbf{e}}_p$ of $\mathbf{S}$:

1. Let $\boldsymbol{\Lambda}$ be the diagonal matrix of eigenvalues $\lambda_1, \ldots, \lambda_p$ of $\boldsymbol{\Sigma}$, then $\sqrt{n}\,(\hat{\boldsymbol{\lambda}} - \boldsymbol{\lambda})$ is approximately $N_p(\mathbf{0}, 2\boldsymbol{\Lambda}^2)$.

2. Let
$$\mathbf{E}_i = \lambda_i \sum_{\substack{k=1 \\ k \neq i}}^{p} \frac{\lambda_k}{(\lambda_k - \lambda_i)^2}\, \mathbf{e}_k \mathbf{e}_k'$$

then $\sqrt{n}\,(\hat{\mathbf{e}}_i - \mathbf{e}_i)$ is approximately $N_p(\mathbf{0}, \mathbf{E}_i)$.

3. Each $\hat{\lambda}_i$ is distributed independently of the elements of the associated $\hat{\mathbf{e}}_i$.

Result 1 implies that, for n large, the $\hat{\lambda}_i$ are independently distributed. Moreover, $\hat{\lambda}_i$ has an approximate $N(\lambda_i, 2\lambda_i^2/n)$ distribution. Using this normal distribution, we obtain $P[|\hat{\lambda}_i - \lambda_i| \leq z(\alpha/2)\lambda_i\sqrt{2/n}] = 1 - \alpha$. A large sample $100(1 - \alpha)\%$ confidence interval for λ_i is thus provided by

$$\frac{\hat{\lambda}_i}{(1 + z(\alpha/2)\sqrt{2/n})} \leq \lambda_i \leq \frac{\hat{\lambda}_i}{(1 - z(\alpha/2)\sqrt{2/n})} \tag{8-33}$$

where $z(\alpha/2)$ is the upper $100(\alpha/2)$th percentile of a standard normal distribution. Bonferroni-type simultaneous $100(1 - \alpha)\%$ intervals for m λ_i's are obtained by replacing $z(\alpha/2)$ with $z(\alpha/2m)$. (See Section 5.4.)

Result 2 implies that the $\hat{\mathbf{e}}_i$'s are normally distributed about the corresponding $\mathbf{e}_i$'s for large samples. The elements of each $\hat{\mathbf{e}}_i$ are correlated, and the correlation depends to a large extent on the separation of the eigenvalues $\lambda_1, \lambda_2, \ldots, \lambda_p$ (which is unknown) and the sample size n. Approximate standard errors for the coefficients $\hat{e}_{ik}$ are given by the square roots of the diagonal elements of $(1/n)\hat{\mathbf{E}}_i$ where $\hat{\mathbf{E}}_i$ is derived from $\mathbf{E}_i$ by substituting $\hat{\lambda}_i$'s for the λ_i's and $\hat{\mathbf{e}}_i$'s for the $\mathbf{e}_i$'s.

Example 8.8 (Constructing a confidence interval for λ_1) We shall obtain a 95% confidence interval for λ_1, the variance of the first population principal component, using the stock price data listed in Table 8.4 in the Exercises.

Assume that the stock rates of return represent independent drawings from an $N_5(\boldsymbol{\mu}, \boldsymbol{\Sigma})$ population, where $\boldsymbol{\Sigma}$ is positive definite with distinct eigenvalues $\lambda_1 > \lambda_2 > \cdots > \lambda_5 > 0$. Since $n = 103$ is large, we can use (8-33) with $i = 1$ to construct a 95% confidence interval for λ_1. From Exercise 8.10, $\hat{\lambda}_1 = .0014$ and in addition, $z(.025) = 1.96$. Therefore, with 95% confidence,

$$\frac{.0014}{\left(1 + 1.96\sqrt{\frac{2}{103}}\right)} \le \lambda_1 \le \frac{.0014}{\left(1 - 1.96\sqrt{\frac{2}{103}}\right)} \quad \text{or} \quad .0011 \le \lambda_1 \le .0019 \quad \blacksquare$$

Whenever an eigenvalue is large, such as 100 or even 1000, the intervals generated by (8-33) can be quite wide, for reasonable confidence levels, even though n is fairly large. In general, the confidence interval gets wider at the same rate that $\hat{\lambda}_i$ gets larger. Consequently, some care must be exercised in dropping or retaining principal components based on an examination of the $\hat{\lambda}_i$'s.

Testing for the Equal Correlation Structure

The special correlation structure $\text{Cov}(X_i, X_k) = \sqrt{\sigma_{ii}\sigma_{kk}}\,\rho$, or $\text{Corr}(X_i, X_k) = \rho$, all $i \ne k$, is one important structure in which the eigenvalues of $\boldsymbol{\Sigma}$ are not distinct and the previous results do not apply.

To test for this structure, let

$$H_0: \underset{(p \times p)}{\boldsymbol{\rho}} = \boldsymbol{\rho}_0 = \begin{bmatrix} 1 & \rho & \cdots & \rho \\ \rho & 1 & \cdots & \rho \\ \vdots & \vdots & \ddots & \vdots \\ \rho & \rho & \cdots & 1 \end{bmatrix}$$

and

$$H_1: \boldsymbol{\rho} \ne \boldsymbol{\rho}_0$$

A test of H_0 versus H_1 may be based on a likelihood ratio statistic, but Lawley [14] has demonstrated that an equivalent test procedure can be constructed from the off-diagonal elements of $\mathbf{R}$.

Lawley's procedure requires the quantities

$$\bar{r}_k = \frac{1}{p-1} \sum_{\substack{i=1 \\ i \neq k}}^{p} r_{ik} \quad k = 1, 2, \dots, p; \quad \bar{r} = \frac{2}{p(p-1)} \sum_{i<k} \sum r_{ik}$$

$$\hat{\gamma} = \frac{(p-1)^2[1 - (1-\bar{r})^2]}{p - (p-2)(1-\bar{r})^2} \tag{8-34}$$

It is evident that $\bar{r}_k$ is the average of the off-diagonal elements in the kth column (or row) of $\mathbf{R}$ and $\bar{r}$ is the overall average of the off-diagonal elements.

The large sample approximate α-level test is to reject H_0 in favor of H_1 if

$$T = \frac{(n-1)}{(1-\bar{r})^2} \left[\sum_{i<k} \sum (r_{ik} - \bar{r})^2 - \hat{\gamma} \sum_{k=1}^{p} (\bar{r}_k - \bar{r})^2 \right] > \chi^2_{(p+1)(p-2)/2}(\alpha) \tag{8-35}$$

where $\chi^2_{(p+1)(p-2)/2}(\alpha)$ is the upper (100α)th percentile of a chi-square distribution with $(p+1)(p-2)/2$ d.f.

Example 8.9 (Testing for equicorrelation structure) From Example 8.6, the sample correlation matrix constructed from the $n = 150$ post-birth weights of female mice is

$$\mathbf{R} = \begin{bmatrix} 1.0 & .7501 & .6329 & .6363 \\ .7501 & 1.0 & .6925 & .7386 \\ .6329 & .6925 & 1.0 & .6625 \\ .6363 & .7386 & .6625 & 1.0 \end{bmatrix}$$

We shall use this correlation matrix to illustrate the large sample test in (8-35).

Here $p = 4$, and we set

$$H_0: \boldsymbol{\rho} = \boldsymbol{\rho}_0 = \begin{bmatrix} 1 & \rho & \rho & \rho \\ \rho & 1 & \rho & \rho \\ \rho & \rho & 1 & \rho \\ \rho & \rho & \rho & 1 \end{bmatrix}$$

$$H_1: \boldsymbol{\rho} \neq \boldsymbol{\rho}_0$$

Using (8-34) and (8-35), we obtain

$$\bar{r}_1 = \frac{1}{3}(.7501 + .6329 + .6363) = .6731, \quad \bar{r}_2 = .7271,$$

$$\bar{r}_3 = .6626, \quad \bar{r}_4 = .6791$$

$$\bar{r} = \frac{2}{4(3)}(.7501 + .6329 + .6363 + .6925 + .7386 + .6625) = .6855$$

$$\sum_{i<k} \sum (r_{ik} - \bar{r})^2 = (.7501 - .6855)^2$$

$$+ (.6329 - .6855)^2 + \cdots + (.6625 - .6855)^2$$

$$= .01277$$

$$\sum_{k=1}^{4} (\bar{r}_k - \bar{r})^2 = (.6731 - .6855)^2 + \cdots + (.6791 - .6855)^2 = .00245$$

$$\hat{\gamma} = \frac{(4-1)^2[1-(1-.6855)^2]}{4-(4-2)(1-.6855)^2} = 2.1329$$

and

$$T = \frac{(150-1)}{(1-.6855)^2}[.01277 - (2.1329)(.00245)] = 11.4$$

Since $(p+1)(p-2)/2 = 5(2)/2 = 5$, the 5% critical value for the test in (8-35) is $\chi_5^2(.05) = 11.07$. The value of our test statistic is approximately equal to the large sample 5% critical point, so the evidence against H_0 (equal correlations) is strong, but not overwhelming.

As we saw in Example 8.6, the smallest eigenvalues $\hat{\lambda}_2$, $\hat{\lambda}_3$, and $\hat{\lambda}_4$ are slightly different, with $\hat{\lambda}_4$ being somewhat smaller than the other two. Consequently, with the large sample size in this problem, small differences from the equal correlation structure show up as statistically significant. ■

Assuming a multivariate normal population, a large sample test that all variables are independent (all the off-diagonal elements of Σ are zero) is contained in Exercise 8.9.

8.6 Monitoring Quality with Principal Components

In Section 5.6, we introduced multivariate control charts, including the quality ellipse and the T^2 chart. Today, with electronic and other automated methods of data collection, it is not uncommon for data to be collected on 10 or 20 process variables. Major chemical and drug companies report measuring over 100 process variables, including temperature, pressure, concentration, and weight, at various positions along the production process. Even with 10 variables to monitor, there are 45 pairs for which to create quality ellipses. Clearly, another approach is required to both visually display important quantities and still have the sensitivity to detect special causes of variation.

Checking a Given Set of Measurements for Stability

Let $X_1, X_2, \ldots, X_n$ be a random sample from a multivariate normal distribution with mean μ and covariance matrix Σ. We consider the first two sample principal components, $\hat{y}_{j1} = \hat{e}_1'(x_j - \bar{x})$ and $\hat{y}_{j2} = \hat{e}_2'(x_j - \bar{x})$. Additional principal components could be considered, but two are easier to inspect visually and, of any two components, the first two explain the largest cumulative proportion of the total sample variance.

If a process is stable over time, so that the measured characteristics are influenced only by variations in common causes, then the values of the first two principal components should be stable. Conversely, if the principal components remain stable over time, the common effects that influence the process are likely to remain constant. To monitor quality using principal components, we consider a two-part procedure. The first part of the procedure is to construct an ellipse format chart for the pairs of values $(\hat{y}_{j1}, \hat{y}_{j2})$ for $j = 1, 2, \ldots, n$.

By (8-20), the sample variance of the first principal component $\hat{y}_1$ is given by the largest eigenvalue $\hat{\lambda}_1$, and the sample variance of the second principal component $\hat{y}_2$ is the second-largest eigenvalue $\hat{\lambda}_2$. The two sample components are uncorrelated, so the quality ellipse for n large (see Section 5.6) reduces to the collection of pairs of possible values $(\hat{y}_1, \hat{y}_2)$ such that

$$\frac{\hat{y}_1^2}{\hat{\lambda}_1} + \frac{\hat{y}_2^2}{\hat{\lambda}_2} \leq \chi_2^2(\alpha) \tag{8-36}$$

Example 8.10 (An ellipse format chart based on the first two principal components)
Refer to the police department overtime data given in Table 5.8. Table 8.1 contains the five normalized eigenvectors and eigenvalues of the sample covariance matrix **S**.

The first two sample components explain 82% of the total variance.
The sample values for all five components are displayed in Table 8.2.

Table 8.1 Eigenvectors and Eigenvalues of Sample Covariance Matrix for Police Department Data

Variable	$\hat{e}_1$	$\hat{e}_2$	$\hat{e}_3$	$\hat{e}_4$	$\hat{e}_5$
Appearances overtime (x_1)	.046	−.048	.629	−.643	.432
Extraordinary event (x_2)	.039	.985	−.077	−.151	−.007
Holdover hours (x_3)	−.658	.107	.582	.250	−.392
COA hours (x_4)	.734	.069	.503	.397	−.213
Meeting hours (x_5)	−.155	.107	.081	.586	.784
$\hat{\lambda}_i$	2,770,226	1,429,206	628,129	221,138	99,824

Table 8.2 Values of the Principal Components for the Police Department Data

Period	$\hat{y}_{j1}$	$\hat{y}_{j2}$	$\hat{y}_{j3}$	$\hat{y}_{j4}$	$\hat{y}_{j5}$
1	2044.9	588.2	425.8	−189.1	−209.8
2	−2143.7	−686.2	883.6	−565.9	−441.5
3	−177.8	−464.6	707.5	736.3	38.2
4	−2186.2	450.5	−184.0	443.7	−325.3
5	−878.6	−545.7	115.7	296.4	437.5
6	563.2	−1045.4	281.2	620.5	142.7
7	403.1	66.8	340.6	−135.5	521.2
8	−1988.9	−801.8	−1437.3	−148.8	61.6
9	132.8	563.7	125.3	68.2	611.5
10	−2787.3	−213.4	7.8	169.4	−202.3
11	283.4	3936.9	−0.9	276.2	−159.6
12	761.6	256.0	−2153.6	−418.8	28.2
13	−498.3	244.7	966.5	−1142.3	182.6
14	2366.2	−1193.7	−165.5	270.6	−344.9
15	1917.8	−782.0	−82.9	−196.8	−89.9
16	2187.7	−373.8	170.1	−84.1	−250.2

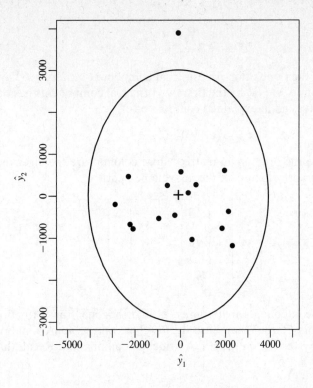

Figure 8.7 The 95% control ellipse based on the first two principal components of overtime hours.

Let us construct a 95% ellipse format chart using the first two sample principal components and plot the 16 pairs of component values in Table 8.2.

Although $n = 16$ is not large, we use $\chi_2^2(.05) = 5.99$, and the ellipse becomes

$$\frac{\hat{y}_1^2}{\hat{\lambda}_1} + \frac{\hat{y}_2^2}{\hat{\lambda}_2} \le 5.99$$

This ellipse centered at $(0,0)$, is shown in Figure 8.7, along with the data.

One point is out of control, because the second principal component for this point has a large value. Scanning Table 8.2, we see that this is the value 3936.9 for period 11. According to the entries of $\hat{\mathbf{e}}_2$ in Table 8.1, the second principal component is essentially extraordinary event overtime hours. The principal component approach has led us to the same conclusion we came to in Example 5.9. ∎

In the event that special causes are likely to produce shocks to the system, the second part of our two-part procedure—that is, a second chart—is required. This chart is created from the information in the principal components not involved in the ellipse format chart.

Consider the deviation vector $\mathbf{X} - \boldsymbol{\mu}$, and assume that $\mathbf{X}$ is distributed as $N_p(\boldsymbol{\mu}, \boldsymbol{\Sigma})$. Even without the normal assumption, $\mathbf{X}_j - \boldsymbol{\mu}$ can be expressed as the sum of its projections on the eigenvectors of $\boldsymbol{\Sigma}$

$$\mathbf{X} - \boldsymbol{\mu} = (\mathbf{X} - \boldsymbol{\mu})'\mathbf{e}_1\mathbf{e}_1 + (\mathbf{X} - \boldsymbol{\mu})'\mathbf{e}_2\mathbf{e}_2$$
$$+ (\mathbf{X} - \boldsymbol{\mu})'\mathbf{e}_3\mathbf{e}_3 + \cdots + (\mathbf{X} - \boldsymbol{\mu})'\mathbf{e}_p\mathbf{e}_p$$

or

$$\mathbf{X} - \boldsymbol{\mu} = Y_1\mathbf{e}_1 + Y_2\mathbf{e}_2 + Y_3\mathbf{e}_3 + \cdots + Y_p\mathbf{e}_p \tag{8-37}$$

where $Y_i = (\mathbf{X} - \boldsymbol{\mu})'\mathbf{e}_i$ is the population ith principal component centered to have mean 0. The approximation to $\mathbf{X} - \boldsymbol{\mu}$ by the first two principal components has the form $Y_1\mathbf{e}_1 + Y_2\mathbf{e}_2$. This leaves an unexplained component of

$$\mathbf{X} - \boldsymbol{\mu} - Y_1\mathbf{e}_1 - Y_2\mathbf{e}_2$$

Let $\mathbf{E} = [\mathbf{e}_1, \mathbf{e}_2, \ldots, \mathbf{e}_p]$ be the orthogonal matrix whose columns are the eigenvectors of $\boldsymbol{\Sigma}$. The orthogonal transformation of the unexplained part,

$$\mathbf{E}'(\mathbf{X} - \boldsymbol{\mu} - Y_1\mathbf{e}_1 - Y_2\mathbf{e}_2) = \begin{bmatrix} Y_1 \\ Y_2 \\ Y_3 \\ \vdots \\ Y_p \end{bmatrix} - \begin{bmatrix} Y_1 \\ 0 \\ 0 \\ \vdots \\ 0 \end{bmatrix} - \begin{bmatrix} 0 \\ Y_2 \\ 0 \\ \vdots \\ 0 \end{bmatrix} = \begin{bmatrix} 0 \\ 0 \\ Y_3 \\ \vdots \\ Y_p \end{bmatrix} = \begin{bmatrix} 0 \\ \mathbf{Y}_{(2)} \end{bmatrix}$$

so the last $p - 2$ principal components are obtained as an orthogonal transformation of the approximation errors. Rather than base the T^2 chart on the approximation errors, we can, equivalently, base it on these last principal components. Recall that

$$\text{Var}\,(Y_i) = \lambda_i \quad \text{for} \quad i = 1, 2, \ldots, p$$

and $\text{Cov}\,(Y_i, Y_k) = 0$ for $i \neq k$. Consequently, the statistic $\mathbf{Y}'_{(2)}\boldsymbol{\Sigma}^{-1}_{\mathbf{Y}_{(2)}, \mathbf{Y}_{(2)}}\mathbf{Y}_{(2)}$, based on the last $p - 2$ population principal components, becomes

$$\frac{Y_3^2}{\lambda_3} + \frac{Y_4^2}{\lambda_4} + \cdots + \frac{Y_p^2}{\lambda_p} \tag{8-38}$$

This is just the sum of the squares of $p - 2$ independent standard normal variables, $\lambda_k^{-1/2}Y_k$, and so has a chi-square distribution with $p - 2$ degrees of freedom.

In terms of the sample data, the principal components and eigenvalues must be estimated. Because the coefficients of the linear combinations $\hat{\mathbf{e}}_i$ are also estimates, the principal components do not have a normal distribution even when the population is normal. However, it is customary to create a T^2-chart based on the statistic

$$T_j^2 = \frac{\hat{y}_{j3}^2}{\hat{\lambda}_3} + \frac{\hat{y}_{j4}^2}{\hat{\lambda}_4} + \cdots + \frac{\hat{y}_{jp}^2}{\hat{\lambda}_p}$$

which involves the estimated eigenvalues and vectors. Further, it is usual to appeal to the large sample approximation described by (8-38) and set the upper control limit of the T^2-chart as $\text{UCL} = c^2 = \chi^2_{p-2}(\alpha)$.

This T^2-statistic is based on high-dimensional data. For example, when $p = 20$ variables are measured, it uses the information in the 18-dimensional space perpendicular to the first two eigenvectors $\hat{\mathbf{e}}_1$ and $\hat{\mathbf{e}}_2$. Still, this T^2 based on the unexplained variation in the original observations is reported as highly effective in picking up special causes of variation.

Example 8.11 (A T^2-chart for the unexplained [orthogonal] overtime hours)
Consider the quality control analysis of the police department overtime hours in
Example 8.10. The first part of the quality monitoring procedure, the quality ellipse
based on the first two principal components, was shown in Figure 8.7. To illustrate
the second step of the two-step monitoring procedure, we create the chart for the
other principal components.

Since $p = 5$, this chart is based on $5 - 2 = 3$ dimensions, and the upper control
limit is $\chi_3^2(.05) = 7.81$. Using the eigenvalues and the values of the principal components, given in Example 8.10, we plot the time sequence of values

$$T_j^2 = \frac{\hat{y}_{j3}^2}{\hat{\lambda}_3} + \frac{\hat{y}_{j4}^2}{\hat{\lambda}_4} + \frac{\hat{y}_{j5}^2}{\hat{\lambda}_5}$$

where the first value is $T^2 = .891$ and so on. The T^2-chart is shown in Figure 8.8.

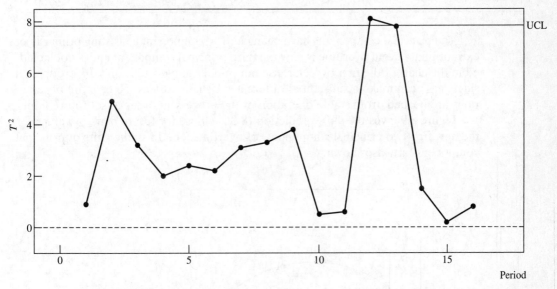

Figure 8.8 A T^2-chart based on the last three principal components of overtime hours.

Since points 12 and 13 exceed or are near the upper control limit, something has
happened during these periods. We note that they are just beyond the period in
which the extraordinary event overtime hours peaked.

From Table 8.2, $\hat{y}_{3j}$ is large in period 12, and from Table 8.1, the large coefficients
in $\mathbf{e}_3$ belong to legal appearances, holdover, and COA hours. Was there some adjusting of these other categories following the period extraordinary hours peaked? ■

Controlling Future Values

Previously, we considered checking whether a given series of multivariate observations was stable by considering separately the first two principal components and
then the last $p - 2$. Because the chi-square distribution was used to approximate
the UCL of the T^2-chart and the critical distance for the ellipse format chart, no further modifications are necessary for monitoring future values.

Example 8.12 (Control ellipse for future principal components) In Example 8.10, we determined that case 11 was out of control. We drop this point and recalculate the eigenvalues and eigenvectors based on the covariance of the remaining 15 observations. The results are shown in Table 8.3.

Table 8.3 Eigenvectors and Eigenvalues from the 15 Stable Observations

	$\hat{e}_1$	$\hat{e}_2$	$\hat{e}_3$	$\hat{e}_4$	$\hat{e}_5$
Appearances overtime (x_1)	.049	.629	.304	.479	.530
Extraordinary event (x_2)	.007	−.078	.939	−.260	−.212
Holdover hours (x_3)	−.662	.582	−.089	−.158	−.437
COA hours (x_4)	.731	.503	−.123	−.336	−.291
Meeting hours (x_5)	−.159	.081	−.058	−.752	.632
$\hat{\lambda}_i$	2,964,749.9	672,995.1	396,596.5	194,401.0	92,760.3

The principal components have changed. The component consisting primarily of extraordinary event overtime is now the third principal component and is not included in the chart of the first two. Because our initial sample size is only 16, dropping a single case can make a substantial difference. Usually, at least 50 or more observations are needed, from stable operation of the process, in order to set future limits.

Figure 8.9 gives the 99% prediction (8-36) ellipse for future pairs of values for the new first two principal components of overtime. The 15 stable pairs of principal components are also shown. ■

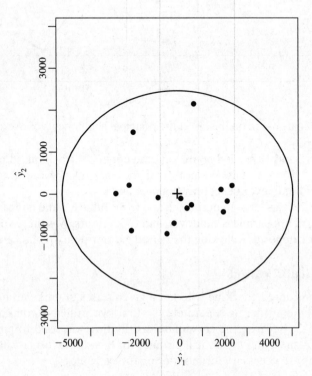

Figure 8.9 A 99% ellipse format chart for the first two principal components of future values of overtime.

In some applications of multivariate control in the chemical and pharmaceutical industries, more than 100 variables are monitored simultaneously. These include numerous process variables as well as quality variables. Typically, the space orthogonal to the first few principal components has a dimension greater than 100 and some of the eigenvalues are very small. An alternative approach (see [13]) to constructing a control chart, that avoids the difficulty caused by dividing a small squared principal component by a very small eigenvalue, has been successfully applied. To implement this approach, we proceed as follows.

For each stable observation, take the sum of squares of its unexplained component

$$d_{Uj}^2 = (\mathbf{x}_j - \bar{\mathbf{x}} - \hat{y}_{j1}\hat{\mathbf{e}}_1 - \hat{y}_{j2}\mathbf{e}_2)'(\mathbf{x}_j - \bar{\mathbf{x}} - \hat{y}_{j1}\hat{\mathbf{e}}_1 - \hat{y}_{j2}\hat{\mathbf{e}}_2)$$

Note that, by inserting $\hat{\mathbf{E}}\hat{\mathbf{E}}' = \mathbf{I}$, we also have

$$d_{Uj}^2 = (\mathbf{x}_j - \bar{\mathbf{x}} - \hat{y}_{j1}\hat{\mathbf{e}}_1 - \hat{y}_{j2}\hat{\mathbf{e}}_2)'\hat{\mathbf{E}}\hat{\mathbf{E}}'(\mathbf{x}_j - \bar{\mathbf{x}} - \hat{y}_{j1}\hat{\mathbf{e}}_1 - \hat{y}_{j2}\hat{\mathbf{e}}_2) = \sum_{k=3}^{p} \hat{y}_{jk}^2$$

which is just the sum of squares of the neglected principal components.

Using either form, the d_{Uj}^2 are plotted versus j to create a control chart. The lower limit of the chart is 0 and the upper limit is set by approximating the distribution of d_{Uj}^2 as the distribution of a constant c times a chi-square random variable with ν degrees of freedom.

For the chi-square approximation, the constant c and degrees of freedom ν are chosen to match the sample mean and variance of the d_{Uj}^2, $j = 1, 2, \ldots, n$. In particular, we set

$$\overline{d_U^2} = \frac{1}{n}\sum_{j=1}^{n} d_{Uj}^2 = c\,\nu$$

$$s_{d^2}^2 = \frac{1}{n-1}\sum_{j=1}^{n}(d_{Uj}^2 - \overline{d_U^2})^2 = 2c^2\nu$$

and determine

$$c = \frac{s_{d^2}^2}{2\overline{d_U^2}} \quad \text{and} \quad \nu = 2\frac{(\overline{d_U^2})^2}{s_{d^2}^2}$$

The upper control limit is then $c\chi_\nu^2(\alpha)$, where $\alpha = .05$ or $.01$.

8A

THE GEOMETRY OF THE SAMPLE PRINCIPAL COMPONENT APPROXIMATION

In this supplement, we shall present interpretations for approximations to the data based on the first r sample principal components. The interpretations of both the p-dimensional scatter plot and the n-dimensional representation rely on the algebraic result that follows. We consider approximations of the form $\underset{(n \times p)}{\mathbf{A}} = [\mathbf{a}_1, \mathbf{a}_2, \dots, \mathbf{a}_n]'$ to the mean corrected data matrix

$$[\mathbf{x}_1 - \bar{\mathbf{x}}, \mathbf{x}_2 - \bar{\mathbf{x}}, \dots, \mathbf{x}_n - \bar{\mathbf{x}}]'$$

The error of approximation is quantified as the sum of the np squared errors

$$\sum_{j=1}^{n} (\mathbf{x}_j - \bar{\mathbf{x}} - \mathbf{a}_j)'(\mathbf{x}_j - \bar{\mathbf{x}} - \mathbf{a}_j) = \sum_{j=1}^{n} \sum_{i=1}^{p} (x_{ji} - \bar{x}_i - a_{ji})^2 \qquad (8A\text{-}1)$$

Result 8A.I Let $\underset{(n \times p)}{\mathbf{A}}$ be any matrix with rank(A) $\le r < \min(p, n)$. Let $\hat{\mathbf{E}}_r = [\hat{\mathbf{e}}_1, \hat{\mathbf{e}}_2, \dots, \hat{\mathbf{e}}_r]$, where $\hat{\mathbf{e}}_i$ is the ith eigenvector of $\mathbf{S}$. The error of approximation sum of squares in (8A-1) is minimized by the choice

$$\hat{\mathbf{A}} = \begin{bmatrix} (\mathbf{x}_1 - \bar{\mathbf{x}})' \\ (\mathbf{x}_2 - \bar{\mathbf{x}})' \\ \vdots \\ (\mathbf{x}_n - \bar{\mathbf{x}})' \end{bmatrix} \hat{\mathbf{E}}_r \hat{\mathbf{E}}_r' = [\hat{\mathbf{y}}_1, \hat{\mathbf{y}}_2, \dots, \hat{\mathbf{y}}_r] \hat{\mathbf{E}}_r'$$

so the jth column of its transpose $\hat{\mathbf{A}}'$ is

$$\hat{\mathbf{a}}_j = \hat{y}_{j1} \hat{\mathbf{e}}_1 + \hat{y}_{j2} \hat{\mathbf{e}}_2 + \cdots + \hat{y}_{jr} \hat{\mathbf{e}}_r$$

where

$$[\hat{y}_{j1}, \hat{y}_{j2}, \ldots, \hat{y}_{jr}]' = [\hat{\mathbf{e}}_1'(\mathbf{x}_j - \bar{\mathbf{x}}), \hat{\mathbf{e}}_2'(\mathbf{x}_j - \bar{\mathbf{x}}), \ldots, \hat{\mathbf{e}}_r'(\mathbf{x}_j - \bar{\mathbf{x}})]'$$

are the values of the first r sample principal components for the jth unit. Moreover,

$$\sum_{j=1}^{n} (\mathbf{x}_j - \bar{\mathbf{x}} - \hat{\mathbf{a}}_j)'(\mathbf{x}_j - \bar{\mathbf{x}} - \hat{\mathbf{a}}_j) = (n-1)(\hat{\lambda}_{r+1} + \cdots + \hat{\lambda}_p)$$

where $\hat{\lambda}_{r+1} \geq \cdots \geq \hat{\lambda}_p$ are the smallest eigenvalues of $\mathbf{S}$.

Proof. Consider first any $\mathbf{A}$ whose transpose $\mathbf{A}'$ has columns $\mathbf{a}_j$ that are a linear combination of a *fixed* set of r perpendicular vectors $\mathbf{u}_1, \mathbf{u}_2, \ldots, \mathbf{u}_r$, so that $\mathbf{U} = [\mathbf{u}_1, \mathbf{u}_2, \ldots, \mathbf{u}_r]$ satisfies $\mathbf{U}'\mathbf{U} = \mathbf{I}$. For fixed $\mathbf{U}$, $\mathbf{x}_j - \bar{\mathbf{x}}$ is best approximated by its projection on the space spanned by $\mathbf{u}_1, \mathbf{u}_2, \ldots, \mathbf{u}_r$ (see Result 2A.3), or

$$(\mathbf{x}_j - \bar{\mathbf{x}})'\mathbf{u}_1\mathbf{u}_1 + (\mathbf{x}_j - \bar{\mathbf{x}})'\mathbf{u}_2\mathbf{u}_2 + \cdots + (\mathbf{x}_j - \bar{\mathbf{x}})'\mathbf{u}_r\mathbf{u}_r$$

$$= [\mathbf{u}_1, \mathbf{u}_2, \ldots, \mathbf{u}_r]\begin{bmatrix} \mathbf{u}_1'(\mathbf{x}_j - \bar{\mathbf{x}}) \\ \mathbf{u}_2'(\mathbf{x}_j - \bar{\mathbf{x}}) \\ \vdots \\ \mathbf{u}_r'(\mathbf{x}_j - \bar{\mathbf{x}}) \end{bmatrix} = \mathbf{U}\mathbf{U}'(\mathbf{x}_j - \bar{\mathbf{x}}) \qquad (8A\text{-}2)$$

This follows because, for an arbitrary vector $\mathbf{b}_j$,

$$\mathbf{x}_j - \bar{\mathbf{x}} - \mathbf{U}\mathbf{b}_j = \mathbf{x}_j - \bar{\mathbf{x}} - \mathbf{U}\mathbf{U}'(\mathbf{x}_j - \bar{\mathbf{x}}) + \mathbf{U}\mathbf{U}'(\mathbf{x}_j - \bar{\mathbf{x}}) - \mathbf{U}\mathbf{b}_j$$
$$= (\mathbf{I} - \mathbf{U}\mathbf{U}')(\mathbf{x}_j - \bar{\mathbf{x}}) + \mathbf{U}(\mathbf{U}'(\mathbf{x}_j - \bar{\mathbf{x}}) - \mathbf{b}_j)$$

so the error sum of squares is

$$(\mathbf{x}_j - \bar{\mathbf{x}} - \mathbf{U}\mathbf{b}_j)'(\mathbf{x}_j - \bar{\mathbf{x}} - \mathbf{U}\mathbf{b}_j) = (\mathbf{x}_j - \bar{\mathbf{x}})'(\mathbf{I} - \mathbf{U}\mathbf{U}')(\mathbf{x}_j - \bar{\mathbf{x}}) + 0$$
$$+ (\mathbf{U}'(\mathbf{x}_j - \bar{\mathbf{x}}) - \mathbf{b}_j)'(\mathbf{U}'(\mathbf{x}_j - \bar{\mathbf{x}}) - \mathbf{b}_j)$$

where the cross product vanishes because $(\mathbf{I} - \mathbf{U}\mathbf{U}')\mathbf{U} = \mathbf{U} - \mathbf{U}\mathbf{U}'\mathbf{U} = \mathbf{U} - \mathbf{U} = \mathbf{0}$. The last term is positive unless $\mathbf{b}_j$ is chosen so that $\mathbf{b}_j = \mathbf{U}'(\mathbf{x}_j - \bar{\mathbf{x}})$ and $\mathbf{U}\mathbf{b}_j = \mathbf{U}\mathbf{U}'(\mathbf{x}_j - \bar{\mathbf{x}})$ is the projection of $\mathbf{x}_j - \bar{\mathbf{x}}$ on the plane.

Further, with the choice $\mathbf{a}_j = \mathbf{U}\mathbf{b}_j = \mathbf{U}\mathbf{U}'(\mathbf{x}_j - \bar{\mathbf{x}})$, (8A-1) becomes

$$\sum_{j=1}^{n} (\mathbf{x}_j - \bar{\mathbf{x}} - \mathbf{U}\mathbf{U}'(\mathbf{x}_j - \bar{\mathbf{x}}))'(\mathbf{x}_j - \bar{\mathbf{x}} - \mathbf{U}\mathbf{U}'(\mathbf{x}_j - \bar{\mathbf{x}}))$$

$$= \sum_{j=1}^{n} (\mathbf{x}_j - \bar{\mathbf{x}})'(\mathbf{I} - \mathbf{U}\mathbf{U}')(\mathbf{x}_j - \bar{\mathbf{x}})$$

$$= \sum_{j=1}^{n} (\mathbf{x}_j - \bar{\mathbf{x}})'(\mathbf{x}_j - \bar{\mathbf{x}}) - \sum_{j=1}^{n} (\mathbf{x}_j - \bar{\mathbf{x}})'\mathbf{U}\mathbf{U}'(\mathbf{x}_j - \bar{\mathbf{x}}) \qquad (8A\text{-}3)$$

We are now in a position to minimize the error over choices of $\mathbf{U}$ by maximizing the last term in (8A-3). By the properties of trace (see Result 2A.12),

$$\sum_{j=1}^{n} (\mathbf{x}_j - \bar{\mathbf{x}})'\mathbf{U}\mathbf{U}'(\mathbf{x}_j - \bar{\mathbf{x}}) = \sum_{j=1}^{n} \text{tr}[(\mathbf{x}_j - \bar{\mathbf{x}})'\mathbf{U}\mathbf{U}'(\mathbf{x}_j - \bar{\mathbf{x}})]$$

$$= \sum_{j=1}^{n} \text{tr}[\mathbf{U}\mathbf{U}'(\mathbf{x}_j - \bar{\mathbf{x}})(\mathbf{x}_j - \bar{\mathbf{x}})']$$

$$= (n-1)\,\text{tr}[\mathbf{U}\mathbf{U}'\mathbf{S}] = (n-1)\,\text{tr}[\mathbf{U}'\mathbf{S}\mathbf{U}] \qquad (8A\text{-}4)$$

That is, the best choice for $\mathbf{U}$ maximizes the sum of the diagonal elements of $\mathbf{U}'\mathbf{S}\mathbf{U}$. From (8-19), selecting $\mathbf{u}_1$ to maximize $\mathbf{u}_1'\mathbf{S}\mathbf{u}_1$, the first diagonal element of $\mathbf{U}'\mathbf{S}\mathbf{U}$, gives $\mathbf{u}_1 = \hat{\mathbf{e}}_1$. For $\mathbf{u}_2$ perpendicular to $\hat{\mathbf{e}}_1$, $\mathbf{u}_2'\mathbf{S}\mathbf{u}_2$ is maximized by $\hat{\mathbf{e}}_2$. [See (2-52).] Continuing, we find that $\hat{\mathbf{U}} = [\hat{\mathbf{e}}_1, \hat{\mathbf{e}}_2, \ldots, \hat{\mathbf{e}}_r] = \hat{\mathbf{E}}_r$ and $\hat{\mathbf{A}}' = \hat{\mathbf{E}}_r\hat{\mathbf{E}}_r'[\mathbf{x}_1 - \bar{\mathbf{x}}, \mathbf{x}_2 - \bar{\mathbf{x}}, \ldots, \mathbf{x}_n - \bar{\mathbf{x}}]$, as asserted.

With this choice the ith diagonal element of $\hat{\mathbf{U}}'\mathbf{S}\hat{\mathbf{U}}$ is $\hat{\mathbf{e}}_i'\mathbf{S}\hat{\mathbf{e}}_i = \hat{\mathbf{e}}_i'(\hat{\lambda}_i\hat{\mathbf{e}}_i) = \hat{\lambda}_i$ so $\mathrm{tr}[\hat{\mathbf{U}}'\mathbf{S}\hat{\mathbf{U}}] = \hat{\lambda}_1 + \hat{\lambda}_2 + \cdots + \hat{\lambda}_r$. Also, $\sum_{j=1}^{n} (\mathbf{x}_j - \bar{\mathbf{x}})'(\mathbf{x}_j - \bar{\mathbf{x}}) = \mathrm{tr}\left[\sum_{j=1}^{n} (\mathbf{x}_j - \bar{\mathbf{x}})(\mathbf{x}_j - \bar{\mathbf{x}})'\right]$ $= (n-1)\,\mathrm{tr}(\mathbf{S}) = (n-1)(\hat{\lambda}_1 + \hat{\lambda}_2 + \cdots + \hat{\lambda}_p)$. Let $\mathbf{U} = \hat{\mathbf{U}}$ in (8A-3), and the error bound follows. ∎

The p-Dimensional Geometrical Interpretation

The geometrical interpretations involve the determination of best approximating planes to the p-dimensional scatter plot. The plane through the origin, determined by $\mathbf{u}_1, \mathbf{u}_2, \ldots, \mathbf{u}_r$, consists of all points $\mathbf{x}$ with

$$\mathbf{x} = b_1\mathbf{u}_1 + b_2\mathbf{u}_2 + \cdots + b_r\mathbf{u}_r = \mathbf{U}\mathbf{b}, \qquad \text{for some } \mathbf{b}$$

This plane, translated to pass through $\mathbf{a}$, becomes $\mathbf{a} + \mathbf{U}\mathbf{b}$ for some $\mathbf{b}$.

We want to select the r-dimensional plane $\mathbf{a} + \mathbf{U}\mathbf{b}$ that *minimizes the sum of squared distances* $\sum_{j=1}^{n} d_j^2$ *between the observations* $\mathbf{x}_j$ *and the plane*. If $\mathbf{x}_j$ is approximated by $\mathbf{a} + \mathbf{U}\mathbf{b}_j$ with $\sum_{j=1}^{n} \mathbf{b}_j = \mathbf{0}$,[5] then

$$\sum_{j=1}^{n} (\mathbf{x}_j - \mathbf{a} - \mathbf{U}\mathbf{b}_j)'(\mathbf{x}_j - \mathbf{a} - \mathbf{U}\mathbf{b}_j)$$

$$= \sum_{j=1}^{n} (\mathbf{x}_j - \bar{\mathbf{x}} - \mathbf{U}\mathbf{b}_j + \bar{\mathbf{x}} - \mathbf{a})'(\mathbf{x}_j - \bar{\mathbf{x}} - \mathbf{U}\mathbf{b}_j + \bar{\mathbf{x}} - \mathbf{a})$$

$$= \sum_{j=1}^{n} (\mathbf{x}_j - \bar{\mathbf{x}} - \mathbf{U}\mathbf{b}_j)'(\mathbf{x}_j - \bar{\mathbf{x}} - \mathbf{U}\mathbf{b}_j) + n(\bar{\mathbf{x}} - \mathbf{a})'(\bar{\mathbf{x}} - \mathbf{a})$$

$$\geq \sum_{j=1}^{n} (\mathbf{x}_j - \bar{\mathbf{x}} - \hat{\mathbf{E}}_r\hat{\mathbf{E}}_r'(\mathbf{x}_j - \bar{\mathbf{x}}))'(\mathbf{x}_j - \bar{\mathbf{x}} - \hat{\mathbf{E}}_r\hat{\mathbf{E}}_r'(\mathbf{x}_j - \bar{\mathbf{x}}))$$

by Result 8A.1, since $[\mathbf{U}\mathbf{b}_1, \ldots, \mathbf{U}\mathbf{b}_n] = \mathbf{A}'$ has rank $(\mathbf{A}) \leq r$. The lower bound is reached by taking $\mathbf{a} = \bar{\mathbf{x}}$, so the plane passes through the sample mean. This plane is determined by $\hat{\mathbf{e}}_1, \hat{\mathbf{e}}_2, \ldots, \hat{\mathbf{e}}_r$. The coefficients of $\hat{\mathbf{e}}_k$ are $\hat{\mathbf{e}}_k'(\mathbf{x}_j - \bar{\mathbf{x}}) = \hat{y}_{jk}$, the kth sample principal component evaluated at the jth observation.

The approximating plane interpretation of sample principal components is illustrated in Figure 8.10.

An alternative interpretation can be given. The investigator places a plane through $\bar{\mathbf{x}}$ and moves it about to obtain the *largest spread* among the shadows of the

---

[5] If $\sum_{j=1}^{n} \mathbf{b}_j = n\bar{\mathbf{b}} \neq \mathbf{0}$, use $\mathbf{a} + \mathbf{U}\mathbf{b}_j = (\mathbf{a} + \mathbf{U}\bar{\mathbf{b}}) + \mathbf{U}(\mathbf{b}_j - \bar{\mathbf{b}}) = \mathbf{a}^* + \mathbf{U}\mathbf{b}_j^*$.

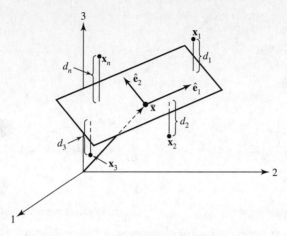

Figure 8.10 The $r = 2$-dimensional plane that approximates the scatter plot by minimizing $\sum\limits_{j=1}^{n} d_j^2$.

observations. From (8A-2), the projection of the deviation $\mathbf{x}_j - \bar{\mathbf{x}}$ on the plane $\mathbf{Ub}$ is $\mathbf{v}_j = \mathbf{UU}'(\mathbf{x}_j - \bar{\mathbf{x}})$. Now, $\bar{\mathbf{v}} = \mathbf{0}$ and the *sum of the squared lengths of the projection deviations*

$$\sum_{j=1}^{n} \mathbf{v}_j'\mathbf{v}_j = \sum_{j=1}^{n} (\mathbf{x}_j - \bar{\mathbf{x}})'\mathbf{UU}'(\mathbf{x}_j - \bar{\mathbf{x}}) = (n-1)\,\mathrm{tr}[\mathbf{U}'\mathbf{SU}]$$

is maximized by $\mathbf{U} = \hat{\mathbf{E}}$. Also, since $\bar{\mathbf{v}} = \mathbf{0}$,

$$(n-1)\mathbf{S}_{\mathbf{v}} = \sum_{j=1}^{n} (\mathbf{v}_j - \bar{\mathbf{v}})(\mathbf{v}_j - \bar{\mathbf{v}})' = \sum_{j=1}^{n} \mathbf{v}_j\mathbf{v}_j'$$

and this plane also maximizes the total variance

$$\mathrm{tr}\,(\mathbf{S}_{\mathbf{v}}) = \frac{1}{(n-1)}\,\mathrm{tr}\left[\sum_{j=1}^{n} \mathbf{v}_j\mathbf{v}_j'\right] = \frac{1}{(n-1)}\,\mathrm{tr}\left[\sum_{j=1}^{n} \mathbf{v}_j'\mathbf{v}_j\right]$$

The n-Dimensional Geometrical Interpretation

Let us now consider, by columns, the approximation of the mean-centered data matrix by $\mathbf{A}$. For $r = 1$, the ith column $[x_{1i} - \bar{x}_i, x_{2i} - \bar{x}_i, \ldots, x_{ni} - \bar{x}_i]'$ is approximated by a multiple $c_i\mathbf{b}'$ of a fixed vector $\mathbf{b}' = [b_1, b_2, \ldots, b_n]$. The square of the length of the error of approximation is

$$L_i^2 = \sum_{j=1}^{n} (x_{ji} - \bar{x}_i - c_i b_j)^2$$

Considering $\underset{(n \times p)}{\mathbf{A}}$ to be of rank one, we conclude from Result 8A.1 that

$$\hat{\mathbf{A}} = \begin{bmatrix} \hat{\mathbf{e}}_1\hat{\mathbf{e}}_1'(\mathbf{x}_1 - \bar{\mathbf{x}}) \\ \hat{\mathbf{e}}_1\hat{\mathbf{e}}_1'(\mathbf{x}_2 - \bar{\mathbf{x}}) \\ \vdots \\ \hat{\mathbf{e}}_1\hat{\mathbf{e}}_1'(\mathbf{x}_n - \bar{\mathbf{x}}) \end{bmatrix} = \begin{bmatrix} \hat{y}_{11} \\ \hat{y}_{21} \\ \vdots \\ \hat{y}_{1n} \end{bmatrix}\hat{\mathbf{e}}_1'$$

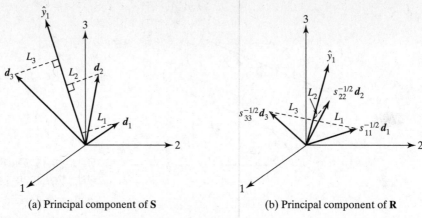

(a) Principal component of **S**   (b) Principal component of **R**

Figure 8.11 The first sample principal component, $\hat{y}_1$, minimizes the sum of the squares of the distances, L_i^2, from the deviation vectors, $d_i' = [x_{1i} - \bar{x}_i, x_{2i} - \bar{x}_i, \ldots, x_{ni} - \bar{x}_i]$, to a line.

minimizes the sum of squared lengths $\sum\limits_{i=1}^{p} L_i^2$. That is, the best direction is determined by the vector of values of the first principal component. This is illustrated in Figure 8.11(a). Note that the longer deviation vectors (the larger s_{ii}'s) have the most influence on the minimization of $\sum\limits_{i=1}^{p} L_i^2$.

If the variables are first standardized, the resulting vector $[(x_{1i} - \bar{x}_i)/\sqrt{s_{ii}}, (x_{2i} - \bar{x}_i)/\sqrt{s_{ii}}, \ldots, (x_{ni} - \bar{x}_i)/\sqrt{s_{ii}}]$ has length $n - 1$ for all variables, and each vector exerts equal influence on the choice of direction. [See Figure 8.11(b).]

In either case, the vector **b** is moved around in n-space to minimize the sum of the squares of the distances $\sum\limits_{i=1}^{p} L_i^2$. In the former case L_i^2 is the squared distance between $[x_{1i} - \bar{x}_i, x_{2i} - \bar{x}_i, \ldots, x_{ni} - \bar{x}_i]'$ and its projection on the line determined by **b**. The second principal component minimizes the same quantity among all vectors perpendicular to the first choice.

Exercises

8.1. Determine the population principal components Y_1 and Y_2 for the covariance matrix

$$\Sigma = \begin{bmatrix} 5 & 2 \\ 2 & 2 \end{bmatrix}$$

Also, calculate the proportion of the total population variance explained by the first principal component.

8.2. Convert the covariance matrix in Exercise 8.1 to a correlation matrix $\boldsymbol{\rho}$.

(a) Determine the principal components Y_1 and Y_2 from $\boldsymbol{\rho}$ and compute the proportion of total population variance explained by Y_1.

(b) Compare the components calculated in Part a with those obtained in Exercise 8.1. Are they the same? Should they be?

(c) Compute the correlations ρ_{Y_1,Z_1}, ρ_{Y_1,Z_2}, and ρ_{Y_2,Z_1}.

8.3. Let

$$\Sigma = \begin{bmatrix} 2 & 0 & 0 \\ 0 & 4 & 0 \\ 0 & 0 & 4 \end{bmatrix}$$

Determine the principal components Y_1, Y_2, and Y_3. What can you say about the eigenvectors (and principal components) associated with eigenvalues that are not distinct?

8.4. Find the principal components and the proportion of the total population variance explained by each when the covariance matrix is

$$\Sigma = \begin{bmatrix} \sigma^2 & \sigma^2\rho & 0 \\ \sigma^2\rho & \sigma^2 & \sigma^2\rho \\ 0 & \sigma^2\rho & \sigma^2 \end{bmatrix}, \quad -\frac{1}{\sqrt{2}} < \rho < \frac{1}{\sqrt{2}}$$

8.5. (a) Find the eigenvalues of the correlation matrix

$$\boldsymbol{\rho} = \begin{bmatrix} 1 & \rho & \rho \\ \rho & 1 & \rho \\ \rho & \rho & 1 \end{bmatrix}$$

Are your results consistent with (8-16) and (8-17)?

(b) Verify the eigenvalue–eigenvector pairs for the $p \times p$ matrix $\boldsymbol{\rho}$ given in (8-15).

8.6. Data on $x_1 = $ sales and $x_2 = $ profits for the 10 largest companies in the world were listed in Exercise 1.4 of Chapter 1. From Example 4.12

$$\bar{\mathbf{x}} = \begin{bmatrix} 155.60 \\ 14.70 \end{bmatrix}, \quad \mathbf{S} = \begin{bmatrix} 7476.45 & 303.62 \\ 303.62 & 26.19 \end{bmatrix}$$

(a) Determine the sample principal components and their variances for these data. (You may need the quadratic formula to solve for the eigenvalues of $\mathbf{S}$.)

(b) Find the proportion of the total sample variance explained by $\hat{y}_1$.

(c) Sketch the constant density ellipse $(\mathbf{x} - \bar{\mathbf{x}})'\mathbf{S}^{-1}(\mathbf{x} - \bar{\mathbf{x}}) = 1.4$, and indicate the principal components $\hat{y}_1$ and $\hat{y}_2$ on your graph.

(d) Compute the correlation coefficients $r_{\hat{y}_1, x_k}$, $k = 1, 2$. What interpretation, if any, can you give to the first principal component?

8.7. Convert the covariance matrix $\mathbf{S}$ in Exercise 8.6 to a sample correlation matrix $\mathbf{R}$.

(a) Find the sample principal components $\hat{y}_1$, $\hat{y}_2$ and their variances.

(b) Compute the proportion of the total sample variance explained by $\hat{y}_1$.

(c) Compute the correlation coefficients $r_{\hat{y}_1, z_k}$, $k = 1, 2$. Interpret $\hat{y}_1$.

(d) Compare the components obtained in Part a with those obtained in Exercise 8.6(a). Given the original data displayed in Exercise 1.4, do you feel that it is better to determine principal components from the sample covariance matrix or sample correlation matrix? Explain.

8.8. Use the results in Example 8.5.

(a) Compute the correlations r_{y_i, z_k} for $i = 1, 2$ and $k = 1, 2, \ldots, 5$. Do these correlations reinforce the interpretations given to the first two components? Explain.

(b) Test the hypothesis

$$H_0: \boldsymbol{\rho} = \boldsymbol{\rho}_0 = \begin{bmatrix} 1 & \rho & \rho & \rho & \rho \\ \rho & 1 & \rho & \rho & \rho \\ \rho & \rho & 1 & \rho & \rho \\ \rho & \rho & \rho & 1 & \rho \\ \rho & \rho & \rho & \rho & 1 \end{bmatrix}$$

versus

$$H_1: \boldsymbol{\rho} \neq \boldsymbol{\rho}_0$$

at the 5% level of significance. List any assumptions required in carrying out this test.

8.9. *(A test that all variables are independent.)*

(a) Consider that the normal theory likelihood ratio test of H_0: $\boldsymbol{\Sigma}$ is the diagonal matrix

$$\begin{bmatrix} \sigma_{11} & 0 & \cdots & 0 \\ 0 & \sigma_{22} & \cdots & 0 \\ \vdots & \vdots & \ddots & \vdots \\ 0 & 0 & \cdots & \sigma_{pp} \end{bmatrix}, \quad \sigma_{ii} > 0$$

Show that the test is as follows: Reject H_0 if

$$\Lambda = \frac{|\mathbf{S}|^{n/2}}{\displaystyle\prod_{i=1}^{p} s_{ii}^{n/2}} = |\mathbf{R}|^{n/2} < c$$

For a large sample size, $-2 \ln \Lambda$ is approximately $\chi^2_{p(p-1)/2}$. Bartlett [3] suggests that the test statistic $-2[1 - (2p + 11)/6n] \ln \Lambda$ be used in place of $-2 \ln \Lambda$. This results in an improved chi-square approximation. The large sample α critical point is $\chi^2_{p(p-1)/2}(\alpha)$. Note that testing $\boldsymbol{\Sigma} = \boldsymbol{\Sigma}_0$ is the same as testing $\boldsymbol{\rho} = \mathbf{I}$.

(b) Show that the likelihood ratio test of H_0: $\boldsymbol{\Sigma} = \sigma^2 \mathbf{I}$ rejects H_0 if

$$\Lambda = \frac{|\mathbf{S}|^{n/2}}{(\text{tr}(\mathbf{S})/p)^{np/2}} = \left[\frac{\displaystyle\prod_{i=1}^{p} \hat{\lambda}_i}{\left(\dfrac{1}{p} \displaystyle\sum_{i=1}^{p} \hat{\lambda}_i\right)^{p}} \right]^{n/2} = \left[\frac{\text{geometric mean } \hat{\lambda}_i}{\text{arithmetic mean } \hat{\lambda}_i} \right]^{np/2} < c$$

For a large sample size, Bartlett [3] suggests that

$$-2[1 - (2p^2 + p + 2)/6pn] \ln \Lambda$$

is approximately $\chi^2_{(p+2)(p-1)/2}$. Thus, the large sample α critical point is $\chi^2_{(p+2)(p-1)/2}(\alpha)$. This test is called a *sphericity test*, because the constant density contours are spheres when $\boldsymbol{\Sigma} = \sigma^2 \mathbf{I}$.

Hint:

(a) $\max\limits_{\mu,\Sigma} L(\mu,\Sigma)$ is given by (5-10), and $\max L(\mu,\Sigma_0)$ is the product of the univariate

likelihoods, $\max\limits_{\mu_i \sigma_{ii}} (2\pi)^{-n/2} \sigma_{ii}^{-n/2} \exp\left[-\sum\limits_{j=1}^{n}(x_{ji}-\mu_i)^2/2\sigma_{ii}\right]$. Hence $\hat{\mu}_i = n^{-1}\sum\limits_{j=1}^{n} x_{ji}$

and $\hat{\sigma}_{ii} = (1/n)\sum\limits_{j=1}^{n}(x_{ji}-\bar{x}_i)^2$. The divisor n cancels in Λ, so S may be used.

(b) Verify $\hat{\sigma}^2 = \left[\sum\limits_{j=1}^{n}(x_{j1}-\bar{x}_1)^2 + \cdots + \sum\limits_{j=1}^{n}(x_{jp}-\bar{x}_p)^2\right]\Big/ np$ under H_0. Again,

the divisors n cancel in the statistic, so S may be used. Use Result 5.2 to calculate the chi-square degrees of freedom.

The following exercises require the use of a computer.

8.10. The weekly rates of return for five stocks listed on the New York Stock Exchange are given in Table 8.4. (See the stock-price data on the following website: www.prenhall.com/statistics.)

(a) Construct the sample covariance matrix S, and find the sample principal components in (8-20). (Note that the sample mean vector $\bar{x}$ is displayed in Example 8.5.)

(b) Determine the proportion of the total sample variance explained by the first three principal components. Interpret these components.

(c) Construct Bonferroni simultaneous 90% confidence intervals for the variances $\lambda_1, \lambda_2,$ and λ_3 of the first three population components $Y_1, Y_2,$ and Y_3.

(d) Given the results in Parts a–c, do you feel that the stock rates-of-return data can be summarized in fewer than five dimensions? Explain.

Table 8.4 Stock-Price Data (Weekly Rate Of Return)

Week	J P Morgan	Citibank	Wells Fargo	Royal Dutch Shell	Exxon Mobil
1	0.01303	−0.00784	−0.00319	−0.04477	0.00522
2	0.00849	0.01669	−0.00621	0.01196	0.01349
3	−0.01792	−0.00864	0.01004	0	−0.00614
4	0.02156	−0.00349	0.01744	−0.02859	−0.00695
5	0.01082	0.00372	−0.01013	0.02919	0.04098
6	0.01017	−0.01220	−0.00838	0.01371	0.00299
7	0.01113	0.02800	0.00807	0.03054	0.00323
8	0.04848	−0.00515	0.01825	0.00633	0.00768
9	−0.03449	−0.01380	−0.00805	−0.02990	−0.01081
10	−0.00466	0.02099	−0.00608	−0.02039	−0.01267
⋮	⋮	⋮	⋮	⋮	⋮
94	0.03732	0.03593	0.02528	0.05819	0.01697
95	0.02380	0.00311	−0.00688	0.01225	0.02817
96	0.02568	0.05253	0.04070	−0.03166	−0.01885
97	−0.00606	0.00863	0.00584	0.04456	0.03059
98	0.02174	0.02296	0.02920	0.00844	0.03193
99	0.00337	−0.01531	−0.02382	−0.00167	−0.01723
100	0.00336	0.00290	−0.00305	−0.00122	−0.00970
101	0.01701	0.00951	0.01820	−0.01618	−0.00756
102	0.01039	−0.00266	0.00443	−0.00248	−0.01645
103	−0.01279	−0.01437	−0.01874	−0.00498	−0.01637

8.11. Consider the census-tract data listed in Table 8.5. Suppose the observations on X_5 = median value home were recorded in ten thousands, rather than hundred thousands, of dollars; that is, multiply all the numbers listed in the sixth column of the table by 10.

(a) Construct the sample covariance matrix $\mathbf{S}$ for the census-tract data when X_5 = median value home is recorded in ten thousands of dollars. (Note that this covariance matrix can be obtained from the covariance matrix given in Example 8.3 by multiplying the off-diagonal elements in the fifth column and row by 10 and the diagonal element s_{55} by 100. Why?)

(b) Obtain the eigenvalue–eigenvector pairs and the first two sample principal components for the covariance matrix in Part a.

(c) Compute the proportion of total variance explained by the first two principal components obtained in Part b. Calculate the correlation coefficients, r_{y_i, x_k}, and interpret these components if possible. Compare your results with the results in Example 8.3. What can you say about the effects of this change in scale on the principal components?

8.12. Consider the air-pollution data listed in Table 1.5. Your job is to summarize these data in fewer than $p = 7$ dimensions if possible. Conduct a principal component analysis of the data using both the covariance matrix $\mathbf{S}$ and the correlation matrix $\mathbf{R}$. What have you learned? Does it make any difference which matrix is chosen for analysis? Can the data be summarized in three or fewer dimensions? Can you interpret the principal components?

Table 8.5 Census-tract Data

Tract	Total population (thousands)	Professional degree (percent)	Employed age over 16 (percent)	Government employment (percent)	Median home value ($100,000)
1	2.67	5.71	69.02	30.3	1.48
2	2.25	4.37	72.98	43.3	1.44
3	3.12	10.27	64.94	32.0	2.11
4	5.14	7.44	71.29	24.5	1.85
5	5.54	9.25	74.94	31.0	2.23
6	5.04	4.84	53.61	48.2	1.60
7	3.14	4.82	67.00	37.6	1.52
8	2.43	2.40	67.20	36.8	1.40
9	5.38	4.30	83.03	19.7	2.07
10	7.34	2.73	72.60	24.5	1.42
⋮	⋮	⋮	⋮	⋮	⋮
52	7.25	1.16	78.52	23.6	1.50
53	5.44	2.93	73.59	22.3	1.65
54	5.83	4.47	77.33	26.2	2.16
55	3.74	2.26	79.70	20.2	1.58
56	9.21	2.36	74.58	21.8	1.72
57	2.14	6.30	86.54	17.4	2.80
58	6.62	4.79	78.84	20.0	2.33
59	4.24	5.82	71.39	27.1	1.69
60	4.72	4.71	78.01	20.6	1.55
61	6.48	4.93	74.23	20.9	1.98

Note: Observations from adjacent census tracts are likely to be correlated. That is, these 61 observations may not constitute a random sample. Complete data set available at www.prenhall.com/statistics.

8.13. In the radiotherapy data listed in Table 1.7 (see also the radiotherapy data on the website www.prenhall.com/statistics), the $n = 98$ observations on $p = 6$ variables represent patients' reactions to radiotherapy.

(a) Obtain the covariance and correlation matrices $\mathbf{S}$ and $\mathbf{R}$ for these data.

(b) Pick one of the matrices $\mathbf{S}$ or $\mathbf{R}$ (justify your choice), and determine the eigenvalues and eigenvectors. Prepare a table showing, in decreasing order of size, the percent that each eigenvalue contributes to the total sample variance.

(c) Given the results in Part b, decide on the number of important sample principal components. Is it possible to summarize the radiotherapy data with a single reaction-index component? Explain.

(d) Prepare a table of the correlation coefficients between each principal component you decide to retain and the original variables. If possible, interpret the components.

8.14. Perform a principal component analysis using the sample covariance matrix of the sweat data given in Example 5.2. Construct a Q–Q plot for each of the important principal components. Are there any suspect observations? Explain.

8.15. The four sample standard deviations for the postbirth weights discussed in Example 8.6 are

$$\sqrt{s_{11}} = 32.9909, \quad \sqrt{s_{22}} = 33.5918, \quad \sqrt{s_{33}} = 36.5534, \quad \text{and} \quad \sqrt{s_{44}} = 37.3517$$

Use these and the correlations given in Example 8.6 to construct the sample covariance matrix $\mathbf{S}$. Perform a principal component analysis using $\mathbf{S}$.

8.16. Over a period of five years in the 1990s, yearly samples of fishermen on 28 lakes in Wisconsin were asked to report the time they spent fishing and how many of each type of game fish they caught. Their responses were then converted to a catch rate per hour for

$$x_1 = \text{Bluegill} \qquad x_2 = \text{Black crappie} \quad x_3 = \text{Smallmouth bass}$$

$$x_4 = \text{Largemouth bass} \quad x_5 = \text{Walleye} \qquad x_6 = \text{Northern pike}$$

The estimated correlation matrix (courtesy of Jodi Barnet)

$$\mathbf{R} = \begin{bmatrix} 1 & .4919 & .2636 & .4653 & -.2277 & .0652 \\ .4919 & 1 & .3127 & .3506 & -.1917 & .2045 \\ .2635 & .3127 & 1 & .4108 & .0647 & .2493 \\ .4653 & .3506 & .4108 & 1 & -.2249 & .2293 \\ -.2277 & -.1917 & .0647 & -.2249 & 1 & -.2144 \\ .0652 & .2045 & .2493 & .2293 & -.2144 & 1 \end{bmatrix}$$

is based on a sample of about 120. (There were a few missing values.)

Fish caught by the same fisherman live alongside of each other, so the data should provide some evidence on how the fish group. The first four fish belong to the centrarchids, the most plentiful family. The walleye is the most popular fish to eat.

(a) Comment on the pattern of correlation within the centrarchid family x_1 through x_4. Does the walleye appear to group with the other fish?

(b) Perform a principal component analysis using only x_1 through x_4. Interpret your results.

(c) Perform a principal component analysis using all six variables. Interpret your results.

8.17. Using the data on bone mineral content in Table 1.8, perform a principal component analysis of **S**.

8.18. The data on national track records for women are listed in Table 1.9.

p. 44
data: T1-9

HW
8.18 − 8.21
are linked

(a) Obtain the sample correlation matrix **R** for these data, and determine its eigenvalues and eigenvectors.

(b) Determine the first two principal components for the standardized variables. Prepare a table showing the correlations of the standardized variables with the components, and the cumulative percentage of the total (standardized) sample variance explained by the two components.

(c) Interpret the two principal components obtained in Part b. (Note that the first component is essentially a normalized unit vector and might measure the athletic excellence of a given nation. The second component might measure the relative strength of a nation at the various running distances.)

(d) Rank the nations based on their score on the first principal component. Does this ranking correspond with your inituitive notion of athletic excellence for the various countries?

8.19. Refer to Exercise 8.18. Convert the national track records for women in Table 1.9 to speeds measured in meters per second. Notice that the records for 800 m, 1500 m, 3000 m, and the marathon are given in minutes. The marathon is 26.2 miles, or 42,195 meters, long. Perform a principal components analysis using the covariance matrix **S** of the speed data. Compare the results with the results in Exercise 8.18. Do your interpretations of the components differ? If the nations are ranked on the basis of their score on the first principal component, does the subsequent ranking differ from that in Exercise 8.18? Which analysis do you prefer? Why?

8.20. The data on national track records for men are listed in Table 8.6. (See also the data on national track records for men on the website www.prenhall.com/statistics) Repeat the principal component analysis outlined in Exercise 8.18 for the men. Are the results consistent with those obtained from the women's data?

8.21. Refer to Exercise 8.20. Convert the national track records for men in Table 8.6 to speeds measured in meters per second. Notice that the records for 800 m, 1500 m, 5000 m, 10,000 m and the marathon are given in minutes. The marathon is 26.2 miles, or 42,195 meters, long. Perform a principal component analysis using the covariance matrix **S** of the speed data. Compare the results with the results in Exercise 8.20. Which analysis do you prefer? Why?

8.22. Consider the data on bulls in Table 1.10. Utilizing the seven variables YrHgt, FtFrBody, PrctFFB, Frame, BkFat, SaleHt, and SaleWt, perform a principal component analysis using the covariance matrix **S** and the correlation matrix **R**. Your analysis should include the following:

(a) Determine the appropriate number of components to effectively summarize the sample variability. Construct a scree plot to aid your determination.

(b) Interpret the sample principal components.

(c) Do you think it is possible to develop a "body size" or "body configuration" index from the data on the seven variables above? Explain.

(d) Using the values for the first two principal components, plot the data in a two-dimensional space with $\hat{y}_1$ along the vertical axis and $\hat{y}_2$ along the horizontal axis. Can you distinguish groups representing the three breeds of cattle? Are there any outliers?

(e) Construct a Q–Q plot using the first principal component. Interpret the plot.

Table 8.6 National Track Records for Men   now 8 variables

Country	100 m (s)	200 m (s)	400 m (s)	800 m (min)	1500 m (min)	5000 m (min)	10,000 m (min)	Marathon (min)
Argentina	10.23	20.37	46.18	1.77	3.68	13.33	27.65	129.57
Australia	9.93	20.06	44.38	1.74	3.53	12.93	27.53	127.51
Austria	10.15	20.45	45.80	1.77	3.58	13.26	27.72	132.22
Belgium	10.14	20.19	45.02	1.73	3.57	12.83	26.87	127.20
Bermuda	10.27	20.30	45.26	1.79	3.70	14.64	30.49	146.37
Brazil	10.00	19.89	44.29	1.70	3.57	13.48	28.13	126.05
Canada	9.84	20.17	44.72	1.75	3.53	13.23	27.60	130.09
Chile	10.10	20.15	45.92	1.76	3.65	13.39	28.09	132.19
China	10.17	20.42	45.25	1.77	3.61	13.42	28.17	129.18
Columbia	10.29	20.85	45.84	1.80	3.72	13.49	27.88	131.17
Cook Islands	10.97	22.46	51.40	1.94	4.24	16.70	35.38	171.26
Costa Rica	10.32	20.96	46.42	1.87	3.84	13.75	28.81	133.23
Czech Republic	10.24	20.61	45.77	1.75	3.58	13.42	27.80	131.57
Denmark	10.29	20.52	45.89	1.69	3.52	13.42	27.91	129.43
DominicanRepublic	10.16	20.65	44.90	1.81	3.73	14.31	30.43	146.00
Finland	10.21	20.47	45.49	1.74	3.61	13.27	27.52	131.15
France	10.02	20.16	44.64	1.72	3.48	12.98	27.38	126.36
Germany	10.06	20.23	44.33	1.73	3.53	12.91	27.36	128.47
Great Britain	9.87	19.94	44.36	1.70	3.49	13.01	27.30	127.13
Greece	10.11	19.85	45.57	1.75	3.61	13.48	28.12	132.04
Guatemala	10.32	21.09	48.44	1.82	3.74	13.98	29.34	132.53
Hungary	10.08	20.11	45.43	1.76	3.59	13.45	28.03	132.10
India	10.33	20.73	45.48	1.76	3.63	13.50	28.81	132.00
Indonesia	10.20	20.93	46.37	1.83	3.77	14.21	29.65	139.18
Ireland	10.35	20.54	45.58	1.75	3.56	13.07	27.78	129.15
Israel	10.20	20.89	46.59	1.80	3.70	13.66	28.72	134.21
Italy	10.01	19.72	45.26	1.73	3.35	13.09	27.28	127.29
Japan	10.00	20.03	44.78	1.77	3.62	13.22	27.58	126.16
Kenya	10.28	20.43	44.18	1.70	3.44	12.66	26.46	124.55
Korea, South	10.34	20.41	45.37	1.74	3.64	13.84	28.51	127.20
Korea, North	10.60	21.23	46.95	1.82	3.77	13.90	28.45	129.26
Luxembourg	10.41	20.77	47.90	1.76	3.67	13.64	28.77	134.03
Malaysia	10.30	20.92	46.41	1.79	3.76	14.11	29.50	149.27
Mauritius	10.13	20.06	44.69	1.80	3.83	14.15	29.84	143.07
Mexico	10.21	20.40	44.31	1.78	3.63	13.13	27.14	127.19
Myanmar(Burma)	10.64	21.52	48.63	1.80	3.80	14.19	29.62	139.57
Netherlands	10.19	20.19	45.68	1.73	3.55	13.22	27.44	128.31
New Zealand	10.11	20.42	46.09	1.74	3.54	13.21	27.70	128.59
Norway	10.08	20.17	46.11	1.71	3.62	13.11	27.54	130.17
Papua New Guinea	10.40	21.18	46.77	1.80	4.00	14.72	31.36	148.13
Philippines	10.57	21.43	45.57	1.80	3.82	13.97	29.04	138.44
Poland	10.00	19.98	44.62	1.72	3.59	13.29	27.89	129.23
Portugal	9.86	20.12	46.11	1.75	3.50	13.05	27.21	126.36
Romania	10.21	20.75	45.77	1.76	3.57	13.25	27.67	132.30
Russia	10.11	20.23	44.60	1.71	3.54	13.20	27.90	129.16
Samoa	10.78	21.86	49.98	1.94	4.01	16.28	34.71	161.50
Singapore	10.37	21.14	47.60	1.84	3.86	14.96	31.32	144.22
Spain	10.17	20.59	44.96	1.73	3.48	13.04	27.24	127.23
Sweden	10.18	20.43	45.54	1.76	3.61	13.29	27.93	130.38
Switzerland	10.16	20.41	44.99	1.71	3.53	13.13	27.90	129.56
Taiwan	10.36	20.81	46.72	1.79	3.77	13.91	29.20	134.35
Thailand	10.23	20.69	46.05	1.81	3.77	14.25	29.67	139.33
Turkey	10.38	21.04	46.63	1.78	3.59	13.45	28.33	130.25
U.S.A.	9.78	19.32	43.18	1.71	3.46	12.97	27.23	125.38

Source: *IAAF/ATES Track and Field Statistics Handbook* for the Helsinki 2005 Olympics. Courtesy of Ottavio Castellini.

8.23. A naturalist for the Alaska Fish and Game Department studies grizzly bears with the goal of maintaining a healthy population. Measurements on $n = 61$ bears provided the following summary statistics:

Variable	Weight (kg)	Body length (cm)	Neck (cm)	Girth (cm)	Head length (cm)	Head width (cm)
Sample mean $\bar{x}$	95.52	164.38	55.69	93.39	17.98	31.13

Covariance matrix

$$S = \begin{bmatrix} 3266.46 & 1343.97 & 731.54 & 1175.50 & 162.68 & 238.37 \\ 1343.97 & 721.91 & 324.25 & 537.35 & 80.17 & 117.73 \\ 731.54 & 324.25 & 179.28 & 281.17 & 39.15 & 56.80 \\ 1175.50 & 537.35 & 281.17 & 474.98 & 63.73 & 94.85 \\ 162.68 & 80.17 & 39.15 & 63.73 & 9.95 & 13.88 \\ 238.37 & 117.73 & 56.80 & 94.85 & 13.88 & 21.26 \end{bmatrix}$$

(a) Perform a principal component analysis using the covariance matrix. Can the data be effectively summarized in fewer than six dimensions?

(b) Perform a principal component analysis using the correlation matrix.

(c) Comment on the similarities and differences between the two analyses.

8.24. Refer to Example 8.10 and the data in Table 5.8, page 240. Add the variable $x_6 =$ regular overtime hours whose values are (read across)

6187	7336	6988	6964	8425	6778	5922	7307
7679	8259	10954	9353	6291	4969	4825	6019

and redo Example 8.10.

8.25. Refer to the police overtime hours data in Example 8.10. Construct an alternate control chart, based on the sum of squares d_{Uj}^2, to monitor the unexplained variation in the original observations summarized by the additional principal components.

8.26. Consider the psychological profile data in Table 4.6. Using the five variables, Indep, Supp, Benev, Conform and Leader, performs a principal component analysis using the covariance matrix S and the correlation matrix R. Your analysis should include the following:

(a) Determine the appropriate number of components to effectively summarize the variability. Construct a scree plot to aid in your determination.

(b) Interpret the sample principal components.

(c) Using the values for the first two principal components, plot the data in a two-dimensional space with $\hat{y}_1$ along the vertical axis and $\hat{y}_2$ along the horizontal axis. Can you distinguish groups representing the two socioeconomic levels and/or the two genders? Are there any outliers?

(d) Construct a 95% confidence interval for λ_1, the variance of the first population principal component from the covariance matrix.

8.27. The pulp and paper properties data is given in Table 7.7. Using the four paper variables, BL (breaking length), EM (elastic modulus), SF (Stress at failure) and BS (burst strength), perform a principal component analysis using the covariance matrix S and the correlation matrix R. Your analysis should include the following:

(a) Determine the appropriate number of components to effectively summarize the variability. Construct a scree plot to aid in your determination.

(b) Interpret the sample principal components.

(c) Do you think it it is possible to develop a "paper strength" index that effectively contains the information in the four paper variables? Explain.

(d) Using the values for the first two principal components, plot the data in a two-dimensional space with $\hat{y}_1$ along the vertical axis and $\hat{y}_2$ along the horizontal axis. Identify any outliers in this data set.

8.28. Survey data were collected as part of a study to assess options for enhancing food security through the sustainable use of natural resources in the Sikasso region of Mali (West Africa). A total of $n = 76$ farmers were surveyed and observations on the nine variables

$x_1 = $ Family (total number of individuals in household)

$x_2 = $ DistRd (distance in kilometers to nearest passable road)

$x_3 = $ Cotton (hectares of cotton planted in year 2000)

$x_4 = $ Maize (hectares of maize planted in year 2000)

$x_5 = $ Sorg (hectares of sorghum planted in year 2000)

$x_6 = $ Millet (hectares of millet planted in year 2000)

$x_7 = $ Bull (total number of bullocks or draft animals)

$x_8 = $ Cattle (total); $x_9 = $ Goats (total)

were recorded. The data are listed in Table 8.7 and on the website www.prenhall.com/statistics

(a) Construct two-dimensional scatterplots of Family versus DistRd, and DistRd versus Cattle. Remove any obvious outliers from the data set.

Table 8.7 Mali Family Farm Data

Family	DistRD	Cotton	Maize	Sorg	Millet	Bull	Cattle	Goats
12	80	1.5	1.00	3.0	.25	2	0	1
54	8	6.0	4.00	0	1.00	6	32	5
11	13	.5	1.00	0	0	0	0	0
21	13	2.0	2.50	1.0	0	1	0	5
61	30	3.0	5.00	0	0	4	21	0
20	70	0	2.00	3.0	0	2	0	3
29	35	1.5	2.00	0	0	0	0	0
29	35	2.0	3.00	2.0	0	0	0	0
57	9	5.0	5.00	0	0	4	5	2
23	33	2.0	2.00	1.0	0	2	1	7
⋮	⋮	⋮	⋮	⋮	⋮	⋮	⋮	⋮
20	0	1.5	1.00	3.0	0	1	6	0
27	41	1.1	.25	1.5	1.50	0	3	1
18	500	2.0	1.00	1.5	.50	1	0	0
30	19	2.0	2.00	4.0	1.00	2	0	5
77	18	8.0	4.00	6.0	4.00	6	8	6
21	500	5.0	1.00	3.0	4.00	1	0	5
13	100	.5	.50	0	1.00	0	0	4
24	100	2.0	3.00	0	.50	3	14	10
29	90	2.0	1.50	1.5	1.50	2	0	2
57	90	10.0	7.00	0	1.50	7	8	7

Source: Data courtesy of Jay Angerer.

(b) Perform a principal component analysis using the correlation matrix **R**. Determine the number of components to effectively summarize the variability. Use the proportion of variation explained and a scree plot to aid in your determination.

(c) Interpret the first five principal components. Can you identify, for example, a "farm size" component? A, perhaps, "goats and distance to road" component?

8.29. Refer to Exercise 5.28. Using the covariance matrix **S** for the first 30 cases of car body assembly data, obtain the sample principal components.

(a) Construct a 95% ellipse format chart using the first two principal components $\hat{y}_1$ and $\hat{y}_2$. Identify the car locations that appear to be out of control.

(b) Construct an alternative control chart, based on the sum of squares $d^2_{U\,j}$, to monitor the variation in the original observations summarized by the remaining four principal components. Interpret this chart.

References

1. Anderson, T. W. *An Introduction to Multivariate Statistical Analysis* (3rd ed.). New York: John Wiley, 2003.

2. Anderson, T. W. "Asymptotic Theory for Principal Components Analysis." *Annals of Mathematical Statistics*, **34** (1963), 122–148.

3. Bartlett, M. S. "A Note on Multiplying Factors for Various Chi-Squared Approximations." *Journal of the Royal Statistical Society (B)*, **16** (1954), 296–298.

4. Dawkins, B. "Multivariate Analysis of National Track Records." *The American Statistician*, **43** (1989), 110–115.

5. Girschick, M. A. "On the Sampling Theory of Roots of Determinantal Equations." *Annals of Mathematical Statistics*, **10** (1939), 203–224.

6. Hotelling, H. "Analysis of a Complex of Statistical Variables into Principal Components." *Journal of Educational Psychology*, **24** (1933), 417–441, 498–520.

7. Hotelling, H. "The Most Predictable Criterion." *Journal of Educational Psychology*, **26** (1935), 139–142.

8. Hotelling, H. "Simplified Calculation of Principal Components." *Psychometrika*, **1** (1936), 27–35.

9. Hotelling, H. "Relations between Two Sets of Variates." *Biometrika*, **28** (1936), 321–377.

10. Jolicoeur, P. "The Multivariate Generalization of the Allometry Equation." *Biometrics*, **19** (1963), 497–499.

11. Jolicoeur, P., and J. E. Mosimann. "Size and Shape Variation in the Painted Turtle: A Principal Component Analysis." *Growth*, **24** (1960), 339–354.

12. King, B. "Market and Industry Factors in Stock Price Behavior." *Journal of Business*, **39** (1966), 139–190.

13. Kourti, T., and J. McGregor, "Multivariate SPC Methods for Process and Product Monitoring," *Journal of Quality Technology*, **28** (1996), 409–428.

14. Lawley, D. N. "On Testing a Set of Correlation Coefficients for Equality." *Annals of Mathematical Statistics*, **34** (1963), 149–151.

15. Rao, C. R. *Linear Statistical Inference and Its Applications* (2nd ed.). New York: Wiley-Interscience, 2002.

16. Rencher, A. C. "Interpretation of Canonical Discriminant Functions, Canonical Variates and Principal Components." *The American Statistician*, **46** (1992), 217–225.

Chapter

9

FACTOR ANALYSIS AND INFERENCE FOR STRUCTURED COVARIANCE MATRICES

The goal is to generate a linear model based on factors F that cannot be measured themselves, but can be postulated from observations in measured variables X.

PCA would help find these interpretations, but factor analysis alone can let you find these interpretations (Factors) and generate a model to predict observations

9.1 Introduction

Factor analysis has provoked rather turbulent controversy throughout its history. Its modern beginnings lie in the early-20th-century attempts of Karl Pearson, Charles Spearman, and others to define and measure intelligence. Because of this early association with constructs such as intelligence, factor analysis was nurtured and developed primarily by scientists interested in psychometrics. Arguments over the psychological interpretations of several early studies and the lack of powerful computing facilities impeded its initial development as a statistical method. The advent of high-speed computers has generated a renewed interest in the theoretical and computational aspects of factor analysis. Most of the original techniques have been abandoned and early controversies resolved in the wake of recent developments. It is still true, however, that each application of the technique must be examined on its own merits to determine its success.

The essential purpose of factor analysis is to describe, if possible, the covariance relationships among many variables in terms of a few underlying, but unobservable, random quantities called *factors*. Basically, the factor model is motivated by the following argument: Suppose variables can be grouped by their correlations. That is, suppose all variables within a particular group are highly correlated among themselves, but have relatively small correlations with variables in a different group. Then it is conceivable that each group of variables represents a single underlying construct, or factor, that is responsible for the observed correlations. For example, correlations from the group of test scores in classics, French, English, mathematics, and music collected by Spearman suggested an underlying "intelligence" factor. A second group of variables, representing physical-fitness scores, if available, might correspond to another factor. It is this type of structure that factor analysis seeks to confirm.

intelligence factor postulates: people have a 'verbal intelligence' plus a 'mathematical intelligence' and with a FA model you can predict how well someone will do in subjects requiring elements of both based upon your individual scores in these factors

481

Similarity/
difference
between
FA /PCA

Factor analysis can be considered an extension of principal component analysis. Both can be viewed as attempts to approximate the covariance matrix Σ. However, the approximation based on the factor analysis model is more elaborate. The primary question in factor analysis is whether the data are consistent with a prescribed structure.

9.2 The Orthogonal Factor Model

The observable random vector $\mathbf{X}$, with p components, has mean $\boldsymbol{\mu}$ and covariance matrix Σ. The factor model postulates that $\mathbf{X}$ is linearly dependent upon a few unobservable random variables $F_1, F_2, \ldots, F_m$, called _common factors_, and p additional sources of variation $\varepsilon_1, \varepsilon_2, \ldots, \varepsilon_p$, called _errors_ or, sometimes, _specific factors_.[1] In particular, the factor analysis model is

$$
\begin{aligned}
X_1 - \mu_1 &= \ell_{11}F_1 + \ell_{12}F_2 + \cdots + \ell_{1m}F_m + \varepsilon_1 \\
X_2 - \mu_2 &= \ell_{21}F_1 + \ell_{22}F_2 + \cdots + \ell_{2m}F_m + \varepsilon_2 \\
&\vdots \qquad\qquad\qquad\qquad\qquad \vdots \\
X_p - \mu_p &= \ell_{p1}F_1 + \ell_{p2}F_2 + \cdots + \ell_{pm}F_m + \varepsilon_p
\end{aligned}
\tag{9-1}
$$

X not Y

or, in matrix notation,

$$
\underset{(p\times1)}{\mathbf{X} - \boldsymbol{\mu}} = \underset{(p\times m)(m\times1)}{\mathbf{L}\quad\mathbf{F}} + \underset{(p\times1)}{\boldsymbol{\varepsilon}}
\tag{9-2}
$$

The coefficient ℓ_{ij} is called the _loading_ of the ith variable on the jth factor, so the matrix $\mathbf{L}$ is the _matrix of factor loadings_. Note that the ith specific factor ε_i is associated only with the ith response X_i. The p deviations $X_1 - \mu_1, X_2 - \mu_2, \ldots, X_p - \mu_p$ are expressed in terms of $p + m$ random variables $F_1, F_2, \ldots, F_m, \varepsilon_1, \varepsilon_2, \ldots, \varepsilon_p$ which are _unobservable_. This distinguishes the factor model of (9-2) from the multivariate regression model in (7-23), in which the independent variables [whose position is occupied by $\mathbf{F}$ in (9-2)] can be observed. dif to linear regression

With so many unobservable quantities, a direct verification of the factor model from observations on $X_1, X_2, \ldots, X_p$ is hopeless. However, with some additional assumptions about the random vectors $\mathbf{F}$ and $\boldsymbol{\varepsilon}$, the model in (9-2) implies certain covariance relationships, which can be checked.

We assume that　　　each factor is unique i.e. uncorrelated with another F

$$
\underset{(m\times1)}{E(\mathbf{F}) = \mathbf{0}}, \qquad \underset{(m\times m)}{\mathrm{Cov}(\mathbf{F}) = E[\mathbf{F}\mathbf{F}'] = \mathbf{I}}
$$

$$
\underset{(p\times1)}{E(\boldsymbol{\varepsilon}) = \mathbf{0}}, \qquad \underset{(p\times p)}{\mathrm{Cov}(\boldsymbol{\varepsilon}) = E[\boldsymbol{\varepsilon}\boldsymbol{\varepsilon}'] = \boldsymbol{\Psi}} = \begin{bmatrix} \psi_1 & 0 & \cdots & 0 \\ 0 & \psi_2 & \cdots & 0 \\ \vdots & \vdots & \ddots & \vdots \\ 0 & 0 & \cdots & \psi_p \end{bmatrix}
\tag{9-3}
$$

[1] As Maxwell [12] points out, in many investigations the ε_i tend to be combinations of measurement error and factors that are uniquely associated with the individual variables.

and that $\mathbf{F}$ and $\boldsymbol{\varepsilon}$ are independent, so

$$\text{Cov}(\boldsymbol{\varepsilon}, \mathbf{F}) = E(\boldsymbol{\varepsilon}\mathbf{F}') = \underset{(p \times m)}{\mathbf{0}}$$

These assumptions and the relation in (9-2) constitute the *orthogonal factor model.*[2]

Orthogonal Factor Model with m Common Factors

a linear model to predict observations

$$\underset{(p \times 1)}{\mathbf{X}} = \underset{(p \times 1)}{\boldsymbol{\mu}} + \underset{(p \times m)(m \times 1)}{\mathbf{L} \quad \mathbf{F}} + \underset{(p \times 1)}{\boldsymbol{\varepsilon}}$$

$\mu_i = $ *mean* of variable i

$\varepsilon_i = i$th *specific factor*   (9-4)

$F_j = j$th *common factor*

$\ell_{ij} = $ *loading* of the ith variable on the jth factor

The unobservable random vectors $\mathbf{F}$ and $\boldsymbol{\varepsilon}$ satisfy the following conditions:

$\mathbf{F}$ and $\boldsymbol{\varepsilon}$ are independent

$E(\mathbf{F}) = \mathbf{0}, \text{Cov}(\mathbf{F}) = \mathbf{I}$

$E(\boldsymbol{\varepsilon}) = \mathbf{0}, \text{Cov}(\boldsymbol{\varepsilon}) = \boldsymbol{\Psi}$, where $\boldsymbol{\Psi}$ is a diagonal matrix

The orthogonal factor model implies a covariance structure for $\mathbf{X}$. From the model in (9-4),

$$(\mathbf{X} - \boldsymbol{\mu})(\mathbf{X} - \boldsymbol{\mu})' = (\mathbf{LF} + \boldsymbol{\varepsilon})(\mathbf{LF} + \boldsymbol{\varepsilon})'$$
$$= (\mathbf{LF} + \boldsymbol{\varepsilon})((\mathbf{LF})' + \boldsymbol{\varepsilon}')$$
$$= \mathbf{LF}(\mathbf{LF})' + \boldsymbol{\varepsilon}(\mathbf{LF})' + \mathbf{LF}\boldsymbol{\varepsilon}' + \boldsymbol{\varepsilon}\boldsymbol{\varepsilon}'$$

so that

$$\boldsymbol{\Sigma} = \text{Cov}(\mathbf{X}) = E(\mathbf{X} - \boldsymbol{\mu})(\mathbf{X} - \boldsymbol{\mu})'$$

$E(FF') = \text{cov}(F) = I$

$$= \mathbf{L}E(\mathbf{FF}')\mathbf{L}' + E(\boldsymbol{\varepsilon}\mathbf{F}')\mathbf{L}' + \mathbf{L}E(\mathbf{F}\boldsymbol{\varepsilon}') + E(\boldsymbol{\varepsilon}\boldsymbol{\varepsilon}')$$
$$= \mathbf{LL}' + \boldsymbol{\Psi}$$

0 covar   0 covar Ψ specific var

according to (9-3). Also by independence, $\text{Cov}(\boldsymbol{\varepsilon}, \mathbf{F}) = E(\boldsymbol{\varepsilon}, \mathbf{F}') = \mathbf{0}$

Also, by the model in (9-4), $(\mathbf{X} - \boldsymbol{\mu})\mathbf{F}' = (\mathbf{LF} + \boldsymbol{\varepsilon})\mathbf{F}' = \mathbf{LFF}' + \boldsymbol{\varepsilon}\mathbf{F}'$, so $\text{Cov}(\mathbf{X}, \mathbf{F}) = E(\mathbf{X} - \boldsymbol{\mu})\mathbf{F}' = \mathbf{L}E(\mathbf{FF}') + E(\boldsymbol{\varepsilon}\mathbf{F}') = \mathbf{L}$.

[2]Allowing the factors $\mathbf{F}$ to be correlated so that $\text{Cov}(\mathbf{F})$ is *not* diagonal gives the oblique factor model. The oblique model presents some additional estimation difficulties and will not be discussed in this book. (See [10].)

Covariance Structure for the Orthogonal Factor Model

1. $\text{Cov}(\mathbf{X}) = \mathbf{LL'} + \mathbf{\Psi}$

or

$$\text{Var}(X_i) = \ell_{i1}^2 + \cdots + \ell_{im}^2 + \psi_i$$

$$\text{Cov}(X_i, X_k) = \ell_{i1}\ell_{k1} + \cdots + \ell_{im}\ell_{km}$$

(9-5)

2. $\text{Cov}(\mathbf{X}, \mathbf{F}) = \mathbf{L}$

or

$$\text{Cov}(X_i, F_j) = \ell_{ij}$$

linear model

The model $\mathbf{X} - \boldsymbol{\mu} = \mathbf{LF} + \boldsymbol{\varepsilon}$ is *linear* in the common factors. If the p responses $\mathbf{X}$ are, in fact, related to underlying factors, but the relationship is nonlinear, such as in $X_1 - \mu_1 = \ell_{11}F_1F_3 + \varepsilon_1$, $X_2 - \mu_2 = \ell_{21}F_2F_3 + \varepsilon_2$, and so forth, then the covariance structure $\mathbf{LL'} + \mathbf{\Psi}$ given by (9-5) may not be adequate. The very important assumption of linearity is inherent in the formulation of the traditional factor model.

That portion of the variance of the ith variable contributed by the m common factors is called the ith *communality*. That portion of $\text{Var}(X_i) = \sigma_{ii}$ due to the specific factor is often called the *uniqueness*, or *specific variance*. Denoting the ith communality by h_i^2, we see from (9-5) that

$$\underbrace{\sigma_{ii}}_{\text{Var}(X_i)} = \underbrace{\ell_{i1}^2 + \ell_{i2}^2 + \cdots + \ell_{im}^2}_{\text{communality}} + \underbrace{\psi_i}_{\text{specific variance}}$$

or

$$h_i^2 = \ell_{i1}^2 + \ell_{i2}^2 + \cdots + \ell_{im}^2$$

(9-6)

and

$$\sigma_{ii} = h_i^2 + \psi_i, \qquad i = 1, 2, \ldots, p$$

The ith communality is the sum of squares of the loadings of the ith variable on the m common factors.

Example 9.1 (Verifying the relation $\Sigma = \mathbf{LL'} + \mathbf{\Psi}$ for two factors) Consider the covariance matrix

$$\Sigma = \begin{bmatrix} 19 & 30 & 2 & 12 \\ 30 & 57 & 5 & 23 \\ 2 & 5 & 38 & 47 \\ 12 & 23 & 47 & 68 \end{bmatrix}$$

The equality

$$\begin{bmatrix} 19 & 30 & 2 & 12 \\ 30 & 57 & 5 & 23 \\ 2 & 5 & 38 & 47 \\ 12 & 23 & 47 & 68 \end{bmatrix} = \begin{bmatrix} 4 & 1 \\ 7 & 2 \\ -1 & 6 \\ 1 & 8 \end{bmatrix} \begin{bmatrix} 4 & 7 & -1 & 1 \\ 1 & 2 & 6 & 8 \end{bmatrix} + \begin{bmatrix} 2 & 0 & 0 & 0 \\ 0 & 4 & 0 & 0 \\ 0 & 0 & 1 & 0 \\ 0 & 0 & 0 & 3 \end{bmatrix}$$

or

$$\Sigma = LL' + \Psi$$

may be verified by matrix algebra. Therefore, Σ has the structure produced by an $m = 2$ orthogonal factor model. Since

$$L = \begin{bmatrix} \ell_{11} & \ell_{12} \\ \ell_{21} & \ell_{22} \\ \ell_{31} & \ell_{32} \\ \ell_{41} & \ell_{42} \end{bmatrix} = \begin{bmatrix} 4 & 1 \\ 7 & 2 \\ -1 & 6 \\ 1 & 8 \end{bmatrix},$$

$$\Psi = \begin{bmatrix} \psi_1 & 0 & 0 & 0 \\ 0 & \psi_2 & 0 & 0 \\ 0 & 0 & \psi_3 & 0 \\ 0 & 0 & 0 & \psi_4 \end{bmatrix} = \begin{bmatrix} 2 & 0 & 0 & 0 \\ 0 & 4 & 0 & 0 \\ 0 & 0 & 1 & 0 \\ 0 & 0 & 0 & 3 \end{bmatrix}$$

specific variance (due to the specific factor)

the communality of X_1 is, from (9-6),

$$h_1^2 = \ell_{11}^2 + \ell_{12}^2 = 4^2 + 1^2 = 17$$

Communality: variance from a variable contributed by all factors

and the variance of X_1 can be decomposed as

$$\sigma_{11} = (\ell_{11}^2 + \ell_{12}^2) + \psi_1 = h_1^2 + \psi_1$$

or

$$\underbrace{19}_{\text{variance}} = \underbrace{4^2 + 1^2}_{\text{communality}} + \underbrace{2}_{\substack{\text{specific} \\ \text{variance}}} = 17 + 2$$

A similar breakdown occurs for the other variables. ∎

The factor model assumes that the $p + p(p - 1)/2 = p(p + 1)/2$ variances and covariances for X can be reproduced from the pm factor loadings ℓ_{ij} and the p specific variances ψ_i. When $m = p$, any covariance matrix Σ can be reproduced exactly as LL' [see (9-11)], so Ψ can be the zero matrix. However, it is when m is small relative to p that factor analysis is most useful. In this case, the factor model provides a "simple" explanation of the covariation in X with fewer parameters than the $p(p + 1)/2$ parameters in Σ. For example, if X contains $p = 12$ variables, and the factor model in (9-4) with $m = 2$ is appropriate, then the $p(p + 1)/2 = 12(13)/2 = 78$ elements of Σ are described in terms of the $mp + p = 12(2) + 12 = 36$ parameters ℓ_{ij} and ψ_i of the factor model.

Unfortunately for the factor analyst, most covariance matrices cannot be factored as $\mathbf{LL'} + \mathbf{\Psi}$, where the number of factors m is much less than p. The following example demonstrates one of the problems that can arise when attempting to determine the parameters ℓ_{ij} and ψ_i from the variances and covariances of the observable variables.

Example 9.2 (Nonexistence of a proper solution) Let $p = 3$ and $m = 1$, and suppose the random variables X_1, X_2, and X_3 have the positive definite covariance matrix

$$\mathbf{\Sigma} = \begin{bmatrix} 1 & .9 & .7 \\ .9 & 1 & .4 \\ .7 & .4 & 1 \end{bmatrix}$$

Using the factor model in (9-4), we obtain

$$X_1 - \mu_1 = \ell_{11}F_1 + \varepsilon_1$$
$$X_2 - \mu_2 = \ell_{21}F_1 + \varepsilon_2$$
$$X_3 - \mu_3 = \ell_{31}F_1 + \varepsilon_3$$

The covariance structure in (9-5) implies that

$$\mathbf{\Sigma} = \mathbf{LL'} + \mathbf{\Psi}$$

or

$$1 = \ell_{11}^2 + \psi_1 \qquad .90 = \ell_{11}\ell_{21} \qquad .70 = \ell_{11}\ell_{31}$$
$$1 = \ell_{21}^2 + \psi_2 \qquad .40 = \ell_{21}\ell_{31}$$
$$1 = \ell_{31}^2 + \psi_3$$

The pair of equations

$$.70 = \ell_{11}\ell_{31}$$
$$.40 = \ell_{21}\ell_{31}$$

implies that

$$\ell_{21} = \left(\frac{.40}{.70}\right)\ell_{11}$$

Substituting this result for ℓ_{21} in the equation

$$.90 = \ell_{11}\ell_{21}$$

yields $\ell_{11}^2 = 1.575$, or $\ell_{11} = \pm 1.255$. Since $\text{Var}(F_1) = 1$ (by assumption) and $\text{Var}(X_1) = 1$, $\ell_{11} = \text{Cov}(X_1, F_1) = \text{Corr}(X_1, F_1)$. Now, a correlation coefficient cannot be greater than unity (in absolute value), so, from this point of view, $|\ell_{11}| = 1.255$ is too large. Also, the equation

$$1 = \ell_{11}^2 + \psi_1, \quad \text{or} \quad \psi_1 = 1 - \ell_{11}^2$$

gives

$$\psi_1 = 1 - 1.575 = -.575$$

which is unsatisfactory, since it gives a negative value for $\text{Var}\,(\varepsilon_1) = \psi_1$.

Thus, for this example with $m = 1$, it is possible to get a unique numerical solution to the equations $\boldsymbol{\Sigma} = \mathbf{LL'} + \boldsymbol{\Psi}$. However, the solution is not consistent with the statistical interpretation of the coefficients, so it is not a proper solution. ∎

When $m > 1$, there is always some inherent ambiguity associated with the factor model. To see this, let $\mathbf{T}$ be any $m \times m$ orthogonal matrix, so that $\mathbf{TT'} = \mathbf{T'T} = \mathbf{I}$. Then the expression in (9-2) can be written

$$\mathbf{X} - \boldsymbol{\mu} = \mathbf{LF} + \boldsymbol{\varepsilon} = \mathbf{LTT'F} + \boldsymbol{\varepsilon} = \mathbf{L^*F^*} + \boldsymbol{\varepsilon} \tag{9-7}$$

where

$$\mathbf{L^*} = \mathbf{LT} \quad \text{and} \quad \mathbf{F^*} = \mathbf{T'F}$$

Since

$$E(\mathbf{F^*}) = \mathbf{T'}E(\mathbf{F}) = \mathbf{0}$$

and

$$\text{Cov}\,(\mathbf{F^*}) = \mathbf{T'}\,\text{Cov}\,(\mathbf{F})\mathbf{T} = \mathbf{T'T} = \underset{(m \times m)}{\mathbf{I}}$$

it is impossible, on the basis of observations on $\mathbf{X}$, to distinguish the loadings $\mathbf{L}$ from the loadings $\mathbf{L^*}$. That is, the factors $\mathbf{F}$ and $\mathbf{F^*} = \mathbf{T'F}$ have the same statistical properties, and even though the loadings $\mathbf{L^*}$ are, in general, different from the loadings $\mathbf{L}$, they both generate the same covariance matrix $\boldsymbol{\Sigma}$. That is,

$$\boldsymbol{\Sigma} = \mathbf{LL'} + \boldsymbol{\Psi} = \mathbf{LTT'L'} + \boldsymbol{\Psi} = (\mathbf{L^*})(\mathbf{L^*})' + \boldsymbol{\Psi} \tag{9-8}$$

This ambiguity provides the rationale for "factor rotation," since orthogonal matrices correspond to rotations (and reflections) of the coordinate system for $\mathbf{X}$.

> Factor loadings $\mathbf{L}$ are determined only up to an orthogonal matrix $\mathbf{T}$. Thus, the loadings
>
> $$\mathbf{L^*} = \mathbf{LT} \quad \text{and} \quad \mathbf{L} \tag{9-9}$$
>
> both give the same representation. The communalities, given by the diagonal elements of $\mathbf{LL'} = (\mathbf{L^*})(\mathbf{L^*})'$ are also unaffected by the choice of $\mathbf{T}$.

The analysis of the factor model proceeds by imposing conditions that allow one to uniquely estimate $\mathbf{L}$ and $\boldsymbol{\Psi}$. The loading matrix is then rotated (multiplied by an orthogonal matrix), where the rotation is determined by some "ease-of-interpretation" criterion. Once the loadings and specific variances are obtained, factors are identified, and estimated values for the factors themselves (called *factor scores*) are frequently constructed.

9.3 Methods of Estimation

Given observations $\mathbf{x}_1, \mathbf{x}_2, \ldots, \mathbf{x}_n$ on p generally correlated variables, factor analysis seeks to answer the question, Does the factor model of (9-4), with a small number of factors, adequately represent the data? In essence, we tackle this statistical model-building problem by trying to verify the covariance relationship in (9-5).

The sample covariance matrix $\mathbf{S}$ is an estimator of the unknown population covariance matrix $\mathbf{\Sigma}$. If the off-diagonal elements of $\mathbf{S}$ are small or those of the sample correlation matrix $\mathbf{R}$ essentially zero, the variables are not related, and a factor analysis will not prove useful. In these circumstances, the *specific* factors play the dominant role, whereas the major aim of factor analysis is to determine a few important common factors.

i.e Cross Correlations exist

If $\mathbf{\Sigma}$ appears to deviate significantly from a diagonal matrix, then a factor model can be entertained, and the initial problem is one of estimating the factor loadings ℓ_{ij} and specific variances ψ_i. We shall consider two of the most popular methods of parameter estimation, the *principal component* (and the related *principal factor*) *method* and the *maximum likelihood method*. The solution from either method can be rotated in order to simplify the interpretation of factors, as described in Section 9.4. It is always prudent to try more than one method of solution; if the factor model is appropriate for the problem at hand, the solutions should be consistent with one another.

Current estimation and rotation methods require iterative calculations that must be done on a computer. Several computer programs are now available for this purpose.

The Principal Component (and Principal Factor) Method

The spectral decomposition of (2-16) provides us with one factoring of the covariance matrix $\mathbf{\Sigma}$. Let $\mathbf{\Sigma}$ have eigenvalue–eigenvector pairs $(\lambda_i, \mathbf{e}_i)$ with $\lambda_1 \geq \lambda_2 \geq \cdots \geq \lambda_p \geq 0$. Then

$$\mathbf{\Sigma} = \lambda_1 \mathbf{e}_1 \mathbf{e}_1' + \lambda_2 \mathbf{e}_2 \mathbf{e}_2' + \cdots + \lambda_p \mathbf{e}_p \mathbf{e}_p'$$

$\sqrt{\lambda} \cdot \sqrt{\lambda}$

$$= [\sqrt{\lambda_1}\, \mathbf{e}_1 \mid \sqrt{\lambda_2}\, \mathbf{e}_2 \mid \cdots \mid \sqrt{\lambda_p}\, \mathbf{e}_p] \begin{bmatrix} \sqrt{\lambda_1}\, \mathbf{e}_1' \\ \hline \sqrt{\lambda_2}\, \mathbf{e}_2' \\ \hline \vdots \\ \hline \sqrt{\lambda_p}\, \mathbf{e}_p' \end{bmatrix} \qquad (9\text{-}10)$$

This fits the prescribed covariance structure for the factor analysis model having as many factors as variables ($m = p$) and specific variances $\psi_i = 0$ for all i. The loading matrix has jth column given by $\sqrt{\lambda_j}\, \mathbf{e}_j$. That is, we can write

$$\underset{(p \times p)}{\mathbf{\Sigma}} = \underset{(p \times p)(p \times p)}{\mathbf{L}\ \mathbf{L}'} + \underset{(p \times p)}{\mathbf{0}} = \mathbf{L}\mathbf{L}' \qquad (9\text{-}11)$$

Apart from the scale factor $\sqrt{\lambda_j}$, the factor loadings on the jth factor are the coefficients for the jth principal component of the population.

Although the factor analysis representation of $\mathbf{\Sigma}$ in (9-11) is exact, it is not particularly useful: It employs as many common factors as there are variables and does not allow for any variation in the specific factors $\boldsymbol{\varepsilon}$ in (9-4). We prefer models that explain the covariance structure in terms of just a few common factors. One approach, when the last $p - m$ eigenvalues are small, is to neglect the contribution of $\lambda_{m+1}\mathbf{e}_{m+1}\mathbf{e}'_{m+1} + \cdots + \lambda_p\mathbf{e}_p\mathbf{e}'_p$ to $\boldsymbol{\Sigma}$ in (9-10). Neglecting this contribution, we obtain the approximation

$$\boldsymbol{\Sigma} \doteq [\sqrt{\lambda_1}\,\mathbf{e}_1 \;\vdots\; \sqrt{\lambda_2}\,\mathbf{e}_2 \;\vdots\; \cdots \;\vdots\; \sqrt{\lambda_m}\,\mathbf{e}_m] \begin{bmatrix} \sqrt{\lambda_1}\,\mathbf{e}'_1 \\ \hline \sqrt{\lambda_2}\,\mathbf{e}'_2 \\ \vdots \\ \hline \sqrt{\lambda_m}\,\mathbf{e}'_m \end{bmatrix} = \underset{(p\times m)\ (m\times p)}{\mathbf{L} \quad \mathbf{L}'} \qquad (9\text{-}12)$$

The approximate representation in (9-12) assumes that the specific factors $\boldsymbol{\varepsilon}$ in (9-4) are of minor importance and can also be ignored in the factoring of $\boldsymbol{\Sigma}$. If specific factors are included in the model, their variances may be taken to be the diagonal elements of $\boldsymbol{\Sigma} - \mathbf{LL}'$, where $\mathbf{LL}'$ is as defined in (9-12).

Allowing for specific factors, we find that the approximation becomes

$$\boldsymbol{\Sigma} \doteq \mathbf{LL}' + \boldsymbol{\Psi}$$

$$= [\sqrt{\lambda_1}\,\mathbf{e}_1 \;\vdots\; \sqrt{\lambda_2}\,\mathbf{e}_2 \;\vdots\; \cdots \;\vdots\; \sqrt{\lambda_m}\,\mathbf{e}_m] \begin{bmatrix} \sqrt{\lambda_1}\,\mathbf{e}'_1 \\ \hline \sqrt{\lambda_2}\,\mathbf{e}'_2 \\ \vdots \\ \hline \sqrt{\lambda_m}\,\mathbf{e}'_m \end{bmatrix} + \begin{bmatrix} \psi_1 & 0 & \cdots & 0 \\ 0 & \psi_2 & \cdots & 0 \\ \vdots & \vdots & \ddots & \vdots \\ 0 & 0 & \cdots & \psi_p \end{bmatrix} \qquad (9\text{-}13)$$

where $\psi_i = \sigma_{ii} - \sum_{j=1}^{m} \ell_{ij}^2$ for $i = 1, 2, \ldots, p$.

To apply this approach to a data set $\mathbf{x}_1, \mathbf{x}_2, \ldots, \mathbf{x}_n$, it is customary first to center the observations by subtracting the sample mean $\bar{\mathbf{x}}$. The centered observations

$$\mathbf{x}_j - \bar{\mathbf{x}} = \begin{bmatrix} x_{j1} \\ x_{j2} \\ \vdots \\ x_{jp} \end{bmatrix} - \begin{bmatrix} \bar{x}_1 \\ \bar{x}_2 \\ \vdots \\ \bar{x}_p \end{bmatrix} = \begin{bmatrix} x_{j1} - \bar{x}_1 \\ x_{j2} - \bar{x}_2 \\ \vdots \\ x_{jp} - \bar{x}_p \end{bmatrix} \qquad j = 1, 2, \ldots, n \qquad (9\text{-}14)$$

have the same sample covariance matrix $\mathbf{S}$ as the original observations.

In cases in which the units of the variables are not commensurate, it is usually desirable to work with the standardized variables

Standardize variables, of course

$$\mathbf{z}_j = \begin{bmatrix} \dfrac{(x_{j1} - \bar{x}_1)}{\sqrt{s_{11}}} \\[2mm] \dfrac{(x_{j2} - \bar{x}_2)}{\sqrt{s_{22}}} \\[2mm] \vdots \\[1mm] \dfrac{(x_{jp} - \bar{x}_p)}{\sqrt{s_{pp}}} \end{bmatrix} \qquad j = 1, 2, \ldots, n$$

whose sample covariance matrix is the sample correlation matrix $\mathbf{R}$ of the observations $\mathbf{x}_1, \mathbf{x}_2, \ldots, \mathbf{x}_n$. Standardization avoids the problems of having one variable with large variance unduly influencing the determination of factor loadings.

The representation in (9-13), when applied to the sample covariance matrix $\mathbf{S}$ or the sample correlation matrix $\mathbf{R}$, is known as the *principal component solution*. The name follows from the fact that the factor loadings are the scaled coefficients of the first few sample principal components. (See Chapter 8.)

Principal Component Solution of the Factor Model

The principal component factor analysis of the sample covariance matrix $\mathbf{S}$ is specified in terms of its eigenvalue–eigenvector pairs $(\hat{\lambda}_1, \hat{\mathbf{e}}_1)$, $(\hat{\lambda}_2, \hat{\mathbf{e}}_2), \ldots$, $(\hat{\lambda}_p, \hat{\mathbf{e}}_p)$, where $\hat{\lambda}_1 \geq \hat{\lambda}_2 \geq \cdots \geq \hat{\lambda}_p$. Let $m < p$ be the number of common factors. Then the matrix of estimated factor loadings $\{\tilde{\ell}_{ij}\}$ is given by

$$\tilde{\mathbf{L}} = \left[\sqrt{\hat{\lambda}_1}\, \hat{\mathbf{e}}_1 \mid \sqrt{\hat{\lambda}_2}\, \hat{\mathbf{e}}_2 \mid \cdots \mid \sqrt{\hat{\lambda}_m}\, \hat{\mathbf{e}}_m \right] \tag{9-15}$$

The estimated specific variances are provided by the diagonal elements of the matrix $\mathbf{S} - \tilde{\mathbf{L}}\tilde{\mathbf{L}}'$, so

$$\tilde{\boldsymbol{\Psi}} = \begin{bmatrix} \tilde{\psi}_1 & 0 & \cdots & 0 \\ 0 & \tilde{\psi}_2 & \cdots & 0 \\ \vdots & \vdots & \ddots & \vdots \\ 0 & 0 & \cdots & \tilde{\psi}_p \end{bmatrix} \quad \text{with} \quad \tilde{\psi}_i = s_{ii} - \sum_{j=1}^{m} \tilde{\ell}_{ij}^2 \tag{9-16}$$

Communalities are estimated as

$$\tilde{h}_i^2 = \tilde{\ell}_{i\,1}^2 + \tilde{\ell}_{i\,2}^2 + \cdots + \tilde{\ell}_{im}^2 \tag{9-17}$$

The principal component factor analysis of the sample correlation matrix is obtained by starting with $\mathbf{R}$ in place of $\mathbf{S}$.

For the principal component solution, the estimated loadings for a given factor do not change as the number of factors is increased. For example, if $m = 1$, $\tilde{\mathbf{L}} = \left[\sqrt{\hat{\lambda}_1}\, \hat{\mathbf{e}}_1 \right]$, and if $m = 2$, $\tilde{\mathbf{L}} = \left[\sqrt{\hat{\lambda}_1}\, \hat{\mathbf{e}}_1 \mid \sqrt{\hat{\lambda}_2}\, \hat{\mathbf{e}}_2 \right]$, where $(\hat{\lambda}_1, \hat{\mathbf{e}}_1)$ and $(\hat{\lambda}_2, \hat{\mathbf{e}}_2)$ are the first two eigenvalue–eigenvector pairs for $\mathbf{S}$ (or $\mathbf{R}$).

By the definition of $\tilde{\psi}_i$, the diagonal elements of $\mathbf{S}$ are equal to the diagonal elements of $\tilde{\mathbf{L}}\tilde{\mathbf{L}}' + \tilde{\boldsymbol{\Psi}}$. However, the off-diagonal elements of $\mathbf{S}$ are not usually reproduced by $\tilde{\mathbf{L}}\tilde{\mathbf{L}}' + \tilde{\boldsymbol{\Psi}}$. How, then, do we select the number of factors m?

If the number of common factors is not determined by a priori considerations, such as by theory or the work of other researchers, the choice of m can be based on the estimated eigenvalues in much the same manner as with principal components. Consider the *residual matrix*

$$\mathbf{S} - (\tilde{\mathbf{L}}\tilde{\mathbf{L}}' + \tilde{\boldsymbol{\Psi}}) \tag{9-18}$$

resulting from the approximation of $\mathbf{S}$ by the principal component solution. The diagonal elements are zero, and if the other elements are also small, we may subjectively take the m factor model to be appropriate. Analytically, we have (see Exercise 9.5)

$$\text{Sum of squared entries of } (\mathbf{S} - (\tilde{\mathbf{L}}\tilde{\mathbf{L}}' + \tilde{\boldsymbol{\Psi}})) \leq \hat{\lambda}_{m+1}^2 + \cdots + \hat{\lambda}_p^2 \tag{9-19}$$

Consequently, a small value for the sum of the squares of the neglected eigenvalues implies a small value for the sum of the squared errors of approximation.

Ideally, the contributions of the first few factors to the sample variances of the variables should be large. The contribution to the sample variance s_{ii} from the first common factor is $\tilde{\ell}_{i1}^2$. The contribution to the *total* sample variance, $s_{11} + s_{22} + \cdots + s_{pp} = \operatorname{tr}(\mathbf{S})$, from the first common factor is then

$$\tilde{\ell}_{11}^2 + \tilde{\ell}_{21}^2 + \cdots + \tilde{\ell}_{p1}^2 = \left(\sqrt{\hat{\lambda}_1}\,\hat{\mathbf{e}}_1\right)'\left(\sqrt{\hat{\lambda}_1}\,\hat{\mathbf{e}}_1\right) = \hat{\lambda}_1$$

since the eigenvector $\hat{\mathbf{e}}_1$ has unit length. In general,

$$\left(\begin{array}{l}\text{Proportion of total}\\ \text{sample variance}\\ \text{due to } j\text{th factor}\end{array}\right) = \begin{cases} \dfrac{\hat{\lambda}_j}{s_{11} + s_{22} + \cdots + s_{pp}} & \text{for a factor analysis of } \mathbf{S} \\[2ex] \dfrac{\hat{\lambda}_j}{p} & \text{for a factor analysis of } \mathbf{R} \end{cases} \tag{9-20}$$

Criterion (9-20) is frequently used as a heuristic device for determining the appropriate number of common factors. The number of common factors retained in the model is increased until a "suitable proportion" of the total sample variance has been explained.

Another convention, frequently encountered in packaged computer programs, is to set m equal to the number of eigenvalues of $\mathbf{R}$ greater than one if the sample correlation matrix is factored, or equal to the number of positive eigenvalues of $\mathbf{S}$ if the sample covariance matrix is factored. These rules of thumb should not be applied indiscriminately. For example, $m = p$ if the rule for $\mathbf{S}$ is obeyed, since all the eigenvalues are expected to be positive for large sample sizes. The best approach is to retain few rather than many factors, assuming that they provide a satisfactory interpretation of the data and yield a satisfactory fit to $\mathbf{S}$ or $\mathbf{R}$.

Example 9.3 (Factor analysis of consumer-preference data) In a consumer-preference study, a random sample of customers were asked to rate several attributes of a new product. The responses, on a 7-point semantic differential scale, were tabulated and the attribute correlation matrix constructed. The correlation matrix is presented next:

Attribute (Variable)		1	2	3	4	5
Taste	1	1.00	.02	(.96)	.42	.01
Good buy for money	2	.02	1.00	.13	.71	(.85)
Flavor	3	.96	.13	1.00	.50	.11
Suitable for snack	4	.42	.71	.50	1.00	(.79)
Provides lots of energy	5	.01	.85	.11	.79	1.00

It is clear from the circled entries in the correlation matrix that variables 1 and 3 and variables 2 and 5 form groups. Variable 4 is "closer" to the $(2,5)$ group than the $(1,3)$ group. Given these results and the small number of variables, we might expect that the apparent linear relationships between the variables can be explained in terms of, at most, two or three common factors.

The first two eigenvalues, $\hat{\lambda}_1 = 2.85$ and $\hat{\lambda}_2 = 1.81$, of **R** are the only eigenvalues greater than unity. Moreover, $m = 2$ common factors will account for a cumulative proportion

$$\frac{\hat{\lambda}_1 + \hat{\lambda}_2}{p} = \frac{2.85 + 1.81}{5} = .93$$

of the total (standardized) sample variance. The estimated factor loadings, communalities, and specific variances, obtained using (9-15), (9-16), and (9-17), are given in Table 9.1.

Table 9.1

Variable	Estimated factor loadings $\tilde{\ell}_{ij} = \sqrt{\hat{\lambda}_i}\,\hat{e}_{ij}$		Communalities $\tilde{h}_i^2$	Specific variances $\tilde{\psi}_i = 1 - \tilde{h}_i^2$
	F_1	F_2		
1. Taste	.56	.82	.98	.02
2. Good buy for money	.78	−.53	.88	.12
3. Flavor	.65	.75	.98	.02
4. Suitable for snack	.94	−.10	.89	.11
5. Provides lots of energy	.80	−.54	.93	.07
Eigenvalues	2.85	1.81		
Cumulative proportion of total (standardized) sample variance	.571	.932		

Now,

$$\tilde{\mathbf{L}}\tilde{\mathbf{L}}' + \tilde{\boldsymbol{\Psi}} = \begin{bmatrix} .56 & .82 \\ .78 & -.53 \\ .65 & .75 \\ .94 & -.10 \\ .80 & -.54 \end{bmatrix} \begin{bmatrix} .56 & .78 & .65 & .94 & .80 \\ .82 & -.53 & .75 & -.10 & -.54 \end{bmatrix}$$

$$+ \begin{bmatrix} .02 & 0 & 0 & 0 & 0 \\ 0 & .12 & 0 & 0 & 0 \\ 0 & 0 & .02 & 0 & 0 \\ 0 & 0 & 0 & .11 & 0 \\ 0 & 0 & 0 & 0 & .07 \end{bmatrix} = \begin{bmatrix} 1.00 & .01 & .97 & .44 & .00 \\ & 1.00 & .11 & .79 & .91 \\ & & 1.00 & .53 & .11 \\ & & & 1.00 & .81 \\ & & & & 1.00 \end{bmatrix}$$

nearly reproduces the correlation matrix $\mathbf{R}$. Thus, on a purely descriptive basis, we would judge a two-factor model with the factor loadings displayed in Table 9.1 as providing a good fit to the data. The communalities (.98, .88, .98, .89, .93) indicate that the two factors account for a large percentage of the sample variance of each variable.

We shall not interpret the factors at this point. As we noted in Section 9.2, the factors (and loadings) are unique up to an orthogonal rotation. A rotation of the factors often reveals a simple structure and aids interpretation. We shall consider this example again (see Example 9.9 and Panel 9.1) after factor rotation has been discussed.

∎

Example 9.4 (Factor analysis of stock-price data) Stock-price data consisting of $n = 103$ weekly rates of return on $p = 5$ stocks were introduced in Example 8.5. In that example, the first two sample principal components were obtained from $\mathbf{R}$. Taking $m = 1$ and $m = 2$, we can easily obtain principal component solutions to the orthogonal factor model. Specifically, the estimated factor loadings are the sample principal component coefficients (eigenvectors of $\mathbf{R}$), scaled by the square root of the corresponding eigenvalues. The estimated factor loadings, communalities, specific variances, and proportion of total (standardized) sample variance explained by each factor for the $m = 1$ and $m = 2$ factor solutions are available in Table 9.2. The communalities are given by (9-17). So, for example, with $m = 2$, $\tilde{h}_1^2 = \tilde{\ell}_{11}^2 + \tilde{\ell}_{12}^2 = (.732)^2 + (-.437)^2 = .73$.

Table 9.2

Variable	One-factor solution		Two-factor solution		
	Estimated factor loadings	Specific variances	Estimated factor loadings		Specific variances
	F_1	$\tilde{\psi}_i = 1 - \tilde{h}_i^2$	F_1	F_2	$\tilde{\psi}_i = 1 - \tilde{h}_i^2$
1. J P Morgan	.732	.46	.732	−.437	.27
2. Citibank	.831	.31	.831	−.280	.23
3. Wells Fargo	.726	.47	.726	−.374	.33
4. Royal Dutch Shell	.605	.63	.605	.694	.15
5. ExxonMobil	.563	.68	.563	.719	.17
Cumulative proportion of total (standardized) sample variance explained	.487		.487	.769	

The residual matrix corresponding to the solution for $m = 2$ factors is

$$\mathbf{R} - \tilde{\mathbf{L}}\tilde{\mathbf{L}}' - \tilde{\mathbf{\Psi}} = \begin{bmatrix} 0 & -.099 & -.185 & -.025 & .056 \\ -.099 & 0 & -.134 & .014 & -.054 \\ -.185 & -.134 & 0 & .003 & .006 \\ -.025 & .014 & .003 & 0 & -.156 \\ .056 & -.054 & .006 & -.156 & 0 \end{bmatrix}$$

The proportion of the total variance explained by the two-factor solution is appreciably larger than that for the one-factor solution. However, for $m = 2$, $\widetilde{\mathbf{L}}\widetilde{\mathbf{L}}'$ produces numbers that are, in general, larger than the sample correlations. This is particularly true for r_{13}.

It seems fairly clear that the first factor, F_1, represents general economic conditions and might be called a *market factor*. All of the stocks load highly on this factor, and the loadings are about equal. The second factor contrasts the banking stocks with the oil stocks. (The banks have relatively large negative loadings, and the oils have large positive loadings, on the factor.) Thus, F_2 seems to differentiate stocks in different industries and might be called an *industry factor*. To summarize, rates of return appear to be determined by general market conditions and activities that are unique to the different industries, as well as a residual or firm specific factor. This is essentially the conclusion reached by an examination of the sample principal components in Example 8.5. ∎

A Modified Approach—the Principal Factor Solution

A modification of the principal component approach is sometimes considered. We describe the reasoning in terms of a factor analysis of $\mathbf{R}$, although the procedure is also appropriate for $\mathbf{S}$. If the factor model $\boldsymbol{\rho} = \mathbf{LL}' + \boldsymbol{\Psi}$ is correctly specified, the m *common* factors should account for the *off-diagonal* elements of $\boldsymbol{\rho}$, as well as the *communality portions* of the diagonal elements

$$\rho_{ii} = 1 = h_i^2 + \psi_i$$

If the specific factor contribution ψ_i is removed from the diagonal or, equivalently, the 1 replaced by h_i^2, the resulting matrix is $\boldsymbol{\rho} - \boldsymbol{\Psi} = \mathbf{LL}'$.

Suppose, now, that initial estimates ψ_i^* of the specific variances are available. Then replacing the ith diagonal element of $\mathbf{R}$ by $h_i^{*2} = 1 - \psi_i^*$, we obtain a "reduced" sample correlation matrix

$$\mathbf{R}_r = \begin{bmatrix} h_1^{*2} & r_{12} & \cdots & r_{1p} \\ r_{12} & h_2^{*2} & \cdots & r_{2p} \\ \vdots & \vdots & \ddots & \vdots \\ r_{1p} & r_{2p} & \cdots & h_p^{*2} \end{bmatrix}$$

Now, apart from sampling variation, all of the elements of the reduced sample correlation matrix $\mathbf{R}_r$ should be accounted for by the m common factors. In particular, $\mathbf{R}_r$ is factored as

$$\mathbf{R}_r \doteq \mathbf{L}_r^* \mathbf{L}_r^{*\prime} \tag{9-21}$$

where $\mathbf{L}_r^* = \{\ell_{ij}^*\}$ are the estimated loadings.

The *principal factor method* of factor analysis employs the estimates

$$\mathbf{L}_r^* = \left[\sqrt{\hat{\lambda}_1^*}\, \hat{\mathbf{e}}_1^* \mid \sqrt{\hat{\lambda}_2^*}\, \hat{\mathbf{e}}_2^* \mid \cdots \mid \sqrt{\hat{\lambda}_m^*}\, \hat{\mathbf{e}}_m^* \right]$$

$$\psi_i^* = 1 - \sum_{j=1}^{m} \ell_{ij}^{*2} \tag{9-22}$$

where $(\hat{\lambda}_i^*, \hat{\mathbf{e}}_i^*)$, $i = 1, 2, \ldots, m$ are the (largest) eigenvalue-eigenvector pairs determined from $\mathbf{R}_r$. In turn, the communalities would then be (re)estimated by

$$\tilde{h}_i^{*2} = \sum_{j=1}^{m} \ell_{ij}^{*2} \qquad (9\text{-}23)$$

The principal factor solution can be obtained iteratively, with the communality estimates of (9-23) becoming the initial estimates for the next stage.

In the spirit of the principal component solution, consideration of the estimated eigenvalues $\hat{\lambda}_1^*, \hat{\lambda}_2^*, \ldots, \hat{\lambda}_p^*$ helps determine the number of common factors to retain. An added complication is that now some of the eigenvalues may be negative, due to the use of initial communality estimates. Ideally, we should take the number of common factors equal to the rank of the reduced *population* matrix. Unfortunately, this rank is not always well determined from $\mathbf{R}_r$, and some judgment is necessary.

Although there are many choices for initial estimates of specific variances, the most popular choice, when one is working with a correlation matrix, is $\psi_i^* = 1/r^{ii}$, where r^{ii} is the ith diagonal element of $\mathbf{R}^{-1}$. The initial communality estimates then become

$$h_i^{*2} = 1 - \psi_i^* = 1 - \frac{1}{r^{ii}} \qquad (9\text{-}24)$$

which is equal to the square of the multiple correlation coefficient between X_i and the other $p - 1$ variables. The relation to the multiple correlation coefficient means that h_i^{*2} can be calculated even when $\mathbf{R}$ is not of full rank. For factoring $\mathbf{S}$, the initial specific variance estimates use s^{ii}, the diagonal elements of $\mathbf{S}^{-1}$. Further discussion of these and other initial estimates is contained in [6].

Although the principal component method for $\mathbf{R}$ can be regarded as a principal factor method with *initial* communality estimates of unity, or specific variances equal to zero, the two are philosophically and geometrically different. (See [6].) In practice, however, the two frequently produce comparable factor loadings if the number of variables is large and the number of common factors is small.

We do not pursue the principal factor solution, since, to our minds, the solution methods that have the most to recommend them are the principal component method and the maximum likelihood method, which we discuss next.

The Maximum Likelihood Method

If the common factors $\mathbf{F}$ and the specific factors $\boldsymbol{\varepsilon}$ can be assumed to be normally distributed, then maximum likelihood estimates of the factor loadings and specific variances may be obtained. When $\mathbf{F}_j$ and $\boldsymbol{\varepsilon}_j$ are jointly normal, the observations $\mathbf{X}_j - \boldsymbol{\mu} = \mathbf{L}\mathbf{F}_j + \boldsymbol{\varepsilon}_j$ are then normal, and from (4-16), the likelihood is

$$
\begin{aligned}
L(\boldsymbol{\mu}, \boldsymbol{\Sigma}) &= (2\pi)^{-\frac{np}{2}} |\boldsymbol{\Sigma}|^{-\frac{n}{2}} e^{-\left(\frac{1}{2}\right) \text{tr}\left[\boldsymbol{\Sigma}^{-1}\left(\sum_{j=1}^{n} (\mathbf{x}_j - \bar{\mathbf{x}})(\mathbf{x}_j - \bar{\mathbf{x}})' + n(\bar{\mathbf{x}} - \boldsymbol{\mu})(\bar{\mathbf{x}} - \boldsymbol{\mu})'\right)\right]} \\
&= (2\pi)^{-\frac{(n-1)p}{2}} |\boldsymbol{\Sigma}|^{-\frac{(n-1)}{2}} e^{-\left(\frac{1}{2}\right) \text{tr}\left[\boldsymbol{\Sigma}^{-1}\left(\sum_{j=1}^{n} (\mathbf{x}_j - \bar{\mathbf{x}})(\mathbf{x}_j - \bar{\mathbf{x}})'\right)\right]} \\
&\quad \times (2\pi)^{-\frac{p}{2}} |\boldsymbol{\Sigma}|^{-\frac{1}{2}} e^{-\left(\frac{n}{2}\right)(\bar{\mathbf{x}} - \boldsymbol{\mu})' \boldsymbol{\Sigma}^{-1} (\bar{\mathbf{x}} - \boldsymbol{\mu})}
\end{aligned}
\qquad (9\text{-}25)
$$

which depends on $\mathbf{L}$ and $\mathbf{\Psi}$ through $\mathbf{\Sigma} = \mathbf{LL}' + \mathbf{\Psi}$. This model is still not well defined, because of the multiplicity of choices for $\mathbf{L}$ made possible by orthogonal transformations. It is desirable to make $\mathbf{L}$ well defined by imposing the computationally convenient *uniqueness condition*

$$\mathbf{L}'\mathbf{\Psi}^{-1}\mathbf{L} = \mathbf{\Delta} \qquad \text{a diagonal matrix} \tag{9-26}$$

The maximum likelihood estimates $\hat{\mathbf{L}}$ and $\hat{\mathbf{\Psi}}$ must be obtained by numerical maximization of (9-25). Fortunately, efficient computer programs now exist that enable one to get these estimates rather easily.

We summarize some facts about maximum likelihood estimators and, for now, rely on a computer to perform the numerical details.

Result 9.1. Let $\mathbf{X}_1, \mathbf{X}_2, \ldots, \mathbf{X}_n$ be a random sample from $N_p(\boldsymbol{\mu}, \mathbf{\Sigma})$, where $\mathbf{\Sigma} = \mathbf{LL}' + \mathbf{\Psi}$ is the covariance matrix for the m common factor model of (9-4). The maximum likelihood estimators $\hat{\mathbf{L}}$, $\hat{\mathbf{\Psi}}$, and $\hat{\boldsymbol{\mu}} = \bar{\mathbf{x}}$ maximize (9-25) subject to $\hat{\mathbf{L}}'\hat{\mathbf{\Psi}}^{-1}\hat{\mathbf{L}}$ being diagonal.

The maximum likelihood estimates of the communalities are

$$\hat{h}_i^2 = \hat{\ell}_{i1}^2 + \hat{\ell}_{i2}^2 + \cdots + \hat{\ell}_{im}^2 \qquad \text{for } i = 1, 2, \ldots, p \tag{9-27}$$

so

$$\left(\begin{array}{c} \text{Proportion of total sample} \\ \text{variance due to } j\text{th factor} \end{array} \right) = \frac{\hat{\ell}_{1j}^2 + \hat{\ell}_{2j}^2 + \cdots + \hat{\ell}_{pj}^2}{s_{11} + s_{22} + \cdots + s_{pp}} \tag{9-28}$$

Proof. By the invariance property of maximum likelihood estimates (see Section 4.3), functions of $\mathbf{L}$ and $\mathbf{\Psi}$ are estimated by the same functions of $\hat{\mathbf{L}}$ and $\hat{\mathbf{\Psi}}$. In particular, the communalities $h_i^2 = \ell_{i1}^2 + \cdots + \ell_{im}^2$ have maximum likelihood estimates $\hat{h}_i^2 = \hat{\ell}_{i1}^2 + \cdots + \hat{\ell}_{im}^2$. ∎

If, as in (8-10), the variables are standardized so that $\mathbf{Z} = \mathbf{V}^{-1/2}(\mathbf{X} - \boldsymbol{\mu})$, then the covariance matrix $\boldsymbol{\rho}$ of $\mathbf{Z}$ has the representation

$$\boldsymbol{\rho} = \mathbf{V}^{-1/2}\mathbf{\Sigma}\mathbf{V}^{-1/2} = (\mathbf{V}^{-1/2}\mathbf{L})(\mathbf{V}^{-1/2}\mathbf{L})' + \mathbf{V}^{-1/2}\mathbf{\Psi}\mathbf{V}^{-1/2} \tag{9-29}$$

Thus, $\boldsymbol{\rho}$ has a factorization analogous to (9-5) with loading matrix $\mathbf{L}_z = \mathbf{V}^{-1/2}\mathbf{L}$ and specific variance matrix $\mathbf{\Psi}_z = \mathbf{V}^{-1/2}\mathbf{\Psi}\mathbf{V}^{-1/2}$. By the invariance property of maximum likelihood estimators, the maximum likelihood estimator of $\boldsymbol{\rho}$ is

$$\hat{\boldsymbol{\rho}} = (\hat{\mathbf{V}}^{-1/2}\hat{\mathbf{L}})(\hat{\mathbf{V}}^{-1/2}\hat{\mathbf{L}})' + \hat{\mathbf{V}}^{-1/2}\hat{\mathbf{\Psi}}\hat{\mathbf{V}}^{-1/2}$$

$$= \hat{\mathbf{L}}_z\hat{\mathbf{L}}_z' + \hat{\mathbf{\Psi}}_z \tag{9-30}$$

where $\hat{\mathbf{V}}^{-1/2}$ and $\hat{\mathbf{L}}$ are the maximum likelihood estimators of $\mathbf{V}^{-1/2}$ and $\mathbf{L}$, respectively. (See Supplement 9A.)

As a consequence of the factorization of (9-30), whenever the maximum likelihood analysis pertains to the correlation matrix, we call

$$\hat{h}_i^2 = \hat{\ell}_{i1}^2 + \hat{\ell}_{i2}^2 + \cdots + \hat{\ell}_{im}^2 \qquad i = 1, 2, \ldots, p \tag{9-31}$$

the maximum likelihood estimates of the communalities, and we evaluate the importance of the factors on the basis of

$$\left(\begin{array}{c}\text{Proportion of total (standardized)}\\ \text{sample variance due to } j\text{th factor}\end{array}\right) = \frac{\hat{\ell}_{1j}^2 + \hat{\ell}_{2j}^2 + \cdots + \hat{\ell}_{pj}^2}{p} \qquad (9\text{-}32)$$

To avoid more tedious notations, the preceding $\hat{\ell}_{ij}$'s denote the elements of $\hat{\mathbf{L}}_{\mathbf{z}}$.

Comment. Ordinarily, the observations are standardized, and a sample correlation matrix is factor analyzed. The sample correlation matrix $\mathbf{R}$ is inserted for $[(n-1)/n]\mathbf{S}$ in the likelihood function of (9-25), and the maximum likelihood estimates $\hat{\mathbf{L}}_{\mathbf{z}}$ and $\hat{\mathbf{\Psi}}_{\mathbf{z}}$ are obtained using a computer. Although the likelihood in (9-25) is appropriate for $\mathbf{S}$, not $\mathbf{R}$, surprisingly, this practice is equivalent to obtaining the maximum likelihood estimates $\hat{\mathbf{L}}$ and $\hat{\mathbf{\Psi}}$ based on the sample covariance matrix $\mathbf{S}$, setting $\hat{\mathbf{L}}_{\mathbf{z}} = \hat{\mathbf{V}}^{-1/2}\hat{\mathbf{L}}$ and $\hat{\mathbf{\Psi}}_{\mathbf{z}} = \hat{\mathbf{V}}^{-1/2}\hat{\mathbf{\Psi}}\hat{\mathbf{V}}^{-1/2}$. Here $\hat{\mathbf{V}}^{-1/2}$ is the diagonal matrix with the reciprocal of the sample standard deviations (computed with the divisor $\sqrt{n}$) on the main diagonal.

Going in the other direction, given the estimated loadings $\hat{\mathbf{L}}_{\mathbf{z}}$ and specific variances $\hat{\mathbf{\Psi}}_{\mathbf{z}}$ obtained from $\mathbf{R}$, we find that the resulting maximum likelihood estimates for a factor analysis of the covariance matrix $[(n-1)/n]\mathbf{S}$ are $\hat{\mathbf{L}} = \hat{\mathbf{V}}^{1/2}\hat{\mathbf{L}}_{\mathbf{z}}$ and $\hat{\mathbf{\Psi}} = \hat{\mathbf{V}}^{1/2}\hat{\mathbf{\Psi}}_{\mathbf{z}}\hat{\mathbf{V}}^{1/2}$, or

$$\hat{\ell}_{ij} = \hat{\ell}_{\mathbf{z},ij}\sqrt{\hat{\sigma}_{ii}} \quad \text{and} \quad \hat{\psi}_i = \hat{\psi}_{\mathbf{z},i}\hat{\sigma}_{ii}$$

where $\hat{\sigma}_{ii}$ is the sample variance computed with divisor n. The distinction between divisors can be ignored with principal component solutions. ∎

The equivalence between factoring $\mathbf{S}$ and $\mathbf{R}$ has apparently been confused in many published discussions of factor analysis. (See Supplement 9A.)

Example 9.5 (Factor analysis of stock-price data using the maximum likelihood method)

The stock-price data of Examples 8.5 and 9.4 were reanalyzed assuming an $m = 2$ factor model and using the *maximum likelihood method*. The estimated factor loadings, communalities, specific variances, and proportion of total (standardized) sample variance explained by each factor are in Table 9.3.[3] The corresponding figures for the $m = 2$ factor solution obtained by the *principal component method* (see Example 9.4) are also provided. The communalities corresponding to the maximum likelihood factoring of $\mathbf{R}$ are of the form [see (9-31)] $\hat{h}_i^2 = \hat{\ell}_{i1}^2 + \hat{\ell}_{i2}^2$.

So, for example,

$$\hat{h}_1^2 = (.115)^2 + (.765)^2 = .58$$

---

[3] The maximum likelihood solution leads to a *Heywood case*. For this example, the solution of the likelihood equations give estimated loadings such that a specific variance is negative. The software program obtains a feasible solution by slightly adjusting the loadings so that all specific variance estimates are nonnegative. A Heywood case is suggested here by the .00 value for the specific variance of Royal Dutch Shell.

Table 9.3

	Maximum likelihood			Principal components		
	Estimated factor loadings		Specific variances	Estimated factor loadings		Specific variances
Variable	F_1	F_2	$\hat{\psi}_i = 1 - \hat{h}_i^2$	F_1	F_2	$\tilde{\psi}_i = 1 - \tilde{h}_i^2$
1. J P Morgan	.115	.755	.42	.732	−.437	.27
2. Citibank	.322	.788	.27	.831	−.280	.23
3. Wells Fargo	.182	.652	.54	.726	−.374	.33
4. Royal Dutch Shell	1.000	−.000	.00	.605	.694	.15
5. Texaco	.683	−.032	.53	.563	.719	.17
Cumulative proportion of total (standardized) sample variance explained	.323	.647		.487	.769	

P.C better in this case

The residual matrix is

$$\mathbf{R} - \hat{\mathbf{L}}\hat{\mathbf{L}}' - \hat{\mathbf{\Psi}} = \begin{bmatrix} 0 & .001 & -.002 & .000 & .052 \\ .001 & 0 & .002 & .000 & -.033 \\ -.002 & .002 & 0 & .000 & .001 \\ .000 & .000 & .000 & 0 & .000 \\ .052 & -.033 & .001 & .000 & 0 \end{bmatrix}$$

The elements of $\mathbf{R} - \hat{\mathbf{L}}\hat{\mathbf{L}}' - \hat{\mathbf{\Psi}}$ are much smaller than those of the residual matrix corresponding to the principal component factoring of $\mathbf{R}$ presented in Example 9.4. On this basis, we prefer the maximum likelihood approach and typically feature it in subsequent examples.

The cumulative proportion of the total sample variance explained by the factors is larger for principal component factoring than for maximum likelihood factoring. It is not surprising that this criterion typically favors principal component factoring. Loadings obtained by a principal component factor analysis are related to the principal components, which have, by design, a variance optimizing property. [See the discussion preceding (8-19).]

Focusing attention on the maximum likelihood solution, we see that all variables have positive loadings on F_1. We call this factor the *market factor*, as we did in the principal component solution. The interpretation of the second factor is not as clear as it appeared to be in the principal component solution. The bank stocks have large positive loadings and the oil stocks have negligible loadings on the second factor F_2. From this perspective, the second factor differentiates the bank stocks from the oil stocks and might be called an *industry factor*. Alternatively, the second factor might be simply called a *banking factor*.

The patterns of the initial factor loadings for the maximum likelihood solution are constrained by the uniqueness condition that $\hat{\mathbf{L}}'\hat{\mathbf{\Psi}}^{-1}\hat{\mathbf{L}}$ be a diagonal matrix. Therefore, useful factor patterns are often not revealed until the factors are rotated (see Section 9.4).   ■

Example 9.6 (Factor analysis of Olympic decathlon data) Linden [11] originally conducted a factor analytic study of Olympic decathlon results for all 160 complete starts from the end of World War II until the mid-seventies. Following his approach we examine the $n = 280$ complete starts from 1960 through 2004. The recorded values for each event were standardized and the signs of the timed events changed so that large scores are good for all events. We, too, analyze the correlation matrix, which based on all 280 cases, is

$\mathbf{R} =$

$$
\begin{bmatrix}
1.000 & .6386 & .4752 & .3227 & .5520 & .3262 & .3509 & .4008 & .1821 & -.0352 \\
.6386 & 1.0000 & .4953 & .5668 & .4706 & .3520 & .3998 & .5167 & .3102 & .1012 \\
.4752 & .4953 & 1.0000 & .4357 & .2539 & .2812 & .7926 & .4728 & .4682 & -.0120 \\
.3227 & .5668 & .4357 & 1.0000 & .3449 & .3503 & .3657 & .6040 & .2344 & .2380 \\
.5520 & .4706 & .2539 & .3449 & 1.0000 & .1546 & .2100 & .4213 & .2116 & .4125 \\
.3262 & .3520 & .2812 & .3503 & .1546 & 1.0000 & .2553 & .4163 & .1712 & .0002 \\
.3509 & .3998 & .7926 & .3657 & .2100 & .2553 & 1.0000 & .4036 & .4179 & .0109 \\
.4008 & .5167 & .4728 & .6040 & .4213 & .4163 & .4036 & 1.0000 & .3151 & .2395 \\
.1821 & .3102 & .4682 & .2344 & .2116 & .1712 & .4179 & .3151 & 1.0000 & .0983 \\
-.0352 & .1012 & -.0120 & .2380 & .4125 & .0002 & .0109 & .2395 & .0983 & 1.0000
\end{bmatrix}
$$

From a principal component factor analysis perspective, the first four eigenvalues, 4.21, 1.39, 1.06, .92, of $\mathbf{R}$ suggest a factor solution with $m = 3$ or $m = 4$. A subsequent interpretation, much like Linden's original analysis, reinforces the choice $m = 4$.

In this case, the two solution methods produced very different results. For the principal component factorization, all events except the 1,500-meter run have large positive loading on the first factor. This factor might be labeled *general athletic ability*. Factor 2, which loads heavily on the 400-meter run and 1,500-meter run might be called a *running endurance* factor. The remaining factors cannot be easily interpreted to our minds.

For the maximum likelihood method, the first factor appears to be a *general athletic ability factor* but the loading pattern is not as strong as with principal component factor solution. The second factor is primarily a *strength* factor because shot put and discus load highly on this factor. The third factor is *running endurance* since the 400-meter run and 1,500-meter run have large loadings. Again, the fourth factor is not easily identified, although it may have something to do with jumping ability or *leg strength*. We shall return to an interpretation of the factors in Example 9.11 after a discussion of factor rotation.

The four-factor principal component solution accounts for much of the total (standardized) sample variance, although the estimated specific variances are large in some cases (for example, the javelin). This suggests that some events might require *unique* or specific attributes not required for the other events. The four-factor maximum likelihood solution accounts for less of the total sample

Table 9.4

Variable	Principal component					Maximum likelihood				
	Estimated factor loadings				Specific variances	Estimated factor loadings				Specific variances
	F_1	F_2	F_3	F_4	$\tilde{\psi}_i = 1 - \tilde{h}_i^2$	F_1	F_2	F_3	F_4	$\hat{\psi}_i = 1 - \hat{h}_i^2$
1. 100-m run	.696	.022	−.468	−.416	.12	.993	−.069	−.021	.002	.01
2. Long jump	.793	.075	−.255	−.115	.29	.665	.252	.239	.220	.39
3. Shot put	.771	−.434	.197	−.112	.17	.530	.777	−.141	−.079	.09
4. High jump	.711	.181	.005	.367	.33	.363	.428	.421	.424	.33
5. 400-m run	.605	.549	−.045	−.397	.17	.571	.019	.620	−.305	.20
6. 100 m hurdles	.513	−.083	−.372	.561	.28	.343	.189	.090	.323	.73
7. Discus	.690	−.456	.289	−.078	.23	.402	.718	−.102	−.095	.30
8. Pole vault	.761	.162	.018	.304	.30	.440	.407	.390	.263	.42
9. Javelin	.518	−.252	.519	−.074	.39	.218	.461	.084	−.085	.73
10. 1500-m run	.220	.746	.493	.085	.15	−.016	.091	.609	−.145	.60
Cumulative proportion of total variance explained	.42	.56	.67	.76		.27	.45	.57	.62	

variance, but, as the following residual matrices indicate, the maximum likelihood estimates $\hat{\mathbf{L}}$ and $\hat{\mathbf{\Psi}}$ do a better job of reproducing $\mathbf{R}$ than the principal component estimates $\tilde{\mathbf{L}}$ and $\tilde{\mathbf{\Psi}}$.

Principal component:

$$\mathbf{R} - \tilde{\mathbf{L}}\tilde{\mathbf{L}}' - \tilde{\mathbf{\Psi}} =$$

$$\begin{bmatrix}
0 & -.082 & -.006 & -.021 & -.068 & .031 & -.016 & .003 & .039 & .062 \\
-.082 & 0 & -.046 & .033 & -.107 & -.078 & -.048 & -.059 & .042 & .006 \\
-.006 & -.046 & 0 & .006 & -.010 & -.014 & -.003 & -.013 & -.151 & .055 \\
-.021 & .033 & .006 & 0 & -.038 & -.204 & -.015 & -.078 & -.064 & -.086 \\
-.068 & -.107 & -.010 & -.038 & 0 & .096 & .025 & -.006 & .030 & -.074 \\
.031 & -.078 & -.014 & -.204 & .096 & 0 & .015 & -.124 & .119 & .085 \\
-.016 & -.048 & -.003 & -.015 & .025 & .015 & 0 & -.029 & -.210 & .064 \\
.003 & -.059 & -.013 & -.078 & -.006 & -.124 & -.029 & 0 & -.026 & -.084 \\
.039 & .042 & -.151 & -.064 & .030 & .119 & -.210 & -.026 & 0 & -.078 \\
.062 & .006 & .055 & -.086 & -.074 & .085 & .064 & -.084 & -.078 & 0
\end{bmatrix}$$

Maximum likelihood:

$$\mathbf{R} - \hat{\mathbf{L}}\hat{\mathbf{L}}' - \hat{\mathbf{\Psi}} =$$

$$\begin{bmatrix}
0 & .000 & .000 & -.000 & -.000 & .000 & -.000 & .000 & -.001 & 000 \\
.000 & 0 & -.002 & .023 & .005 & .017 & -.003 & -.030 & .047 & -.024 \\
.000 & -.002 & 0 & .004 & -.000 & -.009 & .000 & -.001 & -.001 & .000 \\
-.000 & .023 & .004 & 0 & -.002 & -.030 & -.004 & -.006 & -.042 & .010 \\
-.000 & .005 & -.001 & -.002 & 0 & -.002 & .001 & .001 & .000 & -.001 \\
.000 & -.017 & -.009 & -.030 & -.002 & 0 & .022 & .069 & .029 & -.019 \\
-.000 & -.003 & .000 & -.004 & .001 & .022 & 0 & -.000 & -.000 & .000 \\
.000 & -.030 & -.001 & -.006 & .001 & .069 & -.000 & 0 & .021 & .011 \\
-.001 & .047 & -.001 & -.042 & .001 & .029 & -.000 & .021 & 0 & -.003 \\
.000 & -.024 & .000 & .010 & -.001 & -.019 & .000 & .011 & -.003 & 0
\end{bmatrix}$$

■

A Large Sample Test for the Number of Common Factors

The assumption of a normal population leads directly to a test of the adequacy of the model. Suppose the m common factor model holds. In this case $\mathbf{\Sigma} = \mathbf{L}\mathbf{L}' + \mathbf{\Psi}$, and testing the adequacy of the m common factor model is equivalent to testing

$$H_0: \underset{(p \times p)}{\mathbf{\Sigma}} = \underset{(p \times m)}{\mathbf{L}} \underset{(m \times p)}{\mathbf{L}'} + \underset{(p \times p)}{\mathbf{\Psi}} \tag{9-33}$$

versus $H_1: \mathbf{\Sigma}$ any other positive definite matrix. When $\mathbf{\Sigma}$ does not have any special form, the maximum of the likelihood function [see (4-18) and Result 4.11 with $\hat{\mathbf{\Sigma}} = ((n-1)/n)\mathbf{S} = \mathbf{S}_n$] is proportional to

$$|\mathbf{S}_n|^{-n/2} e^{-np/2} \tag{9-34}$$

Under H_0, Σ is restricted to have the form of (9-33). In this case, the maximum of the likelihood function [see (9-25) with $\hat{\mu} = \bar{x}$ and $\hat{\Sigma} = \hat{L}\hat{L}' + \hat{\Psi}$, where $\hat{L}$ and $\hat{\Psi}$ are the maximum likelihood estimates of L and Ψ, respectively] is proportional to

$$|\hat{\Sigma}|^{-n/2} \exp\left(-\tfrac{1}{2}\operatorname{tr}\left[\hat{\Sigma}^{-1}\left(\sum_{j=1}^{n}(\mathbf{x}_j - \bar{\mathbf{x}})(\mathbf{x}_j - \bar{\mathbf{x}})'\right)\right]\right)$$

$$= |\hat{L}\hat{L}' + \hat{\Psi}|^{-n/2}\exp\left(-\tfrac{1}{2}n\operatorname{tr}[(\hat{L}\hat{L}' + \hat{\Psi})^{-1}\mathbf{S}_n]\right) \quad (9\text{-}35)$$

Using Result 5.2, (9-34), and (9-35), we find that the likelihood ratio statistic for testing H_0 is

$$-2\ln\Lambda = -2\ln\left[\frac{\text{maximized likelihood under } H_0}{\text{maximized likelihood}}\right]$$

$$\quad (9\text{-}36)$$

$$= -2\ln\left(\frac{|\hat{\Sigma}|}{|\mathbf{S}_n|}\right)^{-n/2} + n\left[\operatorname{tr}(\hat{\Sigma}^{-1}\mathbf{S}_n) - p\right]$$

with degrees of freedom,

$$v - v_0 = \tfrac{1}{2}p(p+1) - [p(m+1) - \tfrac{1}{2}m(m-1)] \quad (9\text{-}37)$$

$$= \tfrac{1}{2}[(p-m)^2 - p - m]$$

Supplement 9A indicates that $\operatorname{tr}(\hat{\Sigma}^{-1}\mathbf{S}_n) - p = 0$ provided that $\hat{\Sigma} = \hat{L}\hat{L}' + \hat{\Psi}$ is the maximum likelihood estimate of $\Sigma = LL' + \Psi$. Thus, we have

$$-2\ln\Lambda = n\ln\left(\frac{|\hat{\Sigma}|}{|\mathbf{S}_n|}\right) \quad (9\text{-}38)$$

Bartlett [3] has shown that the chi-square approximation to the sampling distribution of $-2\ln\Lambda$ can be improved by replacing n in (9-38) with the multiplicative factor $(n - 1 - (2p + 4m + 5)/6)$.

Using Bartlett's correction,[4] we reject H_0 at the α level of significance if

$$(n - 1 - (2p + 4m + 5)/6)\ln\frac{|\hat{L}\hat{L}' + \hat{\Psi}|}{|\mathbf{S}_n|} > \chi^2_{[(p-m)^2 - p - m]/2}(\alpha) \quad (9\text{-}39)$$

provided that n and $n - p$ are large. Since the number of degrees of freedom, $\tfrac{1}{2}[(p-m)^2 - p - m]$, must be positive, it follows that

$$m < \tfrac{1}{2}(2p + 1 - \sqrt{8p+1}) \quad (9\text{-}40)$$

in order to apply the test (9-39).

---

[4] Many factor analysts obtain an approximate maximum likelihood estimate by replacing $\mathbf{S}_n$ with the unbiased estimate $\mathbf{S} = [n/(n-1)]\mathbf{S}_n$ and then minimizing $\ln|\Sigma| + \operatorname{tr}[\Sigma^{-1}\mathbf{S}]$. The dual substitution of $\mathbf{S}$ and the approximate maximum likelihood estimator into the test statistic of (9-39) does not affect its large sample properties.

Comment. In implementing the test in (9-39), we are testing for the adequacy of the m common factor model by comparing the generalized variances $|\hat{\mathbf{L}}\hat{\mathbf{L}}' + \hat{\boldsymbol{\Psi}}|$ and $|\mathbf{S}_n|$. If n is large and m is small relative to p, the hypothesis H_0 will usually be rejected, leading to a retention of more common factors. However, $\hat{\boldsymbol{\Sigma}} = \hat{\mathbf{L}}\hat{\mathbf{L}}' + \hat{\boldsymbol{\Psi}}$ may be close enough to $\mathbf{S}_n$ so that adding more factors does not provide additional insights, even though those factors are "significant." Some judgment must be exercised in the choice of m.

Example 9.7 (Testing for two common factors) The two-factor maximum likelihood analysis of the stock-price data was presented in Example 9.5. The residual matrix there suggests that a two-factor solution may be adequate. Test the hypothesis $H_0: \boldsymbol{\Sigma} = \mathbf{L}\mathbf{L}' + \boldsymbol{\Psi}$, with $m = 2$, at level $\alpha = .05$.

The test statistic in (9-39) is based on the ratio of generalized variances

$$\frac{|\hat{\boldsymbol{\Sigma}}|}{|\mathbf{S}_n|} = \frac{|\hat{\mathbf{L}}\hat{\mathbf{L}}' + \hat{\boldsymbol{\Psi}}|}{|\mathbf{S}_n|}$$

Let $\hat{\mathbf{V}}^{-1/2}$ be the diagonal matrix such that $\hat{\mathbf{V}}^{-1/2}\mathbf{S}_n\hat{\mathbf{V}}^{-1/2} = \mathbf{R}$. By the properties of determinants (see Result 2A.11),

$$|\hat{\mathbf{V}}^{-1/2}||\hat{\mathbf{L}}\hat{\mathbf{L}}' + \hat{\boldsymbol{\Psi}}||\hat{\mathbf{V}}^{-1/2}| = |\hat{\mathbf{V}}^{-1/2}\hat{\mathbf{L}}\hat{\mathbf{L}}'\hat{\mathbf{V}}^{-1/2} + \hat{\mathbf{V}}^{-1/2}\hat{\boldsymbol{\Psi}}\hat{\mathbf{V}}^{-1/2}|$$

and

$$|\hat{\mathbf{V}}^{-1/2}||\mathbf{S}_n||\hat{\mathbf{V}}^{-1/2}| = |\hat{\mathbf{V}}^{-1/2}\mathbf{S}_n\hat{\mathbf{V}}^{-1/2}|$$

Consequently,

$$\frac{|\hat{\boldsymbol{\Sigma}}|}{|\mathbf{S}_n|} = \frac{|\hat{\mathbf{V}}^{-1/2}|}{|\hat{\mathbf{V}}^{-1/2}|} \frac{|\hat{\mathbf{L}}\hat{\mathbf{L}}' + \hat{\boldsymbol{\Psi}}|}{|\mathbf{S}_n|} \frac{|\hat{\mathbf{V}}^{-1/2}|}{|\hat{\mathbf{V}}^{-1/2}|}$$

$$= \frac{|\hat{\mathbf{V}}^{-1/2}\hat{\mathbf{L}}\hat{\mathbf{L}}'\hat{\mathbf{V}}^{-1/2} + \hat{\mathbf{V}}^{-1/2}\hat{\boldsymbol{\Psi}}\hat{\mathbf{V}}^{-1/2}|}{|\hat{\mathbf{V}}^{-1/2}\mathbf{S}_n\hat{\mathbf{V}}^{-1/2}|} \qquad (9\text{-}41)$$

$$= \frac{|\hat{\mathbf{L}}_z\hat{\mathbf{L}}_z' + \hat{\boldsymbol{\Psi}}_z|}{|\mathbf{R}|}$$

by (9-30). From Example 9.5, we determine

$$\frac{|\hat{\mathbf{L}}_z\hat{\mathbf{L}}_z' + \hat{\boldsymbol{\Psi}}_z|}{|\mathbf{R}|} = \frac{\begin{vmatrix} 1.000 & & & & \\ .632 & 1.000 & & & \\ .513 & .572 & 1.000 & & \\ .115 & .322 & .182 & 1.000 & \\ .103 & .246 & .146 & .683 & 1.000 \end{vmatrix}}{\begin{vmatrix} 1.000 & & & & \\ .632 & 1.000 & & & \\ .510 & .574 & 1.000 & & \\ .115 & .322 & .182 & 1.000 & \\ .154 & .213 & .146 & .683 & 1.000 \end{vmatrix}} = \frac{.17898}{.17519} = 1.0216$$

Using Bartlett's correction, we evaluate the test statistic in (9-39):

$$[n - 1 - (2p + 4m + 5)/6] \ln \frac{|\hat{\mathbf{L}}\hat{\mathbf{L}}' + \hat{\mathbf{\Psi}}|}{|\mathbf{S}_n|}$$

$$= \left[103 - 1 - \frac{(10 + 8 + 5)}{6} \right] \ln (1.0216) = 2.10$$

Since $\frac{1}{2}[(p - m)^2 - p - m] = \frac{1}{2}[(5 - 2)^2 - 5 - 2] = 1$, the 5% critical value $\chi_1^2(.05) = 3.84$ is not exceeded, and we fail to reject H_0. We conclude that the data do not contradict a two-factor model. In fact, the observed significance level, or P-value, $P[\chi_1^2 > 2.10] \doteq .15$ implies that H_0 would not be rejected at *any* reasonable level. ∎

Large sample variances and covariances for the maximum likelihood estimates $\hat{\ell}_{ij}, \hat{\psi}_i$ have been derived when these estimates have been determined from the sample covariance matrix $\mathbf{S}$. (See [10].) The expressions are, in general, quite complicated.

9.4 Factor Rotation

As we indicated in Section 9.2, all factor loadings obtained from the initial loadings by an orthogonal transformation have the same ability to reproduce the covariance (or correlation) matrix. [See (9-8).] From matrix algebra, we know that an orthogonal transformation corresponds to a rigid rotation (or reflection) of the coordinate axes. For this reason, an orthogonal transformation of the factor loadings, as well as the implied orthogonal transformation of the factors, is called *factor rotation*.

If $\hat{\mathbf{L}}$ is the $p \times m$ matrix of estimated factor loadings obtained by any method (principal component, maximum likelihood, and so forth) then

$$\hat{\mathbf{L}}^* = \hat{\mathbf{L}}\mathbf{T}, \quad \text{where } \mathbf{T}\mathbf{T}' = \mathbf{T}'\mathbf{T} = \mathbf{I} \quad (9\text{-}42)$$

is a $p \times m$ matrix of "rotated" loadings. Moreover, the estimated covariance (or correlation) matrix remains unchanged, since

$$\hat{\mathbf{L}}\hat{\mathbf{L}}' + \hat{\mathbf{\Psi}} = \hat{\mathbf{L}}\mathbf{T}\mathbf{T}'\hat{\mathbf{L}} + \hat{\mathbf{\Psi}} = \hat{\mathbf{L}}^*\hat{\mathbf{L}}^{*'} + \hat{\mathbf{\Psi}} \quad (9\text{-}43)$$

Equation (9-43) indicates that the residual matrix, $\mathbf{S}_n - \hat{\mathbf{L}}\hat{\mathbf{L}}' - \hat{\mathbf{\Psi}} = \mathbf{S}_n - \hat{\mathbf{L}}^*\hat{\mathbf{L}}^{*'} - \hat{\mathbf{\Psi}}$, remains unchanged. Moreover, the specific variances $\hat{\psi}_i$, and hence the communalities $\hat{h}_i^2$, are unaltered. Thus, from a mathematical viewpoint, it is immaterial whether $\hat{\mathbf{L}}$ or $\hat{\mathbf{L}}^*$ is obtained.

Since the original loadings may not be readily interpretable, it is usual practice to rotate them until a "simpler structure" is achieved. The rationale is very much akin to sharpening the focus of a microscope in order to see the detail more clearly.

Ideally, we should like to see a pattern of loadings such that each variable loads highly on a single factor and has small to moderate loadings on the remaining factors. However, it is not always possible to get this simple structure, although the rotated loadings for the decathlon data discussed in Example 9.11 provide a nearly ideal pattern.

We shall concentrate on graphical and analytical methods for determining an orthogonal rotation to a simple structure. When $m = 2$, or the common factors are considered two at a time, the transformation to a simple structure can frequently be determined graphically. The uncorrelated common factors are regarded as unit vectors along perpendicular coordinate axes. A plot of the pairs of factor loadings $(\hat{\ell}_{i1}, \hat{\ell}_{i2})$ yields p points, each point corresponding to a variable. The coordinate axes can then be visually rotated through an angle—call it ϕ—and the new rotated loadings $\hat{\ell}_{ij}^*$ are determined from the relationships

$$\underset{(p\times2)}{\hat{\mathbf{L}}^*} = \underset{(p\times2)}{\hat{\mathbf{L}}}\ \underset{(2\times2)}{\mathbf{T}} \tag{9-44}$$

where

$$\begin{cases} \mathbf{T} = \begin{bmatrix} \cos\phi & \sin\phi \\ -\sin\phi & \cos\phi \end{bmatrix} & \begin{matrix} \text{clockwise} \\ \text{rotation} \end{matrix} \\[2em] \mathbf{T} = \begin{bmatrix} \cos\phi & -\sin\phi \\ \sin\phi & \cos\phi \end{bmatrix} & \begin{matrix} \text{counterclockwise} \\ \text{rotation} \end{matrix} \end{cases}$$

The relationship in (9-44) is rarely implemented in a two-dimensional graphical analysis. In this situation, clusters of variables are often apparent by eye, and these clusters enable one to identify the common factors without having to inspect the magnitudes of the rotated loadings. On the other hand, for $m > 2$, orientations are not easily visualized, and the magnitudes of the *rotated* loadings must be inspected to find a meaningful interpretation of the original data. The choice of an orthogonal matrix $\mathbf{T}$ that satisfies an *analytical* measure of simple structure will be considered shortly.

Example 9.8 (A first look at factor rotation) Lawley and Maxwell [10] present the sample correlation matrix of examination scores in $p = 6$ subject areas for $n = 220$ male students. The correlation matrix is

$$\mathbf{R} = \begin{array}{cccccc} \text{Gaelic} & \text{English} & \text{History} & \text{Arithmetic} & \text{Algebra} & \text{Geometry} \\ \begin{bmatrix} 1.0 & .439 & .410 & .288 & .329 & .248 \\ & 1.0 & .351 & .354 & .320 & .329 \\ & & 1.0 & .164 & .190 & .181 \\ & & & 1.0 & .595 & .470 \\ & & & & 1.0 & .464 \\ & & & & & 1.0 \end{bmatrix} \end{array}$$

and a maximum likelihood solution for $m = 2$ common factors yields the estimates in Table 9.5.

Table 9.5			
	Estimated factor loadings		Communalities
Variable	F_1	F_2	$\hat{h}_i^2$
1. Gaelic	.553	.429	.490
2. English	.568	.288	.406
3. History	.392	.450	.356
4. Arithmetic	.740	-.273	.623
5. Algebra	.724	-.211	.569
6. Geometry	.595	-.132	.372

All the variables have positive loadings on the first factor. Lawley and Maxwell suggest that this factor reflects the overall response of the students to instruction and might be labeled a *general intelligence* factor. Half the loadings are positive and half are negative on the second factor. A factor with this pattern of loadings is called a *bipolar factor*. (The assignment of negative and positive poles is arbitrary, because the signs of the loadings on a factor can be reversed without affecting the analysis.) This factor is not easily identified, but is such that individuals who get above-average scores on the verbal tests get above-average scores on the factor. Individuals with above-average scores on the mathematical tests get below-average scores on the factor. Perhaps this factor can be classified as a "math-nonmath" factor.

The factor loading pairs $(\hat{\ell}_{i1}, \hat{\ell}_{i2})$ are plotted as points in Figure 9.1. The points are labeled with the numbers of the corresponding variables. Also shown is a clockwise orthogonal rotation of the coordinate axes through an angle of $\phi \doteq 20°$. This angle was chosen so that one of the new axes passes through $(\hat{\ell}_{41}, \hat{\ell}_{42})$. When this is done, all the points fall in the first quadrant (the factor loadings are all positive), and the two distinct clusters of variables are more clearly revealed.

The mathematical test variables load highly on F_1^* and have negligible loadings on F_2^*. The first factor might be called a *mathematical-ability* factor. Similarly, the three verbal test variables have high loadings on F_2^* and moderate to small loadings on F_1^*. The second factor might be labeled a *verbal-ability* factor. The *general-intelligence* factor identified initially is submerged in the factors F_1^* and F_2^*.

The rotated factor loadings obtained from (9-44) with $\phi \doteq 20°$ and the corresponding communality estimates are shown in Table 9.6. The magnitudes of the rotated factor loadings reinforce the interpretation of the factors suggested by Figure 9.1.

The communality estimates are unchanged by the orthogonal rotation, since $\hat{L}\hat{L}' = \hat{L}TT'\hat{L}' = \hat{L}*\hat{L}*'$, and the communalities are the diagonal elements of these matrices.

We point out that Figure 9.1 suggests an *oblique rotation* of the coordinates. One new axis would pass through the cluster $\{1, 2, 3\}$ and the other through the $\{4, 5, 6\}$ group. Oblique rotations are so named because they correspond to a *nonrigid* rotation of coordinate axes leading to new axes that are not perpendicular.

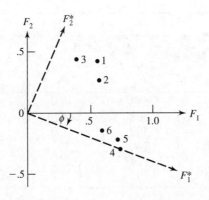

Figure 9.1 Factor rotation for test scores.

Table 9.6

Variable	Estimated rotated factor loadings F_1^*	F_2^*	Communalities $\hat{h}_i^{*2} = \hat{h}_i^2$
1. Gaelic	.369	.594	.490
2. English	.433	.467	.406
3. History	.211	.558	.356
4. Arithmetic	.789	.001	.623
5. Algebra	.752	.054	.568
6. Geometry	.604	.083	.372

It is apparent, however, that the interpretation of the oblique factors for this example would be much the same as that given previously for an orthogonal rotation. ∎

Kaiser [9] has suggested an analytical measure of simple structure known as the *varimax* (or normal varimax) *criterion*. Define $\tilde{\ell}_{ij}^* = \hat{\ell}_{ij}^*/\hat{h}_i$ to be the rotated coefficients scaled by the square root of the communalities. Then the (normal) varimax procedure selects the orthogonal transformation **T** that makes

$$V = \frac{1}{p} \sum_{j=1}^m \left[\sum_{i=1}^p \tilde{\ell}_{ij}^{*4} - \left(\sum_{i=1}^p \tilde{\ell}_{ij}^{*2} \right)^2 \Big/ p \right] \tag{9-45}$$

as large as possible.

Scaling the rotated coefficients $\hat{\ell}_{ij}^*$ has the effect of giving variables with small communalities relatively more weight in the determination of simple structure. After the transformation **T** is determined, the loadings $\tilde{\ell}_{ij}^*$ are multiplied by $\hat{h}_i$ so that the original communalities are preserved.

Although (9-45) looks rather forbidding, it has a simple interpretation. In words,

$$V \propto \sum_{j=1}^m \left(\begin{array}{c} \text{variance of squares of (scaled) loadings for} \\ j\text{th factor} \end{array} \right) \tag{9-46}$$

Effectively, maximizing V corresponds to "spreading out" the squares of the loadings on each factor as much as possible. Therefore, we hope to find groups of large and negligible coefficients in any *column* of the rotated loadings matrix $\hat{\mathbf{L}}^*$.

Computing algorithms exist for maximizing V, and most popular factor analysis computer programs (for example, the statistical software packages SAS, SPSS, BMDP, and MINITAB) provide varimax rotations. As might be expected, varimax rotations of factor loadings obtained by different solution methods (principal components, maximum likelihood, and so forth) will not, in general, coincide. Also, the pattern of rotated loadings may change considerably if additional common factors are included in the rotation. If a dominant single factor exists, it will generally be obscured by any orthogonal rotation. By contrast, it can always be held fixed and the remaining factors rotated.

Example 9.9 (Rotated loadings for the consumer-preference data) Let us return to the marketing data discussed in Example 9.3. The original factor loadings (obtained by the principal component method), the communalities, and the (varimax) rotated factor loadings are shown in Table 9.7. (See the SAS statistical software output in Panel 9.1.)

Table 9.7

Variable	Estimated factor loadings F_1	Estimated factor loadings F_2	Rotated estimated factor loadings F_1^*	Rotated estimated factor loadings F_2^*	Communalities $\widetilde{h}_i^2$
1. Taste	.56	.82	.02	⑨⑨	.98
2. Good buy for money	.78	−.52	㉙④	−.01	.88
3. Flavor	.65	.75	.13	㉙⑧	.98
4. Suitable for snack	.94	−.10	⑧④	.43	.89
5. Provides lots of energy	.80	−.54	⑨⑦	−.02	.93
Cumulative proportion of total (standardized) sample variance explained	.571	.932	.507	.932	

It is clear that variables 2, 4, and 5 define factor 1 (high loadings on factor 1, small or negligible loadings on factor 2), while variables 1 and 3 define factor 2 (high loadings on factor 2, small or negligible loadings on factor 1). Variable 4 is most closely aligned with factor 1, although it has aspects of the trait represented by factor 2. We might call factor 1 a *nutritional* factor and factor 2 a *taste* factor.

The factor loadings for the variables are pictured with respect to the original and (varimax) rotated factor axes in Figure 9.2. ∎

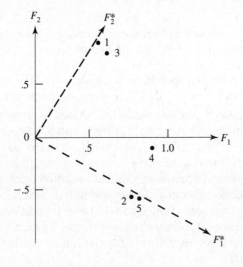

Figure 9.2 Factor rotation for hypothetical marketing data.

PANEL 9.1 SAS ANALYSIS FOR EXAMPLE 9.9 USING PROC FACTOR.

```
title 'Factor Analysis';
data consumer(type = corr);
_type_='CORR';
input _name_$ taste money flavor snack energy;
cards;
taste       1.00      .      .      .      .
money        .02    1.00     .      .      .
flavor       .96     .13   1.00     .      .
snack        .42     .71    .50   1.00     .
energy       .01     .85    .11    .79   1.00
;
proc factor res data=consumer
    method=prin nfact=2rotate=varimax preplot plot;
    var taste money flavor snack energy;
```

PROGRAM COMMANDS

Initial Factor Method: Principal Components   OUTPUT

Prior Communality Estimates: ONE

Eigenvalues of the Correlation Matrix: Total = 5 Average = 1

	1	2	3	4	5
Eigenvalue	2.853090	1.806332	0.204490	0.102409	0.033677
Difference	1.046758	1.601842	0.102081	0.068732	
Proportion	0.5706	0.3613	0.0409	0.0205	0.0067
Cumulative	0.5706	0.9319	0.9728	0.9933	1.0000

2 factors will be retained by the NFACTOR criterion.

Factor Pattern

	FACTOR1	FACTOR2
TASTE	0.55986	0.81610
MONEY	0.77726	−0.52420
FLAVOR	0.64534	0.74795
SNACK	0.93911	−0.10492
ENERGY	0.79821	−0.54323

Final Communality Estimates: Total = 4.659423

TASTE	MONEY	FLAVOR	SNACK	ENERGY
0.97961	0.878920	0.975883	0.892928	0.932231

(continues on next page)

PANEL 9.1 (*continued*)

Rotation Method: Varimax

Rotated Factor Pattern

	FACTOR1	FACTOR2
TASTE	0.01970	0.98948
MONEY	0.93744	−0.01123
FLAVOR	0.12856	0.97947
SNACK	0.84244	0.42805
ENERGY	0.96539	−0.01563

Variance explained by each factor

FACTOR1	FACTOR2
2.537396	2.122027

Rotation of factor loadings is recommended particularly for loadings obtained by maximum likelihood, since the initial values are constrained to satisfy the uniqueness condition that $\hat{\mathbf{L}}'\hat{\mathbf{\Psi}}^{-1}\hat{\mathbf{L}}$ be a diagonal matrix. This condition is convenient for computational purposes, but may not lead to factors that can easily be interpreted.

Example 9.10 (Rotated loadings for the stock-price data) Table 9.8 shows the initial and rotated maximum likelihood estimates of the factor loadings for the stock-price data of Examples 8.5 and 9.5. An $m = 2$ factor model is assumed. The estimated

Table 9.8

Variable	Maximum likelihood estimates of factor loadings		Rotated estimated factor loadings		Specific variances
	F_1	F_2	F_1^*	F_2^*	$\hat{\psi}_i^2 = 1 - \hat{h}_i^2$
J P Morgan	.115	.755	.763	.024	.42
Citibank	.322	.788	.821	.227	.27
Wells Fargo	.182	.652	.669	.104	.54
Royal Dutch Shell	1.000	−.000	.118	.993	.00
ExxonMobil	.683	.032	.113	.675	.53
Cumulative proportion of total sample variance explained	.323	.647	.346	.647	

specific variances and cumulative proportions of the total (standardized) sample variance explained by each factor are also given.

An interpretation of the factors suggested by the unrotated loadings was presented in Example 9.5. We identified *market* and *industry* factors.

The rotated loadings indicate that the bank stocks (JP Morgan, Citibank, and Wells Fargo) load highly on the first factor, while the oil stocks (Royal Dutch Shell and ExxonMobil) load highly on the second factor. (Although the rotated loadings obtained from the principal component solution are not displayed, the same phenomenon is observed for them.) The two rotated factors, together, differentiate the industries. It is difficult for us to label these factors intelligently. Factor 1 represents those unique economic forces that cause bank stocks to move together. Factor 2 appears to represent economic conditions affecting oil stocks.

As we have noted, a general factor (that is, one on which *all* the variables load highly) tends to be "destroyed after rotation." For this reason, in cases where a general factor is evident, an orthogonal rotation is sometimes performed with the general factor loadings fixed.[5]   ∎

Example 9.11 (Rotated loadings for the Olympic decathlon data) The estimated factor loadings and specific variances for the Olympic decathlon data were presented in Example 9.6. These quantities were derived for an $m = 4$ factor model, using both principal component and maximum likelihood solution methods. The interpretation of all the underlying factors was not immediately evident. A varimax rotation [see (9-45)] was performed to see whether the rotated factor loadings would provide additional insights. The varimax rotated loadings for the $m = 4$ factor solutions are displayed in Table 9.9, along with the specific variances. Apart from the estimated loadings, rotation will affect only the *distribution* of the proportions of the total sample variance explained by each factor. The cumulative proportion of the total sample variance explained for *all* factors does not change.

The rotated factor loadings for both methods of solution point to the same underlying attributes, although factors 1 and 2 are not in the same order. We see that shot put, discus, and javelin load highly on a factor, and, following Linden [11], this factor might be called *explosive arm strength*. Similarly, high jump, 110-meter hurdles, pole vault, and—to some extent—long jump load highly on another factor. Linden labeled this factor *explosive leg strength*. The 100-meter run, 400-meter run, and—again to some extent—the long jump load highly on a third factor. This factor could be called *running speed*. Finally, the 1500-meter run loads heavily and the 400-meter run loads heavily on the fourth factor. Linden called this factor *running endurance*. As he notes, "The basic functions indicated in this study are mainly consistent with the traditional classification of track and field athletics."

---

[5]Some general-purpose factor analysis programs allow one to fix loadings associated with certain factors and to rotate the remaining factors.

Table 9.9

Variable	Principal component						Maximum likelihood				
	Estimated rotated factor loadings, $\tilde{\ell}_{ij}^{*}$				Specific variances		Estimated rotated factor loadings, $\hat{\ell}_{ij}^{*}$				Specific variances
	F_1^*	F_2^*	F_3^*	F_4^*	$\tilde{\psi}_i = 1 - \tilde{h}_i^2$		F_1^*	F_2^*	F_3^*	F_4^*	$\hat{\psi}_i = 1 - \hat{h}_i^2$
100-m run	.182	.885	.205	−.139	.12		.204	.296	.928	−.005	.01
Long jump	.291	.664	.429	.055	.29		.280	.554	.451	.155	.39
Shot put	.819	.302	.252	−.097	.17		.883	.278	.228	−.045	.09
High jump	.267	.221	.683	.293	.33		.254	.739	.057	.242	.33
400-m run	.086	.747	.068	.507	.17		.142	.151	.519	.700	.20
110-m hurdles	.048	.108	.826	−.161	.28		.136	.465	.173	−.033	.73
Discus	.832	.185	.204	−.076	.23		.793	.220	.133	−.009	.30
Pole vault	.324	.278	.656	.293	.30		.314	.613	.169	.279	.42
Javelin	.754	.024	.054	.188	.39		.477	.160	.041	.139	.73
1500-m run	−.002	.019	.075	.921	.15		.001	.110	−.070	.619	.60
Cumulative proportion of total sample variance explained	.22	.43	.62	.76			.20	.37	.51	.62	

Plots of rotated maximum likelihood loadings for factors pairs $(1, 2)$ and $(1, 3)$ are displayed in Figure 9.3 on page 513. The points are generally grouped along the factor axes. Plots of rotated principal component loadings are very similar. ∎

Oblique Rotations

Orthogonal rotations are appropriate for a factor model in which the common factors are assumed to be independent. Many investigators in social sciences consider *oblique* (nonorthogonal) rotations, as well as orthogonal rotations. The former are

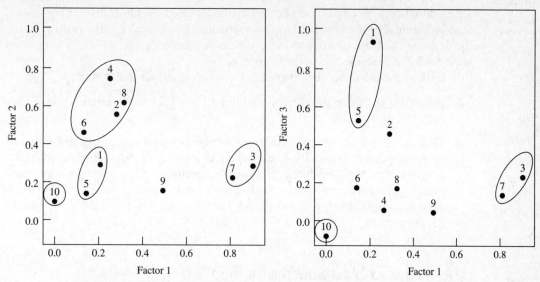

Figure 9.3 Rotated maximum likelihood loadings for factor pairs $(1, 2)$ and $(1, 3)$— decathlon data. (The numbers in the figures correspond to variables.)

often suggested after one views the estimated factor loadings and do not follow from our postulated model. Nevertheless, an oblique rotation is frequently a useful aid in factor analysis.

If we regard the m common factors as coordinate axes, the point with the m coordinates $(\hat{\ell}_{i1}, \hat{\ell}_{i2}, \ldots, \hat{\ell}_{im})$ represents the position of the ith variable in the *factor space*. Assuming that the variables are grouped into nonoverlapping clusters, an orthogonal rotation to a simple structure corresponds to a *rigid* rotation of the coordinate axes such that the axes, after rotation, pass as closely to the clusters as possible. An oblique rotation to a simple structure corresponds to a *nonrigid* rotation of the coordinate system such that the rotated axes (no longer perpendicular) pass (nearly) through the clusters. An oblique rotation seeks to express each variable in terms of a minimum number of factors—preferably, a single factor. Oblique rotations are discussed in several sources (see, for example, [6] or [10]) and will not be pursued in this book.

9.5 Factor Scores

In factor analysis, interest is usually centered on the parameters in the factor model. However, the estimated values of the common factors, called *factor scores*, may also be required. These quantities are often used for diagnostic purposes, as well as inputs to a subsequent analysis.

Factor scores are not estimates of unknown parameters in the usual sense. Rather, they are estimates of values for the unobserved random factor vectors $\mathbf{F}_j$, $j = 1, 2, \ldots, n$. That is, factor scores

$$\hat{\mathbf{f}}_j = \text{estimate of the values } \mathbf{f}_j \text{ attained by } \mathbf{F}_j \ (j\text{th case})$$

The estimation situation is complicated by the fact that the unobserved quantities $\mathbf{f}_j$ and $\boldsymbol{\varepsilon}_j$ outnumber the observed $\mathbf{x}_j$. To overcome this difficulty, some rather heuristic, but reasoned, approaches to the problem of estimating factor values have been advanced. We describe two of these approaches.

Both of the factor score approaches have two elements in common:

1. They treat the estimated factor loadings $\hat{\ell}_{ij}$ and specific variances $\hat{\psi}_i$ as if they were the true values.

2. They involve linear transformations of the original data, perhaps centered or standardized. Typically, the estimated *rotated* loadings, rather than the original estimated loadings, are used to compute factor scores. The computational formulas, as given in this section, do not change when rotated loadings are substituted for unrotated loadings, so we will not differentiate between them.

The Weighted Least Squares Method

Suppose first that the mean vector $\boldsymbol{\mu}$, the factor loadings $\mathbf{L}$, and the specific variance $\boldsymbol{\Psi}$ are known for the factor model

$$\underset{(p \times 1)}{\mathbf{X}} - \underset{(p \times 1)}{\boldsymbol{\mu}} = \underset{(p \times m)}{\mathbf{L}} \underset{(m \times 1)}{\mathbf{F}} + \underset{(p \times 1)}{\boldsymbol{\varepsilon}}$$

Further, regard the specific factors $\boldsymbol{\varepsilon}' = [\varepsilon_1, \ \varepsilon_2, \ldots, \varepsilon_p]$ as errors. Since $\mathrm{Var}(\varepsilon_i) = \psi_i$, $i = 1, 2, \ldots, p$, need not be equal, Bartlett [2] has suggested that weighted least squares be used to estimate the common factor values.

The sum of the squares of the errors, weighted by the reciprocal of their variances, is

$$\sum_{i=1}^{p} \frac{\varepsilon_i^2}{\psi_i} = \boldsymbol{\varepsilon}' \boldsymbol{\Psi}^{-1} \boldsymbol{\varepsilon} = (\mathbf{x} - \boldsymbol{\mu} - \mathbf{Lf})' \boldsymbol{\Psi}^{-1} (\mathbf{x} - \boldsymbol{\mu} - \mathbf{Lf}) \qquad (9\text{-}47)$$

Bartlett proposed choosing the estimates $\hat{\mathbf{f}}$ of $\mathbf{f}$ to minimize (9-47). The solution (see Exercise 7.3) is

$$\hat{\mathbf{f}} = (\mathbf{L}' \boldsymbol{\Psi}^{-1} \mathbf{L})^{-1} \mathbf{L}' \boldsymbol{\Psi}^{-1} (\mathbf{x} - \boldsymbol{\mu}) \qquad (9\text{-}48)$$

Motivated by (9-48), we take the estimates $\hat{\mathbf{L}}$, $\hat{\boldsymbol{\Psi}}$, and $\hat{\boldsymbol{\mu}} = \bar{\mathbf{x}}$ as the true values and obtain the factor scores for the jth case as

$$\hat{\mathbf{f}}_j = (\hat{\mathbf{L}}' \hat{\boldsymbol{\Psi}}^{-1} \hat{\mathbf{L}})^{-1} \hat{\mathbf{L}}' \hat{\boldsymbol{\Psi}}^{-1} (\mathbf{x}_j - \bar{\mathbf{x}}) \qquad (9\text{-}49)$$

When $\hat{\mathbf{L}}$ and $\hat{\boldsymbol{\Psi}}$ are determined by the maximum likelihood method, these estimates must satisfy the uniqueness condition, $\hat{\mathbf{L}}' \hat{\boldsymbol{\Psi}}^{-1} \hat{\mathbf{L}} = \hat{\boldsymbol{\Delta}}$, a diagonal matrix. We then have the following:

Factor Scores Obtained by Weighted Least Squares from the Maximum Likelihood Estimates

$$\hat{\mathbf{f}}_j = (\hat{\mathbf{L}}'\hat{\boldsymbol{\Psi}}^{-1}\hat{\mathbf{L}})^{-1}\hat{\mathbf{L}}'\hat{\boldsymbol{\Psi}}^{-1}(\mathbf{x}_j - \hat{\boldsymbol{\mu}})$$

$$= \hat{\boldsymbol{\Delta}}^{-1}\hat{\mathbf{L}}'\hat{\boldsymbol{\Psi}}^{-1}(\mathbf{x}_j - \bar{\mathbf{x}}), \qquad j = 1, 2, \ldots, n$$

or, if the correlation matrix is factored $\hspace{3cm}$ (9-50)

$$\hat{\mathbf{f}}_j = (\hat{\mathbf{L}}_z'\hat{\boldsymbol{\Psi}}_z^{-1}\hat{\mathbf{L}}_z)^{-1}\hat{\mathbf{L}}_z'\hat{\boldsymbol{\Psi}}_z^{-1}\mathbf{z}_j$$

$$= \hat{\boldsymbol{\Delta}}_z^{-1}\hat{\mathbf{L}}_z'\hat{\boldsymbol{\Psi}}_z^{-1}\mathbf{z}_j, \qquad j = 1, 2, \ldots, n$$

where $\mathbf{z}_j = \mathbf{D}^{-1/2}(\mathbf{x}_j - \bar{\mathbf{x}})$, as in (8-25), and $\hat{\boldsymbol{\rho}} = \hat{\mathbf{L}}_z\hat{\mathbf{L}}_z' + \hat{\boldsymbol{\Psi}}_z$.

The factor scores generated by (9-50) have sample mean vector $\mathbf{0}$ and zero sample covariances. (See Exercise 9.16.)

If rotated loadings $\hat{\mathbf{L}}^* = \hat{\mathbf{L}}\mathbf{T}$ are used in place of the original loadings in (9-50), the subsequent factor scores, $\hat{\mathbf{f}}_j^*$, are related to $\hat{\mathbf{f}}_j$ by $\hat{\mathbf{f}}_j^* = \mathbf{T}'\hat{\mathbf{f}}_j$, $j = 1, 2, \ldots, n$.

Comment. If the factor loadings are estimated by the principal component method, it is customary to generate factor scores using an unweighted (ordinary) least squares procedure. Implicitly, this amounts to assuming that the ψ_i are equal or nearly equal. The factor scores are then

$$\hat{\mathbf{f}}_j = (\tilde{\mathbf{L}}'\tilde{\mathbf{L}})^{-1}\tilde{\mathbf{L}}'(\mathbf{x}_j - \bar{\mathbf{x}})$$

or

$$\hat{\mathbf{f}}_j = (\tilde{\mathbf{L}}_z'\tilde{\mathbf{L}}_z)^{-1}\tilde{\mathbf{L}}_z'\mathbf{z}_j$$

for standardized data. Since $\tilde{\mathbf{L}} = \left[\sqrt{\hat{\lambda}_1}\,\hat{\mathbf{e}}_1 \;\vdots\; \sqrt{\hat{\lambda}_2}\,\hat{\mathbf{e}}_2 \;\vdots\; \cdots \;\vdots\; \sqrt{\hat{\lambda}_m}\,\hat{\mathbf{e}}_m \right]$ [see (9-15)], we have

$$\hat{\mathbf{f}}_j = \begin{bmatrix} \dfrac{1}{\sqrt{\hat{\lambda}_1}}\hat{\mathbf{e}}_1'(\mathbf{x}_j - \bar{\mathbf{x}}) \\[2mm] \dfrac{1}{\sqrt{\hat{\lambda}_2}}\hat{\mathbf{e}}_2'(\mathbf{x}_j - \bar{\mathbf{x}}) \\[1mm] \vdots \\[1mm] \dfrac{1}{\sqrt{\hat{\lambda}_m}}\hat{\mathbf{e}}_m'(\mathbf{x}_j - \bar{\mathbf{x}}) \end{bmatrix} \qquad (9\text{-}51)$$

For these factor scores,

$$\frac{1}{n}\sum_{j=1}^{n}\hat{\mathbf{f}}_j = \mathbf{0} \qquad \text{(sample mean)}$$

and

$$\frac{1}{n-1}\sum_{j=1}^{n}\hat{\mathbf{f}}_j\hat{\mathbf{f}}_j' = \mathbf{I} \qquad \text{(sample covariance)}$$

Comparing (9-51) with (8-21), we see that the $\hat{\mathbf{f}}_j$ are nothing more than the first m (scaled) principal components, evaluated at $\mathbf{x}_j$.

The Regression Method

Starting again with the original factor model $\mathbf{X} - \boldsymbol{\mu} = \mathbf{LF} + \boldsymbol{\varepsilon}$, we initially treat the loadings matrix $\mathbf{L}$ and specific variance matrix $\boldsymbol{\Psi}$ as known. When the common factors $\mathbf{F}$ and the specific factors (or errors) $\boldsymbol{\varepsilon}$ are jointly normally distributed with means and covariances given by (9-3), the linear combination $\mathbf{X} - \boldsymbol{\mu} = \mathbf{LF} + \boldsymbol{\varepsilon}$ has an $N_p(\mathbf{0}, \mathbf{LL}' + \boldsymbol{\Psi})$ distribution. (See Result 4.3.) Moreover, the joint distribution of $(\mathbf{X} - \boldsymbol{\mu})$ and $\mathbf{F}$ is $N_{m+p}(\mathbf{0}, \boldsymbol{\Sigma}*)$, where

$$
\underset{(m+p)\times(m+p)}{\boldsymbol{\Sigma}^*} = \left[\begin{array}{c|c} \underset{(p\times p)}{\boldsymbol{\Sigma} = \mathbf{LL}' + \boldsymbol{\Psi}} & \underset{(p\times m)}{\mathbf{L}} \\ \hline \underset{(m\times p)}{\mathbf{L}'} & \underset{(m\times m)}{\mathbf{I}} \end{array}\right] \tag{9-52}
$$

and $\mathbf{0}$ is an $(m + p) \times 1$ vector of zeros. Using Result 4.6, we find that the conditional distribution of $\mathbf{F}|\mathbf{x}$ is multivariate normal with

$$
\text{mean} = E(\mathbf{F}|\mathbf{x}) = \mathbf{L}'\boldsymbol{\Sigma}^{-1}(\mathbf{x} - \boldsymbol{\mu}) = \mathbf{L}'(\mathbf{LL}' + \boldsymbol{\Psi})^{-1}(\mathbf{x} - \boldsymbol{\mu}) \tag{9-53}
$$

and

$$
\text{covariance} = \text{Cov}(\mathbf{F}|\mathbf{x}) = \mathbf{I} - \mathbf{L}'\boldsymbol{\Sigma}^{-1}\mathbf{L} = \mathbf{I} - \mathbf{L}'(\mathbf{LL}' + \boldsymbol{\Psi})^{-1}\mathbf{L} \tag{9-54}
$$

The quantities $\mathbf{L}'(\mathbf{LL}' + \boldsymbol{\Psi})^{-1}$ in (9-53) are the coefficients in a (multivariate) regression of the factors on the variables. Estimates of these coefficients produce factor scores that are analogous to the estimates of the conditional mean values in multivariate regression analysis. (See Chapter 7.) Consequently, given any vector of observations $\mathbf{x}_j$, and taking the maximum likelihood estimates $\hat{\mathbf{L}}$ and $\hat{\boldsymbol{\Psi}}$ as the true values, we see that the jth factor score vector is given by

$$
\hat{\mathbf{f}}_j = \hat{\mathbf{L}}'\hat{\boldsymbol{\Sigma}}^{-1}(\mathbf{x}_j - \bar{\mathbf{x}}) = \hat{\mathbf{L}}'(\hat{\mathbf{L}}\hat{\mathbf{L}}' + \hat{\boldsymbol{\Psi}})^{-1}(\mathbf{x}_j - \bar{\mathbf{x}}), \qquad j = 1, 2, \ldots, n \tag{9-55}
$$

The calculation of $\hat{\mathbf{f}}_j$ in (9-55) can be simplified by using the matrix identity (see Exercise 9.6)

$$
\underset{(m\times p)}{\hat{\mathbf{L}}'} \underset{(p\times p)}{(\hat{\mathbf{L}}\hat{\mathbf{L}}' + \hat{\boldsymbol{\Psi}})^{-1}} = \underset{(m\times m)}{(\mathbf{I} + \hat{\mathbf{L}}'\hat{\boldsymbol{\Psi}}^{-1}\hat{\mathbf{L}})^{-1}} \underset{(m\times p)}{\hat{\mathbf{L}}'} \underset{(p\times p)}{\hat{\boldsymbol{\Psi}}^{-1}} \tag{9-56}
$$

This identity allows us to compare the factor scores in (9-55), generated by the regression argument, with those generated by the weighted least squares procedure [see (9-50)]. Temporarily, we denote the former by $\hat{\mathbf{f}}_j^R$ and the latter by $\hat{\mathbf{f}}_j^{LS}$. Then, using (9-56), we obtain

$$
\hat{\mathbf{f}}_j^{LS} = (\hat{\mathbf{L}}'\hat{\boldsymbol{\Psi}}^{-1}\hat{\mathbf{L}})^{-1}(\mathbf{I} + \hat{\mathbf{L}}'\hat{\boldsymbol{\Psi}}^{-1}\hat{\mathbf{L}})\mathbf{f}_j^R = (\mathbf{I} + (\hat{\mathbf{L}}'\hat{\boldsymbol{\Psi}}^{-1}\hat{\mathbf{L}})^{-1})\mathbf{f}_j^R \tag{9-57}
$$

For maximum likelihood estimates $(\hat{\mathbf{L}}'\hat{\boldsymbol{\Psi}}^{-1}\hat{\mathbf{L}})^{-1} = \hat{\boldsymbol{\Delta}}^{-1}$ and if the elements of this diagonal matrix are close to zero, the regression and generalized least squares methods will give nearly the same factor scores.

In an attempt to reduce the effects of a (possibly) incorrect determination of the number of factors, practitioners tend to calculate the factor scores in (9-55) by using $\mathbf{S}$ (the original sample covariance matrix) instead of $\hat{\boldsymbol{\Sigma}} = \hat{\mathbf{L}}\hat{\mathbf{L}}' + \hat{\boldsymbol{\Psi}}$. We then have the following:

Factor Scores Obtained by Regression

$$\hat{\mathbf{f}}_j = \hat{\mathbf{L}}'\mathbf{S}^{-1}(\mathbf{x}_j - \bar{\mathbf{x}}), \qquad j = 1, 2, \dots, n$$

or, if a correlation matrix is factored,   (9-58)

$$\hat{\mathbf{f}}_j = \hat{\mathbf{L}}_z'\mathbf{R}^{-1}\mathbf{z}_j, \qquad j = 1, 2, \dots, n$$

where, see (8-25),

$$\mathbf{z}_j = \mathbf{D}^{-1/2}(\mathbf{x}_j - \bar{\mathbf{x}}) \quad \text{and} \quad \hat{\boldsymbol{\rho}} = \hat{\mathbf{L}}_z\hat{\mathbf{L}}_z' + \hat{\boldsymbol{\Psi}}_z$$

Again, if rotated loadings $\hat{\mathbf{L}}^* = \hat{\mathbf{L}}\mathbf{T}$ are used in place of the original loadings in (9-58), the subsequent factor scores $\hat{\mathbf{f}}_j^*$ are related to $\hat{\mathbf{f}}_j$ by

$$\hat{\mathbf{f}}_j^* = \mathbf{T}'\hat{\mathbf{f}}_j, \qquad j = 1, 2, \dots, n$$

A numerical measure of agreement between the factor scores generated from two *different* calculation methods is provided by the sample correlation coefficient between scores on the same factor. Of the methods presented, none is recommended as uniformly superior.

Example 9.12 (Computing factor scores) We shall illustrate the computation of factor scores by the least squares and regression methods using the stock-price data discussed in Example 9.10. A maximum likelihood solution from $\mathbf{R}$ gave the estimated rotated loadings and specific variances

$$\hat{\mathbf{L}}_z^* = \begin{bmatrix} .763 & .024 \\ .821 & .227 \\ .669 & .104 \\ .118 & .993 \\ .113 & .675 \end{bmatrix} \quad \text{and} \quad \hat{\boldsymbol{\Psi}}_z = \begin{bmatrix} .42 & 0 & 0 & 0 & 0 \\ 0 & .27 & 0 & 0 & 0 \\ 0 & 0 & .54 & 0 & 0 \\ 0 & 0 & 0 & .00 & 0 \\ 0 & 0 & 0 & 0 & .53 \end{bmatrix}$$

The vector of standardized observations,

$$\mathbf{z}' = [.50, -1.40, -.20, -.70, 1.40]$$

yields the following scores on factors 1 and 2:

Weighted least squares (9-50):[6]

$$\hat{\mathbf{f}} = (\hat{\mathbf{L}}_z^{*\prime} \hat{\mathbf{\Psi}}_z^{-1} \hat{\mathbf{L}}_z^{*})^{-1} \hat{\mathbf{L}}_z^{*\prime} \hat{\mathbf{\Psi}}_z^{-1} \mathbf{z} = \begin{bmatrix} -.61 \\ -.61 \end{bmatrix}$$

Regression (9-58):

$$\hat{\mathbf{f}} = \hat{\mathbf{L}}_z^{*\prime} \mathbf{R}^{-1} \mathbf{z} = \begin{bmatrix} .331 & .526 & .221 & -.137 & .011 \\ -.040 & -.063 & -.026 & 1.023 & -.001 \end{bmatrix} \begin{bmatrix} .50 \\ -1.40 \\ -.20 \\ -.70 \\ 1.40 \end{bmatrix} = \begin{bmatrix} -.50 \\ -.64 \end{bmatrix}$$

In this case, the two methods produce very similar results. All of the regression factor scores, obtained using (9-58), are plotted in Figure 9.4. ■

Comment. Factor scores with a rather pleasing intuitive property can be constructed very simply. Group the variables with high (say, greater than .40 in absolute value) loadings on a factor. The scores for factor 1 are then formed by summing the (standardized) observed values of the variables in the group, combined according to the sign of the loadings. The factor scores for factor 2 are the sums of the standardized observations corresponding to variables with high loadings

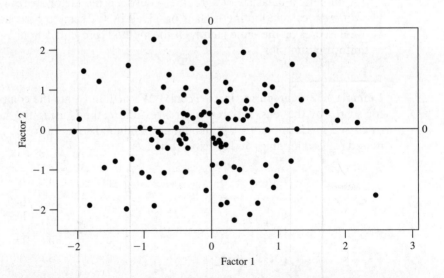

Figure 9.4 Factor scores using (9-58) for factors 1 and 2 of the stock-price data (maximum likelihood estimates of the factor loadings).

---

[6] In order to calculate the weighted least squares factor scores, .00 in the fourth diagonal position of $\hat{\mathbf{\Psi}}_z$ was set to .01 so that this matrix could be inverted.

The sample correlation matrix

$$\mathbf{R} = \begin{bmatrix} 1.000 & .505 & .569 & .602 & .621 & .603 \\ .505 & 1.000 & .422 & .467 & .482 & .450 \\ .569 & .422 & 1.000 & .926 & .877 & .878 \\ .602 & .467 & .926 & 1.000 & .874 & .894 \\ .621 & .482 & .877 & .874 & 1.000 & .937 \\ .603 & .450 & .878 & .894 & .937 & 1.000 \end{bmatrix}$$

was factor analyzed by the principal component and maximum likelihood methods for an $m = 3$ factor model. The results are given in Table 9.10.[7]

Table 9.10 Factor Analysis of Chicken-Bone Data

Principal Component

Variable	Estimated factor loadings			Rotated estimated loadings			$\widetilde{\psi}_i$
	F_1	F_2	F_3	F_1^*	F_2^*	F_3^*	
1. Skull length	.741	.350	.573	.355	.244	.902	.00
2. Skull breadth	.604	.720	−.340	.235	.949	.211	.00
3. Femur length	.929	−.233	−.075	.921	.164	.218	.08
4. Tibia length	.943	−.175	−.067	.904	.212	.252	.08
5. Humerus length	.948	−.143	−.045	.888	.228	.283	.08
6. Ulna length	.945	−.189	−.047	.908	.192	.264	.07
Cumulative proportion of total (standardized) sample variance explained	.743	.873	.950	.576	.763	.950	

Maximum Likelihood

Variable	Estimated factor loadings			Rotated estimated loadings			$\widehat{\psi}$
	F_1	F_2	F_3	F_1^*	F_2^*	F_3^*	
1. Skull length	.602	.214	.286	.467	.506	.128	.51
2. Skull breadth	.467	.177	.652	.211	.792	.050	.33
3. Femur length	.926	.145	−.057	.890	.289	.084	.12
4. Tibia length	1.000	.000	−.000	.936	.345	−.073	.00
5. Humerus length	.874	.463	−.012	.831	.362	.396	.02
6. Ulna length	.894	.336	−.039	.857	.325	.272	.09
Cumulative proportion of total (standardized) sample variance explained	.667	.738	.823	.559	.779	.823	

[7] Notice the estimated specific variance of .00 for tibia length in the maximum likelihood solution. This suggests that maximizing the likelihood function may produce a Heywood case. Readers attempting to replicate our results should try the Hey(wood) option if SAS or similar software is used.

After rotation, the two methods of solution appear to give somewhat different results. Focusing our attention on the principal component method and the cumulative proportion of the total sample variance explained, we see that a three-factor solution appears to be warranted. The third factor explains a "significant" amount of additional sample variation. The first factor appears to be a *body-size* factor dominated by wing and leg dimensions. The second and third factors, collectively, represent skull dimensions and might be given the same names as the variables, *skull breadth* and *skull length*, respectively.

The rotated maximum likelihood factor loadings are consistent with those generated by the principal component method for the first factor, but not for factors 2 and 3. For the maximum likelihood method, the second factor appears to represent head size. The meaning of the third factor is unclear, and it is probably not needed.

Further support for retaining three or fewer factors is provided by the residual matrix obtained from the maximum likelihood estimates:

$$\mathbf{R} - \hat{\mathbf{L}}_z\hat{\mathbf{L}}_z' - \hat{\mathbf{\Psi}}_z = \begin{bmatrix} .000 & & & & & \\ -.000 & .000 & & & & \\ -.003 & .001 & .000 & & & \\ .000 & .000 & .000 & .000 & & \\ -.001 & .000 & .000 & .000 & .000 & \\ .004 & -.001 & -.001 & .000 & -.000 & .000 \end{bmatrix}$$

All of the entries in this matrix are very small. We shall pursue the $m = 3$ factor model in this example. An $m = 2$ factor model is considered in Exercise 9.10.

Factor scores for factors 1 and 2 produced from (9-58) with the rotated maximum likelihood estimates are plotted in Figure 9.5. Plots of this kind allow us to identify observations that, for one reason or another, are not consistent with the remaining observations. Potential outliers are circled in the figure.

It is also of interest to plot pairs of factor scores obtained using the principal component and maximum likelihood estimates of factor loadings. For the chicken-bone data, plots of pairs of factor scores are given in Figure 9.6 on pages 524–526. If the loadings on a particular factor agree, the pairs of scores should cluster tightly about the 45° line through the origin. Sets of loadings that do not agree will produce factor scores that deviate from this pattern. If the latter occurs, it is usually associated with the last factors and may suggest that the number of factors is too large. That is, the last factors are not meaningful. This seems to be the case with the third factor in the chicken-bone data, as indicated by Plot (c) in Figure 9.6.

Plots of pairs of factor scores using estimated loadings from two solution methods are also good tools for detecting outliers. If the sets of loadings for a factor tend to agree, outliers will appear as points in the neighborhood of the 45° line, but far from the origin and the cluster of the remaining points. It is clear from Plot (b) in Figure 9.6 that one of the 276 observations is not consistent with the others. It has an unusually large F_2-score. When this point, $[39.1, 39.3, 75.7, 115, 73.4, 69.1]$, was removed and the analysis repeated, the loadings were not altered appreciably.

When the data set is large, it should be divided into two (roughly) equal sets, and a factor analysis should be performed on each half. The results of these analyses can be compared with each other and with the analysis for the full data set to

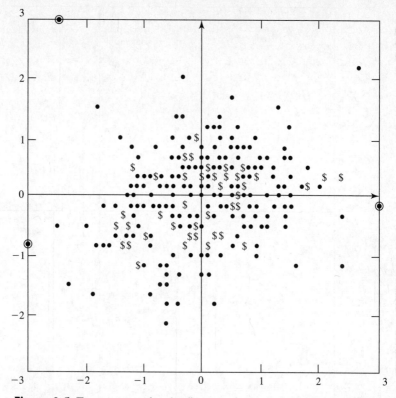

Figure 9.5 Factor scores for the first two factors of chicken-bone data.

test the stability of the solution. If the results are consistent with one another, confidence in the solution is increased.

The chicken-bone data were divided into two sets of $n_1 = 137$ and $n_2 = 139$ observations, respectively. The resulting sample correlation matrices were

$$\mathbf{R}_1 = \begin{bmatrix} 1.000 & & & & & \\ .696 & 1.000 & & & & \\ .588 & .540 & 1.000 & & & \\ .639 & .575 & .901 & 1.000 & & \\ .694 & .606 & .844 & .835 & 1.000 & \\ .660 & .584 & .866 & .863 & .931 & 1.000 \end{bmatrix}$$

and

$$\mathbf{R}_2 = \begin{bmatrix} 1.000 & & & & & \\ .366 & 1.000 & & & & \\ .572 & .352 & 1.000 & & & \\ .587 & .406 & .950 & 1.000 & & \\ .587 & .420 & .909 & .911 & 1.000 & \\ .598 & .386 & .894 & .927 & .940 & 1.000 \end{bmatrix}$$

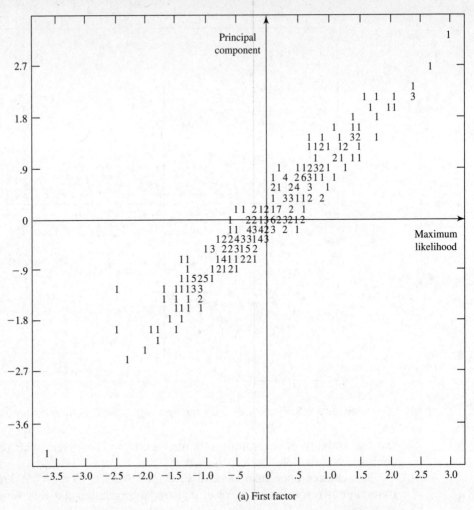

Figure 9.6 Pairs of factor scores for the chicken-bone data. (Loadings are estimated by principal component and maximum likelihood methods.)

The rotated estimated loadings, specific variances, and proportion of the total (standardized) sample variance explained for a principal component solution of an $m = 3$ factor model are given in Table 9.11 on page 525.

The results for the two halves of the chicken-bone measurements are very similar. Factors F_2^* and F_3^* interchange with respect to their labels, skull length and skull breadth, but they collectively seem to represent *head size*. The first factor, F_1^*, again appears to be a *body-size* factor dominated by leg and wing dimensions. These are the same interpretations we gave to the results from a principal component factor analysis of the entire set of data. The solution is remarkably stable, and we can be fairly confident that the large loadings are "real." As we have pointed out however, three factors are probably too many. A one- or two-factor model is surely sufficient for the chicken-bone data, and you are encouraged to repeat the analyses here with fewer factors and alternative solution methods. (See Exercise 9.10.) ∎

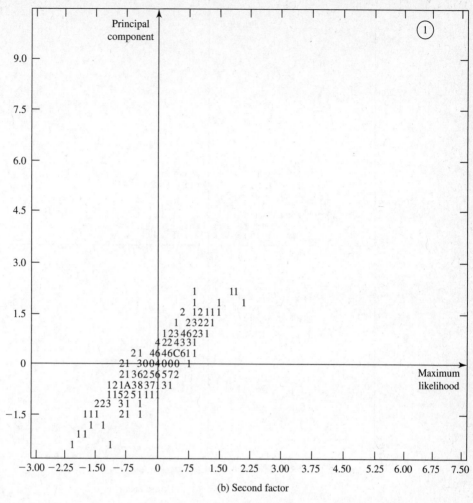

Figure 9.6
(*continued*)

(b) Second factor

Table 9.11

Variable	First set $(n_1 = 137$ observations) Rotated estimated factor loadings				Second set $(n_2 = 139$ observations) Rotated estimated factor loadings			
	F_1^*	F_2^*	F_3^*	$\widetilde{\psi}_i$	F_1^*	F_2^*	F_3^*	$\widetilde{\psi}_i$
1. Skull length	.360	.361	.853	.01	.352	.921	.167	.00
2. Skull breadth	.303	.899	.312	.00	.203	.145	.968	.00
3. Femur length	.914	.238	.175	.08	.930	.239	.130	.06
4. Tibia length	.877	.270	.242	.10	.925	.248	.187	.05
5. Humerus length	.830	.247	.395	.11	.912	.252	.208	.06
6. Ulna length	.871	.231	.332	.08	.914	.272	.168	.06
Cumulative proportion of total (standardized) sample variance explained	.546	.743	.940		.593	.780	.962	

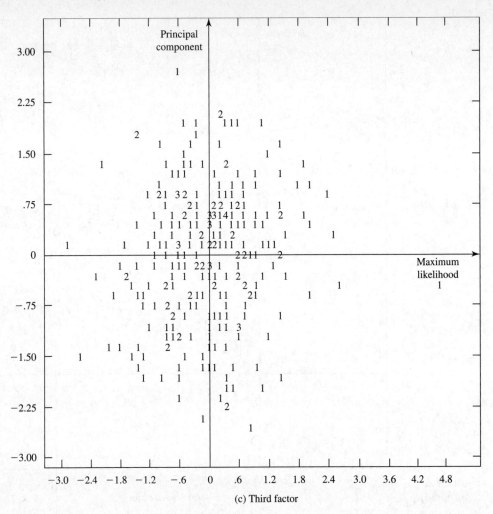

(c) Third factor

Figure 9.6 (*continued*)

Factor analysis has a tremendous intuitive appeal for the behavioral and social sciences. In these areas, it is natural to regard multivariate observations on animal and human processes as manifestations of underlying unobservable "traits." Factor analysis provides a way of explaining the observed variability in behavior in terms of these traits.

Still, when all is said and done, factor analysis remains very subjective. Our examples, in common with most published sources, consist of situations in which the factor analysis model provides reasonable explanations in terms of a few interpretable factors. In practice, the vast majority of attempted factor analyses do not yield such clear-cut results. Unfortunately, the criterion for judging the quality of any factor analysis has not been well quantified. Rather, that quality seems to depend on a

WOW criterion

If, while scrutinizing the factor analysis, the investigator can shout "Wow, I understand these factors," the application is deemed successful.

9A

SOME COMPUTATIONAL DETAILS FOR MAXIMUM LIKELIHOOD ESTIMATION

Although a simple analytical expression cannot be obtained for the maximum likelihood estimators $\hat{\mathbf{L}}$ and $\hat{\boldsymbol{\Psi}}$, they can be shown to satisfy certain equations. Not surprisingly, the conditions are stated in terms of the maximum likelihood estimator $\mathbf{S}_n = (1/n) \sum_{j=1}^{n} (\mathbf{X}_j - \overline{\mathbf{X}})(\mathbf{X}_j - \overline{\mathbf{X}})'$ of an unstructured covariance matrix. Some factor analysts employ the usual sample covariance $\mathbf{S}$, but still use the title *maximum likelihood* to refer to resulting estimates. This modification, referenced in Footnote 4 of this chapter, amounts to employing the likelihood obtained from the Wishart distribution of $\sum_{j=1}^{n} (\mathbf{X}_j - \overline{\mathbf{X}})(\mathbf{X}_j - \overline{\mathbf{X}})'$ and ignoring the minor contribution due to the normal density for $\overline{\mathbf{X}}$. The factor analysis of $\mathbf{R}$ is, of course, unaffected by the choice of $\mathbf{S}_n$ or $\mathbf{S}$, since they both produce the same correlation matrix.

Result 9A.1. Let $\mathbf{x}_1, \mathbf{x}_2, \dots, \mathbf{x}_n$ be a random sample from a normal population. The maximum likelihood estimates $\hat{\mathbf{L}}$ and $\hat{\boldsymbol{\Psi}}$ are obtained by maximizing (9-25) subject to the uniqueness condition in (9-26). They satisfy

$$(\hat{\boldsymbol{\Psi}}^{-1/2}\mathbf{S}_n\hat{\boldsymbol{\Psi}}^{-1/2})(\hat{\boldsymbol{\Psi}}^{-1/2}\hat{\mathbf{L}}) = (\hat{\boldsymbol{\Psi}}^{-1/2}\hat{\mathbf{L}})(\mathbf{I} + \hat{\boldsymbol{\Delta}}) \qquad (9\text{A-}1)$$

so the jth column of $\hat{\boldsymbol{\Psi}}^{-1/2}\hat{\mathbf{L}}$ is the (nonnormalized) eigenvector of $\hat{\boldsymbol{\Psi}}^{-1/2}\mathbf{S}_n\hat{\boldsymbol{\Psi}}^{-1/2}$ corresponding to eigenvalue $1 + \hat{\Delta}_i$. Here

$$\mathbf{S}_n = n^{-1} \sum_{j=1}^{n} (\mathbf{x}_j - \overline{\mathbf{x}})(\mathbf{x}_j - \overline{\mathbf{x}})' = n^{-1}(n-1)\mathbf{S} \quad \text{and} \quad \hat{\Delta}_1 \geq \hat{\Delta}_2 \geq \cdots \geq \hat{\Delta}_m$$

Also, at convergence,

$$\hat{\psi}_i = i\text{th diagonal element of } \mathbf{S}_n - \hat{\mathbf{L}}\hat{\mathbf{L}}' \tag{9A-2}$$

and

$$\text{tr}(\hat{\boldsymbol{\Sigma}}^{-1}\mathbf{S}_n) = p$$

We avoid the details of the proof. However, it is evident that $\hat{\boldsymbol{\mu}} = \bar{\mathbf{x}}$ and a consideration of the log-likelihood leads to the maximization of $-(n/2)[\ln|\boldsymbol{\Sigma}| + \text{tr}(\boldsymbol{\Sigma}^{-1}\mathbf{S}_n)]$ over $\mathbf{L}$ and $\boldsymbol{\Psi}$. Equivalently, since $\mathbf{S}_n$ and p are constant with respect to the maximization, we minimize

$$h(\hat{\boldsymbol{\mu}}, \boldsymbol{\Psi}, \mathbf{L}) = \ln|\boldsymbol{\Sigma}| - \ln|\mathbf{S}_n| + \text{tr}(\boldsymbol{\Sigma}^{-1}\mathbf{S}_n) - p \tag{9A-3}$$

subject to $\mathbf{L}'\boldsymbol{\Psi}^{-1}\mathbf{L} = \boldsymbol{\Delta}$, a diagonal matrix. ∎

Comment. Lawley and Maxwell [10], along with many others who do factor analysis, use the unbiased estimate $\mathbf{S}$ of the covariance matrix instead of the maximum likelihood estimate $\mathbf{S}_n$. Now, $(n-1)\mathbf{S}$ has, for normal data, a Wishart distribution. [See (4-21) and (4-23).] If we ignore the contribution to the likelihood in (9-25) from the second term involving $(\boldsymbol{\mu} - \bar{\mathbf{x}})$, then maximizing the reduced likelihood over $\mathbf{L}$ and $\boldsymbol{\Psi}$ is equivalent to maximizing the Wishart likelihood

$$\text{Likelihood} \propto |\boldsymbol{\Sigma}|^{-(n-1)/2} e^{-[(n-1)/2]\,\text{tr}[\boldsymbol{\Sigma}^{-1}\mathbf{S}]}$$

over $\mathbf{L}$ and $\boldsymbol{\Psi}$. Equivalently, we can minimize

$$\ln|\boldsymbol{\Sigma}| + \text{tr}(\boldsymbol{\Sigma}^{-1}\mathbf{S})$$

or, as in (9A-3),

$$\ln|\boldsymbol{\Sigma}| + \text{tr}(\boldsymbol{\Sigma}^{-1}\mathbf{S}) - \ln|\mathbf{S}| - p$$

Under these conditions, Result (9A-1) holds with $\mathbf{S}$ in place of $\mathbf{S}_n$. Also, for large n, $\mathbf{S}$ and $\mathbf{S}_n$ are almost identical, and the corresponding maximum likelihood estimates, $\hat{\mathbf{L}}$ and $\hat{\boldsymbol{\Psi}}$, would be similar. For testing the factor model [see (9-39)], $|\hat{\mathbf{L}}\hat{\mathbf{L}}' + \hat{\boldsymbol{\Psi}}|$ should be compared with $|\mathbf{S}_n|$ if the actual likelihood of (9-25) is employed, and $|\hat{\mathbf{L}}\hat{\mathbf{L}}' + \hat{\boldsymbol{\Psi}}|$ should be compared with $|\mathbf{S}|$ if the foregoing Wishart likelihood is used to derive $\hat{\mathbf{L}}$ and $\hat{\boldsymbol{\Psi}}$.

Recommended Computational Scheme

For $m > 1$, the condition $\mathbf{L}'\boldsymbol{\Psi}^{-1}\mathbf{L} = \boldsymbol{\Delta}$ effectively imposes $m(m-1)/2$ constraints on the elements of $\mathbf{L}$ and $\boldsymbol{\Psi}$, and the likelihood equations are solved, subject to these contraints, in an iterative fashion. One procedure is the following:

1. Compute initial estimates of the specific variances $\psi_1, \psi_2, \ldots, \psi_p$. Jöreskog [8] suggests setting

$$\hat{\psi}_i = \left(1 - \frac{1}{2} \cdot \frac{m}{p}\right)\left(\frac{1}{s^{ii}}\right) \tag{9A-4}$$

where s^{ii} is the ith diagonal element of $\mathbf{S}^{-1}$.

2. Given $\hat{\boldsymbol{\Psi}}$, compute the first m distinct eigenvalues, $\hat{\lambda}_1 > \hat{\lambda}_2 > \cdots > \hat{\lambda}_m > 1$, and corresponding eigenvectors, $\hat{\mathbf{e}}_1, \hat{\mathbf{e}}_2, \ldots, \hat{\mathbf{e}}_m$, of the "uniqueness-rescaled" covariance matrix

$$\mathbf{S}^* = \hat{\boldsymbol{\Psi}}^{-1/2}\mathbf{S}_n\hat{\boldsymbol{\Psi}}^{-1/2} \qquad (9A\text{-}5)$$

Let $\hat{\mathbf{E}} = [\hat{\mathbf{e}}_1 \mathbin{\vdots} \hat{\mathbf{e}}_2 \mathbin{\vdots} \cdots \mathbin{\vdots} \hat{\mathbf{e}}_m]$ be the $p \times m$ matrix of *normalized* eigenvectors and $\hat{\boldsymbol{\Lambda}} = \mathrm{diag}[\hat{\lambda}_1, \hat{\lambda}_2, \ldots, \hat{\lambda}_m]$ be the $m \times m$ diagonal matrix of eigenvalues. From (9A-1), $\hat{\boldsymbol{\Lambda}} = \mathbf{I} + \hat{\boldsymbol{\Delta}}$ and $\hat{\mathbf{E}} = \hat{\boldsymbol{\Psi}}^{-1/2}\hat{\mathbf{L}}\hat{\boldsymbol{\Delta}}^{-1/2}$. Thus, we obtain the estimates

$$\hat{\mathbf{L}} = \hat{\boldsymbol{\Psi}}^{1/2}\hat{\mathbf{E}}\hat{\boldsymbol{\Delta}}^{1/2} = \hat{\boldsymbol{\Psi}}^{1/2}\hat{\mathbf{E}}(\hat{\boldsymbol{\Lambda}} - \mathbf{I})^{1/2} \qquad (9A\text{-}6)$$

3. Substitute $\hat{\mathbf{L}}$ obtained in (9A-6) into the likelihood function (9A-3), and minimize the result with respect to $\hat{\psi}_1, \hat{\psi}_2, \ldots, \hat{\psi}_p$. A numerical search routine must be used. The values $\hat{\psi}_1, \hat{\psi}_2, \ldots, \hat{\psi}_p$ obtained from this minimization are employed at Step (2) to create a new $\hat{\mathbf{L}}$. Steps (2) and (3) are repeated until convergence—that is, until the differences between successive values of $\hat{\ell}_{ij}$ and $\hat{\psi}_i$ are negligible.

Comment. It often happens that the objective function in (9A-3) has a relative minimum corresponding to *negative* values for some $\hat{\psi}_i$. This solution is clearly inadmissible and is said to be improper, or a *Heywood case*. For most packaged computer programs, negative $\hat{\psi}_i$, if they occur on a particular iteration, are changed to small positive numbers before proceeding with the next step.

Maximum Likelihood Estimators of $\boldsymbol{\rho} = \mathbf{L}_z\mathbf{L}_z' + \boldsymbol{\Psi}_z$

When $\boldsymbol{\Sigma}$ has the factor analysis structure $\boldsymbol{\Sigma} = \mathbf{LL}' + \boldsymbol{\Psi}$, $\boldsymbol{\rho}$ can be factored as $\boldsymbol{\rho} = \mathbf{V}^{-1/2}\boldsymbol{\Sigma}\mathbf{V}^{-1/2} = (\mathbf{V}^{-1/2}\mathbf{L})(\mathbf{V}^{-1/2}\mathbf{L})' + \mathbf{V}^{-1/2}\boldsymbol{\Psi}\mathbf{V}^{-1/2} = \mathbf{L}_z\mathbf{L}_z' + \boldsymbol{\Psi}_z$. The loading matrix for the standardized variables is $\mathbf{L}_z = \mathbf{V}^{-1/2}\mathbf{L}$, and the corresponding specific variance matrix is $\boldsymbol{\Psi}_z = \mathbf{V}^{-1/2}\boldsymbol{\Psi}\mathbf{V}^{-1/2}$, where $\mathbf{V}^{-1/2}$ is the diagonal matrix with ith diagonal element $\sigma_{ii}^{-1/2}$. If $\mathbf{R}$ is substituted for $\mathbf{S}_n$ in the objective function of (9A-3), the investigator minimizes

$$\ln\left(\frac{|\mathbf{L}_z\mathbf{L}_z' + \boldsymbol{\Psi}_z|}{|\mathbf{R}|}\right) + \mathrm{tr}[(\mathbf{L}_z\mathbf{L}_z' + \boldsymbol{\Psi}_z)^{-1}\mathbf{R}] - p \qquad (9A\text{-}7)$$

Introducing the diagonal matrix $\hat{\mathbf{V}}^{1/2}$, whose ith diagonal element is the square root of the ith diagonal element of $\mathbf{S}_n$, we can write the objective function in (9A-7) as

$$\ln\left(\frac{|\hat{\mathbf{V}}^{1/2}||\mathbf{L}_z\mathbf{L}_z' + \boldsymbol{\Psi}_z||\hat{\mathbf{V}}^{1/2}|}{|\hat{\mathbf{V}}^{1/2}||\mathbf{R}||\hat{\mathbf{V}}^{1/2}|}\right) + \mathrm{tr}[(\mathbf{L}_z\mathbf{L}_z' + \boldsymbol{\Psi}_z)^{-1}\hat{\mathbf{V}}^{-1/2}\hat{\mathbf{V}}^{1/2}\mathbf{R}\hat{\mathbf{V}}^{1/2}\hat{\mathbf{V}}^{-1/2}] - p$$

$$= \ln\left(\frac{|(\hat{\mathbf{V}}^{1/2}\mathbf{L}_z)(\hat{\mathbf{V}}^{1/2}\mathbf{L}_z)' + \hat{\mathbf{V}}^{1/2}\boldsymbol{\Psi}_z\hat{\mathbf{V}}^{1/2}|}{|\mathbf{S}_n|}\right)$$

$$\quad + \mathrm{tr}[((\hat{\mathbf{V}}^{1/2}\mathbf{L}_z)(\hat{\mathbf{V}}^{1/2}\mathbf{L}_z)' + \hat{\mathbf{V}}^{1/2}\boldsymbol{\Psi}_z\hat{\mathbf{V}}^{1/2})^{-1}\mathbf{S}_n] - p$$

$$\geq \ln\left(\frac{|\hat{\mathbf{L}}\hat{\mathbf{L}}' + \hat{\boldsymbol{\Psi}}|}{|\mathbf{S}_n|}\right) + \mathrm{tr}[(\hat{\mathbf{L}}\hat{\mathbf{L}}' + \hat{\boldsymbol{\Psi}})^{-1}\mathbf{S}_n] - p \qquad (9A\text{-}8)$$

The last inequality follows because the maximum likelihood estimates $\hat{\mathbf{L}}$ and $\hat{\boldsymbol{\Psi}}$ minimize the objective function (9A-3). [Equality holds in (9A-8) for $\hat{\mathbf{L}}_z = \hat{\mathbf{V}}^{-1/2}\hat{\mathbf{L}}$ and $\hat{\boldsymbol{\Psi}}_z = \hat{\mathbf{V}}^{-1/2}\hat{\boldsymbol{\Psi}}\hat{\mathbf{V}}^{-1/2}$.] Therefore, minimizing (9A-7) over $\mathbf{L}_z$ and $\boldsymbol{\Psi}_z$ is equivalent to obtaining $\hat{\mathbf{L}}$ and $\hat{\boldsymbol{\Psi}}$ from $\mathbf{S}_n$ and estimating $\mathbf{L}_z = \mathbf{V}^{-1/2}\mathbf{L}$ by $\hat{\mathbf{L}}_z = \hat{\mathbf{V}}^{-1/2}\hat{\mathbf{L}}$ and $\boldsymbol{\Psi}_z = \mathbf{V}^{-1/2}\boldsymbol{\Psi}\mathbf{V}^{-1/2}$ by $\hat{\boldsymbol{\Psi}}_z = \hat{\mathbf{V}}^{-1/2}\hat{\boldsymbol{\Psi}}\hat{\mathbf{V}}^{-1/2}$. The rationale for the latter procedure comes from the invariance property of maximum likelihood estimators. [See (4-20).]

Exercises

9.1. Show that the covariance matrix

$$\boldsymbol{\rho} = \begin{bmatrix} 1.0 & .63 & .45 \\ .63 & 1.0 & .35 \\ .45 & .35 & 1.0 \end{bmatrix}$$

for the $p = 3$ standardized random variables Z_1, Z_2, and Z_3 can be generated by the $m = 1$ factor model

$$Z_1 = .9F_1 + \varepsilon_1$$
$$Z_2 = .7F_1 + \varepsilon_2$$
$$Z_3 = .5F_1 + \varepsilon_3$$

where $\mathrm{Var}(F_1) = 1$, $\mathrm{Cov}(\boldsymbol{\varepsilon}, F_1) = \mathbf{0}$, and

$$\boldsymbol{\Psi} = \mathrm{Cov}(\boldsymbol{\varepsilon}) = \begin{bmatrix} .19 & 0 & 0 \\ 0 & .51 & 0 \\ 0 & 0 & .75 \end{bmatrix}$$

That is, write $\boldsymbol{\rho}$ in the form $\boldsymbol{\rho} = \mathbf{LL}' + \boldsymbol{\Psi}$.

9.2. Use the information in Exercise 9.1.

(a) Calculate communalities h_i^2, $i = 1, 2, 3$, and interpret these quantities.

(b) Calculate $\mathrm{Corr}(Z_i, F_1)$ for $i = 1, 2, 3$. Which variable might carry the greatest weight in "naming" the common factor? Why?

9.3. The eigenvalues and eigenvectors of the correlation matrix $\boldsymbol{\rho}$ in Exercise 9.1 are

$$\lambda_1 = 1.96, \qquad \mathbf{e}_1' = [.625, .593, .507]$$
$$\lambda_2 = .68, \qquad \mathbf{e}_2' = [-.219, -.491, .843]$$
$$\lambda_3 = .36, \qquad \mathbf{e}_3' = [.749, -.638, -.177]$$

(a) Assuming an $m = 1$ factor model, calculate the loading matrix $\mathbf{L}$ and matrix of specific variances $\boldsymbol{\Psi}$ using the principal component solution method. Compare the results with those in Exercise 9.1.

(b) What proportion of the total population variance is explained by the first common factor?

9.4. Given $\boldsymbol{\rho}$ and $\boldsymbol{\Psi}$ in Exercise 9.1 and an $m = 1$ factor model, calculate the reduced correlation matrix $\widetilde{\boldsymbol{\rho}} = \boldsymbol{\rho} - \boldsymbol{\Psi}$ and the principal factor solution for the loading matrix $\mathbf{L}$. Is the result consistent with the information in Exercise 9.1? Should it be?

9.5. Establish the inequality (9-19).

Hint: Since $\mathbf{S} - \widetilde{\mathbf{L}}\widetilde{\mathbf{L}}' - \widetilde{\boldsymbol{\Psi}}$ has zeros on the diagonal,

(sum of squared entries of $\mathbf{S} - \widetilde{\mathbf{L}}\widetilde{\mathbf{L}}' - \widetilde{\boldsymbol{\Psi}}$) $\leq$ (sum of squared entries of $\mathbf{S} - \widetilde{\mathbf{L}}\widetilde{\mathbf{L}}'$)

Now, $\mathbf{S} - \widetilde{\mathbf{L}}\widetilde{\mathbf{L}}' = \hat{\lambda}_{m+1}\hat{\mathbf{e}}_{m+1}\hat{\mathbf{e}}'_{m+1} + \cdots + \hat{\lambda}_p\hat{\mathbf{e}}_p\hat{\mathbf{e}}'_p = \hat{\mathbf{P}}_{(2)}\hat{\boldsymbol{\Lambda}}_{(2)}\hat{\mathbf{P}}'_{(2)}$, where $\hat{\mathbf{P}}_{(2)} = [\hat{\mathbf{e}}_{m+1} \vdots \cdots \vdots \hat{\mathbf{e}}_p]$ and $\hat{\boldsymbol{\Lambda}}_{(2)}$ is the diagonal matrix with elements $\hat{\lambda}_{m+1}, \ldots, \hat{\lambda}_p$.

Use (sum of squared entries of $\mathbf{A}$) = tr $\mathbf{A}\mathbf{A}'$ and tr $[\hat{\mathbf{P}}_{(2)}\hat{\boldsymbol{\Lambda}}_{(2)}\hat{\boldsymbol{\Lambda}}_{(2)}\hat{\mathbf{P}}'_{(2)}] = \text{tr}\,[\hat{\boldsymbol{\Lambda}}_{(2)}\hat{\boldsymbol{\Lambda}}_{(2)}]$.

9.6. Verify the following matrix identities.

(a) $(\mathbf{I} + \mathbf{L}'\boldsymbol{\Psi}^{-1}\mathbf{L})^{-1}\mathbf{L}'\boldsymbol{\Psi}^{-1}\mathbf{L} = \mathbf{I} - (\mathbf{I} + \mathbf{L}'\boldsymbol{\Psi}^{-1}\mathbf{L})^{-1}$

Hint: Premultiply both sides by $(\mathbf{I} + \mathbf{L}'\boldsymbol{\Psi}^{-1}\mathbf{L})$.

(b) $(\mathbf{L}\mathbf{L}' + \boldsymbol{\Psi})^{-1} = \boldsymbol{\Psi}^{-1} - \boldsymbol{\Psi}^{-1}\mathbf{L}(\mathbf{I} + \mathbf{L}'\boldsymbol{\Psi}^{-1}\mathbf{L})^{-1}\mathbf{L}'\boldsymbol{\Psi}^{-1}$

Hint: Postmultiply both sides by $(\mathbf{L}\mathbf{L}' + \boldsymbol{\Psi})$ and use (a).

(c) $\mathbf{L}'(\mathbf{L}\mathbf{L}' + \boldsymbol{\Psi})^{-1} = (\mathbf{I} + \mathbf{L}'\boldsymbol{\Psi}^{-1}\mathbf{L})^{-1}\mathbf{L}'\boldsymbol{\Psi}^{-1}$

Hint: Postmultiply the result in (b) by $\mathbf{L}$, use (a), and take the transpose, noting that $(\mathbf{L}\mathbf{L}' + \boldsymbol{\Psi})^{-1}$, $\boldsymbol{\Psi}^{-1}$, and $(\mathbf{I} + \mathbf{L}'\boldsymbol{\Psi}^{-1}\mathbf{L})^{-1}$ are symmetric matrices.

9.7. *(The factor model parameterization need not be unique.)* Let the factor model with $p = 2$ and $m = 1$ prevail. Show that

$$\sigma_{11} = \ell_{11}^2 + \psi_1, \qquad \sigma_{12} = \sigma_{21} = \ell_{11}\ell_{21}$$
$$\sigma_{22} = \ell_{21}^2 + \psi_2$$

and, for given σ_{11}, σ_{22}, and σ_{12}, there is an infinity of choices for $\mathbf{L}$ and $\boldsymbol{\Psi}$.

9.8. *(Unique but improper solution: Heywood case.)*
Consider an $m = 1$ factor model for the population with covariance matrix

$$\boldsymbol{\Sigma} = \begin{bmatrix} 1 & .4 & .9 \\ .4 & 1 & .7 \\ .9 & .7 & 1 \end{bmatrix}$$

Show that there is a unique choice of $\mathbf{L}$ and $\boldsymbol{\Psi}$ with $\boldsymbol{\Sigma} = \mathbf{L}\mathbf{L}' + \boldsymbol{\Psi}$, but that $\psi_3 < 0$, so the choice is not admissible.

9.9. In a study of liquor preference in France, Stoetzel [14] collected preference rankings of $p = 9$ liquor types from $n = 1442$ individuals. A factor analysis of the 9×9 sample correlation matrix of rank orderings gave the following estimated loadings:

	Estimated factor loadings		
Variable (X_1)	F_1	F_2	F_3
Liquors	.64	.02	.16
Kirsch	.50	−.06	−.10
Mirabelle	.46	−.24	−.19
Rum	.17	.74	.97 *
Marc	−.29	.66	−.39
Whiskey	−.29	−.08	.09
Calvados	−.49	.20	−.04
Cognac	−.52	−.03	.42
Armagnac	−.60	−.17	.14

*This figure is too high. It exceeds the maximum value of .64, as a result of an approximation method for obtaining the estimated factor loadings used by Stoetzel.

Given these results, Stoetzel concluded the following: The major principle of liquor preference in France is the distinction between sweet and strong liquors. The second motivating element is price, which can be understood by remembering that liquor is both an expensive commodity and an item of conspicuous consumption. Except in the case of the two most popular and least expensive items (rum and marc), this second factor plays a much smaller role in producing preference judgments. The third factor concerns the sociological and primarily the regional, variability of the judgments. (See [14], p. 11.)

(a) Given what you know about the various liquors involved, does Stoetzel's interpretation seem reasonable?

(b) Plot the loading pairs for the first two factors. Conduct a graphical orthogonal rotation of the factor axes. Generate approximate rotated loadings. Interpret the rotated loadings for the first two factors. Does your interpretation agree with Stoetzel's interpretation of these factors from the unrotated loadings? Explain.

9.10. The correlation matrix for chicken-bone measurements (see Example 9.14) is

$$
\begin{bmatrix}
1.000 & & & & & \\
.505 & 1.000 & & & & \\
.569 & .422 & 1.000 & & & \\
.602 & .467 & .926 & 1.000 & & \\
.621 & .482 & .877 & .874 & 1.000 & \\
.603 & .450 & .878 & .894 & .937 & 1.000
\end{bmatrix}
$$

The following estimated factor loadings were extracted by the maximum likelihood procedure:

	Estimated factor loadings		Varimax rotated estimated factor loadings	
Variable	F_1	F_2	F_1^*	F_2^*
1. Skull length	.602	.200	.484	.411
2. Skull breadth	.467	.154	.375	.319
3. Femur length	.926	.143	.603	.717
4. Tibia length	1.000	.000	.519	.855
5. Humerus length	.874	.476	.861	.499
6. Ulna length	.894	.327	.744	.594

Using the *unrotated* estimated factor loadings, obtain the maximum likelihood estimates of the following.

(a) The specific variances.

(b) The communalities.

(c) The proportion of variance explained by each factor.

(d) The residual matrix $\mathbf{R} - \hat{\mathbf{L}}_z\hat{\mathbf{L}}_z' - \hat{\mathbf{\Psi}}_z$.

9.11. Refer to Exercise 9.10. Compute the value of the varimax criterion using both unrotated and rotated estimated factor loadings. Comment on the results.

9.12. The *covariance* matrix for the logarithms of turtle measurements (see Example 8.4) is

$$
\mathbf{S} = 10^{-3}
\begin{bmatrix}
11.072 & & \\
8.019 & 6.417 & \\
8.160 & 6.005 & 6.773
\end{bmatrix}
$$

The following maximum likelihood estimates of the factor loadings for an $m = 1$ model were obtained:

Variable	Estimated factor loadings F_1
1. ln(length)	.1022
2. ln(width)	.0752
3. ln(height)	.0765

Using the estimated factor loadings, obtain the maximum likelihood estimates of each of the following.

(a) Specific variances.

(b) Communalities.

(c) Proportion of variance explained by the factor.

(d) The residual matrix $\mathbf{S}_n - \hat{\mathbf{L}}\hat{\mathbf{L}}' - \hat{\mathbf{\Psi}}$.

   Hint: Convert $\mathbf{S}$ to $\mathbf{S}_n$.

9.13. Refer to Exercise 9.12. Compute the test statistic in (9-39). Indicate why a test of $H_0: \mathbf{\Sigma} = \mathbf{LL}' + \mathbf{\Psi}$ (with $m = 1$) versus $H_1: \mathbf{\Sigma}$ unrestricted cannot be carried out for this example. [See (9-40).]

9.14. The maximum likelihood factor loading estimates are given in (9A-6) by

$$\hat{\mathbf{L}} = \hat{\mathbf{\Psi}}^{1/2}\hat{\mathbf{E}}\hat{\mathbf{\Delta}}^{1/2}$$

Verify, for this choice, that

$$\hat{\mathbf{L}}'\hat{\mathbf{\Psi}}^{-1}\hat{\mathbf{L}} = \hat{\mathbf{\Delta}}$$

where $\hat{\mathbf{\Delta}} = \hat{\mathbf{\Lambda}} - \mathbf{I}$ is a diagonal matrix.

9.15. Hirschey and Wichern [7] investigate the consistency, determinants, and uses of accounting and market-value measures of profitability. As part of their study, a factor analysis of accounting profit measures and market estimates of economic profits was conducted. The correlation matrix of accounting historical, accounting replacement, and market-value measures of profitability for a sample of firms operating in 1977 is as follows:

Variable	HRA	HRE	HRS	RRA	RRE	RRS	Q	REV
Historical return on assets, HRA	1.000							
Historical return on equity, HRE	.738	1.000						
Historical return on sales, HRS	.731	.520	1.000					
Replacement return on assets, RRA	.828	.688	.652	1.000				
Replacement return on equity, RRE	.681	.831	.513	.887	1.000			
Replacement return on sales, RRS	.712	.543	.826	.867	.692	1.000		
Market Q ratio, Q	.625	.322	.579	.639	.419	.608	1.000	
Market relative excess value, REV	.604	.303	.617	.563	.352	.610	.937	1.000

The following rotated principal component estimates of factor loadings for an $m = 3$ factor model were obtained:

	Estimated factor loadings		
Variable	F_1	F_2	F_3
Historical return on assets	.433	.612	.499
Historical return on equity	.125	.892	.234
Historical return on sales	.296	.238	.887
Replacement return on assets	.406	.708	.483
Replacement return on equity	.198	.895	.283
Replacement return on sales	.331	.414	.789
Market Q ratio	.928	.160	.294
Market relative excess value	.910	.079	.355
Cumulative proportion of total variance explained	.287	.628	.908

(a) Using the estimated factor loadings, determine the specific variances and communalities.

(b) Determine the residual matrix, $\mathbf{R} - \hat{\mathbf{L}}_z\hat{\mathbf{L}}_z' - \hat{\mathbf{\Psi}}_z$. Given this information and the cumulative proportion of total variance explained in the preceding table, does an $m = 3$ factor model appear appropriate for these data?

(c) Assuming that estimated loadings less than .4 are small, interpret the three factors. Does it appear, for example, that market-value measures provide evidence of profitability distinct from that provided by accounting measures? Can you separate accounting historical measures of profitability from accounting replacement measures?

9.16. Verify that factor scores constructed according to (9-50) have sample mean vector $\mathbf{0}$ and zero sample covariances.

9.17. Refer to Example 9.12. Using the information in this example, evaluate $(\hat{\mathbf{L}}_z'\hat{\mathbf{\Psi}}_z^{-1}\hat{\mathbf{L}}_z)^{-1}$.

Note: Set the fourth diagonal element of $\hat{\mathbf{\Psi}}_z$ to .01 so that $\hat{\mathbf{\Psi}}_z^{-1}$ can be determined. Will the regression and generalized least squares methods for constructing factors scores for standardized stock price observations give nearly the same results? *Hint:* See equation (9-57) and the discussion following it.

The following exercises require the use of a computer.

9.18. Refer to Exercise 8.16 concerning the numbers of fish caught.

(a) Using only the measurements $x_1 - x_4$, obtain the principal component solution for factor models with $m = 1$ and $m = 2$.

(b) Using only the measurements $x_1 - x_4$, obtain the maximum likelihood solution for factor models with $m = 1$ and $m = 2$.

(c) Rotate your solutions in Parts (a) and (b). Compare the solutions and comment on them. Interpret each factor.

(d) Perform a factor analysis using the measurements $x_1 - x_6$. Determine a reasonable number of factors m, and compare the principal component and maximum likelihood solutions after rotation. Interpret the factors.

HW2 **9.19.** A firm is attempting to evaluate the quality of its sales staff and is trying to find an examination or series of tests that may reveal the potential for good performance in sales.

The firm has selected a random sample of 50 sales people and has evaluated each on 3 measures of performance: growth of sales, profitability of sales, and new-account sales. These measures have been converted to a scale, on which 100 indicates "average" performance. Each of the 50 individuals took each of 4 tests, which purported to measure creativity, mechanical reasoning, abstract reasoning, and mathematical ability, respectively. The $n = 50$ observations on $p = 7$ variables are listed in Table 9.12 on page 536.

(a) Assume an orthogonal factor model for the *standardized variables* $Z_i = (X_i - \mu_i)/\sqrt{\sigma_{ii}}$, $i = 1, 2, \ldots, 7$. Obtain either the principal component solution or the maximum likelihood solution for $m = 2$ and $m = 3$ common factors.

(b) Given your solution in (a), obtain the rotated loadings for $m = 2$ and $m = 3$. Compare the two sets of rotated loadings. Interpret the $m = 2$ and $m = 3$ factor solutions.

(c) List the estimated communalities, specific variances, and $\hat{\mathbf{L}}\hat{\mathbf{L}}' + \hat{\mathbf{\Psi}}$ for the $m = 2$ and $m = 3$ solutions. Compare the results. Which choice of m do you prefer at this point? Why?

(d) Conduct a test of $H_0: \mathbf{\Sigma} = \mathbf{LL}' + \mathbf{\Psi}$ versus $H_1: \mathbf{\Sigma} \neq \mathbf{LL}' + \mathbf{\Psi}$ for both $m = 2$ and $m = 3$ at the $\alpha = .01$ level. With these results and those in Parts b and c, which choice of m appears to be the best?

(e) Suppose a new salesperson, selected at random, obtains the test scores $\mathbf{x}' = [x_1, x_2, \ldots, x_7] = [110, 98, 105, 15, 18, 12, 35]$. Calculate the salesperson's factor score using the weighted least squares method and the regression method.

✳ *Note:* The components of $\mathbf{x}$ must be standardized using the sample means and variances calculated from the original data.

9.20. Using the air-pollution variables X_1, X_2, X_5, and X_6 given in Table 1.5, generate the sample *covariance* matrix.

(a) Obtain the principal component solution to a factor model with $m = 1$ and $m = 2$.

(b) Find the maximum likelihood estimates of $\mathbf{L}$ and $\mathbf{\Psi}$ for $m = 1$ and $m = 2$.

(c) Compare the factorization obtained by the principal component and maximum likelihood methods.

9.21. Perform a varimax rotation of both $m = 2$ solutions in Exercise 9.20. Interpret the results. Are the principal component and maximum likelihood solutions consistent with each other?

9.22. Refer to Exercise 9.20.

(a) Calculate the factor scores from the $m = 2$ maximum likelihood estimates by (i) weighted least squares in (9-50) and (ii) the regression approach of (9-58).

(b) Find the factor scores from the principal component solution, using (9-51).

(c) Compare the three sets of factor scores.

9.23. Repeat Exercise 9.20, starting from the sample *correlation* matrix. Interpret the factors for the $m = 1$ and $m = 2$ solutions. Does it make a difference if $\mathbf{R}$, rather than $\mathbf{S}$, is factored? Explain.

9.24. Perform a factor analysis of the census-tract data in Table 8.5. Start with $\mathbf{R}$ and obtain both the maximum likelihood and principal component solutions. Comment on your choice of m. Your analysis should include factor rotation and the computation of factor scores.

9.25. Perform a factor analysis of the "stiffness" measurements given in Table 4.3 and discussed in Example 4.14. Compute factor scores, and check for outliers in the data. Use the sample covariance matrix $\mathbf{S}$.

HW2

Table 9.12 Salespeople Data

	Index of:			Score on:			
Salesperson	Sales growth (x_1)	Sales profit-ability (x_2)	New-account sales (x_3)	Creativity test (x_4)	Mechanical reasoning test (x_5)	Abstract reasoning test (x_6)	Mathe-matics test (x_7)
1	93.0	96.0	97.8	09	12	09	20
2	88.8	91.8	96.8	07	10	10	15
3	95.0	100.3	99.0	08	12	09	26
4	101.3	103.8	106.8	13	14	12	29
5	102.0	107.8	103.0	10	15	12	32
6	95.8	97.5	99.3	10	14	11	21
7	95.5	99.5	99.0	09	12	09	25
8	110.8	122.0	115.3	18	20	15	51
9	102.8	108.3	103.8	10	17	13	31
10	106.8	120.5	102.0	14	18	11	39
11	103.3	109.8	104.0	12	17	12	32
12	99.5	111.8	100.3	10	18	08	31
13	103.5	112.5	107.0	16	17	11	34
14	99.5	105.5	102.3	08	10	11	34
15	100.0	107.0	102.8	13	10	08	34
16	81.5	93.5	95.0	07	09	05	16
17	101.3	105.3	102.8	11	12	11	32
18	103.3	110.8	103.5	11	14	11	35
19	95.3	104.3	103.0	05	14	13	30
20	99.5	105.3	106.3	17	17	11	27
21	88.5	95.3	95.8	10	12	07	15
22	99.3	115.0	104.3	05	11	11	42
23	87.5	92.5	95.8	09	09	07	16
24	105.3	114.0	105.3	12	15	12	37
25	107.0	121.0	109.0	16	19	12	39
26	93.3	102.0	97.8	10	15	07	23
27	106.8	118.0	107.3	14	16	12	39
28	106.8	120.0	104.8	10	16	11	49
29	92.3	90.8	99.8	08	10	13	17
30	106.3	121.0	104.5	09	17	11	44
31	106.0	119.5	110.5	18	15	10	43
32	88.3	92.8	96.8	13	11	08	10
33	96.0	103.3	100.5	07	15	11	27
34	94.3	94.5	99.0	10	12	11	19
35	106.5	121.5	110.5	18	17	10	42
36	106.5	115.5	107.0	08	13	14	47
37	92.0	99.5	103.5	18	16	08	18
38	102.0	99.8	103.3	13	12	14	28
39	108.3	122.3	108.5	15	19	12	41
40	106.8	119.0	106.8	14	20	12	37
41	102.5	109.3	103.8	09	17	13	32
42	92.5	102.5	99.3	13	15	06	23
43	102.8	113.8	106.8	17	20	10	32
44	83.3	87.3	96.3	01	05	09	15
45	94.8	101.8	99.8	07	16	11	24
46	103.5	112.0	110.8	18	13	12	37
47	89.5	96.0	97.3	07	15	11	14
48	84.3	89.8	94.3	08	08	08	09
49	104.3	109.5	106.5	14	12	12	36
50	106.0	118.5	105.0	12	16	11	39

Exercises 537

9.26. Consider the mice-weight data in Example 8.6. Start with the sample *covariance* matrix. (See Exercise 8.15 for $\sqrt{s_{ii}}$.)

(a) Obtain the principal component solution to the factor model with $m = 1$ and $m = 2$.

(b) Find the maximum likelihood estimates of the loadings and specific variances for $m = 1$ and $m = 2$.

(c) Perform a varimax rotation of the solutions in Parts a and b.

9.27. Repeat Exercise 9.26 by factoring $\mathbf{R}$ instead of the sample covariance matrix $\mathbf{S}$. Also, for the mouse with standardized weights $[.8, -.2, -.6, 1.5]$, obtain the factor scores using the maximum likelihood estimates of the loadings and Equation (9-58).

9.28. Perform a factor analysis of the national track records for women given in Table 1.9. Use the sample covariance matrix $\mathbf{S}$ and interpret the factors. Compute factor scores, and check for outliers in the data. Repeat the analysis with the sample correlation matrix $\mathbf{R}$. Does it make a difference if $\mathbf{R}$, rather than $\mathbf{S}$, is factored? Explain.

9.29. Refer to Exercise 9.28. Convert the national track records for women to speeds measured in meters per second. (See Exercise 8.19.) Perform a factor analysis of the speed data. Use the sample covariance matrix $\mathbf{S}$ and interpret the factors. Compute factor scores, and check for outliers in the data. Repeat the analysis with the sample correlation matrix $\mathbf{R}$. Does it make a difference if $\mathbf{R}$, rather than $\mathbf{S}$, is factored? Explain. Compare your results with the results in Exercise 9.28. Which analysis do you prefer? Why?

9.30. Perform a factor analysis of the national track records for men given in Table 8.6. Repeat the steps given in Exercise 9.28. Is the appropriate factor model for the men's data different from the one for the women's data? If not, are the interpretations of the factors roughly the same? If the models are different, explain the differences.

9.31. Refer to Exercise 9.30. Convert the national track records for men to speeds measured in meters per second. (See Exercise 8.21.) Perform a factor analysis of the speed data. Use the sample covariance matrix $\mathbf{S}$ and interpret the factors. Compute factor scores, and check for outliers in the data. Repeat the analysis with the sample correlation matrix $\mathbf{R}$. Does it make a difference if $\mathbf{R}$, rather than $\mathbf{S}$, is factored? Explain. Compare your results with the results in Exercise 9.30. Which analysis do you prefer? Why?

9.32. Perform a factor analysis of the data on bulls given in Table 1.10. Use the seven variables YrHgt, FtFrBody, PrctFFB, Frame, BkFat, SaleHt, and SaleWt. Factor the sample covariance matrix $\mathbf{S}$ and interpret the factors. Compute factor scores, and check for outliers. Repeat the analysis with the sample correlation matrix $\mathbf{R}$. Compare the results obtained from $\mathbf{S}$ with the results from $\mathbf{R}$. Does it make a difference if $\mathbf{R}$, rather than $\mathbf{S}$, is factored? Explain.

9.33. Perform a factor analysis of the psychological profile data in Table 4.6. Use the sample correlation matrix $\mathbf{R}$ constructed from measurements on the five variables, Indep, Supp, Benev, Conform and Leader. Obtain both the principal component and maximum likelihood solutions for $m = 2$ and $m = 3$ factors. Can you interpret the factors? Your analysis should include factor rotation and the computation of factor scores.

Note: Be aware that a maximum likelihood solution may result in a Heywood case.

9.34. The pulp and paper properties data are given in Table 7.7. Perform a factor analysis using observations on the four paper property variables, BL, EM, SF, and BS and the sample correlation matrix $\mathbf{R}$. Can the information in these data be summarized by a single factor? If so, can you interpret the factor? Try both the principal component and maximum likelihood solution methods. Repeat this analysis with the sample covariance matrix $\mathbf{S}$. Does your interpretation of the factor(s) change if $\mathbf{S}$ rather than $\mathbf{R}$ is factored?

9.35. Repeat Exercise 9.34 using observations on the pulp fiber characteristic variables AFL, LFF, FFF, and ZST. Can these data be summarized by a single factor? Explain.

9.36. Factor analyze the Mali family farm data in Table 8.7. Use the sample correlation matrix **R**. Try both the principal component and maximum likelihood solution methods for $m = 3, 4$, and 5 factors. Can you interpret the factors? Justify your choice of m. Your analysis should include factor rotation and the computation of factor scores. Can you identify any outliers in these data?

References

1. Anderson, T. W. *An Introduction to Multivariate Statistical Analysis* (3rd ed.). New York: John Wiley, 2003.

2. Bartlett, M. S. "The Statistical Conception of Mental Factors." *British Journal of Psychology,* **28** (1937), 97–104.

3. Bartlett, M. S. "A Note on Multiplying Factors for Various Chi-Squared Approximations." *Journal of the Royal Statistical Society (B)* **16** (1954), 296–298.

4. Dixon, W. S. *Statistical Software Manual to Accompany BMDP Release 7/version 7.0* (paperback). Berkeley, CA: University of California Press, 1992.

5. Dunn, L. C. "The Effect of Inbreeding on the Bones of the Fowl." *Storrs Agricultural Experimental Station Bulletin,* **52** (1928), 1–112.

6. Harmon, H. H. *Modern Factor Analysis* (3rd ed.). Chicago: The University of Chicago Press, 1976.

7. Hirschey, M., and D. W. Wichern. "Accounting and Market-Value Measures of Profitability: Consistency, Determinants and Uses." *Journal of Business and Economic Statistics,* **2**, no. 4 (1984), 375–383.

8. Joreskog, K. G. "Factor Analysis by Least Squares and Maximum Likelihood." In *Statistical Methods for Digital Computers,* edited by K. Enslein, A. Ralston, and H. S. Wilf. New York: John Wiley, 1975.

9. Kaiser, H.F. "The Varimax Criterion for Analytic Rotation in Factor Analysis." *Psychometrika,* **23** (1958), 187–200.

10. Lawley, D. N., and A. E. Maxwell. *Factor Analysis as a Statistical Method* (2nd ed.). New York: American Elsevier Publishing Co., 1971.

11. Linden, M. "A Factor Analytic Study of Olympic Decathlon Data." *Research Quarterly,* **48**, no. 3 (1977), 562–568.

12. Maxwell, A. E. *Multivariate Analysis in Behavioral Research.* London: Chapman and Hall, 1977.

13. Morrison, D. F. *Multivariate Statistical Methods* (4th ed.). Belmont, CA: Brooks/Cole Thompson Learning, 2005.

14. Stoetzel, J. "A Factor Analysis of Liquor Preference." *Journal of Advertising Research,* **1** (1960), 7–11.

15. Wright, S. "The Interpretation of Multivariate Systems." In *Statistics and Mathematics in Biology,* edited by O. Kempthorne and others. Ames, IA: Iowa State University Press, 1954, 11–33.

Chapter 10

CANONICAL CORRELATION ANALYSIS

10.1 Introduction

Canonical correlation analysis seeks to identify and quantify the associations between *two sets* of variables. H. Hotelling ([5], [6]), who initially developed the technique, provided the example of relating arithmetic speed and arithmetic power to reading speed and reading power. (See Exercise 10.9.) Other examples include relating governmental policy variables with economic goal variables and relating college "performance" variables with precollege "achievement" variables.

Canonical correlation analysis focuses on the correlation between a *linear combination* of the variables in one set and a *linear combination* of the variables in another set. The idea is first to determine the pair of linear combinations having the largest correlation. Next, we determine the pair of linear combinations having the largest correlation among all pairs uncorrelated with the initially selected pair, and so on. The pairs of linear combinations are called the *canonical variables*, and their correlations are called *canonical correlations*.

The canonical correlations measure the strength of association between the two sets of variables. The maximization aspect of the technique represents an attempt to concentrate a high-dimensional relationship between two sets of variables into a few pairs of canonical variables.

10.2 Canonical Variates and Canonical Correlations

We shall be interested in measures of association between two groups of variables. The first group, of p variables, is represented by the $(p \times 1)$ random vector $\mathbf{X}^{(1)}$. The second group, of q variables, is represented by the $(q \times 1)$ random vector $\mathbf{X}^{(2)}$. We assume, in the theoretical development, that $\mathbf{X}^{(1)}$ represents the *smaller* set, so that $p \leq q$.

For the random vectors $\mathbf{X}^{(1)}$ and $\mathbf{X}^{(2)}$, let

$$E(\mathbf{X}^{(1)}) = \boldsymbol{\mu}^{(1)}; \qquad \operatorname{Cov}(\mathbf{X}^{(1)}) = \boldsymbol{\Sigma}_{11}$$

$$E(\mathbf{X}^{(2)}) = \boldsymbol{\mu}^{(2)}; \qquad \operatorname{Cov}(\mathbf{X}^{(2)}) = \boldsymbol{\Sigma}_{22} \tag{10-1}$$

$$\operatorname{Cov}(\mathbf{X}^{(1)}, \mathbf{X}^{(2)}) = \boldsymbol{\Sigma}_{12} = \boldsymbol{\Sigma}_{21}'$$

It will be convenient to consider $\mathbf{X}^{(1)}$ and $\mathbf{X}^{(2)}$ jointly, so, using results (2-38) through (2-40) and (10-1), we find that the random vector

$$\underset{((p+q)\times 1)}{\mathbf{X}} = \left[\begin{array}{c} \mathbf{X}^{(1)} \\ \hline \mathbf{X}^{(2)} \end{array} \right] = \left[\begin{array}{c} X_1^{(1)} \\ X_2^{(1)} \\ \vdots \\ X_p^{(1)} \\ \hline X_1^{(2)} \\ X_2^{(2)} \\ \vdots \\ X_q^{(2)} \end{array} \right] \tag{10-2}$$

has mean vector

$$\underset{((p+q)\times 1)}{\boldsymbol{\mu}} = E(\mathbf{X}) = \left[\begin{array}{c} E(\mathbf{X}^{(1)}) \\ \hline E(\mathbf{X}^{(2)}) \end{array} \right] = \left[\begin{array}{c} \boldsymbol{\mu}^{(1)} \\ \hline \boldsymbol{\mu}^{(2)} \end{array} \right] \tag{10-3}$$

and covariance matrix

$$\underset{(p+q)\times(p+q)}{\boldsymbol{\Sigma}} = E(\mathbf{X} - \boldsymbol{\mu})(\mathbf{X} - \boldsymbol{\mu})'$$

$$= \left[\begin{array}{c|c} E(\mathbf{X}^{(1)} - \boldsymbol{\mu}^{(1)})(\mathbf{X}^{(1)} - \boldsymbol{\mu}^{(1)})' & E(\mathbf{X}^{(1)} - \boldsymbol{\mu}^{(1)})(\mathbf{X}^{(2)} \quad \boldsymbol{\mu}^{(2)})' \\ \hline E(\mathbf{X}^{(2)} - \boldsymbol{\mu}^{(2)})(\mathbf{X}^{(1)} - \boldsymbol{\mu}^{(1)})' & E(\mathbf{X}^{(2)} - \boldsymbol{\mu}^{(2)})(\mathbf{X}^{(2)} - \boldsymbol{\mu}^{(2)})' \end{array} \right]$$

$$= \left[\begin{array}{c|c} \underset{(p\times p)}{\boldsymbol{\Sigma}_{11}} & \underset{(p\times q)}{\boldsymbol{\Sigma}_{12}} \\ \hline \underset{(q\times p)}{\boldsymbol{\Sigma}_{21}} & \underset{(q\times q)}{\boldsymbol{\Sigma}_{22}} \end{array} \right] \tag{10-4}$$

The covariances between pairs of variables from different sets—one variable from $\mathbf{X}^{(1)}$, one variable from $\mathbf{X}^{(2)}$—are contained in $\boldsymbol{\Sigma}_{12}$ or, equivalently, in $\boldsymbol{\Sigma}_{21}$. That is, the pq elements of $\boldsymbol{\Sigma}_{12}$ measure the association between the two sets. When p and q are relatively large, interpreting the elements of $\boldsymbol{\Sigma}_{12}$ collectively is ordinarily hopeless. Moreover, it is often linear combinations of variables that are interesting and useful for predictive or comparative purposes. The main task of canonical correlation analysis is to summarize the associations between the $\mathbf{X}^{(1)}$ and $\mathbf{X}^{(2)}$ sets in terms of a *few* carefully chosen covariances (or correlations) rather than the pq covariances in $\boldsymbol{\Sigma}_{12}$.

Linear combinations provide simple summary measures of a set of variables. Set

$$U = \mathbf{a}'\mathbf{X}^{(1)}$$
$$V = \mathbf{b}'\mathbf{X}^{(2)} \tag{10-5}$$

for some pair of coefficient vectors $\mathbf{a}$ and $\mathbf{b}$. Then, using (10-5) and (2–45), we obtain

$$\text{Var}\,(U) = \mathbf{a}'\,\text{Cov}\,(\mathbf{X}^{(1)})\,\mathbf{a} = \mathbf{a}'\boldsymbol{\Sigma}_{11}\mathbf{a}$$
$$\text{Var}\,(V) = \mathbf{b}'\,\text{Cov}\,(\mathbf{X}^{(2)})\,\mathbf{b} = \mathbf{b}'\boldsymbol{\Sigma}_{22}\mathbf{b} \tag{10-6}$$
$$\text{Cov}\,(U, V) = \mathbf{a}'\,\text{Cov}\,(\mathbf{X}^{(1)}, \mathbf{X}^{(2)})\,\mathbf{b} = \mathbf{a}'\boldsymbol{\Sigma}_{12}\mathbf{b}$$

We shall seek coefficient vectors $\mathbf{a}$ and $\mathbf{b}$ such that

$$\text{Corr}\,(U, V) = \frac{\mathbf{a}'\boldsymbol{\Sigma}_{12}\mathbf{b}}{\sqrt{\mathbf{a}'\boldsymbol{\Sigma}_{11}\mathbf{a}}\ \sqrt{\mathbf{b}'\boldsymbol{\Sigma}_{22}\mathbf{b}}} \tag{10-7}$$

is as large as possible.

We define the following:

The *first pair of canonical variables*, or *first canonical variate pair*, is the pair of linear combinations U_1, V_1 having unit variances, which maximize the correlation (10-7);

The *second pair of canonical variables*, or *second canonical variate pair*, is the pair of linear combinations U_2, V_2 having unit variances, which maximize the correlation (10-7) among all choices that are uncorrelated with the first pair of canonical variables.

At the kth step,

The *kth pair of canonical variables*, or *kth canonical variate pair*, is the pair of linear combinations U_k, V_k having unit variances, which maximize the correlation (10-7) among all choices uncorrelated with the previous $k - 1$ canonical variable pairs.

The correlation between the kth pair of canonical variables is called the *kth canonical correlation*.

The following result gives the necessary details for obtaining the canonical variables and their correlations.

Result 10.1. Suppose $p \le q$ and let the random vectors $\underset{(p\times 1)}{\mathbf{X}^{(1)}}$ and $\underset{(q\times 1)}{\mathbf{X}^{(2)}}$ have $\text{Cov}\,(\mathbf{X}^{(1)}) = \underset{(p\times p)}{\boldsymbol{\Sigma}_{11}}$, $\text{Cov}\,(\mathbf{X}^{(2)}) = \underset{(q\times q)}{\boldsymbol{\Sigma}_{22}}$ and $\text{Cov}\,(\mathbf{X}^{(1)}, \mathbf{X}^{(2)}) = \underset{(p\times q)}{\boldsymbol{\Sigma}_{12}}$, where $\boldsymbol{\Sigma}$ has full rank. For coefficient vectors $\underset{(p\times 1)}{\mathbf{a}}$ and $\underset{(q\times 1)}{\mathbf{b}}$, form the linear combinations $U = \mathbf{a}'\mathbf{X}^{(1)}$ and $V = \mathbf{b}'\mathbf{X}^{(2)}$. Then

$$\max_{\mathbf{a}, \mathbf{b}} \text{Corr}\,(U, V) = \rho_1^*$$

attained by the linear combinations (first canonical variate pair)

$$U_1 = \underbrace{\mathbf{e}_1'\boldsymbol{\Sigma}_{11}^{-1/2}}_{\mathbf{a}_1'}\mathbf{X}^{(1)} \quad \text{and} \quad V_1 = \underbrace{\mathbf{f}_1'\boldsymbol{\Sigma}_{22}^{-1/2}}_{\mathbf{b}_1'}\mathbf{X}^{(2)}$$

The kth pair of canonical variates, $k = 2, 3, \ldots, p$,

$$U_k = \mathbf{e}'_k \boldsymbol{\Sigma}_{11}^{-1/2} \mathbf{X}^{(1)} \qquad V_k = \mathbf{f}'_k \boldsymbol{\Sigma}_{22}^{-1/2} \mathbf{X}^{(2)}$$

maximizes

$$\text{Corr}\,(U_k, V_k) = \rho_k^*$$

among those linear combinations uncorrelated with the preceding $1, 2, \ldots, k - 1$ canonical variables.

Here $\rho_1^{*2} \geq \rho_2^{*2} \geq \cdots \geq \rho_p^{*2}$ are the eigenvalues of $\boldsymbol{\Sigma}_{11}^{-1/2}\boldsymbol{\Sigma}_{12}\boldsymbol{\Sigma}_{22}^{-1}\boldsymbol{\Sigma}_{21}\boldsymbol{\Sigma}_{11}^{-1/2}$, and $\mathbf{e}_1, \mathbf{e}_2, \ldots, \mathbf{e}_p$ are the associated $(p \times 1)$ eigenvectors. [The quantities $\rho_1^{*2}, \rho_2^{*2}, \ldots, \rho_p^{*2}$ are also the p largest eigenvalues of the matrix $\boldsymbol{\Sigma}_{22}^{-1/2}\boldsymbol{\Sigma}_{21}\boldsymbol{\Sigma}_{11}^{-1}\boldsymbol{\Sigma}_{12}\boldsymbol{\Sigma}_{22}^{-1/2}$ with corresponding $(q \times 1)$ eigenvectors $\mathbf{f}_1, \mathbf{f}_2, \ldots, \mathbf{f}_p$. Each $\mathbf{f}_i$ is proportional to $\boldsymbol{\Sigma}_{22}^{-1/2}\boldsymbol{\Sigma}_{21}\boldsymbol{\Sigma}_{11}^{-1/2}\mathbf{e}_i$.]

The canonical variates have the properties

$$\text{Var}\,(U_k) = \text{Var}\,(V_k) = 1$$
$$\text{Cov}\,(U_k, U_\ell) = \text{Corr}\,(U_k, U_\ell) = 0 \quad k \neq \ell$$
$$\text{Cov}\,(V_k, V_\ell) = \text{Corr}\,(V_k, V_\ell) = 0 \quad k \neq \ell$$
$$\text{Cov}\,(U_k, V_\ell) = \text{Corr}\,(U_k, V_\ell) = 0 \quad k \neq \ell$$

for $k, \ell = 1, 2, \ldots, p$.

Proof. (See website: www.prenhall.com/statistics)   ∎

If the original variables are standardized with $\mathbf{Z}^{(1)} = [Z_1^{(1)}, Z_2^{(1)}, \ldots, Z_p^{(1)}]'$ and $\mathbf{Z}^{(2)} = [Z_1^{(2)}, Z_2^{(2)}, \ldots, Z_q^{(2)}]'$, from first principles, the canonical variates are of the form

$$U_k = \mathbf{a}'_k \mathbf{Z}^{(1)} = \mathbf{e}'_k \boldsymbol{\rho}_{11}^{-1/2} \mathbf{Z}^{(1)}$$
$$V_k = \mathbf{b}'_k \mathbf{Z}^{(2)} = \mathbf{f}'_k \boldsymbol{\rho}_{22}^{-1/2} \mathbf{Z}^{(2)} \tag{10-8}$$

Here, $\text{Cov}\,(\mathbf{Z}^{(1)}) = \boldsymbol{\rho}_{11}$, $\text{Cov}\,(\mathbf{Z}^{(2)}) = \boldsymbol{\rho}_{22}$, $\text{Cov}\,(\mathbf{Z}^{(1)}, \mathbf{Z}^{(2)}) = \boldsymbol{\rho}_{12} = \boldsymbol{\rho}'_{21}$, and $\mathbf{e}_k$ and $\mathbf{f}_k$ are the eigenvectors of $\boldsymbol{\rho}_{11}^{-1/2}\boldsymbol{\rho}_{12}\boldsymbol{\rho}_{22}^{-1}\boldsymbol{\rho}_{21}\boldsymbol{\rho}_{11}^{-1/2}$ and $\boldsymbol{\rho}_{22}^{-1/2}\boldsymbol{\rho}_{21}\boldsymbol{\rho}_{11}^{-1}\boldsymbol{\rho}_{12}\boldsymbol{\rho}_{22}^{-1/2}$, respectively. The canonical correlations, ρ_k^*, satisfy

$$\text{Corr}\,(U_k, V_k) = \rho_k^*, \qquad k = 1, 2, \ldots, p \tag{10-9}$$

where $\rho_1^{*2} \geq \rho_2^{*2} \geq \cdots \geq \rho_p^{*2}$ are the nonzero eigenvalues of the matrix $\boldsymbol{\rho}_{11}^{-1/2}\boldsymbol{\rho}_{12}\boldsymbol{\rho}_{22}^{-1}\boldsymbol{\rho}_{21}\boldsymbol{\rho}_{11}^{-1/2}$ (or, equivalently, the largest eigenvalues of $\boldsymbol{\rho}_{22}^{-1/2}\boldsymbol{\rho}_{21}\boldsymbol{\rho}_{11}^{-1}\boldsymbol{\rho}_{12}\boldsymbol{\rho}_{22}^{-1/2}$).

Comment. Notice that

$$\mathbf{a}'_k(\mathbf{X}^{(1)} - \boldsymbol{\mu}^{(1)}) = a_{k1}(X_1^{(1)} - \mu_1^{(1)}) + a_{k2}(X_2^{(1)} - \mu_2^{(1)})$$
$$+ \cdots + a_{kp}(X_p^{(1)} - \mu_p^{(1)})$$
$$= a_{k1}\sqrt{\sigma_{11}}\,\frac{(X_1^{(1)} - \mu_1^{(1)})}{\sqrt{\sigma_{11}}} + a_{k2}\sqrt{\sigma_{22}}\,\frac{(X_2^{(1)} - \mu_2^{(1)})}{\sqrt{\sigma_{22}}}$$
$$+ \cdots + a_{kp}\sqrt{\sigma_{pp}}\,\frac{(X_p^{(1)} - \mu_p^{(1)})}{\sqrt{\sigma_{pp}}}$$

where $\mathrm{Var}(X_i^{(1)}) = \sigma_{ii}, i = 1, 2, \ldots, p$. Therefore, the canonical coefficients for the standardized variables, $Z_i^{(1)} = (X_i^{(1)} - \mu_i^{(1)})/\sqrt{\sigma_{ii}}$, are simply related to the canonical coefficients attached to the original variables $X_i^{(1)}$. Specifically, if $\mathbf{a}_k'$ is the coefficient vector for the kth canonical variate U_k, then $\mathbf{a}_k'\mathbf{V}_{11}^{1/2}$ is the coefficient vector for the kth canonical variate constructed from the standardized variables $\mathbf{Z}^{(1)}$. Here $\mathbf{V}_{11}^{1/2}$ is the diagonal matrix with ith diagonal element $\sqrt{\sigma_{ii}}$. Similarly, $\mathbf{b}_k'\mathbf{V}_{22}^{1/2}$ is the coefficient vector for the canonical variate constructed from the set of standardized variables $\mathbf{Z}^{(2)}$. In this case $\mathbf{V}_{22}^{1/2}$ is the diagonal matrix with ith diagonal element $\sqrt{\sigma_{ii}} = \sqrt{\mathrm{Var}(X_i^{(2)})}$. The canonical correlations are *unchanged* by the standardization. However, the choice of the coefficient vectors $\mathbf{a}_k$, $\mathbf{b}_k$ will not be unique if $\rho_k^{*2} = \rho_{k+1}^{*2}$.

The relationship between the canonical coefficients of the standardized variables and the canonical coefficients of the original variables follows from the special structure of the matrix [see also (10–11)]

$$\mathbf{\Sigma}_{11}^{-1/2}\mathbf{\Sigma}_{12}\mathbf{\Sigma}_{22}^{-1}\mathbf{\Sigma}_{21}\mathbf{\Sigma}_{11}^{-1/2} \quad \text{or} \quad \boldsymbol{\rho}_{11}^{-1/2}\boldsymbol{\rho}_{12}\boldsymbol{\rho}_{22}^{-1}\boldsymbol{\rho}_{21}\boldsymbol{\rho}_{11}^{-1/2}$$

and, in this book, is unique to canonical correlation analysis. For example, in principal component analysis, if $\mathbf{a}_k'$ is the coefficient vector for the kth principal component obtained from $\mathbf{\Sigma}$, then $\mathbf{a}_k'(\mathbf{X} - \boldsymbol{\mu}) = \mathbf{a}_k'\mathbf{V}^{1/2}\mathbf{Z}$, but we cannot infer that $\mathbf{a}_k'\mathbf{V}^{1/2}$ is the coefficient vector for the kth principal component derived from $\boldsymbol{\rho}$.

Example 10.1 (Calculating canonical variates and canonical correlations for standardized variables) Suppose $\mathbf{Z}^{(1)} = [Z_1^{(1)}, Z_2^{(1)}]'$ are standardized variables and $\mathbf{Z}^{(2)} = [Z_1^{(2)}, Z_2^{(2)}]'$ are also standardized variables. Let $\mathbf{Z} = [\mathbf{Z}^{(1)}, \mathbf{Z}^{(2)}]'$ and

$$\mathrm{Cov}(\mathbf{Z}) = \begin{bmatrix} \boldsymbol{\rho}_{11} & \boldsymbol{\rho}_{12} \\ \hline \boldsymbol{\rho}_{21} & \boldsymbol{\rho}_{22} \end{bmatrix} = \begin{bmatrix} 1.0 & .4 & .5 & .6 \\ .4 & 1.0 & .3 & .4 \\ \hline .5 & .3 & 1.0 & .2 \\ .6 & .4 & .2 & 1.0 \end{bmatrix}$$

Then

$$\boldsymbol{\rho}_{11}^{-1/2} = \begin{bmatrix} 1.0681 & -.2229 \\ -.2229 & 1.0681 \end{bmatrix}$$

$$\boldsymbol{\rho}_{22}^{-1} = \begin{bmatrix} 1.0417 & -.2083 \\ -.2083 & 1.0417 \end{bmatrix}$$

and

$$\boldsymbol{\rho}_{11}^{-1/2}\boldsymbol{\rho}_{12}\boldsymbol{\rho}_{22}^{-1}\boldsymbol{\rho}_{21}\boldsymbol{\rho}_{11}^{-1/2} = \begin{bmatrix} .4371 & .2178 \\ .2178 & .1096 \end{bmatrix}$$

The eigenvalues, ρ_1^{*2}, ρ_2^{*2}, of $\boldsymbol{\rho}_{11}^{-1/2}\boldsymbol{\rho}_{12}\boldsymbol{\rho}_{22}^{-1}\boldsymbol{\rho}_{21}\boldsymbol{\rho}_{11}^{-1/2}$ are obtained from

$$0 = \begin{vmatrix} .4371 - \lambda & .2178 \\ .2178 & .1096 - \lambda \end{vmatrix} = (.4371 - \lambda)(.1096 - \lambda) - (2.178)^2$$

$$= \lambda^2 - .5467\lambda + .0005$$

yielding $\rho_1^{*2} = .5458$ and $\rho_2^{*2} = .0009$. The eigenvector $\mathbf{e}_1$ follows from the vector equation

$$\begin{bmatrix} .4371 & .2178 \\ .2178 & .1096 \end{bmatrix} \mathbf{e}_1 = (.5458)\mathbf{e}_1$$

Thus, $\mathbf{e}_1' = [.8947, .4466]$ and

$$\mathbf{a}_1 = \boldsymbol{\rho}_{11}^{-1/2}\mathbf{e}_1 = \begin{bmatrix} .8561 \\ .2776 \end{bmatrix}$$

From Result 10.1, $\mathbf{f}_1 \propto \boldsymbol{\rho}_{22}^{-1/2}\boldsymbol{\rho}_{21}\boldsymbol{\rho}_{11}^{-1/2}\mathbf{e}_1$ and $\mathbf{b}_1 = \boldsymbol{\rho}_{22}^{-1/2}\mathbf{f}_1$. Consequently,

$$\mathbf{b}_1 \propto \boldsymbol{\rho}_{22}^{-1}\boldsymbol{\rho}_{21}\mathbf{a}_1 = \begin{bmatrix} .3959 & .2292 \\ .5209 & .3542 \end{bmatrix}\begin{bmatrix} .8561 \\ .2776 \end{bmatrix} = \begin{bmatrix} .4026 \\ .5443 \end{bmatrix}$$

We must scale $\mathbf{b}_1$ so that

$$\text{Var}(V_1) = \text{Var}(\mathbf{b}_1'\mathbf{Z}^{(2)}) = \mathbf{b}_1'\boldsymbol{\rho}_{22}\mathbf{b}_1 = 1$$

The vector $[.4026, .5443]'$ gives

$$[.4026, .5443]\begin{bmatrix} 1.0 & .2 \\ .2 & 1.0 \end{bmatrix}\begin{bmatrix} .4026 \\ .5443 \end{bmatrix} = .5460$$

Using $\sqrt{.5460} = .7389$, we take

$$\mathbf{b}_1 = \frac{1}{.7389}\begin{bmatrix} .4026 \\ .5443 \end{bmatrix} = \begin{bmatrix} .5448 \\ .7366 \end{bmatrix}$$

The first pair of canonical variates is

$$U_1 = \mathbf{a}_1'\mathbf{Z}^{(1)} = .86Z_1^{(1)} + .28Z_2^{(1)}$$
$$V_1 = \mathbf{b}_1'\mathbf{Z}^{(2)} = .54Z_1^{(2)} + .74Z_2^{(2)}$$

and their canonical correlation is

$$\rho_1^* = \sqrt{\rho_1^{*2}} = \sqrt{.5458} = .74$$

This is the largest correlation possible between linear combinations of variables from the $\mathbf{Z}^{(1)}$ and $\mathbf{Z}^{(2)}$ sets.

The second canonical correlation, $\rho_2^* = \sqrt{.0009} = .03$, is very small, and consequently, the second pair of canonical variates, although uncorrelated with members of the first pair, conveys very little information about the association between sets. (The calculation of the second pair of canonical variates is considered in Exercise 10.5.)

We note that U_1 and V_1, apart from a scale change, are not much different from the pair

$$\widetilde{U}_1 = \mathbf{a}'\mathbf{Z}^{(1)} = [3, 1]\begin{bmatrix} Z_1^{(1)} \\ Z_2^{(1)} \end{bmatrix} = 3Z_1^{(1)} + Z_2^{(1)}$$

$$\widetilde{V}_1 = \mathbf{b}'\mathbf{Z}^{(2)} = [1, 1]\begin{bmatrix} Z_1^{(2)} \\ Z_2^{(2)} \end{bmatrix} = Z_1^{(2)} + Z_2^{(2)}$$

For these variates,

$$\text{Var}(\widetilde{U}_1) = \mathbf{a}'\boldsymbol{\rho}_{11}\mathbf{a} = 12.4$$

$$\text{Var}(\widetilde{V}_1) = \mathbf{b}'\boldsymbol{\rho}_{22}\mathbf{b} = 2.4$$

$$\text{Cov}(\widetilde{U}_1, \widetilde{V}_1) = \mathbf{a}'\boldsymbol{\rho}_{12}\mathbf{b} = 4.0$$

and

$$\text{Corr}(\widetilde{U}_1, \widetilde{V}_1) = \frac{4.0}{\sqrt{12.4}\,\sqrt{2.4}} = .73$$

The correlation between the rather simple and, perhaps, easily interpretable linear combinations $\widetilde{U}_1$, $\widetilde{V}_1$ is almost the maximum value $\rho_1^* = .74$. ∎

The procedure for obtaining the canonical variates presented in Result 10.1 has certain advantages. The symmetric matrices, whose eigenvectors determine the canonical coefficients, are readily handled by computer routines. Moreover, writing the coefficient vectors as $\mathbf{a}_k = \boldsymbol{\Sigma}_{11}^{-1/2}\mathbf{e}_k$ and $\mathbf{b}_k = \boldsymbol{\Sigma}_{22}^{-1/2}\mathbf{f}_k$ facilitates analytic descriptions and their geometric interpretations. To ease the computational burden, many people prefer to get the canonical correlations from the eigenvalue equation

$$|\boldsymbol{\Sigma}_{11}^{-1}\boldsymbol{\Sigma}_{12}\boldsymbol{\Sigma}_{22}^{-1}\boldsymbol{\Sigma}_{21} - \rho^{*2}\mathbf{I}| = 0 \tag{10-10}$$

The coefficient vectors $\mathbf{a}$ and $\mathbf{b}$ follow directly from the eigenvector equations

$$\boldsymbol{\Sigma}_{11}^{-1}\boldsymbol{\Sigma}_{12}\boldsymbol{\Sigma}_{22}^{-1}\boldsymbol{\Sigma}_{21}\mathbf{a} = \rho^{*2}\mathbf{a}$$

$$\boldsymbol{\Sigma}_{22}^{-1}\boldsymbol{\Sigma}_{21}\boldsymbol{\Sigma}_{11}^{-1}\boldsymbol{\Sigma}_{12}\mathbf{b} = \rho^{*2}\mathbf{b} \tag{10-11}$$

The matrices $\boldsymbol{\Sigma}_{11}^{-1}\boldsymbol{\Sigma}_{12}\boldsymbol{\Sigma}_{22}^{-1}\boldsymbol{\Sigma}_{21}$ and $\boldsymbol{\Sigma}_{22}^{-1}\boldsymbol{\Sigma}_{21}\boldsymbol{\Sigma}_{11}^{-1}\boldsymbol{\Sigma}_{12}$ are, in general, not symmetric. (See Exercise 10.4 for more details.)

10.3 Interpreting the Population Canonical Variables

Canonical variables are, in general, artificial. That is, they have no physical meaning. If the original variables $\mathbf{X}^{(1)}$ and $\mathbf{X}^{(2)}$ are used, the canonical coefficients $\mathbf{a}$ and $\mathbf{b}$ have units proportional to those of the $\mathbf{X}^{(1)}$ and $\mathbf{X}^{(2)}$ sets. If the original variables are standardized to have zero means and unit variances, the canonical coefficients have no units of measurement, and they must be interpreted in terms of the standardized variables.

Result 10.1 gives the technical definitions of the canonical variables and canonical correlations. In this section, we concentrate on interpreting these quantities.

Identifying the Canonical Variables

Even though the canonical variables are artificial, they can often be "identified" in terms of the subject-matter variables. Many times this identification is aided by computing the correlations between the canonical variates and the original variables. These correlations, however, must be interpreted with caution. They provide only univariate information, in the sense that they do not indicate how the original variables contribute *jointly* to the canonical analyses. (See, for example, [11].)

For this reason, many investigators prefer to assess the contributions of the original variables directly from the standardized coefficients (10-8).

Let $\underset{(p\times p)}{\mathbf{A}} = [\mathbf{a}_1, \mathbf{a}_2, \ldots, \mathbf{a}_p]'$ and $\underset{(q\times q)}{\mathbf{B}} = [\mathbf{b}_1, \mathbf{b}_2, \ldots, \mathbf{b}_q]'$, so that the vectors of canonical variables are

$$\underset{(p\times 1)}{\mathbf{U}} = \mathbf{A}\mathbf{X}^{(1)} \qquad \underset{(q\times 1)}{\mathbf{V}} = \mathbf{B}\mathbf{X}^{(2)} \tag{10-12}$$

where we are primarily interested in the first p canonical variables in $\mathbf{V}$. Then

$$\text{Cov}(\mathbf{U}, \mathbf{X}^{(1)}) = \text{Cov}(\mathbf{A}\mathbf{X}^{(1)}, \mathbf{X}^{(1)}) = \mathbf{A}\boldsymbol{\Sigma}_{11} \tag{10-13}$$

Because $\text{Var}(U_i) = 1$, $\text{Corr}(U_i, X_k^{(1)})$ is obtained by dividing $\text{Cov}(U_i, X_k^{(1)})$ by $\sqrt{\text{Var}(X_k^{(1)})} = \sigma_{kk}^{1/2}$. Equivalently, $\text{Corr}(U_i, X_k^{(1)}) = \text{Cov}(U_i, \sigma_{kk}^{-1/2} X_k^{(1)})$. Introducing the $(p \times p)$ diagonal matrix $\mathbf{V}_{11}^{-1/2}$ with kth diagonal element $\sigma_{kk}^{-1/2}$, we have, in matrix terms,

$$\underset{(p\times p)}{\boldsymbol{\rho}_{\mathbf{U},\mathbf{X}^{(1)}}} = \text{Corr}(\mathbf{U}, \mathbf{X}^{(1)}) = \text{Cov}(\mathbf{U}, \mathbf{V}_{11}^{-1/2}\mathbf{X}^{(1)}) = \text{Cov}(\mathbf{A}\mathbf{X}^{(1)}, \mathbf{V}_{11}^{-1/2}\mathbf{X}^{(1)})$$

$$= \mathbf{A}\boldsymbol{\Sigma}_{11}\mathbf{V}_{11}^{-1/2}$$

Similar calculations for the pairs $(\mathbf{U}, \mathbf{X}^{(2)})$, $(\mathbf{V}, \mathbf{X}^{(2)})$ and $(\mathbf{V}, \mathbf{X}^{(1)})$ yield

$$\underset{(p\times p)}{\boldsymbol{\rho}_{\mathbf{U},\mathbf{X}^{(1)}}} = \mathbf{A}\boldsymbol{\Sigma}_{11}\mathbf{V}_{11}^{-1/2} \qquad \underset{(q\times q)}{\boldsymbol{\rho}_{\mathbf{V},\mathbf{X}^{(2)}}} = \mathbf{B}\boldsymbol{\Sigma}_{22}\mathbf{V}_{22}^{-1/2}$$

$$\underset{(p\times q)}{\boldsymbol{\rho}_{\mathbf{U},\mathbf{X}^{(2)}}} = \mathbf{A}\boldsymbol{\Sigma}_{12}\mathbf{V}_{22}^{-1/2} \qquad \underset{(q\times p)}{\boldsymbol{\rho}_{\mathbf{V},\mathbf{X}^{(1)}}} = \mathbf{B}\boldsymbol{\Sigma}_{21}\mathbf{V}_{11}^{-1/2} \tag{10-14}$$

where $\mathbf{V}_{22}^{-1/2}$ is the $(q \times q)$ diagonal matrix with ith diagonal element $[\text{Var}(X_i^{(2)})]$.

Canonical variables derived from standardized variables are sometimes interpreted by computing the correlations. Thus,

$$\boldsymbol{\rho}_{\mathbf{U},\mathbf{Z}^{(1)}} = \mathbf{A}_z\boldsymbol{\rho}_{11} \qquad \boldsymbol{\rho}_{\mathbf{V},\mathbf{Z}^{(2)}} = \mathbf{B}_z\boldsymbol{\rho}_{22}$$

$$\boldsymbol{\rho}_{\mathbf{U},\mathbf{Z}^{(2)}} = \mathbf{A}_z\boldsymbol{\rho}_{12} \qquad \boldsymbol{\rho}_{\mathbf{V},\mathbf{Z}^{(1)}} = \mathbf{B}_z\boldsymbol{\rho}_{21} \tag{10-15}$$

where $\underset{(p\times p)}{\mathbf{A}_z}$ and $\underset{(q\times q)}{\mathbf{B}_z}$ are the matrices whose rows contain the canonical coefficients for the $\mathbf{Z}^{(1)}$ and $\mathbf{Z}^{(2)}$ sets, respectively. The correlations in the matrices displayed in (10–15) have the *same* numerical values as those appearing in (10–14); that is, $\boldsymbol{\rho}_{\mathbf{U},\mathbf{X}^{(1)}} = \boldsymbol{\rho}_{\mathbf{U},\mathbf{Z}^{(1)}}$, and so forth. This follows because, for example, $\boldsymbol{\rho}_{\mathbf{U},\mathbf{X}^{(1)}} = \mathbf{A}\boldsymbol{\Sigma}_{11}\mathbf{V}_{11}^{-1/2} = \mathbf{A}\mathbf{V}_{11}^{1/2}\mathbf{V}_{11}^{-1/2}\boldsymbol{\Sigma}_{11}\mathbf{V}_{11}^{-1/2} = \mathbf{A}_z\boldsymbol{\rho}_{11} = \boldsymbol{\rho}_{\mathbf{U},\mathbf{Z}^{(1)}}$. The correlations are unaffected by the standardization.

Example 10.2 (Computing correlations between canonical variates and their component variables) Compute the correlations between the first pair of canonical variates and their component variables for the situation considered in Example 10.1.

The variables in Example 10.1 are already standardized, so equation (10–15) is applicable. For the standardized variables,

$$\boldsymbol{\rho}_{11} = \begin{bmatrix} 1.0 & .4 \\ .4 & 1.0 \end{bmatrix} \qquad \boldsymbol{\rho}_{22} = \begin{bmatrix} 1.0 & .2 \\ .2 & 1.0 \end{bmatrix}$$

and

$$\boldsymbol{\rho}_{12} = \begin{bmatrix} .5 & .6 \\ .3 & .4 \end{bmatrix}$$

With $p = 1$,

$$\mathbf{A_z} = [.86, .28] \qquad \mathbf{B_z} = [.54, .74]$$

so

$$\boldsymbol{\rho}_{U_1, \mathbf{z}^{(1)}} = \mathbf{A_z}\boldsymbol{\rho}_{11} = [.86, .28] \begin{bmatrix} 1.0 & .4 \\ .4 & 1.0 \end{bmatrix} = [.97, .62]$$

and

$$\boldsymbol{\rho}_{V_1, \mathbf{z}^{(2)}} = \mathbf{B_z}\boldsymbol{\rho}_{22} = [.54, .74] \begin{bmatrix} 1.0 & .2 \\ .2 & 1.0 \end{bmatrix} = [.69, .85]$$

We conclude that, of the two variables in the set $\mathbf{Z}^{(1)}$, the first is most closely associated with the canonical variate U_1. Of the two variables in the set $\mathbf{Z}^{(2)}$, the second is most closely associated with V_1. In this case, the correlations reinforce the information supplied by the standardized coefficients $\mathbf{A_z}$ and $\mathbf{B_z}$. However, the correlations elevate the relative importance of $Z_2^{(1)}$ in the first set and $Z_1^{(2)}$ in the second set because they ignore the contribution of the remaining variable in each set.

From (10-15), we also obtain the correlations

$$\boldsymbol{\rho}_{U_1, \mathbf{z}^{(2)}} = \mathbf{A_z}\boldsymbol{\rho}_{12} = [.86, .28] \begin{bmatrix} .5 & .6 \\ .3 & .4 \end{bmatrix} = [.51, .63]$$

and

$$\boldsymbol{\rho}_{V_1, \mathbf{z}^{(1)}} = \mathbf{B_z}\boldsymbol{\rho}_{21} = \mathbf{B_z}\boldsymbol{\rho}'_{12} = [.54, .74] \begin{bmatrix} .5 & .3 \\ .6 & .4 \end{bmatrix} = [.71, .46]$$

Later, in our discussion of the sample canonical variates, we shall comment on the interpretation of these last correlations. ∎

The correlations $\boldsymbol{\rho}_{\mathbf{U}, \mathbf{X}^{(1)}}$ and $\boldsymbol{\rho}_{\mathbf{V}, \mathbf{X}^{(2)}}$ can help supply meanings for the canonical variates. The spirit is the same as in principal component analysis when the correlations between the principal components and their associated variables may provide subject-matter interpretations for the components.

Canonical Correlations as Generalizations of Other Correlation Coefficients

First, the canonical correlation generalizes the correlation between two variables. When $\mathbf{X}^{(1)}$ and $\mathbf{X}^{(2)}$ each consist of a single variable, so that $p = q = 1$,

$$|\text{Corr}(X_1^{(1)}, X_1^{(2)})| = |\text{Corr}(aX_1^{(1)}, bX_1^{(2)})| \qquad \text{for all } a, b \neq 0$$

Therefore, the "canonical variates" $U_1 = X_1^{(1)}$ and $V_1 = X_1^{(2)}$ have correlation $\rho_1^* = |\text{Corr}(X_1^{(1)}, X_1^{(2)})|$. When $\mathbf{X}^{(1)}$ and $\mathbf{X}^{(2)}$ have more components, setting $\mathbf{a}' = [0, \ldots, 0, 1, 0, \ldots, 0]$ with 1 in the ith position and $\mathbf{b}' = [0, \ldots, 0, 1, 0, \ldots, 0]$ with 1 in the kth position yields

$$
\begin{aligned}
|\text{Corr}(X_i^{(1)}, X_k^{(2)})| &= |\text{Corr}(\mathbf{a}'\mathbf{X}^{(1)}, \mathbf{b}'\mathbf{X}^{(2)})| \\
&\leq \max_{\mathbf{a},\mathbf{b}} \text{Corr}(\mathbf{a}'\mathbf{X}^{(1)}, \mathbf{b}'\mathbf{X}^{(2)}) = \rho_1^*
\end{aligned}
\tag{10-16}
$$

That is, the first canonical correlation is larger than the absolute value of any entry in $\boldsymbol{\rho}_{12} = \mathbf{V}_{11}^{-1/2} \boldsymbol{\Sigma}_{12} \mathbf{V}_{22}^{-1/2}$.

Second, the multiple correlation coefficient $\rho_{1(\mathbf{X}^{(2)})}$ [see (7-48)] is a special case of a canonical correlation when $\mathbf{X}^{(1)}$ has the single element $X_1^{(1)}(p = 1)$. Recall that

$$
\rho_{1(\mathbf{X}^{(2)})} = \max_{\mathbf{b}} \text{Corr}(X_1^{(1)}, \mathbf{b}'\mathbf{X}^{(2)}) = \rho_1^* \qquad \text{for} \quad p = 1
\tag{10-17}
$$

When $p > 1$, ρ_1^* is larger than each of the multiple correlations of $X_i^{(1)}$ with $\mathbf{X}^{(2)}$ or the multiple correlations of $X_i^{(2)}$ with $\mathbf{X}^{(1)}$.

Finally, we note that

$$
\rho_{U_k(\mathbf{X}^{(2)})} = \max_{\mathbf{b}} \text{Corr}(U_k, \mathbf{b}'\mathbf{X}^{(2)}) = \text{Corr}(U_k, V_k) = \rho_k^*,
\tag{10-18}
$$
$$
k = 1, 2, \ldots, p
$$

from the proof of Result 10.1 (see website: www.prenhall.com/statistics). Similarly,

$$
\rho_{V_k(\mathbf{X}^{(1)})} = \max_{\mathbf{a}} \text{Corr}(\mathbf{a}'\mathbf{X}^{(1)}, V_k) = \text{Corr}(U_k, V_k) = \rho_k^*,
\tag{10-19}
$$
$$
k = 1, 2, \ldots, p
$$

That is, the canonical correlations are also the multiple correlation coefficients of U_k with $\mathbf{X}^{(2)}$ or the multiple correlation coefficients of V_k with $\mathbf{X}^{(1)}$.

Because of its multiple correlation coefficient interpretation, the kth *squared* canonical correlation ρ_k^{*2} is the proportion of the variance of canonical variate U_k "explained" by the set $\mathbf{X}^{(2)}$. It is also the proportion of the variance of canonical variate V_k "explained" by the set $\mathbf{X}^{(1)}$. Therefore, ρ_k^{*2} is often called the *shared variance* between the two sets $\mathbf{X}^{(1)}$ and $\mathbf{X}^{(2)}$. The largest value, ρ_1^{*2}, is sometimes regarded as a measure of set "overlap."

The First r Canonical Variables as a Summary of Variability

The change of coordinates from $\mathbf{X}^{(1)}$ to $\mathbf{U} = \mathbf{A}\mathbf{X}^{(1)}$ and from $\mathbf{X}^{(2)}$ to $\mathbf{V} = \mathbf{B}\mathbf{X}^{(2)}$ is chosen to maximize $\text{Corr}(U_1, V_1)$ and, successively, $\text{Corr}(U_i, V_i)$, where (U_i, V_i) have zero correlation with the previous pairs $(U_1, V_1), (U_2, V_2), \ldots, (U_{i-1}, V_{i-1})$. Correlation between the sets $\mathbf{X}^{(1)}$ and $\mathbf{X}^{(2)}$ has been isolated in the pairs of canonical variables

By design, the coefficient vectors $\mathbf{a}_i, \mathbf{b}_i$ are selected to maximize correlations, not necessarily to provide variables that (approximately) account for the subset covariances $\boldsymbol{\Sigma}_{11}$ and $\boldsymbol{\Sigma}_{22}$. When the first few pairs of canonical variables provide poor summaries of the variability in $\boldsymbol{\Sigma}_{11}$ and $\boldsymbol{\Sigma}_{22}$, it is not clear how a high canonical correlation should be interpreted.

Example 10.3 (Canonical correlation as a poor summary of variability) Consider the covariance matrix

$$\text{Cov}\left(\begin{bmatrix} X_1^{(1)} \\ X_2^{(1)} \\ \hline X_1^{(2)} \\ X_2^{(2)} \end{bmatrix}\right) = \begin{bmatrix} \mathbf{\Sigma}_{11} & \mathbf{\Sigma}_{12} \\ \hline \mathbf{\Sigma}_{21} & \mathbf{\Sigma}_{22} \end{bmatrix} = \begin{bmatrix} 100 & 0 & 0 & 0 \\ 0 & 1 & .95 & 0 \\ \hline 0 & .95 & 1 & 0 \\ 0 & 0 & 0 & 100 \end{bmatrix}$$

The reader may verify (see Exercise 10.1) that the first pair of canonical variates $U_1 = X_2^{(1)}$ and $V_1 = X_1^{(2)}$ has correlation

$$\rho_1^* = \text{Corr}(U_1, V_1) = .95$$

Yet $U_1 = X_2^{(1)}$ provides a very poor summary of the variability in the first set. Most of the variability in this set is in $X_1^{(1)}$, which is uncorrelated with U_1. The same situation is true for $V_1 = X_1^{(2)}$ in the second set. ■

A Geometrical Interpretation of the Population Canonical Correlation Analysis

A geometrical interpretation of the procedure for selecting canonical variables provides some valuable insights into the nature of a canonical correlation analysis.

The transformation

$$\mathbf{U} = \mathbf{A}\mathbf{X}^{(1)}$$

from $\mathbf{X}^{(1)}$ to $\mathbf{U}$ gives

$$\text{Cov}(\mathbf{U}) = \mathbf{A}\mathbf{\Sigma}_{11}\mathbf{A}' = \mathbf{I}$$

From Result 10.1 and (2-22), $\mathbf{A} = \mathbf{E}'\mathbf{\Sigma}_{11}^{-1/2} = \mathbf{E}'\mathbf{P}_1\mathbf{\Lambda}_1^{-1/2}\mathbf{P}_1'$ where $\mathbf{E}'$ is an orthogonal matrix with row $\mathbf{e}_i'$, and $\mathbf{\Sigma}_{11} = \mathbf{P}_1\mathbf{\Lambda}_1\mathbf{P}_1'$. Now, $\mathbf{P}_1'\mathbf{X}^{(1)}$ is the set of principal components derived from $\mathbf{X}^{(1)}$ alone. The matrix $\mathbf{\Lambda}_1^{-1/2}\mathbf{P}_1'\mathbf{X}^{(1)}$ has ith row $(1/\sqrt{\lambda_i})\mathbf{p}_i'\mathbf{X}^{(1)}$, which is the ith principal component scaled to have unit variance. That is,

$$\text{Cov}(\mathbf{\Lambda}_1^{-1/2}\mathbf{P}_1'\mathbf{X}^{(1)}) = \mathbf{\Lambda}_1^{-1/2}\mathbf{P}_1'\mathbf{\Sigma}_{11}\mathbf{P}_1\mathbf{\Lambda}_1^{-1/2} = \mathbf{\Lambda}_1^{-1/2}\mathbf{P}_1'\mathbf{P}_1\mathbf{\Lambda}_1\mathbf{P}_1'\mathbf{P}_1\mathbf{\Lambda}_1^{-1/2}$$

$$= \mathbf{\Lambda}_1^{-1/2}\mathbf{\Lambda}_1\mathbf{\Lambda}_1^{-1/2} = \mathbf{I}$$

Consequently, $\mathbf{U} = \mathbf{A}\mathbf{X}^{(1)} = \mathbf{E}'\mathbf{P}_1\mathbf{\Lambda}_1^{-1/2}\mathbf{P}_1'\mathbf{X}^{(1)}$ can be interpreted as (1) a transformation of $\mathbf{X}^{(1)}$ to uncorrelated standardized principal components, followed by (2) a rigid (orthogonal) rotation $\mathbf{P}_1$ determined by $\mathbf{\Sigma}_{11}$ and then (3) another rotation $\mathbf{E}'$ determined from the full covariance matrix $\mathbf{\Sigma}$. A similar interpretation applies to $\mathbf{V} = \mathbf{B}\mathbf{X}^{(2)}$.

10.4 The Sample Canonical Variates and Sample Canonical Correlations

A random sample of n observations on each of the $(p + q)$ variables $\mathbf{X}^{(1)}, \mathbf{X}^{(2)}$ can be assembled into the $n \times (p + q)$ data matrix

$$\mathbf{X} = \left[\mathbf{X}^{(1)} \mid \mathbf{X}^{(2)}\right]$$

$$= \begin{bmatrix} x_{11}^{(1)} & x_{12}^{(1)} & \cdots & x_{1p}^{(1)} & x_{11}^{(2)} & x_{12}^{(2)} & \cdots & x_{1q}^{(2)} \\ x_{21}^{(1)} & x_{22}^{(1)} & \cdots & x_{2p}^{(1)} & x_{21}^{(2)} & x_{22}^{(2)} & \cdots & x_{2q}^{(2)} \\ \vdots & \vdots & & \vdots & \vdots & \vdots & & \vdots \\ x_{n1}^{(1)} & x_{n2}^{(1)} & \cdots & x_{np}^{(1)} & x_{n1}^{(2)} & x_{n2}^{(2)} & \cdots & x_{nq}^{(2)} \end{bmatrix} = \begin{bmatrix} \mathbf{x}_1^{(1)\prime} & \mathbf{x}_1^{(2)\prime} \\ \vdots & \vdots \\ \mathbf{x}_n^{(1)\prime} & \mathbf{x}_n^{(2)\prime} \end{bmatrix} \quad (10\text{-}20)$$

The vector of sample means can be organized as

$$\underset{(p+q)\times 1}{\bar{\mathbf{x}}} = \begin{bmatrix} \bar{\mathbf{x}}^{(1)} \\ \hline \bar{\mathbf{x}}^{(2)} \end{bmatrix} \quad \text{where} \quad \bar{\mathbf{x}}^{(1)} = \frac{1}{n} \sum_{j=1}^{n} \mathbf{x}_j^{(1)}$$

$$\bar{\mathbf{x}}^{(2)} = \frac{1}{n} \sum_{j=1}^{n} \mathbf{x}_j^{(2)} \quad (10\text{-}21)$$

Similarly, the sample covariance matrix can be arranged analogous to the representation (10-4). Thus,

$$\underset{(p+q)\times(p+q)}{\mathbf{S}} = \begin{bmatrix} \underset{(p\times p)}{\mathbf{S}_{11}} & \underset{(p\times q)}{\mathbf{S}_{12}} \\ \hline \underset{(q\times p)}{\mathbf{S}_{21}} & \underset{(q\times q)}{\mathbf{S}_{22}} \end{bmatrix}$$

where

$$\mathbf{S}_{kl} = \frac{1}{n-1} \sum_{j=1}^{n} \left(\mathbf{x}_j^{(k)} - \bar{\mathbf{x}}^{(k)}\right)\left(\mathbf{x}_j^{(l)} - \bar{\mathbf{x}}^{(l)}\right)', \qquad k, l = 1, 2 \quad (10\text{-}22)$$

The linear combinations

$$\hat{U} = \hat{\mathbf{a}}'\mathbf{x}^{(1)}; \qquad \hat{V} = \hat{\mathbf{b}}'\mathbf{x}^{(2)} \quad (10\text{-}23)$$

have sample correlation [see (3-36)]

$$r_{\hat{U}, \hat{V}} = \frac{\hat{\mathbf{a}}'\mathbf{S}_{12}\hat{\mathbf{b}}}{\sqrt{\hat{\mathbf{a}}'\mathbf{S}_{11}\hat{\mathbf{a}}} \; \sqrt{\hat{\mathbf{b}}'\mathbf{S}_{22}\hat{\mathbf{b}}}} \quad (10\text{-}24)$$

The *first pair of sample canonical variates* is the pair of linear combinations $\hat{U}_1, \hat{V}_1$ having unit sample variances that maximize the ratio (10-24).

In general, the *kth pair of sample canonical variates* is the pair of linear combinations $\hat{U}_k, \hat{V}_k$ having unit sample variances that maximize the ratio (10-24) among those linear combinations uncorrelated with the previous $k - 1$ sample canonical variates.

The sample correlation between $\hat{U}_k$ and $\hat{V}_k$ is called the *kth sample canonical correlation*.

The sample canonical variates and the sample canonical correlations can be obtained from the sample covariance matrices $\mathbf{S}_{11}$, $\mathbf{S}_{12} = \mathbf{S}_{21}'$, and $\mathbf{S}_{22}$ in a manner consistent with the population case described in Result 10.1.

Result 10.2. Let $\widehat{\rho}_1^{*2} \geq \widehat{\rho}_2^{*2} \geq \cdots \geq \widehat{\rho}_p^{*2}$ be the p ordered eigenvalues of $S_{11}^{-1/2}S_{12}S_{22}^{-1}S_{21}S_{11}^{-1/2}$ with corresponding eigenvectors $\hat{e}_1, \hat{e}_2, \ldots, \hat{e}_p$, where the S_{kl} are defined in (10-22) and $p \leq q$. Let $\hat{f}_1, \hat{f}_2, \ldots, \hat{f}_p$ be the eigenvectors of $S_{22}^{-1/2}S_{21}S_{11}^{-1}S_{12}S_{22}^{-1/2}$, where the first p $\hat{f}$'s may be obtained from $\hat{f}_k = (1/\widehat{\rho}_k^*)S_{22}^{-1/2}S_{21}S_{11}^{-1/2}\hat{e}_k$, $k = 1, 2, \ldots, p$. Then the kth sample canonical variate pair[1] is

$$\hat{U}_k = \underbrace{\hat{e}_k' S_{11}^{-1/2}}_{\hat{a}_k'} x^{(1)} \qquad\qquad \hat{V}_k = \underbrace{\hat{f}_k' S_{22}^{-1/2}}_{\hat{b}_k'} x^{(2)}$$

where $x^{(1)}$ and $x^{(2)}$ are the values of the variables $X^{(1)}$ and $X^{(2)}$ for a particular experimental unit. Also, the first sample canonical variate pair has the maximum sample correlation

$$r_{\hat{U}_1, \hat{V}_1} = \widehat{\rho}_1^*$$

and for the kth pair,

$$r_{\hat{U}_k, \hat{V}_k} = \widehat{\rho}_k^*$$

is the largest possible correlation among linear combinations uncorrelated with the preceding $k - 1$ sample canonical variates.

The quantities $\widehat{\rho}_1^*, \widehat{\rho}_2^*, \ldots, \widehat{\rho}_p^*$ are the sample canonical correlations.[2]

Proof. The proof of this result follows the proof of Result 10.1, with S_{kl} substituted for $\Sigma_{kl}, k, l = 1, 2.$ ∎

The sample canonical variates have unit sample variances

$$s_{\hat{U}_k, \hat{U}_k} = s_{\hat{V}_k, \hat{V}_k} = 1 \qquad (10\text{-}25)$$

and their sample correlations are

$$r_{\hat{U}_k, \hat{U}_\ell} = r_{\hat{V}_k, \hat{V}_\ell} = 0, \qquad k \neq \ell$$
$$r_{\hat{U}_k, \hat{V}_\ell} = 0, \qquad k \neq \ell \qquad (10\text{-}26)$$

The interpretation of $\hat{U}_k$, $\hat{V}_k$ is often aided by computing the sample correlations between the canonical variates and the variables in the sets $X^{(1)}$ and $X^{(2)}$. We define the matrices

$$\underset{(p\times p)}{\hat{A}} = [\hat{a}_1, \hat{a}_2, \ldots, \hat{a}_p]' \qquad \underset{(q\times q)}{\hat{B}} = [\hat{b}_1, \hat{b}_2, \ldots, \hat{b}_q]' \qquad (10\text{-}27)$$

whose *rows* are the coefficient vectors for the sample canonical variates.[3] Analogous to (10-12), we have

$$\underset{(p\times 1)}{\hat{U}} = \hat{A} x^{(1)} \qquad \underset{(q\times 1)}{\hat{V}} = \hat{B} x^{(2)} \qquad (10\text{-}28)$$

---

[1] When the distribution is normal, the maximum likelihood method can be employed using $\hat{\Sigma} = S_n$ in place of S. The sample canonical correlations $\widehat{\rho}_k^*$ are, therefore, the maximum likelihood estimates of ρ_k^* and $\sqrt{n/(n-1)}\,\hat{a}_k$, $\sqrt{n/(n-1)}\,\hat{b}_k$ are the maximum likelihood estimates of a_k and b_k, respectively.

[2] If $p > \text{rank}(S_{12}) = p_1$, the nonzero sample canonical correlations are $\widehat{\rho}_1^*, \ldots, \widehat{\rho}_{p_1}^*$.

[3] The vectors $\hat{b}_{p_1+1} = S_{22}^{-1/2}\hat{f}_{p_1+1}, \hat{b}_{p_1+2} = S_{22}^{-1/2}\hat{f}_{p_1+2}, \ldots, \hat{b}_q = S_{22}^{-1/2}\hat{f}_q$ are determined from a choice of the last $q - p_1$ mutually orthogonal eigenvectors $\hat{f}$ associated with the *zero* eigenvalue of $S_{22}^{-1/2}S_{21}S_{11}^{-1}S_{12}S_{22}^{-1/2}$.

and we can define

$$\mathbf{R}_{\hat{\mathbf{U}}, \mathbf{x}^{(1)}} = \text{matrix of sample correlations of } \hat{\mathbf{U}} \text{ with } \mathbf{x}^{(1)}$$

$$\mathbf{R}_{\hat{\mathbf{V}}, \mathbf{x}^{(2)}} = \text{matrix of sample correlations of } \hat{\mathbf{V}} \text{ with } \mathbf{x}^{(2)}$$

$$\mathbf{R}_{\hat{\mathbf{U}}, \mathbf{x}^{(2)}} = \text{matrix of sample correlations of } \hat{\mathbf{U}} \text{ with } \mathbf{x}^{(2)}$$

$$\mathbf{R}_{\hat{\mathbf{V}}, \mathbf{x}^{(1)}} = \text{matrix of sample correlations of } \hat{\mathbf{V}} \text{ with } \mathbf{x}^{(1)}$$

Corresponding to (10-19), we have

$$\begin{aligned}
\mathbf{R}_{\hat{\mathbf{U}}, \mathbf{x}^{(1)}} &= \hat{\mathbf{A}} \mathbf{S}_{11} \mathbf{D}_{11}^{-1/2} \\
\mathbf{R}_{\hat{\mathbf{V}}, \mathbf{x}^{(2)}} &= \hat{\mathbf{B}} \mathbf{S}_{22} \mathbf{D}_{22}^{-1/2} \\
\mathbf{R}_{\hat{\mathbf{U}}, \mathbf{x}^{(2)}} &= \hat{\mathbf{A}} \mathbf{S}_{12} \mathbf{D}_{22}^{-1/2} \\
\mathbf{R}_{\hat{\mathbf{V}}, \mathbf{x}^{(1)}} &= \hat{\mathbf{B}} \mathbf{S}_{21} \mathbf{D}_{11}^{-1/2}
\end{aligned} \tag{10-29}$$

where $\mathbf{D}_{11}^{-1/2}$ is the $(p \times p)$ diagonal matrix with ith diagonal element (sample $\text{var}(x_i^{(1)}))^{-1/2}$ and $\mathbf{D}_{22}^{-1/2}$ is the $(q \times q)$ diagonal matrix with ith diagonal element (sample $\text{var}(x_i^{(2)}))^{-1/2}$.

Comment. If the observations are standardized [see (8-25)], the data matrix becomes

$$\mathbf{Z} = \left[\mathbf{Z}^{(1)} \vdots \mathbf{Z}^{(2)} \right] = \begin{bmatrix} \mathbf{z}_1^{(1)\prime} & \vdots & \mathbf{z}_1^{(2)\prime} \\ \vdots & & \vdots \\ \mathbf{z}_n^{(1)\prime} & \vdots & \mathbf{z}_n^{(2)\prime} \end{bmatrix}$$

and the sample canonical variates become

$$\underset{(p \times 1)}{\hat{\mathbf{U}}} = \hat{\mathbf{A}}_\mathbf{z} \mathbf{z}^{(1)} \qquad \underset{(q \times 1)}{\hat{\mathbf{V}}} = \hat{\mathbf{B}}_\mathbf{z} \mathbf{z}^{(2)} \tag{10-30}$$

where $\hat{\mathbf{A}}_\mathbf{z} = \hat{\mathbf{A}} \mathbf{D}_{11}^{1/2}$ and $\hat{\mathbf{B}}_\mathbf{z} = \hat{\mathbf{B}} \mathbf{D}_{22}^{1/2}$. The sample canonical correlations are unaffected by the standardization. The correlations displayed in (10–29) remain unchanged and may be calculated, for standardized observations, by substituting $\hat{\mathbf{A}}_\mathbf{z}$ for $\hat{\mathbf{A}}$, $\hat{\mathbf{B}}_\mathbf{z}$ for $\hat{\mathbf{B}}$, and $\mathbf{R}$ for $\mathbf{S}$. Note that $\mathbf{D}_{11}^{-1/2} = \underset{(p \times p)}{\mathbf{I}}$ and $\mathbf{D}_{22}^{-1/2} = \underset{(q \times q)}{\mathbf{I}}$ for standardized observations.

Example 10.4 (Canonical correlation analysis of the chicken-bone data) In Example 9.14, data consisting of bone and skull measurements of white leghorn fowl were described. From this example, the chicken-bone measurements for

$$\text{Head } (\mathbf{X}^{(1)}): \quad \begin{cases} X_1^{(1)} = \text{skull length} \\ X_2^{(1)} = \text{skull breadth} \end{cases}$$

$$\text{Leg } (\mathbf{X}^{(2)}): \quad \begin{cases} X_1^{(2)} = \text{femur length} \\ X_2^{(2)} = \text{tibia length} \end{cases}$$

have the sample correlation matrix

$$\mathbf{R} = \begin{bmatrix} \mathbf{R}_{11} & \mathbf{R}_{12} \\ \hline \mathbf{R}_{21} & \mathbf{R}_{22} \end{bmatrix} = \begin{bmatrix} 1.0 & .505 & .569 & .602 \\ .505 & 1.0 & .422 & .467 \\ \hline .569 & .422 & 1.0 & .926 \\ .602 & .467 & .926 & 1.0 \end{bmatrix}$$

A canonical correlation analysis of the head and leg sets of variables using $\mathbf{R}$ produces the two canonical correlations and corresponding pairs of variables

$$\widehat{\rho_1^*} = .631 \qquad \begin{aligned} \hat{U}_1 &= .781 z_1^{(1)} + .345 z_2^{(1)} \\ \hat{V}_1 &= .060 z_1^{(2)} + .944 z_2^{(2)} \end{aligned}$$

and

$$\widehat{\rho_2^*} = .057 \qquad \begin{aligned} \hat{U}_2 &= -.856 z_1^{(1)} + 1.106 z_2^{(1)} \\ \hat{V}_2 &= -2.648 z_1^{(2)} + 2.475 z_2^{(2)} \end{aligned}$$

Here $z_i^{(1)}$, $i = 1, 2$ and $z_i^{(2)}$, $i = 1, 2$ are the standardized data values for sets 1 and 2, respectively. The preceding results were taken from the SAS statistical software output shown in Panel 10.1. In addition, the correlations of the original variables with the canonical variables are highlighted in that panel. ∎

Example 10.5 (Canonical correlation analysis of job satisfaction) As part of a larger study of the effects of organizational structure on "job satisfaction," Dunham [4] investigated the extent to which measures of job satisfaction are related to job characteristics. Using a survey instrument, Dunham obtained measurements of $p = 5$ job characteristics and $q = 7$ job satisfaction variables for $n = 784$ executives from the corporate branch of a large retail merchandising corporation. Are measures of job satisfaction associated with job characteristics? The answer may have implications for job design.

PANEL 10.1   SAS ANALYSIS FOR EXAMPLE 10.4 USING PROC CANCORR.

```
title 'Canonical Correlation Analysis';
data skull (type = corr);
_type_ = 'CORR';
input _name_$ x1 x2 x3 x4;
cards;
x1   1.0      .        .        .
x2   .505    1.0       .        .
x3   .569    .422    1.0        .
x4   .602    .467    .926     1.0
;
proc cancorr data = skull vprefix = head wprefix = leg;
    var x1 x2; with x3 x4;
```

PROGRAM COMMANDS

(continues on next page)

PANEL 10.1   (*continued*)

Canonical Correlation Analysis

	Canonical Correlation	Adjusted Canonical Correlation	Approx Standard Error	Squared Canonical Correlation
1	0.631085	0.628291	0.036286	0.398268
2	0.056794		0.060108	0.003226

Raw Canonical Coefficient for the 'VAR' Variables

	HEAD1	HEAD2
X1	0.7807924389	−0.855973184
X2	0.3445068301	1.1061835145

OUTPUT

Raw Canonical Coefficient for the 'WITH' Variables

	LEG1	LEG2
X3	0.0602508775	−2.648156338
X4	0.943948961	2.4749388913

Canonical Structure

Correlations Between the 'VAR' Variables and Their Canonical Variables

	HEAD1	HEAD2	
X1	0.9548	−0.2974	(see 10-29)
X2	0.7388	0.6739	

Correlations Between the 'WITH' Variables and Their Canonical Variables

	LEG1	LEG2	
X3	0.9343	−0.3564	(see 10-29)
X4	0.9997	0.0227	

Correlations Between the 'VAR' Variables
and the Canonical Variables of the 'WITH' Variables

	LEG1	LEG2	
X1	0.6025	−0.0169	(see 10-29)
X2	0.4663	0.0383	

Correlations Between the 'WITH' Variables
and the Canonical Variables of the 'VAR' Variables

	HEAD1	HEAD2	
X3	0.5897	−0.0202	(see 10-29)
X4	0.6309	0.0013	

The original job characteristic variables, $\mathbf{X}^{(1)}$, and job satisfaction variables, $\mathbf{X}^{(2)}$, were respectively defined as

$$\mathbf{X}^{(1)} = \begin{bmatrix} X_1^{(1)} \\ X_2^{(1)} \\ X_3^{(1)} \\ X_4^{(1)} \\ X_5^{(1)} \end{bmatrix} = \begin{bmatrix} \text{feedback} \\ \text{task significance} \\ \text{task variety} \\ \text{task identity} \\ \text{autonomy} \end{bmatrix}$$

$$\mathbf{X}^{(2)} = \begin{bmatrix} X_1^{(2)} \\ X_2^{(2)} \\ X_3^{(2)} \\ X_4^{(2)} \\ X_5^{(2)} \\ X_6^{(2)} \\ X_7^{(2)} \end{bmatrix} = \begin{bmatrix} \text{supervisor satisfaction} \\ \text{career-future satisfaction} \\ \text{financial satisfaction} \\ \text{workload satisfaction} \\ \text{company identification} \\ \text{kind-of-work-satisfaction} \\ \text{general satisfaction} \end{bmatrix}$$

Responses for variables $\mathbf{X}^{(1)}$ and $\mathbf{X}^{(2)}$ were recorded on a scale and then standardized. The sample correlation matrix based on 784 responses is

$$\mathbf{R} = \begin{bmatrix} \mathbf{R}_{11} & \mathbf{R}_{12} \\ \mathbf{R}_{21} & \mathbf{R}_{22} \end{bmatrix}$$

$$= \begin{bmatrix}
1.0 & & & & & .33 & .32 & .20 & .19 & .30 & .37 & .21 \\
.49 & 1.0 & & & & .30 & .21 & .16 & .08 & .27 & .35 & .20 \\
.53 & .57 & 1.0 & & & .31 & .23 & .14 & .07 & .24 & .37 & .18 \\
.49 & .46 & .48 & 1.0 & & .24 & .22 & .12 & .19 & .21 & .29 & .16 \\
.51 & .53 & .57 & .57 & 1.0 & .38 & .32 & .17 & .23 & .32 & .36 & .27 \\
.33 & .30 & .31 & .24 & .38 & 1.0 & & & & & & \\
.32 & .21 & .23 & .22 & .32 & .43 & 1.0 & & & & & \\
.20 & .16 & .14 & .12 & .17 & .27 & .33 & 1.0 & & & & \\
.19 & .08 & .07 & .19 & .23 & .24 & .26 & .25 & 1.0 & & & \\
.30 & .27 & .24 & .21 & .32 & .34 & .54 & .46 & .28 & 1.0 & & \\
.37 & .35 & .37 & .29 & .36 & .37 & .32 & .29 & .30 & .35 & 1.0 & \\
.21 & .20 & .18 & .16 & .27 & .40 & .58 & .45 & .27 & .59 & .31 & 1.0
\end{bmatrix}$$

The $\min(p, q) = \min(5, 7) = 5$ sample canonical correlations and the sample canonical variate coefficient vectors (from Dunham [4]) are displayed in the following table:

Canonical Variate Coefficients and Canonical Correlations

	Standardized variables						
	$z_1^{(1)}$	$z_2^{(1)}$	$z_3^{(1)}$	$z_4^{(1)}$	$z_5^{(1)}$	$\hat{\rho}_1^*$	
$\hat{\mathbf{a}}_1'$:	.42	.21	.17	−.02	.44	.55	
$\hat{\mathbf{a}}_2'$:	−.30	.65	.85	−.29	−.81	.23	
$\hat{\mathbf{a}}_3'$:	−.86	.47	−.19	−.49	.95	.12	
$\hat{\mathbf{a}}_4'$:	.76	−.06	−.12	−1.14	−.25	.08	
$\hat{\mathbf{a}}_5'$:	.27	1.01	−1.04	.16	.32	.05	

	Standardized variables						
	$z_1^{(2)}$	$z_2^{(2)}$	$z_3^{(2)}$	$z_4^{(2)}$	$z_5^{(2)}$	$z_6^{(2)}$	$z_7^{(2)}$
$\hat{\mathbf{b}}_1'$:	.42	.22	−.03	.01	.29	.52	−.12
$\hat{\mathbf{b}}_2'$:	.03	−.42	.08	−.91	.14	.59	−.02
$\hat{\mathbf{b}}_3'$:	.58	−.76	−.41	−.07	.19	−.43	.92
$\hat{\mathbf{b}}_4'$:	.23	.49	.52	−.47	.34	−.69	−.37
$\hat{\mathbf{b}}_5'$:	−.52	−.63	.41	.21	.76	.02	.10

For example, the first sample canonical variate pair is

$$\hat{U}_1 = .42z_1^{(1)} + .21z_2^{(1)} + .17z_3^{(1)} - .02z_4^{(1)} + .44z_5^{(1)}$$

$$\hat{V}_1 = .42z_1^{(2)} + .22z_2^{(2)} - .03z_3^{(2)} + .01z_4^{(2)} + .29z_5^{(2)} + .52z_6^{(2)} - .12z_7^{(2)}$$

with sample canonical correlation $\widehat{\rho_1^*} = .55$.

According to the coefficients, $\hat{U}_1$ is primarily a feedback and autonomy variable, while $\hat{V}_1$ represents supervisor, career-future, and kind-of-work satisfaction, along with company identification.

To provide interpretations for $\hat{U}_1$ and $\hat{V}_1$, the sample correlations between $\hat{U}_1$ and its component variables and between $\hat{V}_1$ and its component variables were computed. Also, the following table shows the sample correlations between variables in one set and the first sample canonical variate of the other set. These correlations can be calculated using (10-29).

Sample Correlations Between Original Variables and Canonical Variables

$\mathbf{X}^{(1)}$ variables	Sample canonical variates $\hat{U}_1$	$\hat{V}_1$	$\mathbf{X}^{(2)}$ variables	Sample canonical variates $\hat{U}_1$	$\hat{V}_1$
1. Feedback	.83	.46	1. Supervisor satisfaction	.42	.75
2. Task significance	.74	.41	2. Career-future satisfaction	.35	.65
3. Task variety	.75	.42	3. Financial satisfaction	.21	.39
4. Task identity	.62	.34	4. Workload satisfaction	.21	.37
5. Autonomy	.85	.48	5. Company identification	.36	.65
			6. Kind-of-work satisfaction	.44	.80
			7. General satisfaction	.28	.50

All five job characteristic variables have roughly the same correlations with the first canonical variate $\hat{U}_1$. From this standpoint, $\hat{U}_1$ might be interpreted as a job characteristic "index." This differs from the preferred interpretation, based on coefficients, where the task variables are not important.

The other member of the first canonical variate pair, $\hat{V}_1$, seems to be representing, primarily, supervisor satisfaction, career-future satisfaction, company identification, and kind-of-work satisfaction. As the variables suggest, $\hat{V}_1$ might be regarded as a job satisfaction–company identification index. This agrees with the preceding interpretation based on the canonical coefficients of the $z_i^{(2)}$'s. The sample correlation between the two indices $\hat{U}_1$ and $\hat{V}_1$ is $\widehat{\rho_1^*} = .55$. There appears to be some overlap between job characteristics and job satisfaction. We explore this issue further in Example 10.7. ∎

Scatter plots of the first $(\hat{U}_1, \hat{V}_1)$ pair may reveal atypical observations $\mathbf{x}_j$ requiring further study. If the canonical correlations $\widehat{\rho_2^*}, \widehat{\rho_3^*}, \ldots$ are also moderately large, scatter plots of the pairs $(\hat{U}_2, \hat{V}_2), (\hat{U}_3, \hat{V}_3), \ldots$ may also be helpful in this respect. Many analysts suggest plotting "significant" canonical variates against their component variables as an aid in subject-matter interpretation. These plots reinforce the correlation coefficients in (10-29).

If the sample size is large, it is often desirable to split the sample in half. The first half of the sample can be used to construct and evaluate the sample canonical variates and canonical correlations. The results can then be "validated" with the remaining observations. The change (if any) in the nature of the canonical analysis will provide an indication of the sampling variability and the stability of the conclusions.

10.5 Additional Sample Descriptive Measures

If the canonical variates are "good" summaries of their respective sets of variables, then the associations between variables can be described in terms of the canonical variates and their correlations. It is useful to have summary measures of the extent to which the canonical variates account for the variation in their respective sets. It is also useful, on occasion, to calculate the proportion of variance in one set of variables explained by the canonical variates of the other set.

Matrices of Errors of Approximations

Given the matrices $\hat{\mathbf{A}}$ and $\hat{\mathbf{B}}$ defined in (10-27), let $\hat{\mathbf{a}}^{(i)}$ and $\hat{\mathbf{b}}^{(i)}$ denote the ith column of $\hat{\mathbf{A}}^{-1}$ and $\hat{\mathbf{B}}^{-1}$, respectively. Since $\hat{\mathbf{U}} = \hat{\mathbf{A}}\mathbf{x}^{(1)}$ and $\hat{\mathbf{V}} = \hat{\mathbf{B}}\mathbf{x}^{(2)}$ we can write

$$\underset{(p\times 1)}{\mathbf{x}^{(1)}} = \underset{(p\times p)}{\hat{\mathbf{A}}^{-1}} \underset{(p\times 1)}{\hat{\mathbf{U}}} \qquad \underset{(q\times 1)}{\mathbf{x}^{(2)}} = \underset{(q\times q)}{\hat{\mathbf{B}}^{-1}} \underset{(q\times 1)}{\hat{\mathbf{V}}} \qquad (10\text{-}31)$$

Because sample $\text{Cov}(\hat{\mathbf{U}}, \hat{\mathbf{V}}) = \hat{\mathbf{A}}\mathbf{S}_{12}\hat{\mathbf{B}}'$, sample $\text{Cov}(\hat{\mathbf{U}}) = \hat{\mathbf{A}}\mathbf{S}_{11}\hat{\mathbf{A}}' = \underset{(p\times p)}{\mathbf{I}}$, and sample $\text{Cov}(\hat{\mathbf{V}}) = \hat{\mathbf{B}}\mathbf{S}_{22}\hat{\mathbf{B}}' = \underset{(q\times q)}{\mathbf{I}}$,

$$\mathbf{S}_{12} = \hat{\mathbf{A}}^{-1} \begin{bmatrix} \widehat{\rho_1^*} & 0 & \cdots & 0 \\ 0 & \widehat{\rho_2^*} & \cdots & 0 \\ \vdots & \vdots & \ddots & \vdots \\ 0 & 0 & \cdots & \widehat{\rho_p^*} \end{bmatrix} \mathbf{0} \, (\hat{\mathbf{B}}^{-1})' = \widehat{\rho_1^*}\hat{\mathbf{a}}^{(1)}\hat{\mathbf{b}}^{(1)\prime} + \widehat{\rho_2^*}\hat{\mathbf{a}}^{(2)}\hat{\mathbf{b}}^{(2)\prime}$$

$$+ \cdots + \widehat{\rho_p^*}\hat{\mathbf{a}}^{(p)}\hat{\mathbf{b}}^{(p)\prime} \quad (10\text{-}32)$$

$$\mathbf{S}_{11} = (\hat{\mathbf{A}}^{-1})(\hat{\mathbf{A}}^{-1})' = \hat{\mathbf{a}}^{(1)}\hat{\mathbf{a}}^{(1)\prime} + \hat{\mathbf{a}}^{(2)}\hat{\mathbf{a}}^{(2)\prime} + \cdots + \hat{\mathbf{a}}^{(p)}\hat{\mathbf{a}}^{(p)\prime}$$

$$\mathbf{S}_{22} = (\hat{\mathbf{B}}^{-1})(\hat{\mathbf{B}}^{-1})' = \hat{\mathbf{b}}^{(1)}\hat{\mathbf{b}}^{(1)\prime} + \hat{\mathbf{b}}^{(2)}\hat{\mathbf{b}}^{(2)\prime} + \cdots + \hat{\mathbf{b}}^{(q)}\hat{\mathbf{b}}^{(q)\prime}$$

Since $\mathbf{x}^{(1)} = \hat{\mathbf{A}}^{-1}\hat{\mathbf{U}}$ and $\hat{\mathbf{U}}$ has sample covariance $\mathbf{I}$, the first r columns of $\hat{\mathbf{A}}^{-1}$ contain the sample covariances of the first r canonical variates $\hat{U}_1, \hat{U}_2, \ldots, \hat{U}_r$ with their component variables $X_1^{(1)}, X_2^{(1)}, \ldots, X_p^{(1)}$. Similarly, the first r columns of $\hat{\mathbf{B}}^{-1}$ contain the sample covariances of $\hat{V}_1, \hat{V}_2, \ldots, \hat{V}_r$ with their component variables.

If only the first r canonical pairs are used, so that for instance,

$$\widetilde{\mathbf{x}}^{(1)} = [\hat{\mathbf{a}}^{(1)} \;\vdots\; \hat{\mathbf{a}}^{(2)} \;\vdots\; \cdots \;\vdots\; \hat{\mathbf{a}}^{(r)}] \begin{bmatrix} \hat{U}_1 \\ \hat{U}_2 \\ \vdots \\ \hat{U}_r \end{bmatrix}$$

and (10-33)

$$\widetilde{\mathbf{x}}^{(2)} = [\hat{\mathbf{b}}^{(1)} \;\vdots\; \hat{\mathbf{b}}^{(2)} \;\vdots\; \cdots \;\vdots\; \hat{\mathbf{b}}^{(r)}] \begin{bmatrix} \hat{V}_1 \\ \hat{V}_2 \\ \vdots \\ \hat{V}_r \end{bmatrix}$$

then $\mathbf{S}_{12}$ is approximated by sample $\text{Cov}(\widetilde{\mathbf{x}}^{(1)}, \widetilde{\mathbf{x}}^{(2)})$.

Continuing, we see that the *matrices of errors of approximation* are

$$\mathbf{S}_{11} - (\hat{\mathbf{a}}^{(1)}\hat{\mathbf{a}}^{(1)\prime} + \hat{\mathbf{a}}^{(2)}\hat{\mathbf{a}}^{(2)\prime} + \cdots + \hat{\mathbf{a}}^{(r)}\hat{\mathbf{a}}^{(r)\prime}) = \hat{\mathbf{a}}^{(r+1)}\hat{\mathbf{a}}^{(r+1)\prime} + \cdots + \hat{\mathbf{a}}^{(p)}\hat{\mathbf{a}}^{(p)\prime}$$

$$\mathbf{S}_{22} - (\hat{\mathbf{b}}^{(1)}\hat{\mathbf{b}}^{(1)\prime} + \hat{\mathbf{b}}^{(2)}\hat{\mathbf{b}}^{(2)\prime} + \cdots + \hat{\mathbf{b}}^{(r)}\hat{\mathbf{b}}^{(r)\prime}) = \hat{\mathbf{b}}^{(r+1)}\hat{\mathbf{b}}^{(r+1)\prime} + \cdots + \hat{\mathbf{b}}^{(q)}\hat{\mathbf{b}}^{(q)\prime}$$

$$\mathbf{S}_{12} - (\widehat{\rho_1^*}\hat{\mathbf{a}}^{(1)}\hat{\mathbf{b}}^{(1)\prime} + \widehat{\rho_2^*}\hat{\mathbf{a}}^{(2)}\hat{\mathbf{b}}^{(2)\prime} + \cdots + \widehat{\rho_r^*}\hat{\mathbf{a}}^{(r)}\hat{\mathbf{b}}^{(r)\prime})$$

$$= \widehat{\rho_{r+1}^*}\hat{\mathbf{a}}^{(r+1)}\hat{\mathbf{b}}^{(r+1)\prime} + \cdots + \widehat{\rho_p^*}\hat{\mathbf{a}}^{(p)}\hat{\mathbf{b}}^{(p)\prime}$$

(10-34)

The approximation error matrices (10-34) may be interpreted as descriptive summaries of how well the first r sample canonical variates reproduce the sample covariance matrices. Patterns of large entries in the rows and/or columns of the approximation error matrices indicate a poor "fit" to the corresponding variable(s).

Ordinarily, the first r variates do a better job of reproducing the elements of $\mathbf{S}_{12} = \mathbf{S}_{21}'$ than the elements of $\mathbf{S}_{11}$ or $\mathbf{S}_{22}$. Mathematically, this occurs because the residual matrix in the former case is directly related to the smallest $p - r$ sample canonical correlations. These correlations are usually all close to zero. On the other hand, the residual matrices associated with the approximations to the matrices $\mathbf{S}_{11}$ and $\mathbf{S}_{22}$ depend only on the last $p - r$ and $q - r$ coefficient vectors. The elements in these vectors may be relatively large, and hence, the residual matrices can have "large" entries.

For standardized observations, $\mathbf{R}_{kl}$ replaces $\mathbf{S}_{kl}$ and $\hat{\mathbf{a}}_{\mathbf{z}}^{(k)}$, $\hat{\mathbf{b}}_{\mathbf{z}}^{(l)}$ replace $\hat{\mathbf{a}}^{(k)}$, $\hat{\mathbf{b}}^{(l)}$ in (10-34).

Example 10.6 (Calculating matrices of errors of approximation) In Example 10.4, we obtained the canonical correlations between the two head and the two leg variables for white leghorn fowl. Starting with the sample correlation matrix

$$\mathbf{R} = \left[\begin{array}{c|c} \mathbf{R}_{11} & \mathbf{R}_{12} \\ \hline \mathbf{R}_{21} & \mathbf{R}_{22} \end{array}\right] = \left[\begin{array}{cc|cc} 1.0 & .505 & .569 & .602 \\ .505 & 1.0 & .422 & .467 \\ \hline .569 & .422 & 1.0 & .926 \\ .602 & .467 & .926 & 1.0 \end{array}\right]$$

we obtained the two sets of canonical correlations and variables

$$\widehat{\rho_1^*} = .631 \qquad \begin{array}{l} \hat{U}_1 = .781z_1^{(1)} + .345z_2^{(1)} \\ \hat{V}_1 = .060z_1^{(2)} + .944z_2^{(2)} \end{array}$$

and

$$\widehat{\rho_2^*} = .057 \qquad \begin{array}{l} \hat{U}_2 = -.856z_1^{(1)} + 1.106z_2^{(1)} \\ \hat{V}_2 = -2.648z_1^{(2)} + 2.475z_2^{(2)} \end{array}$$

where $z_i^{(1)}$, $i = 1, 2$ and $z_i^{(2)}$, $i = 1, 2$ are the standardized data values for sets 1 and 2, respectively.

We first calculate (see Panel 10.1)

$$\hat{\mathbf{A}}_z^{-1} = \begin{bmatrix} .781 & .345 \\ -.856 & 1.106 \end{bmatrix}^{-1} = \begin{bmatrix} .9548 & -.2974 \\ .7388 & .6739 \end{bmatrix}$$

$$\hat{\mathbf{B}}_z^{-1} = \begin{bmatrix} .9343 & -.3564 \\ .9997 & .0227 \end{bmatrix}$$

Consequently, the matrices of errors of approximation created by using only the first canonical pair are

$$\mathbf{R}_{12} - \text{sample Cov}(\tilde{\mathbf{z}}^{(1)}, \tilde{\mathbf{z}}^{(2)}) = (.057) \begin{bmatrix} -.2974 \\ .6739 \end{bmatrix} [-.3564 \quad .0227]$$

$$= \begin{bmatrix} .006 & -.000 \\ -.014 & .001 \end{bmatrix}$$

$$\mathbf{R}_{11} - \text{sample Cov}(\tilde{\mathbf{z}}^{(1)}) = \begin{bmatrix} -.2974 \\ .6739 \end{bmatrix} [-.2974 \quad .6739]$$

$$= \begin{bmatrix} .088 & -.200 \\ -.200 & .454 \end{bmatrix}$$

$$\mathbf{R}_{22} - \text{sample Cov}(\tilde{\mathbf{z}}^{(2)}) = \begin{bmatrix} -.3564 \\ .0227 \end{bmatrix} [-.3564 \quad .0227]$$

$$= \begin{bmatrix} .127 & -.008 \\ -.008 & .001 \end{bmatrix}$$

where $\tilde{\mathbf{z}}^{(1)}, \tilde{\mathbf{z}}^{(2)}$ are given by (10-33) with $r = 1$ and $\hat{\mathbf{a}}_z^{(1)}, \hat{\mathbf{b}}_z^{(1)}$ replace $\hat{\mathbf{a}}^{(1)}, \hat{\mathbf{b}}^{(1)}$, respectively.

We see that the first pair of canonical variables effectively summarizes (repro-duces) the intraset correlations in $\mathbf{R}_{12}$. However, the individual variates are not particularly effective summaries of the sampling variability in the original $\mathbf{z}^{(1)}$ and $\mathbf{z}^{(2)}$ sets, respectively. This is especially true for $\hat{U}_1$. ∎

Proportions of Explained Sample Variance

When the observations are standardized, the sample covariance matrices $\mathbf{S}_{kl}$ are correlation matrices $\mathbf{R}_{kl}$. The canonical coefficient vectors are the *rows* of the matrices $\hat{\mathbf{A}}_\mathbf{z}$ and $\hat{\mathbf{B}}_\mathbf{z}$ and the *columns* of $\hat{\mathbf{A}}_\mathbf{z}^{-1}$ and $\hat{\mathbf{B}}_\mathbf{z}^{-1}$ are the sample correlations between the canonical variates and their component variables.

Specifically,

$$\text{sample Cov}(\mathbf{z}^{(1)}, \hat{\mathbf{U}}) = \text{sample Cov}(\hat{\mathbf{A}}_\mathbf{z}^{-1}\hat{\mathbf{U}}, \hat{\mathbf{U}}) = \hat{\mathbf{A}}_\mathbf{z}^{-1}$$

and

$$\text{sample Cov}(\mathbf{z}^{(2)}, \hat{\mathbf{V}}) = \text{sample Cov}(\hat{\mathbf{B}}_\mathbf{z}^{-1}\hat{\mathbf{V}}, \hat{\mathbf{V}}) = \hat{\mathbf{B}}_\mathbf{z}^{-1}$$

so

$$\hat{\mathbf{A}}_\mathbf{z}^{-1} = [\hat{\mathbf{a}}_\mathbf{z}^{(1)}, \hat{\mathbf{a}}_\mathbf{z}^{(2)}, \ldots, \hat{\mathbf{a}}_\mathbf{z}^{(p)}] = \begin{bmatrix} r_{\hat{U}_1, z_1^{(1)}} & r_{\hat{U}_2, z_1^{(1)}} & \cdots & r_{\hat{U}_p, z_1^{(1)}} \\ r_{\hat{U}_1, z_2^{(1)}} & r_{\hat{U}_2, z_2^{(1)}} & \cdots & r_{\hat{U}_p, z_2^{(1)}} \\ \vdots & \vdots & \ddots & \vdots \\ r_{\hat{U}_1, z_p^{(1)}} & r_{\hat{U}_2, z_p^{(1)}} & \cdots & r_{\hat{U}_p, z_p^{(1)}} \end{bmatrix}$$

$$\hat{\mathbf{B}}_\mathbf{z}^{-1} = [\hat{\mathbf{b}}_\mathbf{z}^{(1)}, \hat{\mathbf{b}}_\mathbf{z}^{(2)}, \ldots, \hat{\mathbf{b}}_\mathbf{z}^{(q)}] = \begin{bmatrix} r_{\hat{V}_1, z_1^{(2)}} & r_{\hat{V}_2, z_1^{(2)}} & \cdots & r_{\hat{V}_q, z_1^{(2)}} \\ r_{\hat{V}_1, z_2^{(2)}} & r_{\hat{V}_2, z_2^{(2)}} & \cdots & r_{\hat{V}_q, z_2^{(2)}} \\ \vdots & \vdots & \ddots & \vdots \\ r_{\hat{V}_1, z_q^{(2)}} & r_{\hat{V}_2, z_q^{(2)}} & \cdots & r_{\hat{V}_q, z_q^{(2)}} \end{bmatrix} \qquad (10\text{-}35)$$

where $r_{\hat{U}_i, z_k^{(1)}}$ and $r_{\hat{V}_i, z_k^{(2)}}$ are the sample correlation coefficients between the quantities with subscripts.

Using (10-32) with standardized observations, we obtain

Total (standardized) sample variance in first set

$$= \text{tr}(\mathbf{R}_{11}) = \text{tr}(\hat{\mathbf{a}}_\mathbf{z}^{(1)}\hat{\mathbf{a}}_\mathbf{z}^{(1)\prime} + \hat{\mathbf{a}}_\mathbf{z}^{(2)}\hat{\mathbf{a}}_\mathbf{z}^{(2)\prime} + \cdots + \hat{\mathbf{a}}_\mathbf{z}^{(p)}\hat{\mathbf{a}}_\mathbf{z}^{(p)\prime}) = p \qquad (10\text{-}36a)$$

Total (standardized) sample variance in second set

$$= \text{tr}(\mathbf{R}_{22}) = \text{tr}(\hat{\mathbf{b}}_\mathbf{z}^{(1)}\hat{\mathbf{b}}_\mathbf{z}^{(1)\prime} + \hat{\mathbf{b}}_\mathbf{z}^{(2)}\hat{\mathbf{b}}_\mathbf{z}^{(2)\prime} + \cdots + \hat{\mathbf{b}}_\mathbf{z}^{(q)}\hat{\mathbf{b}}_\mathbf{z}^{(q)\prime}) = q \qquad (10\text{-}36b)$$

Since the correlations in the first $r < p$ columns of $\hat{\mathbf{A}}_\mathbf{z}^{-1}$ and $\hat{\mathbf{B}}_\mathbf{z}^{-1}$ involve only the sample canonical variates $\hat{U}_1, \hat{U}_2, \ldots, \hat{U}_r$ and $\hat{V}_1, \hat{V}_2, \ldots, \hat{V}_r$, respectively, we define the contributions of the first r canonical variates to the total (standardized) sample variances as

$$\text{tr}(\hat{\mathbf{a}}_{\mathbf{z}}^{(1)}\hat{\mathbf{a}}_{\mathbf{z}}^{(1)\prime} + \hat{\mathbf{a}}_{\mathbf{z}}^{(2)}\hat{\mathbf{a}}_{\mathbf{z}}^{(2)\prime} + \cdots + \hat{\mathbf{a}}_{\mathbf{z}}^{(r)}\hat{\mathbf{a}}_{\mathbf{z}}^{(r)\prime}) = \sum_{i=1}^{r}\sum_{k=1}^{p} r_{\hat{U}_i, z_k^{(1)}}^2$$

and

$$\text{tr}(\hat{\mathbf{b}}_{\mathbf{z}}^{(1)}\hat{\mathbf{b}}_{\mathbf{z}}^{(1)\prime} + \hat{\mathbf{b}}_{\mathbf{z}}^{(2)}\hat{\mathbf{b}}_{\mathbf{z}}^{(2)\prime} + \cdots + \hat{\mathbf{b}}_{\mathbf{z}}^{(r)}\hat{\mathbf{b}}_{\mathbf{z}}^{(r)\prime}) = \sum_{i=1}^{r}\sum_{k=1}^{p} r_{\hat{V}_i, z_k^{(2)}}^2$$

The *proportions* of total (standardized) sample variances "explained by" the first r canonical variates then become

$$R_{\mathbf{z}^{(1)}|\hat{U}_1,\hat{U}_2,\dots,\hat{U}_r}^2 = \begin{pmatrix} \text{proportion of total standardized} \\ \text{sample variance in first set} \\ \text{explained by } \hat{U}_1,\hat{U}_2,\dots,\hat{U}_r \end{pmatrix}$$

$$= \frac{\text{tr}(\hat{\mathbf{a}}_{\mathbf{z}}^{(1)}\hat{\mathbf{a}}_{\mathbf{z}}^{(1)\prime} + \cdots + \hat{\mathbf{a}}_{\mathbf{z}}^{(r)}\hat{\mathbf{a}}_{\mathbf{z}}^{(r)\prime})}{\text{tr}(\mathbf{R}_{11})}$$

$$= \frac{\sum_{i=1}^{r}\sum_{k=1}^{p} r_{\hat{U}_i, z_k^{(1)}}^2}{p} \tag{10-37}$$

and

$$R_{\mathbf{z}^{(2)}|\hat{V}_1,\hat{V}_2,\dots,\hat{V}_r}^2 = \begin{pmatrix} \text{proportion of total standardized} \\ \text{sample variance in second set} \\ \text{explained by } \hat{V}_1,\hat{V}_2,\dots,\hat{V}_r \end{pmatrix}$$

$$= \frac{\text{tr}(\hat{\mathbf{b}}_{\mathbf{z}}^{(1)}\hat{\mathbf{b}}_{\mathbf{z}}^{(1)\prime} + \cdots + \hat{\mathbf{b}}_{\mathbf{z}}^{(r)}\hat{\mathbf{b}}_{\mathbf{z}}^{(r)\prime})}{\text{tr}(\mathbf{R}_{22})}$$

$$= \frac{\sum_{i=1}^{r}\sum_{k=1}^{q} r_{\hat{V}_i, z_k^{(2)}}^2}{q}$$

Descriptive measures (10-37) provide some indication of how well the canonical variates represent their respective sets. They provide single-number descriptions of the matrices of errors. In particular,

$$\frac{1}{p}\text{tr}[\mathbf{R}_{11} - \hat{\mathbf{a}}_{\mathbf{z}}^{(1)}\hat{\mathbf{a}}_{\mathbf{z}}^{(1)\prime} - \hat{\mathbf{a}}_{\mathbf{z}}^{(2)}\hat{\mathbf{a}}_{\mathbf{z}}^{(2)\prime} - \cdots - \hat{\mathbf{a}}_{\mathbf{z}}^{(r)}\hat{\mathbf{a}}_{\mathbf{z}}^{(r)\prime}] = 1 - R_{\mathbf{z}^{(1)}|\hat{U}_1,\hat{U}_2,\dots,\hat{U}_r}^2$$

$$\frac{1}{q}\text{tr}[\mathbf{R}_{22} - \hat{\mathbf{b}}_{\mathbf{z}}^{(1)}\hat{\mathbf{b}}_{\mathbf{z}}^{(1)\prime} - \hat{\mathbf{b}}_{\mathbf{z}}^{(2)}\hat{\mathbf{b}}_{\mathbf{z}}^{(2)\prime} - \cdots - \hat{\mathbf{b}}_{\mathbf{z}}^{(r)}\hat{\mathbf{b}}_{\mathbf{z}}^{(r)\prime}] = 1 - R_{\mathbf{z}^{(2)}|\hat{V}_1,\hat{V}_2,\dots,\hat{V}_r}^2$$

according to (10-36) and (10-37).

Example 10.7 (Calculating proportions of sample variance explained by canonical variates) Consider the job characteristic–job satisfaction data discussed in Example 10.5. Using the table of sample correlation coefficients presented in that example, we find that

$$R^2_{\mathbf{z}^{(1)}|\hat{U}_1} = \frac{1}{5} \sum_{k=1}^{5} r^2_{\hat{U}_1, z_k^{(1)}} = \frac{1}{5}[(.83)^2 + (.74)^2 + \cdots + (.85)^2] = .58$$

$$R^2_{\mathbf{z}^{(2)}|\hat{V}_1} = \frac{1}{7} \sum_{k=1}^{7} r^2_{\hat{V}_1, z_k^{(2)}} = \frac{1}{7}[(.75)^2 + (.65)^2 + \cdots + (.50)^2] = .37$$

The first sample canonical variate $\hat{U}_1$ of the job characteristics set accounts for 58% of the set's total sample variance. The first sample canonical variate $\hat{V}_1$ of the job satisfaction set explains 37% of the set's total sample variance. We might thus infer that $\hat{U}_1$ is a "better" representative of its set than $\hat{V}_1$ is of its set. The interested reader may wish to see how well $\hat{U}_1$ and $\hat{V}_1$ reproduce the correlation matrices $\mathbf{R}_{11}$ and $\mathbf{R}_{22}$, respectively. [See (10-29).] ∎

10.6 Large Sample Inferences

When $\Sigma_{12} = \mathbf{0}$, $\mathbf{a}'\mathbf{X}^{(1)}$ and $\mathbf{b}'\mathbf{X}^{(2)}$ have covariance $\mathbf{a}'\Sigma_{12}\mathbf{b} = 0$ for all vectors $\mathbf{a}$ and $\mathbf{b}$. Consequently, all the canonical correlations must be zero, and there is no point in pursuing a canonical correlation analysis. The next result provides a way of testing $\Sigma_{12} = \mathbf{0}$, for large samples.

Result 10.3. Let

$$\mathbf{X}_j = \begin{bmatrix} \mathbf{X}_j^{(1)} \\ \hline \mathbf{X}_j^{(2)} \end{bmatrix}, \qquad j = 1, 2, \ldots, n$$

be a random sample from an $N_{p+q}(\boldsymbol{\mu}, \Sigma)$ population with

$$\Sigma = \begin{bmatrix} \Sigma_{11} & \Sigma_{12} \\ {\scriptstyle(p\times p)} & {\scriptstyle(p\times q)} \\ \hline \Sigma_{21} & \Sigma_{22} \\ {\scriptstyle(q\times p)} & {\scriptstyle(q\times q)} \end{bmatrix}$$

Then the likelihood ratio test of $H_0: \underset{(p\times q)}{\Sigma_{12}} = \underset{}{\mathbf{0}}$ versus $H_1: \underset{(p\times q)}{\Sigma_{12}} \neq \mathbf{0}$ rejects H_0 for large values of

$$-2 \ln \Lambda = n \ln \left(\frac{|\mathbf{S}_{11}||\mathbf{S}_{22}|}{|\mathbf{S}|} \right) = -n \ln \prod_{i=1}^{p} (1 - \widehat{\rho_i^{*2}}) \qquad (10\text{-}38)$$

where

$$\mathbf{S} = \left[\begin{array}{c|c} \mathbf{S}_{11} & \mathbf{S}_{12} \\ \hline \mathbf{S}_{21} & \mathbf{S}_{22} \end{array}\right]$$

is the unbiased estimator of Σ. For large n, the test statistic (10-38) is approximately distributed as a chi-square random variable with pq d.f.

Proof. See Kshirsagar [8]. ∎

The likelihood ratio statistic (10-38) compares the sample generalized variance under H_0, namely,

$$\begin{vmatrix} \mathbf{S}_{11} & \mathbf{0} \\ \mathbf{0}' & \mathbf{S}_{22} \end{vmatrix} = |\mathbf{S}_{11}||\mathbf{S}_{22}|$$

with the unrestricted generalized variance $|\mathbf{S}|$.

Bartlett [3] suggests replacing the multiplicative factor n in the likelihood ratio statistic with the factor $n - 1 - \frac{1}{2}(p + q + 1)$ to improve the χ^2 approximation to the sampling distribution of $-2 \ln \Lambda$. Thus, for n and $n - (p + q)$ large, we

Reject $H_0: \Sigma_{12} = \mathbf{0}$ $(\rho_1^* = \rho_2^* = \cdots = \rho_p^* = 0)$ at significance level α if

$$-\left(n - 1 - \frac{1}{2}(p + q + 1)\right) \ln \prod_{i=1}^{p} (1 - \widehat{\rho_i^{*2}}) > \chi_{pq}^2(\alpha) \tag{10-39}$$

where $\chi_{pq}^2(\alpha)$ is the upper (100α)th percentile of a chi-square distribution with pq d.f.

If the null hypothesis $H_0: \Sigma_{12} = \mathbf{0}$ $(\rho_1^* = \rho_2^* = \cdots = \rho_p^* = 0)$ is rejected, it is natural to examine the "significance" of the individual canonical correlations. Since the canonical correlations are ordered from the largest to the smallest, we can begin by assuming that the first canonical correlation is nonzero and the remaining $p - 1$ canonical correlations are zero. If this hypothesis is rejected, we assume that the first two canonical correlations are nonzero, but the remaining $p - 2$ canonical correlations are zero, and so forth.

Let the implied sequence of hypotheses be

$$H_0^k: \rho_1^* \neq 0, \rho_2^* \neq 0, \ldots, \rho_k^* \neq 0, \rho_{k+1}^* = \cdots = \rho_p^* = 0$$

$$\tag{10-40}$$

$$H_1^k: \rho_i^* \neq 0, \text{ for some } i \geq k + 1$$

Bartlett [2] has argued that the kth hypothesis in (10-40) can be tested by the likelihood ratio criterion. Specifically,

Reject $H_0^{(k)}$ at significance level α if

$$-\left(n - 1 - \frac{1}{2}(p + q + 1)\right) \ln \prod_{i=k+1}^{p} (1 - \widehat{\rho_i^{*2}}) > \chi^2_{(p-k)(q-k)}(\alpha) \qquad (10\text{-}41)$$

where $\chi^2_{(p-k)(q-k)}(\alpha)$ is the upper (100α)th percentile of a chi-square distribution with $(p - k)(q - k)$ d.f. We point out that the test statistic in (10-41) involves $\prod_{i=k+1}^{p} (1 - \widehat{\rho_i^{*2}})$, the "residual" after the first k sample canonical correlations have been removed from the total criterion $\Lambda^{2/n} = \prod_{i=1}^{p} (1 - \widehat{\rho_i^{*2}})$.

If the members of the sequence H_0, $H_0^{(1)}$, $H_0^{(2)}$, and so forth, are tested one at a time until $H_0^{(k)}$ is not rejected for some k, the overall significance level is not α and, in fact, would be difficult to determine. Another defect of this procedure is the tendency it induces to conclude that a null hypothesis is correct simply because it is not rejected.

To summarize, the overall test of significance in Result 10.3 is useful for multivariate normal data. The sequential tests implied by (10-41) should be interpreted with caution and are, perhaps, best regarded as rough guides for selecting the number of important canonical variates.

Example 10.8 (Testing the significance of the canonical correlations for the job satisfaction data) Test the significance of the canonical correlations exhibited by the job characteristics–job satisfaction data introduced in Example 10.5.

All the test statistics of immediate interest are summarized in the table on page 566. From Example 10.5, $n = 784$, $p = 5$, $q = 7$, $\widehat{\rho_1^*} = .55$, $\widehat{\rho_2^*} = .23$, $\widehat{\rho_3^*} = .12$, $\widehat{\rho_4^*} = .08$, and $\widehat{\rho_5^*} = .05$.

Assuming multivariate normal data, we find that the first two canonical correlations, ρ_1^* and ρ_2^*, appear to be nonzero, although with the very large sample size, small deviations from zero will show up as statistically significant. From a practical point of view, the second (and subsequent) sample canonical correlations can probably be ignored, since (1) they are reasonably small in magnitude and (2) the corresponding canonical variates explain *very* little of the sample variation in the variable sets $\mathbf{X}^{(1)}$ and $\mathbf{X}^{(2)}$. ∎

The distribution theory associated with the sample canonical correlations and the sample canonical variate coefficients is extremely complex (apart from the $p = 1$ and $q = 1$ situations), even in the null case, $\boldsymbol{\Sigma}_{12} = \mathbf{0}$. The reader interested in the distribution theory is referred to Kshirsagar [8].

Test Results

Null hypothesis	Observed test statistic (Barlett correction)	Degrees of freedom	Upper 1% point of χ^2 distribution	Conclusion
1. $H_0: \boldsymbol{\Sigma}_{12} = \mathbf{0}$ (all $\rho_i^* = 0$)	$-\left(n - 1 - \dfrac{1}{2}(p + q + 1)\right)\ln\displaystyle\prod_{i=1}^{5}(1 - \widehat{\rho}_i^{*2})$ $= -\left(784 - 1 - \dfrac{1}{2}(5 + 7 + 1)\right)\ln(.6453)$ $= 340.1$	$pq = 5(7) = 35$	$\chi^2_{35}(.01) = 57$	Reject H_0.
2. $H_0^{(1)}: \rho_1^* \neq 0,$ $\rho_2^* = \cdots = \rho_5^* = 0$	$-\left(n - 1 - \dfrac{1}{2}(p + q + 1)\right)\ln\displaystyle\prod_{i=2}^{5}(1 - \widehat{\rho}_i^{*2})$ $= 60.4$	$(p - 1)(q - 1) = 24$	$\chi^2_{24}(.01) = 42.98$	Reject H_0.
3. $H_0^{(2)}: \rho_1^* \neq 0, \rho_2^* \neq 0,$ $\rho_3^* = \cdots = \rho_5^* = 0$	$-\left(n - 1 - \dfrac{1}{2}(p + q + 1)\right)\ln\displaystyle\prod_{i=3}^{5}(1 - \widehat{\rho}_i^{*2})$ $= 18.2$	$(p - 2)(q - 2) = 15$	$\chi^2_{15}(.01) = 30.58$	Do not reject H_0.

Exercises

10.1. Consider the covariance matrix given in Example 10.3:

$$\text{Cov}\left(\begin{bmatrix} X_1^{(1)} \\ X_2^{(1)} \\ \hline X_1^{(2)} \\ X_2^{(2)} \end{bmatrix}\right) = \begin{bmatrix} \boldsymbol{\Sigma}_{11} & \boldsymbol{\Sigma}_{12} \\ \hline \boldsymbol{\Sigma}_{21} & \boldsymbol{\Sigma}_{22} \end{bmatrix} = \begin{bmatrix} 100 & 0 & 0 & 0 \\ 0 & 1 & .95 & 0 \\ \hline 0 & .95 & 1 & 0 \\ 0 & 0 & 0 & 100 \end{bmatrix}$$

Verify that the first pair of canonical variates are $U_1 = X_2^{(1)}$, $V_1 = X_1^{(2)}$ with canonical correlation $\rho_1^* = .95$.

10.2. The (2×1) random vectors $\mathbf{X}^{(1)}$ and $\mathbf{X}^{(2)}$ have the joint mean vector and joint covariance matrix

$$\boldsymbol{\mu} = \begin{bmatrix} \boldsymbol{\mu}^{(1)} \\ \hline \boldsymbol{\mu}^{(2)} \end{bmatrix} = \begin{bmatrix} -3 \\ 2 \\ \hline 0 \\ 1 \end{bmatrix};$$

$$\boldsymbol{\Sigma} = \begin{bmatrix} \boldsymbol{\Sigma}_{11} & \boldsymbol{\Sigma}_{12} \\ \hline \boldsymbol{\Sigma}_{21} & \boldsymbol{\Sigma}_{22} \end{bmatrix} = \begin{bmatrix} 8 & 2 & 3 & 1 \\ 2 & 5 & -1 & 3 \\ \hline 3 & -1 & 6 & -2 \\ 1 & 3 & -2 & 7 \end{bmatrix}$$

(a) Calculate the canonical correlations ρ_1^*, ρ_2^*.

(b) Determine the canonical variate pairs (U_1, V_1) and (U_2, V_2).

(c) Let $\mathbf{U} = [U_1, U_2]'$ and $\mathbf{V} = [V_1, V_2]'$. From first principles, evaluate

$$E\left(\begin{bmatrix} \mathbf{U} \\ \hline \mathbf{V} \end{bmatrix}\right) \quad \text{and} \quad \text{Cov}\left(\begin{bmatrix} \mathbf{U} \\ \hline \mathbf{V} \end{bmatrix}\right) = \begin{bmatrix} \boldsymbol{\Sigma}_{UU} & \boldsymbol{\Sigma}_{UV} \\ \hline \boldsymbol{\Sigma}_{VU} & \boldsymbol{\Sigma}_{VV} \end{bmatrix}$$

Compare your results with the properties in Result 10.1.

10.3. Let $\mathbf{Z}^{(1)} = \mathbf{V}_{11}^{-1/2}(\mathbf{X}^{(1)} - \boldsymbol{\mu}^{(1)})$ and $\mathbf{Z}^{(2)} = \mathbf{V}_{22}^{-1/2}(\mathbf{X}^{(2)} - \boldsymbol{\mu}^{(2)})$ be two sets of standardized variables. If $\rho_1^*, \rho_2^*, \ldots, \rho_p^*$ are the canonical correlations for the $\mathbf{X}^{(1)}, \mathbf{X}^{(2)}$ sets and $(U_i, V_i) = (\mathbf{a}_i'\mathbf{X}^{(1)}, \mathbf{b}_i'\mathbf{X}^{(2)})$, $i = 1, 2, \ldots, p$, are the associated canonical variates, determine the canonical correlations and canonical variates for the $\mathbf{Z}^{(1)}, \mathbf{Z}^{(2)}$ sets. That is, express the canonical correlations and canonical variate coefficient vectors for the $\mathbf{Z}^{(1)}$, $\mathbf{Z}^{(2)}$ sets in terms of those for the $\mathbf{X}^{(1)}, \mathbf{X}^{(2)}$ sets.

10.4. *(Alternative calculation of canonical correlations and variates.)* Show that, if λ_i is an eigenvalue of $\boldsymbol{\Sigma}_{11}^{-1/2}\boldsymbol{\Sigma}_{12}\boldsymbol{\Sigma}_{22}^{-1}\boldsymbol{\Sigma}_{21}\boldsymbol{\Sigma}_{11}^{-1/2}$ with associated eigenvector $\mathbf{e}_i$, then λ_i is also an eigenvalue of $\boldsymbol{\Sigma}_{11}^{-1}\boldsymbol{\Sigma}_{12}\boldsymbol{\Sigma}_{22}^{-1}\boldsymbol{\Sigma}_{21}$ with eigenvector $\boldsymbol{\Sigma}_{11}^{-1/2}\mathbf{e}_i$.

Hint: $|\boldsymbol{\Sigma}_{11}^{-1/2}\boldsymbol{\Sigma}_{12}\boldsymbol{\Sigma}_{22}^{-1}\boldsymbol{\Sigma}_{21}\boldsymbol{\Sigma}_{11}^{-1/2} - \lambda_i\mathbf{I}| = 0$ implies that

$$0 = |\boldsymbol{\Sigma}_{11}^{-1/2}||\boldsymbol{\Sigma}_{11}^{-1/2}\boldsymbol{\Sigma}_{12}\boldsymbol{\Sigma}_{22}^{-1}\boldsymbol{\Sigma}_{21}\boldsymbol{\Sigma}_{11}^{-1/2} - \lambda_i\mathbf{I}||\boldsymbol{\Sigma}_{11}^{1/2}|$$

$$= |\boldsymbol{\Sigma}_{11}^{-1}\boldsymbol{\Sigma}_{12}\boldsymbol{\Sigma}_{22}^{-1}\boldsymbol{\Sigma}_{21} - \lambda_i\mathbf{I}|$$

10.5. Use the information in Example 10.1.

(a) Find the eigenvalues of $\Sigma_{11}^{-1}\Sigma_{12}\Sigma_{22}^{-1}\Sigma_{21}$ and verify that these eigenvalues are the same as the eigenvalues of $\Sigma_{11}^{-1/2}\Sigma_{12}\Sigma_{22}^{-1}\Sigma_{21}\Sigma_{11}^{-1/2}$.

(b) Determine the second pair of canonical variates (U_2, V_2) and verify, from first principles, that their correlation is the second canonical correlation $\rho_2^* = .03$.

10.6. Show that the canonical correlations are invariant under nonsingular linear transformations of the $\mathbf{X}^{(1)}, \mathbf{X}^{(2)}$ variables of the form $\underset{(p\times p)}{\mathbf{C}}\ \underset{(p\times 1)}{\mathbf{X}^{(1)}}$ and $\underset{(q\times q)}{\mathbf{D}}\ \underset{(q\times 1)}{\mathbf{X}^{(2)}}$.

Hint: Consider $\text{Cov}\left(\left[\dfrac{\mathbf{CX}^{(1)}}{\mathbf{DX}^{(2)}}\right]\right) = \left[\begin{array}{c|c} \mathbf{C}\Sigma_{11}\mathbf{C}' & \mathbf{C}\Sigma_{12}\mathbf{D}' \\ \hline \mathbf{D}\Sigma_{21}\mathbf{C}' & \mathbf{D}\Sigma_{22}\mathbf{D}' \end{array}\right]$. Consider any linear combination $\mathbf{a}_1'(\mathbf{CX}^{(1)}) = \mathbf{a}'\mathbf{X}^{(1)}$ with $\mathbf{a}' = \mathbf{a}_1'\mathbf{C}$. Similarly, consider $\mathbf{b}_1'(\mathbf{DX}^{(2)}) = \mathbf{b}'\mathbf{X}^{(2)}$ with $\mathbf{b}' = \mathbf{b}_1'\mathbf{D}$. The choices $\mathbf{a}_1' = \mathbf{e}'\Sigma_{11}^{-1/2}\mathbf{C}^{-1}$ and $\mathbf{b}_1' = \mathbf{f}'\Sigma_{22}^{-1/2}\mathbf{D}^{-1}$ give the maximum correlation.

10.7. Let $\boldsymbol{\rho}_{12} = \begin{bmatrix} \rho & \rho \\ \rho & \rho \end{bmatrix}$ and $\boldsymbol{\rho}_{11} = \boldsymbol{\rho}_{22} = \begin{bmatrix} 1 & \rho \\ \rho & 1 \end{bmatrix}$, corresponding to the equal correlation structure where $\mathbf{X}^{(1)}$ and $\mathbf{X}^{(2)}$ each have two components.

(a) Determine the canonical variates corresponding to the nonzero canonical correlation.

(b) Generalize the results in Part a to the case where $\mathbf{X}^{(1)}$ has p components and $\mathbf{X}^{(2)}$ has $q \geq p$ components.

Hint: $\boldsymbol{\rho}_{12} = \rho \mathbf{1}\mathbf{1}'$, where $\mathbf{1}$ is a $(p \times 1)$ column vector of 1's and $\mathbf{1}'$ is a $(q \times 1)$ row vector of 1's. Note that $\boldsymbol{\rho}_{11}\mathbf{1} = [1 + (p-1)\rho]\mathbf{1}$ so $\boldsymbol{\rho}_{11}^{-1/2}\mathbf{1} = [1 + (p-1)\rho]^{-1/2}\mathbf{1}$.

10.8. *(Correlation for angular measurement.)* Some observations, such as wind direction, are in the form of angles. An angle θ_2 can be represented as the pair $\mathbf{X}^{(2)} = [\cos(\theta_2), \sin(\theta_2)]'$.

(a) Show that $\mathbf{b}'\mathbf{X}^{(2)} = \sqrt{b_1^2 + b_2^2}\cos(\theta_2 - \beta)$ where $b_1/\sqrt{b_1^2 + b_2^2} = \cos(\beta)$ and $b_2/\sqrt{b_1^2 + b_2^2} = \sin(\beta)$.

Hint: $\cos(\theta_2 - \beta) = \cos(\theta_2)\cos(\beta) + \sin(\theta_2)\sin(\beta)$.

(b) Let $\mathbf{X}^{(1)}$ have a single component $X_1^{(1)}$. Show that the single canonical correlation is $\rho_1^* = \underset{\beta}{\max}\ \text{Corr}(X_1^{(1)}, \cos(\theta_2 - \beta))$. Selecting the canonical variable V_1 amounts to selecting a new origin β for the angle θ_2. (See Johnson and Wehrly [7].)

(c) Let $X_1^{(1)}$ be ozone (in parts per million) and $\theta_2 = $ wind direction measured from the north. Nineteen observations made in downtown Milwaukee, Wisconsin, give the sample correlation matrix

$$\mathbf{R} = \left[\begin{array}{c|c} \mathbf{R}_{11} & \mathbf{R}_{12} \\ \hline \mathbf{R}_{21} & \mathbf{R}_{22} \end{array}\right] = \begin{array}{ccc} & \text{ozone} \quad \cos(\theta_2) \quad \sin(\theta_2) \\ \left[\begin{array}{c|cc} 1.0 & .166 & .694 \\ \hline .166 & 1.0 & -.051 \\ .694 & -.051 & 1.0 \end{array}\right] \end{array}$$

Find the sample canonical correlation $\widehat{\rho_1^*}$ and the canonical variate $\hat{V}_1$ representing the new origin $\hat{\beta}$.

(d) Suppose $\mathbf{X}^{(1)}$ is also angular measurements of the form $\mathbf{X}^{(1)} = [\cos(\theta_1), \sin(\theta_1)]'$. Then $\mathbf{a}'\mathbf{X}^{(1)} = \sqrt{a_1^2 + a_2^2}\cos(\theta_1 - \alpha)$. Show that

$$\rho_1^* = \underset{\alpha, \beta}{\max}\ \text{Corr}(\cos(\theta_1 - \alpha), \cos(\theta_2 - \beta))$$

(e) Twenty-one observations on the 6:00 A.M. and noon wind directions give the correlation matrix

$$
\mathbf{R} = \begin{array}{c} \begin{array}{cccc} \cos(\theta_1) & \sin(\theta_1) & \cos(\theta_2) & \sin(\theta_2) \end{array} \\ \left[\begin{array}{cc|cc} 1.0 & -.291 & .440 & .372 \\ -.291 & 1.0 & -.205 & .243 \\ \hline .440 & -.205 & 1.0 & .181 \\ .372 & .243 & .181 & 1.0 \end{array}\right] \end{array}
$$

Find the sample canonical correlation $\widehat{\rho_1^*}$ and $\hat{U}_1, \hat{V}_1$.

The following exercises may require a computer.

10.9. H. Hotelling [5] reports that $n = 140$ seventh-grade children received four tests on $X_1^{(1)} = $ reading speed, $X_2^{(1)} = $ reading power, $X_1^{(2)} = $ arithmetic speed, and $X_2^{(2)} = $ arithmetic power. The correlations for performance are

$$
\mathbf{R} = \left[\begin{array}{c|c} \mathbf{R}_{11} & \mathbf{R}_{12} \\ \hline \mathbf{R}_{21} & \mathbf{R}_{22} \end{array}\right] = \left[\begin{array}{cc|cc} 1.0 & .6328 & .2412 & .0586 \\ .6328 & 1.0 & -.0553 & .0655 \\ \hline .2412 & -.0553 & 1.0 & .4248 \\ .0586 & .0655 & .4248 & 1.0 \end{array}\right]
$$

(a) Find all the sample canonical correlations and the sample canonical variates.

(b) Stating any assumptions you make, test the hypotheses

$$
\begin{aligned}
H_0 &: \boldsymbol{\Sigma}_{12} = \boldsymbol{\rho}_{12} = \mathbf{0} \qquad (\rho_1^* = \rho_2^* = 0) \\
H_1 &: \boldsymbol{\Sigma}_{12} = \boldsymbol{\rho}_{12} \neq \mathbf{0}
\end{aligned}
$$

at the $\alpha = .05$ level of significance. If H_0 is rejected, test

$$
\begin{aligned}
H_0^{(1)} &: \rho_1^* \neq 0, \rho_2^* = 0 \\
H_1^{(1)} &: \rho_2^* \neq 0
\end{aligned}
$$

with a significance level of $\alpha = .05$. Does reading ability (as measured by the two tests) correlate with arithmetic ability (as measured by the two tests)? Discuss.

(c) Evaluate the matrices of approximation errors for $\mathbf{R}_{11}$, $\mathbf{R}_{22}$, and $\mathbf{R}_{12}$ determined by the first sample canonical variate pair $\hat{U}_1, \hat{V}_1$.

10.10. In a study of poverty, crime, and deterrence, Parker and Smith [10] report certain summary crime statistics in various states for the years 1970 and 1973. A portion of their sample correlation matrix is

$$
\mathbf{R} = \left[\begin{array}{c|c} \mathbf{R}_{11} & \mathbf{R}_{12} \\ \hline \mathbf{R}_{21} & \mathbf{R}_{22} \end{array}\right] = \left[\begin{array}{cc|cc} 1.0 & .615 & -.111 & -.266 \\ .615 & 1.0 & -.195 & -.085 \\ \hline -.111 & -.195 & 1.0 & -.269 \\ -.266 & -.085 & -.269 & 1.0 \end{array}\right]
$$

The variables are

$X_1^{(1)} = $ 1973 nonprimary homicides

$X_2^{(1)} = $ 1973 primary homicides (homicides involving family or acquaintances)

$X_1^{(2)} = $ 1970 severity of punishment (median months served)

$X_2^{(2)} = $ 1970 certainty of punishment (number of admissions to prison divided by number of homicides)

(a) Find the sample canonical correlations.

(b) Determine the first canonical pair $\hat{U}_1$, $\hat{V}_1$ and interpret these quantities.

10.11. Example 8.5 presents the correlation matrix obtained from $n = 103$ successive weekly rates of return for five stocks. Perform a canonical correlation analysis with $\mathbf{X}^{(1)} = [X_1^{(1)}, X_2^{(1)}, X_3^{(1)}]'$, the rates of return for the banks, and $\mathbf{X}^{(2)} = [X_1^{(2)}, X_2^{(2)}]'$, the rates of return for the oil companies.

10.12. A random sample of $n = 70$ families will be surveyed to determine the association between certain "demographic" variables and certain "consumption" variables. Let

$$
\begin{array}{ll}
\text{Criterion} & \begin{cases} X_1^{(1)} = \text{annual frequency of dining at a restaurant} \\ X_2^{(1)} = \text{annual frequency of attending movies} \end{cases} \\[2em]
\text{set} & \\[1em]
\text{Predictor} & \begin{cases} X_1^{(2)} = \text{age of head of household} \\ X_2^{(2)} = \text{annual family income} \\ X_3^{(2)} = \text{educational level of head of household} \end{cases} \\
\text{set} &
\end{array}
$$

Suppose 70 observations on the preceding variables give the sample correlation matrix

$$
\mathbf{R} = \begin{bmatrix} \mathbf{R}_{11} & \mathbf{R}_{12} \\ \hline \mathbf{R}_{21} & \mathbf{R}_{22} \end{bmatrix} = \begin{bmatrix}
1.0 & & & & \\
.80 & 1.0 & & & \\
\hline
.26 & .33 & 1.0 & & \\
.67 & .59 & .37 & 1.0 & \\
.34 & .34 & .21 & .35 & 1.0
\end{bmatrix}
$$

(a) Determine the sample canonical correlations, and test the hypothesis $H_0: \Sigma_{12} = \mathbf{0}$ (or, equivalently, $\boldsymbol{\rho}_{12} = \mathbf{0}$) at the $\alpha = .05$ level. If H_0 is rejected, test for the significance ($\alpha = .05$) of the first canonical correlation.

(b) Using standardized variables, construct the canonical variates corresponding to the "significant" canonical correlation(s).

(c) Using the results in Parts a and b, prepare a table showing the canonical variate coefficients (for "significant" canonical correlations) and the sample correlations of the canonical variates with their component variables.

(d) Given the information in (c), interpret the canonical variates.

(e) Do the demographic variables have something to say about the consumption variables? Do the consumption variables provide much information about the demographic variables?

10.13. Waugh [12] provides information about $n = 138$ samples of Canadian hard red spring wheat and the flour made from the samples. The $p = 5$ wheat measurements (in standardized form) were

$$
\begin{aligned}
z_1^{(1)} &= \text{kernel texture} \\
z_2^{(1)} &= \text{test weight} \\
z_3^{(1)} &= \text{damaged kernels} \\
z_4^{(1)} &= \text{foreign material} \\
z_5^{(1)} &= \text{crude protein in the wheat}
\end{aligned}
$$

The $q = 4$ (standardized) flour measurements were

$$z_1^{(2)} = \text{wheat per barrel of flour}$$

$$z_2^{(2)} = \text{ash in flour}$$

$$z_3^{(2)} = \text{crude protein in flour}$$

$$z_4^{(2)} = \text{gluten quality index}$$

The sample correlation matrix was

$$\mathbf{R} = \begin{bmatrix} \mathbf{R}_{11} & \mathbf{R}_{12} \\ \hline \mathbf{R}_{21} & \mathbf{R}_{22} \end{bmatrix}$$

$$= \begin{bmatrix}
1.0 & & & & & & & & \\
.754 & 1.0 & & & & & & & \\
-.690 & -.712 & 1.0 & & & & & & \\
-.446 & -.515 & .323 & 1.0 & & & & & \\
.692 & .412 & -.444 & -.334 & 1.0 & & & & \\
-.605 & -.722 & .737 & .527 & -.383 & 1.0 & & & \\
-.479 & -.419 & .361 & .461 & -.505 & .251 & 1.0 & & \\
.780 & .542 & -.546 & -.393 & .737 & -.490 & -.434 & 1.0 & \\
-.152 & -.102 & .172 & -.019 & -.148 & .250 & -.079 & -.163 & 1.0
\end{bmatrix}$$

(a) Find the sample canonical variates corresponding to significant (at the $\alpha = .01$ level) canonical correlations.

(b) Interpret the first sample canonical variates $\hat{U}_1$, $\hat{V}_1$. Do they in some sense represent the overall quality of the wheat and flour, respectively?

(c) What proportion of the total sample variance of the first set $\mathbf{Z}^{(1)}$ is explained by the canonical variate $\hat{U}_1$? What proportion of the total sample variance of the $\mathbf{Z}^{(2)}$ set is explained by the canonical variate $\hat{V}_1$? Discuss your answers.

10.14. Consider the correlation matrix of profitability measures given in Exercise 9.15. Let $\mathbf{X}^{(1)} = [X_1^{(1)}, X_2^{(1)}, \ldots, X_6^{(1)}]'$ be the vector of variables representing accounting measures of profitability, and let $\mathbf{X}^{(2)} = [X_1^{(2)}, X_2^{(2)}]'$ be the vector of variables representing the two market measures of profitability. Partition the sample correlation matrix accordingly, and perform a canonical correlation analysis. Specifically,

(a) Determine the first sample canonical variates $\hat{U}_1$, $\hat{V}_1$ and their correlation. Interpret these canonical variates.

(b) Let $\mathbf{Z}^{(1)}$ and $\mathbf{Z}^{(2)}$ be the sets of standardized variables corresponding to $\mathbf{X}^{(1)}$ and $\mathbf{X}^{(2)}$, respectively. What proportion of the total sample variance of $\mathbf{Z}^{(1)}$ is explained by the canonical variate $\hat{U}_1$? What proportion of the total sample variance of $\mathbf{Z}^{(2)}$ is explained by the canonical variate $\hat{V}_1$? Discuss your answers.

10.15. Observations on four measures of stiffness are given in Table 4.3 and discussed in Example 4.14. Use the data in the table to construct the sample covariance matrix $\mathbf{S}$. Let $\mathbf{X}^{(1)} = [X_1^{(1)}, X_2^{(1)}]'$ be the vector of variables representing the dynamic measures of stiffness (shock wave, vibration), and let $\mathbf{X}^{(2)} = [X_1^{(2)}, X_2^{(2)}]'$ be the vector of variables representing the static measures of stiffness. Perform a canonical correlation analysis of these data.

10.16. Andrews and Herzberg [1] give data obtained from a study of a comparison of nondiabetic and diabetic patients. Three primary variables,

$$X_1^{(1)} = \text{glucose intolerance}$$
$$X_2^{(1)} = \text{insulin response to oral glucose}$$
$$X_3^{(1)} = \text{insulin resistance}$$

and two secondary variables,

$$X_1^{(2)} = \text{relative weight}$$
$$X_2^{(2)} = \text{fasting plasma glucose}$$

were measured. The data for $n = 46$ nondiabetic patients yield the covariance matrix

$$\mathbf{S} = \begin{bmatrix} \mathbf{S}_{11} & \mathbf{S}_{12} \\ \hline \mathbf{S}_{21} & \mathbf{S}_{22} \end{bmatrix} = \begin{bmatrix} 1106.000 & 396.700 & 108.400 & .787 & 26.230 \\ 396.700 & 2382.000 & 1143.000 & -.214 & -23.960 \\ 108.400 & 1143.000 & 2136.000 & 2.189 & -20.840 \\ \hline .787 & -.214 & 2.189 & .016 & .216 \\ 26.230 & -23.960 & -20.840 & .216 & 70.560 \end{bmatrix}$$

Determine the sample canonical variates and their correlations. Interpret these quantities. Are the first canonical variates good summary measures of their respective sets of variables? Explain. Test for the significance of the canonical relations with $\alpha = .05$.

10.17. Data concerning a person's desire to smoke and psychological and physical state were collected for $n = 110$ subjects. The data were responses, coded 1 to 5, to each of 12 questions (variables). The four standardized measurements related to the desire to smoke are defined as

$$z_1^{(1)} = \text{smoking 1 (first wording)}$$
$$z_2^{(1)} = \text{smoking 2 (second wording)}$$
$$z_3^{(1)} = \text{smoking 3 (third wording)}$$
$$z_4^{(1)} = \text{smoking 4 (fourth wording)}$$

The eight standardized measurements related to the psychological and physical state are given by

$$z_1^{(2)} = \text{concentration}$$
$$z_2^{(2)} = \text{annoyance}$$
$$z_3^{(2)} = \text{sleepiness}$$
$$z_4^{(2)} = \text{tenseness}$$
$$z_5^{(2)} = \text{alertness}$$
$$z_6^{(2)} = \text{irritability}$$
$$z_7^{(2)} = \text{tiredness}$$
$$z_8^{(2)} = \text{contentedness}$$

The correlation matrix constructed from the data is

$$\mathbf{R} = \begin{bmatrix} \mathbf{R}_{11} & \mathbf{R}_{12} \\ \hline \mathbf{R}_{21} & \mathbf{R}_{22} \end{bmatrix}$$

where

$$\mathbf{R}_{11} = \begin{bmatrix} 1.000 & .785 & .810 & .775 \\ .785 & 1.000 & .816 & .813 \\ .810 & .816 & 1.000 & .845 \\ .775 & .813 & .845 & 1.000 \end{bmatrix}$$

$$\mathbf{R}_{12} = \mathbf{R}_{21}' = \begin{bmatrix} .086 & .144 & .140 & .222 & .101 & .189 & .199 & .239 \\ .200 & .119 & .211 & .301 & .223 & .221 & .274 & .235 \\ .041 & .060 & .126 & .120 & .039 & .108 & .139 & .100 \\ .228 & .122 & .277 & .214 & .201 & .156 & .271 & .171 \end{bmatrix}$$

$$\mathbf{R}_{22} = \begin{bmatrix} 1.000 & .562 & .457 & .579 & .802 & .595 & .512 & .492 \\ .562 & 1.000 & .360 & .705 & .578 & .796 & .413 & .739 \\ .457 & .360 & 1.000 & .273 & .606 & .337 & .798 & .240 \\ .579 & .705 & .273 & 1.000 & .594 & .725 & .364 & .711 \\ .802 & .578 & .606 & .594 & 1.000 & .605 & .698 & .605 \\ .595 & .796 & .337 & .725 & .605 & 1.000 & .428 & .697 \\ .512 & .413 & .798 & .364 & .698 & .428 & 1.000 & .394 \\ .492 & .739 & .240 & .711 & .605 & .697 & .394 & 1.000 \end{bmatrix}$$

Determine the sample canonical variates and their correlations. Interpret these quantities. Are the first canonical variates good summary measures of their respective sets of variables? Explain.

10.18. The data in Table 7.7 contain measurements on characteristics of pulp fibers and the paper made from them. To correspond with the notation in this chapter, let the paper characteristics be

$$x_1^{(1)} = \text{breaking length}$$
$$x_2^{(1)} = \text{elastic modulus}$$
$$x_3^{(1)} = \text{stress at failure}$$
$$x_4^{(1)} = \text{burst strength}$$

and the pulp fiber characteristics be

$$x_1^{(2)} = \text{arithmetic fiber length}$$
$$x_2^{(2)} = \text{long fiber fraction}$$
$$x_3^{(2)} = \text{fine fiber fraction}$$
$$x_4^{(2)} = \text{zero span tensile}$$

Determine the sample canonical variates and their correlations. Are the first canonical variates good summary measures of their respective sets of variables? Explain. Test for the significance of the canonical relations with $\alpha = .05$. Interpret the significant canonical variables.

10.19. Refer to the correlation matrix for the Olympic decathlon results in Example 9.6. Obtain the canonical correlations between the results for the running speed events (100-meter run, 400-meter run, long jump) and the arm strength events (discus, javelin, shot put). Recall that the signs of standardized running events values were reversed so that large scores are best for all events.

References

1. Andrews, D.F., and A. M. Herzberg. *Data*. New York: Springer-Verlag, 1985.

2. Bartlett, M.S. "Further Aspects of the Theory of Multiple Regression." *Proceedings of the Cambridge Philosophical Society*, **34** (1938), 33–40.

3. Bartlett, M. S. "A Note on Tests of Significance in Multivariate Analysis." *Proceedings of the Cambridge Philosophical Society*, **35** (1939), 180–185.

4. Dunham, R.B. "Reaction to Job Characteristics: Moderating Effects of the Organization." *Academy of Management Journal*, **20**, no. 1 (1977), 42–65.

5. Hotelling, H. "The Most Predictable Criterion." *Journal of Educational Psychology*, **26** (1935), 139–142.

6. Hotelling, H. "Relations between Two Sets of Variables." *Biometrika*, **28** (1936), 321–377.

7. Johnson, R. A., and T. Wehrly. "Measures and Models for Angular Correlation and Angular-Linear Correlation." *Journal of the Royal Statistical Society (B)*, **39** (1977), 222–229.

8. Kshirsagar, A. M. *Multivariate Analysis*. New York: Marcel Dekker, Inc., 1972.

9. Lawley, D. N. "Tests of Significance in Canonical Analysis." *Biometrika*, **46** (1959), 59–66.

10. Parker, R. N., and M. D. Smith. "Deterrence, Poverty, and Type of Homicide." *American Journal of Sociology*, **85** (1979), 614–624.

11. Rencher, A. C. "Interpretation of Canonical Discriminant Functions, Canonical Variates and Principal Components." *The American Statistician*, **46** (1992), 217–225.

12. Waugh, F. W. "Regression between Sets of Variates." *Econometrica*, **10** (1942), 290–310.

Chapter 11

DISCRIMINATION AND CLASSIFICATION

11.1 Introduction

Discrimination and classification are multivariate techniques concerned with *separating* distinct sets of objects (or observations) and with *allocating* new objects (observations) to previously defined groups. Discriminant analysis is rather exploratory in nature. As a separative procedure, it is often employed on a one-time basis in order to investigate observed differences when causal relationships are not well understood. Classification procedures are less exploratory in the sense that they lead to well-defined rules, which can be used for assigning new objects. Classification ordinarily requires more problem structure than discrimination does.

Thus, the immediate goals of discrimination and classification, respectively, are as follows:

Goal 1. To describe, either graphically (in three or fewer dimensions) or algebraically, the differential features of objects (observations) from several known collections (populations). We try to find "discriminants" whose numerical values are such that the collections are separated as much as possible.

Goal 2. To sort objects (observations) into two or more labeled classes. The emphasis is on deriving a rule that can be used to optimally assign *new* objects to the labeled classes.

We shall follow convention and use the term *discrimination* to refer to Goal 1. This terminology was introduced by R. A. Fisher [10] in the first modern treatment of separative problems. A more descriptive term for this goal, however, is *separation*. We shall refer to the second goal as *classification* or *allocation*.

A function that separates objects may sometimes serve as an allocator, and, conversely, a rule that allocates objects may suggest a discriminatory procedure. In practice, Goals 1 and 2 frequently overlap, and the distinction between separation and allocation becomes blurred.

11.2 Separation and Classification for Two Populations

To fix ideas, let us list situations in which one may be interested in (1) separating two classes of objects or (2) assigning a new object to one of two classes (or both). It is convenient to label the classes π_1 and π_2. The objects are ordinarily separated or classified on the basis of measurements on, for instance, p associated random variables $\mathbf{X}' = [X_1, X_2, \ldots, X_p]$. The observed values of $\mathbf{X}$ differ to some extent from one class to the other.[1] We can think of the totality of values from the first class as being the population of $\mathbf{x}$ values for π_1 and those from the second class as the population of $\mathbf{x}$ values for π_2. These two populations can then be described by probability density functions $f_1(\mathbf{x})$ and $f_2(\mathbf{x})$, and consequently, we can talk of assigning observations to populations or objects to classes interchangeably.

You may recall that some of the examples of the following separation–classification situations were introduced in Chapter 1.

Populations π_1 and π_2	Measured variables $\mathbf{X}$
1. Solvent and distressed property-liability insurance companies.	Total assets, cost of stocks and bonds, market value of stocks and bonds, loss expenses, surplus, amount of premiums written.
2. Nonulcer dyspeptics (those with upset stomach problems) and controls ("normal").	Measures of anxiety, dependence, guilt, perfectionism.
3. *Federalist Papers* written by James Madison and those written by Alexander Hamilton.	Frequencies of different words and lengths of sentences.
4. Two species of chickweed.	Sepal and petal length, petal cleft depth, bract length, scarious tip length, pollen diameter.
5. Purchasers of a new product and laggards (those "slow" to purchase).	Education, income, family size, amount of previous brand switching.
6. Successful or unsuccessful (fail to graduate) college students.	Entrance examination scores, high school grade-point average, number of high school activities.
7. Males and females.	Anthropological measurements, like circumference and volume on ancient skulls.
8. Good and poor credit risks.	Income, age, number of credit cards, family size.
9. Alcoholics and nonalcoholics.	Activity of monoamine oxidase enzyme, activity of adenylate cyclase enzyme.

We see from item 5, for example, that objects (consumers) are to be separated into two labeled classes ("purchasers" and "laggards") on the basis of observed values of presumably relevant variables (education, income, and so forth). In the terminology of *observation* and *population*, we want to identify an observation of

---

[1]If the values of $\mathbf{X}$ were not very different for objects in π_1 and π_2, there would be no problem; that is, the classes would be indistinguishable, and new objects could be assigned to either class indiscriminately.

the form $\mathbf{x}' = [x_1(\text{education}),\ x_2(\text{income}),\ x_3(\text{family size}),\ x_4(\text{amount of brand switching})]$ as population π_1, purchasers, or population π_2, laggards.

At this point, we shall concentrate on classification for two populations, returning to separation in Section 11.3.

Allocation or classification rules are usually developed from "learning" samples. Measured characteristics of randomly selected objects *known* to come from each of the two populations are examined for differences. Essentially, the set of all possible sample outcomes is divided into two regions, R_1 and R_2, such that if a *new* observation falls in R_1, it is allocated to population π_1, and if it falls in R_2, we allocate it to population π_2. Thus, one set of observed values favors π_1, while the other set of values favors π_2.

You may wonder at this point how it is we *know* that some observations belong to a particular population, but we are unsure about others. (This, of course, is what makes classification a problem!) Several conditions can give rise to this apparent anomaly (see [20]):

1. *Incomplete knowledge of future performance.*

   Examples: In the past, extreme values of certain financial variables were observed 2 years prior to a firm's subsequent bankruptcy. Classifying another firm as *sound* or *distressed* on the basis of observed values of these leading indicators may allow the officers to take corrective action, if necessary, before it is too late.

   A medical school applications office might want to classify an applicant as *likely to become M.D.* or *unlikely to become M.D.* on the basis of test scores and other college records. Here the actual determination can be made only at the end of several years of training.

2. *"Perfect" information requires destroying the object.*

   Example: The lifetime of a calculator battery is determined by using it until it fails, and the strength of a piece of lumber is obtained by loading it until it breaks. Failed products cannot be sold. One would like to classify products as *good* or *bad* (not meeting specifications) on the basis of certain preliminary measurements.

3. *Unavailable or expensive information.*

   Examples: It is assumed that certain of the *Federalist Papers* were written by James Madison or Alexander Hamilton because they signed them. Others of the *Papers*, however, were unsigned and it is of interest to determine which of the two men wrote the unsigned *Papers*. Clearly, we cannot ask them. Word frequencies and sentence lengths may help classify the disputed *Papers*.

   Many medical problems can be identified conclusively only by conducting an expensive operation. Usually, one would like to diagnose an illness from easily observed, yet potentially fallible, external symptoms. This approach helps avoid needless—and expensive—operations.

It should be clear from these examples that classification rules cannot usually provide an error-free method of assignment. This is because there may not be a clear distinction between the measured characteristics of the populations; that is, the groups may overlap. It is then possible, for example, to incorrectly classify a π_2 object as belonging to π_1 or a π_1 object as belonging to π_2.

Example 11.1 (Discriminating owners from nonowners of riding mowers) Consider two groups in a city: π_1, riding-mower owners, and π_2, those without riding mowers—that is, nonowners. In order to identify the best sales prospects for an intensive sales campaign, a riding-mower manufacturer is interested in classifying families as prospective owners or nonowners on the basis of $x_1 = $ income and $x_2 = $ lot size. Random samples of $n_1 = 12$ current owners and $n_2 = 12$ current nonowners yield the values in Table 11.1.

Table 11.1

π_1: Riding-mower owners		π_2: Nonowners	
x_1 (Income in $1000s)	x_2 (Lot size in 1000 ft^2)	x_1 (Income in $1000s)	x_2 (Lot size in 1000 ft^2)
90.0	18.4	105.0	19.6
115.5	16.8	82.8	20.8
94.8	21.6	94.8	17.2
91.5	20.8	73.2	20.4
117.0	23.6	114.0	17.6
140.1	19.2	79.2	17.6
138.0	17.6	89.4	16.0
112.8	22.4	96.0	18.4
99.0	20.0	77.4	16.4
123.0	20.8	63.0	18.8
81.0	22.0	81.0	14.0
111.0	20.0	93.0	14.8

These data are plotted in Figure 11.1. We see that riding-mower owners tend to have larger incomes and bigger lots than nonowners, although income seems to be a better "discriminator" than lot size. On the other hand, there is some overlap between the two groups. If, for example, we were to allocate those values of (x_1, x_2) that fall into region R_1 (as determined by the solid line in the figure) to π_1, mower owners, and those (x_1, x_2) values which fall into R_2 to π_2, nonowners, we would make some mistakes. Some riding-mower owners would be incorrectly classified as nonowners and, conversely, some nonowners as owners. The idea is to create a rule (regions R_1 and R_2) that minimizes the chances of making these mistakes. (See Exercise 11.2.) ∎

A good classification procedure should result in few misclassifications. In other words, the chances, or probabilities, of misclassification should be small. As we shall see, there are additional features that an "optimal" classification rule should possess.

It may be that one class or population has a greater likelihood of occurrence than another because one of the two populations is relatively much larger than the other. For example, there tend to be more financially sound firms than bankrupt firms. As another example, one species of chickweed may be more prevalent than another. An optimal classification rule should take these "prior probabilities of occurrence" into account. If we really believe that the (prior) probability of a financially distressed and ultimately bankrupted firm is very small, then one should

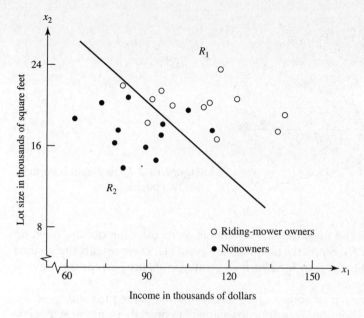

Figure 11.1 Income and lot size for riding-mower owners and nonowners.

classify a randomly selected firm as nonbankrupt unless the data overwhelmingly favors bankruptcy.

Another aspect of classification is <u>cost.</u> Suppose that classifying a π_1 object as belonging to π_2 represents a more serious error than classifying a π_2 object as belonging to π_1. Then one should be cautious about making the former assignment. As an example, failing to diagnose a potentially fatal illness is substantially more "costly" than concluding that the disease is present when, in fact, it is not. An optimal classification procedure should, whenever possible, account for the costs associated with misclassification.

Let $f_1(\mathbf{x})$ and $f_2(\mathbf{x})$ be the probability density functions associated with the $p \times 1$ vector random variable $\mathbf{X}$ for the populations π_1 and π_2, respectively. An object with associated measurements $\mathbf{x}$ *must* be assigned to either π_1 or π_2. Let Ω be the sample space—that is, the collection of all possible observations $\mathbf{x}$. Let R_1 be that set of $\mathbf{x}$ values for which we classify objects as π_1 and $R_2 = \Omega - R_1$ be the remaining $\mathbf{x}$ values for which we classify objects as π_2. Since every object must be assigned to one and only one of the two populations, the sets R_1 and R_2 are mutually exclusive and exhaustive. For $p = 2$, we might have a case like the one pictured in Figure 11.2.

The conditional probability, $P(2 \mid 1)$, of classifying an object as π_2 when, in fact, it is from π_1 is

$$P(2 \mid 1) = P(\mathbf{X} \in R_2 \mid \pi_1) = \int_{R_2 = \Omega - R_1} f_1(\mathbf{x}) \, d\mathbf{x} \qquad (11\text{-}1)$$

Similarly, the conditional probability, $P(1 \mid 2)$, of classifying an object as π_1 when it is really from π_2 is

$$P(1 \mid 2) = P(\mathbf{X} \in R_1 \mid \pi_2) = \int_{R_1} f_2(\mathbf{x}) \, d\mathbf{x} \qquad (11\text{-}2)$$

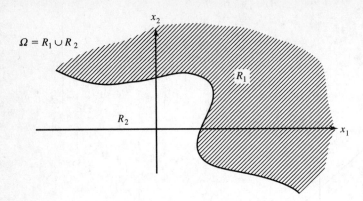

$\Omega = R_1 \cup R_2$

Figure 11.2 Classification regions for two populations.

The integral sign in (11-1) represents the volume formed by the density function $f_1(\mathbf{x})$ over the region R_2. Similarly, the integral sign in (11-2) represents the volume formed by $f_2(\mathbf{x})$ over the region R_1. This is illustrated in Figure 11.3 for the univariate case, $p = 1$.

Let p_1 be the *prior* probability of π_1 and p_2 be the *prior* probability of π_2, where $p_1 + p_2 = 1$. Then the overall probabilities of correctly or incorrectly classifying objects can be derived as the product of the prior and conditional classification probabilities:

$P(\text{observation is correctly classified as } \pi_1) = P(\text{observation comes from } \pi_1$
$\qquad\qquad\qquad\qquad\qquad\qquad\qquad\qquad \text{and is correctly classified as } \pi_1)$
$$= P(\mathbf{X} \in R_1 | \pi_1)P(\pi_1) = P(1|1)p_1$$

$P(\text{observation is misclassified as } \pi_1) = P(\text{observation comes from } \pi_2$
$\qquad\qquad\qquad\qquad\qquad\qquad\qquad\qquad \text{and is misclassified as } \pi_1)$
$$= P(\mathbf{X} \in R_1 | \pi_2)P(\pi_2) = P(1|2)p_2$$

$P(\text{observation is correctly classified as } \pi_2) = P(\text{observation comes from } \pi_2$
$\qquad\qquad\qquad\qquad\qquad\qquad\qquad\qquad \text{and is correctly classified as } \pi_2)$
$$= P(\mathbf{X} \in R_2 | \pi_2)P(\pi_2) = P(2|2)p_2$$

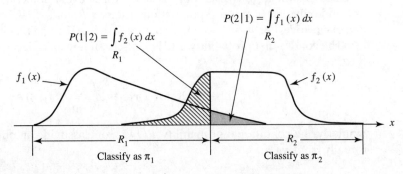

$$P(1|2) = \int_{R_1} f_2(x)\, dx$$

$$P(2|1) = \int_{R_2} f_1(x)\, dx$$

$f_1(x)$ $\qquad$ $f_2(x)$

R_1 $\qquad$ R_2

Classify as π_1 $\qquad\qquad$ Classify as π_2

Figure 11.3 Misclassification probabilities for hypothetical classification regions when $p = 1$.

$$P(\text{observation is misclassified as } \pi_2) = P(\text{observation comes from } \pi_1$$
$$\text{and is misclassified as } \pi_2)$$
$$= P(\mathbf{X} \in R_2 | \pi_1)P(\pi_1) = P(2|1)p_1$$

$$(11\text{-}3)$$

Classification schemes are often evaluated in terms of their misclassification probabilities (see Section 11.4), but this ignores misclassification cost. For example, even a seemingly small probability such as $.06 = P(2|1)$ may be too large if the cost of making an incorrect assignment to π_2 is extremely high. A rule that ignores costs may cause problems.

The costs of misclassification can be defined by a cost matrix:

		Classify as:	
		π_1	π_2
True population:	π_1	0	$c(2|1)$
	π_2	$c(1|2)$	0

$$(11\text{-}4)$$

The costs are (1) zero for correct classification, (2) $c(1|2)$ when an observation from π_2 is incorrectly classified as π_1, and (3) $c(2|1)$ when a π_1 observation is incorrectly classified as π_2.

For any rule, the average, or *expected cost of misclassification* (ECM) is provided by multiplying the off-diagonal entries in (11-4) by their probabilities of occurrence, obtained from (11-3). Consequently,

$$\text{ECM} = c(2|1)P(2|1)p_1 + c(1|2)P(1|2)p_2 \qquad (11\text{-}5)$$

optimization A reasonable classification rule should have an ECM as small, or nearly as small, as possible.

Result 11.1. The regions R_1 and R_2 that minimize the ECM are defined by the values $\mathbf{x}$ for which the following inequalities hold:

$$R_1: \quad \frac{f_1(\mathbf{x})}{f_2(\mathbf{x})} \geq \left(\frac{c(1|2)}{c(2|1)}\right)\left(\frac{p_2}{p_1}\right)$$

$$\left(\begin{array}{c}\text{density}\\\text{ratio}\end{array}\right) \geq \left(\begin{array}{c}\text{cost}\\\text{ratio}\end{array}\right)\left(\begin{array}{c}\text{prior}\\\text{probability}\\\text{ratio}\end{array}\right) \qquad (11\text{-}6)$$

$$R_2: \quad \frac{f_1(\mathbf{x})}{f_2(\mathbf{x})} < \left(\frac{c(1|2)}{c(2|1)}\right)\left(\frac{p_2}{p_1}\right)$$

$$\left(\begin{array}{c}\text{density}\\\text{ratio}\end{array}\right) < \left(\begin{array}{c}\text{cost}\\\text{ratio}\end{array}\right)\left(\begin{array}{c}\text{prior}\\\text{probability}\\\text{ratio}\end{array}\right)$$

Proof. See Exercise 11.3. ∎

It is clear from (11-6) that the implementation of the minimum ECM rule requires (1) the density function ratio evaluated at a new observation $\mathbf{x}_0$, (2) the cost ratio, and (3) the prior probability ratio. The appearance of ratios in the definition of the optimal classification regions is significant. Often, it is much easier to specify the ratios than their component parts.

For example, it may be difficult to specify the costs (in appropriate units) of classifying a student as college material when, in fact, he or she is not and classifying a student as not college material, when, in fact, he or she is. The cost to taxpayers of educating a college dropout for 2 years, for instance, can be roughly assessed. The cost to the university and society of not educating a capable student is more difficult to determine. However, it may be that a realistic number for the ratio of these mis-classification costs can be obtained. Whatever the units of measurement, not admitting a prospective college graduate may be five times more costly, over a suitable time horizon, than admitting an eventual dropout. In this case, the cost ratio is five.

It is interesting to consider the classification regions defined in (11-6) for some special cases.

Special Cases of Minimum Expected Cost Regions

(a) $p_2/p_1 = 1$ (equal prior probabilities)

$$R_1: \quad \frac{f_1(\mathbf{x})}{f_2(\mathbf{x})} \geq \frac{c(1|2)}{c(2|1)} \qquad R_2: \quad \frac{f_1(\mathbf{x})}{f_2(\mathbf{x})} < \frac{c(1|2)}{c(2|1)}$$

(b) $c(1|2)/c(2|1) = 1$ (equal misclassification costs)

$$R_1: \quad \frac{f_1(\mathbf{x})}{f_2(\mathbf{x})} \geq \frac{p_2}{p_1} \qquad R_2: \quad \frac{f_1(\mathbf{x})}{f_2(\mathbf{x})} < \frac{p_2}{p_1} \qquad (11\text{-}7)$$

(c) $p_2/p_1 = c(1|2)/c(2|1) = 1$ or $p_2/p_1 = 1/(c(1|2)/c(2|1))$
    (equal prior probabilities and equal misclassification costs)

$$R_1: \quad \frac{f_1(\mathbf{x})}{f_2(\mathbf{x})} \geq 1 \qquad R_2: \quad \frac{f_1(\mathbf{x})}{f_2(\mathbf{x})} < 1$$

When the prior probabilities are unknown, they are often taken to be equal, and the minimum ECM rule involves comparing the ratio of the population densities to the ratio of the appropriate misclassification costs. If the misclassification cost ratio is indeterminate, it is usually taken to be unity, and the population density ratio is compared with the ratio of the prior probabilities. (Note that the prior probabilities are in the reverse order of the densities.) Finally, when both the prior probability and misclassification cost ratios are unity, or one ratio is the reciprocal of the other, the optimal classification regions are determined simply by comparing the values of the density functions. In this case, if $\mathbf{x}_0$ is a new observation and $f_1(\mathbf{x}_0)/f_2(\mathbf{x}_0) \geq 1$—that is, $f_1(\mathbf{x}_0) \geq f_2(\mathbf{x}_0)$—we assign $\mathbf{x}_0$ to π_1. On the other hand, if $f_1(\mathbf{x}_0)/f_2(\mathbf{x}_0) < 1$, or $f_1(\mathbf{x}_0) < f_2(\mathbf{x}_0)$, we assign $\mathbf{x}_0$ to π_2.

It is common practice to arbitrarily use case (c) in (11-7) for classification. This is tantamount to assuming equal prior probabilities and equal misclassification costs for the minimum ECM rule.[2]

---

[2]This is the justification generally provided. It is also equivalent to assuming the prior probability ratio to be the reciprocal of the misclassification cost ratio.

Example 11.2 (Classifying a new observation into one of the two populations) A researcher has enough data available to estimate the density functions $f_1(\mathbf{x})$ and $f_2(\mathbf{x})$ associated with populations π_1 and π_2, respectively. Suppose $c(2|1) = 5$ units and $c(1|2) = 10$ units. In addition, it is known that about 20% of *all* objects (for which the measurements $\mathbf{x}$ can be recorded) belong to π_2. Thus, the prior probabilities are $p_1 = .8$ and $p_2 = .2$.

Given the prior probabilities and costs of misclassification, we can use (11-6) to derive the classification regions R_1 and R_2. Specifically, we have

$$R_1: \quad \frac{f_1(\mathbf{x})}{f_2(\mathbf{x})} \geq \left(\frac{10}{5}\right)\left(\frac{.2}{.8}\right) = .5$$

$$R_2: \quad \frac{f_1(\mathbf{x})}{f_2(\mathbf{x})} < \left(\frac{10}{5}\right)\left(\frac{.2}{.8}\right) = .5$$

Suppose the density functions evaluated at a new observation $\mathbf{x}_0$ give $f_1(\mathbf{x}_0) = .3$ and $f_2(\mathbf{x}_0) = .4$. Do we classify the new observation as π_1 or π_2? To answer the question, we form the ratio

$$\frac{f_1(\mathbf{x}_0)}{f_2(\mathbf{x}_0)} = \frac{.3}{.4} = .75$$

and compare it with .5 obtained before. Since

$$\frac{f_1(\mathbf{x}_0)}{f_2(\mathbf{x}_0)} = .75 > \left(\frac{c(1|2)}{c(2|1)}\right)\left(\frac{p_2}{p_1}\right) = .5$$

we find that $\mathbf{x}_0 \in R_1$ and classify it as belonging to π_1. ∎

Criteria other than the expected cost of misclassification can be used to derive "optimal" classification procedures. For example, one might ignore the costs of misclassification and choose R_1 and R_2 to minimize the *total probability of misclassification* (TPM):

TPM = P(misclassifying a π_1 observation *or* misclassifying a π_2 observation)

$\quad = P$(observation comes from π_1 and is misclassified)

$\quad\quad + P$(observation comes from π_2 and is misclassified)

$$= p_1 \int_{R_2} f_1(\mathbf{x})\, d\mathbf{x} + p_2 \int_{R_1} f_2(\mathbf{x})\, d\mathbf{x} \qquad (11\text{-}8)$$

Mathematically, this problem is equivalent to minimizing the expected cost of misclassification when the costs of misclassification are equal. Consequently, the optimal regions in this case are given by (b) in (11-7).

We could also allocate a new observation $\mathbf{x}_0$ to the population with the largest "posterior" probability $P(\pi_i | \mathbf{x}_0)$. By Bayes's rule, the posterior probabilities are

$$P(\pi_1 | \mathbf{x}_0) = \frac{P(\pi_1 \text{ occurs and we observe } \mathbf{x}_0)}{P(\text{we observe } \mathbf{x}_0)}$$

$$= \frac{P(\text{we observe } \mathbf{x}_0 | \pi_1) P(\pi_1)}{P(\text{we observe } \mathbf{x}_0 | \pi_1) P(\pi_1) + P(\text{we observe } \mathbf{x}_0 | \pi_2) P(\pi_2)}$$

$$= \frac{p_1 f_1(\mathbf{x}_0)}{p_1 f_1(\mathbf{x}_0) + p_2 f_2(\mathbf{x}_0)}$$

$$P(\pi_2 | \mathbf{x}_0) = 1 - P(\pi_1 | \mathbf{x}_0) = \frac{p_2 f_2(\mathbf{x}_0)}{p_1 f_1(\mathbf{x}_0) + p_2 f_2(\mathbf{x}_0)} \tag{11-9}$$

Classifying an observation $\mathbf{x}_0$ as π_1 when $P(\pi_1 | \mathbf{x}_0) > P(\pi_2 | \mathbf{x}_0)$ is equivalent to using the (b) rule for total probability of misclassification in (11-7) because the denominators in (11-9) are the same. However, computing the probabilities of the populations π_1 and π_2 after observing $\mathbf{x}_0$ (hence the name *posterior* probabilities) is frequently useful for purposes of identifying the less clear-cut assignments.

11.3 Classification with Two Multivariate Normal Populations

Classification procedures based on normal populations predominate in statistical practice because of their simplicity and reasonably high efficiency across a wide variety of population models. We now assume that $f_1(\mathbf{x})$ and $f_2(\mathbf{x})$ are multivariate normal densities, the first with mean vector $\boldsymbol{\mu}_1$ and covariance matrix $\boldsymbol{\Sigma}_1$ and the second with mean vector $\boldsymbol{\mu}_2$ and covariance matrix $\boldsymbol{\Sigma}_2$.

The special case of equal covariance matrices leads to a particularly simple linear classification statistic.

Classification of Normal Populations When $\boldsymbol{\Sigma}_1 = \boldsymbol{\Sigma}_2 = \boldsymbol{\Sigma}$

Suppose that the joint densities of $\mathbf{X}' = [X_1, X_2, \ldots, X_p]$ for populations π_1 and π_2 are given by

$$f_i(\mathbf{x}) = \frac{1}{(2\pi)^{p/2} |\boldsymbol{\Sigma}|^{1/2}} \exp\left[-\frac{1}{2}(\mathbf{x} - \boldsymbol{\mu}_i)' \boldsymbol{\Sigma}^{-1}(\mathbf{x} - \boldsymbol{\mu}_i)\right] \quad \text{for } i = 1, 2 \tag{11-10}$$

Suppose also that the population parameters $\boldsymbol{\mu}_1, \boldsymbol{\mu}_2$, and $\boldsymbol{\Sigma}$ are known. Then, after cancellation of the terms $(2\pi)^{p/2} |\boldsymbol{\Sigma}|^{1/2}$ the minimum ECM regions in (11-6) become

$$R_1: \quad \exp\left[-\frac{1}{2}(\mathbf{x} - \boldsymbol{\mu}_1)' \boldsymbol{\Sigma}^{-1}(\mathbf{x} - \boldsymbol{\mu}_1) + \frac{1}{2}(\mathbf{x} - \boldsymbol{\mu}_2)' \boldsymbol{\Sigma}^{-1}(\mathbf{x} - \boldsymbol{\mu}_2)\right]$$

$$\geq \left(\frac{c(1|2)}{c(2|1)}\right)\left(\frac{p_2}{p_1}\right)$$

$$R_2: \quad \exp\left[-\frac{1}{2}(\mathbf{x} - \boldsymbol{\mu}_1)' \boldsymbol{\Sigma}^{-1}(\mathbf{x} - \boldsymbol{\mu}_1) + \frac{1}{2}(\mathbf{x} - \boldsymbol{\mu}_2)' \boldsymbol{\Sigma}^{-1}(\mathbf{x} - \boldsymbol{\mu}_2)\right]$$

$$< \left(\frac{c(1|2)}{c(2|1)}\right)\left(\frac{p_2}{p_1}\right) \tag{11-11}$$

Given these regions R_1 and R_2, we can construct the classification rule given in the following result.

Result 11.2. Let the populations π_1 and π_2 be described by multivariate normal densities of the form (11-10). Then the allocation rule that minimizes the ECM is as follows:

Allocate $\mathbf{x}_0$ to π_1 if

$$(\boldsymbol{\mu}_1 - \boldsymbol{\mu}_2)'\boldsymbol{\Sigma}^{-1}\mathbf{x}_0 - \frac{1}{2}(\boldsymbol{\mu}_1 - \boldsymbol{\mu}_2)'\boldsymbol{\Sigma}^{-1}(\boldsymbol{\mu}_1 + \boldsymbol{\mu}_2) \geq \ln\left[\left(\frac{c(1|2)}{c(2|1)}\right)\left(\frac{p_2}{p_1}\right)\right] \quad (11\text{-}12)$$

Allocate $\mathbf{x}_0$ to π_2 otherwise.

Proof. Since the quantities in (11-11) are nonnegative for all $\mathbf{x}$, we can take their natural logarithms and preserve the order of the inequalities. Moreover (see Exercise 11.5),

$$-\frac{1}{2}(\mathbf{x} - \boldsymbol{\mu}_1)'\boldsymbol{\Sigma}^{-1}(\mathbf{x} - \boldsymbol{\mu}_1) + \frac{1}{2}(\mathbf{x} - \boldsymbol{\mu}_2)'\boldsymbol{\Sigma}^{-1}(\mathbf{x} - \boldsymbol{\mu}_2)$$

$$= (\boldsymbol{\mu}_1 - \boldsymbol{\mu}_2)'\boldsymbol{\Sigma}^{-1}\mathbf{x} - \frac{1}{2}(\boldsymbol{\mu}_1 - \boldsymbol{\mu}_2)'\boldsymbol{\Sigma}^{-1}(\boldsymbol{\mu}_1 + \boldsymbol{\mu}_2) \quad (11\text{-}13)$$

and, consequently,

$$R_1: \quad (\boldsymbol{\mu}_1 - \boldsymbol{\mu}_2)'\boldsymbol{\Sigma}^{-1}\mathbf{x} - \frac{1}{2}(\boldsymbol{\mu}_1 - \boldsymbol{\mu}_2)'\boldsymbol{\Sigma}^{-1}(\boldsymbol{\mu}_1 + \boldsymbol{\mu}_2) \geq \ln\left[\left(\frac{c(1|2)}{c(2|1)}\right)\left(\frac{p_2}{p_1}\right)\right]$$

$$R_2: \quad (\boldsymbol{\mu}_1 - \boldsymbol{\mu}_2)'\boldsymbol{\Sigma}^{-1}\mathbf{x} - \frac{1}{2}(\boldsymbol{\mu}_1 - \boldsymbol{\mu}_2)'\boldsymbol{\Sigma}^{-1}(\boldsymbol{\mu}_1 + \boldsymbol{\mu}_2) < \ln\left[\left(\frac{c(1|2)}{c(2|1)}\right)\left(\frac{p_2}{p_1}\right)\right]$$

$$(11\text{-}14)$$

The minimum ECM classification rule follows. ∎

In most practical situations, the population quantities $\boldsymbol{\mu}_1, \boldsymbol{\mu}_2$, and $\boldsymbol{\Sigma}$ are unknown, so the rule (11-12) must be modified. Wald [31] and Anderson [2] have suggested replacing the population parameters by their sample counterparts.

Suppose, then, that we have n_1 observations of the multivariate random variable $\mathbf{X}' = [X_1, X_2, \ldots, X_p]$ from π_1 and n_2 measurements of this quantity from π_2, with $n_1 + n_2 - 2 \geq p$. Then the respective data matrices are

$$\underset{(n_1 \times p)}{\mathbf{X}_1} = \begin{bmatrix} \mathbf{x}'_{11} \\ \mathbf{x}'_{12} \\ \vdots \\ \mathbf{x}'_{1n_1} \end{bmatrix}$$

$$(11\text{-}15)$$

$$\underset{(n_2 \times p)}{\mathbf{X}_2} = \begin{bmatrix} \mathbf{x}'_{21} \\ \mathbf{x}'_{22} \\ \vdots \\ \mathbf{x}'_{2n_2} \end{bmatrix}$$

From these data matrices, the sample mean vectors and covariance matrices are determined by $p = $ # of variables

$$\underset{(p\times 1)}{\bar{\mathbf{x}}_1} = \frac{1}{n_1} \sum_{j=1}^{n_1} \mathbf{x}_{1j}, \quad \underset{(p\times p)}{\mathbf{S}_1} = \frac{1}{n_1 - 1} \sum_{j=1}^{n_1} (\mathbf{x}_{1j} - \bar{\mathbf{x}}_1)(\mathbf{x}_{1j} - \bar{\mathbf{x}}_1)'$$

$$\underset{(p\times 1)}{\bar{\mathbf{x}}_2} = \frac{1}{n_2} \sum_{j=1}^{n_2} \mathbf{x}_{2j}, \quad \underset{(p\times p)}{\mathbf{S}_2} = \frac{1}{n_2 - 1} \sum_{j=1}^{n_2} (\mathbf{x}_{2j} - \bar{\mathbf{x}}_2)(\mathbf{x}_{2j} - \bar{\mathbf{x}}_2)'$$

$$(11\text{-}16)$$

Since it is assumed that the parent populations have the same covariance matrix Σ, the sample covariance matrices $\mathbf{S}_1$ and $\mathbf{S}_2$ are combined (pooled) to derive a single, unbiased estimate of Σ as in (6-21). In particular, the weighted average

$$\mathbf{S}_{\text{pooled}} = \left[\frac{n_1 - 1}{(n_1 - 1) + (n_2 - 1)} \right] \mathbf{S}_1 + \left[\frac{n_2 - 1}{(n_1 - 1) + (n_2 - 1)} \right] \mathbf{S}_2 \qquad (11\text{-}17)$$

is an unbiased estimate of Σ if the data matrices $\mathbf{X}_1$ and $\mathbf{X}_2$ contain *random* samples from the populations π_1 and π_2, respectively.

Substituting $\bar{\mathbf{x}}_1$ for $\boldsymbol{\mu}_1$, $\bar{\mathbf{x}}_2$ for $\boldsymbol{\mu}_2$, and $\mathbf{S}_{\text{pooled}}$ for Σ in (11-12) gives the "sample" classification rule:

The Estimated Minimum ECM Rule for Two Normal Populations

Allocate $\mathbf{x}_0$ to π_1 if

$$(\bar{\mathbf{x}}_1 - \bar{\mathbf{x}}_2)' \mathbf{S}_{\text{pooled}}^{-1} \mathbf{x}_0 - \frac{1}{2} (\bar{\mathbf{x}}_1 - \bar{\mathbf{x}}_2)' \mathbf{S}_{\text{pooled}}^{-1} (\bar{\mathbf{x}}_1 + \bar{\mathbf{x}}_2) \geq \ln\left[\left(\frac{c(1|2)}{c(2|1)} \right) \left(\frac{p_2}{p_1} \right) \right]$$

$$(11\text{-}18)$$

Allocate $\mathbf{x}_0$ to π_2 otherwise.

If, in (11-18),

$$\left(\frac{c(1|2)}{c(2|1)} \right) \left(\frac{p_2}{p_1} \right) = 1$$

then $\ln(1) = 0$, and the estimated minimum ECM rule for two normal populations amounts to comparing the scalar variable

$$\hat{y} = (\bar{\mathbf{x}}_1 - \bar{\mathbf{x}}_2)' \mathbf{S}_{\text{pooled}}^{-1} \mathbf{x} = \hat{\mathbf{a}}' \mathbf{x} \qquad (11\text{-}19)$$

evaluated at $\mathbf{x}_0$, with the number

$$\hat{m} = \frac{1}{2} (\bar{\mathbf{x}}_1 - \bar{\mathbf{x}}_2)' \mathbf{S}_{\text{pooled}}^{-1} (\bar{\mathbf{x}}_1 + \bar{\mathbf{x}}_2)$$

$$= \frac{1}{2} (\bar{y}_1 + \bar{y}_2) \qquad (11\text{-}20)$$

where

$$\bar{y}_1 = (\bar{\mathbf{x}}_1 - \bar{\mathbf{x}}_2)' \mathbf{S}_{\text{pooled}}^{-1} \bar{\mathbf{x}}_1 = \hat{\mathbf{a}}' \bar{\mathbf{x}}_1$$

and

$$\bar{y}_2 = (\bar{\mathbf{x}}_1 - \bar{\mathbf{x}}_2)' \mathbf{S}_{\text{pooled}}^{-1} \bar{\mathbf{x}}_2 = \hat{\mathbf{a}}' \bar{\mathbf{x}}_2$$

That is, the estimated minimum ECM rule for two normal populations is tantamount to creating two *univariate* populations for the y values by taking an appropriate linear combination of the observations from populations π_1 and π_2 and then assigning a new observation $\mathbf{x}_0$ to π_1 or π_2, depending upon whether $\hat{y}_0 = \hat{\mathbf{a}}'\mathbf{x}_0$ falls to the right or left of the midpoint $\hat{m}$ between the two univariate means $\bar{y}_1$ and $\bar{y}_2$.

Once parameter estimates are inserted for the corresponding unknown population quantities, there is no assurance that the resulting rule will minimize the expected cost of misclassification in a particular application. This is because the optimal rule in (11-12) was derived assuming that the multivariate normal densities $f_1(\mathbf{x})$ and $f_2(\mathbf{x})$ were known completely. Expression (11-18) is simply an estimate of the optimal rule. However, it seems reasonable to expect that it should perform well if the sample sizes are large.[3]

To summarize, if the data appear to be multivariate normal[4], the classification statistic to the left of the inequality in (11-18) can be calculated for each new observation $\mathbf{x}_0$. These observations are classified by comparing the values of the statistic with the value of $\ln[(c(1|2)/c(2|1))(p_2/p_1)]$.

Example 11.3 (Classification with two normal populations—common Σ and equal costs) This example is adapted from a study [4] concerned with the detection of hemophilia A carriers. (See also Exercise 11.32.)

To construct a procedure for detecting potential hemophilia A carriers, blood samples were assayed for two groups of women and measurements on the two variables,

$$X_1 = \log_{10}(\text{AHF activity})$$

$$X_2 = \log_{10}(\text{AHF-like antigen})$$

recorded. ("AHF" denotes antihemophilic factor.) The first group of $n_1 = 30$ women were selected from a population of women who did not carry the hemophilia gene. This group was called the *normal* group. The second group of $n_2 = 22$ women was selected from known hemophilia A carriers (daughters of hemophiliacs, mothers with more than one hemophilic son, and mothers with one hemophilic son and other hemophilic relatives). This group was called the *obligatory carriers*. The pairs of observations (x_1, x_2) for the two groups are plotted in Figure 11.4. Also shown are estimated contours containing 50% and 95% of the probability for bivariate normal distributions centered at $\bar{\mathbf{x}}_1$ and $\bar{\mathbf{x}}_2$, respectively. Their common covariance matrix was taken as the pooled sample covariance matrix $\mathbf{S}_{\text{pooled}}$. In this example, bivariate normal distributions seem to fit the data fairly well.

The investigators (see [4]) provide the information

$$\bar{\mathbf{x}}_1 = \begin{bmatrix} -.0065 \\ -.0390 \end{bmatrix}, \qquad \bar{\mathbf{x}}_2 = \begin{bmatrix} -.2483 \\ .0262 \end{bmatrix}$$

---

[3] As the sample sizes increase, $\bar{\mathbf{x}}_1$, $\bar{\mathbf{x}}_2$, and $\mathbf{S}_{\text{pooled}}$ become, with probability approaching 1, indistinguishable from $\boldsymbol{\mu}_1$, $\boldsymbol{\mu}_2$, and Σ, respectively [see (4-26) and (4-27)].

[4] At the very least, the marginal frequency distributions of the observations on each variable can be checked for normality. This must be done for the samples from both populations. Often, some variables must be transformed in order to make them more "normal looking." (See Sections 4.6 and 4.8.)

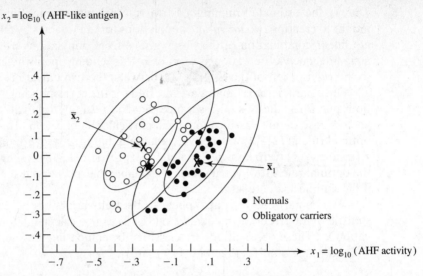

Figure 11.4 Scatter plots of $[\log_{10}(\text{AHF activity}), \log_{10}(\text{AHF-like antigen})]$ for the normal group and obligatory hemophilia A carriers.

and

$$\mathbf{S}_{\text{pooled}}^{-1} = \begin{bmatrix} 131.158 & -90.423 \\ -90.423 & 108.147 \end{bmatrix}$$

Therefore, the equal costs and equal priors discriminant function [see (11-19)] is

$$\hat{y} = \hat{\mathbf{a}}'\mathbf{x} = [\bar{\mathbf{x}}_1 - \bar{\mathbf{x}}_2]'\mathbf{S}_{\text{pooled}}^{-1}\mathbf{x}$$

$$= [.2418 \quad -.0652] \begin{bmatrix} 131.158 & -90.423 \\ -90.423 & 108.147 \end{bmatrix} \begin{bmatrix} x_1 \\ x_2 \end{bmatrix}$$

$$= 37.61 x_1 - 28.92 x_2$$

Moreover,

$$\bar{y}_1 = \hat{\mathbf{a}}'\bar{\mathbf{x}}_1 = [37.61 \quad -28.92] \begin{bmatrix} -.0065 \\ -.0390 \end{bmatrix} = .88$$

$$\bar{y}_2 = \hat{\mathbf{a}}'\bar{\mathbf{x}}_2 = [37.61 \quad -28.92] \begin{bmatrix} -.2483 \\ .0262 \end{bmatrix} = -10.10$$

and the midpoint between these means [see (11-20)] is

$$\hat{m} = \tfrac{1}{2}(\bar{y}_1 + \bar{y}_2) = \tfrac{1}{2}(.88 - 10.10) = -4.61$$

Measurements of AHF activity and AHF-like antigen on a woman who may be a hemophilia A carrier give $x_1 = -.210$ and $x_2 = -.044$. Should this woman be classified as π_1 (normal) or π_2 (obligatory carrier)?

Using (11-18) with equal costs and equal priors so that $\ln(1) = 0$, we obtain

Allocate $\mathbf{x}_0$ to π_1 if $\hat{y}_0 = \hat{\mathbf{a}}'\mathbf{x}_0 \geq \hat{m} = -4.61$

Allocate $\mathbf{x}_0$ to π_2 if $\hat{y}_0 = \hat{\mathbf{a}}'\mathbf{x}_0 < \hat{m} = -4.61$

where $x'_0 = [-.210, -.044]$. Since

$$\hat{y}_0 = \hat{a}'x_0 = [37.61 \quad -28.92] \begin{bmatrix} -.210 \\ -.044 \end{bmatrix} = -6.62 < -4.61$$

we classify the woman as π_2, an obligatory carrier. The new observation is indicated by a star in Figure 11.4. We see that it falls within the estimated .50 probability contour of population π_2 and about on the estimated .95 probability contour of population π_1. Thus, the classification is not clear cut.

Suppose now that the prior probabilities of group membership are known. For example, suppose the blood yielding the foregoing x_1 and x_2 measurements is drawn from the maternal first cousin of a hemophiliac. Then the genetic chance of being a hemophilia A carrier in this case is .25. Consequently, the prior probabilities of group membership are $p_1 = .75$ and $p_2 = .25$. Assuming, somewhat unrealistically, that the costs of misclassification are equal, so that $c(1|2) = c(2|1)$, and using the classification statistic

$$\hat{w} = (\bar{x}_1 - \bar{x}_2)'S^{-1}_{pooled} x_0 - \tfrac{1}{2}(\bar{x}_1 - \bar{x}_2)'S^{-1}_{pooled}(\bar{x}_1 + \bar{x}_2)$$

or $\hat{w} = \hat{a}'x_0 - \hat{m}$ with $x'_0 = [-.210, -.044]$, $\hat{m} = -4.61$, and $\hat{a}'x_0 = -6.62$, we have

$$\hat{w} = -6.62 - (-4.61) = -2.01$$

Applying (11-18), we see that

$$\hat{w} = -2.01 < \ln\left[\frac{p_2}{p_1}\right] = \ln\left[\frac{.25}{.75}\right] = -1.10$$

and we classify the woman as π_2, an obligatory carrier.   ∎

Scaling

The coefficient vector $\hat{a} = S^{-1}_{pooled}(\bar{x}_1 - \bar{x}_2)$ is unique only up to a multiplicative constant, so, for $c \neq 0$, any vector $c\hat{a}$ will also serve as discriminant coefficients.

The vector $\hat{a}$ is frequently "scaled" or "normalized" to ease the interpretation of its elements. Two of the most commonly employed normalizations are

1. Set

$$\hat{a}^* = \frac{\hat{a}}{\sqrt{\hat{a}'\hat{a}}} \tag{11-21}$$

so that $\hat{a}^*$ has unit length.

2. Set

$$\hat{a}^* = \frac{\hat{a}}{\hat{a}_1} \tag{11-22}$$

so that the first element of the new coefficient vector $\hat{a}^*$ is 1.

In both cases, $\hat{a}^*$ is of the form $c\hat{a}$. For normalization (1), $c = (\hat{a}'\hat{a})^{-1/2}$ and for (2), $c = \hat{a}_1^{-1}$.

The magnitudes of $\hat{a}_1^*, \hat{a}_2^*, \ldots, \hat{a}_p^*$ in (11-21) all lie in the interval $[-1, 1]$. In (11-22), $\hat{a}_1^* = 1$ and $\hat{a}_2^*, \ldots, \hat{a}_p^*$ are expressed as multiples of $\hat{a}_1^*$. Constraining the $\hat{a}_i^*$ to the interval $[-1, 1]$ usually facilitates a visual comparison of the coefficients. Similarly, expressing the coefficients as multiples of $\hat{a}_1^*$ allows one to readily assess the relative importance (vis-à-vis X_1) of variables $X_2, \ldots, X_p$ as discriminators.

Normalizing the $\hat{a}_i$'s is recommended only if the X variables have been standardized. If this is not the case, a great deal of care must be exercised in interpreting the results.

Fisher's Approach to Classification with Two Populations

Fisher [10] actually arrived at the linear classification statistic (11-19) using an entirely different argument. Fisher's idea was to transform the multivariate observations $\mathbf{x}$ to univariate observations y such that the y's derived from population π_1 and π_2 were separated as much as possible. Fisher suggested taking linear combinations of $\mathbf{x}$ to create y's because they are simple enough functions of the $\mathbf{x}$ to be handled easily. Fisher's approach does not assume that the populations are normal. It does, however, implicitly assume that the population covariance matrices are equal, because a pooled estimate of the common covariance matrix is used.

A fixed linear combination of the $\mathbf{x}$'s takes the values $y_{11}, y_{12}, \ldots, y_{1n_1}$ for the observations from the first population and the values $y_{21}, y_{22}, \ldots, y_{2n_2}$ for the observations from the second population. The separation of these two sets of univariate y's is assessed in terms of the difference between $\bar{y}_1$ and $\bar{y}_2$. expressed in standard deviation units. That is,

$$\text{separation} = \frac{|\bar{y}_1 - \bar{y}_2|}{s_y}, \quad \text{where } s_y^2 = \frac{\sum_{j=1}^{n_1}(y_{1j} - \bar{y}_1)^2 + \sum_{j=1}^{n_2}(y_{2j} - \bar{y}_2)^2}{n_1 + n_2 - 2}$$

is the pooled estimate of the variance. The objective is to select the linear combination of the $\mathbf{x}$ to achieve maximum separation of the sample means $\bar{y}_1$ and $\bar{y}_2$.

Result 11.3. The linear combination $\hat{y} = \hat{\mathbf{a}}'\mathbf{x} = (\bar{\mathbf{x}}_1 - \bar{\mathbf{x}}_2)'\mathbf{S}_{\text{pooled}}^{-1}\mathbf{x}$ maximizes the ratio

$$\left(\frac{\text{squared distance}}{\text{(sample variance of } y)}\right) = \frac{(\bar{y}_1 - \bar{y}_2)^2}{s_y^2}$$

$$= \frac{(\hat{\mathbf{a}}'\bar{\mathbf{x}}_1 - \hat{\mathbf{a}}'\bar{\mathbf{x}}_2)^2}{\hat{\mathbf{a}}'\mathbf{S}_{\text{pooled}}\,\hat{\mathbf{a}}}$$

$$= \frac{(\hat{\mathbf{a}}'\mathbf{d})^2}{\hat{\mathbf{a}}'\mathbf{S}_{\text{pooled}}\,\hat{\mathbf{a}}} \qquad (11\text{-}23)$$

over all possible coefficient vectors $\hat{\mathbf{a}}$ where $\mathbf{d} = (\bar{\mathbf{x}}_1 - \bar{\mathbf{x}}_2)$. The maximum of the ratio (11-23) is $D^2 = (\bar{\mathbf{x}}_1 - \bar{\mathbf{x}}_2)'\mathbf{S}_{\text{pooled}}^{-1}(\bar{\mathbf{x}}_1 - \bar{\mathbf{x}}_2)$.

Proof. The maximum of the ratio in (11-23) is given by applying (2-50) directly. Thus, setting $\mathbf{d} = (\bar{\mathbf{x}}_1 - \bar{\mathbf{x}}_2)$, we have

$$\max_{\hat{\mathbf{a}}} \frac{(\hat{\mathbf{a}}'\mathbf{d})^2}{\hat{\mathbf{a}}'\mathbf{S}_{pooled}\,\hat{\mathbf{a}}} = \mathbf{d}'\mathbf{S}_{pooled}^{-1}\mathbf{d} = (\bar{\mathbf{x}}_1 - \bar{\mathbf{x}}_2)'\mathbf{S}_{pooled}^{-1}(\bar{\mathbf{x}}_1 - \bar{\mathbf{x}}_2) = D^2$$

where D^2 is the sample squared distance between the two means. ∎

Note that s_y^2 in (11-33) may be calculated as

$$s_y^2 = \frac{\sum_{j=1}^{n_1}(y_{1j} - \bar{y}_1)^2 + \sum_{j=1}^{n_2}(y_{2j} - \bar{y}_2)^2}{n_1 + n_2 - 2} \tag{11-24}$$

with $y_{1j} = \hat{\mathbf{a}}'\mathbf{x}_{1j}$ and $y_{2j} = \hat{\mathbf{a}}'\mathbf{x}_{2j}$.

Example 11.4 (Fisher's linear discriminant for the hemophilia data) Consider the detection of hemophilia A carriers introduced in Example 11.3. Recall that the equal costs and equal priors linear discriminant function was

$$\hat{y} = \hat{\mathbf{a}}'\mathbf{x} = (\bar{\mathbf{x}}_1 - \bar{\mathbf{x}}_2)'\mathbf{S}_{pooled}^{-1}\mathbf{x} = 37.61x_1 - 28.92x_2$$

This linear discriminant function is Fisher's linear function, which maximally separates the two populations, and the maximum separation in the samples is

$$D^2 = (\bar{\mathbf{x}}_1 - \bar{\mathbf{x}}_2)'\mathbf{S}_{pooled}^{-1}(\bar{\mathbf{x}}_1 - \bar{\mathbf{x}}_2)$$

$$= [.2418, \quad -.0652]\begin{bmatrix} 131.158 & -90.423 \\ -90.423 & 108.147 \end{bmatrix}\begin{bmatrix} .2418 \\ -.0652 \end{bmatrix}$$

$$= 10.98$$ ∎

Fisher's solution to the separation problem can also be used to classify new observations.

An Allocation Rule Based on Fisher's Discriminant Function[5]

Allocate $\mathbf{x}_0$ to π_1 if

$$\hat{y}_0 = (\bar{\mathbf{x}}_1 - \bar{\mathbf{x}}_2)'\mathbf{S}_{pooled}^{-1}\mathbf{x}_0$$

$$\geq \hat{m} = \tfrac{1}{2}(\bar{\mathbf{x}}_1 - \bar{\mathbf{x}}_2)'\mathbf{S}_{pooled}^{-1}(\bar{\mathbf{x}}_1 + \bar{\mathbf{x}}_2)$$

or $\tag{11-25}$

$$\hat{y}_0 - \hat{m} \geq 0$$

Allocate $\mathbf{x}_0$ to π_2 if

$$\hat{y}_0 < \hat{m}$$

or

$$\hat{y}_0 - \hat{m} < 0$$

---

[5]We must have $(n_1 + n_2 - 2) \geq p$; otherwise $\mathbf{S}_{pooled}$ is singular, and the usual inverse, $\mathbf{S}_{pooled}^{-1}$, does not exist.

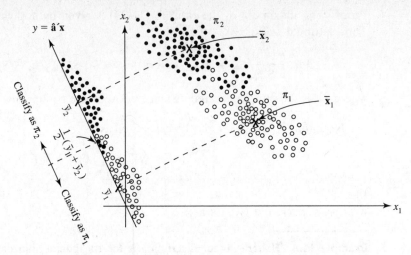

Figure 11.5 A pictorial representation of Fisher's procedure for two populations with $p = 2$.

The procedure (11-23) is illustrated, schematically, for $p = 2$ in Figure 11.5. All points in the scatter plots are projected onto a line in the direction $\hat{\mathbf{a}}$, and this direction is varied until the samples are maximally separated.

Fisher's linear discriminant function in (11-25) was developed under the assumption that the two populations, whatever their form, have a common covariance matrix. Consequently, it may not be surprising that Fisher's method corresponds to a particular case of the minimum expected-cost-of-misclassification rule. The first term, $\hat{y} = (\bar{\mathbf{x}}_1 - \bar{\mathbf{x}}_2)' \mathbf{S}_{\text{pooled}}^{-1} \mathbf{x}$, in the classification rule (11-18) is the linear function obtained by Fisher that maximizes the univariate "between" samples variability relative to the "within" samples variability. [See (11-23).] The entire expression

$$\hat{w} = (\bar{\mathbf{x}}_1 - \bar{\mathbf{x}}_2)' \mathbf{S}_{\text{pooled}}^{-1} \mathbf{x} - \tfrac{1}{2}(\bar{\mathbf{x}}_1 - \bar{\mathbf{x}}_2)' \mathbf{S}_{\text{pooled}}^{-1} (\bar{\mathbf{x}}_1 + \bar{\mathbf{x}}_2)$$

$$= (\bar{\mathbf{x}}_1 - \bar{\mathbf{x}}_2)' \mathbf{S}_{\text{pooled}}^{-1} \left[\mathbf{x} - \tfrac{1}{2}(\bar{\mathbf{x}}_1 + \bar{\mathbf{x}}_2) \right] \tag{11-26}$$

is frequently called *Anderson's classification function (statistic)*. Once again, if $[(c(1|2)/c(2|1))(p_2/p_1)] = 1$, so that $\ln[(c(1|2)/c(2|1))(p_2/p_1)] = 0$, Rule (11-18) is comparable to Rule (11-26), based on Fisher's linear discriminant function. Thus, provided that the two normal populations have the same covariance matrix, Fisher's classification rule is equivalent to the minimum ECM rule with equal prior probabilities and equal costs of misclassification.

Is Classification a Good Idea?

For two populations, the maximum relative separation that can be obtained by considering linear combinations of the multivariate observations is equal to the distance D^2. This is convenient because D^2 can be used, in certain situations, to test whether the population means $\boldsymbol{\mu}_1$ and $\boldsymbol{\mu}_2$ differ significantly. Consequently, a test for differences in mean vectors can be viewed as a test for the "significance" of the separation that can be achieved.

Suppose the populations π_1 and π_2 are multivariate normal *with a common co-variance matrix* $\boldsymbol{\Sigma}$. Then, as in Section 6.3, a test of $H_0: \boldsymbol{\mu}_1 = \boldsymbol{\mu}_2$ versus $H_1: \boldsymbol{\mu}_1 \neq \boldsymbol{\mu}_2$ is accomplished by referring

$$\left(\frac{n_1 + n_2 - p - 1}{(n_1 + n_2 - 2)p} \right) \left(\frac{n_1 n_2}{n_1 + n_2} \right) D^2$$

to an *F*-distribution with $v_1 = p$ and $v_2 = n_1 + n_2 - p - 1$ d.f. If H_0 is rejected, we can conclude that the separation between the two populations π_1 and π_2 is significant.

Comment. Significant separation does not necessarily imply good classification. As we shall see in Section 11.4, the efficacy of a classification procedure can be evaluated independently of any test of separation. By contrast, if the separation is not significant, the search for a useful classification rule will probably prove fruitless.

Classification of Normal Populations When $\boldsymbol{\Sigma}_1 \neq \boldsymbol{\Sigma}_2$

As might be expected, the classification rules are more complicated when the population covariance matrices are unequal.

Consider the multivariate normal densities in (11-10) with $\boldsymbol{\Sigma}_i$, $i = 1, 2$, replacing $\boldsymbol{\Sigma}$. Thus, the covariance matrices, as well as the mean vectors, are different from one another for the two populations. As we have seen, the regions of minimum ECM and minimum total probability of misclassification (TPM) depend on the ratio of the densities, $f_1(\mathbf{x})/f_2(\mathbf{x})$, or, equivalently, the natural logarithm of the density ratio, $\ln[f_1(\mathbf{x})/f_2(\mathbf{x})] = \ln[f_1(\mathbf{x})] - \ln[f_2(\mathbf{x})]$. When the multivariate normal densities have different covariance structures, the terms in the density ratio involving $|\boldsymbol{\Sigma}_i|^{1/2}$ do not cancel as they do when $\boldsymbol{\Sigma}_1 = \boldsymbol{\Sigma}_2$. Moreover, the quadratic forms in the exponents of $f_1(\mathbf{x})$ and $f_2(\mathbf{x})$ do not combine to give the rather simple result in (11-13).

Substituting multivariate normal densities with different covariance matrices into (11-6) gives, after taking natural logarithms and simplifying (see Exercise 11.15), the classification regions

$$R_1: \quad -\frac{1}{2}\mathbf{x}'(\boldsymbol{\Sigma}_1^{-1} - \boldsymbol{\Sigma}_2^{-1})\mathbf{x} + (\boldsymbol{\mu}_1'\boldsymbol{\Sigma}_1^{-1} - \boldsymbol{\mu}_2'\boldsymbol{\Sigma}_2^{-1})\mathbf{x} - k \geq \ln\left[\left(\frac{c(1|2)}{c(2|1)}\right)\left(\frac{p_2}{p_1}\right)\right]$$

$$R_2: \quad -\frac{1}{2}\mathbf{x}'(\boldsymbol{\Sigma}_1^{-1} - \boldsymbol{\Sigma}_2^{-1})\mathbf{x} + (\boldsymbol{\mu}_1'\boldsymbol{\Sigma}_1^{-1} - \boldsymbol{\mu}_2'\boldsymbol{\Sigma}_2^{-1})\mathbf{x} - k < \ln\left[\left(\frac{c(1|2)}{c(2|1)}\right)\left(\frac{p_2}{p_1}\right)\right]$$

if $\Sigma_1 = \Sigma_2 \to$ gives m for LDA

(11-27)

where

determinant

$$k = \frac{1}{2}\ln\left(\frac{|\boldsymbol{\Sigma}_1|}{|\boldsymbol{\Sigma}_2|}\right) + \frac{1}{2}(\boldsymbol{\mu}_1'\boldsymbol{\Sigma}_1^{-1}\boldsymbol{\mu}_1 - \boldsymbol{\mu}_2'\boldsymbol{\Sigma}_2^{-1}\boldsymbol{\mu}_2) \qquad (11\text{-}28)$$

The classification regions are defined by *quadratic* functions of $\mathbf{x}$. When $\boldsymbol{\Sigma}_1 = \boldsymbol{\Sigma}_2$, the quadratic term, $-\frac{1}{2}\mathbf{x}'(\boldsymbol{\Sigma}_1^{-1} - \boldsymbol{\Sigma}_2^{-1})\mathbf{x}$, disappears, and the regions defined by (11-27) reduce to those defined by (11-14).

The classification rule for general multivariate normal populations follows directly from (11-27).

Result 11.4. Let the populations π_1 and π_2 be described by multivariate normal densities with mean vectors and covariance matrices $\boldsymbol{\mu}_1, \boldsymbol{\Sigma}_1$ and $\boldsymbol{\mu}_2, \boldsymbol{\Sigma}_2$, respectively. The allocation rule that minimizes the expected cost of misclassification is given by

Allocate $\mathbf{x}_0$ to π_1 if

$$-\frac{1}{2}\mathbf{x}_0'(\boldsymbol{\Sigma}_1^{-1} - \boldsymbol{\Sigma}_2^{-1})\mathbf{x}_0 + (\boldsymbol{\mu}_1'\boldsymbol{\Sigma}_1^{-1} - \boldsymbol{\mu}_2'\boldsymbol{\Sigma}_2^{-1})\mathbf{x}_0 - k \geq \ln\left[\left(\frac{c(1|2)}{c(2|1)}\right)\left(\frac{p_2}{p_1}\right)\right]$$

Allocate $\mathbf{x}_0$ to π_2 otherwise.

Here k is set out in (11-28). ∎

In practice, the classification rule in Result 11.5 is implemented by substituting the sample quantities $\bar{\mathbf{x}}_1, \bar{\mathbf{x}}_2, \mathbf{S}_1$, and $\mathbf{S}_2$ (see (11-16)) for $\boldsymbol{\mu}_1, \boldsymbol{\mu}_2, \boldsymbol{\Sigma}_1$, and $\boldsymbol{\Sigma}_2$, respectively.[6]

Quadratic Classification Rule
(Normal Populations with Unequal Covariance Matrices)

Allocate $\mathbf{x}_0$ to π_1 if

$$-\frac{1}{2}\underbrace{\mathbf{x}_0'(\mathbf{S}_1^{-1} - \mathbf{S}_2^{-1})\mathbf{x}_0}_{=0 \text{ if } S_1 = S_2} + (\bar{\mathbf{x}}_1'\mathbf{S}_1^{-1} - \bar{\mathbf{x}}_2'\mathbf{S}_2^{-1})\mathbf{x}_0 - k \geq \ln\left[\left(\frac{c(1|2)}{c(2|1)}\right)\left(\frac{p_2}{p_1}\right)\right]$$

(11-29)

Allocate $\mathbf{x}_0$ to π_2 otherwise.

Classification with quadratic functions is rather awkward in more than two dimensions and can lead to some strange results. This is particularly true when the data are not (essentially) multivariate normal.

Figure 11.6(a) shows the equal costs and equal priors rule based on the idealized case of two normal distributions with different variances. This quadratic rule leads to a region R_1 consisting of two disjoint sets of points.

In many applications, the lower tail for the π_1 distribution will be smaller than that prescribed by a normal distribution. Then, as shown in Figure 11.6(b), the lower part of the region R_1, produced by the quadratic procedure, does not line up well with the population distributions and can lead to large error rates. A serious weakness of the quadratic rule is that it is sensitive to departures from normality.

---

[6] The inequalities $n_1 > p$ and $n_2 > p$ must both hold for $\mathbf{S}_1^{-1}$ and $\mathbf{S}_2^{-1}$ to exist. These quantities are used in place of $\boldsymbol{\Sigma}_1^{-1}$ and $\boldsymbol{\Sigma}_2^{-1}$, respectively, in the sample analog (11-29).

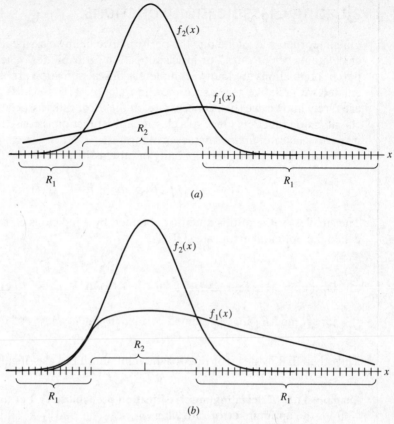

Figure 11.6 Quadratic rules for (a) two normal distribution with unequal variances and (b) two distributions, one of which is nonnormal—rule not appropriate.

If the data are not multivariate normal, two options are available. First, the non-normal data can be transformed to data more nearly normal, and a test for the equality of covariance matrices can be conducted (see Section 6.6) to see whether the linear rule (11-18) or the quadratic rule (11-29) is appropriate. Transformations are discussed in Chapter 4. (The usual tests for covariance homogeneity are greatly affected by nonnormality. The conversion of nonnormal data to normal data must be done before this testing is carried out.)

Second, we can use a linear (or quadratic) rule without worrying about the form of the parent populations and hope that it will work reasonably well. Studies (see [22] and [23]) have shown, however, that there are nonnormal cases where a linear classification function performs poorly, even though the population covariance matrices are the same. The moral is to always check the performance of any classification procedure. At the very least, this should be done with the data sets used to build the classifier. Ideally, there will be enough data available to provide for "training" samples and "validation" samples. The training samples can be used to develop the classification function, and the validation samples can be used to evaluate its performance.

11.4 Evaluating Classification Functions

One important way of judging the performance of any classification procedure is to calculate its "error rates," or misclassification probabilities. When the forms of the parent populations are known completely, misclassification probabilities can be calculated with relative ease, as we show in Example 11.5. Because parent populations are rarely known, we shall concentrate on the error rates associated with the sample classification function. Once this classification function is constructed, a measure of its performance in *future* samples is of interest.

From (11-8), the total probability of misclassification is

$$\text{TPM} = p_1 \int_{R_2} f_1(\mathbf{x}) \, d\mathbf{x} + p_2 \int_{R_1} f_2(\mathbf{x}) \, d\mathbf{x}$$

The smallest value of this quantity, obtained by a judicious choice of R_1 and R_2, is called the optimum error rate (OER).

$$\text{Optimum error rate (OER)} = p_1 \int_{R_2} f_1(\mathbf{x}) \, d\mathbf{x} + p_2 \int_{R_1} f_2(\mathbf{x}) \, d\mathbf{x} \qquad (11\text{-}30)$$

where R_1 and R_2 are determined by case (b) in (11-7).

Thus, the OER is the error rate for the minimum TPM classification rule.

Example 11.5 (Calculating misclassification probabilities) Let us derive an expression for the optimum error rate when $p_1 = p_2 = \frac{1}{2}$ and $f_1(\mathbf{x})$ and $f_2(\mathbf{x})$ are the multivariate normal densities in (11-10).

Now, the minimum ECM and minimum TPM classification rules coincide when $c(1|2) = c(2|1)$. Because the prior probabilities are also equal, the minimum TPM classification regions are defined for normal populations by (11-12), with

$$\ln\left[\left(\frac{c(1|2)}{c(2|1)} \right) \left(\frac{p_2}{p_1} \right) \right] = 0.$$ We find that

$$R_1: \quad (\boldsymbol{\mu}_1 - \boldsymbol{\mu}_2)' \boldsymbol{\Sigma}^{-1} \mathbf{x} - \tfrac{1}{2}(\boldsymbol{\mu}_1 - \boldsymbol{\mu}_2)' \boldsymbol{\Sigma}^{-1}(\boldsymbol{\mu}_1 + \boldsymbol{\mu}_2) \geq 0$$

$$R_2: \quad (\boldsymbol{\mu}_1 - \boldsymbol{\mu}_2)' \boldsymbol{\Sigma}^{-1} \mathbf{x} - \tfrac{1}{2}(\boldsymbol{\mu}_1 - \boldsymbol{\mu}_2)' \boldsymbol{\Sigma}^{-1}(\boldsymbol{\mu}_1 + \boldsymbol{\mu}_2) < 0$$

These sets can be expressed in terms of $y = (\boldsymbol{\mu}_1 - \boldsymbol{\mu}_2)' \boldsymbol{\Sigma}^{-1} \mathbf{x} = \mathbf{a}' \mathbf{x}$ as

$$R_1(y): \quad y \geq \tfrac{1}{2}(\boldsymbol{\mu}_1 - \boldsymbol{\mu}_2)' \boldsymbol{\Sigma}^{-1}(\boldsymbol{\mu}_1 + \boldsymbol{\mu}_2)$$

$$R_2(y): \quad y < \tfrac{1}{2}(\boldsymbol{\mu}_1 - \boldsymbol{\mu}_2)' \boldsymbol{\Sigma}^{-1}(\boldsymbol{\mu}_1 + \boldsymbol{\mu}_2)$$

But Y is a linear combination of normal random variables, so the probability densities of Y, $f_1(y)$ and $f_2(y)$, are univariate normal (see Result 4.2) with means and a variance given by

$$\mu_{1Y} = \mathbf{a}' \boldsymbol{\mu}_1 = (\boldsymbol{\mu}_1 - \boldsymbol{\mu}_2)' \boldsymbol{\Sigma}^{-1} \boldsymbol{\mu}_1$$

$$\mu_{2Y} = \mathbf{a}' \boldsymbol{\mu}_2 = (\boldsymbol{\mu}_1 - \boldsymbol{\mu}_2)' \boldsymbol{\Sigma}^{-1} \boldsymbol{\mu}_2$$

$$\sigma_Y^2 = \mathbf{a}' \boldsymbol{\Sigma} \mathbf{a} = (\boldsymbol{\mu}_1 - \boldsymbol{\mu}_2)' \boldsymbol{\Sigma}^{-1}(\boldsymbol{\mu}_1 - \boldsymbol{\mu}_2) = \Delta^2$$

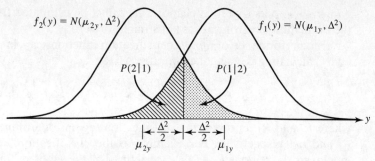

$$f_2(y) = N(\mu_{2y}, \Delta^2) \qquad\qquad f_1(y) = N(\mu_{1y}, \Delta^2)$$

$$P(2|1) \qquad\qquad P(1|2)$$

$$\left|\leftarrow \frac{\Delta^2}{2} \rightarrow\right|\leftarrow \frac{\Delta^2}{2} \rightarrow\right|$$

$$\mu_{2y} \qquad\qquad \mu_{1y}$$

Figure 11.7 The misclassification probabilities based on Y.

Now,

$$\text{TPM} = \tfrac{1}{2} P\,[\text{misclassifying a } \pi_1 \text{ observation as } \pi_2]$$
$$+ \tfrac{1}{2} P\,[\text{misclassifying a } \pi_2 \text{ observation as } \pi_1]$$

But, as shown in Figure 11.7

$$P[\text{misclassifying a } \pi_1 \text{ observation as } \pi_2] = P(2|1)$$

$$= P[Y < \tfrac{1}{2}(\boldsymbol{\mu}_1 - \boldsymbol{\mu}_2)'\boldsymbol{\Sigma}^{-1}(\boldsymbol{\mu}_1 + \boldsymbol{\mu}_2)]$$

$$= P\left(\frac{Y - \mu_{1Y}}{\sigma_Y} < \frac{\tfrac{1}{2}(\boldsymbol{\mu}_1 - \boldsymbol{\mu}_2)'\boldsymbol{\Sigma}^{-1}(\boldsymbol{\mu}_1 + \boldsymbol{\mu}_2) - (\boldsymbol{\mu}_1 - \boldsymbol{\mu}_2)'\boldsymbol{\Sigma}^{-1}\boldsymbol{\mu}_1}{\Delta} \right)$$

$$= P\left(Z < \frac{-\tfrac{1}{2}\Delta^2}{\Delta} \right) = \Phi\left(\frac{-\Delta}{2}\right)$$

where $\Phi\,(\cdot)$ is the cumulative distribution function of a standard normal random variable. Similarly,

$$P[\text{misclassifying a } \pi_2 \text{ observation as } \pi_1]$$

$$= P(1|2) = P[Y \geq \tfrac{1}{2}(\boldsymbol{\mu}_1 - \boldsymbol{\mu}_2)'\boldsymbol{\Sigma}^{-1}(\boldsymbol{\mu}_1 + \boldsymbol{\mu}_2)]$$

$$= P\left(Z \geq \frac{\Delta}{2} \right) = 1 - \Phi\left(\frac{\Delta}{2}\right) = \Phi\left(\frac{-\Delta}{2}\right)$$

Therefore, the optimum error rate is

$$\text{OER} = \text{minimum TPM} = \frac{1}{2}\,\Phi\left(\frac{-\Delta}{2}\right) + \frac{1}{2}\,\Phi\left(\frac{-\Delta}{2}\right) = \Phi\left(\frac{-\Delta}{2}\right) \qquad (11\text{-}31)$$

If, for example, $\Delta^2 = (\boldsymbol{\mu}_1 - \boldsymbol{\mu}_2)'\boldsymbol{\Sigma}^{-1}(\boldsymbol{\mu}_1 - \boldsymbol{\mu}_2) = 2.56$, then $\Delta = \sqrt{2.56} = 1.6$, and, using Table 1 in the appendix, we obtain

$$\text{Minimum TPM} = \Phi\left(\frac{-1.6}{2}\right) = \Phi(-.8) = .2119$$

The optimal classification rule here will incorrectly allocate about 21% of the items to one population or the other. ∎

Example 11.5 illustrates how the optimum error rate can be calculated when the population density functions are known. If, as is usually the case, certain population parameters appearing in allocation rules must be estimated from the sample, then the evaluation of error rates is not straightforward.

The performance of *sample* classification functions can, in principle, be evaluated by calculating the actual error rate (AER),

$$\text{AER} = p_1 \int_{\hat{R}_2} f_1(\mathbf{x}) \, d\mathbf{x} + p_2 \int_{\hat{R}_1} f_2(\mathbf{x}) \, d\mathbf{x} \tag{11-32}$$

where $\hat{R}_1$ and $\hat{R}_2$ represent the classification regions determined by samples of size n_1 and n_2, respectively. For example, if the classification function in (11-18) is employed, the regions $\hat{R}_1$ and $\hat{R}_2$ are defined by the set of $\mathbf{x}$'s for which the following inequalities are satisfied.

$$\hat{R}_1: \quad (\bar{\mathbf{x}}_1 - \bar{\mathbf{x}}_2)' \mathbf{S}_{\text{pooled}}^{-1} \mathbf{x} - \frac{1}{2} (\bar{\mathbf{x}}_1 - \bar{\mathbf{x}}_2)' \mathbf{S}_{\text{pooled}}^{-1} (\bar{\mathbf{x}}_1 + \bar{\mathbf{x}}_2) \geq \ln \left[\left(\frac{c(1|2)}{c(2|1)} \right) \left(\frac{p_2}{p_1} \right) \right]$$

$$\hat{R}_2: \quad (\bar{\mathbf{x}}_1 - \bar{\mathbf{x}}_2)' \mathbf{S}_{\text{pooled}}^{-1} \mathbf{x} - \frac{1}{2} (\bar{\mathbf{x}}_1 - \bar{\mathbf{x}}_2)' \mathbf{S}_{\text{pooled}}^{-1} (\bar{\mathbf{x}}_1 + \bar{\mathbf{x}}_2) < \ln \left[\left(\frac{c(1|2)}{c(2|1)} \right) \left(\frac{p_2}{p_1} \right) \right]$$

The AER indicates how the sample classification function will perform in future samples. Like the optimal error rate, it cannot, in general, be calculated, because it depends on the unknown density functions $f_1(\mathbf{x})$ and $f_2(\mathbf{x})$. However, an estimate of a quantity related to the actual error rate can be calculated, and this estimate will be discussed shortly.

There is a measure of performance that does not depend on the form of the parent populations and that can be calculated for *any* classification procedure. This measure, called the *apparent error rate* (APER), is defined as the fraction of observations in the *training* sample that are misclassified by the sample classification function.

The apparent error rate can be easily calculated from the *confusion matrix*, which shows actual versus predicted group membership. For n_1 observations from π_1 and n_2 observations from π_2, the confusion matrix has the form

Predicted membership

		π_1	π_2		
Actual	π_1	n_{1C}	$n_{1M} = n_1 - n_{1C}$	n_1	(11-33)
membership	π_2	$n_{2M} = n_2 - n_{2C}$	n_{2C}	n_2	

where

$$n_{1C} = \text{number of } \pi_1 \text{ items correctly classified as } \pi_1 \text{ items}$$
$$n_{1M} = \text{number of } \pi_1 \text{ items misclassified as } \pi_2 \text{ items}$$
$$n_{2C} = \text{number of } \pi_2 \text{ items correctly classified}$$
$$n_{2M} = \text{number of } \pi_2 \text{ items misclassified}$$

The apparent error rate is then

$$\text{APER} = \frac{n_{1M} + n_{2M}}{n_1 + n_2} \tag{11-34}$$

which is recognized as the *proportion* of items in the training set that are misclassified.

Example 11.6 (Calculating the apparent error rate) Consider the classification regions R_1 and R_2 shown in Figure 11.1 for the riding-mower data. In this case, observations northeast of the solid line are classified as π_1, mower owners; observations southwest of the solid line are classified as π_2, nonowners. Notice that some observations are misclassified. The confusion matrix is

Predicted membership

		π_1: riding-mower owners	π_2: nonowners	
Actual membership	π_1: riding-mower owners	$n_{1C} = 10$	$n_{1M} = 2$	$n_1 = 12$
	π_2: nonowners	$n_{2M} = 2$	$n_{2C} = 10$	$n_2 = 12$

The apparent error rate, expressed as a percentage, is

$$\text{APER} = \left(\frac{2+2}{12+12}\right)100\% = \left(\frac{4}{24}\right)100\% = 16.7\%$$

The APER is intuitively appealing and easy to calculate. Unfortunately, it tends to underestimate the AER, and the problem does not disappear unless the sample sizes n_1 and n_2 are very large. Essentially, this optimistic estimate occurs because the data used to build the classification function are also used to evaluate it.

Error-rate estimates can be constructed that are better than the apparent error rate, remain relatively easy to calculate, and do not require distributional assumptions. One procedure is to split the total sample into a training sample and a validation sample. The training sample is used to construct the classification function, and the validation sample is used to evaluate it. The error rate is determined by the proportion misclassified in the validation sample. Although this method overcomes the bias problem by not using the same data to both build and judge the classification function, it suffers from two main defects:

(i) It requires large samples.

(ii) The function evaluated is not the function of interest. Ultimately, almost *all* of the data must be used to construct the classification function. If not, valuable information may be lost.

A second approach that seems to work well is called Lachenbruch's "holdout" procedure[7] (see also Lachenbruch and Mickey [24]):

1. Start with the π_1 group of observations. Omit one observation from this group, and develop a classification function based on the remaining $n_1 - 1, n_2$ observations.

2. Classify the "holdout" observation, using the function constructed in Step 1.

[7]Lachenbruch's holdout procedure is sometimes referred to as *jackknifing* or *cross-validation*.

3. Repeat Steps 1 and 2 until all of the π_1 observations are classified. Let $n_{1M}^{(H)}$ be the number of holdout (H) observations misclassified in this group.

4. Repeat Steps 1 through 3 for the π_2 observations. Let $n_{2M}^{(H)}$ be the number of holdout observations misclassified in this group.

Estimates $\hat{P}(2|1)$ and $\hat{P}(1|2)$ of the conditional misclassification probabilities in (11-1) and (11-2) are then given by

$$\hat{P}(2|1) = \frac{n_{1M}^{(H)}}{n_1}$$

$$\hat{P}(1|2) = \frac{n_{2M}^{(H)}}{n_2} \tag{11-35}$$

and the total proportion misclassified, $(n_{1M}^{(H)} + n_{2M}^{(H)})/(n_1 + n_2)$, is, for moderate samples, a nearly unbiased estimate of the *expected* actual error rate, $E(\text{AER})$.

$$\hat{E}(\text{AER}) = \frac{n_{1M}^{(H)} + n_{2M}^{(H)}}{n_1 + n_2} \tag{11-36}$$

Lachenbruch's holdout method is computationally feasible when used in conjunction with the linear classification statistics in (11-18) or (11-19). It is offered as an option in some readily available discriminant analysis computer programs.

Example 11.7 Calculating an estimate of the error rate using the holdout procedure)
We shall illustrate Lachenbruch's holdout procedure and the calculation of error rate estimates for the equal costs and equal priors version of (11-18). Consider the following data matrices and descriptive statistics. (We shall assume that the $n_1 = n_2 = 3$ bivariate observations were selected randomly from two populations π_1 and π_2 with a common covariance matrix.)

$$\mathbf{X}_1 = \begin{bmatrix} 2 & 12 \\ 4 & 10 \\ 3 & 8 \end{bmatrix}; \quad \bar{\mathbf{x}}_1 = \begin{bmatrix} 3 \\ 10 \end{bmatrix}, \quad 2\mathbf{S}_1 = \begin{bmatrix} 2 & -2 \\ -2 & 8 \end{bmatrix}$$

$$\mathbf{X}_2 = \begin{bmatrix} 5 & 7 \\ 3 & 9 \\ 4 & 5 \end{bmatrix}; \quad \bar{\mathbf{x}}_2 = \begin{bmatrix} 4 \\ 7 \end{bmatrix}, \quad 2\mathbf{S}_2 = \begin{bmatrix} 2 & -2 \\ -2 & 8 \end{bmatrix}$$

The pooled covariance matrix is

$$\mathbf{S}_{\text{pooled}} = \frac{1}{4}(2\mathbf{S}_1 + 2\mathbf{S}_2) = \begin{bmatrix} 1 & -1 \\ -1 & 4 \end{bmatrix}$$

Using $\mathbf{S}_{\text{pooled}}$, the rest of the data, and Rule (11-18) with equal costs and equal priors, we may classify the sample observations. You may then verify (see Exercise 11.19) that the confusion matrix is

Classify as:

	π_1	π_2
True population: π_1	2	1
π_2	1	2

and consequently,

$$\text{APER(apparent error rate)} = \frac{2}{6} = .33$$

Holding out the first observation $\mathbf{x}'_H = [2, 12]$ from $\mathbf{X}_1$, we calculate

$$\mathbf{X}_{1H} = \begin{bmatrix} 4 & 10 \\ 3 & 8 \end{bmatrix}; \quad \bar{\mathbf{x}}_{1H} = \begin{bmatrix} 3.5 \\ 9 \end{bmatrix}; \quad \text{and} \quad 1\mathbf{S}_{1H} = \begin{bmatrix} .5 & 1 \\ 1 & 2 \end{bmatrix}$$

The new pooled covariance matrix, $\mathbf{S}_{H,\text{pooled}}$, is

$$\mathbf{S}_{H,\text{pooled}} = \frac{1}{3}[1\mathbf{S}_{1H} + 2\mathbf{S}_2] = \frac{1}{3}\begin{bmatrix} 2.5 & -1 \\ -1 & 10 \end{bmatrix}$$

with inverse[8]

$$\mathbf{S}^{-1}_{H,\text{pooled}} = \frac{1}{8}\begin{bmatrix} 10 & 1 \\ 1 & 2.5 \end{bmatrix}$$

It is computationally quicker to classify the holdout observation $\mathbf{x}_{1H}$ on the basis of its squared distances from the group means $\bar{\mathbf{x}}_{1H}$ and $\bar{\mathbf{x}}_2$. This procedure is equivalent to computing the value of the linear function $\hat{y} = \hat{\mathbf{a}}'_H\mathbf{x}_H = (\bar{\mathbf{x}}_{1H} - \bar{\mathbf{x}}_2)'\mathbf{S}^{-1}_{H,\text{pooled}}\mathbf{x}_H$ and comparing it to the midpoint $\hat{m}_H = \frac{1}{2}(\bar{\mathbf{x}}_{1H} - \bar{\mathbf{x}}_2)'\mathbf{S}^{-1}_{H,\text{pooled}}(\bar{\mathbf{x}}_{1H} + \bar{\mathbf{x}}_2)$. [See (11-19) and (11-20).]

Thus with $\mathbf{x}'_H = [2, 12]$ we have

Squared distance from $\bar{\mathbf{x}}_{1H} = (\mathbf{x}_H - \bar{\mathbf{x}}_{1H})'\mathbf{S}^{-1}_{H,\text{pooled}}(\mathbf{x}_H - \bar{\mathbf{x}}_{1H})$

$$= [2 - 3.5 \quad 12 - 9]\frac{1}{8}\begin{bmatrix} 10 & 1 \\ 1 & 2.5 \end{bmatrix}\begin{bmatrix} 2 & -3.5 \\ 12 & -9 \end{bmatrix} = 4.5$$

Squared distance from $\bar{\mathbf{x}}_2 = (\mathbf{x}_H - \bar{\mathbf{x}}_2)'\mathbf{S}^{-1}_{H,\text{pooled}}(\mathbf{x}_H - \bar{\mathbf{x}}_2)$

$$= [2 - 4 \quad 12 - 7]\frac{1}{8}\begin{bmatrix} 10 & 1 \\ 1 & 2.5 \end{bmatrix}\begin{bmatrix} 2 & -4 \\ 12 & -7 \end{bmatrix} = 10.3$$

Since the distance from $\mathbf{x}_H$ to $\bar{\mathbf{x}}_{1H}$ is smaller than the distance from $\mathbf{x}_H$ to $\bar{\mathbf{x}}_2$, we classify $\mathbf{x}_H$ as a π_1 observation. In this case, the classification is correct.

If $\mathbf{x}'_H = [4, 10]$ is withheld, $\bar{\mathbf{x}}_{1H}$ and $\mathbf{S}^{-1}_{H,\text{pooled}}$ become

$$\bar{\mathbf{x}}_{1H} = \begin{bmatrix} 2.5 \\ 10 \end{bmatrix} \quad \text{and} \quad \mathbf{S}^{-1}_{H,\text{pooled}} = \frac{1}{8}\begin{bmatrix} 16 & 4 \\ 4 & 2.5 \end{bmatrix}$$

---

[8]A matrix identity due to Bartlett [3] allows for the quick calculation of $\mathbf{S}^{-1}_{H,\text{pooled}}$ directly from $\mathbf{S}^{-1}_{\text{pooled}}$. Thus one does not have to recompute the inverse after withholding each observation. (See Exercise 11.20.)

We find that

$$(\mathbf{x}_H - \bar{\mathbf{x}}_{1H})'\mathbf{S}_{H,\text{pooled}}^{-1}(\mathbf{x}_H - \bar{\mathbf{x}}_{1H}) = [4 - 2.5 \quad 10 - 10]\frac{1}{8}\begin{bmatrix} 16 & 4 \\ 4 & 2.5 \end{bmatrix}\begin{bmatrix} 4 - 2.5 \\ 10 - 10 \end{bmatrix}$$

$$= 4.5$$

$$(\mathbf{x}_H - \bar{\mathbf{x}}_2)'\mathbf{S}_{H,\text{pooled}}^{-1}(\mathbf{x}_H - \bar{\mathbf{x}}_2) = [4 - 4 \quad 10 - 7]\frac{1}{8}\begin{bmatrix} 16 & 4 \\ 4 & 2.5 \end{bmatrix}\begin{bmatrix} 4 - 4 \\ 10 - 7 \end{bmatrix}$$

$$= 2.8$$

and consequently, we would incorrectly assign $\mathbf{x}_H' = [4, 10]$ to π_2. Holding out $\mathbf{x}_H' = [3, 8]$ leads to incorrectly assigning this observation to π_2 as well. Thus, $n_{1M}^{(H)} = 2$.

Turning to the second group, suppose $\mathbf{x}_H' = [5, 7]$ is withheld. Then

$$\mathbf{X}_{2H} = \begin{bmatrix} 3 & 9 \\ 4 & 5 \end{bmatrix}; \quad \bar{\mathbf{x}}_{2H} = \begin{bmatrix} 3.5 \\ 7 \end{bmatrix}; \quad \text{and} \quad 1\mathbf{S}_{2H} = \begin{bmatrix} .5 & -2 \\ -2 & 8 \end{bmatrix}$$

The new pooled covariance matrix is

$$\mathbf{S}_{H,\text{pooled}} = \frac{1}{3}[2\mathbf{S}_1 + 1\mathbf{S}_{2H}] = \frac{1}{3}\begin{bmatrix} 2.5 & -4 \\ -4 & 16 \end{bmatrix}$$

with inverse

$$\mathbf{S}_{H,\text{pooled}}^{-1} = \frac{3}{24}\begin{bmatrix} 16 & 4 \\ 4 & 2.5 \end{bmatrix}$$

We find that

$$(\mathbf{x}_H - \bar{\mathbf{x}}_1)'\mathbf{S}_{H,\text{pooled}}^{-1}(\mathbf{x}_H - \bar{\mathbf{x}}_1) = [5 - 3 \quad 7 - 10]\frac{3}{24}\begin{bmatrix} 16 & 4 \\ 4 & 2.5 \end{bmatrix}\begin{bmatrix} 5 - 3 \\ 7 - 10 \end{bmatrix}$$

$$= 4.8$$

$$(\mathbf{x}_H - \bar{\mathbf{x}}_{2H})'\mathbf{S}_{H,\text{pooled}}^{-1}(\mathbf{x}_H - \bar{\mathbf{x}}_{2H}) = [5 - 3.5 \quad 7 - 7]\frac{3}{24}\begin{bmatrix} 16 & 4 \\ 4 & 2.5 \end{bmatrix}\begin{bmatrix} 5 - 3.5 \\ 7 - 7 \end{bmatrix}$$

$$= 4.5$$

and $\mathbf{x}_H' = [5, 7]$ is correctly assigned to π_2.

When $\mathbf{x}_H' = [3, 9]$ is withheld,

$$(\mathbf{x}_H - \bar{\mathbf{x}}_1)'\mathbf{S}_{H,\text{pooled}}^{-1}(\mathbf{x}_H - \bar{\mathbf{x}}_1) = [3 - 3 \quad 9 - 10]\frac{3}{24}\begin{bmatrix} 10 & 1 \\ 1 & 2.5 \end{bmatrix}\begin{bmatrix} 3 - 3 \\ 9 - 10 \end{bmatrix}$$

$$= .3$$

$$(\mathbf{x}_H - \bar{\mathbf{x}}_{2H})'\mathbf{S}_{H,\text{pooled}}^{-1}(\mathbf{x}_H - \bar{\mathbf{x}}_{2H}) = [3 - 4.5 \quad 9 - 6]\frac{3}{24}\begin{bmatrix} 10 & 1 \\ 1 & 2.5 \end{bmatrix}\begin{bmatrix} 3 - 4.5 \\ 9 - 6 \end{bmatrix}$$

$$= 4.5$$

and $\mathbf{x}_H' = [3, 9]$ is incorrectly assigned to π_1. Finally, withholding $\mathbf{x}_H' = [4, 5]$ leads to correctly classifying this observation as π_2. Thus, $n_{2M}^{(H)} = 1$.

An estimate of the expected actual error rate is provided by

$$\hat{E}(\text{AER}) = \frac{n_{1M}^{(H)} + n_{2M}^{(H)}}{n_1 + n_2} = \frac{2 + 1}{3 + 3} = .5$$

Hence, we see that the apparent error rate APER = .33 is an optimistic measure of performance. Of course, in practice, sample sizes are larger than those we have considered here, and the difference between APER and $\hat{E}(\text{AER})$ may not be as large. ∎

If you are interested in pursuing the approaches to estimating classification error rates, see [23].

The next example illustrates a difficulty that can arise when the variance of the discriminant is not the same for both populations.

Example 11.8 (Classifying Alaskan and Canadian salmon) The salmon fishery is a valuable resource for both the United States and Canada. Because it is a limited resource, it must be managed efficiently. Moreover, since more than one country is involved, problems must be solved equitably. That is, Alaskan commercial fishermen cannot catch too many Canadian salmon and vice versa.

These fish have a remarkable life cycle. They are born in freshwater streams and after a year or two swim into the ocean. After a couple of years in salt water, they return to their place of birth to spawn and die. At the time they are about to return as mature fish, they are harvested while still in the ocean. To help regulate catches, samples of fish taken during the harvest must be identified as coming from Alaskan or Canadian waters. The fish carry some information about their birthplace in the growth rings on their scales. Typically, the rings associated with freshwater growth are smaller for the Alaskan-born than for the Canadian-born salmon. Table 11.2 gives the diameters of the growth ring regions, magnified 100 times, where

$$X_1 = \text{diameter of rings for the first-year freshwater growth}$$
$$\text{(hundredths of an inch)}$$

$$X_2 = \text{diameter of rings for the first-year marine growth}$$
$$\text{(hundredths of an inch)}$$

In addition, females are coded as 1 and males are coded as 2.

Training samples of sizes $n_1 = 50$ Alaskan-born and $n_2 = 50$ Canadian-born salmon yield the summary statistics

$$\bar{\mathbf{x}}_1 = \begin{bmatrix} 98.380 \\ 429.660 \end{bmatrix}, \quad \mathbf{S}_1 = \begin{bmatrix} 260.608 & -188.093 \\ -188.093 & 1399.086 \end{bmatrix}$$

$$\bar{\mathbf{x}}_2 = \begin{bmatrix} 137.460 \\ 366.620 \end{bmatrix}, \quad \mathbf{S}_2 = \begin{bmatrix} 326.090 & 133.505 \\ 133.505 & 893.261 \end{bmatrix}$$

Table 11.2 Salmon Data (Growth-Ring Diameters)

	Alaskan			Canadian	
Gender	Freshwater	Marine	Gender	Freshwater	Marine
2	108	368	1	129	420
1	131	355	1	148	371
1	105	469	1	179	407
2	86	506	2	152	381
1	99	402	2	166	377
2	87	423	2	124	389
1	94	440	1	156	419
2	117	489	2	131	345
2	79	432	1	140	362
1	99	403	2	144	345
1	114	428	2	149	393
2	123	372	1	108	330
1	123	372	1	135	355
2	109	420	2	170	386
2	112	394	1	152	301
1	104	407	1	153	397
2	111	422	1	152	301
2	126	423	2	136	438
2	105	434	2	122	306
1	119	474	1	148	383
1	114	396	2	90	385
2	100	470	1	145	337
2	84	399	1	123	364
2	102	429	2	145	376
2	101	469	2	115	354
2	85	444	2	134	383
1	109	397	1	117	355
2	106	442	2	126	345
1	82	431	1	118	379
2	118	381	2	120	369
1	105	388	1	153	403
1	121	403	2	150	354
1	85	451	1	154	390
1	83	453	1	155	349
1	53	427	2	109	325
1	95	411	2	117	344
1	76	442	1	128	400
1	95	426	1	144	403
2	87	402	2	163	370
1	70	397	2	145	355
2	84	511	1	133	375
2	91	469	1	128	383
1	74	451	2	123	349
2	101	474	1	144	373
1	80	398	2	140	388

(continues on next page)

Table 11.2 *(continued)*

	Alaskan			Canadian	
Gender	Freshwater	Marine	Gender	Freshwater	Marine
1	95	433	2	150	339
2	92	404	2	124	341
1	99	481	1	125	346
2	94	491	1	153	352
1	87	480	1	108	339

Gender Key: 1 = female; 2 = male.
Source: Data courtesy of K. A. Jensen and B. Van Alen of the State of Alaska Department of Fish and Game.

The data appear to satisfy the assumption of bivariate normal distributions (see Exercise 11.31), but the covariance matrices may differ. However, to illustrate a point concerning misclassification probabilities, we will use the linear classification procedure.

The classification procedure, using equal costs and equal prior probabilities, yields the holdout estimated error rates

Predicted membership

		π_1: Alaskan	π_2: Canadian
Actual	π_1: Alaskan	44	6
membership	π_2: Canadian	1	49

based on the linear classification function [see (11-19) and (11-20)]

$$\hat{w} = \hat{y} - \hat{m} = -5.54121 - .12839x_1 + .05194x_2$$

There is some difference in the sample standard deviations of $\hat{w}$ for the two populations:

	n	Sample Mean	Sample Standard Deviation
Alaskan	50	4.144	3.253
Canadian	50	−4.147	2.450

Although the overall error rate (7/100, or 7%) is quite low, there is an unfairness here. It is less likely that a Canadian-born salmon will be misclassified as Alaskan born, rather than vice versa. Figure 11.8, which shows the two normal densities for the linear discriminant $\hat{y}$, explains this phenomenon. Use of the

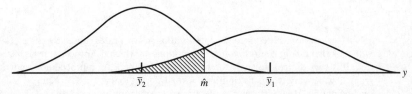

Figure 11.8 Schematic of normal densities for linear discriminant—salmon data.

midpoint between the two sample means does not make the two misclassification probabilities equal. It clearly penalizes the population with the largest variance. Thus, blind adherence to the linear classification procedure can be unwise. ∎

It should be intuitively clear that good classification (low error rates) will depend upon the separation of the populations. The farther apart the groups, the more likely it is that a *useful* classification rule can be developed. This separative goal, alluded to in Section 11.1, is explored further in Section 11.6.

As we shall see, allocation rules appropriate for the case involving equal prior probabilities and equal misclassification costs correspond to functions designed to maximally separate populations. It is in this situation that we begin to lose the distinction between classification and separation.

11.5 Classification with Several Populations

In theory, the generalization of classification procedures from 2 to $g \geq 2$ groups is straightforward. However, not much is known about the properties of the corresponding *sample* classification functions, and in particular, their error rates have not been fully investigated.

The "robustness" of the *two* group linear classification statistics to, for instance, unequal covariances or nonnormal distributions can be studied with computer generated sampling experiments.[9] For more than two populations, this approach does not lead to general conclusions, because the properties depend on where the populations are located, and there are far too many configurations to study conveniently.

As before, our approach in this section will be to develop the theoretically optimal rules and then indicate the modifications required for real-world applications.

The Minimum Expected Cost of Misclassification Method

Let $f_i(\mathbf{x})$ be the density associated with population π_i, $i = 1, 2, \ldots, g$. [For the most part, we shall take $f_i(\mathbf{x})$ to be a multivariate normal density, but this is unnecessary for the development of the general theory.] Let

$$p_i = \text{the prior probability of population } \pi_i, \quad i = 1, 2, \ldots, g$$

$$c(k \mid i) = \text{the cost of allocating an item to } \pi_k \text{ when, in fact, it belongs}$$
$$\text{to } \pi_i, \quad \text{for } k, i = 1, 2, \ldots, g$$

For $k = i$, $c(i \mid i) = 0$. Finally, let R_k be the set of $\mathbf{x}$'s classified as π_k and

$$P(k \mid i) = P(\text{classifying item as } \pi_k \mid \pi_i) = \int_{R_k} f_i(\mathbf{x}) \, d\mathbf{x}$$

for $k, i = 1, 2, \ldots, g$ with $P(i \mid i) = 1 - \sum_{\substack{k=1 \\ k \neq i}}^{g} P(k \mid i)$.

---

[9]Here *robustness* refers to the deterioration in error rates caused by using a classification procedure with data that do not conform to the assumptions on which the procedure was based.

It is very difficult to study the robustness of classification procedures analytically. However, data from a wide variety of distributions with different covariance structures can be easily generated on a computer. The performance of various classification rules can then be evaluated using computer-generated "samples" from these distributions.

The conditional expected cost of misclassifying an $\mathbf{x}$ from π_1 into π_2, or $\pi_3, \ldots,$ or π_g is

$$\text{ECM}(1) = P(2|1)c(2|1) + P(3|1)c(3|1) + \cdots + P(g|1)c(g|1)$$

$$= \sum_{k=2}^{g} P(k|1)c(k|1)$$

This conditional expected cost occurs with prior probability p_1, the probability of π_1.

In a similar manner, we can obtain the conditional expected costs of misclassification $\text{ECM}(2), \ldots, \text{ECM}(g)$. Multiplying each conditional ECM by its prior probability and summing gives the overall ECM:

$$\text{ECM} = p_1 \text{ECM}(1) + p_2 \text{ECM}(2) + \cdots + p_g \text{ECM}(g)$$

$$= p_1 \left(\sum_{k=2}^{g} P(k|1)c(k|1) \right) + p_2 \left(\sum_{\substack{k=1 \\ k \neq 2}}^{g} P(k|2)c(k|2) \right)$$

$$+ \cdots + p_g \left(\sum_{k=1}^{g-1} P(k|g)c(k|g) \right)$$

$$= \sum_{i=1}^{g} p_i \left(\sum_{\substack{k=1 \\ k \neq i}}^{g} P(k|i)c(k|i) \right) \tag{11-37}$$

Determining an optimal classification procedure amounts to choosing the mutually exclusive and exhaustive classification regions $R_1, R_2, \ldots, R_g$ such that (11-37) is a minimum.

Result 11.5. The classification regions that minimize the ECM (11-37) are defined by allocating $\mathbf{x}$ to that population $\pi_k, k = 1, 2, \ldots, g$, for which

$$\sum_{\substack{i=1 \\ i \neq k}}^{g} p_i f_i(\mathbf{x}) c(k|i) \tag{11-38}$$

is smallest. If a tie occurs, $\mathbf{x}$ can be assigned to any of the tied populations.

Proof. See Anderson [2].   ∎

Suppose all the misclassification costs are equal, in which case the minimum expected cost of misclassification rule is the minimum total probability of misclassification rule. (Without loss of generality, we can set all the misclassification costs equal to 1.) Using the argument leading to (11-38), we would allocate $\mathbf{x}$ to that population $\pi_k, k = 1, 2, \ldots, g$, for which

$$\sum_{\substack{i=1 \\ i \neq k}}^{g} p_i f_i(\mathbf{x}) \tag{11-39}$$

is smallest. Now, (11-39) will be smallest when the omitted term, $p_k f_k(\mathbf{x})$, is *largest*. Consequently, when the misclassification costs are the same, the minimum expected cost of misclassification rule has the following rather simple form.

Minimum ECM Classification Rule with Equal Misclassification Costs

Allocate $\mathbf{x}_0$ to π_k if

$$p_k f_k(\mathbf{x}) > p_i f_i(\mathbf{x}) \quad \text{for all } i \neq k \tag{11-40}$$

or, equivalently,

Allocate $\mathbf{x}_0$ to π_k if

$$\ln p_k f_k(\mathbf{x}) > \ln p_i f_i(\mathbf{x}) \quad \text{for all } i \neq k \tag{11-41}$$

It is interesting to note that the classification rule in (11-40) is identical to the one that maximizes the "posterior" probability $P(\pi_k | \mathbf{x}) = P$ (**x** comes from π_k given that **x** was observed), where

$$P(\pi_k | \mathbf{x}) = \frac{p_k f_k(\mathbf{x})}{\displaystyle\sum_{i=1}^{g} p_i f_i(\mathbf{x})} = \frac{(\text{prior}) \times (\text{likelihood})}{\sum [(\text{prior}) \times (\text{likelihood})]} \quad \text{for } k = 1, 2, \ldots, g$$

$$\tag{11-42}$$

Equation (11-42) is the generalization of Equation (11-9) to $g \geq 2$ groups.

You should keep in mind that, in general, the minimum ECM rules have three components: prior probabilities, misclassification costs, and density functions. These components must be specified (or estimated) before the rules can be implemented.

Example 11.9 (Classifying a new observation into one of three known populations) Let us assign an observation $\mathbf{x}_0$ to one of the $g = 3$ populations $\pi_1, \pi_2,$ or π_3, given the following hypothetical prior probabilities, misclassification costs, and density values:

		True population		
		π_1	π_2	π_3
Classify as:	π_1	$c(1\|1) = 0$	$c(1\|2) = 500$	$c(1\|3) = 100$
	π_2	$c(2\|1) = 10$	$c(2\|2) = 0$	$c(2\|3) = 50$
	π_3	$c(3\|1) = 50$	$c(3\|2) = 200$	$c(3\|3) = 0$
Prior probabilities:		$p_1 = .05$	$p_2 = .60$	$p_3 = .35$
Densities at $\mathbf{x}_0$:		$f_1(\mathbf{x}_0) = .01$	$f_2(\mathbf{x}_0) = .85$	$f_3(\mathbf{x}_0) = 2$

We shall use the minimum ECM procedures.

The values of $\sum\limits_{\substack{i=1 \\ i \ne k}}^{3} p_i f_i(\mathbf{x}_0) c(k \mid i)$ [see (11-38)] are

$k = 1$: $p_2 f_2(\mathbf{x}_0) c(1 \mid 2) + p_3 f_3(\mathbf{x}_0) c(1 \mid 3)$
    $= (.60)(.85)(500) + (.35)(2)(100) = 325$

$k = 2$: $p_1 f_1(\mathbf{x}_0) c(2 \mid 1) + p_3 f_3(\mathbf{x}_0) c(2 \mid 3)$
    $= (.05)(.01)(10) + (.35)(2)(50) = 35.055$

$k = 3$: $p_1 f_1(\mathbf{x}_0) c(3 \mid 1) + p_2 f_2(\mathbf{x}_0) c(3 \mid 2)$
    $= (.05)(.01)(50) + (.60)(.85)(200) = 102.025$

Since $\sum\limits_{\substack{i=1 \\ i \ne k}}^{3} p_i f_i(\mathbf{x}_0) c(k \mid i)$ is smallest for $k = 2$, we would allocate $\mathbf{x}_0$ to π_2.

If all costs of misclassification were equal, we would assign $\mathbf{x}_0$ according to (11-40), which requires only the products

$$p_1 f_1(\mathbf{x}_0) = (.05)(.01) = .0005$$
$$p_2 f_2(\mathbf{x}_0) = (.60)(.85) = .510$$
$$p_3 f_3(\mathbf{x}_0) = (.35)(2) = .700$$

Since

$$p_3 f_3(\mathbf{x}_0) = .700 \ge p_i f_i(\mathbf{x}_0), i = 1, 2$$

we should allocate $\mathbf{x}_0$ to π_3. Equivalently, calculating the posterior probabilities [see (11-42)], we obtain

$$P(\pi_1 \mid \mathbf{x}_0) = \frac{p_1 f_1(\mathbf{x}_0)}{\sum\limits_{i=1}^{3} p_i f_i(\mathbf{x}_0)}$$

$$= \frac{(.05)(.01)}{(.05)(.01) + (.60)(.85) + (.35)(2)} = \frac{.0005}{1.2105} = .0004$$

$$P(\pi_2 \mid \mathbf{x}_0) = \frac{p_2 f_2(\mathbf{x}_0)}{\sum\limits_{i=1}^{3} p_i f_i(\mathbf{x}_0)} = \frac{(.60)(.85)}{1.2105} = \frac{.510}{1.2105} = .421$$

$$P(\pi_3 \mid \mathbf{x}_0) = \frac{p_3 f_3(\mathbf{x}_0)}{\sum\limits_{i=1}^{3} p_i f_i(\mathbf{x}_0)} = \frac{(.35)(2)}{1.2105} = \frac{.700}{1.2105} = .578$$

We see that $\mathbf{x}_0$ is allocated to π_3, the population with the largest posterior probability. ∎

Classification with Normal Populations

An important special case occurs when the

$$f_i(\mathbf{x}) = \frac{1}{(2\pi)^{p/2} |\mathbf{\Sigma}_i|^{1/2}} \exp\left[-\frac{1}{2} (\mathbf{x} - \boldsymbol{\mu}_i)' \mathbf{\Sigma}_i^{-1} (\mathbf{x} - \boldsymbol{\mu}_i) \right],$$

$$i = 1, 2, \dots, g \qquad (11\text{-}43)$$

are multivariate normal densities with mean vectors $\boldsymbol{\mu}_i$ and covariance matrices $\boldsymbol{\Sigma}_i$. If, further, $c(i \mid i) = 0$, $c(k \mid i) = 1$, $k \neq i$ (or, equivalently, the misclassification costs are all equal), then (11-41) becomes

Allocate $\mathbf{x}$ to π_k if

$$\ln p_k f_k(\mathbf{x}) = \ln p_k - \left(\frac{p}{2}\right) \ln (2\pi) - \frac{1}{2}\ln |\boldsymbol{\Sigma}_k| - \frac{1}{2}(\mathbf{x} - \boldsymbol{\mu}_k)'\boldsymbol{\Sigma}_k^{-1}(\mathbf{x} - \boldsymbol{\mu}_k)$$

$$= \max_i \ln p_i f_i(\mathbf{x}) \tag{11-44}$$

The constant $(p/2) \ln (2\pi)$ can be ignored in (11-44), since it is the same for all populations. We therefore define the *quadratic discrimination score* for the ith population to be

$$d_i^Q(\mathbf{x}) = -\tfrac{1}{2}\ln |\boldsymbol{\Sigma}_i| - \tfrac{1}{2}(\mathbf{x} - \boldsymbol{\mu}_i)'\boldsymbol{\Sigma}_i^{-1}(\mathbf{x} - \boldsymbol{\mu}_i) + \ln p_i$$

$$i = 1, 2, \ldots, g \tag{11-45}$$

The quadratic score $d_i^Q(\mathbf{x})$ is composed of contributions from the generalized variance $|\boldsymbol{\Sigma}_i|$, the prior probability p_i, and the square of the distance from $\mathbf{x}$ to the population mean $\boldsymbol{\mu}_i$. Note, however, that a different distance function, with a different orientation and size of the constant-distance ellipsoid, must be used for each population.

Using discriminant scores, we find that the classification rule (11-44) becomes the following:

Minimum Total Probability of Misclassification (TPM) Rule for Normal Populations—Unequal $\boldsymbol{\Sigma}_i$

Allocate $\mathbf{x}$ to π_k if

the quadratic score $d_k^Q(\mathbf{x}) = $ largest of $d_1^Q(\mathbf{x}), d_2^Q(\mathbf{x}), \ldots, d_g^Q(\mathbf{x})$ (11-46)

where $d_i^Q(\mathbf{x})$ is given by (11-45).

In practice, the $\boldsymbol{\mu}_i$ and $\boldsymbol{\Sigma}_i$ are unknown, but a training set of correctly classified observations is often available for the construction of estimates. The relevant sample quantities for population π_i are

$$\bar{\mathbf{x}}_i = \text{sample mean vector}$$
$$\mathbf{S}_i = \text{sample covariance matrix}$$

and

$$n_i = \text{sample size}$$

The estimate of the quadratic discrimination score $\hat{d}_i^Q(\mathbf{x})$ is then

$$\hat{d}_i^Q(\mathbf{x}) = -\tfrac{1}{2}\ln |\mathbf{S}_i| - \tfrac{1}{2}(\mathbf{x} - \bar{\mathbf{x}}_i)'\mathbf{S}_i^{-1}(\mathbf{x} - \bar{\mathbf{x}}_i) + \ln p_i, \quad i = 1, 2, \ldots, g \tag{11-47}$$

and the classification rule based on the sample is as follows:

Estimated Minimum (TPM) Rule
for Several Normal Populations—Unequal Σ_i

Allocate $\mathbf{x}$ to π_k if

$$\text{the quadratic score } \hat{d}_k^Q(\mathbf{x}) = \text{largest of } \hat{d}_1^Q(\mathbf{x}), \hat{d}_2^Q(\mathbf{x}), \ldots, \hat{d}_g^Q(\mathbf{x}) \qquad (11\text{-}48)$$

where $\hat{d}_i^Q(\mathbf{x})$ is given by (11-47).

A simplification is possible if the population covariance matrices, Σ_i, are equal. When $\Sigma_i = \Sigma$, for $i = 1, 2, \ldots, g$, the discriminant score in (11-45) becomes

$$d_i^Q(\mathbf{x}) = -\tfrac{1}{2}\ln|\Sigma| - \tfrac{1}{2}\mathbf{x}'\Sigma^{-1}\mathbf{x} + \boldsymbol{\mu}_i'\Sigma^{-1}\mathbf{x} - \tfrac{1}{2}\boldsymbol{\mu}_i'\Sigma^{-1}\boldsymbol{\mu}_i + \ln p_i$$

The first two terms are the same for $d_1^Q(\mathbf{x}), d_2^Q(\mathbf{x}), \ldots, d_g^Q(\mathbf{x})$, and, consequently, they can be ignored for allocative purposes. The remaining terms consist of a constant $c_i = \ln p_i - \tfrac{1}{2}\boldsymbol{\mu}_i'\Sigma^{-1}\boldsymbol{\mu}_i$ and a *linear* combination of the components of $\mathbf{x}$.

Next, define the *linear discriminant score*

$$d_i(\mathbf{x}) = \boldsymbol{\mu}_i'\Sigma^{-1}\mathbf{x} - \tfrac{1}{2}\boldsymbol{\mu}_i'\Sigma^{-1}\boldsymbol{\mu}_i + \ln p_i \qquad (11\text{-}49)$$
$$\text{for } i = 1, 2, \ldots, g$$

An estimate $\hat{d}_i(\mathbf{x})$ of the linear discriminant score $d_i(\mathbf{x})$ is based on the pooled estimate of Σ.

$$\mathbf{S}_{\text{pooled}} = \frac{1}{n_1 + n_2 + \cdots + n_g - g}\left((n_1 - 1)\mathbf{S}_1 + (n_2 - 1)\mathbf{S}_2 + \cdots + (n_g - 1)\mathbf{S}_g\right)$$
$$(11\text{-}50)$$

and is given by

$$\hat{d}_i(\mathbf{x}) = \bar{\mathbf{x}}_i'\mathbf{S}_{\text{pooled}}^{-1}\mathbf{x} - \tfrac{1}{2}\bar{\mathbf{x}}_i'\mathbf{S}_{\text{pooled}}^{-1}\bar{\mathbf{x}}_i + \ln p_i \qquad (11\text{-}51)$$
$$\text{for } i = 1, 2, \ldots, g$$

Consequently, we have the following:

Estimated Minimum TPM Rule
for Equal-Covariance Normal Populations

Allocate $\mathbf{x}$ to π_k if

$$\text{the linear discriminant score } \hat{d}_k(\mathbf{x}) = \text{the largest of } \hat{d}_1(\mathbf{x}), \hat{d}_2(\mathbf{x}), \ldots, \hat{d}_g(\mathbf{x})$$
$$(11\text{-}52)$$

with $\hat{d}_i(\mathbf{x})$ given by (11-51).

Comment. Expression (11-49) is a convenient linear function of $\mathbf{x}$. An equivalent classifier for the equal-covariance case can be obtained from (11-45) by ignoring the constant term, $-\tfrac{1}{2}\ln|\Sigma|$. The result, with sample estimates inserted for unknown population quantities, can then be interpreted in terms of the squared distances

$$D_i^2(\mathbf{x}) = (\mathbf{x} - \bar{\mathbf{x}}_i)'\mathbf{S}_{\text{pooled}}^{-1}(\mathbf{x} - \bar{\mathbf{x}}_i) \qquad (11\text{-}53)$$

from $\mathbf{x}$ to the sample mean vector $\bar{\mathbf{x}}_i$. The allocatory rule is then

> Assign $\mathbf{x}$ to the population π_i for which $-\frac{1}{2}D_i^2(\mathbf{x}) + \ln p_i$ is largest   (11-54)

We see that this rule—or, equivalently, (11-52)—assigns $\mathbf{x}$ to the "closest" population. (The distance measure is penalized by $\ln p_i$.)

If the prior probabilities are unknown, the usual procedure is to set $p_1 = p_2 = \cdots = p_g = 1/g$. An observation is then assigned to the closest population.

Example 11.10 (Calculating sample discriminant scores, assuming a common covariance matrix) Let us calculate the linear discriminant scores based on data from $g = 3$ populations assumed to be bivariate normal with a common covariance matrix.

Random samples from the populations $\pi_1, \pi_2,$ and π_3, along with the sample mean vectors and covariance matrices, are as follows:

$$\pi_1: \quad \mathbf{X}_1 = \begin{bmatrix} -2 & 5 \\ 0 & 3 \\ -1 & 1 \end{bmatrix}, \quad \text{so } n_1 = 3, \quad \bar{\mathbf{x}}_1 = \begin{bmatrix} -1 \\ 3 \end{bmatrix}, \quad \text{and } \mathbf{S}_1 = \begin{bmatrix} 1 & -1 \\ -1 & 4 \end{bmatrix}$$

$$\pi_2: \quad \mathbf{X}_2 = \begin{bmatrix} 0 & 6 \\ 2 & 4 \\ 1 & 2 \end{bmatrix}, \quad \text{so } n_2 = 3, \quad \bar{\mathbf{x}}_2 = \begin{bmatrix} 1 \\ 4 \end{bmatrix}, \quad \text{and } \mathbf{S}_2 = \begin{bmatrix} 1 & -1 \\ -1 & 4 \end{bmatrix}$$

$$\pi_3: \quad \mathbf{X}_3 = \begin{bmatrix} 1 & -2 \\ 0 & 0 \\ -1 & -4 \end{bmatrix}, \quad \text{so } n_3 = 3, \quad \bar{\mathbf{x}}_3 = \begin{bmatrix} 0 \\ -2 \end{bmatrix}, \quad \text{and } \mathbf{S}_3 = \begin{bmatrix} 1 & 1 \\ 1 & 4 \end{bmatrix}$$

Given that $p_1 = p_2 = .25$ and $p_3 = .50$, let us classify the observation $\mathbf{x}_0' = [x_{01}, x_{02}] = [-2 \quad -1]$ according to (11-52). From (11-50),

$$\mathbf{S}_{\text{pooled}} = \frac{3-1}{9-3}\begin{bmatrix} 1 & -1 \\ -1 & 4 \end{bmatrix} + \frac{3-1}{9-3}\begin{bmatrix} 1 & -1 \\ -1 & 4 \end{bmatrix} + \frac{3-1}{9-3}\begin{bmatrix} 1 & 1 \\ 1 & 4 \end{bmatrix}$$

$$= \frac{2}{6}\begin{bmatrix} 1+1+1 & -1-1+1 \\ -1-1+1 & 4+4+4 \end{bmatrix} = \begin{bmatrix} 1 & -\frac{1}{3} \\ -\frac{1}{3} & 4 \end{bmatrix}$$

so

$$\mathbf{S}_{\text{pooled}}^{-1} = \frac{9}{35}\begin{bmatrix} 4 & \frac{1}{3} \\ \frac{1}{3} & 1 \end{bmatrix} = \frac{1}{35}\begin{bmatrix} 36 & 3 \\ 3 & 9 \end{bmatrix}$$

Next,

$$\bar{\mathbf{x}}_1'\mathbf{S}_{\text{pooled}}^{-1} = \begin{bmatrix} -1 & 3 \end{bmatrix}\frac{1}{35}\begin{bmatrix} 36 & 3 \\ 3 & 9 \end{bmatrix} = \frac{1}{35}\begin{bmatrix} -27 & 24 \end{bmatrix}$$

and

$$\bar{\mathbf{x}}_1' \mathbf{S}_{\text{pooled}}^{-1} \bar{\mathbf{x}}_1 = \frac{1}{35} [-27 \quad 24] \begin{bmatrix} -1 \\ 3 \end{bmatrix} = \frac{99}{35}$$

so

$$\hat{d}_1(\mathbf{x}_0) = \ln p_1 + \bar{\mathbf{x}}_1' \mathbf{S}_{\text{pooled}}^{-1} \mathbf{x}_0 - \frac{1}{2} \bar{\mathbf{x}}_1' \mathbf{S}_{\text{pooled}}^{-1} \bar{\mathbf{x}}_1$$

$$= \ln (.25) + \left(\frac{-27}{35} \right) x_{01} + \left(\frac{24}{35} \right) x_{02} - \frac{1}{2} \left(\frac{99}{35} \right)$$

Notice the linear form of $\hat{d}_1(\mathbf{x}_0) = \text{constant} + (\text{constant}) \, x_{01} + (\text{constant}) \, x_{02}$. In a similar manner,

$$\bar{\mathbf{x}}_2' \mathbf{S}_{\text{pooled}}^{-1} = [1 \quad 4] \frac{1}{35} \begin{bmatrix} 36 & 3 \\ 3 & 9 \end{bmatrix} = \frac{1}{35} [48 \quad 39]$$

$$\bar{\mathbf{x}}_2' \mathbf{S}_{\text{pooled}}^{-1} \bar{\mathbf{x}}_2 = \frac{1}{35} [48 \quad 39] \begin{bmatrix} 1 \\ 4 \end{bmatrix} = \frac{204}{35}$$

and

$$\hat{d}_2(\mathbf{x}_0) = \ln (.25) + \left(\frac{48}{35} \right) x_{01} + \left(\frac{39}{35} \right) x_{02} - \frac{1}{2} \left(\frac{204}{35} \right)$$

Finally,

$$\bar{\mathbf{x}}_3' \mathbf{S}_{\text{pooled}}^{-1} = [0 \quad -2] \frac{1}{35} \begin{bmatrix} 36 & 3 \\ 3 & 9 \end{bmatrix} = \frac{1}{35} [-6 \quad -18]$$

$$\bar{\mathbf{x}}_3' \mathbf{S}_{\text{pooled}}^{-1} \bar{\mathbf{x}}_3 = \frac{1}{35} [-6 \quad -18] \begin{bmatrix} 0 \\ -2 \end{bmatrix} = \frac{36}{35}$$

and

$$\hat{d}_3(\mathbf{x}_0) = \ln (.50) + \left(\frac{-6}{35} \right) x_{01} + \left(\frac{-18}{35} \right) x_{02} - \frac{1}{2} \left(\frac{36}{35} \right)$$

Substituting the numerical values $x_{01} = -2$ and $x_{02} = -1$ gives

$$\hat{d}_1(\mathbf{x}_0) = -1.386 + \left(\frac{-27}{35} \right) (-2) + \left(\frac{24}{35} \right) (-1) - \frac{99}{70} = -1.943$$

$$\hat{d}_2(\mathbf{x}_0) = -1.386 + \left(\frac{48}{35} \right) (-2) + \left(\frac{39}{35} \right) (-1) - \frac{204}{70} = -8.158$$

$$\hat{d}_3(\mathbf{x}_0) = -.693 + \left(\frac{-6}{35} \right) (-2) + \left(\frac{-18}{35} \right) (-1) - \frac{36}{70} = -.350$$

Since $\hat{d}_3(\mathbf{x}_0) = -.350$ is the largest discriminant score, we allocate $\mathbf{x}_0$ to π_3. ∎

Example 11.11 (Classifying a potential business-school graduate student) The admission officer of a business school has used an "index" of undergraduate grade point average (GPA) and graduate management aptitude test (GMAT) scores to help decide which applicants should be admitted to the school's graduate programs. Figure 11.9 shows pairs of x_1 = GPA, x_2 = GMAT values for groups of recent applicants who have been categorized as π_1: admit; π_2: do not admit; and π_3: borderline.[10] The data pictured are listed in Table 11.6. (See Exercise 11.29.) These data yield (see the SAS statistical software output in Panel 11.1)

$$n_1 = 31 \qquad\qquad n_2 = 28 \qquad\qquad n_3 = 26$$

$$\bar{\mathbf{x}}_1 = \begin{bmatrix} 3.40 \\ 561.23 \end{bmatrix} \qquad \bar{\mathbf{x}}_2 = \begin{bmatrix} 2.48 \\ 447.07 \end{bmatrix} \qquad \bar{\mathbf{x}}_3 = \begin{bmatrix} 2.99 \\ 446.23 \end{bmatrix}$$

$$\bar{\mathbf{x}} = \begin{bmatrix} 2.97 \\ 488.45 \end{bmatrix} \qquad \mathbf{S}_{\text{pooled}} = \begin{bmatrix} .0361 & -2.0188 \\ -2.0188 & 3655.9011 \end{bmatrix}$$

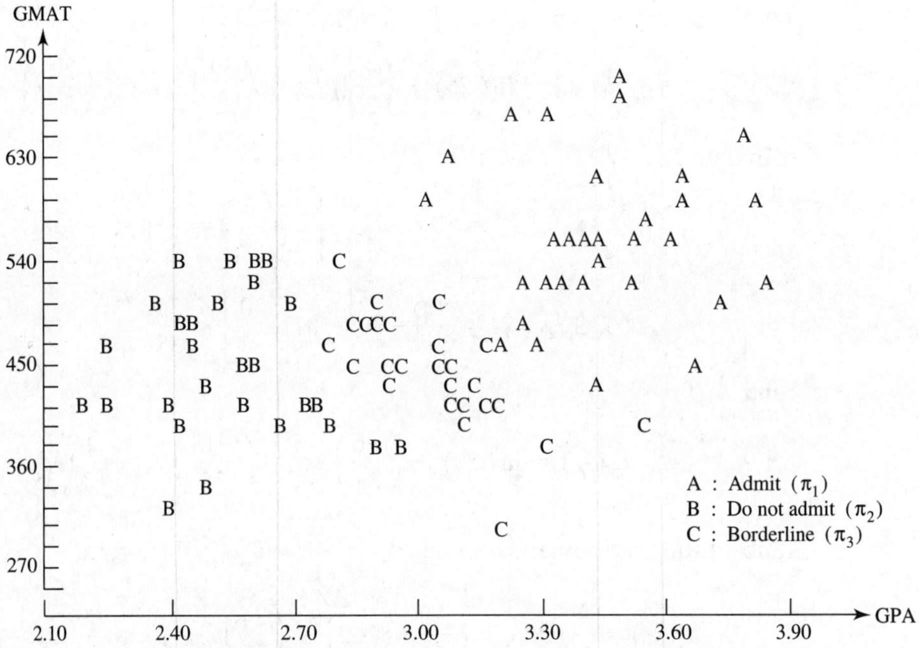

Figure 11.9 Scatter plot of $(x_1 = \text{GPA}, x_2 = \text{GMAT})$ for applicants to a graduate school of business who have been classified as admit, do not admit, or borderline.

[10] In this case, the populations are artificial in the sense that they have been created by the admissions officer. On the other hand, experience has shown that applicants with high GPA and high GMAT scores generally do well in a graduate program; those with low readings on these variables generally experience difficulty.

Suppose a new applicant has an undergraduate GPA of $x_1 = 3.21$ and a GMAT score of $x_2 = 497$. Let us classify this applicant using the rule in (11-54) with equal prior probabilities.

With $\mathbf{x}_0' = [3.21, 497]$, the sample squared distances are

$$D_1^2(\mathbf{x}_0) = (\mathbf{x}_0 - \bar{\mathbf{x}}_1)' \mathbf{S}_{\text{pooled}}^{-1} (\mathbf{x}_0 - \bar{\mathbf{x}}_1)$$

$$= [3.21 - 3.40, \quad 497 - 561.23] \begin{bmatrix} 28.6096 & .0158 \\ .0158 & .0003 \end{bmatrix} \begin{bmatrix} 3.21 - 3.40 \\ 497 - 561.23 \end{bmatrix}$$

$$= 2.58$$

$$D_2^2(\mathbf{x}_0) = (\mathbf{x}_0 - \bar{\mathbf{x}}_2)' \mathbf{S}_{\text{pooled}}^{-1} (\mathbf{x}_0 - \bar{\mathbf{x}}_2) = 17.10$$

$$D_3^2(\mathbf{x}_0) = (\mathbf{x}_0 - \bar{\mathbf{x}}_3)' \mathbf{S}_{\text{pooled}}^{-1} (\mathbf{x}_0 - \bar{\mathbf{x}}_3) = 2.47$$

Since the distance from $\mathbf{x}_0' = [3.21, 497]$ to the group mean $\bar{\mathbf{x}}_3$ is smallest, we assign this applicant to π_3, borderline. ∎

The linear discriminant scores (11-49) can be compared, two at a time. Using these quantities, we see that the condition that $d_k(\mathbf{x})$ is the largest linear discriminant score among $d_1(\mathbf{x}), d_2(\mathbf{x}), \ldots, d_g(\mathbf{x})$ is equivalent to

$$0 \le d_k(\mathbf{x}) - d_i(\mathbf{x})$$

$$= (\boldsymbol{\mu}_k - \boldsymbol{\mu}_i)' \boldsymbol{\Sigma}^{-1} \mathbf{x} - \frac{1}{2} (\boldsymbol{\mu}_k - \boldsymbol{\mu}_i)' \boldsymbol{\Sigma}^{-1} (\boldsymbol{\mu}_k + \boldsymbol{\mu}_i) + \ln\left(\frac{p_k}{p_i}\right)$$

for all $i = 1, 2, \ldots, g$.

PANEL 11.1 SAS ANALYSIS FOR ADMISSION DATA USING PROC DISCRIM.

```
title 'Discriminant Analysis';
data gpa;
infile 'T11-6.dat';
input gpa gmat admit $;                              PROGRAM COMMANDS
proc discrim data = gpa
method = normal pool = yes manova wcov pcov listerr crosslisterr;
priors 'admit' = .3333 'notadmit' = .3333 'border' = .3333;
class admit; var gpa gmat;
```

DISCRIMINANT ANALYSIS

85 Observations	84 DF Total	OUTPUT
2 Variables	82 DF Within Classes	
3 Classes	2 DF Between Classes	

Class Level Information

ADMIT	Frequency	Weight	Proportion	Prior Probability
admit	31	31.0000	0.364706	0.333333
border	26	26.0000	0.305882	0.333333
notadmit	28	28.0000	0.329412	0.333333

(continues on next page)

PANEL 11.1 *(continued)*

DISCRIMINANT ANALYSIS   WITHIN-CLASS COVARIANCE MATRICES

ADMIT = admit   DF = 30

Variable	GPA	GMAT
GPA	0.043558	0.058097
GMAT	0.058097	4618.247312

ADMIT = border   DF = 25

Variable	GPA	GMAT
GPA	0.029692	−5.403846
GMAT	−5.403846	2246.904615

ADMIT = notadmit   DF = 27

Variable	GPA	GMAT
GPA	0.033649	−1.192037
GMAT	−1.192037	3891.253968

> Pooled Within-Class Covariance Matrix   DF = 82

Variable	GPA	GMAT
GPA	0.036068	−2.018759
GMAT	−2.018759	3655.901121

Multivariate Statistics and F Approximations

S = 2   M = −0.5   N = 39.5

Statistic	Value	F	Num DF	Den DF	Pr > F
Wilks' Lambda	0.12637661	73.4257	4	162	0.0001
Pillai's Trace	1.00963002	41.7973	4	164	0.0001
Hotelling-Lawley Trace	5.83665601	116.7331	4	160	0.0001
Roy's Greatest Root	5.64604452	231.4878	2	82	0.0001

NOTE: F Statistic for Roy's Greatest Root is an upper bound.
NOTE: F Statistic for Wilks' Lambda is exact.

DISCRIMINANT ANALYSIS   LINEAR DISCRIMINANT FUNCTION

Constant = $-.5\bar{X}_j' \text{COV}^{-1}\bar{X}_j + \ln \text{PRIOR}_j$ Coefficient Vector = $\text{COV}^{-1}\bar{X}_j$

ADMIT

	admit	border	notadmit
CONSTANT	−241.47030	−178.41437	−134.99753
GPA	106.24991	92.66953	78.08637
GMAT	0.21218	0.17323	0.16541

> Classification Results for Calibration Data: WORK.GPA
> Resubstitution Results using Linear Discriminant Function

Generalized Squared Distance Function:

$$D_j^2(X) = (X - \bar{X}_j)' \text{cov}^{-1}(X - \bar{X}_j)$$

Posterior Probability of Membership in each ADMIT:

$$\Pr(j|X) = \exp(-.5 D_j^2(X))/\underset{k}{\text{SUM}} \exp(-.5 D_k^2(X))$$

PANEL 11.1 *(continued)*

Posterior Probability of Membership in ADMIT:

Obs	From ADMIT	Classified into ADMIT		admit	border	notadmit
2	admit	border	*	0.1202	0.8778	0.0020
3	admit	border	*	0.3654	0.6342	0.0004
24	admit	border	*	0.4766	0.5234	0.0000
31	admit	border	*	0.2964	0.7032	0.0004
58	notadmit	border	*	0.0001	0.7550	0.2450
59	notadmit	border	*	0.0001	0.8673	0.1326
66	border	admit	*	0.5336	0.4664	0.0000

*Misclassified observation

Classification Summary for Calibration Data: WORK.GPA
Cross validation Summary using Linear Discriminant Function

Generalized Squared Distance Function:
$$D_j^2(X) = (X - \bar{X}_{(X)j})' COV_{(X)}^{-1}(X - \bar{X}_{(X)j})$$
Posterior Probability of Membership in each ADMIT:
$$Pr(j|X) = \exp(-.5D_j^2(X))/\underset{k}{SUM} \exp(-.5D_k^2(X))$$

Number of Observations and Percent Classified into ADMIT:

From	ADMIT	admit	border	notadmit	Total
	admit	26	5	0	31
		83.87	16.13	0.00	100.00
	border	1	24	1	26
		3.85	92.31	3.85	100.00
	notadmit	0	2	26	28
		0.00	7.14	92.86	100.00
	Total	27	31	27	85
	Percent	31.76	36.47	31.76	100.00
	Priors	0.3333	0.3333	0.3333	

Error Count Estimates for ADMIT:

	admit	border	notadmit	Total
Rate	0.1613	0.0769	0.0714	0.1032
Priors	0.3333	0.3333	0.3333	

Adding $-\ln(p_k/p_i) = \ln(p_i/p_k)$ to both sides of the preceding inequality gives the alternative form of the classification rule that minimizes the total probability of misclassification. Thus, we

Allocate **x** to π_k if

$$(\boldsymbol{\mu}_k - \boldsymbol{\mu}_i)'\boldsymbol{\Sigma}^{-1}\mathbf{x} - \frac{1}{2}(\boldsymbol{\mu}_k - \boldsymbol{\mu}_i)'\boldsymbol{\Sigma}^{-1}(\boldsymbol{\mu}_k + \boldsymbol{\mu}_i) \geq \ln\left(\frac{p_i}{p_k}\right) \qquad (11\text{-}55)$$

for all $i = 1, 2, \ldots, g$.

Now, denote the left-hand side of (11-55) by $d_{ki}(\mathbf{x})$. Then the conditions in (11-55) define classification regions $R_1, R_2, \ldots, R_g$, which are separated by (hyper) planes. This follows because $d_{ki}(\mathbf{x})$ is a linear combination of the components of $\mathbf{x}$. For example, when $g = 3$, the classification region R_1 consists of all $\mathbf{x}$ satisfying

$$R_1: d_{1i}(\mathbf{x}) \geq \ln\left(\frac{p_i}{p_1}\right) \qquad \text{for } i = 2, 3$$

That is, R_1 consists of those $\mathbf{x}$ for which

$$d_{12}(\mathbf{x}) = (\boldsymbol{\mu}_1 - \boldsymbol{\mu}_2)'\boldsymbol{\Sigma}^{-1}\mathbf{x} - \frac{1}{2}(\boldsymbol{\mu}_1 - \boldsymbol{\mu}_2)'\boldsymbol{\Sigma}^{-1}(\boldsymbol{\mu}_1 + \boldsymbol{\mu}_2) \geq \ln\left(\frac{p_2}{p_1}\right)$$

and, *simultaneously,*

$$d_{13}(\mathbf{x}) = (\boldsymbol{\mu}_1 - \boldsymbol{\mu}_3)'\boldsymbol{\Sigma}^{-1}\mathbf{x} - \frac{1}{2}(\boldsymbol{\mu}_1 - \boldsymbol{\mu}_3)'\boldsymbol{\Sigma}^{-1}(\boldsymbol{\mu}_1 + \boldsymbol{\mu}_3) \geq \ln\left(\frac{p_3}{p_1}\right)$$

Assuming that $\boldsymbol{\mu}_1, \boldsymbol{\mu}_2$, and $\boldsymbol{\mu}_3$ do not lie along a straight line, the equations $d_{12}(\mathbf{x}) = \ln(p_2/p_1)$ and $d_{13}(\mathbf{x}) = \ln(p_3/p_1)$ define two intersecting hyperplanes that delineate R_1 in the p-dimensional variable space. The term $\ln(p_2/p_1)$ places the plane closer to $\boldsymbol{\mu}_1$ than $\boldsymbol{\mu}_2$ if p_2 is greater than p_1. The regions R_1, R_2, and R_3 are shown in Figure 11.10 for the case of two variables. The picture is the same for more variables if we graph the plane that contains the three mean vectors.

The sample version of the alternative form in (11-55) is obtained by substituting $\bar{\mathbf{x}}_i$ for $\boldsymbol{\mu}_i$ and inserting the pooled sample covariance matrix $\mathbf{S}_{\text{pooled}}$ for $\boldsymbol{\Sigma}$. When $\sum_{i=1}^{g} (n_i - 1) \geq p$, so that $\mathbf{S}_{\text{pooled}}^{-1}$ exists, this sample analog becomes

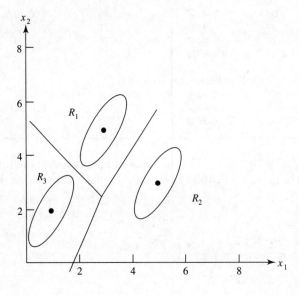

Figure 11.10 The classification regions R_1, R_2, and R_3 for the linear minimum TPM rule $(p_1 = \frac{1}{4}, p_2 = \frac{1}{2}, p_3 = \frac{1}{4})$.

Allocate $\mathbf{x}$ to π_k if

$$\hat{d}_{ki}(\mathbf{x}) = (\bar{\mathbf{x}}_k - \bar{\mathbf{x}}_i)' \mathbf{S}_{\text{pooled}}^{-1} \mathbf{x} - \frac{1}{2}(\bar{\mathbf{x}}_k - \bar{\mathbf{x}}_i)' \mathbf{S}_{\text{pooled}}^{-1}(\bar{\mathbf{x}}_k + \bar{\mathbf{x}}_i)$$

$$\geq \ln\left(\frac{p_i}{p_k}\right) \quad \text{for all } i \neq k \tag{11-56}$$

Given the fixed training set values $\bar{\mathbf{x}}_i$ and $\mathbf{S}_{\text{pooled}}$, $\hat{d}_{ki}(\mathbf{x})$ is a linear function of the components of $\mathbf{x}$. Therefore, the classification regions defined by (11-56)—or, equivalently, by (11-52)—are also bounded by hyperplanes, as in Figure 11.10.

As with the sample linear discriminant rule of (11-52), if the prior probabilities are difficult to assess, they are frequently all taken to be equal. In this case, $\ln(p_i/p_k) = 0$ for all pairs.

Because they employ estimates of population parameters, the sample classification rules (11-48) and (11-52) may no longer be optimal. Their performance, however, can be evaluated using Lachenbruch's holdout procedure. If $n_{iM}^{(H)}$ is the number of misclassified holdout observations in the ith group, $i = 1, 2, \ldots, g$, then an estimate of the expected actual error rate, $E(\text{AER})$, is provided by

$$\hat{E}(\text{AER}) = \frac{\sum_{i=1}^{g} n_{iM}^{(H)}}{\sum_{i=1}^{g} n_i} \tag{11-57}$$

Example 11.12 (Effective classification with fewer variables) In his pioneering work on discriminant functions, Fisher [9] presented an analysis of data collected by Anderson [1] on three species of iris flowers. (See Table 11.5, Exercise 11.27.)

Let the classes be defined as

$$\pi_1 \text{: } Iris\,setosa; \quad \pi_2 \text{: } Iris\,versicolor; \quad \pi_3 \text{: } Iris\,virginica$$

The following four variables were measured from 50 plants of each species.

$$X_1 = \text{sepal length}, \quad X_2 = \text{sepal width}$$
$$X_3 = \text{petal length}, \quad X_4 = \text{petal width}$$

Using all the data in Table 11.5, a linear discriminant analysis produced the confusion matrix

		Predicted membership			
		π_1: Setosa	π_2: Versicolor	π_3: Virginica	Percent correct
Actual membership	π_1: Setosa	50	0	0	100
	π_2: Versicolor	0	48	2	96
	π_3: Virginica	0	1	49	98

The elements in this matrix were generated using the holdout procedure, so (see 11-57)

$$\hat{E}(\text{AER}) = \frac{3}{150} = .02$$

The error rate, 2%, is low.

Often, it is possible to achieve effective classification with fewer variables. It is good practice to try all the variables one at a time, two at a time, three at a time, and so forth, to see how well they classify compared to the discriminant function, which uses all the variables.

If we adopt the holdout estimate of the expected AER as our criterion, we find for the data on irises:

Single variable	Misclassification rate
X_1	.253
X_2	.480
X_3	.053
X_4	.040

Pairs of variables	Misclassification rate
X_1, X_2	.207
X_1, X_3	.040
X_1, X_4	.040
X_2, X_3	.047
X_2, X_4	.040
X_3, X_4	.040

We see that the single variable X_4 = petal width does a very good job of distinguishing the three species of iris. Moreover, very little is gained by including more variables. Box plots of X_4 = petal width are shown in Figure 11.11 for the three species of iris. It is clear from the figure that petal width separates the three groups quite well, with, for example, the petal widths for *Iris setosa* much smaller than the petal widths for *Iris virginica*.

Darroch and Mosimann [6] have suggested that these species of iris may be discriminated on the basis of "shape" or scale-free information alone. Let $Y_1 = X_1/X_2$ be the sepal shape and $Y_2 = X_3/X_4$ be the petal shape. The use of the variables Y_1 and Y_2 for discrimination is explored in Exercise 11.28.

The selection of appropriate variables to use in a discriminant analysis is often difficult. A summary such as the one in this example allows the investigator to make reasonable and simple choices based on the ultimate criteria of how well the procedure classifies its target objects.   ■

Our discussion has tended to emphasize the linear discriminant rule of (11-52) or (11-56), and many commercial computer programs are based upon it. Although the linear discriminant rule has a simple structure, you must remember that it was derived under the rather strong assumptions of multivariate normality and equal covariances. Before implementing a linear classification rule, these tentative

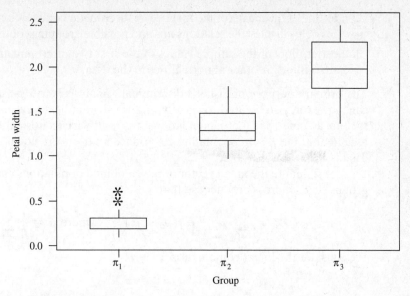

Figure 11.11 Box plots of petal width for the three species of iris.

assumptions should be checked in the order multivariate normality and then equality of covariances. If one or both of these assumptions is violated, improved classification may be possible if the data are first suitably transformed.

The quadratic rules are an alternative to classification with linear discriminant functions. They are appropriate if normality appears to hold, but the assumption of equal covariance matrices is seriously violated. However, the assumption of normality seems to be more critical for quadratic rules than linear rules. If doubt exists as to the appropriateness of a linear or quadratic rule, both rules can be constructed and their error rates examined using Lachenbruch's holdout procedure.

11.6 Fisher's Method for Discriminating among Several Populations

Fisher also proposed an extension of his discriminant method, discussed in Section 11.3, to several populations. The motivation behind the Fisher discriminant analysis is the need to obtain a reasonable representation of the populations that involves only a *few* linear combinations of the observations, such as $\mathbf{a}_1'\mathbf{x}$, $\mathbf{a}_2'\mathbf{x}$, and $\mathbf{a}_3'\mathbf{x}$. His approach has several advantages when one is interested in *separating* several populations for (1) visual inspection or (2) graphical descriptive purposes. It allows for the following:

1. Convenient representations of the g populations that reduce the dimension from a very large number of characteristics to a relatively few linear combinations. Of course, some information—needed for optimal classification—may be lost, unless the population means lie completely in the lower dimensional space selected.

2. Plotting of the means of the first two or three linear combinations (discriminants). This helps display the relationships and possible groupings of the populations.

3. Scatter plots of the sample values of the first two discriminants, which can indicate outliers or other abnormalities in the data.

The primary purpose of Fisher's discriminant analysis is to *separate* populations. It can, however, also be used to classify, and we shall indicate this use. It is not necessary to assume that the g populations are multivariate normal. However, we do assume that the $p \times p$ population covariance matrices are equal and of full rank.[11] That is, $\Sigma_1 = \Sigma_2 = \cdots = \Sigma_g = \Sigma$.

Let $\bar{\boldsymbol{\mu}}$ denote the mean vector of the combined populations and $\mathbf{B}_{\mu}$ the between groups sums of cross products, so that

$$\mathbf{B}_{\mu} = \sum_{i=1}^{g} (\boldsymbol{\mu}_i - \bar{\boldsymbol{\mu}})(\boldsymbol{\mu}_i - \bar{\boldsymbol{\mu}})' \qquad \text{where } \bar{\boldsymbol{\mu}} = \frac{1}{g} \sum_{i=1}^{g} \boldsymbol{\mu}_i \qquad (11\text{-}58)$$

We consider the linear combination

$$Y = \mathbf{a}'\mathbf{X}$$

which has expected value

$$E(Y) = \mathbf{a}' E(\mathbf{X} \mid \pi_i) = \mathbf{a}' \boldsymbol{\mu}_i \qquad \text{for population } \pi_i$$

and variance

$$\text{Var}(Y) = \mathbf{a}' \text{Cov}(\mathbf{X})\mathbf{a} = \mathbf{a}'\Sigma\mathbf{a} \qquad \text{for all populations}$$

Consequently, the expected value $\mu_{iY} = \mathbf{a}' \boldsymbol{\mu}_i$ changes as the population from which $\mathbf{X}$ is selected changes. We first define the overall mean

$$\bar{\mu}_Y = \frac{1}{g} \sum_{i=1}^{g} \mu_{iY} = \frac{1}{g} \sum_{i=1}^{g} \mathbf{a}' \boldsymbol{\mu}_i = \mathbf{a}' \left(\frac{1}{g} \sum_{i=1}^{g} \boldsymbol{\mu}_i \right)$$

$$= \mathbf{a}' \bar{\boldsymbol{\mu}}$$

and form the ratio

$$\frac{\left(\begin{array}{c} \text{sum of squared distances from} \\ \text{populations to overall mean of } Y \end{array} \right)}{(\text{variance of } Y)} = \frac{\sum\limits_{i=1}^{g} (\mu_{iY} - \bar{\mu}_Y)^2}{\sigma_Y^2} = \frac{\sum\limits_{i=1}^{g} (\mathbf{a}' \boldsymbol{\mu}_i - \mathbf{a}' \bar{\boldsymbol{\mu}})^2}{\mathbf{a}'\Sigma\mathbf{a}}$$

$$= \frac{\mathbf{a}' \left(\sum\limits_{i=1}^{g} (\boldsymbol{\mu}_i - \bar{\boldsymbol{\mu}})(\boldsymbol{\mu}_i - \bar{\boldsymbol{\mu}})' \right) \mathbf{a}}{\mathbf{a}'\Sigma\mathbf{a}}$$

or

$$\frac{\sum\limits_{i=1}^{g} (\mu_{iY} - \bar{\mu}_Y)^2}{\sigma_Y^2} = \frac{\mathbf{a}'\mathbf{B}_{\mu}\mathbf{a}}{\mathbf{a}'\Sigma\mathbf{a}} \qquad (11\text{-}59)$$

---

[11] If not, we let $\mathbf{P} = [\mathbf{e}_1, \ldots, \mathbf{e}_q]$ be the eigenvectors of Σ corresponding to nonzero eigenvalues $[\lambda_1, \ldots, \lambda_q]$. Then we replace $\mathbf{X}$ by $\mathbf{P}'\mathbf{X}$ which has a full rank covariance matrix $\mathbf{P}'\Sigma\mathbf{P}$.

The ratio in (11-59) measures the variability *between* the groups of Y-values relative to the common variability *within* groups. We can then select $\mathbf{a}$ to maximize this ratio.

Ordinarily, $\mathbf{\Sigma}$ and the $\boldsymbol{\mu}_i$ are unavailable, but we have a training set consisting of correctly classified observations. Suppose the training set consists of a random sample of size n_i from population π_i, $i = 1, 2, \ldots, g$. Denote the $n_i \times p$ data set, from population π_i, by $\mathbf{X}_i$ and its jth row by $\mathbf{x}'_{ij}$. After first constructing the sample mean vectors

$$\bar{\mathbf{x}}_i = \frac{1}{n_i} \sum_{j=1}^{n_i} \mathbf{x}_{ij}$$

and the covariance matrices $\mathbf{S}_i$, $i = 1, 2, \ldots, g$, we define the "overall average" vector

$$\bar{\mathbf{x}} = \frac{1}{g} \sum_{i=1}^{g} \bar{\mathbf{x}}_i$$

which is the $p \times 1$ vector average of the individual sample averages.

Next, analogous to $\mathbf{B}_{\boldsymbol{\mu}}$ we define the *sample between groups* matrix $\mathbf{B}$. Let

$$\mathbf{B} = \sum_{i=1}^{g} (\bar{\mathbf{x}}_i - \bar{\mathbf{x}})(\bar{\mathbf{x}}_i - \bar{\mathbf{x}})' \tag{11-60}$$

Also, an estimate of $\mathbf{\Sigma}$ is based on the *sample within groups* matrix

$$\mathbf{W} = \sum_{i=1}^{g} (n_i - 1)\mathbf{S}_i = \sum_{i=1}^{g} \sum_{j=1}^{n_i} (\mathbf{x}_{ij} - \bar{\mathbf{x}}_i)(\mathbf{x}_{ij} - \bar{\mathbf{x}}_i)' \tag{11-61}$$

Consequently, $\mathbf{W}/(n_1 + n_2 + \cdots + n_g - g) = \mathbf{S}_{\text{pooled}}$ is the estimate of $\mathbf{\Sigma}$. Before presenting the sample discriminants, we note that $\mathbf{W}$ is the constant $(n_1 + n_2 + \cdots + n_g - g)$ times $\mathbf{S}_{\text{pooled}}$, so the same $\hat{\mathbf{a}}$ that maximizes $\hat{\mathbf{a}}'\mathbf{B}\hat{\mathbf{a}}/\hat{\mathbf{a}}'\mathbf{S}_{\text{pooled}}\hat{\mathbf{a}}$ also maximizes $\hat{\mathbf{a}}'\mathbf{B}\hat{\mathbf{a}}/\hat{\mathbf{a}}'\mathbf{W}\hat{\mathbf{a}}$. Moreover, we can present the optimizing $\hat{\mathbf{a}}$ in the more customary form as eigenvectors $\hat{\mathbf{e}}_i$ of $\mathbf{W}^{-1}\mathbf{B}$, because if $\mathbf{W}^{-1}\mathbf{B}\hat{\mathbf{e}} = \hat{\lambda}\hat{\mathbf{e}}$ then $\mathbf{S}_{\text{pooled}}^{-1}\mathbf{B}\hat{\mathbf{e}} = \hat{\lambda}(n_1 + n_2 + \cdots + n_g - g)\hat{\mathbf{e}}$.

Fisher's Sample Linear Discriminants

Let $\hat{\lambda}_1, \hat{\lambda}_2, \ldots, \hat{\lambda}_s > 0$ denote the $s \le \min(g - 1, p)$ nonzero eigenvalues of $\mathbf{W}^{-1}\mathbf{B}$ and $\hat{\mathbf{e}}_1, \ldots, \hat{\mathbf{e}}_s$ be the corresponding eigenvectors (scaled so that $\hat{\mathbf{e}}'\mathbf{S}_{\text{pooled}}\hat{\mathbf{e}} = 1$). Then the vector of coefficients $\hat{\mathbf{a}}$ that maximizes the ratio

$$\frac{\hat{\mathbf{a}}'\mathbf{B}\hat{\mathbf{a}}}{\hat{\mathbf{a}}'\mathbf{W}\hat{\mathbf{a}}} = \frac{\hat{\mathbf{a}}'\left(\displaystyle\sum_{i=1}^{g} (\bar{\mathbf{x}}_i - \bar{\mathbf{x}})(\bar{\mathbf{x}}_i - \bar{\mathbf{x}})' \right)\hat{\mathbf{a}}}{\hat{\mathbf{a}}'\left[\displaystyle\sum_{i=1}^{g} \sum_{j=1}^{n_i} (\mathbf{x}_{ij} - \bar{\mathbf{x}}_i)(\mathbf{x}_{ij} - \bar{\mathbf{x}}_i)' \right]\hat{\mathbf{a}}} \tag{11-62}$$

is given by $\hat{\mathbf{a}}_1 = \hat{\mathbf{e}}_1$. The linear combination $\hat{\mathbf{a}}_1'\mathbf{x}$ is, called the *sample first discriminant*. The choice $\hat{\mathbf{a}}_2 = \hat{\mathbf{e}}_2$ produces the *sample second discriminant*, $\hat{\mathbf{a}}_2'\mathbf{x}$, and continuing, we obtain $\hat{\mathbf{a}}_k'\mathbf{x} = \hat{\mathbf{e}}_k'\mathbf{x}$, the *sample kth discriminant*, $k \le s$.

Exercise 11.21 outlines the derivation of the Fisher discriminants. The discriminants will not have zero covariance for each random sample $\mathbf{X}_i$. Rather, the condition

$$\hat{\mathbf{a}}_i' \mathbf{S}_{\text{pooled}}\, \hat{\mathbf{a}}_k = \begin{cases} 1 & \text{if } i = k \le s \\ 0 & \text{otherwise} \end{cases} \tag{11-63}$$

will be satisfied. The use of $\mathbf{S}_{\text{pooled}}$ is appropriate because we tentatively assumed that the g population covariance matrices were equal.

Example 11.13 (Calculating Fisher's sample discriminants for three populations) Consider the observations on $p = 2$ variables from $g = 3$ populations given in Example 11.10. Assuming that the populations have a common covariance matrix $\mathbf{\Sigma}$, let us obtain the Fisher discriminants. The data are

$$\pi_1\ (n_1 = 3) \qquad\qquad \pi_2\ (n_2 = 3) \qquad\qquad \pi_3\ (n_3 = 3)$$

$$\mathbf{X}_1 = \begin{bmatrix} -2 & 5 \\ 0 & 3 \\ -1 & 1 \end{bmatrix}; \qquad \mathbf{X}_2 = \begin{bmatrix} 0 & 6 \\ 2 & 4 \\ 1 & 2 \end{bmatrix}; \qquad \mathbf{X}_3 = \begin{bmatrix} 1 & -2 \\ 0 & 0 \\ -1 & -4 \end{bmatrix}$$

In Example 11.10, we found that

$$\bar{\mathbf{x}}_1 = \begin{bmatrix} -1 \\ 3 \end{bmatrix}; \qquad \bar{\mathbf{x}}_2 = \begin{bmatrix} 1 \\ 4 \end{bmatrix}; \qquad \bar{\mathbf{x}}_3 = \begin{bmatrix} 0 \\ -2 \end{bmatrix}$$

so

$$\bar{\mathbf{x}} = \begin{bmatrix} 0 \\ \dfrac{5}{3} \end{bmatrix}; \qquad \mathbf{B} = \sum_{i=1}^{3} (\bar{\mathbf{x}}_i - \bar{\mathbf{x}})(\bar{\mathbf{x}}_i - \bar{\mathbf{x}})' = \begin{bmatrix} 2 & 1 \\ 1 & 62/3 \end{bmatrix}$$

$$\mathbf{W} = \sum_{i=1}^{3} \sum_{j=1}^{n_i} (\mathbf{x}_{ij} - \bar{\mathbf{x}}_i)(\mathbf{x}_{ij} - \bar{\mathbf{x}}_i)' = (n_1 + n_2 + n_3 - 3)\,\mathbf{S}_{\text{pooled}}$$

$$= \begin{bmatrix} 6 & -2 \\ -2 & 24 \end{bmatrix}$$

$$\mathbf{W}^{-1} = \frac{1}{140}\begin{bmatrix} 24 & 2 \\ 2 & 6 \end{bmatrix}; \qquad \mathbf{W}^{-1}\mathbf{B} = \begin{bmatrix} .3571 & .4667 \\ .0714 & .9000 \end{bmatrix}$$

To solve for the $s \le \min(g-1, p) = \min(2, 2) = 2$ nonzero eigenvalues of $\mathbf{W}^{-1}\mathbf{B}$, we must solve

$$|\mathbf{W}^{-1}\mathbf{B} - \lambda\mathbf{I}| = \left| \begin{bmatrix} .3571 - \lambda & .4667 \\ .0714 & .9000 - \lambda \end{bmatrix} \right| = 0$$

or

$$(.3571 - \lambda)(.9000 - \lambda) - (.4667)(.0714) = \lambda^2 - 1.2571\lambda + .2881 = 0$$

Using the quadratic formula, we find that $\hat{\lambda}_1 = .9556$ and $\hat{\lambda}_2 = .3015$. The normalized eigenvectors $\hat{\mathbf{a}}_1$ and $\hat{\mathbf{a}}_2$ are obtained by solving

$$(\mathbf{W}^{-1}\mathbf{B} - \hat{\lambda}_i\mathbf{I})\,\hat{\mathbf{a}}_i = \mathbf{0} \qquad i = 1, 2$$

and scaling the results such that $\hat{\mathbf{a}}_i' \mathbf{S}_{\text{pooled}} \hat{\mathbf{a}}_i = 1$. For example, the solution of

$$(\mathbf{W}^{-1}\mathbf{B} - \hat{\lambda}_1\mathbf{I})\,\hat{\mathbf{a}}_1 = \begin{bmatrix} .3571 - .9556 & .4667 \\ .0714 & .9000 - .9556 \end{bmatrix} \begin{bmatrix} \hat{a}_{11} \\ \hat{a}_{12} \end{bmatrix} = \begin{bmatrix} 0 \\ 0 \end{bmatrix}$$

is, after the normalization $\hat{\mathbf{a}}_1' \mathbf{S}_{\text{pooled}} \hat{\mathbf{a}}_1 = 1$,

$$\hat{\mathbf{a}}_1' = [.386 \quad .495]$$

Similarly,

$$\hat{\mathbf{a}}_2' = [.938 \quad -.112]$$

The two discriminants are

$$\hat{y}_1 = \hat{\mathbf{a}}_1'\mathbf{x} = [.386 \quad .495]\begin{bmatrix} x_1 \\ x_2 \end{bmatrix} = .386x_1 + .495x_2$$

$$\hat{y}_2 = \hat{\mathbf{a}}_2'\mathbf{x} = [.938 \quad -.112]\begin{bmatrix} x_1 \\ x_2 \end{bmatrix} = .938x_1 - .112x_2 \qquad \blacksquare$$

Example 11.14 (Fisher's discriminants for the crude-oil data) Gerrild and Lantz [13] collected crude-oil samples from sandstone in the Elk Hills, California, petroleum reserve. These crude oils can be assigned to one of the three stratigraphic units (populations)

$$\pi_1: \text{Wilhelm sandstone}$$
$$\pi_2: \text{Sub-Mulinia sandstone}$$
$$\pi_3: \text{Upper sandstone}$$

on the basis of their chemistry. For illustrative purposes, we consider only the five variables:

$$X_1 = \text{vanadium (in percent ash)}$$
$$X_2 = \sqrt{\text{iron (in percent ash)}}$$
$$X_3 = \sqrt{\text{beryllium (in percent ash)}}$$
$$X_4 = 1/[\text{saturated hydrocarbons (in percent area)}]$$
$$X_5 = \text{aromatic hydrocarbons (in percent area)}$$

The first three variables are trace elements, and the last two are determined from a segment of the curve produced by a gas chromatograph chemical analysis. Table 11.7 (see Exercise 11.30) gives the values of the five original variables (vanadium, iron, beryllium, saturated hydrocarbons, and aromatic hydrocarbons) for 56 cases whose population assignment was certain.

A computer calculation yields the summary statistics

$$\bar{\mathbf{x}}_1 = \begin{bmatrix} 3.229 \\ 6.587 \\ .303 \\ .150 \\ 11.540 \end{bmatrix}, \quad \bar{\mathbf{x}}_2 = \begin{bmatrix} 4.445 \\ 5.667 \\ .344 \\ .157 \\ 5.484 \end{bmatrix}, \quad \bar{\mathbf{x}}_3 = \begin{bmatrix} 7.226 \\ 4.634 \\ .598 \\ .223 \\ 5.768 \end{bmatrix}, \quad \bar{\mathbf{x}} = \begin{bmatrix} 6.180 \\ 5.081 \\ .511 \\ .201 \\ 6.434 \end{bmatrix}$$

and

$$(n_1 + n_2 + n_3 - 3)\mathbf{S}_{\text{pooled}} = (38 + 11 + 7 - 3)\mathbf{S}_{\text{pooled}}$$

$$= \mathbf{W} = \begin{bmatrix} 187.575 & & & & \\ 1.957 & 41.789 & & & \\ -4.031 & 2.128 & 3.580 & & \\ 1.092 & -.143 & -.284 & .077 & \\ 79.672 & -28.243 & 2.559 & -.996 & 338.023 \end{bmatrix}$$

There are at most $s = \min(g - 1, p) = \min(2, 5) = 2$ positive eigenvalues of $\mathbf{W}^{-1}\mathbf{B}$, and they are 4.354 and .559. The centered Fisher linear discriminants are

$$\hat{y}_1 = .312(x_1 - 6.180) - .710(x_2 - 5.081) + 2.764(x_3 - .511)$$
$$+ 11.809(x_4 - .201) - .235(x_5 - 6.434)$$
$$\hat{y}_2 = .169(x_1 - 6.180) - .245(x_2 - 5.081) - 2.046(x_3 - .511)$$
$$- 24.453(x_4 - .201) - .378(x_5 - 6.434)$$

The separation of the three group means is fully explained in the two-dimensional "discriminant space." The group means and the scatter of the individual observations in the discriminant coordinate system are shown in Figure 11.12. The separation is quite good. ∎

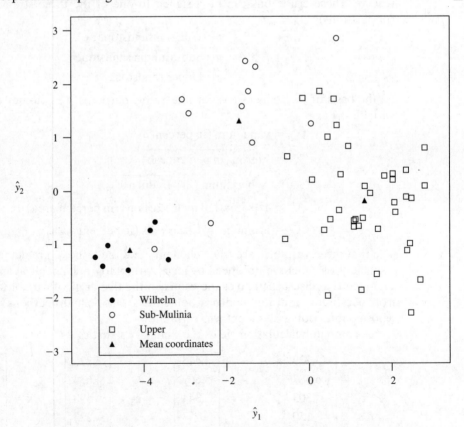

Figure 11.12 Crude-oil samples in discriminant space.

Example 11.15 (Plotting sports data in two-dimensional discriminant space) Investigators interested in sports psychology administered the Minnesota Multiphasic Personality Inventory (MMPI) to 670 letter winners at the University of Wisconsin in Madison. The sports involved and the coefficients in the two discriminant functions are given in Table 11.3.

A plot of the group means using the first two discriminant scores is shown in Figure 11.13. Here the separation on the basis of the MMPI scores is not good, although a test for the equality of means is significant at the 5% level. (This is due to the large sample sizes.)

While the discriminant coefficients suggest that the first discriminant is most closely related to the L and Pa scales, and the second discriminant is most closely associated with the D and Pt scales, we will give the interpretation provided by the investigators.

The first discriminant, which accounted for 34.4% of the common variance, was highly correlated with the Mf scale ($r = -.78$). The second discriminant, which accounted for an additional 18.3% of the variance, was most highly related to scores on the Sc, F, and D scales (r's $= .66, .54$, and $.50$, respectively). The investigators suggest that the first discriminant best represents an interest dimension; the second discriminant reflects psychological adjustment.

Ideally, the standardized discriminant function coefficients should be examined to assess the importance of a variable in the presence of other variables. (See [29].) Correlation coefficients indicate only how each variable by itself distinguishes the groups, ignoring the contributions of the other variables. Unfortunately, in this case, the standardized discriminant coefficients were unavailable.

In general, plots should also be made of other pairs of the first few discriminants. In addition, scatter plots of the discriminant scores for pairs of discriminants can be made for each sport. Under the assumption of multivariate normality, the

Table 11.3				
Sport	Sample size	MMPI Scale	First discriminant	Second discriminant
Football	158	QE	.055	−.098
Basketball	42	L	−.194	.046
Baseball	79	F	−.047	−.099
Crew	61	K	.053	−.017
Fencing	50	Hs	.077	−.076
Golf	28	D	.049	.183
Gymnastics	26	Hy	−.028	.031
Hockey	28	Pd	.001	−.069
Swimming	51	Mf	−.074	−.076
Tennis	31	Pa	.189	.088
Track	52	Pt	.025	−.188
Wrestling	64	Sc	−.046	.088
		Ma	−.103	.053
		Si	.041	.016

Source: W. Morgan and R. W. Johnson.

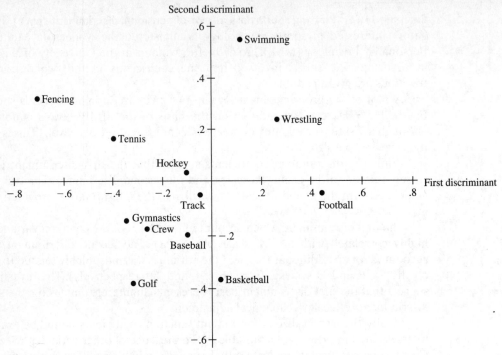

Figure 11.13 The discriminant means $\bar{\mathbf{y}}' = [\bar{y}_1, \bar{y}_2]$ for each sport.

unit ellipse (circle) centered at the discriminant mean vector $\bar{\mathbf{y}}$ should contain approximately a proportion

$$P[(\mathbf{Y} - \boldsymbol{\mu}_Y)'(\mathbf{Y} - \boldsymbol{\mu}_Y) \leq 1] = P[\chi_2^2 \leq 1] = .39$$

of the points when two discriminants are plotted. ∎

Using Fisher's Discriminants to Classify Objects

Fisher's discriminants were derived for the purpose of obtaining a low-dimensional representation of the data that separates the populations as much as possible. Although they were derived from considerations of separation, the discriminants also provide the basis for a classification rule. We first explain the connection in terms of the population discriminants $\mathbf{a}_i' \mathbf{X}$.

Setting

$$Y_k = \mathbf{a}_k' \mathbf{X} = k\text{th discriminant}, \quad k \leq s \tag{11-64}$$

we conclude that

$$\mathbf{Y} = \begin{bmatrix} Y_1 \\ Y_2 \\ \vdots \\ Y_s \end{bmatrix} \text{ has mean vector } \boldsymbol{\mu}_{iY} = \begin{bmatrix} \mu_{iY_1} \\ \vdots \\ \mu_{iY_s} \end{bmatrix} = \begin{bmatrix} \mathbf{a}_1' \boldsymbol{\mu}_i \\ \vdots \\ \mathbf{a}_s' \boldsymbol{\mu}_i \end{bmatrix}$$

under population π_i and covariance matrix $\mathbf{I}$, for all populations. (See Exercise 11.21.)

Because the components of $\mathbf{Y}$ have unit variances and zero covariances, the appropriate measure of squared distance from $\mathbf{Y} = \mathbf{y}$ to $\boldsymbol{\mu}_{iY}$ is

$$(\mathbf{y} - \boldsymbol{\mu}_{iY})'(\mathbf{y} - \boldsymbol{\mu}_{iY}) = \sum_{j=1}^{s} (y_j - \mu_{iY_j})^2$$

A reasonable classification rule is one that assigns $\mathbf{y}$ to population π_k if the square of the distance from $\mathbf{y}$ to $\boldsymbol{\mu}_{kY}$ is smaller than the square of the distance from $\mathbf{y}$ to $\boldsymbol{\mu}_{iY}$ for $i \neq k$.

If only r of the discriminants are used for allocation, the rule is

Allocate $\mathbf{x}$ to π_k if

$$\sum_{j=1}^{r} (y_j - \mu_{kY_j})^2 = \sum_{j=1}^{r} [\mathbf{a}_j'(\mathbf{x} - \boldsymbol{\mu}_k)]^2$$

$$\leq \sum_{j=1}^{r} [\mathbf{a}_j'(\mathbf{x} - \boldsymbol{\mu}_i)]^2 \quad \text{for all } i \neq k \tag{11-65}$$

Before relating this classification procedure to those of Section 11.5, we look more closely at the restriction on the number of discriminants. From Exercise 11.21,

$$s = \text{number of discriminants} = \text{number of nonzero eigenvalues of } \boldsymbol{\Sigma}^{-1}\mathbf{B}_{\mu}$$

$$\text{or of } \boldsymbol{\Sigma}^{-1/2}\mathbf{B}_{\mu}\boldsymbol{\Sigma}^{-1/2}$$

Now, $\boldsymbol{\Sigma}^{-1}\mathbf{B}_{\mu}$ is $p \times p$, so $s \leq p$. Further, the g vectors

$$\boldsymbol{\mu}_1 - \bar{\boldsymbol{\mu}}, \boldsymbol{\mu}_2 - \bar{\boldsymbol{\mu}}, \ldots, \boldsymbol{\mu}_g - \bar{\boldsymbol{\mu}} \tag{11-66}$$

satisfy $(\boldsymbol{\mu}_1 - \bar{\boldsymbol{\mu}}) + (\boldsymbol{\mu}_2 - \bar{\boldsymbol{\mu}}) + \cdots + (\boldsymbol{\mu}_g - \bar{\boldsymbol{\mu}}) = g\bar{\boldsymbol{\mu}} - g\bar{\boldsymbol{\mu}} = \mathbf{0}$. That is, the first difference $\boldsymbol{\mu}_1 - \bar{\boldsymbol{\mu}}$ can be written as a linear combination of the last $g - 1$ differences. Linear combinations of the g vectors in (11-66) determine a hyperplane of dimension $q \leq g - 1$. Taking any vector $\mathbf{e}$ perpendicular to every $\boldsymbol{\mu}_i - \bar{\boldsymbol{\mu}}$, and hence the hyperplane, gives

$$\mathbf{B}_{\mu}\mathbf{e} = \sum_{i=1}^{g} (\boldsymbol{\mu}_i - \bar{\boldsymbol{\mu}})(\boldsymbol{\mu}_i - \bar{\boldsymbol{\mu}})'\mathbf{e} = \sum_{i=1}^{g} (\boldsymbol{\mu}_i - \bar{\boldsymbol{\mu}})0 = \mathbf{0}$$

so

$$\boldsymbol{\Sigma}^{-1}\mathbf{B}_{\mu}\mathbf{e} = 0\mathbf{e}$$

There are $p - q$ orthogonal eigenvectors corresponding to the zero eigenvalue. This implies that there are q or fewer *nonzero* eigenvalues. Since it is always true that $q \leq g - 1$, the number of nonzero eigenvalues s must satisfy $s \leq \min(p, g - 1)$.

Thus, there is no loss of information for discrimination by plotting in two dimensions if the following conditions hold.

Number of variables	Number of populations	Maximum number of discriminants
Any p	$g = 2$	1
Any p	$g = 3$	2
$p = 2$	Any g	2

We now present an important relation between the classification rule (11-65) and the "normal theory" discriminant scores [see (11-49)],

$$d_i(\mathbf{x}) = \boldsymbol{\mu}_i' \boldsymbol{\Sigma}^{-1} \mathbf{x} - \tfrac{1}{2} \boldsymbol{\mu}_i' \boldsymbol{\Sigma}^{-1} \boldsymbol{\mu}_i + \ln p_i$$

or, equivalently,

$$d_i(\mathbf{x}) - \tfrac{1}{2} \mathbf{x}' \boldsymbol{\Sigma}^{-1} \mathbf{x} = -\tfrac{1}{2} (\mathbf{x} - \boldsymbol{\mu}_i)' \boldsymbol{\Sigma}^{-1} (\mathbf{x} - \boldsymbol{\mu}_i) + \ln p_i$$

obtained by adding the same constant $-\tfrac{1}{2} \mathbf{x}' \boldsymbol{\Sigma}^{-1} \mathbf{x}$ to each $d_i(\mathbf{x})$.

Result 11.6. Let $y_j = \mathbf{a}_j' \mathbf{x}$, where $\mathbf{a}_j = \boldsymbol{\Sigma}^{-1/2} \mathbf{e}_j$ and $\mathbf{e}_j$ is an eigenvector of $\boldsymbol{\Sigma}^{-1/2} \mathbf{B}_{\boldsymbol{\mu}} \boldsymbol{\Sigma}^{-1/2}$. Then

$$\sum_{j=1}^{p} (y_j - \mu_{iY_j})^2 = \sum_{j=1}^{p} [\mathbf{a}_j' (\mathbf{x} - \boldsymbol{\mu}_i)]^2 = (\mathbf{x} - \boldsymbol{\mu}_i)' \boldsymbol{\Sigma}^{-1} (\mathbf{x} - \boldsymbol{\mu}_i)$$

$$= -2d_i(\mathbf{x}) + \mathbf{x}' \boldsymbol{\Sigma}^{-1} \mathbf{x} + 2 \ln p_i$$

If $\lambda_1 \geq \cdots \geq \lambda_s > 0 = \lambda_{s+1} = \cdots = \lambda_p$, $\sum_{j=s+1}^{p} (y_j - \mu_{iY_j})^2$ is constant for all populations $i = 1, 2, \ldots, g$ so only the first s discriminants y_j, or $\sum_{j=1}^{s} (y_j - \mu_{iY_j})^2$, contribute to the classification.

Also, if the prior probabilities are such that $p_1 = p_2 = \cdots = p_g = 1/g$, the rule (11-65) with $r = s$ is equivalent to the population version of the minimum TPM rule (11-52).

Proof. The squared distance $(\mathbf{x} - \boldsymbol{\mu}_i)' \boldsymbol{\Sigma}^{-1} (\mathbf{x} - \boldsymbol{\mu}_i) = (\mathbf{x} - \boldsymbol{\mu}_i)' \boldsymbol{\Sigma}^{-1/2} \boldsymbol{\Sigma}^{-1/2} (\mathbf{x} - \boldsymbol{\mu}_i)$ $= (\mathbf{x} - \boldsymbol{\mu}_i)' \boldsymbol{\Sigma}^{-1/2} \mathbf{E} \mathbf{E}' \boldsymbol{\Sigma}^{-1/2} (\mathbf{x} - \boldsymbol{\mu}_i)$, where $\mathbf{E} = [\mathbf{e}_1, \mathbf{e}_2, \ldots, \mathbf{e}_p]$ is the orthogonal matrix whose columns are eigenvectors of $\boldsymbol{\Sigma}^{-1/2} \mathbf{B}_{\boldsymbol{\mu}} \boldsymbol{\Sigma}^{-1/2}$. (See Exercise 11.21.)
Since $\boldsymbol{\Sigma}^{-1/2} \mathbf{e}_i = \mathbf{a}_i$ or $\mathbf{a}_i' = \mathbf{e}_i' \boldsymbol{\Sigma}^{-1/2}$,

$$\mathbf{E}' \boldsymbol{\Sigma}^{-1/2} (\mathbf{x} - \boldsymbol{\mu}_i) = \begin{bmatrix} \mathbf{a}_1' (\mathbf{x} - \boldsymbol{\mu}_i) \\ \mathbf{a}_2' (\mathbf{x} - \boldsymbol{\mu}_i) \\ \vdots \\ \mathbf{a}_p' (\mathbf{x} - \boldsymbol{\mu}_i) \end{bmatrix}$$

and

$$(\mathbf{x} - \boldsymbol{\mu}_i)' \boldsymbol{\Sigma}^{-1/2} \mathbf{E} \mathbf{E}' \boldsymbol{\Sigma}^{-1/2} (\mathbf{x} - \boldsymbol{\mu}_i) = \sum_{j=1}^{p} [\mathbf{a}_j' (\mathbf{x} - \boldsymbol{\mu}_i)]^2$$

Next, each $\mathbf{a}_j = \boldsymbol{\Sigma}^{-1/2} \mathbf{e}_j$, $j > s$, is an (unscaled) eigenvector of $\boldsymbol{\Sigma}^{-1} \mathbf{B}_{\boldsymbol{\mu}}$ with eigenvalue zero. As shown in the discussion following (11-66), $\mathbf{a}_j$ is perpendicular to every $\boldsymbol{\mu}_i - \bar{\boldsymbol{\mu}}$ and hence to $(\boldsymbol{\mu}_k - \bar{\boldsymbol{\mu}}) - (\boldsymbol{\mu}_i - \bar{\boldsymbol{\mu}}) = \boldsymbol{\mu}_k - \boldsymbol{\mu}_i$ for $i, k = 1, 2, \ldots, g$. The condition $0 = \mathbf{a}'_j(\boldsymbol{\mu}_k - \boldsymbol{\mu}_i) = \mu_{kY_j} - \mu_{iY_j}$ implies that $y_j - \mu_{kY_j} = y_j - \mu_{iY_j}$ so $\sum_{j=s+1}^{p} (y_j - \mu_{iY_j})^2$ is constant for all $i = 1, 2, \ldots, g$. Therefore, only the first s discriminants y_j need to be used for classification. ∎

We now state the classification rule based on the first $r \leq s$ sample discriminants.

Fisher's Classification Procedure Based on Sample Discriminants

Allocate $\mathbf{x}$ to π_k if

$$\sum_{j=1}^{r} (\hat{y}_j - \bar{y}_{kj})^2 = \sum_{j=1}^{r} [\hat{\mathbf{a}}'_j(\mathbf{x} - \bar{\mathbf{x}}_k)]^2 \leq \sum_{j=1}^{r} [\hat{\mathbf{a}}'_j(\mathbf{x} - \bar{\mathbf{x}}_i)]^2 \qquad \text{for all } i \neq k$$

(11-67)

where $\hat{\mathbf{a}}_j$ is defined in (11-62), $\bar{y}_{kj} = \hat{\mathbf{a}}'_j\bar{\mathbf{x}}_k$ and $r \leq s$.

When the prior probabilities are such that $p_1 = p_2 = \cdots = p_g = 1/g$ and $r = s$, rule (11-67) is equivalent to rule (11-52), which is based on the largest linear discriminant score. In addition, if $r < s$ discriminants are used for classification, there is a loss of squared distance, or score, of $\sum_{j=r+1}^{p} [\hat{\mathbf{a}}'_j(\mathbf{x} - \bar{\mathbf{x}}_i)]^2$ for each population π_i where $\sum_{j=r+1}^{s} [\hat{\mathbf{a}}'_j(\mathbf{x} - \bar{\mathbf{x}}_i)]^2$ is the part useful for classification.

Example 11.16 (Classifying a new observation with Fisher's discriminants) Let us use the Fisher discriminants

$$\hat{y}_1 = \hat{\mathbf{a}}'_1\mathbf{x} = .386x_1 + .495x_2$$
$$\hat{y}_2 = \hat{\mathbf{a}}'_2\mathbf{x} = .938x_1 - .112x_2$$

from Example 11.13 to classify the new observation $\mathbf{x}'_0 = [1 \quad 3]$ in accordance with (11-67).

Inserting $\mathbf{x}'_0 = [x_{01}, x_{02}] = [1 \quad 3]$, we have

$$\hat{y}_1 = .386x_{01} + .495x_{02} = .386(1) + .495(3) = 1.87$$
$$\hat{y}_2 = .938x_{01} - .112x_{02} = .938(1) - .112(3) = .60$$

Moreover, $\bar{y}_{kj} = \hat{\mathbf{a}}'_j\bar{\mathbf{x}}_k$, so that (see Example 11.13)

$$\bar{y}_{11} = \hat{\mathbf{a}}'_1\bar{\mathbf{x}}_1 = [.386 \quad .495]\begin{bmatrix} -1 \\ 3 \end{bmatrix} = 1.10$$

$$\bar{y}_{12} = \hat{\mathbf{a}}'_2\bar{\mathbf{x}}_1 = [.938 \quad -.112]\begin{bmatrix} -1 \\ 3 \end{bmatrix} = -1.27$$

Similarly,

$$\bar{y}_{21} = \hat{\mathbf{a}}_1' \bar{\mathbf{x}}_2 = 2.37$$

$$\bar{y}_{22} = \hat{\mathbf{a}}_2' \bar{\mathbf{x}}_2 = .49$$

$$\bar{y}_{31} = \hat{\mathbf{a}}_1' \bar{\mathbf{x}}_3 = -.99$$

$$\bar{y}_{32} = \hat{\mathbf{a}}_2' \bar{\mathbf{x}}_3 = .22$$

Finally, the smallest value of

$$\sum_{j=1}^{2} (\hat{y}_j - \bar{y}_{kj})^2 = \sum_{j=1}^{2} [\hat{\mathbf{a}}_j' (\mathbf{x} - \bar{\mathbf{x}}_k)]^2$$

for $k = 1, 2, 3$, must be identified. Using the preceding numbers gives

$$\sum_{j=1}^{2} (\hat{y}_j - \bar{y}_{1j})^2 = (1.87 - 1.10)^2 + (.60 + 1.27)^2 = 4.09$$

$$\sum_{j=1}^{2} (\hat{y}_j - \bar{y}_{2j})^2 = (1.87 - 2.37)^2 + (.60 - .49)^2 = .26$$

$$\sum_{j=1}^{2} (\hat{y}_j - \bar{y}_{3j})^2 = (1.87 + .99)^2 + (.60 - .22)^2 = 8.32$$

Since the minimum of $\sum_{j=1}^{2} (\hat{y}_j - \bar{y}_{kj})^2$ occurs when $k = 2$, we allocate $\mathbf{x}_0$ to population π_2. The situation, in terms of the classifi0ers $\hat{y}_j$, is illustrated schematically in Figure 11.14. ∎

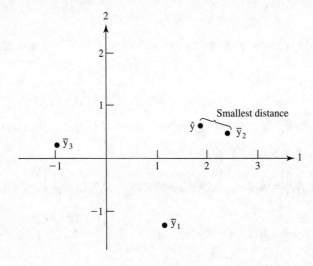

Figure 11.14
The points $\hat{\mathbf{y}}' = [\hat{y}_1, \hat{y}_2]$, $\bar{\mathbf{y}}_1' = [\bar{y}_{11}, \bar{y}_{12}]$, $\bar{\mathbf{y}}_2' = [\bar{y}_{21}, \bar{y}_{22}]$, and $\bar{\mathbf{y}}_3' = [\bar{y}_{31}, \bar{y}_{32}]$ in the classification plane.

Comment. When two linear discriminant functions are used for classification, observations are assigned to populations based on Euclidean distances in the two-dimensional discriminant space.

Up to this point, we have not shown why the first few discriminants are more important than the last few. Their relative importance becomes apparent from their contribution to a numerical measure of spread of the populations. Consider the separatory measure

$$\Delta_S^2 = \sum_{i=1}^{g} (\boldsymbol{\mu}_i - \bar{\boldsymbol{\mu}})' \boldsymbol{\Sigma}^{-1} (\boldsymbol{\mu}_i - \bar{\boldsymbol{\mu}}) \tag{11-68}$$

where

$$\bar{\boldsymbol{\mu}} = \frac{1}{g} \sum_{i=1}^{g} \boldsymbol{\mu}_i$$

and $(\boldsymbol{\mu}_i - \bar{\boldsymbol{\mu}})' \boldsymbol{\Sigma}^{-1} (\boldsymbol{\mu}_i - \bar{\boldsymbol{\mu}})$ is the squared statistical distance from the ith population mean $\boldsymbol{\mu}_i$ to the centroid $\bar{\boldsymbol{\mu}}$. It can be shown (see Exercise 11.22) that $\Delta_S^2 = \lambda_1 + \lambda_2 + \cdots + \lambda_p$ where the $\lambda_1 \geq \lambda_2 \geq \cdots \geq \lambda_s$ are the *nonzero* eigenvalues of $\boldsymbol{\Sigma}^{-1}\mathbf{B}$ (or $\boldsymbol{\Sigma}^{-1/2}\mathbf{B}\boldsymbol{\Sigma}^{-1/2}$) and $\lambda_{s+1}, \ldots, \lambda_p$ are the zero eigenvalues.

The separation given by Δ_S^2 can be reproduced in terms of discriminant means. The first discriminant, $Y_1 = \mathbf{e}_1' \boldsymbol{\Sigma}^{-1/2} \mathbf{X}$ has means $\mu_{iY_1} = \mathbf{e}_1' \boldsymbol{\Sigma}^{-1/2} \boldsymbol{\mu}_i$ and the squared distance $\sum_{i=1}^{g} (\mu_{iY_1} - \bar{\mu}_{Y_1})^2$ of the μ_{iY_1}'s from the central value $\bar{\mu}_{Y_1} = \mathbf{e}_1' \boldsymbol{\Sigma}^{-1/2} \bar{\boldsymbol{\mu}}$ is λ_1. (See Exercise 11.22.) Since Δ_S^2 can also be written as

$$\Delta_S^2 = \lambda_1 + \lambda_2 + \cdots + \lambda_p$$

$$= \sum_{i=1}^{g} (\boldsymbol{\mu}_{iY} - \bar{\boldsymbol{\mu}}_Y)' (\boldsymbol{\mu}_{iY} - \bar{\boldsymbol{\mu}}_Y)$$

$$= \sum_{i=1}^{g} (\mu_{iY_1} - \bar{\mu}_{Y_1})^2 + \sum_{i=1}^{g} (\mu_{iY_2} - \bar{\mu}_{Y_2})^2 + \cdots + \sum_{i=1}^{g} (\mu_{iY_p} - \bar{\mu}_{Y_p})^2$$

it follows that the first discriminant makes the largest single contribution, λ_1, to the separative measure Δ_S^2. In general, the rth discriminant, $Y_r = \mathbf{e}_r' \boldsymbol{\Sigma}^{-1/2} \mathbf{X}$, contributes λ_r to Δ_S^2. If the next $s - r$ eigenvalues (recall that $\lambda_{s+1} = \lambda_{s+2} = \cdots = \lambda_p = 0$) are such that $\lambda_{r+1} + \lambda_{r+2} + \cdots + \lambda_s$ is small compared to $\lambda_1 + \lambda_2 + \cdots + \lambda_r$, then the last discriminants $Y_{r+1}, Y_{r+2}, \ldots, Y_s$ can be neglected without appreciably decreasing the amount of separation.[12]

Not much is known about the efficacy of the allocation rule (11-67). Some insight is provided by computer-generated sampling experiments, and Lachenbruch [23] summarizes its performance in particular cases. The development of the population result in (11-65) required a common covariance matrix $\boldsymbol{\Sigma}$. If this is essentially true and the samples are reasonably large, rule (11-67) should perform fairly well. In any event, its performance can be checked by computing estimated error rates. Specifically, Lachenbruch's estimate of the expected actual error rate given by (11-57) should be calculated.

---

[12] See [18] for further optimal dimension-reducing properties.

11.7 Logistic Regression and Classification

Introduction

The classification functions already discussed are based on quantitative variables. Here we discuss an approach to classification where some or all of the variables are qualitative. This approach is called logistic regression. In its simplest setting, the response variable Y is restricted to two values. For example, Y may be recorded as "male" or "female" or "employed" and "not employed."

 binomial is one example

Even though the response may be a two outcome qualitative variable, we can always code the two cases as 0 and 1. For instance, we can take male $= 0$ and female $= 1$. Then the probability p of 1 is a parameter of interest. It represents the proportion in the population who are coded 1. The mean of the distribution of 0's and 1's is also p since

$$\text{mean} = 0 \times (1 - p) + 1 \times p = p$$

The proportion of 0's is $1 - p$ which is sometimes denoted as q.

The variance of the distribution is

$$\text{variance} = 0^2 \times (1 - p) + 1^2 \times p - p^2 = p(1 - p)$$

var depends on p

It is clear the variance is not constant. For $p = .5$, it equals $.5 \times .5 = .25$ while for $p = .8$, it is $.8 \times .2 = .16$. The variance approaches 0 as p approaches either 0 or 1.

Let the response Y be either 0 or 1. If we were to model the probability of 1 with a single predictor linear model, we would write

$$p = E(Y \mid z) = \beta_0 + \beta_1 z$$

and then add an error term ε. But there are serious drawbacks to this model.

- The predicted values of the response Y could become greater than 1 or less than 0 because the linear expression for its expected value is unbounded.

- One of the assumptions of a regression analysis is that the variance of Y is constant across all values of the predictor variable Z. We have shown this is not the case. Of course, weighted least squares might improve the situation.

We need another approach to introduce predictor variables or covariates $\mathbf{Z}$ into the model (see [26]). Throughout, if the covariates are not fixed by the investigator, the approach is to make the models for $p(z)$ conditional on the observed values of the covariates $\mathbf{Z} = \mathbf{z}$.

The Logit Model

Instead of modeling the probability p directly with a linear model, we first consider the *odds ratio*

$$\text{odds} = \frac{p}{1 - p}$$

which is the ratio of the probability of 1 to the probability of 0. Note, unlike probability, the odds ratio can be greater than 1. If a proportion .8 of persons will get

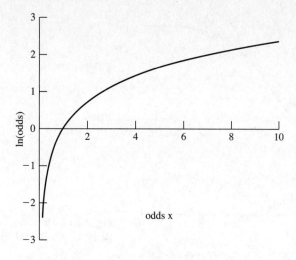

asymmetric, but taking
the natural logarithm will
make it symmetric

Figure 11.15 Natural log of odds ratio.

through customs without their luggage being checked, then $p = .8$ but the odds of not getting checked is $.8/.2 = 4$ or 4 to 1 of not being checked. There is a lack of symmetry here since the odds of being checked are $.2/.8 = 1/4$. Taking the natural logarithms, we find that $\ln(4) = 1.386$ and $\ln(1/4) = -1.386$ are exact opposites.

Consider the natural log function of the odds ratio that is displayed in Figure 11.15. When the odds x are 1, so outcomes 0 and 1 are equally likely, the natural log of x is zero. When the odds x are greater than one, the natural log increases slowly as x increases. However, when the odds x are less than one, the natural log decreases rapidly as x decreases toward zero.

In logistic regression for a binary variable, we model the natural log of the odds ratio, which is called *logit(p)*. Thus

$$logit(p) = \ln(odds) = \ln\left(\frac{p}{1-p}\right) \tag{11-69}$$

The logit is a function of the probability p. In the simplest model, we assume that the logit graphs as a straight line in the predictor variable Z so

$$logit(p) = \ln(odds) = \ln\left(\frac{p}{1-p}\right) = \beta_0 + \beta_1 z \tag{11-70}$$

In other words, the log odds are linear in the predictor variable.

Because it is easier for most people to think in terms of probabilities, we can convert from the logit or log odds to the probability p. By first exponentiating

$$\ln\left(\frac{p}{1-p}\right) = \beta_0 + \beta_1 z$$

we obtain

$$\theta(z) = \frac{p(z)}{1 - p(z)} = \exp(\beta_0 + \beta_1 z)$$

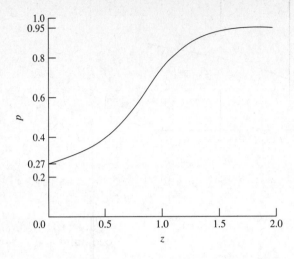

Figure 11.16 Logistic function with $\beta_0 = -1$ and $\beta_1 = 2$.

where $\exp = e = 2.718$ is the base of the natural logarithm. Next solving for $\theta(z)$, we obtain

$$p(z) = \frac{\exp(\beta_0 + \beta_1 z)}{1 + \exp(\beta_0 + \beta_1 z)} \qquad (11\text{-}71)$$

which describes a *logistic curve*. The relation between p and the predictor z is not linear but has an S-shaped graph as illustrated in Figure 11.16 for the case $\beta_0 = -1$ and $\beta_1 = 2$. The value of β_0 gives the value $\exp(\beta_0)/(1 + \exp(\beta_0))$ for p when $z = 0$.

The parameter β_1 in the logistic curve determines how quickly p changes with z but its interpretation is not as simple as in ordinary linear regression because the relation is not linear, either in z or β_1. However, we can exploit the linear relation for log odds.

To summarize, the logistic curve can be written as

$$p(z) = \frac{\exp(\beta_0 + \beta_1 z)}{1 + \exp(\beta_0 + \beta_1 z)} \quad \text{or} \quad p(z) = \frac{1}{1 + \exp(-\beta_0 - \beta_1 z)}$$

Logistic Regression Analysis

Consider the model with several predictor variables. Let $(z_{j1}, z_{j2}, \ldots, z_{jr})$ be the values of the r predictors for the j-th observation. It is customary, as in normal theory linear regression, to set the first entry equal to 1 and $\mathbf{z}_j = [1, z_{j1}, z_{j2}, \ldots, z_{jr}]'$. Conditional on these values, we assume that the observation Y_j is Bernoulli with success probability $p(\mathbf{z_j})$, depending on the values of the covariates. Then

$$P(Y_j = y_j) = p^{y_j}(\mathbf{z}_j)(1 - p(\mathbf{z}_j))^{1-y_j} \qquad \text{for } y_j = 0, 1$$

so

$$E(Y_j) = p(\mathbf{z}_j) \quad \text{and} \quad \text{Var}(Y_j) = p(\mathbf{z}_j)(1 - p(\mathbf{z}_j))$$

It is not the mean that follows a linear model but the natural log of the odds ratio. In particular, we assume the model

$$\ln\left(\frac{p(\mathbf{z})}{1 - p(\mathbf{z})}\right) = \beta_0 + \beta_1 z_1 + \cdots + \beta_r z_r = \boldsymbol{\beta}'\mathbf{z}_j \tag{11-72}$$

where $\boldsymbol{\beta} = [\beta_0, \beta_1, \ldots, \beta_r]'$.

Maximum Likelihood Estimation. Estimates of the β's can be obtained by the method of maximum likelihood. The likelihood L is given by the joint probability distribution evaluated at the observed counts y_j. Hence

$$L(b_0, b_1, \ldots, b_r) = \prod_{j=1}^{n} p^{y_j}(\mathbf{z}_j)(1 - p(\mathbf{z}_j))^{1-y_j}$$

$$= \frac{\prod_{j=1}^{n} e^{y_j(b_0 + b_1 z_{j1} + \ldots + b_r z_{jr})}}{\prod_{j=1}^{n}(1 + e^{b_0 + b_1 z_{j1} + \ldots + b_r z_{jr}})} \tag{11-73}$$

The values of the parameters that maximize the likelihood cannot be expressed in a nice closed form solution as in the normal theory linear models case. Instead they must be determined numerically by starting with an initial guess and iterating to the maximum of the likelihood function. Technically, this procedure is called an iteratively re-weighted least squares method (see [26]).

We denote the numerically obtained values of the maximum likelihood estimates by the vector $\hat{\boldsymbol{\beta}}$.

Confidence Intervals for Parameters. When the sample size is large, $\hat{\boldsymbol{\beta}}$ is approximately normal with mean $\boldsymbol{\beta}$, the prevailing values of the parameters and approximate covariance matrix

$$\widehat{\text{Cov}}(\hat{\boldsymbol{\beta}}) \approx \left[\sum_{j=1}^{n} \hat{p}(\mathbf{z}_j)(1 - \hat{p}(\mathbf{z}_j))\mathbf{z}_j\mathbf{z}_j'\right]^{-1} \tag{11-74}$$

The square roots of the diagonal elements of this matrix are the large sample estimated standard deviations or standard errors (SE) of the estimators $\hat{\beta}_0, \hat{\beta}_1, \ldots, \hat{\beta}_r$ respectively. The large sample 95% confidence interval for β_k is

$$\hat{\beta}_k \pm 1.96\, SE(\hat{\beta}_k) \qquad k = 0, 1, \ldots, r \tag{11-75}$$

The confidence intervals can be used to judge the significance of the individual terms in the model for the logit. Large sample confidence intervals for the logit and for the population proportion $p(\mathbf{z}_j)$ can be constructed as well. See [17] for details.

Likelihood Ratio Tests. For the model with r predictor variables plus the constant, we denote the maximized likelihood by

$$L_{\max} = L(\hat{\beta}_0, \hat{\beta}_1, \ldots, \hat{\beta}_r)$$

If the null hypothesis is $H_0: \beta_k = 0$, numerical calculations again give the maximum likelihood estimate of the reduced model and, in turn, the maximized value of the likelihood

$$L_{\text{max, Reduced}} = L(\hat{\beta}_0, \hat{\beta}_1, \ldots, \hat{\beta}_{k-1}, \hat{\beta}_{k+1}, \ldots, \hat{\beta}_r)$$

When doing logistic regression, it is common to test H_0 using minus twice the log-likelihood ratio

$$-2 \ln \left(\frac{L_{\text{max, Reduced}}}{L_{\text{max}}} \right) \tag{11-76}$$

which, in this context, is called the *deviance*. It is approximately distributed as chi-square with 1 degree of freedom when the reduced model has one fewer predictor variables. H_0 is rejected for a large value of the deviance.

An alternative test for the significance of an individual term in the model for *the logit is due* to Wald (see [17]). The Wald test of $H_0: \beta_k = 0$ uses the test statistic $Z = \hat{\beta}_k/SE(\hat{\beta}_k)$ or its chi-square version Z^2 with 1 degree of freedom. The likelihood ratio test is preferable to the Wald test as the level of this test is typically closer to the nominal α.

Generally, if the null hypothesis specifies a subset of, say, m parameters are simultaneously 0, the deviance is constructed for the implied reduced model and referred to a chi-squared distribution with m degrees of freedom.

When working with individual binary observations Y_i, the residuals

$$\frac{Y_j - \hat{p}(\mathbf{z}_j)}{\sqrt{\hat{p}(\mathbf{z}_j)(1 - \hat{p}(\mathbf{z}_j))}}$$

each can assume only two possible values and are not particularly useful. It is better if they can be grouped into reasonable sets and a total residual calculated for each set. If there are, say, t residuals in each group, sum these residuals and then divide by $\sqrt{t}$ to help keep the variances compatible.

We give additional details on logistic regression and model checking following and application to classification.

Classification

Let the response variable Y be 1 if the observational unit belongs to population 1 and 0 if it belongs to population 2. (The choice of 1 and 0 for response outcomes is arbitrary but convenient. In Example 11.17, we use 1 and 2 as outcomes.) Once a logistic regression function has been established, and using training sets for each of the two populations, we can proceed to classify. Priors and costs are difficult to incorporate into the analysis, so the classification rule becomes

> Assign $\mathbf{z}$ to population 1 if the estimated odds ratio is greater than 1 or

$$\frac{\hat{p}(\mathbf{z})}{1 - \hat{p}(\mathbf{z})} = \exp(\hat{\beta}_0 + \hat{\beta}_1 z_1 + \cdots + \hat{\beta}_r z_r) > 1$$

Equivalently, we have the simple linear discriminant rule

Assign $\mathbf{z}$ to population 1 if the linear discriminant is greater than 0 or

$$\ln \frac{\hat{p}(\mathbf{z})}{1 - \hat{p}(\mathbf{z})} = \hat{\beta}_0 + \hat{\beta}_1 z_1 + \cdots + \hat{\beta}_r z_r > 0 \qquad (11\text{-}77)$$

Example 11.17 (Logistic regression with the salmon data) We introduced the salmon data in Example 11.8 (see Table 11.2). In Example 11.8, we ignored the gender of the salmon when considering the problem of classifying salmon as Alaskan or Canadian based on growth ring measurements. Perhaps better classification is possible if gender is included in the analysis. Panel 11.2 contains the SAS output from a logistic regression analysis of the salmon data. Here the response Y is 1 if Alaskan salmon and 2 if Canadian salmon. The predictor variables (covariates) are gender (1 if female, 2 if male), freshwater growth and marine growth. From the SAS output under Testing the Global Null Hypothesis, the likelihood ratio test result (see 11–76) with the reduced model containing only a β_0 term) is significant at the $< .0001$ level. At least one covariate is required in the linear model for the logit. Examining the significance of individual terms under the heading Analysis of Maximum Likelihood Estimates, we see that the Wald test suggests gender is not significant (p-value $= .7356$). On the other hand, freshwater growth and marine are significant covariates. Gender can be dropped from the model. It is not a useful variable for classification. The logistic regression model can be re-estimated without gender and the resulting function used to classify the salmon as Alaskan or Canadian using rule (11–77).

Turning to the classification problem, but retaining gender, we assign salmon j to population 1, Alaskan, if the linear classifier

$$\boldsymbol{\beta}'\mathbf{z} = 3.5054 + .2816 \text{ gender} + .1264 \text{ freshwater} + .0486 \text{ marine} \geq 0$$

The observations that are misclassified are

Row	Pop	Gender	Freshwater	Marine	Linear Classifier
2	1	1	131	355	3.093
12	1	2	123	372	1.537
13	1	1	123	372	1.255
30	1	2	118	381	0.467
51	2	1	129	420	-0.319
68	2	2	136	438	-0.028
71	2	2	90	385	-3.266

From these misclassifications, the confusion matrix is

		Predicted membership	
		π_1: Alaskan	π_1: Canadian
Actual	π_1: Alaskan	46	4
	π_1: Canadian	3	47

and the apparent error rate, expressed as a percentage is

$$\text{APER} = \frac{4 + 3}{50 + 50} \times 100 = 7\%$$ ∎

When performing a logistic classification, it would be preferable to have an estimate of the misclassification probabilities using the jackknife (holdout) approach but this is not currently available in the major statistical software packages.

We could have continued the analysis in Example 11.17 by dropping gender and using just the freshwater and marine growth measurements. However, when normal distributions with equal covariance matrices prevail, logistic classification can be quite inefficient compared to the normal theory linear classifier (see [7]).

Logistic Regression with Binomial Responses

We now consider a slightly more general case where several runs are made at the same values of the covariates $\mathbf{z}_j$ and there are a total of m different sets where these predictor variables are constant. When n_j independent trials are conducted with the predictor variables $\mathbf{z}_j$, the response Y_j is modeled as a binomial distribution with probability $p(\mathbf{z}_j) = P(\text{Success} \mid \mathbf{z}_j)$.

Because the Y_j are assumed to be independent, the likelihood is the product

$$L(\beta_0, \beta_1, \ldots, \beta_r) = \prod_{j=1}^{m} \binom{n_j}{y_j} p_j^y(\mathbf{z}_j)(1 - p(\mathbf{z}_j))^{n_j - y_j} \tag{11-78}$$

where the probabilities $p(\mathbf{z}_j)$ follow the logit model (11-72)

PANEL 11.2 SAS ANALYSIS FOR SALMON DATA USING PROC LOGISTIC.

```
title 'Logistic Regression and Discrimination';
data salmon;
infile 'T11-2.dat';                                    PROGRAM COMMANDS
input country gender freshwater marine;
proc logistic desc;
model country = gender freshwater marine / expb;
```

OUTPUT

Logistic Regression and Discrimination

The LOGISTIC procedure

Model Information

Model   binary logit

Response Profile

Ordered Value	country	Total Frequency
1	2	50
2	1	50

(continues on next page)

PANEL 11.2 *(continued)*

Probability modeled is country = 2.
Model Fit Statistics

Criterion	Intercept Only	Intercept and Covariates
AIC	140.629	46.674
SC	143.235	57.094
-2 Log L	138.629	38.674

Testing Global Null Hypothesis: BETA = 0

Test	Chi-Square	DF	Pr > ChiSq
Likelihood Ratio	99.9557	3	<.0001
Wald	19.4435	3	0.0002

The LOGISTIC Procedure

Analysis of Maximum Likelihood Estimates

Parameter	DF	Estimate	Standard Error	Wald Chi-Square	Pr > ChiSp	Exp (Est)
Intercept	1	3.5054	6.3935	0.3006	0.5835	33.293
gender	1	0.2816	0.8338	0.1140	0.7356	1.325
freshwater	1	0.1264	0.0357	12.5423	0.0004	1.135
marine	1	−0.0486	0.0146	11.1460	0.0008	0.953

The maximum likelihood estimates $\hat{\boldsymbol{\beta}}$ must be obtained numerically because there is no closed form expression for their computation. When the total sample size is large, the approximate covariance matrix $\widehat{\text{Cov}}(\hat{\boldsymbol{\beta}})$ is

$$\widehat{\text{Cov}}(\hat{\boldsymbol{\beta}}) \approx \left[\sum_{j=1}^{m} n_j \hat{p}(\mathbf{z}_j)(1 - \hat{p}(\mathbf{z}_j)) \mathbf{z}_j \mathbf{z}_j' \right]^{-1} \tag{11-79}$$

and the i-th diagonal element is an estimate of the variance of $\hat{\beta}_{i+1}$. It's square root is an estimate of the large sample standard error $SE\,(\hat{\beta}_{i+1})$.

It can also be shown that a large sample estimate of the variance of the probability $\hat{p}(\mathbf{z}_j)$ is given by

$$\widehat{\text{Var}}(\hat{p}(\mathbf{z}_k)) \approx (\hat{p}(\mathbf{z}_k)(1 - \hat{p}(\mathbf{z}_k))^2 \mathbf{z}_j' \left[\sum_{j=1}^{m} n_j \hat{p}(\mathbf{z}_j)(1 - \hat{p}(\mathbf{z}_j)) \mathbf{z}_j \mathbf{z}_j' \right]^{-1} \mathbf{z}_k$$

Consideration of the interval plus and minus two estimated standard deviations from $\hat{p}(\mathbf{z}_j)$ may suggest observations that are difficult to classify.

Model Checking. Once any model is fit to the data, it is good practice to investigate the adequacy of the fit. The following questions must be addressed.

- Is there any systematic departure from the fitted logistic model?
- Are there any observations that are unusual in that they don't fit the overall pattern of the data (*outliers*)?
- Are there any observations that lead to important changes in the statistical analysis when they are included or excluded (*high influence*)?

If there is no parametric structure to the single trial probabilities $p(\mathbf{z}_j) = P(\text{Success} \mid \mathbf{z}_j)$, each would be estimated using the observed number of successes (1's) y_i in n_i trials. Under this nonparametric model, or saturated model, the contribution to the likelihood for the j-th case is

$$\binom{n_j}{y_j} p^{y_j}(\mathbf{z}_j)(1 - p(\mathbf{z}_j))^{n_j - y_j}$$

which is maximized by the choices $\hat{p}(\mathbf{z}_j) = y_j/n_j$ for $j = 1, 2, \ldots, n$. Here $m = \Sigma n_j$. The resulting value for minus twice the maximized nonparametric (NP) likelihood is

$$-2 \ln L_{\text{max, }NP} = -2 \sum_{j=1}^{m} \left[y_j \ln \left(\frac{y_j}{n_j} \right) + (n_j - y_j) \ln \left(1 - \frac{y_j}{n_j} \right) \right] + 2 \ln \left(\prod_{j=1}^{m} \binom{n_j}{y_j} \right)$$

(11-80)

The last term on the right hand side of (11-80) is common to all models.

We also define a deviance between the nonparametric model and a fitted model having a constant and $r-1$ predicators as minus twice the log-likelihood ratio or

$$G^2 = 2 \sum_{j=1}^{m} \left[y_j \ln \left(\frac{y_j}{\hat{y}_j} \right) + (n_j - y_j) \ln \left(\frac{n_j - y_j}{n_j - \hat{y}_j} \right) \right]$$

(11-81)

where $\hat{y}_j = n_j \hat{p}(\mathbf{z}_j)$ is the fitted number of successes. This is the specific deviance quantity that plays a role similar to that played by the residual (error) sum of squares in the linear models setting.

For large sample sizes, G^2 has approximately a chi square distribution with f degrees of freedom equal to the number of observations, m, minus the number of parameters β estimated.

Notice the deviance for the full model, G^2_{Full}, and the deviance for a reduced model, $G^2_{Reduced}$, lead to a contribution for the extra predictor terms

$$G^2_{Reduced} - G^2_{Full} = -2 \ln \left(\frac{L_{\text{max, }Reduced}}{L_{\text{max}}} \right)$$

(11-82)

This difference is approximately χ^2 with degrees of freedom $df = df_{Reduced} - df_{Full}$. A large value for the difference implies the full model is required.

When m is large, there are too many probabilities to estimate under the nonparametric model and the chi-square approximation cannot be established by existing methods of proof. It is better to rely on likelihood ratio tests of logistic models where a few terms are dropped.

Residuals and Goodness-of-Fit Tests. Residuals can be inspected for patterns that suggest lack of fit of the logit model form and the choice of predictor variables (covariates). In logistic regression residuals are not as well defined as in the multiple regression models discussed in Chapter 7. Three different definitions of residuals are available.

Deviance residuals (d_j):

$$d_j = \pm \sqrt{2\left[y_j \ln\left(\frac{y_j}{n_j\hat{p}(\mathbf{z}_j)}\right) + (n_j - y_j)\ln\left(\frac{n_j - y_j}{n_j(1 - \hat{p}(\mathbf{z}_j))}\right)\right]}$$

where the sign of d_j is the same as that of $y_j - n_j\hat{p}(\mathbf{z}_j)$ and,

$$\text{if } y_j = 0, \text{ then } d_j = -\sqrt{2n_j|\ln(1 - \hat{p}(\mathbf{z}_j))|}$$

$$\text{if } y_j = n_j, \text{ then } d_j = -\sqrt{2n_j|\ln\hat{p}(\mathbf{z}_j)|} \qquad (11\text{-}83)$$

Pearson residuals (r_j): $$r_j = \frac{y_j - n_j\hat{p}(\mathbf{z}_j)}{\sqrt{n_j\hat{p}(\mathbf{z}_j)(1 - \hat{p}(\mathbf{z}_j))}} \qquad (11\text{-}84)$$

Standardized Pearson residuals (r_{sj}): $$r_{sj} = \frac{r_j}{\sqrt{1 - h_{jj}}} \qquad (11\text{-}85)$$

where h_{jj} is the (j,j)th element in the "hat" matrix $\mathbf{H}$ given by equation (11-87). Values larger than about 2.5 suggest lack of fit at the particular $\mathbf{z}_j$.

An overall test of goodness of fit—preferred especially for smaller sample sizes—is provided by Pearson's chi square statistic

$$X^2 = \sum_{j=1}^{m} r_j^2 = \sum_{j=1}^{n} \frac{(y_j - n_j\hat{p}(\mathbf{z}_j))^2}{n_j\hat{p}(\mathbf{z}_j)(1 - \hat{p}(\mathbf{z}_j))} \qquad (11\text{-}86)$$

Notice that the chi square statistic, a single number summary of fit, is the sum of the squares of the Pearson residuals. Inspecting the Pearson residuals themselves allows us to examine the quality of fit over the entire pattern of covariates.

Another goodness-of-fit test due to Hosmer and Lemeshow [17] is only applicable when the proportion of observations with tied covariate patterns is small and all the predictor variables (covariates) are continuous.

Leverage Points and Influential Observations. The logistic regression equivalent of the hat matrix $\mathbf{H}$ contains the estimated probabilities $\hat{p}_k(\mathbf{z}_j)$. The logistic regression version of *leverages* are the diagonal elements h_{jj} of this hat matrix.

$$\mathbf{H} = \mathbf{V}^{-1/2}\mathbf{Z}(\mathbf{Z}'\mathbf{V}^{-1}\mathbf{Z})^{-1}\mathbf{Z}'\mathbf{V}^{-1/2} \qquad (11\text{-}87)$$

where $\mathbf{V}^{-1}$ is the diagonal matrix with (j, j) element $n_j\hat{p}(\mathbf{z}_j)(1 - \hat{p}(\mathbf{z}_j))$, $\mathbf{V}^{-1/2}$ is the diagonal matrix with (j, j) element $\sqrt{n_j\hat{p}(\mathbf{z}_j)(1 - \hat{p}(\mathbf{z}_j))}$.

Besides the leverages given in (11–87), other measures are available. We describe the most common called the *delta beta* or *deletion displacement*. It helps identify observations that, by themselves, have a strong influence on the regression estimates. This change in regression coefficients, when all observations with the same covariate values as the j-th case $\mathbf{z}_j$ are deleted, is quantified as

$$\Delta \beta_j = \frac{r_{sj}^2 \, h_{jj}}{1 - h_{jj}} \tag{11-88}$$

A plot of $\Delta \beta_j$ versus j can be inspected for influential cases.

11.8 Final Comments

Including Qualitative Variables

Our discussion in this chapter assumes that the discriminatory or classificatory variables, $X_1, X_2, \ldots, X_p$ have natural units of measurement. That is, each variable can, in principle, assume any real number, and these numbers can be recorded. Often, a *qualitative* or *categorical variable* may be a useful discriminator (classifier). For example, the presence or absence of a characteristic such as the color red may be a worthwhile classifier. This situation is frequently handled by creating a variable X whose numerical value is 1 if the object possesses the characteristic and zero if the object does not possess the characteristic. The variable is then treated like the measured variables in the usual discrimination and classification procedures.

Except for logistic classification, there is very little theory available to handle the case in which some variables are continuous and some qualitative. Computer simulation experiments (see [22]) indicate that Fisher's linear discriminant function can perform poorly or satisfactorily, depending upon the correlations between the qualitative and continuous variables. As Krzanowski [22] notes, "A low correlation in one population but a high correlation in the other, or a change in the sign of the correlations between the two populations could indicate conditions unfavorable to Fisher's linear discriminant function." This is a troublesome area and one that needs further study.

Classification Trees

An approach to classification completely different from the methods discussed in the previous sections of this chapter has been developed. (See [5].) It is very computer intensive and its implementation is only now becoming widespread. The new approach, called *classification and regression trees* (CART), is closely related to divisive clustering techniques. (See Chapter 12.)

Initially, all objects are considered as a single group. The group is split into two subgroups using, say, high values of a variable for one group and low values for the other. The two subgroups are then each split using the values of a second variable. The splitting process continues until a suitable stopping point is reached. The values of the splitting variables can be ordered or unordered categories. It is this feature that makes the CART procedure so general.

For example, suppose subjects are to be classified as

π_1: heart-attack prone
π_2: not heart-attack prone

on the basis of age, weight, and exercise activity. In this case, the CART procedure can be diagrammed as the tree shown in Figure 11.17. The branches of the tree actually

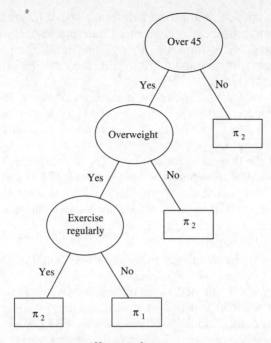

π_1 : Heart-attack prone
π_2 : Not heart-attack prone

Figure 11.17 A classification tree.

correspond to divisions in the sample space. The region R_1, defined as being over 45, being overweight, and undertaking no regular exercise, could be used to classify a subject as π_1: heart-attack prone. The CART procedure would try splitting on different ages, as well as first splitting on weight or on the amount of exercise.

The classification tree that results from using the CART methodology with the Iris data (see Table 11.5), and variables $X_3 =$ petal length (PetLength) and $X_4 =$ petal width (PetWidth), is shown in Figure 11.18. The binary splitting rules are indicated in the figure. For example, the first split occurs at petal length $= 2.45$. Flowers with petal lengths ≤ 2.45 form one group (left), and those with petal lengths > 2.45 form the other group (right).

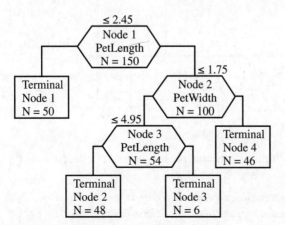

Figure 11.18 A classification tree for the Iris data.

The next split occurs with the right-hand side group (petal length > 2.45) at petal width = 1.75. Flowers with petal widths ≤ 1.75 are put in one group (left), and those with petal widths > 1.75 form the other group (right). The process continues until there is no gain with additional splitting. In this case, the process stops with four terminal nodes (TN).

The binary splits form terminal node rectangles (regions) in the positive quadrant of the X_3, X_4 sample space as shown in Figure 11.19. For example, TN #2 contains those flowers with $2.45 <$ petal lengths ≤ 4.95 and petal widths ≤ 1.75— essentially the Iris Versicolor group.

Since the majority of the flowers in, for example, TN #3 are species Virginica, a new item in this group would be classified as Virginica. That is, TN #3 and TN #4 are both assigned to the Virginica population. We see that CART has correctly classified 50 of 50 of the Setosa flowers, 47 of 50 of the Versicolor flowers, and 49 of 50 of the Virginica flowers. The APER $= \dfrac{4}{150} = .027$. This result is comparable to the result obtained for the linear discriminant analysis using variables X_3 and X_4 discussed in Example 11.12.

The CART methodology is not tied to an underlying population probability distribution of characteristics. Nor is it tied to a particular optimality criterion. In practice, the procedure requires hundreds of objects and, often, many variables. The resulting tree is very complicated. Subjective judgments must be used to prune the tree so that it ends with groups of several objects rather than all single objects. Each terminal group is then assigned to the population holding the majority membership. A new object can then be classified according to its ultimate group.

Breiman, Friedman, Olshen, and Stone [5] have developed special-purpose software for implementing a CART analysis. Also, Loh (see [21] and [25]) has developed improved classification tree software called QUEST[13] and CRUISE.[14] Their programs use several intelligent rules for splitting and usually produces a tree that often separates groups well. CART has been very successful in data mining applications (see Supplement 12A).

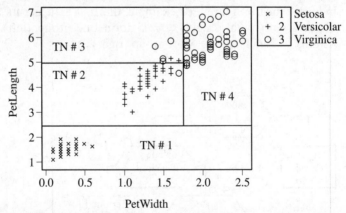

Figure 11.19 Classification tree terminal nodes (regions) in the petal width, petal length sample space.

[13] Available for download at www.stat.wisc.edu/~loh/quest.html
[14] Available for download at www.stat.wisc.edu/~loh/cruise.html

Neural Networks

A *neural network* (NN) is a computer-intensive, algorithmic procedure for transforming inputs into desired outputs using highly connected networks of relatively simple processing units (neurons or nodes). Neural networks are modeled after the neural activity in the human brain. The three essential features, then, of an NN are the basic computing units (neurons or nodes), the network architecture describing the connections between the computing units, and the training algorithm used to find values of the network parameters (weights) for performing a particular task.

The computing units are connected to one another in the sense that the output from one unit can serve as part of the input to another unit. Each computing unit transforms an input to an output using some prespecified function that is typically monotone, but otherwise arbitrary. This function depends on constants (parameters) whose values must be determined with a training set of inputs and outputs.

Network architecture is the organization of computing units and the types of connections permitted. In statistical applications, the computing units are arranged in a series of layers with connections between nodes in different layers, but not between nodes in the same layer. The layer receiving the initial inputs is called the input layer. The final layer is called the output layer. Any layers between the input and output layers are called hidden layers. A simple schematic representation of a multilayer NN is shown in Figure 11.20.

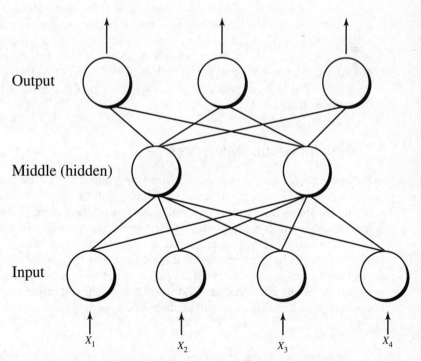

Figure 11.20 A neural network with one hidden layer.

Neural networks can be used for discrimination and classification. When they are so used, the input variables are the measured group characteristics X_1, $X_2, \ldots, X_p$, and the output variables are categorical variables indicating group membership. Current practical experience indicates that properly constructed neural networks perform about as well as logistic regression and the discriminant functions we have discussed in this chapter. Reference [30] contains a good discussion of the use of neural networks in applied statistics.

Selection of Variables

In some applications of discriminant analysis, data are available on a large number of variables. Mucciardi and Gose [27] discuss a discriminant analysis based on 157 variables.[15] In this case, it would obviously be desirable to select a relatively small subset of variables that would contain almost as much information as the original collection. This is the objective of *stepwise discriminant analysis*, and several popular commercial computer programs have such a capability.

If a stepwise discriminant analysis (or any variable selection method) is employed, the results should be interpreted with caution. (See [28].) There is no guarantee that the subset selected is "best," regardless of the criterion used to make the selection. For example, subsets selected on the basis of minimizing the apparent error rate or maximizing "discriminatory power" may perform poorly in future samples. Problems associated with variable-selection procedures are magnified if there are large correlations among the variables or between linear combinations of the variables.

Choosing a subset of variables that seems to be optimal for a *given data set* is especially disturbing if classification is the objective. At the very least, the derived classification function should be evaluated with a validation sample. As Murray [28] suggests, a better idea might be to split the sample into a number of batches and determine the "best" subset for each batch. The number of times a given variable appears in the best subsets provides a measure of the worth of that variable for future classification.

Testing for Group Differences

We have pointed out, in connection with two group classification, that effective allocation is probably not possible unless the populations are well separated. The same is true for the many group situation. Classification is ordinarily not attempted, unless the population mean vectors differ significantly from one another. Assuming that the data are nearly multivariate normal, with a common covariance matrix, MANOVA can be performed to test for differences in the population mean vectors. Although apparent significant differences do not automatically imply effective classification, testing is a necessary first step. If no significant differences are found, constructing classification rules will probably be a waste of time.

---

[15] Imagine the problems of verifying the assumption of 157-variate normality and simultaneously estimating, for example, the 12,403 parameters of the 157×157 presumed common covariance matrix!

Graphics

Sophisticated computer graphics now allow one visually to examine multivariate data in two and three dimensions. Thus, groupings in the variable space for any choice of two or three variables can often be discerned by eye. In this way, potentially important classifying variables are often identified and outlying, or "atypical," observations revealed. Visual displays are important aids in discrimination and classification, and their use is likely to increase as the hardware and associated computer programs become readily available. Frequently, as much can be learned from a visual examination as by a complex numerical analysis.

Practical Considerations Regarding Multivariate Normality

The interplay between the choice of tentative assumptions and the form of the resulting classifier is important. Consider Figure 11.21, which shows the kidney-shaped density contours from two very nonnormal densities. In this case, the normal theory linear (or even quadratic) classification rule will be inadequate compared to another choice. That is, linear discrimination here is inappropriate.

Often discrimination is attempted with a large number of variables, some of which are of the presence–absence, or 0–1, type. In these situations and in others with restricted ranges for the variables, multivariate normality may not be a sensible assumption. As we have seen, classification based on Fisher's linear discriminants can be optimal from a minimum ECM or minimum TPM point of view only when multivariate normality holds. How are we to interpret these quantities when normality is clearly not viable?

In the absence of multivariate normality, Fisher's linear discriminants can be viewed as providing an approximation to the total sample information. The values of the first few discriminants themselves can be checked for normality and rule (11-67) employed. Since the discriminants are linear combinations of a large number of variables, they will often be nearly normal. Of course, one must keep in mind that the first few discriminants are an *incomplete* summary of the original sample information. Classification rules based on this restricted set may perform poorly, while optimal rules derived from all of the sample information may perform well.

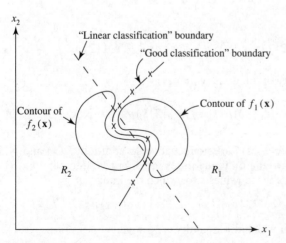

Figure 11.21 Two nonnormal populations for which linear discrimination is inappropriate.

EXERCISES

11.1. Consider the two data sets

$$\mathbf{X}_1 = \begin{bmatrix} 3 & 7 \\ 2 & 4 \\ 4 & 7 \end{bmatrix} \quad \text{and} \quad \mathbf{X}_2 = \begin{bmatrix} 6 & 9 \\ 5 & 7 \\ 4 & 8 \end{bmatrix}$$

for which

$$\bar{\mathbf{x}}_1 = \begin{bmatrix} 3 \\ 6 \end{bmatrix}, \quad \bar{\mathbf{x}}_2 = \begin{bmatrix} 5 \\ 8 \end{bmatrix}$$

and

$$\mathbf{S}_{\text{pooled}} = \begin{bmatrix} 1 & 1 \\ 1 & 2 \end{bmatrix}$$

(a) Calculate the linear discriminant function in (11-19).

(b) Classify the observation $\mathbf{x}_0' = [2 \quad 7]$ as population π_1 or population π_2, using Rule (11-18) with equal priors and equal costs.

11.2. (a) Develop a linear classification function for the data in Example 11.1 using (11-19).

(b) Using the function in (a) and (11-20), construct the "confusion matrix" by classifying the given observations. Compare your classification results with those of Figure 11.1, where the classification regions were determined "by eye." (See Example 11.6.)

(c) Given the results in (b), calculate the apparent error rate (APER).

(d) State any assumptions you make to justify the use of the method in Parts a and b.

11.3. Prove Result 11.1.
Hint: Substituting the integral expressions for $P(2|1)$ and $P(1|2)$ given by (11-1) and (11-2), respectively, into (11-5) yields

$$\text{ECM} = c(2|1)p_1 \int_{R_2} f_1(\mathbf{x}) \, d\mathbf{x} + c(1|2)p_2 \int_{R_1} f_2(\mathbf{x}) d\mathbf{x}$$

Noting that $\Omega = R_1 \cup R_2$, so that the total probability

$$1 = \int_{\Omega} f_1(\mathbf{x}) \, d\mathbf{x} = \int_{R_1} f_1(\mathbf{x}) \, d\mathbf{x} + \int_{R_2} f_1(\mathbf{x}) \, d\mathbf{x}$$

we can write

$$\text{ECM} = c(2|1)p_1 \left[1 - \int_{R_1} f_1(\mathbf{x}) \, d\mathbf{x} \right] + c(1|2)p_2 \int_{R_1} f_2(\mathbf{x}) \, d\mathbf{x}$$

By the additive property of integrals (volumes),

$$\text{ECM} = \int_{R_1} [c(1|2)p_2 f_2(\mathbf{x}) - c(2|1)p_1 f_1(\mathbf{x})] \, d\mathbf{x} + c(2|1)p_1$$

Now, $p_1, p_2, c(1|2)$, and $c(2|1)$ are nonnegative. In addition, $f_1(\mathbf{x})$ and $f_2(\mathbf{x})$ are non-negative for all $\mathbf{x}$ and are the only quantities in ECM that depend on $\mathbf{x}$. Thus, ECM is minimized if R_1 includes those values $\mathbf{x}$ for which the integrand

$$[c(1|2)p_2 f_2(\mathbf{x}) - c(2|1)p_1 f_1(\mathbf{x})] \le 0$$

and excludes those $\mathbf{x}$ for which this quantity is positive.

11.4. A researcher wants to determine a procedure for discriminating between two multivariate populations. The researcher has enough data available to estimate the density functions $f_1(\mathbf{x})$ and $f_2(\mathbf{x})$ associated with populations π_1 and π_2, respectively. Let $c(2|1) = 50$ (this is the cost of assigning items as π_2, given that π_1 is true) and $c(1|2) = 100$.

In addition, it is known that about 20% of all possible items (for which the measurements $\mathbf{x}$ can be recorded) belong to π_2.

(a) Give the minimum ECM rule (in general form) for assigning a new item to one of the two populations.

(b) Measurements recorded on a new item yield the density values $f_1(\mathbf{x}) = .3$ and $f_2(\mathbf{x}) = .5$. Given the preceding information, assign this item to population π_1 or population π_2.

11.5. Show that

$$-\tfrac{1}{2}(\mathbf{x} - \boldsymbol{\mu}_1)'\boldsymbol{\Sigma}^{-1}(\mathbf{x} - \boldsymbol{\mu}_1) + \tfrac{1}{2}(\mathbf{x} - \boldsymbol{\mu}_2)'\boldsymbol{\Sigma}^{-1}(\mathbf{x} - \boldsymbol{\mu}_2)$$
$$= (\boldsymbol{\mu}_1 - \boldsymbol{\mu}_2)'\boldsymbol{\Sigma}^{-1}\mathbf{x} - \tfrac{1}{2}(\boldsymbol{\mu}_1 - \boldsymbol{\mu}_2)'\boldsymbol{\Sigma}^{-1}(\boldsymbol{\mu}_1 + \boldsymbol{\mu}_2)$$

[see Equation (11-13).]

11.6. Consider the linear function $Y = \mathbf{a}'\mathbf{X}$. Let $E(\mathbf{X}) = \boldsymbol{\mu}_1$ and $\text{Cov}(\mathbf{X}) = \boldsymbol{\Sigma}$ if $\mathbf{X}$ belongs to population π_1. Let $E(\mathbf{X}) = \boldsymbol{\mu}_2$ and $\text{Cov}(\mathbf{X}) = \boldsymbol{\Sigma}$ if $\mathbf{X}$ belongs to population π_2. Let $m = \tfrac{1}{2}(\mu_{1Y} + \mu_{2Y}) = \tfrac{1}{2}(\mathbf{a}'\boldsymbol{\mu}_1 + \mathbf{a}'\boldsymbol{\mu}_2)$. Given that $\mathbf{a}' = (\boldsymbol{\mu}_1 - \boldsymbol{\mu}_2)'\boldsymbol{\Sigma}^{-1}$, show each of the following.

(a) $E(\mathbf{a}'\mathbf{X}|\pi_1) - m = \mathbf{a}'\boldsymbol{\mu}_1 - m > 0$

(b) $E(\mathbf{a}'\mathbf{X}|\pi_2) - m = \mathbf{a}'\boldsymbol{\mu}_2 - m < 0$

Hint: Recall that $\boldsymbol{\Sigma}$ is of full rank and is positive definite, so $\boldsymbol{\Sigma}^{-1}$ exists and is positive definite.

11.7. Let $f_1(x) = (1 - |x|)$ for $|x| \le 1$ and $f_2(x) = (1 - |x - .5|)$ for $-.5 \le x \le 1.5$.

(a) Sketch the two densities.

(b) Identify the classification regions when $p_1 = p_2$ and $c(1|2) = c(2|1)$.

(c) Identify the classification regions when $p_1 = .2$ and $c(1|2) = c(2|1)$.

11.8. Refer to Exercise 11.7. Let $f_1(x)$ be the same as in that exercise, but take $f_2(x) = \tfrac{1}{4}(2 - |x - .5|)$ for $-1.5 \le x \le 2.5$.

(a) Sketch the two densities.

(b) Determine the classification regions when $p_1 = p_2$ and $c(1|2) = c(2|1)$.

11.9. For $g = 2$ groups, show that the ratio in (11-59) is proportional to the ratio

$$\left(\frac{\text{squared distance}}{\text{between means of } Y}\right) = \frac{(\mu_{1Y} - \mu_{2Y})^2}{\sigma_Y^2} = \frac{(\mathbf{a}'\boldsymbol{\mu}_1 - \mathbf{a}'\boldsymbol{\mu}_2)^2}{\mathbf{a}'\boldsymbol{\Sigma}\mathbf{a}}$$

$$= \frac{\mathbf{a}'(\boldsymbol{\mu}_1 - \boldsymbol{\mu}_2)(\boldsymbol{\mu}_1 - \boldsymbol{\mu}_2)'\mathbf{a}}{\mathbf{a}'\boldsymbol{\Sigma}\mathbf{a}} = \frac{(\mathbf{a}'\boldsymbol{\delta})^2}{\mathbf{a}'\boldsymbol{\Sigma}\mathbf{a}}$$

where $\boldsymbol{\delta} = (\boldsymbol{\mu}_1 - \boldsymbol{\mu}_2)$ is the difference in mean vectors. This ratio is the population counterpart of (11-23). Show that the ratio is maximized by the linear combination

$$\mathbf{a} = c\boldsymbol{\Sigma}^{-1}\boldsymbol{\delta} = c\boldsymbol{\Sigma}^{-1}(\boldsymbol{\mu}_1 - \boldsymbol{\mu}_2)$$

for any $c \ne 0$.

Hint: Note that $(\boldsymbol{\mu}_i - \bar{\boldsymbol{\mu}})(\boldsymbol{\mu}_i - \bar{\boldsymbol{\mu}})' = \frac{1}{4}(\boldsymbol{\mu}_1 - \boldsymbol{\mu}_2)(\boldsymbol{\mu}_1 - \boldsymbol{\mu}_2)'$ for $i = 1, 2$, where $\bar{\boldsymbol{\mu}} = \frac{1}{2}(\boldsymbol{\mu}_1 + \boldsymbol{\mu}_2)$.

11.10. Suppose that $n_1 = 11$ and $n_2 = 12$ observations are made on two random variables X_1 and X_2, where X_1 and X_2 are assumed to have a bivariate normal distribution with a common covariance matrix $\boldsymbol{\Sigma}$, but possibly different mean vectors $\boldsymbol{\mu}_1$ and $\boldsymbol{\mu}_2$ for the two samples. The sample mean vectors and pooled covariance matrix are

$$\bar{\mathbf{x}}_1 = \begin{bmatrix} -1 \\ -1 \end{bmatrix}; \qquad \bar{\mathbf{x}}_2 = \begin{bmatrix} 2 \\ 1 \end{bmatrix}$$

$$\mathbf{S}_{\text{pooled}} = \begin{bmatrix} 7.3 & -1.1 \\ -1.1 & 4.8 \end{bmatrix}$$

(a) Test for the difference in population mean vectors using Hotelling's two-sample T^2-statistic. Let $\alpha = .10$.

(b) Construct Fisher's (sample) linear discriminant function. [See (11-19) and (11-25).]

(c) Assign the observation $\mathbf{x}_0' = \begin{bmatrix} 0 & 1 \end{bmatrix}$ to either population π_1 or π_2. Assume equal costs and equal prior probabilities.

11.11. Suppose a univariate random variable X has a normal distribution with variance 4. If X is from population π_1, its mean is 10; if it is from population π_2, its mean is 14. Assume equal prior probabilities for the events $A1 = X$ is from population π_1 and $A2 = X$ is from population π_2, and assume that the misclassification costs $c(2|1)$ and $c(1|2)$ are equal (for instance, $10). We decide that we shall allocate (classify) X to population π_1 if $X \leq c$, for some c to be determined, and to population π_2 if $X > c$. Let $B1$ be the event X is classified into population π_1 and $B2$ be the event X is classified into population π_2. Make a table showing the following: $P(B1|A2)$, $P(B2|A1)$, $P(A1 \text{ and } B2)$, $P(A2 \text{ and } B1)$, $P(\text{misclassification})$, and expected cost for various values of c. For what choice of c is expected cost minimized? The table should take the following form:

c	$P(B1|A2)$	$P(B2|A1)$	$P(A1 \text{ and } B2)$	$P(A2 \text{ and } B1)$	$P(\text{error})$	Expected cost
10						
$\vdots$						
14						

What is the value of the minimum expected cost?

11.12. Repeat Exercise 11.11 if the prior probabilities of $A1$ and $A2$ are equal, but $c(2|1) = \$5$ and $c(1|2) = \$15$.

11.13. Repeat Exercise 11.11 if the prior probabilities of $A1$ and $A2$ are $P(A1) = .25$ and $P(A2) = .75$ and the misclassification costs are as in Exercise 11.12.

11.14. Consider the discriminant functions derived in Example 11.3. Normalize $\hat{\mathbf{a}}$ using (11-21) and (11-22). Compute the two midpoints $\hat{m}_1^*$ and $\hat{m}_2^*$ corresponding to the two choices of normalized vectors, say, $\hat{\mathbf{a}}_1^*$ and $\hat{\mathbf{a}}_2^*$. Classify $\mathbf{x}_0' = [-.210, -.044]$ with the function $\hat{y}_0^* = \hat{\mathbf{a}}^{*\prime} \mathbf{x}_0$ for the two cases. Are the results consistent with the classification obtained for the case of equal prior probabilities in Example 11.3? Should they be?

11.15. Derive the expressions in (11-27) from (11-6) when $f_1(\mathbf{x})$ and $f_2(\mathbf{x})$ are multivariate normal densities with means $\boldsymbol{\mu}_1, \boldsymbol{\mu}_2$ and covariances $\boldsymbol{\Sigma}_1, \boldsymbol{\Sigma}_2$, respectively.

11.16. Suppose **x** comes from one of two populations:

π_1: Normal with mean μ_1 and covariance matrix Σ_1

π_2: Normal with mean μ_2 and covariance matrix Σ_2

If the respective density functions are denoted by $f_1(\mathbf{x})$ and $f_2(\mathbf{x})$, find the expression for the quadratic discriminator

$$Q = \ln\left[\frac{f_1(\mathbf{x})}{f_2(\mathbf{x})}\right]$$

If $\Sigma_1 = \Sigma_2 = \Sigma$, for instance, verify that Q becomes

$$(\mu_1 - \mu_2)'\Sigma^{-1}\mathbf{x} - \tfrac{1}{2}(\mu_1 - \mu_2)'\Sigma^{-1}(\mu_1 + \mu_2)$$

11.17. Suppose populations π_1 and π_2 are as follows:

	Population	
	π_1	π_2
Distribution	Normal	Normal
Mean μ	$[10, 15]'$	$[10, 25]'$
Covariance Σ	$\begin{bmatrix} 18 & 12 \\ 12 & 32 \end{bmatrix}$	$\begin{bmatrix} 20 & -7 \\ -7 & 5 \end{bmatrix}$

Assume equal prior probabilities and misclassifications costs of $c(2|1) = \$10$ and $c(1|2) = \$73.89$. Find the posterior probabilities of populations π_1 and π_2, $P(\pi_1|\mathbf{x})$ and $P(\pi_2|\mathbf{x})$, the value of the quadratic discriminator Q in Exercise 11.16, and the classification for each value of **x** in the following table:

x	$P(\pi_1|\mathbf{x})$	$P(\pi_2|\mathbf{x})$	Q	Classification
$[10, 15]'$				
$[12, 17]'$				
$\vdots$				
$[30, 35]'$				

(*Note:* Use an increment of 2 in each coordinate—11 points in all.)

Show each of the following on a graph of the x_1, x_2 plane.

(a) The mean of each population

(b) The ellipse of minimal area with probability .95 of containing **x** for each population

(c) The region R_1 (for population π_1) and the region $\Omega - R_1 = R_2$ (for population π_2)

(d) The 11 points classified in the table

11.18. If **B** is defined as $c(\mu_1 - \mu_2)(\mu_1 - \mu_2)'$ for some constant c, verify that $\mathbf{e} = c\Sigma^{-1}(\mu_1 - \mu_2)$ is in fact an (unscaled) eigenvector of $\Sigma^{-1}\mathbf{B}$, where Σ is a covariance matrix.

11.19. (a) Using the original data sets $\mathbf{X}_1$ and $\mathbf{X}_2$ given in Example 11.7, calculate $\bar{\mathbf{x}}_i$, $\mathbf{S}_i$, $i = 1, 2$, and $\mathbf{S}_{pooled}$, verifying the results provided for these quantities in the example.

(b) Using the calculations in Part a, compute Fisher's linear discriminant function, and use it to classify the sample observations according to Rule (11-25). Verify that the confusion matrix given in Example 11.7 is correct.

(c) Classify the sample observations on the basis of smallest squared distance $D_i^2(\mathbf{x})$ of the observations from the group means $\bar{\mathbf{x}}_1$ and $\bar{\mathbf{x}}_2$. [See (11-54).] Compare the results with those in Part b. Comment.

11.20. The matrix identity (see Bartlett [3])

$$\mathbf{S}_{H,\text{pooled}}^{-1} = \frac{n-3}{n-2}\left(\mathbf{S}_{\text{pooled}}^{-1} + \frac{c_k}{1 - c_k(\mathbf{x}_H - \bar{\mathbf{x}}_k)'\mathbf{S}_{\text{pooled}}^{-1}(\mathbf{x}_H - \bar{\mathbf{x}}_k)}\right.$$

$$\left. \cdot \mathbf{S}_{\text{pooled}}^{-1}(\mathbf{x}_H - \bar{\mathbf{x}}_k)(\mathbf{x}_H - \bar{\mathbf{x}}_k)'\mathbf{S}_{\text{pooled}}^{-1}\right)$$

where

$$c_k = \frac{n_k}{(n_k - 1)(n - 2)}$$

allows the calculation of $\mathbf{S}_{H,\text{pooled}}^{-1}$ from $\mathbf{S}_{\text{pooled}}^{-1}$. Verify this identity using the data from Example 11.7. Specifically, set $n = n_1 + n_2$, $k = 1$, and $\mathbf{x}'_H = [2, 12]$. Calculate $\mathbf{S}_{H,\text{pooled}}^{-1}$ using the full data $\mathbf{S}_{\text{pooled}}^{-1}$ and $\bar{\mathbf{x}}_1$, and compare the result with $\mathbf{S}_{H,\text{pooled}}^{-1}$ in Example 11.7.

11.21. Let $\lambda_1 \geq \lambda_2 \geq \cdots \geq \lambda_s > 0$ denote the $s \leq \min(g - 1, p)$ nonzero eigenvalues of $\mathbf{\Sigma}^{-1}\mathbf{B}_\mu$ and $\mathbf{e}_1, \mathbf{e}_2, \ldots, \mathbf{e}_s$ the corresponding eigenvectors (scaled so that $\mathbf{e}'\mathbf{\Sigma}\mathbf{e} = 1$). Show that the vector of coefficients $\mathbf{a}$ that maximizes the ratio

$$\frac{\mathbf{a}'\mathbf{B}_\mu\mathbf{a}}{\mathbf{a}'\mathbf{\Sigma}\mathbf{a}} = \frac{\mathbf{a}'\left[\sum_{i=1}^{g}(\boldsymbol{\mu}_i - \bar{\boldsymbol{\mu}})(\boldsymbol{\mu}_i - \bar{\boldsymbol{\mu}})'\right]\mathbf{a}}{\mathbf{a}'\mathbf{\Sigma}\mathbf{a}}$$

is given by $\mathbf{a}_1 = \mathbf{e}_1$. The linear combination $\mathbf{a}_1'\mathbf{X}$ is called the *first discriminant*. Show that the value $\mathbf{a}_2 = \mathbf{e}_2$ maximizes the ratio subject to $\text{Cov}(\mathbf{a}_1'\mathbf{X}, \mathbf{a}_2'\mathbf{X}) = 0$. The linear combination $\mathbf{a}_2'\mathbf{X}$ is called the *second discriminant*. Continuing, $\mathbf{a}_k = \mathbf{e}_k$ maximizes the ratio subject to $0 = \text{Cov}(\mathbf{a}_k'\mathbf{X}, \mathbf{a}_i'\mathbf{X})$, $i < k$, and $\mathbf{a}_k'\mathbf{X}$ is called the *k*th *discriminant*. Also, $\text{Var}(\mathbf{a}_i'\mathbf{X}) = 1$, $i = 1, \ldots, s$. [See (11-62) for the sample equivalent.]

Hint: We first convert the maximization problem to one already solved. By the spectral decomposition in (2-20), $\mathbf{\Sigma} = \mathbf{P}'\mathbf{\Lambda}\mathbf{P}$ where $\mathbf{\Lambda}$ is a diagonal matrix with positive elements λ_i. Let $\mathbf{\Lambda}^{1/2}$ denote the diagonal matrix with elements $\sqrt{\lambda_i}$. By (2-22), the symmetric square-root matrix $\mathbf{\Sigma}^{1/2} = \mathbf{P}'\mathbf{\Lambda}^{1/2}\mathbf{P}$ and its inverse $\mathbf{\Sigma}^{-1/2} = \mathbf{P}'\mathbf{\Lambda}^{-1/2}\mathbf{P}$ satisfy $\mathbf{\Sigma}^{1/2}\mathbf{\Sigma}^{1/2} = \mathbf{\Sigma}$, $\mathbf{\Sigma}^{1/2}\mathbf{\Sigma}^{-1/2} = \mathbf{I} = \mathbf{\Sigma}^{-1/2}\mathbf{\Sigma}^{1/2}$ and $\mathbf{\Sigma}^{-1/2}\mathbf{\Sigma}^{-1/2} = \mathbf{\Sigma}^{-1}$. Next, set

$$\mathbf{u} = \mathbf{\Sigma}^{1/2}\mathbf{a}$$

so $\mathbf{u}'\mathbf{u} = \mathbf{a}'\mathbf{\Sigma}^{1/2}\mathbf{\Sigma}^{1/2}\mathbf{a} = \mathbf{a}'\mathbf{\Sigma}\mathbf{a}$ and $\mathbf{u}'\mathbf{\Sigma}^{-1/2}\mathbf{B}_\mu\mathbf{\Sigma}^{-1/2}\mathbf{u} = \mathbf{a}'\mathbf{\Sigma}^{1/2}\mathbf{\Sigma}^{-1/2}\mathbf{B}_\mu\mathbf{\Sigma}^{-1/2}\mathbf{\Sigma}^{1/2}\mathbf{a} = \mathbf{a}'\mathbf{B}_\mu\mathbf{a}$. Consequently, the problem reduces to maximizing

$$\frac{\mathbf{u}'\mathbf{\Sigma}^{-1/2}\mathbf{B}_\mu\mathbf{\Sigma}^{-1/2}\mathbf{u}}{\mathbf{u}'\mathbf{u}}$$

over $\mathbf{u}$. From (2-51), the maximum of this ratio is λ_1, the largest eigenvalue of $\mathbf{\Sigma}^{-1/2}\mathbf{B}_\mu\mathbf{\Sigma}^{-1/2}$. This maximum occurs when $\mathbf{u} = \mathbf{e}_1$, the normalized eigenvector associated with λ_1. Because $\mathbf{e}_1 = \mathbf{u} = \boldsymbol{\Sigma}^{1/2}\mathbf{a}_1$, or $\mathbf{a}_1 = \boldsymbol{\Sigma}^{-1/2}\mathbf{e}_1$, $\text{Var}(\mathbf{a}_1'\mathbf{X}) = \mathbf{a}_1'\boldsymbol{\Sigma}\mathbf{a}_1 = \mathbf{e}_1'\boldsymbol{\Sigma}^{-1/2}\boldsymbol{\Sigma}\boldsymbol{\Sigma}^{-1/2}\mathbf{e}_1 = \mathbf{e}_1'\boldsymbol{\Sigma}^{-1/2}\boldsymbol{\Sigma}^{1/2}\boldsymbol{\Sigma}^{1/2}\boldsymbol{\Sigma}^{-1/2}\mathbf{e}_1 = \mathbf{e}_1'\mathbf{e}_1 = 1$. By (2-52), $\mathbf{u} \perp \mathbf{e}_1$ maximizes the preceding ratio when $\mathbf{u} = \mathbf{e}_2$, the normalized eigenvector corresponding to λ_2. For this choice, $\mathbf{a}_2 = \boldsymbol{\Sigma}^{-1/2}\mathbf{e}_2$, and $\text{Cov}(\mathbf{a}_2'\mathbf{X}, \mathbf{a}_1'\mathbf{X}) = \mathbf{a}_2'\boldsymbol{\Sigma}\mathbf{a}_1 = \mathbf{e}_2'\boldsymbol{\Sigma}^{-1/2}\boldsymbol{\Sigma}\boldsymbol{\Sigma}^{-1/2}\mathbf{e}_1 = \mathbf{e}_2'\mathbf{e}_1 = 0$, since $\mathbf{e}_2 \perp \mathbf{e}_1$. Similarly, $\text{Var}(\mathbf{a}_2'\mathbf{X}) = \mathbf{a}_2'\boldsymbol{\Sigma}\mathbf{a}_2 = \mathbf{e}_2'\mathbf{e}_2 = 1$. Continue in this fashion for the remaining discriminants. Note that if λ and $\mathbf{e}$ are an eigenvalue–eigenvector pair of $\boldsymbol{\Sigma}^{-1/2}\mathbf{B}_\mu\boldsymbol{\Sigma}^{-1/2}$, then

$$\boldsymbol{\Sigma}^{-1/2}\mathbf{B}_\mu\boldsymbol{\Sigma}^{-1/2}\mathbf{e} = \lambda\mathbf{e}$$

and multiplication on the left by $\boldsymbol{\Sigma}^{-1/2}$ gives

$$\boldsymbol{\Sigma}^{-1/2}\boldsymbol{\Sigma}^{-1/2}\mathbf{B}_\mu\boldsymbol{\Sigma}^{-1/2}\mathbf{e} = \lambda\boldsymbol{\Sigma}^{-1/2}\mathbf{e} \quad \text{or} \quad \boldsymbol{\Sigma}^{-1}\mathbf{B}_\mu(\boldsymbol{\Sigma}^{-1/2}\mathbf{e}) = \lambda(\boldsymbol{\Sigma}^{-1/2}\mathbf{e})$$

Thus, $\boldsymbol{\Sigma}^{-1}\mathbf{B}_\mu$ has the same eigenvalues as $\boldsymbol{\Sigma}^{-1/2}\mathbf{B}_\mu\boldsymbol{\Sigma}^{-1/2}$, but the corresponding eigenvector is proportional to $\boldsymbol{\Sigma}^{-1/2}\mathbf{e} = \mathbf{a}$, as asserted.

11.22. Show that $\Delta_S^2 = \lambda_1 + \lambda_2 + \cdots + \lambda_p = \lambda_1 + \lambda_2 + \cdots + \lambda_s$, where $\lambda_1, \lambda_2, \ldots, \lambda_s$ are the nonzero eigenvalues of $\boldsymbol{\Sigma}^{-1}\mathbf{B}_\mu$ (or $\boldsymbol{\Sigma}^{-1/2}\mathbf{B}_\mu\boldsymbol{\Sigma}^{-1/2}$) and Δ_S^2 is given by (11-68). Also, show that $\lambda_1 + \lambda_2 + \cdots + \lambda_r$ is the resulting separation when only the first r discriminants, $Y_1, Y_2, \ldots, Y_r$ are used.

Hint: Let $\mathbf{P}$ be the orthogonal matrix whose ith row $\mathbf{e}_i'$ is the eigenvector of $\boldsymbol{\Sigma}^{-1/2}\mathbf{B}_\mu\boldsymbol{\Sigma}^{-1/2}$ corresponding to the ith largest eigenvalue, $i = 1, 2, \ldots, p$. Consider

$$\mathbf{Y}_{(p\times1)} = \begin{bmatrix} Y_1 \\ \vdots \\ Y_s \\ \vdots \\ Y_p \end{bmatrix} = \begin{bmatrix} \mathbf{e}_1'\boldsymbol{\Sigma}^{-1/2}\mathbf{X} \\ \vdots \\ \mathbf{e}_s'\boldsymbol{\Sigma}^{-1/2}\mathbf{X} \\ \vdots \\ \mathbf{e}_p'\boldsymbol{\Sigma}^{-1/2}\mathbf{X} \end{bmatrix} = \mathbf{P}\boldsymbol{\Sigma}^{-1/2}\mathbf{X}$$

Now, $\boldsymbol{\mu}_{iY} = E(\mathbf{Y}\,|\,\pi_i) = \mathbf{P}\boldsymbol{\Sigma}^{-1/2}\boldsymbol{\mu}_i$ and $\bar{\boldsymbol{\mu}}_Y = \mathbf{P}\boldsymbol{\Sigma}^{-1/2}\bar{\boldsymbol{\mu}}$, so

$$(\boldsymbol{\mu}_{iY} - \bar{\boldsymbol{\mu}}_Y)'(\boldsymbol{\mu}_{iY} - \bar{\boldsymbol{\mu}}_Y) = (\boldsymbol{\mu}_i - \bar{\boldsymbol{\mu}})'\boldsymbol{\Sigma}^{-1/2}\mathbf{P}'\mathbf{P}\boldsymbol{\Sigma}^{-1/2}(\boldsymbol{\mu}_i - \bar{\boldsymbol{\mu}})$$
$$= (\boldsymbol{\mu}_i - \bar{\boldsymbol{\mu}})'\boldsymbol{\Sigma}^{-1}(\boldsymbol{\mu}_i - \bar{\boldsymbol{\mu}})$$

Therefore, $\Delta_S^2 = \sum\limits_{i=1}^{g} (\boldsymbol{\mu}_{iY} - \bar{\boldsymbol{\mu}}_Y)'(\boldsymbol{\mu}_{iY} - \bar{\boldsymbol{\mu}}_Y)$. Using Y_1, we have

$$\sum_{i=1}^{g} (\mu_{iY_1} - \bar{\mu}_{Y_1})^2 = \sum_{i=1}^{g} \mathbf{e}_1'\boldsymbol{\Sigma}^{-1/2}(\boldsymbol{\mu}_i - \bar{\boldsymbol{\mu}})(\boldsymbol{\mu}_i - \bar{\boldsymbol{\mu}})'\boldsymbol{\Sigma}^{-1/2}\mathbf{e}_1$$
$$= \mathbf{e}_1'\boldsymbol{\Sigma}^{-1/2}\mathbf{B}_\mu\boldsymbol{\Sigma}^{-1/2}\mathbf{e}_1 = \lambda_1$$

because $\mathbf{e}_1$ has eigenvalue λ_1. Similarly, Y_2 produces

$$\sum_{i=1}^{g} (\mu_{iY_2} - \bar{\mu}_{Y_2})^2 = \mathbf{e}_2'\boldsymbol{\Sigma}^{-1/2}\mathbf{B}_\mu\boldsymbol{\Sigma}^{-1/2}\mathbf{e}_2 = \lambda_2$$

and Y_p produces

$$\sum_{i=1}^{g} (\mu_{iY_p} - \bar{\mu}_{Y_p})^2 = \mathbf{e}_p'\boldsymbol{\Sigma}^{-1/2}\mathbf{B}_\mu\boldsymbol{\Sigma}^{-1/2}\mathbf{e}_p = \lambda_p$$

Thus,

$$\Delta_S^2 = \sum_{i=1}^{g} (\boldsymbol{\mu}_{iY} - \bar{\boldsymbol{\mu}}_Y)'(\boldsymbol{\mu}_{iY} - \bar{\boldsymbol{\mu}}_Y)$$

$$= \sum_{i=1}^{g} (\mu_{iY_1} - \bar{\mu}_{Y_1})^2 + \sum_{i=1}^{g} (\mu_{iY_2} - \bar{\mu}_{Y_2})^2 + \cdots + \sum_{i=1}^{g} (\mu_{iY_p} - \bar{\mu}_{Y_p})^2$$

$$= \lambda_1 + \lambda_2 + \cdots + \lambda_p = \lambda_1 + \lambda_2 + \cdots + \lambda_s$$

since $\lambda_{s+1} = \cdots = \lambda_p = 0$. If only the first r discriminants are used, their contribution to Δ_S^2 is $\lambda_1 + \lambda_2 + \cdots + \lambda_r$.

The following exercises require the use of a computer.

11.23. Consider the data given in Exercise 1.14.

(a) Check the marginal distributions of the x_i's in both the multiple-sclerosis (MS) group and non-multiple-sclerosis (NMS) group for normality by graphing the corresponding observations as normal probability plots. Suggest appropriate data transformations if the normality assumption is suspect.

(b) Assume that $\Sigma_1 = \Sigma_2 = \Sigma$. Construct Fisher's linear discriminant function. Do all the variables in the discriminant function appear to be important? Discuss your answer. Develop a classification rule assuming equal prior probabilities and equal costs of misclassification.

(c) Using the results in (b), calculate the apparent error rate. If computing resources allow, calculate an estimate of the expected actual error rate using Lachenbruch's holdout procedure. Compare the two error rates.

11.24. Annual financial data are collected for bankrupt firms approximately 2 years prior to their bankruptcy and for financially sound firms at about the same time. The data on four variables, $X_1 = $ CF/TD = (cash flow)/(total debt), $X_2 = $ NI/TA = (net income)/(total assets), $X_3 = $ CA/CL = (current assets)/(current liabilities), and $X_4 = $ CA/NS = (current assets)/(net sales), are given in Table 11.4.

(a) Using a different symbol for each group, plot the data for the pairs of observations (x_1, x_2), (x_1, x_3) and (x_1, x_4). Does it appear as if the data are approximately bivariate normal for any of these pairs of variables?

(b) Using the $n_1 = 21$ pairs of observations (x_1, x_2) for bankrupt firms and the $n_2 = 25$ pairs of observations (x_1, x_2) for nonbankrupt firms, calculate the sample mean vectors $\bar{\mathbf{x}}_1$ and $\bar{\mathbf{x}}_2$ and the sample covariance matrices $\mathbf{S}_1$ and $\mathbf{S}_2$.

(c) Using the results in (b) and assuming that both random samples are from bivariate normal populations, construct the classification rule (11-29) with $p_1 = p_2$ and $c(1|2) = c(2|1)$.

(d) Evaluate the performance of the classification rule developed in (c) by computing the apparent error rate (APER) from (11-34) and the estimated expected actual error rate $\hat{E}$ (AER) from (11-36).

(e) Repeat Parts c and d, assuming that $p_1 = .05$, $p_2 = .95$, and $c(1|2) = c(2|1)$. Is this choice of prior probabilities reasonable? Explain.

(f) Using the results in (b), form the pooled covariance matrix $\mathbf{S}_{pooled}$, and construct Fisher's sample linear discriminant function in (11-19). Use this function to classify the sample observations and evaluate the APER. Is Fisher's linear discriminant function a sensible choice for a classifier in this case? Explain.

(g) Repeat Parts b–e using the observation pairs (x_1, x_3) and (x_1, x_4). Do some variables appear to be better classifiers than others? Explain.

(h) Repeat Parts b–e using observations on all four variables (X_1, X_2, X_3, X_4).

Table 11.4 Bankruptcy Data

Row	$x_1 = \dfrac{CF}{TD}$	$x_2 = \dfrac{NI}{TA}$	$x_3 = \dfrac{CA}{CL}$	$x_4 = \dfrac{CA}{NS}$	Population $\pi_i, i = 1,2$
1	−.45	−.41	1.09	.45	0
2	−.56	−.31	1.51	.16	0
3	.06	.02	1.01	.40	0
4	−.07	−.09	1.45	.26	0
5	−.10	−.09	1.56	.67	0
6	−.14	−.07	.71	.28	0
7	.04	.01	1.50	.71	0
8	−.06	−.06	1.37	.40	0
9	.07	−.01	1.37	.34	0
10	−.13	−.14	1.42	.44	0
11	−.23	−.30	.33	.18	0
12	.07	.02	1.31	.25	0
13	.01	.00	2.15	.70	0
14	−.28	−.23	1.19	.66	0
15	.15	.05	1.88	.27	0
16	.37	.11	1.99	.38	0
17	−.08	−.08	1.51	.42	0
18	.05	.03	1.68	.95	0
19	.01	−.00	1.26	.60	0
20	.12	.11	1.14	.17	0
21	−.28	−.27	1.27	.51	0
1	.51	.10	2.49	.54	1
2	.08	.02	2.01	.53	1
3	.38	.11	3.27	.35	1
4	.19	.05	2.25	.33	1
5	.32	.07	4.24	.63	1
6	.31	.05	4.45	.69	1
7	.12	.05	2.52	.69	1
8	−.02	.02	2.05	.35	1
9	.22	.08	2.35	.40	1
10	.17	.07	1.80	.52	1
11	.15	.05	2.17	.55	1
12	−.10	−.01	2.50	.58	1
13	.14	−.03	.46	.26	1
14	.14	.07	2.61	.52	1
15	.15	.06	2.23	.56	1
16	.16	.05	2.31	.20	1
17	.29	.06	1.84	.38	1
18	.54	.11	2.33	.48	1
19	−.33	−.09	3.01	.47	1
20	.48	.09	1.24	.18	1
21	.56	.11	4.29	.45	1
22	.20	.08	1.99	.30	1
23	.47	.14	2.92	.45	1
24	.17	.04	2.45	.14	1
25	.58	.04	5.06	.13	1

Legend: $\pi_1 = 0$: bankrupt firms; $\pi_2 = 1$: nonbankrupt firms.
Source: 1968, 1969, 1970, 1971, 1972 Moody's Industrial Manuals.

11.25. The annual financial data listed in Table 11.4 have been analyzed by Johnson [19] with a view toward detecting influential observations in a discriminant analysis. Consider variables X_1 = CF/TD and X_3 = CA/CL.

(a) Using the data on variables X_1 and X_3, construct Fisher's linear discriminant function. Use this function to classify the sample observations and evaluate the APER. [See (11-25) and (11-34).] Plot the data and the discriminant line in the (x_1, x_3) coordinate system.

(b) Johnson [19] has argued that the multivariate observations in rows 16 for bankrupt firms and 13 for sound firms are influential. Using the X_1, X_3 data, calculate Fisher's linear discriminant function with *only* data point 16 for bankrupt firms deleted. Repeat this procedure with *only* data point 13 for sound firms deleted. Plot the respective discriminant lines on the scatter in part a, and calculate the APERs, ignoring the deleted point in each case. Does deleting either of these multivariate observations make a difference? (Note that neither of the potentially influential data points is particularly "distant" from the center of its respective scatter.)

11.26. Using the data in Table 11.4, define a binary response variable Z that assumes the value 0 if a firm is bankrupt and 1 if a firm is not bankrupt. Let X = CA/CL, and consider the straight-line regression of Z on X.

(a) Although a binary response variable does not meet the standard regression assumptions, consider using least squares to determine the fitted straight line for the X, Z data. Plot the fitted values for bankrupt firms as a dot diagram on the interval $[0, 1]$. Repeat this procedure for nonbankrupt firms and overlay the two dot diagrams. A reasonable discrimination rule is to predict that a firm will go bankrupt if its fitted value is closer to 0 than to 1. That is, the fitted value is less than .5. Similarly, a firm is predicted to be sound if its fitted value is greater than .5. Use this decision rule to classify the sample firms. Calculate the APER.

(b) Repeat the analysis in Part a using all four variables, $X_1, \ldots, X_4$. Is there any change in the APER? Do data points 16 for bankrupt firms and 13 for nonbankrupt firms stand out as influential?

(c) Perform a logistic regression using all four variables.

11.27. The data in Table 11.5 contain observations on X_2 = sepal width and X_4 = petal width for samples from three species of iris. There are $n_1 = n_2 = n_3 = 50$ observations in each sample.

(a) Plot the data in the (x_2, x_4) variable space. Do the observations for the three groups appear to be bivariate normal?

Table 11.5 Data on Irises

π_1: *Iris setosa*				π_2: *Iris versicolor*				π_3: *Iris virginica*			
Sepal length x_1	Sepal width x_2	Petal length x_3	Petal width x_4	Sepal length x_1	Sepal width x_2	Petal length x_3	Petal width x_4	Sepal length x_1	Sepal width x_2	Petal length x_3	Petal width x_4
5.1	3.5	1.4	0.2	7.0	3.2	4.7	1.4	6.3	3.3	6.0	2.5
4.9	3.0	1.4	0.2	6.4	3.2	4.5	1.5	5.8	2.7	5.1	1.9
4.7	3.2	1.3	0.2	6.9	3.1	4.9	1.5	7.1	3.0	5.9	2.1
4.6	3.1	1.5	0.2	5.5	2.3	4.0	1.3	6.3	2.9	5.6	1.8
5.0	3.6	1.4	0.2	6.5	2.8	4.6	1.5	6.5	3.0	5.8	2.2
5.4	3.9	1.7	0.4	5.7	2.8	4.5	1.3	7.6	3.0	6.6	2.1

(continues on next page)

Table 11.5 *(continued)*

π_1: Iris setosa				π_2: Iris versicolor				π_3: Iris virginica			
Sepal length x_1	Sepal width x_2	Petal length x_3	Petal width x_4	Sepal length x_1	Sepal width x_2	Petal length x_3	Petal width x_4	Sepal length x_1	Sepal width x_2	Petal length x_3	Petal width x_4
4.6	3.4	1.4	0.3	6.3	3.3	4.7	1.6	4.9	2.5	4.5	1.7
5.0	3.4	1.5	0.2	4.9	2.4	3.3	1.0	7.3	2.9	6.3	1.8
4.4	2.9	1.4	0.2	6.6	2.9	4.6	1.3	6.7	2.5	5.8	1.8
4.9	3.1	1.5	0.1	5.2	2.7	3.9	1.4	7.2	3.6	6.1	2.5
5.4	3.7	1.5	0.2	5.0	2.0	3.5	1.0	6.5	3.2	5.1	2.0
4.8	3.4	1.6	0.2	5.9	3.0	4.2	1.5	6.4	2.7	5.3	1.9
4.8	3.0	1.4	0.1	6.0	2.2	4.0	1.0	6.8	3.0	5.5	2.1
4.3	3.0	1.1	0.1	6.1	2.9	4.7	1.4	5.7	2.5	5.0	2.0
5.8	4.0	1.2	0.2	5.6	2.9	3.6	1.3	5.8	2.8	5.1	2.4
5.7	4.4	1.5	0.4	6.7	3.1	4.4	1.4	6.4	3.2	5.3	2.3
5.4	3.9	1.3	0.4	5.6	3.0	4.5	1.5	6.5	3.0	5.5	1.8
5.1	3.5	1.4	0.3	5.8	2.7	4.1	1.0	7.7	3.8	6.7	2.2
5.7	3.8	1.7	0.3	6.2	2.2	4.5	1.5	7.7	2.6	6.9	2.3
5.1	3.8	1.5	0.3	5.6	2.5	3.9	1.1	6.0	2.2	5.0	1.5
5.4	3.4	1.7	0.2	5.9	3.2	4.8	1.8	6.9	3.2	5.7	2.3
5.1	3.7	1.5	0.4	6.1	2.8	4.0	1.3	5.6	2.8	4.9	2.0
4.6	3.6	1.0	0.2	6.3	2.5	4.9	1.5	7.7	2.8	6.7	2.0
5.1	3.3	1.7	0.5	6.1	2.8	4.7	1.2	6.3	2.7	4.9	1.8
4.8	3.4	1.9	0.2	6.4	2.9	4.3	1.3	6.7	3.3	5.7	2.1
5.0	3.0	1.6	0.2	6.6	3.0	4.4	1.4	7.2	3.2	6.0	1.8
5.0	3.4	1.6	0.4	6.8	2.8	4.8	1.4	6.2	2.8	4.8	1.8
5.2	3.5	1.5	0.2	6.7	3.0	5.0	1.7	6.1	3.0	4.9	1.8
5.2	3.4	1.4	0.2	6.0	2.9	4.5	1.5	6.4	2.8	5.6	2.1
4.7	3.2	1.6	0.2	5.7	2.6	3.5	1.0	7.2	3.0	5.8	1.6
4.8	3.1	1.6	0.2	5.5	2.4	3.8	1.1	7.4	2.8	6.1	1.9
5.4	3.4	1.5	0.4	5.5	2.4	3.7	1.0	7.9	3.8	6.4	2.0
5.2	4.1	1.5	0.1	5.8	2.7	3.9	1.2	6.4	2.8	5.6	2.2
5.5	4.2	1.4	0.2	6.0	2.7	5.1	1.6	6.3	2.8	5.1	1.5
4.9	3.1	1.5	0.2	5.4	3.0	4.5	1.5	6.1	2.6	5.6	1.4
5.0	3.2	1.2	0.2	6.0	3.4	4.5	1.6	7.7	3.0	6.1	2.3
5.5	3.5	1.3	0.2	6.7	3.1	4.7	1.5	6.3	3.4	5.6	2.4
4.9	3.6	1.4	0.1	6.3	2.3	4.4	1.3	6.4	3.1	5.5	1.8
4.4	3.0	1.3	0.2	5.6	3.0	4.1	1.3	6.0	3.0	4.8	1.8
5.1	3.4	1.5	0.2	5.5	2.5	4.0	1.3	6.9	3.1	5.4	2.1
5.0	3.5	1.3	0.3	5.5	2.6	4.4	1.2	6.7	3.1	5.6	2.4
4.5	2.3	1.3	0.3	6.1	3.0	4.6	1.4	6.9	3.1	5.1	2.3
4.4	3.2	1.3	0.2	5.8	2.6	4.0	1.2	5.8	2.7	5.1	1.9
5.0	3.5	1.6	0.6	5.0	2.3	3.3	1.0	6.8	3.2	5.9	2.3
5.1	3.8	1.9	0.4	5.6	2.7	4.2	1.3	6.7	3.3	5.7	2.5
4.8	3.0	1.4	0.3	5.7	3.0	4.2	1.2	6.7	3.0	5.2	2.3
5.1	3.8	1.6	0.2	5.7	2.9	4.2	1.3	6.3	2.5	5.0	1.9
4.6	3.2	1.4	0.2	6.2	2.9	4.3	1.3	6.5	3.0	5.2	2.0
5.3	3.7	1.5	0.2	5.1	2.5	3.0	1.1	6.2	3.4	5.4	2.3
5.0	3.3	1.4	0.2	5.7	2.8	4.1	1.3	5.9	3.0	5.1	1.8

Source: Anderson [1].

(b) Assume that the samples are from bivariate normal populations with a common covariance matrix. Test the hypothesis $H_0: \boldsymbol{\mu}_1 = \boldsymbol{\mu}_2 = \boldsymbol{\mu}_3$ versus H_1: at least one $\boldsymbol{\mu}_i$ is different from the others at the $\alpha = .05$ significance level. Is the assumption of a common covariance matrix reasonable in this case? Explain.

(c) Assuming that the populations are bivariate normal, construct the quadratic discriminate scores $\hat{d}_i^Q(\mathbf{x})$ given by (11-47) with $p_1 = p_2 = p_3 = \frac{1}{3}$. Using Rule (11-48), classify the new observation $\mathbf{x}_0' = [3.5 \quad 1.75]$ into population π_1, π_2, or π_3.

(d) Assume that the covariance matrices $\boldsymbol{\Sigma}_i$ are the same for all three bivariate normal populations. Construct the linear discriminate score $\hat{d}_i(\mathbf{x})$ given by (11-51), and use it to assign $\mathbf{x}_0' = [3.5 \quad 1.75]$ to one of the populations $\pi_i, i = 1, 2, 3$ according to (11-52). Take $p_1 = p_2 = p_3 = \frac{1}{3}$. Compare the results in Parts c and d. Which approach do you prefer? Explain.

(e) Assuming equal covariance matrices and bivariate normal populations, and supposing that $p_1 = p_2 = p_3 = \frac{1}{3}$, allocate $\mathbf{x}_0' = [3.5 \quad 1.75]$ to π_1, π_2, or π_3 using Rule (11-56). Compare the result with that in Part d. Delineate the classification regions $\hat{R}_1, \hat{R}_2$, and $\hat{R}_3$ on your graph from Part a determined by the linear functions $\hat{d}_{ki}(\mathbf{x}_0)$ in (11-56).

(f) Using the linear discriminant scores from Part d, classify the sample observations. Calculate the APER and $\hat{E}(\text{AER})$. (To calculate the latter, you should use Lachenbruch's holdout procedure. [See (11-57).])

11.28. Darroch and Mosimann [6] have argued that the three species of iris indicated in Table 11.5 can be discriminated on the basis of "shape" or scale-free information alone. Let $Y_1 = X_1/X_2$ be sepal shape and $Y_2 = X_3/X_4$ be petal shape.

(a) Plot the data in the $(\log Y_1, \log Y_2)$ variable space. Do the observations for the three groups appear to be bivariate normal?

(b) Assuming equal covariance matrices and bivariate normal populations, and supposing that $p_1 = p_2 = p_3 = \frac{1}{3}$, construct the linear discriminant scores $\hat{d}_i(\mathbf{x})$ given by (11-51) using both variables $\log Y_1, \log Y_2$ and each variable individually. Calculate the APERs.

(c) Using the linear discriminant functions from Part b, calculate the holdout estimates of the expected AERs, and fill in the following summary table:

Variable(s)	Misclassification rate
$\log Y_1$	
$\log Y_2$	
$\log Y_1, \log Y_2$	

Compare the preceding misclassification rates with those in the summary tables in Example 11.12. Does it appear as if information on shape alone is an effective discriminator for these species of iris?

(d) Compare the corresponding error rates in Parts b and c. Given the scatter plot in Part a, would you expect these rates to differ much? Explain.

11.29. The GPA and GMAT data alluded to in Example 11.11 are listed in Table 11.6.

(a) Using these data, calculate $\bar{\mathbf{x}}_1, \bar{\mathbf{x}}_2, \bar{\mathbf{x}}_3, \bar{\mathbf{x}}$, and $\mathbf{S}_{\text{pooled}}$ and thus verify the results for these quantities given in Example 11.11.

Table 11.6 Admission Data for Graduate School of Business

	π_1: Admit			π_2: Do not admit			π_3: Borderline	
Applicant no.	GPA (x_1)	GMAT (x_2)	Applicant no.	GPA (x_1)	GMAT (x_2)	Applicant no.	GPA (x_1)	GMAT (x_2)
1	2.96	596	32	2.54	446	60	2.86	494
2	3.14	473	33	2.43	425	61	2.85	496
3	3.22	482	34	2.20	474	62	3.14	419
4	3.29	527	35	2.36	531	63	3.28	371
5	3.69	505	36	2.57	542	64	2.89	447
6	3.46	693	37	2.35	406	65	3.15	313
7	3.03	626	38	2.51	412	66	3.50	402
8	3.19	663	39	2.51	458	67	2.89	485
9	3.63	447	40	2.36	399	68	2.80	444
10	3.59	588	41	2.36	482	69	3.13	416
11	3.30	563	42	2.66	420	70	3.01	471
12	3.40	553	43	2.68	414	71	2.79	490
13	3.50	572	44	2.48	533	72	2.89	431
14	3.78	591	45	2.46	509	73	2.91	446
15	3.44	692	46	2.63	504	74	2.75	546
16	3.48	528	47	2.44	336	75	2.73	467
17	3.47	552	48	2.13	408	76	3.12	463
18	3.35	520	49	2.41	469	77	3.08	440
19	3.39	543	50	2.55	538	78	3.03	419
20	3.28	523	51	2.31	505	79	3.00	509
21	3.21	530	52	2.41	489	80	3.03	438
22	3.58	564	53	2.19	411	81	3.05	399
23	3.33	565	54	2.35	321	82	2.85	483
24	3.40	431	55	2.60	394	83	3.01	453
25	3.38	605	56	2.55	528	84	3.03	414
26	3.26	664	57	2.72	399	85	3.04	446
27	3.60	609	58	2.85	381			
28	3.37	559	59	2.90	384			
29	3.80	521						
30	3.76	646						
31	3.24	467						

(b) Calculate $\mathbf{W}^{-1}$ and $\mathbf{B}$ and the eigenvalues and eigenvectors of $\mathbf{W}^{-1}\mathbf{B}$. Use the linear discriminants derived from these eigenvectors to classify the new observation $\mathbf{x}_0' = [3.21 \quad 497]$ into one of the populations π_1: admit; π_2: not admit; and π_3: borderline. Does the classification agree with that in Example 11.11? Should it? Explain.

11.30. Gerrild and Lantz [13] chemically analyzed crude-oil samples from three zones of sandstone:

π_1: Wilhelm

π_2: Sub-Mulinia

π_3: Upper

The values of the trace elements

$X_1 =$ vanadium (in percent ash)
$X_2 =$ iron (in percent ash)
$X_3 =$ beryllium (in percent ash)

and two measures of hydrocarbons,

$$X_4 = \text{saturated hydrocarbons (in percent area)}$$

$$X_5 = \text{aromatic hydrocarbons (in percent area)}$$

are presented for 56 cases in Table 11.7. The last two measurements are determined from areas under a gas–liquid chromatography curve.

(a) Obtain the estimated minimum TPM rule, assuming normality. Comment on the adequacy of the assumption of normality.

(b) Determine the estimate of $E(\text{AER})$ using Lachenbruch's holdout procedure. Also, give the confusion matrix.

(c) Consider various transformations of the data to normality (see Example 11.14), and repeat Parts a and b.

Table 11.7 Crude-Oil Data					
	x_1	x_2	x_3	x_4	x_5
π_1	3.9	51.0	0.20	7.06	12.19
	2.7	49.0	0.07	7.14	12.23
	2.8	36.0	0.30	7.00	11.30
	3.1	45.0	0.08	7.20	13.01
	3.5	46.0	0.10	7.81	12.63
	3.9	43.0	0.07	6.25	10.42
	2.7	35.0	0.00	5.11	9.00
π_2	5.0	47.0	0.07	7.06	6.10
	3.4	32.0	0.20	5.82	4.69
	1.2	12.0	0.00	5.54	3.15
	8.4	17.0	0.07	6.31	4.55
	4.2	36.0	0.50	9.25	4.95
	4.2	35.0	0.50	5.69	2.22
	3.9	41.0	0.10	5.63	2.94
	3.9	36.0	0.07	6.19	2.27
	7.3	32.0	0.30	8.02	12.92
	4.4	46.0	0.07	7.54	5.76
	3.0	30.0	0.00	5.12	10.77
π_3	6.3	13.0	0.50	4.24	8.27
	1.7	5.6	1.00	5.69	4.64
	7.3	24.0	0.00	4.34	2.99
	7.8	18.0	0.50	3.92	6.09
	7.8	25.0	0.70	5.39	6.20
	7.8	26.0	1.00	5.02	2.50
	9.5	17.0	0.05	3.52	5.71
	7.7	14.0	0.30	4.65	8.63
	11.0	20.0	0.50	4.27	8.40
	8.0	14.0	0.30	4.32	7.87
	8.4	18.0	0.20	4.38	7.98

(continues on next page)

Table 11.7 *(continued)*

x_1	x_2	x_3	x_4	x_5
10.0	18.0	0.10	3.06	7.67
7.3	15.0	0.05	3.76	6.84
9.5	22.0	0.30	3.98	5.02
8.4	15.0	0.20	5.02	10.12
8.4	17.0	0.20	4.42	8.25
9.5	25.0	0.50	4.44	5.95
7.2	22.0	1.00	4.70	3.49
4.0	12.0	0.50	5.71	6.32
6.7	52.0	0.50	4.80	3.20
9.0	27.0	0.30	3.69	3.30
7.8	29.0	1.50	6.72	5.75
4.5	41.0	0.50	3.33	2.27
6.2	34.0	0.70	7.56	6.93
5.6	20.0	0.50	5.07	6.70
9.0	17.0	0.20	4.39	8.33
8.4	20.0	0.10	3.74	3.77
9.5	19.0	0.50	3.72	7.37
9.0	20.0	0.50	5.97	11.17
6.2	16.0	0.05	4.23	4.18
7.3	20.0	0.50	4.39	3.50
3.6	15.0	0.70	7.00	4.82
6.2	34.0	0.07	4.84	2.37
7.3	22.0	0.00	4.13	2.70
4.1	29.0	0.70	5.78	7.76
5.4	29.0	0.20	4.64	2.65
5.0	34.0	0.70	4.21	6.50
6.2	27.0	0.30	3.97	2.97

11.31. Refer to the data on salmon in Table 11.2.

(a) Plot the bivariate data for the two groups of salmon. Are the sizes and orientation of the scatters roughly the same? Do bivariate normal distributions with a common co-variance matrix appear to be viable population models for the Alaskan and Canadian salmon?

(b) Using a linear discriminant function for two normal populations with equal priors and equal costs [see (11-19)], construct dot diagrams of the discriminant scores for the two groups. Does it appear as if the growth ring diameters separate for the two groups reasonably well? Explain.

(c) Repeat the analysis in Example 11.8 for the male and female salmon separately. Is it easier to discriminate Alaskan male salmon from Canadian male salmon than it is to discriminate the females in the two groups? Is gender (male or female) likely to be a useful discriminatory variable?

11.32. Data on hemophilia A carriers, similar to those used in Example 11.3, are listed in Table 11.8 on page 664. (See [15].) Using these data,

(a) Investigate the assumption of bivariate normality for the two groups.

Table 11.8 Hemophilia Data

	Noncarriers (π_1)			Obligatory carriers (π_2)	
Group	$\log_{10}$ (AHF activity)	$\log_{10}$ (AHF antigen)	Group	$\log_{10}$ (AHF activity)	$\log_{10}$ (AHF antigen)
1	−.0056	−.1657	2	−.3478	.1151
1	−.1698	−.1585	2	−.3618	−.2008
1	−.3469	−.1879	2	−.4986	−.0860
1	−.0894	.0064	2	−.5015	−.2984
1	−.1679	.0713	2	−.1326	.0097
1	−.0836	.0106	2	−.6911	−.3390
1	−.1979	−.0005	2	−.3608	.1237
1	−.0762	.0392	2	−.4535	−.1682
1	−.1913	−.2123	2	−.3479	−.1721
1	−.1092	−.1190	2	−.3539	.0722
1	−.5268	−.4773	2	−.4719	−.1079
1	−.0842	.0248	2	−.3610	−.0399
1	−.0225	−.0580	2	−.3226	.1670
1	.0084	.0782	2	−.4319	−.0687
1	−.1827	−.1138	2	−.2734	−.0020
1	.1237	.2140	2	−.5573	.0548
1	−.4702	−.3099	2	−.3755	−.1865
1	−.1519	−.0686	2	−.4950	−.0153
1	.0006	−.1153	2	−.5107	−.2483
1	−.2015	−.0498	2	−.1652	.2132
1	−.1932	−.2293	2	−.2447	−.0407
1	.1507	.0933	2	−.4232	−.0998
1	−.1259	−.0669	2	−.2375	.2876
1	−.1551	−.1232	2	−.2205	.0046
1	−.1952	−.1007	2	−.2154	−.0219
1	.0291	.0442	2	−.3447	.0097
1	−.2228	−.1710	2	−.2540	−.0573
1	−.0997	−.0733	2	−.3778	−.2682
1	−.1972	−.0607	2	−.4046	−.1162
1	−.0867	−.0560	2	−.0639	.1569
			2	−.3351	−.1368
			2	−.0149	.1539
			2	−.0312	.1400
			2	−.1740	−.0776
			2	−.1416	.1642
			2	−.1508	.1137
			2	−.0964	.0531
			2	−.2642	.0867
			2	−.0234	.0804
			2	−.3352	.0875
			2	−.1878	.2510
			2	−.1744	.1892
			2	−.4055	−.2418
			2	−.2444	.1614
			2	−.4784	.0282

Source: See [15].

(b) Obtain the sample linear discriminant function, assuming equal prior probabilities, and estimate the error rate using the holdout procedure.

(c) Classify the following 10 new cases using the discriminant function in Part b.

(d) Repeat Parts a–c, assuming that the prior probability of obligatory carriers (group 2) is $\frac{1}{4}$ and that of noncarriers (group 1) is $\frac{3}{4}$.

New Cases Requiring Classification

Case	$\log_{10}$(AHF activity)	$\log_{10}$(AHF antigen)
1	−.112	−.279
2	−.059	−.068
3	.064	.012
4	−.043	−.052
5	−.050	−.098
6	−.094	−.113
7	−.123	−.143
8	−.011	−.037
9	−.210	−.090
10	−.126	−.019

11.33. Consider the data on bulls in Table 1.10.

(a) Using the variables YrHgt, FtFrBody, PrctFFB, Frame, BkFat, SaleHt, and SaleWt, calculate Fisher's linear discriminants, and classify the bulls as Angus, Hereford, or Simental. Calculate an estimate of $E(\text{AER})$ using the holdout procedure. Classify a bull with characteristics YrHgt = 50, FtFrBody = 1000, PrctFFB = 73, Frame = 7, BkFat = .17, SaleHt = 54, and SaleWt = 1525 as one of the three breeds. Plot the discriminant scores for the bulls in the two-dimensional discriminant space using different plotting symbols to identify the three groups.

(b) Is there a subset of the original seven variables that is almost as good for discriminating among the three breeds? Explore this possibility by computing the estimated $E(\text{AER})$ for various subsets.

11.34. Table 11.9 on pages 666–667 contains data on breakfast cereals produced by three *like Assy 3* different American manufacturers: General Mills (G), Kellogg (K), and Quaker (Q). Assuming multivariate normal data with a common covariance matrix, equal costs, and equal priors, classify the cereal brands according to manufacturer. Compute the estimated $E(\text{AER})$ using the holdout procedure. Interpret the coefficients of the discriminant functions. Does it appear as if some manufacturers are associated with more "nutritional" cereals (high protein, low fat, high fiber, low sugar, and so forth) than others? Plot the cereals in the two-dimensional discriminant space, using different plotting symbols to identify the three manufacturers.

11.35. Table 11.10 on page 668 contains measurements on the gender, age, tail length (mm), and snout to vent length (mm) for Concho Water Snakes.

Define the variables

$$X_1 = \text{Gender}$$
$$X_2 = \text{Age}$$
$$X_3 = \text{TailLength}$$
$$X_4 = \text{SntoVnLength}$$

Table 11.9 Data on Brands of Cereal

Brand	Manufacturer	Calories	Protein	Fat	Sodium	Fiber	Carbohydrates	Sugar	Potassium	Group
1 Apple_Cinnamon_Cheerios	G	110	2	2	180	1.5	10.5	10	70	1
2 Cheerios	G	110	6	2	290	2.0	17.0	1	105	1
3 Cocoa_Puffs	G	110	1	1	180	0.0	12.0	13	55	1
4 Count_Chocula	G	110	1	1	180	0.0	12.0	13	65	1
5 Golden_Grahams	G	110	1	1	280	0.0	15.0	9	45	1
6 Honey_Nut_Cheerios	G	110	3	1	250	1.5	11.5	10	90	1
7 Kix	G	110	2	1	260	0.0	21.0	3	40	1
8 Lucky_Charms	G	110	2	1	180	0.0	12.0	12	55	1
9 Multi_Grain_Cheerios	G	100	2	1	220	2.0	15.0	6	90	1
10 Oatmeal_Raisin_Crisp	G	130	3	2	170	1.5	13.5	10	120	1
11 Raisin_Nut_Bran	G	100	3	2	140	2.5	10.5	8	140	1
12 Total_Corn_Flakes	G	110	2	1	200	0.0	21.0	3	35	1
13 Total_Raisin_Bran	G	140	3	1	190	4.0	15.0	14	230	1
14 Total_Whole_Grain	G	100	3	1	200	3.0	16.0	3	110	1
15 Trix	G	110	1	1	140	0.0	13.0	12	25	1
16 Wheaties	G	100	3	1	200	3.0	17.0	3	110	1
17 Wheaties_Honey_Gold	G	110	2	1	200	1.0	16.0	8	60	1
18 All_Bran	K	70	4	1	260	9.0	7.0	5	320	2
19 Apple_Jacks	K	110	2	0	125	1.0	11.0	14	30	2
20 Corn_Flakes	K	100	2	0	290	1.0	21.0	2	35	2
21 Corn_Pops	K	110	1	0	90	1.0	13.0	12	20	2

continued

22	Cracklin'_Oat_Bran	K	110	3	3	140	4.0	10.0	7	160	2
23	Crispix	K	110	2	0	220	1.0	21.0	3	30	2
24	Froot_Loops	K	110	2	1	125	1.0	11.0	13	30	2
25	Frosted_Flakes	K	110	1	0	200	1.0	14.0	11	25	2
26	Frosted_Mini_Wheats	K	100	3	0	0	3.0	14.0	7	100	2
27	Fruitful_Bran	K	120	3	0	240	5.0	14.0	12	190	2
28	Just_Right_Crunchy_Nuggets	K	110	2	1	170	1.0	17.0	6	60	2
29	Mueslix_Crispy_Blend	K	160	3	2	150	3.0	17.0	13	160	2
30	Nut&Honey_Crunch	K	120	2	1	190	0.0	15.0	9	40	2
31	Nutri-grain_Almond-Raisin	K	140	3	2	220	3.0	21.0	7	130	2
32	Nutri-grain_Wheat	K	90	3	0	170	3.0	18.0	2	90	2
33	Product_19	K	100	3	0	320	1.0	20.0	3	45	2
34	Raisin_Bran	K	120	3	1	210	5.0	14.0	12	240	2
35	Rice_Krispies	K	110	2	0	290	0.0	22.0	3	35	2
36	Smacks	K	110	2	1	70	1.0	9.0	15	40	2
37	Special_K	K	110	6	0	230	1.0	16.0	3	55	2
38	Cap'n'Crunch	Q	120	1	2	220	0.0	12.0	12	35	3
39	Honey_Graham_Ohs	Q	120	1	2	220	1.0	12.0	11	45	3
40	Life	Q	100	4	2	150	2.0	12.0	6	95	3
41	Puffed_Rice	Q	50	1	0	0	0.0	13.0	0	15	3
42	Puffed_Wheat	Q	50	2	0	0	1.0	10.0	0	50	3
43	Quaker_Oatmeal	Q	100	5	2	0	2.7	1.0	1	110	3

Source: Data courtesy of Chad Dacus.

667

Table 11.10 Concho Water Snake Data

	Gender	Age	TailLength	Snto VnLength		Gender	Age	TailLength	Snto VnLength
1	Female	2	127	441	1	Male	2	126	457
2	Female	2	171	455	2	Male	2	128	466
3	Female	2	171	462	3	Male	2	151	466
4	Female	2	164	446	4	Male	2	115	361
5	Female	2	165	463	5	Male	2	138	473
6	Female	2	127	393	6	Male	2	145	477
7	Female	2	162	451	7	Male	3	145	507
8	Female	2	133	376	8	Male	3	145	493
9	Female	2	173	475	9	Male	3	158	558
10	Female	2	145	398	10	Male	3	152	495
11	Female	2	154	435	11	Male	3	159	521
12	Female	3	165	491	12	Male	3	138	487
13	Female	3	178	485	13	Male	3	166	565
14	Female	3	169	477	14	Male	3	168	585
15	Female	3	186	530	15	Male	3	160	550
16	Female	3	170	478	16	Male	4	181	652
17	Female	3	182	511	17	Male	4	185	587
18	Female	3	172	475	18	Male	4	172	606
19	Female	3	182	487	19	Male	4	180	591
20	Female	3	172	454	20	Male	4	205	683
21	Female	3	183	502	21	Male	4	175	625
22	Female	3	170	483	22	Male	4	182	612
23	Female	3	171	477	23	Male	4	185	618
24	Female	3	181	493	24	Male	4	181	613
25	Female	3	167	490	25	Male	4	167	600
26	Female	3	175	493	26	Male	4	167	602
27	Female	3	139	477	27	Male	4	160	596
28	Female	3	183	501	28	Male	4	165	611
29	Female	4	198	537	29	Male	4	173	603
30	Female	4	190	566					
31	Female	4	192	569					
32	Female	4	211	574					
33	Female	4	206	570					
34	Female	4	206	573					
35	Female	4	165	531					
36	Female	4	189	528					
37	Female	4	195	536					

Source: Data courtesy of Raymond J. Carroll.

(a) Plot the data as a scatter plot with tail length (x_3) as the horizontal axis and snout to vent length (x_4) as the vertical axis. Use different plotting symbols for female and male snakes, and different symbols for different ages. Does it appear as if tail length and snout to vent length might usefully discriminate the genders of snakes? The different ages of snakes?

(b) Assuming multivariate normal data with a common covariance matrix, equal priors, and equal costs, classify the Concho Water Snakes according to gender. Compute the estimated $E(\text{AER})$ using the holdout procedure.

(c) Repeat part (b) using age as the groups rather than gender.

(d) Repeat part (b) using only snout to vent length to classify the snakes according to age. Compare the results with those in part (c). Can effective classification be achieved with only a single variable in this case? Explain.

11.36. Refer to Example 11.17. Using logistic regression, refit the salmon data in Table 11.2 with only the covariates freshwater growth and marine growth. Check for the significance of the model and the significance of each individual covariate. Set $\alpha = .05$. Use the fitted function to classify each of the observations in Table 11.2 as Alaskan salmon or Canadian salmon using rule (11-77). Compute the apparent error rate, APER, and compare this error rate with the error rate from the linear classification function discussed in Example 11.8.

References

1. Anderson, E. "The Irises of the Gaspé Peninsula." *Bulletin of the American Iris Society*, **59** (1939), 2–5.

2. Anderson, T. W. *An Introduction to Multivariate Statistical Analysis* (3rd ed.). New York: John Wiley, 2003.

3. Bartlett, M. S. "An Inverse Matrix Adjustment Arising in Discriminant Analysis." *Annals of Mathematical Statistics*, **22** (1951), 107–111.

4. Bouma, B. N., et al. "Evaluation of the Detection Rate of Hemophilia Carriers." *Statistical Methods for Clinical Decision Making*, **7**, no. 2 (1975), 339–350.

5. Breiman, L., J. Friedman, R. Olshen, and C. Stone. *Classification and Regression Trees*. Belmont, CA: Wadsworth, Inc., 1984.

6. Darroch, J. N., and J. E. Mosimann. "Canonical and Principal Components of Shape." *Biometrika*, **72**, no. 1 (1985), 241–252.

7. Efron, B. "The Efficiency of Logistic Regression Compared to Normal Discriminant Analysis." *Journal of the American Statistical Association*, **81** (1975), 321–327.

8. Eisenbeis, R. A. "Pitfalls in the Application of Discriminant Analysis in Business, Finance and Economics." *Journal of Finance*, **32**, no. 3 (1977), 875–900.

9. Fisher, R. A. "The Use of Multiple Measurements in Taxonomic Problems." *Annals of Eugenics*, **7** (1936), 179–188.

10. Fisher, R. A. "The Statistical Utilization of Multiple Measurements." *Annals of Eugenics*, **8** (1938), 376–386.

11. Ganesalingam, S. "Classification and Mixture Approaches to Clustering via Maximum Likelihood." *Applied Statistics*, **38**, no. 3 (1989), 455–466.

12. Geisser, S. "Discrimination, Allocatory and Separatory, Linear Aspects." In *Classification and Clustering*, edited by J. Van Ryzin, pp. 301–330. New York: Academic Press, 1977.

13. Gerrild, P. M., and R. J. Lantz. "Chemical Analysis of 75 Crude Oil Samples from Pliocene Sand Units, Elk Hills Oil Field, California." *U.S. Geological Survey Open-File Report*, 1969.

14. Gnanadesikan, R. *Methods for Statistical Data Analysis of Multivariate Observations* (2nd ed.). New York: Wiley-Interscience, 1997.

15. Habbema, J. D. F., J. Hermans, and K. Van Den Broek. "A Stepwise Discriminant Analysis Program Using Density Estimation." In *Compstat 1974, Proc. Computational Statistics*, pp. 101–110. Vienna: Physica, 1974.

16. Hills, M. "Allocation Rules and Their Error Rates." *Journal of the Royal Statistical Society (B)*, **28** (1966), 1–31.

17. Hosmer, D. W. and S. Lemeshow. *Applied Logistic Regression* (2nd ed.). New York: Wiley-Interscience, 2000.

18. Hudlet, R., and R. A. Johnson. "Linear Discrimination and Some Further Results on Best Lower Dimensional Representations." In *Classification and Clustering*, edited by J. Van Ryzin, pp. 371–394. New York: Academic Press, 1977.

19. Johnson, W. "The Detection of Influential Observations for Allocation, Separation, and the Determination of Probabilities in a Bayesian Framework." *Journal of Business and Economic Statistics*, **5**, no. 3 (1987), 369–381.

20. Kendall, M. G. *Multivariate Analysis*. New York: Hafner Press, 1975.

21. Kim, H. and Loh, W. Y., "Classification Trees with Unbiased Multiway Splits," *Journal of the American Statistical Association*, **96**, (2001), 589–604.

22. Krzanowski, W. J. "The Performance of Fisher's Linear Discriminant Function under Non-Optimal Conditions." *Technometrics*, **19**, no. 2 (1977), 191–200.

23. Lachenbruch, P. A. *Discriminant Analysis*. New York: Hafner Press, 1975.

24. Lachenbruch, P. A., and M. R. Mickey. "Estimation of Error Rates in Discriminant Analysis." *Technometrics*, **10**, no. 1 (1968), 1–11.

25. Loh, W. Y. and Shih, Y. S., "Split Selection Methods for Classification Trees," *Statistica Sinica*, **7**, (1997), 815–840.

26. McCullagh, P., and J. A. Nelder. *Generalized Linear Models* (2nd ed.). London: Chapman and Hall, 1989.

27. Mucciardi, A. N., and E. E. Gose. "A Comparison of Seven Techniques for Choosing Subsets of Pattern Recognition Properties." *IEEE Trans. Computers*, **C20** (1971), 1023–1031.

28. Murray, G. D. "A Cautionary Note on Selection of Variables in Discriminant Analysis." *Applied Statistics*, **26**, no. 3 (1977), 246–250.

29. Rencher, A. C. "Interpretation of Canonical Discriminant Functions, Canonical Variates and Principal Components." *The American Statistician*, **46** (1992), 217–225.

30. Stern, H. S. "Neural Networks in Applied Statistics." *Technometrics*, **38**, (1996), 205–214.

31. Wald, A. "On a Statistical Problem Arising in the Classification of an Individual into One of Two Groups." *Annals of Mathematical Statistics*, **15** (1944), 145–162.

32. Welch, B. L. "Note on Discriminant Functions." *Biometrika*, **31** (1939), 218–220.

Clustering, Distance Methods, and Ordination

12.1 Introduction

Rudimentary, exploratory procedures are often quite helpful in understanding the complex nature of multivariate relationships. For example, throughout this book, we have emphasized the value of data plots. In this chapter, we shall discuss some additional displays based on certain measures of distance and suggested step-by-step rules (algorithms) for grouping objects (variables or items). Searching the data for a structure of "natural" groupings is an important exploratory technique. Groupings can provide an informal means for assessing dimensionality, identifying outliers, and suggesting interesting hypotheses concerning relationships.

Grouping, or clustering, is distinct from the classification methods discussed in the previous chapter. Classification pertains to a *known* number of groups, and the operational objective is to assign new observations to one of these groups. Cluster analysis is a more primitive technique in that no assumptions are made concerning the number of groups or the group structure. Grouping is done on the basis of similarities or distances (dissimilarities). The inputs required are similarity measures or data from which similarities can be computed.

To illustrate the nature of the difficulty in defining a natural grouping, consider sorting the 16 face cards in an ordinary deck of playing cards into clusters of similar objects. Some groupings are illustrated in Figure 12.1. It is immediately clear that meaningful partitions depend on the definition of *similar*.

In most practical applications of cluster analysis, the investigator knows enough about the problem to distinguish "good" groupings from "bad" groupings. Why not enumerate all possible groupings and select the "best" ones for further study?

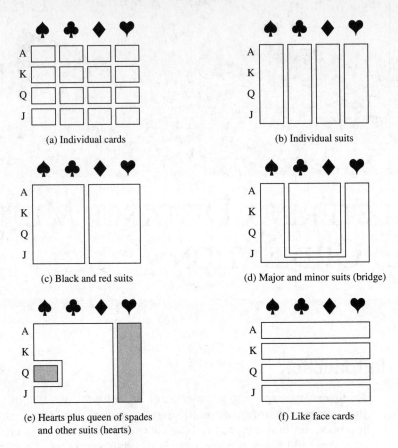

Figure 12.1 Grouping face cards.

For the playing-card example, there is one way to form a *single* group of 16 face cards, there are 32,767 ways to partition the face cards into *two* groups (of varying sizes), there are 7,141,686 ways to sort the face cards into *three* groups (of varying sizes), and so on.[1] Obviously, time constraints make it impossible to determine the best groupings of similar objects from a list of all possible structures. Even fast computers are easily overwhelmed by the typically large number of cases, so one must settle for *algorithms* that search for good, but not necessarily the best, groupings.

To summarize, the basic objective in cluster analysis is to discover natural groupings of the items (or variables). In turn, we must first develop a quantitative scale on which to measure the association (similarity) between objects. Section 12.2 is devoted to a discussion of similarity measures. After that section, we describe a few of the more common algorithms for sorting objects into groups.

---

[1] The number of ways of sorting n objects into k nonempty groups is a Stirling number of the second kind given by $(1/k!) \sum_{j=0}^{k} (-1)^{k-j} \binom{k}{j} j^n$. (See [1].) Adding these numbers for $k = 1, 2, \ldots, n$ groups, we obtain the total number of possible ways to sort n objects into groups.

Even without the precise notion of a natural grouping, we are often able to group objects in two- or three-dimensional plots by eye. Stars and Chernoff faces, discussed in Section 1.4, have been used for this purpose. (See Examples 1.11 and 1.12.) Additional procedures for depicting high-dimensional observations in two dimensions such that similar objects are, in some sense, close to one another are considered in Sections 12.5–12.7.

12.2 Similarity Measures

Most efforts to produce a rather simple group structure from a complex data set require a measure of "closeness," or "similarity." There is often a great deal of subjectivity involved in the choice of a similarity measure. Important considerations include the nature of the variables (discrete, continuous, binary), scales of measurement (nominal, ordinal, interval, ratio), and subject matter knowledge.

When *items* (units or cases) are clustered, proximity is usually indicated by some sort of distance. By contrast, *variables* are usually grouped on the basis of correlation coefficients or like measures of association.

Distances and Similarity Coefficients for Pairs of Items

We discussed the notion of distance in Chapter 1, Section 1.5. Recall that the Euclidean (straight-line) distance between two p-dimensional observations (items) $\mathbf{x}' = [x_1, x_2, \ldots, x_p]$ and $\mathbf{y}' = [y_1, y_2, \ldots, y_p]$ is, from (1-12),

l₂ norm

$$d(\mathbf{x}, \mathbf{y}) = \sqrt{(x_1 - y_1)^2 + (x_2 - y_2)^2 + \cdots + (x_p - y_p)^2}$$

$$= \sqrt{(\mathbf{x} - \mathbf{y})'(\mathbf{x} - \mathbf{y})} \tag{12-1}$$

The statistical distance between the same two observations is of the form [see (1-23)]

$$d(\mathbf{x}, \mathbf{y}) = \sqrt{(\mathbf{x} - \mathbf{y})'\mathbf{A}(\mathbf{x} - \mathbf{y})} \tag{12-2}$$

Ordinarily, $\mathbf{A} = \mathbf{S}^{-1}$, where $\mathbf{S}$ contains the sample variances and covariances. However, without prior knowledge of the distinct groups, these sample quantities cannot be computed. For this reason, Euclidean distance is often preferred for clustering.

Another distance measure is the Minkowski metric

$$d(\mathbf{x}, \mathbf{y}) = \left[\sum_{i=1}^{p} |x_i - y_i|^m \right]^{1/m} \quad \text{any norm} \tag{12-3}$$

For $m = 1$, $d(\mathbf{x}, \mathbf{y})$ measures the "city-block" distance between two points in p dimensions. For $m = 2$, $d(\mathbf{x}, \mathbf{y})$ becomes the Euclidean distance. In general, varying m changes the weight given to larger and smaller differences.

Two additional popular measures of "distance" or dissimilarity are given by the Canberra metric and the Czekanowski coefficient. Both of these measures are defined for nonnegative variables only. We have

Canberra metric:
$$d(\mathbf{x}, \mathbf{y}) = \sum_{i=1}^{p} \frac{|x_i - y_i|}{(x_i + y_i)} \tag{12-4}$$

Czekanowski coefficient:
$$d(\mathbf{x}, \mathbf{y}) = 1 - \frac{2 \sum_{i=1}^{p} \min(x_i, y_i)}{\sum_{i=1}^{p} (x_i + y_i)} \tag{12-5}$$

Whenever possible, it is advisable to use "true" distances—that is, distances satisfying the distance properties of (1-25)—for clustering objects. On the other hand, most clustering algorithms will accept subjectively assigned distance numbers that may not satisfy, for example, the triangle inequality.

When items cannot be represented by meaningful p-dimensional measurements, pairs of items are often compared on the basis of the presence or absence of certain characteristics. Similar items have more characteristics in common than do dissimilar items. The presence or absence of a characteristic can be described mathematically by introducing a *binary variable*, which assumes the value 1 if the characteristic is present and the value 0 if the characteristic is absent. For $p = 5$ binary variables, for instance, the "scores" for two items i and k might be arranged as follows:

	Variables				
	1	2	3	4	5
Item i	1	0	0	1	1
Item k	1	1	0	1	0

In this case, there are two 1–1 matches, one 0–0 match, and two mismatches.

Let x_{ij} be the score (1 or 0) of the jth binary variable on the ith item and x_{kj} be the score (again, 1 or 0) of the jth variable on the kth item, $j = 1, 2, \ldots, p$. Consequently,

$$(x_{ij} - x_{kj})^2 = \begin{cases} 0 & \text{if } x_{ij} = x_{kj} = 1 \quad \text{or} \quad x_{ij} = x_{kj} = 0 \\ 1 & \text{if } x_{ij} \neq x_{kj} \end{cases} \tag{12-6}$$

and the squared Euclidean distance, $\sum_{j=1}^{p} (x_{ij} - x_{kj})^2$, provides a count of the number of mismatches. A large distance corresponds to many mismatches—that is, dissimilar items. From the preceding display, the square of the distance between items i and k would be

$$\sum_{j=1}^{5} (x_{ij} - x_{kj})^2 = (1 - 1)^2 + (0 - 1)^2 + (0 - 0)^2 + (1 - 1)^2 + (1 - 0)^2$$

$$= 2$$

Although a distance based on (12-6) might be used to measure similarity, it suffers from weighting the 1–1 and 0–0 matches equally. In some cases, a 1–1 match is a stronger indication of similarity than a 0–0 match. For instance, in grouping people, the evidence that two persons both read ancient Greek is stronger evidence of similarity than the absence of this ability. Thus, it might be reasonable to discount the 0–0 matches or even disregard them completely. To allow for differential treatment of the 1–1 matches and the 0–0 matches, several schemes for defining similarity coefficients have been suggested.

To introduce these schemes, let us arrange the frequencies of matches and mismatches for items i and k in the form of a contingency table:

	Item k		
	1	0	Totals
Item i 1	a	b	$a + b$
0	c	d	$c + d$
Totals	$a + c$	$b + d$	$p = a + b + c + d$

(12-7)

In this table, a represents the frequency of 1–1 matches, b is the frequency of 1–0 matches, and so forth. Given the foregoing five pairs of binary outcomes, $a = 2$ and $b = c = d = 1$.

Table 12.1 lists common similarity coefficients defined in terms of the frequencies in (12-7). A short rationale follows each definition.

Table 12.1 Similarity Coefficients for Clustering Items*

Coefficient	Rationale
1. $\dfrac{a + d}{p}$	Equal weights for 1–1 matches and 0–0 matches.
2. $\dfrac{2(a + d)}{2(a + d) + b + c}$	Double weight for 1–1 matches and 0–0 matches.
3. $\dfrac{a + d}{a + d + 2(b + c)}$	Double weight for unmatched pairs.
4. $\dfrac{a}{p}$	No 0–0 matches in numerator.
5. $\dfrac{a}{a + b + c}$	No 0–0 matches in numerator or denominator. (The 0–0 matches are treated as irrelevant.)
6. $\dfrac{2a}{2a + b + c}$	No 0–0 matches in numerator or denominator. Double weight for 1–1 matches.
7. $\dfrac{a}{a + 2(b + c)}$	No 0–0 matches in numerator or denominator. Double weight for unmatched pairs.
8. $\dfrac{a}{b + c}$	Ratio of matches to mismatches with 0–0 matches excluded.

*[p binary variables; see (12-7).]

Coefficients 1, 2, and 3 in the table are monotonically related. Suppose coefficient 1 is calculated for two contingency tables, Table I and Table II. Then if $(a_I + d_I)/p \geq (a_{II} + d_{II})/p$, we also have $2(a_I + d_I)/[2(a_I + d_I) + b_I + c_I] \geq 2(a_{II} + d_{II})/[2(a_{II} + d_{II}) + b_{II} + c_{II}]$, and coefficient 3 will be at least as large for Table I as it is for Table II. (See Exercise 12.4.) Coefficients 5, 6, and 7 also retain their relative orders.

Monotonicity is important, because some clustering procedures are not affected if the definition of similarity is changed in a manner that leaves the relative orderings of similarities unchanged. The single linkage and complete linkage hierarchical procedures discussed in Section 12.3 are not affected. For these methods, any choice of the coefficients 1, 2, and 3 in Table 12.1 will produce the same groupings. Similarly, any choice of the coefficients 5, 6, and 7 will yield identical groupings.

Example 12.1 (Calculating the values of a similarity coefficient) Suppose five individuals possess the following characteristics:

	Height	Weight	Eye color	Hair color	Handedness	Gender
Individual 1	68 in	140 lb	green	blond	right	female
Individual 2	73 in	185 lb	brown	brown	right	male
Individual 3	67 in	165 lb	blue	blond	right	male
Individual 4	64 in	120 lb	brown	brown	right	female
Individual 5	76 in	210 lb	brown	brown	left	male

Define six binary variables $X_1, X_2, X_3, X_4, X_5, X_6$ as

$$X_1 = \begin{cases} 1 & \text{height} \geq 72 \text{ in.} \\ 0 & \text{height} < 72 \text{ in.} \end{cases} \qquad X_4 = \begin{cases} 1 & \text{blond hair} \\ 0 & \text{not blond hair} \end{cases}$$

$$X_2 = \begin{cases} 1 & \text{weight} \geq 150 \text{ lb} \\ 0 & \text{weight} < 150 \text{ lb} \end{cases} \qquad X_5 = \begin{cases} 1 & \text{right handed} \\ 0 & \text{left handed} \end{cases}$$

$$X_3 = \begin{cases} 1 & \text{brown eyes} \\ 0 & \text{otherwise} \end{cases} \qquad X_6 = \begin{cases} 1 & \text{female} \\ 0 & \text{male} \end{cases}$$

The scores for individuals 1 and 2 on the $p = 6$ binary variables are

		X_1	X_2	X_3	X_4	X_5	X_6
Individual	1	0	0	0	1	1	1
	2	1	1	1	0	1	0

and the number of matches and mismatches are indicated in the two-way array

		Individual 2		Total
		1	0	
Individual 1	1	1	2	3
	0	3	0	3
	Totals	4	2	6

Employing similarity coefficient 1, which gives equal weight to matches, we compute

$$\frac{a+d}{p} = \frac{1+0}{6} = \frac{1}{6}$$

Continuing with similarity coefficient 1, we calculate the remaining similarity numbers for pairs of individuals. These are displayed in the 5×5 symmetric matrix

$$
\begin{array}{c}
\text{Individual} \\
\begin{array}{ccccc}
1 & 2 & 3 & 4 & 5
\end{array}
\end{array}
$$

$$
\text{Individual}
\begin{array}{c}
1 \\ 2 \\ 3 \\ 4 \\ 5
\end{array}
\begin{bmatrix}
1 & & & & \\
\frac{1}{6} & 1 & & & \\
\frac{4}{6} & \frac{3}{6} & 1 & & \\
\frac{4}{6} & \frac{3}{6} & \frac{2}{6} & 1 & \\
0 & \boxed{\frac{5}{6}} & \frac{2}{6} & \frac{2}{6} & 1
\end{bmatrix}
\qquad Similar = 1
$$

Based on the magnitudes of the similarity coefficient, we should conclude that individuals 2 and 5 are most similar and individuals 1 and 5 are least similar. Other pairs fall between these extremes. If we were to divide the individuals into two relatively homogeneous subgroups on the basis of the similarity numbers, we might form the subgroups (1 3 4) and (2 5).

Note that $X_3 = 0$ implies an absence of brown eyes, so that two people, one with blue eyes and one with green eyes, will yield a 0–0 match. Consequently, it may be inappropriate to use similarity coefficient 1, 2, or 3 because these coefficients give the same weights to 1–1 and 0–0 matches. ∎

We have described the construction of distances and similarities. It is always possible to construct similarities from distances. For example, we might set

$$\widetilde{s}_{ik} = \frac{1}{1 + d_{ik}} \tag{12-8}$$

where $0 < \widetilde{s}_{ik} \leq 1$ is the similarity between items i and k and d_{ik} is the corresponding distance.

However, distances that must satisfy (1-25) cannot always be constructed from similarities. As Gower [11, 12] has shown, this can be done only if the matrix of similarities is nonnegative definite. With the nonnegative definite condition, and with the maximum similarity scaled so that $\widetilde{s}_{ii} = 1$,

$$d_{ik} = \sqrt{2(1 - \widetilde{s}_{ik})} \tag{12-9}$$

has the properties of a distance.

Similarities and Association Measures for Pairs of Variables

Thus far, we have discussed similarity measures for items. In some applications, it is the variables, rather than the items, that must be grouped. Similarity measures for variables often take the form of sample correlation coefficients. Moreover, in some clustering applications, negative correlations are replaced by their absolute values.

When the variables are binary, the data can again be arranged in the form of a contingency table. This time, however, the variables, rather than the items, delineate the categories. For each pair of variables, there are n items categorized in the table. With the usual 0 and 1 coding, the table becomes as follows:

		Variable k			
		1	0	Totals	
Variable i	1	a	b	$a + b$	(12-10)
	0	c	d	$c + d$	
	Totals	$a + c$	$b + d$	$n = a + b + c + d$	

For instance, variable i equals 1 and variable k equals 0 for b of the n items.

The usual product moment correlation formula applied to the binary variables in the contingency table of (12-10) gives (see Exercise 12.3)

$$r = \frac{ad - bc}{[(a + b)(c + d)(a + c)(b + d)]^{1/2}} \tag{12-11}$$

This number can be taken as a measure of the similarity between the two variables.

The correlation coefficient in (12-11) is related to the chi-square statistic $(r^2 = \chi^2/n)$ for testing the independence of two categorical variables. For n fixed, a large similarity (or correlation) is consistent with the presence of dependence.

Given the table in (12-10), measures of association (or similarity) exactly analogous to the ones listed in Table 12.1 can be developed. The only change required is the substitution of n (the number of items) for p (the number of variables).

Concluding Comments on Similarity

To summarize this section, we note that there are many ways to measure the similarity between pairs of objects. It appears that most practitioners use distances [see (12-1) through (12-5)] or the coefficients in Table 12.1 to cluster *items* and correlations to cluster *variables*. However, at times, inputs to clustering algorithms may be simple frequencies.

Example 12.2 (Measuring the similarities of 11 languages) The meanings of words change with the course of history. However, the meaning of the numbers 1, 2, 3, ... represents one conspicuous exception. Thus, a first comparison of languages might be based on the numerals alone. Table 12.2 gives the first 10 numbers in English, Polish, Hungarian, and eight other modern European languages. (Only languages that use the Roman alphabet are considered, and accent marks, cedillas, diereses, etc., are omitted.) A cursory examination of the spelling of the numerals in the table suggests that the first five languages (English, Norwegian, Danish, Dutch, and German) are very much alike. French, Spanish, and Italian are in even closer agreement. Hungarian and Finnish seem to stand by themselves, and Polish has some of the characteristics of the languages in each of the larger subgroups.

Table 12.2 Numerals in 11 Languages

English (E)	Norwegian (N)	Danish (Da)	Dutch (Du)	German (G)	French (Fr)	Spanish (Sp)	Italian (I)	Polish (P)	Hungarian (H)	Finnish (Fi)
one	en	en	een	eins	un	uno	uno	jeden	egy	yksi
two	to	to	twee	zwei	deux	dos	due	dwa	ketto	kaksi
three	tre	tre	drie	drei	trois	tres	tre	trzy	harom	kolme
four	fire	fire	vier	vier	quatre	cuatro	quattro	cztery	negy	neljä
five	fem	fem	vijf	funf	cinq	cinco	cinque	piec	ot	viisi
six	seks	seks	zes	sechs	six	seis	sei	szesc	hat	kuusi
seven	sju	syv	zeven	sieben	sept	siete	sette	siedem	het	seitseman
eight	atte	otte	acht	acht	huit	ocho	otto	osiem	nyolc	kahdeksan
nine	ni	ni	negen	neun	neuf	nueve	nove	dziewiec	kilenc	yhdeksan
ten	ti	ti	tien	zehn	dix	diez	dieci	dziesiec	tiz	kymmenen

679

Table 12.3 Concordant First Letters for Numbers in 11 Languages

	E	N	Da	Du	G	Fr	Sp	I	P	H	Fi
E	10										
N	8	10									
Da	8	9	10								
Du	3	5	4	10							
G	4	6	5	5	10						
Fr	4	4	4	1	3	10					
Sp	4	4	5	1	3	8	10				
I	4	4	5	1	3	9	9	10			
P	3	3	4	0	2	5	7	6	10		
H	1	2	2	2	1	0	0	0	0	10	
Fi	1	1	1	1	1	1	1	1	1	2	10

seen in class

The words for 1 in French, Spanish, and Italian all begin with *u*. For illustrative purposes, we might compare languages by looking at the *first letters* of the numbers. We call the words for the same number in two different languages *concordant* if they have the same first letter and *discordant* if they do not. From Table 12.2, the table of concordances (frequencies of matching first initials) for the numbers 1–10 is given in Table 12.3. We see that English and Norwegian have the same first letter for 8 of the 10 word pairs. The remaining frequencies were calculated in the same manner.

The results in Table 12.3 confirm our initial visual impression of Table 12.2. That is, English, Norwegian, Danish, Dutch, and German seem to form a group. French, Spanish, Italian, and Polish might be grouped together, whereas Hungarian and Finnish appear to stand alone. ■

In our examples so far, we have used our visual impression of similarity or distance measures to form groups. We now discuss less subjective schemes for creating clusters.

12.3 Hierarchical Clustering Methods

We can rarely examine all grouping possibilities, even with the largest and fastest computers. Because of this problem, a wide variety of clustering algorithms have emerged that find "reasonable" clusters without having to look at all configurations.

Hierarchical clustering techniques proceed by either a series of successive mergers or a series of successive divisions. *Agglomerative hierarchical methods* start with the individual objects. Thus, there are initially as many clusters as objects. The most similar objects are first grouped, and these initial groups are merged according to their similarities. Eventually, as the similarity decreases, all subgroups are fused into a single cluster.

Divisive hierarchical methods work in the opposite direction. An initial single group of objects is divided into two subgroups such that the objects in one subgroup are "far from" the objects in the other. These subgroups are then further divided into dissimilar subgroups; the process continues until there are as many subgroups as objects—that is, until each object forms a group.

The results of both agglomerative and divisive methods may be displayed in the form of a two-dimensional diagram known as a *dendrogram*. As we shall see, the dendrogram illustrates the mergers or divisions that have been made at successive levels.

In this section we shall concentrate on agglomerative hierarchical procedures and, in particular, *linkage methods*. Excellent elementary discussions of divisive hierarchical procedures and other agglomerative techniques are available in [3] and [8].

Linkage methods are suitable for clustering items, as well as variables. This is not true for all hierarchical agglomerative procedures. We shall discuss, in turn, *single linkage* (minimum distance or nearest neighbor), *complete linkage* (maximum distance or farthest neighbor), and *average linkage* (average distance). The merging of clusters under the three linkage criteria is illustrated schematically in Figure 12.2.

From the figure, we see that single linkage results when groups are fused according to the distance between their nearest members. Complete linkage occurs when groups are fused according to the distance between their farthest members. For average linkage, groups are fused according to the average distance between pairs of members in the respective sets.

The following are the steps in the agglomerative hierarchical clustering algorithm for grouping N objects (items or variables):

 1. Start with N clusters, each containing a single entity and an $N \times N$ symmetric matrix of distances (or similarities) $\mathbf{D} = \{d_{ik}\}$.

 2. Search the distance matrix for the nearest (most similar) pair of clusters. Let the distance between "most similar" clusters U and V be d_{UV}.

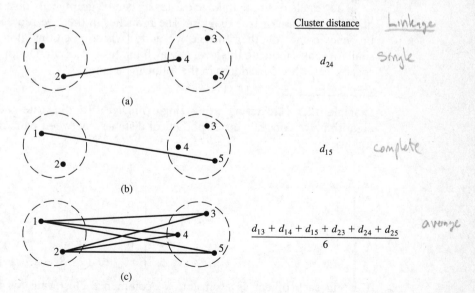

Cluster distance   Linkage

d_{24} single

(a)

d_{15} complete

(b)

$$\frac{d_{13} + d_{14} + d_{15} + d_{23} + d_{24} + d_{25}}{6}$$ average

(c)

Figure 12.2 Intercluster distance (dissimilarity) for (a) single linkage, (b) complete linkage, and (c) average linkage.

3. Merge clusters U and V. Label the newly formed cluster (UV). Update the entries in the distance matrix by (a) deleting the rows and columns corresponding to clusters U and V and (b) adding a row and column giving the distances between cluster (UV) and the remaining clusters.

4. Repeat Steps 2 and 3 a total of $N - 1$ times. (All objects will be in a *single* cluster after the algorithm terminates.) Record the identity of clusters that are merged and the levels (distances or similarities) at which the mergers take place.   (12-12)

The ideas behind any clustering procedure are probably best conveyed through examples, which we shall present after brief discussions of the input and algorithmic components of the linkage methods.

Single Linkage

The inputs to a single linkage algorithm can be distances or similarities between pairs of objects. Groups are formed from the individual entities by merging nearest neighbors, where the term *nearest neighbor* connotes the smallest distance or largest similarity.

Initially, we must find the smallest distance in $\mathbf{D} = \{d_{ik}\}$ and merge the corresponding objects, say, U and V, to get the cluster (UV). For Step 3 of the general algorithm of (12-12), the distances between (UV) and any other cluster W are computed by

$$d_{(UV)W} = \min\{d_{UW}, d_{VW}\} \qquad (12\text{-}13)$$

Here the quantities d_{UW} and d_{VW} are the distances between the nearest neighbors of clusters U and W and clusters V and W, respectively.

The results of single linkage clustering can be graphically displayed in the form of a *dendrogram*, or tree diagram. The branches in the tree represent clusters. The branches come together (merge) at nodes whose positions along a distance (or similarity) axis indicate the level at which the fusions occur. Dendrograms for some specific cases are considered in the following examples.

Example 12.3 (Clustering using single linkage) To illustrate the single linkage algorithm, we consider the hypothetical distances between pairs of five objects as follows:

$$\mathbf{D} = \{d_{ik}\} = \begin{matrix} & \begin{matrix} 1 & \ 2 & \ 3 & \ 4 & \ 5 \end{matrix} \\ \begin{matrix} 1 \\ 2 \\ 3 \\ 4 \\ 5 \end{matrix} & \begin{bmatrix} 0 & & & & \\ 9 & 0 & & & \\ 3 & 7 & 0 & & \\ 6 & 5 & 9 & 0 & \\ 11 & 10 & ② & 8 & 0 \end{bmatrix} \end{matrix}$$

Treating each object as a cluster, we commence clustering by merging the two closest items. Since

$$\min_{i,k}(d_{ik}) = d_{53} = 2$$

objects 5 and 3 are merged to form the cluster (35). To implement the next level of clustering, we need the distances between the cluster (35) and the remaining objects, 1, 2, and 4. The nearest neighbor distances are

$$d_{(35)1} = \min\{d_{31}, d_{51}\} = \min\{3, 11\} = 3$$
$$d_{(35)2} = \min\{d_{32}, d_{52}\} = \min\{7, 10\} = 7$$
$$d_{(35)4} = \min\{d_{34}, d_{54}\} = \min\{9,\ 8\} = 8$$

Deleting the rows and columns of $\mathbf{D}$ corresponding to objects 3 and 5, and adding a row and column for the cluster (35), we obtain the new distance matrix

$$
\begin{array}{cc}
 & \begin{array}{cccc} (35) & 1 & 2 & 4 \end{array} \\
\begin{array}{c} (35) \\ 1 \\ 2 \\ 4 \end{array} &
\left[\begin{array}{cccc}
0 & & & \\
③ & 0 & & \\
7 & 9 & 0 & \\
8 & 6 & 5 & 0
\end{array}\right]
\end{array}
$$

The smallest distance between pairs of clusters is now $d_{(35)1} = 3$, and we merge cluster (1) with cluster (35) to get the next cluster, (135). Calculating

$$d_{(135)2} = \min\{d_{(35)2}, d_{12}\} = \min\{7, 9\} = 7$$
$$d_{(135)4} = \min\{d_{(35)4}, d_{14}\} = \min\{8, 6\} = 6$$

we find that the distance matrix for the next level of clustering is

$$
\begin{array}{cc}
 & \begin{array}{ccc} (135) & 2 & 4 \end{array} \\
\begin{array}{c} (135) \\ 2 \\ 4 \end{array} &
\left[\begin{array}{ccc}
0 & & \\
7 & 0 & \\
6 & ⑤ & 0
\end{array}\right]
\end{array}
$$

The minimum nearest neighbor distance between pairs of clusters is $d_{42} = 5$, and we merge objects 4 and 2 to get the cluster (24).

At this point we have two distinct clusters, (135) and (24). Their nearest neighbor distance is

$$d_{(135)(24)} = \min\{d_{(135)2}, d_{(135)4}\} = \min\{7, 6\} = 6$$

The final distance matrix becomes

$$
\begin{array}{cc}
 & \begin{array}{cc} (135) & (24) \end{array} \\
\begin{array}{c} (135) \\ (24) \end{array} &
\left[\begin{array}{cc}
0 & \\
⑥ & 0
\end{array}\right]
\end{array}
$$

Consequently, clusters (135) and (24) are merged to form a single cluster of all five objects, (12345), when the nearest neighbor distance reaches 6.

The dendrogram picturing the hierarchical clustering just concluded is shown in Figure 12.3. The groupings and the distance levels at which they occur are clearly illustrated by the dendrogram. ∎

In typical applications of hierarchical clustering, the intermediate results—where the objects are sorted into a moderate number of clusters—are of chief interest.

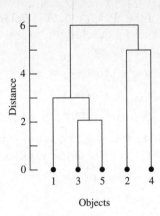

Figure 12.3 Single linkage dendrogram for distances between five objects.

Example 12.4 (Single linkage clustering of 11 languages) Consider the array of concordances in Table 12.3 representing the closeness between the numbers 1–10 in 11 languages. To develop a matrix of distances, we subtract the concordances from the perfect agreement figure of 10 that each language has with itself. The subsequent assignments of distances are

	E	N	Da	Du	G	Fr	Sp	I	P	H	Fi
E	0										
N	2	0									
Da	2	①	0								
Du	7	5	6	0							
G	6	4	5	5	0						
Fr	6	6	6	9	7	0					
Sp	6	6	5	9	7	2	0				
I	6	6	5	9	7	①	①	0			
P	7	7	6	10	8	5	3	4	0		
H	9	8	8	8	9	10	10	10	10	0	
Fi	9	9	9	9	9	9	9	9	9	8	0

We first search for the minimum distance between pairs of languages (clusters). The minimum distance, 1, occurs between Danish and Norwegian, Italian and French, and Italian and Spanish. Numbering the languages in the order in which they appear across the top of the array, we have

$$d_{32} = 1; \qquad d_{86} = 1; \quad \text{and } d_{87} = 1$$

Since $d_{76} = 2$, we can merge only clusters 8 and 6 or clusters 8 and 7. We cannot merge clusters 6, 7, and 8 at level 1. We choose first to merge 6 and 8, and then to update the distance matrix and merge 2 and 3 to obtain the clusters (68) and (23). Subsequent computer calculations produce the dendrogram in Figure 12.4.

From the dendrogram, we see that Norwegian and Danish, and also French and Italian, cluster at the minimum distance (maximum similarity) level. When the allowable distance is increased, English is added to the Norwegian–Danish group,

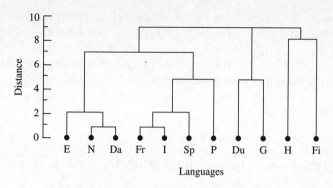

Figure 12.4 Single linkage dendrograms for distances between numbers in 11 languages.

and Spanish merges with the French–Italian group. Notice that Hungarian and Finnish are more similar to each other than to the other clusters of languages. However, these two clusters (languages) do not merge until the distance between nearest neighbors has increased substantially. Finally, all the clusters of languages are merged into a single cluster at the largest nearest neighbor distance, 9. ∎

Since single linkage joins clusters by the shortest link between them, the technique cannot discern poorly separated clusters. [See Figure 12.5(a).] On the other hand, single linkage is one of the few clustering methods that can delineate nonellipsoidal clusters. The tendency of single linkage to pick out long stringlike clusters is known as *chaining*. [See Figure 12.5(b).] Chaining can be misleading if items at opposite ends of the chain are, in fact, quite dissimilar.

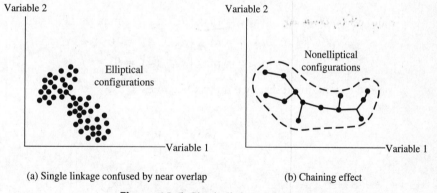

(a) Single linkage confused by near overlap   (b) Chaining effect

Figure 12.5 Single linkage clusters.

The clusters formed by the single linkage method will be unchanged by any assignment of distance (similarity) that gives the same relative orderings as the initial distances (similarities). In particular, any one of a set of similarity coefficients from Table 12.1 that are monotonic to one another will produce the same clustering.

Complete Linkage

Complete linkage clustering proceeds in much the same manner as single linkage clusterings, with one important exception: At each stage, the distance (similarity) between clusters is determined by the distance (similarity) between the two elements, one from each cluster, that are *most distant*. Thus, complete linkage ensures that all items in a cluster are within some maximum distance (or minimum similarity) of each other.

The general agglomerative algorithm again starts by finding the minimum entry in $\mathbf{D} = \{d_{ik}\}$ and merging the corresponding objects, such as U and V, to get cluster (UV). For Step 3 of the general algorithm in (12-12), the distances between (UV) and any other cluster W are computed by

$$d_{(UV)W} = \max\{d_{UW}, d_{VW}\} \tag{12-14}$$

Here d_{UW} and d_{VW} are the distances between the most distant members of clusters U and W and clusters V and W, respectively.

Example 12.5 (Clustering using complete linkage) Let us return to the distance matrix introduced in Example 12.3:

$$\mathbf{D} = \{d_{ik}\} = \begin{array}{c} \\ 1 \\ 2 \\ 3 \\ 4 \\ 5 \end{array} \begin{array}{ccccc} 1 & 2 & 3 & 4 & 5 \\ \left[\begin{array}{ccccc} 0 & & & & \\ 9 & 0 & & & \\ 3 & 7 & 0 & & \\ 6 & 5 & 9 & 0 & \\ 11 & 10 & ② & 8 & 0 \end{array}\right] \end{array}$$

Still take most similar

At the first stage, objects 3 and 5 are merged, since they are most similar. This gives the cluster (35). At stage 2, we compute

only dif is, choose max
$$d_{(35)1} = \underline{\max}\{d_{31}, d_{51}\} = \max\{3, 11\} = 11$$
$$d_{(35)2} = \max\{d_{32}, d_{52}\} = 10$$
$$d_{(35)4} = \max\{d_{34}, d_{54}\} = 9$$

and the modified distance matrix becomes

$$\begin{array}{c} \\ (35) \\ 1 \\ 2 \\ 4 \end{array} \begin{array}{cccc} (35) & 1 & 2 & 4 \\ \left[\begin{array}{cccc} 0 & & & \\ 11 & 0 & & \\ 10 & 9 & 0 & \\ 9 & 6 & ⑤ & 0 \end{array}\right] \end{array}$$

The next merger occurs between the most similar groups, 2 and 4, to give the cluster (24). At stage 3, we have

$$d_{(24)(35)} = \max\{d_{2(35)}, d_{4(35)}\} = \max\{10, 9\} = 10$$
$$d_{(24)1} = \max\{d_{21}, d_{41}\} = 9$$

and the distance matrix

$$\begin{array}{c} \\ (35) \\ (24) \\ 1 \end{array} \begin{array}{ccc} (35) & (24) & 1 \\ \left[\begin{array}{ccc} 0 & & \\ 10 & 0 & \\ 11 & ⑨ & 0 \end{array}\right] \end{array}$$

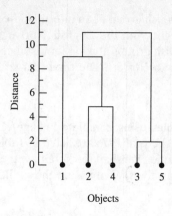

Figure 12.6 Complete linkage dendrogram for distances between five objects.

The next merger produces the cluster (124). At the final stage, the groups (35) and (124) are merged as the single cluster (12345) at level

$$d_{(124)(35)} = \max\{d_{1(35)}, d_{(24)(35)}\} = \max\{11, 10\} = 11$$

The dendrogram is given in Figure 12.6. ∎

Comparing Figures 12.3 and 12.6, we see that the dendrograms for single linkage and complete linkage differ in the allocation of object 1 to previous groups.

Example 12.6 (Complete linkage clustering of 11 languages) In Example 12.4, we presented a distance matrix for numbers in 11 languages. The complete linkage clustering algorithm applied to this distance matrix produces the dendrogram shown in Figure 12.7.

Comparing Figures 12.7 and 12.4, we see that both hierarchical methods yield the English–Norwegian–Danish and the French–Italian–Spanish language groups. Polish is merged with French–Italian–Spanish at an intermediate level. In addition, both methods merge Hungarian and Finnish only at the penultimate stage.

However, the two methods handle German and Dutch differently. Single linkage merges German and Dutch at an intermediate distance, and these two languages remain a cluster until the final merger. Complete linkage merges German

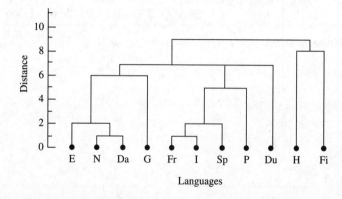

Figure 12.7 Complete linkage dendrogram for distances between numbers in 11 languages.

with the English–Norwegian–Danish group at an intermediate level. Dutch remains a cluster by itself until it is merged with the English–Norwegian–Danish–German and French–Italian–Spanish–Polish groups at a higher distance level. The final complete linkage merger involves two clusters. The final merger in single linkage involves three clusters. ∎

Example 12.7 (Clustering variables using complete linkage) Data collected on 22 U.S. public utility companies for the year 1975 are listed in Table 12.4. (Although it is more interesting to group companies, we shall see here how the complete linkage algorithm can be used to cluster variables.) We measure the similarity between pairs of
group 2 ways

Table 12.4 Public Utility Data (1975)

Company	X_1	X_2	X_3	X_4	X_5	X_6	X_7	X_8
1. Arizona Public Service	1.06	9.2	151	54.4	1.6	9077	0.	.628
2. Boston Edison Co.	.89	10.3	202	57.9	2.2	5088	25.3	1.555
3. Central Louisiana Electric Co.	1.43	15.4	113	53.0	3.4	9212	0.	1.058
4. Commonwealth Edison Co.	1.02	11.2	168	56.0	.3	6423	34.3	.700
5. Consolidated Edison Co. (N.Y.)	1.49	8.8	192	51.2	1.0	3300	15.6	2.044
6. Florida Power & Light Co.	1.32	13.5	111	60.0	-2.2	11127	22.5	1.241
7. Hawaiian Electric Co.	1.22	12.2	175	67.6	2.2	7642	0.	1.652
8. Idaho Power Co.	1.10	9.2	245	57.0	3.3	13082	0.	.309
9. Kentucky Utilities Co.	1.34	13.0	168	60.4	7.2	8406	0.	.862
10. Madison Gas & Electric Co.	1.12	12.4	197	53.0	2.7	6455	39.2	.623
11. Nevada Power Co.	.75	7.5	173	51.5	6.5	17441	0.	.768
12. New England Electric Co.	1.13	10.9	178	62.0	3.7	6154	0.	1.897
13. Northern States Power Co.	1.15	12.7	199	53.7	6.4	7179	50.2	.527
14. Oklahoma Gas & Electric Co.	1.09	12.0	96	49.8	1.4	9673	0.	.588
15. Pacific Gas & Electric Co.	.96	7.6	164	62.2	-0.1	6468	.9	1.400
16. Puget Sound Power & Light Co.	1.16	9.9	252	56.0	9.2	15991	0.	.620
17. San Diego Gas & Electric Co.	.76	6.4	136	61.9	9.0	5714	8.3	1.920
18. The Southern Co.	1.05	12.6	150	56.7	2.7	10140	0.	1.108
19. Texas Utilities Co.	1.16	11.7	104	54.0	-2.1	13507	0.	.636
20. Wisconsin Electric Power Co.	1.20	11.8	148	59.9	3.5	7287	41.1	.702
21. United Illuminating Co.	1.04	8.6	204	61.0	3.5	6650	0.	2.116
22. Virginia Electric & Power Co.	1.07	9.3	174	54.3	5.9	10093	26.6	1.306

KEY: X_1: Fixed-charge coverage ratio (income/debt).
   X_2: Rate of return on capital.
   X_3: Cost per KW capacity in place.
   X_4: Annual load factor.
   X_5: Peak kWh demand growth from 1974 to 1975.
   X_6: Sales (kWh use per year).
   X_7: Percent nuclear.
   X_8: Total fuel costs (cents per kWh).

Source: Data courtesy of H. E. Thompson.

Table 12.5 Correlations Between Pairs of Variables (Public Utility Data)

	X_1	X_2	X_3	X_4	X_5	X_6	X_7	X_8
	1.000							
	.643	1.000						
	−.103	−.348	1.000					
	−.082	−.086	.100	1.000				
	−.259	−.260	.435	.034	1.000			
	−.152	−.010	.028	−.288	.176	1.000		
	.045	.211	.115	−.164	−.019	−.374	1.000	
	−.013	−.328	.005	.486	−.007	−.561	−.185	1.000

variables by the product-moment correlation coefficient. The correlation matrix is given in Table 12.5.

When the sample correlations are used as similarity measures, variables with large negative correlations are regarded as very dissimilar; variables with large positive correlations are regarded as very similar. In this case, the "distance" between clusters is measured as the *smallest* similarity between members of the corresponding clusters. The complete linkage algorithm, applied to the foregoing similarity matrix, yields the dendrogram in Figure 12.8.

We see that variables 1 and 2 (fixed-charge coverage ratio and rate of return on capital), variables 4 and 8 (annual load factor and total fuel costs), and variables 3 and 5 (cost per kilowatt capacity in place and peak kilowatthour demand growth) cluster at intermediate "similarity" levels. Variables 7 (percent nuclear) and 6 (sales) remain by themselves until the final stages. The final merger brings together the (12478) group and the (356) group. ∎

As in single linkage, a "new" assignment of distances (similarities) that have the same relative orderings as the initial distances will not change the configuration of the complete linkage clusters.

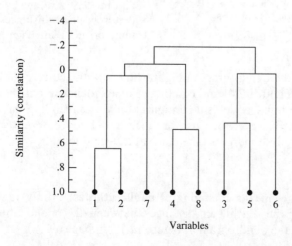

Figure 12.8 Complete linkage dendrogram for similarities among eight utility company variables.

Average Linkage

Average linkage treats the distance between two clusters as the average distance between all pairs of items where one member of a pair belongs to each cluster.

Again, the input to the average linkage algorithm may be distances or similarities, and the method can be used to group objects or variables. The average linkage algorithm proceeds in the manner of the general algorithm of (12-12). We begin by searching the distance matrix $\mathbf{D} = \{d_{ik}\}$ to find the nearest (most similar) objects— for example, U and V. These objects are merged to form the cluster (UV). For Step 3 of the general agglomerative algorithm, the distances between (UV) and the other cluster W are determined by

$$\text{avg:} \qquad d_{(UV)W} = \frac{\sum_i \sum_k d_{ik}}{N_{(UV)} N_W} \qquad (12\text{-}15)$$

where d_{ik} is the distance between object i in the cluster (UV) and object k in the cluster W, and $N_{(UV)}$ and N_W are the number of items in clusters (UV) and W, respectively.

Example 12.8 (Average linkage clustering of 11 languages) The average linkage algorithm was applied to the "distances" between 11 languages given in Example 12.4. The resulting dendrogram is displayed in Figure 12.9.

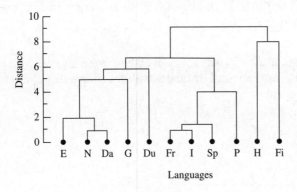

Figure 12.9 Average linkage dendrogram for distances between numbers in 11 languages.

A comparison of the dendrogram in Figure 12.9 with the corresponding single linkage dendrogram (Figure 12.4) and complete linkage dendrogram (Figure 12.7) indicates that average linkage yields a configuration very much like the complete linkage configuration. However, because distance is defined differently for each case, it is not surprising that mergers take place at different levels.   ■

Example 12.9 (Average linkage clustering of public utilities) An average linkage algorithm applied to the Euclidean distances between 22 public utilities (see Table 12.6) produced the dendrogram in Figure 12.10 on page 692.

Table 12.6 Distances Between 22 Utilities

Firm no.	1	2	3	4	5	6	7	8	9	10	11	12	13	14	15	16	17	18	19	20	21	22
1	.00																					
2	3.10	.00																				
3	3.68	4.92	.00																			
4	2.46	2.16	4.11	.00																		
5	4.12	3.85	4.47	4.13	.00																	
6	3.61	4.22	2.99	3.20	4.60	.00																
7	3.90	3.45	4.22	3.97	4.60	3.35	.00															
8	2.74	3.89	4.99	3.69	5.16	4.91	4.36	.00														
9	3.25	3.96	2.75	3.75	4.49	3.73	2.80	3.59	.00													
10	3.10	2.71	3.93	1.49	4.05	3.83	4.51	3.67	3.57	.00												
11	3.49	4.79	5.90	4.86	6.46	6.00	6.00	3.46	5.18	5.08	.00											
12	3.22	2.43	4.03	3.50	3.60	3.74	1.66	4.06	2.74	3.94	5.21	.00										
13	3.96	3.43	4.39	2.58	4.76	4.55	5.01	4.14	3.66	1.41	5.31	4.50	.00									
14	2.11	4.32	2.74	3.23	4.82	3.47	4.91	4.34	3.82	3.61	4.32	4.34	4.39	.00								
15	2.59	2.50	5.16	3.19	4.26	4.07	2.93	3.85	4.11	4.26	4.74	2.33	5.10	4.24	.00							
16	4.03	4.84	5.26	4.97	5.82	5.84	5.04	2.20	3.63	4.53	3.43	4.62	4.41	5.17	5.18	.00						
17	4.40	3.62	6.36	4.89	5.63	6.10	4.58	5.43	4.90	5.48	4.75	3.50	5.61	5.56	3.40	5.56	.00					
18	1.88	2.90	2.72	2.65	4.34	2.85	2.95	3.24	2.43	3.07	3.95	2.45	3.78	2.30	3.00	3.97	4.43	.00				
19	2.41	4.63	3.18	3.46	5.13	2.58	4.52	4.11	4.11	4.13	4.52	4.41	5.01	1.88	4.03	5.23	6.09	2.47	.00			
20	3.17	3.00	3.73	1.82	4.39	2.91	3.54	4.09	2.95	2.05	5.35	3.43	2.23	3.74	3.78	4.82	4.87	2.92	3.90	.00		
21	3.45	2.32	5.09	3.88	3.64	4.63	2.68	3.98	3.74	4.36	4.88	1.38	4.94	4.93	2.10	4.57	3.10	3.19	4.97	4.15	.00	
22	2.51	2.42	4.11	2.58	3.77	4.03	4.00	3.24	3.21	2.56	3.44	3.00	2.74	3.51	3.35	3.46	3.63	2.55	3.97	2.62	3.01	.00

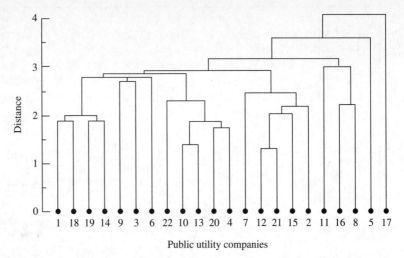

Figure 12.10 Average linkage dendrogram for distances between 22 public utility companies.

Concentrating on the intermediate clusters, we see that the utility companies tend to group according to geographical location. For example, one intermediate cluster contains the firms 1 (Arizona Public Service), 18 (The Southern Company—primarily Georgia and Alabama), 19 (Texas Utilities Company), and 14 (Oklahoma Gas and Electric Company). There are some exceptions. The cluster (7, 12, 21, 15, 2) contains firms on the eastern seaboard and in the far west. On the other hand, all these firms are located near the coasts. Notice that Consolidated Edison Company of New York and San Diego Gas and Electric Company stand by themselves until the final amalgamation stages.

It is, perhaps, not surprising that utility firms with similar locations (or types of locations) cluster. One would expect regulated firms in the same area to use, basically, the same type of fuel(s) for power plants and face common markets. Consequently, types of generation, costs, growth rates, and so forth should be relatively homogeneous among these firms. This is apparently reflected in the hierarchical clustering. ∎

For average linkage clustering, changes in the assignment of distances (similarities) can affect the arrangement of the final configuration of clusters, even though the changes preserve relative orderings.

Ward's Hierarchical Clustering Method

Ward [32] considered hierarchical clustering procedures based on minimizing the 'loss of information' from joining two groups. This method is usually implemented with loss of information taken to be an increase in an error sum of squares criterion,

ESS. First, for a given cluster k, let ESS_k be the sum of the squared deviations of every item in the cluster from the cluster mean (centroid). If there are currently K clusters, define ESS as the sum of the ESS_k or $ESS = ESS_1 + ESS_2 + \ldots + ESS_K$. At each step in the analysis, the union of every possible pair of clusters is considered, and the two clusters whose combination results in the smallest increase in ESS (minimum loss of information) are joined. Initially, each cluster consists of a single item, and, if there are N items, $ESS_k = 0$, $k = 1, 2, \ldots, N$, so $ESS = 0$. At the other extreme, when all the clusters are combined in a single group of N items, the value of ESS is given by

$$ESS = \sum_{j=1}^{N} (\mathbf{x}_j - \bar{\mathbf{x}})'(\mathbf{x}_j - \bar{\mathbf{x}})$$

where $\mathbf{x}_j$ is the multivariate measurement associated with the jth item and $\bar{\mathbf{x}}$ is the mean of all the items.

The results of Ward's method can be displayed as a dendrogram. The vertical axis gives the values of ESS at which the mergers occur.

Ward's method is based on the notion that the clusters of multivariate observations are expected to be roughly elliptically shaped. It is a hierarchical precursor to nonhierarchical clustering methods that optimize some criterion for dividing data into a *given* number of elliptical groups. We discuss nonhierarchical clustering procedures in the next section. Additional discussion of optimization methods of cluster analysis is contained in [8].

Example 12.10 (Clustering pure malt scotch whiskies) Virtually all the world's pure malt Scotch whiskies are produced in Scotland. In one study (see [22]), 68 binary variables were created measuring characteristics of Scotch whiskey that can be broadly classified as color, nose, body, palate, and finish. For example, there were 14 color characteristics (descriptions), including white wine, yellow, very pale, pale, bronze, full amber, red, and so forth. LaPointe and Legendre clustered 109 pure malt Scotch whiskies, each from a different distillery. The investigators were interested in determining the major types of single-malt whiskies, their chief characteristics, and the best representative. In addition, they wanted to know whether the groups produced by the hierarchical clustering procedure corresponded to different geographical regions, since it is known that whiskies are affected by local soil, temperature, and water conditions.

Weighted similarity coefficients $\{s_{ik}\}$ were created from binary variables representing the presence or absence of characteristics. The resulting "distances," defined as $\{d_{ik} = 1 - s_{ik}\}$, were used with Ward's method to group the 109 pure (single-) malt Scotch whiskies. The resulting dendrogram is shown in Figure 12.11. (An average linkage procedure applied to a similarity matrix produced almost exactly the same classification.)

The groups labelled A–L in the figure are the 12 groups of similar Scotches identified by the investigators. A follow-up analysis suggested that these 12 groups have a large geographic component in the sense that Scotches with similar characteristics tend to be produced by distilleries that are located reasonably

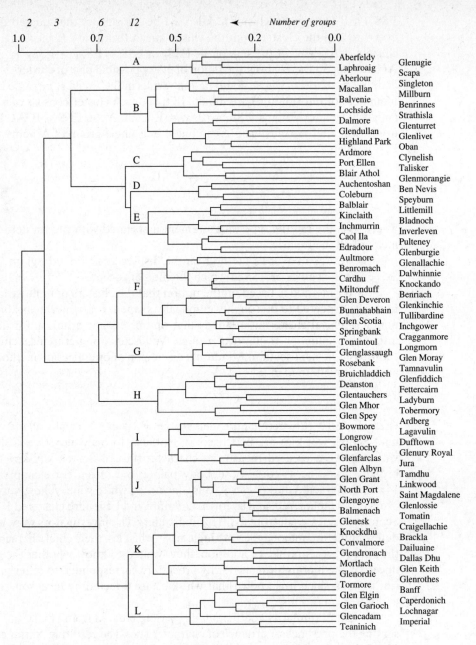

Figure 12.11 A dendrogram for similarities between 109 pure malt Scotch whiskies.

close to one another. Consequently, the investigators concluded, "The relationship with geographic features was demonstrated, supporting the hypothesis that whiskies are affected not only by distillery secrets and traditions but also by factors dependent on region such as water, soil, microclimate, temperature and even air quality." ∎

Final Comments—Hierarchical Procedures

There are many agglomerative hierarchical clustering procedures besides single linkage, complete linkage, and average linkage. However, all the agglomerative procedures follow the basic algorithm of (12-12).

As with most clustering methods, sources of error and variation are not formally considered in hierarchical procedures. This means that a clustering method will be sensitive to outliers, or "noise points."

In hierarchical clustering, there is no provision for a reallocation of objects that may have been "incorrectly" grouped at an early stage. Consequently, the final configuration of clusters should always be carefully examined to see whether it is sensible.

For a particular problem, it is a good idea to try several clustering methods and, within a given method, a couple different ways of assigning distances (similarities). If the outcomes from the several methods are (roughly) consistent with one another, perhaps a case for "natural" groupings can be advanced.

The *stability* of a hierarchical solution can sometimes be checked by applying the clustering algorithm before and after *small* errors (perturbations) have been added to the data units. If the groups are fairly well distinguished, the clusterings before perturbation and after perturbation should agree.

Common values (ties) in the similarity or distance matrix can produce multiple solutions to a hierarchical clustering problem. That is, the dendrograms corresponding to different treatments of the tied similarities (distances) can be different, particularly at the lower levels. This is not an inherent problem of any method; rather, multiple solutions occur for certain kinds of data. Multiple solutions are not necessarily bad, but the user needs to know of their existence so that the groupings (dendrograms) can be properly interpreted and different groupings (dendrograms) compared to assess their overlap. A further discussion of this issue appears in [27].

Some data sets and hierarchical clustering methods can produce *inversions*. (See [27].) An inversion occurs when an object joins an existing cluster at a smaller distance (greater similarity) than that of a previous consolidation. An inversion is represented two different ways in the following diagram:

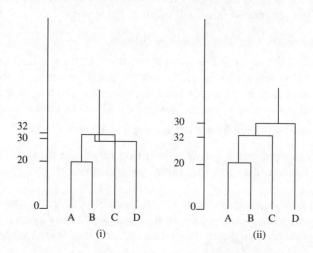

In this example, the clustering method joins A and B at distance 20. At the next step, C is added to the group (AB) at distance 32. Because of the nature of the clustering algorithm, D is added to group (ABC) at distance 30, a smaller distance than the distance at which C joined (AB). In (i) the inversion is indicated by a dendrogram with crossover. In (ii), the inversion is indicated by a dendrogram with a nonmonotonic scale.

Inversions can occur when there is no clear cluster structure and are generally associated with two hierarchical clustering algorithms known as the centroid method and the median method. The hierarchical procedures discussed in this book are not prone to inversions.

12.4 Nonhierarchical Clustering Methods

Nonhierarchical clustering techniques are designed to group *items*, rather than *variables*, into a collection of K clusters. The number of clusters, K, may either be specified in advance or determined as part of the clustering procedure. Because a matrix of distances (similarities) does not have to be determined, and the basic data do not have to be stored during the computer run, nonhierarchical methods can be applied to much larger data sets than can hierarchical techniques.

Nonhierarchical methods start from either (1) an initial partition of items into groups or (2) an initial set of seed points, which will form the nuclei of clusters. Good choices for starting configurations should be free of overt biases. One way to start is to randomly select seed points from among the items or to randomly partition the items into initial groups.

In this section, we discuss one of the more popular nonhierarchical procedures, the K-means method.

K-means Method

MacQueen [25] suggests the term K-means for describing an algorithm of his that assigns each item to the cluster having the nearest centroid (mean). In its simplest version, the process is composed of these three steps:

1. Partition the items into K initial clusters.

2. Proceed through the list of items, assigning an item to the cluster whose centroid (mean) is nearest. (Distance is usually computed using Euclidean distance with either standardized or unstandardized observations.) Recalculate the centroid for the cluster receiving the new item and for the cluster losing the item.

3. Repeat Step 2 until no more reassignments take place.   (12-16)

Rather than starting with a partition of all items into K preliminary groups in Step 1, we could specify K initial centroids (seed points) and then proceed to Step 2.

The final assignment of items to clusters will be, to some extent, dependent upon the initial partition or the initial selection of seed points. Experience suggests that most major changes in assignment occur with the first reallocation step.

Example 12.11 (Clustering using the K-means method) Suppose we measure two variables X_1 and X_2 for each of four items A, B, C, and D. The data are given in the following table:

Item	Observations	
	x_1	x_2
A	5	3
B	−1	1
C	1	−2
D	−3	−2

The objective is to divide these items into $K = 2$ clusters such that the items within a cluster are closer to one another than they are to the items in different clusters. To implement the $K = 2$-means method, we *arbitrarily* partition the items into two clusters, such as (AB) and (CD), and compute the co-ordinates $(\bar{x}_1, \bar{x}_2)$ of the cluster centroid (mean). Thus, at Step 1, we have

Cluster	Coordinates of centroid	
	$\bar{x}_1$	$\bar{x}_2$
(AB)	$\dfrac{5 + (-1)}{2} = 2$	$\dfrac{3 + 1}{2} = 2$
(CD)	$\dfrac{1 + (-3)}{2} = -1$	$\dfrac{-2 + (-2)}{2} = -2$

At Step 2, we compute the Euclidean distance of each item from the group centroids and reassign each item to the nearest group. If an item is moved from the initial configuration, the cluster centroids (means) must be updated before proceeding. The ith coordinate, $i = 1, 2, \ldots, p$, of the centroid is easily updated using the formulas:

$$\bar{x}_{i, new} = \frac{n\bar{x}_i + x_{ji}}{n + 1} \qquad \text{if the jth item is } \textit{added} \text{ to a group}$$

$$\bar{x}_{i, new} = \frac{n\bar{x}_i - x_{ji}}{n - 1} \qquad \text{if the jth item is } \textit{removed} \text{ from a group}$$

Here n is the number of items in the "old" group with centroid $\bar{x}' = (\bar{x}_1, \bar{x}_2, \ldots, \bar{x}_p)$.

Consider the initial clusters (AB) and (CD). The coordinates of the centroids are $(2, 2)$ and $(-1, -2)$ respectively. Suppose item A with coordinates $(5, 3)$ is moved to the (CD) group. The new groups are (B) and (ACD) with updated centroids:

Group (B) $\qquad \bar{x}_{1, new} = \dfrac{2(2) - 5}{2 - 1} = -1 \quad \bar{x}_{2, new} = \dfrac{2(2) - 3}{2 - 1} = 1$, the coordinates of B

Group (ACD) $\bar{x}_{1, new} = \dfrac{2(-1) + 5}{2 + 1} = 1 \quad \bar{x}_{2, new} = \dfrac{2(-2) + 3}{2 + 1} = -.33$

$\left(\left(x-x_2\right)^2+\left(y_1-y_2\right)^2\right.$

Returning to the initial groupings in Step 1, we compute the squared distances

$d^2(A,(AB)) = (5 - 2)^2 + (3 - 2)^2 = 10$ if A is not moved
$d^2(A,(CD)) = (5 + 1)^2 + (3 + 2)^2 = 61$

$d^2(A,(B)) = (5 + 1)^2 + (3 - 1)^2 = 40$ if A is moved to the (CD) group
$d^2(A,(ACD)) = (5 - 1)^2 + (3 + .33)^2 = 27.09$

Since A is closer to the center of (AB) than it is to the center of (ACD), it is not reassigned.

Continuing, we consider reassigning B. We get

$d^2(B,(AB)) = (-1 - 2)^2 + (1 - 2)^2 = 10$ if B is not moved
$d^2(B,(CD)) = (-1 + 1)^2 + (1 + 2)^2 = 9$

$d^2(B,(A))) = (-1-5)^2 + (1 - 3)^2 = 40$ if B is moved to the (CD) group
$d^2(B,(BCD)) = (-1 + 1)^2 + (1 + 1)^2 = 4$

Since B is closer to the center of (BCD) than it is to the center of (AB), B is reassigned to the (CD) group. We now have the clusters (A) and (BCD) with centroid coordinates $(5, 3)$ and $(-1, -1)$ respectively.

We check C for reassignment.

$d^2(C,(A)) = (1 - 5)^2 + (-2 - 3)^2 = 41$ if C is not moved
$d^2(C,(BCD)) = (1 + 1)^2 + (-2 + 1)^2 = 5$

$d^2(C,(AC)) = (1 - 3)^2 + (-2 - .5)^2 = 10.25$ if C is moved to the (A) group
$d^2(C,(BD)) = (1 + 2)^2 + (-2 + .5)^2 = 11.25$

Since C is closer to the center of the BCD group than it is to the center of the AC group, C is not moved. Continuing in this way, we find that no more reassignments take place and the final $K = 2$ clusters are (A) and (BCD).

For the final clusters, we have

	Item			
Cluster	A	B	C	D
A	0	40	41	89
(BCD)	52	4	5	5

Squared distances to group centroids

The within cluster sum of squares (sum of squared distances to centroid) are

Cluster A: 0
Cluster (BCD): $4 + 5 + 5 = 14$

Equivalently, we can determine the $K = 2$ clusters by using the criterion

$$\min E = \sum d^2_{i,\, c(i)}$$

where the minimum is over the number of $K = 2$ clusters and $d^2_{i,\,c(i)}$ is the squared distance of case i from the centroid (mean) of the assigned cluster.

In this example, there are seven possibilities for $K = 2$ clusters:

$A, (BCD)$

$B, (ACD)$

$C, (ABD)$

$D, (ABC)$

$(AB), (CD)$

$(AC), (BD)$
$(AD), (BC)$

For the $A, (BCD)$ pair:

$$A \qquad d^2_{A,\,c(A)} = 0$$

$$(BCD) \qquad d^2_{B,\,c(B)} + d^2_{C,\,c(C)} + d^2_{D,\,c(D)} = 4 + 5 + 5 = 14$$

Consequently, $\sum d^2_{i,\,c(i)} = 0 + 14 = 14$

For the remaining pairs, you may verify that

$$B, (ACD) \quad \sum d^2_{i,\,c(i)} = 48.7$$

$$C, (ABD) \quad \sum d^2_{i,\,c(i)} = 27.7$$

$$D, (ABC) \quad \sum d^2_{i,\,c(i)} = 31.3$$

$$(AB), (CD) \quad \sum d^2_{i,\,c(i)} = 28$$

$$(AC), (BD) \quad \sum d^2_{i,\,c(i)} = 27$$

$$(AD), (BC) \quad \sum d^2_{i,\,c(i)} = 51.3$$

Since the smallest $\sum d^2_{i,\,c(i)}$ occurs for the pair of clusters (A) and (BCD), this is the final partition. ∎

To check the stability of the clustering, it is desirable to rerun the algorithm with a new initial partition. Once clusters are determined, intuitions concerning their interpretations are aided by rearranging the list of items so that those in the first cluster appear first, those in the second cluster appear next, and so forth. A table of the cluster centroids (means) and within-cluster variances also helps to delineate group differences.

Example 12.12 (K-means clustering of public utilities) Let us return to the problem of clustering public utilities using the data in Table 12.4. The K-means algorithm for several choices of K was run. We present a summary of the results for $K = 4$ and $K = 5$. In general, the choice of a particular K is not clear cut and depends upon subject-matter knowledge, as well as data-based appraisals. (Data-based appraisals might include choosing K so as to maximize the between-cluster variability relative to the within-cluster variability. Relevant measures might include $|\mathbf{W}|/|\mathbf{B} + \mathbf{W}|$ [see (6-38)] and $\text{tr}\,(\mathbf{W}^{-1}\mathbf{B})$.) The summary is as follows:

$K = 4$

Cluster	Number of firms	Firms
1	5	Idaho Power Co. (8), Nevada Power Co. (11), Puget Sound Power & Light Co. (16), Virginia Electric & Power Co. (22), Kentucky Utilities Co. (9).
2	6	Central Louisiana Electric Co. (3), Oklahoma Gas & Electric Co. (14), The Southern Co. (18), Texas Utilities Co. (19), Arizona Public Service (1), Florida Power & Light Co. (6).
3	5	New England Electric Co. (12), Pacific Gas & Electric Co. (15), San Diego Gas & Electric Co. (17), United Illuminating Co. (21), Hawaiian Electric Co. (7).
4	6	Consolidated Edison Co. (N.Y.) (5), Boston Edison Co. (2), Madison Gas & Electric Co. (10), Northern States Power Co. (13), Wisconsin Electric Power Co. (20), Commonwealth Edison Co. (4).

Distances between Cluster Centers

$$
\begin{array}{c}
 \\
1 \\
2 \\
3 \\
4
\end{array}
\begin{array}{cccc}
1 & 2 & 3 & 4 \\
\left[\begin{array}{cccc}
0 & & & \\
3.08 & 0 & & \\
3.29 & 3.56 & 0 & \\
3.05 & 2.84 & 3.18 & 0
\end{array}\right]
\end{array}
$$

$K = 5$

Cluster	Number of firms	Firms
1	5	Nevada Power Co. (11), Puget Sound Power & Light Co. (16), Idaho Power Co. (8), Virginia Electric & Power Co. (22), Kentucky Utilities Co. (9).
2	6	Central Louisiana Electric Co. (3), Texas Utilities Co. (19), Oklahoma Gas & Electric Co. (14), The Southern Co. (18), Arizona Public Service (1), Florida Power & Light Co. (6).
3	5	New England Electric Co. (12), Pacific Gas & Electric Co. (15), San Diego Gas & Electric Co. (17), United Illuminating Co. (21), Hawaiian Electric Co. (7).
4	2	Consolidated Edison Co. (N.Y.) (5), Boston Edison Co. (2).
5	4	Commonwealth Edison Co. (4), Madison Gas & Electric Co. (10), Northern States Power Co. (13), Wisconsin Electric Power Co. (20).

Distances between Cluster Centers

$$
\begin{array}{c}
 \quad 1 \qquad 2 \qquad 3 \qquad 4 \qquad 5 \\
\begin{array}{c}
1 \\
2 \\
3 \\
4 \\
5
\end{array}
\left[
\begin{array}{ccccc}
0 & & & & \\
3.08 & 0 & & & \\
3.29 & 3.56 & 0 & & \\
3.63 & 3.46 & 2.63 & 0 & \\
3.18 & 2.99 & 3.81 & 2.89 & 0
\end{array}
\right]
\end{array}
$$

The cluster profiles ($K = 5$) shown in Figure 12.12 order the eight variables according to the ratios of their between-cluster variability to their within-cluster variability. [For univariate F-ratios, see Section 6.4.] We have

$$
F_{\text{nuc}} = \frac{\text{mean square percent nuclear between clusters}}{\text{mean square percent nuclear within clusters}} = \frac{3.335}{.255} = 13.1
$$

so firms within different clusters are widely separated with respect to percent nuclear, but firms within the same cluster show little percent nuclear variation. Fuel costs (FUELC) and annual sales (SALES) also seem to be of some importance in distinguishing the clusters.

Reviewing the firms in the five clusters, it is apparent that the K-means method gives results generally consistent with the average linkage hierarchical method. (See Example 12.9.) Firms with common or compatible geographical locations cluster. Also, the firms in a given cluster seem to be roughly the same in terms of percent nuclear.   ∎

We must caution, as we have throughout the book, that the importance of *individual* variables in clustering must be judged from a multivariate perspective. *All* of the variables (multivariate observations) determine the cluster means and the reassignment of items. In addition, the values of the descriptive statistics measuring the importance of individual variables are functions of the number of clusters and the final configuration of the clusters. On the other hand, descriptive measures can be helpful, after the fact, in assessing the "success" of the clustering procedure.

Final Comments—Nonhierarchical Procedures

There are strong arguments for not fixing the number of clusters, K, in advance, including the following:

1. If two or more seed points inadvertently lie within a single cluster, their resulting clusters will be poorly differentiated.

Cluster profiles—variables are ordered by F-ratio size

Percent nuclear
Total fuel costs
Sales
Cost per KW capacity in place
Annual load factor
Peak kWh demand growth
Rate of return on capital
Fixed-charge coverage ratio

F-ratio
13.1
12.4
8.4
5.1
4.4
2.7
2.4
0.5

Each column describes a cluster.
The cluster number is printed at the mean of each variable.
Dashes indicate one standard deviation above and below mean.

Figure 12.12 Cluster profiles ($K = 5$) for public utility data.

2. The existence of an outlier might produce at least one group with very disperse items.

3. Even if the population is known to consist of K groups, the sampling method may be such that data from the rarest group do not appear in the sample. Forcing the data into K groups would lead to nonsensical clusters.

In cases in which a single run of the algorithm requires the user to specify K, it is always a good idea to rerun the algorithm for several choices.

Discussions of other nonhierarchical clustering procedures are available in [3], [8], and [16].

12.5 Clustering Based on Statistical Models

The popular clustering methods discussed earlier in this chapter, including single linkage, complete linkage, average linkage, Ward's method and K-means clustering, are intuitively reasonable procedures but that is as much as we can say without having a model to explain how the observations were produced. Major advances in clustering methods have been made through the introduction of statistical models that indicate how the collection of ($p \times 1$) measurements $\mathbf{x}_j$, from the N objects, was generated. The most common model is one where cluster k has expected proportion p_k of the objects and the corresponding measurements are generated by a probability density function $f_k(\mathbf{x})$. Then, if there are K clusters, the observation vector for a single object is modeled as arising from the *mixing distribution*

$$f_{Mix}(\mathbf{x}) = \sum_{k=1}^{K} p_k f_k(\mathbf{x})$$

where each $p_k \geq 0$ and $\sum_{k=1}^{K} p_k = 1$. This distribution $f_{\text{Mix}}(\mathbf{x})$ is called a mixture of the K distributions $f_1(\mathbf{x}), f_2(\mathbf{x}), \dots, f_K(\mathbf{x})$ because the observation is generated from the component distribution $f_k(\mathbf{x})$ with probability p_k. The collection of N observation vectors generated from this distribution will be a mixture of observations from the component distributions.

The most common mixture model is a mixture of multivariate normal distributions where the k-th component $f_k(\mathbf{x})$ is the $N_p(\boldsymbol{\mu}_k, \boldsymbol{\Sigma}_k)$ density function.

The normal mixture model for one observation $\mathbf{x}$ is

$$f_{Mix}(\mathbf{x} \mid \boldsymbol{\mu}_1, \boldsymbol{\Sigma}_1, \dots, \boldsymbol{\mu}_K, \boldsymbol{\Sigma}_K)$$

$$= \sum_{k=1}^{K} p_k \frac{1}{(2\pi)^{p/2} |\boldsymbol{\Sigma}_k|^{1/2}} \exp\left(-\frac{1}{2} (\mathbf{x} - \boldsymbol{\mu}_k)' \boldsymbol{\Sigma}_k^{-1} (\mathbf{x} - \boldsymbol{\mu}_k) \right) \qquad (12\text{-}17)$$

Clusters generated by this model are ellipsoidal in shape with the heaviest concentration of observations near the center.

Inferences are based on the likelihood, which for N objects and a fixed number of clusters K, is

$$L(p_1, \ldots, p_K, \boldsymbol{\mu}_1, \boldsymbol{\Sigma}_1, \ldots, \boldsymbol{\mu}_k, \boldsymbol{\Sigma}_K) = \prod_{j=1}^{N} f_{Mix}(\mathbf{x}_j \mid \boldsymbol{\mu}_1, \boldsymbol{\Sigma}_1, \ldots, \boldsymbol{\mu}_K, \boldsymbol{\Sigma}_K)$$

$$= \prod_{j=1}^{N}\left(\sum_{k=1}^{K} p_k \frac{1}{(2\pi)^{p/2} \mid \boldsymbol{\Sigma}_k \mid^{1/2}} \exp\left(-\frac{1}{2}(\mathbf{x}_j - \boldsymbol{\mu}_k)'\boldsymbol{\Sigma}_k^{-1}(\mathbf{x}_j - \boldsymbol{\mu}_k)\right)\right) \quad (12\text{-}18)$$

where the proportions $p_1, \ldots, p_k$, the mean vectors $\boldsymbol{\mu}_1, \ldots, \boldsymbol{\mu}_k$, and the covariance matrices $\boldsymbol{\Sigma}_1, \ldots, \boldsymbol{\Sigma}_k$ are unknown. The measurements for different objects are treated as independent and identically distributed observations from the mixture distribution.

There are typically far too many unknown parameters for making inferences when the number of objects to be clustered is at least moderate. However, certain conclusions can be made regarding situations where a heuristic clustering method should work well. In particular, the likelihood based procedure under the normal mixture model with all $\boldsymbol{\Sigma}_k$ the same multiple of the identity matrix, $\eta\mathbf{I}$, is approximately the same as K-means clustering and Ward's method. To date, no statistical models have been advanced for which the cluster formation procedure is approximately the same as single linkage, complete linkage or average linkage.

Most importantly, under the sequence of mixture models (12-17) for different K, the problems of choosing the number of clusters and choosing an appropriate clustering method has been reduced to the problem of selecting an appropriate statistical model. This is a major advance.

A good approach to selecting a model is to first obtain the maximum likelihood estimates $\hat{p}_1, \ldots, \hat{p}_K, \hat{\boldsymbol{\mu}}_1, \hat{\boldsymbol{\Sigma}}_1, \ldots, \hat{\boldsymbol{\mu}}_K, \hat{\boldsymbol{\Sigma}}_K$ for a fixed number of clusters K. These estimates must be obtained numerically using special purpose software. The resulting value of the maximum of the likelihood

$$L_{\max} = L(\hat{p}_1, \ldots, \hat{p}_K, \hat{\boldsymbol{\mu}}_1, \hat{\boldsymbol{\Sigma}}_1, \ldots, \hat{\boldsymbol{\mu}}_K, \hat{\boldsymbol{\Sigma}}_K)$$

provides the basis for model selection. How do we decide on a reasonable value for the number of clusters K? In order to compare models with different numbers of parameters, a penalty is subtracted from twice the maximized value of the log-likelihood to give

$$-2 \ln L_{\max} - Penalty$$

where the penalty depends on the number of parameters estimated and the number of observations N. Since the probabilities p_k sum to 1, there are only $K - 1$ probabilities that must be estimated, $K \times p$ means and $K \times p(p + 1)/2$ variances and covariances. For the Akaike information criterion (AIC), the penalty is $2N \times$ (number of parameters) so

$$\text{AIC} = 2 \ln L_{\max} - 2N\left(K\frac{1}{2}(p + 1)(p + 2) - 1\right) \quad (12\text{-}19)$$

The Bayesian information criterion (BIC) is similar but uses the logarithm of the number of parameters in the penalty function

$$\text{BIC} = 2 \ln L_{\max} - 2 \ln(N) \left(K \frac{1}{2} (p+1)(p+2) - 1 \right) \qquad (12\text{-}20)$$

There is still occasional difficulty with too many parameters in the mixture model so simple structures are assumed for the $\boldsymbol{\Sigma}_k$. In particular, progressively more complicated structures are allowed as indicated in the following table.

Assumed form for $\boldsymbol{\Sigma}_k$	Total number of parameters	BIC
$\boldsymbol{\Sigma}_k = \eta \, \mathbf{I}$	$K(p+1)$	$\ln L_{\max} - 2\ln(N)K(p+1)$
$\boldsymbol{\Sigma}_k = \eta_k \, \mathbf{I}$	$K(p+2) - 1$	$\ln L_{\max} - 2\ln(N)(K(p+2) - 1)$
$\boldsymbol{\Sigma}_k = \eta_k \, Diag(\lambda_1, \lambda_2, \ldots, \lambda_p)$	$K(p+2) + p - 1$	$\ln L_{\max} - 2\ln(N)(K(p+2) + p - 1)$

Additional structures for the covariance matrices are considered in [6] and [9].

Even for a fixed number of clusters, the estimation of a mixture model is complicated. One current software package, *MCLUST*, available in the R software library, combines hierarchical clustering, the EM algorithm and the BIC criterion to develop an appropriate model for clustering. In the 'E'-step of the EM algorithm, a $(N \times K)$ matrix is created whose jth row contains estimates of the conditional (on the current parameter estimates) probabilities that observation $\mathbf{x}_j$ belongs to cluster $1, 2, \ldots, K$. So, at convergence, the jth observation (object) is assigned to the cluster k for which the conditional probability

$$p(k \mid \mathbf{x}_j) = \hat{p}_j f(\mathbf{x}_j \mid k) / \sum_{i=1}^{K} \hat{p}_i f(\mathbf{x}_i \mid k)$$

of membership is the largest. (See [6] and [9] and the references therein.)

Example 12.13 (A model based clustering of the iris data) Consider the Iris data in Table 11.5. Using *MCLUST* and specifically the *me* function, we first fit the $p = 4$ dimensional normal mixture model restricting the covariance matrices to satisfy $\boldsymbol{\Sigma}_k = \eta_k \, \mathbf{I}$, $k = 1, 2, 3$.

Using the BIC criterion, the software chooses $K = 3$ clusters with estimated centers

$$\boldsymbol{\mu}_1 = \begin{bmatrix} 5.01 \\ 3.43 \\ 1.46 \\ 0.25 \end{bmatrix}, \quad \boldsymbol{\mu}_2 = \begin{bmatrix} 5.90 \\ 2.75 \\ 4.40 \\ 1.43 \end{bmatrix}, \quad \boldsymbol{\mu}_3 = \begin{bmatrix} 6.85 \\ 3.07 \\ 5.73 \\ 2.07 \end{bmatrix},$$

and estimated variance-covariance scale factors $\hat{\eta}_1 = .076$, $\hat{\eta}_2 = .163$ and $\hat{\eta}_3 = .163$. The estimated mixing proportions are $\hat{p}_1 = .3333$, $\hat{p}_2 = .4133$ and $\hat{p}_3 = .2534$. For this solution, BIC $= -853.8$. A matrix plot of the clusters for pairs of variables is shown in Figure 12.13.

Once we have an estimated mixture model, a new object $\mathbf{x}_j$ will be assigned to the cluster for which the conditional probability of membership is the largest (see [9]).

Assuming the $\boldsymbol{\Sigma}_k = \eta_k \, \mathbf{I}$ covariance structure and allowing up to $K = 7$ clusters, the BIC can be increased to BIC $= -705.1$.

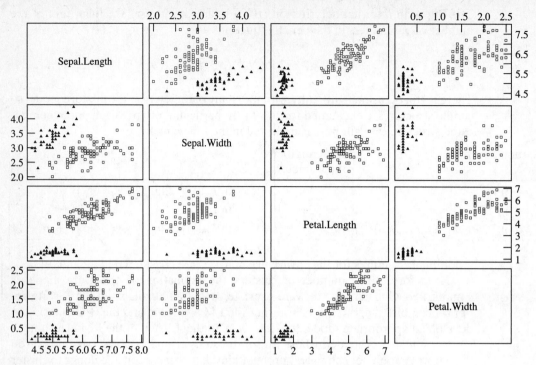

Figure 12.13 Multiple scatter plots of $K = 3$ clusters for Iris data

Finally, using the BIC criterion with up to $K = 9$ groups and several different covariance structures, the best choice is a two group mixture model with unconstrained covariances. The estimated mixing probabilities are $\hat{p}_1 = .3333$ and $\hat{p}_2 = .6667$. The estimated group centers are

$$\mu_1 = \begin{bmatrix} 5.01 \\ 3.43 \\ 1.46 \\ 0.25 \end{bmatrix}, \quad \mu_2 = \begin{bmatrix} 6.26 \\ 2.87 \\ 4.91 \\ 1.68 \end{bmatrix}$$

and the two estimated covariance matrices are

$$\hat{\Sigma}_1 = \begin{bmatrix} .1218 & .0972 & .0160 & .0101 \\ .0972 & .1408 & .0115 & .0091 \\ .0160 & .0115 & .0296 & .0059 \\ .0101 & .0091 & .0059 & .0109 \end{bmatrix} \quad \hat{\Sigma}_2 = \begin{bmatrix} .4530 & .1209 & .4489 & .1655 \\ .1209 & .1096 & .1414 & .0792 \\ .4489 & .1414 & .6748 & .2858 \\ .1655 & .0792 & .2858 & .1786 \end{bmatrix}$$

Essentially, two species of Iris have been put in the same cluster as the projected view of the scatter plot of the sepal measurements in Figure 12.14 shows. ∎

12.6 Multidimensional Scaling

This section begins a discussion of methods for displaying (transformed) multivariate data in low-dimensional space. We have already considered this issue when we

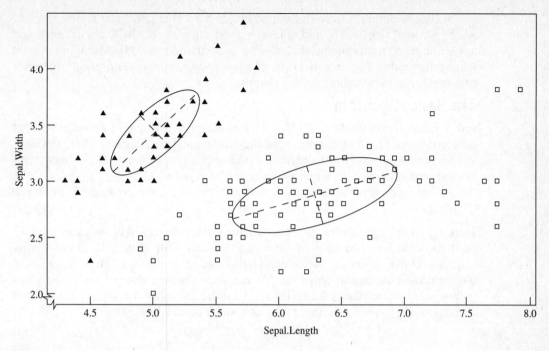

Figure 12.14 Scatter plot of sepal measurements for best model.

discussed plotting scores on, say, the first two principal components or the scores on the first two linear discriminants. The methods we are about to discuss differ from these procedures in the sense that their *primary* objective is to "fit" the original data into a low-dimensional coordinate system such that any distortion caused by a re-duction in dimensionality is minimized. Distortion generally refers to the similarities or dissimilarities (distances) among the original data points. Although Euclidean distance may be used to measure the closeness of points in the final low-dimensional configuration, the notion of similarity or dissimilarity depends upon the underlying technique for its definition. A low-dimensional plot of the kind we are alluding to is called an *ordination* of the data.

Multidimensional scaling techniques deal with the following problem: For a set of observed similarities (or distances) between every pair of N items, find a repre-sentation of the items in few dimensions such that the interitem proximities "nearly match" the original similarities (or distances).

It may not be possible to match exactly the ordering of the original similarities (distances). Consequently, scaling techniques attempt to find configurations in $q \leq N - 1$ dimensions such that the match is as close as possible. The numerical measure of closeness is called the *stress*.

It is possible to arrange the N items in a low-dimensional coordinate system using only the *rank orders* of the $N(N - 1)/2$ original similarities (distances), and not their magnitudes. When only this ordinal information is used to obtain a geometric repre-sentation, the process is called *nonmetric multidimensional scaling*. If the actual magni-tudes of the original similarities (distances) are used to obtain a geometric representation in q dimensions, the process is called *metric multidimensional scaling*. Metric multidimensional scaling is also known as *principal coordinate analysis*.

Scaling techniques were developed by Shepard (see [29] for a review of early work), Kruskal [19, 20, 21], and others. A good summary of the history, theory, and applications of multidimensional scaling is contained in [35]. Multidimensional scaling invariably requires the use of a computer, and several good computer programs are now available for the purpose.

The Basic Algorithm

For N items, there are $M = N(N - 1)/2$ similarities (distances) between pairs of different items. These similarities constitute the basic data. (In cases where the similarities cannot be easily quantified as, for example, the similarity between two colors, the rank orders of the similarities are the basic data.)

Assuming no ties, the similarities can be arranged in a strictly ascending order as

$$s_{i_1 k_1} < s_{i_2 k_2} < \cdots < s_{i_M k_M} \tag{12-21}$$

Here $s_{i_1 k_1}$ is the smallest of the M similarities. The subscript $i_1 k_1$ indicates the pair of items that are least similar—that is, the items with rank 1 in the similarity ordering. Other subscripts are interpreted in the same manner. We want to find a q-dimensional configuration of the N items such that the distances, $d_{ik}^{(q)}$, between pairs of items match the ordering in (12-21). If the distances are laid out in a manner corresponding to that ordering, a perfect match occurs when

$$d_{i_1 k_1}^{(q)} > d_{i_2 k_2}^{(q)} > \cdots > d_{i_M k_M}^{(q)} \tag{12-22}$$

That is, the descending ordering of the distances in q dimensions is exactly analogous to the ascending ordering of the initial similarities. As long as the order in (12-22) is preserved, the magnitudes of the distances are unimportant.

For a given value of q, it may not be possible to find a configuration of points whose pairwise distances are monotonically related to the original similarities. Kruskal [19] proposed a measure of the extent to which a geometrical representation falls short of a perfect match. This measure, the stress, is defined as

$$\text{Stress}\,(q) = \left\{ \frac{\displaystyle\sum_{i<k} \sum (d_{ik}^{(q)} - \hat{d}_{ik}^{(q)})^2}{\displaystyle\sum_{i<k} \sum [d_{ik}^{(q)}]^2} \right\}^{1/2} \tag{12-23}$$

The $\hat{d}_{ik}^{(q)}$'s in the stress formula are numbers known to satisfy (12-22); that is, they are monotonically related to the similarities. The $\hat{d}_{ik}^{(q)}$'s are *not* distances in the sense that they satisfy the usual distance properties of (1-25). They are merely reference numbers used to judge the nonmonotonicity of the observed $d_{ik}^{(q)}$'s.

The idea is to find a representation of the items as points in q-dimensions such that the stress is as small as possible. Kruskal [19] suggests the stress be informally interpreted according to the following guidelines:

Stress	Goodness of fit	
20%	Poor	
10%	Fair	
5%	Good	(12-24)
2.5%	Excellent	
0%	Perfect	

Goodness of fit refers to the monotonic relationship between the similarities and the final distances.

A second measure of discrepancy, introduced by Takane et al. [31], is becoming the preferred criterion. For a given dimension q, this measure, denoted by SStress, replaces the d_{ik}'s and $\hat{d}_{ik}$'s in (12-23) by their squares and is given by

$$\text{SStress} = \left[\frac{\displaystyle\sum_{i<k}\sum (d_{ik}^2 - \hat{d}_{ik}^2)^2}{\displaystyle\sum_{i<k}\sum d_{ik}^4}\right]^{1/2} \tag{12-25}$$

The value of SStress is always between 0 and 1. Any value less than .1 is typically taken to mean that there is a good representation of the objects by the points in the given configuration.

Once items are located in q dimensions, their $q \times 1$ vectors of coordinates can be treated as multivariate observations. For display purposes, it is convenient to represent this q-dimensional scatter plot in terms of its principal component axes. (See Chapter 8.)

We have written the stress measure as a function of q, the number of dimensions for the geometrical representation. For each q, the configuration leading to the minimum stress can be obtained. As q increases, minimum stress will, within rounding error, decrease and will be zero for $q = N - 1$. Beginning with $q = 1$, a plot of these stress (q) numbers versus q can be constructed. The value of q for which this plot begins to level off may be selected as the "best" choice of the dimensionality. That is, we look for an "elbow" in the stress-dimensionality plot.

The entire multidimensional scaling algorithm is summarized in these steps:

1. For N items, obtain the $M = N(N - 1)/2$ similarities (distances) between distinct pairs of items. Order the similarities as in (12-21). (Distances are ordered from largest to smallest.) If similarities (distances) cannot be computed, the rank orders must be specified.

2. Using a trial configuration in q dimensions, determine the interitem distances $d_{ik}^{(q)}$ and numbers $\hat{d}_{ik}^{(q)}$, where the latter satisfy (12-22) and minimize the stress (12-23) or SStress (12-25). (The $\hat{d}_{ik}^{(q)}$ are frequently determined within scaling computer programs using regression methods designed to produce monotonic "fitted" distances.)

3. Using the $\hat{d}_{ik}^{(q)}$'s, move the points around to obtain an improved configuration. (For q fixed, an improved configuration is determined by a general function minimization procedure applied to the stress. In this context, the stress is regarded as a function of the $N \times q$ coordinates of the N items.) A new configuration will have new $d_{ik}^{(q)}$'s new $\hat{d}_{ik}^{(q)}$'s and smaller stress. The process is repeated until the best (minimum stress) representation is obtained.

4. Plot minimum stress (q) versus q and choose the best number of dimensions, q^*, from an examination of this plot. (12-26)

We have assumed that the initial similarity values are symmetric $(s_{ik} = s_{ki})$, that there are no ties, and that there are no missing observations. Kruskal [19, 20] has suggested methods for handling asymmetries, ties, and missing observations. In addition, there are now multidimensional scaling computer programs that will handle not only Euclidean distance, but any distance of the Minkowski type. [See (12-3).]

The next three examples illustrate multidimensional scaling with distances as the initial (dis)similarity measures.

Example 12.14 (Multidimensional scaling of U.S. cities) Table 12.7 displays the airline distances between pairs of selected U.S. cities.

Table 12.7 Airline-Distance Data

	Atlanta (1)	Boston (2)	Cincinnati (3)	Columbus (4)	Dallas (5)	Indianapolis (6)	Little Rock (7)	Los Angeles (8)	Memphis (9)	St. Louis (10)	Spokane (11)	Tampa (12)
(1)	0											
(2)	1068	0										
(3)	461	867	0									
(4)	549	769	107	0								
(5)	805	1819	943	1050	0							
(6)	508	941	108	172	882	0						
(7)	505	1494	618	725	325	562	0					
(8)	2197	3052	2186	2245	1403	2080	1701	0				
(9)	366	1355	502	586	464	436	137	1831	0			
(10)	558	1178	338	409	645	234	353	1848	294	0		
(11)	2467	2747	2067	2131	1891	1959	1988	1227	2042	1820	0	
(12)	467	1379	928	985	1077	975	912	2480	779	1016	2821	0

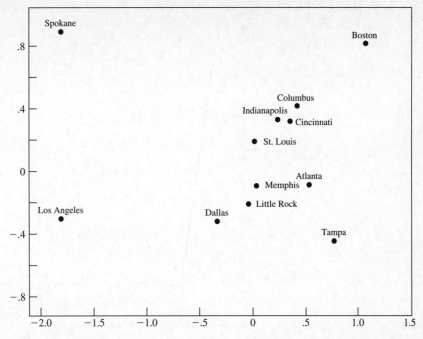

Figure 12.15 A geometrical representation of cities produced by multidimensional scaling.

Since the cities naturally lie in a two-dimensional space (a nearly level part of the curved surface of the earth), it is not surprising that multidimensional scaling with $q = 2$ will locate these items about as they occur on a map. Note that if the distances in the table are ordered from largest to smallest—that is, from a least similar to most similar—the first position is occupied by $d_{\text{Boston, L.A.}} = 3052$.

A multidimensional scaling plot for $q = 2$ dimensions is shown in Figure 12.15. The axes lie along the sample principal components of the scatter plot.

A plot of stress (q) versus q is shown in Figure 12.16 on page 712. Since stress $(1) \times 100\% = 12\%$, a representation of the cities in one dimension (along a single axis) is not unreasonable. The "elbow" of the stress function occurs at $q = 2$. Here stress $(2) \times 100\% = 0.8\%$, and the "fit" is almost perfect.

The plot in Figure 12.16 indicates that $q = 2$ is the best choice for the dimension of the final configuration. Note that the stress actually increases for $q = 3$. This anomaly can occur for extremely small values of stress because of difficulties with the numerical search procedure used to locate the minimum stress. ■

Example 12.15 (Multidimensional scaling of public utilities) Let us try to represent the 22 public utility firms discussed in Example 12.7 as points in a low-dimensional space. The measures of (dis)similarities between pairs of firms are the Euclidean distances listed in Table 12.6. Multidimensional scaling in $q = 1, 2, \ldots, 6$ dimensions produced the stress function shown in Figure 12.17.

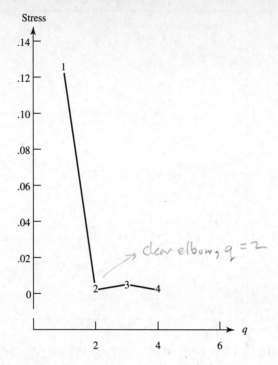

Figure 12.16 Stress function for airline distances between cities.

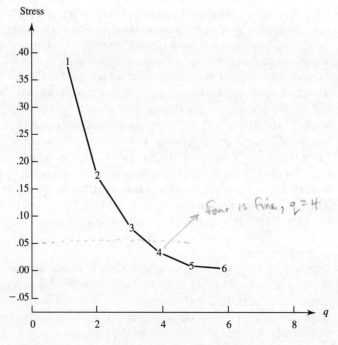

Figure 12.17 Stress function for distances between utilities.

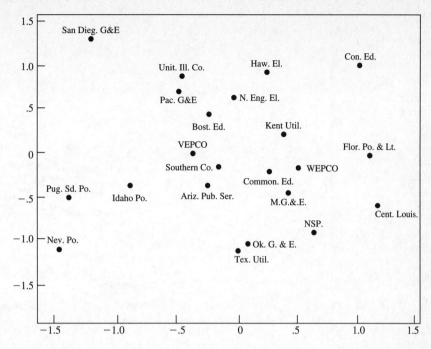

Figure 12.18 A geometrical representation of utilities produced by multidimensional scaling.

The stress function in Figure 12.17 has no sharp elbow. The plot appears to level out at "good" values of stress (less than or equal to 5%) in the neighborhood of $q = 4$. A good four-dimensional representation of the utilities is achievable, but difficult to display. We show a plot of the utility configuration obtained in $q = 2$ dimensions in Figure 12.18. The axes lie along the sample principal components of the final scatter.

Although the stress for two dimensions is rather high (stress (2) $\times$ 100% = 19%), the distances between firms in Figure 12.18 are not wildly inconsistent with the clustering results presented earlier in this chapter. For example, the midwest utilities—Commonwealth Edison, Wisconsin Electric Power (WEPCO), Madison Gas and Electric (MG & E), and Northern States Power (NSP)—are close together (similar). Texas Utilities and Oklahoma Gas and Electric (Ok. G & E) are also very close together (similar). Other utilities tend to group according to geographical locations or similar environments.

The utilities cannot be positioned in two dimensions such that the interutility distances $d_{ik}^{(2)}$ are entirely consistent with the original distances in Table 12.6. More flexibility for positioning the points is required, and this can only be obtained by introducing additional dimensions.   ■

Example 12.16 (Multidimensional scaling of universities) Data related to 25 U.S. universities are given in Table 12.9 on page 729. (See Example 12.19.) These data give the average SAT score of entering freshmen, percent of freshmen in top

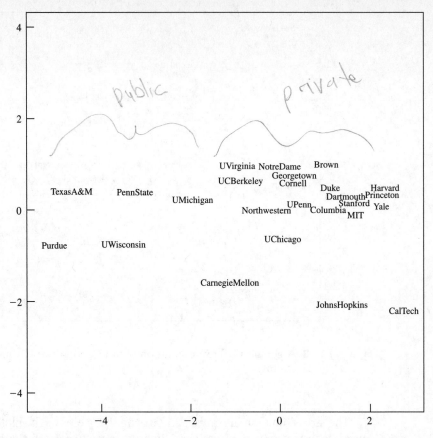

Figure 12.19 A two-dimensional representation of universities produced by metric multidimensional scaling.

10% of high school class, percent of applicants accepted, student–faculty ratio, estimated annual expense, and graduation rate (%). A metric multidimensional scaling algorithm applied to the standardized university data gives the two-dimensional representation shown in Figure 12.19. Notice how the private universities cluster on the right of the plot while the large public universities are, generally, on the left. A nonmetric multidimensional scaling two-dimensional configuration is shown in Figure 12.20. For this example, the metric and nonmetric scaling representations are very similar, with the two dimensional stress value being approximately 10% for both scalings. ∎

Classical metric scaling, or principal coordinate analysis, is equivalent to ploting the principal components. Different software programs choose the signs of the appropriate eigenvectors differently, so at first sight, two solutions may appear to be different. However, the solutions will coincide with a reflection of one or more of the axes. (See [26].)

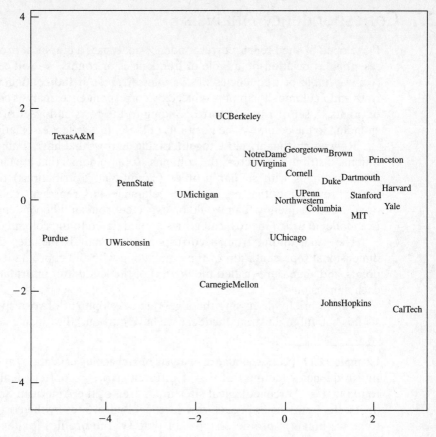

Figure 12.20 A two-dimensional representation of universities produced by nonmetric multidimensional scaling.

To summarize, the key objective of multidimensional scaling procedures is a low-dimensional picture. Whenever multivariate data can be presented graphically in two or three dimensions, visual inspection can greatly aid interpretations.

When the multivariate observations are naturally numerical, and Euclidean distances in p-dimensions, $d_{ik}^{(p)}$, can be computed, we can seek a $q < p$-dimensional representation by minimizing

$$E = \left[\sum_{i<k}\sum (d_{ik}^{(p)} - d_{ik}^{(q)})^2 / d_{ik}^{(p)} \right] \left[\sum_{i<k}\sum d_{ik}^{(p)} \right]^{-1} \qquad (12\text{-}27)$$

In this alternative approach, the Euclidean distances in p and q dimensions are compared directly. Techniques for obtaining low-dimensional representations by minimizing E are called *nonlinear mappings*.

The final goodness of fit of any low-dimensional representation can be depicted graphically by *minimal spanning trees*. (See [16] for a further discussion of these topics.)

12.7 Correspondence Analysis

Developed by the French, correspondence analysis is a graphical procedure for representing associations in a table of frequencies or counts. We will concentrate on a two-way table of frequencies or *contingency table*. If the contingency table has I rows and J columns, the plot produced by correspondence analysis contains two sets of points: A set of I points corresponding to the rows and a set of J points corresponding to the columns. The positions of the points reflect associations.

Row points that are close together indicate rows that have similar profiles (conditional distributions) across the columns. Column points that are close together indicate columns with similar profiles (conditional distributions) down the rows. Finally, row points that are close to column points represent combinations that occur more frequently than would be expected from an independence model—that is, a model in which the row categories are unrelated to the column categories.

The usual output from a correspondence analysis includes the "best" two-dimensional representation of the data, along with the coordinates of the plotted points, and a measure (called the *inertia*) of the amount of information retained in each dimension.

Before briefly discussing the algebraic development of correspondence analysis, it is helpful to illustrate the ideas we have introduced with an example.

Example 12.17 (Correspondence analysis of archaeological data) Table 12.8 contains the frequencies (counts) of $J = 4$ different types of pottery (called potsherds) found at $I = 7$ archaeological sites in an area of the American Southwest. If we divide the frequencies in each row (archaeological site) by the corresponding row total, we obtain a profile of types of pottery. The profiles for the different sites (rows) are shown in a bar graph in Figure 12.21(a). The widths of the bars are proportional to the total row frequencies. In general, the profiles are different; however, the profiles for sites P1 and P2 are similar, as are the profiles for sites P4 and P5.

The archaeological site profile for different types of pottery (columns) are shown in a bar graph in Figure 12.21(b). The site profiles are constructed using the

Table 12.8 Frequencies of Types of Pottery				
	Type			
Site	A	B	C	D	Total
P0	30	10	10	39	89
P1	53	4	16	2	75
P2	73	1	41	1	116
P3	20	6	1	4	31
P4	46	36	37	13	132
P5	45	6	59	10	120
P6	16	28	169	5	218
Total	283	91	333	74	781

Source: Data courtesy of M. J. Tretter.

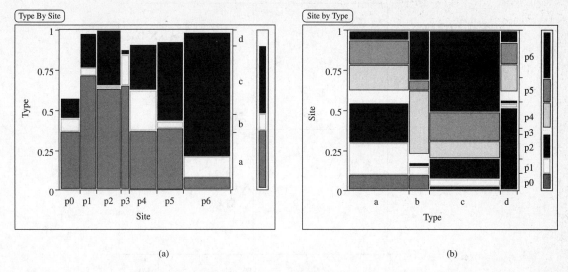

Figure 12.21 Site and pottery type profiles for the data in Table 12.8.

column totals. The bars in the figure appear to be quite different from one another. This suggests that the various types of pottery are not distributed over the archaeological sites in the same way.

The two-dimensional plot from a correspondence analysis[2] of the pottery type–site data is shown in Figure 12.22.

The plot in Figure 12.22 indicates, for example, that sites P1 and P2 have similar pottery type profiles (the two points are close together), and sites P0 and P6 have very different profiles (the points are far apart). The individual points representing the types of pottery are spread out, indicating that their archaeological site profiles are quite different. These findings are consistent with the profiles pictured in Figure 12.21.

Notice that the points P0 and D are quite close together and separated from the remaining points. This indicates that pottery type D tends to be associated, almost exclusively, with site P0. Similarly, pottery type A tends to be associated with site P1 and, to lesser degrees, with sites P2 and P3. Pottery type B is associated with sites P4 and P5, and pottery type C tends to be associated, again, almost exclusively, with site P6. Since the archaeological sites represent different periods, these associations are of considerable interest to archaeologists.

The number $\lambda_1^2 = .28$ at the end of the first coordinate axis in the two-dimensional plot is the *inertia* associated with the first dimension. This inertia is 55% of the total inertia. The inertia associated with the second dimension is $\lambda_2^2 = .17$, and the second dimension accounts for 33% of the total inertia. Together, the two dimensions account for 55% + 33% = 88% of the total inertia. Since, in this case, the data could be exactly represented in three dimensions, relatively little information (variation) is lost by representing the data in the two-dimensional plot of Figure 12.22. Equivalently, we may regard this plot as the best two-dimensional representation of the multidimensional scatter of row points and the multidimensional

[2]The JMP software was used for a correspondence analysis of the data in Table 12.8.

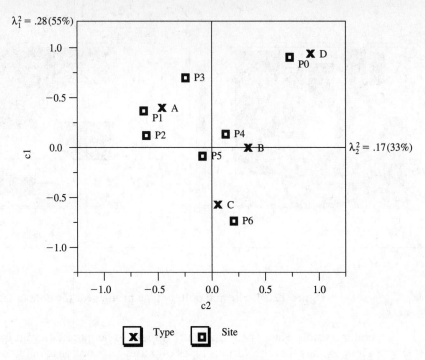

Figure 12.22 A correspondence analysis plot of the pottery type–site data.

scatter of column points. The combined inertia of 88% suggests that the representation "fits" the data well.

In this example, the graphical output from a correspondence analysis shows the nature of the associations in the contingency table quite clearly. ■

Algebraic Development of Correspondence Analysis

To begin, let $\mathbf{X}$, with elements x_{ij}, be an $I \times J$ two-way table of unscaled frequencies or counts. In our discussion we take $I > J$ and assume that $\mathbf{X}$ is of full column rank J. The rows and columns of the contingency table $\mathbf{X}$ correspond to different categories of two different characteristics. As an example, the array of frequencies of different pottery types at different archaeological sites shown in Table 12.8 is a contingency table with $I = 7$ archaeological sites and $J = 4$ pottery types.

If n is the total of the frequencies in the data matrix $\mathbf{X}$, we first construct a matrix of proportions $\mathbf{P} = \{p_{ij}\}$ by dividing each element of $\mathbf{X}$ by n. Hence

$$p_{ij} = \frac{x_{ij}}{n}, \quad i = 1, 2, \ldots, I, \quad j = 1, 2, \ldots, J, \quad \text{or} \quad \underset{(I \times J)}{\mathbf{P}} = \frac{1}{n} \underset{(I \times J)}{\mathbf{X}} \quad (12\text{-}28)$$

The matrix $\mathbf{P}$ is called the *correspondence matrix*.

Next define the vectors of row and column sums $\mathbf{r}$ and $\mathbf{c}$ respectively, and the diagonal matrices $\mathbf{D}_r$ and $\mathbf{D}_c$ with the elements of $\mathbf{r}$ and $\mathbf{c}$ on the diagonals. Thus

$$r_i = \sum_{j=1}^{J} p_{ij} = \sum_{j=1}^{J} \frac{x_{ij}}{n}, \qquad i = 1, 2, \ldots, I, \quad \text{or} \quad \underset{(I \times 1)}{\mathbf{r}} = \underset{(I \times J)(J \times 1)}{\mathbf{P} \, \mathbf{1}_J}$$

$$c_j = \sum_{i=1}^{I} p_{ij} = \sum_{i=1}^{I} \frac{x_{ij}}{n}, \qquad j = 1, 2, \ldots, J, \quad \text{or} \quad \underset{(J \times 1)}{\mathbf{c}} = \underset{(J \times I)(I \times 1)}{\mathbf{P}' \, \mathbf{1}_I}$$

(12-29)

where $\mathbf{1}_J$ is a $J \times 1$ and $\mathbf{1}_I$ is a $I \times 1$ vector of 1's and

$$\mathbf{D}_r = \text{diag}\,(r_1, r_2, \ldots, r_I) \quad \text{and} \quad \mathbf{D}_c = \text{diag}\,(c_1, c_2, \ldots, c_J) \qquad (12\text{-}30)$$

We define the square root matrices

$$\mathbf{D}_r^{1/2} = \text{diag}\,(\sqrt{r_1}, \ldots, \sqrt{r_I}) \qquad \mathbf{D}_r^{-1/2} = \text{diag}\left(\frac{1}{\sqrt{r_1}}, \ldots, \frac{1}{\sqrt{r_I}}\right)$$

$$\mathbf{D}_c^{1/2} = \text{diag}\,(\sqrt{c_1}, \ldots, \sqrt{c_J}) \qquad \mathbf{D}_c^{-1/2} = \text{diag}\left(\frac{1}{\sqrt{c_1}}, \ldots, \frac{1}{\sqrt{c_J}}\right)$$

(12-31)

for scaling purposes.

Correspondence analysis can be formulated as the weighted least squares problem to select $\hat{\mathbf{P}} = \{\hat{p}_{ij}\}$, a matrix of specified reduced rank, to minimize

$$\sum_{i=1}^{I} \sum_{j=1}^{J} \frac{(p_{ij} - \hat{p}_{ij})^2}{r_i c_j} = \text{tr}\,[(\mathbf{D}_r^{-1/2}(\mathbf{P} - \hat{\mathbf{P}})\mathbf{D}_c^{-1/2})\,(\mathbf{D}_r^{-1/2}(\mathbf{P} - \hat{\mathbf{P}})\mathbf{D}_c^{-1/2})'] \quad (12\text{-}32)$$

since $(p_{ij} - \hat{p}_{ij})/\sqrt{r_i c_j}$ is the (i, j) element of $\mathbf{D}_r^{-1/2}(\mathbf{P} - \hat{\mathbf{P}})\mathbf{D}_c^{-1/2}$.

As Result 12.1 demonstrates, the term $\mathbf{rc}'$ is common to the approximation $\hat{\mathbf{P}}$ whatever the $I \times J$ correspondence matrix $\mathbf{P}$. The matrix $\hat{\mathbf{P}} = \mathbf{rc}'$ can be shown to be the best rank 1 approximation to $\mathbf{P}$.

Result 12.1. The term $\mathbf{rc}'$ is common to the approximation $\hat{\mathbf{P}}$ whatever the $I \times J$ correspondence matrix $\mathbf{P}$.

The reduced rank s approximation to $\mathbf{P}$, which minimizes the sum of squares (12-32), is given by

$$\mathbf{P} \doteq \sum_{k=1}^{s} \widetilde{\lambda}_k (\mathbf{D}_r^{1/2}\,\widetilde{\mathbf{u}}_k)\,(\mathbf{D}_c^{1/2}\,\widetilde{\mathbf{v}}_k)' = \mathbf{rc}' + \sum_{k=2}^{s} \widetilde{\lambda}_k (\mathbf{D}_r^{1/2}\,\widetilde{\mathbf{u}}_k)\,(\mathbf{D}_c^{1/2}\,\widetilde{\mathbf{v}}_k)'$$

where the $\widetilde{\lambda}_k$ are the singular values and the $I \times 1$ vectors $\widetilde{\mathbf{u}}_k$ and the $J \times 1$ vectors $\widetilde{\mathbf{v}}_k$ are the corresponding singular vectors of the $I \times J$ matrix $\mathbf{D}_r^{-1/2}\mathbf{P}\mathbf{D}_c^{-1/2}$. The minimum value of (12-32) is $\sum_{k=s+1}^{J} \widetilde{\lambda}_k^2$.

The reduced rank $K > 1$ approximation to $\mathbf{P} - \mathbf{rc}'$ is

$$\mathbf{P} - \mathbf{rc}' \doteq \sum_{k=1}^{K} \lambda_k (\mathbf{D}_r^{1/2}\mathbf{u}_k)\,(\mathbf{D}_c^{1/2}\mathbf{v}_k)' \qquad (12\text{-}33)$$

where the λ_k are the singular values and the $I \times 1$ vectors $\mathbf{u}_k$ and the $J \times 1$ vectors $\mathbf{v}_k$ are the corresponding singular vectors of the $I \times J$ matrix $\mathbf{D}_r^{-1/2}(\mathbf{P} - \mathbf{rc}')\mathbf{D}_c^{-1/2}$. Here $\lambda_k = \widetilde{\lambda}_{k+1}$, $\mathbf{u}_k = \widetilde{\mathbf{u}}_{k+1}$, and $\mathbf{v}_k = \widetilde{\mathbf{v}}_{k+1}$ for $k = 1, \ldots, J - 1$.

Proof. We first consider a scaled version $\mathbf{B} = \mathbf{D}_r^{-1/2}\mathbf{P}\mathbf{D}_c^{-1/2}$ of the correspondence matrix $\mathbf{P}$. According to Result 2A.16, the best low rank $= s$ approximation $\hat{\mathbf{B}}$ to $\mathbf{D}_r^{-1/2}\mathbf{P}\mathbf{D}_c^{-1/2}$ is given by the first s terms in the the singular-value decomposition

$$\mathbf{D}_r^{-1/2}\mathbf{P}\mathbf{D}_c^{-1/2} = \sum_{k=1}^{J} \widetilde{\lambda}_k \widetilde{\mathbf{u}}_k \widetilde{\mathbf{v}}_k' \tag{12-34}$$

where

$$\mathbf{D}_r^{-1/2}\mathbf{P}\mathbf{D}_c^{-1/2}\widetilde{\mathbf{v}}_k = \widetilde{\lambda}_k \widetilde{\mathbf{u}}_k \qquad \widetilde{\mathbf{u}}_k'\mathbf{D}_r^{-1/2}\mathbf{P}\mathbf{D}_c^{-1/2} = \widetilde{\lambda}_k \widetilde{\mathbf{v}}_k' \tag{12-35}$$

and

$$|(\mathbf{D}_r^{-1/2}\mathbf{P}\mathbf{D}_c^{-1/2})(\mathbf{D}_r^{-1/2}\mathbf{P}\mathbf{D}_c^{-1/2})' - \widetilde{\lambda}_k^2\mathbf{I}| = 0 \quad \text{for} \quad k = 1, \ldots, J$$

The approximation to $\mathbf{P}$ is then given by

$$\hat{\mathbf{P}} = \mathbf{D}_r^{1/2}\hat{\mathbf{B}}\mathbf{D}_c^{1/2} \doteq \sum_{k=1}^{s} \widetilde{\lambda}_k (\mathbf{D}_r^{1/2}\widetilde{\mathbf{u}}_k)(\mathbf{D}_c^{1/2}\widetilde{\mathbf{v}}_k)'$$

and, by Result 2A.16, the error of approximation is $\displaystyle\sum_{k=s+1}^{J} \widetilde{\lambda}_k^2$.

Whatever the correspondence matrix $\mathbf{P}$, the term $\mathbf{rc}'$ always provides a (the best) rank one approximation. This corresponds to the assumption of independence of the rows and columns. To see this, let $\widetilde{\mathbf{u}}_1 = \mathbf{D}_r^{1/2}\mathbf{1}_I$ and $\widetilde{\mathbf{v}}_1 = \mathbf{D}_c^{1/2}\mathbf{1}_J$, where $\mathbf{1}_I$ is a $I \times 1$ and $\mathbf{1}_J$ a $J \times 1$ vector of 1's. We verify that (12-35) holds for these choices.

$$\widetilde{\mathbf{u}}_1'(\mathbf{D}_r^{-1/2}\mathbf{P}\mathbf{D}_c^{-1/2}) = (\mathbf{D}_r^{1/2}\mathbf{1}_I)'(\mathbf{D}_r^{-1/2}\mathbf{P}\mathbf{D}_c^{-1/2})$$

$$= \mathbf{1}_I'\mathbf{P}\mathbf{D}_c^{-1/2} = \mathbf{c}'\mathbf{D}_c^{-1/2}$$

$$= [\sqrt{c_1}, \ldots, \sqrt{c_J}] = (\mathbf{D}_c^{1/2}\mathbf{1}_J)' = \widetilde{\mathbf{v}}_1'$$

and

$$(\mathbf{D}_r^{-1/2}\mathbf{P}\mathbf{D}_c^{-1/2})\widetilde{\mathbf{v}}_1 = (\mathbf{D}_r^{-1/2}\mathbf{P}\mathbf{D}_c^{-1/2})(\mathbf{D}_c^{1/2}\mathbf{1}_J)$$

$$= \mathbf{D}_r^{-1/2}\mathbf{P}\mathbf{1}_J = \mathbf{D}_r^{-1/2}\mathbf{r}$$

$$= \begin{bmatrix} \sqrt{r_1} \\ \vdots \\ \sqrt{r_I} \end{bmatrix} = \mathbf{D}_r^{1/2}\mathbf{1}_I = \widetilde{\mathbf{u}}_1$$

That is,

$$(\widetilde{\mathbf{u}}_1, \widetilde{\mathbf{v}}_1) = (\mathbf{D}_r^{1/2}\mathbf{1}_I, \mathbf{D}_c^{1/2}\mathbf{1}_J) \tag{12-36}$$

are singular vectors associated with singular value $\widetilde{\lambda}_1 = 1$. For any correspondence matrix, $\mathbf{P}$, the common term in every expansion is

$$\mathbf{D}_r^{1/2}\mathbf{u}_1\mathbf{v}_1'\mathbf{D}_c^{1/2} = \mathbf{D}_r\mathbf{1}_I\mathbf{1}_J'\mathbf{D}_c = \mathbf{rc}'$$

Therefore, we have established the first approximation and (12-34) can always be expressed as

$$\mathbf{P} = \mathbf{rc}' + \sum_{k=2}^{J} \tilde{\lambda}_k (\mathbf{D}_r^{1/2} \tilde{\mathbf{u}}_k)(\mathbf{D}_c^{1/2} \tilde{\mathbf{v}}_k)'$$

Because of the common term, the problem can be rephrased in terms of $\mathbf{P} - \mathbf{rc}'$ and its scaled version $\mathbf{D}_r^{-1/2}(\mathbf{P} - \mathbf{rc}')\mathbf{D}_c^{-1/2}$. By the orthogonality of the singular vectors of $\mathbf{D}_r^{-1/2}\mathbf{P}\mathbf{D}_c^{-1/2}$, we have $\tilde{\mathbf{u}}_k'(\mathbf{D}_r^{1/2}\mathbf{1}_I) = 0$ and $\tilde{\mathbf{v}}_k'(\mathbf{D}_c^{1/2}\mathbf{1}_J) = 0$, for $k > 1$, so

$$\mathbf{D}_r^{-1/2}(\mathbf{P} - \mathbf{rc}')\mathbf{D}_c^{-1/2} = \sum_{k=2}^{J} \tilde{\lambda}_k \tilde{\mathbf{u}}_k \tilde{\mathbf{v}}_k'$$

is the singular-value decomposition of $\mathbf{D}_r^{-1/2}(\mathbf{P} - \mathbf{rc}')\mathbf{D}_c^{-1/2}$ in terms of the singular values and vectors obtained from $\mathbf{D}_r^{-1/2}\mathbf{P}\mathbf{D}_c^{-1/2}$. Converting to singular values and vectors λ_k, $\mathbf{u}_k$, and $\mathbf{v}_k$ from $\mathbf{D}_r^{-1/2}(\mathbf{P} - \mathbf{rc}')\mathbf{D}_c^{-1/2}$ only amounts to changing k to $k - 1$ so $\lambda_k = \lambda_{k+1}$, $\mathbf{u}_k = \tilde{\mathbf{u}}_{k+1}$, and $\mathbf{v}_k = \tilde{\mathbf{v}}_{k+1}$ for $k = 1, \ldots, J - 1$.

In terms of the singular value decomposition for $\mathbf{D}_r^{-1/2}(\mathbf{P} - \mathbf{rc}')\mathbf{D}_c^{-1/2}$, the expansion for $\mathbf{P} - \mathbf{rc}'$ takes the form

$$\mathbf{P} - \mathbf{rc}' = \sum_{k=1}^{J-1} \lambda_k (\mathbf{D}_r^{1/2}\mathbf{u}_k)(\mathbf{D}_c^{1/2}\mathbf{v}_k)' \qquad (12\text{-}37)$$

The best rank K approximation to $\mathbf{D}_r^{-1/2}(\mathbf{P} - \mathbf{rc}')\mathbf{D}_c^{-1/2}$ is given by $\sum_{k=1}^{K} \lambda_k \mathbf{u}_k \mathbf{v}_k'$. Then, the best approximation to $\mathbf{P} - \mathbf{rc}'$ is

$$\mathbf{P} - \mathbf{rc}' \doteq \sum_{k=1}^{K} \lambda_k (\mathbf{D}_r^{1/2}\mathbf{u}_k)(\mathbf{D}_c^{1/2}\mathbf{v}_k)' \qquad (12\text{-}38)$$

∎

Remark. Note that the vectors $\mathbf{D}_r^{1/2}\mathbf{u}_k$ and $\mathbf{D}_c^{1/2}\mathbf{v}_k$ in the expansion (12-38) of $\mathbf{P} - \mathbf{rc}'$ need not have length 1 but satisfy the scaling

$$(\mathbf{D}_r^{1/2}\mathbf{u}_k)'\mathbf{D}_r^{-1}(\mathbf{D}_r^{1/2}\mathbf{u}_k) = \mathbf{u}_k'\mathbf{u}_k = 1$$

$$(\mathbf{D}_c^{1/2}\mathbf{v}_k)'\mathbf{D}_c^{-1}(\mathbf{D}_c^{1/2}\mathbf{v}_k) = \mathbf{v}_k'\mathbf{v}_k = 1$$

Because of this scaling, the expansions in Result 12.1 have been called a generalized singular-value decomposition.

Let Λ, $\mathbf{U} = [\mathbf{u}_1, \ldots, \mathbf{u}_I]$ and $\mathbf{V} = [\mathbf{v}_1, \ldots, \mathbf{v}_J]$ be the matricies of singular values and vectors obtained from $\mathbf{D}_r^{-1/2}(\mathbf{P} - \mathbf{rc}')\mathbf{D}_c^{-1/2}$. It is usual in correspondence analysis to plot the first two or three columns of $\mathbf{F} = \mathbf{D}_r^{-1}(\mathbf{D}_r^{1/2}\mathbf{U})\Lambda$ and $\mathbf{G} = \mathbf{D}_c^{-1}(\mathbf{D}_c^{1/2}\mathbf{V})\Lambda$ or $\lambda_k \mathbf{D}_r^{-1/2}\mathbf{u}_k$ and $\lambda_k \mathbf{D}_c^{-1/2}\mathbf{v}_k$ for $k = 1$, 2, and maybe 3.

The joint plot of the coordinates in $\mathbf{F}$ and $\mathbf{G}$ is called a *symmetric map* (see Greenacre [13]) since the points representing the rows and columns have the same normalization, or scaling, along the dimensions of the solution. That is, the geometry for the row points is identical to the geometry for the column points.

Example 12.18 (Calculations for correspondence analysis) Consider the 3×2 contingency table

	B1	B2	Total
A1	24	12	36
A2	16	48	64
A3	60	40	100
	100	100	200

The correspondence matrix is

$$
\mathbf{P} = \begin{bmatrix} .12 & .06 \\ .08 & .24 \\ .30 & .20 \end{bmatrix}
$$

with marginal totals $\mathbf{c}' = [.5, .5]$ and $\mathbf{r}' = [.18, .32, .50]$. The negative square root matrices are

$$
\mathbf{D}_r^{-1/2} = \operatorname{diag}\left(\sqrt{2}/.6, \sqrt{2}/.8, \sqrt{2}\right) \qquad \mathbf{D}_c^{-1/2} = \operatorname{diag}\left(\sqrt{2}, \sqrt{2}\right)
$$

Then

$$
\mathbf{P} - \mathbf{rc}' = \begin{bmatrix} .12 & .06 \\ .08 & .24 \\ .30 & .20 \end{bmatrix} - \begin{bmatrix} .18 \\ .32 \\ .50 \end{bmatrix} [.5 \quad .5] = \begin{bmatrix} .03 & -.03 \\ -.08 & .08 \\ .05 & -.05 \end{bmatrix}
$$

The scaled version of this matrix is

$$
\mathbf{A} = \mathbf{D}_r^{-1/2}(\mathbf{P} - \mathbf{rc}')\mathbf{D}_c^{-1/2} = \begin{bmatrix} \dfrac{\sqrt{2}}{.6} & 0 & 0 \\ 0 & \dfrac{\sqrt{2}}{.8} & 0 \\ 0 & 0 & \sqrt{2} \end{bmatrix} \begin{bmatrix} .03 & -.03 \\ -.08 & .08 \\ .05 & -.05 \end{bmatrix} \begin{bmatrix} \sqrt{2} & 0 \\ 0 & \sqrt{2} \end{bmatrix}
$$

$$
= \begin{bmatrix} 0.1 & -0.1 \\ -0.2 & 0.2 \\ 0.1 & -0.1 \end{bmatrix}
$$

Since $I > J$, the square of the singular values and the $\mathbf{v}_i$ are determined from

$$
\mathbf{A}'\mathbf{A} = \begin{bmatrix} .1 & -.2 & .1 \\ -.1 & .2 & -.1 \end{bmatrix} \begin{bmatrix} .1 & -.1 \\ -.2 & .2 \\ .1 & -.1 \end{bmatrix} = \begin{bmatrix} .06 & -.06 \\ -.06 & .06 \end{bmatrix}
$$

It is easily checked that $\lambda_1^2 = .12$, $\lambda_2^2 = 0$, since $J - 1 = 1$, and that

$$
\mathbf{v}_1 = \begin{bmatrix} \dfrac{1}{\sqrt{2}} \\ \dfrac{-1}{\sqrt{2}} \end{bmatrix}
$$

Further,

$$
\mathbf{A}\mathbf{A}' = \begin{bmatrix} .1 & -.1 \\ -.2 & .2 \\ .1 & -.1 \end{bmatrix} \begin{bmatrix} .1 & -.2 & .1 \\ -.1 & .2 & -.1 \end{bmatrix} = \begin{bmatrix} .02 & -.04 & .02 \\ -.04 & .08 & -.04 \\ .02 & -.04 & .02 \end{bmatrix}
$$

A computer calculation confirms that the single nonzero eigenvalue is $\lambda_1^2 = .12$, so that the singular value has absolute value $\lambda_1 = .2\sqrt{3}$ and, as you can easily check,

$$
\mathbf{u}_1 = \begin{bmatrix} \dfrac{1}{\sqrt{6}} \\ -\dfrac{2}{\sqrt{6}} \\ \dfrac{1}{\sqrt{6}} \end{bmatrix}
$$

The expansion of $\mathbf{P} - \mathbf{r}\mathbf{c}'$ is then the single term

$$
\lambda_1 (\mathbf{D}_r^{1/2}\mathbf{u}_1)(\mathbf{D}_c^{1/2}\mathbf{v}_1)'
$$

$$
= \sqrt{.12} \begin{bmatrix} \dfrac{.6}{\sqrt{2}} & 0 & 0 \\ 0 & \dfrac{.8}{\sqrt{2}} & 0 \\ 0 & 0 & \dfrac{1}{\sqrt{2}} \end{bmatrix} \begin{bmatrix} \dfrac{1}{\sqrt{6}} \\ -\dfrac{2}{\sqrt{6}} \\ \dfrac{1}{\sqrt{6}} \end{bmatrix} \begin{bmatrix} \dfrac{1}{\sqrt{2}} & \dfrac{-1}{\sqrt{2}} \end{bmatrix} \begin{bmatrix} \dfrac{1}{\sqrt{2}} & 0 \\ 0 & \dfrac{1}{\sqrt{2}} \end{bmatrix}
$$

$$
= \sqrt{.12} \begin{bmatrix} \dfrac{.3}{\sqrt{3}} \\ -\dfrac{.8}{\sqrt{3}} \\ \dfrac{.5}{\sqrt{3}} \end{bmatrix} \begin{bmatrix} \dfrac{1}{2} & \dfrac{-1}{2} \end{bmatrix} = \begin{bmatrix} .03 & -.03 \\ -.08 & .08 \\ .05 & -.05 \end{bmatrix} \quad \text{check}
$$

There is only one pair of vectors to plot

$$\lambda_1 \mathbf{D}_r^{1/2} \mathbf{u}_1 = \sqrt{.12} \begin{bmatrix} \dfrac{.6}{\sqrt{2}} & 0 & 0 \\ 0 & \dfrac{.8}{\sqrt{2}} & 0 \\ 0 & 0 & \dfrac{1}{\sqrt{2}} \end{bmatrix} \begin{bmatrix} \dfrac{1}{\sqrt{6}} \\ -\dfrac{2}{\sqrt{6}} \\ \dfrac{1}{\sqrt{6}} \end{bmatrix} = \sqrt{.12} \begin{bmatrix} \dfrac{.3}{\sqrt{3}} \\ -\dfrac{.8}{\sqrt{3}} \\ \dfrac{.5}{\sqrt{3}} \end{bmatrix}$$

and

$$\lambda_1 \mathbf{D}_c^{1/2} \mathbf{v}_1 = \sqrt{.12} \begin{bmatrix} \dfrac{1}{\sqrt{2}} & 0 \\ 0 & \dfrac{1}{\sqrt{2}} \end{bmatrix} \begin{bmatrix} \dfrac{1}{\sqrt{2}} \\ \dfrac{-1}{\sqrt{2}} \end{bmatrix} = \sqrt{.12} \begin{bmatrix} \dfrac{1}{2} \\ -\dfrac{1}{2} \end{bmatrix} \qquad \blacksquare$$

There is a second way to define contingency analysis. Following Greenacre [13], we call the preceding approach the *matrix approximation method* and the approach to follow the *profile approximation method*. We illustrate the profile approximation method using the row profiles; however, an analogous solution results if we were to begin with the column profiles.

Algebraically, the row profiles are the rows of the matrix $\mathbf{D}_r^{-1}\mathbf{P}$, and contingency analysis can be defined as the approximation of the row profiles by points in a low-dimensional space. Consider approximating the row profiles by the matrix $\mathbf{P}^*$. Using the square-root matrices $\mathbf{D}_r^{1/2}$ and $\mathbf{D}_c^{1/2}$ defined in (12-31), we can write

$$(\mathbf{D}_r^{-1}\mathbf{P} - \mathbf{P}^*)\mathbf{D}_c^{-1/2} = \mathbf{D}_r^{-1/2}(\mathbf{D}_r^{-1/2}\mathbf{P} - \mathbf{D}_r^{1/2}\mathbf{P}^*)\mathbf{D}_c^{-1/2}$$

and the least squares criterion (12-32) can be written, with $p_{ij}^* = \hat{p}_{ij}/r_i$, as

$$\sum_i \sum_j \frac{(p_{ij} - \hat{p}_{ij})^2}{r_i c_j} = \sum_i r_i \sum_j \frac{(p_{ij}/r_i - p_{ij}^*)^2}{c_j}$$
$$= \text{tr}\big[\mathbf{D}_r^{1/2}\mathbf{D}_r^{1/2}(\mathbf{D}_r^{-1}\mathbf{P} - \mathbf{P}^*)\mathbf{D}_c^{-1/2}\mathbf{D}_c^{-1/2}(\mathbf{D}_r^{-1}\mathbf{P} - \mathbf{P}^*)'\big]$$
$$= \text{tr}\big[\mathbf{D}_r^{1/2}(\mathbf{D}_r^{-1/2}\mathbf{P} - \mathbf{D}_r^{1/2}\mathbf{P}^*)\mathbf{D}_c^{-1/2}\mathbf{D}_c^{-1/2}(\mathbf{D}_r^{-1/2}\mathbf{P} - \mathbf{D}_r^{1/2}\mathbf{P}^*)'\mathbf{D}_r^{-1/2}\big]$$
$$= \text{tr}\big[[(\mathbf{D}_r^{-1/2}\mathbf{P} - \mathbf{D}_r^{1/2}\mathbf{P}^*)\mathbf{D}_c^{-1/2}][(\mathbf{D}_r^{-1/2}\mathbf{P} - \mathbf{D}_r^{1/2}\mathbf{P}^*)\mathbf{D}_c^{-1/2}]'\big] \qquad (12\text{-}39)$$

Minimizing the last expression for the trace in (12-39) is precisely the first minimization problem treated in the proof of Result 12.1. By (12-34), $\mathbf{D}_r^{-1/2}\mathbf{P}\mathbf{D}_c^{-1/2}$ has the singular-value decomposition

$$\mathbf{D}_r^{-1/2}\mathbf{P}\mathbf{D}_c^{-1/2} = \sum_{k=1}^{J} \widetilde{\lambda}_k \widetilde{\mathbf{u}}_k \widetilde{\mathbf{v}}_k' \qquad (12\text{-}40)$$

The best rank K approximation is obtained by using the first K terms of this expansion. Since, by (12-39), we have $\mathbf{D}_r^{-1/2}\mathbf{P}\mathbf{D}_c^{1/2}$ approximated by $\mathbf{D}_r^{1/2}\mathbf{P}^*\mathbf{D}_c^{-1/2}$, we left multiply by $\mathbf{D}_r^{-1/2}$ and right multiply by $\mathbf{D}_c^{1/2}$ to obtain the generalized singular-value decomposition

$$\mathbf{D}_r^{-1}\mathbf{P} = \sum_{k=1}^{J} \tilde{\lambda}_k \mathbf{D}_r^{-1/2}\tilde{\mathbf{u}}_k (\mathbf{D}_c^{1/2}\tilde{\mathbf{v}}_k)' \tag{12-41}$$

where, from (12-36), $(\tilde{\mathbf{u}}_1, \tilde{\mathbf{v}}_1) = (\mathbf{D}_r^{1/2}\mathbf{1}_I, \mathbf{D}_c^{1/2}\mathbf{1}_J)$ are singular vectors associated with singular value $\lambda_1 = 1$. Since $\mathbf{D}_r^{-1/2}(\mathbf{D}_r^{1/2}\mathbf{1}_I) = \mathbf{1}_I$ and $(\mathbf{D}_c^{1/2}\mathbf{1}_J)'\mathbf{D}_c^{1/2} = \mathbf{c}'$, the leading term in the decomposition (12-41) is $\mathbf{1}_I\mathbf{c}'$.

Consequently, in terms of the singular values and vectors from $\mathbf{D}_r^{-1/2}\mathbf{P}\mathbf{D}_c^{-1/2}$, the reduced rank $K < J$ approximation to the row profiles $\mathbf{D}_r^{-1}\mathbf{P}$ is

$$\mathbf{P}^* \doteq \mathbf{1}_I\mathbf{c}' + \sum_{k=2}^{K} \tilde{\lambda}_k \mathbf{D}_r^{-1/2}\tilde{\mathbf{u}}_k (\mathbf{D}_c^{1/2}\tilde{\mathbf{v}}_k)' \tag{12-42}$$

In terms of the singular values and vectors $\lambda_k, \mathbf{u}_k$ and $\mathbf{v}_k$ obtained from $\mathbf{D}_r^{-1/2}(\mathbf{P} - \mathbf{rc}')\mathbf{D}_c^{-1/2}$, we can write

$$\mathbf{P}^* - \mathbf{1}_I\mathbf{c}' \doteq \sum_{k=1}^{K-1} \lambda_k \mathbf{D}_r^{-1/2}\mathbf{u}_k (\mathbf{D}_c^{1/2}\mathbf{v}_k)'$$

(Row profiles for the archaeological data in Table 12.8 are shown in Figure 12.21 on page 717.)

Inertia

Total inertia is a measure of the variation in the count data and is defined as the weighted sum of squares

$$\text{tr}\left[\mathbf{D}_r^{-1/2}(\mathbf{P} - \mathbf{rc}')\mathbf{D}_c^{-1/2}(\mathbf{D}_r^{-1/2}(\mathbf{P} - \mathbf{rc}')\mathbf{D}_c^{-1/2})'\right] = \sum_i \sum_j \frac{(p_{ij} - r_ic_j)^2}{r_ic_j} = \sum_{k=1}^{J-1} \lambda_k^2 \tag{12-43}$$

where the λ_k are the singular values obtained from the singular-value decomposition of $\mathbf{D}_r^{-1/2}(\mathbf{P} - \mathbf{rc}')\mathbf{D}_c^{-1/2}$ (see the proof of Result 12.1).[3]

The inertia associated with the best reduced rank $K < J$ approximation to the centered matrix $\mathbf{P} - \mathbf{rc}'$ (the K-dimensional solution) has inertia $\sum_{k=1}^{K} \lambda_k^2$. The residual inertia (variation) not accounted for by the rank K solution is equal to the sum of squares of the remaining singular values: $\lambda_{K+1}^2 + \lambda_{K+2}^2 + \ldots + \lambda_{J-1}^2$. For plots, the inertia associated with dimension k, λ_k^2, is ordinarily displayed along the kth coordinate axis, as in Figure 12.22 for $k = 1, 2$.

[3]Total inertia is related to the chi-square measure of association in a two-way contingency table,

$\chi^2 = \sum_{i,j} \frac{(O_{ij} - E_{ij})^2}{E_{ij}}$. Here $O_{ij} = x_{ij}$ is the observed frequency and E_{ij} is the expected frequency for

the ijth cell. In our context, if the row variable is independent of (unrelated to) the column variable, $E_{ij} = nr_ic_j$, and

$$\text{Total inertia} = \sum_{i=1}^{I} \sum_{j=1}^{J} \frac{(p_{ij} - r_ic_j)^2}{r_ic_j} = \frac{\chi^2}{n}$$

Interpretation in Two Dimensions

Since the inertia is a measure of the data table's total variation, how do we interpret a large value for the proportion $(\lambda_1^2 + \lambda_2^2)/\sum_{k=1}^{J-1} \lambda_k^2$? Geometrically, we say that the associations in the centered data are well represented by points in a plane, and this best approximating plane accounts for nearly all the variation in the data beyond that accounted for by the rank 1 solution (independence model). Algebraically, we say that the approximation

$$\mathbf{P} - \mathbf{rc}' \doteq \lambda_1 \mathbf{u}_1 \mathbf{v}_1' + \lambda_2 \mathbf{u}_2 \mathbf{v}_2'$$

is very good or, equivalently, that

$$\mathbf{P} \doteq \mathbf{rc}' + \lambda_1 \mathbf{u}_1 \mathbf{v}_1' + \lambda_2 \mathbf{u}_2 \mathbf{v}_2'$$

Final Comments

Correspondence analysis is primarily a graphical technique designed to represent associations in a low-dimensional space. It can be regarded as a scaling method, and can be viewed as a complement to other methods such as multidimensional scaling (Section 12.6) and biplots (Section 12.8). Correspondence analysis also has links to principal component analysis (Chapter 8) and canonical correlation analysis (Chapter 10). The book by Greenacre [14] is one choice for learning more about correspondence analysis.

12.8 Biplots for Viewing Sampling Units and Variables

A *biplot* is a graphical representation of the information in an $n \times p$ data matrix. The *bi-* refers to the two kinds of information contained in a data matrix. The information in the rows pertains to samples or sampling units and that in the columns pertains to variables.

When there are only two variables, scatter plots can represent the information on both the sampling units and the variables in a single diagram. This permits the visual inspection of the position of one sampling unit relative to another and the relative importance of each of the two variables to the position of any unit.

With several variables, one can construct a matrix array of scatter plots, but there is no one single plot of the sampling units. On the other hand, a two-dimensional plot of the sampling units can be obtained by graphing the first two principal components, as in Section 8.4. The idea behind biplots is to add the information about the variables to the principal component graph.

Figure 12.23 gives an example of a biplot for the public utilities data in Table 12.4.

You can see how the companies group together and which variables contribute to their positioning within this representation. For instance, X_4 = annual load factor and X_8 = total fuel costs are primarily responsible for the grouping of the mostly coastal companies in the lower right. The two variables X_1 = fixed-

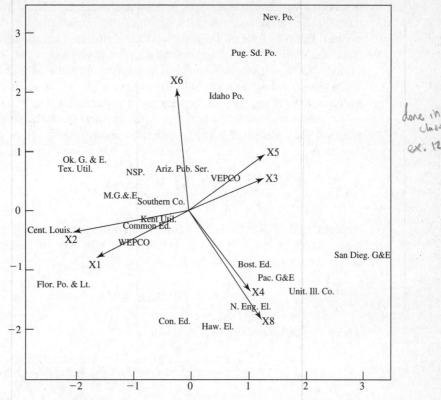

Figure 12.23 A biplot of the data on public utilities.

charge ratio and X_2 = rate of return on capital put the Florida and Louisiana companies together.

Constructing Biplots

The construction of a biplot proceeds from the sample principal components.

According to Result 8A.1, the best two-dimensional approximation to the data matrix $\mathbf{X}$ approximates the jth observation $\mathbf{x}_j$ in terms of the sample values of the first two principal components. In particular,

$$\mathbf{x}_j \doteq \bar{\mathbf{x}} + \hat{y}_{j1}\hat{\mathbf{e}}_1 + \hat{y}_{j2}\hat{\mathbf{e}}_2 \tag{12-44}$$

where $\hat{\mathbf{e}}_1$ and $\hat{\mathbf{e}}_2$ are the first two eigenvectors of $\mathbf{S}$ or, equivalently, of $\mathbf{X}_c'\mathbf{X}_c = (n-1)\mathbf{S}$. Here $\mathbf{X}_c$ denotes the mean corrected data matrix with rows $(\mathbf{x}_j - \bar{\mathbf{x}})'$. The eigenvectors determine a plane, and the coordinates of the jth unit (row) are the pair of values of the principal components, $(\hat{y}_{j1}, \hat{y}_{j2})$.

To include the information on the variables in this plot, we consider the pair of eigenvectors $(\hat{\mathbf{e}}_1, \hat{\mathbf{e}}_2)$. These eigenvectors are the coefficient vectors for the first two sample principal components. Consequently, each row of the matrix $\hat{\mathbf{E}} = [\hat{\mathbf{e}}_1, \hat{\mathbf{e}}_2]$

positions a variable in the graph, and the magnitudes of the coefficients (the coordinates of the variable) show the weightings that variable has in each principal component. The positions of the variables in the plot are indicated by a vector. Usually, statistical computer programs include a multiplier so that the lengths of all of the vectors can be suitably adjusted and plotted on the same axes as the sampling units. Units that are close to a variable likely have high values on that variable. To interpret a new point $\mathbf{x}_0$, we plot its principal components $\hat{\mathbf{E}}'(\mathbf{x}_0 - \bar{\mathbf{x}})$.

A direct approach to obtaining a biplot starts from the singular value decomposition (see Result 2A.15), which first expresses the $n \times p$ mean corrected matrix $\mathbf{X}_c$ as

$$\mathbf{X}_c_{(n \times p)} = \mathbf{U}_{(n \times p)} \ \mathbf{\Lambda}_{(p \times p)} \ \mathbf{V}'_{(p \times p)} \tag{12-45}$$

where $\mathbf{\Lambda} = \text{diag}(\lambda_1, \lambda_2, \ldots, \lambda_p)$ and $\mathbf{V}$ is an orthogonal matrix whose columns are the eigenvectors of $\mathbf{X}'_c \mathbf{X}_c = (n-1)\mathbf{S}$. That is, $\mathbf{V} = \hat{\mathbf{E}} = [\hat{\mathbf{e}}_1, \hat{\mathbf{e}}_2, \ldots, \hat{\mathbf{e}}_p]$. Multiplying (12-45) on the right by $\hat{\mathbf{E}}$, we find

$$\mathbf{X}_c \hat{\mathbf{E}} = \mathbf{U}\mathbf{\Lambda} \tag{12-46}$$

where the jth row of the left-hand side,

$$[(\mathbf{x}_j - \bar{\mathbf{x}})'\hat{\mathbf{e}}_1, (\mathbf{x}_j - \bar{\mathbf{x}})'\hat{\mathbf{e}}_2, \ldots, (\mathbf{x}_j - \bar{\mathbf{x}})'\hat{\mathbf{e}}_p] = [\hat{y}_{j1}, \hat{y}_{j2}, \ldots, \hat{y}_{jp}]$$

is just the value of the principal components for the jth item. That is, $\mathbf{U}\mathbf{\Lambda}$ contains all of the values of the principal components, while $\mathbf{V} = \hat{\mathbf{E}}$ contains the coefficients that define the principal components.

The best rank 2 approximation to $\mathbf{X}_c$ is obtained by replacing $\mathbf{\Lambda}$ by $\mathbf{\Lambda}^* = \text{diag}(\lambda_1, \lambda_2, 0, \ldots, 0)$. This result, called the Eckart–Young theorem, was established in Result 8.A.1. The approximation is then

$$\mathbf{X}_c \doteq \mathbf{U}\mathbf{\Lambda}^*\mathbf{V}' = [\hat{\mathbf{y}}_1, \hat{\mathbf{y}}_2] \begin{bmatrix} \hat{\mathbf{e}}'_1 \\ \hat{\mathbf{e}}'_2 \end{bmatrix} \tag{12-47}$$

where $\hat{\mathbf{y}}_1$ is the $n \times 1$ vector of values of the first principal component and $\hat{\mathbf{y}}_2$ is the $n \times 1$ vector of values of the second principal component.

In the biplot, each *row* of the data matrix, or item, is represented by the point located by the pair of values of the principal components. The ith *column* of the data matrix, or variable, is represented as an arrow from the origin to the point with coordinates (e_{1i}, e_{2i}), the entries in the ith column of the second matrix $[\hat{\mathbf{e}}_1, \hat{\mathbf{e}}_2]'$ in the approximation (12-47). This scale may not be compatible with that of the principal components, so an arbitrary multiplier can be introduced that adjusts all of the vectors by the same amount.

The idea of a biplot, to represent both units and variables in the same plot, extends to canonical correlation analysis, multidimensional scaling, and even more complicated nonlinear techniques. (See [12].)

Example 12.19 (A biplot of universities and their characteristics) Table 12.9 gives the data on some universities for certain variables used to compare or rank major universities. These variables include X_1 = average SAT score of new freshmen, X_2 = percentage of new freshmen in top 10% of high school class, X_3 = percentage of applicants accepted, X_4 = student–faculty ratio, X_5 = estimated annual expenses and X_6 = graduation rate (%).

Because two of the variables, SAT and Expenses, are on a much different scale from that of the other variables, we standardize the data and base our biplot on the matrix of standardized observations z_j. The biplot is given in Figure 12.24 on page 730.

Notice how Cal Tech and Johns Hopkins are off by themselves; the variable Expense is mostly responsible for this positioning. The large state universities in our sample are to the left in the biplot, and most of the private schools are on the right.

Table 12.9 Data on Universities

University	SAT	Top10	Accept	SFRatio	Expenses	Grad
Harvard	14.00	91	14	11	39.525	97
Princeton	13.75	91	14	8	30.220	95
Yale	13.75	95	19	11	43.514	96
Stanford	13.60	90	20	12	36.450	93
MIT	13.80	94	30	10	34.870	91
Duke	13.15	90	30	12	31.585	95
CalTech	14.15	100	25	6	63.575	81
Dartmouth	13.40	89	23	10	32.162	95
Brown	13.10	89	22	13	22.704	94
JohnsHopkins	13.05	75	44	7	58.691	87
UChicago	12.90	75	50	13	38.380	87
UPenn	12.85	80	36	11	27.553	90
Cornell	12.80	83	33	13	21.864	90
Northwestern	12.60	85	39	11	28.052	89
Columbia	13.10	76	24	12	31.510	88
NotreDame	12.55	81	42	13	15.122	94
UVirginia	12.25	77	44	14	13.349	92
Georgetown	12.55	74	24	12	20.126	92
CarnegieMellon	12.60	62	59	9	25.026	72
UMichigan	11.80	65	68	16	15.470	85
UCBerkeley	12.40	95	40	17	15.140	78
UWisconsin	10.85	40	69	15	11.857	71
PennState	10.81	38	54	18	10.185	80
Purdue	10.05	28	90	19	9.066	69
TexasA&M	10.75	49	67	25	8.704	67

Source: *U.S. News & World Report*, September 18, 1995, p. 126.

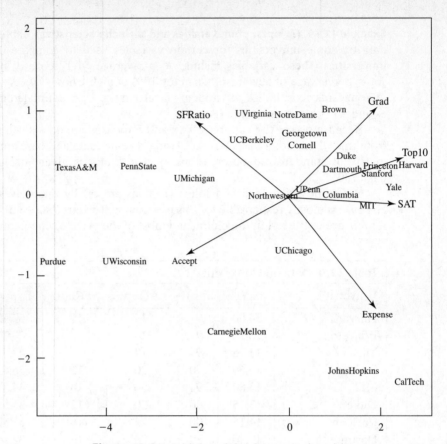

Figure 12.24 A biplot of the data on universities.

Large values for the variables SAT, Top10, and Grad are associated with the private school group. Northwestern lies in the middle of the biplot. ∎

A newer version of the biplot, due to Gower and Hand [12], has some advantages. Their biplot, developed as an extension of the scatter plot, has features that make it easier to interpret.

- The two axes for the principal components are suppressed.
- An axis is constructed for each variable and a scale is attached.

As in the original biplot, the i-th item is located by the corresponding pair of values of the first two principal components

$$(\hat{y}_{1i}, \hat{y}_{2i}) = ((\mathbf{x}_i - \bar{\mathbf{x}})' \hat{\mathbf{e}}_1, (\mathbf{x}_i - \bar{\mathbf{x}})' \hat{\mathbf{e}}_2)$$

where $\hat{\mathbf{e}}_1$ and where $\hat{\mathbf{e}}_2$ are the first two eigenvectors of $\mathbf{S}$. The scales for the principal components are not shown on the graph.

In addition the arrows for the variables in the original biplot are replaced by axes that extend in both directions and that have scales attached. As was the case with the arrows, the axis for the i-the variable is determined by the i-the row of $\hat{\mathbf{E}} = [\hat{\mathbf{e}}_1, \hat{\mathbf{e}}_2]$.

To begin, we let $\mathbf{u}_i$ the vector with 1 in the i-th position and 0's elsewhere. Then an arbitrary $p \times 1$ vector $\mathbf{x}$ can be expressed as

$$\mathbf{x} = \sum_{i=1}^{p} x_i \mathbf{u}_i$$

and, by Definition 2.A.12, its projection onto the space of the first two eigenvectors has coefficient vector

$$\hat{\mathbf{E}}'\mathbf{x} = \sum_{i=1}^{p} x_i (\hat{\mathbf{E}}'\mathbf{u}_i)$$

so the contribution of the i-th variable to the vector sum is $x_i(\hat{\mathbf{E}}'\mathbf{u}_i) = x_i[e_{1i}, e_{2i}]'$. The two entries e_{1i} and e_{2i} in the i-the row of $\hat{\mathbf{E}}$ determine the direction of the axis for the i-th variable.

The projection vector of the sample mean $\bar{\mathbf{x}} = \sum_{i=1}^{p} \bar{x}_i \mathbf{u}_i$

$$\hat{\mathbf{E}}'\bar{\mathbf{x}} = \sum_{i=1}^{p} \bar{x}_i (\hat{\mathbf{E}}'\mathbf{u}_i)$$

is the origin of the biplot. Every $\mathbf{x}$ can also be written as $\mathbf{x} = \bar{\mathbf{x}} + (\mathbf{x} - \bar{\mathbf{x}})$ and its projection vector has two components

$$\sum_{i=1}^{p} \bar{x}_i (\hat{\mathbf{E}}'\mathbf{u}_i) + \sum_{i=1}^{p} (x_i - \bar{x}_i)(\hat{\mathbf{E}}'\mathbf{u}_i) .$$

Starting from the origin, the points in the direction $w[e_{1i}, e_{2i}]'$ are plotted for $w = 0, \pm 1, \pm 2, \ldots$ This provides a scale for the mean centered variable $x_i - \bar{x}_i$. It defines the distance in the biplot for a change of one unit in x_i. But, the origin for the i-th variable corresponds to $w = 0$ because the term $\bar{x}_i(\hat{\mathbf{E}}'\mathbf{u}_i)$ was ignored. The axis label needs to be translated so that the value $\bar{x}_i$ is at the origin of the biplot. Since $\bar{x}_i$ is typically not an integer (or another nice number), an integer (or other nice number) closest to it can be chosen and the scale translated appropriately. Computer software simplifies this somewhat difficult task.

The scale allows us to visually interpolate the position of $x_i[e_{1i}, e_{2i}]'$ in the biplot. The scales predict the values of a variable, not give its exact value, as they are based on a two dimensional approximation.

Example 12.20 (An alternative biplot for the university data) We illustrate this newer biplot with the university data in Table 12.9. The alternative biplot with an axis for each variable is shown in Figure 12.25. Compared with Figure 12.24, the software reversed the direction of the first principal component. Notice, for example, that expenses and student faculty ratio separate Cal Tech and Johns Hopkins from the other universities. Expenses for Cal Tech and Johns Hopkins can be seen to be about 57 thousand a year, and the student faculty ratios are in the single digits. The large state universities, on the right hand side of the plot, have relatively high student faculty ratios, above 20, relatively low SAT scores of entering freshman, and only about 50% or fewer of their entering students in the top 10% of their high school class. The scaled axes on the newer biplot are more informative than the arrows in the original biplot. ∎

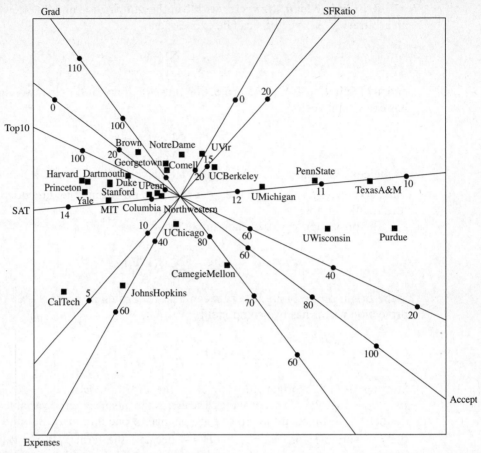

Figure 12.25 An alternative biplot of the data on universities.

See le Roux and Gardner [23] for more examples of this alternative biplot and references to appropriate special purpose statistical software.

12.9 Procrustes Analysis: A Method for Comparing Configurations

Starting with a given $n \times n$ matrix of distances **D**, or similarities **S**, that relate n objects, two or more configurations can be obtained using different techniques. The possible methods include both metric and nonmetric multidimensional scaling. The question naturally arises as to how well the solutions coincide. Figures 12.19 and 12.20 in Example 12.16 respectively give the metric multidimensional scaling (principal coordinate analysis) and nonmetric multidimensional scaling solutions for the data on universities. The two configurations appear to be quite similar, but a quantitative measure would be useful. A numerical comparison of two configurations, obtained by moving one configuration so that it aligns best with the other, is called *Procrustes analysis*, after the innkeeper Procrustes, in Greek mythology, who would either stretch or lop off customers' limbs so they would fit his bed.

Constructing the Procrustes Measure of Agreement

Suppose the $n \times p$ matrix $\mathbf{X}^*$ contains the coordinates of the n points obtained for plotting with technique 1 and the $n \times q$ matrix $\mathbf{Y}^*$ contains the coordinates from technique 2, where $q \leq p$. By adding columns of zeros to $\mathbf{Y}^*$, if necessary, we can assume that $\mathbf{X}^*$ and $\mathbf{Y}^*$ both have the same dimension $n \times p$. To determine how compatible the two configurations are, we move, say, the second configuration to match the first by shifting each point by the same amount and rotating or reflecting the configuration about the coordinate axes.[4]

Mathematically, we translate by a vector $\mathbf{b}$ and multiply by an orthogonal matrix $\mathbf{Q}$ so that the coordinates of the jth point $\mathbf{y}_j$ are transformed to

$$\mathbf{Q}\mathbf{y}_j + \mathbf{b}$$

The vector $\mathbf{b}$ and orthogonal matrix $\mathbf{Q}$ are then varied to order to minimize the sum, over all n points, of squared distances

$$d_j^2(\mathbf{x}_j, \mathbf{Q}\mathbf{y}_j + \mathbf{b}) = (\mathbf{x}_j - \mathbf{Q}\mathbf{y}_j - \mathbf{b})'(\mathbf{x}_j - \mathbf{Q}\mathbf{y}_j - \mathbf{b}) \qquad (12\text{-}48)$$

between $\mathbf{x}_j$ and the transformed coordinates $\mathbf{Q}\mathbf{y}_j + \mathbf{b}$ obtained for the second technique. We take, as a measure of fit, or agreement, between the two configurations, the residual sum of squares

$$PR^2 = \min_{\mathbf{Q},\mathbf{b}} \sum_{j=1}^{n} (\mathbf{x}_j - \mathbf{Q}\mathbf{y}_j - \mathbf{b})'(\mathbf{x}_j - \mathbf{Q}\mathbf{y}_j - \mathbf{b}) \qquad (12\text{-}49)$$

The next result shows how to evaluate this Procrustes residual sum of squares measure of agreement and determines the *Procrustes rotation* of $\mathbf{Y}^*$ relative to $\mathbf{X}^*$.

Result 12.2 Let the $n \times p$ configurations $\mathbf{X}^*$ and $\mathbf{Y}^*$ both be centered so that all columns have mean zero. Then

$$PR^2 = \sum_{j=1}^{n} \mathbf{x}_j'\mathbf{x}_j + \sum_{j=1}^{n} \mathbf{y}_j'\mathbf{y}_j - 2 \sum_{i=1}^{p} \lambda_i$$

$$= \text{tr}[\mathbf{X}^*\mathbf{X}^{*'}] + \text{tr}[\mathbf{Y}^*\mathbf{Y}^{*'}] - 2\,\text{tr}[\mathbf{\Lambda}] \qquad (12\text{-}50)$$

where $\mathbf{\Lambda} = \text{diag}(\lambda_1, \lambda_2, \ldots, \lambda_p)$ and the minimizing transformation is

$$\hat{\mathbf{Q}} = \sum_{i=1}^{p} \mathbf{v}_i\mathbf{u}_i' = \mathbf{V}\mathbf{U}' \qquad \hat{\mathbf{b}} = \mathbf{0} \qquad (12\text{-}51)$$

---

[4] Sibson [30] has proposed a numerical measure of the agreement between two configurations, given by the coefficient

$$\gamma = 1 - \frac{[\text{tr}(\mathbf{Y}^{*'}\mathbf{X}^*\mathbf{X}^{*'}\mathbf{Y}^*)^{1/2}]^2}{\text{tr}(\mathbf{X}^{*'}\mathbf{X}^*)\,\text{tr}(\mathbf{Y}^{*'}\mathbf{Y}^*)}$$

For identical configurations, $\gamma = 0$. If necessary, γ can be computed after a Procrustes analysis has been completed.

Here Λ, $\mathbf{U}$, and $\mathbf{V}$ are obtained from the singular-value decomposition

$$\sum_{j=1}^{n} \mathbf{y}_j \mathbf{x}_j' = \underset{(p \times n)}{\mathbf{Y}^{*\prime}} \underset{(n \times p)}{\mathbf{X}^{*}} = \underset{(p \times p)}{\mathbf{U}} \underset{(p \times p)}{\Lambda} \underset{(p \times p)}{\mathbf{V}'}$$

Proof. Because the configurations are centered to have zero means $\left(\sum_{j=1}^{n} \mathbf{x}_j = \mathbf{0} \right.$ and $\left. \sum_{j=1}^{n} \mathbf{y}_j = \mathbf{0} \right)$, we have

$$\sum_{j=1}^{n} (\mathbf{x}_j - \mathbf{Q}\mathbf{y}_j - \mathbf{b})'(\mathbf{x}_j - \mathbf{Q}\mathbf{y}_j - \mathbf{b}) = \sum_{j=1}^{n} (\mathbf{x}_j - \mathbf{Q}\mathbf{y}_j)'(\mathbf{x}_j - \mathbf{Q}\mathbf{y}_j) + n\mathbf{b}'\mathbf{b}$$

The last term is nonnegative, so the best fit occurs for $\hat{\mathbf{b}} = \mathbf{0}$. Consequently, we need only consider

$$PR^2 = \min_{\mathbf{Q}} \sum_{j=1}^{n} (\mathbf{x}_j - \mathbf{Q}\mathbf{y}_j)'(\mathbf{x}_j - \mathbf{Q}\mathbf{y}_j) = \sum_{j=1}^{n} \mathbf{x}_j'\mathbf{x}_j + \sum_{j=1}^{n} \mathbf{y}_j'\mathbf{y}_j - 2 \max_{\mathbf{Q}} \sum_{j=1}^{n} \mathbf{x}_j'\mathbf{Q}\mathbf{y}_j$$

Using $\mathbf{x}_j'\mathbf{Q}\mathbf{y}_j = \operatorname{tr}[\mathbf{Q}\mathbf{y}_j\mathbf{x}_j']$, we find that the expression being maximized becomes

$$\sum_{j=1}^{n} \mathbf{x}_j'\mathbf{Q}\mathbf{y}_j = \sum_{j=1}^{n} \operatorname{tr}[\mathbf{Q}\mathbf{y}_j\mathbf{x}_j'] = \operatorname{tr}\left[\mathbf{Q} \sum_{j=1}^{n} \mathbf{y}_j\mathbf{x}_j' \right]$$

By the singular-value decomposition,

$$\sum_{j=1}^{n} \mathbf{y}_j \mathbf{x}_j' = \mathbf{Y}^{*\prime}\mathbf{X}^{*} = \mathbf{U}\Lambda\mathbf{V}' = \sum_{j=1}^{p} \lambda_i \mathbf{u}_i \mathbf{v}_i'$$

where $\mathbf{U} = [\mathbf{u}_1, \mathbf{u}_2, \ldots, \mathbf{u}_p]$ and $\mathbf{V} = [\mathbf{v}_1, \mathbf{v}_2, \ldots, \mathbf{v}_p]$ are $p \times p$ orthogonal matrices. Consequently,

$$\sum_{j=1}^{n} \mathbf{x}_j'\mathbf{Q}\mathbf{y}_j = \operatorname{tr}\left[\mathbf{Q}\left(\sum_{i=1}^{p} \lambda_i \mathbf{u}_i \mathbf{v}_i' \right) \right] = \sum_{i=1}^{p} \lambda_i \operatorname{tr}[\mathbf{Q}\mathbf{u}_i \mathbf{v}_i']$$

The variable quantity in the ith term

$$\operatorname{tr}[\mathbf{Q}\mathbf{u}_i \mathbf{v}_i'] = \mathbf{v}_i'\mathbf{Q}\mathbf{u}_i$$

has an upper bound of 1 as can be seen by applying the Cauchy–Schwarz inequality (2–48) with $\mathbf{b} = \mathbf{Q}\mathbf{v}_i$ and $\mathbf{d} = \mathbf{u}_i$. That is, since $\mathbf{Q}$ is orthogonal,

$$\mathbf{v}_i'\mathbf{Q}\mathbf{u}_i \leq \sqrt{\mathbf{v}_i'\mathbf{Q}\mathbf{Q}'\mathbf{v}_i} \sqrt{\mathbf{u}_i'\mathbf{u}_i} = \sqrt{\mathbf{v}_i'\mathbf{v}_i} \times 1 = 1$$

Each of these p terms can be maximized by the same choice $\mathbf{Q} = \mathbf{VU}'$. With this choice,

$$\mathbf{v}_i'\mathbf{Qu}_i = \mathbf{v}_i'\mathbf{VU}'\mathbf{u}_i = [0,\ldots,0,1,0,\ldots,0]\begin{bmatrix} 0 \\ \vdots \\ 0 \\ 1 \\ 0 \\ \vdots \\ 0 \end{bmatrix} = 1$$

Therefore,

$$-2\max_{\mathbf{Q}} \sum_{j=1}^{n} \mathbf{x}_j'\mathbf{Qy}_j = -2(\lambda_1 + \lambda_2 + \cdots + \lambda_p)$$

Finally, we verify that $\mathbf{QQ}' = \mathbf{VU}'\mathbf{UV}' = \mathbf{VI}_p\mathbf{V}' = \mathbf{I}_p$, so $\mathbf{Q}$ is a $p \times p$ orthogonal matrix, as required. ∎

Example 12.21 (Procrustes analysis of the data on universities) Two configurations, produced by metric and nonmetric multidimensional scaling, of data on universities are given Example 12.16. The two configurations appear to be quite close. There is a two-dimensional array of coordinates for each of the two scaling methods. Initially, the sum of squared distances is

$$\sum_{j=1}^{25} (\mathbf{x}_j - \mathbf{y}_j)'(\mathbf{x}_j - \mathbf{y}_j) = 3.862$$

A computer calculation gives

$$\mathbf{U} = \begin{bmatrix} -.9990 & .0448 \\ .0448 & .9990 \end{bmatrix} \qquad \mathbf{V} = \begin{bmatrix} -1.0000 & .0076 \\ .0076 & 1.0000 \end{bmatrix}$$

$$\mathbf{\Lambda} = \begin{bmatrix} 114.9439 & 0.000 \\ 0.000 & 21.3673 \end{bmatrix}$$

According to Result 12.2, to better align these two solutions, we multiply the nonmetric scaling solution by the orthogonal matrix

$$\hat{\mathbf{Q}} = \sum_{i=1}^{2} \mathbf{v}_i\mathbf{u}_i' = \mathbf{VU}' = \begin{bmatrix} .9993 & -.0372 \\ .0372 & .9993 \end{bmatrix}$$

This corresponds to clockwise rotation of the nonmetric solution by about 2 degrees. After rotation, the sum of squared distances, 3.862, is reduced to the Procrustes measure of fit

$$PR^2 = \sum_{j=1}^{25} \mathbf{x}_j'\mathbf{x}_j + \sum_{j=1}^{25} \mathbf{y}_j'\mathbf{y}_j - 2\sum_{j=1}^{2} \lambda_i = 3.673 \qquad ∎$$

Example 12.22 (Procrustes analysis and additional ordinations of data on forests) Data were collected on the populations of eight species of trees growing on ten upland sites in southern Wisconsin. These data are shown in Table 12.10.

The metric, or principal coordinate, solution and nonmetric multidimensional scaling solution are shown in Figures 12.26 and 12.27.

Table 12.10 Wisconsin Forest Data

Tree	Site									
	1	2	3	4	5	6	7	8	9	10
BurOak	9	8	3	5	6	0	5	0	0	0
BlackOak	8	9	8	7	0	0	0	0	0	0
WhiteOak	5	4	9	9	7	7	4	6	0	2
RedOak	3	4	0	6	9	8	7	6	4	3
AmericanElm	2	2	4	5	6	0	5	0	2	5
Basswood	0	0	0	0	2	7	6	6	7	6
Ironwood	0	0	0	0	0	0	7	4	6	5
SugarMaple	0	0	0	0	0	5	4	8	8	9

Source: See [24].

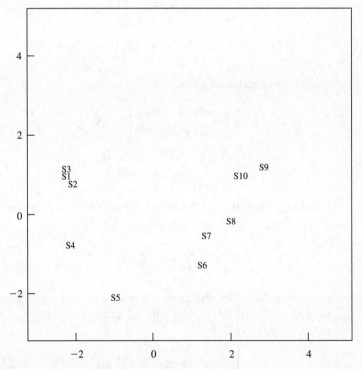

Figure 12.26 Metric multidimensional scaling of the data on forests.

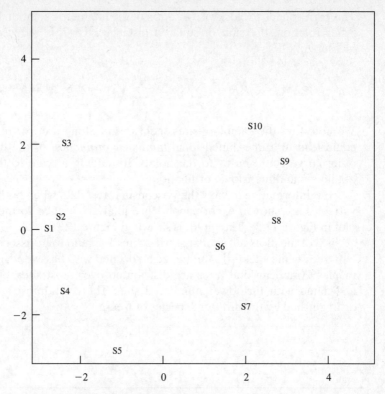

Figure 12.27 Nonmetric multidimensional scaling of the data on forests.

Using the coordinates of the points in Figures 12.26 and 12.27, we obtain the initial sum of squared distances for fit:

$$\sum_{j=1}^{10} (\mathbf{x}_j - \mathbf{y}_j)' (\mathbf{x}_j - \mathbf{y}_j) = 8.547$$

A computer calculation gives

$$\mathbf{U} = \begin{bmatrix} -.9833 & -.1821 \\ -.1821 & .9833 \end{bmatrix} \qquad \mathbf{V} = \begin{bmatrix} -1.0000 & -.0001 \\ -.0001 & 1.0000 \end{bmatrix}$$

$$\mathbf{\Lambda} = \begin{bmatrix} 43.3748 & 0.0000 \\ 0.0000 & 14.9103 \end{bmatrix}$$

According to Result 12.2, to better align these two solutions, we multiply the non-metric scaling solution by the orthogonal matrix

$$\hat{\mathbf{Q}} = \sum_{i=1}^{2} \mathbf{v}_j \mathbf{u}_i' = \mathbf{V}\mathbf{U}' = \begin{bmatrix} .9833 & .1821 \\ -.1821 & .9833 \end{bmatrix}$$

This corresponds to clockwise rotation of the nonmetric solution by about 10 degrees. After rotation, the sum of squared distances, 8.547, is reduced to the Procrustes measure of fit

$$PR^2 = \sum_{j=1}^{10} \mathbf{x}_j' \mathbf{x}_j + \sum_{j=1}^{10} \mathbf{y}_j' \mathbf{y}_j - 2 \sum_{i=1}^{2} \lambda_i = 6.599$$

We note that the sampling sites seem to fall along a curve in both pictures. This could lead to a one-dimensional *nonlinear ordination* of the data. A quadratic or other curve could be fit to the points. By adding a scale to the curve, we would obtain a one-dimensional ordination.

It is informative to view the Wisconsin forest data when both sampling units and variables are shown. A correspondence analysis applied to the data produces the plot in Figure 12.28. The biplot is shown in Figure 12.29.

All of the plots tell similar stories. Sites 1–5 tend to be associated with species of oak trees, while sites 7–10 tend to be associated with basswood, ironwood, and sugar maples. American elm trees are distributed over most sites, but are more closely associated with the lower numbered sites. There is almost a continuum of sites distinguished by the different species of trees. ■

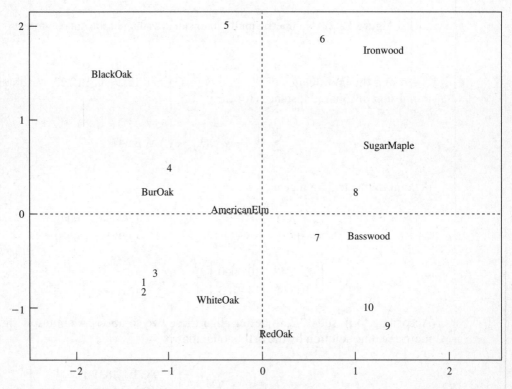

Figure 12.28 The correspondence analysis plot of the data on forests.

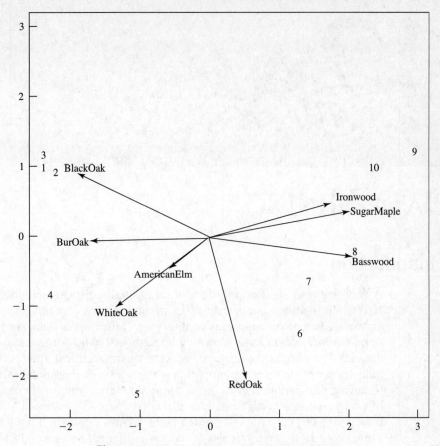

Figure 12.29 The biplot of the data on forests.

DATA MINING

Introduction

A very large sample in applications of traditional statistical methodology may mean 10,000 observations on, perhaps, 50 variables. Today, computer-based repositories known as data warehouses may contain many terabytes of data. For some organizations, corporate data have grown by a factor of 100,000 or more over the last few decades. The telecommunications, banking, pharmaceutical, and (package) shipping industries provide several examples of companies with huge databases. Consider the following illustration. If each of the approximately 17 million books in the Library of Congress contained a megabyte of text (roughly 450 pages) in MS Word format, then typing this collection of printed material into a computer database would consume about 17 terabytes of disk space. United Parcel Service (UPS) has a package-level detail database of about 17 terabytes to track its shipments.

For our purposes, *data mining* refers to the process associated with discovering patterns and relationships in extremely large data sets. That is, data mining is concerned with extracting a few nuggets of knowledge from a relative mountain of numerical information. From a business perspective, the nuggets of knowledge represent *actionable* information that can be exploited for a competitive advantage.

Data mining is not possible without appropriate software and fast computers. Not surprisingly, many of the techniques discussed in this book, along with algorithms developed in the machine learning and artificial intelligence fields, play important roles in data mining. Companies with well-known statistical software packages now offer comprehensive data mining programs.[5] In addition, special purpose programs such as CART have been used successfully in data mining applications.

Data mining has helped to identify new chemical compounds for prescription drugs, detect fraudulent claims and purchases, create and maintain individual customer relationships, design better engines and build appropriate inventories, create better medical procedures, improve process control, and develop effective credit scoring rules.

---

[5]SAS Institute's data mining program is currently called Enterprise Miner. SPSS's data mining program is Clementine.

In traditional statistical applications, sample sizes are relatively small, data are carefully collected, sample results provide a basis for inference, anomalies are treated but are often not of immediate interest, and models are frequently highly structured. In data mining, sample sizes can be huge; data are scattered and historical (routinely recorded), samples are used for training, validation, and testing (no formal inference); anomalies are of interest; and models are often unstructured. Moreover, data preparation—including data collection, assessment and cleaning, and variable definition and selection—is typically an arduous task and represents 60 to 80% of the data mining effort.

Data mining problems can be roughly classified into the following categories:

- Classification (discrete outcomes):

  Who is likely to move to another cellular phone service?

- Prediction (continuous outcomes):

  What is the appropriate appraised value for this house?

- Association/market basket analysis:

  Is skim milk typically purchased with low-fat cottage cheese?

- Clustering:

  Are there groups with similar buying habits?

- Description:

  On Thursdays, grocery store consumers often purchase corn chips and soft drinks together.

Given the nature of data mining problems, it should not be surprising that many of the statistical methods discussed in this book are part of comprehensive data mining software packages. Specifically, regression, discrimination and classification procedures (linear rules, logistic regression, decision trees such as those produced by CART), and clustering algorithms are important data mining tools. Other tools, whose discussion is beyond the scope of this book, include association rules, multivariate adaptive regression splines (MARS), K-nearest neighbor algorithm, neural networks, genetic algorithms, and visualization.[6]

The Data Mining Process

Data mining is a process requiring a sequence of steps. The steps form a strategy that is not unlike the strategy associated with any model building effort. Specifically, data miners must

1. Define the problem and identify objectives.
2. Gather and prepare the appropriate data.
3. Explore the data for suspected associations, unanticipated characteristics, and obvious anomalies to gain understanding.
4. Clean the data and perform any variable transformation that seems appropriate.

[6]For more information on data mining in general and data mining tools in particular, see the references at the end of this chapter.

5. Divide the data into training, validation, and, perhaps, test data sets.

6. Build the model on the training set.

7. Modify the model (if necessary) based on its performance with the validation data.

8. Assess the model by checking its performance on validation or test data. Compare the model outcomes with the initial objectives. Is the model likely to be useful?

9. Use the model.

10. Monitor the model performance. Are the results reliable, cost effective?

In practice, it is typically necessary to repeat one of more of these steps several times until a satisfactory solution is achieved. Data mining software suites such as Enterprise Miner and Clementine are typically organized so that the user can work sequentially through the steps listed and, in fact, can picture them on the screen as a process flow diagram.

Data mining requires a rich collection of tools and algorithms used by a skilled analyst with sound subject matter knowledge (or working with someone with sound subject matter knowledge) to produce acceptable results. Once established, any successful data mining effort is an ongoing exercise. New data must be collected and processed, the model must be updated or a new model developed, and, in general, adjustments made in light of new experience. The cost of a poor data mining effort is high, so careful model construction and evaluation is imperative.

Model Assessment

In the model development stage of data mining, several models may be examined simultaneously. In the example to follow, we briefly discuss the results of applying logistic regression, decision tree methodology, and a neural network to the problem of credit scoring (determining good credit risks) using a publicly available data set known as the German Credit data. Although the data miner can control the model inputs and certain parameters that govern the development of individual models, in most data mining applications there is little formal statistical inference. Models are ordinarily assessed (and compared) by domain experts using descriptive devices such as confusion matrices, summary profit or loss numbers, lift charts, threshold charts, and other, mostly graphical, procedures.

The split of the very large initial data set into training, validation, and testing subsets allows potential models to be assessed with data that were not involved in model development. Thus, the training set is used to build models that are assessed on the validation (holdout) data set. If a model does not perform satisfactorily in the validation phase, it is retrained. Iteration between training and validation continues until satisfactory performance with validation data is achieved. At this point, a trained and validated model is assessed with test data. The test data set is ordinarily used once at the end of the modeling process to ensure an unbiased assessment of model performance. On occasion, the test data step is omitted and the final assessment is done with the validation sample, or by cross-validation.

An important assessment tool is the *lift chart*. Lift charts may be formatted in various ways, but all indicate improvement of the selected procedures (models) over what can be achieved by a baseline activity. The baseline activity often represents a prior conviction or a random assignment. Lift charts are particularly useful for comparing the performance of different models.

Lift is defined as

$$\text{Lift} = \frac{P(\text{result} \mid \text{condition})}{P(\text{result})}$$

If the result is independent of the condition, then Lift = 1. A value of Lift > 1 implies the condition (generally a model or algorithm) leads to a greater probability of the desired result and, hence, the condition is useful and potentially profitable. Different conditions can be compared by comparing their lift charts.

Example 12.23 (A small-scale data mining exercise) A publicly available data set known as the German Credit data[7] contains observations on 20 variables for 1000 past applicants for credit. In addition, the resulting credit rating ("Good" or "Bad") for each applicant was recorded. The objective is to develop a credit scoring rule that can be used to determine if a new applicant is a good credit risk or a bad credit risk based on values for one or more of the 20 explanatory variables. The 20 explanatory variables include CHECKING (checking account status), DURATION (duration of credit in months), HISTORY (credit history), AMOUNT (credit amount), EMPLOYED (present employment since), RESIDENT (present resident since), AGE (age in years), OTHER (other installment debts), INSTALLP (installment rate as % of disposable income), and so forth. Essentially, then, we must develop a function of several variables that allows us to classify a new applicant into one of two categories: Good or Bad.

We will develop a classification procedure using three approaches discussed in Sections 11.7 and 11.8; logistic regression, classification trees, and neural networks. An abbreviated assessment of the three approaches will allow us compare the performance of the three approaches on a validation data set. This data mining exercise is implemented using the general data mining process described earlier and SAS Enterprise Miner software.

In the full credit data set, 70% of the applicants were Good credit risks and 30% of the applicants were Bad credit risks. The initial data were divided into two sets for our purposes, a training set and a validation set. About 60% of the data (581 cases) were allocated to the training set and about 40% of the data (419 cases) were allocated to the validation set. The random sampling scheme employed ensured that each of the training and validation sets contained about 70% Good applicants and about 30% Bad applicants. The applicant credit risk profiles for the data sets follow.

	Credit data	Training data	Validation data
Good:	700	401	299
Bad:	300	180	120
Total:	1000	581	419

---

[7] At the time this supplement was written, the German Credit data were available in a sample data file accompanying SAS Enterprise Miner. Many other publicly available data sets can be downloaded from the following Web site: www.kdnuggets.com.

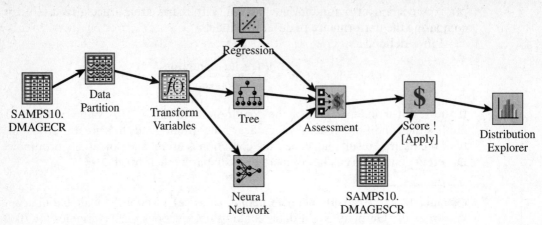

Figure 12.30 The process flow diagram.

Figure 12.30 shows the process flow diagram from the Enterprise Miner screen. The icons in the figure represent various activities in the data mining process. As examples, SAMPS10.DMAGECR contains the data; Data Partition allows the data to be split into training, validation, and testing subsets; Transform Variables, as the name implies, allows one to make variable transformations; the Regression, Tree, and Neural Network icons can each be opened to develop the individual models; and Assessment allows an evaluation of each predictive model in terms of predictive power, lift, profit or loss, and so on, and a comparison of all models.

The best model (with the training set parameters) can be used to score a new selection of applicants without a credit designation (SAMPS10.DMAGESCR). The results of this scoring can be displayed, in various ways, with Distribution Explorer.

For this example, the prior probabilities were set proportional to the data; consequently, $P(\text{Good}) = .7$ and $P(\text{Bad}) = .3$. The cost matrix was initially specified as follows:

		Predicted (Decision)	
		Good (Accept)	Bad (Reject)
Actual	Good	0	$1
	Bad	$5	0

so that it is 5 times as costly to classify a Bad applicant as Good (Accept) as it is to classify a Good applicant as Bad (Reject). In practice, accepting a Good credit risk should result in a profit or, equivalently, a negative cost. To match this formulation more closely, we subtract $1 from the entries in the first row of the cost matrix to obtain the "realistic" cost matrix:

		Predicted (Decision)	
		Good (Accept)	Bad (Reject)
Actual	Good	−$1	0
	Bad	$5	0

This matrix yields the same decisions as the original cost matrix, but the results are easier to interpret relative to the expected cost objective function. For example, after further adjustments, a negative expected cost score may indicate a potential profit so the applicant would be a Good credit risk.

Next, input variables need to be processed (perhaps transformed), models (or algorithms) must be specified, and required parameters must be set in all of the icons in the process flow diagram. Then the process can be executed up to any point in the diagram by clicking on an icon. All previous connected icons are run. For example, clicking on Score executes the process up to and including the Score icon. Results associated with individual icons can then be examined by clicking on the appropriate icon.

We illustrate model assessment using lift charts. These lift charts, available in the Assessment icon, result from one execution of the process flow diagram in Figure 12.30.

Consider the logistic regression classifier. Using the logistic regression function determined with the training data, an expected cost can be computed for each case in the validation set. These expected cost "scores" can then ordered from smallest to largest and partitioned into groups by the 10th, 20th, ..., and 90th percentiles. The first percentile group then contains the 42 (10% of 419) of the applicants with the smallest negative expected costs (largest potential profits), the second percentile group contains the next 42 applicants (next 10%), and so on. (From a classification viewpoint, those applicants with negative expected costs might be classified as Good risks and those with nonnegative expected costs as Bad risks.)

If the model has no predictive power, we would expect, approximately, a uniform distribution of, say, Good credit risks over the percentile groups. That is, we would expect 10% or $.10(299) = 30$ Good credit risks among the 42 applicants in each of the percentile groups.

Once the validation data have been scored, we can count the number of Good credit risks (of the 42 applicants) actually falling in each percentile group. For example, of the 42 applicants in the first percentile group, 40 were actually Good risks for a "captured response rate" of $40/299 = .133$ or 13.3%. In this case, lift for the first percentile group can be calculated as the ratio of the number of Good predicted by the model to the number of Good from a random assignment or

$$\text{Lift} = \frac{40}{30} = 1.33$$

The lift value indicates the model assigns $10/299 = .033$ or 3.3% more Good risks to the first percentile group (largest negative expected cost) than would be assigned by chance.[8]

Lift statistics can be displayed as individual (noncumulative) values or as cumulative values. For example, 40 Good risks also occur in the second percentile group for the logistic regression classifier, and the cumulative risk for the first two percentile groups is

$$\text{Lift} = \frac{40 + 40}{30 + 30} = 1.33$$

---

[8]The lift numbers calculated here differ a bit from the numbers displayed in the lift diagrams to follow because of rounding.

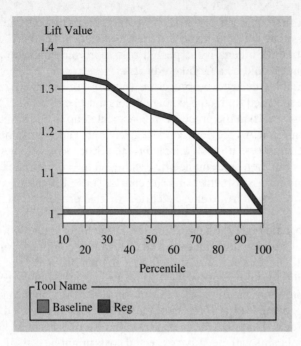

Figure 12.31 Cumulative lift chart for the logistic regression classifier.

The cumulative lift chart for the logistic regression model is displayed in Figure 12.31.

Lift and cumulative lift statistics can be determined for the classification tree tool and for the neural network tool. For each classifier, the entire data set is scored (expected costs computed), applicants ordered from smallest score to largest score and percentile groups created. At this point, the lift calculations follow those outlined for the logistic regression method. The cumulative charts for all three classifiers are shown in Figure 12.32.

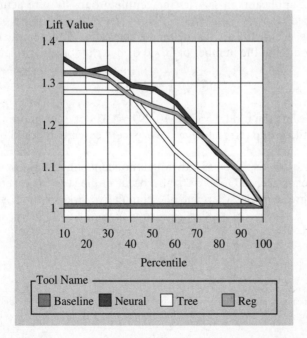

Figure 12.32 Cumulative lift charts for neural network, classification tree, and logistic regression tools.

We see from Figure 12.32 that the neural network and the logistic regression have very similar predictive powers and they both do better, in this case, than the classification tree. The classification tree, in turn, outperforms a random assignment. If this represented the end of the model building and assessment effort, one model would be picked (say, the neural network) to score a new set of applicants (without a credit risk designation) as Good (accept) or Bad (reject).

In the decision flow diagram in Figure 12.30, the SAMPS10.DMAGESCR file contains 75 new applicants. Expected cost scores for these applicants were created using the neural network model. Of the 75 applicants, 33 were classified as Good credit risks (with negative expected costs). ∎

Data mining procedures and software continue to evolve, and it is difficult to predict what the future might bring. Database packages with embedded data mining capabilities, such as SQL Server 2005, represent one evolutionary direction.

Exercises

12.1. Certain characteristics associated with a few recent U.S. presidents are listed in Table 12.11.

Table 12.11

President	Birthplace (region of United States)	Elected first term?	Party	Prior U.S. congressional experience?	Served as vice president?
1. R. Reagan	Midwest	Yes	Republican	No	No
2. J. Carter	South	Yes	Democrat	No	No
3. G. Ford	Midwest	No	Republican	Yes	Yes
4. R. Nixon	West	Yes	Republican	Yes	Yes
5. L. Johnson	South	No	Democrat	Yes	Yes
6. J. Kennedy	East	Yes	Democrat	Yes	No

(a) Introducing appropriate binary variables, calculate similarity coefficient 1 in Table 12.1 for pairs of presidents.

Hint: You may use birthplace as South, non-South.

(b) Proceeding as in Part a, calculate similarity coefficients 2 and 3 in Table 12.1 Verify the monotonicity relation of coefficients 1, 2, and 3 by displaying the order of the 15 similarities for each coefficient.

12.2. Repeat Exercise 12.1 using similarity coefficients 5, 6, and 7 in Table 12.1.

12.3. Show that the sample correlation coefficient [see (12-11)] can be written as

$$r = \frac{ad - bc}{[(a + b)(a + c)(b + d)(c + d)]^{1/2}}$$

for two 0–1 binary variables with the following frequencies:

		Variable 2	
		0	1
Variable 1	0	a	b
	1	c	d

12.4. Show that the monotonicity property holds for the similarity coefficients 1, 2, and 3 in Table 12.1.

Hint: $(b + c) = p - (a + d)$. So, for instance,

$$\frac{a + d}{a + d + 2(b + c)} = \frac{1}{1 + 2[p/(a + d) - 1]}$$

This equation relates coefficients 3 and 1. Find analogous representations for the other pairs.

12.5. Consider the matrix of distances

$$\begin{array}{c} \\ 1 \\ 2 \\ 3 \\ 4 \end{array} \begin{array}{cccc} 1 & 2 & 3 & 4 \\ \left[\begin{array}{cccc} 0 & & & \\ 1 & 0 & & \\ 11 & 2 & 0 & \\ 5 & 3 & 4 & 0 \end{array}\right] \end{array}$$

Cluster the four items using each of the following procedures.

(a) Single linkage hierarchical procedure.

(b) Complete linkage hierarchical procedure.

(c) Average linkage hierarchical procedure.

Draw the dendrograms and compare the results in (a), (b), and (c).

12.6. The distances between pairs of five items are as follows:

$$\begin{array}{c} \\ 1 \\ 2 \\ 3 \\ 4 \\ 5 \end{array} \begin{array}{ccccc} 1 & 2 & 3 & 4 & 5 \\ \left[\begin{array}{ccccc} 0 & & & & \\ 4 & 0 & & & \\ 6 & 9 & 0 & & \\ 1 & 7 & 10 & 0 & \\ 6 & 3 & 5 & 8 & 0 \end{array}\right] \end{array}$$

Cluster the five items using the single linkage, complete linkage, and average linkage hierarchical methods. Draw the dendrograms and compare the results.

12.7. Sample correlations for five stocks were given in Example 8.5. These correlations, rounded to two decimal places, are reproduced as follows:

	JP Morgan	Citibank	Wells Fargo	Royal DutchShell	Exxon Mobil
JP Morgan	1				
Citibank	.63	1			
Wells Fargo	.51	.57	1		
Royal DutchShell	.12	.32	.18	1	
ExxonMobil	.16	.21	.15	.68	1

Treating the sample correlations as similarity measures, cluster the stocks using the single linkage and complete linkage hierarchical procedures. Draw the dendrograms and compare the results.

12.8. Using the distances in Example 12.3, cluster the items using the average linkage hierarchical procedure. Draw the dendrogram. Compare the results with those in Examples 12.3 and 12.5.

12.9. The vocabulary "richness" of a text can be quantitatively described by counting the words used once, the words used twice, and so forth. Based on these counts, a linguist proposed the following distances between chapters of the Old Testament book Lamentations (data courtesy of Y. T. Radday and M. A. Pollatschek):

$$
\begin{array}{c}
\text{Lamentations} \\
\text{chapter}
\end{array}
$$

$$
\begin{array}{c}
\text{Lamentations} \\
\text{chapter}
\end{array}
\begin{array}{c}
\\ 1 \\ 2 \\ 3 \\ 4 \\ 5
\end{array}
\begin{array}{ccccc}
1 & 2 & 3 & 4 & 5 \\
\left[\begin{array}{ccccc}
0 & & & & \\
.76 & 0 & & & \\
2.97 & .80 & 0 & & \\
4.88 & 4.17 & .21 & 0 & \\
3.86 & 1.92 & 1.51 & .51 & 0
\end{array}\right]
\end{array}
$$

Cluster the chapters of Lamentations using the three linkage hierarchical methods we have discussed. Draw the dendrograms and compare the results.

12.10. Use Ward's method to cluster the four items whose measurements on a single variable X are given in the following table.

Item	Measurements x
1	2
2	1
3	5
4	8

(a) Initially, each item is a cluster and we have the clusters

$$\{1\} \quad \{2\} \quad \{3\} \quad \{4\}$$

Show that ESS = 0, as it must.

(b) If we join clusters $\{1\}$ and $\{2\}$, the new cluster $\{12\}$ has

$$\text{ESS}_1 = \sum (x_j - \bar{x})^2 = (2 - 1.5)^2 + (1 - 1.5)^2 = .5$$

and the ESS associated with the grouping $\{12\}$, $\{3\}$, $\{4\}$ is ESS = .5 + 0 + 0 = .5. The *increase* in ESS (loss of information) from the first step to the current step in .5 − 0 = .5. Complete the following table by determining the increase in ESS for all the possibilities at step 2.

Clusters			Increase in ESS
$\{12\}$	$\{3\}$	$\{4\}$	.5
$\{13\}$	$\{2\}$	$\{4\}$	
$\{14\}$	$\{2\}$	$\{3\}$	
$\{1\}$	$\{23\}$	$\{4\}$	
$\{1\}$	$\{24\}$	$\{3\}$	
$\{1\}$	$\{2\}$	$\{34\}$	

(c) Complete the last two algamation steps, and construct the dendrogram showing the values of ESS at which the mergers take place.

12.11. Suppose we measure two variables X_1 and X_2 for four items $A, B, C,$ and D. The data are as follows:

Item	Observations	
	x_1	x_2
A	5	4
B	1	-2
C	-1	1
D	3	1

Use the K-means clustering technique to divide the items into $K = 2$ clusters. Start with the initial groups (AB) and (CD).

12.12. Repeat Example 12.11, starting with the initial groups (AC) and (BD). Compare your solution with the solution in the example. Are they the same? Graph the items in terms of their (x_1, x_2) coordinates, and comment on the solutions.

12.13. Repeat Example 12.11, but start at the bottom of the list of items, and proceed up in the order D, C, B, A. Begin with the initial groups (AB) and (CD). [The first potential reassignment will be based on the distances $d^2(D, (AB))$ and $d^2(D, (CD))$.] Compare your solution with the solution in the example. Are they the same? Should they be the same?

The following exercises require the use of a computer.

12.14. Table 11.9 lists measurements on 8 variables for 43 breakfast cereals.

(a) Using the data in the table, calculate the Euclidean distances between pairs of cereal brands.

(b) Treating the distances calculated in (a) as measures of (dis)similarity, cluster the cereals using the single linkage and complete linkage hierarchical procedures. Construct dendrograms and compare the results.

12.15. Input the data in Table 11.9 into a K-means clustering program. Cluster the cereals into $K = 2, 3,$ and 4 groups. Compare the results with those in Exercise 12.14.

12.16. The national track records data for women are given in Table 1.9.

(a) Using the data in Table 1.9, calculate the Euclidean distances between pairs of countries.

(b) Treating the distances in (a) as measures of (dis)similarity, cluster the countries using the single linkage and complete linkage hierarchical procedures. Construct dendrograms and compare the results.

(c) Input the data in Table 1.9 into a K-means clustering program. Cluster the countries into groups using several values of K. Compare the results with those in Part b.

12.17. Repeat Exercise 12.16 using the national track records data for men given in Table 8.6. Compare the results with those of Exercise 12.16. Explain any differences.

12.18. Table 12.12 gives the road distances between 12 Wisconsin cities and cities in neighboring states. Locate the cities in $q = 1, 2,$ and 3 dimensions using multidimensional scaling. Plot the minimum stress (q) versus q and interpret the graph. Compare the two-dimensional multidimensional scaling configuration with the locations of the cities on a map from an atlas.

12.19. Table 12.13 on page 752 gives the "distances" between certain archaeological sites from different periods, based upon the frequencies of different types of potsherds found at the sites. Given these distances, determine the coordinates of the sites in $q = 3, 4,$ and 5 dimensions using multidimensional scaling. Plot the minimum stress (q) versus q

Table 12.12 Distances Between Cities in Wisconsin and Cities in Neighboring States

	Appleton (1)	Beloit (2)	Fort Atkinson (3)	Madison (4)	Marshfield (5)	Milwaukee (6)	Monroe (7)	Superior (8)	Wausau (9)	Dubuque (10)	St. Paul (11)	Chicago (12)
(1)	0											
(2)	130	0										
(3)	98	33	0									
(4)	102	50	36	0								
(5)	103	185	164	138	0							
(6)	100	73	54	77	184	0						
(7)	149	33	58	47	170	107	0					
(8)	315	377	359	330	219	394	362	0				
(9)	91	186	166	139	45	181	186	223	0			
(10)	196	94	119	95	186	168	61	351	215	0		
(11)	257	304	287	258	161	322	289	162	175	274	0	
(12)	186	97	113	146	276	93	130	467	275	184	395	0

Table 12.13 Distances Between Archaeological Sites

	P1980918 (1)	P1931131 (2)	P1550960 (3)	P1530987 (4)	P1361024 (5)	P1351005 (6)	P1340945 (7)	P1311137 (8)	P1301062 (9)
(1)	0								
(2)	2.202	0							
(3)	1.004	2.025	0						
(4)	1.108	1.943	0.233	0					
(5)	1.122	1.870	0.719	0.541	0				
(6)	0.914	2.070	0.719	0.679	0.539	0			
(7)	0.914	2.186	0.452	0.681	1.102	0.916	0		
(8)	2.056	2.055	1.986	1.990	1.963	2.056	2.027	0	
(9)	1.608	1.722	1.358	1.168	0.681	1.005	1.719	1.991	0

KEY: P1980918 refers to site P198 dated A.D. 0918, P1931131 refers to site P193 dated A.D. 1131, and so forth.

Source: Data Courtesy of M. J. Tretter.

752

and interpret the graph. If possible, locate the sites in two dimensions (the first two principal components) using the coordinates for the $q = 5$-dimensional solution. (Treat the sites as variables.) Noting the periods associated with the sites, interpret the two-dimensional configuration.

12.20. A sample of $n = 1660$ people is cross-classified according to mental health status and socioeconomic status in Table 12.14.

Perform a correspondence analysis of these data. Interpret the results. Can the associations in the data be well represented in one dimension?

12.21. A sample of 901 individuals was cross-classified according to three categories of income and four categories of job satisfaction. The results are given in Table 12.15.

Perform a correspondence analysis of these data. Interpret the results.

12.22. Perform a correspondence analysis of the data on forests listed in Table 12.10, and verify Figure 12.28 given in Example 12.22.

12.23. Construct a biplot of the pottery data in Table 12.8. Interpret the biplot. Is the biplot consistent with the correspondence analysis plot in Figure 12.22? Discuss your answer. (Use the row proportions as a vector of observations at a site.)

12.24. Construct a biplot of the mental health and socioeconomic data in Table 12.14. Interpret the biplot. Is the biplot consistent with the correspondence analysis plot in Exercise 12.20? Discuss your answer. (Use the column proportions as the vector of observations for each status.)

Table 12.14 Mental Health Status and Socioeconomic Status Data

Mental Health Status	Parental Socioeconomic Status				
	A (High)	B	C	D	E (Low)
Well	121	57	72	36	21
Mild symptom formation	188	105	141	97	71
Moderate symptom formation	112	65	77	54	54
Impaired	86	60	94	78	71

Source: Adapted from data in Srole, L., T. S. Langner, S. T. Michael, P. Kirkpatrick, M. K. Opler, and T. A. C. Rennie, *Mental Health in the Metropolis: The Midtown Manhatten Study*, rev. ed. (New York: NYU Press, 1978).

Table 12.15 Income and Job Satisfaction Data

Income	Job Satisfaction			
	Very dissatisfied	Somewhat dissatisfied	Moderately satisfied	Very satisfied
< $ 25,000	42	62	184	207
$25,000–$50,000	13	28	81	113
> $ 50,000	7	18	54	92

Source: Adapted from data in Table 8.2 in Agresti, A., *Categorical Data Analysis* (New York: John Wiley, 1990).

12.25. Using the archaeological data in Table 12.13, determine the two-dimensional metric and nonmetric multidimensional scaling plots. (See Exercise 12.19.) Given the coordinates of the points in each of these plots, perform a Procrustes analysis. Interpret the results.

12.26. Table 8.7 contains the Mali family farm data (see Exercise 8.28). Remove the outliers 25, 34, 69 and 72, leaving at total of $n = 72$ observations in the data set. Treating the Euclidean distances between pairs of farms as a measure of similarity, cluster the farms using average linkage and Ward's method. Construct the dendrograms and compare the results. Do there appear to be several distinct clusters of farms?

12.27. Repeat Exercise 12.26 using standardized observations. Does it make a difference whether standardized or unstandardized observations are used? Explain.

12.28. Using the Mali family farm data in Table 8.7 with the outliers 25, 34, 69 and 72 removed, cluster the farms with the K-means clustering algorithm for $K = 5$ and $K = 6$. Compare the results with those in Exercise 12.26. Is 5 or 6 about the right number of distinct clusters? Discuss.

12.29. Repeat Exercise 12.28 using standardized observations. Does it make a difference whether standardized of unstandardized observations are used? Explain.

12.30. A company wants to do a mail marketing campaign. It costs the company $1 for each item mailed. They have information on 100,000 customers. Create and interpret a cumulative lift chart from the following information.

  Overall Response Rate: Assume we have no model other than the prediction of the overall response rate which is 20%. That is, if all 100,000 customers are contacted (at a cost of $100,000), we will receive around 20,000 positive responses.

  Results of Response Model: A response model predicts who will respond to a marketing campaign. We use the response model to assign a score to all 100,000 customers and predict the positive responses from contacting only the top 10,000 customers, the top 20,000 customers, and so forth. The model predictions are summarized below.

Cost ($)	Total Customers Contacted	Positive Responses
10000	10000	6000
20000	20000	10000
30000	30000	13000
40000	40000	15800
50000	50000	17000
60000	60000	18000
70000	70000	18800
80000	80000	19400
90000	90000	19800
100000	100000	20000

12.31. Consider the crude-oil data in Table 11.7. Transform the data as in Example 11.14. Ignore the known group membership. Using the special purpose software *MCLUST*,

(a) select a mixture model using the BIC criterion allowing for the different covariance structures listed in Section 12.5 and up to $K = 7$ groups.

(b) compare the clustering results for the best model with the known classifications given in Example 11.14. Notice how several clusters correspond to one crude-oil classification.

References

1. Abramowitz, M., and I. A. Stegun, eds. *Handbook of Mathematical Functions*. U.S. Department of Commerce, National Bureau of Standards Applied Mathematical Series. 55, 1964.

2. Adriaans, P., and D. Zantinge. *Data Mining*. Harlow, England: Addison-Wesley, 1996.

3. Anderberg, M. R. *Cluster Analysis for Applications*. New York: Academic Press, 1973.

4. Berry, M. J. A., and G. Linoff. *Data Mining Techniques: For Marketing, Sales and Customer Relationship Management* (2nd ed.) (paperback). New York: John Wiley, 2004.

5. Berthold, M., and D. J. Hand. *Intelligent Data Analysis* (2nd ed.). Berlin, Germany: Springer-Verlag, 2003.

6. Celeux, G., and G. Govaert. "Gaussian Parsimonious Clustering Models." *Pattern Recognition*, **28** (1995), 781–793.

7. Cormack, R. M. "A Review of Classification (with discussion)." *Journal of the Royal Statistical Society (A)*, **134**, no. 3 (1971), 321–367.

8. Everitt, B. S., S. Landau and M. Leese. *Cluster Analysis* (4th ed.). London: Hodder Arnold, 2001.

9. Fraley, C., and A. E. Raftery. "Model-Based Clustering, Discriminant Analysis and Density Estimation." *Journal of the American Statistical Association*, **97** (2002), 611–631.

10. Gower, J. C. "Some Distance Properties of Latent Root and Vector Methods Used in Multivariate Analysis." *Biometrika*, **53** (1966), 325–338.

11. Gower, J. C. "Multivariate Analysis and Multidimensional Geometry." *The Statistician*, **17** (1967), 13–25.

12. Gower, J. C., and D. J. Hand. *Biplots*. London: Chapman and Hall, 1996.

13. Greenacre, M. J. "Correspondence Analysis of Square Asymmetric Matrices," *Applied Statistics*, **49**, (2000) 297–310.

14. Greenacre, M. J. *Theory and Applications of Correspondence Analysis*. London: Academic Press, 1984.

15. Hand, D., H. Mannila, and P. Smyth. *Principles of Data Mining*. Cambridge, MA: MIT Press, 2001.

16. Hartigan, J. A. *Clustering Algorithms*. New York: John Wiley, 1975.

17. Hastie, T. R., R. Tibshirani and J. Friedman. *The Elements of Statistical Learning: Data Mining, Inference and Prediction*. Berlin, Germany: Springer-Verlag, 2001.

18. Kennedy, R. L., L. Lee, B. Van Roy, C. D. Reed, and R. P. Lippmann. *Solving Data Mining Problems Through Pattern Recognition*. Upper Saddle River, NJ: Prentice-Hall, 1997.

19. Kruskal, J. B. "Multidimensional Scaling by Optimizing Goodness of Fit to a Nonmetric Hypothesis." *Psychometrika*, **29**, no. 1 (1964), 1–27.

20. Kruskal, J. B. "Non-metric Multidimensional Scaling: A Numerical Method." *Psychometrika*, **29**, no. 1 (1964), 115–129.

21. Kruskal, J. B., and M. Wish. "Multidimensional Scaling." Sage University Paper Series on Quantitative Applications in the Social Sciences, 07–011. Beverly Hills and London: Sage Publications, 1978.

22. LaPointe, F-J, and P. Legendre. "A Classification of Pure Malt Scotch Whiskies." *Applied Statistics*, **43**, no. 1 (1994), 237–257.

23. le Roux, N. J., and S. Gardner. "Analysing Your Multivariate Data as a Pictorial: A Case for Applying Biplot Methodology." *International Statistical Review*, **73** (2005), 365–387.

24. Ludwig, J. A., and J. F. Reynolds. *Statistical Ecology—a Primer on Methods and Computing.* New York: Wiley-Interscience, 1988.

25. MacQueen, J. B. "Some Methods for Classification and Analysis of Multivariate Observations." *Proceedings of 5th Berkeley Symposium on Mathematical Statistics and Probability*, **1**, Berkeley, CA: University of California Press (1967), 281–297.

26. Mardia, K. V., J. T. Kent, and J. M. Bibby. *Multivariate Analysis* (Paperback). London: Academic Press, 2003.

27. Morgan, B. J. T., and A. P. G. Ray. "Non-uniqueness and Inversions in Cluster Analysis." *Applied Statistics*, **44**, no. 1 (1995), 117–134.

28. Pyle, D. *Data Preparation for Data Mining.* San Francisco: Morgan Kaufmann, 1999.

29. Shepard, R. N. "Multidimensional Scaling, Tree-Fitting, and Clustering." *Science*, **210**, no. 4468 (1980), 390–398.

30. Sibson, R. "Studies in the Robustness of Multidimensional Scaling" *Journal of the Royal Statistical Society (B)*, **40** (1978), 234–238.

31. Takane, Y., F. W. Young, and J. De Leeuw. "Non-metric Individual Differences Multidimensional Scaling: Alternating Least Squares with Optimal Scaling Features." *Psycometrika*, **42** (1977), 7–67.

32. Ward, Jr., J. H. "Hierarchical Grouping to Optimize an Objective Function." *Journal of the American Statistical Association*, **58** (1963), 236–244.

33. Westphal, C., and T. Blaxton. *Data Mining Solutions: Methods and Tools for Solving Real World Problems* (Paperback). New York: John Wiley, 1998.

34. Whitten, I. H., and E. Frank. *Data Mining: Practical Machine Learning Tools and Techniques* (2nd ed.) (Paperback). San Francisco: Morgan Kaufmann, 2005.

35. Young, F. W., and R. M. Hamer. *Multidimensional Scaling: History, Theory, and Applications.* Hillsdale, NJ: Lawrence Erlbaum Associates, Publishers, 1987.

Selected Additional References for Model Based Clustering

Banfield, J. D., and A. E. Raftery. "Model-Based Gaussian and Non-Gaussian Clustering." *Biometrics*, **49** (1993), 803–821.

Biernacki, C., and G. Govaert. "Choosing Models in Model Based Clustering and Discriminant Analysis." *Journal of Statistical Computation and Simulation*, **64** (1999), 49–71.

Celeux, G., and G. Govaert. "A Classification EM Algorithm for Clustering and Two Stochastic Versions." *Computational Statistics and Data Analysis*, **14** (1992), 315–332.

Fraley, C., and A. E. Raftery. "MCLUST: Software for Model Based Cluster Analysis." *Journal of Classification*, **16** (1999), 297–306.

Hastie, T., and R. Tibshirani. "Discriminant Analysis by Gaussian Mixtures." *Journal of the Royal Statistical Society* (B), **58** (1996), 155–176.

McLachlan, G. J., and K. E. Basford. *Mixture Models: Inference and Applications to Clustering.* New York: Marcel Dekker, 1988.

Schwarz, G. "Estimating the Dimension of a Model." *Annals of Statistics*, **6** (1978), 461–464.

Appendix

TABLE 1   STANDARD NORMAL PROBABILITIES

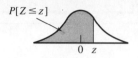

$P[Z \leq z]$

z	.00	.01	.02	.03	.04	.05	.06	.07	.08	.09
.0	.5000	.5040	.5080	.5120	.5160	.5199	.5239	.5279	.5319	.5359
.1	.5398	.5438	.5478	.5517	.5557	.5596	.5636	.5675	.5714	.5753
.2	.5793	.5832	.5871	.5910	.5948	.5987	.6026	.6064	.6103	.6141
.3	.6179	.6217	.6255	.6293	.6331	.6368	.6406	.6443	.6480	.6517
.4	.6554	.6591	.6628	.6664	.6700	.6736	.6772	.6808	.6844	.6879
.5	.6915	.6950	.6985	.7019	.7054	.7088	.7123	.7157	.7190	.7224
.6	.7257	.7291	.7324	.7357	.7389	.7422	.7454	.7486	.7517	.7549
.7	.7580	.7611	.7642	.7673	.7703	.7734	.7764	.7794	.7823	.7852
.8	.7881	.7910	.7939	.7967	.7995	.8023	.8051	.8078	.8106	.8133
.9	.8159	.8186	.8212	.8238	.8264	.8289	.8315	.8340	.8365	.8389
1.0	.8413	.8438	.8461	.8485	.8508	.8531	.8554	.8577	.8599	.8621
1.1	.8643	.8665	.8686	.8708	.8729	.8749	.8770	.8790	.8810	.8830
1.2	.8849	.8869	.8888	.8907	.8925	.8944	.8962	.8980	.8997	.9015
1.3	.9032	.9049	.9066	.9082	.9099	.9115	.9131	.9147	.9162	.9177
1.4	.9192	.9207	.9222	.9236	.9251	.9265	.9279	.9292	.9306	.9319
1.5	.9332	.9345	.9357	.9370	.9382	.9394	.9406	.9418	.9429	.9441
1.6	.9452	.9463	.9474	.9484	.9495	.9505	.9515	.9525	.9535	.9545
1.7	.9554	.9564	.9573	.9582	.9591	.9599	.9608	.9616	.9625	.9633
1.8	.9641	.9649	.9656	.9664	.9671	.9678	.9686	.9693	.9699	.9706
1.9	.9713	.9719	.9726	.9732	.9738	.9744	.9750	.9756	.9761	.9767
2.0	.9772	.9778	.9783	.9788	.9793	.9798	.9803	.9808	.9812	.9817
2.1	.9821	.9826	.9830	.9834	.9838	.9842	.9846	.9850	.9854	.9857
2.2	.9861	.9864	.9868	.9871	.9875	.9878	.9881	.9884	.9887	.9890
2.3	.9893	.9896	.9898	.9901	.9904	.9906	.9909	.9911	.9913	.9916
2.4	.9918	.9920	.9922	.9925	.9927	.9929	.9931	.9932	.9934	.9936
2.5	.9938	.9940	.9941	.9943	.9945	.9946	.9948	.9949	.9951	.9952
2.6	.9953	.9955	.9956	.9957	.9959	.9960	.9961	.9962	.9963	.9964
2.7	.9965	.9966	.9967	.9968	.9969	.9970	.9971	.9972	.9973	.9974
2.8	.9974	.9975	.9976	.9977	.9977	.9978	.9979	.9979	.9980	.9981
2.9	.9981	.9982	.9982	.9983	.9984	.9984	.9985	.9985	.9986	.9986
3.0	.9987	.9987	.9987	.9988	.9988	.9989	.9989	.9989	.9990	.9990
3.1	.9990	.9991	.9991	.9991	.9992	.9992	.9992	.9992	.9993	.9993
3.2	.9993	.9993	.9994	.9994	.9994	.9994	.9994	.9995	.9995	.9995
3.3	.9995	.9995	.9995	.9996	.9996	.9996	.9996	.9996	.9996	.9997
3.4	.9997	.9997	.9997	.9997	.9997	.9997	.9997	.9997	.9997	.9998
3.5	.9998	.9998	.9998	.9998	.9998	.9998	.9998	.9998	.9998	.9998

TABLE 2   STUDENT'S *t*-DISTRIBUTION PERCENTAGE POINTS

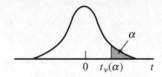

d.f.	α								
ν	.250	.100	.050	.025	.010	.00833	.00625	.005	.0025
1	1.000	3.078	6.314	12.706	31.821	38.190	50.923	63.657	127.321
2	.816	1.886	2.920	4.303	6.965	7.649	8.860	9.925	14.089
3	.765	1.638	2.353	3.182	4.541	4.857	5.392	5.841	7.453
4	.741	1.533	2.132	2.776	3.747	3.961	4.315	4.604	5.598
5	.727	1.476	2.015	2.571	3.365	3.534	3.810	4.032	4.773
6	.718	1.440	1.943	2.447	3.143	3.287	3.521	3.707	4.317
7	.711	1.415	1.895	2.365	2.998	3.128	3.335	3.499	4.029
8	.706	1.397	1.860	2.306	2.896	3.016	3.206	3.355	3.833
9	.703	1.383	1.833	2.262	2.821	2.933	3.111	3.250	3.690
10	.700	1.372	1.812	2.228	2.764	2.870	3.038	3.169	3.581
11	.697	1.363	1.796	2.201	2.718	2.820	2.981	3.106	3.497
12	.695	1.356	1.782	2.179	2.681	2.779	2.934	3.055	3.428
13	.694	1.350	1.771	2.160	2.650	2.746	2.896	3.012	3.372
14	.692	1.345	1.761	2.145	2.624	2.718	2.864	2.977	3.326
15	.691	1.341	1.753	2.131	2.602	2.694	2.837	2.947	3.286
16	.690	1.337	1.746	2.120	2.583	2.673	2.813	2.921	3.252
17	.689	1.333	1.740	2.110	2.567	2.655	2.793	2.898	3.222
18	.688	1.330	1.734	2.101	2.552	2.639	2.775	2.878	3.197
19	.688	1.328	1.729	2.093	2.539	2.625	2.759	2.861	3.174
20	.687	1.325	1.725	2.086	2.528	2.613	2.744	2.845	3.153
21	.686	1.323	1.721	2.080	2.518	2.601	2.732	2.831	3.135
22	.686	1.321	1.717	2.074	2.508	2.591	2.720	2.819	3.119
23	.685	1.319	1.714	2.069	2.500	2.582	2.710	2.807	3.104
24	.685	1.318	1.711	2.064	2.492	2.574	2.700	2.797	3.091
25	.684	1.316	1.708	2.060	2.485	2.566	2.692	2.787	3.078
26	.684	1.315	1.706	2.056	2.479	2.559	2.684	2.779	3.067
27	.684	1.314	1.703	2.052	2.473	2.552	2.676	2.771	3.057
28	.683	1.313	1.701	2.048	2.467	2.546	2.669	2.763	3.047
29	.683	1.311	1.699	2.045	2.462	2.541	2.663	2.756	3.038
30	.683	1.310	1.697	2.042	2.457	2.536	2.657	2.750	3.030
40	.681	1.303	1.684	2.021	2.423	2.499	2.616	2.704	2.971
60	.679	1.296	1.671	2.000	2.390	2.463	2.575	2.660	2.915
120	.677	1.289	1.658	1.980	2.358	2.428	2.536	2.617	2.860
∞	.674	1.282	1.645	1.960	2.326	2.394	2.498	2.576	2.813

TABLE 3 χ^2 DISTRIBUTION PERCENTAGE POINTS

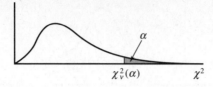

d.f.	α								
ν	.990	.950	.900	.500	.100	.050	.025	.010	.005
1	.0002	.004	.02	.45	2.71	3.84	5.02	6.63	7.88
2	.02	.10	.21	1.39	4.61	5.99	7.38	9.21	10.60
3	.11	.35	.58	2.37	6.25	7.81	9.35	11.34	12.84
4	.30	.71	1.06	3.36	7.78	9.49	11.14	13.28	14.86
5	.55	1.15	1.61	4.35	9.24	11.07	12.83	15.09	16.75
6	.87	1.64	2.20	5.35	10.64	12.59	14.45	16.81	18.55
7	1.24	2.17	2.83	6.35	12.02	14.07	16.01	18.48	20.28
8	1.65	2.73	3.49	7.34	13.36	15.51	17.53	20.09	21.95
9	2.09	3.33	4.17	8.34	14.68	16.92	19.02	21.67	23.59
10	2.56	3.94	4.87	9.34	15.99	18.31	20.48	23.21	25.19
11	3.05	4.57	5.58	10.34	17.28	19.68	21.92	24.72	26.76
12	3.57	5.23	6.30	11.34	18.55	21.03	23.34	26.22	28.30
13	4.11	5.89	7.04	12.34	19.81	22.36	24.74	27.69	29.82
14	4.66	6.57	7.79	13.34	21.06	23.68	26.12	29.14	31.32
15	5.23	7.26	8.55	14.34	22.31	25.00	27.49	30.58	32.80
16	5.81	7.96	9.31	15.34	23.54	26.30	28.85	32.00	34.27
17	6.41	8.67	10.09	16.34	24.77	27.59	30.19	33.41	35.72
18	7.01	9.39	10.86	17.34	25.99	28.87	31.53	34.81	37.16
19	7.63	10.12	11.65	18.34	27.20	30.14	32.85	36.19	38.58
20	8.26	10.85	12.44	19.34	28.41	31.41	34.17	37.57	40.00
21	8.90	11.59	13.24	20.34	29.62	32.67	35.48	38.93	41.40
22	9.54	12.34	14.04	21.34	30.81	33.92	36.78	40.29	42.80
23	10.20	13.09	14.85	22.34	32.01	35.17	38.08	41.64	44.18
24	10.86	13.85	15.66	23.34	33.20	36.42	39.36	42.98	45.56
25	11.52	14.61	16.47	24.34	34.38	37.65	40.65	44.31	46.93
26	12.20	15.38	17.29	25.34	35.56	38.89	41.92	45.64	48.29
27	12.88	16.15	18.11	26.34	36.74	40.11	43.19	46.96	49.64
28	13.56	16.93	18.94	27.34	37.92	41.34	44.46	48.28	50.99
29	14.26	17.71	19.77	28.34	39.09	42.56	45.72	49.59	52.34
30	14.95	18.49	20.60	29.34	40.26	43.77	46.98	50.89	53.67
40	22.16	26.51	29.05	39.34	51.81	55.76	59.34	63.69	66.77
50	29.71	34.76	37.69	49.33	63.17	67.50	71.42	76.15	79.49
60	37.48	43.19	46.46	59.33	74.40	79.08	83.30	88.38	91.95
70	45.44	51.74	55.33	69.33	85.53	90.53	95.02	100.43	104.21
80	53.54	60.39	64.28	79.33	96.58	101.88	106.63	112.33	116.32
90	61.75	69.13	73.29	89.33	107.57	113.15	118.14	124.12	128.30
100	70.06	77.93	82.36	99.33	118.50	124.34	129.56	135.81	140.17

TABLE 4 *F*-DISTRIBUTION PERCENTAGE POINTS ($\alpha = .10$)

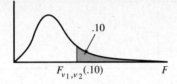

$$F_{v_1, v_2}(.10)$$

v_2 \ v_1	1	2	3	4	5	6	7	8	9	10	12	15	20	25	30	40	60
1	39.86	49.50	53.59	55.83	57.24	58.20	58.91	59.44	59.86	60.19	60.71	61.22	61.74	62.05	62.26	62.53	62.79
2	8.53	9.00	9.16	9.24	9.29	9.33	9.35	9.37	9.38	9.39	9.41	9.42	9.44	9.45	9.46	9.47	9.47
3	5.54	5.46	5.39	5.34	5.31	5.28	5.27	5.25	5.24	5.23	5.22	5.20	5.18	5.17	5.17	5.16	5.15
4	4.54	4.32	4.19	4.11	4.05	4.01	3.98	3.95	3.94	3.92	3.90	3.87	3.84	3.83	3.82	3.80	3.79
5	4.06	3.78	3.62	3.52	3.45	3.40	3.37	3.34	3.32	3.30	3.27	3.24	3.21	3.19	3.17	3.16	3.14
6	3.78	3.46	3.29	3.18	3.11	3.05	3.01	2.98	2.96	2.94	2.90	2.87	2.84	2.81	2.80	2.78	2.76
7	3.59	3.26	3.07	2.96	2.88	2.83	2.78	2.75	2.72	2.70	2.67	2.63	2.59	2.57	2.56	2.54	2.51
8	3.46	3.11	2.92	2.81	2.73	2.67	2.62	2.59	2.56	2.54	2.50	2.46	2.42	2.40	2.38	2.36	2.34
9	3.36	3.01	2.81	2.69	2.61	2.55	2.51	2.47	2.44	2.42	2.38	2.34	2.30	2.27	2.25	2.23	2.21
10	3.29	2.92	2.73	2.61	2.52	2.46	2.41	2.38	2.35	2.32	2.28	2.24	2.20	2.17	2.16	2.13	2.11
11	3.23	2.86	2.66	2.54	2.45	2.39	2.34	2.30	2.27	2.25	2.21	2.17	2.12	2.10	2.08	2.05	2.03
12	3.18	2.81	2.61	2.48	2.39	2.33	2.28	2.24	2.21	2.19	2.15	2.10	2.06	2.03	2.01	1.99	1.96
13	3.14	2.76	2.56	2.43	2.35	2.28	2.23	2.20	2.16	2.14	2.10	2.05	2.01	1.98	1.96	1.93	1.90
14	3.10	2.73	2.52	2.39	2.31	2.24	2.19	2.15	2.12	2.10	2.05	2.01	1.96	1.93	1.91	1.89	1.86
15	3.07	2.70	2.49	2.36	2.27	2.21	2.16	2.12	2.09	2.06	2.02	1.97	1.92	1.89	1.87	1.85	1.82
16	3.05	2.67	2.46	2.33	2.24	2.18	2.13	2.09	2.06	2.03	1.99	1.94	1.89	1.86	1.84	1.81	1.78
17	3.03	2.64	2.44	2.31	2.22	2.15	2.10	2.06	2.03	2.00	1.96	1.91	1.86	1.83	1.81	1.78	1.75
18	3.01	2.62	2.42	2.29	2.20	2.13	2.08	2.04	2.00	1.98	1.93	1.89	1.84	1.80	1.78	1.75	1.72
19	2.99	2.61	2.40	2.27	2.18	2.11	2.06	2.02	1.98	1.96	1.91	1.86	1.81	1.78	1.76	1.73	1.70
20	2.97	2.59	2.38	2.25	2.16	2.09	2.04	2.00	1.96	1.94	1.89	1.84	1.79	1.76	1.74	1.71	1.68
21	2.96	2.57	2.36	2.23	2.14	2.08	2.02	1.98	1.95	1.92	1.87	1.83	1.78	1.74	1.72	1.69	1.66
22	2.95	2.56	2.35	2.22	2.13	2.06	2.01	1.97	1.93	1.90	1.86	1.81	1.76	1.73	1.70	1.67	1.64
23	2.94	2.55	2.34	2.21	2.11	2.05	1.99	1.95	1.92	1.89	1.84	1.80	1.74	1.71	1.69	1.66	1.62
24	2.93	2.54	2.33	2.19	2.10	2.04	1.98	1.94	1.91	1.88	1.83	1.78	1.73	1.70	1.67	1.64	1.61
25	2.92	2.53	2.32	2.18	2.09	2.02	1.97	1.93	1.89	1.87	1.82	1.77	1.72	1.68	1.66	1.63	1.59
26	2.91	2.52	2.31	2.17	2.08	2.01	1.96	1.92	1.88	1.86	1.81	1.76	1.71	1.67	1.65	1.61	1.58
27	2.90	2.51	2.30	2.17	2.07	2.00	1.95	1.91	1.87	1.85	1.80	1.75	1.70	1.66	1.64	1.60	1.57
28	2.89	2.50	2.29	2.16	2.06	2.00	1.94	1.90	1.87	1.84	1.79	1.74	1.69	1.65	1.63	1.59	1.56
29	2.89	2.50	2.28	2.15	2.06	1.99	1.93	1.89	1.86	1.83	1.78	1.73	1.68	1.64	1.62	1.58	1.55
30	2.88	2.49	2.28	2.14	2.05	1.98	1.93	1.88	1.85	1.82	1.77	1.72	1.67	1.63	1.61	1.57	1.54
40	2.84	2.44	2.23	2.09	2.00	1.93	1.87	1.83	1.79	1.76	1.71	1.66	1.61	1.57	1.54	1.51	1.47
60	2.79	2.39	2.18	2.04	1.95	1.87	1.82	1.77	1.74	1.71	1.66	1.60	1.54	1.50	1.48	1.44	1.40
120	2.75	2.35	2.13	1.99	1.90	1.82	1.77	1.72	1.68	1.65	1.60	1.55	1.48	1.45	1.41	1.37	1.32
∞	2.71	2.30	2.08	1.94	1.85	1.77	1.72	1.67	1.63	1.60	1.55	1.49	1.42	1.38	1.34	1.30	1.24

TABLE 5   *F*-DISTRIBUTION PERCENTAGE POINTS ($\alpha = .05$)

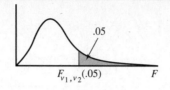

$F_{\nu_1,\nu_2}(.05)$

ν_2 \ ν_1	1	2	3	4	5	6	7	8	9	10	12	15	20	25	30	40	60
1	161.5	199.5	215.7	224.6	230.2	234.0	236.8	238.9	240.5	241.9	243.9	246.0	248.0	249.3	250.1	251.1	252.2
2	18.51	19.00	19.16	19.25	19.30	19.33	19.35	19.37	19.38	19.40	19.41	19.43	19.45	19.46	19.46	19.47	19.48
3	10.13	9.55	9.28	9.12	9.01	8.94	8.89	8.85	8.81	8.79	8.74	8.70	8.66	8.63	8.62	8.59	8.57
4	7.71	6.94	6.59	6.39	6.26	6.16	6.09	6.04	6.00	5.96	5.91	5.86	5.80	5.77	5.75	5.72	5.69
5	6.61	5.79	5.41	5.19	5.05	4.95	4.88	4.82	4.77	4.74	4.68	4.62	4.56	4.52	4.50	4.46	4.43
6	5.99	5.14	4.76	4.53	4.39	4.28	4.21	4.15	4.10	4.06	4.00	3.94	3.87	3.83	3.81	3.77	3.74
7	5.59	4.74	4.35	4.12	3.97	3.87	3.79	3.73	3.68	3.64	3.57	3.51	3.44	3.40	3.38	3.34	3.30
8	5.32	4.46	4.07	3.84	3.69	3.58	3.50	3.44	3.39	3.35	3.28	3.22	3.15	3.11	3.08	3.04	3.01
9	5.12	4.26	3.86	3.63	3.48	3.37	3.29	3.23	3.18	3.14	3.07	3.01	2.94	2.89	2.86	2.83	2.79
10	4.96	4.10	3.71	3.48	3.33	3.22	3.14	3.07	3.02	2.98	2.91	2.85	2.77	2.73	2.70	2.66	2.62
11	4.84	3.98	3.59	3.36	3.20	3.09	3.01	2.95	2.90	2.85	2.79	2.72	2.65	2.60	2.57	2.53	2.49
12	4.75	3.89	3.49	3.26	3.11	3.00	2.91	2.85	2.80	2.75	2.69	2.62	2.54	2.50	2.47	2.43	2.38
13	4.67	3.81	3.41	3.18	3.03	2.92	2.83	2.77	2.71	2.67	2.60	2.53	2.46	2.41	2.38	2.34	2.30
14	4.60	3.74	3.34	3.11	2.96	2.85	2.76	2.70	2.65	2.60	2.53	2.46	2.39	2.34	2.31	2.27	2.22
15	4.54	3.68	3.29	3.06	2.90	2.79	2.71	2.64	2.59	2.54	2.48	2.40	2.33	2.28	2.25	2.20	2.16
16	4.49	3.63	3.24	3.01	2.85	2.74	2.66	2.59	2.54	2.49	2.42	2.35	2.28	2.23	2.19	2.15	2.11
17	4.45	3.59	3.20	2.96	2.81	2.70	2.61	2.55	2.49	2.45	2.38	2.31	2.23	2.18	2.15	2.10	2.06
18	4.41	3.55	3.16	2.93	2.77	2.66	2.58	2.51	2.46	2.41	2.34	2.27	2.19	2.14	2.11	2.06	2.02
19	4.38	3.52	3.13	2.90	2.74	2.63	2.54	2.48	2.42	2.38	2.31	2.23	2.16	2.11	2.07	2.03	1.98
20	4.35	3.49	3.10	2.87	2.71	2.60	2.51	2.45	2.39	2.35	2.28	2.20	2.12	2.07	2.04	1.99	1.95
21	4.32	3.47	3.07	2.84	2.68	2.57	2.49	2.42	2.37	2.32	2.25	2.18	2.10	2.05	2.01	1.96	1.92
22	4.30	3.44	3.05	2.82	2.66	2.55	2.46	2.40	2.34	2.30	2.23	2.15	2.07	2.02	1.98	1.94	1.89
23	4.28	3.42	3.03	2.80	2.64	2.53	2.44	2.37	2.32	2.27	2.20	2.13	2.05	2.00	1.96	1.91	1.86
24	4.26	3.40	3.01	2.78	2.62	2.51	2.42	2.36	2.30	2.25	2.18	2.11	2.03	1.97	1.94	1.89	1.84
25	4.24	3.39	2.99	2.76	2.60	2.49	2.40	2.34	2.28	2.24	2.16	2.09	2.01	1.96	1.92	1.87	1.82
26	4.23	3.37	2.98	2.74	2.59	2.47	2.39	2.32	2.27	2.22	2.15	2.07	1.99	1.94	1.90	1.85	1.80
27	4.21	3.35	2.96	2.73	2.57	2.46	2.37	2.31	2.25	2.20	2.13	2.06	1.97	1.92	1.88	1.84	1.79
28	4.20	3.34	2.95	2.71	2.56	2.45	2.36	2.29	2.24	2.19	2.12	2.04	1.96	1.91	1.87	1.82	1.77
29	4.18	3.33	2.93	2.70	2.55	2.43	2.35	2.28	2.22	2.18	2.10	2.03	1.94	1.89	1.85	1.81	1.75
30	4.17	3.32	2.92	2.69	2.53	2.42	2.33	2.27	2.21	2.16	2.09	2.01	1.93	1.88	1.84	1.79	1.74
40	4.08	3.23	2.84	2.61	2.45	2.34	2.25	2.18	2.12	2.08	2.00	1.92	1.84	1.78	1.74	1.69	1.64
60	4.00	3.15	2.76	2.53	2.37	2.25	2.17	2.10	2.04	1.99	1.92	1.84	1.75	1.69	1.65	1.59	1.53
120	3.92	3.07	2.68	2.45	2.29	2.18	2.09	2.02	1.96	1.91	1.83	1.75	1.66	1.60	1.55	1.50	1.43
∞	3.84	3.00	2.61	2.37	2.21	2.10	2.01	1.94	1.88	1.83	1.75	1.67	1.57	1.51	1.46	1.39	1.32

TABLE 6   *F*-DISTRIBUTION PERCENTAGE POINTS ($\alpha = .01$)

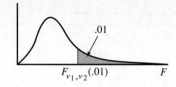

$F_{\nu_1,\nu_2}(.01)$ F

ν_1 ν_2	1	2	3	4	5	6	7	8	9	10	12	15	20	25	30	40	60
1	4052.	5000.	5403.	5625.	5764.	5859.	5928.	5981.	6023.	6056.	6106.	6157.	6209.	6240.	6261.	6287.	6313.
2	98.50	99.00	99.17	99.25	99.30	99.33	99.36	99.37	99.39	99.40	99.42	99.43	99.45	99.46	99.47	99.47	99.48
3	34.12	30.82	29.46	28.71	28.24	27.91	27.67	27.49	27.35	27.23	27.05	26.87	26.69	26.58	26.50	26.41	26.32
4	21.20	18.00	16.69	15.98	15.52	15.21	14.98	14.80	14.66	14.55	14.37	14.20	14.02	13.91	13.84	13.75	13.65
5	16.26	13.27	12.06	11.39	10.97	10.67	10.46	10.29	10.16	10.05	9.89	9.72	9.55	9.45	9.38	9.29	9.20
6	13.75	10.92	9.78	9.15	8.75	8.47	8.26	8.10	7.98	7.87	7.72	7.56	7.40	7.30	7.23	7.14	7.06
7	12.25	9.55	8.45	7.85	7.46	7.19	6.99	6.84	6.72	6.62	6.47	6.31	6.16	6.06	5.99	5.91	5.82
8	11.26	8.65	7.59	7.01	6.63	6.37	6.18	6.03	5.91	5.81	5.67	5.52	5.36	5.26	5.20	5.12	5.03
9	10.56	8.02	6.99	6.42	6.06	5.80	5.61	5.47	5.35	5.26	5.11	4.96	4.81	4.71	4.65	4.57	4.48
10	10.04	7.56	6.55	5.99	5.64	5.39	5.20	5.06	4.94	4.85	4.71	4.56	4.41	4.31	4.25	4.17	4.08
11	9.65	7.21	6.22	5.67	5.32	5.07	4.89	4.74	4.63	4.54	4.40	4.25	4.10	4.01	3.94	3.86	3.78
12	9.33	6.93	5.95	5.41	5.06	4.82	4.64	4.50	4.39	4.30	4.16	4.01	3.86	3.76	3.70	3.62	3.54
13	9.07	6.70	5.74	5.21	4.86	4.62	4.44	4.30	4.19	4.10	3.96	3.82	3.66	3.57	3.51	3.43	3.34
14	8.86	6.51	5.56	5.04	4.69	4.46	4.28	4.14	4.03	3.94	3.80	3.66	3.51	3.41	3.35	3.27	3.18
15	8.68	6.36	5.42	4.89	4.56	4.32	4.14	4.00	3.89	3.80	3.67	3.52	3.37	3.28	3.21	3.13	3.05
16	8.53	6.23	5.29	4.77	4.44	4.20	4.03	3.89	3.78	3.69	3.55	3.41	3.26	3.16	3.10	3.02	2.93
17	8.40	6.11	5.19	4.67	4.34	4.10	3.93	3.79	3.68	3.59	3.46	3.31	3.16	3.07	3.00	2.92	2.83
18	8.29	6.01	5.09	4.58	4.25	4.01	3.84	3.71	3.60	3.51	3.37	3.23	3.08	2.98	2.92	2.84	2.75
19	8.18	5.93	5.01	4.50	4.17	3.94	3.77	3.63	3.52	3.43	3.30	3.15	3.00	2.91	2.84	2.76	2.67
20	8.10	5.85	4.94	4.43	4.10	3.87	3.70	3.56	3.46	3.37	3.23	3.09	2.94	2.84	2.78	2.69	2.61
21	8.02	5.78	4.87	4.37	4.04	3.81	3.64	3.51	3.40	3.31	3.17	3.03	2.88	2.79	2.72	2.64	2.55
22	7.95	5.72	4.82	4.31	3.99	3.76	3.59	3.45	3.35	3.26	3.12	2.98	2.83	2.73	2.67	2.58	2.50
23	7.88	5.66	4.76	4.26	3.94	3.71	3.54	3.41	3.30	3.21	3.07	2.93	2.78	2.69	2.62	2.54	2.45
24	7.82	5.61	4.72	4.22	3.90	3.67	3.50	3.36	3.26	3.17	3.03	2.89	2.74	2.64	2.58	2.49	2.40
25	7.77	5.57	4.68	4.18	3.85	3.63	3.46	3.32	3.22	3.13	2.99	2.85	2.70	2.60	2.54	2.45	2.36
26	7.72	5.53	4.64	4.14	3.82	3.59	3.42	3.29	3.18	3.09	2.96	2.81	2.66	2.57	2.50	2.42	2.33
27	7.68	5.49	4.60	4.11	3.78	3.56	3.39	3.26	3.15	3.06	2.93	2.78	2.63	2.54	2.47	2.38	2.29
28	7.64	5.45	4.57	4.07	3.75	3.53	3.36	3.23	3.12	3.03	2.90	2.75	2.60	2.51	2.44	2.35	2.26
29	7.60	5.42	4.54	4.04	3.73	3.50	3.33	3.20	3.09	3.00	2.87	2.73	2.57	2.48	2.41	2.33	2.23
30	7.56	5.39	4.51	4.02	3.70	3.47	3.30	3.17	3.07	2.98	2.84	2.70	2.55	2.45	2.39	2.30	2.21
40	7.31	5.18	4.31	3.83	3.51	3.29	3.12	2.99	2.89	2.80	2.66	2.52	2.37	2.27	2.20	2.11	2.02
60	7.08	4.98	4.13	3.65	3.34	3.12	2.95	2.82	2.72	2.63	2.50	2.35	2.20	2.10	2.03	1.94	1.84
120	6.85	4.79	3.95	3.48	3.17	2.96	2.79	2.66	2.56	2.47	2.34	2.19	2.03	1.93	1.86	1.76	1.66
∞	6.63	4.61	3.78	3.32	3.02	2.80	2.64	2.51	2.41	2.32	2.18	2.04	1.88	1.78	1.70	1.59	1.47

Data Index

Subject Index